D1712500

Cell Lineage and Fate Determination

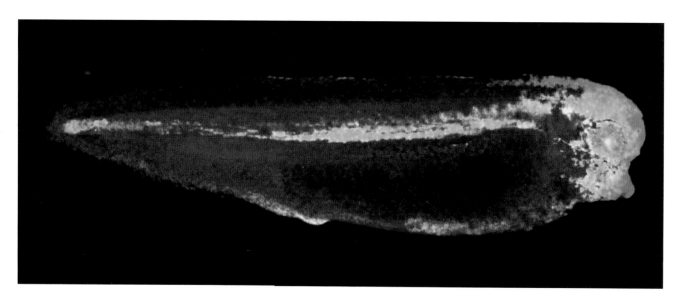

Is your origin your fate? The distribution of progeny from two lineage-labeled 8-cell blastomeres of a *Xenopus* embryo is altered when the embryo is rotated 90 degrees with respect to gravity shortly after fertilization. Red cells are derived from the original V1 blastomere and green cells are derived from the original V2 blastomere. This early rotation alters the direction of cortical rotation, presumably changes the distributions of maternal determinant molecules, and subsequently changes the fates of the blastomeres (see Chapter 20 by Sullivan *et al.*). Courtesy of Dr. Sen Huang, The George Washington University.

CELL LINEAGE AND FATE DETERMINATION

Edited by

SALLY A. MOODY
Department of Anatomy and Cell Biology
Institute for Biomedical Sciences
The George Washington University
Washington, D.C. 20037

ACADEMIC PRESS
San Diego   London   Boston   New York   Sydney   Tokyo   Toronto

Cover photograph: Wild-type embryonic central nervous system of *Drosophila* stained for the pan-neuronal protein Elav (dark blue), the axonal marker BP102 (red), and the Even-skipped transcription factor (yellow, cyan). Courtesy of Julie Broadus.

This book is printed on acid-free paper.

Academic Press
a division of Harcourt Brace & Company
525 B Street, Suite 1900, San Diego, California 92101-4495, USA
http://www.apnet.com

Academic Press
24-28 Oval Road, London NW1 7DX, UK
http://www.hbuk.co.uk/ap/

Library of Congress Catalog Card Number: 98-88229

International Standard Book Number: 0-12-505255-3

PRINTED IN CANADA
98 99 00 01 02 03 FR 9 8 7 6 5 4 3 2 1

Contents

Contributors

Numbers in parentheses indicate the pages on which the authors' contributions begin.

Ruben Adler (463), The Johns Hopkins University School of Medicine, Baltimore, Maryland 21287

Teri Belecky-Adams (463), The Johns Hopkins University School of Medicine, Baltimore, Maryland 21287

Shirley T. Bissen (197), Department of Biology, University of Missouri, St. Louis, St. Louis, Missouri 63121

Bruce Bowerman (97), Institute of Molecular Biology, University of Oregon, Eugene, Oregon 97403

Julie Broadus (273), Howard Hughes Medical Institute, Department of Human Genetics, University of Utah, Salt Lake City, Utah 84112

Margaret Buckingham (617), CNRS URA 1947, Department of Molecular Biology, Institut Pasteur, 75724 Paris, France

R. Andrew Cameron (11), Division of Biology, California Institute of Technology, Pasadena, California 91125

José A. Campos-Ortega (227), Institut für Entwicklungsbiologie, Universität zu Köln, 50931 Köln, Germany

Anne Camus (491), Embryology Unit, Children's Medical Research Institute, Wentworthville, New South Wales 2145, Australia

Richard W. Carthew (235), Department of Biological Sciences, University of Pittsburgh, Pittsburgh, Pennsylvania 15260

James A. Coffman (11), Stowers Institute for Medical Research and Division of Biology, California Institute of Technology, Pasadena, California 91125

Bruce P. Davidson (491), Embryology Unit, Children's Medical Research Institute, Wentworthville, New South Wales 2145, Australia

Igor B. Dawid (289), Laboratory of Molecular Genetics, National Institute of Child Health and Human Development, National Institutes of Health, Bethesda, Maryland 20892

Wolfgang Driever (371), Department of Developmental Biology, University of Freiburg, D-79104 Freiburg, Germany

Judith S. Eisen (415), Institute of Neuroscience, University of Oregon, Eugene, Oregon 97403

Marc Ekker (399), Loeb Institute for Medical Research, Ottawa Civic Hospital, Anatomy and Neurobiology, University of Ottawa, Ontario, Canada K1N 6N5

Ronald E. Ellis (119), Department of Biology, University of Michigan, Ann Arbor, Ann Arbor, Michigan 48109

Scott W. Emmons (139), Department of Molecular Genetics, Albert Einstein College of Medicine, Bronx, New York 10461

Scott E. Fraser (383), Division of Biology, Biological Imaging Center, Beckman Institute, California Institute of Technology, Pasadena, California 91125

Andreas Fritz (399), Institute of Neuroscience, University of Oregon, Eugene, Oregon 97403

Maureen Gannon (583), Department of Cell Biology, Vanderbilt University Medical Center, Nashville, Tennessee 37232

Alex Hajnal (157), Department of Pathology, Division of Cancer Research, University of Zürich, CH-8091 Zürich, Switzerland

William A. Harris (353), Department of Anatomy, University of Cambridge, Cambridge CB2 3DY, United Kingdom

Danielle Hickford (505), Department of Zoology, La Trobe University, Bundoora, Victoria 3083, Australia

Robert K. Ho (399), Department of Molecular Biology, Princeton University, Princeton, New Jersey 08544

Françoise Z. Huang (185), Department of Molecular and Cell Biology, University of California, Berkeley, Berkeley, California 94720

Deborah E. Isaksen (185), Virginia Mason Research Center, Seattle, Washington 98101

Lucille Joly (399), Loeb Institute for Medical Research, Ottawa Civic Hospital, Anatomy and Neurobiology, University of Ottawa, Ontario, Canada K1N 6N5

John P. Kanki (399), Department of Molecular Biology, Princeton University, Princeton, New Jersey 08544

Rachele C. Kauffmann (235), Department of Biological Sciences, University of Pittsburgh, Pittsburgh, Pennsylvania 15260

Daniel S. Kessler (323), Department of Cell and Developmental Biology, University of Pennsylvania School of Medicine, Philadelphia, Pennsylvania 19104

Susan Kladny (235), Department of Biological Sciences, University of Pittsburgh, Pittsburgh, Pennsylvania 15260

William H. Klein (25), Department of Biochemistry and Molecular Biology, M. D. Anderson Cancer Center, University of Texas, Houston, Texas 77030

Paul A. Krieg (341), Department of Zoology, Institute for Cellular and Molecular Biology, University of Texas, Austin, Austin, Texas 78712

Songhui Li (235), Department of Biological Sciences, University of Pittsburgh, Pittsburgh, Pennsylvania 15260

Eric C. Liao (569), Division of Hematology/Oncology, Children's Hospital, Department of Pediatrics, and Howard Hughes Medical Institute, Harvard Medical School, Boston, Massachusetts 02215, and Division of Health Sciences and Technology, Massachusetts Institute of Technology, Cambridge, Massachusetts 02139

Catriona Y. Logan (41), Stanford University, Stanford, California 94305

David R. McClay (41), Duke University, Durham, North Carolina, 27708

Takashi Mikawa (451), Department of Cell Biology, Cornell University Medical College, New York, New York 10021

Sally A. Moody (297, 551), Department of Anatomy and Cell Biology, The George Washington University Medical Center, Washington D.C. 20037

Kathryn B. Moore (297), Department of Anatomy and Cell Biology, The George Washington University Medical Center, Washington D.C. 20037

Maria I. Morasso (553), Section on Vertebrate Development, Laboratory of Molecular Genetics, National Institute of Child Health and Human Development, National Institutes of Health, Bethesda, Maryland 20892

Craig S. Newman (341), Department of Zoology, Institute for Cellular and Molecular Biology, University of Texas, Austin, Austin, Texas 78712

Roger A. Pedersen (477), Reproductive Genetics Unit, Department of Obstetrics, Gynecology, and Reproductive Sciences, University of California, San Francisco, San Francisco, California 94143

Muriel Perron (353), Department of Anatomy, University of Cambridge, Cambridge CB2 3DY, United Kingdom

Victoria E. Prince (399), Department of Molecular Biology, Princeton University, Princeton, New Jersey 08544

David W. Raible (415), Department of Biological Structure, University of Washington, Seattle, Washington 98195

Thomas D. Sargent (553), Section on Vertebrate Development, Laboratory of Molecular Genetics, National Institute of Child Health and Human Development, National Institutes of Health, Bethesda, Maryland 20892

Noriyuki Satoh (59), Department of Zoology, Graduate School of Science, Kyoto University, Kyoto 606-8502, Japan

Gary C. Schoenwolf (429), Department of Neurobiology and Anatomy, University of Utah School of Medicine, Salt Lake City, Utah 84132

Lynne Selwood (505), Department of Zoology, La Trobe University, Bundoora, Victoria 3083, Australia

Marty Shankland (207), Department of Zoology and Institute of Cellular and Molecular Biology, University of Texas, Austin, Austin, Texas 78712

Esther Siegfried (249), Departments of Biology and Biochemistry and Molecular Biology, Pennsylvania State University, University Park, Pennsylvania 16802

Eric P. Spana (273), Department of Genetics, Harvard Medical School, Boston, Massachusetts 02115

Gunther S. Stent (173), Department of Molecular and Cell Biology, University of California, Berkeley, Berkeley, California 94720

Claudio D. Stern (437), Department of Genetics and Development, Columbia University, New York, New York 10032

Andrea Streit (437), Department of Genetics and Development, Columbia University, New York, New York 10032

Steven A. Sullivan (297), Department of Anatomy and Cell Biology, The George Washington University Medical Center, Washington D.C. 20037

Shahragim Tajbakhsh (617), CNRS URA 1947, Department of Molecular Biology, Institut Pasteur, 75724 Paris, France

Patrick P. L. Tam (491), Embryology Unit, Children's Medical Research Institute, Wentworthville, New South Wales 2145, Australia

Christopher A. Walsh (529), Beth Israel Deaconess Medical Center, Harvard Medical School, Boston, Massachusetts 02115

Marcus L. Ware (529), Harvard Medical School, Boston, Massachusetts 02115

David A. Weisblat (185), Department of Molecular and

Cell Biology, University of California, Berkeley, Berkeley, California 94720

Athula H. Wikramanayake (25), Department of Biochemistry and Molecular Biology, M. D. Anderson Cancer Center, University of Texas, Houston, Texas 77030

William B. Wood (77), Department of Molecular, Cellular, and Developmental Biology, University of Colorado, Boulder, Colorado 80309

Gregory A. Wray (3), Department of Ecology and Evolution, State University of New York, Stony Brook, New York 11794

Christopher V. E. Wright (583), Department of Cell Biology, Vanderbilt University Medical Center, Nashville, Tennessee 37232

Magdalena Zernicka-Goetz (521), Wellcome/CRC Institute, University of Cambridge, Cambridge CB2 1QR, United Kingdom

Jianjun Zhang (235), Department of Biological Sciences, University of Pittsburgh, Pittsburgh, Pennsylvania 15260

Leonard I. Zon (569), Division of Hematology/Oncology, Children's Hospital, Department of Pediatrics, and Howard Hughes Medical Institute, Harvard Medical School, Boston, Massachusetts 02215

Foreword: A Historical Perspective on the Study of Cell Lineages and Fate Determination

The study of how embryonic cells attain their final differentiated state has a very long history. We tend to accept the modern concepts of embryological development as obvious and therefore easily derived, but in fact the route of discovery has been long and difficult. The most obvious impediment is the fact that most developmental events occur on a microscopic scale. Thus, they cannot be observed without the aid of specialized devices and techniques that make the events visible, such as microscopes, histological techniques, cell-labeling techniques, and biochemical and molecular assays. One of the purposes of this book is to illustrate that major advances in our understanding of developmental mechanisms have required technological advances that uncover new ways of observing the developmental processes. Another purpose is to illustrate that although our hypotheses are driven by new observations, it has been very important to choose, by design or serendipity, the appropriate animal in which to experimentally test the question. Until recently this has been an impediment to progress given the different sizes, availability, and accessibility of various embryos. Finally, major breakthroughs historically have occurred when scientific disciplines have communicated and influenced each other's ways of thinking about or approaching a problem. Of course, an important purpose of this book is to communicate exciting new discoveries by researchers focused on the development of different animal models and to illustrate in several cases the potential application of recent discoveries in basic research to clinical problems in birth defects.

To accomplish these goals, this book presents eight different animal models whose experimental advantages are discussed in the introduction to each section. Chapters describing research on each animal were chosen for their interdisciplinary use of experimental approaches (biochemistry, cell and explant culture, genetics, lineage tracing, molecular biology, and morphology) to answer key questions about the mechanisms that found cell lineages and determine cell

fates. The final section on tissue determination illustrates how the fields of genetics, developmental biology, and human congenital disease are being cross-fertilized to construct a unifying picture of how distinct tissues arise and provides insights into the developmental causes of malformations.

Below I discuss how the issues of technology, animal models, and cross communication have influenced the progress of embryological research. This is not intended as an exhaustive review, but as a brief historical perspective that touches on some highlights. My apologies to the many very important contributors to our esteemed field that I fail to mention; excellent and thorough accounts can be found in the reference section of this foreword (Adelmann, 1966; Hamburger, 1988; Horder *et al.*, 1986; Jacobson, 1991; Maienschein, 1986; Willier and Oppenheimer, 1964).

What Is Development?

As early as the 4th century BC the Greek scholars debated whether the embryo arises by intrinsic or extrinsic forces. Aristotle made the critical observations that infant animals always grow up to resemble their parents rather than another animal type and that the offspring of parents with amputated limbs are born with completely formed limbs. These observations demonstrated that important information is passed down from parents to create the next generation and that this information does not transmit physical features acquired by later events in life. In a way, this was the first discussion of the genetic inheritance of traits and presaged what we now consider the "old" nature-versus-nurture issue. Aristotle also recognized the advantage of using avians to study embryonic development. He incubated hen's eggs and cracked them open at timed intervals to directly observe development. This led him to formulate the concept that the embryo arises from simple rudiments, rather than by the growth of a tiny, but completely formed, individual. Nearly 2000 years later,

William Harvey, best known for discovering the circulation of the blood, coined the phrase "epigenesis" to describe the process by which the embryo originates from simpler rudiments that gradually increase in complexity to ultimately give rise to the adult form. Nearly 200 years after that, Wilhelm His proved the existence of these rudiments (anlagen) by inventing histological techniques that allowed him to identify the cellular composition of simple embryonic tissues that become more complex with time (Jacobson, 1991).

After Aristotle, the next 2000 years of embryological inquiry were characterized by postulations on the origin of embryos from parental body fluids without direct testing or observation. Without the aid of the magnifying lens it was not possible to understand the cellular nature of the gametes within these fluids, and because mammalian development takes place in the body cavity of the mother, observation of development was difficult. It was Galen, during the 2nd century AD, who correctly recognized that the testes and ovaries are involved in reproduction and that the mammalian fetus develops in the uterus. But the origin and function of the gametes and their contributions to the various body parts could not be ascertained until the cells involved could be seen with the aid of the magnifying lens. Fabricius of Padua, also studying hen's eggs, discovered that eggs originate in the ovary, and he coined the term "ovarium." His student, William Harvey, although a firm believer in the ovum, denied that male semen was involved in embryo formation. The results of an experiment carried out on King Charles I's deer population convinced him that semen merely stimulates the formation of the embryo in the female uterus. Deer were allowed to mate, and the females were sacrificed at regular intervals; nothing resembling an embryo could be seen in the uterus until a month after mating. Shortly later, however, De Graaf performed the same experiment on rabbits; because of their more rapid development and larger ova, he found embryos in the uterine tubes three to four days after mating. He also discovered the mammalian follicle of the ovary, which since has born his name (the Graafian follicle), and correctly reported that the ovum is released from the ovary and travels to the uterus through the uterine tubes. Thus, our understanding of the role of the female gamete was advanced greatly by the use of appropriate animal models. Nearly 200 years later Von Baer showed that the ovum was only a part of the Graafian follicle and observed that the mammalian ovum is a single cell. He had the advantage of much improved microscopes, the development of histological sectioning and staining techniques, and the conceptual framework of the "cell theory" of Schwann and Schleiden.

In all animal species the ovum is a storage unit of the RNA and proteins used by the zygote during the first several hours after fertilization. Thus, it is much larger than the sperm and, not surprisingly, was the first gamete to be "discovered." Although the ancients were well aware that semen had some role in reproduction, they hotly debated whether the embryo derived from semen or was activated by it to grow. It was not until the manufacture of the microscope in the early 1700s by Van Leeuwenhoek that spermatozoa were discovered. For the next century it was argued whether the sperm alone gave rise to the embryo (the "spermatists" of which Leeuwenhoek was one) or whether the ovum alone gave rise to the embryo (the "ovulists"). It was during this debate that the idea of a miniature, preformed man curled up in the head of a sperm (a homunculus) was popularized. A contemporary of De Graaf, Malpighi, used the new microscopic lens to identify the fine structure of the early chick embryo, and his descriptions of the organ rudiments are quite accurate, even by today's standards (Adelmann, 1966). However, Malpighi never observed a completely undeveloped embryo because chick eggs are in the process of gastrulation when they are laid, and he therefore supported the preformationist theory of development.

Thus, as insightful as accurate descriptions of morphology can be, a cellular understanding of fertilization did not come for another century. The important discovery that both gametes are necessary for development to proceed was made possible by using the large frog egg and a new technology to produce high-quality ground lenses. Spallanzani, in the late 1760s, and later Prevost and Dumas, in 1824, experimentally demonstrated this by adding sperm versus filtered seminal fluid to frog eggs. The entry of the sperm into the egg was not described until 1843 by Martin Barry, and the fusion of the female and male pronuclei until 1875 by Oscar Hertwig.

It is useful to think about the historical steps in the process of understanding the most fundamental step in development, fertilization, because it is easy to see that there were certain technological and philosophical tenets that had to be in place for the important discoveries to be made. First, the technology had to be in place for accurate observations to be made. The manufacture of the magnifying lens and its subsequent improvements were critical to visualizing events that happened at the cellular level. Histological preparations did not come into existence until the mid-1800s, making subcellular details appreciable. A favorable philosophical climate also was necessary. It was not until the Renaissance period that European scientists appreciated the need for direct observation and experimentation; reliance on Greek observations made in the previous millennium was no longer sufficient. Finally, more widespread communication of ideas from other fields, e.g., from the cell theory, made it possible to put together a convincing picture of what occurs when sperm meets egg. Without the proper theoretical framework, what sense could be made of the observations? It also is important to appreciate that use of the optimal animal model provided breakthrough observations. In this case, study of some of the largest eggs available in the animal kingdom (avian, frog, and rabbit) was the key to advancing scientific knowledge.

What Drives Developmental Processes?

During the 1800s, as technology, theory, and refined observations were coming together to allow accurate descriptions of the beginnings of embryonic development, a revolution in zoological thought was fomenting. Charles Darwin and others challenged the long-held idea that all extant species originated as is by divine creation. In its stead they presented the idea that species originated by evolution in response to pressures found within the physical world. In the mid-1800s Von Baer's meticulous microscopic descriptions of the early embryonic forms of a large number of vertebrates demonstrated that they shared many early features, as though they progressed through common themes during development. Perhaps we can attribute to him the first recognition that study of one animal model can reveal developmental mechanisms fundamental to most animals. Ernst Haeckel strenuously applied basic concepts of evolutionary theory to explain the developmental commonalities described by Von Baer. He hypothesized that the developmental events of a given animal were driven by its biological (evolutionary) history. He postulated that somewhere in the fossil record there are animal forms whose adult body plan closely resembles each stage of vertebrate development. Although Haeckel's hypotheses were subsequently discredited, mostly because they are not testable, he generated much enthusiasm for studying the forces that drive the development of embryos, and he trained many of the leading scientists who truly began the modern era of developmental biological inquiry.

The father of embryological experimentation was Wilhelm Roux. A student of Haeckel's, he advocated experimental intervention to unlock the secrets of the driving force of development. Influenced by Newtonian physics, he founded the discipline of "developmental mechanics" and established a new scientific journal to highlight studies using these methodologies. First published under the name of *Archiv für Entwicklungsmechanik der Organismen* in 1894–1895, it is still published (until recently as *Roux's Archives of Developmental Biology*) as *Development, Genes and Evolution*. He introduced the concept of studying embryological problems by breaking them down into simple processes that could be analyzed in physical/chemical ways. One of his most famous experiments, published in 1888

(Willier and Oppenheimer, 1964), was to determine whether embryonic cells differentiate according to an intrinsic program or whether they rely on interactions with other cells in the embryo. Using a hot needle, he destroyed one blastomere of the two-cell frog embryo. The remaining cell formed only half an embryo, leading to the conclusion that each cell contains its own, determined developmental program. Although shown later to have an erroneous conclusion, this experiment led the way in establishing experimental intervention as an important scientific tool for developmental biology. Roux's influence is still with us today, as evidenced in all of the chapters presented in this book.

A follower of Roux's, Hans Driesch, applied these same techniques to echinoderm eggs, which could be more easily separated mechanically and chemically, and demonstrated that blastomeres from even later stages could regulate to each form an entire, albeit smaller, embryo. These experiments are particularly relevant to the work presented in this book, because they demonstrated that the fate of a cell depends on its environment—what a blastomere makes when part of a whole embryo is only a subset of its fate potential. Many of the ideas we use today to describe cell fate determination were outlined in his 1894 book *Analytische Theorie der Organischen Entwicklung*. These ideas include the importance of position, induction by cell contact and by chemicals, polarity of the egg by cytoplasmic constituents, and the action of the nucleus being through "ferments," or what we call enzymes. However, Driesch, although a champion of the experimental method, could never reconcile the experiments in separating blastomeres with Roux's favorite analogy of the embryo as a machine. He could not imagine a machine that could reconstruct two complete miniatures if divided into parts. He eventually left science for philosophy because the mechanical, Newtonian world provided only examples that could not explain the complexities of interactions that occur during embryogenesis. I believe that the chapters in this book demonstrate that Driesch was right, not in abandoning experimental embryology but in recognizing that development proceeds in a nonlinear manner. It is a complex matrix of temporal and spatial interactions that are beginning to be unraveled by the use of multidisciplinary techniques and approaches and several animal models, as illustrated in this book.

Cell Lineage Analyses

The decade before the start of the 20th century marked the beginnings of cell lineage analyses. Evolutionary theory postulated that cells must contain "particles" responsible for the transmission of physical traits from one generation to the next over the millennia. Could evidence for the heredity of traits be found within the zygote by new cytological means? The Naples Marine Station and the Marine Biological Laboratory at Woods Hole were the major sites for seminal discoveries of how single-cell zygotes are transformed into adult body plans. The study of cell lineage began with Whitman's study (1878) of the clepsine worm (see the leech section), in which he traced the progeny of individual blastomeres to the principal organs of the body (Jacobson, 1991). Soon after, Wilson, Conklin, Lillie, Morgan, and others made similar observations on a number of marine worms and echinoderms (see the section on sea urchins and ascidians). These animals were ideal because they are transparent, cleave in reproducible patterns, and contain small numbers of cells and because the partitioning of cytoplasmic inclusions such as pigment granules can be observed as development progresses. These researchers demonstrated that every individual cell makes a unique and definable contribution to the final body form, that some cells are set aside to create certain tissues or organs, and that cytoplasmic inclusions might determine cell fate. They also demonstrated that in order to test mechanisms that establish cell fate, one needs to know the complete and detailed fate map of the normal embryo and the lineages by which it arises. These predecessors profoundly influenced those who in recent years created the very detailed fate maps for the modern repertoire of experimental animals presented in this book. However, although very accurate lineage maps could be constructed 100 years ago simply by using microscopy, cytology, and keen observation, the development of several new techniques was required before this approach could be applied to many of the animal models presented in this book. These techniques include differential interference contrast microscopy (*Caenorhabditis elegans* and mammals), genetic markers (fly), intracellular labeling techniques (leech, frog, fish, and mammals), and replication-deficient tagged retroviruses (chick and mammals). These only came into use about 25 years ago, but now allow lineage and fate analyses at the single-cell level in almost any animal.

Genetic Analyses

Using the new cytological techniques, many researchers during the mid-1800s described the nucleus and chromosomes in cells, but the functions of these were unknown. Theodor Boveri, using marine worms and echinoderms, made significant contributions to this field with his detailed descriptions of the role of centrosomes, spindles, asters, and the nucleus (Willier and Oppenheimer, 1964). He demonstrated that each parent furnishes equivalent groups of chromosomes and that chromatin is diminished in the germ line under the influence of part of the cytoplasm. He demonstrated by cytological means that each chromosome is an independently inherited entity and that an improper number or

combination of chromosomes causes developmental aberrations. Although this work provided support for the idea that each chromosome carries different information, as predicted by Mendel's experiments and evolutionary theory, many embryologists rejected the idea of heritable particles because it sounded too "preformationist." Even T. H. Morgan, who founded the field of *Drosophila* genetics with the discovery of the linkage of eye color to sex, at first rejected the theory of Mendelian inheritance as an affront to the fundamental concept of epigenesis. However, rapid breakthroughs in genetics, as a result of convenient mutagenesis and rapid generation times of the fruit fly, later led Morgan and collaborators to embrace the concept of discrete intracellular particles that direct heritable traits, which they called "genes" (Morgan, 1926). Interestingly, these advances in "heredity" divorced Morgan from his embryological upbringing and investigation of the epigenetic regulation of developmental processes. This resulted in a 50- to 60-year division between the fields of genetics and developmental biology (see Allen in Horder *et al.,* 1986). Twelve years ago, G. E. Allen predicted that the recent advances in molecular genetics would soon bring these two cousin fields together again. The presentations in this book demonstrate that this prediction is being realized.

Tissue Inductions

At the beginning of the 20th century, experimental embryology was one of the most exciting fields to study (Hamburger, 1988). Asking cause-and-effect questions about living embryos was innovative and opened biology to a whole new field of ideas and experimental approaches. But there were limitations. The animals that could be studied had to be large, hardy, easy to obtain, and relatively easy to manipulate. Questions were limited to those concerned with tissue interactions and autonomy of fate and were answered by deletion of parts, addition of parts, transplantation of parts, and explant culture. Ross Harrison is credited with developing animal tissue culture to explore the cellular nature of neuritic outgrowth, but when he was nominated for this achievement the Nobel Prize committee found it of very limited importance. Of course, this was a major technical advance and is now a fundamental experimental technique in nearly all fields of biological inquiry. The burgeoning field of genetics soon split away from embryology and had little impact on developmental research until recent decades (see above section and chapters in this book). A wealth of information, including the Nobel Prize-winning discovery of neural induction and the "organizer" (Spemann, 1967; Hamburger, 1988), was gleaned about tissue and cellular interactions during the first 30 years of the 20th century. This led many biochemists (notably Needham, Waddington, and J. Brachet) to enthusiastically try to discover the chemical nature of these "inductions." These efforts established the field of chemical embryology (Brachet, 1950; Horder *et al.*, 1986), but although the journals of the 1930–1950s are filled with descriptions of embryonic metabolism, this period was one of few advances in the subcellular mechanisms of the inductive interactions that had been described by experimental embryologists. The methodologies at hand were not up to the challenge. Only the most abundant molecules could be detected and the molecular nature of the known compounds often was not understood. For example, although J. Brachet meticulously described temporal and spatial changes in the distribution of an intriguingly new compound, ribonucleic acid, its function in protein synthesis was completely unknown.

And Today?

"The main task of the future will be to find a common ground between experimental embryology and chemical embryology since it is only the fusion of these two disciplines which will permit a penetration into the mysteries of development."

(J. Brachet, p. 470)

We are again in an exciting, innovative period of embryological research, in which our basic understanding of biological and evolutionary processes is being expanded and translated into an understanding of human disease and congenital malformations. To what can we attribute this revitalization of developmental biology, and how have the recent, major advances in our understanding of cell lineage and fate determination that are presented in this book been achieved? Certainly, the advances in our understanding of basic biochemical, cell biological, and molecular processes over the past 50 years have created the common ground Brachet envisioned. What are some of the recent innovations that have specifically impacted developmental biology?

(1) New methodologies for cell lineage tracing were developed in the 1970s. Just as Spemann and Mangold could not have proven the inductive power of the "organizer" without transplanting darkly pigmented tissue from one species into a lightly pigmented host, modern experimenters cannot discern cell-by-cell responses to molecular manipulations without being able to identify which cells carry the mutation or the ectopic gene product. Cell lineage techniques were achieved only a very short while ago in the leech (Weisblat *et al.*, 1978), frog (Jacobson and Hirose, 1978), zebrafish (Kimmel and Law, 1985), and mouse (Pedersen *et al.*, 1986).

(2) Screening the genetic animal models for developmental abnormalities and subsequent molecular cloning of the responsible genes identified important

regulatory molecules present in such low abundance that they otherwise would not have been identified.

(3) Homologous cloning techniques have had a major impact on our understanding of the universality of developmental principles at a molecular level. As genes are being identified in those animals best suited to genetic analyses (flies and worms), homologous relatives are being cloned in animals in which experimental embryological manipulations are easy (frogs and chicks). These techniques are currently being exploited to understand the relationship between evolution and development and someday may address Haeckel's idea of the interrelationship between these two processes.

(4) The development of whole embryo and *in vitro* expression assays to study gene function during development, especially in the frog and the chick, has elucidated the molecular pathways of cell–cell interactions.

(5) Advances and improvements in microscopy have allowed cellular and molecular interactions to be studied in the embryo as developmental events are happening.

(6) Many other methods have become more and more "micro" (microinjection, microsurgery, microextraction of nucleic acids, microgel electrophoresis, PCR, etc.), allowing smaller embryos to be studied in the same way as the traditional larger ones and making analyses proceed more quickly and with less starting material.

(7) Transgenic approaches for the deletion, overexpression, and misexpression of genes have profoundly influenced our understanding of gene function during development.

(8) The development of cell markers, both of proteins and of nucleic acids, has significantly enhanced the detail and molecular level of our observations after experimental manipulations.

(9) The increased communication and overlap of techniques between specialized fields (e.g., developmental biology, neuroscience, genetics, biochemistry, molecular biology, and immunology) have resulted in a broader, integrative approach to the questions we ask and the answers we receive.

(10) The demonstration that the interesting problems of basic embryology and developmental biology can have a direct and profound impact on clinical problems has provided an impetus for bidirectional translation between clinical and basic sciences.

As I stated at the beginning of this foreword, advances in our understanding of development rely on creating new techniques, using the appropriate animal models to ask the right questions, and doing the work in the right theoretical framework. The authors of the chapters presented in this book have clearly benefited from these recent advances and have uncovered significant insights into the mechanisms of cell fate determination by combining classical approaches with modern molecular, biochemical, genetic, and morphological approaches. When Viktor Hamburger published a history of the organizer experiments in 1988, I was disappointed to read in his preface that with the success of molecular biological approaches in developmental biology, "old-style experimental embryology was doomed; it attained the status of a 'classical' science, an epithet which gave it a venerable aura to some, but had a disparaging flavor for others." I was saddened because I knew what he had written was true. The new students were enthralled with cloning and misexpressing genes and impatient with the thought of having to learn the microsurgical approaches and morphological analyses that had served the field so well. But, as the following chapters will show, this attitude was only temporary. Although molecular genetics is an integral and important part of the developmental biologist's technical repertoire, we have reached the point where combining these techniques with the classical approaches provides the most significant breakthroughs in our understanding of the developmental processes underlying cell fate determination. From ascidians to zebrafish, you will find examples in this book of the modern and the classical approaches performed side by side to reveal how cell lineages and final phenotypic choices are made during embryogenesis. And breakthroughs made with one animal powerfully influence the thought on and approaches taken with others. We no longer do genetics in one animal and transplantations in another nor consider the mouse the only reasonable model for human disease and congenital malformations. It is an exciting time to be an embryologist, experimental, molecular, and morphological. You will read this excitement in what follows, 41 chapters of new avenues of research in eight of the most popular animal models used in developmental biology.

Acknowledgments

I thank Steven Klein for many helpful comments and sharing his knowledge about the history of embryology. I am grateful to all the members of my laboratory for being understanding when my attention was diverted to this book. Finally, I thank Craig Panner for his advice and patience throughout this project.

SALLY A. MOODY

References

Adelmann, H. B. (1966). Marcello Malpighi and the Evolution of Embryology. Cornell University Press, Ithaca, NY.

Brachet, J. (1950). Chemical Embryology. Interscience Publishers, New York, NY.

Hamburger, V. (1988). The Heritage of Experimental Embryology: Hans Spemann and the Organizer. Oxford University Press, New York, NY.

Horder, T. J., Witkowski, J. A., and Wylie, C. C. (1986). A History of Embryology. Cambridge University Press, New York, NY.

Jacobson, M. (1991). Developmental Neurobiology, 3rd ed. Chapter 1. Plenum Press, New York, NY.

Jacobson, M., and Hirose, G. (1978). Origin of the retina from both sides of the embryonic brain: A contribution to the problem of crossing at the optic chiasma. *Science* **202**, 637–639.

Kimmel, C. B., and Law, R. D. (1985). Cell lineage of zebrafish blastomeres. I. Cleavage pattern and cytoplasmic bridges between cells. *Dev. Biol.* **108**, 78–85.

Maienschein, J. (1986). Defining Biology: Lectures from the 1890s. Harvard University Press, Cambridge, MA.

Morgan, T. H. (1926). The Theory of the Gene. Yale University Press, New Haven, CT.

Pedersen, R. A., Wu, K., and Balakier, H. (1986). Origin of the inner cell mass in mouse embryos: Cell lineage analysis by microinjection. *Dev. Biol.* **11**, 581–595.

Spemann, H. (1967). Embryonic Development and Induction. Hafner Publishing Co., New York, NY.

Weisblat, D. A., Sawyer, R. T., and Stent, G. S. (1978). Cell lineage analysis by intracellular injection of a tracer enzyme. *Science* **202**, 1295–1298.

Willier, B. H., and Oppenheimer, J. M. (1964). Foundations of Experimental Embryology. Prentice-Hall, Englewood Cliffs, NJ.

I

Sea Urchins and Ascidians

1

Introduction to Sea Urchins

GREGORY A. WRAY
Department of Ecology and Evolution
State University of New York
Stony Brook, New York 11794

I. Introduction

Some of the earliest experimental studies of fate specification were carried out by Driesch and Fiedler in the 1890s, using sea urchin embryos (Driesch, 1892). These pioneering studies were followed by extensive experimental analyses of axial and cell fate specification by Boveri, Child, von Ubisch, Morgan, Hörstadius, and others (for a historical perspective, see Ernst, 1997). Both sea urchin and ascidian embryos provide important experimental advantages for studying animal development today, particularly for experimental studies of cell fate specification and morphogenesis. Chapters 2–4 review mechanisms of fate specification for a variety of cell types in sea urchin embryos; Chapter 5 reviews these issues in ascidian embryos. Together, these chapters illustrate the power of an investigative approach that combines experimental embryology, modern molecular methods, and the ability to visualize and record events in live embryos. In this chapter, I will focus on sea urchin development. Ascidian development is reviewed in Chapter 5 by Satoh.

II. Utility for Studies of Cell Lineage and Fate Specification

Sea urchin embryos offer several important experimental benefits to investigators interested in developmental mechanisms. These features are complementary to benefits of the genetically oriented model species that are commonly used for developmental studies (see following sections on *Caenorhabditis elegans*, fly, and mouse).

First, sea urchin embryos can be cultured easily, inexpensively, and in large quantities. In particular, it is simple to rear large numbers of synchronously developing embryos and larvae. This makes possible biochemical, molecular, and cellular techniques that would be difficult or costly in many other model systems. The rapid development of sea urchins to larvae containing fully differentiated cells means that the consequences of experimental manipulations can be assayed within a day or so.

Second, sea urchin embryos are physically robust and easily manipulated. Most commonly studied animal embryos permit simple experimental manipulations such as single-cell ablation and microinjection. Sea urchin embryos permit a much wider range of informative experimental procedures, including transplantation of single or multiple blastomeres, assembly of "artificial embryos" from virtually any combination of blastomeres, isolation of particular cell types or germ layers en masse, production of interspecies chimeras, and bisections of eggs, embryos, and larvae (e.g., Hörstadius, 1973; Ettensohn, 1990; Ransick and Davidson, 1993; Armstrong and McClay, 1994; McClay and Logan, 1996). Many of these procedures are difficult or impossible to carry out in other animal model systems, with the exception of amphibians. Following even severe experimental manipulations, sea urchin embryos typically continue to develop, allowing one to assay effects on cell movements and differentiation.

Third, the embryos and larvae of sea urchins are transparent and anatomically simple. Together, transparency and rapid development allow the consequences of experimental manipulations to be documented and followed over time in living embryos. The larvae of sea urchins are simple anatomically, initially containing about 2000 cells, comprising relatively few cell types (Okazaki, 1975; Cameron and Davidson, 1991). Molecular markers exist for many cell types, allowing one to assay the differentiated state of individual cells in the absence of overt phenotypic signs (Coffman and Davidson, 1992).

Fourth, echinoderms are relatively close to chordates phylogenetically. Both classic anatomical comparisons (Brusca and Brusca, 1990) and more recent molecular studies (e.g., Turbeville *et al.*, 1994) confirm that echinoderms, hemichordates, and chordates are very closely related. This means that developmental studies of echinoderms will be crucial for deciphering the evolutionary origin of important chordate innovations, such as the notochord and hollow dorsal nerve tube (Holland and Garcia-Fernàndez, 1996; Lacalli, 1996).

The most important limitation of sea urchins as an experimental system is the impracticality of carrying out genetic screens and Mendelian genetics. These limitations are less important than they once seemed, now that it is possible to isolate readily the orthologues of genes originally identified as developmentally important in other taxa. A powerful arsenal of diagnostic and experimental procedures has greatly expanded the range of molecular genetic approaches that are possible using echinoderm embryos, including the ease with which it is possible to visualize reporter gene expression in any cell of a living embryo (e.g., Arnone *et al.*, 1997) and the ability to carry out direct experimental manipulations of gene expression (e.g., Franks *et al.*, 1988; Maxson and Tan, 1994).

III. Overview of Sea Urchin Development

The following brief overview of development in sea urchins (Fig. 1) is oriented towards providing a background for the following three chapters. More comprehensive and detailed reviews of sea urchin development include Okazaki (1975), Ettensohn and Ingersoll (1992), Hardin (1994), and Wray (1997).

A. Geometry of Cleavage Divisions

Cleavage divisions in all echinoderms are highly ordered and stereotypic (reviewed in Cameron and Davidson, 1991; Wray, 1993). All divisions are holoblastic in sea urchin embryos. The first three mitoses are equal and oriented perpendicular to each other: the first two cleavage planes are parallel to the animal–vegetal axis and the third bisects it. The resulting 8-cell embryo is composed of four animal blastomeres whose progeny are ectodermal and four vegetal blastomeres which contribute to all three germ layers (Hörstadius, 1939; Cameron et al., 1987). In most sea urchins, fourth cleavage is longitudinal in the animal tier, and equatorial and unequal in the vegetal tier. The resulting 16-cell embryo is com-

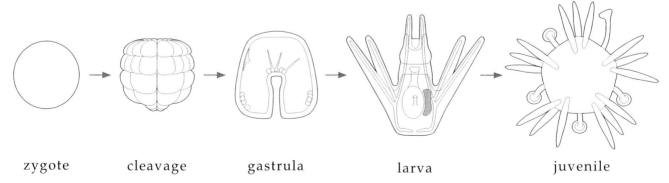

| zygote | cleavage | gastrula | larva | juvenile |

Figure 1 Development of sea urchins. Cleavage divisions in sea urchins are highly stereotypic in orientation and timing, and blastomeres have predictable fates (the 60-cell stage is shown). Differential gene expression and morphogenesis commence soon after cleavage ends. By the end of gastrulation, several fully differentiated cell types are already present. The swimming, feeding larva is called a pluteus (Latin for easel). Larvae are transparent and relatively simple morphologically. An imaginal rudiment arises on the left side of the larvae (shown in gray), and contains most of the cells that will comprise the adult. Metamorphosis results in a juvenile with 5-fold radial symmetry.

posed of eight mesomeres, four macromeres, and four micromeres (Fig. 2). The micromeres play a central role in early inductive interactions, as discussed later. The mesomeres and macromeres divide equally but in perpendicular orientations during fifth and sixth cleavages, resulting in an_1, an_2, veg_1, and veg_2 tiers of eight cells each in the 60-cell embryo (Fig. 2). Blastomeres in the an and veg tiers divide several more times before the end of cleavage. Meanwhile, the micromeres divide unequally again during fifth cleavage, producing four large and four small micromeres (Fig. 2). The large micromeres divide three more times, and the small micromeres twice, before the end of cleavage.

B. Embryonic Cell Fates

The fates of early blastomeres in sea urchin embryos have been traced in detail (Hörstadius, 1939; Pehrson and Cohen, 1986; Cameron *et al.,* 1987, 1989, 1991; Ruffins and Ettensohn, 1996; Cameron and Davidson, 1997; Logan and McClay, 1997). These studies have shown that many, but not all, blastomeres have specified fates in 60-cell embryos. Few cell fates are committed at this point, as evident from the extraordinary plasticity of fates that can be evoked from experimental manipulations of the embryo; but specification is evident from the single cell type (or set of related cell types) to which a particular blastomere consistently gives rise in larvae.

The primary embryonic cell lineages are diagrammed in Fig. 3. All larval cell types are specified polyclonally, typically from at least four founder cells. The animal half of the embryo gives rise exclusively to ectoderm and ectodermally derived neurons. The vegetal half of the embryo produces some ectoderm, but most descendant cells are mesodermal and endodermal. Although the micromere-derived mesodermal cell fates

are specified by fifth cleavage, those derived from veg_1 and veg_2 are not specified until after seventh cleavage. Some variation among individual embryos is apparent in the origin of some larval cell types (Ruffins and Ettensohn, 1996) and in the precise location of the boundary between endoderm and vegetal ectoderm (Logan and McClay, 1997).

C. Midblastula Transition and Activation of the Zygotic Genome

The rapid, ordered divisions of cleavage last for about 9 to 13 cell cycles, depending on the species (Masuda and Sato, 1984; Cameron and Davidson, 1991). An abrupt slowing of cell divisions marks the midblastula transition, after which the position and orientation of cell divisions become unpredictable. Soon after cleavage ends, the first zygotic transcripts begin to accumulate (Davidson, 1986; Kingsley *et al.,* 1993). Even the earliest localized domains of zygotic gene expression are delimited by cell lineages (Chapter 2). Numerous molecular markers for specified and differentiated cell types are available (Coffman and Davidson, 1992) and have been used to interpret the effects of various experimental manipulations.

D. Gastrulation and Early Morphogenesis

The first morphogenetic movements begin a few hours after cleavage ends (Ettensohn and Ingersoll, 1992). The onset of gastrulation, in the broad sense of establishing three germ layers, occurs when the descendants of the large micromeres, the primary mesenchyme cells, ingress from the vegetal plate into the blastocoel. After a brief period of migration, they differentiate into skeletogenic cells, form syncitia, and begin to synthesize the calcitic spicules of the larval skeleton (Fig. 1). Invagina-

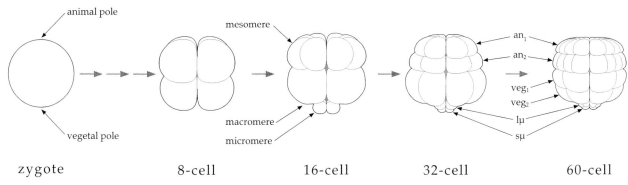

Figure 2 Geometry of early cleavages. Many crucial cell interactions occur between fourth and sixth cleavage in sea urchins. Fourth cleavage (16-cell) establishes three size classes of blastomeres. The vegetal-most cells, the micromeres, are autonomously specified, and initiate a series of inductive interactions with the other blastomere tiers. Cell interactions continue through the next two cell cycles. The resulting 60-cell embryo is a highly ordered mosaic of specified (but not committed) cell fates (see Fig. 3 for details). Additional cell interactions and fate specifications occur beyond sixth cleavage, particularly within the veg_2 descendants (which comprise all of the endoderm and much of the mesoderm) and within the ectoderm (to establish the ciliated band and nervous system of the larva). Abbreviations: $l\mu$, large micromeres; $s\mu$, small micromeres.

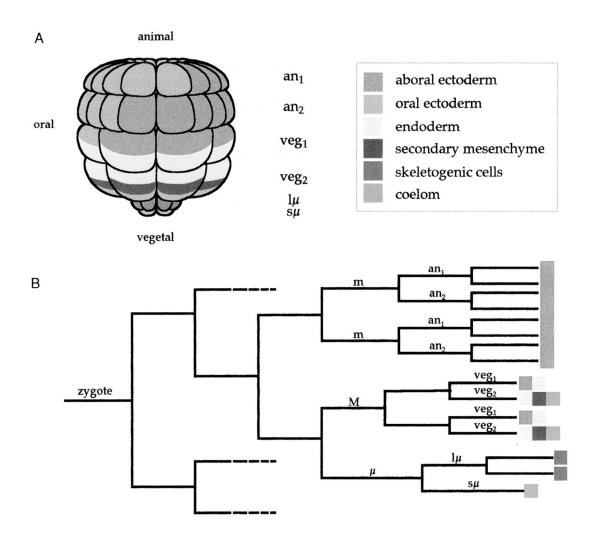

Figure 3 Fate map and embryonic cell lineages. (A) Fate map of the 60-cell embryo of *Strongylocentrotus purpuratus,* showing the left side. Blastomere fates are segregated along the animal–vegetal axis and, to a lesser extent, along the oral–aboral axis. (B) Primary embryonic cell lineages of S. *purpuratus.* For simplicity, only one quadrant of the embryo is shown beyond second cleavage; the other quadrants differ primarily in partitioning of oral versus aboral ectoderm and within the veg_2 lineages. Beyond fourth cleavage, the timing of cell divisions differs between tiers of blastomeres, with smaller cells dividing more slowly. All differentiated larval cell types are specified polyclonally. Coelom arises from two completely independent sources: the small micromeres and some veg_2 descendants. Note that "secondary mesenchyme" is a heterogeneous population of cells that gives rise to myocytes, coelomocytes, and pigmented cells. For additional details, particularly within the complex vegetal plate region, see Cameron and Davidson (1991, 1997), Cameron *et al.* (1991), Ruffins and Ettensohn (1996), and Logan and McClay (1997). Abbreviations: M, macromere; m, mesomere; μ, micromere; $l\mu$, large micromeres; $s\mu$, small micromeres.

tion of the vegetal plate begins a few hours after the primary mesenchyme cells ingress, and establishes the archenteron. Cell rearrangements and shape changes rapidly elongate the archenteron. Secondary mesenchyme cells ingress from the tip of the archenteron as it elongates, giving rise to pigmented cells, muscle cells, and coelomocytes. Toward the end of gastrulation, the tip of the archenteron flattens and produces a pair of lateral coelomic pouches; these bud off and somewhat later each divides into three pouches. Shortly after the coelomic pouches form, the archenteron contacts the oral ectoderm and the larval mouth forms. Meanwhile, the ciliated band forms as a tract of thickened ectodermal cells with unusually long cilia encircling the oral field.

E. Larval Morphology

The larvae of sea urchins are bilaterally symmetrical and bear no resemblance to the radially symmetrical adults to which they give rise (Fig. 1). Larvae of most species are planktotrophic and must feed for several days or weeks before accumulating sufficient mass to complete metamorphosis. Much of their anatomy is geared towards the efficient capture of phytoplankton (Hart and Strathmann, 1995). The larval skeleton supports "arms," around which runs the ciliated band that is used for locomotion and phytoplankton capture. A network of several dozen neurons, mostly associated with the ciliated band (Bisgrove and Burke, 1987), coordinates feeding and swimming behaviors. Mesenchymally derived pigmented cells are embedded in the ectoderm, and ceolomocytes prowl the blastocoel. A simple tripartite gut and associated muscles comprise the digestive system. The coeloms are quiescent in early larvae, but begin to grow in mid-larvae and eventually contribute substantially to the adult.

F. Preparations for Metamorphosis

Long before overt metamorphosis, populations of ectodermal, coelomic, and mesenchymal cells begin to proliferate on the left side of the larval body. Together, these cells comprise the imaginal adult rudiment, which gives rise to much of the cellular mass of the postmetamorphic juvenile. No cell lineage analyses have traced the fates of marked cells into the adult rudiment or across metamorphosis [except in the case of a species with highly modified development (Wray and Raff, 1990)]. It is clear from anatomical studies that many, but not all, adult structures derive from the rudiment (Bury, 1895). Extensive cell death and remodeling of larval tissues accompany metamorphosis (Cameron and Hinegardner, 1978; Burke, 1982).

IV. Axial and Cell Fate Specification

Extensive experimental evidence demonstrates that many differentiated fates in sea urchin embryos are specified on a cell-by-cell basis in early embryos, that early blastomeres have consistent fates in larvae, and that most cell fates are specified conditionally (reviewed in Hörstadius, 1973; Wilt, 1987; Davidson, 1989; see also Chapter 2). This overall strategy of cell fate specification is phylogenetically widespread within the Metazoa (e.g., present in molluscs, ascidians, and nematodes), and is clearly ancestral to the atypical strategies used by insect and chordate embryos (Davidson, 1990).

A. Embryonic Axes

The first experimental demonstration of a consistent spatial relationship between the polar bodies, cleavage planes, and body axes in any animal came from work on sea urchins done by Boveri in 1901. It is now known that the animal–vegetal axis in sea urchins is both specified and committed prior to fertilization. Equatorial egg bisections demonstrate that a maternally synthesized component is sequestered in the vegetal half of the egg which is essential for the correct development of posterior structures in larvae (Hörstadius, 1973; Maruyama *et al.*, 1985). In contrast, the dorsoventral axis is not committed until well after fertilization in most sea urchin species (Hörstadius, 1973; McClay and Logan, 1996). Microinjections and surface marking indicate that dorsoventral specification has occurred by the beginning of first cleavage in some species (Cameron *et al.*, 1989), but a few cleavage divisions later in others

(Kominami, 1988; Henry *et al.*, 1992). The mechanism of dorsoventral axis specification is not well understood in sea urchins.

B. Autonomous Specification of the Micromeres

The first blastomeres to be specified in sea urchin embryos are the micromeres; these are also the only cells to be autonomously specified. If micromeres are isolated in tissue culture or transplanted to ectopic regions of the embryo, they will differentiate into their normal skeletogenic fate (Hörstadius, 1973; Okazaki, 1975; Ransick and Davidson, 1993). It is not known whether the coelomic fate of the small micromeres (which are micromere descendants) is likewise autonomously specified, as unambiguous molecular or morphological markers of coelomic cells are lacking. It is likely that a maternally sequestered molecule is responsible for the autonomous specification of micromeres, but its identity remains elusive. Micromeres play a central role in early cell fate specification in sea urchin embryos, initiating a cascade of cell–cell interactions that occurs over the next few cleavage divisions (Wilt, 1987; Davidson, 1989; Cameron and Davidson, 1991). However, if micromeres are removed from an embryo, macromeres produce some progeny with a skeletogenic fate and a more or less normal larva results (Hörstadius, 1939). Even mesomeres, which are the blastomeres farthest from the micromeres in the embryo, can produce skeletogenic cells if isolated and recombined (Henry *et al.*, 1989) or treated with lithium (Livingston and Wilt, 1990).

C. Conditional Specification of Other Cell Lineages

In contrast to micromeres, the fates of mesomeres and macromeres are not committed in the early embryo. Indeed, both are capable of producing cells belonging to all three germ layers under a variety of experimental conditions (e.g., Hörstadius, 1973; Henry *et al.*, 1989), a competence that exceeds their normal fate. Extensive cell–cell interactions occur between the fourth and seventh cleavage divisions, establishing most of the major cell types of the larva. However, the timing of some fate specifications is not well known. This is particularly true of the heterogeneous population of cells comprising the vegetal plate region and of the larval neurons. Conditional specification within sea urchin embryos includes both instructive and repressive cell–cell interactions (Davidson, 1989; Henry *et al.*, 1989; Ransick and Davidson, 1993). Detailed discussions of how cell fates are established within the mesomere- and macromere-derived lineages appear in Chapters 3 (ectodermal) and 4 (mesodermal and endodermal fates), and are not covered here.

D. Evolution of Cell Lineage and Specification Mechanisms

Some of the features of axial and cell fate specification just discussed have changed during the evolutionary history of sea urchins (reviewed in Wray, 1993). Examples include: (1) changes in the relative position of the dorsoventral axis and the first two cleavage planes (Henry *et al.*, 1992); (2) commitment of the dorsoventral axis prior to first cleavage in *Heliocidaris erythrogramma,* a species with nonfeeding larvae (Henry and Raff, 1989); (3) loss of the unequal cleavage divisions that produce the micromeres in several sea urchin species with nonfeeding larvae (Wray, 1996) and variable micromere numbers and sizes in the cidaroid *Eucidaris tribuloides* (Schroeder, 1981); and (4) extensive changes in the timing and order of cell fate specification in *H. erythrogramma* (Wray and Raff, 1990). Some of these evolutionary changes have probably evolved in response to particular life-history modifications (Wray, 1996).

E. Adult Axes and Cell Fates

The mechanisms by which the body axes of the radially symmetrical adults of echinoderms are specified have not been investigated. Where data are available, however, it is clear that both the oral–aboral axis and the radial orientation of the adult body axes have a consistent relationship to the larval body. This suggests that spatial information established within the early embryo is used to specify the axes of the adult body. No experimental studies have examined the fate of larval cells in post-metamorphic juveniles. From anatomical studies (e.g., Bury, 1895; Cameron and Hinegardner, 1978; Burke, 1982), however, it seems likely that some differentiated larval cells are carried over into the adult while others must be specified following metamorphosis.

Acknowledgments

Thanks to Jon Stone for helpful comments.

References

Armstrong, N., and McClay, D. R. (1994). Skeletal pattern is specified autonomously by the primary mesenchyme cells in sea urchin embryos. Dev. Biol. 162, 329–338.

Arnone, M. I., Bogarad, L. D., Collazo, A., Kirchhamer, C. V., Cameron, R. A., Rast, J. P., Gregorians, A., and Davidson, E. H. (1997). Green fluorescent protein in the sea urchin: new experimental approaches to transcriptional regulatory analysis in embryos and larvae. *Development (Cambridge, UK)* **124,** 4649–4659.

Bisgrove, B. W., and Burke, R. D. (1987). Development of the nervous system of the pluteus larva of *Strongylocentrotus droebachiensis. Cell Tissue Res.* **248,** 335–343.

Boveri, T. (1901). Über die Polarität von Ovocyte, Ei und Larve des *Stronylocentrotus lividus. Zool. Jahrb. (Abt. Anat.)* **14,** 630–653.

Brusca, R. C., and Brusca, G. J. (1990). "Invertebrates." Sinauer, Sunderland, Massachusetts.

Burke, R. D. (1982). Echinoid metamorphosis: Retraction and resorption of larval tissues. *In* "International Echinoderms Conference, Tampa Bay" (J. M. Lawrence, ed.), pp. 513–518. A. A. Balkema, Amsterdam.

Bury, H. (1895). The metamorphosis of echinoderms. *Q. J. Microsc. Sci.* **29,** 45–135.

Cameron, R. A., and Davidson, E. H. (1991). Cell type specification during sea urchin development. *Trends Genet.* 7, 212–218.

Cameron, R. A., and Davidson, E. H. (1997). LiCl perturbs ectodermal veg₁ lineage allocations in *Strongylocentrotus purpuratus* embryos. *Dev. Biol.* **187,** 236–239.

Cameron, R. A., and Hinegardner, R. T. (1978). Early events in sea urchin metamorphosis, description and analysis. *J. Morphol.* **157,** 21–32.

Cameron, R. A., Hough-Evans, B., Britten, R. J., and Davidson, E. H. (1987). Lineage and fate of each blastomere of the eight-cell sea urchin embryo. *Genes Dev.* **1,** 75–84.

Cameron, R. A., Fraser, S. E., Britten, R. J., and Davidson, E. H. (1989). The oral–aboral axis of a sea urchin embryo is specified by first cleavage. *Development (Cambridge, UK)* **106,** 641–647.

Cameron, R. A., Fraser, S. E., Britten, R. J., and Davidson, E. H. (1991). Macromere fates during sea urchin development. *Development (Cambridge, UK)* **113,** 1085–1091.

Coffman, J. A., and Davidson, E. H. (1992). Expression of spatially regulated genes in the sea urchin embryo. *Curr. Opin. Genet. Dev.* **2,** 260–268.

Davidson, E. H. (1986). "Gene Activity in Early Development." Academic Press, Orlando, Florida.

Davidson, E. H. (1989). Lineage-specific gene expression and the regulative capacities of the sea urchin embryo: A proposed mechanism. *Development (Cambridge, UK)* **105,** 421–445.

Davidson, E. H. (1990). How embryos work: A comparative view of diverse modes of cell fate specification. *Development (Cambridge, UK)* **108,** 365–389.

Driesch, H. (1892). The potency of the first two cleavage cells in echinoderm development: Experimental production of double and partial formations. *Reprinted in* "Foundations of Experimental Embryology" (B. H. Willier and J. M. Oppenheimer, eds.). Hafner, New York.

Ernst, S. G. (1997). A century of sea urchin development. *Am. Zool.* **37,** 250–259.

Ettensohn, C. A. (1990). Cell interactions in the sea urchin embryo studied by fluorescence photoablation. *Science* **248,** 1115–1118.

Ettensohn, C. A., and Ingersoll, E. P. (1992). Morphogenesis of the sea urchin embryo. *In* "Morphogenesis: An Analysis of the Development of Biological Form" (E. F. Rossomondo and S. Alexander, eds.), pp. 189–262. Dekker, New York.

Franks, R. R., Hough-Evans, B. R., Britten, R. J., and David-son, E. H. (1988). Spatially deranged though temporally correct expression of a *Strongylocentrotus purpuratus* actin gene fusion in transgenic embryos of a different sea urchin family. *Genes Dev.* **2**, 2–12.

Hardin, J. (1994). The sea urchin. *In* "Embryos: Color Atlas of Development" (J. Bard, ed.), pp. 37–54. Mosby-Year Book Europe, London.

Hart, M. W., and Strathmann, R. R. (1995). Mechanisms and rates of suspension feeding. *In* "Larval Ecology of Marine Invetebrates" (L. R. McEdward, ed.), pp. 193–222. CRC Press, Boca Raton, Florida.

Henry, J. J., and Raff, R. A. (1990). Evolutionary change in the process of dorsoventral axis determination in the direct developing sea urchin, *Heliocidaris erythrogramma. Dev. Biol.* **141**, 55–69.

Henry, J. J., Amemiya, S., Wray, G. A., and Raff, R. A. (1989). Early inductive interactions are involved in restricting cell fates of mesomeres in sea urchin embryos. *Dev. Biol.* **136**, 140–153.

Henry, J. J., Klueg, K. M., and Raff, R. A. (1992). Evolutionary dissociation between cleavage, cell lineage and embryonic axes in sea urchin embryos. *Development (Cambridge, UK)* **114**, 931–938.

Holland, P. W. H., and Garcia-Fernàndez, J. (1996). *Hox* genes and chordate evolution. *Dev. Biol.* **173**, 382–395.

Hörstadius, S. (1939). The mechanics of sea urchin development, studied by operative methods. *Biol. Rev.* **14**, 132–179.

Hörstadius, S. (1973). "Experimental Embryology of Echinoderms." Oxford Univ. Press (Claredon), London.

Kingsley, P. D., Angerer, L. M., and Angerer, R. C. (1993). Major temporal and spatial patterns of gene expression during differentiation of the sea urchin embryo. *Dev. Biol.* **155**, 216–234.

Kominami, T. (1988). Determination of dorso–ventral axis in early embryos of the sea urchin, *Hemicentrotus pulcherrimus. Dev. Biol.* **127**, 187–196.

Lacalli, T. C. (1996). Landmarks and subdomains in the larval brain of *Branchiostoma*: vertebrate homologs and invertebrate antecedents. *Isr. J. Zool.* **42**, S131–S146.

Livingston, B. T., and Wilt, F. H. (1990). Range and stability of cell fate determination in isolated sea urchin blastomeres. *Development (Cambridge, UK)* **108**, 403–410.

Logan, C. Y., and McClay, D. R. (1997). The allocation of early blastomeres to the ectoderm and endoderm is variable in the sea urchin embryo. *Development (Cambridge, UK)* **124**, 2213–2223.

McClay D. R., and Logan, C. Y. (1996). Regulative capacity of the archenteron during gastrulation in the sea urchin. *Development (Cambridge, UK)* **122**, 607–616.

Maruyama, Y. K., Nakaseko, Y., and Yagi, S. (1985). Localization of cytoplasmic determinants responsible for primary mesenchyme formation and gastrulation in the unfertilized eggs of the sea urchin *Hemicentrotus pulcherrimus. J. Exp. Zool.* **236**, 155–163.

Masuda, M., and Sato, H. (1984). Asynchronization of cell division is concurrently related with ciliogenesis in sea urchin blastulae. *Dev. Growth Differ.* **26**, 281–294.

Maxson, R., and Tan, H. (1994). Promoter analysis meets pattern formation: Transcriptional regulatory genes in sea urchin embryogenesis. *Curr. Opin. Genet. Dev.* **4**, 678–684.

Okazaki, K. (1975). Spicule formation by isolated micromeres of the sea urchin embryo. *Am. Zool.* **15**, 567–581.

Pehrson, J. R. and Cohen, L. H. (1986). The fate of the small micromeres in sea urchin development. *Dev. Biol.* **113**, 522–536.

Ransick, A., and Davidson, E. H. (1993). A complete second gut induced by transplanted micromeres in the sea urchin embryo. *Science* **259**, 1134–1138.

Ruffins, S., and Ettensohn, C. A. (1996). A fate map of the vegetal plate of the sea urchin (*Lytechinus variegatus*) mesenchyme blastula. *Development (Cambridge, UK)* **122**, 253–264.

Schroeder, T. E. (1981). Development of a "primitive" sea urchin (*Eucidaris tribuloides*): Irregularities in the hyaline layer, micromeres, and primary mesenchyme. *Biol. Bull.* **161**, 141–151.

Turbeville, J. M., Schultz, J. R., and Raff, R. A. (1994). Deuterostome phylogeny and the sister group of the chordates: Evidence from molecules and morphology. *Mol. Biol. Evol.* **11**, 648–655.

Wilt, F. H. (1987). Determination and morphogenesis in the sea urchin embryo. *Development (Cambridge, UK)* **100**, 559–575.

Wray, G. A. (1993). The evolution of cell lineage in echinoderms. *Am. Zool.* **34**, 353–363.

Wray, G. A. (1996). Parallel evolution of nonfeeding larvae in echinoids. *Syst. Biol.* **45**, 308–322.

Wray, G. A. (1997). Echinoderms. *In* "Embryology: Constructing the Organism" (S. F. Gilbert and A. M. Raunio, eds.), pp. 309–329. Sinauer, Sunderland, Massachusetts.

Wray, G. A., and Raff, R. A. (1990). Novel origins of lineage founder cells in the direct-developing sea urchin *Heliocidaris erythrogramma. Dev. Biol.* **141**, 41–54.

2

Gene Expression and Early Cell Fate Specification in Embryos of the Purple Sea Urchin (*Strongylocentrotus purpuratus*)

R. Andrew Cameron[1]
James A. Coffman[1,2]
[1]Division of Biology and [2]Stowers Institute of Medical Research
California Institute of Technology
Pasadena, California 91125

I. Introduction

Sea urchin embryos possess a distinctive suite of key developmental features that are shared with a diverse group of invertebrate taxa (Davidson 1990, 1991). The unfertilized egg possesses a maternally derived animal–vegetal (A/V) polarity (Boveri, 1901) and the early cleavages follow an invariant pattern that is oriented with respect to the A/V axis. As a result there is a predictable relationship between cell lineage and fate during early embryogenesis (Hörstadius, 1939; Cameron *et al.*, 1987). The earliest cell specification events occur during cleavage and therefore the early founder cells for embryonic territories arise before large-scale cell movements begin. A large body of experimental embryological investigations demonstrate that both autonomous and conditional specification mechanisms are employed to establish patterns of spatially restricted gene activity in sea urchin embryos (Davidson, 1989). Transcription is already active in pronuclei at the time of pronuclear fusion, and blastomere nuclei attain peak levels of transcription by the fifth cleavage (see Davidson, 1986, for

review). Several examples of spatially restricted zygotic transcription have been detected in fifth cleavage embryos (Reynolds *et al.*, 1992, DiBernardo *et al.*, 1995, Wang *et al.*, 1996), and numerous territory specific zygotic genes are expressed in the early blastula (Lee *et al.*, 1992; reviewed in Coffman and Davidson, 1992).

Spatially distinct territories are established relatively early in sea urchin embryogenesis. The four main territories initially specified by blastula stage comprise the skeletogenic territory, the vegetal plate (or endomesodermal) territory, and the oral and aboral ectoderm territories. The cells of a fifth territory, the small micromeres, cease to divide early in embryogenesis, but contribute to the coelomic pouches that give rise to the adult rudiment later in development. Founder cells for each territory have been identified and territory specific genes have been cloned. The focus of our research is the cellular and molecular basis of the initial specification processes that occur in normal embryogenesis. Cell lineage analyses, experimental embryological investigations, and the detection of localized patterns of zygotic gene expression have each helped to refine our understanding. Here we discuss our current view of the initial specification processes that establish early territory-specific gene expression in the sea urchin embryo, with an emphasis on studies of the purple sea urchin, *Strongylocentrotus purpuratus*. We start by considering cell lineage relationships, to introduce the context in which initial specification processes proceed. Next we address what is definitively known about the specification process as it occurs in the normal embryo. The final section reviews what we have learned about how individual territory specific genes process information provided through trans-acting gene regulatory factors.

II. The Relationship between Cell Lineage and Cell Specification Processes

A. Normal Cleavage and Early Lineage

The developmental mode discussed below is representative of those Euechinoidea (the regular sea urchins) which develop indirectly by producing a feeding pluteus larva. The embryos undergo three consecutive orthogonal cleavages (Fig. 1A). The first two cleavages are longitudinal and pass through the animal vegetal pole, producing four equal blastomeres, each of which contains an equivalent mass of cytoplasm. The third cleavage is then equatorial and divides the embryo into more or less equal quartets of animal and vegetal blastomeres. The fourth cleavage is unequal and subequatorial in the vegetal quartet, giving rise to micromeres and macromeres, while the cells of the animal quartet divide equally and longitudinally, producing eight mesomeres. From this point onward, the timing of cleavage divisions becomes more asynchronous and the mesomere derived lineages soon lose precise orthogonal positioning of cleavage planes (Summers *et al.*, 1993). In the micromere lineage, the mitotic cycle lengthens and the micromeres divide unequally, giving rise to the small micromeres and the large micromeres. Two cleavage divisions of the macromeres, one longitudinal and the next equatorial, yield two tiers each containing eight cells (Fig. 1A). The "veg_1" tier is closer to the embryonic equator and the "veg_2" tier is closer to the vegetal pole. Classical embryologists maintained that the veg_2 cells give rise to the entire archenteron, while the veg_1 cells formed the anal plate ectoderm and the mesomeres gave rise to all of the remaining ectodermal regions (see Hörstadius, 1939, 1973, for review). Recent lineage analyses specifically focused on the macromere lineages have provided details that significantly modify this classical view (see Section II,C and Chapter 4 by Logan and McClay; also Ransick and Davidson, 1998). Cleavage proceeds fur-

ther until about 9 divisions have occurred and the embryo consists of about 500 cells arranged in a hollow sphere that is one cell layer thick. Although this blastula still appears largely homogeneous and with little differentiation of form, its organization includes cell populations that are already specified as endoderm, mesoderm, and oral and aboral ectoderm, that is, at this stage cells within each of these territories are expressing characteristic sets of genes (Fig. 1B). Following cleavage the embryo engages in morphogenetic processes that lead to the formation of a pluteus larva containing about 1800 cells and one dozen cell types.

The progeny of the animal quartet of the 8-cell embryo form contiguous tracts of cells in the ectoderm of the larva (Cameron *et al.*, 1987; Cameron and Davidson, 1991). The highly reproducible clonal allocations to ectodermal cell types demonstrates not only the absence of cell mixing in the ectodermal layer, but also that both the cleavage divisions and the distortions of the spherical blastula to form the pyramidal pluteus follow highly regular patterns (Cameron and Davidson, 1991). Using this knowledge and the labeling patterns obtained from injecting single blastomeres in 2-cell embryos, it was shown that oral–aboral axis specification occurs with respect to the first cleavage plane (Cameron *et al.*, 1989). Subsequent lineage tracing reveal that all of the descendants of some 32-cell stage animal blastomeres lie entirely within a single ectodermal territory (Cameron *et al.*, 1990). Yet the ciliated band, a structure that separates the oral and aboral ectoderm, receives variable contributions from adjacent blastomeres of the 8-cell embryo (Cameron *et al.*, 1993). Thus the ultimate position of this boundary is not a function of cell lineage.

There are two insights that derive from analyses of cleavage and early lineages. The first relates to the timing of cell specification. Although initial specification of embryonic territories is well underway by the sixth cleavage stage, all cells are not specified until some time later. Indeed, the progeny of discrete sets of sixth cleavage blastomeres will come to lie in two different territories (Fig. 1B). Specifically, progeny of the veg_1 lineage become both ectoderm and endoderm, while the Nl lineages contribute oral and aboral ectoderm. The second insight has to do with the mode of specification. Because territory boundaries that form relatively late are variable in position from embryo to embryo and do not conform to lineage boundaries, local cell signaling processes must play a role in these specification events. Early specification events may employ cell signaling or they may make use of the invariant cleavage pattern to segregate regional asymmetries of the egg into specific cells. Thus the division of the egg by cleavage accomplishes two things: it segregates asymmetries into specific cells and sets up cell membranes across which cell signaling can occur.

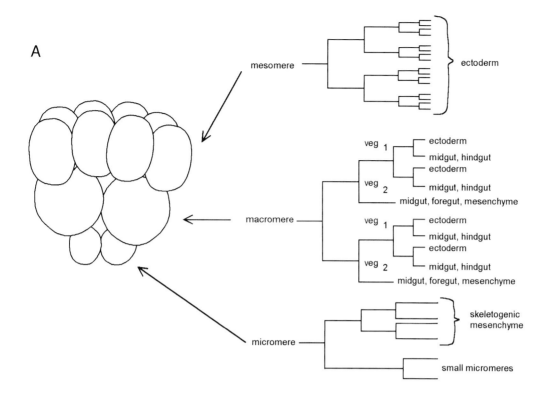

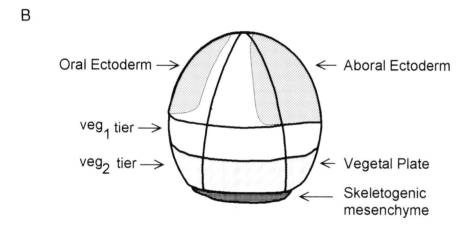

Figure 1 The cell lineage and cell fates for early territorial specification in the purple sea urchin, *Strongylocentrotus purpuratus*. (A) The sixteen cell stage embryo and the lineages that derive from each the three types of cells (mesomeres, macromeres, and micromeres). The cell types that arise from each sublineage are labeled. (B) A schematic diagram of the embryo at the end of cleavage. The borders that separate the contributions from early blastomeres are lined: The longitudinal lines represent the first and second cleavage; the equatorial line represents the third cleavage; the latitudinal line separating the skeletogenic mesenchyme represents the fourth and the line that separates the veg$_1$ and veg$_2$ occurs at the sixth cleavage. The regions occupied by specific early blastomere progeny that have been shown by lineage tracing to contribute to a territory (founder cells) are represented by filled areas. The exact boundary between the two ectoderms and the one that occurs in the progeny of the veg$_1$ to separate endoderm and ectoderm are depicted as white since the exact time of their specification is later.

B. Autonomous Specification and the Skeletogenic Territory

The large micromeres which result from the second of two consecutive unequal divisions at the vegetal pole are the founder cells of the skeletogenic territory (Fig. 1B) and they are the first to be specified. That the large micromeres are irreversibly committed to the skeletogenic pathway from the time they are formed is demonstrated by their unalterable pathway of differentiation in standard embryological tests of explantation (Okazaki, 1975) or transplantation (Hörstadius, 1939). The cells of a normally specified skeletogenic lineage display a stereotyped cleavage pattern and rhythm, as well as a powerful inductive capacity (Hörstadius, 1939; Ransick and Davidson, 1993, 1995). By the blastula stage the expression of territory specific genes is detectable in large micromere descendants (reviewed in Coffman and Davidson, 1992). Soon thereafter a nearly simultaneous blastocoelar ingression of this entire lineage occurs and the cells, now known as the skeletogenic or "primary" mesenchyme cells (PMCs) will eventually secrete the spicules of the embryo. While the early program of the skeletogenic territory requires no interactions with other cells, later aspects of PMC morphogenesis, such as position in the blastocoel and pattern of spiculogenesis, do require external signals and environmental cues (reviewed by Harkey, 1983; Armstrong and McClay, 1994; Guss and Ettensohn, 1997).

C. Specification of the Vegetal Plate Territory

The vegetal plate territory derives its name from the flattened and thickened vegetal portion of the blastula wall that is most clearly visualized just prior to the initiation of gastrulation. The vegetal plate territory contributes all of the endoderm and that part of the larval mesoderm that is not the PMCs. In this sense it is appropriate to consider it to be an endomesodermal territory. The cells which give rise to the morphological structure called the vegetal plate (VP) are derived from the veg_2 lineages (Hörstadius, 1939; Fig. 1B), and until recently it was believed that these were the only lineages that contributed cells to the endomesoderm.

The expression of $Endo16$ in S. purpuratus embryos fulfills the criteria of a territory specific gene (Davidson, 1989). Its expression is clearly restricted to the vegetal plate at the mesenchyme blastula stage and it marks the cells that will participate in gastrulation (Nocente-McGrath et al., 1989; Ransick et al., 1993). Recently, examination by whole mount in situ hybridization for $Endo16$ transcripts in combination with the tracing of veg_2 and veg_1 lineages (Ransick and Davidson, 1998) showed that $Endo16$ mRNA occurs exclusively in the veg_2 lineages at the mesenchyme blastula stage, and remains restricted to that domain during the early phase of gastrulation.

The initial specification of the endomesodermal founder cells involves an inductive signal from the micromeres, which are the immediate neighbors of the vegetal plate founder cells. The molecular nature of the inducing signal and the immediate response pathway(s) are unknown. The sensitive period for this interaction has been shown by micromere deletion experiments (Ransick and Davidson, 1995). The number of cells expressing $Endo16$ in the early blastula stage exceeds that reached in the absence of micromeres when the contact between micromeres and macromere daughters continues up to the fifth cleavage. Contact with the mi-

cromeres prolonged through the sixth cleavage cycle results in a nearly normal number of veg_2 derived cells with $Endo16$ expression at the blastula stage. The veg_2 lineage possesses an inherent capacity to activate the pathway of vegetal plate specification, since embryos from which the micromeres are deleted will often form normal larvae successfully but only after a significant delay in initiating gastrulation. Embryos that develop with the progeny of a single micromere left in place show normal levels of $Endo16$ expression and gastrulate in synchrony with controls (Ransick and Davidson, 1995). Thus it is likely that the veg_2 cells normally signal between themselves to maintain the effect of the micromere signal. While in the absence of the micromeres, veg_2 cells can activate the VP specification pathway, this is a relatively inefficient route to vegetal plate specification compared to the strong induction by micromeres.

A second phase of cellular allocation to the endomesodermal territory occurs later in development. Recent lineage tracing studies have shown that this allocation consists of a large portion of the endoderm and is derived from the veg_1 lineages (Logan and McClay, 1997; Ransick and Davidson, 1998). Thus, the specification of the endomesodermal cells is considerably more complex than simply the segregation of the veg_2 lineage. The involvement of veg_1 derived cells in endoderm formation was closely followed in both Lytechinus variegatus and S. purpuratus using a lipid soluble fluorescent dye loaded into the membranes of individual blastomeres (Logan and McClay, 1997; Ransick and Davidson, 1998). In both species, veg_1 derived cells contribute significantly to the endoderm, comprising the hindgut and a portion of the midgut. Interestingly, the lineage analyses show that in both species the early phase of gastrulation involves only the cells of the veg_2 lineages. In other words, the traditionally recognized vegetal plate does behave as a cohesive unit during the processes of vegetal plate buckling and primary archenteron invagination. Veg_1 progeny do not directly participate in these early morphogenetic cell movements. They do not move through the blastopore region until the archenteron begins its elongation phase, which is roughly at midgastrula stage. Movements of cells that elongate the prospective hindgut region continue for many hours after the archenteron reaches its oral field target. Endodermal portions of veg_1 clones become arranged in narrow linear tracts that stretch from the midgut to the anus. The final boundary between endoderm and ectoderm, marked by the anus, forms without specific relation to any particular early cleavage plane.

While early lineage relationships are very similar among the species studied during the early phase of gastrulation, species differences become evident later when veg_1 lineages participate. The pattern of recruitment of veg_1 cells to the endoderm in L. variegatus is highly variable and shows no bias with respect to the oral–aboral axis (Logan and McClay, 1997). In S. *purpuratus* there is some embryo to embryo variability in the proportion of veg_1 cells that become endoderm, but most significantly, there is consistently a greater contribution to endoderm from the oral quadrant veg_1 lineages (Ransick and Davidson, 1998). It is worth noting that the specific veg_1 allocations to the endoderm in S. *purpuratus* and *L. variegatus* correlate to the pathways of archenteron elongation followed by each species. Intraspecies variability and interspecies differences in recruitment of veg_1-derived cells to the endoderm both support the view that the late specification of veg_1-derived cells is a conditional event not correlated with cell lineage.

The expression of *Endo16* subsequently spreads to the veg_1 lineages in a pattern consistent with the contribution of these lineages to endoderm. Specifically, *Endo16* is detected in the portion of veg_1 clones surrounding the blastopore, with the larger recruitment of cells observed on the oral side of the embryo. All *Endo16* positive veg_1-derived cells then participate in gastrulation. The late expression of *Endo16* in veg_1 descendants confirms their contribution to the vegetal plate territory, and shows that the endodermal precursors are specified in two phases: an early phase that involves uniform response by the entire veg_2 lineage and a later phase that is independent of lineage per se. These cells are specified on an individual basis as gastrulation proceeds. An important part of endodermal specification is the formation of the endoderm–ectoderm boundary during gastrulation. Early establishment of this boundary might place constraints on recruitment of cells to the endoderm. In the aboral ectoderm, for example, the territory-specific genes *Spec1* and *CyIIIa* are expressed as early as the mesenchyme blastula stage. The early expression of these genes raises the possibility that prior ectodermal specification of veg_1-derived cells in the aboral quadrant has already occurred and thus limits recruitment of cells to the endoderm on this side of the embryo.

The discovery of immediate early zygotic responses in the veg_2 lineage could shed light on the specification process. The earliest zygotic response to be identified thus far is SpKrox1, a Kruppel-like zinc-finger transcription factor. SpKrox1 mRNA can already be detected in veg_2 cells in the sixth cleavage embryo and expression is restricted to the vegetal plate in the blastula (Wang *et al.*, 1996). If SpKrox1 expression is an immediate downstream response to micromere inductive signaling, it would provide an excellent starting point for determining the signaling pathway that transduces the inductive signal.

It is likely that the vegetal plate founder cells are first specified as an endomesodermal field, then subsequent specification processes occur sequentially in embryogenesis to subdivide the vegetal plate into a central mesodermal domain of secondary mesenchyme cells (SMC) surrounded by an endodermal domain (Cameron *et al.*, 1991; Ruffins and Ettensohn, 1996). Because the central mesodermal region of the vegetal plate contacts the micromere lineage directly, micromere inductive signaling may play a role in this specification pathway as well. The growing list of molecules that mark cells within this mesodermal domain (Miller *et al.*, 1996; Harada *et al.*, 1995; Martinez and Davidson, 1997) make a test of this hypothesis a realistic possibility. Notch receptor expression has been localized in the vegetal region of blastula stage embryos in a pattern that suggests a link to specification of the mesodermal domain in the vegetal plate (Sherwood and McClay, 1997). However, a causal relationship has not yet been established between this interesting pattern and any known cell fate specification process in normal development.

D. Specification of the Oral–Aboral Axis and Ectodermal Territories

The oral–aboral (O/A) axis is the second embryonic axis and it is nearly congruent with the anterior–posterior axis of the larva. However, this axis is not detectable morphologically until the early gastrula stage. Only then do the first morphogenetic processes produce O/A asymmetries, such as the bilateral arrangement of the oral clusters of PMCs or directed tilting of the invaginating archenteron. By the midgastrula stage cell shape changes occur which begin to flatten the oral ectoderm, while the aboral ectoderm expands as it becomes a squamous epithelium. By the end of gastrulation a distinct boundary between the ectodermal territories is demarcated by the ciliated band.

Lineage tracing studies show the entire ectoderm derives from the animal quartet blastomeres of the 8-cell embryo, except for the anal plate aboral ectoderm which is veg_1-derived (Fig. 1B). In S. *purpuratus*, where the O/A axis is specified early, the two opposite 8-cell stage animal blastomeres No and Na give rise to clones that reside entirely within the oral or aboral ectoderm, respectively (Cameron *et al.*, 1987). The other two animal blastomeres (right and left Nl) yield clones with both oral and aboral ectoderm progeny, since they lie across the region where the boundary between the ectodermal territories (i.e., the ciliated band) arises. Although clearly demonstrating the early segregation of pure oral and aboral clones, lineage tracing studies alone can not address the timing or mechanism of ectodermal specification events.

Despite the absence of early morphological O/A markers, there are indications for a mechanism that establishes an O/A polarity during early cleavage stages. As early as the 8-cell stage, the prospective oral side exhibits an elevated level of cytochrome oxidase activity (reviewed in Czihak, 1971). In both S. *purpuratus* and *Lytechinus pictus,* the orientations of the early cleavage planes bear a consistent relationship to the O/A axis (Cameron *et al.,* 1991; Henry *et al.,* 1992). It has to be pointed out, however, that this cleavage relationship tends to be species-specific, and no relationship between the first two cleavages and the O/A axis can be found in *Paracentrotus lividus* or *Hemicentrotus pulcherrimus* (Hörstadius, 1939; Kominami, 1988). Furthermore, the initial polarity is labile and easily overridden by blastomere dissociation even in those species where it is established early. Two examples of early molecular level O/A asymmetries have been recently described. One involves the asymmetric localization of a maternal mRNA (SpCOUP) in eggs and embryos of S. *purpuratus* (Vlahou *et al.,* 1996), and the other the transient asymmetric zygotic expression of a homeodomain transcription factor (PlHbox12) in early cleavage stage *P. lividus* embryos (DiBernardo *et al.,* 1994, 1995). However, the relationship of these patterns to specification of the O/A axis is not known. When animal quartets of 8-cell embryos are isolated intact and cultured, the resulting epithelial balls lack mesoderm or endoderm, but display distinct morphological polarity along the O/A axis (Wikramanyake and Klein, 1997). Although some spatially restricted gene expression also occurs in these isolates, an additional vegetal influence is required to obtain full differentiation of oral and aboral ectoderm. The molecular basis of the O/A polarity is not understood. That labile asymmetries and multiple pathways exist from the earliest cleavages offers the simplest explanation for the establishment of the O/A axis.

Territory specific zygotic expression patterns can be used as the cornerstones of detailed investigation into the mechanisms of cell specification, because they identify and distinguish cell type primordia on a molecular level. It follows that a detailed description of the spatial and temporal patterns of expression of such markers is crucial. The best studied aboral ectoderm specific markers, *CyIIIa* actin and *Spec1,* are known to have spatially restricted expression patterns in early blastula stage embryos. Unfortunately, there has not been a detailed examination of the lineages expressing these markers, particularly in the early phase of expression that occurs during the later cleavage stages. It is fundamentally important to determine the manner in which these aboral specific genes turn on with respect to the early lineage clones. For example, if the expression pattern is initially in a few cells and spreads to surrounding cells gradually, that implies a different mechanism of specification than if the initial expression domain includes all Na derived cells. Such a detailed spatiotemporal analysis of the domain of expression of *Endo16* did reveal important insights into the specification of the vegetal plate territory (see earlier discussion).

III. Spatial Regulation of Gene Expression

The molecular consequence of the cellular specification events discussed above is the regionalized expression of specific sets of genes. Specific gene expression patterns define each of the territories of the early sea urchin embryo. In the following, we address the manner in which spatially distinct patterns of gene expression are generated in the embryo by considering the mechanisms by which individual genes in the sea urchin embryo process the information provided by the higher level processes of cell specification discussed above.

A. Cis-Regulatory Systems of Spatially Regulated Genes: Lessons from Detailed Analyses of *CyIIIa* and *Endo16*

One of the most thoroughly analyzed developmental cis-regulatory systems is that which controls transcription of the *CyIIIa* actin gene, a specific marker of aboral ectoderm in embryos of S. *purpuratus* (see Coffman *et al.,* 1996, 1997; Kirchhamer and Davidson, 1996, and references therein). Given at first glance the relatively simple pattern of *CyIIIa* expression, one might be surprised at the complexity of its 2.3-kb 5' regulatory domain (Fig. 2A). *In vitro* studies have revealed over 20 specific protein binding sites which engage at least 8 different DNA-binding proteins, most of which have been cloned (Coffman *et al.,* 1997, and references therein). Remarkably, each of these binding sites functions in the regulation of *CyIIIa*.CAT reporter genes, as revealed both by *in vivo* competition analysis (Franks *et al.,* 1990; Hough-Evans *et al.,* 1990), and by target site deletion or mutation (Wang *et al.,* 1995a; Kirchhamer and Davidson, 1996; Coffman *et al.,* 1996, 1997). The *CyIIIa* regulatory domain is organized into functional modules (Kirchhamer and Davidson, 1996; Kirchhamer *et al.,* 1996). It has recently become apparent that the functional units of many cis-regulatory systems are not the individual target sites (and associated transcription factors) themselves, but clusters of target sites for transcription factors whose interactions fulfill some specific spatial or temporal function in regulating transcription (Kirchhamer *et al.,* 1996). Sometimes these clusters contain target sites for both activators and repressors. The repressors often function within their respective modules to specifically interfere with the activators (Gray and Levine, 1996). Clusters of transcription factor target sites that function as a unit have been variously termed "enhancers" (Gray and Levine, 1996, and references therein) or "modules" (Arnone and Davidson, 1997, and references therein). Studies of *CyIIIa* have revealed two distinct regulatory modules, a proxi-

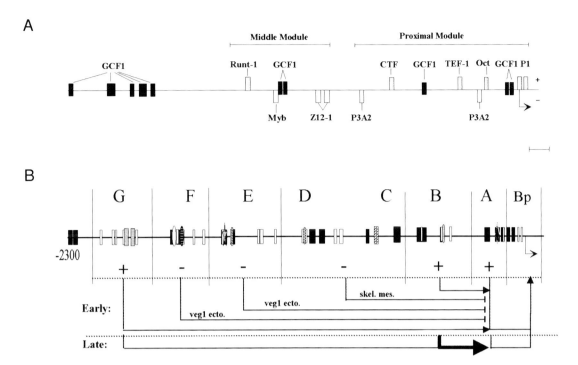

Figure 2 The cis-regulatory systems of *CyIIIa* and *Endo16*. (A) Linear representation of the *CyIIIa* cis-regulatory system (modified from Coffman *et al.,* 1997). Specific target sites for transcription factors are shown as boxes. Shaded boxes above the line represent sites of interaction known to stimulate transcription (+), while open boxes below the line represent sites of interaction known to repress transcription (−). The black boxes that cross the line represent binding sites for the ubiquitous DNA looping factor SpGCF1 (Zeller *et al.,* 1995b,c). Except for the P1 factor, which has been partially purified, clones have been obtained for each of the factors depicted. For simplicity, the prefix "Sp" has been omitted from each of the transcription factor designations (see Coffman *et al.,* 1997, and references therein). Brackets above the line show the approximate extent of the two *CyIIIa* regulatory modules (the proximal and middle modules). Scale bar represents 100 bp. The bent arrow denotes the transcription start site. (B) Linear representation of the *Endo16* cis-regulatory system (modified from Kirchhamer *et al.,* 1996). Specific target sites for transcription factors are shown as symbols along the line (Yuh *et al.,* 1996); open boxes represent sites for the ubiquitous looping factor SpGCF1 (Zeller *et al.,* 1995b,c). The regulatory domain is organized as six modules (G, F, E, D–C, B, A), the approximate boundaries of which are shown by the vertical lines. Positively acting modules are denoted by "+", negatively acting modules by "−" below the line. The scale is the same as in (A). The bent arrow represents the transcription start site. The information flow, communicated through module A to the basal promoter (Bp), is depicted by lines with arrowheads (for positive interactions) or bars (for negative interactions) at their termini. The thick arrow under module B is meant to emphasize the dominant positive regulatory role that this module exercises late in development. For more details refer to Yuh and Davidson (1996), and Yuh *et al.* (1996, 1998).

mal module that mediates early activation of the gene during initial (cleavage stage) territorial specification, and a middle module that is required for normal levels of transcription during terminal differentiation of the aboral ectoderm. In addition to activator sites, both the proximal and middle modules also contain repressor sites required for spatial localization of *CyIIIa* transcription to the aboral ectoderm (Fig. 2A; Kirchhamer and Davidson, 1996). The best-understood example is the middle module, which contains sites for SpRunt-1, required for normal levels of transcriptional activation in the postgastrula stage embryo (Coffman *et al.,* 1996), and SpMyb, required to confine that activation to the aboral ectoderm (Coffman *et al.,* 1997). Finally, both the proximal and middle modules, as well as the distal region of the *CyIIIa* regulatory domain contain multiple target sites for SpGCF1, a transcription factor which has been shown to loop DNA *in vitro* (Zeller *et al.,* 1995 b,c). These sites are required for maximal expres-

sion of the gene (Flytzanis *et al.,* 1987; Franks *et al.,* 1990; Kirchhamer and Davidson, 1996; Coffman *et al.,* 1997).

An even more complex regulatory system is found in the *Endo16* gene, which encodes an endoderm-specific transcript that is initially expressed throughout the vegetal plate and invaginating archenteron, but later becomes localized to the midgut (Nocente-McGrath *et al.,* 1989; Ransick *et al.,* 1993). As in the case of *CyIIIa*, the regulatory domain of *Endo16* contains a large number of specific protein binding sites which are organized as six functionally distinct modules (Fig. 2B). Recent studies by Yuh *et al.,* (1996, 1998) demonstrate that module A functions as an integrator for the entire system. Early in development, during initial vegetal plate specification, module A serves as a stand-alone positive regulatory element. Later it functions as an amplifier of positive regulatory information from modules B and G. In addition, module A is required to deliver negative regulatory information from modules E and F (which prevent ectopic activation in the veg$_1$ ectoderm) and CD (which prevents ectopic activation in the skeletogenic mesenchyme) to the basal promoter (Bp). It has been shown that these negatively acting modules are converted to positively acting modules when embryos are grown in the presence of LiCl (a teratogen which causes expansion of the endodermal territory at the expense of ectoderm), and that these positive functions also are integrated by module A (Yuh *et al.,* 1998). Since LiCl is thought to disrupt signal transduction pathways (Berridge, 1995; Klein and Mclton; 1996), this evidence suggests that the negative regulatory modules of *Endo16* are responsive to cell interactions in the embryo. Specific target sites for an as yet uncharacterized factor (nominally termed CG), in both module A and proximal to the Bp, appear to be required for specifically communicating the information integrated by module A to the Bp (Yuh *et al.,* 1996). Thus, if the SV40 promoter is substituted for the endogenous *Endo16* Bp and its associated CG target sites, each *Endo16* regulatory module functions independently, and the integrative function of module A is lost (Yuh *et al.,* 1996). Finally, late in development the situation is simplified somewhat, as module B becomes the dominant regulatory element in *Endo16,* by itself controlling stomach-specific expression with only positive regulators (Yuh and Davidson, 1996). Interestingly, the unique target site within module B responsible for its activation function also is found in the regulatory domain of the cytoskeletal actin gene *CyIIa,* wherein it also drives midgut-specific expression late in development (Arnone *et al.,* 1998). This latter fact underscores a common theme in the regulation of complex patterns of gene expression. That is, early in the development of a *conditionally* specified territory, spatially restricted gene expression depends on multiple positive and negative regulators, reflecting the complexity in intercellular signals that must be processed by the gene. Later, after cells have become specified and begin to differentiate to their terminal state, control of transcription no longer needs such a complex information processing system. At this stage only a few positive regulators provided by spatially restricted, zygotic gene expression are required (Arnone and Davidson, 1997; Arnone *et al.,* 1998).

B. The Regulation of Gene Expression in Autonomously Specified Cells: *SM50*

A number of spatially regulated genes of the sea urchin embryo have relatively simple regulatory systems, consisting of a single module. Many of these are expressed in autonomously specified territories. One example of such a gene is *SM50,* which is expressed exclusively in the micromere lineage (Benson *et al.,* 1987; Sucov *et al.,* 1987), which as discussed above only has potential

to develop into primary (skeletogenic) mesenchyme cells (PMCs). Makabe *et al.* (1995) showed that within the *SM50* regulatory domain exists a positively acting target site, the "C-element," that is absolutely required for PMC-specific expression of reporter constructs. This element binds a factor of the cut-domain family of homeodomain proteins (K. Makabe, 1996, unpublished result). Significantly, no ectopic expression of *SM50* reporter genes was ever detected, suggesting that this gene only utilizes positive regulation. Since the other positive regulators required for PMC-specific expression of *SM50* are ubiquitous (as inferred from their role in the regulation of various other genes such as *CyIIIa*), it is likely that the factor that binds the C-element is either localized to the micromere lineage, or is specifically activated there (Makabe *et al.,* 1995). A similarly simple situation has been shown to exist for a gene whose spatial pattern of expression is essentially complementary to that of *SM50*, in that it is expressed everywhere but the vegetal-most region of the embryo: the "very early blastula" (VEB) gene *SpHE* (Wei *et al.,* 1997, and references therein). As in the case of *SM50,* only positively acting target sites exist within the relatively short (several hundred bp) spatial regulatory domain of this gene. Interestingly, there is very little overlap between the activators that regulate expression of *SpHE* and those that regulate *SpAN*, another VEB gene with essentially the same expression pattern as *SpHE* (Kozlowski *et al.,* 1996). Thus, there are in this case at least two completely different regulatory systems (though both are perhaps tied into the same higher level mechanism, such as cytoskeletal organization) that generate essentially the same spatial pattern of expression.

C. Studies of Cloned Transcription Factors of the Sea Urchin Embryo: Emerging Themes

The function of genetic cis-regulatory systems is to process cellular information, which in the nucleus takes the form of DNA-binding proteins whose interactions regulate transcription. In the last few years a growing number of transcriptional regulatory proteins of the sea urchin embryo have been cloned. Several of these, listed in Table I, were initially identified by virtue of their specific interaction with a target sequence of known regulatory function in a downstream structural gene. Cloning of complementary DNAs (cDNAs) that encode transcription factors facilitates several kinds of analysis that would not otherwise be possible. A big advantage of this approach is that purified recombinant proteins are much more easily obtained than are purified native proteins, facilitating production of high-quality antibodies. This allows immunologic detection of transcription factors *in situ* in the embryo, and of transcription factor variants separated by electrophoresis. These two tech-

TABLE I

Cloned Transcription Factors with Known Target Genes in Embryos of *Strongylocentrotus purpuratus*

Factor name	Family	Known target gene(s)	Function in regulating target gene	Maternal protein?	Zygotically transcribed?	References
SpP3A2	Unique[a]	*CyIIIa, SM50*	Spatial repressor	Yes	Yes	Calzone *et al.* (1991); Cutting *et al.* (1990); Zeller *et al.* (1995a)
SpZ12-1	Zinc finger	*CyIIIa*	Spatial repressor	Yes	Yes	Wang *et al.* (1995a,b)
SpMyb	Myb domain	*CyIIIa*	Spatial repressor	N.D.[b]	Yes	Coffman *et al.* (1997)
SpRunt-1	Runt domain	*CyIIIa*	Activator	Yes	Yes	Coffman *et al.* (1996)
SpGCF	Unique	*SpAn, CyIIIa, CyIIa, Endo16*	DNA looping	Yes	Yes	Kozlowski *et al.* (1996); Zeller *et al.*(1995b,c); Coffman *et al.* (1997); Arnone *et al.* (1998)
SpZ2-1	Zinc finger	*CyIIIa*	Repressor	Yes	Yes	Höög *et al.* (1991); Cutting *et al.* (1990); Zeller *et al.* (1995a)
SpOtx	Homeo domain	*SpHE, Spec2a, Endo16*	Activator	Yes	Yes	Wei *et al.* (1995); Gan *et al.* (1995); Mao *et al.* (1996); Yuh *et al.* (1998)
SpUSF	bHLH	*Spec1, Spec2a, Spec2c, SpU6*	Activator[c]	N.D.	N.D.	Tomlinson *et al.* (1990); Kozlowski *et al.*(1991); Li *et al.* (1994)
SpCOUP-TF	SH receptor	*CyIIIb*[d]	Spatial repressor[d]	Yes	Yes	Chan *et al.* (1992); Niemeyer and Flytzanis (1993); Vlahou *et al.* (1996)
SpSHR2	SH receptor	*CyIIIb*[d]	N.D.[d]	Yes	N.D.	Kontrogianni-Konstantopoulos *et al.* (1996)
SpOct	POU homeo domain	*CyIIIa α Histone H2B*	Activator	Yes	Yes	Char *et al.* (1993)
SpSSAP	RRM[e]	Late Histone *H1*	Activator	Yes[f]	Yes	DeAngelo *et al.* (1993, 1995)

[a]This protein belongs to a family of transcription factors that include the product of the *Drosophila erect wing* gene (DeSimone and White, 1993) and the human protein NRF-1 (Virbasius *et al.*, 1993)

[b]N.D., Not determined.

[c]While mutation of the USF binding sites in the Spec regulatory domain do not effect the level of expression of injected reporter constructs (Tomlinson *et al.*, 1990), a USF site in the U6 snRNA gene of S. *purpuratus* is required for transcription of the gene *in vitro* (Li *et al.*, 1994).

[d]The hormone-response element in the regulatory domain of *CyIIIb* has been shown to be required for spatial regulation of transcription. While both SpCOUP-TF and SpSHR2 have the same target site specificity, the expression pattern of SpCOUP-TF suggests that it is the spatial repressor that acts through this site.

[e]The DNA-binding domain of this protein is homologous to the conserved RNA-recognition motif (RRM) found in certain RNA-binding proteins (DeAngelo *et al.*, 1995).

[f]This is based on the low level of DNA-binding activity detected in egg extracts by electrophoretic mobility shift assays (DeAngelo *et al.*, 1993).

niques have revealed some interesting generalities that apply to transcription factors of the sea urchin embryo. The first is that many of the transcription factors involved in early embryogenesis are present in the egg as maternal proteins (Wang *et al.*, 1995b; Zeller *et al.*, 1995a; J. A. Coffman, 1996, unpublished results). Where analyzed, these proteins are globally distributed in the early embryo (Mao *et al.*, 1996; Chuang *et al.*, 1996; C. N. Flytzanis, 1997, personal communication; J. A. Coffman, 1996, unpublished results). This generality is also supported by a recent study of DNA-binding activities that interact with regulatory target sequences of *CyIIIa*: all nine of the activities tested were present in cytoplasm of unfertilized eggs (Calzone *et al.*, 1997).

The second generality is that maternal transcriptional regulatory proteins are subject to extensive posttranslational modification during early embryogenesis, some (but not all) of which represents phosphorylation (Harrington *et al.*, 1997). These two generalities are entirely consistent with the proposal that the processes of initial territorial specification in the sea urchin embryo result in regionalized modification of the structure, and thus presumably the activity, of maternally supplied transcriptional regulatory proteins (Davidson, 1989; Coffman and Davidson, 1994). A totally unexplored area of research for the future is thus the nature and origin of the posttranslational modifications that affect the activity of maternal transcription factors of the sea urchin embryo.

While initial territorial specification appears to depend largely on maternal transcription factors that are globally distributed but locally activated, later development clearly utilizes zygotic transcription factors that are themselves spatially localized at the transcriptional level. To list several examples, localized zygotic transcripts of SpOtx (Li *et al.*, 1997), SpMyb (Coffman *et al.*, 1997), SpHox8 (Angerer *et al.*, 1989), SpHmx (Martinez and Davidson, 1997), SpCOUP-TF (Vlahou *et al.*, 1996), SpFkh1 (Luke *et al.*, 1997), and SpKrox1 (Wang *et al.*, 1996) have been observed in postgastrula stage embryos of *S. purpuratus*. Thus, one of the results of initial regionalization of the embryo during cleavage is the differential spatial activation of genes encoding transcription factors required for further territorial specification. Indeed, this process probably begins early during cleavage, as demonstrated in *P. lividus* by homeobox gene PlHbox12, which is transiently expressed beginning at the 4-cell stage with maximal expression at the 60-cell stage and declining sharply thereafter (DiBernardo *et al.*, 1994, 1995). Transcripts of this gene are expressed on only one side of the embryo along the prospective oral–aboral axis (DiBernardo *et al.*, 1995). Studies of the cis-regulatory system of this gene, and perhaps others like it that have yet to be discovered, will provide insight into the earliest events whereby intercellular signaling affects territory specific activation of genes that encode regulatory proteins required for subsequent regional specification.

D. Conclusions and Prospects

The data briefly reviewed in this section point to two obvious directions for future research on the molecular biology of cell fate specification in early sea urchin development. First, the mother lode of information that can be mined from the detailed analysis of cis-regulatory systems should be obvious from the few examples discussed here. Cis-regulatory systems represent a developmental nexus, wherein environmental information provided by the cell (transcription factors active in the nucleus) is processed into gene activity (transcription). Analysis of genetic cis-regulatory systems gives direct access to the transcription factors functioning in the regulation of gene expression. Second, the available data suggest that the earliest processes that lead to territory-specific gene activity in the sea urchin embryo involve cellular mechanisms that modulate the activity of maternally synthesized transcription factors, probably through covalent posttranslational modification. Elucidation of the biochemical nature of the mechanisms that carry out these modifications, and their role in processes of cell specification, will be a major challenge for the future.

Acknowledgments

The original research from the laboratory of Eric Davidson which is reviewed in this chapter was supported by National Institute of Child Health and Human Development, National Science Foundation Program in Developmental Mechanisms, Office of Naval Research, and the Stowers Institute for Medical Research. We thank Eric Davidson and Andrew Ransick for many helpful discussions and critical reviews of the manuscript.

References

Angerer, L. M., Dolecki, G. J., Gagnon, M. L., Lum, R., Wang, G., Yang, Q., Humphreys, T., and Angerer, R. C. (1989). Progressively restricted expression of a homeo box gene within the aboral ectoderm of developing sea-urchin embryos. *Genes Dev.* **3,** 370–383.

Armstrong, N., and McClay, D. R. (1994). Skeletal pattern is specified autonomously by the primary mesenchyme cells in sea-urchin embryos. *Dev. Biol.* **162,** 329–338.

Arnone, M., and Davidson, E. H. (1997). The hardwiring of development: Organization and function of genomic regulatory systems. *Development* (*Cambridge, UK*) **124,** 1851–1864.

Arnone, M., Martin, E., and Davidson, E. H. (1998). Cis-regulation of cell type specification: A single compact element controls the complex expression of the *CYIIa* gene in sea urchin embryos. *Development* (*Cambridge, UK*) **125,** 1381–1395.

Benson, S. C., Sucov, H. M., Stephens, L., Davidson, E. H., and Wilt, F. (1987). A lineage-specific gene encoding a major matrix protein of the sea urchin embryo spicule. I. Authentication of the cloned gene and its developmental expression. *Dev. Biol.* **120,** 499–506.

Berridge, M. J. (1995). Calcium signalling and cell proliferation. *BioEssays* **17,** 491–500.

Boveri, T. (1901). Die Polaritat von Ovocyte, Ei und Larve des *Strongylocentrotus lividus*. *Zool. Jahrb. Abt. Anat. Ontog. Tiere* **14,** 630–653.

Calzone, F. J., Höög, C., Teplow, D. B., Cutting, A. E., Zeller, R. W., Britten, R. J., and Davidson, E. H. (1991). Gene regulatory factors of the sea urchin embryo. I. Purification by affinity chromatography and cloning of P3A2, a novel DNA binding protein. *Development* (*Cambridge, UK*) **112,** 335–350.

Calzone, F. J., Grainger, J., Coffman, J. A., and Davidson, E. H. (1997). Extensive maternal representation of DNA-binding proteins that interact with regulatory target sites of the *Strongylocentrotus purpuratus CyIIIa* gene. *Mol. Marine Biol. Biotechnol.* **6,** 79–83.

Cameron, R. A., and Davidson, E. H. (1991). Cell type specification during sea urchin development. *Trends Genet.* **7,** 212–218.

Cameron, R. A., Hough-Evans, B. R., Britten, R. J., and

Davidson, E. H. (1987). Lineage and fate of each blastomere of the eight-cell sea urchin embryo. *Genes Dev.* **1**, 75–84.

Cameron, R. A., Fraser, S. E., Britten, R. J., and Davidson, E. H. (1989). The oral–aboral axis of a sea urchin embryo is specified by first cleavage. *Development (Cambridge, UK)* **106**, 641–647.

Cameron, R. A., Fraser, S. E., Britten, R. J., and Davidson, E. H. (1990). Segregation of oral from aboral ectoderm precursors is completed at 5th cleavage in the embryogenesis of *Strongylocentrotus purpuratus*. *Dev. Biol.* **137**, 77–85.

Cameron, R. A., Fraser, S., Britten, R. J., and Davidson, E. H. (1991). Macromere cell fates during sea urchin development. *Development (Cambridge, UK)* **113**, 1085–1091.

Cameron, R. A., Britten, R. J., and Davidson, E. H. (1993). The embryonic ciliated band of the sea urchin, *Strongylocentrotus purpuratus* derives from both oral and aboral ectoderm. *Dev. Biol.* **160**, 369–376.

Chan, S. M., Xu, N. D., Niemeyer, C. C., Bone, J. R., and Flytzanis, C. N. (1992). SpCOUP-TF—A sea-urchin member of the steroid thyroid-hormone receptor family. *Proc. Natl. Acad. Sci. U.S.A.* **89**, 10568–10572.

Char, B. R., Bell, J. R., Dovala, J., Coffman, J. A., Harrington, M. G., Becerra, J. C., Davidson, E. H., Calzone, F. J., and Maxson, R. (1993). SpOct, a gene encoding the major octamer binding protein in sea urchin embryos: Expression profile, evolutionary relationships, and DNA binding of expressed protein. *Dev. Biol.* **158**, 350–363.

Chuang, C.-K., Wikramanayake, A. H., Mao, C.-A., Li, X., and Klein, W. H. (1996). Transient appearance of *Strongylocentrotus purpuratus* Otx in micromere nuclei: Cytoplasmic retention of SpOtx possibly mediated through an alpha–actinin interaction. *Dev. Genet.* **19**, 231–237.

Coffman, J. A., and Davidson, E. H. (1992). Expression of spatially regulated genes in the sea urchin embryo. *Curr. Opin. Genet. Dev.* **2**, 260–268.

Coffman, J. A., and Davidson, E. H. (1994). Regulation of gene expression in the sea urchin embryo. *J. Marine Biol. Assoc., U.K.* **74**, 17–26.

Coffman, J. A., Kirchhamer, C. V., Harrington, M. G., and Davidson, E. H. (1996). SpRunt-1, a new member of the runt-domain family of transcription factors, is a positive regulator of the aboral ectoderm-specific CyIIIa gene in sea urchin embryos. *Dev. Biol.*, **174**, 43–54.

Coffman, J. A., Kirchhamer, C. V., Harrington, M. G., and Davidson, E. H. (1997). SpMyb functions as an intramodular repressor to regulate spatial expression of CyIIIa in sea urchin embryos. *Development (Cambridge, UK)* **124**, 4717–4727.

Cutting, A. E., Höög, C., Calzone, F. J., Britten, R. J., and Davidson, E. H. (1990). Rare maternal mRNAs code for regulatory proteins that control lineage specific gene expression in the sea urchin embryo. *Proc. Natl. Acad. Sci. U.S.A.* **87**, 7953–7957.

Czihak, G. (1971). Echinoids. *In* "Experimental Embryology of Marine and Freshwater Invertebrates." (G. Reverberi, ed.), Chap. 12, pp. 363–482. North-Holland Publ., Amsterdam.

Davidson, E. H. (1986). "Gene Activity in Early Development," 3rd Ed. Academic Press, New York.

Davidson, E. H. (1989). Lineage-specific gene expression and the regulative capacities of the sea urchin embryo: A proposed mechanism. *Development (Cambridge, UK)* **105**, 421–445.

Davidson, E. H. (1990). How embryos work: A comparative view of diverse modes of cell fate specification. *Development (Cambridge, UK)* **108**, 365–389.

Davidson, E. H. (1991). Spatial mechanisms of gene regulation in metazoan embryos. *Development (Cambridge, UK)* **113**, 1–26.

DeAngelo, D. J., DeFalco, J., and Childs, G. (1993). Purification and characterization of the stage-specific embryonic enhancer-binding protein SSAP-1. *Mol. Cell. Biol.* **13**, 1746–1758.

DeAngelo, D. J., DeFalco, J., Rybacki, L., and Childs, G. (1995). Embryonic enhancer-binding protein ssap contains a novel DNA-binding domain which has homology to several RNA-binding proteins *Mol. Cell. Biol.* **15**, 1254–1264.

DeSimone, S. M., and White, K. (1993). The *Drosophila erect wing* gene, which is important for both neuronal and muscle development, encodes a protein which is similar to the sea-urchin P3A2 DNA-binding protein. *Mol. Cell. Biol.* **13**, 3641–3649.

DiBernardo, M., Russo, R., Oliveri, P., Melfi, R., and Spinelli, G. (1994). Expression of homeobox-containing genes in the sea urchin (*Paracentrotus lividus*) embryo. *Genetica (The Hague)* **94**, 141–150.

DiBernardo, M., Russo, R., Oliveri, P., Melfi, R., and Spinelli, G. (1995). Homeobox-containing gene transiently expressed in a spatially restricted pattern in the early sea urchin embryo. *Proc. Natl. Acad. Sci. U.S.A.* **92**, 8180–8184.

Flytzanis, C. N., Britten, R. J., and Davidson, E. H. (1987). Ontogenic activation of a fusion gene introduced into sea urchin eggs. *Proc. Natl. Acad. Sci. U.S.A.* **84**, 151–155.

Franks, R. R., Anderson, R., Moore, J. G., Hough-Evans, B. R., Britten, R. J., and Davidson, E. H. (1990). Competitive titration in living sea urchin embryos of regulatory factors required for expression of the CyIIIa actin gene. *Development (Cambridge, UK)* **110**, 31–40.

Gan, L., Mao, C. A., Wikramanayake, A., Angerer, L. M., Angerer, R. C., and Klein, W. H. (1995). An orthodenticle-related protein from *Strongylocentrotus purpuratus*. *Dev. Biol.* **167**, 517–528.

Gray, S., and Levine, M. (1996). Transcriptional repression in development. *Curr. Opin. Cell Biol.* **8**, 358–364.

Guss, K. A., and Ettensohn, C. A. (1997). Skeletal morphogenesis in the sea urchin embryo—regulation of primary mesenchyme gene-expression and skeletal rod growth by ectoderm–derived cues. *Development (Cambridge, UK)* **124**, 1899–1908.

Harada, Y., Yasuo, H., and Satoh, N. (1995). A sea urchin homologue of the chordate Brachyury (T) gene is expressed in the secondary mesenchyme founder cells. *Development (Cambridge, UK)* **121**, 2747–2754.

Harkey, A. H. (1983). Determination and differentiation of micromeres in the sea urchin embryo. *In* "Time, Space and Pattern in Embryonic Development" (R. Raff and W. Jeffery, eds), pp. 131–155. Alan R. Liss, New York.

Harrington, M. G., Coffman, J. A., and Davidson, E. H. (1997). Covalent variation is a general property of transcription factors in the sea urchin embryo. *Mol. Marine Biol. Biotechnol.* **6**, 153–162.

Henry, J. J., Kleug, K. M., and Raff, R. A. (1992). Evolutionary dissociation between cleavage, cell lineage and embryonic axes in sea urchin embryos. *Development (Cambridge, UK)* **114**, 931–938.

Höög, C., Calzone, F. J., Cutting, A. E., Britten, R. J., and Davidson, E. H. (1991). Gene regulatory factors of the sea urchin embryo. II. Two dissimilar proteins, P3A1 and P3A2, bind to the same target sites that are required for early territorial gene expression. *Development (Cambridge, UK)* **112**, 351–364.

Hough-Evans, B. R., Franks, R. R., Zeller, R. W., Britten, R. J., and Davidson, E. H. (1990). Negative spatial regulation of the lineage specific *CyIIIa* actin gene in the sea urchin embryo. *Development (Cambridge, UK)* **110**, 41–50.

Hörstadius, S. (1939). The mechanics of sea urchin development, studied by operative methods. *Biol. Rev. Cambridge Philos. Soc.* **14**, 132–179.

Hörstadius, S. (1973). "Experimental Embryology of Echinoderms." Oxford Univ. Press (Clarendon), Oxford.

Kirchhamer, C. V., and Davidson, E. H. (1996). Spatial and temporal information processing in the sea urchin embryo: Modular and intramodular organization of the *CyIIIa* gene *cis*-regulatory system. *Development (Cambridge, UK)* **122**, 333–348.

Kirchhamer, C. V., Yuh, C. H., and Davidson, E. H. (1996). Modular cis-regulatory organization of developmentally expressed genes: Two genes transcribed territorially in the sea urchin embryo, and additional examples. *Proc. Natl. Acad. Sci. U.S.A.* **93**, 9322–9328.

Klein, P. S., and Melton, D. A. (1996). A molecular mechanism for the effect of lithium on development. *Proc. Natl. Acad. Sci. U.S.A.* **93**, 8455–8459.

Kominami, T. (1988). Determination of dorso-ventral axis in early embryos of the sea urchin, *Hemicentrotus pulcherrimus*. *Dev. Biol.* **127**, 187–196.

Kontrogianni-Konstantopoulos, A., Vlahou, A., Vu, D., and Flytzanis, C. N. (1996). A novel sea-urchin nuclear receptor encoded by alternatively spliced maternal RNAs. *Dev. Biol.* **177**, 371–382.

Kozlowski, M. T., Gan, L., Venuti, J. M., Sawadogo, M., and Klein, W. H. (1991). Sea urchin USF—a helix-loop-helix protein active in embryonic ectoderm cells. *Dev. Biol.* **148**, 625–630.

Kozlowski, D. J., Gagnon, M. L., Marchant, J. K., Reynolds, S. D., Angerer, L. M., and Angerer, R. C. (1996). Characterization of a span promoter sufficient to mediate correct spatial regulation along the animal–vegetal axis of the sea-urchin embryo. *Dev. Biol.* **176**, 95–107.

Lee, J. J., Calzone, F. J., and Davidson, E. H. (1992). Modulation of sea urchin actin mRNA prevalence during embryogenesis: Nuclear synthesis and decay rate measurements of transcripts from five different genes. *Dev. Biol.* **149**, 415–431.

Li, J. M., Parsons, R. A., and Marzluff, W. F. (1994). Transcription of the sea urchin U6 gene *in vitro* requires a TATA-like box, a proximal sequence element, and sea-urchin USF, which binds an essential E-box. *Mol. Cell. Biol.* **14**, 2191–2200.

Li, X. T., Chuang, C. K., Mao, C. A., Angerer, L. M., and Klein, W. H. (1997). 2 Otx proteins generated from multiple transcripts of a single-gene in *Strongylocentrotus purpuratus Dev. Biol.* **187**, 253–266.

Logan, C. Y., and McClay, D. R. (1997). The allocation of early blastomeres to the ectoderm and endoderm is variable in the sea urchin embryo. *Development (Cambridge, UK)* **124**, 2213–2223.

Luke, N. H., Killian, C. E., and Livingston, B. T. (1997). SpFkh1 encodes a transcription factor implicated in gut formation during sea urchin development. *Dev. Growth Differ.* **39**, 285–294.

Makabe, K. W., Kirchhamer, C. V., Britten, R. J. and Davidson, E. H. (1995). Cis-regulatory control of the *SM50* gene, an early marker of skeletogenic lineage specification in the sea urchin embryo. *Development (Cambridge, UK)* **121**, 1957–1970.

Mao, C. A., Wikramanayake, A. H., Gan, L., Chuang, C. K., Summers, R. G., and Klein, W. H. (1996). Altering cell fates in sea urchin embryos by overexpressing SpOtx, an orthodenticle-related protein. *Development (Cambridge, UK)* **122**, 1489–1498.

Martinez, P., and Davidson, E. H. (1997). *SpHmx*, a sea urchin homeobox gene expressed in embryonic pigment cells. *Dev. Biol.* **181**, 213–222.

Miller, R. N., Dalamagas, D. G., Kingsley, P. D., and Ettensohn, C. A. (1996). Expression of S9 and actin *CyIIa* messenger-RNAs reveals dorsoventral polarity and mesodermal sublineages in the vegetal plate of the sea urchin embryo. *Mech. Dev.* **60**, 3–12.

Niemeyer, C. C. and Flytzanis, C. N. (1993). Upstream elements involved in the embryonic regulation of the sea urchin *CyIIIb* actin gene—temporal and spatial specific interactions at a single cis-acting element. *Dev. Biol.* **156**, 293–302.

Nocente-McGrath, C., Brenner, C. A., and Ernst, S. G. (1989). Endo16, a lineage-specific protein of the sea urchin embryo, is first expressed just prior to gastrulation. *Dev. Biol.* **136**, 264–272.

Okazaki, K. (1975). Spicule formation by isolated micromeres of the sea urchin embryo. *Am. Zool.* **15**, 567–581.

Ransick, A., and Davidson, E. H. (1993). A complete second gut induced by transplanted micromeres in the sea urchin embryo. *Science* **259**, 1134–1138.

Ransick, A., and Davidson, E. H. (1995). Micromeres are required for normal vegetal plate specification in sea urchin embryos. *Development (Cambridge, UK)* **121**, 3215–3222.

Ransick, A., and Davidson, E. H. (1998). Late specification of veg$_1$ lineages to endodermal fate in the sea urchin embryo. *Dev. Biol.* **195**, 38–48.

Ransick, A., Ernst, S., Britten, R. J., and Davidson, E. H. (1993). Whole mount *in situ* hybridization shows Endo16 to be a marker for the vegetal plate territory in sea urchin embryos. *Mech. Dev.* **42**, 117–124.

Reynolds, S. D., Angerer, L. M., Palis, J., Nasir, A., and Angerer, R. C. (1992). Early mRNAs, spatially restricted along the animal–vegetal axis of sea urchin embryos, include one encoding a protein related to tolloid and BMP-1. *Development (Cambridge, UK)* **114**, 769–786.

Ruffins, S. W., and Ettensohn, C. A., (1996). A fate map of the vegetal plate of the sea urchin (*Lytechinus variegatus*) mesenchyme blastula. *Development (Cambridge, UK)* **122**, 253–263.

Sherwood, D., and McClay, D. (1997). Identification and localization of a sea-urchin notch homolog—insights into vegetal plate regionalization and notch receptor regulation. *Development (Cambridge, UK)* **124**, 3363–3374.

Sucov, H. M., Benson, S., Robinson, J. J., Britten, R. J., Wilt, F., and Davidson, E. H. (1987). A lineage-specific gene encoding a major matrix protein of the sea urchin embryo spicule. II. Structure of the gene and derived sequence of the protein. *Dev. Biol.* **120**, 507–519.

Summers, R. G., Morrill, J. B., Leith, A., Marko, M., Piston, D. W., and Stonebraker, A. T. (1993). A stereometric analysis of karyokinesis, cytokinesis and cell arrangements during and following fourth cleavage period in the sea urchin, *Lytechinus variegatus. Dev. Growth Differ.* **35**, 41–57.

Tomlinson, C. R., Kozlowski, M. T., and Klein, W. H. (1990). Ectoderm nuclei from sea urchin embryos contain a Spec-DNA binding-protein similar to the vertebrate transcription factor USF. *Development (Cambridge, UK)* **110**, 259–272.

Virbasius, C. M. A., Virbasius, J. V., and Scarpulla, R. C. (1993). NRF-1, an activator involved in nuclear–mitochondrial interactions, utilizes a new DNA-binding domain conserved in a family of developmental regulators. *Genes Dev.* **7**, 2431–2445.

Vlahou, A., Gonzalez-Rimbau, M., and Flytzanis, C. N. (1996). Maternal mRNA encoding the steroid hormone receptor SpCOUP-TF is localized in sea urchin eggs. *Development (Cambridge, UK)* **122**, 521–526.

Wang, D. G.-W., Kirchhamer, C. V., Britten, R. J., and Davidson, E. H. (1995a). SpZ12-1, a negative regulator required for spatial control of the territory-specific *CyIIIa* gene in the sea urchin embryo. *Development (Cambridge, UK)* **121**, 1111–1122.

Wang, D. G.-W., Britten, R. J., and Davidson, E. H. (1995b). Maternal and embryonic provenance of a sea urchin embryo transcription factor, SpZ12-1. *Mol. Marine Biol. Biotechnol.* **4**, 148–153.

Wang, W., Wikramanayake, A. H., Gonzalez-Rimbau, M., Vlahou, A., Flytzanis, C. N., and Klein, W. H. (1996). Very early and transient vegetal plate expression of SpKrox1, a Kruppel/Krox gene from *Strongylocentrotus purpuratus. Mech. Dev.* **60**, 185–195.

Wei, Z., Angerer, L. M., Gagnon, M. L., and Angerer, R. C. (1995). Characterization of the SpHE promoter that is spatially regulated along the animal vegetal axis of the sea urchin embryo. *Dev. Biol.* **171**, 195–211.

Wei, Z., Angerer, L. M., and Angerer, R. C. (1997). Multiple positive cis elements regulate the asymmetric expression of the SpHE gene along the sea urchin embryo animal–vegetal axis. *Dev. Biol.* **187**, 71–78.

Wikramanayake, A. H., and Klein, W. H. (1997). Multiple signaling events specify ectoderm and pattern the oral–aboral axis in the sea urchin embryo. *Development (Cambridge, UK)* **124**, 13–20.

Yuh, C.-H., and Davidson, E. H. (1996). Modular cis-regulatory organization of *Endo16*, a gut-specific gene of the sea urchin embryo. *Development (Cambridge, UK)* **122**, 1069–1082.

Yuh, C.-H., Moore, J. G., and Davidson, E. H. (1996). Quantitative functional interrelations within the cis-regulatory system of the S. *purpuratus Endo16* gene. *Development (Cambridge, UK)* **122**, 4045–4056.

Yuh, C.-H., Bolouri, H., and Davidson, E. H. (1998). Genomic cis-regulatory logic: Functional analysis and computational model of a sea urchin gene control system. *Science* **279**, 1896–1902.

Zeller, R. W., Britten, R. J., and Davidson, E. H. (1995a). Developmental utilization of SpP3A1 and SpP3A2: Two proteins which recognize the same DNA target site in several sea urchin gene regulatory regions. *Dev. Biol.* **170**, 75–82.

Zeller, R. W., Coffman, J. A., Harrington, M. G., Britten, R. J., and Davidson, E. H. (1995b). SpGCF1, a sea urchin embryo DNA binding protein, exists as five nested variants encoded by a single mRNA. *Dev. Biol.* **169**, 713–727.

Zeller, R. W., Griffith, J. D., Moore, J. G., Kirchhamer, C. V., Britten, R. J., and Davidson, E. H. (1995c). A multimerizing transcription factor of sea urchin embryos capable of looping DNA. *Proc. Natl. Acad. Sci. U.S.A.* **92**, 2989–2993.

3

Otx, β-Catenin, and the Specification of Ectodermal Cell Fates in the Sea Urchin Embryo

Athula H. Wikramanayake
William H. Klein
Department of Biochemistry and Molecular Biology
University of Texas M. D. Anderson Cancer Center
Houston, Texas 77030

I. Introduction

Understanding how individual cell types arise during embryogenesis is at the heart of contemporary developmental biology. In fact, theoretical arguments have been recently put forward suggesting that cell fate specification events supersede those of axial patterning and may be indicative of a more primordial step in the evolution of embryo development (Davidson et al., 1995). In sea urchins, the fates of all the major cell types are specified very early in embryogenesis and it is the establishment of these individual cell types, rather than reiterative patterning along a body axis, which is critical for embryonic development. Later developmental events, on the other hand, probably require axial patterning; in particular, the development of the adult rudiment within the growing sea urchin larva appears to depend on homologs of genes associated with axial patterning in insects and vertebrates (Panganiban et al., 1997; Lowe and Wray, 1997). In this chapter, we will review the current status of the specification events leading to the differentiation of ectodermal cell types in the sea urchin embryo. Section II describes the embryological and molecular features that make ectodermal cell fate specification an attractive experimental system. In Section III, we discuss transcription factors, particularly members of the Otx family, thought to be involved in directing ectoderm-specific gene expression and more generally ectoderm specification and differentiation. Section IV presents the current evidence linking extracellular signaling events with ectoderm-specific gene expression. The prevailing model that the activities of uniformly distributed transcription factors are regionally modulated by external signaling pathways will be evaluated in light of new experiments. Section V discusses our current knowledge on cell–cell interactions and vegetal cell signaling associated with ectoderm specification and patterning. We will attempt to place these signaling events in the context of cell fate decisions along the entire animal–vegetal axis. Finally, in Section VI we compare the relevance of recent findings on sea urchin development to other organisms and consider how evolutionary mechanisms influence cell fate-specification processes.

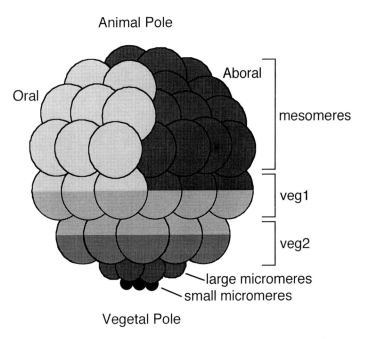

Figure 1 Fate map of S. *purpuratus* 60-cell stage embryo. [Adapted from Logan and McClay (1997a), © Company of Biologists, Ltd.] The figure depicts the approximate veg$_2$ boundary between mesoderm and endoderm, and the veg$_1$ boundary between endoderm and ectoderm, as well as the five embryonic territories: oral ectoderm, aboral ectoderm, vegetal plate, large micromeres, and small micromeres.

II. Ectodermal Territories and Cell Types

A. Descriptive Embryology and Cell Lineage Analysis

By the sixth cleavage (60-cell stage), the *Strongylocentrotus purpuratus*[1] embryo has been partitioned into five embryonic territories: small micromeres, large micromeres, vegetal plate, oral ectoderm, and aboral ectoderm (Fig. 1; Cameron and Davidson, 1991). These territories, readily defined by invariant cell lineages and expression patterns of cell type-specific genes, underlie the importance of regional specification as a critical step in sea urchin development. The formation of the ectodermal territories is of obvious importance for normal development. However, until the advent of modern lineage tracing techniques and the isolation of molecular markers, knowledge of ectodermal cell lineages was ambiguous, and the oral–aboral axis, which mainly reflects the polarity of oral and aboral ectoderm, was largely ignored. This was because the embryonic ectoderm has few distinguishing morphological characteristics and mapping the fate of ectodermal cells in normal or experimentally perturbed embryos by classic techniques was difficult. It is now evident that specification

of cell types along the oral–aboral axis is closely integrated with cell specification events along the animal–vegetal axis (Davidson, 1989).

The oral ectoderm territory is derived from the progeny of seven founder cells, which are segregated by the sixth cleavage (Cameron and Davidson, 1991). At the pluteus stage, the oral ectoderm consists of a facial epithelium and neurons associated with the mouth and ciliated band. This latter structure is a ciliated columnar epithelium that marks the border of aboral and oral ectoderm and probably arises through an inductive interaction at the oral–aboral interface (Cameron *et al.*, 1990). The aboral ectoderm territory is derived from eleven founder cells which have segregated by the sixth cleavage (Cameron and Davidson, 1991). The differentiated aboral ectoderm of the pluteus is made up of approximately 450 ciliated squamous epithelial cells that are in close contact with each other through septate junctions.

Recent cell lineage-tracing experiments indicate that the establishment of the vegetal plate and ectoderm territories is somewhat more complex than the conventional view described in the preceding paragraphs. Although it was previously thought that the entire veg$_1$ tier at the 60-cell stage gave rise solely to ectoderm, it is now generally accepted that the more vegetal descendants of veg$_1$ give rise to endoderm (Fig. 1; Logan and McClay, 1997a,b; Ransick and Davidson, 1998). Since the veg$_1$ tier of blastomeres divide after the sixth cleavage, those veg$_1$ descendants destined to become ectoderm will contribute to oral and aboral ectoderm

---

[1] Several sea urchin species are discussed in this chapter and it should be noted that many aspects of development can show significant variation from one species to another. These differences are pointed out where relevant.

territories later than previously believed. *In situ* hybridization patterns with *Spec1*, an aboral ectoderm-specific mRNA, and *Endo16*, an endoderm-specific transcript, are consistent with the idea that veg$_1$-derived cells within the aboral ectoderm territory arise at later times than other cells within the territory and that a well-defined ectoderm–endoderm boundary between veg$_1$ descendants is set up by the early blastula stage (Logan and McClay, 1997a,b; Ransick and Davidson, 1998; see also Chapter 2 by Cameron and Coffman; and Chapter 4 by Logan and McClay).

The ability to accurately predict the cell lineages which will give rise to the oral and aboral ectoderm territories provides an important prerequisite for uncovering the mechanisms by which cell type specification occurs. Along with the well-defined cell-fate map, progress has also been made on the identification and characterization of genes expressed specifically in ectoderm territories and cell types. Some of these genes have been particularly useful as cell differentiation markers and as sources for cis-regulatory elements and trans-factors associated with ectoderm-specific transcriptional control.

B. Ectoderm-Specific Genes

Numerous genes have been identified in sea urchins whose expression patterns are restricted at some stage in development to ectodermal lineages. For the purposes of this review, these genes are grouped into five categories: early aboral ectoderm-specific, late aboral ectoderm-restricted, late oral ectoderm/foregut-restricted, very early blastula (VEB), and *COUP-TF/Hbox12*.

1. Early Aboral Ectoderm-Specific Genes

The first genes shown to be expressed exclusively in aboral ectoderm cells were the *Spec1* and *Spec2* genes of S. *purpuratus* (Lynn *et al.*, 1983). This eight-member gene family encodes a group of calcium-binding proteins believed to function as calcium ion buffers in aboral ectoderm cells (Klein *et al.*, 1991). The homologs of the *Spec1* and *Spec2* genes from *Lytechinus pictus* (*LpS1*) and *Lythechinus variegatus* (*LvS1*) have also been cloned (Klein *et al.*, 1991). The *Spec* genes and their *Lytechinus* counterparts are activated at the late cleavage–early blastula stage, shortly after the lineages giving rise to aboral ectoderm are established. These genes have proven to be extremely valuable as markers for aboral ectoderm specification and differentiation. Although the *Spec* mRNAs accumulate exclusively in aboral ectoderm progenitors, experiments monitoring the spatial distribution of *Spec* pre-mRNAs

suggests that aboral ectoderm specificity may be post-transcriptionally regulated, at least at the midgastrula stage (Gagnon *et al.*, 1992). However, reporter genes fused to regulatory sequences from the *Spec2a* gene show aboral ectoderm-specific expression throughout embryo development (Gan *et al.*, 1990a).

The other prominent aboral ectoderm-specific gene is *CyIIIa actin*. This gene is part of a six-member gene family encoding cytoplasmic and muscle actins (Cox *et al.*, 1986). In S. *purpuratus*, each member of the actin gene family is expressed in a distinct temporal and spatial pattern, although these expression patterns are not conserved in other sea urchin species (Fang and Brandhorst, 1996). *CyIIIa actin* is activated at virtually the same time as the *Spec* genes suggesting that common cis-regulatory elements are associated with the temporal and spatial expression of these genes. However, comparisons between *CyIIIa actin* and *Spec2a* transcriptional control regions show little similarity and it appears that these genes are largely regulated by different transcription factors (Mao *et al.*, 1994; Kirchhamer and Davidson, 1996).

2. Late Aboral Ectoderm-Restricted Genes

In contrast to genes whose mRNAs accumulate exclusively in aboral ectoderm cell lineages, several zygotically expressed genes have been isolated which are initially expressed more uniformly throughout the embryo then gradually restrict to aboral ectoderm lineages after the mesenchyme blastula stage. These genes include *HpArs*, which encodes an arylsulfatase from *Hemicentrotus pulcherrimus* (Akasaka *et al.*, 1990), *Hox8/Hbox1*, a member of the sea urchin Hox cluster (Angerer *et al.*, 1989), and *SpMTA*, a metallothionein gene from S. *purpuratus* (Nemer *et al.*, 1995). *Hox8/Hbox1* is expressed in a subset of aboral ectoderm cells at the pluteus stage, the only gene known to be restricted in this manner (Angerer *et al.*, 1989). The expression patterns of the late aboral ectoderm-restricted genes indicate that in addition to early gene activation events, late mechanisms must exist for generating aboral ectoderm-specific gene expression. In support of the idea for separable early and late events, a "module" that specifically regulates the late expression of *CyIIIa actin* has been identified in the *CyIIIa actin* transcriptional control region (Kirchhamer and Davidson, 1996). It is clear, however, that the aboral ectoderm cells are committed to their fate many hours before these late expression events.

3. Late Oral Ectoderm/Foregut-Restricted Genes

In contrast to aboral ectoderm-specific genes, there are no reported examples of genes strictly associated with oral ectoderm-specific expression, although it is likely that these exist, particularly for neurogenic progenitors. However, a large class of genes has been characterized with uniform expression patterns between the early cleavage and mesenchyme blastula stages followed by a gradual restriction to the oral ectoderm and foregut at later times (Kingsley *et al.*, 1993). One of these genes encodes the Ecto V antigen, a secreted glycoprotein that serves as a useful marker for oral ectoderm (Coffman and McClay, 1990). Ecto V is widely used to distinguish oral ectoderm because it is present in relatively low amounts in eggs and cleavage stage embryos and accumulates several fold during embryogenesis specifically in the oral ectoderm and foregut. It is unclear why so many sea urchin embryonic genes have this oral ectoderm/foregut expression pattern, but it may be related to the fact that the oral ectoderm and foregut cells are the most rapidly proliferating cells between the blastula and pluteus stages. The accumulations of these gene products are thought to anticipate the later events of larval growth and development (Kingsley *et al.*, 1993).

4. VEB Genes

Several genes have been identified that are transiently expressed in a spatially restricted pattern between the 16-cell and hatching blastula stages (Reynolds *et al.*, 1992; Ghiglione *et al.*, 1993). These genes are expressed in an asymmetric pattern along the animal–vegetal axis but their expression along the oral–aboral axis is uniform. VEB transcripts are the earliest mRNAs currently known to reveal asymmetry along the animal–vegetal axis although their expression patterns are not consistent with their involvement in any known cell type- or territory-specification process. Interestingly, the VEB genes are expressed along the animal–vegetal axis with a variable vegetal boarder that can be shifted upwards by the vegetalizing agent, lithium chloride (Ghiglione *et al.*, 1996). Thus, the VEB mRNAs accumulate in all cells of the presumptive ectoderm as well as in variable portions of presumptive endoderm and mesenchyme. Because of their early activation and spatial expression patterns, the VEB genes are likely to be regulated by maternal factors that are prelocalized in the egg cytoplasm along the animal–vegetal axis. As would be predicted by this hypothesis, the VEB genes are activated in a cell autonomous fashion; when embryos are dissociated at the two-cell stage and the cultured dividing cells are kept dissociated, VEB genes are expressed at the appropriate times and levels (Reynolds *et al.*, 1992; Ghiglione *et al.*, 1993).

The function of at least one of the VEB genes appears to involve cell-fate specification along the animal–vegetal axis. Two well-described VEB genes encode metalloendoproteases: one is the hatching enzyme

(termed *SpHe* in *S. purpuratus* and HE in *Paracentrotus lividus*) and the other (*SpAn* or *BP10*) is similar to *BMP-1* and *Tolloid* (Reynolds *et al.*, 1992; Ghiglione *et al.*, 1993, 1994). Recent experiments suggest that the SpAn protease functions to activate a sea urchin BMB-4 homolog, which appears to suppress endoderm and oral ectoderm and enhance aboral ectoderm formation when overexpressed in sea urchin embryos by mRNA injection (Angerer and Angerer, 1997).

5. COUP-TF/Hbox12

The nuclear factor *SpCOUP-TF* is the highly conserved sea urchin homolog of the COUP family of transcription factors. *SpCOUP-TF* mRNA is asymmetrically distributed in the egg and early blastomeres lateral to the animal–vegetal axis and at a 45° angle to the oral–aboral axis (Vlahou *et al.*, 1996). The embryonic expression of the *SpCOUP-TF* gene is spatially restricted in the oral ectoderm of the early embryo and at later stages in the neurogenic region of the ciliated band. SpCOUP-TF protein is distributed uniformly in eggs but following fertilization, translation of *SpCOUP-TF* mRNA leads to SpCOUP-TF protein accumulation on one side of the embryo (Vlahou *et al.*, 1998). Flytzanis and co-workers have suggested that SpCOUP-TF is involved in specification events of neurogenic cells in later embryogenesis but the significance of the maternal spatial expression pattern of *SpCOUP-TF* is not certain since it does not coincide with the oral ectoderm territory where neurons are derived (Vlahou *et al.*, 1997). A similar asymmetric expression pattern has been reported for a paired homeobox-containing gene from *P. lividus*, *PlHbox12* (DiBernardo *et al.*, 1995). The *Hbox12* gene is transiently expressed between the 4- and 120-cell stages and, as with *SpCOUP-TF*, is restricted to one side of the embryo lateral to the animal–vegetal axis. The expression patterns of *PlHbox12* suggest a role in early ectoderm specification events, but direct evidence for such a role must still be obtained. Because *PlHbox12* and *SpCOUP-TF* have been isolated from different sea urchin species, it is not clear if they have overlapping spatial expression patterns or if each gene defines a unique region along the animal–vegetal axis.

III. Role of Transcription Factors in Aboral Ectoderm Formation

A. Cis-Regulatory Elements and Trans-Factors

The identification of aboral ectoderm-specific genes affords the opportunity to characterize corresponding transcriptional control regions and isolate transcription factors involved in aboral ectoderm-specific gene ex pression. This approach has been particularly useful in the sea urchin system where moving "upstream" from gene control regions to transcription factors is not limited by availability of material and should lead ultimately to the mechanisms by which the transcription factors are able to achieve aboral ectoderm-specific transcription. The underlying assumption in these experiments is that the events that lead to aboral ectoderm-specific gene activation are intimately related to cell fate specification.

Mapping of DNA elements onto the transcriptional control regions of five aboral ectoderm-specific genes, *CyIIIa actin*, *CyIIIb actin*, *HpArs*, *SpMTA*, and *Spec2a*, has been reported (Kirchhamer and Davidson, 1996; Xu *et al.*, 1996; Iuchi *et al.*, 1995; Nemer *et al.*, 1995; Mao *et al.*, 1994). Some important conclusions can be drawn from these analyses. First, cis-regulatory elements associated with aboral ectoderm gene expression are generally dissimilar, both within and among sea urchin species, and the comparisons indicate that there are multiple ways to achieve aboral ectoderm-specific gene expression. Notable exceptions exist however, including a conserved cassette of DNA elements found upstream of *CyIIIa actin* and within the first intron of *SpMTA* (Nemer *et al.*, 1993, 1995), and similar Otx binding sites associated with the transcriptional enhancers of *Spec2a* and *HpArs* (Mao *et al.*, 1994; Sakamoto *et al.*, 1997). Second, a complex network of interacting enhancers (termed modules by Davidson and co-workers) each containing multiple DNA-regulatory elements act in concert to confer the observed spatial and temporal expression patterns on the aboral ectoderm genes. This is best documented in the case of *CyIIIa actin*, although the other genes that have been less extensively analyzed appear to fit this model (Kirchhamer *et al.*, 1996). Embryo expression experiments with transcriptional control regions from aboral ectoderm-expressed genes clearly oppose the simple scheme whereby a single enhancer with one or a few DNA elements can confer the appropriate temporal and spatial expression patterns. It is becoming increasingly clear that a multitude of transcription factors and associated proteins are involved in aboral ectoderm-specific gene expression. Third, both positive and negative control elements work in combination to confer temporal and spatial specificity. For example, for *CyIIIa actin*, a positive DNA element that binds to the transcription factor SpRunt-1 is required for late expression in all but endodermal cells of the embryo, while two negative DNA elements, one binding to SpMyb and the other to SpP3AII, are required to repress expression in the oral ectoderm and mesenchyme cell lineages (Kirchhamer and Davidson, 1996; Coffman *et al.*, 1996, 1997; Zeller *et al.*, 1995; Chapter 2 by Cameron and Coffman). Similarly, a cluster of four Otx sites within the *Spec2a* control region is essential for expression in aboral ectoderm

and mesenchyme cells while a region further upstream of these sites is required to silence expression in mesenchyme lineages (Gan *et al.*, 1990a; Mao *et al.*, 1994).

B. Otx and Cell Fate Specification

1. Otx and Aboral Ectoderm

Characterization of the *Spec2a* transcriptional control region led to the identification of an 188-bp enhancer [the repeat-spacer-repeat (RSR) enhancer] which is both necessary and sufficient for reporter gene activation in aboral ectoderm and mesenchymal cells (Gan *et al.*, 1990a,b). The RSR enhancer is conserved within the 5' region of all *Spec* genes and an additional 50 to 100 RSR sequences are interspersed throughout the *S. purpuratus* genome (Hardin *et al.*, 1988). Four redundant TAATCC/T sites were shown to be required for the RSR enhancer activity and to be bound by members of the Otx family of homeobox-containing transcription factors (Mao *et al.*, 1994; Gan *et al.*, 1995; Li *et al.*, 1997). Otx proteins have a strong preference for TAATCC/T sequence motifs due to the presence of a lysine in position 50 of the third (DNA recognition) helix of the homeobox (Wilson *et al.*, 1993). Members of this family are defined by their highly conserved homeoboxes and essential roles in anterior cephalic patterning and other axial patterning events in insects and vertebrates (Finkelstein and Boncinelli, 1994). In *S. purpuratus*, there are two Otx forms, SpOtx(α) and SpOtx(β) which differ in their N-terminal regions but are identical in their homeobox and C-terminal regions (Li *et al.*, 1997). SpOtx(α) and SpOtx(β) are generated by differential promoter utilization and alternative splicing of a single *SpOtx* gene. Thus, α- and β-specific exons are responsible for the different N termini found in the two SpOtx forms. The single *S. purpuratus SpOtx* gene encoding two proteins contrasts with the vertebrate *Otx1* and *Otx2* genes which each encode a single protein (Simeone *et al.*, 1992). Sequence comparisons show that the N-terminal region of SpOtx(β) has significant similarity to the corresponding regions of Otx1 and Otx2 while the N-terminal region of SpOtx(α) is completely distinct (Li *et al.*, 1997). A likely model for *Otx* gene evolution in deuterostomes is that a single copy of *Otx* existed in the genome of a deuterostome ancestor of sea urchins and vertebrates. Sometime after the split of chordate and echinoderm lineages, the *Otx* gene was duplicated in the chordate lineage giving rise to *Otx1* and *Otx2* while the α-specific exon was acquired by the *Otx* gene as a separate event in the echinoderm lineage.

The two SpOtx forms have distinct temporal and spatial expression patterns and may have different functions (Li *et al.*, 1997). SpOtx(α) transcripts accumulate in all cells during cleavage stages and are gradually con-

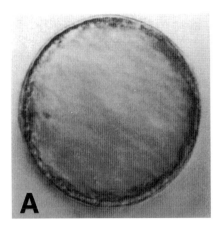
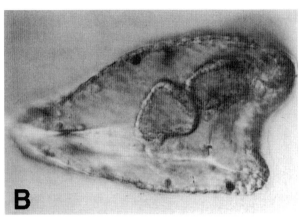

Figure 2 Alteration in cell fate by overexpression of SpOtx(α). S. *purpuratus* eggs were fertilized and injected with 2 pl of a 2 mg/ml solution containing SpOtx(α) transcripts. Embryos were collected after 3 days of culturing. (A) An embryoid showing the most severe defect, a hollow ball of aboral ectoderm cells. (B) A noninjected control embryo; surface oral ectoderm is on the right. The foregut and midgut, dark pigment cells, and skeletal rods derived from primary mesenchyme cells can be seen within the blastocoel. These elements are missing in the SpOtx(α)-injected embryoids.

centrated to oral ectoderm and vegetal plate territories during gastrulation as transcript levels decline. In contrast, *SpOtx(β)* transcripts begin to accumulate at the blastula stage primarily in ectoderm and later are largely restricted to oral ectoderm and vegetal plate territories where they remain at maximal levels throughout embryogenesis. These patterns imply that SpOtx(α) has an early function and SpOtx(β) a late function in sea urchin development. The SpOtx(α) protein is found in the cytoplasm and at the nuclear membrane of eggs and cleaving blastomeres until the 60-cell stage when it begins to translocate into nuclei (Mao *et al.*, 1996). By the 120-cell stage, SpOtx(α) is in the nuclei of all cells of the embryo. This is coincident with the time of *Spec2a* activation and supports the hypothesis that SpOtx(α) is involved in the initial temporal activation of *Spec2a* and other *Spec* genes. SpOtx(β) protein accumulates later than SpOtx(α) and appears to be mostly nuclear. Both forms of SpOtx are present in the nucleus at the end of the cleavage stage although the SpOtx(α) is the predominant form (C.-K. Chuang and W. H. Klein, 1996, unpublished results). Because SpOtx(α) and SpOtx(β) bind to the Otx sites on *Spec2a* with equal affinity (Li *et al.*, 1997), it is possible that either or both forms are involved in *Spec2a* gene activation. However, the uniform nuclear distribution of the SpOtx proteins at the time when *Spec2a* is activated implies that neither SpOtx(α) nor SpOtx(β) can be solely responsible for conferring aboral ectoderm-specific expression.

Dramatic evidence that SpOtx forms are involved in aboral ectoderm differentiation comes from overexpression studies. When large amounts of *SpOtx(α)* or *SpOtx(β)* mRNAs are injected into sea urchin eggs, the embryoids that develop are swimming balls of aboral ectoderm cells, as evidenced by the enhancement of aboral ectoderm-specific gene expression and the corre-

sponding loss of oral ectoderm, mesenchyme, and endoderm-specific gene expression (Fig. 2; Mao *et al.*, 1996). No such morphological defects are observed when the homeobox, N- or C-terminal transcriptional activation domains (TADs) from SpOtx(α) are deleted, indicating that the effects are due to SpOtx functioning as a transcription factor. Overexpressing *SpOtx(α)* extinguishes endogenous *SpOtx(β)* expression but overexpressing *SpOtx(β)* has no effect on *SpOtx(α)* expression, suggesting that SpOtx(α) may repress *SpOtx(β)* expression at early times in normal embryogenesis (Li *et al.*, 1997). These experiments demonstrate that when overexpressed, either form of SpOtx can direct cells toward the aboral ectoderm differentiation program by altering the normal fate of nonaboral ectoderm cells.

One likely interpretation for these results is that SpOtx[2] interacts with another factor to prevent it from functioning as an activator of aboral ectoderm gene expression in nonaboral ectoderm cells. In *Xenopus*, Otx2 requires an unidentified factor to restrict cement gland gene expression to the anterior regions of the embryo (Gammill and Sive, 1997). Producing large amounts of SpOtx in nonaboral ectoderm cells would overcome the inhibitory effects of the putative factor. In the unperturbed embryo, vegetal signaling mechanisms could disrupt the interaction between SpOtx and the hypothesized inhibitory factor specifically in aboral ectoderm cells, allowing SpOtx to participate in the activation of *Spec2a*.

Otx forms from *H. pulcherrimus* have been reported by Akasaka and co-workers (Sakamoto *et al.*, 1997). The aboral ectoderm-restricted *HpArs* gene con-

---

[2] For simplicity, we occasionally refer to SpOtx(α) and SpOtx(β) collectively as SpOtx, realizing that there are likely to be functional differences between the two forms.

tains a 229-bp enhancer in its first intron that is required for the activation of *HpArs* at the blastula stage but does not appear to be involved in aboral ectoderm-specific expression, at least not initially (Iuchi *et al.,* 1995). As discussed above, the *HpArs* gene is first expressed in all embryonic cells but by the gastrula stage is restricted to aboral ectoderm (Akasaka *et al.,* 1990). The *HpArs* enhancer contains two tandemly repeated Otx sites that are required for enhancer activity (Sakamoto *et al.,* 1997). Two Otx forms, HpOtx$_E$ and HpOtx$_L$, which are virtually identical to SpOtx(α) and SpOtx(β) in their sequence, gene organization, and temporal expression patterns, bind to the Otx sites on the *HpArs* enhancer and either or both are likely to be involved in the temporal activation of *HpArs*. The fact that *HpArs* expression is initially uniform throughout the embryo and is subsequently restricted to aboral ectoderm cells, suggests that in *H. pulcherrimus*, mechanisms which modulate the spatial activity of HpOtx forms may not exist or may be delayed relative to similar mechanisms in *S. purpuratus*. When overexpressed by mRNA injection, both forms of HpOtx produce squamous epithelial balls with suppression of endodermal and mesenchymal cell types, but unlike what is observed for SpOtx overexpression, oral ectoderm as well as aboral ectoderm markers are expressed in these embryoids (K. Akasaka, 1997, personal communication).

2. Otx and Endoderm

Analysis of the transcriptional control region of the *Endo16* gene from *S. purpuratus* reveals that a single Otx-binding site located within a region denoted module A is essential for the initial expression of *Endo16* in the vegetal plate (Yuh and Davidson, 1996; Yuh et al., 1996; see Chapter 2 by Cameron and Coffman). SpOtx(α) and SpOtx(β) bind to the *Endo16* Otx site with equal affinities, comparable to the Otx sites of *Spec2a* (X. Li and W. H. Klein, 1997, unpublished results). Apparently, SpOtx proteins are required for two disparate gene activation events at the early blastula stage; the activation of the *Endo16* gene in the vegetal plate and the activation of the *Spec2a* gene in progenitors of aboral ectoderm. DNA elements interacting with the Otx sites within the *Endo16* and *Spec2a* control regions in combination with the cellular environment may allow SpOtx to discriminate between the two genes in the different embryonic territories. For example, module A of *Endo16* contains binding sites for several other DNA-binding proteins besides Otx, including several redundant elements termed CG sites (Yuh and Davidson, 1996). These CG sites are also present on the *Endo16* basal promoter and match the consensus binding site for the transcription factor LEF1/Tcf [(A/T)(A/T)CAAAG] (Brannon *et al.,* 1997). The CG sites are required for maximal transcription from the *Endo16* control region (Yuh *et al.,* 1996). LEF1/Tcf, which functions as a heterodimer with β-

catenin, may only be functional in vegetal plate cells where β-catenin is nuclear (see Section V). Thus, vegetal plate-specific expression of *Endo16* could result from a synergistic interaction of SpOtx and LEF1/Tcf-β-catenin on the *Endo16* control region. Within the *Endo16* control region, SpOtx may function only weakly without LEF1/Tcf-β-catenin, although this might not be the case for the *Spec2a* control region which has multiple Otx sites and no LEF1/Tcf sites.

IV. Linking Vegetal Signaling to Aboral Ectoderm Gene Activation

By focusing on the specification of the aboral ectoderm cell type and the activation of the *Spec2a* gene, which is expressed specifically in progenitors of this cell type, some preliminary conclusions can be drawn about the mechanisms of cell fate specification in sea urchin development. In Section III.B.1, we summarized the experiments which implicate SpOtx in *Spec2a* gene activation. However, SpOtx is uniformly distributed in the embryo when *Spec2a* is activated, and in addition, also is required for *Endo16* expression in vegetal plate cells. It is now clear that many transcription factors involved in cell type-specific gene activation in the sea urchin embryo are not restricted to the region where cell type-specific expression is occurring (e.g., Coffman and Davidson, 1994; Mao *et al.,* 1996). The unfertilized egg contains significant levels of several transcription factors and their mRNAs to be used later for cell type-specific expression, as is the case for SpOtx(α). Davidson has suggested that for conditional cell fate-specification to occur, cell–cell interactions must initiate a signaling cascade which ultimately leads to the regional activation or inhibition of uniformly distributed transcription factors (Davidson, 1989). Such specification events would be predicted to modulate the activity of SpOtx in different regions of the embryo.

Is SpOtx responsible for aboral ectoderm-specific expression of *Spec2a*? It is clear that the Otx sites within the RSR enhancer are the major positive cis-regulatory elements associated with the *Spec2a* control region (Mao *et al.,* 1994). Nevertheless, it might be that spatial repression in nonaboral ectoderm cells is the chief means of aboral ectoderm-specific *Spec2a* expression. At least one spatial repression element exists upstream of the *Spec2a* RSR enhancer since deleting this DNA from the *Spec2a* control region causes ectopic expression in mesenchyme cells (Gan *et al.,* 1990b). However, there is no evidence for spatial repressor elements within the RSR enhancer, which by itself is capable of directing reporter gene expression in aboral ectoderm and mesenchyme cells when placed in front of a minimal promoter (Mao *et al.*, 1994). Importantly, the RSR enchancer and subfragments containing one or more Otx sites are incapable of directing expression to endoderm cells. The simplest interpretation of these results is that SpOtx cannot function to activate *Spec2a* in endoderm progenitor cells even though it is present in the nuclei of these cells and can activate *Endo16*. While the basis of mesenchyme cell expression is not clear, it is probably due to the presence of a non-Otx element within the enhancer that can be inhibited by the upstream negative mesenchyme element (Mao *et al.*, 1994).

It is particularly striking that excess SpOtx in sea urchin embryos suppresses nonaboral ectoderm cell fates and enhances aboral ectoderm, suggesting that excess SpOtx is overcoming an inhibitory effect in nonaboral ectoderm progenitor cells. SpOtx plays an important role in the activation of the *Endo16* gene in the vegetal plate. It is possible that in these cells excess SpOtx overcomes the normal endoderm differentiation program, which may rely on LEF1/Tcf-β-catenin interactions with SpOtx as discussed above. In the unperturbed embryo, the ratio of SpOtx in complex with LEF1/Tcf-β-catenin versus free SpOtx might be a critical indicator of endoderm versus ectoderm differentiation.

In general, the activity of SpOtx could be modulated by signals sent from the more vegetal veg$_1$ descendants at the ectoderm–endoderm boundary to more animal tiers of cells. Since activated β-catenin induces aboral ectoderm-specific gene expression when introduced into animal halves, the aboral ectoderm-inducing signals are likely to be initiated by the β-catenin signaling pathway. This pathway cannot be directly responsible for aboral ectoderm gene activation however, because β-catenin is not normally localized to the nucleus in ectoderm progenitor cells (Chapter 4 by Logan and McClay). While the basis of the hypothesized regional modulation of SpOtx is not known, phosphorylated forms of SpOtx and HpOtx have been reported (Sakamoto *et al.*, 1997; C.-K. Chuang and W. H. Klein, 1996, unpublished results).

V. Vegetal Signaling in Ectoderm Patterning and Cell Type Specification

A. Using Lithium as a Tool to Study Ectoderm Specification

In indirect developing sea urchins, most of the ectoderm is derived from the animal half of the embryo. The animal half is segregated at the third (equatorial) cleavage and consists of the an$_1$ and an$_2$ tiers of the 60-cell stage embryo (Fig. 1). A vegetal contribution to ectoderm comes from the more animal region of the veg$_1$

tier and these cells are committed to an ectodermal fate later in development than an$_1$ and an$_2$ derived ectoderm (Logan and McClay, 1997a,b; Ransick and Davidson, 1998). Historically, there have been differing views on the mechanisms through which ectoderm cells achieve their respective fates. Hörstadius, in his remarkable embryological studies showed that isolated animal halves develop into hollow, ciliated epithelial balls that were termed "dauerblastulae." He interpreted these results in terms of the double gradient model and concluded that these embryoids were programmed to follow an animal or ectodermal fate by "animalizing" morphogens localized to the animal pole (Hörstadius, 1973). Cameron and Davidson (1991), on the other hand, attempted to place these results in the context of Davidson's vegetal signaling model and described these embryoids as being "arrested in development" due to deprivation of inductive signals from vegetal cells. Recent experimental studies, however, support a model where both autonomous as well as nonautonomous mechanisms play a role in ectoderm specification and patterning in sea urchin embryos (Wikramanayake *et al.*, 1995; Wikramanayake and Klein, 1997).

Attempts to study ectoderm specification in sea urchin embryos have been hampered by the lack of methods to easily isolate large numbers of animal halves and by the lack of molecular markers to distinguish different ectodermal cell types. These shortcomings have been overcome with the development of a batch method to isolate large numbers of animal halves from 8-cell embryos and with the development of molecular markers to identify various ectodermal cell types. Animal halves made using this new method form the typical epithelial balls that have a cuboidal epithelium that tapers off into a squamous epithelium similar to those animal halves made using microsurgical methods. Hyperextended cilia are seen on the surface of the cuboidal epithelium and shorter motile cilia are produced by the squamous cells. As reported by Hörstadius (1939) and others, these embryoids do not normally form ectoderm-derived structures such as the ciliary band and the stomodeum (Fig. 3; Wikramanayake *et al.*, 1995; Wikramanayake and Klein, 1997). The oral ectoderm marker Ecto V is globally expressed in animal half-derived embryoids suggesting that in the absence of vegetal cells all mesomeres follow an oral ectoderm fate. These animal halves also produce serotonin-expressing cells that normally originate from the oral ectoderm in the undisturbed embryo. In these explants the serotonergic cells are always associated with the cuboidal epithelium revealing that in addition to the morphological polarity of the epithelia, a molecular polarity is present in the animal halves as well. A further molecular polarity is revealed when animal halves are stained with an antibody that recognizes a subset of embryonic neurons in pluteus stage embryos. This antibody reveals a large num-

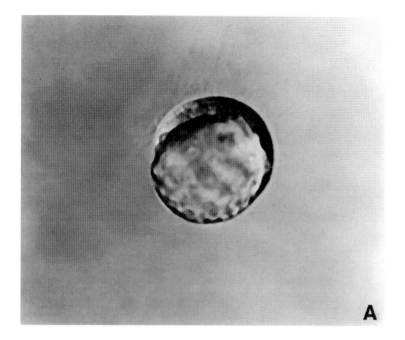

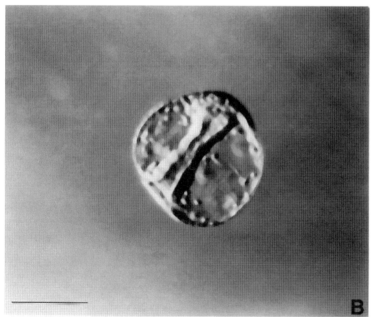

Figure 3 Effect of lithium chloride on *L. pictus* animal halves.(A) An embryoid derived from an animal half isolated from an 8-cell stage embryo. (B) An embryoid from an animal half treated with 25 m*M* LiCl.

ber of neuronlike processes that are produced by the cuboidal epithelium but not by the squamous epithelium of animal halves (Wikramanayake and Klein, 1997). Therefore, animal halves isolated at the 8-cell stage display a variety of oral ectoderm characteristics and appear to have an inherent polarity demonstrating that by as early as the 8-cell stage the four mesomeres are not an equivalence group.

Animal halves made from 8-cell *Lytechinus* embryos do not express aboral ectoderm-specific genes, but these genes can be activated by either recombining the explants with vegetal cells or by treating them with lithium chloride. In addition to the induction of aboral ectoderm genes, treating animal halves with 25 m*M* lithium chloride also results in the induction of endoderm in these explants (Wikramanayake *et al.*, 1995). It was originally believed that induction of aboral ectoderm genes was through secondary signaling from these vegetal cell types. However, subsequent experiments using lower concentrations of lithium, which does not induce endoderm, showed that correct patterning of ectodermal explants along the oral–aboral axis could be obtained with this treatment. That is, aboral ectoderm genes are expressed in the squamous epithelium, the Ecto V antigen is restricted to the cuboidal epithelium and there is an induction of a ciliary band at the boundary between the cuboidal and squamous epithelia. In addition, a stomodeum is induced in the cuboidal epithelium (Fig. 4; Wikramanayake and Klein, 1997). These results reveal that by morphological as well as molecular criteria, the cuboidal epithelium corresponds to oral ectoderm.

The morphological effects of the low lithium treatment are very similar to the effects seen by Hörstadius (1973) when he recombined the veg_1 tier with isolated animal halves. Hörstadius (1973) also showed that when he treated animal halves with lithium, separated the an_1 and an_2 tiers and then recombined these tiers with untreated animal halves, only the animal half recombined with the an_2 tier formed a pluteus larva. In these experiments lithium was differentially activating factors in an_2 and not in the an_1 tier. Thus, one explanation for the low lithium effects on isolated animal halves described above is that lithium was inducing a veg_1-like cell type in an_2 descendants. Signaling from these induced cells would then mimic veg_1 signaling shown in the experiments by Hörstadius. Hörstadius (1973) reported that these an_1, an_2, veg_1 explants did not develop any endoderm, except in embryos that displayed a subequatorial cleavage. These results must be reevaluated in view of the recent observations that the veg_1 tier contributes to endoderm (Logan and McClay, 1997a; Ransick and Davidson, 1998). In addition, Logan and McClay (1997b) have also reported that an_1, an_2, veg_1 explants from 60-cell embryos do form endoderm but that it is poorly patterned.

In *S. purpuratus*, all the observations reported for the *Lytechinus sp.* hold with a notable exception. Animal halves that are made from this species express the Ecto V antigen on all cells and these explants do not form ciliary bands and stomodea. But in sharp contrast to animal halves from *Lytechinus sp.*, the aboral ectoderm-specific *Spec1* gene is expressed at the same time as control embryos and is localized to the squamous cells (Wikramanayake *et al.*, 1995). Thus, these squamous cells express genes specific to aboral as well as to oral ectoderm. It is likely, but not proven, that in this species the signal to specify the aboral ectoderm is transmitted at or prior to the third cleavage or that the isolation protocol artificially stimulates aboral ectoderm-specific gene expression. If the signal to specify the aboral ectoderm is transmitted at the third cleavage, the signal must come from the four vegetal cells, since at this time the micromeres have not yet segregated.

The studies described above have led to a model for ectoderm specification and ectoderm patterning that invokes localized maternal determinants as well as multiple signaling events. According to this model, oral ec-

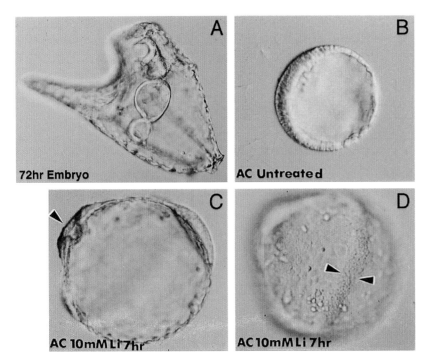

Figure 4 Ectoderm patterning of *L. pictus* animal halves treated with lithium chloride. (A) A pluteus larva at 3 days postfertilization. (B) An animal half isolated from an 8-cell stage embryo following 3 days in culture. (C) An animal half cultured for 3 days following exposure to 10 m*M* LiCl. Arrowhead points to the stomodeum. (D) The same embryo as in C at a different plane of focus. The arrowheads point to the ciliary band.

toderm in sea urchins is autonomously specified by inherited maternal determinants while aboral ectoderm specification requires signaling from vegetal blastomeres. Aboral ectoderm genes are not expressed in prospective oral ectoderm cells because inherited maternal components allow these cells to be refractory to this vegetal signaling. A suppressive signal that is transmitted from vegetal blastomeres represses the expression of oral ectoderm-specific genes such as Ecto V in aboral ectoderm cells. Presumably, this suppressive signal does not affect the oral ectoderm genes being expressed in the prospective oral ectoderm region due to factors in these cells making them refractory to vegetal signals (Wikramanayake and Klein, 1997). Localized maternal factors in oral ectoderm cells have not been identified, but several lines of evidence suggest that they exist and function in the manner described above. As described earlier, the morphology of isolated animal halves clearly show that the four animal-half blastomeres are not equivalent by as early as the 8-cell stage. The cuboidal cells produce serotonergic cells in the absence of vegetal signaling and in addition produce processes reminiscent of neurons. When explants are treated with lithium, the ion should freely enter all the mesomeres, yet aboral ectoderm genes are only expressed in the squamous cells suggesting that their activation is repressed in prospective oral ectoderm cells. Finally, when Ecto V expression is suppressed by lithium treatment, the suppression is only from the squamous cells and it always restricts to the cuboidal cells (Wikramanayake and Klein, 1997). Evidence supporting an early segregation of oral blastomeres comes from classic studies, which have shown that by the 16-cell stage, the future oral blastomeres already manifest metabolic differences from the future aboral blastomeres (Child, 1936; Czihak, 1963).

B. Signaling Molecules That Specify and Pattern the Ectoderm

While there is a large, growing body of literature on the proteins that bind transcriptional control regions of lineage-specific genes and on how these factors control the spatial and temporal expression of these genes in sea urchin embryos, virtually nothing is known about the signaling molecules that may regulate the activity of these factors in distinct cell types. An understanding of these signaling factors is essential for testing the validity of existing models for pattern formation along the animal–vegetal axis. Recent work in several labs has focused on attempting to identify the signaling molecules that mediate territorial specification in the early sea urchin embryo. An advance in these studies has come from the recent identification of a molecular target of lithium in cells. As described earlier lithium has power-

ful morphogenetic effects on embryos and isolated animal halves and for many years it was thought that these effects were due to its inhibition of components of the inositol trisphosphate signaling pathway (Berridge et al., 1989). However, it has been increasingly apparent that lithium affecting this pathway alone was insufficient to explain all its effects on embryos. Several labs have now shown that lithium is a powerful uncompetitive inhibitor of glycogen synthase kinase-3 (GSK-3) which is a negative regulator of the widely conserved Wnt/wingless signaling pathway (Klein and Melton, 1996; Stambolic et al., 1996; Hedgepeth et al., 1997). Inhibition of GSK-3 by lithium leads to the stabilization of β-catenin, thus mimicking Wnt signaling in lithium treated cells. β-Catenin is seen in the nuclei of vegetal cells of early cleavage stage sea urchin embryos putting it in the right place at the right time to play a role in specification of these vegetal cells (Miller and Moon, 1996; see Chapter 4 by Logan and McClay). Overexpression of an activated β-catenin molecule by injecting RNA into fertilized eggs or by ectopically expressing activated β-catenin in animal halves has shown that this protein has concentration-dependent effects that match the fates of cells along the animal–vegetal axis. That is, β-catenin promotes mesodermal, endodermal, and aboral ectodermal fates in a concentration-dependent fashion. Blocking the signaling pool of β-catenin by overexpressing the cell adhesion molecule C-Cadherin leads to the loss of mesodermal, endodermal, and aboral ectodermal cell fates. Since β-catenin is not seen in nuclei of animal blastomeres, this suggests that β-catenin has a more global role in generating pattern along the entire animal–vegetal axis (Wikramanayake et al., 1998). Because many of the effects of injecting β-catenin RNA can be reproduced by injection of Wnt RNAs (Wikramanayake et al., 1998) it is thought that a Wnt signaling pathway is present and functional in vegetal cells to mediate pattern formation along the animal–vegetal axis. Since nuclear β-catenin is not seen in the mesomere derivatives, activation of aboral ectoderm-specific genes and normal patterning of ectoderm may be initiated by a non-cell-autonomous signal that is induced in vegetal cells by β-catenin.

Non-cell-autonomous signals from the vegetal cells that specify the ectoderm are unknown at this time. However, a series of studies by Tomlinson and colleagues have implicated an important role for the extracellular matrix (ECM) and a platelet derived growth factor-like (PDGF-like) signaling pathway in this process (Ramachandran et al., 1995, 1997). These studies were motivated by the observation that inhibition of collagen deposition by treating cleavage stage embryos with the lathrytic factor β-aminopropionitrile (BAPN), inhibited the transcriptional activation of aboral ectoderm-specific genes (Wessel et al., 1989). Subsequent studies showed that the effects of BAPN could be rescued by treating embryos with PDGF-BB and tumor growth factor α (TGFα) (Ramachandran *et al.,* 1995). These studies indicated that a growth factor pathway, dependent on an intact ECM was necessary for aboral ectoderm specification. This hypothesis was further strengthened by the observation that expressing a dominant/negative human PDGF receptor (PDGF-R) in sea urchin embryos resulted in complete inhibition of aboral ectoderm formation (Ramachandran *et al.,* 1997). In addition, expression of this truncated receptor also resulted in the ectopic expression of the oral ectoderm specific protein Ecto V in all ectoderm cells. This is very similar to the effect seen when animal halves are made from 8-cell *Lytechinus* embryos and to the effect seen with the depletion of β-catenin in early embryos leading to the speculation that a PDGF-like signaling pathway may specify aboral ectoderm in *Lytechinus sp.* It is possible that a PDGF-like signal is produced in the vegetal plate and this molecule then signals to the more animal tiers as envisaged by Davidson (1989). An intact ECM may be necessary for presenting the signaling factors to the ectodermal cells. Unfortunately, neither a PDGF homolog nor a PDGF-R homolog have been isolated from sea urchin embryos as yet, making it difficult to ascribe a definite function for this growth factor pathway in ectoderm specification or differentiation. Cloning of sea urchin PDGF-like molecules or identifying downstream targets of β-catenin signaling may provide further insight into the factors produced by the vegetal plate that in turn signal the ectoderm territories in the animal pole.

It is now clear that a β-catenin/Wnt signaling pathway plays an indispensable role in patterning the sea urchin animal–vegetal axis. In other systems, notably in *Drosophila* the Wnt pathway acts in opposition to the BMP-4/Dpp signaling pathway. In the *Drosophila* leg imaginal disc, Dpp and wingless are expressed in nonoverlapping patterns in the anterior boundary. In this system it has been shown very clearly that these two pathways act antagonistically to one another but where they are at the highest concentration they synergistically induce a distinct set of genes that further pattern the developing leg (Jiang and Struhl, 1996; Brook and Cohen, 1996; Lecuit and Cohen, 1997). As described in Section II, a sea urchin BMP-4 homolog, which belongs to the VEB class will suppress endoderm formation when overexpressed by RNA injection (Angerer and Angerer, 1997). Perhaps this suppression of vegetal cells is by the suppression of the Wnt pathway in vegetal cells. Similarly, overexpression of β-catenin suppresses ectoderm differentiation perhaps by suppression of the BMP-4 pathway. The expression patterns of VEB genes and nuclear β-catenin are largely nonoverlapping along the animal–vegetal axis. It is possible that as in other systems these opposing pathways act antagonistically and cooperatively to pattern the sea urchin

animal–vegetal axis. For example, β-catenin promotes vegetal cell fates while BMP-4 signaling opposes this pathway in ectoderm cells. These two pathway appear to interact at the veg_1 tier. As shown by Logan and McClay (1997a), the segregation of ectoderm and endoderm in this tier takes place gradually and eventually forms the ectoderm–endoderm boundary. Perhaps this boundary is formed by the interaction of these two antagonistic pathways. One would predict that the interaction of these pathways at the veg_1 tier would produce distinct patterns of gene expression in this tier or its descendants and currently at least one such candidate gene has been identified. *Univin* is a TGF-β related signaling molecule that is initially expressed throughout the embryo and gradually restricts to what appears to be the an_2, veg_1 tiers by late cleavage (Stenzel *et al.,* 1994). A function has not been ascribed to this gene in sea urchin development, but it is tempting to speculate that its restricted pattern of expression is due to its dependence on both BMP-4 and Wnt signaling. It would be interesting to test the function of *Univin* by expressing it in isolated animal halves and to examine its expression pattern in embryos overexpressing BPM-4 or β-catenin.

VI. Recruitment of Conserved Factors and Pathways for Novel Uses in Sea Urchins

Sea urchin embryos have a distinct mode of development which requires novel use of conserved regulatory systems. Because of their position as primitive deuterostomes, sea urchin development offers a particularly interesting comparison with vertebrates. The family of Otx transcription factors are exemplary. Unlike vertebrates and insects, sea urchin embryos do not have anterior cephalic structures and SpOtx has clearly been recruited for other roles in sea urchin embryogenesis than head formation. SpOtx may function in axial patterning in later development since it is found in adult tissues in reiterating patterns which reflect the pentagonal body plan (Panganiban *et al.,* 1997; Lowe and Wray, 1997). Although SpOtx may have a role in axis patterning in the metamorphosing larva and adult sea urchin, it has nevertheless been co-opted for other functions than anterior cephalic or axial patterning associated with Otx forms in vertebrates and flies. The role of SpOtx in post-larval sea urchin development might reflect changes in the basic body plan that occurred during the evolution of echinoderms (Davidson *et al.,* 1995; Davidson, 1997). Similarly, SpOtx has also been recruited for highly specialized roles during sea urchin embryogenesis which have evolved sometime after the last common ancestor to vertebrates and sea urchins. This illustrates the evolutionary flexibility associated with Otx proteins.

The generation of new sets of target genes for Otx proteins over evolutionary time by creating TAATCC/T motifs in the control regions of non-Otx genes could readily lead to novel regulatory patterns and diversity in embryo development.

Conserved signaling pathways are also put to different use in sea urchin development. The β-catenin/Wnt pathway in sea urchin embryos works to pattern cell types along the animal–vegetal axis. What was the original function of the β-catenin/Wnt pathway? Many marine invertebrate larvae have a predetermined animal–vegetal axis with embryological properties much like those of sea urchins (Davidson *et al.*, 1995). A recent model proposes that indirect developing embryos, mostly marine invertebrates, reflect a primal mode of embryogenesis which was used by an early metazoan ancestor (Davidson *et al.*, 1995). Is it possible that β-catenin/Wnt signaling along the animal–vegetal axis is a universal feature of marine invertebrate development? The original function of this ubiquitous signaling pathway might have been to specify cell types via a vegetal signaling center in an ancient ancestral embryo. Other roles for β-catenin/Wnt signaling could be derived from this original function.

References

Akasaka, K., Ueda, T., Higashinakagawa, T., Yamada, K., and Shimada, H. (1990). Spatial pattern of arylsulfatase mRNA expression in sea urchin embryos. *Dev. Growth Differ.* **32**, 9–13.

Angerer, R. C., and Angerer, L. M. (1997). Fate specification along the sea urchin embryo animal–vegetal axis. *Biol. Bull.* **192**, 175–177.

Angerer, L. M., Dolecki, G. J., Gagnon, M. L., Lum, R., Wong, G., Yang, Q., Humphreys, T., and Angerer, R. C. (1989). Progressively restricted expression of a homeo box gene within the aboral ectoderm of developing sea urchin embryos. *Genes Dev.* **3**, 370–383.

Berridge, M. J., Downes, C. P. and Hanley, M. R. (1989). Neural and developmental actions of lithium: a unifying hypothesis. *Cell (Cambridge, Mass.)* **59**, 411–419.

Brannon, M., Gomperts, M., Sumoy, L., Moon, R. T., and Kimelman, D. (1997). A β-catenin/XTcf-3 complex binds to the siamois promoter to regulate dorsal axis specification in *Xenopus. Genes Dev.* **11**, 2359–2370.

Brook, W. J., and Cohen, S. M. (1996). Antagonistic interactions between wingless and decapentaplegic responsible for dorsal–ventral pattern in the *Drosophila* leg. *Science* **273**, 1373–1377.

Cameron, R. A., and Davidson, E. H. (1991). Cell type specification during sea urchin development. *Trends Genet.* **7**, 212–218.

Cameron, R. A., Fraser, S. E., Britten, R. J., and Davidson, E. H. (1990). Segragation of oral from aboral ectoderm precursors is completed at fifth cleavage in the embryogenesis of *Strongylocentrotus purpuratus. Dev. Biol.* **137**, 77–85.

Child, C. M. (1936). Differential reduction of vital dyes in the early development of echinoderms. *Wilhelm Roux'Arch. Entwicklungsmech. Org.* **135**, 426–456.

Coffman, J. A., and Davidson, E. H. (1994). Regulation of gene expression in the sea urchin embryo. *J. Mar. Biol. Assoc. U.K.* **74**, 17–26.

Coffman, J. A., Kirchhamer, C. V., Harrington, M. G., and Davidson, E. H. (1996). SpRunt-1, a new member of the runt domain family of transcription factors, is a positive regulator of the aboral ectoderm-specific *CyIIIa* gene in sea urchin embryos. *Dev. Biol.* **174**, 43–54.

Coffman, J. A., and McClay, D. R. (1990). A hyaline layer protein that becomes localized to the oral ectoderm and foregut of sea urchin embryos. *Dev. Biol.* **140**, 93–104.

Coffman, J. A., Kirchhamer, C. V., Harrington, M. G., and Davidson, E. H. (1997). SpMyb functions as an intramodular repressor to regulate spatial expression of CyIIIa in sea urchin embryos. *Development (Cambridge, UK)* **124**, 4717–4727.

Cox, K. H., Angerer, L. M., Lee, J. J., Davidson, E. H., and Angerer, R. C. (1986). Cell lineage-specific programs of expression of multiple actin genes during sea urchin embryogenesis. *J. Mol. Biol.* **188**, 159–172.

Czihak, G. (1963). Entwicklungsphysiologische untersuchungen an echiniden (verteilung und be deutung der cytochromeoxydase). *Wilhelm Roux' Arch. Entwicklunsmech. Org.* **154**, 272–292.

Davidson, E. H. (1989). Lineage-specific gene expression and the regulative capacities of the sea urchin embryo: A proposed mechanism. *Development (Cambridge, UK)* **105**, 421–445.

Davidson, E. H. (1997). Insights from the echinoderms. *Nature, (London)* **389**, 679–680.

Davidson, E. H., Peterson, K. J., and Cameron, R. A. (1995). Origin of bilaterian body plans: Evolution of developmental regulatory mechanisms. *Science* **270**, 1319–1325.

DiBernardo, M., Russo, R., Oliveri, P., Melfi, R., and Spinelli, G. (1995). Homeobox-containing gene transiently expressed in a spatially restricted pattern in the early sea urchin embryo. *Proc. Natl. Acad. Sci. U.S.A.* **92**, 8180–8184.

Fang, H., and Brandhorst, B. P. (1996). Expression of the actin gene family in embryos of the sea urchin *Lytechinus pictus. Dev. Biol.* **173**, 306–317.

Finkelstein, R., and Boncinelli, E. (1994). From fly head to mammalian forebrain: The story of otd and Otx. *Trends Genet.* **10**, 310–315.

Gagnon, M. L., Angerer, L. M., and Angerer, R. C. (1992). Posttranscriptional regulation of ectoderm-specific gene expression in early sea urchin embryos. *Development (Cambridge, UK)* **114**, 457–467.

Gammill, L. S., and Sive, H. (1997). Identification of Otx2 target genes and restrictions in ectodermal competence during *Xenopus* cement gland formation. *Development (Cambridge, UK)* **124**, 471–481.

Gan, L., Wessel, G. M., and Klein, W. H. (1990a). Regulatory elements from the related Spec genes of *Strongylocentrotus purpuratus* yield different spatial patterns with a *lacZ* reporter gene. *Dev. Biol.* **142**, 346–359.

Gan, L., Zhang, W., and Klein, W. H. (1990b). Repetitive

DNA sequences linked to the sea urchin *Spec* genes contain transcriptional enhancer-like elements. *Dev. Biol.* **139**, 186–196.

Gan, L., Mao, C.-A., Wikramanayake, A., Angerer, L. M., Angerer, R. C., and Klein, W. H. (1995). An orthodenticle-related protein from *Strongylocentrotus purpuratus. Dev. Biol.* **167**, 517–528.

Ghiglione, C., Lhomond, G., Lepage, T., and Gache, C. (1993). Cell-autonomous expression and position-dependent repression by Li⁺ of two zygotic genes during sea urchin development. *EMBO J.* **12**, 87–96.

Ghiglione, C., Lhomond, G., Lepage, T., and Gache, C. (1994). Structure of the sea urchin hatching enzyme gene. *Eur. J. Biochem.* **219**, 845–854.

Ghiglione, C., Emilyfenouil, F., Chang, P., and Gache, C. (1996). Early gene expression along the animal–vegetal axis in sea urchin embryoids and embryos. *Development (Cambridge, UK)* **122**, 3067–3074.

Hardin, P. E., Angerer, L. M., Hardin, S. H., Angerer, R. C., and Klein, W. H. (1988). The *Spec2* genes of *Strongylocentrotus purpuratus*: Structure and differential expression in embryonic aboral ectoderm cells. *J. Mol. Biol.* **202**, 417–431.

Hedgepeth, C. M., Conrad, L. J., Zhang, J., Huang, H. C., Lee, V. M., and Klein, P. S. (1997). Activation of the Wnt signaling pathway: A molecular mechanism for lithium action. *Dev. Biol.* **185**, 82–91.

Hörstadius, S. (1939). The mechanisms of sea urchin development studied by operative methods. *Biol. Rev. Cambridge Philos. Soc.* **14**, 132–179.

Hörstadius, S. (1973). "Experimental Embryology of Echinoderms." Oxford Univ. Press (Clarendon), Oxford.

Iuchi, Y., Morokuma, J., Akasaka, K., and Shimada, H. (1995). Detection and characterization of the cis-element in the first intron of the *Ars* gene in the sea urchin. *Dev. Growth. Differ.* **37**, 373–378.

Jiang, J., and Struhl, G. (1996). Complementary and mutually exclusive activities of decapentaplegic and wingless organize axial patterning during *Drosophila* leg development. *Cell (Cambridge, Mass.)* **86**, 401–409.

Kingsley, P. D., Angerer, L. M., and Angerer, R. C. (1993). Major temporal and spatial patterns of gene expression during differentiation of the sea urchin embryo. *Dev. Biol.* **155**, 216–234.

Kirchhamer, C. V., and Davidson, E. H. (1996). Spatial and temporal information processing in the sea urchin embryo: Modular and intramodular organization of the *CyIIIa* gene cis-regulatory system. *Development (Cambridge, UK)* **122**, 333–348.

Kirchhamer, C. V., Yuh, C.-H., and Davidson, E. H. (1996). Modular cis-regulatory organization of developmentally expressed genes: Two genes transcribed territorially in the sea urchin embryo, and additional examples. *Proc. Natl. Acad. Sci. U.S.A.* **93**, 9322–9328.

Klein, P. S., and Melton, D. A. (1996). A molecular mechanism for the effect of lithium on development. *Proc. Natl. Acad. Sci. U.S.A.* **93**, 8455–8459.

Klein, W. H., Xiang, M., and Wessel, G. M. (1991). Spec proteins: Calcium binding proteins in the embryonic ectoderm of sea urchins. *In* "Novel Calcium Biding Proteins: Structures, Principles and Chemical Relevance"

(C. W. Heizmann, ed.), pp. 465–479. Springer-Verlag, Heidelberg.

Lecuit, T., and Cohen, S. M. (1997). Proximal–distal axis formation in the *Drosophila leg. Nature (London)* **388**, 139–145.

Li, X., Chuang, C.-K., Mao, C.-A., Angerer, L. M., and Klein, W. H. (1997). Two Otx proteins generated from multiple transcripts of a single gene in *Strongylocentrotus purpuratus. Dev. Biol.* **187**, 253–266.

Logan, C. Y., and McClay, D. R. (1997a). The allocation of early blastomeres to the ectoderm and endoderm is variable in the sea urchin embryo. *Development (Cambridge, UK)* **124**, 2213–2223.

Logan, C. Y., and McClay, D. R. (1997b). Cell fate specification of the endoderm in the sea urchin embryo. *Dev. Biol.* **186**, B202.

Lowe, C. J., and Wray, G. A. (1997). Radical alterations in the roles of homeobox genes during echinoderm evolution. *Nature (London)* **389**, 718–721.

Lynn, D. A., Angerer, L. M., Bruskin, A. M., Klein, W. H., and Angerer, R. C. (1983). Localization of a family of mRNAs in a single cell type and its precursors in sea urchin embryos. *Proc. Natl. Acad. Sci. U.S.A.* **80**, 2656–2660.

Mao, C.-A., Gan, L., and Klein, W. H. (1994). Multiple Otx binding sites required for expression of the *Strongylocentrotus purpuratus Spec2a* gene. *Dev. Biol.* **165**, 229–242.

Mao, C.-A., Wikramanayake, A. H., Gan, L., Chuang, C.-K., Summers, R. G., and Klein, W. H. (1996). Altering cell fates in sea urchin embryos by overexpressing SpOtx, an orthodenticle-related protein. *Development (Cambridge, UK)* **122**, 1489–1498.

Miller, J. R., and Moon, R. T. (1996). Signal transduction through β-catenin and specification of cell fate during embryogenesis. *Genes Dev.* **10**, 2527–2539.

Nemer, M., Bai, G., and Stuebing, E. W. (1993). Highly identical cassettes of gene regulatory elements, genomically repetitive and present in RNA. *Proc. Natl. Acad. Sci. U.S.A.* **90**, 10851–10855.

Nemer, M., Stuebing, E. W., Bai, G., and Parker, H. R. (1995). Spatial regulation of SpMTA metallothionein gene expression in sea urchin embryos by a regulatory cassette in intron 1. *Mech. Dev.* **50**, 131–137.

Panganiban, G., Irvine, S. M., Lowe, C., Roehl, H., Corley, L. S., Sherbon, B., Grenier, J. K., Fallon, J. F., Kimble, J., Walker, M., Wray, G. A., Swalla, B. J., Martindale, M. Q., and Carroll, S. B. (1997). The origin and evolution of animal appendages. *Proc. Natl. Acad. Sci. U.S.A.* **94**, 5162–5166.

Penton, A., and Hoffman, F. M. (1996). Decapentaplegic restricts the domain of wingless during *Drosophila* limb patterning. *Nature (London)* **382**, 162–164.

Ramachandran, R. K., Govindarajan, V., Seid, C. A., Patil, S., and Tomlinson, C. R. (1995). Role for platelet-derived growth factor-like and epidermal growth factor-like signaling pathways in gastrulation and spiculogenesis in the *Lytechinus* sea urchin embryo. *Dev. Dyn.* **204**, 77–88.

Ramachandran, R. K., Wikramanayake, A. H., Uzman, J. A., Govindarajan, V., and Tomlinson, C. R. (1997). Disruption of gastrulation and oral–aboral ectoderm differentiation in the *Lytechinus pictus* embryo by a dominant/negative PDGF receptor. *Development (Cambridge, UK)* **124**, 2355–2364.

Ransick, A., and Davidson, E. H. (1998). Late specification of veg$_1$ lineages to endodermal fate in the sea urchin embryo. *Dev. Biol.* **195**, 38–48.

Reynolds, S. D., Angerer, L. M., Palis, J., Nasir, A., and Angerer, R. C. (1992). Early mRNAs, spatially restricted along the animal–vegetal axis of sea urchin embryos, include one encoding a protein related to tolloid and BMB-1. *Development (Cambridge, UK)* **114**, 769–786.

Sakamoto, N., Akasaka, K., Mitsunaga-Nakatsubo, K., Takata, K., Nishitani, T., and Shimada, H. (1997). Two isoforms of orthodenticle–related proteins (HpOtx) bind to the enhancer element of sea urchin arylsulfatase gene. *Dev. Biol.* **181**, 284–295.

Simeone, A., Acampora, D., Guilisano, M., Stornaluolo, A., and Boncinelli, E. (1992). Nested expression domains of four homeobox genes in the developing rostral brain. *Nature (London)* **358**, 687–690.

Stambolic, V., Ruel, L., and Woodgett, J. R. (1996). Lithium inhibits glycogen synthase kinase-3 activity and mimics wingless signaling in intact cells. *Curr. Biol.* **6**, 1664–1668.

Stenzel, P., Angerer, L. M., Smith, B. J., Angerer, R. C., and Wylie, W. (1994). The *univin* gene encodes a member of the transforming growth factor-beta superfamily with restricted expression in the sea urchin embryo. *Dev. Biol.* **166**, 149–158.

Vlahou, A., Gonzalez-Rimbau, M., and Flytzanis, C. N. (1996). Maternal mRNA encoding the orphan steroid receptor SpCOUP-TF is localized in sea urchin eggs. *Development (Cambridge, UK)* **122**, 521–526.

Vlahou, A., Schultz, C. P., and Flytzanis, C. N. (1998). Cell cycle dependent intracellular movements and asymmetric distribution of the SpCOUP-TF nuclear receptor in early embryonic development. Submitted.

Wessel, G. M., Zhang, W., Tomlinson, C. R., Lennarz, W. J., and Klein, W. H. (1989). Transcription of the *Spec1*-like gene of *Lytechinus* is selectively inhibited in response to disruption of the extracellular matrix. *Development (Cambridge, UK)* **106**, 355–365.

Wikramanayake, A. H., and Klein, W. H. (1997). Multiple signaling events specify ectoderm and pattern the oral–aboral axis in the sea urchin embryo. *Development (Cambridge, UK)* **124**, 13–20.

Wikramanayake, A. H., Brandhorst, B. P., and Klein, W. H. (1995). Autonomous and nonautonomous differentiation of ectoderm in different sea urchin species. *Development (Cambridge, UK)* **121**, 1497–1505.

Wikramanayake, A. H., Huang, L., and Klein, W. H. (1998). β catenin is essential for patterning the maternally specified animal-vegetal axis in the sea urchin embryo. *Proc. Natl. Acad. Sci. U.S.A.* **95**, in press.

Wilson, D., Sheng, G., Lecuit, T., Dostatni, N., and Desplan, C. (1993). Cooperative dimerization of paired class homeo domains on DNA. *Genes Dev.* **7**, 2120–2134.

Xu, N. D., Niemeyer, C. C., Gonzalez-Rimbau, M., Bogosian, E. A., and Flytzanis, C. N. (1996). Distal cis-acting elements restrict expression of the *CyIIIb actin* gene in the aboral ectoderm of the sea urchin embryo. *Mech. Dev.* **60**, 151–162.

Yuh, C.-H., and Davidson, E. H. (1996). Modular cis-regulatory organization of *Endo16*, a gut-specific gene of the sea urchin embryo. *Development (Cambridge, UK)* **122**, 1069–1082.

Yuh, C.-H., Moore, J. G., and Davidson, E. H. (1996). Quantitative functional interrelations within the cis-regulatory system of the *S. purpuratus Endo16* gene. *Development (Cambridge, UK)* **122**, 4045–4056.

Zeller, R. W., Britten, R. J., and Davidson, E. H. (1995). Developmental utilization of SpP3A1 and SpP3A2: Two proteins which recognize the same DNA target site in several sea urchin gene regulatory regions. *Dev. Biol.* **170**, 75–82.

4

Lineages That Give Rise to Endoderm and Mesoderm in the Sea Urchin Embryo

Catriona Y. Logan
Stanford University
Stanford, California 94305

David R. McClay
DCMB Group
Duke University
Durham, North Carolina 27708

I. Introduction

During development of the sea urchin, the ectoderm, the mesoderm, and the endoderm arise from distinct regions along the animal–vegetal axis (Fig. 1; see reviews in Cameron and Davidson, 1991; Logan and McClay, 1997; and Chapter 2 by Cameron and Coffman). The ectoderm forms from the animal half of the embryo, whereas the endoderm and mesoderm are produced from the vegetal hemisphere. Endodermal cells differentiate into foregut, midgut, and hindgut tissues, which correspond to the esophagus, stomach, and intestine, respectively. The mesoderm generates two classes of mesenchyme cells, the skeletogenic primary mesenchyme cells (PMCs) and the secondary mesenchyme cells (SMCs), which are specified further as four subpopulations. These are the pigment cells, the muscles cells of the esophagus, unpigmented blastocoelar cells, and coelomic sac cells which form pouches on either side of the foregut (reviewed by Ettensohn, 1992). The

various endodermal and mesodermal cell types originate from distinct positions within the embryo and become identifiable either morphologically or by marker expression at specific times during development, suggesting that their specification and differentiation involve a precisely coordinated series of developmental mechanisms.

Several models have attempted to explain the specification of cell identities in the sea urchin embryo. Early models proposed that the interplay between two opposing gradients emanating from the animal and vegetal poles establish cell identities along the animal–vegetal axis (A–V) (reviewed in Hörstadius, 1973; Runnström, 1975). More recent models have suggested that the specification of different cell identities along the animal–vegetal axis is initiated by signals emanating from the vegetal pole (Davidson, 1989; Wilt, 1987). Davidson's model (1989) has proposed that a preformed A–V asymmetry is progressively transformed into distinct territories of gene expression by a cascade of cell–cell signaling events that emanates from the mi-

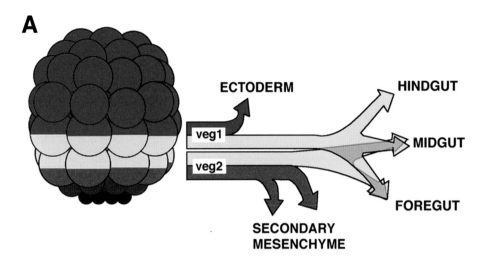

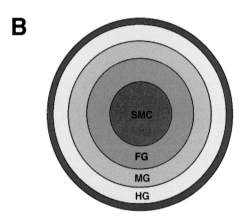

Figure 1 Fate maps of the sea urchin embryo (*Lytechinus variegatus*) at the 60-cell (A) and at the mesenchyme blastula stages (B). (A) Schematic of a 60-cell stage embryo viewed from the side, showing the contribution of early blastomeres to different cell types within the embryo. The animal pole is up, the vegetal pole is down. At this stage, the embryo is arranged as distinct tiers of cells; two of these tiers are called veg1 and veg2. Cell marking studies have shown that descendants of both veg1 and veg2 cells will form endoderm; veg1 progeny contribute to all three regions of the gut and veg2 progeny contribute to the foregut and midgut. In addition, a subset of veg1 cells contribute to the ectoderm, and a portion of veg2 progeny form secondary mesenchyme cells (SMCs). Colors represent different tissues: blue, ectoderm; yellow, endoderm; and red, mesoderm. The light red marks the prospective secondary mesenchyme cells (SMCs) and the dark red marks skeletogenic, primary mesenchyme cells (PMCs). The small dark brown cells are the small micromeres which will eventually contribute to the coelomic sacs. (B) A fate map of a mesenchyme blastula stage embryo viewed from the vegetal pole. Cell marking studies by Ruffins and Ettensohn (1996) have shown that prospective SMC, foregut (FG), midgut (MG), and hindgut (HG) cells are arranged as concentric rings within the vegetal plate. The different populations have been color coded (blue, ectoderm; yellows and oranges, endoderm; red, SMCs). (Adapted from Ruffins and Ettensohn, 1996, © Company of Biologists, Ltd.)

cromeres and that occurs across cleavage boundaries. Cell identities within these territories, in turn, are thought to be specified further as distinct cell and tissue types by additional cell–cell signaling interactions (Davidson, 1993; Chapter 2 by Cameron and Coffman).

The focus of our laboratory has been to examine how cell type diversity arises during development and to understand the mechanisms that promote the formation of distinct tissues and refine their boundaries. In particular, we have focused on the specification of the endoderm and secondary mesoderm and have examined their development by asking three main questions: (1) From where do the endoderm and mesoderm arise? (2) How are cells specified to form endoderm and mesoderm and how are the limits of these tissues decided? (3) How are further subdivisions and patterning within the endoderm and mesoderm accomplished?

Different definitions of specification have been applied depending on whether the experimental test of specification involves microsurgical or molecular analyses (for some examples, see Slack, 1991; Davidson, 1989). In this review, since we describe both molecular and embryological data, we have adapted a definition of specification from Davidson (1989) but define the term somewhat more loosely, as the process(es) that lead to the acquisition of distinct cell identities. One way to assess whether a cell is specified at a given time is to ask whether it expresses any particular markers of gene expression, whether its progeny, if labeled, will give rise to only one cell type, or whether its progeny will form what they are normally fated to make if cultured in a neutral environment. In all cases, it is important to remember that these readouts only suggest that the cell of interest is somehow different from its neighbors and that such difference(s) may be changed if the cell is placed in a new environment. Hence, our use of the term specification will imply that the fate choices open to a cell may still be flexible or conditional on its local environment, and we use the term commitment when the specification state of a cell becomes irreversible.

II. From Where Do the Endoderm and Mesoderm Arise?

Many cell labeling studies have been performed on sea urchin embryos during the past century, and new data have both confirmed and improved on previous fate maps as novel lineage tracers have become available (reviewed in Cameron and Davidson, 1991; see Hörstadius, 1939 and references therein). Recently, fate maps of 60-cell stage and mesenchyme blastula stage embryos have provided new information about how tissue precursors are distributed along the animal–vegetal axis (Ruffins and Ettensohn, 1996; Logan and McClay, 1997). Since these types of studies reveal where prospective cell and tissue types reside in unperturbed embryos, the fate maps provide new tools for experimentally examining how the endoderm and mesoderm are specified during development.

A. The 60-Cell Stage Embryo

The early cleavages of the sea urchin embryo are highly stereotyped and produce blastomeres that are easily distinguished by their size and location along the animal–vegetal axis. This predictability has allowed cells to be identified reliably and marked with lineage tracers to assess their fates in the embryo. It has been well established that the micromeres produce the skeletogenic, primary mesenchyme cells and that the mesomeres give rise to the ectoderm (reviewed in Cameron *et al.*, 1991). The macromeres form several different cell types which include SMCs, endoderm, and ectoderm. At the 60-cell stage, a horizontal division of the macromeres produces two tiers of eight cells each, called veg1 and veg2 which can be identified by their relative positions along the animal–vegetal axis (Hörstadius, 1939). Previous fate mapping studies had suggested that the allocation of cells to the ectoderm and endoderm was completed by the 60-cell stage when the veg1 and veg2 tiers were formed (reviewed in Hörstadius, 1939; Hörstadius,

1973 and references cited therein). By allocation, we mean that all of the progeny of a single labeled cell contribute to only one cell or tissue type. Since veg2 cells were thought to form the archenteron (endoderm and SMCs), and veg1 cells were thought to form only ectoderm, the ectoderm–endoderm boundary was predicted to coincide with the veg1/veg2 cleavage plane.

A recent study examining the fates of vegetal blastomeres at the 60-cell stage in embryos of the sea urchin species *Lytechinus variegatus* has revealed that both veg1 and veg2 cells contribute to the gut and that the ectoderm–endoderm boundary lies within veg1-derived clones (Fig. 1; Logan and McClay, 1997). These data demonstrate that the sixth cleavage does not necessarily accompany the segregation of the ectoderm and endoderm to distinct fates and that the ectoderm–endoderm boundary can be shifted more towards the animal pole than was predicted from previous analyses (Logan and McClay, 1997). Furthermore, the more animal positioning of the ectoderm–endoderm boundary is not simply due to the fact that the ectoderm and endoderm are allocated to different tissues by a later horizontal cleavage; cell labeling studies performed at the eighth division (the next horizontal cleavage) showed that a single labeled veg1 descendant segregates progeny to both ectoderm and endoderm in a variable manner. These data suggest that the actual position of veg1 progeny along the A–V axis is a better predictor of cell fate than the orientation or timing of specific cleavages. Consistent with this observation, when the position and numbers of DiI-labeled veg1-derived progeny were mapped in the anal ectoderm, clones contributing to ectoderm at similar locations in different embryos showed variable shapes, were variably distributed between the ectoderm and endoderm, and contained different numbers of cells. Cell marking studies of mesomeres in unperturbed 32-cell stage embryos also revealed that mesomeres can contribute descendants to the endoderm (16% of labeled cases; Logan and McClay, 1997). Hence, the allocation of cells to the ectoderm and endoderm and the positioning of the ectoderm–endoderm boundary is surprisingly imprecise and appears to bear no relationship to the early cleavage boundaries in *L. variegatus* embryos. Since cell labeling studies in *Strongylocentrotus purpuratus* also show that both veg1 and veg2 cells form gut tissues (Ransick and Davidson, 1995), the contribution of both veg1 and veg2 blastomeres to the gut may be a property that is common to many sea urchin species.

Cell labeling studies demonstrate that the allocation of veg1 and veg2 cells to distinct subregions within the gut is also highly variable (Logan and McClay, 1997). Veg2 cells contribute to the secondary mesoderm, the foregut, and the midgut, and veg1 cells can contribute to all three parts of the gut and the anal ectoderm (Fig. 1). When the labeling patterns of veg1 and veg2 descendants were compared among different embryos, each labeled veg1- or veg2-derived clone formed a different range of tissues within the archenteron, even when the clones were found in similar locations. Thus, although cell marking studies are not an experimental test of the mechanisms involved in allocating cells to distinct fates, the observation that early cleavage planes do not demarcate either germ layer boundaries or the segregation of endodermal cells to particular regions of the gut predicts that cell–cell signaling after the 60-cell stage plays an important role in assigning cells their distinct fates. Also implied from these results is that cell–cell signaling continues to refine cell identities at germ layer and tissue boundaries.

B. The Mesenchyme Blastula Stage Embryo

Following the early cleavage stages, the embryo hatches out of its fertilization envelope and initiates gastrulation movements. Gastrulation is marked by ingression of the PMCs into the blastocoel and archenteron formation which involves a buckling of the vegetal epithelium and subsequent cell rearrangements (Gustafson and Wolpert, 1967; Ettensohn, 1984, 1985; Hardin and Cheng, 1986). If the segregation of the ectoderm, endoderm, and mesoderm is not complete until after the early cleavage stages, when are cells allocated to distinct tissues? One possibility is that prospective endodermal and mesodermal populations only become segregated to distinct fates during gastrulation and are intermixed within the vegetal plate until the cell rearrangements of archenteron elongation bring them into register. Alternatively, prospective endodermal and mesodermal populations are arranged in a highly predictable pattern and are segregated prior to formation of the archenteron. To distinguish between these possibilities, Ruffins and Ettensohn (1996) produced a fate map of the vegetal plate by marking single cells with DiI(C18) at the mesenchyme blastula stage. Their results demonstrate that the prospective endoderm and mesoderm are already segregated as distinct cell populations that are arranged circumferentially in an orderly manner within the vegetal epithelium by the mesenchyme blastula stage (Fig. 1B).

The vegetal plate fate map is of high resolution and predicts 66 prospective SMCs in the vegetal plate center surrounded by a ring of 155 prospective endodermal cells. In addition, a dorsal–ventral bias in the distribution of SMCs is revealed by the analysis; presumptive pigment cells lie towards the dorsal side of the vegetal plate and presumptive blastocoelar cells are found on the ventral side of the vegetal plate. Finally, the prospective foregut, midgut, and hindgut are arranged as concentric rings within the endoderm, and each labeled endodermal precursor cell is fated to form only one cell type (Fig. 1B). These data offer the possi-

bility that cells of the vegetal plate can be at least conditionally specified to distinct fates prior to invagination of the archenteron. They also indicate that developmental events occurring between the late cleavage (~240 cells) and mesenchyme blastula stages (~1000 cells) segregate cells as distinct populations prior to gastrulation.

Analyses of molecular expression patterns support the hypothesis that cells are at least conditionally specified by the mesenchyme blastula stage, since several markers are found in specific subpopulations within the vegetal plate at this time. For example, the sea urchin Notch receptor is localized at the mesenchyme blastula stage on the endodermal side of the SMC–endoderm boundary (Sherwood and McClay, 1997). The CyIIa actin mRNA is expressed predominantly in the ventral region of the vegetal plate prior to archenteron formation, providing molecular confirmation of a dorsal–ventral asymmetry in this region of the embryo (Miller *et al.*, 1996). Finally, sea urchin β-catenin is found in the nuclei of the cells that lie on the endodermal side of the ectoderm–endoderm boundary prior to invagination of the vegetal plate (D. R. Sherwood and D. R. McClay, 1997; Logan *et al.*, 1998). Together, these data show that the vegetal plate is subdivided into cell lineage and gene expression domains even prior to gastrulation and the overt differentiation of the cells.

Together, the early and late maps shown in Fig. 1 indicate that a progressive allocation of cells to distinct germ layers and subpopulations within tissues occurs during early development and is largely complete by the mesenchyme blastula stage. The next question that arises, then, is how the endoderm and mesoderm are specified within the embryo, and how cell identities within the endoderm and mesoderm and at tissue boundaries become refined and committed over time.

III. Specifying Tissues and Defining Tissue Boundaries

Cell marking studies provide valuable information regarding the normal distribution of prospective cell types, but they do not provide clear mechanistic insights into how those cell types are directed toward particular fates. To understand the processes that specify different cell identities, experimental approaches are necessary. In the sea urchin, the specification of the endoderm and mesoderm have been experimentally analyzed in two general ways. Classic embryological studies have examined cell fate specification processes by assessing the formation of particular tissues following various microsurgical or pharmacological perturbations (for examples, see Hörstadius, 1973; Khaner and Wilt, 1991;

Nocente-McGrath *et al.*, 1991; Hardin *et al.*, 1992). More recent approaches have included analyses of RNA and protein expression patterns, and examination of upstream regulatory regions of tissue specific genes to understand how cells become distinct on a molecular level (for some recent examples, see Ransick *et al.*, 1993; Wikramanayake *et al.*, 1995; Yuh and Davidson, 1996; Sherwood and McClay, 1997). Both molecular and embryological analyses have generally supported the hypothesis that the early embryo is an assemblage of distinct territories of gene expression that are subsequently refined by cell–cell interactions (Davidson, 1989, 1993). One of the goals of our laboratory has been to identify the cellular interactions that play a role in specifying the endoderm and mesoderm and in assigning identities within these tissues.

A. How Are Cells Specified to Form Endoderm and How Are Its Limits Decided?

1. Specification of the Endoderm Involves Vegetally Derived Cell–Cell Signaling

The specification of any tissue during development must involve mechanisms that both allocate cells to that particular tissue and that define the limits of the tissue. In the sea urchin embryo, specification of the endoderm produces the larval gut and the ectoderm–endoderm junction which is an important site of tissue differentiation and patterning. Cells at the ectoderm–endoderm boundary eventually form the anal sphincter which consists of thickened intestinal epithelial cells (Burke, 1981) that are recognized by antibodies to the SUM-1 transcription factor (Venuti *et al.*, 1993) and myosin heavy chain (Wessel *et al.*, 1990). In addition, cells near the ectoderm–endoderm junction mark where PMCs will form a subequatorial ring and initiate spiculogenesis, to produce the final form of the larva (reviewed in Okazaki, 1975).

Positioning of the ectoderm–endoderm boundary in the sea urchin is at least partially dependent on the processes that specify the endoderm itself, since the manipulation of cell signals that induce the endoderm also affects the positioning of the ectoderm–endoderm junction. In the normal embryo, micromere-derived signals appear to trigger a sequence of signaling events from the vegetal pole that promote endoderm formation and influence the positioning of the ectoderm–endoderm boundary. For example, micromeres transplanted to the animal pole can induce a secondary archenteron (Ransick and Davidson, 1993; reviewed in Hörstadius, 1973). Endoderm formation by host tissues also is accompanied by the formation of constrictions within the gut, a new ectoderm–endoderm boundary, and ectopic

sites of skeletogenesis. Similarly, by reaggregating mesomeres with micromeres (Khaner and Wilt, 1991) or by recombining micromeres with animal blastomeres obtained from 16- to 60-cell stage embryos (reviewed in Hörstadius, 1939), it has been shown that micromere signaling can respecify prospective ectodermal cells as endoderm and produce plutei with well developed tripartite gut tissues.

The use of molecular markers has permitted the role of micromere-derived cues to be examined in greater detail and has revealed that at least one of the effects of micromere-derived signaling may be to trigger the initial specification of the vegetal plate territory (Ransick and Davidson, 1993, 1995; Chapter 2 by Cameron and Coffman). The vegetal plate territory can be identified by expression of Endo16 and is thought to be comprised mostly of descendants of the veg2 tier (Ransick *et al.*, 1993). Micromere transplantation induces ectopic Endo16 expression (Ransick and Davidson, 1993), and micromere deletion reduces the number of cells that express Endo16 and delays the initiation of archenteron formation, although it does not abolish development of the gut itself (Ransick and Davidson, 1995). Descendants of the vegetal plate territory also form a thickened columnar epithelium that buckles during the initial stages of archenteron formation. This morphologically identifiable structure can be induced at the animal pole following micromere transplantation (Ransick and Davidson, 1993) and its formation is delayed if micromeres are deleted from the vegetal pole (Ransick and Davidson, 1995). Therefore, one function of the micromeres may be to induce the vegetal plate territory; in turn, this induction may bring about subsequent morphological changes that accompany endoderm specification.

2. Does the Vegetal Plate Territory Specify veg1-Derived Endoderm?

Although the micromeres may induce the vegetal plate territory, the endoderm is formed by both veg1 and veg2 cells, suggesting that processes in addition to specification of the vegetal plate territory are necessary to produce endoderm from veg1 cells. Several different mechanisms can account for this process. First, micromeres may signal to the macromeres between the 16- and 32-cell stages such that when the veg1 and veg2 tiers are formed at the 60-cell stage, both cell layers are already specified to assume an endodermal fate. Alternatively, micromeres may secrete long-range diffusible signals that can reach both veg1 and veg2 cells and direct veg1 cells toward forming endoderm. There is no evidence at present to support either of these hypotheses, although additional molecular markers may reveal that micromere signaling affects both veg1 and veg2 cells.

Another possible mechanism is that veg1 cells are induced to form endoderm by the underlying veg2 tier, and results of blastomere recombinations performed at the 60-cell stage suggest that this type of mechanism may contribute to endoderm formation by veg1 blastomeres (Logan and McClay, 1998; reviewed in Hörstadius, 1973). For example, veg1 cells of normal embryos form anal ectoderm and also always contribute to the endoderm (Fig. 1; Logan and McClay, 1997). When mesomeres and the veg1 tier at the 60-cell stage are recombined, two types of embroids result; the majority of cases form endoderm that resembles small archenteronlike structures, and the remaining cases form only ectoderm (Fig. 2a; Logan and McClay, 1998). One interpretation of this result is that veg1 cells are normally specified to form endoderm by underlying veg2 cells and

will form only ectoderm if veg2 cells are missing; embryoids resulting from mesomere/veg1 recombinants lacking endoderm are rare because the transfer of veg2-derived signals to veg1 cells is extremely rapid and cannot be prevented quickly enough by the microsurgery.

Consistent with the idea that veg2 cells possess inductive abilities, when mesomere/veg2 recombinants are made at the 60-cell stage and a single mesomere overlaying the veg2 tier is labeled with DiI so that mesomere descendants can be followed, the progeny of the labeled mesomere are recruited into the endoderm in 100% of the recombinants (Fig. 2b; Logan and McClay, 1998). Since mesomeres rarely form gut tissues in unperturbed embryos (Logan and McClay, 1997),

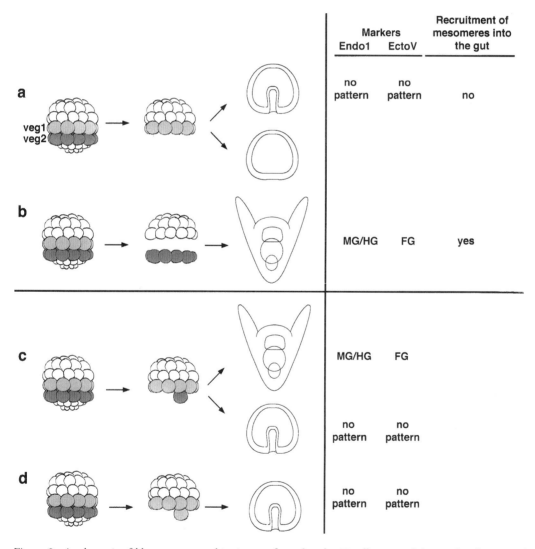

Figure 2 A schematic of blastomere recombinations performed at the 60-cell stage and the results of assaying for region-specific gut marker expression or changes in cell fate based on cell lineage analyses. (a) Mesomeres recombined with the veg1 tier produce embryos that possess endoderm in most cases; a few consist of only ectoderm. Gut markers exhibit no region-specific expression and mesomeres are not recruited into the gut. (b) Mesomeres recombined with the veg2 tier produce embryos that resemble normal plutei. Region-specific gut markers are expressed appropriately, and the mesomeres which usually make mostly ectoderm are recruited into the endoderm. (c) The addition of a single veg2 cell to a mesomere/veg1 recombinant can promote the formation of three-part gut tissues in some cases; these embryos exhibit appropriate region-specific marker expression. (d) The addition of an additional veg1 cell to a mesomere/veg1 recombinant does not result in the development of embryoids with three-part gut tissues.

these results demonstrate that veg2 cells produce a signal(s) that induces the overlying mesomeres in mesomere/veg2 recombinants to form gut tissues. Labeled mesomeres in mesomere/veg1 recombinants are not recruited to form endoderm, suggesting that the inductive capabilities are limited to the veg2 tier (Fig. 2b; Logan and McClay, 1998; see also Khaner and Wilt, 1991) and to micromeres (Ransick and Davidson, 1993; reviewed in Hörstadius, 1973) during the early cleavage stages. Hörstadius (reviewed in 1973) previously demonstrated that mesomeres form gut tissues in macromere/mesomere recombinants and he concluded that the veg2 cells were the source of the inductive signals. Therefore, since the normal neighbors of veg2 cells are the veg1 tier and not the mesomeres, veg2 cells may induce veg1 cells to form endoderm in the unperturbed embryo.

Because the recombination experiments involve the elimination of either veg1 or veg2 cells once they are formed, it cannot be unambiguously determined whether endoderm formation by veg1 cells *requires* the presence of veg2 cells. However, these data suggest that veg2 signaling may play some role in either directly inducing veg1 cells to form endoderm or in modulating their capacity to differentiate as endoderm. When molecules involved in endoderm and mesoderm specification are defined in more detail, the mechanisms that specify veg1-derived endoderm can be examined more directly.

Taken together, these results suggest that endoderm formation in the normal embryo and positioning of the ectoderm–endoderm junction involve a micromere-derived signal which specifies the vegetal plate territory in veg2 cells. This territory, in turn, may produce inductive signals that both instruct the overlying veg1 progeny to form gut tissues and provide positional cues that define the limits of the endoderm (Fig. 3a).

3. Veg2 Cells Pattern Veg1-Derived Endoderm

Additional evidence to support the hypothesis that there is a transfer of signals from the veg2 to veg1 tier comes from blastomere recombination experiments made at the 60-cell stage which show that regionalization of the gut as foregut, midgut, and hindgut also relies, at least in part, on veg2-derived cues (Fig. 2; Logan and McClay, 1998). As described above, when mesomeres were recombined with either the veg1 tier or veg2 tier at the 60-cell stage, both types of recombinants usually formed endoderm (Fig. 2a,b; Logan and McClay, 1998; see also Khaner and Wilt, 1991). However, veg1-derived endoderm in mesomere/veg1 combinations made morphologically poorly patterned gut tissues whereas veg2/mesomere recombinants resembled normal plutei. All recombinants produced embryoids

that expressed the gut-specific marker, Endo1 and the oral ectoderm/foregut marker, EctoV. Endo1 expression in mesomere/veg2 recombinants appeared normal and was localized to the midgut and hindgut as found in unmanipulated embryos. Similarly, EctoV also was expressed normally in mesomere/veg2 combinations. However, Endo1 expression in mesomere/veg1 recombinants was found throughout the archenteronlike tissue without exhibiting any region-specific expression. EctoV expression in mesomere/veg1 recombinants was present on the oral ectoderm but was missing completely or restricted to a small patch at the archenteron tip. Therefore, although both veg1 and veg2 cells can form endoderm by the 60-cell stage, veg1 cells in veg1/mesomere recombinants do not form foregut tissues and clear constrictions in the absence of veg2 cells, suggesting that additional signals from veg2 cells are required for proper patterning of veg1-derived endoderm.

To test directly whether the patterning of veg1-derived endoderm in normal embryos requires veg2-derived signals, mesomere/veg1 recombinants were cultured in the presence of a single veg2 cell (Fig. 2c). Single veg1 or veg2 blastomeres in these recombinants were labeled with DiI in order to follow their progeny. Roughly 20% of resulting embryoids formed gut tissues that consisted of three distinct parts containing veg1-derived cells in all three regions of the gut, suggesting that veg2 cells can pattern veg1-derived endoderm. The regionalization of the endoderm was not due to the addition of an extra blastomere, since the addition of a single veg1 cell to mesomere/veg1 recombinants did not produce any embryoids with three-part gut tissues (Fig. 2d). These results demonstrate that veg2 cells are a source of cell–cell signals that can provide patterning information to veg1-derived endoderm. Hence, in the normal embryo, veg2-derived cues may both induce endoderm and promote regionalization of the gut into three distinct parts.

4. Are Maternal Determinants Involved in Cell Fate Specification of the Endoderm?

Although cell–cell interactions appear to specify the endoderm, there is also some evidence to suggest that cell-autonomous mechanisms play a role in this process (Fig. 3b). First, the ability of mesomere caps isolated from 16-cell embryos to form endoderm is correlated with the position of the third cleavage plane (Henry et al., 1989). Normally, the third division is horizontal and separates the embryo into animal and vegetal halves. Mesomeres isolated from embryos that naturally undergo a subequatorial cleavage can contribute progeny to the endoderm at a much higher frequency than do those obtained from equatorially or supraequatorially cleaving embryos (Henry et al., 1989). The proposed explanation is that maternal determinants,

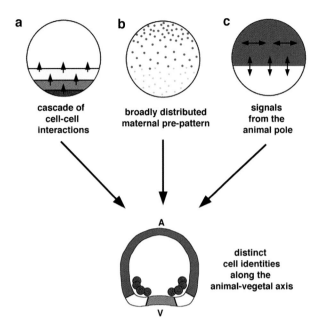

Figure 3 Three types of mechanisms that might function in patterning the embryo along the animal–vegetal axis. (a) A tier-to-tier cascade of inductive signals may emanate from the vegetal–most pole, subdividing the embryo into distinct cell populations along the animal–vegetal axis (reviewed in Davidson, 1989). (b) Maternally provided factors may be broadly distributed along the entire animal–vegetal axis and specify regional identity along the animal–vegetal axis. (c) Signaling among mesomeres or signals from the animal hemisphere may influence cell identities near the ectoderm–endoderm boundary. Note that these three possible mechanisms are not necessarily mutually exclusive and also may not be completely independent from each other. For example, it is possible that broadly distributed maternal factors (b) or a cascade of inductive interactions (a) might promote signaling among animal blastomeres at a later stage (c).

broadly distributed throughout the entire vegetal half of the embryo, are more likely to be inherited by mesomeres if cleavage occurs subequatorially and that these factors can bias cells toward assuming an endodermal identity. Experimental shifting of the third cleavage plane towards the vegetal pole with chemical agents also results in endoderm formation by isolated animal halves (Kitajima and Okazaki, 1980). In this case, the animal blastomeres were separated from the vegetal blastomeres at the 8-cell stage prior to production of the micromeres, precluding the possibility that micromere-derived cues already had determined the position of the ectoderm–endoderm junction. These observations suggest an inherent bias in endoderm-forming potential along the animal–vegetal axis which may act independently of micromere-derived signaling.

Comparisons between the effects of lithium chloride treatment and micromere signaling on the specification of cell fates along the animal–vegetal axis also suggest that the embryo may become broadly regionalized along the animal–vegetal axis by several different, independent mechanisms that may act combinatorially.

Lithium chloride is a vegetalizing agent in the sea urchin embryo which has been shown to increase the amount of secondary mesoderm and endoderm at the expense of the ectoderm (Nocente-McGrath *et al.*, 1991; reviewed in Hörstadius, 1973). Cell labeling experiments have demonstrated directly that the ectoderm–endoderm junction is shifted towards the veg1/mesomere boundary following exposure to lithium chloride (Cameron and Davidson, 1997). Similarly, isolated animal halves which usually only form ectoderm can differentiate to form endoderm when treated with LiCl (reviewed in Hörstadius, 1973; Livingston and Wilt, 1990; Wikramanayake *et al.*, 1995; Chapter 3 by Wikramanayake and Klein). Although this effect appears to be similar to the consequences of micromere-derived signaling, it has been shown that lithium treatment and micromeres have distinct and synergistic effects on endoderm and mesoderm specification (Livingston and Wilt, 1990). Lithium is thought to affect signaling pathways by perturbing inositol triphosphate metabolism or by affecting the activity the serine/threonine kinase GSK-3 which is a component of the Wnt signaling pathway (Berridge *et al.*, 1989; Klein and Melton, 1996). The sensitivity of endoderm formation to lithium treatment suggests that the cell signaling mechanisms triggered by lithium stimulate both the specification of gut tissues and the positioning of the ectoderm–endoderm junction.

Evidence to suggest that the lithium-sensitive signaling machinery in the sea urchin embryo acts independently of micromere-derived cues come from studies on two genes, hatching enzyme (HE), and β-catenin. HE is a zygotic gene that is normally broadly expressed in a domain that is excluded from the vegetal-most regions of the embryo (Lepage *et al.*, 1992; Reynolds *et al.*, 1992). It continues to be expressed when cells are dissociated, indicating that regulation of its transcription is cell autonomous (Reynolds *et al.*, 1992; Ghiglione *et al.*, 1993). In the species *Paracentrotus lividus*, the boundary of hatching enzyme expression is shifted towards the animal pole when embryos are treated with lithium chloride (Ghiglione *et al.*, 1993, 1996), consistent with the ability of lithium to vegetalize the embryo at the expense of ectoderm. However, transcription of HE is not responsive to transplantation of micromeres to the animal pole, demonstrating that its expression pattern is not regulated by micromere-derived cues (Ghiglione *et al.*, 1996). This result suggests that control of HE expression in *P. lividus* is influenced by a mechanism(s) that imparts positional information along the animal–vegetal axis but that acts independently of the micromeres. Similarly, sea urchin β-catenin protein is asymmetrically localized; during cleavage it is found vegetally in nuclei of the micromeres and the veg2 tier (Logan *et al.*, 1998). This localization of β-catenin is lithium chloride responsive, because its expression domain expands into the veg1 tier following exposure to lithium. However, nuclear accumulation of β-catenin in veg2 cells is not prevented if micromeres are deleted from the vegetal pole and cannot be induced in mesomeres if micromeres are transplanted to the animal pole. Thus, the machinery that promotes nuclear β-catenin localization also appears to be regulated independently of micromere-derived signals. Together, these data indicate that hatching enzyme and β-catenin are markers that reveal a broadly distributed asymmetry along the animal–vegetal axis. This asymmetry may act in conjunction with micromere-derived cues to initiate signaling events that will specify the endoderm and that will determine the position of the ectoderm–endoderm junction.

5. Do Signals from Animal Blastomeres Influence the Position of the Ectoderm–Endoderm Junction?

Although both cell-autonomous and cell signaling mechanisms appear to be involved in specifying vegetal cell fates within the sea urchin embryo, animally derived signals may also influence the positioning of the endoderm and the placement of the ectoderm–endoderm junction (Fig. 3c). For example, blastomere isolation experiments by Henry *et al.* (1989) have demonstrated that animal blastomeres can interact with each other and reinforce their ectodermal identity. Isolated mesomere pairs that are placed in culture have a greater tendency to form endoderm than do intact animal caps, whereas larger numbers of mesomeres undergo interactions that restrict their ability to form endodermal tissues. Although vegetally acting mechanisms may play a role in specifying the endoderm in the normal embryo, these data suggest that cell fates near the ectoderm–endoderm junction may also be influenced by signals from the mesomeres.

One molecule that may be involved in animally derived cell–cell signaling in the sea urchin embryo is BMP-4, a member of the transforming growth factor-β (TGF-β) family of growth factors. Overexpression or misexpression of *Lytechinus pictus* or *Xenopus* BMP-4 in sea urchin embryos leads to animalized phenotypes in which embryos form ectoderm while endoderm and mesoderm differentiation are suppressed (L. M. Angerer, D. Oleksyn, L. Dale, and R. C. Angerer, 1997, personal communication). Previous studies have shown that *decapentaplegic*, the *Drosophila* homolog of BMP-4 is involved in dorsal–ventral patterning and requires the metalloendoprotease, *tolloid*, in order to function during *Drosphila* embryogenesis (Shimell *et al.*, 1991; Ferguson and Anderson, 1992). More recent studies in both *Drosophila* and *Xenopus* have demonstrated that metalloproteases like *tolloid/xolloid* activate BMP-4 by cleaving BMP-4 antagonists such as *sog* in *Drosophila*

(Marqués *et al.*, 1997) or its homolog, *chordin*, in *Xenopus* (Piccolo *et al.*, 1997). Interestingly, in the sea urchin, a candidate metalloendoprotease that may affect the function of sea urchin BMP-4 is SpAN (Reynolds *et al.*, 1992; reviewed in Angerer and Angerer, 1997). SpAN resembles tolloid in protein sequence (Reynolds *et al.*, 1992) and is localized to the animal 85% of blastula stage embryos (Reynolds *et al.*, 1992). In addition, SpAN can animalize sea urchin embryos when overexpressed (D. Oleksyn and L. Angerer, personal communication), suggesting that it may participate in mechanisms that influence cell fate specification events along the animal–vegetal axis. Although a sea urchin "*chordin*" has not yet been identified, these data suggest that similar to *Xenopus* and *Drosphila*, a mechanism involving BMP-4 and a tolloidlike metalloprotease will be utilized in the sea urchin, perhaps to mediate signaling events that refine cell identities at the ectoderm–endoderm boundary.

B. Commitment of Endodermal Cells and Stabilization of the Ectoderm–Endoderm Boundary

Cell dissociation studies combined with assays of marker expression have indicated that the irreversible commitment of cells to an endodermal fate occurs between the late blastula and the early mesenchyme blastula stages (Chen and Wessel, 1996). When expression of the endoderm-specific markers Endo1 and LvN1.2 was examined following dissociation of embryonic cells at different times during development, embryos dissociated after the late blastula stages exhibited autonomous, endoderm-specific marker expression whereas those dissociated prior to late blastula stage did not. Hence, the late blastula stage marks a change in the specification state of prospective endodermal cells in that they acquire an autonomous capacity to form endoderm at this time. These data predict that, in normal embryos, both the commitment of endodermal cells and the processes that define the ectoderm–endoderm boundary are completed between the late blastula and mesenchyme blastula stages.

Consistent with this idea, there appears to be a stable boundary at the junction between the ectoderm and endoderm by the mesenchyme blastula stage. Deletion of large amounts of presumptive gut tissue from the vegetal half of the embryo prior to invagination of the vegetal plate can give embryos that lack endoderm completely (Fig. 4a,b; McClay and Logan, 1996). The simplest interpretation of this result is that the entire presumptive endodermal field has been deleted by the operation, and that the remaining ectodermal cells cannot be respecified because the fates of veg1 progeny at the ectoderm–endoderm boundary are already committed. Further, the average number of ectodermal veg1 progeny in embryos following the removal of endo-dermal tissues does not dif-

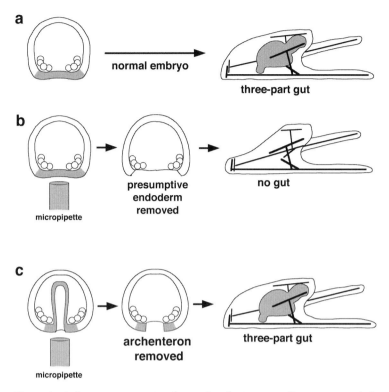

Figure 4 A diagram representing the results of microsurgically removing endodermal tissues (shaded) from gastrulating embryos. Embryos are viewed from the side; the animal pole is up and the vegetal pole is down. (a) An unmanipulated embryo forms a normal, three-part gut (shaded). (b) When the entire presumptive endoderm is removed at the mesenchyme blastula stage, the missing gut tissues are not replaced showing that the remaining tissues cannot form endoderm. (c) When only a portion of the gut is removed as seen when the archenteron is deleted from a mid-to-late gastrula stage embryo, the remaining endodermal tissues form a new three-part gut, demonstrating that the remaining endodermal cells can be respecified to new fates within the gut.

fer significantly from that of unmanipulated embryos, suggesting that there is no large-scale conversion of veg1-derived ectoderm to endodermal fates (Logan, 1997). Thus, the cell–cell interactions that specify the endoderm and that position the ectoderm–endoderm junction must occur during early development such that the ectoderm and endoderm become distinct populations by the mesenchyme blastula stage.

The endoderm eventually becomes subdivided into the foregut, the midgut, and the hindgut. These three regions are functionally distinct and can be distinguished at the pluteus stage both morphologically and by region-specific markers. The constrictions between different gut tissues are marked by antibodies to myosin heavy chain (Wessel *et al.*, 1990). In addition, the Endo16 marker (Nocente-McGrath *et al.*, 1989) shows dynamic changes in expression before eventually becoming confined to the midgut. Endo1 and LvN1.2 eventually mark both the midgut and hindgut (Wessel *et al.*, 1989; Wessel and McClay, 1985), and the EctoV antibody recognizes an epitope on both oral ectoderm and foregut cells (Coffman and McClay, 1990).

In contrast to endodermal fate and cell identities at the ectoderm–endoderm boundary which appear to be irreversibly stabilized by the mesenchyme blastula

stage, the patterning of the gut into different subregions and the establishment of their boundaries appear to be extremely plastic for much of gastrulation (McClay and Logan, 1996). The transplantation of small endodermal pieces to new positions within the gut throughout gastrulation results in changes in region-specific marker expression in donor tissues that are appropriate for the host position, suggesting that subregions of the endoderm can be respecified when placed in new contexts. Deletion of portions of the archenteron or the entire archenteron results in replacement of all three parts of the gut, demonstrating that the remaining endodermal cells can be repatterned even when large portions are removed (Fig. 4c). The prolonged plasticity within the gut appears to continue until all morphogenetic movements are complete. These data indicate that once the endoderm has been specified, further regionalization of the gut involves local cell–cell interactions that persist throughout archenteron formation.

Additional evidence to support the observation that patterning of the gut involves local cell interactions comes from embryological and pharmacological studies. In normal embryos, the archenteron contacts that blastocoel roof at the animal pole and subsequently becomes constricted into three parts (Hardin and McClay,

1990; Gustafson and Kinnander, 1960). Lithium chloride or sodium dodecyl-sulfate (SDS) treatment can produce exogastrulae in which the archenteron projects outwards from the embryo instead of protruding into the blastocoel, preventing the archenteron from contacting the blastocoel roof (Hardin and Cheng, 1986; Coffman and McClay, 1990; also reviewed in Hörstadius, 1973). In these cases, the archenteron nevertheless becomes subdivided and expresses gut markers appropriately (Coffman and McClay, 1990; Ransick *et al.,* 1993), demonstrating that the processes which pattern the gut do not require cues from the blastocoel roof target. In addition, archenteron tissues that have been isolated with a micropipette from gastrulating embryos will form constrictions when placed in culture, suggesting that the gut also does not require contact with the rest of the embryo (D. R. McClay, 1996, unpublished observation). Hence, although endoderm may be committed by the mesenchyme blastula stage, further patterning of the gut appears to involve cell signaling mechanisms that persist throughout gastrulation, that partition the gut into subregions, and that refine cell identities at their boundaries.

C. How Are Cells Specified to Form the Secondary Mesenchyme Cells and How Are the Limits of This Population Decided?

1. Specification of Secondary Mesenchyme Cells

The secondary mesenchyme cells develop from the center of the vegetal plate and differentiate into four distinct subpopulations. The pigment cells are most easily identified by their red color once they differentiate in the ectoderm (Gibson and Burke, 1985; Gustafson and Wolpert, 1967; Ryberg and Lundgren, 1979) and by staining with molecular markers such as SP1 (Gibson and Burke, 1985) and SpHmx (Martinez and Davidson, 1997). Blastocoelar cells are a multipolar, unpigmented, migratory SMC population whose function is unclear at the present time (Tamboline and Burke, 1992). They are recognized by an antibody called Sp12 which reveals that they reside in the blastocoel and form a fibrous network of cell processes near the gut and the skeletal rods (Tamboline and Burke, 1989; 1992). Coelomic sacs form as two pouches on either side of the foregut endoderm (Gustafson and Wolpert, 1963), and the circumesophageal muscle cells arise relatively late during development after the coelomic sacs have formed. Once differentiated, muscle cells are recognizable morphologically (Ishimoda-Takagi and Sato, 1984) and functionally by their contractions of the esophagus. In addition, they stain positively with antibodies to actin (Cox *et al.,* 1986; Burke and Alvarez, 1988), tropomyosin (Ishi-

moda-Takagi and Sato, 1984), myosin heavy chain (Wessel *et al.,* 1990), and Sum-1, a basic helix–loop–helix-containing transcription factor (Venuti *et al.,* 1993).

Available evidence suggests that the specification of some SMCs may involve micromere-derived signaling. For example, micromeres recombined with mesomeres induce pigment cells (Khaner and Wilt, 1991), suggesting that pigment cell specification in normal embryos may involve micromere-derived cues. An additional behavior exhibited by normal SMCs is the capacity to convert to a skeletogenic fate if PMCs are experimentally eliminated (Ettensohn and McClay, 1988; reviewed in Ettensohn, 1992). Studies in the sand dollar have demonstrated that the SMCs induced in micromere/mesomere recombinants possess the ability to convert to skeletogenic mesenchyme when PMCs are removed (Minokawa *et al.,* 1997). Hence, micromere-derived signaling can induce isolated mesomeres to form SMCs, and these SMCs can exhibit behaviors mimicking those of SMCs in normal embryos. Whether micromeres induce all other SMC cell types is not known at this time, but these data suggest that micromere-derived cues may be one source of signals involved in SMC specification.

2. How Are Further Subdivisions within the Secondary Mesenchyme Cells Accomplished?

By the mesenchyme blastula stage, both cell marking studies (Ruffins and Ettensohn, 1996) and the localization of Notch protein to apical membranes of endodermal cells surrounding the presumptive SMCs (Sherwood and McClay, 1997), demonstrate that SMCs are clearly distinct from neighboring cells at this time. Within the prospective SMCs, many subtypes also may be at least conditionally specified by this stage, since cells of specific SMC subpopulations are distributed in predictable patterns within the vegetal epithelium (Ruffins and Ettensohn, 1996). Additionally, the expression pattern of the Sp1 marker demonstrates that at least a subset of prospective pigment cells are molecularly distinct from other SMCs (Gibson and Burke, 1985), and the expression pattern of CyIIa mRNA suggests that there is diversity within the SMCs prior to archenteron formation (Miller *et al.,* 1996). Cell marking studies performed between hatching and the mesenchyme blastula stage have demonstrated a gradual restriction in the range of SMC subtypes that can be formed by individually labeled cells in the vegetal plate (Ruffins and Ettensohn, 1993), indicating a progressive allocation of cells to different fates over time. These data are consistent with the hypothesis that, similar to the endoderm, cell–cell signaling processes regulate cell fate specification among the prospective SMCs.

3. Cell–Cell Interactions during Gastrulation

The mechanisms that regulate SMC specification to distinct fates within the embryo have been difficult to examine because the lack of early molecular markers and the intermixing of SMC precursors within the vegetal plate (Ruffins and Ettensohn, 1996) have made the SMCs relatively inaccessible to experimental analyses. However, several studies have demonstrated that cell interactions among the endoderm, PMCs, and other SMC subpopulations that occur during gastrulation may regulate the final differentiation of some SMC subtypes. For example, cell–cell interactions between the PMCs and SMCs play a role in the final specification of SMCs, particularly in the differentiation of the pigment cells. As mentioned above, the experimental deletion of PMCs triggers a conversion of SMCs to a PMC fate (Ettensohn and McClay, 1988), demonstrating that PMCs normally provide a suppressive signal(s) that prevents SMCs from assuming a skeletogenic identity. High resolution video microscopy techniques have demonstrated interactions between thin filopodia on PMCs and lamellipodia of SMCs, suggesting that the cell–cell signaling between PMCs and SMCs during gastrulation may be mediated via direct cell–cell contact (Ettensohn, 1992; Miller *et al.*, 1995). Studies by Ettensohn and Ruffins involving careful cell counts of SMC subpopulations following the removal of the PMCs suggest that the specific SMC population that converts is a subset of prospective pigment cells (Ettensohn and Ruffins, 1993). Hence, cell–cell interactions occur between at

least the prospective pigment cells and PMCs, and play an important role in deciding whether these SMCs will form skeleton or pigment-forming mesenchyme.

SMCs also may interact with the endoderm during gastrulation to specify coelomic sac tissues. When veg1 cells are labeled with DiI at the 60-cell stage and the archenteron is removed with a micropipette at the midgastrula stage, veg1-derived endoderm can differentiate into coelomic sacs (Fig. 5; Logan, 1997). Similarly, veg1 progeny in recombinants made at the 60-cell stage containing mesomeres, veg1 cells, and only a single veg2 cell can form coelomic sacs, indicating that veg1 cells can differentiate into coelomic sac tissues if a large portion of the SMCs or veg2-derived tissues are missing (Logan and McClay, manuscript in preparation). Ettensohn (1992) has raised the question of whether coelomic sacs truly represent an SMC-derived population, because although they are protrusive, they remain continuously associated with the epithelium of the endoderm unlike the other SMC subtypes (Gustafson and Wolpert, 1963). If we include the coelomic pouches as an SMC subpopulation, however, this result suggests that cell–cell interactions must occur across the SMC–endoderm boundary. Although the nature of the intercellular interactions between endoderm and SMCs is not known, one possible mechanism that may account for coelomic sac formation by endoderm is that similar to the repressive effects of PMCs on SMC conversion to a skeletogenic fate, the prospective coelomic sac cells or other veg2-derived tissues may exert a suppressive effect on conversion of veg1-derived endoderm

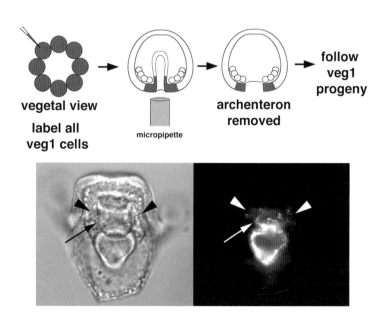

Figure 5 Veg1 descendants can form coelomic sac tissues when the gut is deleted during gastrulation. In this experiment, all veg1 cells of a 60-cell stage embryo were labeled (top left) and the gut was removed during the midgastrula stage. The labeled progeny of veg1 cells were followed to assess their fates after gut deletions. Although veg1 cells normally only form endoderm and ectoderm (Logan and McClay, 1997), veg1 progeny also can form coelomic sac tissues following microsurgical removal of the archenteron. (Arrow points to the foregut, arrowheads point to the coelomic sacs.)

to a coelomic pouch fate. Further studies will be necessary to understand this interaction in more detail.

Molecular evidence consistent with the observation that cell–cell interactions can occur between the endoderm and SMCs comes from studies on the sea urchin Notch receptor. Notch is normally downregulated during the mesenchyme blastula stage in the region where the prospective SMCs reside and this Notch-deficient region persists during early gastrulation (Sherwood and McClay, 1997). If the archenteron tip is microsurgically deleted during gastrulation and the Notch-deficient SMC population is removed, Notch protein is found throughout the remaining tissue immediately after the operation. By 1.5 hr following surgery, however, the newly forming tip of the archenteron downregulates Notch protein again. Although the specific SMC cell types replaced by converting endoderm are unclear, the downregulation of Notch in prospective endoderm following microsurgery suggests that an endoderm to SMC respecification has occurred.

There are still many questions that are unresolved regarding interactions among SMCs, between the SMCs and PMCs, and between the SMCs and endoderm. For example, the deletion of gut tissues produces embryos that possess circumesophageal muscle cells and blastocoelar cells (McClay and Logan, 1996). However, although these cell types are formed, their numbers are reduced (C. Y. Logan, 1995, unpublished observation), and it is unclear whether these cells arise from the respecification of remaining cells into SMCs, or whether they simply reflect the continued differentiation of remaining muscle and blastocoelar cells that were prespecified prior to archenteron removal. In addition, PMC removals followed by gut or vegetal plate deletions result in both PMC and gut replacement (McClay and Logan, 1996; C.Y. Logan, 1995, unpublished observations). Since it has been argued that the normal SMC population that converts to replace missing PMCs are presumptive pigment cells (Ettensohn and Ruffins, 1993), the replacement of PMCs and endoderm in PMC-less/gut-less or vegetal plate-less embryos would seem to result from a respecification of endoderm to SMCs (possibly presumptive pigment cells) which might, in turn, convert to PMCs. However, pigment cells do not appear to be respecified following the removal of gut or vegetal plate tissues (McClay and Logan, 1996; Logan, 1997). How, then, can we explain the replacement of PMCs in following both PMC and gut or vegetal plate deletions, since in these embryos, the presumptive pigment cells that might normally convert to PMCs have been eliminated? Clearly, we do not yet fully understand the complex interactions that occur among the SMCs and between the SMCs and other tissues. Unlike the conditional specification of other cells within the archenteron, pigment cells appear to be restricted during early development (Logan, 1997) even

though they retain an ability to interact with PMCs during gastrulation. Additional SMC-specific markers will provide useful tools to examine SMC formation, and further studies will allow us to understand the extent and nature of cell–cell interactions that assign specific identities within the SMCs.

IV. Future Directions

It is clear from the evidence in the foregoing discussion that multiple cell–cell signaling events regulate cell fate specification and patterning during development of the sea urchin embryo. As proposed by Davidson (1989), sequential cell–cell interactions initiated by the micromeres are involved in specifying germ layers, in refining the boundaries between tissues, and in specifying sublineages within them (Chapter 2 by Cameron and Coffman; Chapter 3 by Wikramanayake and Klein). At the same time, there is evidence to suggest that tissue specification involves maternally encoded information that acts independently of micromeres. Hence, our future task will be to understand the relative roles of micromere-dependent and micromere-independent processes in specifying cell identities along the A–V axis and how these mechanisms eventually regulate the final three-dimensional organization of the embryo. Given this background, one of the general areas in which we feel that the sea urchin will make a significant contribution in the future is in the area of early axial specification.

To gain a deeper understanding of how early axial specification is regulated during development, the characterization of additional molecular markers and secreted signaling factors will be crucial. A recent theme that has emerged from molecular studies is that embryos of different organisms use similar sets of molecules to specify and pattern tissues even when the developmental processes they regulate appear distinct. For example, the sea urchin embryo, the *Xenopus,* and zebrafish embryos differ greatly in size, yolk content, how their axes are arranged, and where gastrulation is initiated. In sea urchins, the micromeres and veg2-derived tissues of the vegetal pole appear to be potent sources of signaling activities (Logan and McClay, 1998; Benink *et al.,* 1997; also reviewed in Hörstadius, 1973) that organize the embryo along the animal–vegetal axis. In vertebrate embryos, prospective anterior endoderm and dorsal axial mesoderm are sources of inductive cell signals that specify cell fates along the dorsal–ventral axis (Kimmelman *et al.,* 1992; Shih and Keller, 1992; Chapter 19 by Dawid; Chapter 21 by Kessler). Nuclear β-catenin has been observed in vegetal cells of the sea urchin (Logan *et al.,* 1998), and transcription factors such as brachyury and a member of the forkhead family have been localized to veg2 and veg1 progeny (Harada *et al.*, 1995, 1996). Similarly, β-catenin has been observed in dorsal nuclei of both *Xenopus* and zebrafish embryos (Schneider *et al.*, 1996; Larabell *et al.*, 1997), and cells in this region also express brachyury (Wilkinson et al., 1990; Smith *et al.*, 1991; Schulte-Merker *et al.*, 1992), and HNF-3β, a forkhead family member (Ruiz i Altaba and Jessell, 1992). Although the organizing centers for axial specification are found in different locations within these embryos, similar sets of molecules appear to be utilized in order to create a region of active signaling or of morphogenetic activity. In addition, several growth factors and their receptors known to play important roles in cell–cell signaling of other organisms (for reviews, see Artavanis-Tsakonas *et al.*, 1995; Hogan, 1995; Moon *et al.*, 1997; Siegfried and Perrimon, 1994) have also been cloned in sea urchins. These include Notch (Sherwood and McClay, 1997), TGF-β family members such as univin (Stenzel *et al.*, 1994) and BMP-4 (C. Y. Logan and D. R. McClay, 1994, unpublished; L. M. Angerer, D. Oleksyn, L. Dale, and R. C. Angerer, 1997, personal communication), and Wnt homologs (Sidow, 1992). Although not enough data are available at the present time to fully understand their functions and to place these factors in a coherent network of signaling pathways, these molecules will provide useful starting points for understanding how cell fates are specified within the sea urchin embryo. Eventually, understanding the similarities and differences in the molecular regulation of cell fate specification along embryonic axes of diverse organisms may provide novel insights into how these molecules function and how they are deployed in ways that are unique to the organization of various embryos.

With an increasing understanding of the mechanisms that specify distinct cell identities during development, how cell fate choices are coordinated with later morphogenetic movements to effect three-dimensional changes within the embryo will also become an important aspect of axial development. The cell movements during gastrulation result in dramatic rearrangements of vegetal cell populations (Ettensohn, 1984, 1985; Fink and McClay, 1985; Hardin and Cheng, 1986; Hardin, 1989; Burke *et al.*, 1991; Lane *et al.*, 1993) which are coordinated with both cell fate decisions and cytokinesis (Nislow and Morrill, 1988). Morphogenetic events such a convergent-extension and ingression have been documented in a wide variety of species and not only have implications for understanding developmental processes but also for understanding cell motility and metastasis in cancer. Therefore, elucidating how gene expression and cell behavioral changes are coordinated during development will aid in comprehending the processes that regulate axis formation. The optical clarity of the sea urchin embryo allows all cell types to be easily visualized, unlike many less transparent organisms. Moreover, dyes, antibodies, and tagged molecules

provide experimental tools for following cells. Combining both molecular and embryological approaches in the sea urchin to study both cell behavior and cell fate specification will eventually allow one to examine how morphogenesis and specification events are coordinately regulated during development.

The simplicity and transparency of the sea urchin embryo provide new opportunities to experimentally explore axial and tissue specification mechanisms and their relationships to later patterning and morphogenesis. In addition, they allow questions to be asked and interpreted in ways that are not as easily feasible in other organisms. By asking how endoderm and mesoderm arise in both sea urchins and other organisms, it will be possible to understand how early axial specification leads to tissue specification and the generation of embryonic form.

Acknowledgments

The authors thank Dr. Jeff Miller for extensive discussions and insightful comments. We are also extremely grateful to Drs. Chuck Ettensohn, Lynne Angerer, Bob Angerer, and Fred Wilt for valuable feedback, and many thoughtful suggestions. We also thank Lynne and Bob Angerer for sharing their unpublished data.

References

Angerer, R. C., and Angerer, L. M. (1997). Fate specification along the sea urchin embryo animal–vegetal axis. *Biol. Bull.* **192**, 175–177.

Artavanis-Tsakonas, S., Matsuno, K., and Fortini, M. E. (1995). Notch signaling. *Science* **268**, 225–232.

Benink, H., Wray, G., and Hardin, J. (1997). Archenteron precursor cells can organize secondary axial structures in the sea urchin embryo. *Development* (*Cambridge, UK*) **124**, 3461–3470.

Berridge, M. J., Downes, C. P., and Hanley, M. R. (1989). Neural and developmental actions of lithium: A unifying hypothesis. *Cell* **59**, 411–419.

Burke, R. D. (1981). Structure of the digestive tract of the pluteus larva of *Dendraster excentricus* (Echinodermata: Echinoida). *Zoomorphology* **98**, 209–225.

Burke, R. D., and Alvarez, C.M. (1988). Development of the esophageal muscles in embryos of the sea urchin *Strongylocentrotus purpuratus*. *Cell Tissue Res.* **252**, 411–417.

Burke, R. D., Myers, R. L., Sexton, T. L., and Jackson, C. (1991). Cell movements during the initial phase of gastrulation in the sea urchin embryo. *Dev. Biol.* **146**, 542–557.

Cameron, R. A., and Davidson, E. H. (1991). Cell type specification during sea urchin development. *Trends Genet.* **7**, 212–218.

Cameron, R. A., and Davidson, E. H. (1997). LiCl perturbs ectodermal veg₁ lineage allocations in *Strongylocentrotus purpuratus* embryos. *Dev. Biol.* **187**, 236–239.

Cameron, R. A., Fraser, S. E., Britten, R. J., and Davidson, E. H. (1991). Macromere cell fates during sea urchin development. *Development (Cambridge, UK)* **113**, 1085–1091.

Chen, S. W., and Wessel, G. M. (1996). Endoderm differentiation in vitro identifies a transitional period for endoderm ontogeny in the sea urchin embryo. *Dev. Biol.* **175**, 57–65.

Coffman, J. A., and McClay, D. R. (1990). A hyaline layer protein that becomes localized to the oral ectoderm and foregut of sea urchin embryos. *Dev. Biol.* **140**, 93–104.

Cox, K. H., Angerer, L. M., Lee, J. J., Davidson, E. H., and Angerer, R.C. (1986). Cell lineage-specific programs of expression of multiple actin genes during sea urchin embryogenesis. *J. Mol. Biol.* **188**, 159–172.

Davidson, E. H. (1989). Lineage-specific gene expression and the regulative capacities of the sea urchin embryo: A proposed mechanism. *Development (Cambridge, UK)* **105**, 421–445.

Davidson, E. H. (1993). Later embryogenesis: Regulatory circuitry in morphogenetic fields. *Development (Cambridge, UK)* **118**, 665–690.

Ettensohn, C. A. (1984). Primary invagination of the vegetal plate during sea urchin gastrulation. *Am. Zool.* **24**, 571–588.

Ettensohn, C. A. (1985). Gastrulation in the sea urchin embryo is accompanied by rearrangement of invaginating epithelial cells. *Dev. Biol.* **112**, 383–390.

Ettensohn, C. A. (1992). Cell interactions and mesodermal cell fates in the sea urchin embryo. *Development (Suppl.)*, 43–51.

Ettensohn, C. A., and McClay, D. R. (1988). Cell lineage conversion in the sea urchin embryo. *Dev. Biol.* **125**, 396–409.

Ettensohn, C. A., and Ruffins, S. W. (1993). Mesodermal cell interactions in the sea urchin embryo: Properties of skeletogenic secondary mesenchyme cells. *Development (Cambridge, UK)* **117**, 1275–1285.

Ferguson, E. L., and Anderson, K. V. (1992). Localized enhancement and repression of the activity of the TGF-beta family member, decapentaplegic, is necessary for dorsal–ventral pattern formation in *Drosophila* embryo. *Development (Cambridge, UK)* **114**, 583–597.

Fink, R. D., and McClay, D. R. (1985). Three cell recognition changes accompany the ingression of sea urchin primary mesenchyme cells. *Dev. Biol.* **107**, 66–74.

Ghiglione, C., Lhomond, G., Lepage, T., and Gache, C. (1993). Cell-autonomous expression and position-dependent repression by Li⁺ of two zygotic genes during sea urchin early development. *EMBO J.* **12**, 87–96.

Ghiglione, C., Emily-Fenouil, F., Chang, P., and Gache, C. (1996). Early gene expression along the animal–vegetal axis in sea urchin embryoids and grafted embryos. *Development (Cambridge, UK)* **122**, 3067–3074.

Gibson, A. W., and Burke, R. D. (1985). The origin of pigment cells in the sea urchin *Strongylocentrotus purpuratus*. *Dev. Biol.* **107**, 414–419.

Gustafson, T., and Wolpert, L. (1963). Studies on the cellular basis of morphogenesis in the sea urchin embryo: Formation of the coelom, the mouth, and the primary pore canal. *Exp. Cell Res.* **29**, 561–582.

Gustafson, T., and Wolpert, L. (1967). Cellular movement and contact in sea urchin morphogenesis. *Biol. Rev. Cambridge Philos. Soc.* **42**, 442–498.

Gustafson, W., and Kinnander, H. (1960). Cellular mechanisms in morphogenesis of the sea urchin gastrula: The oral contact. *Exp. Cell Res.* **21**, 361–373.

Harada, Y., Yasuo, H., and Satoh, N. (1995). A sea urchin homologue of the chordate Brachyury (T) gene is expressed in the secondary mesenchyme founder cells. *Development (Cambridge, UK)* **121**, 2747–2754.

Harada, Y., Akasaka, K., Shimada, H., Peterson, K. J., Davidson, E. H., and Satoh, N. (1996). Spatial expression of a forkhead homologue in the sea urchin embryo. *Mech. Dev.* **60**, 163–173.

Hardin, J. (1989). Local shifts in position and polarized motility drive cell rearrangement during sea urchin gastrulation. *Dev. Biol.* **136**, 430–445.

Hardin, J. D., and Cheng, L. Y. (1986). The mechanisms and mechanics of archenteron elongation during sea urchin gastulation. *Dev. Biol.* **115**, 490–501.

Hardin, J., and McClay, D. R. (1990). Target recognition by the archenteron during sea urchin gastrulation. *Dev. Biol.* **142**, 86–102.

Hardin, J., Coffman, J. A., Black, S. D., and McClay, D. R. (1992). Commitment along the dorsoventral axis of the sea urchin embryo is altered in response to NiCl₂. *Development (Cambridge, UK)* **116**, 671–685.

Henry, J. H., Amemiya, S., Wray, G. A., and Raff, R. A. (1989). Early inductive interactions are involved in restricting cell fates of mesomeres in sea urchin embryos. *Dev. Biol.* **136**, 140–153.

Hogan, B. L. M. (1995). The TGF-β-related signalling system in mouse development. *Semin. Dev. Biol.* 257–265.

Hörstadius, S. (1939). The mechanics of sea urchin development, studies by operative methods. *Biol. Rev. Cambridge Philos. Soc.* **14**, 132–179.

Hörstadius, S. (1973). "Experimental Embryology of Echinoderms." Oxford Univ. Press (Clarendon), Oxford.

Ishimoda-Takagi, T., and Sato, H. (1984). Evidence for the involvement of muscle tropomyosin in the contractile elements of the coelom–esophagus complex in sea urchin embryos. *Dev. Biol.* **105**, 365–376.

Khaner, O., and Wilt, F. (1991). Interactions of different vegetal cells with mesomeres during early stages of sea urchin development. *Development (Cambridge, UK)* **112**, 881–890.

Kimmelman, D., Christian, J. L., and Moon, R. T. (1992). Synergistic principles of development: Overlapping patterning systems in *Xenopus* mesoderm induction. *Development (Cambridge, UK)* **116**, 1–9.

Kitajima, T., and Okazaki, K. (1980). Spicule formation *in vitro* by the descendants of precocious micromere formed at the 8-cell stage of sea urchin embryo. *Dev. Growth Differ.* **22**, 265–279.

Klein, P. S., and Melton, D. A. (1996). A molecular mechanism for the effect of lithium on development. *Proc. Natl. Acad. Sci. U.S.A.* **93**, 8455–8459.

Lane, M. C., Koehl, M. A. R., and Keller, R. (1993). A role for regulated secretion of the apical extracellular matrix during epithelial invagination in the sea urchin. *Development (Cambridge, UK)* **117**, 1049–1060.

Larabell, C. A., Torres, M., Rowning, B. A., Yost, C., Miller, J. R., Wu, M., Kimelman, D., and Moon, R. T. (1997). Establishment of the dorsoventral axis in *Xenopus* embryos is presaged by early asymmetries in β-catenin that are modulated by the wnt signaling pathway. *J. Cell Biol.* **136**, 1123–1136.

Lepage, T., Sardet, C., and Gache, C. (1992). Spatial expression of the hatching enzyme gene in the sea urchin embryo. *Dev. Biol.* **150**, 23–32.

Livingston, B. T., and Wilt, F. H. (1990). Range and stability of cell fate determination in isolated sea urchin blastomeres. *Development (Cambridge, UK)* **108**, 403–410.

Logan, C. Y. (1997). "Specification of the Endoderm and Mesoderm in the Sea Urchin Embryo." Duke University, Durham, North Carolina.

Logan, C. Y., and McClay, D. R. (1997). The allocation of early blastomeres to the ectoderm and endoderm is variable in the sea urchin embryo. *Development (Cambridge, UK)* **124**, 2213–2223.

Logan, C. Y., and McClay, D. R. (1998). Manuscript in preparation.

Logan, C. Y., Miller, J., Ferkowicz, M., and McClay, D. R. (1998). Developmental Regulation of β-catenin distribution and the establishment of cell fates along the animal–vegetal axis in the sea urchin embryo. Submitted to *Development.*

Marqués, G., Musacchio, M., Schimell, M. J., Wünnenberg-Stapleton, K., Cho, K. W. Y., and O'Connor, M. B. (1997). Production of a DPP activity gradient in the early *Drosophila* embryo through the opposing actions of the SOG and TLD proteins. *Cell (Cambridge, Mass.)* **91**, 417–426.

Martinez, P., and Davidson, E. (1997). SpHmx, a sea urchin homeobox gene expressed in embryonic pigment cells. *Dev. Biol.* **181**, 213–222.

McClay, D. R., and Logan, C. Y. (1996). Regulative capacity of the archenteron during gastrulation in the sea urchin. *Development (Cambridge, UK)* **122**, 607–616.

Miller, J., Fraser, S. E., and McClay, D. (1995). Dynamics of thin filopodia during sea urchin gastrulation. *Development (Cambridge, UK)* **121**, 2501–2511.

Miller, R. N., Dalamagas, D. G., Kingsley, P. D., and Ettensohn, C. A. (1996). Expressin of S9 and actin CyIIa mRNAs reveals dorso-ventral polarity and mesodermal sublineages in the vegetal plate of the sea urchin embryo. *Mech. Dev.* **60**, 3–12.

Minokawa, T., Hamaguchi, Y., and Amemiya, S. (1997). Skeletogenic potential of induced secondary mesenchyme cells derived from the presumptive ectoderm in echinoid embryos. *Dev. Genes Evol.* **206**, 472–476.

Moon, R. T., Brown, J. D., and Torres, M. (1997). Wnts modulate cell fate and behavior during vertebrate development. *Trends Genet.* **13**, 157–162.

Nislow, C., and Morrill, J.B. (1988). Regionalized cell division during sea urchin gastrulation contributes to archenteron formation and is correlated with the establish-ment of larval symmetry. *Dev. Growth Differ.* **30**, 483–499.

Nocente-McGrath, C., Brenner, C. A., and Ernst, S. G. (1989). Endo16, a lineage-specific protein of the sea urchin embryo, is first expressed just prior to gastrulation. *Dev. Biol.* **136**, 264–272.

Nocente-McGrath, C., McIsaac, R., and Ernst, S. G. (1991). Altered cell fate in LiCl-treated sea urchin embryos. *Dev. Biol.* **147**, 445–450.

Okazaki, K. (1975). Normal development to metamorphosis. *In* "The Sea Urchin Embryo" (G. Czihak, ed.), pp. 177–232. Springer-Verlag, Berlin.

Piccolo, S., Agius, E., Lu, B., Goodman, S., Dale, L., and DeRobertis, E. M. (1997). Cleavage of Chordin by Xolloid metalloprotease suggests a role for proteolytic processing in the regulation of Spemann organizer activity. *Cell (Cambridge, Mass.)* **91**, 407–416.

Ransick, A., and Davidson, E. H. (1993). A complete second gut induced by transplanted micromeres in the sea urchin embryo. *Science* **259**, 1134–1138.

Ransick, A., and Davidson, E. H. (1995). Micromeres are required for normal vegetal plate specification in sea urchin embryos. *Development (Cambridge, UK)* **121**, 3215–3222.

Ransick, A., Ernst, S., Britten, R. J., and Davidson, E. H. (1993). Whole mount *in situ* hybridization shows Endo16 to be a marker for the vegetal plate territory in sea urchin embryos. *Mech. Dev.* **42**, 117–124.

Reynolds, S. D., Angerer, L. M., Palis, J., Nasir, A., and Angerer, R. C. (1992). Early mRNAs, spatially restricted along the animal–vegetal axis of sea urchin embryos, include one encoding a protein related to tolloid and BMP-1. *Development (Cambridge, UK)* **114**, 769–786.

Ruffins, S. W., and Ettensohn, C. A. (1993). A clonal analysis of secondary mesenchyme cell fates in the sea urchin embryo. *Dev. Biol.* **160**, 285–288.

Ruffins, S. W., and Ettensohn, C. A. (1996). A fate map of the vegetal plate of the sea urchin (*Lytechinus variegatus*) mesenchyme blastula. *Development (Cambridge, UK)* **122**, 253–263.

Ruiz i Altaba, A., and Jessell, T. M. (1992). Pintallavis, a gene expressed in the organizer and midline cell of frog embryos: Involvement in the development of the neural axis. *Development (Cambridge, UK)* **116**, 81–93.

Runnström, J. (1975). Integrating factors. *In* "The Sea Urchin Embryo: Biochemistry and Morphogenesis" (G. Czihak, ed.), pp. 646–670. Springer-Verlag, Berlin.

Ryberg, E., and Lundgren, B. (1979). Some aspects on pigment cell distribution and function in the developing echinopluteus of *Psammechinus miliaris*. *Dev. Growth Differ.* **21**, 129–140.

Schneider, S., Steinbeisser, H., Warga, R. M., and Hausen, P. (1996). β-Catenin translocation into nuclei demarcates the dorsalizing centers in frog and fish embryos. *Mech. Dev.* **57**, 191–198.

Schulte-Merker, S., Ho, R. K., Herrmann, B. G., and Nüsslein-Volhard, C. (1992). The protein product of the zebrafish homologue of the mouse T gene is expressed in nuclei of the germ ring and the notochord of the early embryo. *Development (Cambridge, UK)* **116**, 1021–1032.

Sherwood, D. R. (1997), The Identification and function of a Notch receptor in the sea urchin embryo. Duke University, Durham, North Carolina.

Sherwood, D. R., and McClay, D. R. (1997). Identification and localization of a sea urchin Notch homologue: Insights into vegetal plate regionalization and Notch receptor regulation. *Development (Cambridge, UK)* **124**, 3363–3374.

Shih, J., and Keller, R. (1992). The epithelium of the dorsal marginal zone of *Xenopus* has organizer properties. *Development (Cambridge, UK)* **116**, 887–899.

Shimell, M. J., Ferguson, E. L., Childs, S. R., and O'Connor, M. (1991). The *Drosophila* dorsal-ventral patterning gene tolloid is related to human bone morphogenetic protein 1. *Cell (Cambridge, Mass.)* **67**, 469–481.

Sidow, A. (1992). Diversification of the Wnt gene family on the ancestral lineage of vertebrates. *Proc. Natl. Acad. Sci. U.S.A.* **89**, 5098–5112.

Siegfried, E., and Perrimon, N. (1994). *Drosophila* Wingless: A paradigm for the function and mechanism of Wnt signaling. *Bioessays* **16**, 395–404.

Slack, J. M. W. (1991). "From Egg to Embryo." Cambridge, Univ. Press, Cambridge.

Smith, J. C., Price, B. M. J., Green, J. B. A., Weigel, D., and Herrmann, B. G. (1991). Expression of a *Xenopus* homolog of Brachyury (T) is an immediate-early response to mesoderm induction. *Cell (Cambridge, Mass.)* **67**, 79–87.

Stenzel, P., Angerer, L. M., Smith, B. J., Angerer, R. C., and Vale, W.W. (1994). The univin gene encodes a member of the transforming growth factor-beta superfamily with restricted expression in the sea urchin embryo. *Dev. Biol.* **166**, 149–158.

Tamboline, C. R., and Burke, R. D. (1989). Ontogeny and characterization of mesenchyme antigens of the sea urchin embryo. *Dev. Biol.* **136**, 75–86.

Tamboline, C. R., and Burke, R. D. (1992). Secondary mesenchyme of the sea urchin embryo: Ontogeny of blastocoelar cells. *J. Exp. Zool.* **262**, 51–60.

Venuti, J. M., Gan, L., Kozlowski, M. T., and Klein, W. H. (1993). Developmental potential of muscle cell progenitors and the myogenic factor SUM-1 in the sea urchin embryo. *Mech. Dev.* **41**, 3–14.

Wessel, G. M., and McClay, D. R. (1985). Sequential expression of germ-layer specific molecules in the sea urchin embryo. *Dev. Biol.* **111**, 451–463.

Wessel, G. M., Goldberg, L., Lennarz, W. J., and Klein, W. H. (1989). Gastrulation in the sea urchin is accompanied by the accumulation of an endoderm-specific mRNA. *Dev. Biol.* **136**, 526–536.

Wessel, G. M., Zhang, W., and Klein, W. H. (1990). Myosin heavy chain accumulates in dissimilar cell types of the macromere lineage in the sea urchin embryo. *Dev. Biol.* **140**, 447–454.

Wikramanayake, A. H., Brandhorst, B. P., and Klein, W. H. (1995). Autonomous and nonautonomous differentiation of ectoderm in different sea urchin species. *Development (Cambridge, UK)* **121**, 1497–1505.

Wilkinson, D. G., Bhatt, S., and Herrmann, B. G. (1990). Expression pattern of the mouse T gene and its role in mesoderm formation. *Nature (London)* **343**, 657–659.

Wilt, F. H. (1987). Determination and morphogenesis in the sea urchin embryo. *Development (Cambridge, UK)* **100**, 559–575.

Yuh, C., and Davidson, E. H. (1996). Modular cis-regulatory organization of Endo16, a gut-specific gene of the sea urchin embryo. *Development (Cambridge, UK)* **122**, 1069–1082.

5

Cell Fate Determination in the Ascidian Embryo

NORIYUKI SATOH
Department of Zoology
Graduate School of Science
Kyoto University
Kyoto 606-8502, Japan

I. Introduction

Ascidians (subphylum Urochordata or Tunicata, class Ascidiacea), or sea squirts, are marine animals ubiquitous throughout the world. They are sessile and specialized for filter feeding. Descriptions of ascidian embryogenesis and cell lineages first appeared in the late nineteenth century. Ascidians were originally thought to be molluscs. The description by Kowalevsky (1866) of the occurrence of ascidian tadpole-type larvae with notochord and nerve cord shifted this animal group to a new and proper taxonomic position, close to vertebrates. The descriptive work of van Beneden and Julin (1884) was the first demonstration in the history of embryology of the relationship between the egg axis and larval body plan. In addition, in 1887, the French biologist Laurent Chabry carried out blastomere-destruction experiments with ascidian eggs, the first such effort in the history of embryology. Chabry destroyed one blastomere of a 2-cell *Ascidiella* embryo and found that the remaining blastomere continued to cleave as if it were half of the whole embryo and eventually formed a half-larva instead of a complete dwarf larva. After obtaining a similar result using a 4-cell embryo, he reached the conclusion that ascidian embryos could not compensate for missing parts, and thus the developmental pattern was "mosaic."

In 1905, Conklin described in elaborate detail the cell lineages of ascidian embryos by direct observation of *Cynthia partita* living materials (Conklin, 1905a). His observations were facilitated by the presence of natural colored cytoplasms in the egg that became segregated into specific blastomeres during cleavage. Subsequently, Conklin's work constituted a milestone in the study of ascidian embryogenesis. His study was confirmed later by Ortolani (1955, 1957), Nishida and Satoh (1983, 1985), and Nishida (1987).

Along with the well-described lineage, many experimental studies were conducted to elucidate cellular mechanisms involved in developmental fate determination of ascidian embryonic cells, in particular by Italian embryologists around 1940s and 1950s (reviewed by Reverberi, 1971). The gate for the modern era of ascidian embryology was open by Whittaker (1973) with his work on cleavage-arrested embryos, in which he showed very beautifully the existence and segregation of muscle determinants into muscle-lineage cells. More recently, Nishida has greatly contributed to our understanding of the cellular mechanisms of developmental fate determination in the ascidian embryo (reviewed by Satoh, 1994; Nishida, 1997).

II. General Description of Ascidian Embryogenesis

Matured eggs of ascidians cease at the metaphase of first meiotic division. Fertilization reinitiates the meiotic division, protruding the first and second polar bodies. Fertilization also evokes a dynamic rearrangement of the egg cytoplasm called ooplasmic segregation. The first phase of ooplasmic segregation involves rapid movement of the peripheral cytoplasm including the myoplasm (the cytoplasm which is segregated into muscle lineage) to form a transient cap near the vegetal pole of the egg. During the second phase, the myoplasm shifts from the vegetal-pole region to a new position near the subequatorial zone of the egg and forms a crescent, which is a landmark of the posterior side of the future embryo.

The cleavage pattern is invariant and essentially the same in eggs of all ascidian species. Cleavage is bilaterally symmetrical; the first cleavage plane establishes the only plane of symmetry in the embryo, and the right-half side of the embryo is the mirror image of the left-half side (Fig. 1). Because every division of the embryonic cells is recognizable, developmental stages of early ascidian embryos are called the 8-, 16-, 32-, 64-, and 110-cell stages, instead of the morula and blastula stages (Fig. 1). Gastrulation initiates around the 118-cell stage. It involves epibolic movements of ectodermal cells and migration of endodermal and mesodermal cells inside the embryo. Neurulation is accomplished by folding of the presumptive neural cells, as in vertebrate embryos. Then, the tailbud embryo is formed, which eventually develops into a tadpole larva.

The ascidian tadpole consists of about 2600 cells that form several distinct types of tissues (Table I, Fig. 1; Satoh, 1994). The tadpole is organized into a trunk and tail (Fig. 1K–M). The trunk contains a dorsal central nervous system (CNS) with two sensory organs (otolith and ocellus), endoderm, mesenchyme and trunk lateral cells. The tail contains a notochord flanked dorsally by the nerve cord, ventrally by endodermal strand, and bilaterally by three rows of muscle cells (Fig. 1K–M). The entire surface of the larva is covered by an epidermis. The nonfeeding larva is a dispersal phase. During metamorphosis, the larval tail is retracted and disintegrated, and adult tissues and organs differentiate from precursors in the trunk (Hirano and Nishida, 1997).

Recent studies support the ascidian tadpole prototype for the ancestral chordate (reviewed by Satoh and Jeffery, 1995). In addition, recent molecular cloning succeeded in the isolation of cDNA clones for genes that are expressed specifically in certain types of tissues (reviewed by Satoh *et al.*, 1996a). The expression of

these genes are used as molecular markers for cell differentiation, and are summarized in Table I.

III. Nomenclature of the Blastomeres, Lineage, and Restriction of Developmental Fates

Each blastomere of the early ascidian embryo is distinguishable, and they are named according to the nomenclature of Conklin (1905a), such as a4.2, b5.3, A6.1, and B7.4 (Fig. 1). The letter "a" denotes descendents of two (or pair of) anterior animal blastomeres of the 8-cell embryo; "b" shows those of the posterior animal blastomeres; "A", those of the anterior vegetal; and "B", those of the posterior vegetal. The first numerical digit denotes cell generation, counting the unsegmented eggs as the first. The second digit gives the cell its own number, which doubles at each division (e.g., B7.4 divides into B8.7 and B8.8). Cells that lie near the vegetal pole are assigned the lower number. Underlining is used to indicate the blastomeres on the right side of the bilaterally symmetrical embryo. The uncleaved zygote is named A1. Blastomeres of the 2-cell embryo are called the left AB2 and the right AB2. The anterior and posterior blastomeres of the 4-cell embryo are named A3 and B3 pair, respectively (Fig. 1A).

As shown in Fig. 2, the complete cell lineage has been described up to the early gastrula stage (Conklin, 1905a; Ortolani, 1957; Nishida, 1987; Nicol and Meinertzhagen, 1988). The lineages of epidermis, CNS, notochord, and muscle are documented. Cell lineage analyses by intracellular marking of embryonic cells with a tracer enzyme revealed that the lineage is invariant except for two cases; the bilateral pair of a 8.25 and a8.25 give rise to the otolith and ocellus in complementary manner (see section V.B), and the bilateral pair of b8.17 and b8.17 give rise to the endodermal strand and nerve cord in the same manner (Nishida, 1987).

During ascidian embryogenesis, as in other animals, the developmental fates of embryonic cells are gradually restricted to give rise to a single type of tissue. The fate restriction in ascidian embryos takes place relatively early (Fig. 1). As early as the 16-cell stage, a pair of epidermis-restricted cells (b5.4 pair) appear in the animal hemisphere (Fig. 1C). Then, at the 32-cell stage, a pair of cells (A6.1 pair) in the vegetal hemisphere become restricted to endoderm (Fig. 1F). At the 64-cell stage, blastomeres appear that are restricted to their various individual fates: notochord, muscle, nerve cord, mesenchyme, and trunk lateral cells (Fig. 1G, H).

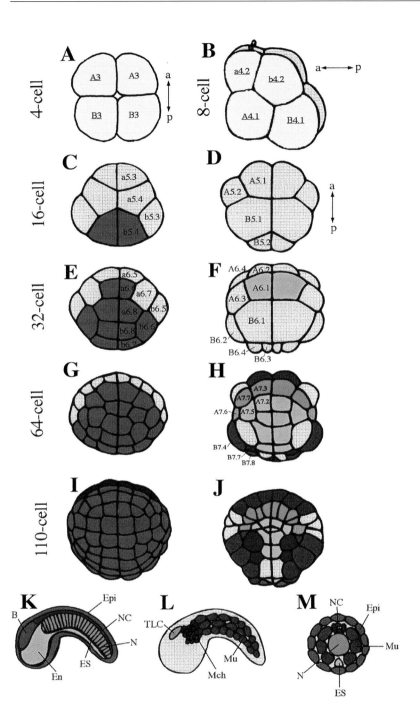

Figure 1 Cleavage pattern, nomenclature of blastomeres, lineage, and gradual restriction of developmental fates of ascidian embryonic cells. Blastomeres are colored when the developmental fate is restricted to give rise to cells of a single type of tissue. (A) A 4-cell embryo, animal pole view. Anterior is up and posterior is down. The name of each blastomere is indicated. (B) An 8-cell embryo, lateral view. Animal pole is up and vegetal pole is down. Anterior is to the left and posterior is to the right. (C, D) A 16-cell embryo, viewed from animal (C) and vegetal pole (D). Anterior is up and posterior is down. (E, F) A 32-cell embryo, animal (E) and vegetal (F) views, respectively. (G, H) A 64-cell embryo, animal (G) and vegetal (H) views, respectively. (I, J) A 110-cell embryos, animal (I) and vegetal (J) views, respectively. (K, L, M) Schematic drawings showing tissues and organs of the tailbud embryo. Midsagittal section (K) and sagittal section (L) of the embryo, and transverse section of the tail (M). B, brain (marked by purple); En, endoderm (yellow); Epi, epidermis (green); ES, endodermal strand (yellow); Mch, mesenchyme (dark green); Mu, muscle (red); N, notochord (pink); NC, nerve cord (light purple); TLC, trunk lateral cells (light blue). (Original drawing by Dr. H. Nishida, modified.)

TABLE I

Cell-Types, Specification Pattern, and Specific Genes in Embryos of the Ascidian *Halocynthia roretzi*

Cell type	Number of constituting cells	Specification pattern	Specific gene	References
Epidermis	800	Autonomous	*HrEpiA~H*	Ueki *et al.* (1994)
CNS	~265	Conditional	*TuNa1*	Okamura *et al.* (1994)
			HrTBB2	Miya and Satoh (1997)
Sensory Pigment cells	2	Conditional	Tyrosinase gene	Sato *et al.* (1997)
Nerve cord	~65	Conditional[a]	—	
Endoderm	~500	Autonomous	Alkaline phosphatase gene	H. Nishida (personal communication, 1998)
Mesenchyme	~900	Autonomous[a]	*HrCA1*	Araki *et al.* (1996)
Trunk lateral cells	~32	Conditional	Specific genes	Takahashi *et al.* (1997)
Notochord	40		*As-T*	Yasuo and Satoh (1993, 1994)
A-line	(32)	Conditional		
B-line	(8)	Conditional[a]		
Muscle	42			
B-line	(28)	Autonomous	[*HrMA4* (actin gene)	Kusakabe *et al.* (1992)
A- and b-line	(10, 4)	Conditional[a]	[*HrMHC* (myosin HC gene)	Araki and Satoh (1996)

[a]Not fully determined.

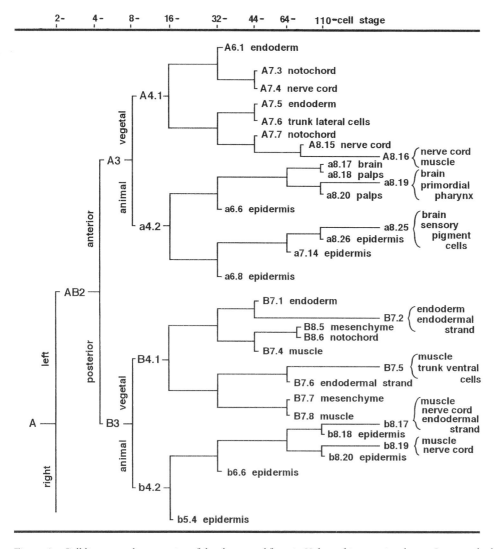

Figure 2   Cell lineage and segregation of developmental fates in *Halocynthia roretzi* embryos. Because the lineage is bilaterally symmetrical, only the left half of the embryo is shown. Divisions of cells that are tissue restricted are abbreviated. (From Nishida, 1987.)

IV. Autonomous Specification of Embryonic Cells

Reflecting such an early fate restriction, the ascidian embryo shows a highly determinate mode of development. In many cell types, if precursor cells are lost, they cannot be replaced by other cells during development. Blastomeres isolated from early embryos show a high potential for self-differentiation, dependent on prelocalized maternal factors. In addition to classic descriptive and experimental studies (reviewed by Reverberi, 1971), recent studies have provided convincing evidence for maternal factors or determinants responsible for differentiation of muscle (Nishida, 1992; Marikawa *et al.,* 1994), endoderm (Nishida, 1993), and epidermis (Nishida, 1994a), factors for the establishment of anteroposterior axis of the embryo (Nishida, 1994b), and those for initiation of gastrulation (Jeffery, 1990; Nishida, 1996). In particular, the posterior-vegetal cytoplasm of the fertilized egg or the so-called myoplasm contains muscle determinants, factors for the anteroposterior axis establishment, and those for initiation of gastrulation. Here, we will focus on the autonomous specification of muscle, endoderm, and epidermis.

A. Muscle

1. Lineage

During ascidian embryogenesis, unicellular and striated muscle cells are formed on each side of the tail of the tadpole larva (Fig. 1L, M). The number of muscle cells differs among species; 36 and 42 muscle cells are formed in *Ciona intestinalis* and *Halocynthia roretzi,* respectively (Nishida, 1987). This difference is due to the number of b4.2-derived muscle cells. In an *H. roretzi* embryo, of 42 muscle cells, 28 of the anterior and middle part of the tail are derived from the B-line, 4 of the posterior part of the tail from the A-line, and 10 of the caudal tip from the b-line blastomeres (Fig. 2; Nishida, 1987). Because the specification pattern of B-line muscle cells differs from that of A- and b-lines, the B-line is sometimes called the primary lineage, while the A- and b-lines are called the secondary lineage (Meedel *et al.,* 1987).

The developmental potential that allows the B4.1 to give rise to muscle is inherited by the B5.1 and B5.2 of the 16-cell embryos, and then B6.2, B6.3, and B6.4 of the 32-cell embryo (Fig. 2). At the 64-cell stage, B7.4, B7.5, and B7.8 are the B-line muscle lineage, and the developmental fate of B7.4 and B7.8 is restricted to muscle (Figs. 1 and 2). Blastomere B7.4 gives rise to a clone of 8 muscle cells after 3 subsequent divisions, whereas B7.8 forms a clone of 4 muscle cells after subsequent divisions. Previously it was thought that B7.5 had the potential to give rise to two muscle cells and two

endoderm cells (Nishida, 1987). A recent reexamination showed that the latter cells are not endoderm cells but trunk ventral cells that give rise to adult body-wall muscle (Hirano and Nishida, 1997; Fig. 2).

Development of muscle cells from the A-line precursor is through the A5.2 of the 16-cell embryo, the A6.4 of the 32-cell embryo, the A7.8 of the 64-cell embryo, and then the A8.16 of the 110-cell embryo (Fig. 2). However, the developmental fate of the A8.6 is not restricted yet; it has the potential to form two muscle cells and several nerve-cord cells (Nishida, 1987), although in *Ciona* the A9.31 forms only muscle (Nicol and Meinertzhagen, 1988).

The b-line muscle precursor cells in *H. roretzi* are the b5.3 of the 16-cell embryo, the b6.5 of the 32-cell embryo, the b7.9 and b7.10 of the 64-cell embryo, and the b8.17 and b8.19 of the 110-cell embryo (Fig. 2). In this species, the b8.17 has the potential to form three muscle cells, as well as nerve cord and endodermal strand, whereas b8.19 gives rise to two muscle cells, as well as epidermis, nerve cord, and brain stem cells. On the other hand, in *C. intestinalis*, only the b7.9 has muscle fate; it gives rise to two muscle cells, in addition to epidermis, nerve cord, and endodermal strand.

2. Specification Mechanisms

The mechanisms of ascidian larval-muscle differentiation have intensively been investigated since Conklin described the segregation of a yellow pigmented cytoplasm (the myoplasm) into muscle cells of *Cynthia* embryos (Conklin, 1905b). Presumptive muscle cells isolated from early embryos differentiate autonomously into muscle cells (e.g., Nishida, 1992). Muscle differentiation occurs even if cleavage of the early embryo is arrested with cytochalasin B, and expression of a muscle differentiation marker, acetylcholinesterase (AChE), in embryos arrested at the 1-, 2-, 4-, 8-, 16-, 32-, and 64-cell stages, precisely follows the primary-lineage muscle blastomeres (Whittaker, 1973). In addition, even if the B4.1 quadrant is isolated and arrested immediately after isolation, this quandrant develops the muscle-specific marker antigen (Nishikata *et al.,* 1987). Therefore, the B4.1 is a self-sustained system in regard to muscle differentiation.

Whittaker (1980) altered the position of the third cleavage furrow so that the myoplasm is partitioned into four blastomeres instead of two (B4.1 pair) in the normal 8-cell embryo. When cleavage of myoplasm-redistributed embryos was arrested, they eventually developed AChE in three or four blastomeres. Deno and Satoh (1984) transplanted the myoplasm of *Halocynthia* eggs into the secondary lineage A4.1 cell, and induced expression of AChE in the A4.1 cell of cleavage-arrested 8-cell embryos. More recently, convincing evidence for the presence of muscle determinants has been offered by Nishida (1992). He manually isolated the B4.1 cell from *Halocynthia* 8-cell embryos and divided the cell into fragments that were with and without the myoplasm. The enucleated myoplasm was fused with nonmuscle lineage a4.2 cells and the fusion products were allowed to develop into partial embryos. Nearly all of the partial embryos produced markers of muscle differentiation, while none of those from the fusion of a4.2 cells with enucleated B4.1 fragments without myoplasm produced such markers. Nishida (1992) and Yamada and Nishida (1996) extended this type of experiment with unfertilized eggs, fertilized eggs, and early embryos. As shown in Fig. 3, they were able to determine the segregation pattern of the cytoplasmic region that has the potential to promote muscle differentiation. Therefore, it can be concluded that the primary-lineage muscle cells differentiate autonomously mediated by cytoplasmic determinants. The specification mechanism of the secondary lineage cells remain to be elucidated (see Satoh, 1994).

3. *Molecular Identification of Maternal Factors*

The molecular identification of localized maternal factors including muscle determinants, the elucidation of the machinery responsible for the localization, and the exploration of the mode of action of the localized factors are therefore key research subjects for the elucidation of the cell fate determination of ascidian embryos.

Jeffery (1985) isolated the yellow myoplasm from *Styela plicata* eggs and found that the myoplasm fraction contains at least 15 polypeptides that are unde-

tectable in the other cytoplasmic fraction. Nishikata *et al.* (1987) produced monoclonal antibodies that specifically recognize components of the myoplasm of *C. intestinalis* eggs. One of the antigens, named myoplasmin-C1, is a single 40-kDa polypeptide of the cortex of the myoplasm. The myoplasmin-C1 is implicated in muscle differentiation, because injection of its antibody into fertilized eggs partially blocks the development of AChE activity (Nishikata *et al.*, 1987). Isolation and characterization of myoplasmin-C1 cDNA clones suggested that myoplasmin-C1 is a cytoskeletal component of the myoplasm and that it may play a role in anchorage and segregation of the determinants (Nishikata and Wada, 1996). Recently, using cDNA probes synthesized from the yellow myoplasm and supernatant fractions isolated from *Styela clava* eggs, Swalla and Jeffery (1995, 1996) succeeded in the isolation of cDNA clones for the *yellow crescent* (*YC*) RNA and *PCNA* (proliferating cell nuclear antigen) mRNA, the former being localized in the myoplasm and the latter in the ectoplasm.

Eggs of *Ciona savignyi* have brownish myoplasm. Centrifugation of unfertilized eggs yielded four types of fragments: a large nucleated red fragment and small enucleated black, clear, and brown fragments (Marikawa *et al.*, 1994). When inseminated, only red fragments cleave and develop, but they form so-called permanent blastulae in which only epidermal cell differentiation is evident. However, when red fragments are fused with black fragments and fusion products are fertilized, nearly all of the fusion products develop muscle cells (Marikawa *et al.*, 1994). In addition, the ability of black fragments to promote muscle cell differentia-

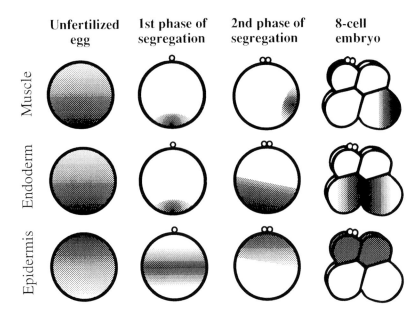

Figure 3 Distribution of cytoplasmic determinants for muscle (top), endoderm (middle), and epidermis (bottom). Shaded areas represent location of cytoplasmic determinants. Animal pole is up and vegetal pole is down. Anterior is to the left and posterior is to the right. (From Nishida, 1997.)

tion is evident when the fragments are fused with non-muscle lineage a4.2 cell. In contrast, clear and brown fragments have no such abilities. Therefore, the muscle determinants appear to be preferentially separated into black fragments.

Irradiation of black fragments with UV light diminishes the ability of promotion of muscle differentiation of the fragments (Marikawa *et al.*, 1995). The effective wavelength (250–275 nm) suggested that maternal mRNAs are one of the UV targets. This was proven by a poly(A)$^+$ RNA injection experiment. Black fragments which were first UV-irradiated then injected with poly(A)$^+$ RNA of intact black fragments recover the muscle differentiation-promoting activity (Marikawa *et al.*, 1995). Injection of poly(A)$^+$ RNAs of black fragments into red fragments, however, did not induce muscle differentiation. These results suggest that muscle determinants are comprised of not only maternal mRNAs but also factors other than mRNAs.

Yoshida *et al.* (1996) took advantage of this experimental system to isolate cDNA clones for mRNAs specific to the black fragments, and they succeeded in the isolation of a novel maternal gene, *posterior end mark*

(*pem*). As shown in Fig. 4, the *pem* transcript is initially distributed in the peripheral cytoplasm of the unfertilized egg (Fig. 4A). After fertilization, *pem* transcript is concentrated in the posterior-vegetal cytoplasm of the fertilized egg (Fig. 4C), and later the distribution of the transcript marks the posterior end of developing embryos (Fig. 4D–F). Because overexpression of PEM by microinjection of synthetic capped mRNA into fertilized eggs resulted in the development of larvae with deficiencies of the anterior-most adhesive organ, dorsal brain, and sensory pigment cells, *pem* may be involved in pattern formation of the embryo, although its activity is not directly associated with muscle specification (Yoshida *et al.*, 1996, 1997).

Another approach has been adopted to elucidate maternal genes with localized mRNAs. Satou and Satoh (1997) constructed a cDNA library of *C. savignyi* fertilized egg mRNAs subtracted with gastrula mRNAs. By examining the localization of the corresponding mRNAs of randomly selected clones by whole-mount *in situ* hybridization, they isolated five independent genes with localized mRNAs. Interestingly, all of the five genes showed a distribution pattern of the maternal mRNA

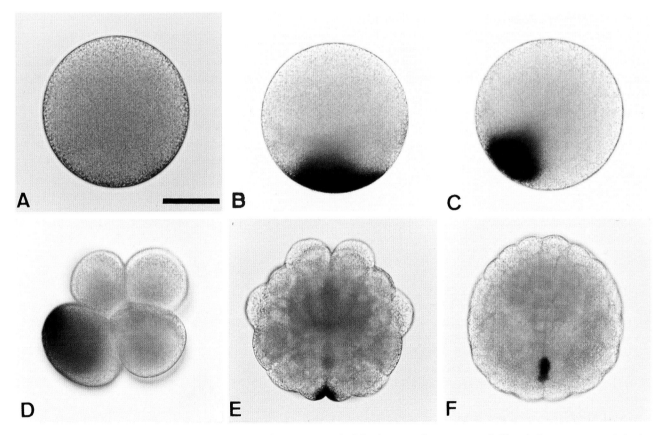

Figure 4 Distribution of *pem* maternal mRNA marks the posterior end of developing embryos, as revealed by whole-mount *in situ* hybridization. (A) An unfertilized egg, lateral view. Animal pole is up and vegetal pole is down. Scale bar represents 50 μm for all panels. (B) A fertilized egg after completion of the first phase of ooplasmic segregation, lateral view. (C) A fertilized egg after completion of the second phase of ooplasmic segregation, lateral view. (D) An 8-cell embryo, lateral view. (E) A 32-cell embryo, vegetal pole view. (F) An early gastrula, vegetal pole view. (From Yoshida *et al.*, 1996, © Company of Biologists, Ltd.)

very similar to that of *pem*, and therefore named *posterior end mark 2* (*pem-2*), *pem-3*, *pem-4*, *pem-5*, and *pem-6* (Satou and Satoh, 1997). The predicted amino acid sequence suggested that PEM-2 contains a signal for nuclear localization, a src homology 3 (SH3) domain and a consensus sequence of the CDC24 family guanine nucleotide dissociation stimulators (GDSs). PEM-4 has a signal for nuclear localization and three C2H2-type zinc finger motifs, but PEM-5 and PEM-6 show no similarity to known proteins. Functions of these maternal messages, however, were not deduced, since injection of synthetic mRNAs into fertilized eggs failed to induce any distinct effects on the embryogenesis (Satou and Satoh, 1997). Interestingly, *pem-3* encodes a probable RNA-binding protein with two KH domains and is a candidate homolog of *Caenorhabditis elegans mex-3* (*muscle excess*) (Draper *et al.*, 1996). These molecular approaches may disclose maternal genes that are involved in autonomous specification of ascidian larval muscle cells.

4. Regulation of Muscle-Specific Gene Expression

In *H. roretzi*, five muscle actin genes form a cluster (*HrMA2/4* cluster) within a 30-kb region of the genome (Kusakabe *et al.*, 1992). Examination of timing of the gene expression by whole-mount *in situ* hybridization revealed that zygotic transcripts of *HrMA4a* are first evident in B6.2 (the progenitor of B7.4) at the 32-cell stage, in B7.8 at the 64-cell stage, and in B7.5 around the 76-cell stage, respectively, suggesting that the transcription of this gene is initiated prior to the developmental fate restriction in the B7.4-sublineage (Satou *et al.*, 1995). The presence of the actin transcripts in the 32-cell embryos was confirmed by means of reverse transcriptase–polymerase chain reaction (RT–PCR) (Satou *et al.*, 1995). The *HrMHC1*, an ascidian myosin heavy chain gene, is expressed in the same manner as *HrMA4a*; namely, zygotic transcripts of *HrMHC1* are evident in the B-line muscle precursor cells as early as the 32-cell stage (Satou *et al.*, 1995).

The 5′ upstream region close to the transcription start site of *HrMA4a* contains several consensus sequences, which include a TATA box at −30, an E-box at −71, a CArG box at −116, and a cluster of three E-boxes between −150 and −190 (Kusakabe *et al.*, 1992). When deletion constructs of this region fused with a bacterial gene *lacZ* and were microinjected into fertilized eggs, the reporter gene was expressed in muscle cells of the tailbud embryo (Hikosaka *et al.*, 1994). Analyses of the deletion constructs suggested that the 103-bp upstream region, but not the 72-bp upstream region, is sufficient for the appropriate spatial expression of the reporter gene. Mutations in the proximal E-box sequence did not disturb the muscle-specific expression of the reporter gene (Hikosaka *et al.*, 1994; Satou and Satoh, 1996).

Vertebrate myogenic basic helix-loop-helix (bHLH) proteins (MyoD, myogenin, myf-5, and MRF4) are able to induce the expression of muscle structural genes in nonmuscle cells (e.g., Weintraub *et al.*, 1989; Olson and Klein, 1994). An ascidian homolog of vertebrate myogenic bHLH genes is interesting with respect to muscle determinants in ascidian eggs, and therefore a homolog, designated *AMD1*, was characterized (Araki *et al.*, 1994). However, zygotic transcripts of *AMD1* are first detected in the primary lineage of the 64-cell embryos, a little later than the detection of *HrMA4* and *HrMHC1* zygotic transcripts. Together with the result that mutations in the proximal E-box of the 5′ upstream of *HrMA4a* did not affect the muscle-specific expression of the reporter gene, it is unlikely that AMD1 itself is the muscle determinant.

This type of approach, namely going upstream of the genetic cascade by analyzing transcriptional control mechanisms of muscle-specific structural genes, may also contribute to the disclosure of the molecular nature of muscle determinants (Satoh *et al.*, 1996b).

B. Endoderm

1. Lineage

The endoderm tissue of an ascidian larva is usually subdivided into endoderm in the trunk region and an endodermal strand in the tail region (Fig. 1K). This classification is arbitrary and does not necessarily mean that there are structural and/or functional differences between them. The endoderm cells are rich in yolk granules, which may provide the embryos with nutrients. Because ascidian tadpoles are nonfeeding and in dispersal phase during their life history, endoderm differentiation is not so conspicuous in embryogenesis.

In *H. roretzi*, all endodermal cells, except for one in the caudal-tip region, are derived from the vegetal blastomeres. The developmental potential of the A4.1 to form endoderm is further segregated into the A5.1 and A5.2 of the 16-cell embryo, then the A6.1 and A6.3 of the 32-cell embryo (Fig. 2). As early as the 32-cell stage, the A6.1 becomes restricted to endoderm (Fig. 1F). At the 64-cell stage, not only the A7.1 and A7.2 but also the A7.5 become endoderm-restricted (Fig. 1H). On the other hand, the B-line presumptive endoderm cells are the B5.1 and B5.2 of the 16-cell embryo, the B6.1 and B6.3 of the 32-cell embryo, and the B7.1, B7.2, and B7.6 of the 64-cell embryo (Fig. 2). The B7.1 gives rise to only endoderm, the B7.2 to endoderm and endodermal strand, and the B7.6 to endodermal strand only.

The number of cells constituting the endodermal tissue of a newly hatched larva is estimated to be about 500 (Monroy, 1979). Because the B7.6 gives rise to one or two endodermal-strand cells of the posterior part of the tail, most of the endodermal cells are derived from the five pairs, A7.1, A7.2, A7.5, B7.1, and B7.2. Each of these primordial cells may divide five or six times to form a clone of 32 or 64 cells, in order to form the endodermal tissue consisting of about 500 cells.

2. Specification Mechanisms

Endoderm cells of the ascidian embryo express alkaline phosphatase (AP). Histochemical detection of this enzymatic activity has been used to monitor differentiation of this cell type. Although *H. roretzi* larvae express AP only in the endodermal tissues (endoderm and endodermal strand), *C. intestinalis* larvae express AP not only in the endodermal tissue but also in B-line notochord cells in the tip of the tail. This suggests a close relationship or similar property between the endoderm and B-line notochord cells. It has long been debated whether development of AP activity requires nuclear activity (Whittaker, 1977; Meedel and Whittaker, 1983; Bates and Jeffery, 1987). Recent isolation of *Halocynthia* AP protein (Kumano *et al.*, 1996) may contribute to the isolation of cDNA clone for the AP gene, with which this problem must be reexamined.

Endoderm precursors isolated from early embryos develop AP activity autonomously. Experiments of cleavage-arrested embryos suggested that endoderm determinants are segregated into endoderm-lineage blastomeres (Whittaker, 1977). As in the case of muscle determinants mentioned above, fusion of blastomeres and cytoplasmic fragments revealed the presence and distribution of endoderm determinants in eggs and early embryos (Nishida, 1993; Yamada and Nishida, 1996). As shown in Fig. 3, endoderm determinants are widely distributed in the unfertilized *Halocynthia* eggs, in the form of a gradient, with maximum activity at the vegetal pole. After fertilization, endoderm determinants move toward the vegetal pole of the egg during the first phase of segregation, and extend in the equatorial direction prior to the first cleavage. These determinants are partitioned into endoderm-lineage cells during cleavage.

A developmental gene that is implicated in the autonomous differentiation of endoderm cells is *fork head/HNF-3*. The cDNA clones for the ascidian *fork head/HNF-3* have been isolated from *C. intestinalis* (Corbo *et al.*, 1997a), *Molgula oculata* (Olsen and Jeffery, 1997), and *H. roretzi* (Shimauchi *et al.*, 1997). The ascidian *fork head/HNF-3* gene begins to express as early as the 16-cell stage in blastomeres of the lineages of endoderm, notochord, and nerve cord. Inhibition of possible functions of *Molgula fork head/HNF-3* by treatment with antisense oligonucleotides resulted in incomplete gastrulation without movement of endoderm and notochord cells inside the embryo (Olsen and Jeffery, 1997). In *H. roretzi*, it has been shown that *HrHNF3-1* is expressed in blastomeres that were continuously dis-sociated from the first division until the 16-cell stage (Shimauchi *et al.*, 1997). This very early and autonomous expression of *HrHNF3-1* strongly suggests that the transcription of this gene is directly regulated by maternal factors or endoderm determinants.

C. Epidermis

1. Lineage

The outer surface of an ascidian tadpole is composed of a single layer of epidermal cells. The number of epidermal cells of the mid-tailbud embryo is 800. During later embryogenesis, the epidermal cells produce a transparent larval tunic, which is sometimes used as a differentiation marker of the cells.

The epidermis is derived entirely from blastomeres of the animal hemisphere (a- and b-lines) (Figs. 1 and 2). As early as the 16-cell stage, the developmental fate of the b5.4 pair becomes restricted to give rise to epidermis (Fig. 1C). Through its derivatives of the b6.7 and b6.8 pairs of the 32-cell embryo, the b7.13, b7.14, b7.15, and b7.16 pairs of the 64-cell embryo, the b5.4 pair give rise to 256 epidermal cells of the ventral part of the tail of the mid-tailbud embryo, because each of the b7.13, b7.14, b7.15, and b7.16 pairs gives rise to a clone of 64 epidermal cells.

The three pairs of a6.6, a6.8, and b6.6 become epidermis-restricted at the 32-cell stage (Fig. 1E). The first two pairs give rise to a total of 256 epidermal cells of the lateral and ventral regions of the head of the mid-tailbud embryo after 6 subsequent divisions, whereas the b6.6 pair form 128 epidermal cells of the lateral part of the tail. At the 64-cell stage, the a7.14 pair also becomes restricted to give rise to 64 epidermal cells of the dorsal part of the head (Fig. 1G). In addition, epidermis of the dorsal part of the trunk and tail is derived from the b8.20 and b8.18 pairs, which are derivatives of b5.3, b6.5, b7.9, and b7.10. Each of these pairs forms 32 epidermal cells. Furthermore, 32 epidermal cells of the anterior dorsal part of the head are derived from the 8.26 pair. The total number of epidermal cells is therefore 800.

2. Specification Mechanisms

Reflecting their early restriction of developmental fate, the presumptive epidermal cells of early embryos showed a high potential of self-differentiation (e.g., Ueki *et al.*, 1994). This autonomy has been suggested to be dependent on determinants localized at the ectoplasm (Conklin, 1905b; Nishida, 1994a). As in the case of muscle and endoderm determinants mentioned above, fusion of blastomeres and cytoplasmic fragments revealed the presence and distribution of epidermis determinants in eggs and early embryos (Nishida, 1994a; Yamada and Nishida, 1996). As shown in Fig. 3, epidermis determinants are widely distributed in the egg cytoplasm along a gradient, concentrated in the animal pole region. After fertilization, these determinants concentrate once in the equatorial region, and then move into the animal hemisphere. The epidermal determinants are then inherited by epidermal-lineage cells of the animal hemisphere (Fig. 3).

Taking advantage of the fact that cleavage-arrested *H. roretzi* fertilized eggs express only epidermis differentiation markers, nine different cDNA clones were isolated and shown to be specific to cleavage-arrested one-celled embryos (Ueki *et al.*, 1991). Transcripts corresponding to each of the clones were present in normal tailbud embryos, and their distribution was restricted to epidermal cells (Ueki *et al.*, 1994). The timing of the accumulation and spatial distribution pattern of transcripts of these epidermis-specific genes, *HrEpiA-H*, are not identical, however, and their spatiotemporal expression patterns are categorized into several types (Ueki *et al.*, 1994; Ishida *et al.*, 1996).

The predicted amino acid sequence of *HrEpiB* is similar to UDP glucose-4-epimerases and 3β-hydroxysteroid dehydrogenase/isomerases, and *HrEpiD* is an ascidian homolog of the *SEC61* gene of mammals and yeast (Ueki and Satoh, 1994). Since all of these genes are specifically expressed in epidermal cells, they may share several common cis-regulatory elements required for proper spatial expression. In addition, they may have common and specific cis-elements required for appropriate temporal expression patterns. The putative cis-regulatory elements of *HrEpiB* and *HrEpiD* were studied by microinjecting fusion gene constructs (Ueki and Satoh, 1995). The 5′ upstream region from −345 to +200 of *HrEpiB* and that from −166 to +108 of *HrEpiD* are sufficient for the epidermis-specific expression of the reporter gene. The 5′ flanking regions of the genes share several common sequence motifs (Ueki and Satoh, 1995).

V. Cellular Interactions Are Important for Conditional Specification

As mentioned above, muscle, endoderm, and epidermis of ascidian embryos are specified autonomously dependent on prelocalized maternal factors. In addition, recent studies revealed that cell–cell interactions play crucial roles in differentiation of certain cell-types, including the central nervous system, sensory pigment cells, notochord, and trunk lateral cells (Table I).

A. The Central Nervous System

1. Lineage

The CNS of the ascidian larva is formed from the neural plate, which rolls into a neural tube. According to elaborate descriptive studies by Nicol and Meinertzhagen (1988, 1991), the complement of the *C. intestinalis* larval CNS numbers some 330 cells, comprising approximately 215 cells in the sensory vesicle or the brain, 50 cells in the visceral or tail ganglion or the brain stem, and about 65 ciliated ependymal cells of the caudal nerve cord or spinal cord. In addition, another 40–50 cells constitute the neurohypophysis that will give rise to the adult nervous system.

Cells of the CNS are derived from a4.2 and A4.1-cell pairs of the 8-cell embryo, and their differentiation is accomplished by serial interactions of a4.2-progenitor cells with other cells (reviewed by Okamura *et al.*, 1993). According to the description by Nishida on the CNS of *H. roretzi* embryos (Nishida, 1987), the anterior and posterior parts of the brain are derived from the a8.17, the ventral part is from the a8.19, and the dorsal and lateral parts are from the a8.25 (Fig. 2). Most of the brain stem is derived from the A7.4, but the dorsal part is from the b8.19. The nerve cord consists of four rows of cells, a dorsal row, a ventral row, and two lateral rows. The dorsal row is derived from the b8.19 pair, the ventral row is from the A7.4 pair, and the two lateral rows originate mainly from the A8.15 pairs (Fig. 2).

2. Specification Mechanisms

In 1939, Rose conducted a series of microsurgical experiments to define the cell interactions required for the sensory pigment cell development in the ascidian larval brain. Results of his experiments suggested that (1) the presumptive neural cells do not differentiate autonomously, (2) an inductive interaction emanating from descendent of the A4.1 is required for progeny a4.2 to form the brain and sensory pigment cells, and (3) the b4.2, which in normal development does not form these tissues, develops the potential to form at least the sensory pigment cells when combined with the A4.1. These results have been confirmed by others (Reverberi and Minganti, 1946; Reverberi *et al.*, 1960). Nishida and Satoh (1989) and Nishida (1991) also examined cellular mechanisms of the CNS induction by isolating various sets of blastomeres from different stages of *Halocynthia* embryos. Results of their experiments indicated that (1) the potential of A4.1 to induce the CNS is retained by the middle gastrula stage, (2) the potential of b4.2 to form the sensory pigment cells disappeared by the 64-cell stage, and (3) signals for the CNS induction emanate from the nerve cord cells.

TuNa I is a gene for a putative neural-specific sodium channel protein in *H. roretzi* embryos (Okamura *et al.*, 1994). This voltage-activated sodium current undergoes a change in kinetics during embryogenesis of *H. roretzi*. *TuNa I* transcription is first detected at late gastrula stage, and the gene is expressed in neural cells of the tailbud embryo (Okamura *et al.*, 1994). When a single a4.2 was dissociated from the 8-cell embryo, its subsequent division was arrested with cytochalasin B, and the cell was cultured in isolation, it differentiated exclusively into an epidermal cell judging from membrane excitability. However, when the same blastomere was cultured in contact with a single A4.1, it displayed neural differentiation (Okado and Takahashi, 1988). Neural induction in this simple two-cell system was confirmed by *TuNa I* expression; *TuNa I* is expressed in isolated a4.2 cell cultured in contact with a single A4.1 (Okamura *et al.*, 1994). These results suggest cell contact, prior to neurulation, is required to activate the *TuNa I* gene in cells of the neural lineage.

A developmental gene associated with the neuronal specification is an ascidian homolog (*HrBMPb*) of *BMP-2/4* (Miya *et al.*, 1997). Overexpression of *HrBMPb* in *Xenopus* eggs caused ventralization of the frog embryo, suggesting conservative function of *BMP-2/4* between ascidians and vertebrates. However, overexpression of *HrBMPb* did not cause any alteration in mesodermal specification in the ascidian embryo. Instead, *HrBMPb* overexpression inhibited neuronal differentiation and induced epidermal cells in the ascidian embryo.

B. Sensory Pigment Cells

1. Lineage

The ascidian tadpole has two major sensory organs within the brain vesicle; the anterior otolith and the posterior ocellus. The otolith is composed of a single cell which contains melanin pigment granules as its major constituent, whereas the ocellus is composed of a large melanocyte and several lens cells. The developmental fate to give rise to pigment cells is segregated along the following cell lineage; a4.2, a5.3, a6.7, a7.13, and a8.25 (Fig. 2). The developmental potential of the a8.25 pair at the 110-cell stage is not restricted to pigment cells, because it contains brain potential as well. Lineage analysis showed that the right and left a8.25 cells give rise to the two kind of pigment cells in a complementary manner (i.e., if the otolith is derived from the left a8.25, then the ocellus pigment cells is from the right a8.25, and vice versa) (Nishida, 1987).

2. Specification Mechanisms

A group of embryonic cells that share a common developmental potential is referred to as an "equivalence group" (Kimble *et al.*, 1979). The otolith and ocellus pigment cells belong to such a group, and specification of pigment cells requires rather complex inductive signals emanating from the adjacent cells (Nishida and Satoh, 1989; Nishida, 1991). The decision to develop into either an ocellus or otolith is likely to be irreversibly determined by the middle tailbud stage (Nishida and Satoh, 1989). When either the right or left pigment-

lineage blastomere is destroyed prior to the early tailbud stage, most of the operated embryos develop only an ocellus, suggesting that the ocellus is a dominant fate. However, if the ablation is done at the middle and late tailbud stages, some of operated embryos develop an otolith. During the early tailbud stage the bilaterally positioned pigment precursors become situated along the median plane when the neural tube closes. During the following several hours, the determination of the pigment precursors to give rise to otolith may take place. It has been suggested that the fates are specified by positional cues along the anteroposterior axis of the embryo (Nishida and Satoh, 1989).

Tyrosinase is a key enzyme in melanin biosynthesis. A cDNA encoding tyrosinase has been isolated from an *H. roretzi* tailbud cDNA library (Sato *et al.*, 1997). Expression of the gene is detected first in two pigment precursor cells positioned in the neural plate of early neurulae, and later in two melanin-containing pigment cells within the brain of late tailbud embryos. Its expression pattern correlates well with the appearance of tyrosinase enzyme activity in developing brain. In addition, recent molecular cloning of an ascidian *Notch* homolog showed that this gene is expressed in neuronal cells including cells destined to form pigment cells (Hori *et al.*, 1997). These two genes are therefore involved in the determinative and differentiation events in the sensory pigment cell equivalence group of the ascidian embryo.

C. Notochord

1. Lineage

The ascidian notochord is a larval organ which is disintegrated during metamorphosis. It consists of 40 cells aligned in single file along the center of the larval tail (Fig. 1K, L). The cells become vacuolated during elongation of the tail, and the notochord as a whole is enclosed within a notochordal sheath.

The notochord is derived from 32 A-line and 8 B-line blastomeres (Figs. 1 and 2). The A-line notochord potential is inherited by the A5.1 and A5.2 of the 16-cell embryo, by the A6.2 and A6.4 of the 32-cell embryo, and then by the A7.3 and A7.7 of the 64-cell embryo; the A7.3 and A7.7 become restricted to give rise to notochord and divide three times to form 32 notochord cells at the anterior and middle part of the tail (Fig. 1H). In addition, the B5.1 of the 16-cell embryo, the B6.2 of the 32-cell embryo, the B7.3 of the 64-cell embryo, and the B8.6 of the 110-cell embryo are the presumptive notochord cells (Fig. 2). The B8.6 pair become restricted to notochord, giving rise to 8 notochord cells in the posterior part of the tail after 2 divisions. Therefore, every notochord cell has a history of 9 divisions from the zygote stage until the ultimate differentiation.

2. Specification Mechanisms

The results of earlier studies suggested that notochord cells differentiate autonomously without cellular interaction (Reverberi and Minganti, 1946; Nishikata and Satoh, 1990). However, recent experiments of isolation and recombination of blastomeres at various times of development revealed that the differentiation of A-line notochord cells of *H. roretzi* is induced during a very short time window at the 32-cell stage (Nakatani and Nishida, 1994). Inductive signals emanate from the neighboring endoderm cells and from the neighboring notochord cells as well. The induction of notochord cells in ascidian embryos differs from that of amphibian embryos, because the ascidian animal hemisphere cells do not have the competence to form notochord (Nakatani and Nishida, 1994). One candidate molecule for the inductive signals is bFGF (Nakatani *et al.*, 1996), and thus inhibition of signal transduction pathways including Ras resulted in the failure of notochord cell differentiation (Nakatani and Nishida, 1997), although it should be determined whether Ras acts upstream of the *Brachyury* gene or a target of the *Brachyury* (see below). On the other hand, the induction of B-line notochord cells is not fully understood yet.

Interestingly, immediately after induction, the developmental fate of the A-line presumptive notochord blastomeres is restricted to give rise to notochord at the 64-cell stage, and then the *H. roretzi Brachyury* gene (*As-T*) begins to be expressed in the primordial notochord cells (Yasuo and Satoh, 1993, 1994). *As-T* expression is characteristic not only because it is expressed exclusively in the notochord cells but also because the timing of the gene expression coincides with that of the developmental fate restriction in both A- and B-line presumptive cells (Yasuo and Satoh, 1993, 1994). In addition, overexpression of As-T by microinjection of synthetic capped mRNA into fertilized eggs resulted in the notochord differentiation without any inductional influence (Yasuo and Satoh, 1998). Ectopic differentiation of notochord in blastomeres of, at least, endoderm, mesenchyme, and nerve cord lineages also was promoted by overexpression of As-T (Yasuo and Satoh, 1998). Therefore, *As-T* may be a kind of master control gene for notochord differentiation in the ascidian embryo.

The *Brachyury* homolog (*CiBra*) also has been isolated from *C. intestinalis* (Corbo *et al.*, 1997b). The expression pattern of *CiBra* is a little different from that of *As-T*, because *CiBra* begins to be expressed at the 64-cell stage not only in the A-line primordial cells (A7.3 and A7.7) but also in the B-line presumptive cells (B7.3). A fusion gene was constructed in which an approximately 3.5-kb 5′ flanking region of *CiBra* gene was ligated to GFP or lacZ with the nuclear localization signal. Injection of the fusion construct as well as various deletion constructs demonstrated a minimal, 435-bp enhancer from the promoter region that mediates the notochord-restricted expression of the reporter genes (Corbo *et al.*, 1997b). Interestingly, this enhancer contains a negative control region that excludes *CiBra* expression from inappropriate embryonic lineages, including trunk mesenchyme and tail muscle (Corbo *et al.*, 1997b).

Recently, T-box genes have been recognized as a novel family of transcriptional factors that appear to play a crucial role in the development of various animal groups (Herrmann and Kispert, 1994). In relation to T-box genes, it should be mentioned that the *H. roretzi* genome contains at least three additional T-box genes, an ascidian homolog of *Drosophila optomoter blind* (*omb*) (Pflugfelder *et al.*, 1992), As-T2 (Yasuo *et al.*, 1996), and maternal As-T3 (Takada *et al.*, submitted, 1998). The As-T2 gene encodes a diverged T-domain protein. As-T2 is mainly expressed in involuting primary muscle cells during gastrulation and at the tip of the tail in the tailbud embryo. Interestingly, the combined pattern of As-T and As-T2 expression appears to be equivalent to that of a single mouse *T-box* gene.

D. Trunk Lateral Cells

1. Lineage

The A7.6 cell of the *H. roretzi* 64-cell embryo gives rise to a cluster of about 16 trunk lateral cells (TLCs) at the late tailbud stage (Nishida, 1987). As shown in Fig. 2, the embryonic precursors of the TLCs are the A4.1 of the 8-cell embryo, the A5.2 of the 16-cell embryo, the A6.3 of the 32-cell embryo, and the A7.6 of the 64-cell embryo. The A7.6 fate is restricted (Fig. 1H), and then A7.6 divides 4 times to form 16 TLCs. TLCs give rise to adult blood cells after metamorphosis (Hirano and Nishida, 1997).

2. Specification Mechanisms

Although the lineage is simple, the specification mechanism for TLCs seems rather complex. A TLC-specific antigen recognized by a monoclonal antibody is expressed from the middle of the tailbud stage onward, which has been used as a marker for TLC differentiation (Mita-Miyazawa *et al.*, 1987). When the antigen expression was examined in partial embryos, the A4.1 quarter-embryos rarely developed the antigen (Nishikata and Satoh, 1991).

This experimental result was reexamined by Kawaminami and Nishida (1997). When prospective TLC blastomeres were isolated from embryos before the 16-cell stage, they failed to express the TLC-specific antigen. Isolates after the 32-cell stage, however, autonomously expressed the antigen. Results of experiments involving coisolation and recombination of blastomeres at the 16-cell stage showed that an inductive influence emanating from cells of the animal hemisphere (presumptive epidermis blastomeres) is required for TLC formation. The inductive interaction takes place at the 16-cell stage. The inducing activity is distributed widely in the animal hemisphere. By contrast, only presumptive TLC blastomeres are competent to be induced to form TLCs.

VI. Conclusions

As discussed above, the ascidian embryo has several advantages as an experimental system to study the genetic circuitry underlying the developmental fate determination of embryonic cells. An advantage is that the ascidian tadpole consists of a very small number of distinct types of tissues, and the cell lineage is completely described up to the gastrula stage. We are able to identify every blastomere in early embryos, and thus are able to analyze mechanisms of cell fate determination at the single cell level. This advantage is particularly pronounced when we will determine the expression of specific genes. Namely, we could assess the timing of gene expression with respect to that of developmental fate restrictions and in relation to the developmental potentials of blastomeres. These situations seem less feasible in vertebrate embryos in which lineage is not rigidly determined.

The ascidian embryo has been most intensively studied with respect to localized egg cytoplasmic determinants. Recent studies revealed the presence of determinants for differentiation of muscle, endoderm, and epidermis. Larval muscle cells, for example, provide an experimental system with which to explore the entire genetic program from the beginning of muscle determinants until the terminal differentiation of specific gene expression. As was discussed here, various approaches to understand the molecular nature of muscle determinants may disclose the developmentally important maternal factors in the near future.

Recent studies support the ascidian tadpole prototype for the ancestral chordate. Therefore, molecular developmental studies of ascidian embryos should facilitate our understanding of the evolutionary aspects of the genetic circuitry among chordates or deuterostomes. Altogether, ascidian embryos may be one of the most appropriate experimental systems to explore the mechanisms of cell lineage and cell fate determination.

Acknowledgments

I thank Dr. Hiroki Nishida for providing figures. The work from my own laboratory of the Kyoto University was supported by a Grant-in-Aid for Specially Promoted Research (07102012) from the Ministry of Education, Science, Sports and Culture of Japan to N.S.

References

Araki, I., and Satoh, N. (1996). Cis-regulatory elements conserved in the proximal promoter region of an ascidian embryonic muscle myosin heavy-chain gene. *Dev. Genes Evol.* **206**, 54–63.

Araki, I., Saiga, H., Makabe, K. W., and Satoh, N. (1994). Expression of AMD1, a gene for a MyoD1-related factor in the ascidian *Halocynthia roretzi. Roux's Arch. Dev. Biol.* **203**, 320–327.

Araki, I., Tagawa, K., Kusakabe, T., and Satoh, N. (1996). Predominant expression of a cytoskeletal actin gene in mesenchyme cells during embryogenesis of the ascidian *Halocynthia roretzi. Dev. Growth Differ.* **38**, 401–411.

Bates, W. R., and Jeffery, W. R. (1987). Alkaline phosphatase expression in ascidian egg fragments and andromerogons. *Dev. Biol.* **119**, 382–389.

Chabry, L. (1887). Contribution a l'embryologie normale et teratologique des Ascidies simples. *J. Anat. Physiol. (Paris)* **23**, 167–319.

Conklin, E. G. (1905a). The organization and cell lineage of the acidian egg. *J. Acad. Nat. Sci. (Philadelphia)* **13**, 1–119.

Conklin, E. G. (1905b). Organ forming substances in the eggs of ascidians. *Biol. Bull.* **8**, 205–230.

Corbo, J. C., Erives, A., Di Gregorio, A., Chang, A., and Levine, M. (1997a). Dorsoventral patterning of the vertebrate neural tube is conserved in a protochordate. *Development (Cambridge, UK)* **124**, 2335–2344.

Corbo, J. C., Levine, M., and Zeller, R. W. (1997b). Characterization of a notochord-specific enhancer from the *Brachyury* promoter region of the ascidian, *Ciona intestinalis. Development (Cambridge, UK)* **124**, 589–602.

Deno, T., and Satoh, N. (1984). Studies on the cytoplasmic determinant for muscle cell differentiation in ascidian embryos: an attempt at transplantation of the myoplasm. *Dev. Growth Differ.* **26**, 43–48.

Draper, B. W., Mello, C. C., Bowerman, B., Hardin, J., and Priess, J. R. (1996). MEX-3 is a KH domain protein that regulates blastomere identity in early *C. elegans* embryos. *Cell (Cambridge, Mass.)* **87**, 205–216.

Herrmann, B. G., and Kispert, A. (1994). The *T* genes in embryogenesis. *Trends Genet.* **10**, 280–286.

Hikosaka, A., Kusakabe, T., and Satoh, N. (1994). Short upstream sequences associated with the muscle-specific expression of an actin gene in ascidian embryos. *Dev. Biol.* **166**, 763–769.

Hirano, T., and Nishida, H. (1997). Developmental fates of larval tissues after metamorphosis in ascidian *Halocynthia roretzi.* I. Origin of mesodermal tissues of the juvenile. *Dev. Biol.* **192**, 199–210.

Hori, S., Saitoh, T., Matsumoto, M., Makabe, K. W., and Nishida, H. (1997). *Notch* homologue from *Halocynthia roretzi* is preferentially expressed in the central nervous system during ascidian embryogenesis. *Dev. Genes Evol.* **207**, 371–380.

Ishida, K., Ueki, T., and Satoh, N. (1996). Spatio-temporal expression patterns of eight epidermis-specific genes in the ascidian embryo. *Zool. Sci.* **13**, 699–709.

Jeffery, W. R. (1985). Identification of proteins and mRNAs in isolated yellow crescent of ascidian eggs. *J. Embryol. Exp. Morphol.* **89**, 275–287.

Jeffery, W. R. (1990). Ultraviolet irradiation during ooplasmic segregation prevents gastrulation, sensory cell induction, and axis formation in the ascidian embryo. *Dev. Biol.* **140**, 388–400.

Kawaminami, S., and Nishida, H. (1997). Induction of trunk lateral cells, the blood cell precursors, during ascidian embryogenesis. *Dev. Biol.* **181**, 14–20.

Kimble, J., Sulston, J., and White, J. (1979). Regulative development in the postembryonic lineage of *Caenorhabditis elegans.* In "Cell Lineage, Stem Cells and Cell Determination" (N. L. Douarin, ed.), pp. 59–68. Elsevier, Amsterdam.

Kowalevsky, A. (1866). Entwicklungsgeschichte der einfachen Ascidien. *Mem. Acad. St. Petersbourg, Ser. 7* **10**, 1–19.

Kumano, G., Yokosawa, H., and Nishida, H. (1996). Biochemical evidence for membrane-bound endoderm-specific alkaline phosphatase in larvae of the ascidian, *Halocynthia roretzi. Eur. J. Biochem.* **240**, 485–489.

Kusakabe, T., Makabe, K. W., and Satoh, N. (1992). Tunicate muscle actin genes: Structure and organization as a gene cluster. *J. Mol. Biol.* **227**, 955–960.

Marikawa, Y., Yoshida, S., and Satoh, N. (1994). Development of egg fragments of the ascidian *Ciona savignyi*: The cytoplasmic factors responsible for muscle differentiation are separated into a specific fragment. *Dev. Biol.* **162**, 134–142

Marikawa, Y., Yoshida, S., and Satoh, N. (1995). Muscle determinants in the ascidian egg are inactivated by UV irradiation and the inactivation is partially rescued by injection of maternal mRNAs. *Roux's Arch. Dev. Biol.* **204**, 180–186.

Meedel, T. H., and Whittaker, J. R. (1983). Development of translationally active mRNA for larval muscle acetylcholinesterase during ascidian embryogenesis. *Proc. Natl. Acad. Sci. U.S.A.* **80**, 4761–4765.

Meedel, T. H., Crowther, R. J., and Whittaker, J. R. (1987). Determinative properties of muscle lineages in ascidian embryos. *Development (Cambridge, UK)* **100**, 245–260.

Mita-Miyazawa, I., Nishikata, T., and Satoh, N. (1987). Cell- and tissue-specific monoclonal antibodies in eggs and embryos of the ascidian *Halocynthia roretzi. Development (Cambridge, UK)* **99**, 151–162.

Miya, T., and Satoh, N. (1997). Isolation and characterization of cDNA clones for b-tubulin genes as a molecular marker for neural cell differentiation in the ascidian embryo. *Int. J. Dev. Biol.* **41**, 551–557.

Miya, T., Morita, K., Suzuki, A., Ueno, N., and Satoh, N. (1997). Functional analysis of an ascidian homologue of vertebrate *Bmp-2/Bmp-4* suggests its role in the inhibition of neural fate specification. *Development (Cambridge, UK)* **124**, 5149–5159.

Monroy, A. (1979). Introductory remarks on the segregation of cell lines in the embryo. In "Cell Lineage, Stem Cells and Cell Determination" (N. L. Douarin, ed.), pp. 3–13. Elsevier, Amsterdam.

Nakatani, Y., and Nishida, H. (1994). Induction of notochord during ascidian embryogenesis. *Dev. Biol.* **166**, 289–299.

Nakatani, Y., and Nishida, H. (1997). Ras is an essential component for notochord formation during ascidian embryogenesis. *Mech. Dev.* **68**, 81–89.

Nakatani, Y., Yasuo, H., Satoh, N., and Nishida, H. (1996). Basic fibroblast growth factor induces notochord formation and the expression of *As-T,* a *Brachyury* homolog, during ascidian embryogenesis. *Development (Cambridge, UK)* **122**, 2023–2031.

Nicol, D. R., and Meinertzhagen, I. A. (1988). Development of the central nervous system of the larva of the ascidian, *Ciona intestinalis* L. II. Neural plate morphogenesis and cell lineages during neurulation. *Dev. Biol.* **130**, 737–766.

Nicol, D. R., and Meinertzhagen, I. A. (1991). Cell counts and maps in the larval central nervous system of the ascidian *Ciona intestinalis* (L.). *J. Comp. Neurol.* **309**, 415–429.

Nishida, H. (1987). Cell lineage analysis in ascidian embryos by intracellular injection of a tracer enzyme. III. Up to the tissue restricted stage. *Dev. Biol.* **121**, 526–541.

Nishida, H. (1991). Induction of brain and sensory pigment cells in the ascidian embryo analyzed by experiments with isolated blastomeres. *Development (Cambridge, UK)* **112**, 389–395.

Nishida, H. (1992). Regionality of egg cytoplasm that promotes muscle differentiation in embryo of the ascidian, *Halocynthia roretzi. Development (Cambridge, UK)* **116**, 521–529.

Nishida, H. (1993). Localized regions of egg cytoplasm that promote expression of endoderm-specific alkaline phosphatase in embryos of the ascidian *Halocynthia roretzi. Development (Cambridge, UK)* **118**, 1–7.

Nishida, H. (1994a). Localization of egg cytoplasm that promotes differentiation to epidermis in embryos of the ascidian *Halocynthia roretzi. Development (Cambridge, UK)* **120**, 235–243.

Nishida, H. (1994b). Localization of determinants for formation of the anterior-posterior axis in eggs of the ascidian *Halocynthia roretzi. Development (Cambridge, UK)* **120**, 3093–3104.

Nishida, H. (1996). Vegetal egg cytoplasm promotes gastrulation and is responsible for specification of vegetal blastomeres in embryos of the ascidian *Halocynthia roretzi. Development (Cambridge, UK)* **122**, 1271–1279.

Nishida, H. (1997). Cell fate specification by localized cytoplasmic determinants and cell interactions in ascidian embryos. *Int. Rev. Cytol.* **176**, 245–306.

Nishida, H., and Satoh, N. (1983). Cell lineage analysis in ascidian embryos by intracellular injection of a tracer enzyme. I. Up to the eight-cell stage. *Dev. Biol.* **99**, 382–394.

Nishida, H., and Satoh, N. (1985). Cell lineage analysis in ascidian embryos by intracellular injection of a tracer enzyme. II. The 16- and 32-cell stages. *Dev. Biol.* **101**, 440–454.

Nishida, H., and Satoh, N. (1989). Determination and regulation in the pigment cell lineage of the ascidian embryo. *Dev. Biol.* **132**, 355–367.

Nishikata, T., and Satoh, N. (1990). Specification of notochord cells in the ascidian embryo analysis with a specific monoclonal antibody. *Cell Differ. Dev.* **30**, 43–53.

Nishikata, T., and Satoh, N. (1991). Expression of an antigen specific for trunk lateral cells in quarter embryos of the ascidian, *Halocynthia roretzi. J. Exp. Zool.* **258**, 344–352.

Nishikata, T., and Wada, M. (1996). Molecular characterization of myoplasmin-C1: A cytoskeletal component localized in the myoplasm of the ascidian egg. *Dev. Genes Evol.* **206**, 72–79.

Nishikata, T., Mita-Miyazawa, I., Deno, T., and Satoh, N. (1987). Muscle cell differentiation in ascidian embryos analysed with a tissue-specific monoclonal antibody. *Development (Cambridge, UK)* **99**, 163–171.

Okado, H., and Takahashi, K. (1988). A simple "neural induction" model with two interacting cleavage-arrested ascidian blastomeres. *Proc. Natl. Acad. Sci. U.S.A.* **85**, 6197–6201.

Okamura, Y., Okado, H., and Takahashi, K. (1993). The ascidian embryo as a prototype of vertebrate neurogenesis. *BioEssays* **15**, 723–730.

Okamura, Y., Ono, F., Okagaki, R., Chong, J. A., and Mandel, G. (1994). Neural expression of a sodium channel gene requires cell-specific interactions. *Neuron* **13**, 937–948.

Olsen, C. L., and Jeffery, W. R. (1997). A *forkhead* gene related to *HNF-3β* is required for gastrulation and axis formation in the ascidian embryo. *Development (Cambridge, UK)* **124**, 3609–3619.

Olson, E. N., and Klein, W. H. (1994). bHLH factors in muscle development: Dead lines and commitments, what to leave in and what to leave out. *Genes Dev.* **8**, 1–8.

Ortolani, G. (1955). The presumptive territory of the mesoderm in the ascidian germ. *Experientia* **11**, 445–446.

Ortolani, G. (1957). Il territorio precoce della corda nelle Ascidie. *Acta Embryol. Morphol. Exp.* **1**, 33–36.

Pflugfelder, G. O., Roth, H., and Poeck, B. (1992). A homology domain shared between *Drosophila* optomoter-blind and mouse *Brachyury* is involved in DNA binding. *Biochem. Biophy. Res. Commun.* **186**, 918–925.

Reverberi, G. (1971). Ascidians. *In* "Experimental Embryology of Marine and Fresh-Water Invertebrates" (G. Reverberi, ed.), pp. 507–550. North-Holland Publ., Amsterdam.

Reverberi, G., and Minganti, A. (1946). Fenomeni di evocazione nello sviluppo dell'uovo di Ascidie. Risultati dell'indagine sperimentale sull'uovo di *Ascidiella aspersa* e di *Ascidia malaca* allo stadio di 8 blastomeri. *Pubbl. Stn. Zool. Napoli* **20**, 199–252.

Reverberi, G., Ortolani, G., and Farinella-Ferruzza, N. (1960). The causal formation of the brain in the ascidian larva. *Acta Embryol. Morphol. Exp.* **3**, 296–336.

Rose, S. M. (1939). Embryonic induction in the Ascidia. *Biol. Bull.* **76**, 216–232.

Sato, S., Masuya, H., Numakunai, T., Satoh, N., Ikeo, K., Gojobori, T., Tamura, K., Ide, H., Takeuchi, T., and Yamamoto, H. (1997). Ascidian tyrosinase gene: Its unique structure and expression in the developing brain. *Dev. Dyn.* **208**, 363–374.

Satoh, N. (1994). "Developmental Biology of Ascidians." Cambridge Univ. Press, New York.

Satoh, N., and Jeffery, W. R. (1995). Chasing tails in ascidians: Developmental insights into the origin and evolution of chordates. *Trends Genet.* **11**, 354–359.

Satoh, N., Makabe, K. W., Katsuyama, Y., Wada, S., and Saiga, H. (1996a). The ascidian embryo: An experimental system for studying genetic circuitry for embryonic cell specification and morphogenesis. *Dev. Growth Differ.* **38**, 325–340.

Satoh, N., Araki, I., and Satou, Y. (1996b). An intrinsic genetic program for autonomous differentiation of muscle cells in the ascidian embryo. *Proc. Natl. Acad. Sci. U.S.A.* **93**, 9315–9321.

Satou, Y., and Satoh, N. (1996). Two cis-regulatory elements are essential for the muscle-specific expression of an actin gene in the ascidian embryo. *Dev. Growth Differ.* **38**, 565–573.

Satou, Y., and Satoh, N. (1997). *posterior end mark 2 (pem-2)*, *pem-4, pem-5* and *pem-6*: Maternal genes with localized mRNA in the ascidian embryo. *Dev. Biol.* **192**, 467–481.

Satou, Y., Kusakabe, T., Araki, I., and Satoh, N. (1995). Timing of initiation of muscle-specific gene expression in the ascidian embryo precedes that of developmental fate restriction in lineage cells. *Dev. Growth Differ.* **37**, 319–327.

Shimauchi, Y., Yasuo, H., and Satoh, N. (1997). Autonomy of ascidian *fork head/HNF-3* gene expression. *Mech. Dev.* **69**, 143–154.

Swalla, B. J., and Jeffery, W. R. (1995). A maternal RNA localized in the yellow crescent is segregated to the larval muscle cells during ascidian development. *Dev. Biol.* **170**, 353–364.

Swalla, B. J., and Jeffery, W. R. (1996). PCNA mRNA Has a 3′ UTR antisense to yellow crescent RNA and is localized in ascidian eggs and embryos. *Dev. Biol.* **178**, 23–34.

Takahashi, H., Ishida, K., Makabe, K. W., and Satoh, N. (1997). Isolation of cDNA clones for genes that are expressed in the tail region of the ascidian tailbud embryo. *Int. J. Dev. Biol.* **41**, 691–698.

Ueki, T., and Satoh, N. (1994). An ascidian homolog of *SEC61* is expressed predominantly in epidermal cells of the embryo. *Dev. Biol.* **165**, 185–192.

Ueki, T., and Satoh, N. (1995). Sequence motifs shared by the 5′ flanking regions of two epidermis-specific genes in the ascidian embryo. *Dev. Growth Differ.* **37**, 597–604.

Ueki, T., Makabe, K. W., and Satoh, N. (1991). Isolation of cDNA clones for epidermis-specific genes of the ascidian embryo. *Dev. Growth Differ.* **33**, 579–586.

Ueki, T., Yoshida, S., Marikawa, Y., and Satoh, N. (1994). Autonomy of expression of epidermis-specific genes in the ascidian embryo. *Dev. Biol.* **164**, 207–218.

van Beneden, E., and Julin, C. H. (1884). La segmentation chez les ascidens dans ses rapportes avec l'organization de la larve. *Arch. Biol.* **5**, 111–126.

Weintraub, H., Tapscott, S. J., Davis, R. L., Thayer, M. J., Adam, M. A., Lassar, A. B., and Miller, A. D. (1989). Activation of muscle-specific genes in pigment, nerve, fat, liver, and fibroblast cell lines by forced expression of MyoD. *Proc. Natl. Acad. Sci. U.S.A.* **86,** 5434–5438.

Whittaker, J. R. (1973). Segregation during ascidian embryogenesis of egg cytoplasmic information for tissue-specific enzyme development. *Proc. Natl. Acad. Sci. U.S.A.* **70,** 2096–2100.

Whittaker, J. R. (1977). Segregation during cleavage of a factor determining endodermal alkaline phosphatase development in ascidian embryos. *J. Exp. Zool.* **202,** 139–154.

Whittaker, J. R. (1980). Acetylcholinesterase development in extra cells caused by changing the distribution of myoplasm in ascidian embryos. *J. Embryol. Exp. Morphol.* **55,** 343–354.

Yamada, A., and Nishida, H. (1996). Distribution of cytoplasmic determinants in unfertilized eggs of the ascidian *Halocynthia roretzi. Dev. Genes Evol.* **206,** 297–304.

Yasuo, H., and Satoh, N. (1993). Function of vertebrate *T* gene. *Nature (London)* **364,** 582–583.

Yasuo, H., and Satoh, N. (1994). An ascidian homolog of the mouse *Brachyury (T)* gene is expressed exclusively in notochord cells at the fate restricted stage. *Dev. Growth Differ.* **36,** 9–18.

Yasuo, H., and Satoh, N. (1998). Conservation of the developmental role of *Brachyury* in notochord formation in a urochordate, the ascidian *Halocynthia roretzi. Dev. Biol.* in press.

Yasuo, H., Kobayashi, M., Shimauchi, Y., and Satoh, N. (1996). The ascidian genome contains another T-domain gene that is expressed in differentiating muscle and the tip of the tail of the embryo. *Dev. Biol.* **180,** 773–779.

Yoshida, S., Marikawa, Y., and Satoh, N. (1996). *posterior end mark,* A novel maternal gene encoding a localized factor in the ascidian embryo. *Development (Cambridge, UK)* **122,** 2005–2012.

Yoshida, S., Satou, Y., and Satoh, N. (1997). Maternal genes with localized mRNA and pattern formation of the ascidian embryo. Cold Spring Harbor Symp. *Quant. Biol.,* **62,** 89–95.

II

Nematode

6

Cell Lineages in Caenorhabditis elegans Development

WILLIAM B. WOOD
Department of Molecular, Cellular, and Developmental Biology
University of Colorado, Boulder
Boulder, Colorado 80309

The roundworm *Caenorhabditis elegans* is one of the simplest metazoans, with a small fixed number of somatic cells (959), a small genome (100 Mb) which will soon be completely sequenced, a short life cycle (3 days), transparent embryos, larvae and adults accessible to light microscopy, and several convenient features for genetic analysis that have allowed mutational identification of many genes affecting all aspects of development, including the cell lineage. The normal cell lineage proved to be invariant and possible to determine completely from zygote to adulthood by direct observation using Nomarski optics. Early lineage studies revealed that analogous cells in different parts of the animal usually arise by homologous sublineages and that many of

these lineages appear to be autonomously programmed. These findings suggested that lineal patterns might somehow be directly related to establishment of the resulting cell fates. Molecular genetic analysis of mutations that affect lineage patterns has provided insights into how these patterns are controlled. In particular, recent study of mutations that alter polarities of anterior–posterior cleavages in early embryos as well as polarities of later lineages has revealed that components of the Wnt signaling pathway are important in controlling these aspects of the lineage. We can speculate that the asymmetric expression of transcription factors and the regulation of cell division orientations under control of Wnt-related signals may account for the correspondence between homologous lineages and analogous cell fates and may help us to understand the "hieroglyphics" of cell lineage patterns.

I. Introduction to *Caenorhabditis elegans*

A. Why *C. elegans*?

Partly because of the easy accessibility of its cell lineage, the free-living soil nematode *Caenorhabditis elegans* was chosen by Sydney Brenner in the 1960s for a concerted genetic, ultrastructural, and molecular attack on the problems of animal development and neurobiology (Brenner, 1974, 1988). *C. elegans* is transparent throughout its life cycle, so that cell divisions at all stages of development can be observed in living animals by light microscopy, most conveniently using Nomarski differential-interference-contrast optics. Moreover, it is one of the few metazoans whose development involves a fixed, relatively small number of cell divisions that are essentially invariant with regard to timing and spatial orientation. Adults of the two sexes, hermaphrodites and males, each about 1 mm in length, are comprised of only 959 and 1041 somatic cells, respectively. Because of these properties, it was possible for John Sulston and others to determine first, the complete postembryonic

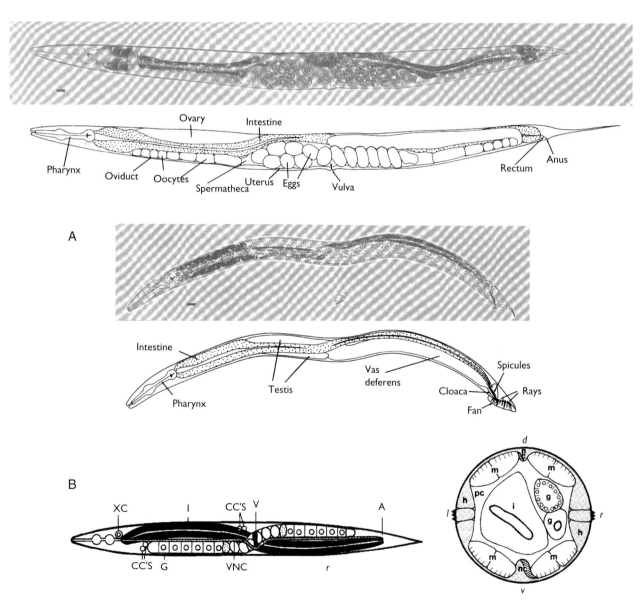

Figure 1 (A) Bright-field micrographs and diagrams of the *C. elegans* adult hermaphrodite (above) and male (below) (modified from Sulston and Horvitz, 1977, with permission). (B) Diagrams of the adult hermaphrodite, showing a ventral view and a cross section through the anterior of the animal. In the ventral view: XC, excretory cell; CC's, coelomocytes; G, gonad; I, intestine; VNC, ventral nerve cord; V, vulva; A, anus; *l,r,* left, right, respectively. In the cross section: *l,r,d,v* left, right, dorsal, ventral, respectively; alae (not labeled) are seen as lateral protrusions from the cuticle; nc, nerve cords; m, muscle; h, hypodermis; g, gonad; i, intestine. (Adapted from Wood and Kershaw, 1991.)

cell lineage (Sulston and Horvitz, 1977; Kimble and Hirsh, 1979; Sulston *et al.*, 1980), and later, the complete embryonic cell lineage (Sulston *et al.*, 1983). Despite its small number of cells, *C. elegans* includes all the common invertebrate tissue types, and its development has many conserved mechanisms in common with other metazoans. Because the ancestry of every cell from zygote to adult is known in both sexes, *C. elegans* provides an excellent experimental organism with which to study the genetic control of cell lineage as well as the significance of cell lineage patterns in cell fate determination, differentiation, and morphogenesis.

B. Anatomy

Light micrographs and diagrams of the adult hermaphrodite and male are shown in Fig. 1. The basic body plan of *C. elegans,* like that of all nematodes, can be simply described as a tube within a tube. The outer tube is an epidermal layer of cells (hypodermis), which secretes a tough, flexible collagenous cuticle covering the exterior of the animal. Associated with the inside of this tube are neurons and body-wall muscles. The inner tube is the digestive tract, including a muscular pharynx as well as the intestine, which runs from the posterior end of the pharynx to the anus. Between the two tubes is a space called the pseudocoelom, which is occupied by the gonad in adults. The pharynx is used to ingest and crush bacteria as well as to pump in fluids; nematodes lack skeletal components and rely on hydrostatic pressure to maintain body shape.

C. Genome

The genome of *C. elegans* is correspondingly simple and will soon become the first metazoan genome to be completely sequenced. It is distributed among six roughly equal-sized chromosomes: five autosomes (I–V) and a sex chromosome X (hermaphrodites have two, males one). The genome is predicted from sequence analysis to include about 15,000 functional genes. Over 2000 of these have been mutationally identified and genetically analyzed to provide a genetic map that is extensively anchored to the physical map and the nucleotide sequence, facilitating positional cloning. Genetic and genomic data are publicly available in "A *C. elegans* Database" (ACeDB), which can be accessed at *http:// elegans.swmed.edu/.*

D. Life Cycle

Caenorhabitis elegans embryos develop inside a chitinous eggshell, which forms from within the zygote shortly after fertilization, and are expelled from the hermaphrodite uterus through the vulva at gastrulation stage. After hatching as first stage (L1) juveniles (commonly referred to as larvae), the worms grow through three more larval stages (L2–L4), separated by molts at which a new cuticle forms and the old one is shed (Fig. 2). The L4's molt to sexually mature adults. Under adverse conditions, such as overcrowding and limited food, L2's can molt to an alternative form of the L3, called a dauer (enduring) larva, also called simply a

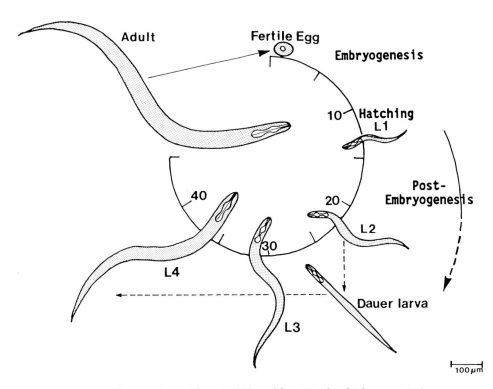

Figure 2 Diagram of the *C. elegans* life cycle. (Adapted from Wood and Johnson, 1994.)

dauer, which is resistant to environmental stress, does not feed, and can survive for several months, much longer than the normal laboratory life span of about 3 weeks. If conditions improve, the dauers molt to L4 and resume the normal life cycle.

II. Embryology

A. Establishment of Polarities

Fertilization occurs as oocytes pass from the oviduct into the spermatheca (Fig. 1A) and there encounter the ameboid sperm of either the hermaphrodite itself or a male, if the hermaphrodite has mated. Sperm entry establishes the future posterior pole of the embryo (Goldstein and Hird, 1996), initiates eggshell formation, and sets in motion a series of events that establish several anterior–posterior (A–P) polarities as the zygote cleaves unequally to form a larger anterior AB cell and a smaller posterior P_1 cell (see Fig. 3 and see Chapter 7 by Bowerman).

Dorsal–ventral (D–V) polarity appears to be established at second cleavage, as AB divides equally and transversely, and one of its daughters is constrained by the eggshell to move posteriorly, defining the future dorsal side of the embryo (Fig. 3). Cells in *C. elegans* lineages are named according to their founder cell (e.g., AB) or a later precursor cell and the spatial relationships of the daughters at each subsequent division; therefore, the posterior AB daughter defining the dorsal side is named ABp and the anterior daughter ABa.

Left–right (L–R) asymmetry of the embryo is established with a normally invariant handedness during third cleavage as ABa and ABp divide transversely in an L–R direction with a slight clockwise skewing of the spindles (as viewed dorsally), so that the daughter cells ABal and ABpl on the left are slightly anterior to their respective sisters ABar and ABpr on the right (Wood, 1991). As a result, the subsequent A–P cleavages of the ventral EMS cell and the posterior P_2 cell are skewed as well, so that the 8-cell embryo is markedly L–R asymmetric, always with the handedness shown in Fig. 3. This asymmetry persists until much later in embryogenesis, when bilateral symmetry is gradually regained as the result of cell-lineage differences on the left and right (see Section IV). However, several L–R asymmetries are retained in the adult (Fig. 1B), the handedness of which is dictated by the initial embryonic asymmetry (Wood, 1991).

During the early cleavages, successive divisions of the germ-line (P) lineage give rise to the six so-called founder cells from which the three germ layers and major tissue lineages of the embryo are derived (Fig. 4): AB (ectoderm, pharynx), MS (mesoderm, pharynx), E (en-

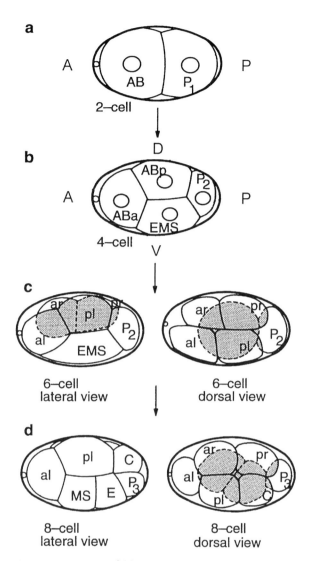

Figure 3 Diagram of blastomere arrangements following the first three cleavages of the *C. elegans* embryo.

doderm), C (ectoderm, mesoderm), D (mesoderm), and P_4 (germ line).

B. Gastrulation

A simple gastrulation process (reviewed by Bucher and Seydoux, 1994) is initiated at the 26-cell stage after all 6 founder cells have been generated, when Ea and Ep leave the ventral surface and migrate interiorly into an inconspicuous blastocoel, followed by descendants of MS, D, P_4, and finally mesoblasts from the C and AB lineages. Although only 53 cells move to the interior during gastrulation, the result is a typical triploblastic embryo with a gut rudiment of E-derived cells near the center, a dorsal–lateral sheet of ectoderm derived from AB and C, and mesodermal and pharyngeal precursors in between. Gastrulation terminates with a process of epiboly as the ectodermal sheet extends ventrally to cover the mesodermal cells still on the surface and then

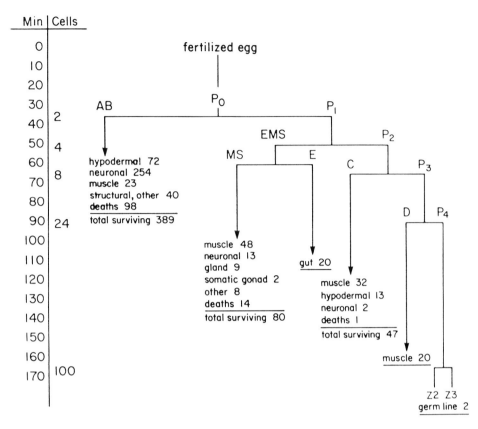

Figure 4 Cell lineages in the early *C. elegans* embryo; ancestry and fates of the embryonic founder cells. (From Wood, 1988.)

closes the ventral cleft (Williams-Masson *et al.*, 1997; L. Edgar and W. B. Wood, unpublished). At this point, about 270 min into embryogenesis, the embryo is still spheroidal and has about 360 cells.

C. Morphogenesis

Following a last round of divisions to produce about 600 cells, cell division largely ceases and a process of morphogenesis begins, during which the spheroidal embryo is transformed into a vermiform shape. Hypodermal cells coalesce into seven extensive hypodermal syncytia, which surround the embryo as shown in Fig. 5A, and tight junctions form between them. The seam cells remain as a lateral row of mononucleate hypodermal cells on each side of the animal and later secrete the specialized cuticular structures known as alae (Fig. 1B). Circumferential bands of actin filaments then form through the syncytia and literally squeeze the spheroidal embryo into a worm that is eventually almost four times as long as the eggshell (Priess and Hirsh, 1986). Characteristic successive stages in morphogenesis are often referred to as comma, lima bean, 1½-fold, 2-fold, and pretzel (see Fig. 5B). A collagenous cuticle, secreted by the hypodermal cells as elongation is completed, stabilizes the vermiform shape and prepares the L1 for hatching, which is accomplished with the help of

a secreted enzyme that digests the eggshell from the inside.

D. Postembryonic Development

The L1 hermaphrodite hatches with 558 cells (560 in the male) which comprise a hypodermis, functional pharynx, gut, excretory system, body musculature, and rudimentary nervous system, in addition to a few other specialized cells including a gonad primordium of two somatic and two germ-line cells. Most of the larval cells, born during the first half of embryogenesis, divide no further (see Fig. 6), but 55 larval blast cells undergo further divisions to produce the adult reproductive structures, with associated elaborations of the musculature and nervous system. In the hermaphrodite, the two somatic cells of the gonad primordium become the distal tip cells, which lead the growth and morphogenesis of the somatic gonad. The germ cells inside proliferate mitotically during gonad growth, differentiate into spermatocytes during L4, and then switch to oocyte production at the molt to adulthood. These processes are described in Chapter 8 by Ellis. Concomitantly, specific ventral hypodermal cells are induced by a signal from the anchor cell in the developing gonad to form the vulva, as descendants of the M mesoblast produce the egg-laying muscles. The vulva

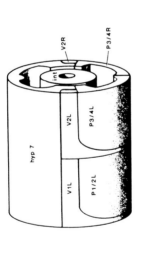

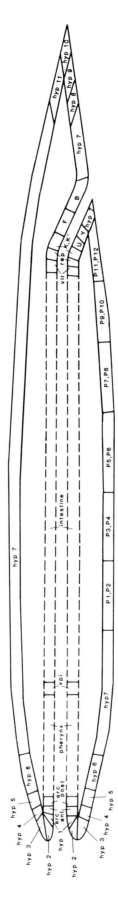

Figure 5 (A) Schematic diagram of a longitudinal section of the L1 larva, showing arrangement of the hypodermal cells. Upper diagram shows the three-dimensional organization of a region of the central body. In the lower diagram, commas indicate two cells meeting in the plane of the drawing. hyp7 is a large syncytium, which enlarges by cell fusion and spreads over most of the ventral region during post-embryonic development. (From Sulston *et al.*, 1983, with permission). (B) Three stages in embryonic morphogenesis. Scanning electron micrographs of embryos with the eggshells removed at the 1½-fold, 2-fold, and pretzel stages are shown in panels a, c, and e, respectively. Panels b, d, and f show embryos at stages corresponding to those on the left that have been fixed and stained with an antibody that recognizes the tight junctions between hypodermal cells. (From Priess and Hirsh, 1986, with permission.)

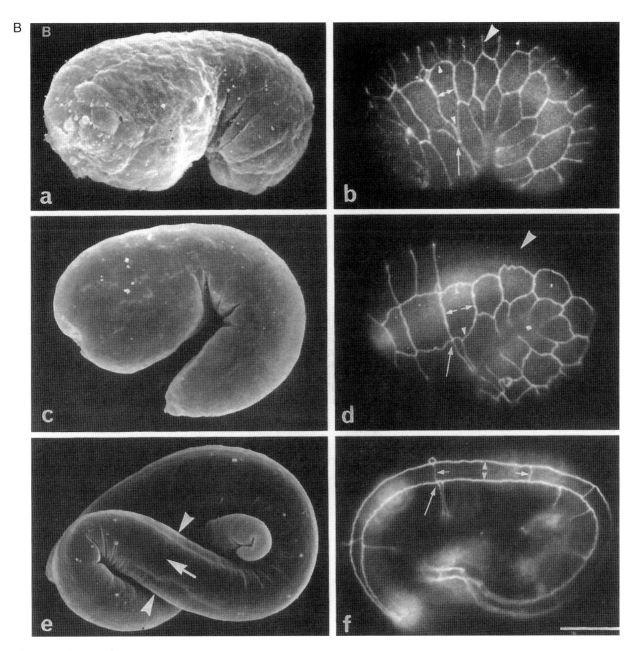

Figure 5 *Continued*

lineages and their control are described in Chapter 10 by Hajnal.

In the male, the gonad primordium develops differently, with the linker cell leading growth first anteriorly and then posteriorly to form the single-armed male gonad and attach it to the cloaca. The germ cells proliferate inside and begin differentiation into spermatocytes during L4. Concomitantly, the male tail acquires specialized sex muscles from descendants of the M cell and undergoes a complex process of morphogenesis at the posterior end to form the male mating structures: the fan, sensory rays, hook, and spicules. Male tail morphogenesis is described in Chapter 9 by Emmons. Each of these processes involves specialized cell lineages, as

described in the following section and subsequent chapters.

III. Properties of Cell Lineages

A. Methods

Cell lineages in *C. elegans* embryos and larvae are generally determined by direct observation of living specimens, using Nomarski optics to note the position of each nucleus and watch the cell divide, and then recording the positions of the daughter nuclei. For later embryos in particular, this method of analysis becomes

Zygote

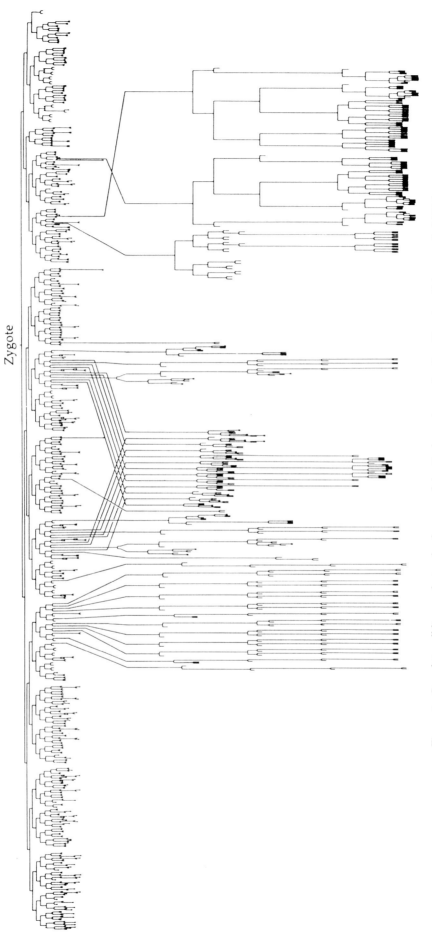

Figure 6 Complete cell lineage during development of the *C. elegans* hermaphrodite. (Based on Sulston *et al.*, 1983.)

extremely difficult as the number of cells increases, and only the remarkable invariance of the process allowed Sulston and co-workers (1983) to reconstruct the entire embryonic lineage from observations on hundreds of wild-type embryos over the course of several years. Embryonic lineage analysis has recently been greatly simplified by development of multiple-focal-plane time-lapse video recording systems ("4D microscopes"), which can automatically record a through-focus series of Nomarski images every half minute or less during embryogenesis and store these as digital files, providing a complete archival record of the development of an embryo in three dimensions over time. Subsequently, successive images in a single focal plane can be played back under control of the operator to allow reconstruction of cell lineages (Fire, 1994; Thomas *et al.*, 1996). This method is particularly useful for mutant analysis, where it is often desirable to reconstruct the complete lineage of a single defective embryo.

B. Cell Lineage Hieroglyphics

The monumental publications of the postembryonic (Sulston and Horvitz, 1977) and embryonic lineages (Sulston *et al.*, 1983) presented developmental biologists with an array of reproducible cell division patterns that have been tantalizing to think of as possibly decipherable pictograms (Fig. 6). Does the structure of the lineage itself have meaning? What, if any, is the developmental significance of its essential invariance from one embryo to the next, of the apparently different logic of different founder-cell lineages, and, most intriguingly, of the many precisely repeated lineal patterns (sublineages) within several of these lineages? How did the lineage evolve?

C. Generalizations

Several generalizations emerged from the initial descriptions of the lineage (Sulston and Horvitz, 1997; Sulston *et al.*, 1983). Its invariance is clearly important for establishment of the body plan. Only 12 cells migrate significant distances during embryonic development, apart from the more restricted movements of gastrulation; the rest are placed close to their final relative positions in the process of cell division.

Founder-cell lineages do not correlate consistently with germ layers. Although E and D lineages are exclusively endodermal and mesodermal, respectively (Fig. 4), the other three somatic founder cells (AB, MS, C) contribute both ectodermal and mesodermal derivatives. Founder-cell lineages also do not correlate consistently with tissue types. Although progeny of the E (gut), D (muscle), and P_4 (germ line) cells are limited to single tissues, the other three founder cell lineages each contribute to more than one. Conversely, several tissues include cells of more than one founder cell lineage; for example, body wall muscle is derived from progeny of the MS, C, D, and AB founder cells in different parts of the body.

Besides the various cell types that are generated in the lineage, programmed cell death (apoptosis) is a common cell fate. About 1 out of 6 cells generated in embryogenesis dies. Cell deaths, like other cell fates, are invariant features of specific lineages, occurring autonomously in fated cells shortly after their birth. The mechanism of this apoptosis is described in Chapter 8 by Ellis.

D. Clonal and Stem-Cell Patterns

An intriguing question since the first lineages were described has been the relationship between observed lineal patterns and the specification of cell fates. This question can be explored by examining the ancestries of developmentally equivalent (analogous) cells. In the simplest cases, analogous cells are generated by clonal patterns, in which all progeny of a founder cell have the same general fate. In this manner, the intestine, the germ line, and some of the body-wall muscle are generated by divisions of the E, P_4, and D cells, respectively. Simple stem-cell patterns, in which one daughter at each division is like the parent and the other daughter is of a second type, can give rise to several cells of the second type. Such parental reiteration patterns could have evolved simply from clonal patterns. Modified stem-cell patterns exhibiting grandparental reiteration are also encountered a few times in the lineage, and can give rise to several cells of two or three types.

E. Repeated Sublineages

The above mechanisms generate analogous cells in a restricted area. However, analogous cells are often found at different places in the animal; for example, motor neurons of a particular class occur at intervals along the entire length of the ventral nerve cord. These cells are generated by repeated sublineages that generate groups of several kinds of neurons. The general rule in these sublineages is that lineally homologous cells have analogous fates. For example, descendants of the P cells in the ventral cord execute a postembryonic sublineage that is repeated 13 times along the length of the animal to give each time (with some variations; see below) the same 5 types of motor neurons and a hypodermal cell (Fig. 7). This observation raises the question of whether the lineal pattern itself might be relevant to the generation of these fates.

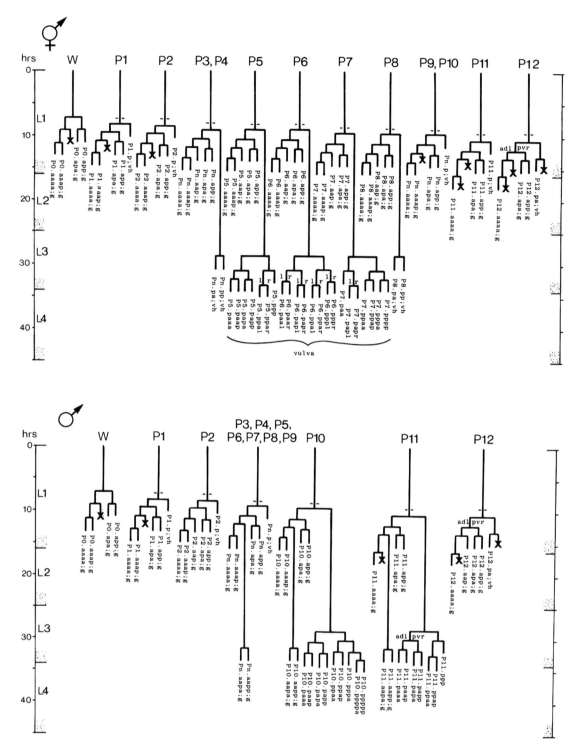

Figure 7 Postembryonic cell lineages in the ventral nerve cord and hypodermis of hermaphrodites (above) and males (below). (Modified from Sulston and Horvitz, 1977, with permission.)

F. Modified Sublineages

Repeated sublineages, such as those in the ventral cord, are often similar but not identical; that is, region-specific modifications cause production of somewhat different sets of cells at different locations (Fig. 7). Programmed cell deaths provide a mechanism for modifying a sublineage to prevent production of a certain cell type. And near the center of the hermaphrodite, the fates of three hypodermal Pnp cells (P5p, P6p, and P7p) in the ventral cord lineages are modified to form the vulva in response to an inductive signal from the gonadal anchor cell, by an inductive mechanism that is quite well understood (see Chapter 10 by Hajnal). Mod-

ified sublineages represent exceptions to the above general rule, since lineal homologues may no longer have analogous fates. Such sublineages, like segments in insects, may have evolved by duplication and subsequent modification.

A curious manifestation of such modification is seen on comparison of homologous lineages on the two sides of the embryo. As mentioned above, the early embryo is grossly L–R asymmetric, so that bilateral symmetry must be superimposed by compensating differences in left and right lineages, first pointed out by Sulston (Sulston *et al.,* 1983; Sulston, 1983). These differences are most marked in the anterior AB lineages, as shown in Fig. 8 (Wood and Kershaw, 1991; Wood, 1998).

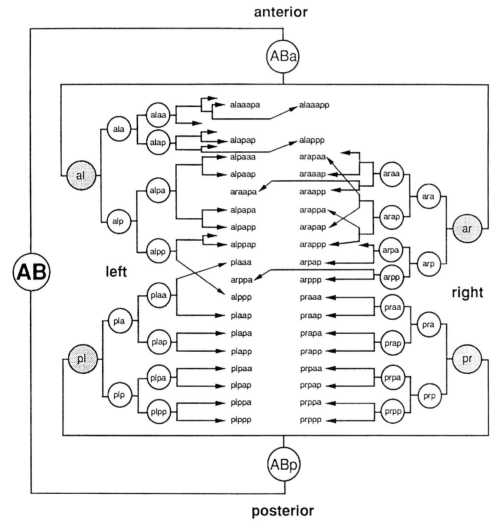

Figure 8 Left–right differences in the AB lineages. The diagram shows the lineal relationships among 18 pairs of contralaterally analogous cells in the AB lineage. Pairs of cells shown opposite each other in the center of the diagram are analogous in both position and fate: The members of each pair occupy approximately equivalent positions on the left and right sides of the embryo at the time they are born (32-AB-cell to 128-AB-cell stages), and they give rise to nearly or completely identical numbers and types of progeny cells during subsequent development. Their ancestors, back to the 4-AB-cell stage, are represented in their approximate relative A–P positions, with cells on the left more anterior than their lineal homologues on the right. Note that analogues among ABp descendants in the seven most posterior pairs are also lineal homologues, whereas the remaining more anterior analogues all have different lineal ancestries on the two sides of the embryo (as can be seen from their cell names as well as from the lineage diagram). Four analogues are generated by divisions that cross the midline. (Based on Sulston *et al.,* 1983; Sulston, 1983; from Wood and Kershaw, 1991, with permission.)

IV. Mechanisms of Cell Fate Determination

A. Cell Interactions

In *C. elegans,* with the complete cell lineage providing the ultimate fate map, the developmental fate of a cell at any stage can be defined in terms of its progeny and their descendant lineages. This developmental fate must be determined in response either to external signals, generally originating from other cells, or to intrinsic factors, or to a combination of both. Although it is of course a fallacy to equate invariance of the cell lineage with any particular mechanism of determination, this invariance and the equivalence of lineal homologues seemed to fit earlier views that nematode development was largely "mosaic" in character, meaning that cell fates were determined autonomously by intrinsic determinants partitioned asymmetrically at certain cell divisions. However, following early experiments indicating cell interactions in the modification of some postembryonic lineages (e.g., Kimble, 1981), a variety of studies too numerous to describe here have shown that many inductive effects are involved in *C. elegans* development, not only postembryonically but also in the early embryo (e.g., Priess and Thomson, 1987; Wood, 1991; Hutter and Schnabel, 1994, 1995; for review see Chapter 7 by Bowerman). Much of this evidence was obtained by observing embryonic development following micromanipulation or laser ablation of specific cells.

B. Lineal Programming by Autonomous Mechanisms

The first laser ablation experiments failed to reveal early inductive interactions, because technical problems prevented destruction of the large early blastomeres without killing the embryo. However, successful experiments on embryos beyond the 28-cell stage indicated that many of the cell fates generated in later embryonic lineages appear to be "lineally programmed," that is, ablation of a variety of single cells had no apparent effect on the subsequent lineage patterns of other cells (Sulston *et al.,* 1983). Similar, apparently autonomous programming has been observed postembryonically, for example, among the descendants of vulval precursor cells following the initial inductive events (see Chapter 10 by Hajnal). Such programming is not necessarily cell-autonomous and may involve incestuous cell interactions between close relatives in the same sublineage, but it appears to be lineally autonomous. The mechanisms for this mode of fate determination are only beginning to be understood (see following Section V and Chapter 7 by Bowerman). They are being elucidated by

analysis of mutations causing specific defects in lineage patterns.

V. Genetic and Molecular Control of Cell Lineages and Cell Fates

A. Screens for Lineage-Defective Mutants

The ability of a *C. elegans* hermaphrodite to lay eggs is not essential for fertility, since progeny will simply develop and hatch in the uterus, commit "endomatricide" on the parent (which becomes an easily recognizable "bag of worms") and eventually escape from the parental cuticle. Horvitz and collaborators took advantage of this property to carry out powerful large-scale screens for nonlethal mutations that cause failure of, or abnormalities in vulval development or egg laying, or both (Horvitz and Sulston, 1980). The resulting mutants are still providing a rich resource for studying the genetic control of postembryonic lineage patterns and cell fate determination. Many of the mutations resulted in postembryonic lineage defects, identifying genes designated *lin* (lineage-defective) that are apparently involved in control of lineal patterns (e.g., Horvitz *et al.,* 1984; Horvitz and Sternberg, 1991). Many of the mutations were *homoeotic,* resulting in specific cells mistakenly adopting the normal fates of other cells. And several of the identified *lin* genes behave as *switch genes,* in that loss-of-function (*lf*) and gain-of-function (*gf*) mutations cause opposite cell-fate transformations, suggesting that these genes encode components of determinative regulatory pathways for cell-fate specification. The *lin* genes (and others with different names that also affect cell lineages) can be grouped into the following several classes based on their mutant phenotypes.

B. Timing Genes Identified by Heterochronic Mutants

One class of mutations causing alteration of lineal patterns turned out to affect temporal rather than spatial cues for determining cell fates. These so-called heterochronic mutations affect the timing of cell lineages, causing normal lineages to be executed at an abnormal time during larval development. For example, *lin-4(lf)* and in *lin-14(gf)* mutations cause retarded lineage execution: that is, lineages normally executed only in early larval stages are reiterated in later larval stages. By contrast, gene *lin-14(lf)* mutations cause precocious lineage execution: lineages normally executed late are executed too early. These and several other heterochronic genes subsequently identified affect other aspects of larval development besides cell lineages, such as type of cuticle synthesized. Therefore, they appear to control the timing mechanism that specifies different stages of larval development. The mechanism of this control, which is beyond the scope of this chapter, is beginning to be understood (for review see Ambros, 1997).

C. Control of Differences between Repeated Sublineages

As discussed above and shown in Fig. 7, repeated sublineages are often nonidentical; the differences between them result in somewhat different sets of cells at different points along the body axis. Mutations that eliminate differences between similar sublineages identify genes that modify the lineally autonomous specification of cell fates within sublineages.

A few embryonic and several postembryonic lineages are sexually dimorphic, as seen from the example in Fig. 7. These differences are controlled by the sex-determining genes including *her-1,* encoding a normally male-specific ligand which acts through a novel signal transduction pathway to negatively regulate the transcription factor encoded by *tra-1,* which is required for hermaprodite development (reviewed by Cline and Meyer, 1996). *her-1(lf)* mutations nonautonomously feminize male-specific lineages (Hunter and Wood, 1992), and *tra-1(lf)* mutations autonomously masculinize hermaphrodite-specific lineages (Hunter and Wood, 1990).

Also on a global level, genes in the Hox cluster appear responsible for region-specific modifications along the anterior–posterior axis (Clark *et al.,* 1993; Salser *et al.,* 1993). For example, lineally homologous cells in the ten P-cell sublineages of the ventral cord normally become VC neurons in the central six sublineages and undergo programmed cell death in the other four. A *lin-39(lf)* mutation causes these cells to die in all ten sublineages. The mechanism by which normal *lin-39* function prevents these cell deaths in the central sublineages is unknown (for review see Ruvkun, 1997).

On a more local level, paracrine inductive effects or juxtacrine signaling, or a combination of both, can reprogram specific members of an *equivalence group* of cells with apparently equivalent developmental potential (often lineally homologous cells from repeated sublineages). One of the best understood examples is assignment of the anchor-cell fate to one of two lineal homologues in the somatic gonad lineages, by competitive juxtacrine signaling between the two cells, mediated by the Notch like receptor encoded by *lin-12* and the Delta/Serrate-like cell-surface ligand encoded by *lag-2* (Wilkinson *et al.,* 1994). Another is the induction of vulval formation in the P5p–P8p equivalence group of the ventral cord, in response to a *lin-3*-encoded inductive signal from the gonadal anchor cell as well as lateral *lin-12*-mediated signals between the P cells (Chapter 10 by Hajnal). A final comes from the 4-cell embryo, where different fates are conferred on the developmentally equivalent ABa and ABp cells by a Delta/Serrate-like signal encoded by *apx-1* in the P$_2$ cell acting through the Notch-like receptor encoded by *glp-1* in the neighboring ABp cell (Mello *et al.,* 1994). Additional *glp-1*-mediated signals occur later during cleavage stage to specify fate differences between subsequently arising cells in the AB lineage (see Chapter 7 by Bowerman). Mutations that affect any of these signals can eliminate the lineage differences they normally promote.

D. Control of Lineage Polarities

Most sublineage patterns are asymmetric and exhibit a consistent A–P polarity, suggesting the existence of global polarizing signals along the A–P axis. One of a few exceptions is the apparent mirror symmetry (actually rotational) of the vulval sublineages on either side of the central P6p cell in hermaphrodites (Fig. 7), suggesting that another polarizing signal could emanate from the center of the developing vulva. In *lin-44(lf)* mutants, the normal A–P polarities in several posterior postembryonic sublineages are reversed. The finding that *lin-44* encodes a homologue of the *Drosophila* Wingless and vertebrate Wnt proteins, produced by hypodermal cells near the tip of the tail, suggests that this ligand could act as one A–P polarizing signal. Its absence apparently allows a second weaker and more anterior polarizing signal to reverse the affected lineal patterns (Herman *et al.,* 1995; see also Ruvkun, 1997).

Additional evidence for the role of Wnt signaling in controlling the polarity of unequal embryonic divisions is accumulating rapidly. Mutations that result in failure of the P$_2$-induced polarization of EMS required for production of MS and E have identified several genes designated *mom* (more mesoderm), whose impairment leads instead to production of two MS-like cells (Thorpe *et al.,* 1997; Rocheleau *et al.,* 1997; see Chapter 7 by Bowerman, for review). Several of these genes encode homologues of Wnt pathway components. For example, *mom-2* encodes a Wnt homologue, and *mom-5* encodes a Frizzled-2 (putative Wnt receptor) homologue. RNA interference (RNAi) experiments have shown that the product of *wrm-1,* a homologue of vertebrate β-catenin (*Drosophila* Armadillo), as well as the product of *apr-1,* a homologue of the vertebrate adenomatous polyposis coli (APC) gene are also required for this polarization (Thorpe *et al.,* 1997; Rocheleau *et al.,* 1997; see Chapter 7 by Bowerman, for review). RNAi, in which gene-specific RNAs are injected into the hermaphrodite gonad to produce defective progeny (Guo and Kemphues, 1996), is only beginning to be understood (Fire *et al.,* 1998) but has proven to be a reliable method for phenocopying most maternal-effect mutations in *C. elegans* (Thorpe *et al.,* 1997; Rocheleau *et al.,* 1997; see Chapter 7 by Bowerman, for review). Inactivation of a gene by this method is indicated below as, for example, *wrm-1(RNAi)*. The *pop-1* gene (posterior pharynx-defective), in which *lf* mutations cause the opposite phenotype of producing two E-like cells and no MS (Lin *et al.*, 1995), encodes a homologue of the vertebrate transcription factor Lef-1, which is regulated by the Wnt pathway. In double mutants, *pop-1(lf)* mutations are epistatic to *mom(lf)* mutations and *wrm-1(RNAi)*, indicating that POP-1 also acts downstream of the Wnt pathway elements.

Two recent papers likely to become landmarks in understanding cell lineages suggest that Wnt signaling may be involved in polarization not only of unequal divisions, but of most or all A–P divisions during embryonic and postembryonic development. In one of these papers, Lin *et al.*, (1998) show that following many A–P divisions throughout development, the POP-1 transcription factor appears by immunostaining to be substantially more concentrated in the nucleus of the anterior daughter than in that of the posterior daughter. (Exceptions include the P-cell descendants that form the vulva, which exhibit mirror symmetry around the center of the P6p lineage; R. Lin, personal communication). Inactivation of the *wrm-1* (β-catenin) gene by *wrm-1(RNAi)*, or elimination of both the Wnt homologue MOM-2 and the APC homologue APR-1 by *apr-1(RNAi)* in a *mom-2(lf)* background, eliminates all the observed asymmetries in POP-1 staining. The *mom-2(lf)* mutation alone eliminates POP-1 asymmetry between MS and E, but not between daughters of A–P divisions in the AB lineage, consistent with findings in vertebrates that APC and Wnt may be involved in alternative controls of β-catenin interaction with Lef-1. A *pop-1(lf)* mutation or *pop-1(RNAi)* causes many embryonic anterior-to-posterior cell-fate transformations between daughter cells of A–P divisions in both AB and EMS lineages. The asymmetric staining of POP-1 (probably resulting from posttranslational modification) between MS and E requires the polarizing contact of EMS with P$_2$. However, once EMS has been polarized in a wild-type embryo, P$_2$ can be removed without affecting the subsequent asymmetric appearance of higher POP-1 staining in the anterior daughters of both E and MS; that is, the generation of A–P asymmetry appears to be self propagating once initiated. Although the mechanisms of initiating polarization and of POP-1 control remain unclear for the present, the authors conclude that asymmetric function of POP-1 in the daughters of A–P divisions in response to Wnt signaling could play a major role in cell fate determination throughout development (see Section VI). In the other paper, Kaletta *et al.*, (1997) show that mutations in a gene called *lit-1* (lack of intestine) cause multiple posterior-to-anterior cell fate transformations between daughter cells of A–P divisions throughout embryogenesis, and that *pop-1* is epistatic to *lit-1*. They conclude that *lit-1* may control *pop-1* in

many binary cell-fate decisions. Although the molecular identity of *lit-1* has not yet been reported, it seems likely also to encode a Wnt pathway component that is required for POP-1 asymmetry.

E. Lineage-Autonomous Controls

In most postembryonic cell divisions, the fates of daughter cells are different from that of the mother cell. Mutations in *unc-86* reveal underlying stem-cell patterns in divisions that normally produce two daughters different from the mother and from each other (A → B,C). In *unc-86(lf)* mutant animals these divisions, in several different lineages, instead produce one daughter cell with the same fate as the mother, so that the lineage is reiterated (A → B,A → B,A, etc.). The UNC-86 gene product is a POU-domain transcription factor and, therefore, likely to act cell-autonomously (Finney and Ruvkun 1990; see also Ruvkun, 1997).

In many postembryonic lineages, sister cells at a given division acquire different fates that appear to be autonomously controlled. Mutations (*lf*) in the *lin-26*, *lin-11*, and *lin-17* genes eliminate such differences, causing cell divisions normally of the type A → B,C in several different sublineages to be converted to the type A → B,B. *lin-17* encodes a Frizzled (Wnt receptor) homologue, consistent with involvement of Wnt signaling in dictating these fate differences (Sawa *et al.*, 1996). The *lin-11* gene encodes a LIM-domain transcription factor (Freyd *et al.*, 1990), and *lin-26* encodes a zinc-finger transcription factor (Labouesse *et al.*, 1996); their roles in fate specification are not clear.

F. Control of Spindle Orientations

Mutations that affect spindle orientations have been found primarily in screens for mutational defects in early embryogenesis. As mentioned previously, the orientations of cell divisions, perhaps especially in the early embryonic cleavages, are crucial to normal morphogenesis. The normal cycle of centrosomal duplication and migration to opposite sides of the nucleus during prophase dictates that in a series of proliferative divisions, the spindle orientation in a given cell cycle will be orthogonal to the orientation in the previous cell cycle, unless the spindle is actively reoriented at the start of mitosis. High-resolution light microscopy, laser ablation, and drug experiments indicate that such a reorientation occurs in the germ-line cells P$_1$, P$_2$, and P$_3$ prior to their unequal, A–P divisions in the early embryo. The mechanism involves actin-containing cortical plaques in these cells, positioned near the midbody from the previous cleavage, which connect via astral microtubules to the poles of the spindle and actively rotate it by 90° during prophase (Hyman and White, 1987; Hyman, 1989; reviewed in White and Strome, 1996). An apparently similar reorientation occurs prior to the unequal A–P cleavage of EMS to form the MS and E founder cells.

Mutations in some of the *par* genes affect the AB and P_1 spindle orientations, leading to grossly disorganized and nonviable embryos; however, these mutations also affect partitioning of many cell components during first cleavage that are essential for normal patterning. Experiments with *par* double mutants have shown that the reorientation machinery normally exists in both cells at the 2-cell stage, but that the activity of the *par-3* gene product prevents it from acting in the AB cell. The *par-2* gene product restricts PAR-3 activity to the AB cell. Both are novel proteins (Cheng *et al.*, 1995). None of the known mutations are defective in the reorientation process itself for the P_1 cell; however, the *mes-1* gene may be required specifically for reorientation in P_2 and P_3 (Hird *et al.*, 1996). A mutation in the *spn-1* gene, which appears to affect preferentially the orientations of the ABa and ABp spindles that determine embryonic handedness at third cleavage, can result in viable embryos with all the normal L–R lineal asymmetries reversed (D. Bergmann, L. Rose, and W. B. Wood, unpublished). None of the mutations in any of these genes appear to affect spindle orientations later in development.

VI. Understanding Cell Lineage Hieroglyphics

A. Cell Lineage Functions

The cell lineage must: (1) generate cells of the correct developmental potential at the (2) correct positions and (3) correct times during embryonic and postembryonic development. Current knowledge of these three functions is considered in reverse order below.

B. Control of Cell Cycling

The generally invariant timing of cell divisions throughout development and the widely varying times between divisions in different branches of the cell lineage indicates precise developmental control of entry into and exit from the cell cycle. Apart from some understanding of the heterochronic genes that are responsible for overall timekeeping during larval development, relatively little is known about this aspect of the lineage. The *cul-1* gene (formerly *lin-19*), is one of a conserved family of CDC53-like genes, also found in yeast and humans and designated cullins. This gene and *lin-23*, which encodes a CDC4 homologue, appear to be important for cell-

cycle exit (Kipreos *et al.*, 1996); *cul-1(lf)* and *lin-23(lf)* mutations result in hyperplasia of all tissues in the embryo. Application of the recently developed RNAi method to determine the results of functionally silencing *C. elegans* homologues of known cell-cycle genes could rapidly advance understanding of developmental cell-cycle controls. It will be of particular interest to learn how cell cycling is terminated in most lineages at midembryogenesis and allowed to resume in only 55 blast cells during larval development.

C. Control of Spindle Orientations

Correct positioning of daughter cells at each division requires control of successive spindle orientations. This control can be exerted in at least two ways, following, for example, an A–P division. First, the spindle orientation for a subsequent transverse division (L–R or D–V) will be dictated by the plane in which the centrosomes after duplication undergo their 90° circumferential migration to opposite poles on the nuclear membrane, where they will form the poles of the new spindle. How this plane of centrosomal migration is chosen remains completely obscure, but could involve controls on centriolar orientations before or during their duplication. Second, if the subsequent division is again to be A–P, the new spindle must presumably be reoriented after its formation by a mechanism such as that described in the previous section (unless there are also unknown controls that can hold one daughter centrosome in place while the other migrates 180°). So far, reorientation has been documented only in the first few germ-line cleavages and the unequal cleavage of EMS. However, some such mechanism must be used subsequently in equal cleavages as well, since many of these do not show orthogonal spindle orientations in successive divisions. In fact, most divisions during development are oriented in the A–P direction, as are perhaps all the determinative divisions (producing daughters with different fates). The cleavages of both MS and E, for example, are A–P, parallel to the cleavage orientation of their parent EMS. How these many somatic reorientations might be controlled is unknown, but it seems plausible that the mechanisms could be related to those controlling cell polarization.

D. Establishment of Cell Polarities

Polarization of cells had been inferred from the A–P asymmetry of many lineage patterns, but it was previously visible only in unequal divisions or those for which internal markers were available, almost exclusively in the early embryo. However, the recent discovery of widespread POP-1 staining asymmetry indicates that polarization may occur in most if not all cells that undergo A–P divisions throughout development (Lin *et al.*,

1998). This finding, and the demonstration that the apparent POP-1 asymmetry is somehow controlled by Wnt signaling, is an important first step in understanding the meaning of cell lineage hieroglyphics.

E. The Relevance of Lineal Patterns to Cell Fate Determination

The invariance of the cell lineage means that each differentiated cell is the product of the same series of spatially and temporally controlled cell divisions in every animal. Is this simply a correlation, or are specific lineal patterns somehow important in generating specific sets of cell types? Some features of the lineage have long suggested that these patterns may be important. One such feature is the observation emphasized above, that in the generation of similar sets of cells by repeated sublineages in widely separated regions of the animal, analogous cells are lineal homologues, with generally only minor variations that can often be explained by local cell interactions or superimposition of regional controls. Another is that the terminal divisions of some lineage branches give rise to one or two generations of daughter cells that undergo apoptosis, while the surviving daughter or granddaughter differentiates. An example is the production of a VA motor neuron by the postembryonic

P11 and P12 lineages in the hermaphrodite ventral nerve cord (Fig. 7; enlarged in Fig. 9A). Why not save these divisions and instead allow the grandparent (P11aa) to become a VA neuron, unless the divisions themselves are somehow part of the program that specifies the fate of the remaining cell?

The recent analyses of POP-1 accumulation, Wnt signaling, and *lit-1* mutants suggest that lineage patterns are in fact functional units of cell fate determination involving successive binary choices in A–P divisions (Kaletta *et al.,* 1997; Lin *et al.,* 1998). Lin *et al.,* (1998) have proposed a simple, plausible model for how such a unit might function (Fig. 9B). It assumes that A–P polarity is first established by Wnt signaling, but can then be propagated such that the apparent level of POP-1 is reduced in the posterior daughter at each division. The authors propose that a factor inhibiting this reduction is initially localized and activated in the anterior cortex of a precursor cell by Wnt signaling, so that it is inherited by the anterior daughter. The posterior daughter then recognizes its own anterior pole, perhaps by the presence of the midbody, and again localizes and activates the inhibitory factor in the surrounding cortex, thereby itself becoming polarized and allowing propagation of the POP-1 asymmetry to its daughters in the subsequent division. POP-1 is a transcription factor, which

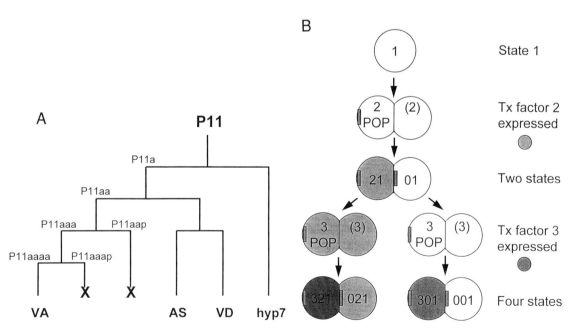

Figure 9 The postembryonic P11 cell lineage in the hermaprodite ventral nerve cord. Programmed cell deaths are indicated by X. Instead of cell P11aa simply being specified to become a VA motor neuron, this cell undergoes two divisions in which the posterior cell is eliminated, so that the VA cell produced in this lineage is Pnaaaa as in all the other ventral cord Pn lineages (see Fig. 7). (B) A model for propagation of anterior–posterior polarity and sequential binary determination of cell fate differences by lineal programming in a series of anterior–posterior cell divisions. A precursor cell in State 1 is polarized by Wnt signaling so that an inhibitor of POP-1 degradation is localized at the anterior pole. The next division triggers expression in both daughters of Transcription (Tx) Factor 2, which is active only in combination with POP-1 and therefore only in the anterior daughter, which consequently adopts a new determination state, 21. The posterior daughter then also becomes polarized, perhaps by recruitment of the degradation inhibitor to the midbody, so that when the following division triggers expression of Transcription Factor 3, it is again active only in the anterior daughters, which consequently adopt the new states 321 and 301 to give each of the four granddaughters a different determination state. (Adapted from Lin *et al.,* 1998.)

probably interacts with others to confer transcriptional specificity that could dictate cell fate. If a new interacting transcription factor is synthesized symmetrically at each division in a sublineage, but is activated only in the anterior daughter by POP-1 interaction, then different cell types could be combinatorially specified by a succession of temporally but not spatially regulated factors interacting with spatially regulated POP-1 (Lin *et al.,* 1998).

We can add to this model the proposal that some of the same machinery could be involved in reorientation of spindles to achieve the successive A–P divisions that are required by the model and observed in most lineages. If such reorientation involves cortical attachment sites as shown for the early blastomeres, then the midbody again could provide a convenient reference point in both anterior and posterior daughters for initiating connections to the spindle poles.

In the context of such a model, lineage patterns take on new developmental meaning. Is this phenomenon limited to nematodes and related organisms with few cells and invariant patterns of cell division throughout development? Although importance of lineal programming may be most easily demonstrated in *C. elegans* because of its small number of cells, the answer is most certainly no. Stereotyped embryonic lineages in sea urchin, ascidians, and leeches may function similarly (see other chapters in this book). In *Drosophila,* there are clear examples of invariant lineal patterns, for example, in later development of the peripheral and central nervous systems, which first provided elegant examples of asymmetrically distributed proteins as indicators of polarity in somatic determinative cell divisions (Rhyu *et al.,* 1994; Knoblich *et al.,* 1995; Kraut *et al.,* 1996; Shen *et al.,* 1997; Chapter 18 by Broadus and Spana). In organisms with larger numbers of cells, it is more difficult to find invariant bits of cell lineage. However, it seems plausible that the mechanisms of lineal programming used so extensively by nematodes will turn out to be generally employed, if only in isolated situations interspersed with other modes of cell proliferation and determination in the development of larger animals.

Acknowledgments

I am grateful to R. Lin and J. Priess for communication of unpublished results. Research from my laboratory was supported by NIH Grants HD-11762, HD-14958, and HD-29397.

References

Ambros, V. (1997). Heterochronic genes. *In* "C. elegans II" (D. L. Riddle, T. Blumenthal, B. J. Meyer, and J. R. Priess, eds.), pp. 501–518. Cold Spring Harbor Laboratory, Cold Spring Harbor, New York.

Brenner, S. (1974). The genetics of *Caenorhabditis elegans, Genetics* **77,** 71–94.

Brenner, S. (1988). Foreword. *In* "The Nematode *Caenorhabditis elegans*" (W. B. Wood, ed.), pp. ix–xiii. Cold Spring Harbor Laboratory, Cold Spring Harbor, New York.

Bucher, E. A., and Seydoux, G. (1994). Gastrulation in the nematode *Caenorhabditis elegans. Semin. Dev. Biol.* **5,** 121–130.

Cheng, N., Kirby, C., and Kemphues, K. (1995). Control of cleavage spindle orientation in *Caenorhabditis elegans:* The role of the genes *par-2* and *par-3. Genetics* **139,** 549–559.

Clark, S., Chisholm, A., and Horvitz, H. (1993). Control of the cell fates in the central body region of *C. elegans* by the homeobox gene *lin-39. Cell (Cambridge, Mass.)* **74,** 43–55.

Cline, T. W., and Meyer, B. J. (1996). Vive la difference: Males vs. females in flies vs. worms. *Annu. Rev. Genet.* **30,** 637–702.

Finney, M., and Ruvkun, G. (1990). The *unc-86* gene product couples cell lineage and cell identity in *C. elegans. Cell (Cambridge, Mass.)* **63,** 895–905.

Fire, A. (1994). A 4-dimensional digital image archiving system for cell lineage tracing and retrospective embryology. *Comput. Appl Biosci.* **10,** 443–447.

Fire, A., Xu, S., Montgomery, M. K., Kostas, S. A., Driver, S. E., and Mello, C. C. (1998). Potent and specific genetic interference by double-stranded RNA in *Caenorhabditis elegans. Nature (London)* **391,** 806–811.

Freyd, G., Kim, S., and Horvitz, H. (1990). Novel cysteine-rich motif and homeodomain in the product of the *Caenorhabditis elegans* cell lineage gene lin-11. *Nature (London)* **344,** 876–879.

Goldstein, B., and Hird, S. N. (1996). Specification of the anteroposterior axis in *Caenorhabditis elegans. Development (Cambridge, UK)* **122,** 1467–1474.

Guo, S., and Kemphues, K. (1996). A non-muscle myosin required for embryonic polarity in *Caenorhabditis elegans. Nature (London)* **382,** 455–458.

Herman, M., Vassilieva, L., Horvitz, R., Shaw, J., and Herman, R. (1995). The *C. elegans* gene *lin-44,* which controls the polarity of certain asymmetric cell divisions, encodes a Wnt protein and acts cell nonautonomously. *Cell (Cambridge, Mass.)* **83,** 101–110.

Hird, S., Paulsen, J., and Strome, S. (1996). Segregation of germ granules in living *Caenorhabditis elegans* embryos: Cell-type-specific mechanisms for cytoplasmic localisation. *Development (Cambridge, UK)* **122,** 1303–1312.

Horvitz, H., and Sternberg, P. (1991). Multiple intercellular signalling systems control the development of the *Caenorhabditis elegans* vulva. *Nature (London)* **351,** 535–541.

Horvitz, H., Sternberg, P., Greenwald, I., Fixsen, W., and Ellis, H. (1984). Mutations that affect neural cell lineages and cell fates during the development of the nematode *Caenorhabditis elegans. Cold Spring Harbor Symp. Quant. Biol.* **48,** 453–463.

Horvitz, R., and Sulston, J. (1980). Isolation and genetic characterization of cell-lineage mutants of the nematode C. elegans. *Genetics* **96**, 435–454.

Hunter, C., and Wood, W. (1990). The *tra-1* gene determines sexual phenotype cell autonomously in *C. elegans*. *Cell (Cambridge, Mass.)* **63**, 1193–1204.

Hunter, C., and Wood, W. (1992). Evidence from mosaic analysis of the masculinizing gene *her-1* for cell interactions in *C. elegans* sex determination. *Nature (London)* **355**, 551–555.

Hutter, H., and Schnabel, R. (1994). *glp-1* And inductions establishing embryonic axes in *C. elegans*. *Development (Cambridge, UK)* **120**, 2051–2064.

Hutter, H., and Schnabel, R. (1995). Specification of anterior–posterior differences within the AB lineage in the *C. elegans* embryo: A polarising induction. *Development (Cambridge, UK)* **121**, 1559–1568.

Hyman, A. (1989). Centrosome movement in the early divisions of *Caenorhabditis elegans:* A cortical site determining centrosome position. *J. Cell Biol.* **109**, 1185–1193.

Hyman, A., and White, J. (1987). Determination of cell division axes in the early embryogenesis of *Caenorhabditis elegans*. *J. Cell Biol.* **105**, 2123–2135.

Kaletta, T., Schnabel, H., and Schnabel, R. (1997). Binary specification of the embryonic lineage in *Caenorhabditis elegans*. *Nature (London)* **390**, 294–298.

Kimble, J. (1981). Alterations in cell lineage following laser ablation of cells in the somatic gonad of *Caenorhabditis elegans*. *Dev. Biol.* **87**, 286–300.

Kimble, J., and Hirsh, D. (1979). The postembryonic cell lineages of the hemaphrodite and male gonads in *Caenorhabditis elegans*. *Dev. Biol.* **70**, 396–417.

Kipreos, E. T., Lander, L. E., Wing, J. P., He, W. W., and Hedgecock, E. M. (1996). *cul-1* Is required for cell cycle exit in *C. elegans* and identifies a novel gene family. *Cell (Cambridge, Mass.)* **85**, 829–839.

Knoblich, J., Jan, L., and Jan, Y. (1995). Asymmetric segregation of Numb and Prospero during cell division. *Nature (London)* **377**, 624–627.

Kraut, R., Chia, W., Jan, L., Jan, Y., and Knoblich, J. (1996). Role of *inscuteable* in orienting asymmetric cell divisions in *Drosophila*. *Nature (London)* **383**, 50–55.

Labouesse, M., Hartwieg, E., and Horvitz, H. R. (1996). The *Caenorhabditis elegans* LIN-26 protein is required to specify and/or maintain all non-neuronal ectodermal cell fates. *Development (Cambridge, UK)* **122**, 2579–2588.

Lin, R., Thompson, S., and Priess, J. (1995). *pop-1* Encodes an Hmg box protein required for the specification of a mesoderm precursor in early *C. elegans* embryos. *Cell (Cambridge, Mass.)* **83**, 599–609.

Lin, R., Hill, R. J., and Priess, J. R. (1998). POP-1 and anterior–posterior fate decisions in *C. elegans* embryos. *Cell (Cambridge, Mass.)* **92**, 229–239.

Mello, C., Draper, B., and Priess, J. (1994). The maternal genes *apx-1* and *glp-1* and establishment of dorsal–ventral polarity in the early *C. elegans* embryo. *Cell (Cambridge, Mass.)* **77**, 95–106.

Priess, J. R., and Hirsh, D. I. (1986). *Caenorhabditis elegans* morphogenesis: The role of the cytoskeleton in elongation of the embryo. *Dev. Biol.* **117**, 156–173.

Priess, J., and Thomson, J. N. (1987). Cellular interactions in early *Caenorhabditis elegans* embryos. *Cell (Cambridge, Mass.)* **48**, 241–250.

Rhyu, M., Jan, L., and Jan, Y. (1994). A symmetric distribution of numb protein during division of the sensory organ precursor cell confers distinct fates to daughter cells. *Cell (Cambridge, Mass.)* **76**, 477–491.

Rocheleau, C. E., Downs, W. D., Lin, R., Wittman, C., Bei, Y., Cha, Y.-H., Ali, M., Priess, J., and Mello, C. C. (1997). Wnt signaling and an APC-related gene specify endoderm in early *C. elegans* embryos. *Cell (Cambridge, Mass.)* **90**, 5567.

Ruvkun, G. (1997). Patterning of the nervous system. *In* "C. elegans II" (D. L. Riddle, T. Blumenthal, B. J. Meyer, and J. R. Priess, eds.), pp. 543–581. Cold Spring Harbor Laboratory, Cold Spring Harbor, New York.

Salser, S. J., Loer, C. M., and Kenyon, C. (1993). Multiple HOM-C gene interactions specify cell fates in the nematode central nervous system. *Genes Dev.* **7**, 1714–1724.

Sawa, H., Lobel, L., and Horvitz, H. R. (1996). The *Caenorhabditis elegans* gene *lin-17*, which is required for certain asymmetric cell divisions, encodes a putative seven-transmembrane protein similar to the *Drosophila* Frizzled protein. *Genes Dev.* **10**, 2189–2197.

Shen, C., Jan, L. Y., and Jan, Y. N. (1997). Miranda is required for the asymmetric localization of Prospero during mitosis in *Drosophila*. *Cell (Cambridge, Mass.)* **90**, 449–458.

Sulston, J. (1983). Neuronal cell lineages in the nematode C. elegans. *Cold Spring Harbor Symp. Quant. Biol.* **48**, 443–452.

Sulston, J., and Horvitz, H. (1977). Post-embryonic cell lineage of the nematode *Caenorhabditis elegans*. *Dev. Biol.* **56**, 110–156.

Sulston, J., Albertson, D., and Thomson, J. (1980). The *Caenorhabditis elegans* male: Postembryonic development of nongonadal structures. *Dev. Biol.* **78**, 542–576.

Sulston, J., Schierenberg, E., White, J., and Thomson, J. (1983). The embryonic cell lineage of the nematode *Caenorhabditis elegans*. *Dev. Biol.* **100**, 64–119.

Thomas, C., DeVries, P., Hardin, J., and White, J. (1996). Four-dimensional imaging: Computer visualization of 3D movements in living specimens. *Science* **273**, 603–607.

Thorpe, C. J., Schlesinger, A., Carter, J. C., and Bowerman, B. (1997). Wnt signaling polarizes an early *C. elegans* blastomere to distinguish endoderm from mesoderm. *Cell (Cambridge, Mass.)* **90**, 695–705.

White, J., and Strome, S. (1996). Cleavage plane specification in *C. elegans*: How to divide the spoils. *Cell (Cambridge, Mass.)* **84**, 195–198.

Wilkinson, H., Fitzgerald, K., and Greenwald, I. (1994). Reciprocal changes in expression of the receptor *lin-12* and its ligand *lag-2* prior to commitment in a *C. elegans* cell fate decision. *Cell (Cambridge, Mass.)* **79**, 1187–1198.

Williams-Masson, E. M., Malik, A. N., and Hardin, J. (1997). An actin-mediated two-step mechanism is required for ventral enclosure of the *C. elegans* hypodermis. *Development (Cambridge, UK)* **124**, 2889–2901.

Wood, W. B. (1988). Embryology. *In* "The nematode *Caenorhabditis elegans*" (W. B. Wood, ed.), pp. 215–242. Cold Spring Harbor Laboratory, Cold Spring Harbor, New York.

Wood, W. B. (1991). Evidence from reversal of handedness in *C. elegans* embryos for early cell interactions determining cell fates. *Nature (London)* **349**, 536–538.

Wood, W. B. (1998). Handed asymmetry in nematodes. *Semin. Cell Dev. Biol.* **9**, 53–60.

Wood, W., and Johnson, T. (1994). Stopping the clock. *Curr. Biol.* **4**, 151–153.

Wood, W., and Kershaw, D. (1991). Handed asymmetry, handedness reversal and mechanisms of cell fate determination in nematode embryos. *In* "Biological Asymmetry and Handedness" (G. R. Bock, and J. Marsh, eds), pp. 143–164. Ciba Foundation Symposia, Vol. 162. Wiley, Chichester.

7

Maternal Control of Polarity and Patterning during Embryogenesis in the Nematode Caenorhabditis elegans

BRUCE BOWERMAN
Institute of Molecular Biology
University of Oregon
Eugene, Oregon 97403

I. Polarization of the 1-Cell Stage Zygote and the Establishment of an Anterior–Posterior Body Axis

Genetic screens for recessive, maternal-effect, embryonic-lethal mutations have identified many key loci that regulate pattern formation in the cellularized *Caenorhabditis elegans* embryo (Table I). Before discussing the functional requirements for the maternal genes identified thus far, it is useful to review first some of the experimental embryology that has set the stage for a genetic and molecular understanding of the mechanisms that control pattern formation during embryogenesis in *C. elegans*.

A. Sperm Entry and the Microfilament-Dependent Establishment of an a–p Axis

Sperm entry appears to provide the initial asymmetric cue that polarizes the *C. elegans* zygote and thereby determines the anterior–posterior body axis (Goldstein and Hird, 1996). After entering an oocyte, the sperm pronucleus and its accompanying centriole appear to generate a cytoplasmic flux that pushes the male pronucleus and centriole to the nearest end of the oblong embryo, with that end becoming the posterior pole (Fig. 1). If an oocyte has polarity, it apparently can be overridden by sperm entry.

    After fertilization, three anterior–posterior (a–p) asymmetries become evident during the first cell cycle

TABLE I
Mutationally Identified Maternal Loci in *Caenorhabditis elegans*: Gene Names and Molecular Identities[a]

Gene	Name	Molecular identity
apx-1	anterior pharynx-defective	Delta-like transmembrane ptn, putative GLP-1 ligand
glp-1	germline proliferation-defective	Notch-like transmembrane ptn, putative receptor for APX-1
let-99	lethal	?
lit-1	loss of intestine	?
mes-1	maternal-effect sterile	?
mex-1	muscle excess	TIS-11 like Zn^{2+} finger ptn
mex-3	muscle excess	Two KH domains, putative RNA binding ptn
mom-1	more mesoderm	Porcupine-like, putative ER transmembrane ptn required for Wnt processing/section
mom-2	more mesoderm	Wnt-like; putative secreted glycoprotein ligand
mom-3	more mesoderm	?
mom-4	more mesoderm	?
mom-5	more mesoderm	Frizzled-like; putative 7-pass transmembrane ptn and Wnt receptor
pal-1	posterior alae-defective	Caudal-like homeodomain ptn, putative transcription factor
par-1	partitioning-defective	Putative ser/thr kinase, binds a nonmuscle conventional myosin
par-2	partitioning-defective	Novel, ATP binding site
par-3	partitioning-defective	Three PDZ domains
par-4	partitioning-defective	putative ser/thr kinase
par-5	partitioning-defective	?
par-6	partitioning-defective	?
pie-1	pharynx and intestine excess	TIS-11-like Zn^{2+} finger ptn, putative transcription factor
pop-1	posterior pharynx-defective	Single HMG domain ptn, putative transcription factor

[a]See text for references.

(Fig. 1). The first asymmetry is the cytoplasmic flux that occurs posteriorly shortly after fertilization and and appears to be initiated by the sperm pronucleus and centriole. Time-lapse videomicroscopy studies have shown that cytoplasmic yolk droplets flow anteriorly in the cortical cytoplasm, while more internal cytoplasm flows posteriorly (Hird and White, 1993). This cytoplasmic flux occurs predominantly in the posterior half of the 1-cell zygote, and similar cytoplasmic fluxes may occur in later blastomeres. A second asymmetry appears to depend on the cytoplasmic flux generated by sperm entry: the posterior localization of cytoplasmic structures called P-granules. Although their function remains largely unknown, P-granules are ribonucleoprotein complexes present specifically in germline precursors and may specify germline fate (Strome and Wood, 1983; Seydoux and Fire, 1994; Draper *et al.*, 1996; Guedes and Priess, 1996; Mello *et al.*, 1996). P-granules intially are present throughout the cytoplasm of the oocyte and the 1-cell zygote, but after fertlization they are actively segregated to the cortical cytoplasm at the posterior pole before the first embryonic mitosis (Strome and Wood, 1983; Wolf *et al.*, 1983; Hird *et al.*, 1996). A third prominent asymmetry in the 1-cell zygote is the posterior displacement of the first mitotic spindle, which results in the production of two daughters with very different fates: a smaller posterior blastomere called P_1, and a larger anterior blastomere called AB

(Sulston *et al.*, 1983; Albertson, 1984). The mosaic nature of pattern formation is evident on completion of the first cleavage: if P_1 and AB are separated by experimental manipulation, they each are capable of producing only partial embryos (Priess and Thomson, 1987).

Intriguingly, the asymmetric positioning of the first mitotic spindle, the localization of P-granules, and the posterior flux of cytoplasm all require functional actin microfilaments. Disruption of microfilaments by treatment with cytochalasin D prevents the cytoplasmic flux from occuring, although disruption of microtubules with nocodazole treatment does not (Hird and White, 1993). In addition, treating 1-cell stage wild-type embryos with brief pulses of cytochalasin D prevents P-granule segregation (Hill and Strome, 1988, 1990) and disrupts the posterior positioning of the first mitotic spindle (Hill and Strome, 1990). Moreover, the time period during which cytochalasin treatment can disrupt spindle positioning and P-granule localization corresponds precisely to the time of cytyoplasmic flux (Hill and Strome, 1990; Hird *et al.*, 1996; Kemphues and Strome, 1997). These results suggest that after fertilization, actin-dependent processes generate a cytoplasmic flux required for localizing P-granules, and possibly for posteriorly displacing the first mitotic spindle, establishing early differences between the anterior and posterior poles of the zygote that presumably are related to the different developmental potentials of P_1 and AB.

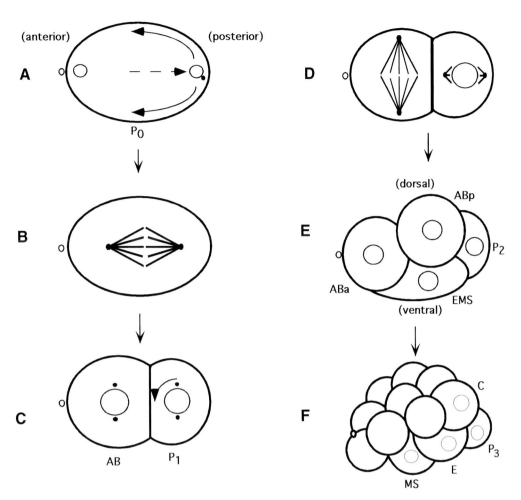

Figure 1 Early asymmetries and blastomere names in the C. *elegans* embryo at the 1-cell (A, B), 2-cell (C, D), 4-cell stages (E), and 12-cell (F) stages. (A) The position of the sperm pronucleus (large open circle) and its accompanying centriole (small filled circle) define the posterior pole of the zygote and initiate a cytoplasmic flux in the posterior half of the zygote (curved and dashed arrows). One of two polar bodies (small open circle) is usually present at the anterior pole, near the oocyte pronucleus (large open circle). The localization of P-granules to the posterior cortex of P_0 is not shown (see text). (B) After pronuclear congression (not shown), the first mitotic spindle becomes displaced slightly posteriorly. (C) The first embryonic cleavage produces a smaller posterior daughter called P_1 and a larger anterior daughter called AB. AB is the first founder cell to be born. Both mitotic spindles in the 2-cell stage embryo initially set up transversely (small filled circles), but the P_1 spindle rotates before the completion of mitosis to lie along the long axis (curved arrow). (D) P_1 divides slightly after AB, again with a posteriorly displaced spindle, while AB divides transversely and equally. (E) In a 4-cell stage embryo, the two daughters of AB, called ABa and ABp, are of equal size and initially have equivalent developmental potential. The two P_1 daughters, called P_2 and EMS, are of different size and are born with different fates. The dorsal–ventral axis is defined by the position of ABp (dorsal) and EMS (ventral). (F) Three more somatic founder cells are present at the 12-cell stage: C, E, and MS. P_3 divides to produce the remaining somatic founder cell, D, and the germline progenitor P_4 at the 24-cell stage (not shown). Nuclei are indicated by open circles in blastomeres but omitted in AB descendants at the 12-cell stage.

B. The *par* Genes and Polarization of the Embryonic Actin Cytoskeleton

Some insight into how regulation of the actin cytoskeleton polarizes the 1-cell stage C. *elegans* zygote on fertilization has come from the identification of six maternally expressed *par* genes, *par-1* through *par-6* (Kemphues *et al.*, 1988; Watts *et al.*, 1996; for recent reviews of the *par* genes, see Guo and Kemphues, 1996a; Kemphues and Strome, 1997). Mutational inactivation of the *par* genes results in defects similar to those observed after treating 1-cell stage wild-type em-

bryos with pulses of cytochalasin (Kemphues and Strome, 1997). The cytoplasmic flux triggered by sperm entry appears defective to some extent in *par-1* through *par-4* mutants, with *par-1* and *par-4* perhaps exhibiting less severe defects (Kirby *et al.*, 1990), and all the *par* mutants are defective in the proper segregation of P-granules (Kemphues *et al.*, 1988; Kirby *et al.*, 1990; Kemphues and Strome, 1997). The first mitotic spindle fails to move posteriorly in all but *par-4* mutant embryos, producing equal-sized anterior and posterior 2-cell stage daughters. Furthermore, in all six *par* mutants the 2-cell stage blastomeres divide synchronously and equally, in contrast to the asynchronous and unequal divisions in wild-type embryos. Finally, *par* mutants exhibit a characteristic defect in the orientations of 2-cell stage mitotic spindle axes (Kemphues *et al.*, 1988; Kemphues and Strome, 1997; see Fig. 2). In 2-cell stage *par-1* mutant embryos, the posterior and anterior blastomeres have longitudinally and transversely oriented spindles, respectively, as in wild-type embryos, although both divide synchronously. In 2-cell stage *par-2* and *par-5* mutant embryos, the spindles divide transversely and synchronously, and in *par-3* and *par-6* mutants, both divide longitudinally and synchronously.

Because *par* mutants are defective in cytoplasmic flux, P-granule localization, and spindle positioning (the same processes disrupted by treating wild-type embryos with cytochalasin) the *par* genes may regulate pattern formation at least in part by interacting with and perhaps polarizing the actin cytoskeleton in response to sperm entry (Guo and Kemphues, 1996a; Kemphues and Strome, 1997). Consistent with this hypothesis, studies have shown that the C-terminus of PAR-1 binds a conventional nonmuscle myosin called NMY-2 that itself is required for proper polarization of the early embryo. Inactivation of NMY-2 results in production of

embryos with transversely oriented 2-cell stage mitotic spindles, similar to the defects observed in *par-2* and *par-5* mutant embryos (Guo and Kemphues, 1996b).

How the ability of PAR-1 to bind myosin relates to the *par-2*-like spindle orientation defect caused by inactivation of NMY-2 is not known. But additional insights into how regulation of the actin cytoskeleton affects polarity are likely to come from the identification of 17 oocyte proteins that bind filamentous actin as determined by F-actin affinity column chromatography (Aroian *et al.*, 1997). Antibodies to three such proteins show distinct localization patterns in the early embryo. CABP1 localizes to the actin-rich cortex throughout all early blastomeres. CAPB14 is remarkably dynamic, cycling from the nucleus during prophase, to the cortex during metaphase, and to the cleavage furrow during cytokinesis. CABP11 is localized to the cortex but only in the anterior part of the embryo, indicating that asymmetries involving proteins that physically interact with actin microfilaments exist as early as the time of pronuclear congression in the 1-cell zygote. Intriguingly, cortical localization of CABP11 requires *par-3* function. With the genetic and biochemical methods now available for identifying and inactivating gene products in C.

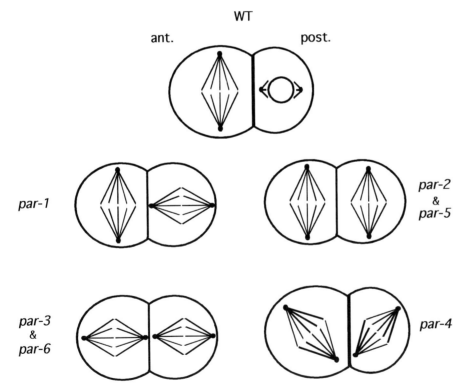

Figure 2 2-cell stage mitotic spindle orientations in wild-type (WT) and in *par* mutant embryos. In the WT embryo (top), the larger anterior blastomere divides transversely before the posterior blastomere divides longitudinally. In *par-1* mutant embryos, the posterior and anterior blastomeres divide with normal spindle orientations but synchronously. In *par-2* and *par-5* mutant embryos, both blastomeres divide transversely and synchronously. In *par-3* and *par-6* mutant embryos, both divide longitudinally and synchronously. In *par-4* mutant embryos, both blastomeres divide with random orientations, one arbitrary example being shown. In all but *par-4* mutant embryos, the first cleavage produces two blastomeres roughly equal in size. In *par-4* mutants, the first cleavage produces a smaller posterior and larger anterior blastomere, as in wild-type embryos, but both blastomeres subsequently divide synchronously.

elegans (Aroian *et al.,* 1997; Jansen *et al.,* 1997; Han, 1997), the discovery of important mechanistic links between the PAR proteins, the actin cytoskeleton, and embryonic polarity in *C. elegans* appear close at hand.

C. Polarized Cortical Distributions of the PAR Proteins

The genes *par-1, par-2, par-3,* and *par-4* have been molecularly cloned. The encoded proteins are cytoplasmic but substantially enriched at the cytoplasmic cortex (Fig. 3). PAR-1 contains a predicted N-terminal ser/thr kinase domain, and a C-terminal domain that interacts with the nonmuscle conventional myosin NMY-2 (Guo and Kemphues, 1995, 1996b). Before fertilization PAR-1 is not polarized in distribution. But after fertilization, by the time the maternal and paternal pronuclei meet, PAR-1 is present at the cortex only posteriorly in the 1-cell zygote. PAR-2 is a protein of unknown function

PAR-1, PAR-2 & PAR-3

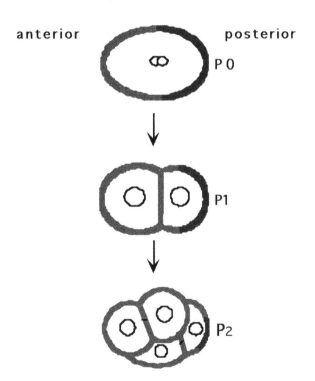

Figure 3 Polarized distributions of PAR-1, PAR-2, and PAR-3 cortically localized proteins. PAR-1 and PAR-2 both are enriched in the posterior part of cytoplasmic cortex by late in the 1-cell stage (red). PAR-3 also is enriched cortically but only in the anterior portion of P_0 (blue). PAR-1, PAR-2, and PAR-3 continue to show a polarized distribution in the germline precursors, P_1 and P_2, at the 2-cell and 4-cell stages of embryogenesis, respectively. This polarized distribution is also maintained in the P_3 daughter of P_2 (not shown). PAR-1 and PAR-2 are absent in somatic blastomeres, while PAR-3 is enriched cortically throughout the entire cortex of all somatic blastomeres. All three PAR proteins are also present cytoplasmically throughout the embryo (not shown).

with a putative ATP binding site and a zinc binding domain of the "RING finger" class (Levitan *et al.,* 1994). Like PAR-1, PAR-2 is enriched in the cortex but only in the posterior part of the zygote by late in the 1-cell stage (Boyd *et al.,* 1996). PAR-3 contains three PDZ domains that presumably mediate protein–protein interactions. PAR-3 is present cortically only in the anterior part of the 1-cell zygote—its posterior boundary roughly coincides with the anterior boundary of PAR-1 and PAR-2 (Etemad-Moghadam and Kemphues, 1995). PAR-4 also contains a ser/thr kinase domain, different from that of PAR-1. PAR-4 is unique in having a nonpolarized distribution, being localized to the cortex throughout the zygote and in all early blastomeres (K. Kemphues, personal communication). The polarized distributions of PAR-1, PAR-2, and PAR-3 are inherited by the germline precursors P_1, P_2, and P_3, but not by P_4, the final germline progenitor (Guo and Kemphues, 1995; Kemphues and Strome, 1997). While it is not known if the PAR proteins are required for maintaining polarity in germline precursors, the temperature sensitive (ts) periods for ts alleles of *par-2* and *par-4* are over by the end of the 1-cell stage, consistent with *par* functions being required only during the first zygotic cell cycle (Kemphues and Strome, 1997).

D. Interactions of the *par* Genes

The distributions of the PAR proteins in *par* mutant embryos indicate that interactions among the *par* genes are important for regulating embryonic polarity. Four interactions have been noted. First, in *par-2* mutant embryos, PAR-1 is present cytoplasmically but with no enrichment at the cortex (Boyd *et al.,* 1996). Also, PAR-2 and PAR-3 depend on each other for their polarized distributions (Etemad-Moghadam and Kemphues, 1995; Boyd *et al.,* 1996; Kemphues and Strome, 1997). In *par-2* mutants, cortical PAR-3 extends posteriorly, and in *par-3* mutants cortical PAR-2 extends anteriorly. Finally, the cortical localization of PAR-3 requires *par-6* function (Watts *et al.,* 1996). None of the other PAR proteins require *par-4* function for their polarized distribution, and PAR-4 appears normal in all other *par* mutants (Kemphues and Strome, 1997).

Genetic studies, together with the polarized distributions of the PAR proteins just described, suggest that the only function of *par-2* may be to limit the posterior extension of cortical PAR-3. This conclusion is based on the observation that eliminating one copy of *par-6* rescues *par-2* mutant embryos: *par-6(−)/par-6(+)*; *par-2 (−)/par-2(−)* mothers produce viable embryos (Watts *et al.,* 1996). As cortical PAR-3 in *par-2* embryos forms a gradient that fades posteriorly, one possible explanation for the suppression of *par-2* is that reducing the dose of *par-6* reduces the posterior extension of PAR-3 sufficiently for normal development to occur (Watts *et al.,*

1996). Finally, it should be noted that there is little or no evidence that the cortical enrichment of the PAR proteins is important for their function—cortical localization could represent a consequence of their function rather than a prerequisite.

E. *par-2* and *par-3* Interact to Pattern Mitotic Spindle Orientations

The observations that the proper cortical localization of some PAR proteins require *par* gene functions suggest that some of the *par* genes interact to regulate polarity. Evidence for *par* genes functioning in a linear pathway has come from studies of *par-2* and *par-3* and their requirements for proper orientation of the mitotic spindles in P_1 and AB (Kemphues *et al.*, 1988; Cheng *et al.*, 1995). In wild-type 2-cell stage embryos, both P_1 and AB initially set up transversely oriented mitotic spindles. AB continues to divide transversely, slightly ahead of P_1 in timing. Just before P_1 divides, though, its mitotic spindle rotates to lie along the longitudinal axis (Hyman and White, 1987; Hyman, 1989; Waddle *et al.*, 1994). In *par-2* mutants, both the P_1 and AB spindles orient transversely, while they both orient longitudinally in *par-3* mutants (Kemphues *et al.*, 1988). Because both spindles orient longitudinally in *par-2;par-3* double mutants, neither *par-2* nor *par-3* are required for spindle rotation (Cheng *et al.*, 1995). Rather, *par-3* is required to prevent spindle rotation in AB, while *par-2* appears to prevent *par-3* from functioning in P_1, thereby permitting some other process to rotate the P_1 spindle.

Because *par-2* restricts cortical PAR-3 mostly to the AB blastomere, it has been proposed that cortical PAR-3 may prevent spindle rotation in AB by interacting with astral microtubules (Etemad-Moghadam and Kemphues, 1995; Guo and Kemphues, 1996a; Kemphues and Strome, 1997). In this model, the presence of cortical PAR-3 only at the anterior of P_1 presumably would be insufficient to prevent spindle rotation. Mutations in *par-2* result in PAR-3 being present throughout the cortex in both P_1 and AB, consistent with PAR-3 mediating a stabilization of the transverse spindles in both 2-cell stage blastomeres in *par-2* mutant embryos. However, mutations recently were identified in another maternal gene named *let-99*. In about 50% of *let-99* embryos, the AB spindle orients longitudinally and the P_1 spindle orients transversely, the opposite of the wild-type pattern (Rose and Kemphues, 1998). Remarkably PAR-1, PAR-2, and PAR-3 all show a normal polarized distribution in *let-99* mutant embryos (Rose and Kemphues, 1998). Therefore spindle rotation in AB is not prevented by PAR-3, and a transverse P_1 spindle can be stabilized without altering the distribution of PAR-3, indicating that in *let-99* mutant embryos PAR-3 is neither necessary nor sufficient to prevent rotation. Alternatively, *par-*

2 and *par-3* might polarize the distribution of other factors, resulting in only P_1 acquiring the machinery necessary for spindle capture and rotation.

F. Independent Functions of the *par* Genes

While the interactions of *par-2* and *par-3* indicate that some *par* genes function in linear pathways, the *par* genes more often appear to act independently of each other to regulate pattern formation. This conclusion is based in part on the dissimilar phenotypes of the *par* mutants (Kemphues *et al.*, 1988; Bowerman *et al.*, 1997), and on the different distributions of four regulatory proteins in *par-1*, *par-2*, *par-3*, and *par-4* mutant embryos (Crittenden *et al.*, 1996; Bowerman *et al.*, 1997). Two of these regulatory proteins, GLP-1 and MEX-3, are present at high levels only in anterior blastomeres at the 2-cell and 4-cell stages. Two others, SKN-1 and PAL-1, are present at high levels only in posterior blastomeres at the 4-cell stage (Fig. 4). No correlation is seen in how the distributions of these four proteins respond to mutations in the *par* genes. For example, mutations in *par-1* result in SKN-1 being present in all 4-cell stage blastomeres but also result in a complete absence of PAL-1. Mutations in *par-2* do not affect MEX-3 distribution, but GLP-1 is present in all 4-cell stage blastomeres in many *par-2* mutants. Thus mutations in the *par* genes uncouple the mechanisms that localize different regulatory molecules. Perhaps the *par* genes act to polarize the distribution of intermediate factors that ultimately act more specifically to restrict regulators like GLP-1, MEX-1, PAL-1, and SKN-1 to specific blastomeres (see below for further discussion of these four regulatory factors).

G. Homologs of the *par* Genes: Cell Polarity in Yeast and in Mammalian Epithelia

Homologs of *par-1* have been identified in yeast and in mammals (Levin *et al.*, 1987; Levin and Bishop, 1990; Drewes *et al.*, 1997; Bohm *et al.*, 1997). The *Schizosaccharomyces pombe* gene *kin1*$^+$ is required for polarized cell growth (Levin and Bishop, 1990). Mammalian homologs of *par-1*, called MARK and mPAR-1 proteins, phosphorylate microtubule associated proteins and may regulate microtubule structure and function (Drewes *et al.*, 1997). Intriguingly, mPAR-1 proteins are localized to the basolateral cortex in epithelial cells, and overexpression of a dominant negative form of mPAR-1 causes a loss of lateral adhesion and a disruption of polarity in cultured epithelial cells (Bohm *et al.*, 1997). Moreover, another emerging family of proteins that include *Drosophila discs large* and the tight junction proteins

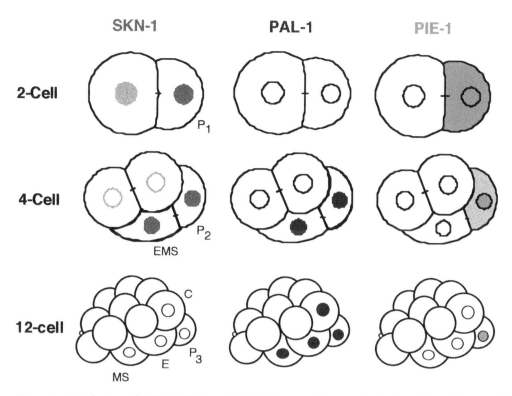

Figure 4 Distributions of SKN-1, PAL-1, and PIE-1, three putative transcriptional regulators that specify the fates of P_1 descendants in the early embryo. SKN-1 accumulates to higher levels in P_1 than in AB, persists in P_1 descendants until the 8-cell stage (not shown), and is undetectable by the 12-cell stage. PAL-1 is detectable first at the 4-cell stage, only in P_2 and EMS, and persists at high levels in P_1 descendants until the 12-cell stage and beyond. PIE-1 is present in the cytoplasm and nucleus at the 2-cell stage but becomes localized to the nuclei of germline precursors in subsequent stages. PIE-1 is also associated with P-granules in germline precursors (not shown). Nuclei are indicated by open or filled circles but are omitted in AB descendants at the 12-cell stage.

Z0-1 and Z0-2 share significant similarity with PAR-3 and also contain PDZ domains (Kurzchalia and Hartmann, 1996). In epithelial cells, Z0-1 and Z0-2 are restricted to the tight junctions, abutting the basolateral distribution of mPAR-1 (Bohm *et al.*, 1997). The involvement of these related kinases in regulating polarity in yeast and in mammalian epithelial cells suggest that the functions of the *par* genes are likely to be of general importance in studies of cell polarity and regulation of the cytoskeleton.

H. The *mes-1* Gene and a Reversal of Polarity during Germline Development

The maternal gene *mes-1* may play an important role in the regulation of cell polarity in germline precursors in later stages of embryogenesis. Early studies of *C. elegans* embryogenesis first suggested that germline precursors undergo a polarity reversal beginning with the division of P_2 (Schierenberg, 1987). If P_1 is extruded from the eggshell and allowed to divide, P_2 is born posteriorly. Subsequently, P_2 divides to produce a smaller P_3 anteriorly, and P_3 makes a smaller P_4 anteriorly.

When constrained at the narrow end of a rigid eggshell, the germline precursors P_3 and P_4 are forced to adopt more ventral and anterior positions relative to their somatic sisters even though their parents divide along a reversed a–p axis relative to P_1 and P_0. The earliest detectable defect in *mes-1* mutants occurs when P_2 divides and the mitotic spindle fails to orient properly (Strome *et al.*, 1995; Hird *et al.*, 1996). The defective positioning of the P_2 spindle in *mes-1* mutant embryos may reflect a specific role for *mes-1* in the reversal of germline polarity, perhaps after the roles of the *par* genes are complete.

II. Specifying the Fates of Individual Blastomeres in the Early *Caenorhabditis elegans* Embryo

A. The Generation of Founder Cells

Before discussing maternal genes that act more specifically than the *par* genes to control pattern formation, an overview of the early embryonic cells, or blastomeres, and the characteristic cell types they each produce is necessary (Sulston *et al.*, 1983; see Fig. 1). The 1-cell zygote, called P_0, divides to make a smaller posterior blastomere P_1 and a larger anterior blastomere AB. AB then divides before P_1 to produce two daughters of equal size, ABa and ABp, that subsequently divide synchronously. Based on the roughly synchronous divisions of its descendants, AB is considered the first founder cell. P_1 and its descendants undergo a series of asymmetric divisions to produce blastomeres differing in size and in the timing of their cell cycles. These early unequal cleavages generate a group of six so-called founder cells, born from the 2-cell to the 24-cell stage. Five founder cells called AB, MS, E, C, and D produce somatic cells, whereas P_4 is the germline progenitor. The descendants of each founder cell exhibit somewhat synchronous cell cycle times and in sum produce the 558 surviving cells that form a hatched larva. Founder cells descendants are named according to their position at birth relative to their sisters. For example, ABal is the left hand daughter of ABa, and ABa is the anterior daughter of AB.

Three founder cells produce descendants that all share roughly the same fate. P_4 divides to make Z2 and Z3, the two embryonic progenitors of the germline that subsequently proliferate during larval development, D produces only body wall muscle cells, and E makes all of the intestinal cells. The remaining founder cells produce more complex patterns of cell fate. For example, MS generates body wall muscle and the somatic gonad in addition to several cell types that form the posterior part of the pharynx, a neuromuscular organ used for feeding. C produces body wall muscle cells and most of the dorsal epidermis. ABa makes many neurons, some epidermal cells, and the anterior half of the pharynx. ABp also makes many neurons and some epidermal cells, and a number of specialized cell types.

B. Defining the Body Axes

As described earlier, the position of sperm entry determines the a–p body axis in *C. elegans* (Goldstein and Hird, 1996) However, a dorsal–ventral axis is not evident until P_1 and AB are nearly done dividing (Sultson *et al.*, 1983; Priess and Thomson, 1987; see Fig. 1). As the AB spindle elongates, it becomes longer than the eggshell is wide and slants to one side, forcing P_1 to slant in a complementary fashion. This sequence of events results in ABa being the anterior-most 4-cell stage blastomere, while ABp becomes the dorsal-most blastomere. The asymmetric division of P_1 produces the smaller germline progenitor P_2, the posterior-most blastomere in a 4-cell stage embryo, and EMS, the ventral-most blastomere. Thus the dorsal–ventral axis is defined

by the positions of ABp and EMS. Left–right differences become apparent at the 8-cell stage, when for unknown reasons the left-side daughters of ABa and ABp adopt positions more anterior than the right-side daughters (Sulston *et al.*, 1983; Wood, 1991).

C. Extrinsic and Intrinsic Mechanisms Controlling the Specification of Blastomere Identities

Experimental manipulations of wild-type embryos indicate that pattern formation in the *C. elegans* embryo relies extensively on both cell intrinsic ("mosaic") and cell extrinsic ("regulative") mechanisms. The first cleavage of the embryo produces two blastomeres, AB and P_1, that appear to be born with substantially different fates. Their different developmental potentials can be demonstrated by separating them on birth and examining the cell types they each produce in isolation (Priess and Thomson, 1987). If AB is physically removed from a 2-cell stage embryo, P_1 still produces many if not all of the cell types it normally makes, suggesting that P_1 inherits a largely intrinsic ability to develop and that cell signals from AB descendants play at most a minor role in P_1 patterning. In contrast, if P_1 is removed and AB left to develop in isolation, the AB descendants fail to make many cell types they normally produce, such as the anterior pharyngeal cells made by ABa in an intact embryo. Thus AB development appears to depend extensively on cell signals from P_1 descendants, but presumably intrinsic factors that act in AB and its descendants also are important for proper development.

The different relative importance of cell intrinsic and extrinsic mechanisms for patterning the fates of P_1 and AB descendants was demonstrated most clearly by one experiment that has shaped much of the mechanistic thinking about embryogenesis in *C. elegans*. If one uses a micromanipulator to switch the positions of ABa and ABp just as they are born, a normal embryo results with an expected reversal in left–right asymmetry (Priess and Thomson, 1987). This simple experiment showed that ABa and ABp are born with equivalent developmental potential and suggested that cell–cell interactions must distinguish their fates. However, after switching the positions of P_2 and EMS, each P_1 daughter appeared to produce its normal complement of cell types in the wrong place, resulting in morphologically abnormal embryos that failed to hatch (Priess and Thomson, 1987). Thus both cell autonomous and cell nonautonomous mechanisms pattern the early embryo, with AB descendants apparently relying more on cell signaling and P_1 descendants relying more on the asymmetric segregation of intrinsic development potential during early cleavages.

III. Maternal Genes That Specify the Identities of P₁ Descendants

1. skn-1, pal-1, and pie-1 Function

The three maternal genes *skn-1*, *pal-1*, and *pie-1* encode putative transcriptional regulators that appear to act intrinsically to regulate initiation of the genetic programs that define the fates of the four P_1-derived somatic founder cells MS, E, C, and D (Bowerman *et al.*, 1992a, 1993; Hunter and Kenyon, 1996; Mello *et al.*, 1992, 1996). This view is based in part on the observation that in *skn-1;pal-1* double mutant embryos, P_1 frequently fails to produce any differentiated somatic cell types and instead produces many small and apparently undifferentiated descendants (Hunter and Kenyon, 1996). For the *C. elegans* genes discussed in this review, no other double mutant combination examined thus far results in a lack of differentiation; all other mutants instead exhibit trans-fating. As described below, *pie-1* appears to separate the functions of *skn-1* and *pal-1* in space and time and may specify germline fate in P_4, the remaining founder cell derived from P_1.

a. skn-1 Function and EMS Fate.
The *skn-1* mutant embryos lack the endoderm normally made by E and the mesoderm normally made by MS. Instead, E and MS in *skn-1* mutant embryos each produce epidermis (skin) and body wall muscle, a fate similar to that of the P_2 daughter C (Bowerman *et al.*, 1992a). In the absence of SKN-1, the PAL-1 present in E and MS appears to respecify E and MS to adopt C-like fates (Bowerman *et al.*, 1992a; Hunter and Kenyon, 1996). As mentioned above, in the absence of *skn-1* and *pal-1* function, very few or no differentiated somatic cell fates are produce by either P_2 or EMS, suggesting that SKN-1 and PAL-1 act very early to specify the identities of P_1-derived somatic founder cells. Other blastomeres in *skn-1* mutant embryos appear to develop completely normally, with the exception that ABa descendants fail to produce pharyngeal cells (Bowerman *et al.*, 1992a). However, the lack of ABa-derived pharyngeal cells is an indirect consequence of the requirement for *skn-1* function in EMS (Shelton and Bowerman, 1996). In wild-type embryos, MS signals ABa descendants to produce pharyngeal cells at about the 12-cell stage (Priess *et al.*, 1987; Hutter and Schnabel, 1994; Mango *et al.*, 1994a). In the absence of *skn-1* function, MS is incapable of signaling, but a wild-type MS can induce *skn-1* mutant ABa descendants to produce pharyngeal cells (Shelton and Bowerman, 1996). Thus *skn-1* appears to specify EMS fate with respect both to the cell fate patterns it produces and to the ability of MS to signal ABa.

b. pal-1 Function and P₂ Fate.
The *pal-1* gene was first identified by partial loss of function mutations that affect cell fate patterning during larval development but are not lethal (Waring and Kenyon, 1990, 1991). The PAL-1 protein, though, is expressed maternally in P_1 descendants beginning at the 4-cell stage in P_2 and EMS, and elimination of maternal *pal-1* function results in P_2 failing to produce the epidermal and body wall muscle cells normally made by its descendants C and D. P_2 in *pal-1* mutant embryos appears to produce germline normally, but the remaining P_2 descendants appear small and undifferentiated (Hunter and Kenyon, 1996).

c. pie-1 Function and Germline Fate.
Mutational inactivation of *pie-1* results in P_2 adopting a fate nearly identical to that of EMS, producing excess pharynx and intestine (Mello *et al.*, 1992). Because *pie-1; skn-1* double mutant embryos resemble *skn-1* mutant embryos, *pie-1* must act to block the function of the SKN-1 present in P_2. However, P_2 still fails to produce germline in *pie-1; skn-1* double mutant embryos, indicating that *pie-1* also is required to specify germline (Mello *et al.*, 1992). Recent studies of *pie-1* indicate that it may specify germline by acting as a general repressor of transcription, maintaining the germline progenitors in a transcriptionally silent or inert state (Mello *et al.*, 1996; Seydoux *et al.*, 1996; Seydoux and Dunn, 1997).

2. SKN-1, PAL-1, and PIE: Protein Identities and Distributions

At the 4-cell stage, SKN-1 and PAL-1 both are present in the nuclei of P_2 and EMS with little or no protein detectable in ABa and ABp (Bowerman *et al.*, 1993; Hunter and Kenyon, 1996; see Fig. 4). PIE-1 is present only in P_2, in its cytoplasm and nucleus (Mello *et al.*, 1996; see Fig. 4). SKN-1 is a sequence-specific DNA binding protein with a DNA binding domain related to basic regions of bZIP transcription factors (Bowerman *et al.*, 1992a; Blackwell *et al.*, 1994). However, SKN-1 lacks a leucine zipper motif, which in bZIP proteins is essential for dimerization and DNA binding. Instead, SKN-1 terminates immediately after its C-terminal basic region and binds DNA as a monomer (Blackwell *et al.*, 1994; Carroll *et al.*, 1997). By late in the 2-cell stage, SKN-1 accumulates to substantially higher levels in the nucleus of P_1 than in AB, and the levels peak midway through the 4-cell stage in EMS and P_2. At the 8-cell stage, SKN-1 is present at lower levels in all four P_1 descendants and undetectable in AB descendants; it is undetectable in any blastomere by the 12-cell stage. PAL-1 is a homeodomain protein most similar in sequence to *Drosophila* Caudal, which like PAL-1 in *C. elegans* is required for posterior patterning

(Hunter and Kenyon, 1996). PAL-1 is first detectable at the 4-cell stage in the nuclei of P_2 and EMS and remains present in all P_1 descendants until well after the 12-cell stage, when SKN-1 is undetectable. Finally, PIE-1 is a Zn^{2+} finger protein that exhibits remarkable localization properties in the early embryo (Mello *et al.*, 1996). At the 2-cell stage PIE-1 is present in the cytoplasm and nucleus of P_1. During mitosis, it localizes to P_1 centrosomes. As P_1 finishes dividing, PIE-1 leaves the centrosomes, disappearing from the somatic daughter EMS but transiting to the nucleus in P_2. At each division of a germline precursor, PIE-1 localizes to centrosomes and then returns to the nucleus of the germline progenitor as mitosis ends. Finally, PIE-1 also localizes to P-granules in germline precursors, but not to P-granules in oocytes.

The temporal and spatial regulation of SKN-1, PAL-1, and PIE-1 expression may be largely responsible for segregating the activities of SKN-1 and PAL-1 (Hunter and Kenyon, 1996). Even though both PAL-1 and SKN-1 are present at high levels in P_2 and EMS, each acts in only one of these two P_1 descendants to specify somatic founder cell fates. By appearing at high levels before PAL-1, SKN-1 may predominate at the 4-cell stage to specify EMS identity. At the same time, PIE-1 prevents SKN-1 from functioning in P_2 as part of its more general germline repressor function. By the time P_2 divides to produce a C daughter free of PIE-1, SKN-1 is barely detectable and PAL-1 is present at high levels. Perhaps PAL-1 can then override the fading levels of SKN-1 to specify C identity and, after P_3 dividies, D identity. It is possible that as yet unidentified genes may provide additional means to regulate the time and place of SKN-1 and PAL-1 function. For example, a factor localized to EMS could serve to block PAL-1 function in EMS, much as PIE-1 blocks SKN-1 in P_2. Finally, as E, MS, C, and D all have very different fates, other genes must be necessary to distinguish E from MS, and C from D.

3. Spatial Regulation of SKN-1, PIE-1, and PAL-1 Expression

Because the maternal mRNAs for *pal-1*, *pie-1*, and *skn-1* are distributed uniformly throughout early embryos (Seydoux and Fire, 1994; Hunter and Kenyon, 1996; Mello *et al.*, 1996), either translational regulation or differences in protein stability must account for their localized expression. The maternal genes *par-1*, *par-3*, and *mex-1* are required for the proper localization of SKN-1 (Bowerman *et al.*, 1993, 1997), *mex-1* is required for proper localization of PIE-1 (Guedes and Priess, 1996), and *par-1*, *par-3*, *par-4*, and *mex-3* are required for the proper localization of PAL-1 (Hunter and Kenyon, 1996; Bowerman *et al.*, 1997).

a. SKN-1 Localization. In both *par-1* and *par-3* mutant embryos, SKN-1 is present at roughly equal levels in all 4-cell stage blastomeres, and all 4-cell stage blastomeres produce SKN-1-dependent pharyngeal cells and body wall muscle cells (Bowerman *et al.* 1993, 1997). Because of the early polarity and cleavage defects in *par* mutant embryos, it seems likely that mislocalization of SKN-1 is an indirect effect of *par* gene functions. The gene products that might mediate these interactions remain unidentified, however, and it is still possible that the effects are direct.

In many respects, *mex-1* mutant embryos resemble *par* mutant embryos but appear to have less extensive losses of a–p polarity, suggesting that the effect on SKN-1 expression might be more direct than for *par-1* and *par-3* (Mello *et al.*, 1992; Schnabel *et al.*, 1996). In some *mex-1* mutant embryos the first embryonic cleavage produces two equally sized blastomeres, and P-granule partitioning is partially defective throughout the early cleavages of all mutant embryos. In *mex-1* mutants, the four granddaughters of AB often adopt fates similar to MS (Mello *et al.*, 1992), although their fates appear to be a mosaic of AB and MS lineages (Schnabel *et al.*, 1996). Thus transformations in blastomere identities in the early *C. elegans* are by no means absolute and can instead be partial and mosaic. Consistent with the MS-like fates produced by AB, SKN-1 protein is mislocalized in *mex-1* mutant embryos, accumulating to high levels in ABa and ABp at the 4-cell stage (Bowerman *et al.*, 1993). Thus *mex-1*, like *par-1* and *par-3*, is required for restricting the accumulation of high levels of SKN-1 to P_1 descendants. However, other blastomeres also exhibit fate transformations in *mex-1* mutant embryos, indicating that *mex-1* either is required for multiple processes or perhaps only indirectly regulates SKN-1 distribution. In some *mex-1* mutant embryos, E fails to produce gut and instead adopts a C-like fate. Moreoever, P_3 in *mex-1* mutants divides equally to produce two D-like daughters, in contrast to a wild-type P_3 which divides unequally to make the smaller germline precursor P_4 and the body wall muscle precursor D. Thus, *mex-1* mutant embryos lack germline, and they produce excess body wall muscle from both P_1 and AB descendants. To summarize, some *mex-1* mutant embryos exhibit extensive defects in the development of both P_1 and AB descendants. However, in many *mex-1* mutant embryos, aside from the equal cleavage of P_3 to produce two D-like fates, P_1 descendants appear to develop normally. Thus, while *mex-1* mutant embryos resemble *par* mutants in many respects, most *mex-1* mutant embryos have less extensive defects in a–p polarity than do most *par* mutant embryos.

b. Localization of MEX-1 Protein. Surprisingly, molecular analysis has shown that *mex-1* encodes a protein with a Zn^{2+} finger domain related to the Zn^{2+} finger domain in PIE-1 (Guedes and Priess, 1996). The spacing of cysteine and histidine residues found in both *mex-1* and *pie-1* also is found in vertebrate TIS-11 genes, which are expressed early in response to treatment of fibroblasts with triphorbol esters but are of unknown function (Varnum *et al.*, 1989; Dubois *et al.*, 1990; Guedes and Priess, 1996). Like PIE-1, MEX-1 is present only in germline precursors and is associated with P-granules. However, neither PIE-1 nor MEX-1 are associated with P-granules in the maternal germline, either due to their absence or perhaps due to masking by other proteins. In either case, these results indicate that P-granules are dynamic structures, raising the interesting possibility that P-granules load and perhaps unload different factors at different times during the early cleavages that generate founder cells.

Unlike PIE-1, MEX-1 is cytoplasmic and does not localize to either centrosomes or to the nucleus as does PIE-1 (see above). Furthermore, while *pie-1* mutant embryos resemble *mex-1* mutants in that P_3 divides equally to produce D-like blastomeres, the phenotypes of *pie-1* and *mex-1* mutant embryos otherwise are very different. Intriguingly, *mex-1* is required to restrict PIE-1 to the germline when P_3 divides: in *mex-1* mutant embryos, PIE-1 is mislocalized to and represses the transcription of zygotic genes in both P_3 daughters, indicating that in addition to being structurally related these genes interact (Guedes and Priess, 1996). How the localization of MEX-1 and its similarity to PIE-1 relates to the mutant phenotype of *mex-1* mutant embryos remains rather mysterious, and the functions of TIS-11-like Zn^{2+} finger domains in general remain largely unknown. Studies of their roles in the control of embryonic patterning in C. *elegans* promises to shed new light on the functions of this recently discovered subfamily of Zn^{2+} finger proteins.

c. mex-3 Function and the Regulation of PAL-1 Expression. The *mex-3* mutant embryos resemble the *mex-1* mutants in that the AB granddaughters produce a large excess of body wall muscle. Instead of adopting MS-like fates, as occurs in *mex-1* mutants, AB descendants in *mex-3* mutant embryos adopt C-like fates (Draper *et al.*, 1996). As described above, *pal-1* is required for specifying C fate, and in wild-type embryos PAL-1 is present only in P_2 and EMS at the 4-cell stage. In *mex-3* mutant embryos, PAL-1 is evenly distributed in all 4-cell stage blastomeres, suggesting that PAL-1 acts ectopically in *mex-3* mutants to specify C-like fates in the granddaughters of AB (Hunter and Kenyon, 1996). Consistent with this hypothesis, AB descendants in *mex-3; pal-1* double mutant embryos do not produce the excess body wall muscle cells made in *mex-3* mutants (Draper *et al.*, 1996; Hunter and Kenyon, 1996). Finally, *mex-3* is unique among mutants with equal P_3 cleavages (*pie-1, mex-1, mes-1,* and *mex-3*) in that both

P_3 daughters adopt a germline fate (Draper *et al.*, 1996). In all the other mutants, both P_3 daughters adopt a D-like fate and produce excess body wall muscle (see above).

The MEX-3 protein contains two KH domains, which are involved in RNA binding and in protein–protein interactions (Draper *et al.*, 1996; Chen *et al.*, 1997). MEX-3 is present at higher levels in the cytoplasm of AB than in P_1 at the beginning of the 2-cell stage, and at higher levels in the daughters of AB than in the daughters of P_1 at the 4-cell stage (Draper *et al.*, 1996). Moreover, *pal-1* mRNA is present throughout the early embryo, and the 3′UTR of *pal-1* mRNA is necessary and sufficient to localize translation of a lacZ reporter RNA to P_1 descendants (Hunter and Kenyon, 1996). Thus it has been proposed that MEX-3 directly regulates PAL-1 expression by acting as a translational repressor in AB, perhaps binding 3′UTR sequences in *pal-1* mRNA (Draper *et al.*, 1996; Hunter and Kenyon, 1996). In later stage embryos, MEX-3 is present at very low levels only in P_2 descendants and is undetectable in other blastomeres.

d. Linking the Establishment of a–p Polarity to the Specification of Blastomere Identities. Studies of *mex-3* and *pal-1* provide one example of a genetic pathway that may connect the more general polarity specification functions of the *par* genes with the localized function of more specifically acting regulators of blastomere identity. MEX-3, normally present at higher levels in ABa and ABp than in P_2 and EMS, is present at high levels in all 4-cell stage blastomeres in *par-1* mutant embryos (Draper *et al.*, 1996). Consistent with the proposed role of MEX-3 as a translational repressor of *pal-1* in ABa and ABp in wild-type embryos, high levels of MEX-3 throughout *par-1* embryos correlates with a complete loss of PAL-1 protein (Hunter and Kenyon, 1996). Thus mislocalization of MEX-3 to P_1 descendants causes ectopic repression of PAL-1 expression in *par-1* mutant embryos. Moreover, eliminating *mex-3* function in *par-1* mutant embryos restores PAL-1 expression and function in all 4-cell stage blastomeres (Draper *et al.*, 1996; Hunter and Kenyon, 1996). Finally, the localization of PAR-1, a putative ser/thr kinase, to the posterior cortex of the 1-cell stage wild-type zygote correlates with the lower levels of *mex-3* mRNA and protein in P_1 and in P_1's daughters. Thus the putative PAR-1 kinase may negatively regulate *mex-3* in the posterior part of the zygote, restricting the translation of *pal-1* to posterior blastomeres. Although PAR-4 is present at the cortex both anteriorly and posteriorly at the 1-cell stage, mutations in *par-4* also result in high levels of MEX-3 in all 4-cell stage blastomeres and a corresponding loss of PAL-1 expression (Bowerman *et al.*, 1997). Thus both PAR-1 and PAR-4 are required to restrict high levels of MEX-3 to AB descendants, with the localization of cortical PAR-1 to the posterior cortex correlating with the lower levels of MEX-3 in posterior blastomeres.

Although a simple model in which *par-1* and *par-4* restrict high levels of MEX-3 to AB descendants, and thereby limit PAL-1 expression to posterior blastomeres is appealing (Fig. 5a), more recent studies of *par-3* mutant embryos indicate that the regulation of PAL-1 expression by *par-1, par-4,* and *mex-3* may be more complex (Bowerman *et al.,* 1997; see Fig. 5b). As in *par-1* mutants, MEX-3 is expressed at high levels in all 4-cell stage blastomeres in *par-3* mutant embryos. However, in contrast to *par-1* mutant embryos, PAL-1 usually is expressed at normal levels and sometimes is even mislocalized in *par-3* mutants. One interpretation of these results is that *par-1* and *par-4* act independently of *mex-3* to derepress the translation of *pal-1* mRNA, with *par-3* acting to restrict these functions of *par-1* and *par-4* to posterior blastomeres.

Whatever mechanism restricts high levels of MEX-3 to anterior blastomeres, regulating the distribution of the maternal *mex-3* mRNA may be important for limiting MEX-3 protein levels posteriorly. *mex-3* is unique among the maternal genes discussed here in showing an enrichment of mRNA in AB (Draper *et al.,* 1996). At the beginning of the 1-cell stage, *mex-3* mRNA is distributed evenly but fades to lower levels posteriorly by the time the zygote divides. During the 2-cell and 4-cell stages, higher levels of *mex-3* mRNA are present in AB and its daughters than in P_1 and its daughters. Subsequently, *mex-3* mRNA levels fade to below detectable levels except in the germline, an observation true of many maternal mRNAs in *C. elegans* (Seydoux and Fire, 1994).

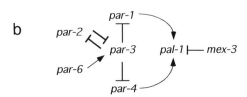

Figure 5 Linking the establishment of cell polarity in the 1-cell zygote to the specification of individual blastomere identities: *par, mex-3,* and *pal-1* interactions. (a) Simple pathway proposed for the regulation of PAL-1 expression by PAR-1 and MEX-3 (see text). (b) More complex network of gene function that attempts to account for all known functional interactions and protein localization data known for the genes shown (see text). The *par* genes appear to function as part of a complex network of overlapping but sometimes distinct activities that somehow regulate the distribution of more specifically acting gene products such as *mex-3* and *pal-1.*

4. Cell Interactions and EMS Fate Specification

The simple fates of most founder cells descended from P_1, the inability of P_2 and EMS to replace each other after having their positions interchanged, and the apparent ability of P_1 to develop normally after removal of AB all suggest that P_1 develops using largely cell autonomous mechanisms. However, none of these observations rule out important roles for cell interactions among P_1 descendants. Indeed, a simple and informative set of experiments in 1987 suggested that only germline precursors in the early embryo have an intrinsic polarity, and that all somatic founder cells might require signals from germline precursors to develop properly (Schierenberg, 1987). Five years later, it was shown conclusively that the proper development of the somatic 4-cell stage blastomere EMS requires a polarizing signal from its neighbor and sister, the germline precursor P_2 (Goldstein, 1992). Normally EMS divides to make one daughter, MS, that produces mesoderm, and another daughter, E, that produces all of the endoderm (Sulston *et al.,* 1983). A signal from P_2 polarizes EMS such that its daughter born next to P_2 produces endoderm. In the absence of P_2 signaling, EMS instead divides to make two MS-like daughters (Goldstein, 1992, 1995a). Thus the specification of E fate, the earliest-born founder cell to produce only a single cell-type, requires an induction. Signals from P_2 not only polarize gut potential to one side of EMS but also orient the EMS mitotic spindle: move P_2 to a different position on EMS, and the EMS spindle will rotate to "point" toward P_2, producing an E daughter close to P_2 and an MS daughter away from P_2. Because the timing of the gut polarization signal and the timing of the mitotic spindle orientation signal appear different, it has been suggested that two different signals from P_2 influence EMS fate (Goldstein, 1995b).

a. Specifying Endoderm by Downregulating an HMG Domain Protein in E. The *pop-1* gene was the first maternal gene identified that is required specifically for distinguishing the fates of E and MS (Lin *et al.,* 1995). In *pop-1* mutant embryos, MS adopts an E-like fate, resulting in an excess of intestine at the expense of MS-derived mesoderm, a phenotype opposite to that caused by elimination of P_2 signaling. The POP-1 protein is a putative transcription factor that contains a single HMG domain and is present in all nuclei at the 4-cell stage. However, when EMS divides, nuclear POP-1 levels remain high in MS but drop to nearly undetectable levels in E (Fig. 6), and eliminating POP-1 from both E and MS in *pop-1* mutant embryos results in both EMS daughters adopting E fates. Therefore downregulation of nuclear POP-1 appears to permit the specification of endoderm fate (Lin *et al.,* 1995). Presumably the polarization of EMS by P_2 results in the

POP-1

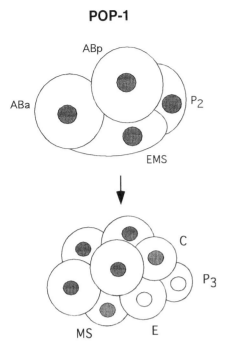

Figure 6   POP-1 protein distribution in 4-cell stage and 8-cell stage wild-type embryos. POP-1 contains a single HMG domain and is present in all nuclei at the 4-cell stage. At the 8-cell stage, EMS divides to produce one daughter, MS, with high levels of nuclear POP-1, and one daughter E, with low or undetectable levels of nuclear POP-1. In *pop-1* mutant embryos, both EMS daughters adopt E-like fates (see text). POP-1 also is present at low or undetectable levels in the P_3 daughter of P_2, but P_3 develops normally in *pop-1* mutant embryos.

differential segregation of regulatory factor(s) to the daughters of EMS such that nuclear POP-1 is present only in MS. If so, mutations in genes required for P_2 signaling should result in EMS producing two daughters that both have high levels of POP-1 and adopt MS fates. Recently, mutations in five such genes have been identified.

b. Identification of the mom Genes.   Mutationally identified genes required for the polarizing induction of gut potential in EMS are named *mom-1* through *mom-5*. In most *mom* mutant embryos, EMS produces two MS-like daughters, resulting in excess mesoderm at the expense of all endoderm (Rocheleau *et al.*, 1997; Thorpe *et al.*, 1997). Genetic mosaic analyses have shown that *mom-1*, *mom-2*, and *mom-3* are required in P_2 for signaling but not in EMS for responding, while *mom-4* is required in EMS for responding but not in P_2 for signaling (Thorpe *et al.*, 1997). The penetrance of the gut defect in *mom-5* mutant embryos is too low to permit a conclusive mosaic analysis of its requirement in P_2 versus EMS. Indeed, all mutant alleles of the *mom* genes show incomplete penetrance for the gut defect. For example, even strong loss of function mutations in *mom-2* result in only about 75% of the mutant embryos

lacking endoderm. In essentially all cases, mutations in the *mom* genes result in a fully penetrant defect in morphogenesis, a poorly understood and likely complex process that converts the round ball of cells made by embryonic cleavages into a long, thin worm. Thus the *mom* genes also participate in other processes in addition to polarizing gut potential in EMS.

c. Molecular Analysis of the mom Genes: Wnt Signaling Specifies Endoderm in C. elegans.   The sequences for three of the five *mom* genes are known (Rocheleau *et al.*, 1997; Thorpe *et al.*, 1997). All three are predicted to encode components of the widely conserved Wnt signal transduction pathway, which is required for many different cell signaling processes that pattern the bodies of both invertebrates and vertebrates (Fig. 7). *mom-1* is related to *Drosophila porcupine*, *mom-2* to *Drosopohila wingless/Wnt*, and *mom-5* to *Drosophila frizzled*. Porcupine is a multipass transmembrane protein localized to the endoplasmic reticulum and required in signaling cells for proper glycosylation and secretion of Wingless, a secreted glycoprotein that is the ligand for the Wnt/wingless signal transduction pathway (Kadowaki *et al.*, 1995). *mom-5* encodes a Frizzled homolog and therefore is likely to be a Wnt receptor based on biochemical studies in *Drosopihila* (Bhanot *et al.*, 1996). In addition to the mutationally identified *mom* genes, reverse genetics techniques have implicated two other *C. elegans* genes related to Wnt pathway components that, based on work in *Drosophila*, function in responding cells (Rocheleau *et al.*, 1997). Inactivation of a nematode β-catenin gene called *wrm-1* (*armadillo* in *Drosophila*) results in a fully penetrant Mom phenotype. Inactivation of the closest nematode relative of the human gene APC (for adenomatous polyposis coli tumor suppressor gene) results in a partially penetrant Mom phenotype, providing the first genetic evidence outside of human cancer cells that APC is a component, or regulator, of Wnt pathway signal transduction. This latter gene has been named *apr-1* for APC-related. The observation that, except for *wrm-1*, all the Mom phenotypes are only partially penetrant suggests that more than one pathway may converge on *wrm-1* during the polarization of gut potential. The *mom-3* and *mom-4* genes have yet to be molecularly cloned but may encode novel components of the Wnt pathway, or components of parallel pathway(s) that perhaps converge on *wrm-1*. While much remains to be learned about the function of the Wnt pathway in the *C. elegans* embryo, studies of the *mom* genes already have provided novel insights into this critical developmental pathway that also is implicated in human colon carcinoma, one of the leading causes of cancer death in the United States (for review, see Miller and Moon, 1996; Gumbiner, 1997; Han, 1997).

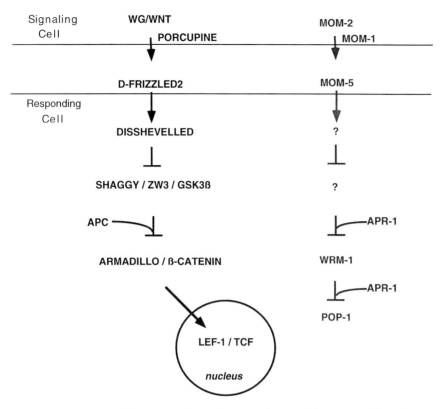

Figure 7   A summary of the Wnt signal transduction pathway and the corresponding identities of *mom-1, mom-2,* and *mom-5.* The *mom-3* gene has not been cloned but is required in P₂ for signaling; *mom-4* has not been cloned but is required in EMS for responding to the polarizing signal from P₂. Wnt signaling in other systems activates HMG domain proteins such as LEF-1 and TCF. In the *C. elegans* embryo, Wnt signaling downregulates the HMG domain protein POP-1 (see text). In human colon carcinoma cell lines, APC is required to downregulate cytoplasmic levels of β-catenin. In *C. elegans,* eliminating the maternal function of the APC-related gene *apr-1* mutant results in a weak Mom phenotype, suggesting that *apr-1* is required positively in the Wnt pathway to downregulate POP-1. Therefore *apr-1* is shown as having a positive function downstream of the negative regulatory function identified in human colon carcinoma cell lines (see text).

d. mom-5/Frizzled: The Receptor for mom-2/Wnt? Although biochemical studies suggest that the seven-pass transmembrane protein D-Frizzled2 functions as a receptor for Wnt ligands, genetic evidence in support of this conclusion is still lacking (Bhanot *et al.,* 1996). The identification of the Frizzled related gene *mom-5* in the *mom/Wnt* pathway promises to further our understanding of this important issue. The current data suggest that if *mom-2/Wnt* and *mom-5/Wnt* encode a ligand and receptor, their relationship may not be simple. First, the penetrance of the gut defect for strong loss of function mutations in the two loci are extremely different (Rocheleau *et al.,* 1997; Thorpe *et al.,* 1997). While about 75% of *mom-2* mutant embryos lack gut, only about 5% of *mom-5* mutant embryos lack gut, even though the mutations in both cases are severe and may be null. Surprisingly, in three different allelic combinations *mom-2;mom-5* double mutant embryos exhibit the weaker *mom-5*-like phenotype, with only about 8% of the double mutants lacking gut (Rocheleau *et al.,* 1997; C. Thorpe and B. Bowerman, unpublished data). These results suggest that *mom-5* acts downstream of *mom-2* but not in a positive, linear pathway. One interpretation

is that in a *mom-2* mutant embryo, *mom-5*(+) function negatively regulates a polarity that specifies endoderm, and that *mom-2*(+) function overcomes this negative regulation. Further experiments using genetic mosaics, and molecular studies of the two gene products, promise to shed more light on these early steps in the Wnt signal transduction pathway.

e. Wnt Signaling and the Downregulation of an HMG Domain Protein.   Studies in *Drosophila* and in *Xenopus* clearly indicate that the final step of the Wnt pathway is the activation of HMG domain proteins, such as Lef-1 and Tcf-1 in vertebrates and *pangolin* in *Drosophila* (for a review, see Nusse, 1997). Briefly, APC and the constitutively active kinase GSK-3ß (for glycogen synthase kinase; called *shaggy* and *zeste-white3* in *Drosophila*) maintain low levels of cytoplasmic β-catenin by promoting its degradation. Wnt signaling appears to stabilize cytoplasmic pools of β-catenin by blocking the function of GSK-3ß. Stabilizing β-catenin allows it to associate with an HMG domain protein and transit to the nucleus as a complex, with β-catenin providing an activation domain and the HMG domain protein providing DNA binding activity and target specificity. As described above, though, in *C. elegans* the eventual consequence of gut polarization is the downregulation of the HMG domain protein POP-1 in E, which permits endoderm fate. In *mom* mutant embryos, downregulation of POP-1 is lost and POP-1 levels remain high in both EMS daughters. Furthermore, in double mutant analyses the Pop-1 phenotype is epistatic to the Mom phenotype for all components of the *mom/Wnt* pathway tested, consistent with *pop-1* functioning downstream of the *mom* genes and serving as the ultimate target of *mom/Wnt* signaling (Rocheleau *et al.*, 1997; Thorpe *et al.*, 1997). How Wnt signaling through *wrm-1*/β-catenin can result in downregulation of POP-1 instead of activation remains to be determined, but these results clearly indicate that the Wnt pathway is capable of more diverse outputs than previously known.

f. Wnt Signaling and Polarization of the Cytoskeleton. Because β-catenin is associated with cell adhesion junctions, and because *wingless* mutants exhibit defects in bristle orientation, it has long been suspected that Wnt signaling might influence cell polarity in addition to influencing patterns of cell fate across a field of tissue (Cox *et al.*, 1996). Studies of the *mom/Wnt* pathway provide strong evidence that Wnt signaling directly influences cytoskeletal polarity in individual cells, and even suggest that the cytoskeleton may be the primary target of Wnt signaling in EMS. Two lines of evidence support this assertion. First, the polarizing induction of endoderm occurs before EMS divides and does not influence the fate of both EMS daughters, only the E daughter (Goldstein, 1992, 1995a). The simplest model is that P2 signaling does not regulate gene expression in EMS but rather results in the daughter of EMS closest to P2 inheriting a cytoplasm that mediates downregulation of POP-1. Exposing EMS in a 4-cell stage embryo with chemicals that depolymerize either microtubules or microfilaments prevents expression of gut markers, while similar treatments after EMS divides do not, providing some evidence that polarization of the cytoskeleton may segregate different cytoplasmic factors to the daughters of EMS (Goldstein, 1995a). More compelling support for *mom/Wnt* signaling polarizing the cytoskeleton comes from the observation that *mom-1, mom-2, mom-3,* and *mom*-5 mutant embryos each exhibit a highly penetrant mitotic spindle orientation defect in an 8-cell stage blastomere called ABar (Rocheleau *et al.*, 1997; Thorpe *et al.*, 1997). Furthermore, some *mom-1;mom-2* double mutant embryos exhibit mitotic spindle orientation defects in all somatic blastomeres (Thorpe *et al.*, 1997). Because the proper orientation of mitotic spindles in the early embryo appears not to require gene transcription (Powell-Coffman *et al.*, 1996), the cytoskeleton appears to be a direct target of *mom/Wnt* sig-

naling. Although experimental manipulations of wild-type embryos suggest that two different signals from P2 polarize gut potential and orient the mitotic spindle in EMS (Goldstein, 1995b), the *mom* genes are required for both processes. It will be interesting to determine at what point the gut polarization and spindle orientation pathways diverge in EMS, and if some outputs of Wnt signaling branch away from the Wnt pathway as currently defined prior to *wrm-1*/β-catenin.

g. lit-1 and a–p Polarity. Recently, another gene required for gut polarization has been identified genetically (Kaletta *et al.*, 1997). Eliminating the maternal function of *lit-1*, as for the *mom* genes, transforms E into an MS-like fate. However, detailed analysis of the cell lineages produced by *lit-1* mutant embryos has shown than all somatic founder cells but D exhibit lineage defects in which both the anterior and posterior daughters of cells that divide along the a–p axis adopt the anterior fate. These transformations are observed in lineages that appear to develop autonomously, and in lineages that require polarizing inductions. Transformations are seen as early as the third round of cell division in the embryo and as late as the eigth round. The *lit-1* gene has yet to be identified molecularly, but the mutant phenotype suggests that it plays a fundamental role in specifying binary switches in a–p fate throughout embryonic development. At least some of the *lit-1* defects appear not to be present in *mom* mutant embryos (Thorpe *et al.*, 1997), raising the possibility that *lit-1* may represent a third system for specifying a–p polarity. In addition to *lit-1*, the *par* genes polarize the 1-cell zygote along the a–p axis, and the Wnt pathway is required for a–p polarization of gut potential and perhaps for additional polarizations during morphogenesis (see above). The widespread conservation of some *par* genes and of the Wnt pathway in other organisms make it seem likely that *lit-1* also will prove to be of general relevance to studies of cell polarity. An alternative view is that *lit-1* may represent a relatively phylum-specific mode of regulation that in part accounts for the invariance of the embryonic lineage in *C. elegans*. A fundamental issue for future studies of pattern formation during embryogenesis in *C. elegans* will be to determine if and how these three different a–p polarity systems interact.

IV. Cell Interactions and the Maternal Genes That Specify the Fates of AB Descendants

Although the invariant lineage of the *C. elegans* embryo initially led many to assume that exclusively cell intrinsic mechanisms would control cell fate patterning in *C.*

elegans, the finding that ABa and ABp are born with equivalent developmental potential made it clear that cell interactions are critical for patterning a substantial proportion of the lineages produced during embryogenesis (Priess and Thomson, 1987). We now understand that the lineage is invariant not because purely mosaic mechanisms pattern the embryo, but rather because the nematode embryo is simple enough for highly reproducible cell contacts and interactions to occur. Since the equivalence of ABa and ABp were made apparent, genetic studies have begun to identify some of the genes that mediate cell signals responsible for distinguishing the fates of ABa and ABp.

A. The *glp-1* Gene, the Notch Pathway, and Inductive Signaling in the Early Embryo

Soon after the demonstration that ABa and ABp are born equivalent, the maternal gene *glp-1* was mutationally identified and shown to be required for a cell interaction that contributes to making ABa and ABp different (Priess *et al.,* 1987). At the 12-cell stage in wild-type embryos, the MS blastomere signals the two granddaughters of ABa that touch MS to adopt fates that include the production of cells forming the anterior part of the pharynx (Priess *et al.,* 1987; Hutter and Schnabel, 1994; Mango *et al.,* 1994a). In *glp-1* mutant

embryos, the MS induction of pharyngeal cells does not occur (Priess *et al.,* 1987), and genetic mosaic studies indicate that *glp-1* is required in the responding cells (Shelton and Bowerman, 1996). *glp-1* encodes a member of the Notch family of transmembrane receptors, which participate in many cell interactions that pattern cell fates during vertebrate and invertebrate development (Austin and Kimble, 1987, 1989; Yochem and Greenwald, 1989). Antibodies to GLP-1 show that this putative receptor is present at the cell surface of ABa and ABp at the 4-cell stage and persists at high levels in their descendants until the 28-cell stage (Evans *et al.,* 1994; see Fig. 8). Thus GLP-1 is present on the responding cells at the appropriate time to act as the receptor for the 12-cell stage signal from MS that induces production of pharyngeal cells. Lower levels of GLP-1 are present in E and MS, and GLP-1 also is required to receive a signal required for MS to produce body wall muscle (Schnabel, 1994).

The finding that MS induces ABa to produce pharyngeal cells raised an important question: If ABa and ABp are born equivalent, why does the MS signal influence only ABa descendants and not ABp descendants that touch MS? Because P_2 touches only ABp and not ABa, it was a logical candidate for sending a signal to break the initial equivalence of ABa and ABp at the 4-cell stage, perhaps rendering ABp descendants insensi-

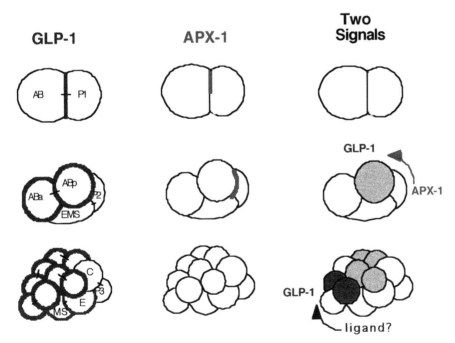

Figure 8   GLP-1 and APX-1 distributions in 2-cell, 4-cell and 12-cell stage embryos. GLP-1 (red) is first detected at the P_1–AB boundary at the 2-cell stage, and on the surfaces of AB descendants at the 4-cell and 12-cell stages. APX-1 (blue) is first detected at the interface of P_1 and AB, but only in about one half of the boundary. The mechanisms controlling this early APX-1 asymmetry, and its functional significance, remain unknown. APX-1 is present at the interface of P_2 and ABp at the 4-cell stage but fades to undetectable levels by the 12-cell stage. APX-1 and GLP-1 are required for a cell–cell interaction at the 4-cell stage that specifies ABp identity (light shading) and breaks the initial equivalence of ABa and ABp. GLP-1 and an unknown ligand produced by MS are required for a second set of cell–cell interactions that specifies the fates of two granddaughters of ABa at the 12-cell stage that touch MS (dark shading).

tive to the subsequent signal from MS. A variety of experimental manipulations provided evidence for such an interaction. (i) If P_2 in a wild-type embryo is killed with a laser microbeam just as it is born, ABp sometimes fails to produce intestinal–rectal valve cells, an ABp-specific cell type (Bowerman et al., 1992b). (ii) Removal of P_2 within 5 minutes of its birth also results in an absence of intestinal–rectal valve cells and moreover results in the production of excess pharyngeal cells, consistent with both ABa and ABp descendants producing pharynx in response to MS signaling (Mango et al., 1994b). (iii) If P_1 is prevented from dividing such that a large undivided P_1 touches both ABa and ABp, extra intestinal–rectal valve cells are made and no pharyngeal cells are induced (Mello et al., 1994). (iv) If P_1 and AB divide without an intact eggshell forcing their spindles to slant diagonally, P_2 fails to contact either AB daughter, in which case no intestinal valve cells and large numbers of pharyngeal cells are produced (Mello et al., 1994). In sum, these experiments indicate that a signal from P_2 at the 4-cell stage is required for the production of ABp-specific cell-types, and for preventing the production of pharyngeal cells by ABp in response to a signal from MS at the 12-cell stage.

B. A Putative Ligand for glp-1, and a Sequence of Two glp-1-Dependent Signals

Genetic evidence for a P_2 signal specifying ABp fate came both from the identification of the maternal gene apx-1, and from elegant studies of ts glp-1 alleles (Hutter and Schnabel, 1994; Mango et al., 1994b; Mello et al., 1994). Mutations in apx-1 result in ABp adopting a fate much like that of ABa: ABp in apx-1 mutant embryos fails to produce ABp-specific cell types, and the ABp granddaughters that touch MS produce extra anterior pharyngeal cells in response to MS signaling at the 12-cell stage (Mango et al., 1994b; Mello et al., 1994). Consistent with apx-1 being required for P_2 signaling, it encodes a transmembrane protein related to the Delta family of Notch ligands, and the APX-1 protein is produced in P_2 at the 4-cell stage, localized to the interface of P_2 and ABp (Mello et al., 1994; Mickey et al., 1996; see Fig. 8). The localization of APX-1 to the interface of P_2 and ABp could be due simply to clustering of APX-1 by its receptor on ABp. However, APX-1 also is present at the 2-cell stage at the interface of P_1 and AB, but only on one side of the embryo. Thus APX-1 localization may be polarized in some manner that cannot be explained by contacts with neighboring blastomeres.

The identity of APX-1 as a Delta family member suggests that GLP-1, the Notch family member present on the surface of ABa and ABp at the 4-cell stage, might act as the receptor for APX-1. Temperature shift experiments using a conditional allele of glp-1 showed that GLP-1 does function as the receptor both for the 4-cell stage signal involving apx-1 that specifies ABp identity, and for the 12-cell stage signal from MS that induces pharyngeal cell production by ABa descendants (Hutter and Schnabel, 1994; Mello et al., 1994). If glp-1 function is blocked by raising embryos at the restrictive temperature, but then restored shortly after the 4-cell stage by shifting to permissive temperature, the P_2 signal is blocked and ABp develops just as it does in apx-1 mutant embryos, failing to produce ABp-specific cell types and instead making excess anterior pharyngeal cells in response to the MS inductive signal at the 12-cell stage. If glp-1 function is blocked at both the 4-cell and 12-cell stages, neither ABp-specific cell types nor anterior pharyngeal cells are produced; instead all AB descendants adopts one of two "uninduced" ABa fates, that of a wild-type ABala or ABarp blastomere. These results showed that GLP-1 is the receptor both for the 4-cell stage signal from P_2 and for the 12-cell stage signal from MS (Fig. 8). While the ligand for the MS signal remains unknown, MS requires SKN-1 function to express signaling activity, and the MS ligand presumably is another Delta family member (Shelton and Bowerman, 1996).

The specific responses by AB descendants to the P_2 and MS signals (ABp fate versus induced ABa granddaughter fates) appear to depend not on the specific nature of the ligand or signaling blastomere, but rather on time-dependent differences in the response of AB descendants to activation of the GLP-1 receptor. By using isolated blastomeres placed in contact with each other in culture medium, it is possible to induce 12-cell stage AB descendants to produce pharyngeal cells either by placing them in contact with MS, the normal signaling blastomere from a 12-cell stage embryo, or by placing them in contact with P_2 to form a "heterochronically" chimaeric partial embryo in vitro (Shelton and Bowerman, 1996). Pharyngeal induction by P_2, but not by MS, requires apx-1 function, indicating that APX-1 in P_2 can substitute for the MS signal to induce 12-cell stage AB descendants to produce pharyngeal cells. Factors that presumably change with time in AB descendants to mediate the different responses to GLP-1 signaling at the 4-cell and 12-cell stages have yet to be identified, as do any autonomously acting factors responsible for specifying the uninduced fates of 12-cell stage AB descendants. Finally, other signals have been identified that serve to specify the eight different fates of the AB descendants present in a 12-cell stage embryo (Hutter and Schnabel, 1995). At least some of these later signals also appear to involve the function of zygotically expressed Delta and Notch family members (Moskowitz and Rothman, 1996).

V. Concluding remarks

Until recently, the fruit fly *Drosophila melanogaster* was the only metazoan used for large-scale genetic screens to study early steps in pattern formation (St. Johnston and Nusslein-Volhard, 1992). More recently, genetic screens in *C. elegans* and in *Arabidopsis thaliana* have begun to broaden our view of early development by identifying additional regulatory loci in another animal and in a plant embryo (this chapter; Jurgens, 1995). Mechanistic comparisons of pattern formation in these and other early embryos may reveal not only how different life forms develop but perhaps also how they evolve.

For a comparison of early development, the insect and nematode embryos are impressively different. The *Drosophila* embryo is a 500 μm-long syncitium in which a peripheral monolayer of nuclei share a common cytoplasm until completion of the 13th round of mitosis (St. Johnston and Nusslein-Volhard, 1992). The diffusion of transcriptional and translational regulators from localized sources forms morphogenetic gradients of positional information that pattern large fields of nuclei. In dramatic contrast, the 50-μm-long *C. elegans* embryo is completely cellularized, and the early events that control patterning must negotiate the plasma membranes that partition all nuclei (Sulston *et al.*, 1983). Another notable difference between fly and worm embryos is the relative importance of pattern formation during oogenesis versus after fertilization. In an early insect embryo, many of the events that establish anterior–posterior and dorsal–ventral asymmetry begin during oogenesis, with fertilization activating previously localized regulators (St. Johnston and Nusslein-Volhard, 1992; Grunert and St. Johnston, 1996). In *C. elegans*, the body axes form not during oogenesis but sequentially during embryogenesis, with sperm entry establishing an anterior–posterior axis (Goldstein and Hird, 1996).

One largely unmet challenge in developmental biology is to understand how and why early embryos begin development in such remarkably different ways. As one route to a broader understanding, the genetics and sequenced genome of *C. elegans* have provided rapid access to the molecules and mechanisms that initiate pattern formation in this cellularized animal embryo. Recent developments in the ability to assign functions to genes identified by sequence (Han, 1997; Jansen, *et al.*, 1997), and more sophisticated genetic screens for additional regulatory loci, promise to provide further insight into the mechanisms that control pattern formation in this elegantly simple embryo.

References

Albertson, D. G. (1984). Formation of the first cleavage spindle in nematode embryos. *Dev. Biol.* **101,** 61–72.

Aroian, R. V., Field, C., Pruliere, G., Kenyon, C., and Alberts, B. M. (1997). Isolation of actin-associated proteins from *Caenorhabditis elegans* oocytes and their localization in the early embryo. *EMBO J.* **16.**

Austin, J., and Kimble, J. (1987). *glp-1* is required in the germ line for regulation of the decision between mitosis and meiosis in *C. elegans*. *Cell (Cambridge, Mass.)* **51,** 589–599.

Austin, J., and Kimble, J. (1989). Transcript analysis of *glp-1* and *lin-12,* homologous genes required for cell interactions during development of *C. elegans*. *Cell (Cambridge, Mass.)* **58,** 565–571.

Bhanot, P. Brink, M., Samos, C. H., Hsieh, J.-C., Wang, Y., Macke, J. P., Andrew, D., Nathans, J., and Nusse, R. (1996). A new member of the frizzled family from *Drosophila* functions as a Wingless receptor. *Nature (London)* **382,** 225–230.

Blackwell, T. K., Bowerman, B., Priess, J. R., and Weintraub, H. (1994). Incorporation of homeodomain and bZIP elements into a DNA binding domain by *C. elegans* SKN-1 protein. *Science* **266,** 621–628.

Bohm, H., Brinkmann, V., Drab, M., Henske, A., and Kurzchalia, T. V. (1997). Mammalian homologues of *C. elegans* PAR-1 are asymmetrically localized in epithelial cells and may influence their polarity. *Curr. Biol.* **7,** 603–606.

Bowerman, B., Eaton, B. A., and Priess, J. R. (1992a). *skn-1,* a maternally expressed gene required to specify the fate of ventral blastomeres in the early *C. elegans* embryo. *Cell (Cambridge, Mass.)* **68,** 1061–1075.

Bowerman, B., Tax, F. E., Thomas, J. H., and Priess, J. R. (1992b). Cell interactions involved in the development of the bilaterally symmetrical intestinal valve cells during embryogenesis in *Caenorhabditis elegans*. *Development (Cambridge, Mass.)* **116,** 1113–1122.

Bowerman, B., Draper, B. W., Mello, C. C., and Priess, J. R. (1993). The maternal gene *skn-1* encodes a protein that is distributed unequally in early *C. elegans* embryos. *Cell (Cambridge, Mass.)* **74,** 443–452.

Bowerman, B., Ingram, M. K., and Hunter, C. P. (1997). The maternal *par* genes and the segregation of cell fate specification activities in early *C. elegans* embryos. *Development (Cambridge, UK)* **124,** 3815–3826.

Boyd, L., Guo, S., Levitan, D., Stinchcomb, D. T, and Kemphues, K. J. (1996). PAR-2 is asymmetrically distributed and promotes association of P granules and PAR-1 with the cortex in *C. elegans* embryos. *Development (Cambridge, UK)* **122,** 3075–3084.

Carroll, A. S., Gilbert, D. E., Liu, X., Cheung, J. W., Michnowicz, J. E., Wagner, G., Ellenberger, T. E., and Blackwell, T. K. (1997). SKN-1 domain folding and basic region monomer stabilization upon DNA binding. *Genes Dev.* **11,** 2222–2238.

Chen, T., Damaj, B. B., Herrera, C., Lasko, P., and Richard, S. (1997). Self-association of the single-KH-domain family members GRP33, GLD-1, and Qk1: role of the KH domain. *Mol. Cell Biol.* **17,** 5707–5718.

Cheng, N. N., Kirby, C. M., and Kemphues, K. J. (1995). Control of cleavage spindle orientation in *Caenorhabditis elegans:* The role of the genes *par-2* and *par-3*. *Genetics* **139,** 549–559.

Cox, R. T., Kirkpatrick, C., and Peifer, M. (1996). Armadillo is required for adherens junction assembly, cell polarity, and morphogenesis during *Drosophila* embryogenesis. *J. Cell Biol.* **134**, 133–148.

Crittenden, S. L., Rudel, D., Binder, J., Evans, T. C., and Kimble, J. (1996). Genes required for GLP-1 asymmetry in the early *C. elegans* embryo. *Dev. Biol.* **181**, 36–46.

Draper, B. W., Mello, C. C., Bowerman, B., Hardin, J., and Priess, J. R. (1996). The maternal gene *mex-3* encodes a KH domain protein and regulates blastomere identity in early *C. elegans* embryos. *Cell (Cambridge, Mass.)* **87**, 205–216.

Drewes, G., Ebneth, A., Preuss, E., Mandelkow, E.-M., and Mandelkow, E. (1997). MARK, a novel family of protein kinases that phosphorylate microtubule-associated proteins and trigger microtubule disruption. *Cell (Cambridge, Mass.)* **89**, 297–308.

Dubois, R. N., McLane, M., Ryder, K., Lau, L. F., and Nathans, D. (1990). A growth factor-inducible nuclear protein with a novel cysteine/histidine repetitive sequence. *J. Biol. Chem.* **265**, 19185–19191.

Etemad-Moghadam, S. G., and Kemphues, K. J. (1995). Asymmetrically distributed PAR-3 protein contributes to cell polarity and spindle alignment in early *C. elegans* embryos. *Cell (Cambridge, Mass.)* **83**, 743–752.

Evans, T. C., Crittenden, S. L., Kodoyianni, V., and Kimble, J. (1994). Translational control of maternal *glp-1* mRNA establishes an asymmetry in the *C. elegans* embryo. *Cell (Cambridge, Mass.)* **77**, 183–194.

Goldstein, B. (1992). Induction of gut in *Caenorhabditis elegans* embryos. *Nature (London)* **357**, 255–257.

Goldstein, B. (1995a). An analysis of the response to gut induction in the *C. elegans* embryo. *Development (Cambridge, UK)* **121**, 1221–1236.

Goldstein, B. (1995b). Cell contacts orient some cell division axes in the *Caenorhabditis elegans* embryo. *J. Cell Biol.* **129**, 1071–1080.

Goldstein, B., and Hird, S. N. (1996). Specification of the anteroposterior axis in *Caenorhabditis elegans*. *Development (Cambridge, UK)* **122**, 1467–1474.

Grunert, S., and St. Johnston, D. (1996). RNA localization and the development of asymmetry during *Drosophila* oogenesis. *Curr. Opin. Genet. Dev.* **6**, 395–402.

Guedes, S., and Priess, J. R. (1996). The *C. elegans* MEX-1 protein is present in germline blastomeres and is a P granule component. *Development (Cambridge, UK)* **124**, 731–739.

Gumbiner, B. (1997). Carcinogenesis: A balance between β-catenin and APC. *Curr. Biol.* **7**, 443–448.

Guo, S., and Kemphues, K. J. (1995). *par-1*, a gene required for establishing polarity in *C. elegans* embryos, encodes a putative ser/thr kinase that is asymmetrically distributed. *Cell (Cambridge, Mass.)* **81**, 611–620.

Guo, S., and Kemphues, K. J. (1996a). Molecular genetics of asymmetric cleavage in the early *C. elegans* embryo. *Curr. Opin. Genet. and Dev.* **6**, 408–415.

Guo, S., and Kemphues, K. J. (1996b). A non-muscle myosin required for embryonic polarity in *Caenorhabditis elegans*. *Nature (London)* **382**, 455–458.

Han, M. (1997). Gut reaction to Wnt signaling in worms. *Cell (Cambridge, Mass.)* **90**, 581–584.

Hill, D. P., and Strome, S. (1988). An analysis of the role of microfilaments in the establishment and maintenance of asymmetry in *Caenorhabditis elegans* zygotes. *Dev. Biol.* **125**, 15–84.

Hill, D. P., and Strome, S. (1990). Brief cytochalasin-induced disruption of microfilaments during a critical interval in 1-cell *C. elegans* embryos alters the partitioning of developmental instructions to the 2-cell embryo. (*Cambridge, UK*) **108**, 159–172.

Hird, S. N., and White, J. G. (1993). Cortical and cytoplasmic flow polarity in early embryonic cells of *Caenorhabditis elegans*. *J. Cell Biol.* **121**, 1343–1355.

Hird, S. N., Paulsen, J. E., and Strome, S. (1996). Segregation of germ granules in living *Caenorhabditis elegans* embryos: Cell-type-specific mechanisms of cytoplasmic localisation. *Development (Cambridge, UK)* **124**, 1303–1312.

Hunter, C. P., and Kenyon, C. (1996). Spatial and temporal controls target *pal-1* blastomere-specification activity to a single blastomere lineage in *C. elegans* embryos. *Cell (Cambridge, Mass.)* 87, 217–226.

Hutter, H., and Schnabel, R. (1994). *glp-1* and inductions establishing embryonic axes in *C. elegans*. *Development (Cambridge, UK)* **120**, 2051–2064.

Hutter, H., and Schnabel, R. (1995). Specification of anterior–posterior differences within the AB lineage in the *C. elegans* embryo: A polarising induction. *Development (Cambridge, UK)* **121**, 1559–1568.

Hyman, A. A. (1989). Centrosome movement in the early divisions of *Caenorhabditis elegans*: A cortical site determining centrosome position. *J. Cell Biol.* **109**, 1185–1194.

Hyman, A. A., and White, J. G. (1987). Determination of cell division axes in the early embryogenesis of *Caenorhabditis elegans*. *J. Cell Biol.* **105**, 2123–2135.

Jansen, G., Hazendonk, E., Thijssen, K. L., and Plasterk, R. H. A. (1997). Reverse genetics by chemical mutagenesis in *Caenorhabditis elegans*. *Nat. Genet.* **17**, 119–121.

Jurgens, G. (1995). Axis formation in plant embryogenesis: Cues and clues. *Cell (Cambridge, Mass.)* **81**, 467–470.

Kadowaki, T., Wilder, E., Klingensmith, J., Zachary, K., and Perrimon, N. (1995). The segment polarity gene *porcupine* encodes a putative multitransmembrane protein involved in *Wingless* processing. *Genes Dev.* **10**, 3116–3128.

Kaletta, T., Schnabel, H., and Schnabel, R. (1997). Binary specification of the embryonic lineage in *Caenorhabditis elegans*. *Nature (London)* **390**, 294–298.

Kemphues, K. J., and Strome, S. (1997). Fertilization and establishment of polarity in the embryo. "*C. elegans* II," pp. 335–360. Cold Spring Harbor Laboratory, Plainview, New York. 335–360.

Kemphues, K. J., Priess, J. R., Morton, D. G., and Cheng, N. S. (1988). Identification of genes required for cytoplasmic localization in early *C. elegans* embryos. *Cell (Cambridge, Mass.)* **52**, 311–320.

Kirby, C., Kusch, M., and Kemphues, K. J. (1990). Mutations in the *par* genes of *Caenorhabditis elegans* affect cytoplasmic reorganization during the first cell cycle. *Dev. Biol.* **142**, 203–215.

Kurzchalia, T., and Hartmann, E. (1996). Are there similarities between the polarization of the *C. elegans* embryo and of an epithelial cell? *Trends Cell Biol.* **6**, 131–132.

Levin, D. E., and Bishop, J. M. (1990). A putative protein kinase gene (kin-1[+]) is important for growth polarity in *Schizosaccharomyces pombe*. *Proc. Natl. Acad. Sci. U.S.A.* **87**, 8272–8276.

Levin, D. E., Hammond, C. I., Ralston, R. O., and Bishop, J. M. (1987). Two yeast genes that encode unusual protein kinases. *Proc. Natl. Acad. Sci. U.S.A.* **84**, 6035–6039.

Levitan, D. J., Boyd, L., Mello, C. C., Kemphues, K. J., and Stinchcomb, D. T. (1994). *par-2*, a gene required for blastomere asymmetry in *Caenorhabditis elegans*, encodes zinc-finger and ATP-binding motifs. *Proc. Natl. Acad. Sci. U.S.A.* **91**, 6108–6112.

Lin, R., Thompson, S., and Priess, J. R. (1995). *pop-1* encodes an HMG box protein required for the specification of a mesoderm precursor in early *C. elegans* embryos. *Cell (Cambridge, Mass.)* **83**, 599–609.

Mango, S. E., Lambie, E. J., and Kimble, J. (1994a). The *pha-4* gene is required to generate the pharyngeal primordium of *Caenorhabditis elegans*. *Development (Cambridge, UK)* **120**, 3019–3031.

Mango, S. E., Thorpe, C. J., Martin, P. R., Chamberlain, S. H., and Bowerman, B. (1994b). Two maternal genes, *apx-1* and *pie-1*, are required to distinguish the fates of equivalent blastomeres in early *C. elegans* embryos. *Development (Cambridge, UK)* **120**, 2305–2315.

Mello, C. C., Draper, B. W., Krause, M., Weintraub, H., and Priess, J. R. (1992). The *pie-1* and *mex-1* genes and maternal control of blastomere identity in early *C. elegans* embryos. *Cell (Cambridge, Mass.)* **70**, 163–76.

Mello, C. C., Draper, B. W., and Priess, J. R. (1994). The maternal genes *apx-1* and *glp-1* and establishment of dorsal–ventral polarity in the early *C. elegans* embryo. *Cell (Cambridge, Mass.)* **77**, 95–106.

Mello, C. C., Schubert, C., Draper, B. W., Zhang, W., Lobel, R., and Priess, J. R. (1996). The PIE-1 protein and germline specification in *C. elegans* embryos. *Nature (London)* **382**, 710–712.

Mickey, K. M., Mello, C. C., Montgomery, M. K., Fire, A., and Priess, J. R. (1996). An inductive interaction in 4-cell stage *C. elegans* embryos involves APX-1 expression in the signalling cell. *Development (Cambridge, UK)* **121**, 1791–1798.

Miller, J. R., and Moon, R. T. (1996). Signal transduction through β-catenin and specification of cell fate during embryogenesis. *Genes Dev.* **10**, 2527–2539.

Moskowitz, I. P. G., and Rothman, J. H. (1996). *lin-12* and *glp-1* are required zygotically for early embryonic cellular interactions and are regulated by maternal GLP-1 signaling in *C. elegans*. *Development (Cambridge, UK)* **122**, 4105–4117.

Nusse, R. (1997). A versatile transcriptional effector of Wingless signaling. *Cell (Cambridge, Mass.)* **89**, 321–323.

Powell-Coffman, J. A., Knight, J., and Wood, W. B. (1996). Onset of *C. elegans* gastrulation is bocked by inhibition of embryonic transcription with an RNA polymerase antisense RNA. *Dev. Biol.* **178**, 472–483.

Priess, J. R., and Thomson, J. N. (1987). Cellular interactions in early *C. elegans* embryos. *Cell (Cambridge, Mass.)* **34**, 85–100.

Priess, J. R., Schnabel, H., and Schnabel, R. (1987). The *glp-1* locus and cellular interactions in early *C. elegans* embryos. *Cell (Cambridge, Mass.)* **51**, 601–611.

Rocheleau, C. E., Downs, W. D., Lin, R., Wittmann, C., Bei, Y., Cha, Y.-H., Ali, M., Priess, J. R., and Mello, C. C. (1997). Wnt signaling and an APC-related gene specify endoderm in early *C. elegans* embryos. *Cell (Cambridge, Mass.)* **90**, 707–716.

Rose, L. S., and Kemphues, K. (1998). The *let-99* gene is required for proper spindle orientation during cleavage of the *C. elegans* embryo. *Development (Cambridge, UK)* **125**, 1337–1346.

Schierenberg, E. (1987). Reversal of cellular polarity and early cell–cell interactions in the embryo of *Caenorhabditis elegans*. *Dev. Biol.* **122**, 452–463.

Schnabel, R. (1994). Autonomoy and nonautonomy in cell fate specification of muscle in the *C. elegans* embryo: A reciprocal induction. *Science* **263**, 1449–1452.

Schnabel, R., Weigner, C., Hutter, H., Feichtinger, R., and Schnabel, H. (1996). *mex-1* and the general partitioning of cell fate in the early *C. elegans* embryo. *Mech. Dev.* **54**, 133–147.

Seydoux, G., and Dunn, M. A. (1997). Transcriptionally repressed germ cells lack a subpopulation of phosphorylated RNA polymerase II in early embryos of *Caenorhabditis elegans* and *Drosophila melanogaster*. *Development (Cambridge, UK)* **124**, 2191–2201.

Seydoux, G., and Fire, A. (1994). Soma-germline asymmetry in the distribution of embryonic RNAs in *C. elegans*. *Development (Cambridge, UK)* **120**, 2823–2834.

Seydoux, G., Mello, C. C., Pettitt, J., Wood, W. B., Priess, J. R., and Fire, A. (1996). Repression of gene expression in the embryonic germ lineage of *C. elegans*. *Nature (London)* **382**, 713–716.

Shelton, C. A., and Bowerman, B. (1996). Time-dependent responses to *glp-1*-mediated inductions in early *C. elegans* embryos. *Development (Cambridge, UK)* **122**, 2043–2050.

St. Johnston, D., and Nusslein-Volhard, C. (1992). The origin or pattern and polarity in the *Drosophila* embryo. *Cell (Cambridge, Mass.)* **68**, 201–219.

Strome, S., and Wood, W. B. (1983). Generation of asymmetry and segregation of germ-line granules in early *C. elegans* embryos. *Cell (Cambridge, Mass.)* **35**, 15–25.

Strome, S., Martin, P. R., Schierenberg, E., and Paulsen, J. (1995). Transformation of the germ line into muscle in mes-1 mutant embryos of *Caenorhabditis elegans*. *Development (Cambridge, UK)* **121**, 2961–2972.

Sulston, J. E., Schierenberg, E., White, J. G., and Thomson, J. N. (1983). The embryonic cell lineage of the nematode *Caenorhabditis elegans*. *Dev. Biol.* **100**, 64–119.

Thorpe, C. J., Schlesinger, A., Carter, J. C., and Bowerman, B. (1997). Wnt signaling polarizes an early C. elegans blastomere to distinguish endoderm from mesoderm. *Cell (Cambridge, Mass.)* **90**, 695–705.

Varnum, B. C., Lim, R. W., Sukhatme, V. P., and Herschman, H. R. (1989). Nucleotide sequence of a cDNA encoding

TIS11, a message induced in Swiss 3T3 cells by the tumor promoter tetradecanoyl phorbol acetate. *Oncogene* **4**, 119–120.

Waddle, J. A., Cooper, J. A., and Waterston, R. H. (1994). Transient localized accumulation of actin in *Caenorhabditis elegans* blastomeres with oriented asymmetric divisions. *Development (Cambridge, UK)* **120**, 2317–2328.

Waring, D. A., and Kenyon, C. (1990). Selective silencing of cell communication influences anteroposterior pattern formation in *C. elegans*. *Cell (Cambridge, Mass.)* **60**, 123–131.

Waring, D. A., and Kenyon, C. (1991). Regulation of cellular responsiveness to inductive signals in the developing *C. elegans* nervous system. *Nature (London)* **350**, 712–715.

Watts, J. L., Etemad-Moghadam, B., Guo, S., Boyd, L.,

Draper, B. W., Mello, C. C., Priess, J. R., and Kemphues, K. J. (1996). *par-6*, a gene involved in the establishment of asymmetry in early *C. elegans* embryos, mediates the asymmetric localization of PAR-3. *Development (Cambridge, UK)* **122**, 3133–3140.

Wolf, N., Priess, J. R., and Hirsh, D. (1983). Segregation of germline granules in early embryos of *C. elegans*: An electron microscopic analysis. *J. Exp. Morphol.* **73**, 297–306.

Wood, W. B. (1991). Evidence from reversal of handedness in *C. elegans* embryos for early cell interactions determining cell fates. *Nature (London)* **349**, 536–538.

Yochem, J., and Greenwald, I. (1989). *glp-1* and *lin-12*, genes implicated in distinct cell–cell interactions in *C. elegans*, encode similar transmembrane proteins. *Cell (Cambridge, Mass.)* **58**, 553–563.

8

Sex and Death in the Caenorhabditis elegans Germ Line

RONALD E. ELLIS
Department of Biology
University of Michigan, Ann Arbor
Ann Arbor, Michigan 48109

I. Introduction

The germ line must have been one of the first animal tissues to evolve, since germ cells are essential for reproduction. This role in reproduction imposes several requirements on differentiating germ cells. First, the cell cycle must be regulated so that some germ cells undergo mitosis and others initiate meiosis, the specialized form of division required to form gametes. Second, the fate of differentiating germ cells must be controlled, so that males produce sperm and females produce oocytes. This dichotomy is the most widespread form of sexual dimorphism in the animal kingdom, and could provide a model for how other sexual fates are regulated. Third, oocytes are extremely complex cells, which in many animals require nurse cells or other accessories to aid their development. In such animals, the germ line must produce and eliminate these helper cells. Several lines of research have led to a picture of how each of these cell-fate decisions is made in the nematode *Caenorhabditis elegans.*

II. Development and Structure of the Germ Line

In *C. elegans,* as in most higher animals, the germ line is set aside early in development (Sulston *et al.,* 1983). By the 24-cell stage, it consists of a single founder cell, which is marked by the presence of germ granules (also known as P-granules). These granules resemble those found in several other types of animals, and are specifically segregated to the germ line during the first set of embryonic cell divisions (Strome and Wood, 1983; Hird *et al.,* 1996). Maternal PIE-1 protein is also segregated to the germ line during these divisions, and is essential both for specifying the identify of the founding germ cell, and for repressing transcription in the embryonic germ line (Mello *et al.,* 1992, 1996; Seydoux *et al.,* 1996). The role these factors and associated genes play in establishing the germ line has been reviewed by Schedl (1997), Schnabel and Priess (1997), Kemphues and Strome (1997) and Bowerman (Chapter 7).

    The founding germ cell generates two daughters,

Z2 and Z3, which remain quiescent during the rest of embryogenesis. Early in larval development these cells begin to divide, a process that continues for the rest of the life of the animal (Hirsh *et al.,* 1976; Klass *et al.,* 1976). At the beginning of the fourth larval stage the most proximal cells enter meiosis, and rapidly progress to the pachytene stage of prophase I. They remain at this stage for several hours, after which they resume differentiation and form gametes. The production of either sperm or oocytes continues throughout adulthood, so that the creation of new germ cells by mitosis is balanced by their loss through meiosis. Although the mechanisms that control these cell fates are similar in males and hermaphrodites, they are not identical, and must be considered separately.

In *C. elegans,* XX animals develop as self-fertile hermaphrodites; these are essentially females that produce some sperm before beginning oogenesis. The germ cells in hermaphrodites are distributed among two symmetrical arms of the gonad (Fig. 1). The cells at the distal end (farthest from the vulva) undergo mitosis, more proximal cells enter meiosis, and those near the bend in the gonad begin final differentiation. In L4 hermaphrodites, differentiating germ cells become sperm. In adult hermaphrodites they either become oocytes or undergo programmed cell deaths. Gumienny and colleagues (1998) suggest that the doomed cells might produce materials that aid the development of nearby oocytes. This process would not require secretion, since most germ cells retain an opening to a central core of cytoplasm, and are thus part of a large syncytium (Fig. 1). Germ cells only separate from this syncytium late in the process of differentiation. (For simplicity, throughout this review I will use the term germ cell to mean a nucleus and its associated cytoplasm and membranes, whether or not it is part of the syncytium).

Animals with a single X chromosome develop as males. In males, the germ cells are arranged in a single reflexed testis (Fig. 2). The cells at the distal end undergo mitosis, more proximal cells enter meiosis, and those midway past the bend in the gonad begin spermatogenesis. The sperm are stored in the seminal vesicle, and pass out through the vas deferens during mating.

III. Regulation of the Cell Cycle in Germ Cells

A. The Decision between Mitosis and Meiosis

Because the adult ovotestes and the male testis resemble long tubes, with mitotic germ cells at one end and fully differentiated gametes at the other, one can find germ cells at any stage of differentiation by scanning along these tubes. The first cell-fate decision occurs near the distal end in both sexes, at the point where slightly more proximal germ cells enter meiosis, while their more distal neighbors undergo mitosis. This choice raises two important questions. First, how is the decision to divide by mitosis or meiosis coordinated, so that each germ cell adopts the correct fate? Second, does implementing this decision in the germ line require unique genes to help regulate the cell cycle and cell division?

1. The Distal Tip Cells Promote Mitosis Instead of Meiosis

In *C. elegans,* two somatic cells are found near the mitotic germ cells at the distal end of the gonad, and are thus known as the distal tip cells. One distal tip cell is present at the end of each ovotestis (Fig. 1), and two distal tip cells lie adjacent to each other at the end of the male testis (Fig. 2). If the distal tip cells are killed with a laser microbeam at any time during development or adulthood, germ cells cease mitosis and enter meiosis (Kimble and White, 1981). Thus, the distal tip cells provide a signal that controls this fate in nearby germ cells. By contrast, killing the cells that generate the rest of the somatic gonad does not force all germ cells to undergo either mitosis or meiosis (McCarter *et al.,* 1997).

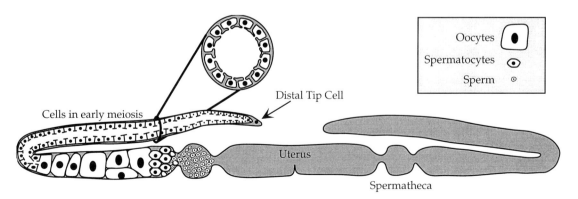

Figure 1   Structure of the adult hermaphrodite gonad and germ line.

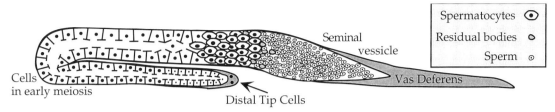

Figure 2 Structure of the adult male gonad and germ line.

This result indicates that the other somatic cells are not essential for regulating this cell-fate decision, although they are necessary for the development of a large and healthy germ line.

2. The glp-1 Signaling Pathway

The *glp-1* gene was identified by screening for mutants in which germ cells undergo meiosis rather than mitosis. Recessive mutations that inactivate *glp-1* cause germ cells to enter meiosis early in larval development, after only two or three rounds of mitotic division (Austin and Kimble, 1987). A dominant mutation that causes GLP-1 to be constitutively active has the opposite effect—it causes ectopic germ cells to undergo mitosis (Berry *et al.*, 1997). Three lines of evidence suggest that *glp-1* encodes a receptor that allows germ cells to transduce a signal from the distal tip cells. First, analysis of genetic mosaic animals shows that wild-type *glp-1* is needed in the germ line, but not in the distal tip cells (Austin and Kimble, 1987). Second, the sequence of *glp-1* suggests that it encodes a transmembrane receptor, similar in structure to the C. *elegans* LIN-12 protein and Notch of *Drosophila* (Austin and Kimble, 1989; Yochem and Greenwald, 1989). Third, antibody staining reveals that GLP-1 is located in the membranes of mitotic germ cells, but is not found in distal tip cells (Crittenden *et al.*, 1994).

Screens for additional mutations that cause a Glp-1 phenotype did not identify other genes in this pathway. However, mutations in *lag-1* or *lag-2* appeared to prevent signaling by both GLP-1 and the related protein LIN-12, suggesting that they encode components common to both signal transduction pathways (Lambie and Kimble, 1991; reviewed by Kimble and Simpson, 1997). This hypothesis was confirmed by experiments that showed the *lag-2* gene encodes the ligand for GLP-1. The predicted *lag-2* product was shown to be a small transmembrane protein similar in structure to Delta of *Drosophila*, which acts as a ligand for Notch (Henderson *et al.*, 1994; Tax *et al.*, 1994). Furthermore, *lag-2* mRNA was detected in the distal tip cells, but not in the germ line (Henderson *et al.*, 1994). Finally, mature LAG-2 protein was observed both on the surface of the distal tip cells, and colocalized with GLP-1 in complexes within the germ line (Henderson *et al.*, 1994).

How does the signaling process select which germ cells should undergo mitosis? Two results suggest that LAG-2 remains on the surface of the distal tip cells prior to its interaction with GLP-1. First, secretion of LAG-2 is not necessary for interaction with GLP-1, since antibody staining reveals that the intracellular portion of LAG-2 is taken up by germ cells (Henderson *et al.*, 1994). Second, the *lag-2(q477)* mutation, which results in the production of a truncated LAG-2 protein containing only the extracellular domain, causes a complete loss of gene activity (Henderson *et al.*, 1994, 1997). Some experiments involving the expression of recombinant proteins in transgenic worms suggest that a secreted ligand is capable of activating the GLP-1 receptor, and can even induce ectopic germ cells to undergo mitosis (Fitzgerald and Greenwald, 1995; Henderson *et al.*, 1997). This effect could be due to diffusion of the secreted LAG-2, so that it stimulates distant germ cells, or perhaps to its increased stability or activity. There is a striking correlation between the region of the germ line that contains mitotic cells which express GLP-1 (Crittenden *et al.*, 1994) and the region in contact with processes extended by the distal tip cell (Fitzgerald and Greenwald, 1995), so perhaps direct contact with the distal tip cell is required for signal transduction. If so, the intracellular message transmitted by GLP-1 must not diffuse very far through the germ line syncytium.

What are the components of this intracellular message? The *lag-1* gene encodes a protein that is similar to the transcription factor *Suppressor of hairless* from *Drosophila*, which acts downstream of the receptor Notch (Christensen *et al.*, 1996). Furthermore, epistasis tests using constitutively active GLP-1 support the pathway LAG-2 → GLP-1 → LAG-1 (Berry *et al.*, 1997). Thus, LAG-1 might transmit the signal from GLP-1 to target genes in the nucleus. Colocalization assays and immunoprecipitation experiments indicate that LAG-1 can interact with GLP-1 in two distinct ways: a nonspecific interaction involving the ANK domain, which contains six ankyrin repeats, and a specific interaction with the RAM domain, located near the cell membrane (Roehl *et al.*, 1996). Expression of the RAM domain has no detectable effect on GLP-1 signaling (Roehl *et al.*, 1996), but expression of the ANK domain mimics active GLP-1 in the regulation of several somatic cell fates (Roehl and Kimble, 1993). Although these experiments have not been repeated in the germ line, they suggest that the ANK domain of GLP-1 might interact with LAG-1 to regulate transcription. The analysis of *glp-1* mutations also indicates that the ANK repeats play a crucial role in the function of the protein. Many loss-of-function (*lf*) mutations map to the ANK repeats (Kodoyianni *et al.*, 1992). Furthermore, all intragenic suppressors of the two temperature-sensitive *glp-1* alleles map to these repeats (Lissemore *et al.*, 1993).

What are the targets of LAG-1, and how do these genes control which pattern of cell division is adopted? *In vitro* studies have shown that LAG-1 is capable of binding DNA with the sequence (A/G)TGGGAA (Christensen *et al.*, 1996). Many copies of this target sequence are found in the *glp-1*, *lag-1*, and *lin-12* genes, suggesting that LAG-1 exerts positive feedback on components of the signaling pathway (Christensen *et al.*, 1996). Two observations support the existence of a feedback mechanism. First, several weak *glp-1* mutations have an all-or-nothing effect on development of the germ line; in these mutants one ovotestis often produces large numbers of germ cells, while the other ovotestis produces very few (Kodoyianni *et al.*, 1992). Second, *glp-1* mRNA is only found in the mitotic region of the male testis, where GLP-1 protein is active (Crittenden *et al.*, 1994). However, transcriptional regulation of *glp-1* cannot explain all aspects of its expression pattern, because in hermaphrodites the transcript is found throughout the germ line, and translational regulation helps restrict GLP-1 protein to mitotic cells (Crittenden *et al.*, 1994; Evans *et al.*, 1994).

So far, no other targets of LAG-1 are known. However, germ cells in which both *gld-1* and *gld-2* are inactive appear to remain permanently in the mitotic cell cycle, even if *glp-1* or *lag-1* is also defective (Kadyk and Kimble, 1998). Thus, the two *gld* genes act downstream of *lag-1* to control entry into meiosis. Furthermore, a screen for mutations that enhance weak alleles of *glp-1* recovered, in addition to five new alleles of *lag-1*, mutations in *glp-4* and the new genes *ego-1* through *ego-5* (Qiao *et al.*, 1995). Some of these new genes might be targets of LAG-1. This possibility is consistent with epistasis tests, which suggest that *ego-1* and *ego-3* act downstream of active GLP-1 (Qiao *et al.*, 1995). However, most of these new genes have pleiotropic effects on the development of the germ line that are difficult to interpret, and their molecular identities are not yet known. Finally, several genes that might interact with LAG-2, GLP-1, or LAG-1 were identified by mutations that suppress *glp-1* temperature-sensitive alleles, or that enhance the formation of tumors in *glp-1(oz112gfoz120)* animals (*sog* mutations, Maine and Kimble, 1993; *teg* mutations, L. Berry, and T. Schedl,

personal communication). So far, none of the *sog* or *teg* mutations that have been tested were able to suppress null alleles of *glp-1*. Thus, they probably act by enhancing the activity of the GLP-1 signal transduction pathway.

B. Control of the Cell Cycle in Germ Cells

To learn what factors are needed for germ cells to progress through the cell cycle, several groups have screened for mutants that develop into sterile adults with small germ lines. The genes they have identified fall into two groups. First, several genes affect the divisions of both germ cells and somatic cells. For example, recessive mutations in the *lin-5* gene prevent most postembryonic cells from completing mitosis (Albertson *et al.*, 1978; Sulston and Horvitz, 1981). Because some germ cells divide in these animals, and these cells occasionally differentiate as sperm, it is not clear if the germ line defects are a direct result of the *lin-5* mutation, or are caused indirectly by the failure of the somatic gonad to develop properly. Second, two genes appear to be required specifically for cell division in the germ line. Recessive mutations in the *glp-3* and *glp-4* genes prevent most germ cells from undergoing either mitosis or meiosis, but do not affect somatic development (Beanan and Strome, 1992; Kadyk *et al.*, 1997). The *glp-4* mutants also produce some oocytes; these are very irregular in morphology, and unable to develop into viable embryos after fertilization (Beanan and Strome, 1992; Qiao *et al.*, 1995). It is not clear if this defect in oogenesis defines a second function for GLP-4, or if these oocytes are produced because small amounts of GLP-4 activity allow partial advancement through the meiotic cell cycle.

C. Progression through Meiosis

Once germ cells have entered meiosis, they rapidly progress through the initial stages of prophase I, take several hours to complete pachytene, and then resume development. This resumption requires the activity of several genes of the *ras* signal transduction pathway (Church *et al.*, 1995), including *let-60* RAS, *lin-45* RAF, *mek-2* MAPKK, and *mpk-1* MAP kinase. If any of these genes is inactive, germ cells arrest at the pachytene stage and do not differentiate further. Analysis of genetic mosaics shows that MAP kinase must be present in germ cells, rather than the somatic gonad, to allow progression through meiosis (Church *et al.*, 1995). Perhaps this MAP kinase phosphorylates proteins that regulate progression through the cell cycle. Three *pex* genes have been identified that are needed specifically for progression of germ cells through meiosis (T. Schedl and E. Lambie, personal communication), but how they interact with the genes of the *ras* pathway is not yet understood.

The fact that the *ras* pathway is involved suggests that progression through meiosis is controlled by an extra-cellular signal. If so, the source of the signal, the ligand, and the receptor remain unknown. If the sheath and spermathecal cells that surround the hermaphrodite gonad are eliminated by a laser early in development, the fraction of germ cells in pachytene increases dramatically, though some still manage to progress through meiosis and differentiate as sperm or oocytes (McCarter *et al.*, 1997). Thus, the sheath or spermathecal cells might signal nearby germ cells to resume meiosis. However, it remains possible that these somatic cells only nurture and support the germ cells, and that unhealthy germ cells cannot exit pachytene. Furthermore, in the male testis there are no somatic cells in an appropriate position to provide this signal.

In hermaphrodites, the *gld-1* gene also plays a critical role regulating the progression of oocytes through the pachytene stage of prophase I (Francis *et al.*, 1995a,b). Inactivation of GLD-1 by a mutation causes germ cells to return to mitosis after entering this stage, instead of completing meiosis and differentiating as oocytes, but does not affect germ cells that are committed to forming sperm. Molecular analysis reveals that *gld-1* encodes a cytoplasmic protein with a potential KH domain, suggesting that GLD-1 might bind RNA; furthermore, mutations in this KH domain inactivate GLD-1 (Jones and Schedl, 1995). Perhaps GLD-1 represses translation of mRNAs that accumulate in the germ line for packaging into oocytes, and whose products normally promote mitosis.

IV. Control of Sexual Fate in Germ Cells

Caenorhabditis elegans has two sexes. *XO* individuals develop as males, and *XX* individuals become hermaphrodites. To understand how the sexual fate of germ cells is regulated, we need to address two distinct questions. First, what genes directly control spermatogenesis and

oogenesis, and how are they controlled by the other sex-determination genes? Second, what regulatory stratagems allow an *XX* animal, which is undergoing female development in most tissues, to produce sperm for a brief period of time? To address these questions we must first consider how the sexual identity of the rest of the animal is regulated.

A. Sex Determination in the Soma

In *C. elegans*, the initial signal that specifies sex is the ratio of X chromosomes to autosomes (Madl and Herman, 1979). If there are two X chromosomes this ratio equals 1.0, and the animal develops as a hermaphrodite. By contrast, a single X chromosome yields a ratio of 0.5, causing the animal to develop as a male. The $X:Autosome$ ratio acts by controlling a sophisticated regulatory hierarchy (Figs. 3 and 4). The entire process can be divided into four steps: (a) initial assessment of sexual identity, (b) communication between cells, (c) interpretation of this signal, and (d) specification of cells to adopt appropriate male or female fates.

1. Initial Determination of Sexual Identity

The primary signal that controls sexual identity is present in all cells—the ratio of X chromosomes to autosomes (Madl and Herman, 1979). While no autosomal elements have been identified so far, genetic studies have defined four X chromosome regions that contain signal elements (Akerib and Meyer, 1994; I. Carmi, J. Kopczynski, and B. J. Meyer, personal communication; Hodgkin and Albertson, 1995). One of these four regions causes both feminization and lethality if multiple, transgenic copies are placed in *XO* individuals (Hodgkin *et al.*, 1994). Genetic and molecular analysis revealed that the gene *fox-1* which encodes a putative RNA-binding protein, corresponds to the signal element in this region (Nicoll *et al.*, 1997). A second region contains the transcription factor *sex-1* (I. Carmi, J. Kopczynski, and B. J. Meyer, personal communication). Both genes appear to act by regulating *xol-1*, which

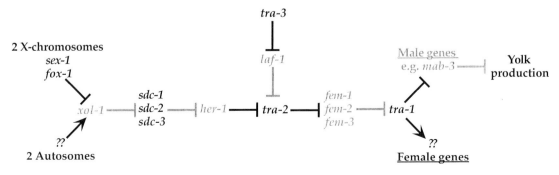

Figure 3　Sex determination in the hermaphrodite soma.

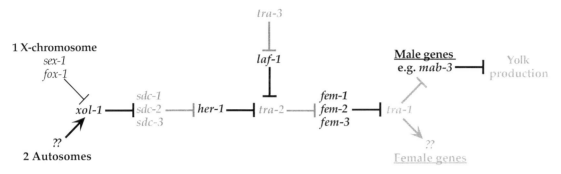

Figure 4 Sex determination in the male soma.

functions as a switch controlling both sex determination and dosage compensation (Miller *et al.*, 1988; Rhind *et al.*, 1995; Nicoll *et al.*, 1997). When the *X : Autosome* ratio is low, active *xol-1* directs male development. By contrast, when the *X : Autosome* ratio is high, *xol-1* is inactive, and female development and dosage compensation ensue. The *xol-1* gene encodes a novel protein which somehow represses *sdc-1, sdc-2,* and *sdc-3* (reviewed by Meyer, 1997). Both *sdc-1* and *sdc-3* encode transcription factors (Nonet and Meyer, 1991; Klein and Meyer, 1993), and appear to regulate sexual identity by repressing transcription of the *her-1* gene. The *sdc* genes also control the process of dosage compensation in *XX* individuals (reviewed by Meyer, 1997). Surprisingly, mutations that alter the activity of the dosage compensation process also can influence an animal's sexual identity (DeLong *et al.*, 1993). This feedback mechanism might be caused by the location of several sex-determination genes on the X chromosome.

2. *Communication between Cells*

The *her-1* gene encodes two transcripts, which are expressed in *XO*, but not *XX* animals (Trent *et al.*, 1991). Two experiments elucidate how this regulation occurs. First, the known gain-of-function (*gf*) mutations alter a single site in the P1 promotor, causing expression of both transcripts in *XO* and *XX* individuals (Trent *et al.*, 1988; Perry *et al.*, 1994). Second, transgenic *XX* worms that contain multiple copies of the P2 promotor are often sickly and partially transformed into males; these worms resemble *sdc* mutants, which are defective both for sex determination and dosage compensation (N. Rajwans and M. Perry, personal communication; W. Li, A. Streit, B. Robertson, and B. Wood, personal communication). These observations support the hypothesis that transcription of *her-1* is repressed in *XX* individuals, probably by the products of the *sdc* genes. This repression appears to involve both the P1 and P2 promotors.

Analysis of genetic mosaic individuals indicates that HER-1 does not act cell autonomously, but instead functions in a signal transduction process (Hunter and Wood, 1992). Two results indicate that HER-1 is prob-

ably a secreted protein that helps to coordinate sexual identity in the developing worm. First, the larger *her-1* transcript encodes a small protein with a signal sequence; this protein is sufficient to rescue *her-1* mutant animals (Perry *et al.*, 1993). Second, ectopic expression of *her-1* in the body wall muscles of *XX* animals can cause the rest of the worm to undergo male development (Perry *et al.*, 1993).

The likely receptor for HER-1 is a product of the *tra-2* gene, TRA-2A. Genetic studies indicate that *tra-2* activity is repressed by *her-1* in males (Hodgkin, 1980). Sequence analysis indicates that TRA-2A has nine transmembrane domains and a signal sequence, and thus functions as a membrane protein (Kuwabara *et al.*, 1992). Furthermore, dominant mutations in the putative extracellular domain of TRA-2A render this protein insensitive to regulation by HER-1 (Hodgkin and Albertson, 1995; Kuwabara, 1996). Two experiments indicate that high levels of *tra-2* activity cause female development. First, using constructs driven by a heat shock promoter, Kuwabara and Kimble (1995) showed that expression of TRA-2A in *XO* animals causes female development. Second, expression of TRA-2B, a protein encoded by a second *tra-2* transcript, also causes female development. Since TRA-2B resembles the intracellular domain of TRA-2A, this domain is sufficient to cause cells to become female.

In addition to the interaction of TRA-2A with HER-1, the activity of the *tra-2* gene is subject to other controls in the soma. Northern analyses suggest that the amount of *tra-2* transcripts is subject to feedback regulation by the *tra-1* gene (Okkema and Kimble, 1991). Furthermore, the translation of *tra-2* mRNA appears to be regulated in a sex-specific manner that involves the *tra-3* and *laf-1* genes (see below, and Hodgkin, 1980; Goodwin *et al.*, 1993, 1997).

3. *Interpretation of the Signal*

The signal transduced by TRA-2A appears to regulate the activity of three genes required for male development: *fem-1, fem-2,* and *fem-3* (Doniach and Hodgkin, 1984; Kimble *et al.*, 1984; Hodgkin, 1986).

Each gene has been cloned, but how they promote male development is not clear. The *fem-1* gene is expressed widely in both sexes, and its product contains ankyrin repeats, which might mediate interactions with other proteins (Spence *et al.*, 1990; Gaudet *et al.*, 1996). Northern analysis indicates that *fem-2* also is expressed in both sexes (Pilgrim *et al.*, 1995). FEM-2 shows homology to protein phosphatase 2C, and mutations that abolish phosphatase activity inactivate FEM-2 (Pilgrim *et al.*, 1995; Chin-Sang and Spence, 1996). Furthermore, studies using the yeast two-hybrid system or immunoprecipitation indicate that FEM-2 can bind FEM-3 (Chin-Sang and Spence, 1996). Expression of the major *fem-3* transcript is restricted to males and the germ line of hermaphrodites (Rosenquist and Kimble, 1988). However, this difference is unlikely to regulate sexual development, since the *fem-3* transcripts in young embryos are derived from the mother, and are found at the same level in both XX and XO animals (Ahringer *et al.*, 1992). Sequence analysis indicates that FEM-3 is a novel protein (Ahringer *et al.*, 1992), and studies using both immunoprecipitation and the yeast two-hybrid system suggest that FEM-3 can bind not only FEM-2 (Chin-Sang and Spence, 1996), but also TRA-2A (A. Spence and P. Kuwabara, personal communication).

By examining double mutants, Hodgkin (1986) showed that *tra-1* acts downstream of the *fem* genes in the soma. Combining these results, one possible model is that FEM-3 is inactivated by binding to TRA-2A; this would allow *tra-1* to direct female development. However, when TRA-2A is inactive, FEM-1, FEM-2, and FEM-3 are free to turn off *tra-1*, resulting in male development (Fig. 4). Gain-of-function (*gf*) mutations in TRA-1A map to a domain near the amino-terminus of the protein (de Bono *et al.*, 1995). If this regulatory model is correct, these *gf* mutations might define a domain needed for repression of TRA-1A by the FEM proteins.

4. Specification of Male and Female Cell Fates

Epistasis experiments indicate that *tra-1* acts at the end of the sex-determination pathway in the some

(Hodgkin, 1980, 1986), and analyses of genetic mosaic animals show that *tra-1* acts cell-autonomously (Hunter and Wood, 1990). These results are consistent with the hypothesis that *tra-1* encodes a switch that controls cell fate in response to the other sex-determination proteins. How does this switch work? Molecular analysis shows that the major product of *tra-1*, TRA-1A, contains five Zinc-fingers; these fingers resemble those found in the human oncogenes *GLI* and *GLI3*, and in the *Cubitus interruptus* protein of *Drosophila* (Zarkower and Hodgkin, 1992). Direct biochemical studies indicate that TRA-1A binds DNA, and that its target sequences are very similar to those favored by the *GLI* proteins (Zarkower and Hodgkin, 1993). These results suggest that TRA-1A might promote transcription of genes required for female fates, and repress genes required for male fates. The *tra-1* gene also encodes a smaller transcript, whose product does not appear to bind DNA (Zarkower and Hodgkin, 1992, 1993). The function of this smaller protein remains unclear. To date, no somatic genes have been shown to be direct targets of TRA-1A.

What genes regulate the sexually dimorphic fates of individual male or female cells? In the soma, the best candidate is the *mab-3* gene, which is needed to prevent yolk production in the male intestine, and also for proper development of the male tail (Shen and Hodgkin, 1988). Although there is no evidence for a direct interaction, the level of *mab-3* transcripts is controlled by activity of the *tra-1* gene (Raymond *et al.*, 1998). Sequence analysis indicates *mab-3* shows similarity to the *Drosophila doublesex* gene, which is a transcriptional regulator of many sexual fates in files, and which controls yolk production (Raymond *et al.*, 1998). This is the first indication that portions of the sex-determination pathways have been conserved between nematodes and insects. Since there are potential MAB-3 binding sites in the promotors of the *C. elegans* vitellogenin genes (which encode the yolk proteins), MAB-3 is probably the terminal regulatory gene for this sexual trait (W. Yi and D. Zarkower, personal communication).

Two other genes might regulate the sexual fates of individual somatic tissues. First, genetic tests indicate

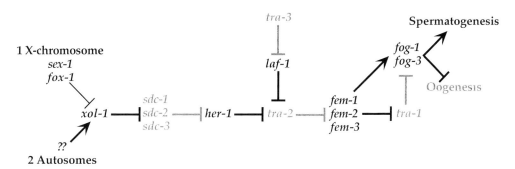

Figure 5　A model for sex determination in the male germ line.

that mutations in *laf-1* can partially feminize the male tail and germ line (E. Goodwin and J. Kimble, personal communication), so perhaps the *laf-1* gene is a target of TRA-1A. Second, dominant mutations in *egl-1* appear to alter the sexual identity of the HSN neurons (Trent *et al.*, 1983; Ellis and Horvitz, 1986), causing them to die in hermaphrodites like they normally do in males. However, the null phenotype of *egl-1* is not known, so it is not clear if the function of *egl-1* is to regulate the sexual fate of these nerve cells.

B. Sex Determination in the Germ Line

Three factors appear responsible for differences between the control of sexual fate in germ cells and in the soma. First, unique genes seem to directly control spermatogenesis and oogenesis, but not any somatic fates. Second, several genes are required in hermaphrodites to allow the production of sperm late in larval development, and the subsequent switch to oogenesis. Third, the production and storage of maternal RNAs for packaging into oocytes requires special regulatory mechanisms to prevent them from influencing germ cell fates.

1. Terminal Regulation of Sexual Fate in Germ Cells

While examining *tra-1 fem* double mutants, Hodgkin (1986) discovered that these animals develop male bodies, but that all germ cells differentiate as oocytes. This result showed that *fem-1*, *fem-2*, and *fem-3* are essential for spermatogenesis. However, it also raised an unexpected question—why does *tra-1* appear to act upstream of the *fem* genes in the germ line, but downstream of them in the soma?

The analysis of *tra-1* mutants raised an additional question about the function of this gene in the germ line (Hodgkin, 1987; Schedl *et al.*, 1989). Gain-of-function mutations in *tra-1* not only cause XX and XO animals to develop female bodies, but also cause them to produce oocytes instead of sperm, suggesting that *tra-1* prevents spermatogenesis and promotes oogenesis. Surprisingly, many *lf* mutations in *tra-1* also cause XX and XO animals to produce oocytes, after an initial period of sperm production. This result implies that *tra-1* is needed to maintain spermatogenesis. Thus, *tra-1* might have two separate functions in germ cells.

Genes that directly control whether germ cells differentiate as sperm or as oocytes could provide the key to disentangling the actions of *tra-1*, *fem-1*, *fem-2*, and *fem-3* in germ cells. Two candidates exist, *fog-1* and *fog-3* (Barton and Kimble, 1990; Ellis and Kimble, 1995). Mutations in these genes cause all germ cells to differentiate as oocytes rather than sperm, but do not affect other aspects of sexual development. Furthermore, mutations in either *fog-1* or *fog-3* are epistatic to mutations

in other sex-determination genes, suggesting that both are absolutely required for spermatogenesis to occur. These results are consistent with models in which germ cells differentiate as sperm if *fog-1* and *fog-3* are active, but as oocytes if either gene is inactive. Molecular studies involving *fog-3* are now testing this model.

a. The fog-3 Gene Appears to Be Regulated by TRA-1A. Northern analyses indicate that transcription of *fog-3* is regulated by the other sex-determination genes (Chen and Ellis, 1998). Transcript levels are high in *fem-3(gf)* mutants, which make only sperm, and low in *fem-1* loss of function (*lf*) mutants, which produce only oocytes. The level of *fog-3* transcripts also is correlated with spermatogenesis in wild-type animals—transcripts are found at high levels in larval hermaphrodites, larval males, and adult males, which all produce sperm, but are not found in adult hermaphrodites, which produce oocytes. In theory, these differences might be caused either by changing levels of transcription or by changes in the stability of the *fog-3* message. Sequence analysis suggests that this expression pattern might involve three sites in the fog-3 promotor. Each site has the sequence (T)TTTCnnnnTGGGTGGTC, which *in vitro* tests show binds TRA-1A with high affinity (Zarkower and Hodgkin, 1993). Thus, the fact that high levels of TRA-1A activity prevent spermatogenesis and promote oogenesis could be due to repression of *fog-3* transcription by TRA-1A binding to these three sites. Since lower levels of TRA-1A activity promote spermatogenesis, it is possible that the occupation of a subset of these sites by TRA-1A promotes transcription. A combination of *in vivo* and *in vitro* assays are probing this model.

The second result to emerge from the molecular analysis is that *fog-3* encodes a product with weak similarity to the Tob proteins of mammals (Chen *et al.*, 1998). The first 113 amino acids of FOG-3 show 28% identity and 48% similarity to the Tob proteins, which have been shown to bind the *erbB-2* receptor tyrosine kinase (Matsuda *et al.*, 1996; Yoshida *et al.*, 1997). Six of the seven characterized missense mutations affect this domain. Although it is not known if this domain mediates binding of FOG-3 to another protein, expression of constructs encoding both this portion of FOG-3 and the middle of the protein (but not the carboxyl-terminus) causes a dominant *fog-3* mutant phenotype. This result implies that these domains bind either wild-type FOG-3 or some other protein required for spermatogenesis (Chen *et al.*, 1998). The carboxyl-terminal half of FOG-3 is novel in sequence, but contains short stretches that are also homologous to the Tob proteins.

Because the three *fem* genes are required for germ cells to become sperm, they must act with or positively regulate *fog-1* and *fog-3*, in addition to repressing TRA-1A. How this interaction occurs remains unknown. One possibility is that FEM-2, which is homologous to protein phosphatase 2C, activates the FOG-1 or FOG-3 proteins. There are several potential phosphorylation sites in the FOG-3 protein, which might be targets for FEM-2 regulation (Chen *et al.*, 1998). Activation of FOG-3 by FEM-2 could explain why mutations in the *fem* genes are epistatic to *tra-1* mutations in the germ line.

b. fog-1 Also Is Required for Spermatogenesis. Genetic studies indicate that *fog-1* also is required for germ cells to differentiate as sperm (Barton and Kimble, 1990; Ellis and Kimble, 1995). Tests of the severity of *fog-1* alleles in trans to the weak mutation *fog-1(q253ts)* indicate that many alleles have a dominant negative effect (R. Ellis, unpublished data). This result suggests that FOG-1 is likely to bind either to itself or to another protein required for spermatogenesis, and that the dominant negative mutations specify a FOG-1 protein that retains this binding ability, but cannot direct spermatogenesis. Since the studies of transgenic animals described above indicate similar behavior for FOG-3, one attractive model is that FOG-1 and FOG-3 must bind each other to function.

How FOG-1, FOG-3, and the FEM proteins promote germ cells to differentiate as sperm rather than as oocytes remains a mystery. However, many genes required for spermatogenesis are expressed only in developing spermatocytes, suggesting that their transcription is regulated (e.g., L'Hernault and Arduengo, 1992; Varkey *et al.*, 1995). Thus, one possibility is that the FOG and FEM proteins activate transcription of genes involved in spermatogenesis, and repress transcription of genes for oogenesis.

c. Negative Regulation of TRA-1A in Germ Cells. The weak gain-of-function allele *tra-1(e2271)* causes all germ cells to differentiate as oocytes, but does not affect somatic cells (Hodgkin, 1987). In addition, several other *tra-1* alleles show a similar *gf* activity in the hermaphrodite germ line, if their transcripts are stabilized by a second mutation in one the *smg* genes (Zarkower *et al.*, 1994). Sequence analysis indicates that these mutations might define a site in the C-terminal region of TRA-1A that is negatively regulated in germ cells.

2. Regulation of Spermatogenesis in Hermaphrodites

By examining double mutants, Schedl and Kimble (1988) showed that the relative activities of the *tra-2* and *fem-3* genes play a crucial role in controlling germ cell fate. Increasing *fem-3* activity causes XX animals to produce only sperm, whereas increasing *tra-2* activity causes them to produce only oocytes. However, when both *fem-3* and *tra-2* activity are increased in the same individual, germ cells differentiate normally. These experiments, combined with studies showing that *tra-2*

acts upstream of *fem-3* (Hodgkin, 1986), support the following model. For hermaphrodites to produce sperm during larval development, the level of TRA-2A activity must be low, to prevent the inactivation of FEM-3, which is required for spermatogenesis. For these individuals to begin producing oocytes as adults, the level of TRA-2A relative to FEM-3 must increase, causing the inactivation of FEM-3.

Two results show that these changes in the relative activities of TRA-2A and FEM-3 cannot be achieved by the temporary production of the HER-1 ligand in XX animals. First *her-1(lf)* worms are able to produce sperm (Hodgkin, 1980). Second, the *her-1* transcript is not found in XX individuals (Trent *et al.*, 1991). Several alternative mechanisms that do help regulate spermatogenesis in hermaphrodites have been identified, although the relative importance of each is not yet known.

a. The fog-2 Gene Promotes Spermatogenesis. Genetic screens have identified several kinds of mutations that prevent spermatogenesis in hermaphrodites, but have little or no effect on males. Some of these mutations define the *fog-2* gene, which is required for spermatogenesis in hermaphrodites but not in males (Schedl and Kimble, 1988). Mutations in *fog-2* are suppressed by mutations that inactivate TRA-2A, or by mutations that cause production of extra FEM-3 protein. Although these results do not indicate if FOG-2 acts by repressing TRA-2A, or by directly promoting FEM-3 activity, the gene recently has been cloned, and molecular tests of FOG-2 function should soon be possible (B. Clifford and T. Schedl, personal communication).

b. The gld-1 Gene Promotes Spermatogenesis. The *gld-1* gene is required for several different aspects of germ line development, and is absolutely required for oogenesis (Francis *et al.*, 1995a,b). Surprisingly, mutations that inactivate *gld-1* also prevent germ cells in hermaphrodites from differentiating as sperm. Instead, these cells appear to begin oogenesis, since they contain several transcripts that are found in oocytes but are not produced in sperm (Jones *et al.*, 1996). However, shortly after the pachytene phase of meiosis I, these germ cells relapse into mitosis, and do not differentiate further (Francis *et al.*, 1995a). Because *gld-1(lf)* mutations do not prevent spermatogenesis in males or in *fem-3(gf)* hermaphrodites, GLD-1 might act upstream of the *fem* genes to control the sexual identity of germ cells. Recent work by B. Clifford and T. Schedl indicates that GLD-1 binds FOG-2, so perhaps these genes act together to promote spermatogenesis in XX individuals (personal communication).

Two additional classes of *gld-1* mutations affect sex determination in germ cells. The *gld-1(Mog)* alleles are semidominant mutations that cause all germ cells to differentiate as sperm. These mutations promote spermatogenesis in hermaphrodites, as well as in some mutant males (Ellis and Kimble, 1995; Francis *et al.*, 1995a). By contrast, the *gld-1(Fog)* alleles are semidominant mutations that promote all germ cells to differentiate as oocytes (Francis *et al.*, 1995a). These mutations also affect both sexes. Although the molecular lesions responsible for these mutations are known (Jones and Schedl, 1995), it is not clear how these changes affect the regulation of germ cell fate. One possibility is that they alter GLD-1 binding to certain key mRNAs that control sex determination in germ cells. Alternatively, they might alter the normal regulation of hermaphrodite spermatogenesis by GLD-1 and FOG-2.

c. Translational Repression of tra-2 Promotes Spermatogenesis. Two kinds of mutations in the *tra-2* gene prevent spermatogenesis in hermaphrodites. One class consists of *gf* mutations such as *e2020*, which prevents spermatogenesis, but does not affect the development of a female body (Doniach, 1986; Schedl and Kimble, 1988). These mutations have only minor effects on male development, and *tra-2(gf)* XO animals are essentially fertile males. Sequence analysis reveals that each of the *tra-2(gf)* mutations is located in the 3'-untranslated region (3'-UTR) of the messenger RNA, in an area that contains two 28-nucleotide direct repeat elements (DREs, Goodwin *et al.*, 1993). Furthermore, a deletion that eliminates both direct repeats increases the fraction of *tra-2* message bound to multiple ribosomes, and thus probably increases the rate of translation. These *gf* mutations should therefore cause increased production of the normal TRA-2A protein. Presumably, this extra protein inactivates FEM-3 in XX individuals, causing all germ cells to develop as oocytes. By contrast, the effect of extra TRA-2A protein in males is small, probably because HER-1 is present in XO individuals, and can inactivate TRA-2A.

How is translation of the *tra-2* mRNA repressed? Biochemical assays suggest that a protein from worm extracts can bind to the direct repeat elements (Goodwin *et al.*, 1993). Experiments using a reporter gene fused to the *tra-2* 3'-UTR suggest that this repression is controlled by the *laf-1* gene (Goodwin *et al.*, 1997). It is not known if LAF-1 binds directly to these messages, or if it acts by regulating a separate binding protein. Because *laf-1* is required for viability, it must control other genes besides *tra-2*, but the nature of these targets is still a mystery. The regulation of mRNA translation by the DREs appears to be conserved, not only in the *tra-2* gene of related nematodes, but even in the control of the *GLI* gene of mammals (Jan *et al.*, 1997).

The *tra-3* gene behaves as an accessory to *tra-2* in genetic assays (Hodgkin, 1980; Schedl *et al.*, 1989). Sequence analysis reveals that *tra-3* encodes a member of the calpain protease family (Barnes and Hodgkin, 1996), and genetic tests suggest that *tra-3* might control

sexual fate by repressing *laf-1* (Goodwin *et al.*, 1997). One attractive model is that TRA-3 promotes female development by cleaving LAF-1, thereby increasing translation of the *tra-2* message and the production of TRA-2A.

d. A Site on the Intracellular Portion of TRA-2A is Required for Spermatogenesis. In addition to the *tra-2(gf)* alleles, a second class of *tra-2* mutations prevents hermaphrodite germ cells from developing as sperm. These mutations include alleles such as *e1941*. Mutations in this class are known as *mixomorphic* alleles, because they not only prevent spermatogenesis in XX individuals, but also partially masculinize their bodies (Doniach, 1986; Schedl and Kimble, 1988; P. Kuwabara and J. Kimble, personal communication). This result implies that the *mixomorphic* alleles have two different effects on TRA-2A. First, they slightly impair the function of the protein causing partial masculinization of XX individuals. Second, they might damage a site needed to downregulate TRA-2A in hermaphrodite germ cells; this would allow the mutant TRA-2A to be constitutively active, causing all germ cells to differentiate as oocytes. Sequence analysis of the *tra-2(mx)* mutations shows that each maps within a 22-amino acid region of the intracellular domain of TRA-2A (P. Kuwabara, P. Okkema, and J. Kimble, personal communication).

e. Translational Repression of fem-3 Promotes Oogenesis. Translational control also plays an important role in controlling *fem-3* activity in germ cells. The major *fem-3* transcript is found in both sexes, and appears to be expressed at high levels in the germ line of L4 larvae and adults (Rosenquist and Kimble, 1988; Ahringer *et al.*, 1992). This transcript encodes a novel protein, which genetic studies indicate is essential both for spermatogenesis, and for repression of TRA-1A in the soma (Hodgkin, 1986). Several mutations that alter the 3'UTR of *fem-3* appear to result in a gain of function, since they cause all germ cells to differentiate as sperm (Barton *et al.*, 1987; Ahringer and Kimble, 1991). This result suggests that an inhibitor of translation might bind to the wild-type message, and that the mutant mRNAs are not sensitive to this inhibition. Recently, the *fbf-1* and *fbf-2* genes have been detected by screening for proteins that bind the *fem-3* 3'-UTR (Zhang *et al.*, 1997). Injection of animals with antisense RNA that should inactivate these two genes yields progeny that resemble *fem-3(gf)* mutants. These results suggests that the FBF proteins must inhibit translation of *fem-3* mRNA for hermaphrodite germ cells to begin oogenesis.

The *fbf-1* and *fbf-2* genes were not identified in screens for XX individuals that produce oocytes instead of sperm, perhaps because they are functionally redundant. However, these screens did uncover six *mog* genes, each of which is needed for hermaphrodites to begin oogenesis (Graham and Kimble, 1993; Graham *et al.*, 1993). Using a reporter gene fused to the *fem-3* 3'-UTR, M. Gallegos and J. Kimble have shown that a mutation in any of these *mog* genes leads to increased activity of the reporter protein, suggesting that the *mog* genes help regulate translation of *fem-3* (personal communication). Mutations in the *mog* genes also have other effects on the germ line, since individuals homozygous for one of these mutations produce many fewer sperm than do *fem-3(gf)* animals. Furthermore, if any of the *mog* mutants produces oocytes because of a second mutation in *fog-1*, *fog-3* or one of the *fem* genes, the embryos are not viable (Graham and Kimble, 1993; Graham *et al.*, 1993). The *mog-1* and *mog-5* genes have been cloned, and each shows similarity to RNA helicases (A. Puoti and J. Kimble, personal communication). Perhaps the six *mog* genes play a general role in the regulation of maternal transcripts in the hermaphrodite germ line.

f. A Model for the Regulation of Germ Cell Fate. Although each of the components described above is required for hermaphrodites to produce sperm during larval development, and then switch to oogenesis during adulthood, it is not clear which ones are directly responsible for this switch. For example, activity of the other proteins might be constant, and changing levels of FOG-2 activity might be the force that allows first spermatogenesis and then oogenesis. Alternatively, FOG-2 levels might be constant, and changing levels of the FBF proteins might cause this switch in germ cell fate. Two kinds of experiments should help resolve this question in the next few years. First, analysis of the expression patterns of these regulatory genes in the germ line could indicate which ones are constant and which vary. Second, study of the expression or activity of downstream genes such as *fog-1* and *fog-3* in a variety of double mutants might resolve the order in which these regulatory genes act to control germ cell fates. For simplicity, in the model presented in Figs. 6 and 7 the

activities of each of the regulatory genes discussed above is depicted as varying between larval and adult hermaphrodites.

3. Regulation of Maternal mRNAs in the Hermaphrodite Germ Line

In adult hermaphrodites, a large variety of mRNAs are produced for packaging into oocytes. Many of these mRNAs have been shown to diffuse toward the proximal end of the germ line syncytium (Jones *et al.*, 1996). Because some of these messages encode proteins that regulate germ cell fates (such as GLP-1 or FEM-3), their translation must be carefully regulated. A variety of experiments have shown that the translation of several of these mRNAs is controlled in the germ line by elements in the 3-UTR of each message. Examples include the regulation of *tra-2*, *fem-3*, and *glp-1* (Ahringer and Kimble, 1991; Ahringer *et al.*, 1992; Goodwin *et al.*, 1993; Evans *et al.*, 1994). Proteins that are likely to bind RNA in the germ line have also been identified. For example, the FBF proteins have been shown to bind to and regulate translation of *fem-3* mRNAs (Zhang *et al.*, 1997). In addition, sequence analysis suggests that the GLD-1 protein is likely to bind RNAs, and thus might play a role in the control of maternal messages. Finally, the *mog-1* and *mog-5* genes appear to encode RNA helicases. How these regulatory factors interact to control the translation of crucial mRNAs in the hermaphrodite germ line, and how this action influences cell fate, have broad implications for understanding oogenesis and the control of mRNA translation in other animals (e.g., Chapter 20 by Sullivan *et al.*).

V. Programmed Cell Death in the Germ Line

After germ cells have entered meiosis and begun to differentiate as sperm or oocytes, some face a final cell fate

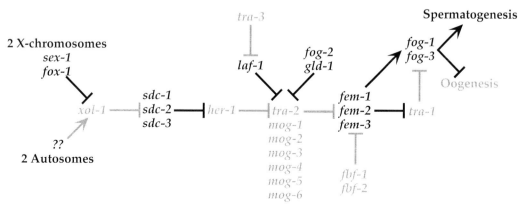

Figure 6   A model for sex determination in the germ line of larval hermaphrodites.

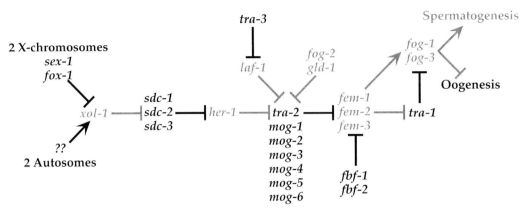

Figure 7   A model for sex determination in the germ line of adult hermaphrodites.

decision. Direct observation reveals that many cells that appear committed to oogenesis choose instead to undergo programmed cell deaths (Gumienny *et al.*, 1998). These suicides are not observed during the process of spermatogenesis. One attractive theory is that these dying germ cells function as nurse cells, producing material for use in developing oocytes. Because they resemble developing oocytes in all other respects, it is likely that the only cell fate decision required to control this process involves determining which of these differentiating germ cells will become oocytes, and which will die.

   Two factors have combined to make C. *elegans* a leading system for studying programmed cell death (Ellis *et al.*, 1991a; Hengartner, 1997). First, these deaths can be studied in living animals (Sulston and Horvitz, 1977; Sulston *et al.*, 1983). In these animals, the dying cells rapidly condense in size and become highly refractile, so that they appear as bright disks when viewed using Nomarski optics, and then disappear. Electron micrographs reveal that neighbors engulf and degrade the dying cells (Robertson and Thomson, 1982). Second, many mutations that affect programmed deaths are viable and easy to identify (Hedgecock *et al.*, 1983; Ellis and Horvitz, 1986). Use of mutants to dissect the process of cell death in the worm has defined four major steps: the decision to die, suicide, engulfment of the

corpse, and degradation (Fig. 8). Molecular analysis has revealed that several proteins which play a central role in cell death in nematodes are part of a conserved program present in most animals.

A. Cell Suicide

The *ced-3* and *ced-4* genes are essential for cells in C. *elegans* to undergo programmed deaths (Ellis and Horvitz, 1986). Analysis of genetic mosaic animals indicates that each gene is likely to act within cells, so these deaths are essentially suicides (Yuan and Horvitz, 1990). This conclusion is supported by experiments in which *ced-3* and *ced-4* were ectopically expressed in the touch neurons, causing these normally healthy cells to undergo programmed death (Shaham and Horvitz, 1996a). In addition to their role in the soma, both *ced-3* and *ced-4* are required for germ cells to die (Gumienny *et al.*, 1998).

   The *ced-3* gene encodes a member of a new family of cysteine proteases, the caspases (Yuan *et al.*, 1993), all of which resemble interleukin-1β-converting enzyme (ICE). The effect that mutations in *ced-3* have on cell death is directly correlated with their effect on protease activity, which supports the hypothesis that this activity is critical for CED-3 to function (Yuan *et al.*, 1993; Xue and Horvitz, 1995). *In vitro* assays indi-

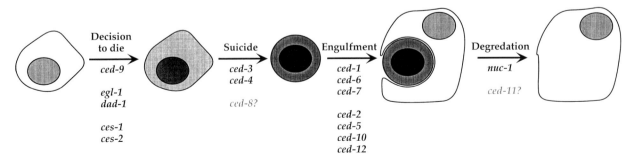

Figure 8   The major steps in programmed cell death in C. *elegans*.

cate that CED-3 cleaves target sequences similar to those of other caspases, and that its targets most closely resemble those of the mammalian caspase CPP32 (Xue *et al.*, 1996). By screening proteins produced from pools of *Xenopus* cDNAs for cleavage by CED-3, D. Xue, K. Lustig, M. Kirschner, and H. R. Horvitz have identified at least four potential targets (personal communication). It is not yet clear which of these target proteins are critical for programmed cell death, nor how inactivation of such targets causes death.

Just as is observed with ICE, the initial *ced-3* product is a proenzyme, which must be cleaved to form an active cysteine protease (Xue and Horvitz, 1995; Xue *et al.*, 1996). Although ectopic expression of *ced-3* is sufficient to kill cells that lack *ced-4* activity, the reverse is not true (Shaham and Horvitz, 1996a). Thus, CED-4 might promote cell death by helping to activate CED-3.

How does CED-4 function? Molecular analysis reveals that *ced-4* generates two transcripts; only the major transcript is needed to promote cell death, and the other appears to protect cells from programmed deaths (Shaham and Horvitz, 1996b). The major transcript encodes a protein of 549 amino acids that induces cell death when ectopically expressed; I will refer to this protein as CED-4 (Yuan and Horvitz, 1992; Shaham and Horvitz, 1996a). Immunoprecipitation experiments and the yeast two-hybrid assay indicate that CED-4 can bind CED-3 (Chinnaiyan *et al.*, 1997; Irmler *et al.*, 1997; Seshagiri and Miller, 1997; Wu *et al.*, 1997). When these two proteins are coexpressed in tissue culture, CED-4 activates the protease activity of CED-3, and stimulates the induction of programmed cell death (Chinnaiyan *et al.*, 1997; Seshagiri and Miller, 1997; Wu *et al.*, 1997). Surprisingly, expression of CED-4 in fission yeast without coexpression of CED-3 causes the chromosomes to condense and the cells to die, raising the possibility that CED-4 might be able to promote cell death in the absence of CED-3; so far, no experiments have replicated this activity in other organisms (James *et al.*, 1997). Recently, mammalian homologues of CED-4 have been shown to bind to and activate caspases, suggesting that the role CED-4 plays in activating CED-3 has been conserved during evolution (Zou *et al.*, 1997).

One other gene in *C. elegans* might play a direct role in killing cells—*ced-8*. Mutations in *ced-8* were first identified because they cause cell corpses to accumulate late in embryogenesis (Ellis *et al.*, 1991b). This accumulation appears to be due to a general slow down of the process of cell death during embryogenesis, suggesting that CED-8 might be needed to aid CED-3 and CED-4 in this process (G. Stanfield and H. R. Horvitz, personal communication). However, analysis of mutants suggests that *ced-8* is not required for germ line deaths to occur (Gumienny *et al.*, 1998). How the CED-8 protein functions is unknown.

B. Decision to Die

CED-3 and CED-4 are capable of causing cell death, and appear to be widely expressed in *C. elegans* (Yuan and Horvitz, 1992; Yuan *et al.*, 1993; Shaham and Horvitz, 1996a). How are their activities regulated to ensure that only appropriate cells die? Two major steps appear to be involved. First, the *ced-9* gene encodes a protein that protects most cells in the animal from undergoing programmed deaths (see Fig. 9; Hengartner *et al.*, 1992; Hengartner and Horvitz, 1994a). Second, genes that regulate the fates of individual cells influence the relative activities of *ced-3, ced-4,* and *ced-9.*

The *ced-9* gene is required to prevent programmed cell deaths—recessive mutations that cause a loss of *ced-9* function cause many extra cells to die (Hengartner *et al.*, 1992), whereas expression of *ced-9* under the control of a heat-shock promoter prevents many cell deaths (Hengartner and Horvitz, 1994a). The CED-9 protein is 280 amino acids long, and shows 49% similarity to the protein Bcl-2, which can prevent programmed cell death in many mammalian tissues (Hengartner and Horvitz, 1994a). Both proteins have a carboxyl-terminal hydrophobic tail; this tail has been shown to localize Bcl-2 to membranes. The activity of these two proteins also has been conserved, since expression of Bcl-2 in nematodes prevents cells from undergoing programmed deaths, and suppresses the effects of a *ced-9(lf)* mutation (Vaux *et al.*, 1992; Hengartner and Horvitz, 1994a). Surprisingly, a *ced-9* mutation in a site required for Bcl-2 to form heterodimers with related proteins prevents all programmed cell deaths in somatic tissues (Hengartner and Horvitz, 1994b), but has no effect on cell death in the germ line (Gumienny *et al.*, 1998).

How does CED-9 prevent programmed cell deaths? Biochemical studies reveal that CED-9 can bind to CED-4, and that this physical association inhibits CED-4 activity (Chinnaiyan *et al.*, 1997; James *et al.*, 1997; Seshagiri and Miller, 1997; Spector *et al.*, 1997; Wu *et al.*, 1997). Furthermore, when CED-9 is coexpressed with CED-4 in tissue culture cells, it causes CED-4 to be relocalized from the cytosol to intracellular membranes and the perinuclear region, where CED-9 is found (Wu *et al.*, 1997). These results support models in which CED-4 can bind to and activate CED-3, causing death, whereas a CED-9–CED-4

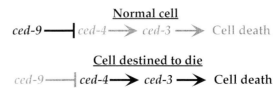

Figure 9 The central genes in the cell death pathway.

complex cannot activate CED-3, preventing death (Fig. 9). *In vitro* assays also show that CED-9 is a substrate for the CED-3 protease, and that mutations which alter these cleavage sites in CED-9 lower its ability to protect cells from death (Xue and Horvitz, 1997). Thus, CED-9 might also act as a competitive inhibitor of the CED-3 protease. This type of competitive inhibition appears to be the manner by which the viral protein p35 inhibits programmed cell death in *C. elegans* (Sugimoto *et al.,* 1994; Xue and Horvitz, 1995). So far, the relative importance of these two CED-9 activities is not known.

For cell death to occur, the activity of CED-9 relative to CED-4 and CED-3 must be carefully regulated. One gene that might be involved in this process is *dad-1*; overexpression of *dad-1*, driven by a heat-shock promoter, prevents most cell deaths in *C. elegans* (Sugimoto *et al.,* 1995). The DAD-1 protein was originally identified in mammals, and appears to encode a subunit of oligosaccharyltransferase. Homologues have been identified in many animals, so it might represent a fourth component of the biochemical machinery involved in all programmed cell deaths.

There are also likely to be genes that regulate the programmed deaths of specific cells in response to unique developmental or environmental cues. One example is the activation of cell death in response to signal transduction by Fas in mammals (Chinnaiyan *et al.,* 1995; Muzio *et al.,* 1996). In nematodes, the only well-characterized genes that act in this manner control the deaths of specific cells in the pharynx (Ellis and Horvitz, 1991; Metzstein *et al.,* 1996). Mutations in these genes do not affect the deaths of germ cells (Gumienny *et al.,* 1998). However, mutations that appear to specifically affect germ cell death have recently been identified (M. Hengartner, personal communication). For germ cells to die they must (1) be committed to oogenesis, and (2) have passed through the pachytene stage of meiosis (Gumienny *et al.,* 1998). Thus, genes such as *fog-1* and *fog-3*, which control the sexual identity of germ cells, and genes of the *ras* pathway, which control progression through meiosis, might also regulate the cell death genes.

C. Engulfment

Once cells have died, their corpses are engulfed by nearby cells. This process can begin even before the cell is completely dead (Robertson and Thomson, 1982), but is not required for death to occur (Hedgecock *et al.,* 1983). At least seven genes are involved in the process of engulfment; *ced-1, ced-6,* and *ced-7* appear to act as one group, and *ced-2, ced-5, ced-10,* and *ced-12* control a redundant process (Hedgecock *et al.,* 1983; Ellis *et al.,* 1991b; M. Hengartner and T. Schedl, personal communication). A mutation in any of these genes hinders the engulfment of dead germ cells (Gumienny *et al.,* 1998).

Electron micrographs reveal that, in these *ced* mutants, many dying germ cells are not engulfed by the sheath cells of the somatic gonad, which normally carry out this task. Thus, these genes are likely to control either the recognition of corpses by the sheath cells, or the beginning of the engulfment process (Gumienny *et al.,* 1998).

Molecular analysis reveals that CED-5 is similar to the human protein DOCK180, which interacts with CRK to regulate cytoskeletal proteins and cell structure (Chun and Horvitz, 1998). The *ced-7* gene encodes a protein similar to ATP-binding cassette transporters (Y. W. Chun and H. R. Horvitz, personal communication). Perhaps CED-7 controls the transport of molecules required for engulfment. Luciani and Chimini (1996) have identified an ATP-binding cassette transporter in the mouse that appears necessary for the engulfment of dead cells by macrophages, so the role these transporters play in engulfment might have been conserved during animal evolution. Finally, sequence analysis indicates that CED-6 contains a phosphotyrosine binding domain, and might act as an adaptor protein (Q. Liu and M. O. Hengartner, personal communication). The CED-6 protein appears to function in engulfing cells rather than cell corpses, and thus might help transduce a signal from the corpse that promotes engulfment.

4. Degradation

Once cell corpses are engulfed, their remains are rapidly degraded. This process takes place within the engulfing cell, since mutations that block engulfment also block degradation (Hedgecock *et al.,* 1983). One gene is known to act in this process: *nuc-1* encodes or controls a nuclease that degrades the DNA of cell corpses (Sulston, 1976; Hedgecock *et al.,* 1983; Hevelone and Hartman, 1988). In the gonad, mutations in *nuc-1* cause pycnotic nuclei from dead germ cells to accumulate in vacuoles of the engulfing sheath cells (Gumienny *et al.,* 1998). One additional gene might control degradation—mutations in the *ced-11* gene affect the appearance of cell corpses (G. Stanfield and H. R. Horvitz, personal communication). However, the CED-11 protein appears novel, and its role in cell death is not yet understood.

VI. Conclusions

The research discussed above has produced a rough portrait of the mechanisms that control cell fate in the developing germ line. First, the decision of some germ cells to remain in mitosis and proliferate is controlled by a signal from the nearby distal tip cells; this signal is mediated by proteins of the LAG-2/GLP-1 signal transduction pathway. Second, whether germ cells form sperm or oocytes is controlled by the sex-determination genes; these act through the HER-1/TRA-2 signal transduction process to control TRA-1A, which might in turn directly regulate transcription of *fog-3*. For germ cells to form sperm instead of oocytes, *fog-3* and *fog-1* must be active, and require assistance from the three *fem* genes. Third, some germ cells that have begun oogenesis do not form oocytes, but instead choose to die. Prior to death, these cells might function as nurse cells for their compatriots. The decisions made during these three processes specify what fate each germ cell will adopt, both during development, and in the mature animal.

Many important questions remain. Perhaps the most essential involve unresolved details of each regulatory process. For example, it is not known how GLP-1 and LAG-1 regulate transcription, nor which genes they control in the germ line, nor what accessory proteins aid in this signal transduction process. Several potential partners and targets have been identified in genetic screens, and should provide interesting subjects for investigation. Furthermore, although the molecular analysis of *fog-3* has provided some clues as to how the sex-determination genes regulate germ cell fate, many questions persist. What is the nature of the FOG-1 protein, and how does it function? Do FOG-1 and FOG-3 regulate transcription of genes required for spermatogenesis or oogenesis, as one might suspect from the existence of many sperm-specific and oocyte-specific transcripts? How do the FEM proteins interact with FOG-1 and FOG-3 to promote spermatogenesis? In addition to these questions, several new genes have been identified that might act in this process, and their roles are not yet understood. These include suppressors of the *fog-1(q253ts)* mutation (S. Jin and R. Ellis, unpublished results) and the homologue of the *Drosophila mago nashi* gene (W. Li and B. Wood, personal communication). Finally, the role of programmed cell death in the germ line has only recently been described. Its purpose in germ cells remains speculative, and we do not yet know how the cell death genes are regulated, so that some germ cells live and others die. Several candidate mutations might soon shed light on this subject.

The manner in which these regulatory processes interact must also be considered. Germ cells do not differentiate as sperm or oocytes prior to meiosis, and do not undergo programmed deaths prior to beginning oogenesis, so information from some fate decisions might play a vital role in making others. Furthermore, a crucial point in the regulation of cell fate occurs during the pachytene phase of meiosis I. Prior to the completion of this phase, germ cells are still capable of returning to mitosis, and neither programmed cell death nor most aspects of sexual differentiation occur. Exit from this arrest is controlled by genes of the *ras* pathway, which might therefore play an important role in helping determine the fates of developing germ cells. It is not known

yet if *ras* and its partners act in response to a signal from the somatic gonad, but interactions between the soma and germ line play a central role in the regulation of mitosis and of sexual fate, and are required for the engulfment of dying germ cells.

For germ cells, deciding how to differentiate is only the beginning of a long developmental process. *C. elegans* provides an excellent model for understanding how spermatogenesis and oogenesis occur, and several groups are pursuing these questions (For reviews, see L'Hernault, 1997; Schedl, 1997). An important task will be determining how the genes involved in spermatogenesis and oogenesis are regulated during the cell fate decisions described above. Recent evidence suggests that the maturation of oocytes depends on an interaction with the sheath cells of the somatic gonad (Iwasaki *et al.*, 1996; Rose *et al.*, 1997), which implies that some crucial decisions about differentiation are made after the initial determination of cell fate.

Finally, it is essential to place this research in a broader biological and evolutionary perspective. It is possible that some of these regulatory mechanisms have been broadly conserved during evolution; this is clearly the case for the role that *ced-3*, *ced-4*, and *ced-9* play controlling programmed cell deaths. Whether it will also prove true for the direct control of the cell cycle, spermatogenesis, and oogenesis, and for the genes that regulate germ cell deaths specifically, remains to be seen. Furthermore, some of these regulatory proteins have important homologues in other animals, which control very different developmental processes. For example, the *glp-1* gene is homologous to *lin-12* in *C. elegans*, *Notch* in *Drosophila*, and several vertebrate genes, and the zinc-fingers of TRA-1A show similarity to *Cubitus interruptus*, GLI and GLI3. Study of the *C. elegans* germ line should therefore help elucidate the processes by which regulatory proteins are recruited from one task to another during evolution.

Acknowledgments

I thank Betsy Goodwin, Michael Hengartner, Judith Kimble, Tim Schedl, Bill Wood, and Dave Zarkower for critical reading of the manuscript, and numerous colleagues for communicating results prior to publication. In addition, I thank Bob Horvitz, Patricia Kuwabara, and Barbara Meyer for specific comments and suggestions. My work has been supported by grants from the American Cancer Society and the March of Dimes.

References

Ahringer, J., and Kimble, J. (1991). Control of the sperm–oocyte switch in *Caenorhabditis elegans* hermaphrodites by the *fem-3* 3' untranslated region. *Nature (London)* **349**, 346–348.

Ahringer, J., Rosenquist, T. A., Lawson, D. N., and Kimble, J. (1992). The *Caenorhabditis elegans* sex determining gene *fem-3* is regulated posttranscriptionally. *EMBO J.* **11**, 2303–2310.

Akerib, C. C., and Meyer, B. J. (1994). Identification of X chromosome regions in *Caenorhabditis elegans* that contain sex-determination signal elements. *Genetics* **138**, 1105–1125.

Albertson, D. G., Sulston, J. E., and White, J. G. (1978). Cell cycling and DNA replication in a mutant blocked in cell division in the nematode *Caenorhabditis elegans*. *Dev. Biol.* **63**, 165–178.

Austin, J., and Kimble, J. (1987). *glp-1* is required in the germ line for regulation of the decision between mitosis and meiosis in *C. elegans*. *Cell* **51**, 589–599.

Austin, J., and Kimble, J. (1989). Transcript analysis of *glp-1* and *lin-12*, homologous genes required for cell interactions during development of *C. elegans*. *Cell* (*Cambridge, Mass.*) **58**, 565–571.

Barnes, T. M., and Hodgkin, J. (1996). The *tra-3* sex determination gene of *Caenorhabditis elegans* encodes a member of the calpain regulatory protease family. *EMBO J.* **15**, 4477–4484.

Barton, M. K., and Kimble, J. (1990). *fog-1*, a regulatory gene required for specification of spermatogenesis in the germ line of *Caenorhabditis elegans*. *Genetics* **125**, 29–39.

Barton, M. K., Schedl, T. B., and Kimble, J. (1987). Gain-of-function mutations of *fem-3*, a sex-determination gene in *Caenorhabditis elegans*. *Genetics* **115**, 107–119.

Beanan, M. J., and Strome, S. (1992). Characterization of a germ-line proliferation mutation in C. *elegans*. *Development* (*Cambridge, UK*) **116**, 755–766.

Berry, L. W., Westlund, B., and Schedl, T. (1997). Germ-line tumor formation caused by activation of *glp-1*, a *Caenorhabditis elegans* member of the Notch family of receptors. *Development* (*Cambridge, UK*) **124**, 925–936.

Chen, P.-J., and Ellis, R. E. (1998). The *tra-1* gene regulates germ cell fate in *C. elegans* by controlling transcription of *fog-3*. (Manuscript in preparation).

Chen, P.-J., Kimble, J., and Ellis, R. E. (1998). FOG-3, which controls sexual fate in *C. elegans* germ cells, is a novel member of the Tob family of antiproliferative proteins. Submitted.

Chin-Sang, I. D., and A. M. Spence, (1996). *Caenorhabditis elegans* sex-determining protein FEM-2 is a protein phosphatase that promotes male development and interacts directly with FEM-3. *Genes Dev.* **10**, 2314–2325.

Chinnaiyan, A. M., O'Rourke, K., Tewari, M., and Dixit, V. M. (1995). FADD, a novel death domain-containing protein, interacts with the death domain of Fas and initiates apoptosis. *Cell* (*Cambridge, Mass.*) **81**, 505–512.

Chinnaiyan, A. M., O'Rourke, K., Lane, B. R., and Dixit, V. M. (1997). Interaction of CED-4 with CED-3 and CED-9: A molecular framework for cell death. *Science* **275**, 1122–1126.

Christensen, S., Kodoyianni, V., Bosenberg, M., Friedman, L., and Kimble, J. (1996). *lag-1*, a gene required for *lin-12* and *glp-1* signaling in *Caenorhabditis elegans*, is homologous to human CBF1 and *Drosophila* Su(H). *Development* (*Cambridge, UK*) **122**, 1373–1383.

Chun, Y. W., and Horvitz, H. R. (1998). *C. elegans* phagocytosis and cell-migration protein CED-5 is similar to human DOCK 180. *Nature* (*London*) **392**, 501–504.

Church, D. L., Guan, K. L., and Lambie, E. J. (1995). Three genes of the MAP kinase cascade, *mek-2, mpk-1/sur-1* and *let-60 ras*, are required for meiotic cell cycle progression in *Caenorhabditis elegans*. *Development* (*Cambridge, UK*) **121**, 2525–2535.

Crittenden, S. L., Troemel, E. R., Evans, T. C., and Kimble, J. (1994). GLP-1 is localized to the mitotic region of the *C. elegans* germ line. *Development* (*Cambridge, UK*) **120**, 2901–2911.

de Bono, M., Zarkower, D., and Hodgkin, J. (1995). Dominant feminizing mutations implicate protein–protein interactions as the main mode of regulation of the nematode sex-determining gene *tra-1*. *Genes Dev.* **9**, 155–167.

DeLong, L., Plenefisch, J. D., Klein, R. D., and Meyer, B. J. (1993). Feedback control of sex determination by dosage compensation revealed through *Caenorhabditis elegans sdc-3* mutations. *Genetics* **133**, 875–896.

Doniach, T. (1986). Activity of the sex-determining gene *tra-2* is modulated to allow spermatogenesis in the *C. elegans* hermaphrodite. *Genetics* **114**, 53–76.

Doniach, T., and Hodgkin, J. (1984). A sex-determining gene, *fem-1*, required for both male and hermaphrodite development in *Caenorhabditis elegans*. *Dev. Biol.* **106**, 223–235.

Ellis, H. M., and Horvitz, H. R. (1986). Genetic control of programmed cell death in the nematode *C. elegans*. *Cell* (*Cambridge, Mass.*) **44**, 817–829.

Ellis, R. E., and Horvitz, H. R. (1991). Two *C. elegans* genes control the programmed deaths of specific cells in the pharynx. *Development* (*Cambridge, UK*) **112**, 591–603.

Ellis, R. E., and Kimble, J. (1995). The *fog-3* gene and regulation of cell fate in the germ line of *Caenorhabditis elegans*. *Genetics* **139**, 561–577.

Ellis, R. E., Yuan, J. Y., and Horvitz, H. R. (1991a). Mechanisms and functions of cell death. *Ann. Rev. Cell Biol.* **7**, 663–698.

Ellis, R. E., Jacobson, D. M., and Horvitz, H. R. (1991b). Genes required for the engulfment of cell corpses during programmed cell death in *Caenorhabditis elegans*. *Genetics* **129**, 79–94.

Evans, T. C., Crittenden, S. L., Kodoyianni, V., and Kimble, J. (1994). Translational control of maternal *glp-1* mRNA establishes an asymmetry in the *C. elegans* embryo. *Cell* (*Cambridge, Mass.*) **77**, 183–194.

Fitzgerald, K., and Greenwald, I. (1995). Interchangeability of *Caenorhabditis elegans* DSL proteins and intrinsic signalling activity of their extracellular domains *in vivo*. *Development* (*Cambridge, UK*) **121**, 4275–4282.

Francis, R., Barton, M. K., Kimble, J., and Schedl, T. (1995a). *gld-1*, a tumor suppressor gene required for oocyte development in *Caenorhabditis elegans*. *Genetics* **139**, 579–606.

Francis, R., Maine, E., and Schedl, T. (1995b). Analysis of the multiple roles of *gld-1* in germline development: Interactions with the sex determination cascade and the *glp-1* signaling pathway. *Genetics* **139**, 607–630.

Gaudet, J., VanderElst, I., and Spence, A. M. (1996). Posttranscriptional regulation of sex determination in *Caenorhabditis elegans*: Widespread expression of the sex-determining gene *fem-1* in both sexes. *Mol. Biol. Cell* **7**, 1107–1121.

Goodwin, E. B., Okkema, P. G., Evans, T. C., and Kimble, J. (1993). Translational regulation of *tra-2* by its 3′ untranslated region controls sexual identity in *C. elegans*. *Cell (Cambridge, Mass.)* **75**, 329–339.

Goodwin, E. B., Hofstra, K., Hurney, C. A., Mango, S., and Kimble, J. (1997). A genetic pathway for regulation of *tra-2* translation. *Development (Cambridge, UK)* **124**, 749–758.

Graham, P. L., and Kimble, J. (1993). The *mog-1* gene is required for the switch from spermatogenesis to oogenesis in *Caenorhabditis elegans*. *Genetics* **133**, 919–931.

Graham, P. L., Schedl, T., and Kimble, J. (1993). More *mog* genes that influence the switch from spermatogenesis to oogenesis in the hermaphrodite germ line of *Caenorhabditis elegans*. *Dev. Genet.* **14**, 471–484.

Gumienny, T. L., Lambie, E., Hartwieg, E., Horvitz, H. R., and Hengartner, M. O. (1998). Elimination of nurse cells by programmed cell death in the *C. elegans* hermaphrodite germ line. Manuscript in preparation.

Hedgecock, E. M., Sulston, J. E., and Thomson, J. N. (1983). Mutations affecting programmed cell deaths in the nematode *Caenorhabditis elegans*. *Science* **220**, 1277–1279.

Henderson, S. T., Gao, D., Lambie, E. J., and Kimble, J. (1994). *lag-2* may encode a signaling ligand for the GLP-1 and LIN-12 receptors of *C. elegans*. *Development (Cambridge, UK)* **120**, 2913–2924.

Henderson, S. T., Gao, D., Christensen, S., and Kimble, J. (1997). Functional domains of LAG-2, a putative signaling ligand for LIN-12 and GLP-1 receptors in *Caenorhabditis elegans*. *Mol. Biol. Cell* **8**, 1751–1762.

Hengartner, M. O. (1997). Cell death. *In* "*C. elegans* II" (D. L. Riddle, T. Blumenthal, B. J. Meyer, and J. R. Priess, eds.), pp. 383–416. Cold Spring Harbor Laboratory, Plainview, New York.

Hengartner, M. O., and Horvitz, H. R. (1994a). *C. elegans* cell survival gene *ced-9* encodes a functional homolog of the mammalian proto-oncogene *bcl-2*. *Cell (Cambridge, Mass.)* **76**, 665–676.

Hengartner, M. O., and Horvitz, H. R. (1994b). Activation of *C. elegans* cell death protein CED-9 by an amino-acid substitution in a domain conserved in *Bcl-2*. *Nature (London)* **369**, 318–320.

Hengartner, M. O., Ellis, R. E., and Horvitz, H. R. (1992). *Caenorhabditis elegans* gene *ced-9* protects cells from programmed cell death. *Nature (London)* **356**, 494–499.

Hevelone, J., and Hartman, P. S. (1988). An endonuclease from *Caenorhabditis elegans*: Partial purification and characterization. *Biochem. Genet.* **26**, 447–461.

Hird, S. N., Paulsen, J. E., and Strome, S. (1996). Segregation of germ granules in living *Caenorhabditis elegans* embryos: Cell-type-specific mechanisms for cytoplasmic localisation. *Development (Cambridge, UK)* **122**, 1303–1312.

Hirsh, D., Oppenheim, D., and Klass, M. (1976). Develop-

ment of the reproductive system of *Caenorhabditis elegans*. *Dev. Biol.* **49**, 200–219.

Hodgkin, J. (1980). More sex-determination mutants of *Caenorhabditis elegans*. *Genetics* **96**, 649–664.

Hodgkin, J. (1986). Sex determination in the nematode *C. elegans*: Analysis of *tra-3* suppressors and characterization of *fem* genes. *Genetics* **114**, 15–52.

Hodgkin, J. (1987). A genetic analysis of the sex-determining gene, *tra-1*, in the nematode *Caenorhabditis elegans*. *Genes Dev.* **1**, 731–745.

Hodgkin, J., and Albertson, D. G. (1995). Isolation of dominant XO-feminizing mutations in *Caenorhabditis elegans*: New regulatory *tra* alleles and an X chromosome duplication with implications for primary sex determination. *Genetics* **141**, 527–542.

Hodgkin, J., Zellan, J. D., and Albertson, D. G. (1994). Identification of a candidate primary sex determination locus, *fox-1*, on the X chromosome of *Caenorhabditis elegans*. *Development (Cambridge, UK)* **120**, 3681–3689.

Hunter, C. P., and Wood, W. B. (1990). The *tra-1* gene determines sexual phenotype cell-autonomously in *C. elegans*. *Cell (Cambridge, Mass.)* **63**, 1193–1204.

Hunter, C. P., and Wood, W. B. (1992). Evidence from mosaic analysis of the masculinizing gene *her-1* for cell interactions in *C. elegans* sex determination. *Nature (London)* **355**, 551–555.

Irmler, M., Hofmann, K., Vaux, D., and Tschopp, J. (1997). Direct physical interaction between the *Caenorhabditis elegans* 'death proteins' CED-3 and CED-4. *FEBS Lett.* **406**, 189–190.

Iwasaki, K., McCarter, J., Francis, R., and Schedl, T. (1996). *emo-1*, a *Caenorhabditis elegans* Sec61p gamma homologue, is required for oocyte development and ovulation. *J. Cell Biol.* **134**, 699–714.

James, C., Gschmeissner, S., Fraser, A., and Evan, G. I. (1997). CED-4 induces chromatin condensation in *Schizosaccharomyces pombe* and is inhibited by direct physical association with CED-9. *Curr. Biol.* **7**, 246–252.

Jan, E., Yoon, J., Walterhouse, D., Iannaccone, P., and Goodwin, E. (1997). Conservation of the *C. elegans tra-2* 3′-UTR translational control. *EMBO J.* **16**, 6301–6313.

Jones, A. R., and Schedl, T. (1995). Mutations in *gld-1*, a female germ cell-specific tumor suppressor gene in *Caenorhabditis elegans*, affect a conserved domain also found in Src-associated protein Sam68. *Genes Dev.* **9**, 1491–1504.

Jones, A. R., Francis, R., and Schedl, T. (1996). GLD-1, a cytoplasmic protein essential for oocyte differentiation, shows stage- and sex-specific expression during *Caenorhabditis elegans* germline development. *Dev. Biol.* **180**, 165–183.

Kadyk, L. C., and Kimble, J. (1998). Genetic regulation of entry into meiosis in *C. elegans*. *Development* **125**, 1803–1813.

Kadyk, L. C., Lambie, E. J., and Kimble, J. (1997). *glp-3* is required for mitosis and meiosis in the *Caenorhabditis elegans* germ line. *Genetics* **145**, 111–121.

Kemphues, K. J., and Strome, S. (1997). Fertilization and establishment of polarity in the embryo. *In* "*C. elegans* II" (D. L. Riddle, T. Blumenthal, B. J. Meyer, and J. R.

Priess, eds.), pp. 335–360. Cold Spring Harbor Laboratory, Plainview, New York.

Kimble, J., and Simpson, P. (1997). The LIN-12/Notch signaling pathway and its regulation. *Ann. Rev. Cell Dev. Biol.* **13**, 333–361.

Kimble, J. E., and White, J. G. (1981). On the control of germ cell development in *Caenorhabditis elegans. Dev. Biol.* **81**, 208–219.

Kimble, J., Edgar, L., and Hirsh, D. (1984). Specification of male development in *Caenorhabditis elegans:* The *fem* genes. *Dev. Biol.* **105**, 234–239.

Klass, M., Wolf, N., and Hirsh, D. (1976). Development of the male reproductive system and sexual transformation in the nematode *Caenorhabditis elegans. Dev. Biol.* **52**, 1–18.

Klein, R. D., and Meyer, B. J. (1993). Independent domains of the Sdc-3 protein control sex determination and dosage compensation in *C. elegans. Cell (Cambridge, Mass.)* **72**, 349–364.

Kodoyianni, V., Maine, E. M., and Kimble, J. (1992). Molecular basis of loss-of-function mutations in the *glp-1* gene of *Caenorhabditis elegans. Mol. Biol. Cell* **3**, 1199–1213.

Kuwabara, P. E. (1996). A novel regulatory mutation in the *C. elegans* sex determination gene *tra-2* defines a candidate ligand/receptor interaction site. *Development (Cambridge, UK)* **122**, 2089–2098.

Kuwabara, P. E., and Kimble, J. (1995). A predicted membrane protein, TRA-2A, directs hermaphrodite development in *Caenorhabditis elegans. Development (Cambridge, UK)* **121**, 2995–3004.

Kuwabara, P. E., Okkema, P. G., and Kimble, J. (1992). *tra-2* encodes a membrane protein and may mediate cell communication in the *Caenorhabditis elegans* sex determination pathway. *Mol. Biol. Cell* **3**, 461–473.

L'Hernault, S. (1997). Spermatogenesis. In "*C. elegans* II" (D. L. Riddle, T. Blumenthal, B. J. Meyer, and J. R. Priess, eds.) pp. 271–294. Cold Spring Harbor Laboratory, Plainview, New York.

L'Hernault, S. W., and Arduengo, P. M. (1992). Mutation of a putative sperm membrane protein in *Caenorhabditis elegans* prevents sperm differentiation but not its associated meiotic divisions. *J. Cell Biol.* **119**, 55–68.

Lambie, E. J., and Kimble, J. (1991). Two homologous regulatory genes, *lin-12* and *glp-1,* have overlapping functions. *Development (Cambridge, UK)* **112**, 231–240.

Lissemore, J. L., Currie, P. D., Turk, C. M., and Maine, E. M. (1993). Intragenic dominant suppressors of *glp-1,* a gene essential for cell-signaling in *Caenorhabditis elegans,* support a role for cdc10/SW16/ankyrin motifs in GLP-1 function. *Genetics* **135**, 1023–1034.

Luciani, M. F., and Chimini, G. (1996). The ATP binding cassette transporter ABC1, is required for the engulfment of corpses generated by apoptotic cell death. *EMBO J.* **15**, 226–235.

McCarter, J., Bartlett, B., Dang, T., and Schedl, T. (1997). Soma-germ cell interactions in *Caenorhabditis elegans:* Multiple events of hermaphrodite germline development require the somatic sheath and spermathecal lineages. *Dev. Biol.* **181**, 121–143.

Madl, J. E., and Herman, R. K. (1979). Polyploids and sex determination in *Caenorhabditis elegans. Genetics* **93**, 393–402.

Maine, E. M., and Kimble, J. (1993). Suppressors of *glp-1,* a gene required for cell communication during development in *Caenorhabditis elegans,* define a set of interacting genes. *Genetics* **135**, 1011–1022.

Matsuda, S., Kawamura-Tsuzuku, J., Ohsugi, M., Yoshida, M., Emi, M., Nakamura, Y., Onda, M., Yoshida, Y., Nishiyama, A., and Yamamoto, T. (1996). Tob, a novel protein that interacts with p185erbB2, is associated with anti-proliferative activity. *Oncogene* **12**, 705–713.

Mello, C. C., Draper, B. W., Krause, M., Weintraub, H., and Priess, J. R. (1992). The *pie-1* and *mex-1* genes and maternal control of blastomere identity in early *C. elegans* embryos. *Cell (Cambridge, Mass.)* **70**, 163–176.

Mello, C. C., Schubert, C., Draper, B., Zhang, W., Lobel, R., and Priess, J. R. (1996). The PIE-1 protein and germline specification in *C. elegans* embryos. *Nature (London)* **382**, 710–712.

Metzstein, M. M., Hengartner, M. O., Tsung, N., Ellis, R. E., and Horvitz, H. R. (1996). Transcriptional regulator of programmed cell death encoded by *Caenorhabditis elegans* gene *ces-2. Nature (London)* **382**, 545–547.

Meyer, B. J. (1997). Sex determination and X chromosome dosage compensation. In "*C. elegans* II" (D. L. Riddle, T. Blumenthal, B. J. Meyer, and J. R. Priess, eds.), pp. 209–240. Cold Spring Harbor Laboratory, Plainview, New York.

Miller, L. M., Plenefisch, J. D., Casson, L. P., and Meyer, B. J. (1988). *xol-1:* A gene that controls the male modes of both sex determination and X chromosome dosage compensation in *C. elegans. Cell (Cambridge, Mass.)* **55**, 167–183.

Muzio, M., Chinnaiyan, A. M., Kischkel, F. C., O'Rourke, K., Shevchenko, A., Ni, J., Scaffidi, C., Bretz, J. D., Zhang, M., Gentz, R., Mann, M., Krammer, P. H., Peter, M. E., and Dixit, V. M. (1996). FLICE, a novel FADD-homologous ICE/CED-3-like protease, is recruited to the CD95 (Fas/APO-1) death-inducing signaling complex. *Cell (Cambridge, Mass.)* **85**, 817–827.

Nicoll, M., Akerib, C. C., and Meyer, B. J. (1997). X-chromosome-counting mechanisms that determine nematode sex. *Nature (London)* **388**, 200–204.

Nonet, M. L., and Meyer, B. J. (1991). Early aspects of *Caenorhabditis elegans* sex determination and dosage compensation are regulated by a zinc-finger protein. *Nature (London)* **351**, 65–68.

Okkema, P. G., and Kimble, J. (1991). Molecular analysis of *tra-2,* a sex determining gene in *C. elegans. EMBO J.* **10**, 171–176.

Perry, M. D., Li, W., Trent, C., Robertson, B., Fire, A., Hageman, J. M., and Wood, W. B. (1993). Molecular characterization of the *her-1* gene suggests a direct role in cell signaling during *Caenorhabditis elegans* sex determination. *Genes Dev.* **7**, 216–228.

Perry, M. D., Trent, C., Robertson, B., Chamblin, C., and Wood, W. B. (1994). Sequenced alleles of the *Caenorhabditis elegans* sex-determining gene *her-1* include a novel class of conditional promoter mutations. *Genetics* **138**, 317–327.

Pilgrim, D., McGregor, A., Jackle, P., Johnson, T., and Hansen, D. (1995). The *C. elegans* sex-determining gene *fem-2* encodes a putative protein phosphatase. *Mol. Biol. Cell* **6**, 1159–1171.

Qiao, L., Lissemore, J. L., Shu, P., Smardon, A., Gelber, M. B., and Maine, E. M. (1995). Enhancers of *glp-1*, a gene required for cell-signaling in *Caenorhabditis elegans* define a set of genes required for germline development. *Genetics* **141**, 551–569.

Raymond, C. S., Shamu, C. E., Shen, M. M., Seifert, K., Hirsch, B., Hodgkin, J., and Zarkower, D. (1998). Evidence for evolutionary conservation of sex determining genes. *Nature* (*London*) **391**, 691–695.

Rhind, N. R., Miller, L. M., Kopczynski, J. B., and Meyer, B. J. (1995). *xol-1* acts as an early switch in the *C. elegans* male/hermaphrodite decision. *Cell* (*Cambridge, Mass.*) **80**, 71–82.

Robertson, A., and Thomson, N. (1982). Morphology of programmed cell death in the ventral nerve cord of *Caenorhabditis elegans* larvae. *J. Embryol. Exp. Morphol.* **67**, 89–100.

Roehl, H., and Kimble, J. (1993). Control of cell fate in *C. elegans* by a GLP-1 peptide consisting primarily of ankyrin repeats. *Nature* (*London*) **364**, 632–635.

Roehl, H., Bosenberg, M., Blelloch, R., and Kimble, J. (1996). Roles of the RAM and ANK domains in signaling by the *C. elegans* GLP-1 receptor. *EMBO J.* **15**, 7002–7012.

Rose, K. L., Winfrey, V. P., Hoffman, L. H., Hall, D. H., Furuta, T., and Greenstein, D. (1997). The POU gene *ceh-18* promotes gonadal sheath cell differentiation and function required for meiotic maturation and ovulation in *Caenorhabditis elegans*. *Dev. Biol.* **192**, 59–77.

Rosenquist, T. A., and Kimble, J. (1988). Molecular cloning and transcript analysis of *fem-3*, a sex-determination gene in *Caenorhabditis elegans*. *Genes Dev.* **2**, 606–616.

Schedl, T. (1997). Developmental genetics of the germ line. In "*C. elegans* II" (D. L. Riddle, T. Blumenthal, B. J. Meyer, and J. R. Priess, eds.), pp. 241–270. Cold Spring Harbor Laboratory, Plainview, New York.

Schedl, T., and Kimble, J. (1988). *fog-2*, a germ-line-specific sex determination gene required for hermaphrodite spermatogenesis in *Caenorhabditis elegans*. *Genetics* **119**, 43–61.

Schedl, T., Graham, P. L., Barton, M. K., and Kimble, J. (1989). Analysis of the role of *tra-1* in germline sex determination in the nematode *Caenorhabditis elegans*. *Genetics* **123**, 755–769.

Schnabel, R., and Priess, J. R. (1997). Specification of cell fates in the early embryo. In "*C. elegans* II" (D. L. Riddle, T. Blumental, B. J. Meyer, and J. R. Priess, eds.), pp. 361–382. Cold Spring Harbor Laboratory, Plainview, New York.

Seshagiri, S., and Miller, L. K. (1997). *Caenorhabditis elegans* CED-4 stimulates CED-3 processing and CED-3-induced apoptosis. *Curr. Biol.* **7**, 455–460.

Seydoux, G., Mello, C. C., Pettitt, J., Wood, W. B., Priess, J. R., and Fire, A. (1996). Repression of gene expression in the embryonic germ lineage of *C. elegans*. *Nature* (*London*) **382**, 713–716.

Shaham, S., and Horvitz, H. R. (1996a). Developing *Caenorhabditis elegans* neurons may contain both cell-

death protective and killer activities. *Genes Dev.* **10**, 578–591.

Shaham, S., and Horvitz, H. R. (1996b). An alternatively spliced *C. elegans ced-4* RNA encodes a novel cell death inhibitor. *Cell* (*Cambridge, Mass.*) **86**, 201–208.

Shen, M. M., and Hodgkin, J. (1988). *mab-3*, a gene required for sex-specific yolk protein expression and a male-specific lineage in *C. elegans*. *Cell* (*Cambridge, Mass.*) **54**, 1019–1031.

Spector, M. S., Desnoyers, S., Hoeppner, D. J., and Hengartner, M. O. (1997). Interaction between the *C. elegans* cell-death regulators CED-9 and CED-4. *Nature* (*London*) **385**, 653–656.

Spence, A. M., Coulson, A., and Hodgkin, J. (1990). The product of *fem-1*, a nematode sex-determining gene, contains a motif found in cell cycle control proteins and receptors for cell–cell interactions. *Cell* (*Cambridge, Mass.*) **60**, 981–990.

Strome, S., and Wood, W. B. (1983). Generation of asymmetry and segregation of germ-line granules in early *C. elegans* embryos. *Cell* (*Cambridge, Mass.*) **35**, 15–25.

Sugimoto, A., Friesen, P. D., and Rothman, J. H. (1994). Baculovirus p35 prevents developmentally programmed cell death and rescues a *ced-9* mutant in the nematode *Caenorhabditis elegans*. *EMBO J.* **13**, 2023–2028.

Sugimoto, A., Hozak, R. R., Nakashima, T., Nishimoto, T., and Rothman, J. H. (1995). *dad-1*, an endogenous programmed cell death suppressor in *Caenorhabditis elegans* and vertebrates. *EMBO J.* **14**, 4434–4441.

Sulston, J. E. (1976). Post-embryonic development in the ventral cord of *Caenorhabditis elegans*. *Philos. Trans. R. Soc. London Ser. B: Biol. Sci.* **275**, 287–297.

Sulston, J. E., and Horvitz, H. R. (1977). Post-embryonic cell lineages of the nematode, *Caenorhabditis elegans*. *Dev. Biol.* **56**, 110–156.

Sulston, J. E., and Horvitz, H. R. (1981). Abnormal cell lineages in mutants of the nematode *Caenorhabditis elegans*. *Dev. Biol.* **82**, 41–55.

Sulston, J. E., Schierenberg, E., White, J. G., and Thomson, J. N. (1983). The embryonic cell lineage of the nematode *Caenorhabditis elegans*. *Dev. Biol.* **100**, 64–119.

Tax, F. E., Yeargers, J. J., and Thomas, J. H. (1994). Sequence of *C. elegans lag-2* reveals a cell-signalling domain shared with Delta and Serrate of *Drosophila*. *Nature* (*London*) **368**, 150–154.

Trent, C., Tsung, N., and Horvitz, H. R. (1983). Egg-laying defective mutants of the nematode *Caenorhabditis elegans*. *Genetics* **104**, 619–647.

Trent, C., Wood, W. B., and Horvitz, H. R. (1988). A novel dominant transformer allele of the sex-determining gene *her-1* of *Caenorhabditis elegans*. *Genetics* **120**, 145–157.

Trent, C., Purnell, B., Gavinski, S., Hageman, I., Chamblin, C., and Wood, W. B. (1991). Sex-specific transcriptional regulation of the *C. elegans* sex-determining gene *her-1*. *Mech. Dev.* **34**, 43–55.

Varkey, J. P., Muhlrad, P. J., Minniti, A. N., Do, B., and Ward, S. (1995). The *Caenorhabditis elegans spe-26* gene is necessary to form spermatids and encodes a protein similar to the actin-associated proteins kelch and scruin. *Genes Dev.* **9**, 1074–1086.

Vaux, D. L., Weissman, I. L., and Kim, S. K. (1992). Prevention of programmed cell death in *Caenorhabditis elegans* by human bcl-2. *Science* **258**, 1955–1957.

Wu, D., Wallen, H. D., Inohara, N., and Nunez, G. (1997). Interaction and regulation of the *Caenorhabditis elegans* death protease CED-3 by CED-4 and CED-9. *J. Biol. Chem.* **272**, 21449–21454.

Wu, D., Wallen, H. D., and Nunez, G. (1997). Interaction and regulation of subcellular localization of CED-4 by CED-9. *Science* **275**, 1126–1129.

Xue, D., and Horvitz, H. R. (1995). Inhibition of the *Caenorhabditis elegans* cell-death protease CED-3 by a CED-3 cleavage site in baculovirus p35 protein. *Nature (London)* **377**, 248–251.

Xue, D., and Horvitz, H. R. (1997). *Caenorhabditis elegans* CED-9 protein is a bifunctional cell-death inhibitor. *Nature (London)* **390**, 305–308.

Xue, D., Shaham, S., and Horvitz, H. R. (1996). The *Caenorhabditis elegans* cell-death protein CED-3 is a cysteine protease with substrate specificities similar to those of the human CPP32 protease. *Genes Dev.* **10**, 1073–1083.

Yochem, J., and Greenwald, I. (1989). *glp-1* and *lin-12*, genes implicated in distinct cell–cell interactions in *C. elegans*, encode similar transmembrane proteins. *Cell (Cambridge, Mass.)* **58**, 553–563.

Yoshida, Y., Matsuda, S., and Yamamoto, T. (1997). Cloning and characterization of the mouse *tob* gene. *Gene* **191**, 109–113.

Yuan, J. Y., and Horvitz, H. R. (1990). The *Caenorhabditis elegans* genes *ced-3* and *ced-4* act cell autonomously to cause programmed cell death. *Dev. Biol.* **138**, 33–41.

Yuan, J., and Horvitz, H. R. (1992). The *Caenorhabditis elegans* cell death gene *ced-4* encodes a novel protein and is expressed during the period of extensive programmed cell death. *Development (Cambridge, UK)* **116**, 309–320.

Yuan, J., Shaham, S., Ledoux, S., Ellis, H. M., and Horvitz, H. R. (1993). The *C. elegans* cell death gene *ced-3* encodes a protein similar to mammalian interleukin-1 beta-converting enzyme. *Cell (Cambridge, Mass.)* **75**, 641–652.

Zarkower, D., and Hodgkin, J. (1992). Molecular analysis of the *C. elegans* sex-determining gene *tra-1*: A gene encoding two zinc finger proteins. *Cell (Cambridge, Mass.)* **70**, 237–249.

Zarkower, D., and Hodgkin, J. (1993). Zinc fingers in sex determination: Only one of the two *C. elegans* Tra-1 proteins binds DNA *in vitro. Nucleic Acids Res.* **21**, 3691–3698.

Zarkower, D., de Bono, M., Aronoff, R., and Hodgkin, J. (1994). Regulatory rearrangements and smg-sensitive alleles of the *C. elegans* sex-determining gene *tra-1*. *Dev. Genet.* **15**, 240–250.

Zhang, B., Gallegos, M., Puoti, A., Durkin, E., Fields, S., Kimble, J., and Wickens, M. (1997). A conserved RNA-binding protein that regulates sexual fates in the *C. elegans* hermaphrodite germ line. *Nature (London)* **390**, 477–484.

Zou, H., Henzel, W. J., Liu, X., Lutschg, A., and Wang, X. (1997). Apaf-1, a human protein homologous to *C. elegans* CED-4, participates in cytochrome *c*-dependent activation of caspase-3. *Cell (Cambridge, Mass.)* **90**, 405–413.

9

Cell Fate Determination in Caenorhabditis elegans Ray Development

Scott W. Emmons
Department of Molecular Genetics
Albert Einstein College of Medicine
Bronx, New York 10461

I. Introduction

The specialized genital apparatus in the tail of male nematodes includes a set of simple, three-celled sensory structures, in some species forming genital papillae and in other species forming extended "rays," that have become the focus of basic studies. Research on the male rays of *Caenorhabditis elegans*, which is facilitated by their cellular simplicity and genetic accessibility, addresses several fundamental questions in metazoan development. Wild-type ray number and morphology is dependent on a cell fate specification process during postembryonic cell lineages. Specification of cell fates is guided by universally significant transcription factors of the *Hox, Pax,* and *achaete/scute* families. Transforming growth factor β (TGFβ) and Wnt signaling pathways are involved in ways yet to be fully established. After the lineage has generated cells of the correct types, these cells assemble into precise sensory organs through establishment of specific heterotypic cellular associations, including extension and targeting of axons of sensory neu-

rons to specific postsynaptic partners. Because nematode male genitalia vary from species to species, comparative analysis of ray development provides an opportunity to define the molecular changes underlying the evolution of morphology. This article reviews the cytological and molecular genetic investigations that have laid a foundation for what it is hoped might one day be a complete description of ray development, revealing how one particular biological structure is encoded in the nucleotide sequence of DNA.

In *C. elegans*, there are nine pairs of rays projecting from the posterior body inside an acellular cuticular webbing known as the fan (Fig. 1). The rays function during copulation to allow the male to establish and maintain contact with the hermaphrodite (Fig. 2). Because in *C. elegans* the hermaphroditic sex can self-fertilize internally, copulation is not necessary for strain propagation. This property has facilitated the isolation of large numbers of mutations affecting many aspects of the male genitalia and the specialized male nervous system (reviewed in Emmons and Sternberg, 1997). Muta-

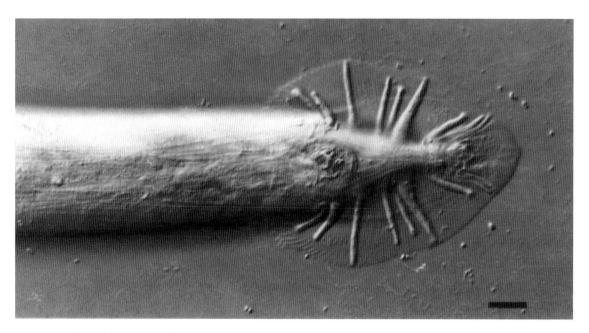

Figure 1 The *C. elegans* male tail, ventral view, showing acellular fan and rays. The ray pairs are numbered 1–9 starting with the most anterior pair. Each ray forms at a reproducible position; the tips of rays 1, 5, and 7 open on the dorsal surface of the fan, the tips of rays 2, 4, and 8 open on the ventral surface of the fan, rays 3 and 9 open at the margin of the fan. Ray 6 has no or only a tiny opening and a different, conical morphology. Nomarski optics. Scale bar: 10 μm.

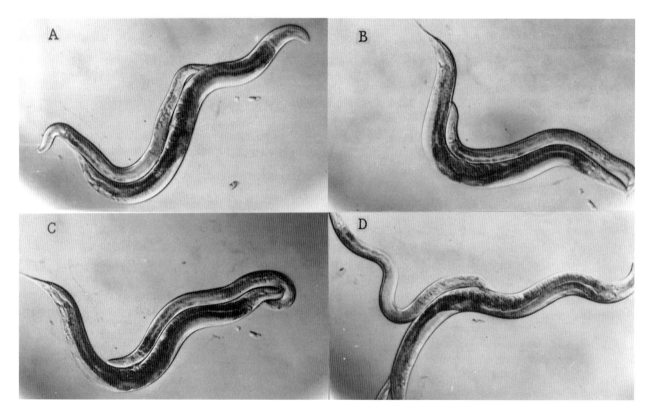

Figure 2 Copulation. (A) A male (the smaller of the two animals) places his tail against the hermaphrodite and swims backward, maintaining contact, in search of the vulva. (B and C) If the male reaches the end of the hermaphrodite, he executes a sharp ventral bend and searches the opposite side. (D) When the vulva is located, the male inserts spicules and transfers sperm.

tions affecting the rays have been easy to isolate and have formed the basis for a study of cell fate determination in the postembryonic cell lineages leading to the rays.

The three cells that comprise each ray, together with an epidermal cell, are a clone derived from a single ray precursor cell. The ray precursor cell, known as an Rn cell (where n = 1–9 corresponding to each of the nine pairs of rays), divides three times during the third and fourth larval stages, following a stereotyped cell lineage pattern known as the ray sublineage (Fig. 3). One set of questions raised by a consideration of ray development concerns the nature of the genetic subprogram that is presumed to underlie this ray sublineage. How does this subprogram resemble and how does it differ from the many other neuronal sublineages involved in generation of the C. elegans nervous system, which are similar to the ray sublineage in many ways but which generate neurons of different types? How are these C. elegans subprograms related to those that underlie development of sensory structures in other groups, for example in insects (Lawrence, 1966; Bodmer et al., 1989)? Is it possible that classes of neuronal subprograms in species of different phyla are related by being descended from a particular ancestral genetic subprogram that generated a sensory structure or other subset of neurons in a basal metazoan?

A second question that arises is how are cells specified to express the Rn cell fate? In C. elegans there are nine Rn cells on each side of the tail, but in other species the number can be less or greater than this. In C. elegans, mutants and ectopic gene expression studies demonstrate that there are epidermal cells in other parts of the body or at other times during development that can be caused to adopt the Rn cell fate and execute the ray subprogram. Such mutations define genes whose ac-

tivities are regulated to specify the number and position of the Rn cells.

A third question concerns the origin of differences among the rays. Although they arise from repetition of a subprogram and contain cells that are differentiated into the same three cell types, the nine ray pairs are not identical. They form at different fixed sites in the fan, they may have different morphology and ultrastructure, and they may express different neurotransmitters. Much of the work on ray development has focused on this third aspect. How serially homologous structures are given distinct identities is a question that recurs throughout biology. Studies of the genetic basis of this phenomenon in C. elegans ray development have shown that patterning the rays comes about through the action of some of the same transcription factors and intercellular signaling systems that pattern structures such as insect segments, the vertebrate nervous system, or vertebrate limbs. The C. elegans rays offer an exceptional opportunity to define the components of these universal patterning systems and to study how they interact in a regulatory hierarchy to assign cells their individual and specific fates.

II. The Rays: Structure, Cellular Composition, and Function

A ray is a tubular protrusion of the large hypodermal syncytium which covers most of the C. elegans body. This tubular protrusion encloses the processes of three cells: two sensory neurons and a support cell called the structural cell (Fig. 4). The rays extend at the end of the L4 larval stage during a brief period when anterior movement of cells results in remodeling of the tail region and generation of the fan (Sulston et al., 1980). Extension of each ray occurs under the influence of the structural cell: if the structural cell is ablated no ray forms, whereas the two ray neurons may be ablated without effect on ray formation or morphology (Sulston et al., 1980; Zhang and Emmons, 1995). How the structural cell causes the hypodermis to form this extended structure is not known.

In the mature male, the cell bodies of the three ray cells are located in a bilateral pair of lumbar ganglia anterior of the anus (Fig. 5). The sensory neurons extend axons through commissures into the preanal ganglion, where they synapse onto one set of identified interneurons (EF interneurons) and probably additional as yet unidentified targets (Sulston et al., 1980). The sensory dendrites extend from the lumbar ganglion, together with the process of the structural cell, into the ray. At the end of the ray the structural cell surrounds and holds the dendritic endings of the neurons (Fig. 4c). The two ray sensory neurons, termed RnA and RnB

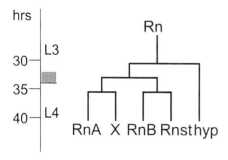

Figure 3 The ray sublineage. Each of the 18 ray precursor cells, Rn (n = 1–9 corresponding to the 9 ray pairs), divides during the L3 and L4 larval stages as shown, giving rise to 4 cell types: RnA, an A-type sensory neuron; RnB, a B-type sensory neuron; Rnst, a support cell; hyp, a hypodermal cell. A fifth cell undergoes programmed cell death, ×. Divisions are generally along the anteroposterior body axis. Hours of postembryonic development are given to the left, along with an indication of the larval stages; the shaded interval is the period of lethargus before the molt.

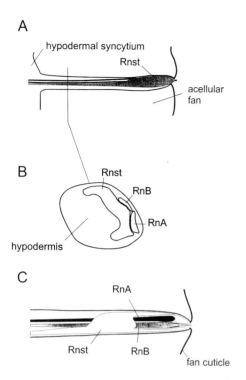

Figure 4 Structure of a ray. Reconstruction of the ultrastructure of a typical ray from electronmicrographs of serial sections (Chow *et al.*, 1995). RnA, A-type sensory neuron; RnB, B-type sensory neuron; Rnst, ray structural cell. The cross section in B shows that the bulk of a ray is hypodermis. The cutaway in C shows how the structural cell holds the tip of the B-type neuron at an opening, and the tip of the A-type neuron inside the ray tip.

neurons, differ from each other in the ultrastructure and position of their endings within the ray tip, and in several additional ultrastructural features. All of the rays have similar structure except ray 6. Ray 6 has a more conical shape due to an increased amount of hypodermis within the ray, has a closed tip, and has a B-type neuron of slightly different ultrastructure (Sulston *et al.*, 1980; Chow *et al.*, 1995).

The functions of the rays in mating have been explored by cell ablation studies (Liu and Sternberg, 1995). Most of the ray tips are open to the exterior, ex-

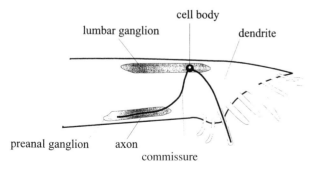

Figure 5 A ray sensory neuron. The cell bodies of the three ray cells, RnA, RnB, and Rnst, reside in the lumbar ganglia. Processes extend into the rays and through commissures into the preanal ganglion, where postsynaptic partners are encountered (Sulston *et al.*, 1980).

posing the dendritic endings of the RnB neurons. This suggests that the rays are chemosensory, allowing the male to detect chemical moieties on the surface of the hermaphrodite cuticle. A chemosensory role also is suggested by evidence for expression of a chemoreceptor in at least one of the rays (Troemel *et al.*, 1995). There also is the possibility that the rays are mechanosensory and detect movement of the fan when it opens as a male encounters a hermaphrodite with his tail, or closes if a male reaches the end of a hermaphrodite or loses contact with her. Therefore each ray may contain one chemosensory neuron and one mechanosensory neuron. However, this is speculation, and the actual modalities of the rays remain unknown.

III. Development of the Rays from Rn Cells

Each of the 18 rays develops from a single cell (Sulston and Horvitz, 1977). These 18 ray precursor cells, or Rn cells, are hypodermal cells lying in a domain of the hypodermis known as the seam. Most of the *C. elegans* hypodermis consists of a large syncytium made up of fused hypodermal cells. The seam is a row of unfused cells; there is one seam on each side of the animal. Seam cells divide in a stem cell pattern during larval development to provide additional cells that fuse with the hypodermal syncytium (see Fig. 8). The seam also generates components of the peripheral nervous system by providing neuroblasts at various points. It is therefore a neuroepithelium similar to those which give rise to components of the nervous systems of other animals. Each neuroblast divides a small number of times to provide the neurons and support cell or cells of a sensory structure.

The 18 Rn cells comprise all but one of the most posterior seam cells on each side of the male tail at the end of the L3 larval stage (Fig. 6, see also Fig. 8). In late L3 and early L4 each Rn cell divides in a stereotyped pattern known as the ray sublineage (Fig. 3). At the first division, which is in the plane of the epidermis, a ray neuroblast and a hypodermal cell are generated. Subsequently, the ray neuroblast divides twice, also in the plane of the epidermis, giving rise to the cells of one ray and a cell that undergoes programmed cell death. Thus the ray cells are born among the epidermal cells in the posterior region, and are joined to them by adherens junctions (Baird *et al.*, 1991) (see Fig. 10). The cell bodies of the differentiating ray neurons and structural cell subsequently migrate out of this epidermal layer to take up their adult positions in the lumbar ganglia, leaving behind processes anchored in the epidermal layer at the sites of their birth.

Specification of the nine bilateral pairs of Rn cells is a crucial step in the specialization of the male tail and

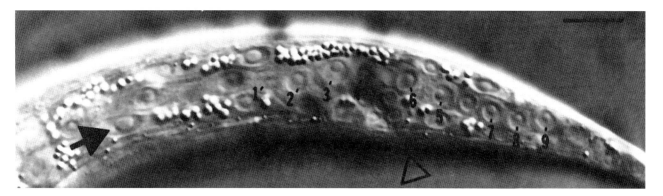

Figure 6   Ray precursor (Rn) cells. Nomarski photomicrograph of the lateral posterior hypodermis of an L3 larval male. The nuclei, with visible nucleoli, of the nine Rn cells are indicated. For some cells, the outline of the cell is also visible (e.g., R7). The closed arrow indicates the nucleus of V6.a (see Fig. 8); the open triangle shows the position of the anus. Scale bar: 10 μm.

a focal point of the regulatory pathways that govern development of the fan and rays. Each Rn cell is largely autonomous. The regulatory pathway ensures that nine Rn cells are generated and that these lie at the posterior end of the seam. If spatial regulation is disrupted, Rn cells can be generated at more anterior positions. Such ectopic Rn cells divide at the appropriate time following the same stereotyped pattern, and give rise to differentiated ray cells. In some genetic backgrounds even a small fan and extended ray are formed (H. Zhang and S. W. Emmons, unpublished observations). Thus local signals in the tail are not necessary for generation and differentiation of a ray.

Other elements of the regulatory pathway ensure that each Rn cell generates a ray of a specific type. The rays differ from each other in their positions, morphology, neurotransmitter usage, and probably other aspects yet to be defined. These characteristics appear to be fixed at the time or soon after an Rn cell is generated. Understanding how these nonequivalent states of the Rn cells are created is a central problem in understanding ray development.

IV. An *achaete/scute* Family Transcription Factor Is Necessary for the Ray Neuroblast Cell Fate

In its first division, each Rn cell generates a ray neuroblast, which divides twice to give the two neurons and support cell of a ray. Expression of the neuroblast fate requires the function of a basic helix-loop-helix (bHLH) transcription factor, the product of the *lin-32* gene (Zhao and Emmons, 1995). The most similar known gene in another organism is the *atonal* gene of *Drosophila,* a member of the *achaete/scute* family, which acts in development of chordatonal sensory organs and the eye (Jarman *et al.,* 1993, 1994). Transcription fac-

tors of the bHLH class play a role in defining tissue type in many organisms. Those related to the *Drosophila* genes *achaete* and *scute* define neuroblasts or neurons.

In *lin-32* loss-of-function mutants, the ray neuroblast fate is not expressed. Instead, the anterior daughters of the Rn cells appear to remain in the same cell state as their mothers; that is, they appear to remain as Rn cells. Like an Rn cell, they divide, producing another Rn.p cell and yet another seam cell. On the other hand, if *lin-32* is ectopically expressed in the anterior seam, it will cause cells that normally are hypodermal seam cells to become neuroblasts (Zhao and Emmons, 1995). Therefore LIN-32 is in some cells both necessary and sufficient for the ray neuroblast cell fate.

As expected from these observations, in the wild-type male seam, *lin-32* is transcribed only in the 18 posterior seam cells that express the ray sublineage (Zhao, 1995) (Fig. 7). However, *lin-32* is expressed in other cells throughout the body as well, and its expression is necessary during both embryonic and postembryonic development for a number of other neuroblast cell fates. Therefore *lin-32* expression alone does not define a ray neuroblast. There is a requirement for one or more additional functions, as yet undefined, which together with *lin-32* result in development of this particular sensillum.

Although LIN-32 seems to be required for the ray neuroblast fate, that is the fate of the Rn.a cell, *lin-32* is first transcribed one cell generation earlier, in the Rn cell (Zhao, 1995). Furthermore, a *lin-32* reporter gene, after turning on in Rn, remains on in both Rn daughters. Thus the gene is transcribed in three cells, Rn, Rn.a, and Rn.p, only one of which becomes a ray neuroblast. These three cells must differ from each other in some respect, such that LIN-32 acts only in the Rn.a cell. We will see that activity of a regulatory transcription factor in only a subset of the cells in which it is found recurs frequently. We will return later to the significance of this observation.

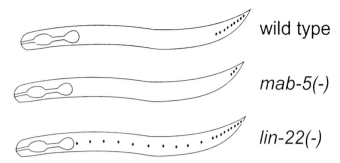

wild type

mab-5(-)

lin-22(-)

Figure 7 Expression of *lin-32*. Diagrammatic indication of the staining pattern observed with a *lin-32::lacZ* reporter gene in an L3 male (Zhao, 1995). In wild type, the nine ray precursor cells express the reporter. In *mab-5*(−), only the ray precursor cells descended from the blast cell T, R7, R8, and R9, express the reporter. In *mab-5* mutants rays 1–6 are not generated. In *lin-22*(−), (*lin-22* is a homolog of *Drosophila hairy* and *Enhancer of split*), *mab-5* is expressed in anterior seam cells, which results in expression of *lin-32* and of the ray sublineage by these cells (see text).

Target genes of the LIN-32 transcription factor have not yet been identified. As LIN-32 is not present after the second division of the ray sublineage, it probably does not directly activate neuronal or structural cell differentiation genes. Its targets must be an intermediate set of genes that are part of the ray genetic subprogram.

V. *Hox* Genes Regulate *lin-32* Transcription

Because the action of LIN-32 is both necessary and sufficient for a seam cell to generate a ray, the regulatory pathway that governs *lin-32* expression determines how many rays the seam will form as well as their position in the body. This regulation appears to be brought about through transcriptional regulation of *lin-32* by *Hox* genes. Two *Hox* genes that are in part responsible for expression of *lin-32* are *mab-5* and *egl-5*. *mab-5* is the *C. elegans* homolog of *Drosophila Antennapedia,* and *egl-5* is the *C. elegans* homolog of *Drosophila Abdominal-B* (see Ruvkun, 1997, for review of *C. elegans Hox* genes).

For the "V rays" (the rays generated by the blast cells V5 and V6, that is rays 1–6), there is a correspondence between MAB-5 and EGL-5 expression, on the one hand, and *lin-32* transcription on the other (Fig. 8). That is, seam cells expressing either of these two proteins transcribe *lin-32,* whereas those that do not, in the anterior and midbody, fail to transcribe *lin-32.* Therefore the *lin-32* promoter could be a direct target of *mab-5* and *egl-5* transcriptional activation, but binding studies between MAB-5 or EGL-5 and the *lin-32* promoter have not been carried out to verify this supposition. In order to turn on *lin-32* in response to *mab-5* or *egl-5,* a seam cell must also be in an L3 male. Therefore addi-

tional factors must be present to potentiate a seam cell to the action of these *Hox* genes. [What regulatory factors may turn on *lin-32* transcription in the three branches of the T lineage, leading to rays 7–9, are not known. One gene, *mab-19,* has been described that is required specifically for generation of rays by the T lineage (Sutherlin and Emmons, 1994). However, the molecular nature of the MAB-19 gene product is unknown, and it is not known whether it acts via regulation of *lin-32.*]

The roles of *mab-5* and *egl-5* in regulating *lin-32* can be demonstrated in various genetic backgrounds. In a *mab-5* null mutant, *lin-32* is not expressed in R1–R6 and rays 1–6 are not formed (Fig. 7) (Kenyon, 1986; Zhao, 1995). The requirement for *mab-5* could in part be through a direct interaction with the *lin-32* promoter, but it also appears to come about in part as a result of *mab-5* activation of *egl-5* transcription during L1 (Salser and Kenyon, 1996; Ferreira *et al.,* 1998). In an *egl-5* null mutant, all the rays are generated except ray 6.

For *mab-5,* expression is not only necessary for expression of the ray sublineage, it is also sufficient. Ectopic expression of *mab-5* in anterior seam cells causes activation of *lin-32* and ectopic generation of the ray sublineage (Salser and Kenyon, 1996). This is particularly evident in a *lin-22* mutant, where MAB-5 is expressed throughout the anterior seam and the ray sublineage is expressed by seam cells all along the body (Horvitz *et al.,* 1983; Wrischnik and Kenyon, 1997) (see below) (Fig. 7). Therefore spatial regulation of *lin-32* is dependent on spatial regulation of *mab-5.* The role of the *Hox* genes may be to mediate the action of spatial patterning signals, thereby bringing spatial patterning information to bear on the *lin-32* promoter. Regulation of the expression of *mab-5* and *egl-5,* including the role of spatial signals, is discussed further below.

Why do *mab-5* and *egl-5,* which are expressed much more widely, turn on *lin-32* transcription only in seam cells, and only at a certain developmental stage? What properties of L3 seam cells potentiate *lin-32* for activation by these two *Hox* genes? No doubt additional transcription factors present in L3 seam cells act combinatorially with the *Hox* genes to potentiate their action at the *lin-32* promoter. The nature of these additional factors is not yet known.

VI. The Rays and Ray Neuroblasts Have Unique Identities

Although they are generated from identical cell sublineages by the action of *lin-32* and have similar structures, the rays are not all alike. Differences between the rays in the characteristics of their constituent neurons,

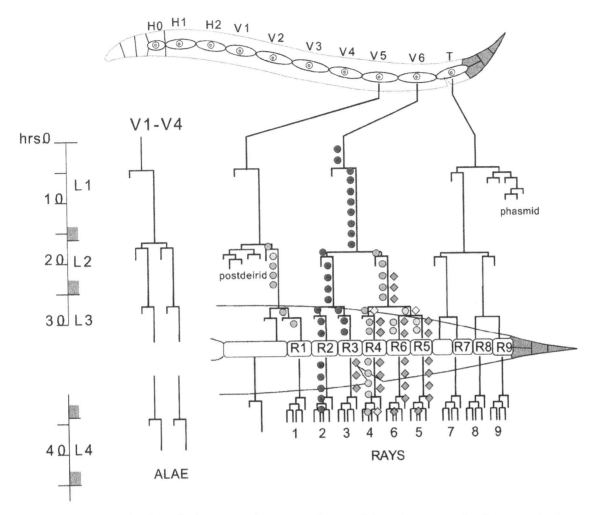

Figure 8 Expression of *mab-5* and *egl-5* in postembryonic seam lineages of the male; MAB-5: red and orange circles (orange indicates weaker antibody staining); EGL-5: green diamonds. At the top is shown the hypodermal structure of an L1 larva before postembryonic cell division has occurred. The seam cells are labeled. Seam cells V1–V6 and T divide at the times and during the larval stages shown to the left. Short, unlabeled lineage branches indicate cells that fuse with the hypodermal syncytium, *hyp7*. Three types of sensory structures are generated: postdeirid, phasmid, and rays. V1–V4 descendants, as well as V5.pppapp, generate ridges in the adult cuticle known as alae. Data for MAB-5 are from Salser and Kenyon (1996); data for EGL-5 are from this laboratory (Ferreira *et al.*, 1998). Shaded cells in the tail tip express the Wnt gene *lin-44* (Herman *et al.*, 1995).

and in location in the fan, probably result in differences in ray functions essential for optimal male mating (Liu and Sternberg, 1995). Thus the rays are like serially homologous structures or segments found in many organisms; they have a similar developmental origin and share many characteristics, but they have diverged from each other in ways that allow specialization in their functions.

One known difference between the ray neurons is in monoamine neurotransmitter expression. The A-type neurons of rays 5, 7, and 9 uniquely contain dopamine (Sulston *et al.*, 1975). One of the two neurons in rays 1, 3, and 9 uniquely express serotonin (it is not known whether it is the A-type or B-type neuron) (Loer and Kenyon, 1993). What other neurotransmitters are expressed by ray neurons is unknown. The B-type neuron of ray 6 has a differing ultrastructure from the B-type

neurons in the other rays (Sulston *et al.*, 1980), and ray 6 itself has a slightly different morphology, apparently due to an increased size of the hypodermal component (Chow *et al.*, 1995). Unlike the nervous system of the *C. elegans* hermaphrodite, the nervous system of the male has only been partially reconstructed (Sulston *et al.*, 1980). Thus the connections and ultrastructural characteristics of the many male-specific neurons found in the tail, including the ray neurons and support cells, are not fully known. It seems likely, in view of their other differing properties, that when the axonal projections of the ray neurons are determined, they may be found to have differing postsynaptic partners.

A further difference between the rays is revealed by the properties of certain mutants that affect ray development. When visual screens were carried out for mutations affecting tail morphology, a striking class of mutants was found that resulted in fused rays (Baird *et al.*, 1991) (Fig. 9). Fused rays contain the constituent cells of two or more rays assembled somewhat as if they were a single large ray (Chow *et al.*, 1995). Furthermore, in these mutants, whether or not they fuse, rays frequently form at incorrect positions in the fan (Zhang and Emmons, 1995). Mutations that cause fused rays appear to define genes that govern expression of the cell recognition and adhesion functions that control the assembly of ray cells with other cells, both other ray cells and the surrounding hypodermal cells that position the rays. These cell recognition and adhesion functions must differ in some way from ray to ray, thereby allowing independent ray assembly and defining the hypodermal location of each ray.

The spectrum of phenotypes found in some fused ray mutants can be interpreted as resulting from transformations in cell identities at the level of the ray neuroblasts. In these mutants, a number of independent properties of the cells generated by one ray neuroblast are all transformed to properties normally associated with the cells generated by a different ray neuroblast.

This has been most thoroughly demonstrated for mutations in the gene *mab-21* (Chow *et al.*, 1995). Mutations in this gene, which encodes a novel protein of unknown cellular function with a conserved human counterpart (Margolis *et al.*, 1996), transform the morphology of ray 6, its position and assembly characteristics, the ultrastructure of its tip, and the ultrastructure of its B-type neuron, all to those associated normally with the corresponding cells of the ray 4 sublineage. Mutations in the gene *mab-18*, which encodes a *C. elegans* homolog of the *Pax-6* transcription factor, and which is also required for the identity of ray 6 (see below), were used to show that it is the structural cell that determines the position and morphology of ray 6: in a *mab-18* background the properties of ray 6 were transformed to those of ray 4 even when the ray 6 neurons were ablated by a laser microbeam (Zhang and Emmons, 1995). Mutations in several genes that encode components of a TGFβ signaling pathway result both in fusions affecting rays 5, 7, and 9, and also in the loss of dopamine expression by these rays (Savage *et al.*, 1996; R. Lints and S. W. Emmons, unpublished observations).

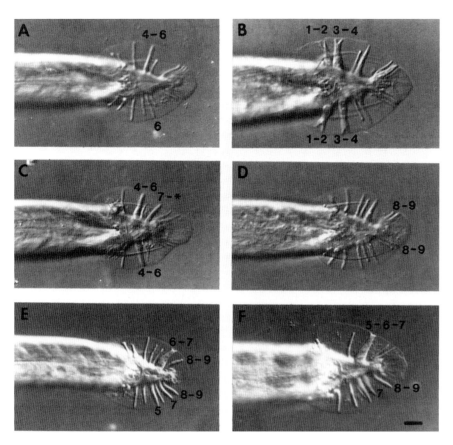

Figure 9 Fused ray mutants. Nomarski photomicrographs, ventral views, of male tails of six mutants (Baird *et al.*, 1991). Affected rays are indicated. (A) *mab-18*, encodes a *Pax-6* homolog (Zhang and Emmons, 1995); (B) *mab-20*, encodes a semaphorin homolog (J. Culotti, personal communication); (C) *mab-21*, encodes a novel protein with a conserved human homolog (Chow *et al.*, 1995; Margolis *et al.*, 1996); (D) *bx28*, an uncharacterized double mutant (S. W. Emmons, unpublished results); (E) *sma-2*, encodes a Smad protein (Savage *et al.*, 1996); (F) *sma-3*, encodes a Smad protein (Savage *et al.*, 1996).

Thus mutations in *mab-21*, *mab-18*, and TGFβ pathway genes affect the properties of both the neurons and the structural cell of one or more ray sublineages. These genes may be thought of as ray identity genes acting as part of a regulatory hierarchy that patterns the rays by assigning different identities to ray neuroblasts.

VII. Nonequivalence of the Rn.p Cells, Hypodermal Cells Generated by Rn Cells

Like the rays themselves, the hypodermal cells generated at the first divisions of the Rn cells are nonequivalent. Their differing properties may be important for positioning the rays. These hypodermal cells, known as Rn.p cells, enlarge to cover the lateral surfaces of the tail in the regions where the rays and fan will form. Rn.p cells are joined to each other and to the surrounding hypodermal syncytium by adherens junctions, resulting in a detailed pattern of cell boundaries on the surface (Fig. 10a). The three clonally related cells destined to form each ray lie at specific sites defined by these adjacent cell boundaries (Fig. 10b). Possibly, ray cell-hypodermal cell recognition and adhesion functions position the ray cells at precise points in this network (Baird *et al.*, 1991). In other nematode species, the ray cells lie at different points in the network, and the network itself varies, suggesting that changes in cell recognition and adhesion functions, or the regulatory mechanisms that govern their expression, are the source of evolutionary variability in ray position (Fitch and Emmons, 1995).

After enlarging, the anterior five Rn.p cells, R1.p–R5.p, fuse together to generate a small syncytium known as the tail seam or SET, whereas R6.p to R9.p remain unfused until a still later time, when they fuse with hyp7, the hypodermal syncytium (Sulston *et al.*, 1980) (Fig. 10c,d). The SET appears to define a domain on the surface that will become the dorsal surface of the fan. This domain may contribute in some way to generation of the fan, but this has not been studied directly.

It appears that ray identities and Rn.p identities are coordinately determined. In two cases that have been examined, mutants that transform properties of a ray also transform the properties of the corresponding Rn.p cell. In both *mab-18* and *mab-21* mutants, which result in a morphological and ultrastructural transformation in the identity of ray 6 to that of ray 4, R6.p also is transformed and fuses with the SET like R4.p (Chow *et al.*, 1995; Zhang and Emmons, 1995). Therefore these two genes act in both branches of the ray sublineage, or else they act within the Rn cell itself to coordinately transform both daughter cells.

VIII. Rn Cells Have Unique Identities

Since both the rays and the Rn.p cells have unique properties, it is possible that the characteristics of both lineage branches leading from an Rn cell are defined by a single mechanism acting at the level of the Rn cell. In this case each Rn cell differs from the others in some property or properties that determine the characteristics of the cells it generates. Thus, two possibly independent pattern formation systems act in generation of the rays: one causes the Rn cell to express the ray sublineage, and the second, overlying the first, generates unique ray and Rn.p cell characteristics. Mutations that only transform ray and Rn.p cell properties, without affecting expression of the ray sublineage and generation of the ray, define genetic functions that act in the second pathway and support the notion that the two pathways are at least partly independent.

Several lines of evidence support the notion of unique Rn cell identities. The first is that, as discussed above, some mutants simultaneously alter the properties of cells of both daughter branches of Rn cell division. Additional evidence for unique Rn cell identities has been provided by cell ablation studies. These studies have shown that the differences between the rays in expression of unique morphogenetic functions are fixed by the time the Rn cells are born. No interactions between neighboring seam cells or positional signals along the body axis appear to be necessary by the Rn cell stage for an Rn cell to generate a ray of the expected type (Chow and Emmons, 1994). Thus, an Rn cell is preprogrammed to autonomously generate a particular ray.

Analysis of mutants also suggested that ray identity is set at the level of the Rn cell. *Hox* gene mutations, described further below, result in transformations in ray identities that were interpreted as being transformations in cellular identities at the level of the Rn cells (Chow and Emmons, 1994). Temperature shift (ts) experiments with a ts allele of a TGFβ receptor showed that the requirement for this pathway in defining the identities of rays 5, 7, and 9 occurred during the Rn cell stage (S. E. Baird, personal communication) (see below).

Thus, the apparent autonomous ability of each Rn cell to generate a particular ray, the effects of mutations on cells in multiple lineage branches, and the time of action of the TGFβ pathway, all suggest that ray identity is set within the Rn cell. Molecular differences between Rn cells in expression of regulatory transcription factors, described further below, clearly show that these cells are not equivalent. It is necessary to understand how these molecular differences between the Rn cells are established in order to understand how different rays are specified.

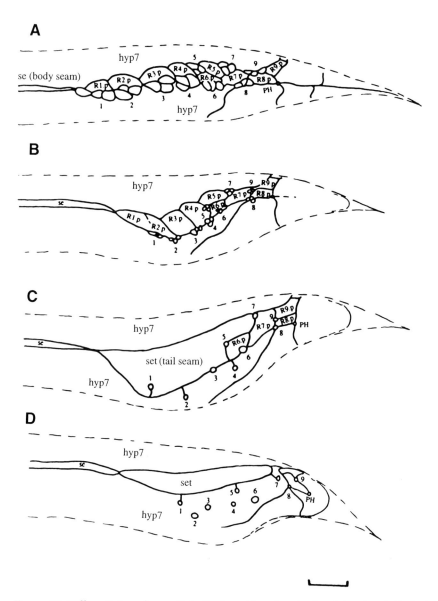

Figure 10 Differentiation of ray cells in the lateral hypodermis of an L4 male tail (dashed outline). Hypodermal cell boundaries (solid lines) have been visualized, at successive, approximately 1-hour intervals, by staining adherens junctions with the monoclonal antibody MH27. Most of the body is covered by a syncytium (hyp7). (A) After the ray sublineages are completed, the ray cells and associated Rn.p cells are arrayed in nine clusters. (B) Ray cell bodies migrate out of the epidermal layer, leaving processes anchored at the surface (clusters of small circles). (C) R1.p–R5.p fuse, creating the tail seam (SET). Ray cells further differentiate, leaving only the structural cell joined to hypodermal cells. The positions of the structural cells at this stage determine the positions where the rays will form. (D) The positions of the structural cells remains fixed as retraction of the SET and morphogenesis of the tail begins. Scale bar: 10 μm.

IX. Rn Cell Identities Are Determined by a Lineage Mechanism

Cells in a line or in a two-dimensional field may be patterned by nearest-neighbor interactions, like cells of the *Drosophila* retina (Tomlinson, 1988; Freeman, 1997), or by a positional signal emanating from a source (St. Johnston and Nusslein-Volhard, 1992; Neumann and Cohen, 1997). Rn cells appear to be patterned by neither of these mechanisms (Chow and Emmons, 1994).

Cell ablation studies demonstrated that each ray type could still be generated after removal of the cells leading to other ray types. This rules out the possibility that one ray type induces another the way one *Drosophila* photoreceptor cell induces another. In addition, Rn cells displaced to different positions along the body axis generated the same ray type rather than the ray type that would normally be generated at the new position. This appears to rule out some kind of positional signal along the anteroposterior body axis. Thus Rn cells were found to have fixed characteristics independent of their neighbors or their axial positions.

If each Rn cell has a characteristic identity, then it follows that the parents of the Rn cells, which generate Rn cells of particular identities, must themselves have characteristic identities. Applying the same reasoning one cell generation earlier, it becomes evident that every cell in the lineage must have a unique identity. The fixed and discrete nature of these presumptive intermediate cell identities was confirmed by ablation studies (Sulston and White, 1980; Chow and Emmons, 1994). For most of the intermediate cells, the wild-type characteristics of unablated cells were not affected by removal of their neighbors. For example, if V6.pap was ablated, V6.ppp still generated rays 4, 5, and 6, whereas if V6.ppp was ablated, V6.pap still generated rays 2 and 3 (see Fig. 8). Thus V6.pap and V6.ppp differ in some fixed characteristic such that they generate different cell lineages and rays of differing identities. Sometimes in such studies the fate of a cell can be transformed. For example, if V6 is ablated, V5 can take V6 identity and generate rays 2–6 instead of ray 1. Such transformations indicate that there are periods when the fate of a cell is influenced by signals. But eventually this fate becomes fixed and cannot be influenced by alterations in the cellular environment.

In this progressive specification of alternative cell states through the cell lineage, the postembryonic cell lineages leading to the rays are not unlike the cell lineages leading to the founder cells of the early *C. elegans* embryo (Chapter 7 by Bowerman). Most likely this paradigm of progressive cell states also operates in embryos of animals without fixed cell lineages. What general molecular mechanisms underlie this process are of great interest.

X. *Hox* Genes Act in the Rn Cell Identity Pathway

Two key players in specification of the intermediate cell states of the postembryonic seam lineages leading to the rays are the *Hox* genes *mab-5* and *egl-5*. Not only do *mab-5* and *egl-5* regulate ray formation by regulating the expression of *lin-32*, they also play a pivotal role in assignment of Rn cell identities. Thus, like many examples of serially repeated but divergent structures in both *C. elegans* and other animals, rays are made distinct from one another through the action of *Hox* genes.

The roles of *mab-5* and *egl-5* in the ray identity pathway were first made apparent by gene dosage studies and by the phenotypes of unusual, non-null mutant alleles (Chow and Emmons, 1994). An increase and decrease in the number of gene copies resulted in transformations and fusions of rays similar to those seen in fused ray mutants. Specifically, a decrease in the number of *mab-5* gene copies, as well as a hypomorphic mutation, resulted in transformation of ray 4 to ray 3, and

a decrease in the number of *egl-5* gene copies resulted in transformation of ray 6 to ray 4. In a gain-of-function *mab-5* mutant, ray 1 was transformed to ray 2 and ray 3 was transformed to ray 4.

Normally, the requirement for a gene in a given process is best assessed in a null mutant background. The requirement for *mab-5* in specifying ray identity is obscured in a null mutant background because the essential role of this gene in activating *lin-32* and generation of the ray sublineage is epistatic to specification of ray identity. For *egl-5* as well, assessment of ray identity in a null mutant is made difficult by absence of tail morphogenesis and the consequent failure of extension of the rays. However, in a mosaic background in which *egl-5(0)* rays did form, the V6 rays were fused together (Chisholm, 1991). Furthermore, in *egl-5(0)* the tips of the 4 V6 rays (rays 2–5, ray 6 is absent) form next to each other at the normal position of ray 2 (Chisholm, 1991). Therefore, in *egl-5(0)*, all the V6 rays may take the identity of ray 2.

The genes *mab-5* and *egl-5* are subject to a detailed pattern of regulation that places them in the Rn cells that genetically require them and supports the notion that ray identity may be specified by some kind of *Hox* "code" (Fig. 8) (Salser and Kenyon, 1996; Ferreira *et al.*, 1998). MAB-5 is present during expression of the ray 2 and ray 4 sublineages and absent from the ray 1 and ray 3 sublineages. If *mab-5* is ectopically expressed from a heat-shock-driven transgene, the identities of ray 1 and ray 3 are transformed respectively to ray 2 and ray 4. Therefore *mab-5* expression, initially on, must be turned off in the ray 1 and ray 3 branches to allow the ray 1 and ray 3 identities to be expressed. Conversely, EGL-5 is present in R5 and R6 and their sublineages, and less strongly in R3 and R4 and their sublineages. This pattern is consistent with the observation that decreased *egl-5* gene dosage caused transformation of ray 6 to ray 4.

As a possible *Hox* gene "code" for the rays these data suggest the following:

	ray 1	ray 2	ray 3	ray 4	ray 5	ray 6
mab-5	OFF	ON	OFF	ON	OFF	OFF
egl-5	OFF	OFF	ON	ON	ON	ON

Such a model is provisional and certainly incomplete. Two aspects of the *mab-5* and *egl-5* expression pattern are not yet clearly established: the level and duration of expression of *egl-5* in ray 3, and whether there is simultaneous, overlapping expression of *mab-5* and *egl-5* in ray 4. Furthermore, this proposed code as it stands does not distinguish between rays 3, 5, and 6. No doubt additional genes play a role. One, for example, might be expected to act in ray 1 in a capacity similar to that of *mab-5* and *egl-5* in the other V rays.

XI. Establishment of the *mab*-5 and *egl*-5 Expression Patterns

Ongoing studies are aimed at determining how the intricately regulated pattern of *mab*-5 and *egl*-5 expression, necessary for establishing the identities of the V rays, is brought about. These highly regulated patterns convey a different impression of the action of these two *Hox* genes from the traditional view of *Hox* genes as indelible markers of positional information (for discussion see Emmons, 1996). As we have seen, in the seam lineages *mab*-5 and *egl*-5 do not remain on as continuous cell markers, but are turned on and off in a pattern that is necessary for specification of the wild-type morphology of the adult.

Several genes that regulate *mab*-5 and *egl*-5 in the seam have been identified in genetic studies. Expression of *mab*-5 in the V6 lineage requires action of *pal-1* (Waring and Kenyon, 1990). *pal-1* encodes a homeobox transcription factor that is the *C. elegans* homolog of *Drosophila* caudal (Waring and Kenyon, 1991). *pal-1* plays a role in specification of early blastomere fates in the embryo (Chapter 7 by Bowerman). Its expression pattern in the seam lineages has not yet been defined, but it is required cell-autonomously in V6 at the time *mab*-5 expression begins, and hence *mab*-5 could be a transcriptional target (Waring and Kenyon, 1991).

What gene or genes are responsible for turning on *mab*-5 expression in the V5 lineage are not known. Whatever this activity is, it appears to be regulated by the product of the *lin-22* gene. *lin-22* encodes a bHLH transcription factor that is the *C. elegans* homolog of *Drosophila hairy* or *enhancer of split* (Wrischnik and Kenyon, 1997). In a *lin-22* mutant, anterior seam cells V1–V4 adopt a V5-like fate and generate a lineage identical to V5 (Horvitz *et al.*, 1983; Kenyon, 1986). All of these V5-like lineages are abnormal in that they produce two rays instead of one. Both ectopic V5-like lineages and extra rays are the result of ectopic activation of *mab*-5 in anterior seam cells or lineage branches (Wrischnik and Kenyon, 1997). Thus a wild-type function of *lin-22* is to prevent expression of *mab*-5 in certain cells.

The expression pattern of *mab*-5 in the V5 lineage is influenced by signals regulated by cell contacts. For example, if V5.p looses contact with either of its anterior or posterior neighbor cells in the seam (as a result of their being ablated), *mab*-5 is turned on in both lineage branches leading from V5.p and as a consequence the postdeirid neuroblast is not generated, with V5.pa remaining as a seam cell (Austin and Kenyon, 1994).

The complex expression patterns of *mab*-5 and *egl*-5 transcription are in part the result of cross-regulatory interactions between them. Expression of *egl*-5 in the

V6 lineage is lost in a *mab*-5(0) background (Ferreira *et al.*, 1998). As mentioned above, a requirement for *mab*-5 transcription occurs in late L1, before appearance of EGL-5 in V6.ppp, and MAB-5 protein is present in V6.ppp. Hence the *egl*-5 promoter could be a transcriptional target of MAB-5. Later, *egl*-5 may repress *mab*-5. A negative regulatory influence of *egl*-5 on *mab*-5 could explain why MAB-5 protein is lost from just those Rn cell lineage branches where EGL-5 remains on (Fig. 8). Such a negative interaction also could explain the results of gene dosage studies in which an added *egl*-5 wild-type gene copy increased the penetrance of the haploinsufficient phenotype of *mab*-5 (Chow and Emmons, 1994). It could also explain the generation of ray 5 in an *egl*-5(0) mutant: *mab*-5 may be ectopically activated in this lineage in *egl*-5(0). Repression of *mab*-5 expression by *egl*-5 has previously been observed in other tail lineages (Salser *et al.*, 1993).

XII. The *Caenorhabditis elegans* Pax-6 Homolog Is Required for the Identity of Ray 6

As mentioned above, *mab*-5 and *egl*-5 alone cannot account for the complete pattern of the V rays. Additional input comes from the *C. elegans* Pax-6 homolog. The *C. elegans* Pax-6 locus is required specifically for the identity of ray 6 and R6.p. Among the rays, ray 6 has a unique overall morphology and a distinct tip with a tiny or absent opening (Sulston *et al.*, 1980; Chow *et al.*, 1995). This morphology, together with the ray 6 position and assembly independent of ray 4, are all dependent on the function of one isoform of the *C. elegans* Pax-6 homolog, defined by *mab-18* mutations (Zhang and Emmons, 1995). In *mab-18* mutants, ray 6 takes the identity of ray 4 and usually fuses with it (Fig. 9A), and R6.p fuses with the tail seam (SET), like R4.p.

PAX-6 is a highly conserved transcription factor present in probably all metazoans. It has two DNA binding domains (an N-terminal PAIRED domain and a C terminal homeodomain) and a C-terminal transcriptional activation domain. The *C. elegans* gene is typical in structure (Fig. 11). It is transcribed from at least two promoters, one lying upstream of the entire gene, and one lying within an intron separating the PAIRED and homeodomains.

It is an isoform containing only the homeodomain, translated from transcripts apparently initiated within the intron, that is required for ray 6 identity and constitutes the *mab-18* genetic locus (Zhang and Emmons, 1995). Mutations within the PAIRED domain, which define a complementing genetic locus, *vab-3*, result in

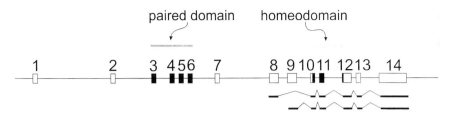

Figure 11 Structure of the *C. elegans Pax-6* gene. The *C. elegans* gene is typical of *Pax-6* in other organisms, having two potential DNA-binding domains: a PAIRED domain and a homeodomain. Mutations in the PAIRED domain define the genetic locus *vab-3* and result in disorganization of head sensory structures, as well as other structures (Chisholm and Horvitz, 1995). Mutations that affect expression of transcripts that contain only the homeodomain define the complementing genetic locus *mab-18* (Zhang and Emmons, 1995). *mab-18* transcripts (two example cDNAs are shown) appear to initiate from a promoter lying within the intron separating exons 7 and 8.

deformities of the body, primarily in the head, and disorganization of head sensory structures (Lewis and Hodgkin, 1977; Chisholm and Horvitz, 1995). *vab-3* mutations have no effect on ray 6. Thus the two *Pax-6* functions are independent, although both utilize the homeodomain portion of the gene.

The expression pattern of *mab-18* is consistent with a function within ray 6 (Fig. 12) (Zhang and Emmons, 1995; Zhang *et al.*, 1998). Expression studies with both reporter genes and antibody staining show that the gene is expressed in cells of ray 6 throughout adulthood. Therefore, MAB-18 may function in expression and maintenance of the terminal differentiation phenotypes of ray 6 cells. As both *mab-18* and *vab-3* mutations cause disorganization of sensory structures, among the terminal differentiation genes controlled by *C. elegans Pax-6* could be cellular recognition and adhesion proteins required in the nervous system. A similar conclusion has been drawn from studies of vertebrate *Pax-6* (Stoykova *et al.*, 1997). Expression of *mab-18* in V6.ppppa (normally R6) is absent in an *egl-5* mutant; the *mab-18* promoter could be a target of this *Hox* gene.

Regulation of *mab-18* expression to ray 6 cells takes place in several steps. In the first step, the gene is transcribed and translated in three ray precursor cells: R6, R7, and R8 (Fig. 12). Interestingly, the protein product, detected by staining with anti-MAB-18 antibodies, is initially present in the cytoplasm, and appears to be relatively excluded or possibly absent from the nucleus. Cytoplasmic staining persists throughout expression of these sublineages. After all cell divisions are completed, staining disappears from the cytoplasm and appears in the nuclei of ray 6 cells, while it disappears entirely from the ray 7 and ray 8 cells. Apparently, MAB-18 translocates to the nucleus in ray 6 cells, but is degraded in ray 7 and ray 8 cells. Finally, expression of *mab-18* appears to become autoactivating, as suggested by the observation that a *mab-18* reporter gene is not expressed in *mab-18* mutant adults.

XIII. Two Signaling Pathways Play a Role in Specification of Ray Identity

As we have seen, a cascade of transcription factor gene interactions appears to underlie the specification of cell fates in the postembryonic seam lineages leading to the rays (Fig. 13). Genetic data indicate that each transcription factor in the cascade is necessary for transcriptional activation of the next in the cascade. Expression data or mosaic studies are consistent with interaction of each transcription factor at the promoter of the next gene, but such interaction has not yet been demonstrated. However, such putative interactions are insufficient to explain the transcriptional pattern of each target gene. The target genes are transcribed only in a subset of the cells where the activating transcription factor is present, and only at a significantly later time than the first appearance of the transcription factor. Thus, PAL-1 is expressed in early embryonic blastomeres but turns on *mab-5* much later during postembryonic development (Hunter and Kenyon, 1996). MAB-5 is present in the V6 cell and in all branches of the V6 lineage until the Rn cell stage, yet it turns on *egl-5* only in V6.ppp. After an interval, *egl-5* causes expression of *mab-18* in V6.ppppa but not in its sister cell V6.ppppp. *mab-5* and *egl-5* activate *lin-32* only in Rn cells, and *lin-32* activates the neurogenic pathway only in Rn.a. For all of these transcription factors, activity at particular target promoters is limited to a subset of the cells where the transcription factor is present.

In order to explain this observation, it is necessary to postulate additional cell and stage-specific transcriptional cofactors or other types of posttranscriptional regulatory events that potentiate the actions of the *Hox* and bHLH transcription factors in certain cells. Transcriptional cofactors may be supplied by the signal transduction pathways associated with intercellular signaling systems. Two such systems are known to affect differentiation of seam cells in the tail: the TGFβ pathway and the Wnt pathway.

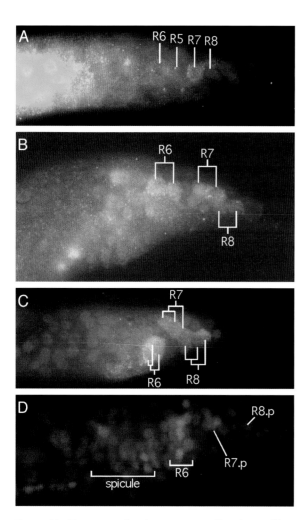

Figure 12 Expression of *mab-18* in seam cells and ray sublineages is regulated at the level of nuclear entry (Zhang *et al.,* 1998). MAB-18, a *Pax-6* isoform containing only the homeodomain, was detected in fixed, whole-mount nematodes by immunofluorescent staining with a polyclonal antibody. The figure shows merged photomicrographs of L4 male tails (anterior to left) in which *Pax-6* antibody staining is green, DAPI staining of nuclear DNA is red, and overlap of red and green appears yellow. Cytoplasmic localization of MAB-18 is shown by green cells with red nuclei. This can be seen in (A) R6-8; (B) R6-8.a and R6-8.p; and (C) R6-8.aa, R6-8.ap, and R6-8.p. (D) After the ray sublineage is complete, staining becomes nuclear in ray 6 cells (four yellow nuclei) and disappears from ray 7 and ray 8 cells, persisting slightly longer in R7.p and R8.p. Additional yellow nuclei anterior of ray 6 cells are internal nuclei of spicule cells which express the paired-domain containing isoform. Green at the left in A is background.

Genes in the TGFβ pathway required for ray pattern formation are shown in Fig. 14. Mutations in each of these genes cause transformations and ray fusions involving rays 4 through 9 (Fig. 9) (Baird *et al.,* 1991; Savage *et al.,* 1996; R. Padgett, personal communication; Y. Suzuki and W. Wood, personal communication), as well as loss of dopamine expression in rays 5, 7, and 9 (R. Lints and S. W. Emmons, unpublished observations). The observation that mutations in the *sma* genes shared a ray morphology phenotype, as well as a small body size phenotype, with mutations in *daf-4*, previously shown to encode a TGFβ Type II receptor homolog, was pivotal in the discovery of the Smad proteins (Savage *et al.,* 1996; Heldin *et al.,* 1997).

By mosaic analysis and by study of a temperature sensitive mutation in the DAF-4 receptor, it has been shown that the TGFβ pathway must be activated within Rn cells for correct specification of ray identity (Savage *et al.,* 1996; S. E. Baird, personal communication). The source of the ligand is unknown. The location above the seam of the rays primarily affected (rays 5, 7, and 9, see Fig. 10) suggests a dorsal source. However, in the strongest mutant backgrounds additional rays are affected, and the source may be more ubiquitous.

A second signaling pathway known to be required in the specification of cell fates in the seam lineages is the Wnt pathway. A Wnt ligand, the product of the *lin-44* gene, is expressed by the hypodermal cells that form the tip of the tail (see Fig. 8) (Herman *et al.,* 1995). In *lin-44* mutants, the polarity of cell fate specification in the postembryonic lineage of the seam cell T, as well as in other male tail postembryonic lineages, is often reversed (Herman and Horvitz, 1994). The receptor for this action of LIN-44 appears to be the product of the *lin-17* gene, a gene related to *Drosophila frizzled* (Sawa *et al.,* 1996). In *lin-17* mutants, polarity of these lineages is lost (Sternberg and Horvitz, 1988). A wider role of the Wnt pathway in patterning cell fates in the seam is suggested by the alternating seam cell expression pattern of POP-1, a transcription factor regulated by Wnt (Lin *et al.,* 1998).

XIV. Summary and Prospects

Metazoan development presents the remarkable phenomenon of a program of cell proliferation during

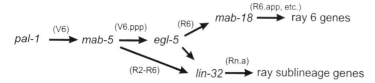

Figure 13 A functional transcription factor cascade in the postembryonic seam lineages. Each gene has been shown genetically to be required for expression of the next gene in the cascade. Interactions between proteins and promoters have not been studied. Activation takes place in the cell shown in parenthesis.

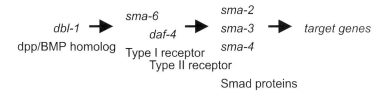

Figure 14 The TGFβ pathway in the male tail. For review of the TGFβ pathway see Heldin *et al.* (1997). Each *C. elegans* gene is related to corresponding genes in *Drosophila* and vertebrates, shown below. The ligand, product of the *dbl-1* gene, is related to *Drosophila* dpp (*decapentaplegic*) and vertebrate BMP (bone morphogenetic protein) (Y. Suzuki and W. Wood, personal communication.) Smad proteins transduce the signal from the activated receptor to the nucleus.

which cells repeatedly change their transcriptional states. When the process is over, a great variety of different cell types has been generated, each cell type expressing a specialized subset of genes. Most of these differentiating cells already lie in the vicinity of the cells of other types, in different transcriptional states, with which they must interact to generate the tissues, organs, limbs, and so forth that constitute the functioning animal.

We have seen how this process operates in the postembryonic lineages leading to the rays. As the lineage progresses, cells of increasingly restricted potential are generated in the correct relative positions until terminal cells differentiate and join with their nearest neighbors to generate adult structures. A cascade of transcription factors appears to explain the progressive specification of cell fates in the seam lineages (Fig. 13). Each transcription factor activates the next in the cascade. However, this occurs only after a significant interval and only in a subset of the cells in which the transcription factor is found. Therefore additional transcriptional cofactors or coactivators, or other posttranscriptional regulatory events, must intervene to potentiate the action of each transcription factor on its target gene. For one of the transcription factors we have described, MAB-18, posttranscriptional activation appears to be by regulation of nuclear entry. For MAB-5 and EGL-5, potentiation of their activity on target promoters is by a different mechanism or mechanisms. Possibly their activity is triggered when an essential cofactor is provided by an intercellular signal, such as the TGFβ or Wnt signals. Such a mechanism has been demonstrated to occur in activation of *Drosophila Hox* genes (Mann and Abu-Shaar, 1996; Rieckhof *et al.*, 1997).

The genetic screens that have lead to the identification of genes involved in ray development are far from saturated. New screens are underway that should result in the discovery of additional components of the cell fate specification pathway (H. Zhang and S. W. Emmons, unpublished). Of particular interest will be genes that regulate expression of *Hox* genes. Understanding how expression and action of these genes is regulated, particularly by intercellular signals such as TGFβ and Wnt signals, is of central importance in developmental biology.

Hox genes act by selective activation of target promoters. We have suggested that among the downstream targets in the postembryonic seam lineages are other *Hox* genes or other types of transcription factor genes. Promoter studies are now necessary to confirm such interactions and to identify the additional transcription factors that bind together with *Hox* proteins.

Eventually, differentiation genes must be activated to give mature cells their distinctive characteristics and functions. Studies of such differentiation genes have been initiated with analysis of the dopamine biosynthetic enzyme tyrosine hydroxylase (R. Lints and S. W. Emmons, unpublished experiments). Recently, the exciting discovery that the fused ray gene *mab-20* (Fig. 9B) encodes a semaphorin homolog demonstrates that important differentiation genes may be identified by the types of genetic screens that have been carried out so far, and that these genes too may be conserved in other animals (Baird *et al.*, 1991; Kolodkin, 1996; J. Culotti, personal communication). By studying additional mutants affecting the rays, it should be possible to understand how specific properties of individual neurons are determined. When we know how a regulatory cell fate specification pathway leads to expression of differentiation genes by specific cells, we will have reached a significant understanding of the developmental process. Hopefully, the postembryonic development of the *C. elegans* male rays is simple enough to allow us to achieve such a satisfying result.

Acknowledgments

I thank the members of my laboratory for their comments and discussions. The author is the Siegfried Ullmann Professor of Molecular Genetics. Work in the author's laboratory is funded by the U.S. National Institutes of Health and the National Science Foundation. Some nematode strains were received from the *Caenorhabditis* Genetics Center, which is funded by a contract from the NIH National Center for Research Resources.

References

Austin, J., and Kenyon, C. (1994). Cell contact regulates neuroblast formation in the *Caenorhabditis elegans* lateral epidermis. *Development (Cambridge, UK)* **120**, 313–324.

Baird, S. E., Fitch, D. H. A., Kassem, I., and Emmons, S. W. (1991). Pattern formation in the nematode epidermis: Determination of the spatial arrangement of peripheral sense organs in the *C. elegans* male tail. *Development (Cambridge, UK)* **113**, 515–526.

Bodmer, R., Carretto, R., and Jan, Y. N. (1989). Neurogenesis of the peripheral nervous system in *Drosophila* embryos: DNA replication patterns and cell lineages. *Neuron* **3**, 21–32.

Chisholm, A. (1991). Control of cell fate in the tail region of *C. elegans* by the gene *egl-5*. *Development (Cambridge, UK)* **111**, 921–932.

Chisholm, A. D., and Horvitz, H. R. (1995). Patterning of the *Caenorhabditis elegans* head region by the *Pax-6* family member *vab-3*. *Nature (London)* **377**, 52–55.

Chow, K. L., and Emmons, S. W. (1994). HOM-C/*Hox* genes and four interacting loci determine the morphogenetic properties of single cells in the nematode male tail. *Development (Cambridge, UK)* **120**, 2579–2593.

Chow, K. L., Hall, D., and Emmons, S. W. (1995). The *mab-21* gene of *C. elegans* encodes a novel protein required for choice of alternate cell fates. *Development (Cambridge, UK)* **121**, 3615–3626.

Emmons, S. W. (1996). Simple worms, complex genes (News and Views). *Nature (London)* **382**, 301–302.

Emmons, S. W., and Sternberg, P. W. (1997). Male development and mating behavior. *In* "*C. elegans* II." (D. L. Riddle, T. Blumenthal, B. J. Meyer, and J. R. Priess, eds.) pp. 295–334. Cold Spring Harbor Laboratory, Cold Spring Harbor, New York.

Ferreira, H. B., Zhang, Y., Zhao, C., and Emmons, S. W. (1998). Patterning of *C. elegans* posterior structures by the *Abdominal-B* homolog, *egl-5*. Submitted.

Fitch, D. H. A., and Emmons, S. W. (1995). Variable cell positions and cell contacts underlie morphological evolution of the rays in the male tails of nematodes related to *Caenorhabditis elegans*. *Dev. Biol.* **170**, 564–582.

Freeman, M. (1997). Cell determination strategies in the developing *Drosophila* eye. *Development (Cambridge, UK)* **124**, 261–270.

Heldin, C., Miyazono, K., and ten Dijke, P. (1997). TGF-β signalling from cell membrane to nucleus through SMAD proteins. *Nature (London)* **390**, 465–471.

Herman, M. A., and Horvitz, H. R. (1994). The *Caenorhabditis elegans* gene *lin-44* controls the polarity of asymmetric cell divisions. *Development (Cambridge, UK)* **120**, 1035–1047.

Herman, M. A., Vassilieva, L. L., Horvitz, H. R., Shaw, J. E., and Herman, R. K. (1995). The *C. elegans* gene *lin-44*, which controls the polarity of certain asymmetric cell divisions, encodes a Wnt protein and acts cell nonautonomously. *Cell (Cambridge, Mass.)* **83**, 101–110.

Horvitz, H. R., Sternberg, P. W., Greenwald, I. S., Fixsen, W., and Ellis, H. M. (1983). Mutations that affect neural cell lineages and cell fates during the development of *Caenorhabditis elegans*. *Cold Spring Harbor Symp. Quant. Biol.* **48**, 453–463.

Hunter, C. P., and Kenyon, C. (1996). Spatial and temporal controls target *pal-1* blastomere-specification activity to a single blastomere lineage in *C. elegans* embryos. *Cell (Cambridge, Mass.)* **87**, 217–226.

Jarman, A. P., Grau, Y., Jan, L. Y., and Jan, Y. N. (1993). *atonal* is a proneural gene that directs chordotonal organ formation in the *Drosophila* peripheral nervous system. *Cell (Cambridge, Mass.)* **73**, 1307–1321.

Jarman, A. P., Grell, E. H., Ackerman, L., Jan, L. Y., and Jan, Y. N. (1994). *atonal* is the proneural gene for *Drosophila* photoreceptors. *Nature (London)* **369**, 398–400.

Kenyon, C. (1986). A gene involved in the development of the posterior body region of *C. elegans*. *Cell (Cambridge, Mass.)* **46**, 477–487.

Kolodkin, A. L. (1996). Semaphorins—mediators of repulsive growth cone guidance. *Trends Cell Biol.* **6**, 15–22.

Lawrence, P. A. (1966). Development and determination of hairs and bristles in the milkweed bug, *Oncopeltus fasciatus* (Lygacidae, Hemiptera). *J. Cell Sci.* **1**, 475–498.

Lewis, J. A., and Hodgkin, J. A. (1977). Specific neuroanatomical changes in chemosensory mutants of the nematode *Caenorhabditis elegans*. *J. Comp. Neurol.* **172**, 489–509.

Lin, R., Hill, R. J., and Priess, J. R. (1998). POP-1 and anterior–posterior fate decisions in *C. elegans* embryos. *Cell* **92**, 229–239.

Liu, K. S., and Sternberg, P. W. (1995). Sensory regulation of male mating behavior in *Caenorhabditis elegans*. *Neuron* **14**, 1–20.

Loer, C. M., and Kenyon, C. J. (1993). Serotonin-deficient mutants and male mating behavior in the nematode *Caenorhabditis elegans*. *J. Neurosci.* **13**, 5407–5417.

Mann, R. S., and Abu-Shaar, M. (1996). Nuclear import of the homeodomain protein Extradenticle in response to Wg and Dpp signalling. *Nature (London)* **383**, 630–633.

Margolis, R. L., Stine, O. C., McInnis, M. G., Ranen, N. G., Rubinsztein, D. C., Leggo, J., Brando, L. V. J., Kidwai, A. S., Soev, S. J., Breschel, T. S., Callahan, C., Simpson, S. G., DePaulo, J. R., McMahon, F. J., Jain, S., Paykel, E. S., Walsh, C., DeLisi, L. E., Crow, T. J., Torrey, E. F., Ashworth, R. G., Macke, J. P., Nathans, J., and Ross, C. A. (1996). cDNA cloning of a human homologue of the *Caenorhabditis elegans* cell fate-determining gene *mab-21*: Expression, chromosomal localization and analysis of a highly polymorphic $(CAG)_n$ trinucleotide repeat. *Hum. Mol. Genet.* **5**, 607–616.

Neumann, C., and Cohen, S. (1997). Morphogens and pattern formation. *BioEssays* **19**, 721–729.

Rieckhof, G. E., Casares, F., Ryook H. D., Abu-Shaar, M., and Mann, R. S. (1997). Nuclear translocation of extradenticle requires *homothorax*, which encodes and extradenticle-related homeodomain protein. *Cell (Cambridge, Mass.)* **91**, 171–183.

Ruvkun, G. (1997). Patterning the nervous system. *In* "*C. elegans* II" (D. L. Riddle, T. Blumenthal, B. J. Meyer, and J. R. Priess, eds.) pp. 295–334. Cold Spring Harbor Laboratory, Cold Spring Harbor, New York.

Salser, S. J., and Kenyon, C. (1996). A *C. elegans Hox* gene switches on, off, on and off again to regulate proliferation, differentiation and morphogenesis. *Development (Cambridge, UK)* **122**, 1651–1661.

Salser, S. J., Loer, C. M., and Kenyon, C. (1993). Multiple HOM-C gene interactions specify cell fates in the nematode central nervous system. *Genes Dev.* **7**, 1714–1724.

Savage, C., Das, P., Finelli, A. L., Townsend, S. R., Sun, C.-Y., Baird, S. E., and Padgett, R. W. (1996). *Caenorhabditis elegans* genes *sma-2*, *sma-3*, and *sma-4* define a conserved family of transforming growth factor β pathway components. *Proc. Natl. Acad. Sci. U.S.A.* **93**, 790–794.

Sawa, H., Lobel, L., and Horvitz, H. R. (1996). The *Caenorhabditis elegans* gene *lin-17*, which is required for certain asymmetric cell divisions, encodes a putative seven-transmembrane protein similar to the *Drosophila* Frizzled protein. *Genes Dev.* **10**, 2189–2197.

St. Johnston, D., and Nusslein-Volhard, C. (1992). The origin of pattern and polarity in the *Drosophila* embryo. *Cell* **68**, 201–219.

Sternberg, P. W., and Horvitz, H. R. (1988). *lin-17* mutations of *Caenorhabditis elegans* disrupt certain asymmetric cell divisions. *Dev. Biol.* **130**, 67–73.

Stoykova, A., Gotz, M., Gruss, P., and Price, J. (1997). *Pax-6*-dependent regulatioin of adhesive patterning, *R-cadherin* expression and boundary formation in developing forebrain. *Development (Cambridge, UK)* **124**, 3765–3777.

Sulston, J. E., and Horvitz, H. R. (1977). Post-embryonic cell lineages of the nematode *Caenorhabditis elegans*. *Dev. Biol.* **56**, 111–156.

Sulston, J. E., and White, J. G. (1980). Regulation and cell autonomy during postembryonic development of *Caenorhabditis elegans*. *Dev. Biol.* **78**, 577–597.

Sulston, J., Dew, M., and Brenner, S. (1975). Dopaminergic neurons in the nematode *Caenorhabditis elegans*. *J. Comp. Neurol.* **163**, 215–226.

Sulston, J. E., Albertson, D. G., and Thomson, J. N. (1980). The *Caenorhabditis elegans* male: Postembryonic development of nongonadal structures. *Dev. Biol.* **78**, 542–576.

Sutherlin, M. E., and Emmons, S. W. (1994). Selective lineage specification by *mab-19* during *Caenorhabditis elegans* male peripheral sense organ development. *Genetics* **138**, 675–688.

Tomlinson, A. (1988). Cellular interactions in the developing *Drosophila* eye. *Development (Cambridge, UK)* **104**, 183–193.

Troemel, E. R., Chou, J. H., Dwyer, N. D., Colbert, H. A., and Bargmann, C. I. (1995). Divergent seven transmembrane receptors are candidate chemosensory receptors in *C. elegans*. *Cell (Cambridge, UK)* **83**, 207–218.

Waring, D. A., and Kenyon, C. (1990). Selective silencing of cell communication influences anteroposterior pattern formation in *C. elegans*. *Cell (Cambridge, Mass.)* **60**, 123–131.

Waring, D. A., and Kenyon, C. (1991). Regulation of cellular responsiveness to inductive signals in the developing *C. elegans* nervous system. *Nature (London)* **350**, 712–715.

Wrischnik, L. A. and Kenyon, C. J. (1997). The role of *lin-22*, a *hairy/enhancer of split* homolog, in patterning the peripheral nervous system of *C. elegans*. *Development (Cambridge, UK)* **124**, 2875–2888.

Zhang, Y., and Emmons, S. W. (1995). Specification of sense-organ identity by a *C. elegans Pax-6* homolog. *Nature (London)* **377**, 55–59.

Zhang, Y., Ferreira, H. B., Greenberg, D., Chisholm, A., and Emmons, S. W. (1998). Regulated nuclear entry of the *C. elegans Pax-6* transcription factor. Submitted.

Zhao, C. (1995). Developmental control of peripheral sense organs in *C. elegans* by a transcription factor of the bHLH family. Ph.D. Thesis. Albert Einstein College of Medicine, New York.

Zhao, C., and Emmons, S. W. (1995). A transcription factor controlling development of peripheral sense organs in *C. elegans*. *Nature (London)* **373**, 74–78.

10

Cell Fate Determination and Signal Transduction during Caenorhabditis elegans Vulval Development

ALEX HAJNAL
Department of Pathology
Division of Cancer Research
University of Zürich
CH-8091 Zürich, Switzerland

I. Introduction

A. The *Caenorhabditis elegans* Vulva as a Model for Epithelial Differentiation

The *Caenorhabditis elegans* vulva forms an opening on the ventral side of the hermaphrodite through which the animals can lay their eggs. This simple organ is formed by the relatively small number of 22 epithelial cells during the third and fourth larval stages (Sulston and Horvitz, 1977). Animals lacking a functional vulva are viable and fertile, a fact that has greatly simplified genetic analysis and allowed the isolation of a large number of mutations that perturb normal vulval development (Ferguson and Horvitz, 1985). The power of this genetic approach combined with the ability to follow cell fate decisions in developing animals at the resolution of single cells has made the *C. elegans* vulva an excellent model system to study epithelial differentiation. In particular, the question how multiple intercellular signals are integrated by cells to generate a specific pattern of cell fates in a reproducible manner is key to un-

derstanding the development of multicellular organisms. Vulval development has proved to be an excellent model to study this question at the molecular level. Furthermore, since the specification of the vulval cell fates occurs in a set of epithelial precursor cells, the specific requirements for signaling processes to occur in polarized, asymmetric cells can be investigated. Many of the genes that control vulval development have closely related homologs in vertebrates and *Drosophila*. What we learn about cell fate determination during vulval development is thus likely to be generally applicable to more complex organisms.

In principle, vulval development can be divided into three temporally distinct steps (Ferguson *et al.*, 1987). During the first larval stage, six epithelial vulval precursor cells (VPCs) are generated, and the vulval equivalence group is established. The cells are positioned along the ventral midline along the anterior–posterior axis and acquire the competence to respond to the various patterning signals. During the second and early third larval stages, three of the six VPCs are induced to adopt vulval cell fates by a signal from the anchor cell in the gonad. This inductive signal combined with lateral signals between the VPCs and inhibitory signals from surrounding cells specifies the pattern of cell fates the VPCs will adopt.

After the cell fates have been specified, the VPCs execute stereotype patterns of cell divisions, and their descendants undergo extensive morphogenesis during the third and fourth larval stages. The cells migrate and invaginate to form the ventral opening, and several cell fusions between the differentiated vulval cells occur (Sulston and Horvitz, 1977). Finally, a set of eight muscles (derived from the sex myoblasts) that control the opening of the vulva attach to the differentiated vulval cells (Stern and DeVore, 1994).

The main focus of this review is on the signaling pathways that control the specification of vulval cell fates during the second step, the induction of vulval development (Section III). The first and third steps, the generation of the vulval equivalence group and the execution of vulval cell fates, are briefly discussed in Sections II and IV, respectively.

B. Intercellular Signals Specify Three Different Cell Fates

The vulva is formed by the 22 descendants of P5.p, P6.p, and P7.p, three of the 12 epithelial Pn.p cells that are positioned along the ventral midline of the animal (Sulston and Horvitz, 1977) (Fig. 1A). During wild-type vulval development, the anchor cell in the gonad induces P6.p to go through three rounds of symmetric cell divisions and generate 8 descendants, a lineage termed the primary (1°) cell fate (Kimble, 1981; Sternberg and Horvitz, 1986) (Fig. 2). In response to the inductive anchor cell signal, P6.p sends a lateral signal that induces the secondary (2°) cell fate in the two adjacent VPCs (P5.p and P7.p) and prevents them from adopting the 1° cell fate (Sternberg, 1988). These two cells also go through three rounds of divisions, but in contrast to the 1° lineage, the last round of divisions in the 2° cell lineage is asymmetric and generates only seven descendants (Sternberg and Horvitz, 1986) (Fig. 2). Because of this asymmetry, there exist two possible orientations

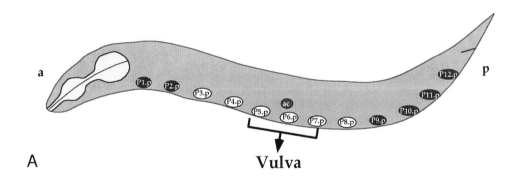

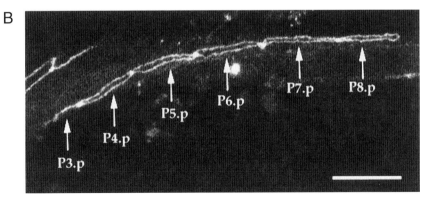

Figure 1 The vulval equivalence group. (A) Location of the 12 Pn.p cells and the anchor cell (ac) in the third larval stage at the time of vulval induction. White circles represent the VPCs (P3.p through P8.p) forming the vulval equivalence group, black circles represent Pn.p cells outside of the vulval equivalence group. The three vulval precursor cells that will generate vulval tissue are indicated with an arrow. Anterior (a) is to the left, posterior (p) to the right and ventral is down. (B) Lateral view of the VPCs in an animal at the third larval stage stained with the MH27 antibody which recognizes a component of the adherens junctions (Waterston, 1988). The white arrows point at the junctions between the six VPCs and the hypodermal skin. Anterior is to the left, posterior to the right. The basal VPC compartment is above, the apical compartment is below the cell junctions. Note that in more anterior or posterior Pn.p cells (P2.p and P9.p), cell junctions are absent. Scale bar: 10 μm.

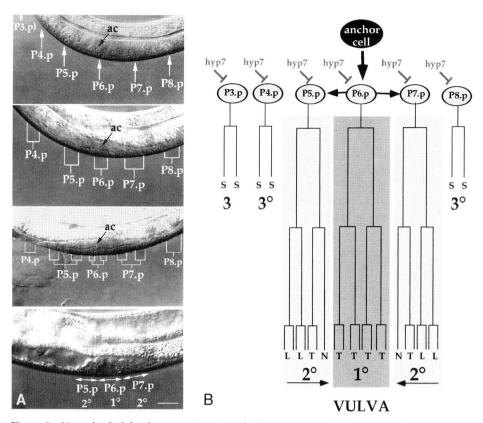

Figure 2 Normal vulval development. (A) Nomarski photomicrographs depicting the different stages of wild-type vulval development. The positions of the nuclei of the anchor cell (ac, where visible), the VPCs, and their descendants are marked. In the animal shown in the bottom panel, vulval morphogenesis has initiated, and only a few of the 22 vulval cells are visible in the focal plane that is shown. Scale bar: 10 μm. (B) The intercellular signals controling vulval cell fate determination are shown together with a schematic representation of the vulval cell lineages. Black arrows indicate the inductive signal from the anchor cell and the lateral signal from P6.p, the gray lines from hyp7 represent the inhibitory signal from the surrounding hyp7 "skin" cell (hypodermis). The cleavage planes during the third round of cell divisions are indicated (Sternberg and Horvitz, 1986): T refers to a division along the left–right axis, L refers to a division along the anterior–posterior axis, and N denotes a cell that did not divide. The primary (1°) cell fate is symmetric and generates eight cells with a TTTT pattern, the secondary (2°) fate is asymmetric and generates seven cells with an LLTN pattern. The black arrows under "2°" indicate the orientation of the 2° fate.

for a 2° cell fate along the anterior–posterior axis, and the orientations of the 2° fates adopted by P5.p and P7.p are opposite. The lateral signal from P6.p combined with the inductive signal from the anchor cell ensures the correct 2°-1°-2° pattern of vulval cell fates (Sternberg and Horvitz, 1989; Koga and Ohshima, 1995; Simske and Kim, 1995).

Finally, an inhibitory signal from the surrounding "skin" (the hyp7 syncytium) prevents induction of VPCs that receive neither inductive nor lateral signals (P3.p, P4.p, and P8.p) (Herman and Hedgecock, 1990). These three distal VPCs go through one round of cell divisions and then fuse with hyp7, a lineage called the tertiary (3°) or uninduced cell fate (Sulston and Horvitz, 1977) (Fig. 2). If the anchor cell is removed by laser ablation, neither inductive nor lateral signals are produced, and the inhibitory hyp7 signal ensures that all six VPCs adopt a 3° fate resulting in a vulvaless phenotype (Kimble, 1981). After the VPCs have integrated these three distinct signals, the pattern of cell fates is irreversibly determined at the beginning of the third larval stage, the

VPCs progress into S-phase and execute their cell fates independently of these intercellular signals (Kimble, 1981).

II. Generation of the Vulval Equivalence Group

A. Six Equivalent Precursor Cells Have the Potential to Adopt a Vulval Cell Fate

Although the vulva is only formed by the descendants of P5.p, P6.p, and P7.p, these three cells belong to a group of six Pn.p cells, P3.p through P8.p, that are equivalent in their developmental potential as they each are competent to adopt a 1°, 2°, or 3° cell fate (Sternberg and Horvitz, 1986) (Fig. 1A). For example, if P5.p–P7.p are ablated using a laser microbeam, P3.p, P4.p, and P8.p will differentiate into 1° or 2° vulval cells instead of adopting the 3° fate (Sternberg and Horvitz, 1989). Similarly, in animals containing mutations that constitutively activate the inductive signaling pathway (Section III,A), P3.p through P8.p (but no other Pn.p cells) adopt induced (1° or 2°) vulval cell fates. P3.p through P8.p therefore form an equivalence group, and they are collectively termed vulval precursor cells (VPCs).

Vulval precursor cells are polarized epithelial cells (Kim, 1995). On their apical side, they secrete components of the cuticle that forms the skin of the animals, and their basal side faces the gonad with the inductive anchor cell. VPC polarity is maintained by specialized adherens junctions (also called belt desmosomes) that separate the apical and basal compartments (Fig. 1B). The six Pn.p cells that do not belong to the vulval equivalence group (P1.p, P2.p, P9.p, P10.p, P11.p, and P12.p) are also born as polarized epithelial cells, but they loose their polarity during the first larval stage, well before vulval induction occurs, and fuse with the surrounding skin (Podbilewicz and White, 1994) (except for P12.p which adopts a tail specific cell fate). VPCs can therefore be distinguished from Pn.p cells outside the equivalence group by the presence of adherens junctions (Fig. 1B).

B. Genes That Specify the Vulval Equivalence Group

A few genes that are required for the generation of the VPCs have been identified in screens for mutations that cause a vulvaless phenotype (Table I). In particular, the *lin-39* gene appears to be a key regulator of VPC identity. *lin-39* encodes a protein with a homeobox domain similar to *Drosophila sex combs reduced/deformed* (*Scr/Dfd*) and belongs to the *C. elegans* HOM-C gene cluster (Clark *et al.*, 1993; Wang *et al.*, 1993). This gene cluster consists of four homeotic genes that control the spatial patterning of the animal along the anterior–posterior axis. *lin-39* specifies the identity of cells in the middle body region where the VPCs are located. *lin-39* is expressed in the VPCs but not in the more anterior or posterior Pn.p cells. In *lin-39* mutants, no vulval equivalence group is specified, and the VPCs adopt the same fate as more anterior or posterior Pn.p cells which is to fuse with the hypodermis during the first larval stage. Mosaic analysis has indicated that *lin-39* functions in a cell autonomous fashion to render the VPCs competent to adopt vulval cell fates and to prevent fusion of the VPCs with the hyp7 cell (Clark *et al.*, 1993). In the posterior VPCs (P7.p and P8.p), *lin-39* appears to be redundant with *mab-5* which encodes a homlog of *Drosophila antennapedia* and specifies cell fates in the more posterior body region (Wang *et al.*, 1993). The targets of *lin-39* are unknown, but it seems likely that *lin-39* might control (directly or indirectly) the transcription of genes that VPCs require to respond to the inductive and lateral

signals (Section III). Expression of *lin-39* itself may be regulated by *bar-1* which encodes a protein related to β-*catenin/armadillo* (D. M. Eisenmann and S. K. Kim, 1996, personal communication) and by *apr-1* which encodes a protein related to the adenomatous polyposis colon (APC) tumor suppressor gene (A. Hajnal and S. K. Kim, 1998, unpublished results).

Unlike *lin-39*, the two genes *unc-83* and *unc-84* are not directly involved in specifying VPC identity. Rather, these two genes are required during the first larval stage for the migration of the P cells, the progenitors of the Pn.p cells, from the lateral position in the animals to the ventral midline where the P cells divide to generate the Pn.p cells (Fixsen, 1985).

C. Heterochronic Genes Control the Timing of Vulval Induction

Although production of the inductive anchor cell signal starts early in the second larval stage, the VPCs do not respond to this signal until later in the third larval stage when the fates of the VPCs are irreversibly determined and they progress into S-phase (Section III,A). Heterochronic genes represent a global "clock" that controls the relative timing of many stage specific events throughout the animal (Ambros and Moss, 1994) (Table I). While the inductive, lateral, and inhibitory signals control the spatial patterning of vulval cell fates, the heterochronic genes regulate the temporal competence of the VPCs by generating a time window during which the VPCs are competent to receive, transduce, and integrate the patterning signals (Ambros and Moss, 1994; Euling and Ambros, 1996). In *lin-14* or *lin-28* loss-of-function mutants, vulval induction occurs precociously during the second larval stage as soon as the anchor cell signal is produced (Ruvkun and Giusto, 1989). In *lin-4* loss-of-function or in *lin-14* gain-of-function mutants, on the other hand, the VPCs never acquire the competence to respond to the inductive signal and no vulva is formed (Lee *et al.*, 1993; Ambros and Moss, 1994). *lin-14* and *lin-28* lengthen the G1 phase of the cell cycle in the VPCs, thereby preventing early vulval induction (Euling and Ambros, 1996), while *lin-4* negatively regulates *lin-14* activity. The heterochronic genes that have so far been analyzed at the molecular level all encode novel proteins or, in the case of *lin-4*, an untranslated RNA that might prevent translation of *lin-14* mRNA (Lee *et al.*, 1993; Wightman *et al.*, 1993). One downstream target of *lin-14* and *lin-28* might be *lin-29* which encodes a putative transcription factor with a Zinc-finger motif (Rougvie and Ambros, 1995). It will be interesting to learn how the heterochronic pathway modulates the length of the cell cycle in the VPCs.

TABLE I

Genes Involved in Vulval Development

Function or signaling pathway	Gene	Gene product
Generation of the equivalence group (II,B)[a]	lin-39	Homeobox protein similar to *Drosophila* Dfd/Scr
	bar-1	β-Catenin/Armadillo related
	apr-1	Related to APC tumor suppressor protein
Heterochronic genes (II, C)	lin-14	Novel
	lin-4	Untranslated RNA
	line-28	Novel
	lin-29	Zinc finger transcription factor
Inductive anchor cell signal (III, A)		
RTK/RAS pathway	lin-3	Epidermal growth factor
	let-23	Epidermal growth factor receptor
	sem-5	GRB2, SH3-SH-2-SH3 adaptor protein
	let-60	RAS
	lin-45	RAF
	mek-2	MEK (MAP kinase kinase)
	mpk-1/sur-1	MAP kinase
New genes in the RTK/RAS pathway	lin-2	Membrane-associated guanylate kinase
	lin-7	PDZ domain
	lin-10	PDZ domains
	ksr-1	Novel ceramide-activated protein kinase
Downstream of MAP kinase	lin-1	ETS, transcription factor
	lin-31	HNF3/forkhead, transcription factor
	lin-25	Novel
	sur-2	Novel
Lateral signaling pathway (III, B)	lin-12	Notch, transmembrane receptor
	emb-5	Chromatin protein?
	sel-1	Novel
	sel-12	Preselinin
Inhibitory genes (III, C)		
Class A synthetic multivulva genes	lin-15A	Novel
	lin-8	—[b]
	lin-38	—
Class B synthetic multivulva genes	lin-15B	Novel
	lin-9	—
	lin-36	Novel
	lin-37	—
Other inhibitory genes	sli-1	c-Cbl proto-oncogene product
	gap-1	GTPase activating protein
	unc-101	AP47 clathrin-associated protein
Execution of vulval cell fates (IV)	lin-11	LIM domain, transcription factor
	lin-17	Fizzled, seven-transmembrane receptor
	lin-18	

[a] The numbers in parentheses in the left column indicate the corresponding section of the text where the genes are discussed.

[b] —, A gene has not been cloned.

III. Cell Fate Specification during Vulval Induction

A. The Anchor Cell Signal Induces the Primary Cell Fate

1. A Conserved Receptor Tyrosine Kinase/RAS/MAP Kinase Pathway

Starting in the second larval stage, the anchor cell expresses the *lin-3* gene which encodes a protein simi- lar to the epidermal growth factor (EGF) and trans- forming growth factor-α (TGF-α) (Hill and Sternberg, 1992). LIN-3 is likely to be the anchor cell signal as *lin-3* mutations cause a vulvaless phenotype similar to ablation of the anchor cell (Fig. 4). The anchor cell signal (LIN-3) is received and transduced by a highly conserved recepter tyrosine kinase (RTK)/RAS pathway (Table I and Fig. 3). *let-23* encodes the putative LIN-3 receptor which is similar to vertebrate EGF receptor tyrosine kinases (Aroian *et al.*, 1990). Reduction-of-function mutations in *let-23* cause a vulvaless phenotype similar to *lin-3*

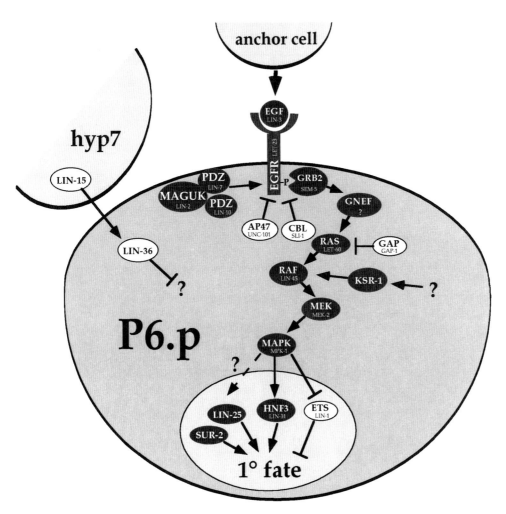

Figure 3   Signaling pathways in P6.p. Proteins in the RTK/RAS signaling pathway, potential nuclear targets of MAP kinase, and some inhibitory proteins are shown. Black circles indicate proteins that promote the 1° cell fate, white circles indicate proteins that repress vulval induction.

mutations (Aroian *et al.*, 1994). Although LET-23 is initially expressed in all six VPCs, mosaic analysis has demonstrated that *let-23* activity is only required in P6.p to induce the 1° fate but not in P5.p and P7.p to induce the 2° cell fate (Koga Ohshima, 1995; Simske and Kim, 1995; Simske *et al.*, 1996). Furthermore, LET-23 expression levels increase in P6.p and decrease in the other VPCs in response to the anchor cell signal (Simske *et al.*, 1996). Taken together, these observations suggest that the anchor cell signal first induces the 1° cell fate in P6.p through the RTK/RAS pathway, and P6.p then sequentially induces the 2° fate in P5.p and P7.p through a *let-23* independent pathway (Section III,B).

Downstream of *let-23*, the signal is transduced by *sem-5* which encodes an adaptor protein similar to vertebrate GRB2 (Clark *et al.*, 1992). In analogy to GRB2, SEM-5 might bind to activated (phosphorylated) LET-23 thereby recruiting a yet unidentified guanine nucleotide exchange factor (GNEF) (Lowenstein *et al.*, 1992; Olivier *et al.*, 1993). GNEF then might bind and

activate LET-60, the *C. elegans* RAS protein (Beitel *et al.*, 1990; Han and Sternberg, 1990). Activation of LET-60 RAS appears to be sufficient to induce vulval development since activating point mutations in *let-60 ras* that are similar to mutations found in *ras* oncogenes allow vulval induction to occur in the absence of the anchor cell signal. Constitutive activation of the RTK/RAS pathway through such a gain-of-function mutation in *let-60 ras* causes all six VPCs to adopt a pattern of alternating 1° and 2° fates resulting in a multivulva phenotype (Fig. 4). A similar multivulva phenotype can be observed if other components of the RTK/RAS pathway (e.g., *let-23*) are constitutively activated (Katz *et al.*, 1996). Similar to *let-23*, mosaic analysis has demonstrated that *let-60 ras* activity is only required in P6.p, supporting the idea that the RTK/RAS pathway is required for specification of the 1° but not the 2° cell fate (Yochem *et al.*, 1997).

Downstream of *let-60 ras*, the signal is transduced through the well known cascade of Ser/Thr kinases which include *lin-45 raf*, *mek-2* MAP kinase kinase, and

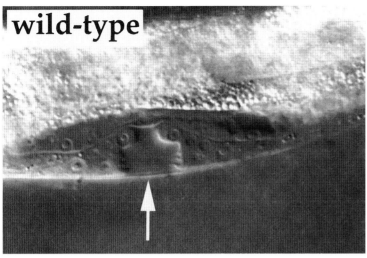

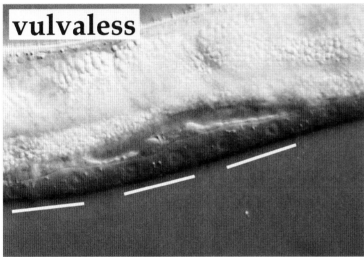

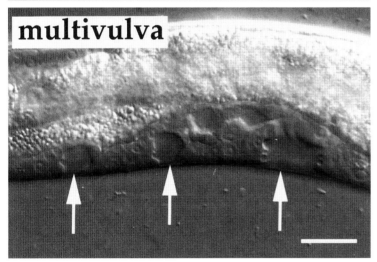

Figure 4   Vulvaless and multivulva phenotypes. Nomarski photomicrographs of a wild-type (top), a vulvaless animal containing a *let-23^{rf}* mutation (middle) and a multivulva animal containing a *let-60gf* mutation (bottom) during the fourth larval stage. Vulval tissue generated from 1° and 2° cells (arrows) and hypodermal tissue generated from 3° cells (bars) are indicated. Scale bar: 10 μm.

mpk-1/sur-1 MAP kinase (Han *et al.*, 1993; Lackner *et al.*, 1994; Wu and Han, 1994; Kornfeld *et al.*, 1995a; Wu *et al.*, 1995). Although constitutive activation of *let-60 ras* or *mpk-1* MAP kinase is sufficient to induce vulval cell fates, this observation does not rule out the possibility that in addition to the RAS pathway, a yet unknown parallel pathway might be activated by LET-23 EGFR in response to the anchor cell signal (Beitel *et al.*, 1990; Han and Sternberg, 1990; M. Lackner and S. K. Kim, 1995, personal communication).

2. New Components of the RTK/RAS Pathway

One attractive aspect is that the genetic screens for mutations that cause defects in vulval development have identified several new components of the RTK/RAS pathway that may not be readily identified using a biochemical approach in other systems. The novel genes discussed in this section all have closely related vertebrate homologs that probably perform analogous functions in the RTK/RAS pathway in these more complex organisms.

a. Receptor Localization by Lin-2, Lin-7, and Lin-10. Null mutations in the three genes *lin-2, lin-7,* and *lin-10* (Fig. 3 and Table I) strongly reduce but do not eliminate vulval induction resulting in a partially penetrant vulvaless phenotype (Ferguson and Horvitz, 1985; Hoskins *et al.*, 1996; Kim and Horvitz, 1990; Simske *et al.*, 1996). In contrast to these three genes, null mutations in other genes that act in the anchor cell pathway (such as *let-23* or *let-60*, Section III,A) completely block the reception of the inductive anchor cell signal. Genetic epistasis experiments have suggested that *lin-2, lin-7,* and *lin-10* most likely act at the level of *let-23* EGF receptor and that they are required for LET-23 to efficiently receive the inductive anchor cell signal (Hoskins *et al.*, 1996; Simske *et al.*, 1996). The three genes all encode proteins containing one or more PDZ (PSD-95/discs-large/ZO-1) domains (Kim, 1995). LIN-2 belongs to the family of membrane-associated guanylate kinases (MAGUK) and contains a single PDZ domain (Hoskins *et al.*, 1996), *lin-7* encodes a protein with a single PDZ (Simske *et al.*, 1996), and *lin-10* a protein with two PDZ domains (Kim and Horvitz, 1990; C. Whitfield and S. K. Kim, 1997, personal communication).

MAGUK proteins are often localized at the sites of cell–cell contacts on the inner side of the plasma membrane, and their PDZ domains have more recently been shown to specifically interact with the C-termini of various transmembrane receptors to direct their subcellular localization (Kim, 1995). In the case of the VPCs, the EGF receptor LET-23 is normally localized to the basal membrane compartment which faces the inducing anchor cell, but in *lin-2, lin-7,* or *lin-10* mutants, LET-23 is mislocalized to the apical cell compartment

(Simske *et al.*, 1996; C. Whitfield and S. K. Kim, 1997, personal communication). LET-23 in the basal cell compartment may exist in a complex together with LIN-2, LIN-7, and LIN-10, and the PDZ domains of these proteins appear to be crucial for complex formation with LET-23 EGFR (S. Kaech and S. K. Kim, 1996, personal communication). Since LIN-3 is secreted by the anchor cell on the basal side of the VPCs and the cell junctions around the VPCs provide an intra- and extracellular barrier between the two cell compartments, LIN-3 is probably unable to reach the apical compartment of the VPCs. Thus, in *lin-2, lin-7,* or *lin-10*-mutants apically mislocalized LET-23 may not be exposed to significant levels of LIN-3 resulting in a vulvaless phenotype.

In addition, basal localization of LET-23 appears to be necessary for sequestering of the anchor cell signal LIN-3 by P6.p (Hajnal *et al.*, 1997). For example, in animals containing a second mutation in an inhibitory gene such as *gap-1* (Section III,C) which allows the VPCs to respond to lower levels of LIN-3, a mutation in *lin-2* may cause a multivulva phenotype due to induction of P3.p, P4.p, or P8.p. One possible explanation for this apparent reversal of the *lin-2* phenotype is that if LET-23 is apically mislocalized in P6.p, unbound LIN-3 might diffuse to the distal VPCs (P3.p, P4.p, and P8.p) and induce them to adopt vulval (1° or 2°) cell fates.

b. ksr-1: Downstream of let-60 ras or in a Parallel Pathway? The *ksr-1* gene was isolated in a genetic screen for mutations that suppress the multivulva phenotype of a gain-of-function mutation in *let-60 ras*. A strong loss-of-function mutation in *ksr-1* only results in a weak vulvaless phenotype, suggesting that *ksr-1* may not be absolutely required for vulval induction but rather positively modulates the activity of the RTK/RAS pathway. *ksr-1* encodes a novel kinase related to RAF kinases, and the mouse homolog of *ksr-1* was more recently shown to be identical to a ceramide-activated protein kinase (Kornfeld *et al.*, 1995b; Sundaram and Han, 1995; Zhang *et al.*, 1997). *ksr-1* might act in a parallel pathway that feeds in or out of the RTK/RAS pathway (Fig. 3). Alternatively, *ksr-1* could be a new component of the RTK/RAS pathway that acts downstream of *let-60 ras* and upstream of *lin-45 raf*. Current genetic data cannot distinguish between these two possibilities and biochemical experiments will be required to clarify the function of *ksr-1*.

3. Downstream of MAP Kinase

Reception of the inductive anchor cell signal by P6.p probably results in the transcriptional induction of many genes that are required for the execution of the 1° cell fate. Induction of the 1° cell fate also involves the production of a lateral signal (Section III,B), and it is conceivable that the gene encoding such a lateral signaling molecule may be a target of the RTK/RAS pathway. Activated MAP kinase might modulate the activity of a number of proteins through phosphorylation on critical Ser/Thr residues. Thus, the inductive anchor cell signal might split into several parallel branches that act downstream of MAP kinase. Of particular interest are the genes that encode candidate transcription factors that may regulate 1° cell fate specific genes.

Genetic epistasis experiments have indicated that the four genes *lin-1* (Beitel *et al.*, 1995), *lin-31* (Miller *et al.*, 1993), *lin-25* (Tuck and Greenwald, 1995), and *sur-2* (Singh and Han, 1995) act downstream of the *mpk-1/sur-1* MAP kinase (Fig. 3 and Table I). The genetic interactions between these four genes suggests that they do not represent a simple linear pathway downstream of MAP kinase but rather function in different branches of the inductive signaling pathway. Two of these four genes, *lin-31* and *lin-1*, encode proteins with similarity to transcription factors. The other two genes, *lin-25* and *sur-2*, encode novel proteins of unknown function. It is presently unclear whether *lin-25* and *sur-2* might be direct target genes for transcription factors such as *lin-1* and *lin-31* or if they function as transcriptional regulators themselves.

Loss of *lin-1* function causes a multivulva phenotype similar to a *let-60 ras* gain-of-function mutation indicating that *lin-1* functions as a negative regulator of vulval induction (Beitel *et al.*, 1995). LIN-1 belongs to the ETS family of transcription factors which are known to serve as substrates for MAP kinases, and LIN-1 indeed has been shown to serve as a high-affinity *in vitro* substrate for Erk MAP kinases (D. Jacobs and K. Kornfeld, 1997, personal communication). Therefore, one possible model is that unphosphorylated LIN-1 might inhibit vulval induction, for example, by binding to the promoter sequences of genes that are necessary for induction or execution of the 1° cell fate thereby preventing their transcription. Reception of the inductive anchor cell signal in P6.p would then lead to phosphorylation of LIN-1 by MAP kinase on residues critical for LIN-1 mediated gene repression and thus relieve inhibition of 1° cell fate induction. A similar, although more complicated, scenario might be true for *lin-31* which encodes a transcription factor homologous to vertebrate HNF3 and *Drosphila forkhead* (Miller *et al.*, 1993). Similar to LIN-1, LIN-31 contains several potential MAP kinase phosphorylation sites and may be phosphorylated on these sites by MAP kinase *in vitro* (P. Tan and S. K. Kim, 1997, personal communication). However, animals lacking *lin-31* display a complete deregulation of all three cell fates (Miller *et al.*, 1993). For example, in a *lin-31* null mutant P5.p, P6.p, and P7.p may adopt the 3° uninduced cell fate instead of the 1° or 2° fates, while P3.p, P4.p, and P8.p may adopt 1° or 2° induced fates instead of the 3° uninduced cell fate. This complex phenotype has suggested a dual role for *lin-31* in inducing (1° and

2°) vulval cell fates in P5.p, P6.p, and P7.p and inhibiting vulval induction in P3.p, P4.p, and P8.p. One possible model to explain this phenotype is that unphosphorylated LIN-31 might act as a repressor in distal VPCs (P3.p, P4.p, and P8.p) while in its phosphorylated form, LIN-31 might function as an activator for genes that promote induced (1° or 2°) vulval cell fates in the proximal VPCs (P5.p, P6.p, and P7.p). Therefore, in animals that produce no LIN-31 protein the inhibitory and activating forms of LIN-31 are both absent resulting in the simultaneous expression of a multivulva and a vulvaless phenotype in the same strain.

Interestingly, production of the lateral signal does not appear to be controlled by *lin-1* since in a *lin-1* loss-of-function mutant, 1° cells are always flanked by 2° cells (Beitel *et al.*, 1995). In contrast, in *lin-25* or *lin-31* mutants VPCs often adopt the 1° cell fate without inducing the 2° fate in the neighboring VPCs (Beitel *et al.*, 1995; Tuck and Greenwald, 1995). Therefore, *lin-25* and *lin-31* might function in a branch of the anchor cell pathway that leads to the production of the lateral signal.

B. A Lateral Signal from P6.p Induces the Secondary Cell Fate in P5.p and P7.p

The anchor cell signal stimulates P6.p to produce a lateral signal that induces the 2° fate in P5.p and P7.p ensuring the correct 2°–1°–2° pattern of vulval cell fates (Section I,B). This lateral signal is received in P5.p and P7.p by *lin-12* which encodes a transmembrane receptor of the Notch family (Greenwald *et al.*, 1983; Yochem *et al.*, 1988). In animals containing a *lin-12* gain-of-function mutation, all six VPCs adopt a 2° cell fate even in the absence of the anchor cell signal (for example after ablation of the anchor cell), and in animals containing a *lin-12* loss-of-function mutations, no VPCs adopt a 2° fate, even if they are next to a 1° cell (Sternberg and Horvitz, 1989; Greenwald and Seydoux, 1990). These observations have suggested that the lateral signal is both necessary and sufficient for the induction of the 2° cell fate. As mentioned above (Section III,A.), mosaic analysis of *let-23* and *let-60* has demonstrated that expression of the 2° cell fate does not require the activity of the RTK/RAS signaling pathway in P5.p and P7.p. In contrast to these observations, other experiment have shown that low levels of LIN-3 may induce a 2° cell fate in the absence of a lateral signal from a 1° cell if a VPC is isolated from its neighbors (Katz *et al.*, 1995). Such a graded anchor cell signal could be a second, functionally redundant mechanism that promotes the 2° cell fate in the absence of a lateral signal (Kenyon, 1995).

Early in *C. elegans* germ line development, LAG-2 which is a member of the DELTA/SERRATE family of proteins, functions as a ligand for LIN-12 (Henderson *et al.*, 1994; Tax *et al.*, 1994). However, *lag-2* does not appear to mediate the induction of the 2° cell fate during vulval induction (S. Kaech and S. K. Kim, 1995, personal communication). In addition to *lag-2*, the *C. elegans* genome sequencing project has identified at least three genes encoding proteins of the Delta/Serrate family, and one of them might encode the yet unidentified lateral signal that is produced by P6.p and induces the 2° cell fate in P5.p and P7.p.

Less is known about the components of the lateral signaling pathway that act downstream of *lin-12* (Table I). The intracellular domain of LIN-12 interacts with EMB-5 (Hubbard *et al.*, 1996) which is similar to a yeast chromatin protein and might transduce the *lin-12* signal into the nucleus of P5.p and P7.p. *sel-12* may also act downstream of *lin-12* and encodes a protein similar to human preselinin which has been implicated in Alzheimer's disease (Levitan and Greenwald, 1995). Finally, *sel-1* functions as an inhibitor of *lin-12* activity and encodes an extracellular protein of unknown function (Grant and Greenwald, 1996).

C. Multiple Inhibitory Signals Prevent Excess Vulval Induction

In addition to the genes that act in the inductive and lateral signaling pathways, several genes that function as negative regulators of vulval development have been isolated by screening for mutations that cause a multivulva phenotype (Fig. 3 and Table I). Interestingly, a single mutation in an inhibitory gene is usually silent, and only animals containing mutations in two or more inhibitors display a multivulva phenotype due to excess vulval differentiation. Thus, vulval induction appears to be repressed through multiple genetically redundant mechanisms that may oppose the activity of the RTK/RAS pathway at different steps. In fact, the default program of a VPC in the absence of any extracellular signals may be to adopt an induced vulval cell fate since in animals containing mutations in inhibitory genes, all six VPCs adopt induced 1° and 2° cell fates even in the absence of the inductive anchor cell signal (see below). The function of the inductive anchor cell signal is therefore to activate the RTK/RAS pathway above a threshold level that overcomes the various inhibitory activities.

1. An Inhibitory Signal from the Hypodermis

The VPCs are surrounded by and in close contact with the hypodermis (hyp7) which forms the "skin" of the animals (Sulston and Horvitz, 1977). A group of genes termed the synthetic multivulva genes generates an inhibitory signal in the hyp7 cell and transduce this signal in the VPCs (Herman and Hedgecock, 1990) (Figs. 2B and 3). These genes fall into two classes (termed class A and B) that may constitute two functionally redundant-signaling pathways (Ferguson and

Horvitz, 1989) (Table I). Single mutant animals containing only a class A or a class B mutation exhibit a wild-type pattern of vulval induction, but in double mutant animals containing loss-of-function mutations in both a class A and a class B gene, all six VPCs express a pattern of alternating 1° and 2° cell fates resulting in a "synthetic" multivulva phenotype. All of the synthetic multivulva genes that have been cloned so far encode novel proteins of unknown function, and they likely belong to a signaling pathway that has not yet been discovered, or may not exist, in other organisms. For example, the *lin-15* locus encodes two novel proteins, LIN-15A and LIN-15B, that are expressed in the nuclei of most hypodermal cells and provide class A and B activity, respectively (Clark *et al.*, 1994; Huang *et al.*, 1994). Animals lacking *lin-15* (class A and B) activity in hyp7 exhibit a multivulva phenotype even if the VPCs are *lin-15(+)* suggesting that *lin-15* may be required for the generation of the inhibitory signal in hyp7 (Herman and Hedgecock, 1990). Interestingly, vulval induction in *lin-15* mutant animals does not depend on the inductive anchor cell signal as *lin-15* mutants that lack the anchor cell exhibit a multivulva phenotype. However, vulval induction in these animals does depend on a basal, signal-independent activity of the RTK/RAS pathway since double mutants between *let-23* EGF receptor and *lin-15* are vulvaless. In contrast to *lin-15*, the class B synthetic multivulva gene, *lin-36* acts cell-autonomously in the VPCs and could be required to receive or transduce the hyp7 inhibitory signal in the VPCs (J. Thomas and H. R. Horvitz, 1994, personal communication). It is currently unknown at what step the synthetic multivulva genes oppose the activity of the RTK/RAS pathway, and biochemical experiments may be required to understand this interesting and novel signaling pathway.

2. Other Inhibitors of the RTK/RAS Pathway

Using genetic screens for mutations that suppress the vulvaless phenotype caused by mutations that reduce but do not eliminate the activity of the RTK/RAS pathway, a few genes that inhibit vulval induction and do not belong to either class of synthetic multivulva genes have been isolated (Fig. 3 and Table I). *sli-1* encodes a protein similar to the mammalian *c-cbl* proto-oncogene product and contains a RING (Zinc) finger domain and a proline rich region that might bind to SH3 domains (Jongeward *et al.*, 1995; Yoon *et al.*, 1995). Vertebrate c-CBL exists in a complex with the EGF receptor and is tyrosine phosphorylated in response to EGF stimulation (Galisteo *et al.*, 1995). These observations suggest that SLI-1 may exist in a complex with LET-23 and negatively regulate transduction of the anchor cell signal by LET-23. The molecular mechanism of this inhibitory activity is presently unclear.

The gene *gap-1* encodes a protein similar to GTPase activating proteins (GAPs) (Hajnal *et al.*, 1997). In analogy to GAPs from other organisms, *gap-1* most likely inhibits the signaling activity of LET-60 RAS by accelerating the hydrolysis of GTP to GDP, thus increasing the levels of GDP-bound, inactive LET-60 RAS. Unlike GAP mutations in most other organisms, loss of *gap-1* function does not result in constitutive activation of the RTK/RAS pathway, as *gap-1* single mutant animals exhibit no vulval defects (Hajnal *et al.*, 1997). Mutations in *gap-1* (and also mutations in *sli-1*) only cause excess vulval induction and a multivulva phenotype when combined with a second mutation in another inhibitor of vulval induction (Yoon *et al.*, 1995; Hajnal *et al.*, 1997). Furthermore, unlike the synthetic multivulva genes, mutations in *sli-1* or *gap-1* do not cause vulval induction independently of the anchor cell signal LIN-3, but they allow the VPCs to respond to lower levels of the anchor cell signal.

Finally, the *unc-101* gene encodes a protein similar to the vertebrate clathrin-associated protein AP47 that inhibits vulval induction (Lee *et al.*, 1994). UNC-101 might negatively regulate the activity of the RTK/RAS pathway by promoting the endocytosis of signaling receptors such as LET-23 or LIN-12.

IV. Execution and Morphogenesis

After vulval cell fates have been specified through the combination of inductive, lateral, and inhibitory signals, the VPCs start to divide in specific patterns according to the cell fates they have adopted (Fig. 2; Greenwald *et al.*, 1983; Sternberg and Horvitz 1986; Euling and Ambros, 1996). This process is called execution of vulval cell fates. Cell ablation experiments have indicated that the execution of vulval cell fates does not require continuing intercellular signals between the VPCs or the anchor cell. For example, if the anchor cell is ablated shortly after the determination step and before the VPCs have started to divide, the VPCs will adopt a normal pattern of vulval cell lineages (Kimble, 1981). Similarly, if one of the two descendants of a VPCs is ablated, the remaining cell will continue to divide according to its previously determined cell fate (Sternberg and Horvitz, 1986). Despite this fact, many genes that control the specification of the vulval cell fates continue to be expressed in the descendants of the VPCs, and they might perform additional functions during the execution phase. For example, the receptor for the anchor cell signal LET-23 continues to be expressed in the 1° descendants of P6.p throughout the fourth larval stage, suggesting that LET-23 might be required for the execution or morphogenesis of 1° vulval cells (Simske *et al.*, 1996). A few genes that specifically affect the execution and not the determination of vulval cell fates are

known (Table I). Mutations in *lin-11* cause a specific defect in the execution of the 2° cell lineage such that P5.p and P7.p generate an LLLL instead of an LLTN and NTLL lineage pattern, respectively (Freyd *et al.*, 1990). *lin-11* encodes a putative transcription factor with a LIM domain that is specifically expressed in the N and T descendants of P5.p and P7.p, suggesting that LIN-11 may define the T and N subset of the 2° lineage. Mutations in *lin-17* or *lin-18* do not alter the 2° lineage pattern, but the polarity of the 2° lineage in P7.p is reversed such that P7.p adopts an LLTN instead of an NTLL lineage (Ferguson *et al.*, 1987; Sternberg and Horvitz, 1988). *lin-17* encodes a protein similar to Frizzled seven-transmembrane receptors which bind and transduce Wingless signals (Sawa *et al.*, 1996). Removal of the gonad before execution of vulval cell fates causes a similar polarity reversal of the 2° fate (Sternberg and Horvitz, 1986), suggesting that a Wingless signal from the gonad may control the polarity of 2° cell fates. This signal could, for example, regulate the expression of genes like *lin-11* in the T and N subset of the 2° lineage.

Morphogenesis of the vulva starts during the last round of cell divisions in the fourth larval stage (Fig. 2A, bottom panel). The vulval cells migrate towards the anchor cell and invaginate to form the ventral opening of the vulva. The anchor cell performs an important role in this process by providing a site of attachment for the central (1°) vulval cells. If the anchor cell is removed after the determination step, the correct number of vulval cells is generated but the cells are unable to undergo proper invagination and fail to attach to the uterus resulting in a nonfunctional vulva (Kimble, 1981). During morphogenesis, cell fusions between vulval cells generate the vulval toroids and a set of muscle cells that control the opening of the vulva attach (J. G. White, E. Southgate, and D. Kershaw, 1990, personal communication; Stern and DeVore, 1994). To study this complex process, genetic screens for mutations that cause vulval defects without altering the vulval lineage have been performed and several genes required for vulval morphogenesis have been isolated. Molecular analysis of these genes will certainly provide insight into the different aspects of vulval morphogenesis.

V. Perspectives

In the 12 years since the original description of a first set of genes that regulate vulval development (Ferguson and Horvitz, 1985) much has been learned about the different signaling pathways that control this process. Thanks to the rapid progress in cloning techniques and thanks to the C. *elegans* genome sequencing project, the molecular analysis of most components of the inductive and a few components of the lateral and inhibitory signaling pathways has been achieved. The coming years are likely to bring the cloning of many more genes and will certainly lead to a more precise description of the lateral and inhibitory signaling pathways. Also the list of genes that control the generation of the vulval equivalence group is getting longer, and it will be interesting to learn how these genes control the expression of the different components of the inductive and lateral signaling pathways. However, the key question of how the inductive, lateral, and inhibitory signals interact with each other and how a cell integrates these opposing inputs it receives is far from being solved. To study this question it will be necessary to define the end points of the different signaling pathways, or in other words, the genes whose transcription is regulated in a cell fate specific manner in order to study their regulation. At the same time, such an analysis should shed light on the little known mechanisms that control the execution of vulval cell fates and vulval morphogenesis.

Acknowledgments

I am grateful to N. Flury, R. Klemenz, and D. Moritz for their critical comments on the manuscript, to S. K. Kim as well as the members of his laboratory for many stimulating discussions and for communicating unpublished results. I also thank N. Wey and H. Nef for their expert assistance with the figures. This work was supported by the Kanton of Zürich.

References

Ambros, V., and Moss, E. G. (1994). Heterochronic genes and the temporal control of *C. elegans* development. *Trends Genet.* **10**, 123–127.

Aroian, R. V., Koga, M., Mendel, J. E., Ohshima, Y., and Sternberg, P. W. (1990). The *let-23* gene necessary for *Caenorhabditis elegans* vulval induction encodes a tyrosine kinase of the EGF receptor subfamily. *Nature (London)* **348**, 693–699.

Aroian, R. V., Lesa, G. M., and Sternberg, P. W. (1994). Mutations in the *Caenorhabditis elegans let-23* EGFR-like gene define elements important for cell-type specificity and function. *EMBO J.* **13**, 360–366.

Beitel, G. J., Clark, S. G., and Horvitz, H. R. (1990). *Caenorhabditis elegans ras* gene *let-60* acts as a switch in the pathway of vulval induction. *Nature (London)* **348**, 503–509.

Beitel, G. J., Tuck, S., Greenwald, I. and Horvitz, H. R. (1995). The *Caenorhabditis elegans* gene *lin-1* encodes an ETS-domain protein and defines a branch of the vulval induction pathway. *Genes Dev.* **9**, 3149–3162.

Clark, S. G., Stern, M. J., and Horvitz, H. R. (1992). The *C. elegans* cell-signalling gene *sem-5* encodes a protein with SH2 and SH3 domains. *Nature (London)* **356**, 340–344.

Clark, S. G., Chisholm, A. D., and Horvitz, H. R. (1993). Control of cell fates in the central body region of *C. elegans* by the homeobox gene *lin-39*. *Cell (Cambridge, Mass.)* **74**, 43–55.

Clark, S. G., Lu, X., and Horvitz, H. R. (1994). The *Caenorhabditis elegans* locus *lin-15*, a negative regulator of a tyrosine kinase signaling pathway, encodes two different proteins. *Genetics* **137**, 987–997.

Euling, S., and Ambros, V. (1996). Heterochronic genes control cell cycle progress and developmental competence of *C. elegans* vulva precursor cells. *Cell* **84**, 667–676.

Ferguson, E. L., and Horvitz, H. R. (1985). Identification and characterization of 22 genes that affect the vulval cell lineages of the nematode *Caenorhabditis elegans*. *Genetics* **110**, 17–72.

Ferguson, E. L., and Horvitz, H. R. (1989). The multivulva phenotype of certain *Caenorhabditis elegans* mutants results from defects in two functionally redundant pathways. *Genetics* **123**, 109–121.

Ferguson, E. L., Sternberg, P. W., and Horvitz, H. R. (1987). A genetic pathway for the specification of the vulval cell lineages of *Caenorhabditis elegans*. *Nature* **326**, 259–267.

Fixsen, W. D. (1985). The genetic control of hypodermal cell lineages during nematode development. Ph.D. thesis, MIT, Cambridge, MA.

Freyd, G., Kim, S. K., and Horvitz, H. R. (1990). Novel cysteine-rich motif and homeodomain in the product of the *Caenorhabditis elegans* cell lineage gene *lin-11*. *Nature (London)* **344**, 876–879.

Galisteo, M. L., Dikic, I., Batzer, A. G., Langdon, W. Y., and Schlessinger, J. (1995). Tyrosine phosphorylation of the *c-cbl* proto-oncogene protein product and association with epidermal growth factor (EGF) receptor upon EGF stimulation. *J. Biol. Chem.* **270**, 20242–20245.

Grant, B., and Greenwald, I. (1996). The *Caenorhabditis elegans sel-1* gene, a negative regulator of *lin-12* and *glp-1*, encodes a predicted extracellular protein. *Genetics* **143**, 237–247.

Greenwald, I., and Seydoux, G. (1990). Analysis of gain-of-function mutations of the *lin-12* gene of *Caenorhabditis elegans*. *Nature (London)* **346**, 197–199.

Greenwald, I. S., Sternberg, P. W., and Horvitz, H. R. (1983). The *lin-12* locus specifies cell fates in *Caenorhabditis elegans*. *Cell (Cambridge, Mass.)* **34**, 435–444.

Hajnal, A., Whitfield, C., and Kim, S. K. (1997). Inhibition of *C. elegans* vulval induction by *gap-1* and cell nonautonomous antagonism by *let-23* receptor tyrosine kinase. *Genes Dev.* **11**, 2715–2728.

Han, M., and Sternberg, P. W. (1990). *let-60*, a gene that specifies cell fates during *C. elegans* vulval induction, encodes a RAS protein. *Cell (Cambridge, Mass.)* **63**, 921–31.

Han, M., Golden, A., Han, Y., and Sternberg, P. W. (1993). *C. elegans lin-45* raf gene participates in *let-60* ras-stimulated vulval differentiation. *Nature (London)* **363**, 133–140.

Henderson, S. T., Gao, D., Lambie, E. J., and Kimble, J. (1994). *lag-2* may encode a signaling ligand for the GLP-1 and LIN-12 receptors of *C. elegans*. *Development (Cambridge, UK)* **120**, 2913–2924.

Herman, R. K., and Hedgecock, E. M. (1990). Limitation of the size of the vulval primordium of *Caenorhabditis elegans* by *lin-15* expression in surrounding hypodermis. *Nature (London)* **348**, 169–171.

Hill, R. J., and Sternberg, P. W. (1992). The gene *lin-3* encodes an inductive signal for vulval development in *C. elegans*. *Nature (London)* **358**, 470–476.

Hoskins, R., Hajnal, A. F., Harp, S. A., and Kim, S. K. (1996). The *C. elegans* vulval induction gene *lin-2* encodes a member of the MAGUK family of cell junction proteins. *Development (Cambridge, UK)* **122**, 97–111.

Huang, L. S., Tzou, P., and Sternberg, P. W. (1994). The *lin-15* locus encodes two negative regulators of *Caenorhabditis elegans* vulval development. *Mol. Biol. Cell* **5**, 395–411.

Hubbard, E. J., Dong, Q., and Greenwald, I. (1996). Evidence for physical and functional association between EMB-5 and LIN-12 in *Caenorhabditis elegans*. *Science* **273**, 112–115.

Jongeward, G. D., Clandinin, T. R., and Sternberg, P. W. (1995). *sli-1*, a negative regulator of *let-23*-mediated signaling in *C. elegans*. *Genetics* **139**, 1553–1566.

Katz, W. S., Hill, R. J., Clandinin, T. R., and Sternberg, P. W. (1995). Different levels of the *C. elegans* growth factor LIN-3 promote distinct vulval precursor fates. *Cell (Cambridge, Mass.)* **82**, 297–307.

Katz, W. S., Lesa, G. M., Yannoukakos, D., Clandinin, T. R., Schlessinger, J., and Sternberg, P. W. (1996). A point mutation in the extracellular domain activates LET-23, the *Caenorhabditis elegans* epidermal growth factor receptor homolog. *Mol. Cell. Biol.* **16**, 529–537.

Kenyon, C. (1995). A perfect vulva every time: Gradients and signaling cascades in *C. elegans Cell (Cambridge, Mass.)* **82**, 171–174.

Kim, S. K. (1995). Tight junctions, membrane-associated guanylate kinases and cell signaling. *Curr. Opin. Cell. Biol.* **7**, 641–649.

Kim, S. K., and Horvitz, H. R. (1990). The *Caenorhabditis elegans* gene *lin-10* is broadly expressed while required specifically for the determination of vulval cell fates. *Genes Dev.* **4**, 357–371.

Kimble, J. (1981). Alterations in cell lineage following laser ablation of cells in the somatic gonad of *Caenorhabditis elegans*. *Dev. Biol.* **87**, 286–300.

Koga, M., and Ohshima, Y. (1995). Mosaic analysis of the *let-23* gene function in vulval induction of *Caenorhabditis elegans*. *Development (Cambridge, UK)* **121**, 2655–2666.

Kornfeld, K., Guan, K. L., and Horvitz, H. R. (1995a). The *Caenorhabditis elegans* gene *mek-2* is required for vulval induction and encodes a protein similar to the protein kinase MEK. *Genes Dev.* **9**, 756–768.

Kornfeld, K., Hom, D. B., and Horvitz, H. R. (1995b). The *ksr-1* gene encodes a novel protein kinase involved in *Ras*-mediated signaling in *C. elegans*. *Cell (Cambridge, Mass.)* **83**, 903–913.

Lackner, M. R., Kornfeld, K., Miller, L. M., Horvitz, H. R., and Kim, S. K. (1994). A MAP kinase homolog, *mpk-1*, is involved in *ras*-mediated induction of vulval cell fates in *Caenorhabditis elegans*. *Genes Dev.* **8**, 160–173.

Lee, J., Jongeward, G. D., and Sternberg, P. W. (1994). *unc-101*, a gene required for many aspects of *Caenorhabditis elegans* development and behavior, encodes a clathrin-associated protein. *Genes Dev.* **8**, 60–73.

Lee, R. C., Feinbaum, R. L., and Ambros, V. (1993). The *C. elegans* heterochronic gene *lin-4* encodes small RNAs with antisense complementarity to *lin-14*. *Cell* **75**, 843–854.

Levitan, D., and Greenwald, I. (1995). Facilitation of *lin-12*-mediated signalling by *sel-12*, a *Caenorhabditis elegans* S182 Alzheimer's disease gene. *Nature (London)* **377**, 351–354.

Lowenstein, E. J., Daly, R. J., Batzer, A. G., Li, W., Margolis, B., Lammers, R., Ullrich, A., Skolnik, E. Y., Bar, S. D., and Schlessinger, J. (1992). The SH2 and SH3 domain-containing protein GRB2 links receptor tyrosine kinases to *ras* signaling. *Cell (Cambridge, Mass.)* **70**, 431–442.

Miller, L. M., Gallegos, M. E., Morisseau, B. A., and Kim, S. K. (1993). *lin-31*, a *Caenorhabditis elegans* HNF-3/fork head transcription factor homolog, specifies three alternative cell fates in vulval development. *Genes Dev.* **7**, 933–947.

Olivier, J. P., Raabe, T., Henkemeyer, M., Dickson, B., Mbamalu, G., Margolis, B., Schlessinger, J., Hafen, E., and Pawson, T. (1993). A *Drosophila* SH2-SH3 adaptor protein implicated in coupling the *sevenless* tyrosine kinase to an activator of *Ras* guanine nucleotide exchange, *Sos*. *Cell (Cambridge, Mass.)* **73**, 179–191.

Podbilewicz, B., and White, J. G. (1994). Cell fusions in the developing epithelia of *C. elegans*. *Dev. Biol.* **161**, 408–424.

Rougvie, A. E., and Ambros, V. (1995). The heterochronic gene *lin-29* encodes a zinc finger protein that controls a terminal differentiation event in *Caenorhabditis elegans*. *Development (Cambridge, UK)* **121**, 2491–2500.

Ruvkun, G., and Giusto, J. (1989). The *Caenorhabditis elegans* heterochronic gene *lin-14* encodes a nuclear protein that forms a temporal developmental switch. *Nature (London)* **338**, 313–319.

Sawa, H., Lobel, L., and Horvitz, H. R. (1996). The *Caenorhabditis elegans* gene *lin-17*, which is required for certain asymmetric cell divisions, encodes a putative seven-transmembrane protein similar to the *Drosophila* Frizzled protein. *Genes Dev.* **10**, 2189–2197.

Simske, J. S., and Kim, S. K. (1995). Sequential signalling during *Caenorhabditis elegans* vulval induction. *Nature (London)* **375**, 142–146.

Simske, J. S., Kaech, S. M., Harp, S. A., and Kim, S. K. (1996). LET-23 receptor localization by the cell junction protein LIN-7 during *C. elegans* vulval induction. *Cell (Cambridge, Mass.)* **85**, 195–204.

Singh, N., and Han, M. (1995). *sur-2*, a novel gene, functions late in the *let-60* ras-mediated signaling pathway during *Caenorhabditis elegans* vulval induction. *Genes Dev.* **9**, 2251–2265.

Stern, M. J., and DeVore, D. L. (1994). Extending and connecting signaling pathways in *C. elegans*. *Dev. Biol.* **166**, 443–459.

Sternberg, P. W. (1988). Lateral inhibition during vulval induction in *Caenorhabditis elegans*. *Nature* **335**, 551–554.

Sternberg, P. W., and Horvitz, H. R. (1986). Pattern formation during vulval development in *C. elegans*. *Cell (Cambridge, Mass.)* **44**, 761–772.

Sternberg, P. W., and Horvitz, H. R. (1988). *lin-17* mutations of *Caenorhabditis elegans* disrupt certain asymmetric cell divisions. *Dev. Biol.* **130,** 67–73.

Sternberg, P. W., and Horvitz, H. R. (1989). The combined action of two intercellular signaling pathways specifies three cell fates during vulval induction in *C. elegans. Cell (Cambridge, Mass.)* **58,** 679–693.

Sulston, J. E., and Horvitz, H. R. (1977). Post-embryonic cell lineages of the nematode, *Caenorhabditis elegans. Dev. Biol.* **56,** 110–156.

Sundaram, M., and Han, M. (1995). The *C. elegans ksr-1* gene encodes a novel *Raf*-related kinase involved in *Ras*-mediated signal transduction. *Cell (Cambridge, Mass.)* **83,** 889–901.

Tax, F. E., Yeargers, J. J., and Thomas, J. H. (1994). Sequence of *C. elegans lag-2* reveals a cell-signalling domain shared with *Delta* and *Serrate* of *Drosophila. Nature (London)* **368,** 150–154.

Tuck, S., and Greenwald, I. (1995). *lin-25*, a gene required for vulval induction in *Caenorhabditis elegans. Genes Dev.* **9,** 341–357.

Wang, B. B., Muller, I. M., Austin, J., Robinson, N. T., Chisholm, A., and Kenyon, C. (1993). A homeotic gene cluster patterns the anteroposterior body axis of *C. elegans. Cell (Cambridge, Mass.)* **74,** 29–42.

Waterston, R. H. (1988). Muscle. *In* "The nematode *Caenorhabditis elegans*" (W. B. Wood, ed.), pp. 281–335. Cold Spring Harbor Laboratory, Cold Spring Harbor, New York.

Wightman, B., Ha, I., and Ruvkun, G. (1993). Posttranscriptional regulation of the heterochronic gene *lin-14* by *lin-4* mediates temporal pattern formation in *C. elegans. Cell* **75,** 855–862.

Wu, Y., and Han, M. (1994). Suppression of activated LET-60 ras protein defines a role of *Caenorhabditis elegans* SUR-1 MAP kinase in vulval differentiation. *Genes Dev.* **8,** 147–159.

Wu, Y., Han, M., and Guan, K. L. (1995). MEK-2, a *Caenorhabditis elegans* MAP kinase kinase, functions in *Ras*-mediated vulval induction and other developmental events. *Genes Dev.* **9,** 742–755.

Yochem, J., Weston, K., and Greenwald, I. (1988). The *Caenorhabditis elegans lin-12* gene encodes a transmembrane protein with overall similarity to *Drosophila* Notch. *Nature (London)* **335,** 547–550.

Yochem, J., Sundaram, M., and Han, M. (1997). *Ras* is required for a limited number of cell fates and not for general proliferation in *C. elegans. Mol. Cell. Biol.* **17,** 2716–2722.

Yoon, C. H., Lee, J., Jongeward, G. D., and Sternberg, P. W. (1995). Similarity of *sli-1*, a regulator of vulval development in *C. elegans,* to the mammalian proto-oncogene *c-cbl. Science* **269,** 1102–1105.

Zhang, Y., Yao, B., Delikat, S., Bayoumy, S., Lin, X. H., Basu, S., McGinley, M., Chan-Hui, P. Y., Lichenstein, H., and Kolesnick, R. (1997). Kinase suppressor of *Ras* is ceramide-activated protein kinase. *Cell (Cambridge, Mass.)* **89,** 63–72.

III

Leech

11

Introduction to the Leech

GUNTHER S. STENT

Department of Molecular and Cell Biology
University of California, Berkeley
Berkeley, California 94720

I. Taxonomy

Leeches are hermaphroditic, bloodsucking worms that form the class of Hirudinea in the phylum Annelida. The class includes two orders: the Gnathobdellidae, which draw blood by biting through the skin of the host with toothed, rasping jaws, and the Rhynchobdellidae, which do so by inserting a proboscis under the skin of the host. In the mid-nineteenth century, when the medicinal employ of leeches for therapeutic bloodletting had reached the peak of its popularity, they came to be also used for basic research in neurobiology and embryology. Because of its ready commercial availability and large size, the gnathobdellid *Hirudo medicinalis* (of the family Hirudinidiae) became (and has remained ever since) the most popular leech species for neurobiological studies. However, for reasons to be detailed further below, the gnathobdellid order is a much less favorable material for embryological investigations than the rhynchobdellid order, on whose family of Glossiphoniidae most work on leech embryogenesis has been carried out. This chapter and the following three chapters focus mainly on the results of studies done with embryos of two of the most popular glossiphoniid species, *Helobdella triserialis* and *Theromyzon rude*.

Helobdella triserialis feeds on the lymph of aquatic snails. It reaches an adult length of 1–2 cm and propagates with an egg-to-egg generation time of about 3 months (Sawyer, 1972). *Theromyzon rude* feeds on the blood of aquatic birds. It reaches an adult length of 2–4 cm and propagates with an egg-to-egg generation time of about 12 months.

II. Gross Anatomy

Despite differences in size, habit, and mode of embryonic development, the anatomy and neurology of adult leeches is fairly constant across the class of Hirudinea. Their tubular body consists of 32 metameric segments and a nonsegmental prostomium. The four rostral segments (designated R1 to R4) are fused. They form the specialized cephalic structures, including pairs of eyes dorsally and a mouth surrounded by a small oral sucker ventrally. The seven caudal segments (designated C1 to C7) are fused as well and form a large rear sucker. Between the fused rostral and fused caudal segments lie 21 unfused, or midbody segments (designated M1 to M21) (Fig. 1A).

The excretory system of leeches consists of paired, metameric nephridia distributed in a species-specific

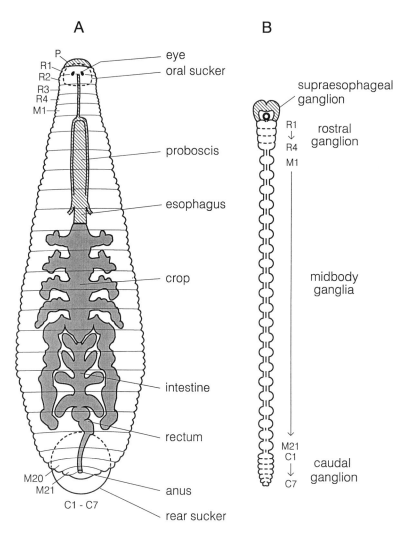

Figure 1 The glossiphoniid leech, *Helobdella triserialis,* shown in dorsal view. Anterior is up. (A) The leech body plan consists of a nonsegmental cephalic domain (hatched) and an elongate trunk composed of 32 serially homologous segments, whose boundaries are marked by transverse lines. The four most rostral segments (R1–R4) join with the nonsegmental prostomium (P) to form the oral sucker, a ventrally oriented concavity surrounding the mouth. They are followed by 21 midbody segments (M1–M21). The seven most caudal segments (C1–C7) form the disk-shaped rear sucker, which extends out past the anus. The digestive tract is subdivided into an unsegmented foregut (hatched) and a segmented midgut (gray) composed of crop, intestine, and rectum. (B) The ventral nerve cord, whose four frontmost and seven hindmost neuromeres are fused to form the compound rostral and caudal ganglia, respectively. The 21 midbody neuromeres remain separate as discrete segmental ganglia. The supraesophageal ganglion joined to the anterior end of the nerve cord is a nonsegmental derivative of the prostomium. (From Shankland and Savage, 1997.)

pattern over most (but not all) of the midbody segments. Individual nephridia excrete urine via a nephridiopore located on the ventral aspect (Harant and Grassé, 1959; Mann, 1962). Segments M5 and M6 are the reproductive segments. The penis and vas deferens lie in segment M5; the sperm are produced and stored in metameric testes distributed over several segments posterior to M5. The bilaterally paired ovaries lie in segment M6; eggs are fertilized internally and laid through the female pore of segment M6.

Below the epidermis lies an outer layer of circular muscle fibers coursing circumferentially and an inner layer of longitudinal muscle fibers coursing longitudi-

nally. The body cavity of the leech is traversed by dorsoventral muscle fibers, which insert into the dorsal body wall at one end and into the ventral body wall at the other.

All three types of muscle fibers are arranged in discrete parallel fascicles, of which there is a fixed number per segment. Contraction of each type of muscle works against the hydrostatic skeleton provided by the fluid-filled leech body tube to effect a characteristic change in body shape: Contraction of the circular fibers causes constriction and lengthening; of the longitudinal fibers, shortening; and of the dorsoventral fibers flattening and lengthening.

III. Nervous System

A. Ventral Nerve Cord

The nervous system of the leech reflects its segmental body plan. The central nervous system (CNS) consists of a ventral nerve cord of 32 segmentally iterated neuromeres (Fig. 1B). The frontmost four neuromeres (belonging to segments R1–R4) are fused. They form the compound rostral ganglion, which is linked at its anterior end to a dorsally situated supraesophageal ganglion via a bilateral pair of circumpharyngeal connective nerves. The supraesophageal ganglion is part of the prostomium, and hence, unlike all other ganglia, is not a segmental organ. The hindmost seven neuromeres (belonging to segments C1–C7) are fused as well and form the compound caudal ganglion. The intervening 21 neuromeres are unfused and form the midbody ganglia (located in segments M1–M21). They are linked via an unpaired, median connective nerve, and two paired, lateral connective nerves.

Each midbody ganglion contains about 200 bilateral pairs of neurons (Macagno, 1980), as well as a few unpaired neurons. Their cell bodies form a cortex around the outer surface of the ganglion and are readily accessible to electrophysiological and histochemical characterization. The ganglionic neurons are monopolar and extend their single axon into a central neuropil, where it branches and makes synaptic contacts. Sensory and effector neurons project their axon branches from the neuropil to peripheral targets via three distinct, bilaterally paired segmental nerves, whose roots emerge from the lateral margin of the segmental ganglion. The axon branches of some interneurons (as well as of some sensory neurons) project via the connective nerves from the neuropil of their ganglion of origin to the neuropils of other ganglia.

The anatomy of the segmental ganglia is highly stereotyped, so that after characterizing a particular neuron in a particular ganglion of a particular specimen according to morphological and physiological criteria, homologous neurons can usually be found on the other side of that same ganglion, in other ganglia of that same specimen, as well as in the ganglia of other specimens of the same species (Muller et al., 1981).

This high degree of neural stereotypy notwithstanding, there do occur some systematic variations in the number of cells among different segmental ganglia within the same nerve cord and among corresponding ganglia in the nerve cords of different leech species. For instance, in H. medicinalis, the ganglia in the two reproductive body segments contain nearly twice as many cells as the other midbody ganglia, while in the glossiphoniid leech, Haementeria ghilianii, the ganglia in the reproductive segment contain only about 5% more cells

as the other midbody ganglia (Macagno, 1980). Moreover, there is some slight variation in the exact number of neurons per ganglion between corresponding ganglia of different specimens of the same species.

B. Identified Cells

About one-quarter of the neurons of the segmental ganglia of the adult H. medicinalis have been identified according to various criteria, including function. Thus, many cells have been classified as sensory, effector, or interneurons, and their connectivity has been elucidated (Nicholls and Baylor, 1968; Baylor and Nicholls, 1969; Stuart, 1970; Nicholls and Purves, 1972; Lent, 1973; Ort et al., 1974; Thompson and Stent, 1976a–c; Friesen et al., 1978; Muller, 1979, Muller et al, 1981; Friesen, 1985; Nusbaum and Kristan, 1986). These surveys have culminated in the description of sensory pathways and of neuronal networks controlling various behaviors, such as body shortening, heartbeat, and swimming (Stent et al., 1978, 1979; Friesen, 1989; Kristan et al., 1988). Despite their phyletic distance from H. medicinalis, the glossiphoniid leech species share with it not only the same general structure of the CNS but even many of the identified neurons (Kramer and Goldman, 1981).

C. Neurotransmitters

The characterization of leech neurons has been extended to the identification of actual or putative neurotransmitters by electrophysiological, pharmacological, autoradiographic, histochemical, and immunohistological techniques. For example, the neurotransmitter released by the identified excitatory motor neurons innervating the body wall muscles is acetylcholine (Kuffler, 1978), while the corresponding identified inhibitory motor neurons release γ-aminobutyric acid (GABA) (Cline, 1986). There may be additional cholinergic and GABAergic neurons in the segmental ganglion, as suggested by the presence of many other cells which, like the excitatory motor neurons, contain choline acetyltransferase (Sargent, 1977) and cholinesterase (Wallace and Gillon, 1982), or which, like the inhibitory motor neurons, have a high affinity uptake system for GABA (Cline, 1983).

Neurons containing monoamine neurotransmitters are present in the leech as well. (Rude, 1969; Stuart et al., 1974). For instance, each midbody ganglion includes three pairs of serotonergic neurons, while three bilateral, segmentally iterated pairs of dopaminergic neurons form part of the peripheral nervous system. (Blair, 1983; Lent et al., 1979; Stuart et al., 1987). In addition to low molecular weight neurotransmitters, several neuropeptide transmitter candidates, such as

FMRFamide, have been identified in the leech nervous system (Kuhlman *et al.*, 1985; Norris and Calabrese, 1990; Shankland and Martindale, 1989).

IV. Morphological Development and Staging

Glossiphoniid leeches lay clutches of yolk-rich eggs, whose diameters are about 0.4 mm and 0.8 mm for *H. triserialis* and *T. rude,* respectively. Each clutch is enclosed in a transparent, soft-walled, saline-filled cocoon, which remains attached to the venter of the brooding parent. The ensuing embryonic development has been divided into 11 stages, beginning with deposition of the internally fertilized egg and ending with exhaustion of the yolk (about 3 and 6 weeks later for *H. triserialis* and *T. rude,* respectively), when the juvenile leech hatches and looks for a host animal in order to take its first meal. The definition of the developmental stages is based on morphological criteria discernible in the living embryo (Fernandez, 1980; Weisblat *et al.*, 1980b; Weisblat, 1981) (Fig. 2; Table I).

Early in stage 1, two polar bodies appear at the animal pole, adumbrating the orientation of the future anterior–posterior embryonic axis. Later, as the egg approaches its first cleavage, the orientation of this axis is buttressed by domains of yolk-free cytoplasm, or teloplasm, which form at the animal and vegetal poles. The first cleavage, yielding cells AB and CD, is meridional and establishes the future dorsal–ventral axis (stage 2). The second cleavage is also meridional; it divides the egg into four macromeres, A, B, C, and D, with D receiving most of the teloplasm (stage 3). The third cleavage is highly unequal, budding off one micromere each from A, B, C, and D (stage 4a); macromere D cleaves equatorially to yield cells designated DNOPQ and DM, while macromeres A, B, and C undergo another round of micromere production, yielding three more micromeres (stage 4b).

Macromeres A, B, and C embody the presumptive endoderm, DNOPQ the presumptive ectoderm, and DM the presumptive mesoderm (Whitman, 1887) (stage 4c). Macromeres A, B, and C each undergo one more round of micromere production, while DM and DNOPQ both cleave equally to yield the right and left cell pairs M and NOPQ, respectively (stage 5). The NOPQ cell pair divides to yield the pairs N and OPQ (stage 6a), whereon the OPQ pair divides to yield pairs OP and Q (stage 6b), and finally the OP pair divides to yield pairs O and P (stage 7). In addition to the micromeres and the macromeres A, B, and C, the embryo now contains the five bilateral cell pairs, M, N, O, P, and Q, designated as teloblasts.

During stage 7, each teloblast carries out a series of 40–100 highly unequal cleavages in the stem cell division mode, generating a bandlet of small primary blast cells. The blast cells and their bandlets produced by the M, N, O, P, and Q teloblasts are designated as m, n, o, p, and q, respectively. The five bandlets produced on either side of the future dorsal midline merge to form two ridges of cells, the right and left germinal bands. The n, o, p, and q ectodermal precursor bandlets lie (side by side, in mediolaterally alphabetical order) on the upper surface of the germinal band ridge, while the m mesodermal precursor bandlet lies under the four ectodermal bandlets (Fig. 2). Between the two germinal bands lies a cluster of cells derived mainly from the micromeres, designated as the micromere cap (Fernandez and Stent, 1980).

With ongoing blast cell production by the five teloblast pairs, the germinal bands migrate circumferentially over the surface of the embryo during stage 8, and then coalesce zipperlike, from the future head rearward along the ventral midline, forming a sheet of cells called germinal plate (Fig. 2). As the cells of the germinal plate proliferate and differentiate to form adult tissues, the plate gradually thickens and expands over the surface of the embryo back into dorsal territory. Even before the germinal bands have started to coalesce, the left and right m bandlets become partitioned into a series of discrete blocks of cells corresponding to paired hemilateral somites. During stage 8 and continuing through stage 9, the coelom arises as a cavity within each hemilateral somite, so that the germinal plate becomes partitioned along its length into a series of tissue blocks, each separated from its anterior and posterior neighbors by transverse septa. Each tissue block corresponds to a future body segment.

Segmentation starts at the front and progresses rearward, and by the time the ganglion of the hindmost segment has formed (end of stage 9), the expanding germinal plate covers about one-third of the ventral surface. When the expanding germinal plate covers about one-half the ventral surface, the embryo hatches from the vitelline membrane. By stage 10, right and left leading edges of the expanding germinal plate meet and coalesce on the dorsal midline, closing the leech body. Meanwhile, formation of the gut is underway. It first appears as a cylinder (filled with yolk provided by macromeres A, B, and C and the remnants of the teloblasts) and then becomes segmented by annular constrictions, which probably correspond to the segmental septa. These constrictions give rise to paired gut lobes, or caeca, in register with the midbody segments. Gut segmentation is completed at body closure (end of stage 10); the embryo now has the general shape of the adult leech. The final steps of morphological development, including maturation of the posterior sucker, occur during stage 11.

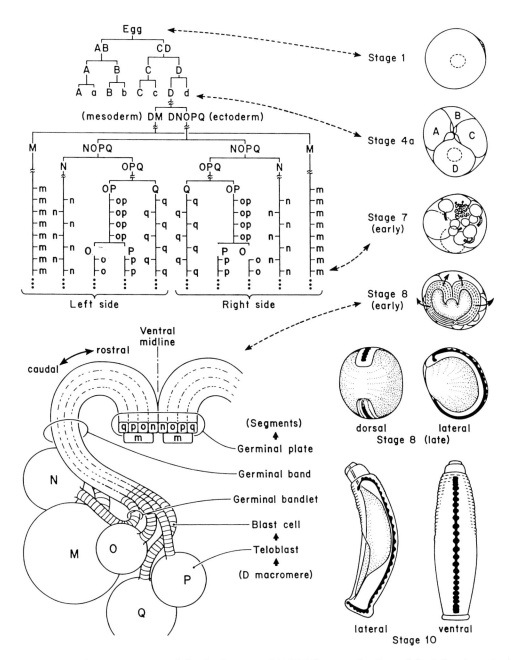

Figure 2 Schematic summary of the development of *Helobdella triserialis*. Upper left: Cell pedigree leading from the uncleaved egg to the macromeres A, B, C, and D, the micromeres a, b, and c, the teloblast pairs M, N, O, P, and Q, and the paired primary blast cell bandlets. Breaks in the lineage indicate nodes at which additional micromeres are produced. The number of op blast cells produced prior to cleavage of the proteloblast OP varies from four to seven. Lower left: Hemilateral disposition of the teloblasts and their primary blast cell bandlets m, n, o, p, and q within the germinal band and germinal plate. Right margin: Diagrammatic view of the embryo at various stages. The dashed circle in the uncleaved egg (stage 1) signifies the teloplasm, which is passed on mainly to the D macromere (stage 4a). In the stage 7 embryo the dashed circle signifies the right M teloblast (which is invisible from the dorsal aspect), while the many small, closed contours in the upper midportion indicate the micromere cap. In the stage 8 (early) embryo, the heart-shaped germinal bands migrate over the surface of the embryo in the direction indicated by the arrows. In the stage 8 (late) embryo the germinal plate is shown to lie on the ventral midline, with the nascent ventral nerve cord and its ganglia and ganglionic primordia indicated in black. In the stage 10 embryo shown, body closure is nearly complete. Here, the stippled areas signify the yolky remnants of other macromeres and teloblasts, now enclosed in the gut of the embryo. The chain of ganglia linked via connective nerves, shown in black, already closely resembles the adult nerve cord. (From Weisblat *et al.*, 1984.)

TABLE I

Stages of Glossiphoniid Embryogenesis

Stage	Start of stage
1. Uncleaved egg (ABCD)	Egg deposition
2. Two blastomeres (AB and CD)	Onset of first cleavage
3. Four blastomeres (A, B, C, and D)	Onset of second cleavage
4a. Micromere quartet (a, b, c, and d)	Onset of micromere cleavage
4b. Macromere quintet (A, B, C, DM, and DNOPQ)	Onset of cleavage of macromere D
4c. M mesoteloblast pair	Onset of cleavage of cell DM
5. NOPQ ectoteloblast precursor pair	Onset of cleavage of cell DNOPQ
6a. N teloblast pair	Onset of cleavage of NOPQ cell pair
6b. Q teloblast pair	Onset of cleavage of OPQ cell pair
7. O and P teloblast pairs; germinal band formation.	Completion of cleavage of OP cell pair
8. Germinal band coalescence	Onset of coalescence of right and left germinal bands
9. Segmentation	Completion of germinal band coalescence
10. Body closure	Appearance of coelomic space in the 32nd segment
11. Yolk exhaustion	Completion of fusion of the lateral edges of the germinal plate along the dorsal midline
Juvenile	Exhaustion of the yolk in the embryonic gut and first feeding

This description of embryogenesis applies to glossiphoniid leeches in general, but not to hirudinid leeches, such as *H. medicinalis*. Hirudinid eggs are much smaller (only about 0.1 mm in diameter) and contain little yolk. They are deposited in a sealed, hard-walled cocoon, which contains an albuminous fluid. The initial stages of hirudinid embryogenesis produce a sac, inappropriately called a "larva," which ingests the albumin via a "mouth" as an exogenous source of organic matter and energy for subsequent development. The embryo forms on the surface of this sac and rapidly increases in size. As with glossiphoniid leeches, embryogenesis in hirudinids proceeds via five pairs of (very small) teloblasts, which produce bandlets of primary blast cells, germinal bands, and a germinal plate. In the hirudinid embryo, germinal bands form on the future ventral (rather than on the future dorsal) surface and coalesce directly, without undergoing the circumferential migration characteristic of the Glossiphoniidae (Schleip, 1936). Development subsequent to formation of the germinal plate follows much the same course in the hirudinids as in the glossiphoniids (Fernandez and Stent, 1983).

V. Developmental Cell Lineage

A. Cell Lineage Tracing

To address one of the fundamental questions of developmental biology, namely, whether the differentiated properties of a cell of an adult animal depend on its line of descent from the fertilized egg or on its interactions with other cells, it is necessary to ascertain the developmental pedigree of the cell. C. O. Whitman (1878, 1887) carried out his pioneering studies on developmental cell lineage in the embryogenesis of glossiphoniid leeches with that question in mind. His method for ascertaining cell lineage relations was following the development of the living embryo under the microscope and keeping track of its successive cleavages.

A century later, Whitman's cell lineage studies on glossiphoniid embryos were refined and extended by devising microinjected cell lineage tracers, to establish the lines of descent of the identified cells of the adult leech (Weisblat *et al.*, 1978, 1980a; Gimlich and Braun, 1985; Stuart *et al.*, 1990). In this procedure, a histologically detectable tracer molecule is injected directly into an identified cell of the embryo early in development. At a later embryonic stage, the cellular distribution of the tracer is observed. Cells containing the tracer are inferred to have descended from the originally injected cell. However, because of the more complicated, indirect early development of hirudinid leeches, few, if any, cell lineage studies have been carried out on them thus far.

To be a candidate for serving as a cell lineage tracer, a molecule has to meet three conditions: (1) It must allow embryonic development to continues normally after its injection; (2) it must remain intact and not be diluted too much by cell growth and multiplication in the developing embryo; and (3) it must not pass through junctions linking embryonic cells and remain confined exclusively to descendants of the injected cell.

Horseradish peroxidase (HRP) was the first molecule to be employed as an intracellular lineage tracer (Weisblat *et al.*, 1978). Soon thereafter covalently linked adducts of large carrier molecules, such as dextrans, and fluorescent dyes, such as rhodamine or fluorescein, came into use (Weisblat *et al.*, 1980a). The rhodamine-dextran tracer is designated RDX and the fluorescein-dextran tracer FDX. To obtain a histologically fixable fluorescent lineage tracer that aldehyde treatment binds to tissue, dextrans to which lysine residues have been bound can be used as the carrier (Gimlich and Braun, 1985; Stuart *et al.*, 1989). The fixable composites of lysinated dextran and rhodamine or fluorescein are designated RDA or FDA.

Lineage tracer injection of individual macromeres and teloblasts showed that the labeled descendants of each form a distinct pattern, which is repeated from segment to segment and is the same from embryo to embryo. Moreover, they confirmed Whitman's early inferences that macromeres A, B, and C are the sources of the endodermal, DNOPQ of the ectodermal, and DM of the mesodermal tissues. Diagrammatic representations of the cellular contributions of the four ectodermal teloblasts to a midbody segment are shown in Fig. 3. It should be noted that the labeling pattern is confined to the same side as that of the injected teloblast (except for cellular processes projected contralaterally by labeled cell somata).

In addition to its utility for tracing cell lineages, fluorescein-labeled tracers can also serve as a specific photosensitizer (Shankland, 1984). On illuminating an FDX-labeled cell at a wavelength of 490 nm, some of the excited fluorescein fluorophores are quenched by transferring the absorbed energy to oxygen molecules in solution, converting them to the highly reactive singlet state. These reactive oxygen molecules, in turn, cause a generalized oxidation of cell constituents, leading to death of the illuminated, labeled cell (Miller and Selver-

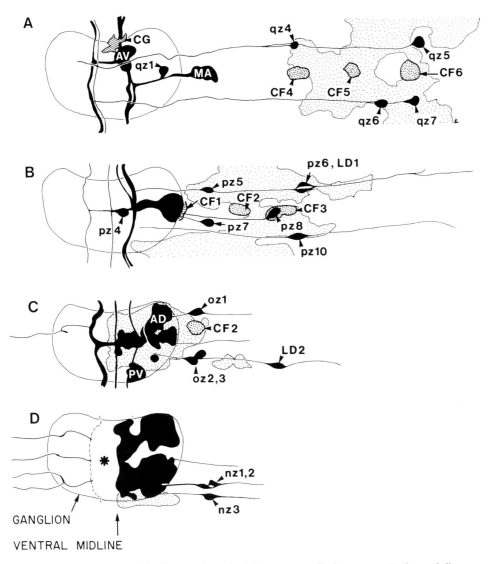

Figure 3 Contributions of the four ectodermal teloblasts to a midbody segment. Each panel illustrates the pattern of tracer-labeled cells observed in one segment of a stage 10 embryo of *T. rude*, following injection at stage 6 or 7 of a lineage tracer into the right Q teloblast (A), right P teloblast (B), right O teloblast (C), or right N teloblast (D). The hatched cell in A is a glioblast in the connective nerve. Labeled squamous epidermis is indicated by light stippling, and labeled cuboidal epidermal cells (or "cell florets") by heavy stippling. In D, fine, labeled processes fill most of the contralateral neuropil (asterisk outlined by dashed line). The right edge of the figure corresponds to the lateral margin of the germinal plate, and thus to the future dorsal midline. Anterior is up. Major labeled axonal tracts are shown, but no attempt has been made to render detailed axonal projection patterns within the CNS. (From Torrence and Stuart, 1986.)

ston, 1979). Because cells are largely transparent to light at 490 nm unless they contain fluorescein, injection of an embryonic progenitor cell with FDX makes it possible to photoablate its progeny selectively, even if they have become intermingled with cells from other lines of descent. Moreover, because the photosensitizer is also a lineage tracer, it automatically provides for visual identification of the cells to be ablated, as well as for direct visualization of their normal fates in unirradiated control embryos.

As detailed in other chapters of this volume, developmental cell lineage analyses were later extended to embryogenesis in animal taxa other than leeches, not only by direct observation, but also by use of such experimental techniques as selective cell ablation, application of extracellular marker particles, and production of chimeras and genetic mosaics.

B. Genealogical Origins of the Segmental Neurons

Because the paired n blast cell bandlets straddle the ventral midline of the germinal plate, Whitman (1887) inferred that the cells of the leech ventral nerve cord are derived from the N teloblast pair. But detailed scrutiny of lineage-tracer-labeled stage 10 embryos shows that here Whitman was mistaken: As seen in Fig. 3, all four ectodermal precursor teloblasts, N, O, P, and Q, contribute some cells to the CNS, some to the peripheral nervous system and some to the epidermis. Even the mesodermal precursor teloblast, M, contributes 3–4 paired neurons to each segmental ganglion, while its main contribution is made to mesenchyme, nephridia, and muscle fibers, the tissues to which embryologists traditionally assign a mesodermal origin. Thus, individual teloblasts generate a mixture of various types of tissue (Weisblat *et al.*, 1984; Kramer and Weisblat, 1985; Weisblat and Shankland, 1985; Torrence and Stuart, 1986).

Descendants of the n blast cell bandlet are found exclusively within the segmental ganglia, except for a few peripheral neurons and a few ventral epidermal cells. By contrast, descendants of the o, p, and q blast cell bandlets contribute substantially fewer cells to the CNS and correspondingly more cells to the peripheral nervous system and the epidermis, including epidermal specializations called cell florets. The stereotyped patterns of peripheral neurons derived from each blast cell bandlet indicates that the peripheral nervous system too comprises individually identifiable neurons, which can be assigned a unique identity based on their position in the body wall and their teloblast of origin (Fig. 3).

The topography of the contribution to the segmental epidermis made by each blast cell bandlet is in general concordance with their relative positions within the germinal plate. Thus, dorsal epidermis derives from the lateral-most q bandlet and ventral epidermis derives mainly from the mediolateral o and p bandlets, while the medial-most n bandlet provides a few epidermal cells on the ventral midline (Weisblat *et al.*, 1980b).

The ganglion cells contributed by any blast cell bandlet form discrete and coherent cell domains. For instance, the n bandlet contributes two transverse slabs of cells in the anterior and posterior regions of the ipsilateral hemiganglion and a longitudinal band of cells adjacent to the midline on the ventral aspect of the ganglion. The cell domains derived from the other bandlets display similarly compact and stereotyped topographies. In fact, these bandlets contribute fewer neurons, allowing some of their progeny to be individually recognized.

C. Kinship Groups

The regular, segmentally iterated patterns of neuronal descent indicate that the developmental cell lineages derived from each blast cell bandlet correspond to five identifiable kinship groups (Stent *et al.*, 1982; Weisblat *et al.*,1980b; Weisblat and Shankland, 1985). On surveying the kinship groups by combined use of fluorescent lineage tracers and the electrophysiological, anatomical, and histochemical techniques used for identifying individual neurons it was found that the members of a given neuronal kinship group do not share a unique set of specific neuronal properties that sets them apart from the members of the other kinship groups, such as functional category (glia or sensory, motor, or interneuron) or type of neurotransmitter (except that all serotonergic neurons belong to the n-bandlet-derived kinship group and that all dopaminergic neurons, as well as some homologous glial and mechanosensory cells, belong to the kinship groups derived from the o and p bandlets.) (Kramer and Weisblat, 1985; Blair, 1983; Stuart *et al.*, 1987).

D. Segmental Founder Cells

The number of primary blast cells derived from each teloblast that contribute to the foundation of one hemisegmental primordium was estimated by two different methods resorting to lineage tracers: one indirect, termed label boundary method (Weisblat *et al.*, 1984; Weisblat and Shankland, 1985) and the other direct, termed double label method (Zackson, 1984). Both methods led to the same estimate, namely that each primary blast cell in the m, o, and p bandlets contributes the entire hemisegmental complement of its kinship group, while in the n and q bandlets two successively born primary blast cells, of which the elder (i.e., anterior) are called n_s and q_s and the younger (i.e., posterior) n_f and q_f, each contributes a specific kinship subgroup to the hemisegmental complements of their respective kinship groups (Bissen and Weisblat, 1987).

Such lineage analyses were later extended to the detailed pedigree of individual cells within each kinship group. First, the initial division pattern of all seven primary hemisegmental founder blast cells were ascertained by observing the caudorostral distribution of second-, third-, and higher-order blast cells in the m, n, o, p, and q cell bandlets that had been labeled by injecting lineage tracers into one of the parent teloblasts (Zackson, 1984; Shankland and Weisblat, 1984; Shankland, 1987a,b; Bissen and Weisblat, 1989; Keleher and Stent, 1990). These studies revealed that each of the seven founder blast cells divides in a sequence that is idiosyncratic and stereotyped with respect to timing, orientation, and asymmetry of cell division. Furthermore, insofar as has been determined, each blast cell line has its own pattern of gene expression (Wedeen and Weisblat, 1991).

Second, the fate of second- and third-order blast cells was ascertained by injecting lineage tracer directly into them. Each higher order blast cell was found to generate a stereotyped subset of the hemisegmental kinship group issuing from its parental primary blast cell. In the majority of cases, each subset still contributed to more than one tissue; for example, higher order o- and p-bandlet blast cells produce both neurons and epidermal cells, whereas higher order m-bandlet blast cells produce both muscle fibers and nonmuscle cells.

One might conclude from the results just described that the mesodermal and ectodermal tissues of each hemisegment arise as clones founded by seven primary blast cells. In fact, the situation is more complicated, as was first inferred from results of the label boundary experiments (Weisblat and Shankland, 1985) and then was demonstrated directly by injecting lineage tracer into individual primary blast cells (Shankland, 1987a,b; Gleizer and Stent, 1993). Each m, n_f, o, p, or q_f blast cell contributes a stereotyped subset of its progeny to each of several successive morphologically defined hemisegments (at least three segments in the case of the m blast cell and two in the cases of the other blast cells). Thus, each morphologically defined hemisegment contains the progeny of not one, but of two or three primary blast cell clones, which interdigitate across segmental borders. By contrast, the contributions of the ns and qs primary blast cells appear to be confined to single hemisegments, that is, do not cross segmental borders.

E. Origin of the Supraesophageal Ganglion

There is one component of the leech CNS that remains unlabeled after tracer injection of any of the five teloblasts, namely the supraesophageal ganglion (Weisblat *et al.*, 1980b). Hence the frontmost cells of the CNS appear to arise from a source other than the blast cells of the germinal bands. The most likely alternative source of this tissue appeared to be some or all of the micromeres that arise during early cleavage from the A, B, C, and D macromeres (Sandig and Dohle, 1988; Bissen and Weisblat, 1989). In agreement with this expectation, direct lineage tracer experiments showed that these micromeres do contribute their progeny to the supraesophageal ganglion (Weisblat *et al.*, 1984). More specifically, the micromeres derived from the C and D macromeres were found to generate bilaterally symmetric sets of supraesophageal neurons, as well as nonneuronal cells, including the prostomial epidermis and muscle cells of the proboscis.

The finding that the supraesophageal ganglion is derived from micromeres rather than from the blast cells of the germinal bands is not without phyletic significance. In polychaetes, another major class of the annelid phylum, the highly developed and complex supraesophageal ganglion is obviously a prostomial organ. Here that ganglion not only lies in a region that is clearly rostral to the metameric body segments, but also arises developmentally as the neural tissue of a nonsegmented larva (itself derived from micromeres) to which the (teloblast-derived) segmental ventral nerve cord is joined later on metamorphosis (Dawydoff, 1959). By contrast, the status of the much less elaborate supraesophageal ganglion of leeches, whose development does not pass through a free-living larval stage endowed with its own nervous system, has long been the subject of controversy. In adult leeches, the supraesophageal ganglion is so far displaced posteriorly that it lies in a region whose body wall actually belongs to the first or second midbody segment. Thus modern lineage tracer methods have confirmed Whitman's (1887) original conjecture that the supraesophageal ganglion of the leech is derived from the micromeres rather than the germinal bands, a conjecture which he later withdrew (Whitman, 1892) and claimed erroneously that frontmost ganglion is a serial homologue of the segmental ganglia rather than a prostomial organ.

VI. Conclusions

As shown by the findings summarized here, leech embryogenesis is highly determinate, in the sense that the genealogical origin of identified cells can be traced, via a sequence of stereotyped cleavages, to the fertilized egg. This determinacy suggests that the developmental fate, that is, characteristic phenotype, of a given cell of the juvenile leech is the product of a series of preordained, somatically heritable restrictive developmental commitments made during successive cleavages in its particular line of descent from a developmentally pluripotent ancestral cell (Stent, 1985).

A. Typologic versus Topographic Hierarchies

It used to be thought that a developmental pathway such as that followed by the leech embryo, in which cleavage-associated commitments leading eventually to subclones of variously differentiated cell phenotypes would proceed according to a stepwise, typologically hierarchic commitment sequence (Slack, 1983). For instance, a cholinergic motor neuron would arise along the following pathway: from an ectodermal clone committed to expression of the characters that distinguish ectoderm from mesoderm to a neural subclone committed to expression of the characters that distinguish nervous tissue from epidermis; from the neural subclone, to a neuronal subclone committed to expression of the characters that distinguish neurons from glia; from the neuronal subclone to a motor neuron subclone committed to expression of the characters that distinguish motor neurons from sensory neurons; and finally from the motor neuron subclone to a cholinergic subclone committed to expression of the gene encoding cholineacetyltransferase (typical of excitatory motor neurons) rather than glutamic acid decarboxylase (typical of inhibitory motor neurons).

The finding that in the leech all the serotonergic neurons are derived from the N teloblast and all the dopaminergic neurons are derived from its OPQ sister cell could be taken as support for the seemingly plausible typologically hierarchic scheme. Thus on division of cell NOPQ, the N daughter would be committed to the generation of serotonergic and the OPQ daughter to the generation of dopaminergic neurons.

But during development of the leech nervous system such typologically hierarchic cases are more the exceptions than the rule. Usually, the phenotypes of identified neurons show little correlation with their membership in any particular kinship group, so that cell lineage relations appear to be typologically arbitrary rather than hierarchic. Indeed, even after a primary blast cell has undergone one or two divisions in the generation of its segmental founder clone, one of its daughter cells may still give rise to a mixed clone of neural and epidermal cells (Shankland and Stent, 1986).

In so far as there is any hierarchy of neurodevelopmental pathways in the leech, it is mainly topographic rather than typologic. In other words, it is the definitive (i.e., postembryonic) position of two cells rather than their differentiated phenotype that tends to be correlated with their genealogical proximity. As is evident in Fig. 3, nearly all of the N-kinship group neurons (including the serotonergic neurons) are destined for the CNS on the ventral midline, while the bulk of the O, P, or Q-kinship group neurons (including the dopaminergics) and epidermal cells are fated for the ventromedial, ventrolateral, or dorsal peripheral nervous system and body wall, respectively (Weisblat *et al.,* 1984). Never-

theless, the topographic hierarchy in kinship relations is far from absolute: Some descendants of the O, P, and Q teloblasts are destined for the CNS on the ventral midline, whither they migrate from their ventrolateral and dorsal sites of birth in the germinal plate at which the cleavage pattern initially deposited them.

It should not be surprising, therefore, that the overall developmental pathways present a mixture of typologically and topographically hierarchic schemes. Any particular developmental pathway probably represents an evolutionary compromise between maximizing the ease of ordering the spatial distribution of the determinants of commitment to a particular differentiated adult phenotype and minimizing the need for migration of differentially committed embryonic cells (Stent, 1985).

B. Determinants of Developmental Commitment

Two kinds of agents are commonly envisaged to contribute to the commitment of embryonic cells to a particular developmental fate under either a typologically or topographically hierarchic mode. One them is a set of intracellular determinants, which would account for the differential commitment of sister cells in terms of their unequal partition in successive cell divisions. For instance, a pluripotent cell might possess two intracellular determinants, a and b, necessary for producing cell types A and B respectively. Commitment of a daughter cell to fate A (and loss of pluripotency) would occur at an asymmetric cell division at which at least one of the daughter cells receives only a, but not b. Under this mechanism cell lineage would play a crucial role in cell commitment by consigning particular subsets of intracellular determinants to particular daughter cells.

The other putative agents of commitment are intercellular determinants, anisotropically distributed throughout the volume of the embryo. Here, a pluripotent embryonic cell would be capable of responding to either of two inducers a or b, provided to it by other cells, which are necessary for producing cell types A or B, respectively. Commitment of the cell to fate A (and loss of its pluripotency) would occur on having responded to the intercellular inducer a. Under this mechanism cell lineage would play a crucial role in cell commitment by placing particular cells at particular sites within the embryo, and hence govern the pattern of their exposure to intercellular inducers. The following three chapters of this section provide some concrete examples of intracellular and intercellular determinants of commitment in the development of leech embryos.

References

Baylor, D. A., and Nicholls, J. G. (1969). Chemical and electrical synaptic connexions between cutaneous mechanoreceptor neurones in the central nervous system of the leech. *J. Physiol. (London)* **203**, 591–609.

Bissen, S. T, and Weisblat, D. A. (1987). Early differences between alternate n blast cells in leech embryo. *J. Neurobiol.* **18**, 251–269.

Bissen, S. T., and Weisblat, D. W. (1989). The durations and compositions of cell cycles in embryos of the leech *Helobdella triserialis. Development (Cambridge, UK)* **106**, 105–118.

Blair, S. S. (1983). Blastomere ablation and the developmental origin of identified monoamine-containing neurons in the leech. *Dev. Biol.* **95**, 65–72.

Cline, H. T. (1983). 3H-GABA uptake selectively labels identifiable neurons in the leech central nervous system. *J. Comp. Neurol.* **215**, 351–358.

Cline, H. T. (1986). Evidence for GABA as a neurotransmitter in the leech. *J. Neurosci.* **6**, 2848–2856.

Dawydoff, C. (1959). Ontogenese des Annelides. *In* "Traité de Zoologie" (P. P. Grassé, ed.), Vol. 5, pp. 594–686. Masson, Paris.

Fernandez, J. (1980). Embryonic development of the glossiphoniid leech *Theromyzon rude*: Structure and development of the germinal bands. *Dev. Biol.* **78**, 407–434.

Fernandez, J., and Stent, G. S. (1983). Embryonic development of the hirudiniid leech *Hirudo medicinalis*. Structure, development and segmentation of the germinal plate. *J. Embryol. Exp. Morphol.* **72**, 71–96.

Friesen, W. O. (1985). Neuronal control of leech swimming movements: Interaction between cell 60 and previously described oscillator neurons. *J. Comp. Physiol.* **A156**, 231–242.

Friesen, W. O. (1989). Neuronal control of leech swimming movements. *In* "Neuronal and Cellular Oscillators" (J. W. Jacklet, ed.), pp. 269–316. Dekker, New York.

Friesen, W. O., Poon M., and Stent, G. S. (1978). Neuronal control of swimming in the medicinal leech. IV. Identification of a network of oscillatory interneurones. *J. Exp. Biol.* **75**, 25–43.

Gimlich, R. L., and Braun, J. (1985). Improved fluorescent compounds for tracing cell lineage. *Dev. Biol.* **109**, 509–514.

Gleizer, L., and Stent, G. S. (1993). Developmental origin of the segmental identity in the leech mesoderm. *Development (Cambridge, UK)* **117**, 177–189.

Harant, H., and Grassé, P. P. (1959). Class des annelides achètes ou Hirudinées ou sangsues. In "Traité de Zoologie" (P. P. Grassé, ed.), Vol. 5. pp. 471–593. Masson, Paris.

Keleher, G. P., and Stent, G. S. (1990). Cell position and developmental fate in leech embryogenesis. *Proc. Natl. Acad. Sci. U.S.A.* **87**, 8457–8461.

Kramer, A. P., and Goldman, J. R. (1981). The nervous system of the glossiphoniid leech *Haementeria ghilianii*. I. Identification of neurons. *J. Comp. Physiol.* **144**, 435–448.

Kramer, A. P., and Weisblat, D. A. (1985). Developmental neural kinship groups in the leech. *J. Neurosci.* **5**, 388–407.

Kristan, W. B., Jr., Wittenberg, G., Nusbaum, M. P., and Stern-Tomlinson, W. (1988). Multifunctional interneurons in behavioral circuits of the medicinal leech. *In*

"Invertebrate Neuroethology" (J. Camhi, ed.), pp. 383–389. Birkhaueser, Basel.

Kuffler, D. P (1978). Neuromuscular transmission in longitudinal muscle of the leech, *Hirudo medicinalis. J. Comp. Physiol.* **124A**, 333–338.

Kuhlman, J. R., Li, C., and Calabrese, R. L. (1985). FMRFamide-like substances in the leech. I. Immunocytochemical localization. *J. Neurosci.* **5**, 2301–2309.

Lent, C. M. (1973). *Science* **179**, 693–696.

Lent, C. M., Ono, J., Keyser, K. T., and Karten, H. J. (1979). Identification of serotonin with vital-stained neurons from leech ganglia. *J. Neurochem.* **32**, 1559–1563.

Macagno, E. R. (1980). The number and distribution of neurons in leech segmental ganglia. *J. Comp. Neurol.* **190**, 283–302.

Mann, K. H. (1962). "Leeches (Hirudinea)." Pergamon, Oxford.

Miller, J. P., and Selverston, A. I. (1979). Rapid killing of single neurons by irradiation of intracellularly injected dye. *Science* **206**, 702–704.

Muller, K. J. (1979). Synapses between neurones in the central nervous system of the leech. *Biol. Rev.* **54**, 99–134.

Muller, K. J., Nicholls, J. G., and Stent, G. S., eds. (1981). "Neurobiology of the Leech." Cold Spring Harbor Laboratory, Cold Spring Harbor, New York.

Nicholls, J. G., and Baylor, D. K. (1968). Specific modalities and receptive fields of sensory neurons in CNS of the leech. *J. Neurophysiol.* **31**, 740–756.

Nicholls, J. G., and Purves, D. (1972). A comparison of chemical and electrical synaptic transmission between single sensory cells and a motoneuron in the central nervous system of the leech. *J. Physiol. (London)* **225**, 637–656.

Norris, B. J., and Calabrese, R. L. (1990). Identification of motor neurons that contain a FMRFamide-like peptide and the effects of FRMFamide on longitudinal muscles of the leech, *Hirudo medicinalis. J. Comp. Neurol.* **266**, 95–111.

Nusbaum, M. P., and Kristan, W. B., Jr. (1986). Swim initiation in the leech by serotonin-containing interneurons, cells 21 and 61. *J. Exp. Biol.* **122**, 277–302.

Ort, C. A., Kristan, W. B., Jr., and Stent, G. S. (1974). Neuronal control of swimming in the medicinal leech. II. Identification and connections of motor neurons. *J. Comp. Physiol.* **94**, 121–154.

Rude, S. (1969). Monoamine-containing neurons in the central nervous system and peripheral nerves of the leech *Hirudo medicinalis. J. Comp. Neurol.* **136**, 349–371.

Sandig, M., and Dohle, W. (1988). The cleavage pattern in the leech *Theromyzon tessulatum* (Hirudinea, Glossiphoniidae). *J. Morphol.* **196**, 217–252.

Sargent, P. B. (1977). Synthesis of acetylcholine by excitatory motoneurons in the central nervous system of the leech. *J. Neurophysiol.* **40**, 453–460.

Sawyer, R. T. (1972). "North American Fresh Water Leeches, Exclusive of the Piscicolidae, with a Key to All Species." Univ. of Illinois Press, Urbana.

Schleip, W. (1936). Ontogenie der Hirudineen. *In* "Klassen und Ordnungen des Tierreich" (H. G Bronn, ed.), Vol. 4, Div. III, Book 4, Part 2, pp. 1–121. Akademische Verlagsgesellschaft, Leipzig.

Shankland, M. (1984). Positional control of supernumerary

blast cell death in the leech embryo. _Nature (London)_ **307**, 541–543.

Shankland, M. (1987a). Differentiation of the O and P cell lines in the embryo of the leech. I. Sequential commitment of blast cell lineages. _Dev. Biol._ **123**, 85–96.

Shankland, M. (1987b). Differentiation of the O and P cell lines in the embryo of the leech. II. Genealogical relationship of descendant pattern elements in alternative developmental pathways. _Dev. Biol._ **123**, 97–107.

Shankland, M., and Martindale, M. Q. (1989). Segmental specificity and lateral asymmetry in the differentiation of developmentally homologous neurons during leech embryogenesis. _Dev. Biol._ **135**, 431–448.

Shankland, M., and Savage, R. M. (1997). Annelids, the segmented worms. _In_ "Embryology" (S. F Gilbert and A. Raunio, eds.), pp. 219–235. Sinauer, Sunderland, Massachusetts.

Shankland M., and Stent, G. S. (1986). Cell lineage and cell interactions in the determination of developmental cell fate. _In_ "Genes, Molecules and Evolution" (J. P. Gustafson, G. L. Stebbins, and F. J. Ayala, eds.), pp. 211–233. Academic Press, Orlando, Florida.

Shankland, M., and Weisblat, D. A. (1984). Stepwise loss of neighbor cell interactions during the positional specification of blast cell fates in the leech embryo. _Dev. Biol._ **106**, 326–342.

Slack, J. M. W. (1983). "From Egg to Embryo." Cambridge Univ. Press, London and New York.

Stent, G. S. (1985). The role of cell lineage in development. _Philos. Trans. R. Soc. London Ser. B_ **312**, 3–19.

Stent, G. S., Kristan, W. B., Jr., Friesen, W. O., Ort, C. A., Poon, M., and Calabrese, R. L. (1978). Neuronal generation of the leech swimming movement. An oscillatory network of neurons driving a locomotory rhythm has been identified. _Science_ **200**, 1348–1357.

Stent, G. S., Thompson W. J., and Calabrese, R. L. (1979). Neural control of heartbeat in the leech and in some other invertebrates. _Physiol. Rev._ **59**, 101–136.

Stent, G. S., Weisblat, D. A., Blair, S., and Zackson, S. L. (1982). Cell lineage in the development of the leech nervous system. _In_ "Neuronal Development" (N. Spitzer, ed.), pp. 1–44. Plenum, New York.

Stuart, A. E. (1970). Physiological and morphological properties of motoneurones in the central nervous system of the leech. _J. Physiol. (London)_ **209**, 627–646.

Stuart, A. E., Hudspeth, A. J., and Hall, Z. W. (1974). Vital staining of specific monoamine-containing cells in the leech nervous system. _Cell Tissue Res._ **153**, 55–61.

Stuart, D. K., Blair, S. S., and Weisblat, D. A. (1987). Cell lineage, cell death, and the developmental origin of identified serotonin- and dopamine-containing neurons in the leech. _J. Neurosci._ **7**, 1107–1122.

Stuart, D. K., Torrence, S. A., and Law, M. I. (1989). Leech neurogenesis I. Positional commitment of neural precursor cells. _Dev. Biol._ **136**, 17–39.

Stuart, D. K., Torrence, S. A., and Stent, G. S. (1990). Mi-

croinjectable probes for tracing cell lineages in development. _In_ "Methods in Neurosciences," Vol. 2, pp. 375–392. Academic Press, San Diego.

Thompson, W. J., and Stent, G. S. (1976a). Neuronal control of heartbeat in the medicinal leech. I. Generation of the vascular constriction rhythm by heart motor neurons. _J. Comp. Physiol._ **111**, 261–279.

Thompson, W. J., and Stent, G. S. (1976b). Neuronal control of heartbeat in the medicinal leech. II. Intersegmental coordination of heart motor neuron activity by heart interneurons. _J. Comp. Physiol._ **111**, 281–307.

Thompson, W. J., and Stent, G. S. (1976c). Neuronal control of heartbeat in the medicinal leech. III. Synaptic relations of the heart interneurons. _J. Comp. Physiol._ **111**, 309–333.

Torrence, S. A., and Stuart, D. K. (1986). Gangliogenesis in leech embryos: Migration of neural precursor cells. _J. Neurosci._ **6**, 2736–2746.

Wallace, B. G., and Gillon, J. W. (1982). Characterization of acetylcholine esterase in individual neurons in the leech central nervous system. _J. Neurosci._ **2**, 1108–1118.

Wedeen, C. J., and Weisblat, D. A. (1991). Segmental expression of an _engrailed_-class gene during early development and neurogenesis in an annelid. _Development (Cambridge, UK)_ **113**, 805–814.

Weisblat, D. A. (1981). Development of the nervous system. _In_ "Neurobiology of the Leech" (K. J. Muller, J. G. Nicholls, and G. S. Stent, eds.), pp. 173–195. Cold Spring Harbor Laboratory, Cold Spring Harbor, New York.

Weisblat, D. A, and Shankland, M. (1985). Cell lineage and segmentation in the leech. _Philos. Trans. R. Soc. London Ser. B_ **312**, 39–56.

Weisblat, D. A., Sawyer, R. T., and Stent, G. S. (1978). Cell lineage analysis by intracellular injection of a tracer enzyme. _Science_ **202**, 1295–1298.

Weisblat, D. A., Zackson, S. L., Blair, S. S., and Young, J. D. (1980a). Cell lineage analysis by intracellular injection of fluorescent tracers. _Science_ **209**, 1538–1541.

Weisblat, D. A., Harper, G., Stent, G. S., and Sawyer, R. T. (1980b). Embryonic cell lineage in the nervous system of the glossiphoniid leech _Helobdella triserialis_. _Dev. Biol._ **76**, 58–78.

Weisblat, D. A., Kim, S. Y., and Stent, G. S. (1984). Embryonic origins of cells in the leech _Helobdella triserialis_. _Dev. Biol._ **104**, 65–85.

Whitman, C. O. (1878). The embryology of Clepsine. _Q. J. Microsc. Sci._ **18**, 215–315.

Whitman, C. O. (1887). A contribution to the history of germ layers in Clepsine. _J. Morphol._ **1**, 105–182.

Whitman, C. O. (1892). The Metamerism of Clepsine. _In_ "Festschrift zum 70. Geburtstage R. Leuckarts," pp. 385–395. Engelmann, Leipzig.

Zackson, S. L. (1984). Cell lineage, cell–cell interaction, and segment formation in the ectoderm of a glossiphoniid leech embryo. _Dev. Biol._ **104**, 143–160.

12

Cell Fate Specification in Glossiphoniid Leech: Macromeres, Micromeres, and Proteloblasts

DAVID A. WEISBLAT
FRANÇOISE Z. HUANG
Department of Molecular and Cell Biology
University of California, Berkeley,
Berkeley, California 94720

DEBORAH E. ISAKSEN
Virginia Mason Research Center
Seattle, Washington 98101

I. Introduction

In this chapter, we focus on three aspects of cell lineage and cell fate determination in the early development of glossiphoniid leech embryos. First, we summarize work on cell fate determination of the D' macromere and the teloblast precursors (DM and DNOPQ proteloblasts) to which it gives rise at fourth cleavage. Second, we describe a stepwise process of cell fusions responsible for forming the syncytial yolk cell (from which the definitive gut epithelium forms) and how this process is regulated in the case of the A‴ and B‴ macromeres. Third, we discuss the diversity of cell lineage patterns and cell fates within the group of 25 micromeres that arise during cleavage from all four quadrants of the embryo. In each of these research areas, our goals are not only to improve our understanding of how leech embryos develop at the cellular and molecular levels, but also to facilitate comparisons between developmental processes in leeches and other animals. Such comparisons should allow us to infer how changes in developmental

processes have given rise to the diversity of body plans seen in modern day animals.

II. Specification of the D Quadrant Lineage

The D blastomere is identifiable as the largest cell in stage 3 embryos. It is also unique in that it contains pools of yolk-deficient cytoplasm at the animal and vegetal ends of the cell. These cytoplasmic domains, referred to as pole plasm in annelids generally and as teloplasm in leeches, are rich in mitochondria and polyadenylated maternal mRNAs (Fernandez and Olea, 1982; Fernandez et al., 1987; Holton et al., 1989, 1994). Teloplasm arises by cytoplasmic rearrangements that occur prior to first cleavage (Fig. 1, stage 1) in a microtubule-dependent process (Astrow et al., 1989). [Curiously, homologous cytoplasmic rearrangements in oligochaete annelids proceed via microfilament-dependent processes (Shimizu, 1982, 1984, 1986).] In

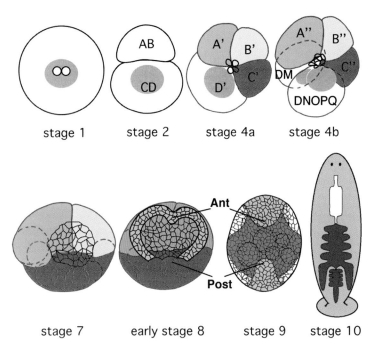

Figure 1   Summary of leech development. Selected stages (Stent *et al.*, 1992) are depicted as seen from the animal pole (prospective dorsal views; posterior toward the bottom). Teloplasm (gray) arises at the animal and vegetal poles prior to first cleavage (stage 1; polar bodies indicated by small white circles). The first cleavage is shown transverse to the A–P axis of the embryo, so that the A and B quadrants remain in apposition at the midline. This results in cells C and D being off axis at stage 4a; the normal symmetry is restored by a counterclockwise shift of the D′ cleavage products (proteloblasts and teloblasts, dotted outlines) and a complementary clockwise shift of macromere C‴ as it comes to envelop most of the proteloblasts and teloblasts (stages 4b–7). Micromeres and their progeny in the epithelium of the provisional integument are shown in dark outline. The A, B, and C quadrant macromeres are shaded blue, yellow, and orange respectively. Macromeres A‴ and B‴ fuse during early stage 8 to form cell A/B (green). By the end of stage 7, the germinal bands (gray) are joined at their anterior (Ant) ends and elongate by addition of blast cells from the teloblasts at their posterior (Post) ends. During stage 8, the germinal bands move ventrovegetally over the surface of the embryo, accompanied by epiboly of the micromere-derived epithelium, and gradually coalesce from anterior to posterior along the ventral midline, forming the germinal plate (gray). By stage 9, germinal plate formation is complete and C‴ has fused with A/B to form cell A/B/C (brown); subsequent fusions of the teloblasts and supernumerary blast cells with A/B/C form the syncytial yolk cell (SYC; purple). By stage 10, the provisional integument has been largely replaced by the definitive body wall (gray) and the definitive gut epithelium has arisen by cellularization of the SYC.

normal development, both animal and vegetal pools of teloplasm are segregated to macromere D′ in the 8-cell embryo (Fig. 1, stages 1–4a), then the vegetal teloplasm moves to merge with the animal teloplasm prior to fourth cleavage; the common pool of teloplasm is divided between cells DNOPQ and DM at fourth cleavage (Fig. 1, stage 4b; Holton *et al.*, 1989).

A variety of experimental results demonstrate that it is the inheritance of teloplasm that causes the D quadrant cells to carry out the unique series of cell divisions during stages 4–6 that lead to the formation of all 10 teloblasts and 16 of the 25 micromeres (Figs. 1 and 2). In one series of experiments (Astrow *et al.*, 1987), gentle centrifugation was used to redistribute

teloplasm in cell CD of the 2-cell embryo. At second cleavage in such embryos, teloplasm was often partitioned into both cells C and D more or less equally, in which case both cells gave rise to teloblasts. In other experiments, zygotes were compressed after teloplasm formation so as to elongate them along the animal–vegetal axis. This operation forces the reorientation of the first cleavage furrow so that both daughter cells inherit teloplasm, in which case both cells give rise to teloblasts (Nelson and Weisblat, 1992). The fact that redistributing teloplasm by either procedure frequently leads to the formation of supernumerary teloblasts suggests that teloplasm acts permissively to release the normal D quadrant cleavage pattern in the macromere.

Figure 2   Partial cell lineage diagram for *Helobdella*, emphasizing micromere production and formation of the gut epithelium. The corresponding developmental stages and hours after zygote deposition are indicated on the time line at left; breaks in the time line denote changes in scale. The macromeres, proteloblast, and teloblasts are indicated in capital letters. The A, B, and C quadrant cells and their fusion products (A/B, A/B/C, and SYC) are also indicated in the colors corresponding to Fig. 1. Lowercase letters denote micromeres (circled) and blast cells. Documented cell fusions are denoted by the merger of the various cell lines. Dotted lines indicate the omission of bilaterally symmetric lineages (M_L and M_R; $NOPQ_L$ and $NOPQ_R$), the continuing production of blast cells from the teloblasts (M, N, O/P, O/P, Q), and uncertainties in the timing of later SYC fusions.

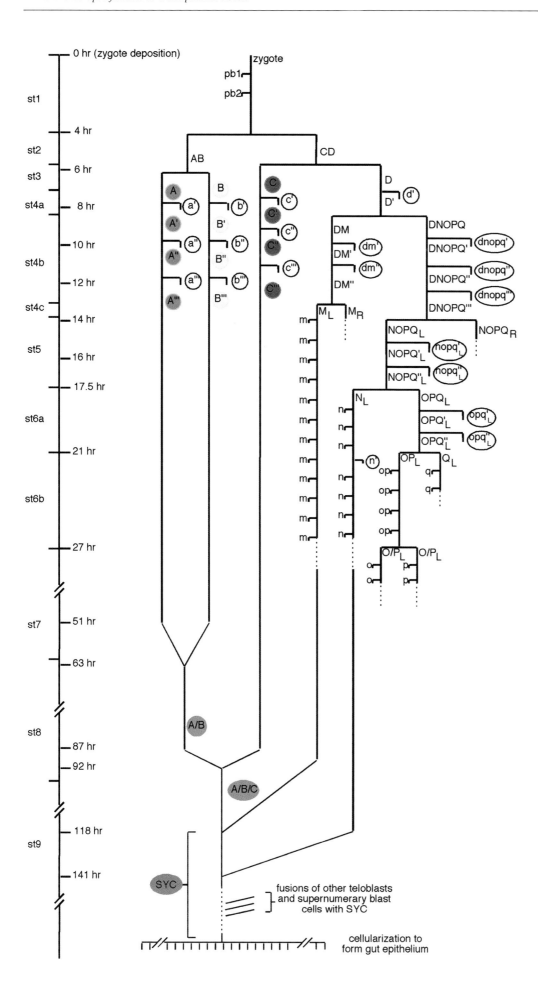

The best information regarding the mechanisms of unequal first and second cleavage in clitellate annelids comes from elegant studies of *Tubifex,* an oligochaete (Ishii and Shimizu, 1997; Shimizu, 1996a,b). It has been shown that the maternal centrosome does not duplicate after meiosis I and II. Therefore, the spindle formed at the first mitotic division contains but a single centrosomal aster (as judged by γ-tubulin immunoreactivity) and the resulting asymmetric spindle results in the establishment of an eccentrically situated cleavage furrow sometime after metaphase. The asymmetry of the second cleavage in *Tubifex* is established by a more traditional mechanism. The centrosome duplicates, giving rise to an amphiastral spindle which undergoes an eccentric translocation, perhaps due to interactions with localized cortical factors as in the nematode *Caenorhabditis elegans* (Hyman, 1989).

III. Cell Fate Determination in the Ectodermal and Mesodermal Proteloblast Lineages

How are the DNOPQ and DM blastomeres, which arise as sister blastomeres at fourth cleavage, assigned their distinct ectoderm and mesodermal fates? According to the model we favor at present (Fig. 3), the ectodermal (DNOPQ) fate is induced in blastomeres that inherit teloplasm in quantities above some threshold quantity and in which the teloplasm interacts at short range with factor(s) localized in cortex derived from the animal region of the zygote. In contrast, the mesodermal (DM) fate is postulated to be a default fate followed by blastomeres that inherit teloplasm but do not undergo the interaction between teloplasm and cortex.

Evidence for this model comes from several types of embryological experiments. First, the fate assignments seem to be on the basis of inherited differences rather than environmental influences. For example, if cell DM or cell DNOPQ is removed from the embryo at the completion of fourth cleavage, the surviving DNOPQ or DM cells follow their normal fates, despite the fact that they occupy equivalent positions in the embryo and appear to make equivalent contacts with micromeres and macromeres. Similarly when the animal and vegetal teloplasms are segregated to distinct cells by equatorial first cleavage planes in response to experimental manipulation, the animal hemisphere cell assumes the ectodermal fate and the vegetal hemisphere cell assumes the mesodermal fate (Nelson and Weisblat, 1992).

Taken in isolation, this latter result could be interpreted as suggesting that the animal and vegetal teloplasms themselves contain distinct ectodermal and mesodermal determinants respectively. But another line of evidence, based on the results of cytoplasm extrusion experiments, indicates that the two pools of teloplasm are equipotent and implicates the animal cortex as critical in determining the ectodermal fate. Either animal or vegetal teloplasm can be selectively extruded from zygotes shortly before first cleavage (Fig. 3; Nelson and Weisblat, 1991). The resultant embryos seem to develop normally through stage 4a, including the vegetal to animal migration of teloplasm in embryos from which the animal teloplasm has been extruded. Embryos from which vegetal teloplasm was extruded make normal complements of meso- and ectoteloblasts and can develop to adulthood. But in embryos from which animal teloplasm was extruded, both proteloblasts assume the mesodermal fate, including the nominal DNOPQ cell!

In these animal-extrusion embryos, we noted that the teloplasm in the nominal DNOPQ cells was separated from the animal cortex by a layer of yolky cytoplasm. Moreover, centrifuging such embryos at the 4-cell stage to force the remaining teloplasm up to the animal pole was able to restore the nominal DNOPQ cell to its normal ectodermal fate (Fig. 3). From these experiments, we concluded first, that both animal and vegetal teloplasm is capable of supporting both ectodermal and mesodermal fates and second, that the ectodermal fate requires that the teloplasm in the DNOPQ cell be at high concentration in the animal end of the cell, presumably in close proximity to factors localized in cortical cytoplasm at the animal pole of the zygote.

An intriguing molecular correlate of the DM-DNOPQ fate differences has emerged from studies of a *nanos*-class gene called *Hro-nos* in the leech *Helobdella robusta.* In *Drosophila, nanos* mRNA is a maternal transcript, localized and translated preferentially at the posterior pole of the zygote, to generate a gradient of protein that is required for the downregulation of the *hunchback* gene activity in the posterior portion of the syncytial blastoderm (Tautz, 1988; Tautz and Pfeifle, 1989). *Hro-nos* also is present as an abundant maternal transcript that decays continuously to much lower levels by the end of cleavage. In contrast, the *Hro-nos* protein is undetectable in the zygote and reaches its peak of expression during stage 4, within the D lineage. Whole-mount immunostaining reveals that, as macromere D′ starts to cleave, *Hro-nos* immunoreactivity is already higher within the prospective DNOPQ cell. Once this cleavage is complete, cell DNOPQ contains significantly higher levels of *Hro-nos* protein than DM or any of the macromeres, as judged both by immunostaining and by Western blots of pooled, dissected blastomeres (Pilon and Weisblat, 1997).

Whether *Hro-nos* plays a role in regulating the distinct fates of DM and DNOPQ as opposed to being a reporter of those distinct fates remains to be determined. In either case, the question of whether it regulates the

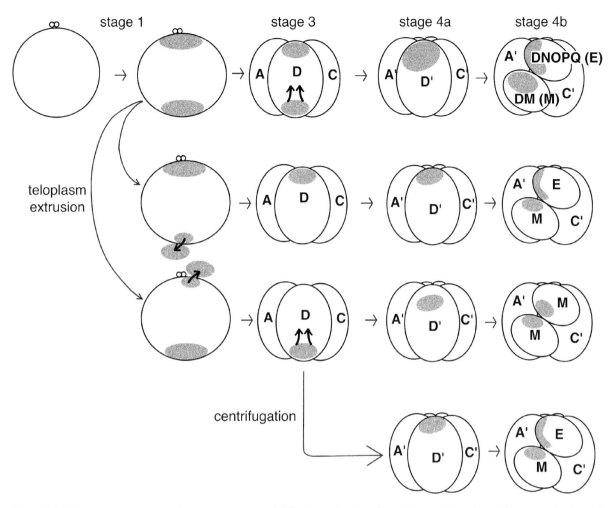

Figure 3 Teloplasm extrusion experiments suggest a model for determination of mesoderm and ectoderm. Diagrammatic views of normal and experimentally perturbed embryos; equatorial views of the prospective or actual D quadrant. In normal development (top row) after two polar bodies (small circles) are extruded at the animal pole (top), cytoplasmic rearrangements lead to accumulation of yolk-deficient cytoplasm (teloplasm; shading) at the animal and vegetal poles of the zygote (stage 1). Unequal cleavages segregate teloplasm to the D blastomere of the 4-cell embryo (stage 3), after which the vegetal teloplasm migrates to the animal pole (arrows). The third cleavage is highly unequal in all four quadrants, producing quartets of animal micromeres and vegetal macromeres (stage 4a). The fourth cleavage in obliquely equatorial in the D quadrant, producing ectodermal (E) and mesodermal (M) precursors, blastomeres DNOPQ and DM, respectively, both of which inherit teloplasm (stage 4b). In embryos from which vegetal teloplasm has been extruded (second row), the remaining teloplasm stays near the animal cortex. The daughters of macromere D′ inherit less teloplasm, but follow their normal fates. But in embryos from which animal teloplasm has been extruded (third row), the vegetal teloplasm fails to reach the animal end of the cell during its normal migration and both cells follow the mesodermal fate. If embryos from which animal teloplasm was extruded are centrifuged at the 4-cell stage, forcing the remaining teloplasm to the animal end of the cell, the normal ectodermal fate is rescued. This suggests that the ectodermal fate of cell DNOPQ is induced by the short-range interaction of teloplasm with factors localized in the animal cortex.

translation of the leech *hunchback*-class gene (Savage and Shankland, 1996) will also be of interest.

IV. Cell–Cell Fusions among Endodermal Precursors

From his pioneering cell lineage studies, Whitman (1878) concluded that the epithelial lining of the gut in glossiphoniid leech arises from multinucleate yolk cells arising from the A, B, and C macromeres. More recently the origins of the gut epithelium from a multinucleate

syncytial yolk cell (SYC) has been conclusively demonstrated using microinjected cell lineage tracers (Nardelli-Haefliger and Shankland, 1993). During stage 9, nuclei accumulate near the surface of the SYC beneath the germinal plate and become cellularized, under the influence of the germinal plate mesoderm. By this means, the definitive gut forms around the remaining yolk, which therefore lies within the lumen of the nascent gut. The crop and intestine form with prominent, segmentally iterated lobes (Fig. 1, stage 10) and express two homeobox containing genes, *Lox10* and *Lox3*, in segmentally iterated stripes corresponding to the peaks and the valleys of the lobes, respectively (Nardelli-Haefliger and Shankland, 1993; Wedeen and Shankland, 1997; Chapter 14 by Shankland). A *Lox3*-class gene is expressed in a similar pattern in the midgut of the jawed leech *Hirudo medicinalis* (Wysocka-Diller *et al.*, 1995).

This mechanism of gut formation raises intriguing parallels with the syncytial blastoderm of *Drosophila* and other insects. In *Drosophila* of course, not only gut epithelia, but also the cells of prospective segmental mesoderm and ectoderm arise essentially simultaneously by the cellularization of the syncytial blastoderm. But in apterygote insects, it seems that syncytial yolk cell nuclei migrate to the periphery of the yolk cell and become cellularized at two separate times in development (reviewed in Anderson, 1973). The first wave gives rise to presumptive mesoderm and ectoderm, and the second gives rise to the midgut epithelium. Perhaps the formation of gut epithelium from a multinucleate syncytium is a shared trait derived from an ancestral protostome?

The formation of the SYC offers another point of comparison between annelids and arthropods. In insects such as *Drosophila*, the SYC arises directly from the zygote by nuclear proliferation without cytokinesis. But in other insects, early development consists of holoblastic cleavage and cells later fuse to form the yolk syncytium (reviewed in Anderson, 1973). A similar process is seen in leech embryos, where the SYC arises through a complex series of selective fusions involving cells from disparate embryonic lineages as well as nuclear proliferations.

That endodermal precursor cells fuse in leech was clearly appreciated by Bychowsky (1921), who reported a gradual loss of visible outlines for the yolk cells, beginning at the anterior end of the prospective midgut. From his drawings, he was referring to embryos at early stage 9, by which time the germinal plate has completely formed. In our laboratory, we have used a sensitive assay to observe cell fusion, namely the visualization of readily detectable lineage tracers that spread from one cell to the next after fusion (Liu *et al.*, 1998). We find that cell fusion in the endodermal lineages occurs much earlier than was apparent to Bychowsky, and that the three macromeres fuse in a stepwise manner to initiate formation of the syncytial yolk cell (Figs. 1 and 2).

The first step is the fusion of macromeres A''' and B''' to form a cell we designate as A/B (Fig. 1). When we injected β-galactosidase into either the A or the B quadrant cell during stages 4–5 and then stained at progressively later times, we first detected diffusion of the enzyme from one cell into the other (as determined by the distribution of the histochemical reaction product) as early as mid-stage 7, roughly 51 hours after zygote deposition (AZD). In *Helobdella*, this is approximately 65

hours before the stage at which Bychowsky inferred fusion by the loss of visible cell outlines within the yolk mass. At about the end of stage 7 (~12 hours later), macromeres A''' and B''' had fused in virtually all embryos in a batch. In rare embryos, all three macromeres seemed to have fused, but we have never seen embryos in which macromere A''' or B''' fuses selectively with C'''. Macromere C''' fuses with A/B within a 5-hour time period at the end of stage 8 (87–92 hours AZD). The resulting cell is designated as A/B/C (Fig. 1).

Schmidt (1939) proposed that the M teloblasts of glossiphoniid leech embryos fused with the macromeres, on the basis of light microscopic examination of sectioned embryos of somewhat indeterminate age. We used the diffusion assay to confirm and extend Schmidt's findings by injecting M or N teloblasts with lineage tracer during stage 6, and then fixing and clearing the embryos at various time points beginning at the end of stage 8. Fusion of M teloblasts with A/B/C occurs during the period 89–118 hours AZD and N teloblast fusion occurs even later (118–141 hours AZD). This delay correlates with the fact that the M teloblasts have produced their full complement of segmental founder cells 35 hours earlier than the N teloblasts. The cell resulting from the fusion of teloblasts with A/B/C is finally designated as the SYC, but there is yet another contribution of nuclei and cytoplasm to the SYC. Near the end of blast cell production, each teloblast generates "supernumerary" blast cells that are not incorporated into the germinal bands or germinal plate. It was originally supposed that these cells die (Shankland, 1984) but it seem more likely that in fact these supernumerary blast cells fuse with the SYC (Desjeux, 1995; Chapter 14 by Shankland). If so, the SYC receives contributions of nuclei (and cytoplasm) from three embryonic lineages: the macromeres, the teloblasts, and the supernumerary blast cells. Whether these different components contribute equally to the definitive gut epithelium and the intralumenal yolk nuclei remains to be determined.

V. Regulation of A'''–B''' Fusion

The gut precursor cell in the B quadrant of the embryo makes continuous and extensive contact with the gut precursors in the A and C quadrants from the time of their birth at stage 3. Yet macromere B''' does not fuse with A''' until early stage 8 much later and with the other SYC contributors later still. Focusing on the fusion of macromeres A''' and B''', we have shown that the fusion process does not occur autonomously. Instead, it is regulated by signals emanating at least in part from the D quadrant of the embryo (Isaksen, 1997; Isaksen *et al.*, 1998).

The first evidence for this conclusion came from cell deletion experiments. When cell C is removed from embryos at during stages 3–4, the remaining A, B, and D quadrants undergo apparently normal subsequent divisions and macromeres A′′′ and B′′′ fuse as in controls. But in embryos from which macromere D′ is removed, A′′′–B′′′ fusion is blocked, suggesting that A′′′–B′′′ fusion requires some sort of signal from the D quadrant. Equivalent results are obtained from intact embryos, in which the C′ or D′macromere is "biochemically arrested" by microinjection of either RNase or the ricin A chain, a powerful protein synthesis inhibitor (Endo and Tsurugi, 1988). Cells injected with either of these reagents round up and undergo at most one further division, but do not lyse (Nelson and Weisblat, 1992; Isaksen et al., 1998).

Using the simpler technique of biochemical arrest allowed us to extend the perturbation experiments to stages of development at which microsurgery is impractical (Fig. 4). If we wait until stage 4b and then arrest cells DM and DNOPQ, fusion is still blocked in almost all of the embryos. However, delaying the arrest until stage 4c and then arresting DM′′ and DNOPQ′′′ yields embryos in which A′′′–B′′′ fusion usually takes place. During this interval, the D quadrant cells have given rise to two dm and three dnopq micromeres that were not injected with ricin in the experiments described above. Are some or all of these micromeres required for the fusion signal? Apparently not, because fusion is observed in a substantial fraction (20–30%) of embryos in

which cell DM and all three dnopq micromeres were arrested by ricin injection, so that cell DNOPQ′′′ is the only D quadrant derivative allowed to continue cleaving. Fusion also is seen in most embryos if only DM is allowed to continue cleaving. Thus, it seems that the capacity to induce fusion is distributed among the D quadrant progeny and that A′′′–B′′′ fusion has gradually become immune to the effects of D quadrant arrest by the end of stage 4.

One interpretation for these observations is that the putative D-derived signal has already been sent by stage 5. Another is that the signal has not yet been sent at that point, but that its transmission is immune to biochemical arrest by stage 5. If the injected ricin A chain acts by inhibiting protein synthesis, one possibility is that the required D lineage signal involves a protein that is synthesized during stage 4, then activated later in a synthesis-independent process. In any case, it is striking that the critical period ends at least 40 hours before fusion actually occurs.

Not surprisingly, A′′′–B′′′ fusion also can be blocked by injecting A or B quadrant cells directly with ricin. But we were surprised to discover that even with this experimental paradigm, fusion becomes immune to ricin with a time course similar to that obtained for D quadrant arrest. That is, arresting either the A/A′ cell or the B/B′cell prior to stage 4b blocks fusion, but fusion occurs if either or both cells are arrested after stage 4c. Curiously,

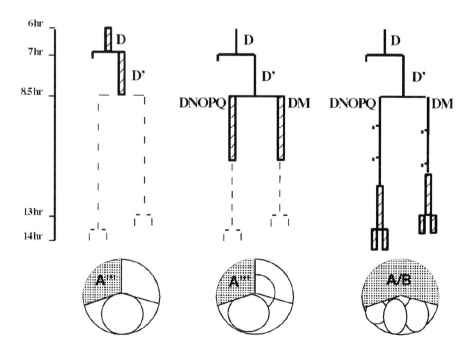

Figure 4 Biochemical arrest of D lineage blastomeres blocks A′′′–B′′′ fusion, with an early critical period. The D quadrant was arrested by microinjection of ricin A chain and was injected into blastomere D, macromere D′, or the descendant proteloblasts and teloblasts during the periods indicated by the hatched bars in the D lineage (top row); dashed lines in the lineage diagram indicate divisions that are blocked by ricin injection; the time line at left is as in Fig. 2. The resultant embryos were scored for fusion at about 75 hours after zygote deposition (bottom row). Note that the cells injected with ricin no longer divide. Biochemical arrest of D, D′, or early DM and DNOPQ lineages block A′′′–B′′′ fusion. Additional biochemical arrest experiments are described in the text.

the A lineage macromere becomes immune to the biochemical arrest several hours prior to B lineage macromeres during the transition period, despite the fact that cleavages in the A and B lineages proceed synchronously and the two cells are generally assumed to be identical. This difference in the behavior of the two cells correlates with the fact that the A quadrant macromere has a much larger area of contact with the D quadrant blastomere and its progeny than does the B quadrant cell, consistent with the notion that the D quadrant signal operates quantitatively and accumulates gradually during the critical stages.

VI. Cell Lineage and Cell Fate Differences among the Micromeres

We designate as micromeres those 25 small cells that arise by highly unequal divisions during cleavage and that are not part of the coherent columns of blast cells produced by the teloblasts (Fig. 1; Sandig and Dohle, 1988; Bissen and Weisblat, 1989). This criterion for identifying cells as micromeres results in some discrepancies in nomenclature with respect to the classic spiralian nomenclature, in which the term is applied to the animal daughter of any macromere division (Table I). For example, the cell we designate as DNOPQ corresponds to micromere 2d in the classic nomenclature and the cell we designate as dm″ corresponds to cell 4D in the classic nomenclature. In any event, an intriguing

problem for students of spiralian embryology is to understand how the early divisions of the embryo, including many of the peculiarities of the micromere-forming divisions are so widely conserved (Dohle, 1998) despite the great variation in the uses to which the cells and their progeny are put in various groups. The contrast between those species that form trochophore larvae and those that undergo direct development is particularly great.

By any criterion other than size, the micromeres appear to be quite a complex population of cells. In glossiphoniid leech embryos, we so far have identified four developmental roles played by various micromeres or their progeny within the "micromere cap" (Fig. 1, stages 4a–7).

1. Superficial cells in the micromere cap form the squamous epithelium of the provisional integument of the embryo. As such, they constitute the epibolizing epithelium that covers the germinal bands and the area dorsal to them during gastrulation (Fig. 1, stages 7–10; Weisblat et al., 1984; Smith and Weisblat, 1994; Smith et al., 1996).

2. The micromere-derived epithelium plays a role in regulating cell fates of blast cells within the germinal bands, including the O-P equivalence group (Ho and Weisblat, 1987; Huang and Weisblat, 1996; F. Z. Huang, unpublished observations).

3. Deep cells in the micromere cap form tissues in the unsegmented anterior of the leech (the prostomium), including the dorsal ganglion and foregut (Weisblat et al., 1984; Nardelli-Haefliger and Shankland, 1993; Smith and Weisblat, 1994).

4. Micromere derivatives also may be required for the normal assembly of the germinal bands, and for initiating germinal band coalescence during gastrulation (Isaksen, 1997).

Although 25 individual micromeres are born during cleavage, it seems unlikely that there are more than 20 different *kinds* of micromeres, because those arising from the left and right NOPQ proteloblasts (nopq′, nopq″, opq′, opq″, and n′) are clearly left–right homologs of one another. Fate mapping experiments (Nardelli-Haefliger and Shankland, 1993; Smith and Weisblat, 1994) indicate that the number of different kinds of micromeres may be reduced further to as few as 14, if we also accept c′ and d′, a′ and b′, a″ and b″, and a‴ and b‴, dnopq′ and dnopq″, and nopq′ and nopq″ as equipotent pairs of cells. Evidence for this idea is that the members of these proposed pairs contribute equivalent numbers of superficial and/or deep cells to equivalent regions of the micromere cap and that the timing and symmetry of their first divisions are similar (Table II; Smith and Weisblat, 1994; Huang *et al.*, 1998). Table II suggests other micromere groupings as well (the entire primary quartet; the secondary and tertiary

TABLE I

Alternative Designations of Micromeres

This chapter [after Bissen and Weisblat (1989)], cells listed in order of birth	Sandig and Dohle (1988)
d′	1d
c′	1c
a′	1a
b′	1b
c″	2c
dnopq′	$2d^1$
dm′	3d
a″	2a
b″	2b
dnopq″	$2d^{21}$
dm″	4D
c‴	3c
a‴	3a
b‴	3b
dnopq	$2d^{221}$
nopq′	t^I
nopq″	t^{II}
opq′	opq^I
opq″	opq^{II}
n′	n^{IV}

TABLE II

Comparison of Early Micromere Fate Maps and Lineage Patterns

| Cell name(s) | Micromere fates (distribution of progeny at end of stage 7) (from Smith and Weisblat, 1994) | | | Micromere lineage features (Haung et al., 1998) | |
	Number of superficial progeny	Number of deep progeny	Age of clone at end of stage 7 (hours)	Age at first division (hours)	Symmetry of cell division
a′, b′	8–9 (each)	11–15 (each)	54	9.5	Unequal
c′, d′	10 (each)	16–20 (each)	43–55	7.5	Unequal
a″, b″	0	5–6 (each)	52	24	Equal
a‴, b‴9	0	6–7 (each)	50	24	Equal
c″	0	7	52	24	Equal
c‴	0	5	50	24	Equal
dnopq′, dnopq″	10–13 (each)	5 (each)	51–52	16	Equal
dm′	0	11	52	16	Equal
dm″	0	15	50	18.5	Equal
dnopq‴	8	0	49	15	Unequal
nopq′$_{L/R}$ nopq″$_{L/R}$,	6 (each)	2–3 (each)	46–47	13.5	Unequal
opq′$_{L/R}$	4 (each)	8 (each)	43	??	??
opq″$_{L/R}$	17 (each)	0	42	9	Equal
n′$_{L/R}$	8 (each)	0	40	11.5	Unequal

trios; and cells dm′ and dm″). But it must be borne in mind that the deep cells may actually have different final fates among the various prostomial cell types (epidermis, neurons, muscle). The definitive fates of the deep micromere progeny in glossiphoniid leeches remain to be determined.

Even with these simplifying assumptions, the micromeres are a heterogeneous set of cells on the basis of their fates in normal development. And these fate differences seem to be maintained in experimentally perturbed embryos. For example, cells dnopq‴ and opq″ both contribute exclusively superficial cells to the micromere cap, but exhibit different rates of cell proliferation; a dnopq‴ has produced only about 8 cells by the time its clone is 49 hours old, while an opq″ clone produces about 17 cells in only 42 hours. Moreover, the rate of proliferation in the dnopq‴ clone is not affected by ablating both opq″ cells, and the epithelium contains significantly fewer epithelial cells after such ablations (Smith and Weisblat, 1994).

Further evidence of the heterogeneity of the micromeres comes from the lineage-specific differences in cell division patterns and cell cycle composition. The early lineages of the various micromere clones have been followed for clonal ages (i.e., the time after the birth of the micromere in question) ranging from 24 to 31.5 hours (Huang et al., 1998) using the combined techniques of cell lineage tracing and carefully timed exposure to an immunohistochemically detectable thymidine analog (bromodeoxyuridine) as used previously by Bissen and Weisblat (1989).

The lineage patterns seen in the primary quartet-micromeres (a′–d′) are noteworthy in that each of these

cells undergoes a series of relatively rapid (2.5–4.5 hour cell cycle) and unequal divisions to generate a column of progeny with significantly longer cell cycle times (greater than 13 hours). It appears that the first cells born from the micromeres are destined for the prostomium, while later cells contribute exclusively to the provisional epithelium (F.-A. Ramirez-Weber, personal communication). This stem cell-like pattern of divisions is not seen for the primary quartet of micromeres in either mollusks (van den Bigelaar, 1971a,b; Damen and Dictus, 1994) or polychaete annelids (Avel et al., 1959).

Among the other micromeres, it appears that each type undergoes an idiosyncratic lineage and that few generalizations can be drawn. Cell divisions may be apparently equal or obviously unequal, in which case the larger daughter cell usually divides prior to the smaller one. Cell cycle times vary from 9 to 24 hours; the most rapid divisions occur in the n′ and opq″ lineages and the slowest in the secondary and tertiary trios micromeres of (a″–c″ and a‴–c‴). No appreciable G1 phase has been detected for any of these cells; S phase varies from 15 minutes to 3 hours and G2 phase varies from 2.5 to 24 hours.

VII. Conclusions

Apart from their intrinsic appeal for a small group of aficionados, the studies of glossiphoniid leech development presented here also merits the attention of those interested in the evolution of developmental mechanisms. Even given the exciting new discoveries of fossilized embryos (Li et al., 1998; Xiao et al., 1998) it seems clear that nothing can replace the comparative analysis of diverse living species to draw inferences about how they arose by changes in the development of ancestral species. Moreover, it seems clear that these comparisons cannot be confined exclusively to those few species that are amenable to genetic analysis if a full picture of evolutionary changes in development is to be obtained (Bolker, 1995; Bolker and Raff, 1997).

A. D Quadrant Specification

The specification of the D quadrant is a central issue in spiralian development. The cellular processes by which the D quadrant of the embryo is set aside from A, B, and C quadrants vary between species. Freeman and Lundelius (1992) argue that the first two cleavages of the ancestral spiralian were equal, with the D quadrant being specified later by inductive interactions between the presumptive D cell and the micromeres. In contrast, Dohle (1998) suggests that a pattern of unequal cleavages, such as those described here for glossiphoniid leeches, is ancestral at least for the annelids. The final resolution of this question will require further studies including the construction of detailed spiralian phylogenies by means of molecular techniques that are independent of the embryonically derived characters we seek to compare.

Understanding the molecular mechanism(s) of D quadrant specification should also be informative. In that regard we are hopeful that identifying and characterizing the expression of the *nanos*-class gene in leech is a useful step. We note, however that the expression of *hro-nos* is first detected only after the first unequal cleavage is already complete, suggesting that there are important upstream factors awaiting discovery. It will be of great interest to see if the distinct mechanisms governing unequal first and second cleavages in *Tubifex* also pertain to leech.

B. Non-D Macromeres

The events leading from the A‴, B‴, and C‴macromeres at stage 4c to the definitive midgut epithelium during stage 9 comprise a catchall of cell biological processes that should certainly qualify for the "Ripley's Believe It or Not" of embryology. The well-regulated fusions between like and unlike cells over long periods of embryogenesis illustrate the use of cellular analyses, in the complex environment of the intact organism, at uncovering phenomena whose existence might not be anticipated from what we know of life in the petri dish. In addition, the description of endoderm formation in leech raises questions about the homologous processes in other spiralians, and suggests comparisons with the formation of the germinal band and gut epithelium from syncytial precursors in arthropods.

C. Micromere Lineages and Beyond

Another intriguing problem for students of spiralian embryology is to understand how the early divisions of the embryo, including many of the peculiarities of the micromere-forming divisions are so widely conserved (Dohle, 1998) despite the great variation in the uses to which the cells and their progeny are put in various groups. The contrast between those that form trochophore larvae and those that undergo direct development is particularly great. In leech, the micromeres seem to play important regulatory roles in early development, in addition to forming the epithelium of the provisional integument and a variety of definitive cell types for the prostomium. As more detailed fate maps become available for the micromeres in leech and in other spiralians, it should be possible to determine the points of homology and divergence in the lineages and fates of these ancient cell homologs.

Clearly the work we have presented is just a scratch at the surface of understanding the mechanisms for the developmental processes described. Moreover, the leech is far from being a "typical" annelid in either its embryological or its adult features. (Can such an animal be found for any taxa, without going back to the very ancestors whose mysteries we seek to fathom?) Among the annelids, there is a remarkable richness in reproductive strategies, embryological processes, and regenerative capacities. Moreover, recent molecular analyses suggest that the Echiura and Pogonophora are actually derived annelids, which would add further variety to the diversity of body plans and embryological processes within this group (McHugh, 1998). There is no shortage of questions for the enterprising embryologist!

References

Anderson, D. T. (1973). "Embryology and Phylogeny in Annelids and Arthropods." Pergamon, Oxford.

Astrow, S. H., Holton, B., and Weisblat, D. A. (1987). Centrifugation redistributes factors determining cleavage patterns in leech embryos. *Dev. Biol.* **120,** 270–283.

Astrow, S. H., Holton, B., and Weisblat, D. A. (1989). Teloplasm formation in a leech *Helobdella triserialis,* is a microtubule-dependent process. *Dev. Biol.* **135,** 306–319.

Avel, M., de Beauchamp, P., Brien, P., Dawydoff, C., Durchon, M., Fauvel, P., Grassé, P.-P., Harant, H., Prenant, M., Roger, J., and Tétry, A. (1959). "Ontogenèse des Annelides, I: Polychètes. Traité de Zoologie" (V. Tome, ed.), pp. 600–602. Masson, Paris.

Bissen, S. T., and Weisblat, D. A. (1989). The durations and compositions of cell cycles in embryos of the leech, *Helobdella triserialis. Development (Cambridge, UK)* **106,** 105–118.

Bolker, J. A. (1995). Model systems in developmental biology. *BioEssays* **17,** 451–455.

Bolker, J. A., and Raff, R. A. (1997). Beyond worms, flies and mice: It's time to widen the scope of developmental biology. *J. NIH Res.* **9**, 35–39.

Bychowsky, A. (1921). Ueber die Entwicklung der Nephridien von *Clepsine sexoculata* Bergmann. *Rev. Suisse Zool.* **29**, 41–131.

Damen, P., and Dictus, W. J. A. G. (1994). Cell lineage of the prototroch of *Patella vulgata*. *Dev. Biol.* **162**, 364–383.

Desjeux, I. (1995). An investigation into the regulation of segment number in the leech. Ph.D. Thesis, Dept. Physiol., University Medical School, Edinburgh, U.K.

Dohle, W. (1998). The ancestral cleavage pattern of clitellates and its evolutionary implications. *Hydrobiologia* (in press).

Endo, Y., and Tsurugi, K. (1988). The RNA N-glycosidase activity of ricin A-chain: The characteristics of the enzymatic activity of ricin A-chain with ribosomes and with rRNA. *J. Biol. Chem.* **263**, 8735–8739.

Fernandez, J., and Olea, N. (1982). Embryonic development of glossiphoniid leeches. *In* "Developmental Biology of Freshwater Invertebrates" F. W. (Harrison and R. R. Cowden, eds.), pp. 317–361. Alan R. Liss, New York.

Fernandez, J., Olea, N., and Matte, C. (1987). Structure and development of the egg of the glossiphoniid leech *Theromyzon rude*: Characterization of developmental stages and structure of the early uncleaved egg. *Development (Cambridge, UK)* **100**, 211–225.

Freeman, G., and Lundelius, J. W. (1992). Evolutionary implications of the mode of D quadrant specification in coelomates with spiral cleavage. *J. Evol. Biol.* **5**, 205–247.

Ho, R. K., and Weisblat, D. A. (1987). A provisional epithelium in leech embryo: Cellular origins and influence on a developmental equivalence group. *Dev. Biol.* **120**, 520–534.

Holton, B., Astrow, S. H., and Weisblat, D. A. (1989). Animal and vegetal teloplasms mix in the early embryo of the leech, *Helobdella triserialis*. *Dev. Biol.* **131**, 182–188.

Holton, B., Wedeen, C. J., Astrow, S. H., and Weisblat, D. A. (1994). Localization of polyadenylated RNAs during teloplasm formation and cleavage in leech embryos. *Roux's Arch. Dev. Biol.* **204**, 46–53.

Huang, F. Z., and Weisblat, D. A. (1996). Cell fate determination in an annelid equivalence group. *Development (Cambridge, UK)* **122**, 1839–1847.

Huang, F. Z., *et al.* (1998). In preparation.

Hyman, A. A. (1989). Centrosome movements in the early divisions of *Caenorhabditis elegans*: A cortical site determining centrosome position. *J. Cell Biol.* **109**, 1185–1193.

Isaksen, D. E. (1997). The identification of a *TGF*-β class member and the regulation of endodermal precursor cell fusion in the leech. Ph.D. Thesis, Dept. Mol. Cell. Biol., University of California, Berkeley.

Isaksen, D. E., Liu, N.-J., and Weisblat, D. A. (1998). Inductive interactions regulate endodermal precursor cell fusion in the leech. Submitted.

Ishii, R., and Shimizu, T. (1997). Equalization of unequal first cleavage in the *Tubifex* egg by introduction of an additional centrosome: Implications for the absence of cor-

tical mechanisms for mitotic spindle asymmetry. *Dev. Biol.* **189**, 49–56.

Li, C.-W., Chen, J.-Y., and Hua, T.-E. (1998). Precambrian sponges with cellular structures. *Science* **279**, 879–882.

Liu, N.-J., Isaksen, D. E., Smith, C. M., and Weisblat, D. A. (1998). Movements and stepwise fusion of endodermal precursor cells in leech. *Dev. Genes Evol.* **208**, 117–127.

McHugh, D. (1998). Phylogenetic hypothesis for annelid relationships based on *elongation factor-1-alpha*, a nuclear coding gene. *Hydrobiologia* (in press).

Nardelli-Haefliger, D., and Shankland, M. (1993). *Lox 10*, a member of the *NK-2* homeobox gene class, is expressed in a segmental pattern in the endoderm and in the cephalic nervous system of the leech *Helobdella*. *Development (Cambridge, UK)* **118**, 877–892.

Nelson, B. H., and Weisblat, D. A. (1991). Conversion of ectoderm to mesoderm by cytoplasmic extrusion in leech embryos. *Science* **253**, 435–438.

Nelson, B. H., and Weisblat, D. A. (1992). Cytoplasmic and cortical determinants interact to specify ectoderm and mesoderm in the leech embryo. *Development (Cambridge, UK)* **115**, 103–115.

Pilon, M., and Weisblat, D. A. (1997). A *nanos* homolog in leech. *Development (Cambridge, UK)* **124**, 1771–1780.

Sandig, M., and Dohle, W. (1988). The cleavage pattern in the leech *Theromyzon tessulatum* (Hirudinea, Glossiphoniidae). *J. Morphol.* **196**, 217–252.

Savage, R. M., and Shankland, M. (1996). Identification and characterization of a *hunchback* orthologue, *Lzf2*, and its expression during leech embryogenesis. *Dev. Biol.* **175**, 205–217.

Shankland, M. (1984). Positional determination of supernumerary blast cell death in the leech embryo. *Nature (London)* **307**, 541–543.

Schmidt, G. A. (1939). Dégénérescence phylogénétique des modes de développement des organes. *Arch. Zool. Exp. Gen.* **81**, 317–370.

Shimizu, T. (1982). Ooplasmic segregation in the *Tubifex* egg: Mode of pole plasm accumulation and possible involvement of microfilaments. *Roux's Arch.* **191**, 246–256.

Shimizu, T. (1984). Dynamics of the actin microfilament system in *Tubifex* during ooplasmic segregation. *Dev. Biol.* **106**, 414–426.

Shimizu, T. (1986). Bipolar segregation of mitochondria, actin network and surface in the *Tubifex* embryo: Role of cortical polarity. *Dev. Biol.* **116**, 241–251.

Shimizu, T. (1996a). The first two cleavages in *Tubifex* involve distinct mechanisms to generate asymmetry in mitotic apparatus. *Hydrobiologia* **334**, 269–276.

Shimizu, T. (1996b). Behaviour of centrosomes in early *Tubifex* embryos: Asymmetric segregation and mitotic cycle-dependent duplication. *Roux's Arch. Dev. Biol.* **205**, 290–299.

Smith, C. M., and Weisblat, D. A. (1994). Micromere fate maps in leech embryos: Lineage-specific differences in rates of cell proliferation. *Development (Cambridge, UK)* **120**, 3427–3438.

Smith, C. M., Lans, D., and Weisblat, D. A. (1996). Cellular mechanisms of epiboly in leech embryos. *Development (Cambridge, UK)* **122**, 1885–1894.

Stent, G. S., Kristan, W. B., Jr., Torrence, S. A., French, K. A., and Weisblat, D. A. (1992). Development of the leech nervous system. *Int. J. Neurobiol.* **33,** 109–193.

Tautz, D. (1988). Regulation of the *Drosophila* segmentation gene *hunchback* by two maternal morphogenetic centres. *Nature (London)* **332,** 281–284.

Tautz, D., and Pfeifle, C. (1989). A non-radioactive *in situ* hybridization method for the localization of specific RNAs in *Drosophila* embryos reveals translational control of the segmentation gene *hunchback*. *Chromosoma* **98,** 81–85.

van den Bigelaar, J. A. M. (1971a). Timing of the phases of the cell cycle with tritiated thymidine and Feulgen cytophotometry during the period of synchronous division up to 49-cell stage in *Lymnea. J. Embryol. Exp. Morphol.* **26,** 351–366.

van den Bigelaar, J. A. M. (1971b). Timing of the phases of the cell cycle during the period of asynchronous divisions up to 49-cell stage in *Lymnea. J. Embryol Exp. Morphol.* **26,** 367–371.

Wedeen, C. J., and Shankland, M. (1997). Mesoderm is required for the formation of a segmented endodermal cell layer in the leech *Helobdella. Dev. Biol.* **191,** 202–214.

Weisblat, D. A., Kim, S. Y., and Stent, G. S. (1984). Embryonic origins of cells in the leech *Helobdella triserialis. Dev. Biol.* **104,** 65–85.

Whitman, C. O. (1878). The embryology of Clepsine. *Q. J. Microsc. Sci.* **18,** 215–315.

Wysocka-Diller, J., Aisemberg, G. O., and Macagno, E. R. (1995). A novel homeobox cluster expressed in repeated structures of the midgut. *Dev. Biol.* **171,** 439–447.

Xiao, S., Zhang, Y., and Knoll, A. H. (1998). Three-dimensional preservation of algae and animal embryos in a Neoproterozoic phosphorite. *Nature (London)* **391,** 553–558.

13

Spatial and Temporal Control of Cell Division during Leech Development

SHIRLEY T. BISSEN
Department of Biology
University of Missouri, St. Louis
St. Louis, Missouri 63121

I. Interactions between Cell Division and Cell Fate

Glossiphoniid leech embryos, like many other embryos, undergo complex patterns of highly regulated cell divisions. Cell division not only produces the large number of cells needed to construct a complex organism, but spatially and temporally regulated cell divisions also contribute to the generation of the many different types of cells needed to give that organism form and function. Proper orientation of the plane of cleavage is necessary for the differential segregation of localized determinants to specific daughter cells and for the optimal positioning of cells in the embryo. The timing of division must be coordinated with other cellular processes, such as cell movements, cell shape changes, and macromolecular syntheses.

This chapter describes the intricate patterns of cell division in *Helobdella triserialis* embryos and details advances in our understanding of the mechanisms that regulate these divisions. First, the requirement for zygotically transcribed gene products to direct the plane of cleavage in a subset of early blastomeres is discussed. Second, the lineage-specific and cell type-specific mechanisms that regulate the timing of cell division are examined.

II. Spatial Regulation of Cell Division

A. Prominent Role of Unequal Cleavage during *Helobdella* Development

Unequal cell cleavages are common during glossiphoniid leech development. In fact, the inequality and asynchrony of cleavage allow individual cells to be identified and named. The first two cleavages are slightly unequal; the four-cell embryo contains cells A, B, and C, which are about equal in size, and cell D, which is larger. These cells exhibit different patterns of cleavage and have different developmental fates. Cells A, B, and C undergo three highly unequal cleavages, each cleavage producing a micromere and a macromere. The resultant macromeres of the A, B, and C lineages eventually fuse and give rise to the endodermal layer of the gut wall (Whitman, 1887; Nardelli-Haefliger and Shankland, 1993).

In contrast, cell D undergoes a complex series of equal and unequal cleavages (see Fig. 1 and Section II,B) to produce 16 additional micromeres, as well as five bilateral pairs of stem cells—the M, N, O, P, and Q teloblasts. The teloblasts are the progenitors of the segmentally repeated mesoderm and ectoderm of the body.

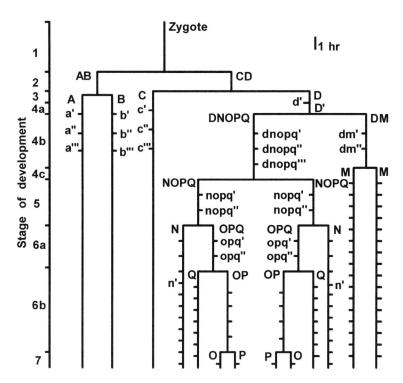

Figure 1 Cleavage pattern in early *Helobdella* embryos, showing the generation of 25 micromeres and 10 teloblasts. The horizontal lines that connect two sister cells, with uppercase names, designate equal or nearly equal cleavages. The short horizontal lines denote highly unequal cleavages, in which micromeres or primary blast cells are produced. The names of micromeres are given [lowercase followed by one or more prime (′) symbols], but not those of the primary blast cells. From Bissen and Smith (1996), courtesy of the Company of Biologists, Ltd. See Chapter 11 by Stent for spatial arrangement of cells.

Each teloblast undergoes several dozen highly unequal divisions in the stem cell mode, producing a longitudinal chain or bandlet of segmental founder cells, the primary blast cells. Each class of primary blast cells undergoes an invariant series of temporally and spatially patterned cell divisions (Zackson, 1984; Shankland, 1987a,b; Bissen and Weisblat, 1989), and ultimately contributes a distinct set of segmentally iterated progeny (Weisblat and Shankland, 1985; Bissen and Weisblat, 1987; Ramírez *et al.*, 1995).

Twenty-five micromeres are generated during these early cleavages. Some micromeres contribute to the nonsegmented cephalic structures, including the supraesophageal ganglion (Weisblat *et al.*, 1984; Nardelli-Haefliger and Shankland, 1993). Others give rise to the epithelial layer of the provisional integument, a temporary body covering that is present during gastrulation (Weisblat *et al.*, 1984; Ho and Weisblat, 1987; Smith and Weisblat, 1994).

B. Pattern of Cleavages of the D Lineage

The unique developmental fate of cell D depends on its selective inheritance of domains of yolk-deficient cytoplasm, or teloplasm. Prior to first cleavage, pools of teloplasm form at the animal and vegetal poles via cytoplas-

mic rearrangements (Fernández *et al.*, 1987; Astrow *et al.*, 1989). The rest of the cytoplasm is filled with yolk particles. Teloplasm is enriched in mitochondria, endoplasmic reticulum, ribosomes (Fernández and Stent, 1980), and polyadenylated RNA (Holton *et al.*, 1994). The differential partitioning of teloplasm to cell D during the first two cleavages directs that cell to undergo the pattern of cleavages leading to the formation of teloblasts. Teloplasm can be redistributed by centrifugation or physical reorientation of the plane of first cleavage, and any blastomere that receives a substantial amount of teloplasm undergoes the pattern of teloblast-producing cleavages normally associated with cell D (Astrow *et al.*, 1987; Nelson and Weisblat, 1992).

On formation of cell D, the vegetal pool of teloplasm migrates to and mixes with the animal pool of teloplasm (Holton *et al.*, 1989). Since the nucleus lies within the domain of teloplasm, the mitotic spindle forms off center. Prior to the highly unequal cleavage of cell D, the peripherally located mitotic spindle aligns with the radial axis of the cell, with one spindle pole aster abutting the cell cortex (Fig. 2A). Consequently, the cleavage furrow forms off center and divides the parental cell such that micromere d′ receives only teloplasm and macromere D′ receives both teloplasm and yolky cytoplasm (Fig. 2B). Macromere D′ then cleaves

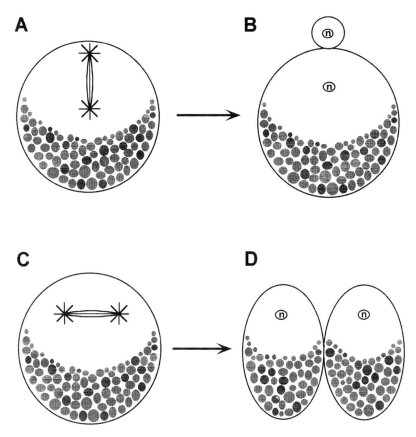

Figure 2 Orientation of mitotic spindle and equality of cleavage. Each large circle or oval represents a cell, which contains cytoplasm rich in yolk particles (small gray circles) and cytoplasm devoid of yolk particles. The mitotic apparatus or nucleus (n) lies off center, within the domain of yolk-deficient cytoplasm. Alignment of the mitotic spindle along the radial axis of a cell (A) produces two cells that differ in size and cytoplasmic composition (B). Orientation of the mitotic spindle perpendicular to the radial axis of a cell (C) yields two cells that are equivalent in size and composition (D).

equally into the ectodermal precursor, cell DNOPQ, and the mesodermal precursor, cell DM. Prior to this cleavage, the eccentrically located mitotic spindle orients perpendicular to the radial axis of cell D′ and the cleavage furrow divides the teloplasm and yolky cytoplasm nearly equally between the two new cells (Fig. 2C,D).

In each of the subsequently formed yolky blastomeres of the D lineage, the teloplasm lies at one side of the cell and the eccentric mitotic spindle orients, in a predictable pattern, either parallel or perpendicular to the radial axis of the cell. In general, each yolky D-derived blastomere undergoes two or three highly unequal cleavages, followed by an equal (or nearly equal) cleavage (Fig. 1). Highly unequal cleavages differentially segregate teloplasm to the micromeres, and equal cleavages distribute teloplasm and yolky cytoplasm to both cells (Fig. 2). On formation of the teloblasts, the pattern of cleavage changes. Teloblasts divide unequally and repeatedly along the same axis. Each teloblast "buds off" several dozen teloplasm-rich primary blast cells that form a coherent bandlet.

C. Specific Unequal Cleavages of D Lineage Require Zygotic Gene Products

The earliest cleavages of *Helobdella* embryos proceed normally in the presence of low concentrations of a transcriptional inhibitor, either α-amanitin or actinomycin D (Bissen and Weisblat, 1991). The unequal cleavages of the A, B, and C lineages occur, as do the first several unequal cleavages of the D lineage (i.e., cells d,′, dnopq′, dnopq″, dnopq‴, dm′, and dm″ are produced). Around the start of stage 5 (formation of the two NOPQ cells), however, the pattern of cleavage is perturbed. Inhibition of transcription converts the unequal, micromere-forming cleavages of the D lineage to equal cleavages (Bissen and Smith, 1996).

In experimental embryos treated with a transcriptional inhibitor, the mitotic spindles of yolky D-derived cells do not align radially, but rather orient perpendicularly. As a consequence, these cells always divide equally, with teloplasm and yolky cytoplasm being distributed to both progeny cells. The resultant embryos, which cease normal development, contain fewer micromere-sized cells and extra yolky D-derived blastomeres (Bissen and Smith, 1996). This requirement for zygotic transcription is specific for a subset of unequal cleavages that occur during the period of teloblast formation. The repeated unequal divisions of the teloblasts, like the earlier unequal cleavages, proceed unhampered in the presence of a transcriptional inhibitor (Bissen and Weisblat, 1991).

D. Spindle Orientation in Unequally Dividing Cells

1. Mechanisms of Spindle Reorientation in Other Organisms

a. Centrosome Movements. The plane of cleavage depends on the position and orientation of the mitotic spindle, since the cleavage furrow bisects the mitotic apparatus. The orientation of the mitotic spindle, in turn, depends on the location of the centrosomes. At the beginning of mitosis, the duplicated centrosomes separate and migrate to opposite sides of the nucleus, where they form the poles of the mitotic spindle. In many proliferating cells, the axis of migration of the centrosomes shifts 90° relative to the previous axis of migration. The mitotic spindle of a cell, therefore, is oriented perpendicular to the axis of the spindle of its parental cell. This predictable pattern of centrosome migration leads to a default pattern of division in which the plane of cleavage occurs at right angles to that of the preceding division. Unequally dividing cells frequently deviate from this pattern, however.

The first cleavage of *Caenorhabditis elegans* embryos is unequal, producing an anterior somatic blastomere and a smaller germline blastomere, which inherits specific cytoplasmic granules and proteins (Strome and Wood, 1983; Bowerman *et al.*, 1993; see Chapter 7 by Bowerman). The posterior germline precursor cell undergoes several more unequal, stem cell-like divisions on the same axis. The centrosomes of the germline blastomere undergo their normal migrations, but the centrosome–nucleus complex undergoes an additional 90° rotation to position the mitotic spindle along the original axis (Hyman and White, 1987).

The neuroblasts of *Drosophila* embryos also divide successively along the same axis, budding off small ganglion mother cells. Specific cell fate determinants (i.e., Numb and Prospero proteins and *prospero* RNA) segregate differentially to the ganglion mother cells (Hirata *et al.*, 1995; Knoblich *et al.*, 1995; Spana *et al.*, 1995; Li *et al.*, 1997; see Chapter 18 by Broadus and Spana). Again, the spindles of the neuroblasts orient repeatedly along the same axis. In this case, however, one centrosome remains in the original position and the other migrates 180° to the opposite side of the nucleus (Spana and Doe, 1995).

b. Physical Alignment. Interactions between the astral microtubules of the spindle and the microfilaments of the cell cortex are critical for spindle alignment. Not only does one pole of the spindle abut the cortex of an unequally dividing cell, but the aster establishes a strong physical linkage with a specific region of the cortex (Conklin, 1917; Kawamura, 1977; Dan, 1979; Dan and Ito, 1984; Dan and Inoue, 1987; Lutz *et al.*, 1988).

In the germline blastomeres of *C. elegans*, laser beam irradiation of the region between the proposed site in the cortex and the leading edge of the rotating mitotic apparatus disrupts spindle reorientation, whereas irradiation of other sites in the cell has no effect (Hyman, 1989). This rotation is sensitive to both microtubule and microfilament inhibitors (Hyman and White, 1987). Furthermore, actin and an actin-capping protein accumulate transiently at a specific cortical site prior to rotation (Waddle *et al.*, 1994).

In *Drosophila* neuroblasts, proper spindle orientation, as well as normal localization of cell fate determinants, requires the protein Inscuteable. In the absence of Inscuteable, the spindle assumes a nearly random orientation, rather than aligning along the previous axis. Inscuteable protein localizes transiently to a specific region of the cortex of neuroblasts in a microfilament dependent manner (Kraut *et al.*, 1996).

2. Possible Roles of Leech Zygotic Gene Products in Spindle Reorientation

The early cleavages of leech embryos follow the default mode in which the plane of cleavage is shifted 90° relative to that of the previous cleavage. Around the start of stage 5, however, some cleavage undergo 45° shifts in orientation (Sandig and Dohle, 1988). The cells in which these shifts occur are the same cells whose spindles do not orient properly after inhibition of zygotic transcription (Bissen and Smith, 1996). Thus, it appears that these 45° shifts in spindle orientation depend on newly transcribed gene products.

It is possible that the requisite gene products function in the yolky D-derived cells to mediate some aspect of the interactions among the cytoskeleton, the centrosomes, or the cell cortex for these special spindle reorientations. Alternatively, these zygotic gene products could function extrinsic to the yolky D-derived cells; for example, cell–cell contacts or signaling mechanisms could help mediate these spindle reorientations (Goldstein, 1995). It is interesting to note that the unequal cleavages of the teloblasts, in which the spindle orients repeatedly along the same axis, do not require newly transcribed gene products, suggesting that their mechanism of spindle reorientation differs from that of these early cells.

III. Temporal Regulation of Cell Division

A. General Characteristics of Embryonic Cell Cycles

In many embryos, the timing of cell division changes during the course of development. The early cleavages, which subdivide the fertilized egg into many cells, are relatively rapid and are regulated by maternal gene products that are stockpiled in the egg. As development proceeds, the cell division cycles lengthen and require zygotic gene products, as cell division becomes coordinated with other developmental events.

The cell division cycles of embryonic cells are generally simpler in composition that those of adult cells. Some embryonic cycles, such as those of early *Drosophila* and *Xenopus* embryos, are extremely simple, comprising back-to-back phases of DNA synthesis (S) and mitosis (M) (Foe and Alberts, 1983; Graham and Morgan, 1966). Most other embryonic cycles have postreplicative gap (G2) phases between the S and M phases (van den Biggelaar, 1971; Edgar *et al.*, 1986; Bissen and Weisblat, 1989; Azzaria and McGhee, 1992). At some point during development, the cycles acquire prereplicative gap (G1) phases (Bissen and Weisblat, 1989; Edgar and O'Farrell, 1990; Frederick and Andrews, 1994) and begin to resemble adult cell cycles, comprising phases of G1, S, G2, and M.

B. Lineage and Cell Specific Differences in Cell Cycle Timing in *Helobdella*

The early cleavages of *Helobdella* embryos are relatively rapid. The cell cycles of the D-derived blastomeres average about 90 minutes whereas those of the A-, B-, and C-derived blastomeres average about 135 minutes (see Fig. 1). While all these cycles have S and M phases of similar lengths, the cycles of the D-derived cells have much shorter G2 phases (Bissen and Weisblat, 1989). Although newly transcribed gene products regulate the equality of cleavage in some of these early cells (see Section II, C), maternal gene products regulate the *timing* of these early cleavages (Bissen and Smith, 1996). Similarly, the iterative 60-minute cycles of the teloblasts have even shorter G2 phases and are under maternal control (Bissen and Weisblat, 1989, 1991).

The rate of cell division slows considerably among the primary blast cells. The cycles of the mesodermal primary blast cells last 10 hours and those of the different classes of ectodermal primary blast cells range from 21 to 33 hours (Zackson, 1984). This increase in cell cycle timing is mainly due to the lengthening of G2 phases (Bissen and Weisblat, 1989). The progeny of the primary blast cells have cell cycles that range from 4.5 to >20 hours (Zackson, 1984; Shankland, 1987a,b; Bissen and Weisblat, 1989). Although some of these cycles acquire G1 phases, differences in cell cycle timing are due to differences in the lengths of G1, S, and G2 phases; only M phase remains constant (Bissen and Weisblat, 1989). Zygotic gene products direct the divisions of the primary blast cells and their progeny; cell division completely stops in these cells in the absence of transcription (Bissen and Weisblat, 1991).

Although different micromeres have different rates of proliferation (Smith and Weisblat, 1994), little is known about the timing or composition of individual micromere cycles. The micromeres are similar to blast cells, however, in that their divisions require zygotic gene products (Bissen and Weisblat, 1991).

C. Molecular Control of Cell Cycle Timing

A family of protein kinase complexes regulates cell cycle progression in all higher eukaryotic cells (reviewed by Morgan, 1995; Nigg, 1995; Elledge, 1996; King *et al.*, 1996). Each complex comprises minimally a kinase catalytic subunit, called cdk for cyclin-dependent kinase, and a cyclin regulatory subunit. Cyclins are synthesized throughout the cell cycle and associate with catalytic subunits to form potentially active complexes. Both stimulatory and inhibitory phosphorylations regulate the activity of these complexes. In some cases, the binding of small regulatory proteins also inhibits activity. Degradation of the associated cyclin subunit inactivates the kinase to promote advancement through the cycle.

Passage through the cell cycle depends on successive activation and inactivation of specific cyclin/cdk complexes. For example, Cyclin D/Cdk4 and Cyclin D/Cdk6 function during G1 phase, Cyclin E/Cdk2 guides the G1 to S transition, and Cyclin A/Cdk2 functions during S phase (Fig. 3). The transition from G2 to M phase requires the sequential activation (and inactivation) of Cyclin A/Cdk1 and Cyclin B/Cdk1. Since pro-

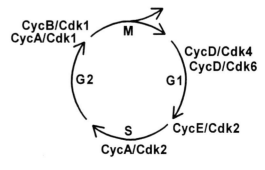

Figure 3 A model of cell cycle regulation. Shown are some of the cyclin-dependent kinase complexes (Cyc/Cdk) whose sequential activation and inactivation guide progression through the cell cycle.

gression through the cycle could be regulated at any of these transitions and any component of the conserved cell cycle control machinery could be limiting, many regulatory mechanisms are possible.

D. Strategies of Cell Cycle Control in Embryonic Cells

The major point of control during most embryonic cell cycles is the initiation of mitosis, or the activation of Cyclin B/Cdk1. Many of the conserved cell cycle regulators, and their regulatory kinases and phosphatases, are maternally supplied and present in excess in most early embryos. Some embryonic cycles, for example, the early abbreviated cycles of *Drosophila* and *Xenopus* embryos, are driven by the accumulation and cyclical destruction of Cyclin B protein (Murray and Kirschner, 1989; Edgar *et al.,* 1994a; Hartley *et al.,* 1996). The levels of Cyclin B rise during interphase due to the continuous translation of maternal transcripts. The accumulation of Cyclin B leads to formation of active Cyclin B/Cdk1 complexes, which are phosphorylated at stimulatory sites but not at inhibitory sites (Fig. 4A). Degradation of Cyclin B by ubiquitin-dependent proteases leads to kinase inactivation and exit from mitosis (Murray *et al.,* 1989; Glotzer *et al.,* 1991).

Cell cycle regulation in other embryonic cells is more complex, in that accumulation of Cyclin B is essential but not sufficient for kinase activity. Cyclin B/Cdk1 complexes are kept inactive by phosphorylation at inhibitory sites on the Cdk1 subunit (Fig. 4B). The phosphatase Cdc25 removes these inhibitory phosphates, leading to kinase activation (Kumagai and Dunphy, 1992; Izumi *et al.,* 1992). At the time of cellularization in *Drosophila* embryos, the cell cycles lengthen

and become asynchronous via the acquisition of G2 phases of different lengths (Foe and Alberts, 1983; Edgar *et al.,* 1986). At this time, zygotically transcribed gene products promote degradation of maternal *cdc25* transcripts and proteins (Edgar and Datar, 1996), leading to inhibitory phosphorylation of Cdk1 subunits and G2 arrest (Edgar *et al.,* 1994a). Bursts of transcription of *cdc25^{string}* during late G2 leads to activation of Cyclin B/Cdk1 and initiation of mitosis in these cells (Edgar and O'Farrell, 1989, 1990). The complex spatiotemporal pattern of *cdc25^{string}* transcription is regulated by products of patterning genes, which bind directly to the promoter of *cdc25^{string}* (Edgar *et al.,* 1994b).

E. Cell Type-Specific Strategies of Control in *Helobdella* Cells

To explore the mechanisms of cell cycle regulation among the different classes of leech cells, we have isolated *Helobdella* homologs of some of the conserved cell cycle control genes. To date, we have examined the patterns of localization and zygotic synthesis of *cdc25, cyclin A,* and *cyclin B3* mRNAs in identified cells of *Helobdella* embryos using whole-mount *in situ* hybridization (Bissen, 1995; Chen and Bissen, 1997; Anderson, 1997).

Maternal *cdc25, cyclin A,* and *cyclin B3* mRNAs are present in the early blastomeres, macromeres, and teloblasts. Treatment of one-cell embryos with a transcriptional inhibitor does not reduce the levels of these mRNAs. These transcripts localize to the yolk-free regions of the cells, that is, the teloplasm of D-derived cells or the cytoplasm surrounding the nuclei of A-, B-, and C-derived cells (Bissen, 1995; Chen and Bissen, 1997; Anderson, 1997).

A

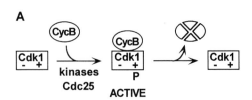

B

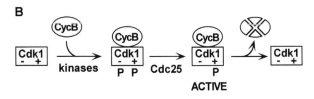

Figure 4 Two different mechanisms of Cyclin B/Cdk1 activation. (A) Accumulation of newly synthesized Cyclin B leads to formation of active Cyclin B/Cdk1 complexes, which are phosphorylated at stimulatory (+) sites. Degradation of Cyclin B inactivates the kinase complex. (B) Newly associated Cyclin B/Cdk1 complexes remain inactive, due to phosphorylation at inhibitory (−) sites. The phosphatase Cdc25 removes the inhibitory phosphates, leading to activation of Cyclin B/Cdk1 kinase.

Although the mechanism of cell cycle control has not been confirmed in these early cells, preliminary data reveal that microinjection of *in vitro* transcribed leech *cdc25* mRNA does not alter the timing of their divisions (S. T. Bissen, 1998, unpublished). Our working hypothesis is that these early cleavages are timed by the accumulation of Cyclin B, translated from maternal messages (see Fig. 4A). The different rates of cell division between the D-derived cells and the A-, B-, and C-derived cells, therefore, would be due to differences in the protein synthetic capability of these cells. The D-derived cells, which contain the ribosome-enriched teloplasm, would be able to synthesize threshold levels of Cyclin B faster than the A-, B-, and C-derived cells. Consequently, the D-derived cells have shorter cell cycles.

The primary blast cells and their progeny contain *cdc25* and *cyclin A* mRNAs, the majority of which are of zygotic origin. The levels of these transcripts are similar in blast cells of all ages (Bissen, 1995; Chen and Bissen, 1997). These data indicate that leech blast cells transcribe *cdc25* constitutively throughout their extended cell cycles, in marked contrast to the postblastoderm cells of *Drosophila*, which transcribe *cdc25* only during late G2 (Edgar and O'Farrell, 1989). Despite this difference in transcriptional regulation, Cdc25 may regulate the timing of division in the blast cells of leech embryos (see Fig. 4B). Introduction of *in vitro* transcribed *cdc25* mRNA accelerates the timing of cell division in the blast cells (S. T. Bissen, 1998, unpublished). Furthermore, our preliminary immunolocalization studies reveal that the levels of Cdc25 protein remain constant throughout the cell cycles of the blast cells, suggesting that the activity of Cdc25 is regulated posttranslationally, for example, via phosphorylation.

Different types of cells in *Helobdella* embryos, therefore, exhibit different strategies of cell cycle control. The relatively rapid cycles of the early blastomeres, macromeres, and teloblasts are driven by maternal regulators (e.g., Cyclin B). The cell cycles of the primary blast cells lengthen and arrest in G2, presumably due to depletion of some maternal regulator. Cell division in the blast cells depends on newly transcribed gene products, of which Cdc25 may be the critical regulator. Since some of these different types of cells coexist in leech embryos and divide concurrently, for example, the first-produced blast cells begin to divide while their parental teloblast is still producing blast cells, these different strategies of cell cycle control operate in parallel. Hence, the timing of cell division is regulated by lineage-specific or cell type-specific mechanisms in *Helobdella* embryos, and not stage-specific mechanisms as in *Drosophila* and *Xenopus* embryos (Edgar *et al.*, 1994a; Hartley *et al.*, 1996).

References

Anderson, J. (1997). Isolation and characterization of a leech *cyclin B3* gene. Master's Thesis, Department of Biology, University of Missouri, St. Louis.

Astrow, S., Holton, B., and Weisblat, D. (1987). Centrifugation redistributes factors determining cleavage patterns in leech embryos. *Dev. Biol.* **120,** 270–283.

Astrow, S. H., Holton, B., and Weisblat, D. A. (1989). Teloplasm formation in a leech, *Helobdella triserialis*, is a microtubule-dependent process. *Dev. Biol.* **135,** 306–319.

Azzaria, M., and McGhee, J. D. (1992). DNA synthesis in the early embryo of the nematode *Ascaris suum. Dev. Biol.* **152,** 89–93.

Bissen, S. T. (1995). Expression of the cell cycle control gene, *cdc25*, is constitutive in the segmental founder cells but is cell cycle-regulated in the micromeres of leech embryos. *Development (Cambridge, UK)* **121,** 3035–3043.

Bissen, S. T., and Smith, C. M. (1996). Unequal cleavage in leech embryos: Zygotic transcription is required for correct spindle orientation in a subset of early blastomeres. *Development (Cambridge, UK)* **122,** 599–606.

Bissen, S. T., and Weisblat, D. A. (1987). Early differences between alternate n blast cells in leech embryos. *J. Neurobiol.* **18,** 251–269.

Bissen, S. T., and Weisblat, D. A. (1989). The durations and compositions of cell cycles in embryos of the leech, *Helobdella triserialis. Development (Cambridge, UK)* **106,** 105–118.

Bissen, S. T., and Weisblat, D. A. (1991). Transcription in leech: mRNA synthesis is required for early cleavages in *Helobdella* embryos. *Dev. Biol.* **146,** 12–23.

Bowerman, B., Draper, B. W., Mello, C. C., and Priess, J. R. (1993). The maternal gene *skn-1* encodes a protein that is distributed unequally in early *C. elegans* embryos. *Cell (Cambridge, Mass.)* **74,** 443–452.

Chen, Y., and Bissen, S. T. (1997). Regulation of *cyclin A* mRNA in leech embryonic stem cells. *Dev. Genes Evol.* **206,** 407–415.

Conklin, E. G. (1917). Effects of centrifugal force on the structure and development of the eggs of *Crepidula. J. Exp. Zool.* **22,** 311–419.

Dan, K. (1979). Studies on unequal cleavage in sea urchins. I. Migration of the nuclei to the vegetal pole. *Dev. Growth Differ.* **21,** 527–535.

Dan, K., and Inoue, S. (1987). Studies on unequal cleavage in molluscs: II. Asymmetric nature of the two asters. *Int. J. Invertebr. Reprod. Dev.* **11,** 335–354.

Dan, K., and Ito, S. (1984). Studies on unequal cleavage in molluscs: I. Nuclear behavior and anchorage of a spindle pole to cortex as revealed by isolation technique. *Dev. Growth Differ.* **26,** 249–262.

Edgar, B. A., and Datar, S. A. (1996). Zygotic degradation of two maternal Cdc25 mRNAs terminates *Drosophila's* early cell cycle program. *Genes Dev.* **10,** 1966–1977.

Edgar, B. A., and O'Farrell, P. H. (1989). Genetic control of cell division patterns in the *Drosophila* embryo. *Cell (Cambridge, Mass.)* **57,** 177–187.

Edgar, B. A., and O'Farrell, P. H. (1990). The three postblastoderm cell cycles of *Drosophila* embryogenesis are regulated in G2 by *string*. *Cell (Cambridge, Mass.)* **62**, 469–480.

Edgar, B. A., Kiehle, C. P., and Schubiger, G. (1986). Cell cycle control by the nucleo-cytoplasmic ratio in early *Drosophila* development. *Cell (Cambridge, Mass.)* **44**, 365–372.

Edgar, B. A., Sprenger, F., Duronio, R. J., Leopold, P., and O'Farrell, P. (1994a). Distinct molecular mechanisms regulate cell cycle timing at successive stages of *Drosophila* embryogenesis. *Genes Dev.* **8**, 440–452.

Edgar, B. A., Lehman, D. A., and O'Farrell, P. H. (1994b). Transcriptional regulation of *string* (*cdc25*): A link between developmental programming and the cell cycle. *Development (Cambridge, UK)* **120**, 3131–3143.

Elledge, S. J. (1996). Cell cycle checkpoints: Preventing an identity crisis. *Science* **274**, 1664–1672.

Fernández, J., and Stent, G. S. (1980). Embryonic development of the glossiphoniid leech *Theromyzon rude*: Structure and development of the germinal bands. *Dev. Biol.* **78**, 407–434.

Fernández, J., Olea, N., and Matte, C. (1987). Structure and development of the egg of the glossiphoniid leech *Theromyzon rude*: Characterization of developmental stages and structure of the early uncleaved egg. *Development (Cambridge, UK)* **100**, 211–225.

Foe, V. E., and Alberts, B. M. (1983). Studies of nuclear and cytoplasmic behavior during the five mitotic cycles that precede gastrulation in *Drosophila* embryogenesis. *J. Cell Sci.* **61**, 31–70.

Frederick, D. L., and Andrews, M. T. (1994). Cell cycle remodeling requires cell–cell interactions in developing *Xenopus* embryos. *J. Exp. Zool.* **270**, 410–416.

Glotzer, M., Murray, A. W., and Kirschner, M. W. (1991). Cyclin is degraded by the ubiquitin pathway. *Nature (London)* **349**, 132–138.

Goldstein, B. (1995). Cell contacts orient some cell division axes in the *Caenorhabditis elegans* embryo. *J. Cell Biol.* **129**, 1071–1080.

Graham, C. F., and Morgan, R. W. (1966). Changes in the cell cycle during early amphibian development. *Dev. Biol.* **14**, 439–460.

Hartley, R. S., Rempel, R. E., and Maller, J. L. (1996). *In vivo* regulation of the early embryonic cell cycly in *Xenopus*. *Dev Biol.* **173**, 408–419.

Hirata, J., Nakagoshi, H., Nabeshima, Y., and Matsuzaki, F. (1995). Asymmetric segration of the homeodomain protein Prospero during *Drosophila* development. *Nature (London)* **377**, 627–630.

Ho, R. K., and Weisblat, D. A. (1987). A provisional epithelium in leech embryo: Cellular origins and influence on a developmental equivalence group. *Dev. Biol.* **120**, 520–534.

Holton, B., Astrow, S. H., and Weisblat, D. A. (1989). Animal and vegetal teloplasms mix in the early embryo of the leech, *Helobdella triserialis*. *Dev. Biol.* **131**, 182–188.

Holton, B., Wedeen, C. J., Astrow, S. H., and Weisblat, D. A. (1994). Localization of polyadenylated RNAs during teloplasm formation and cleavage in leech embryos. *Roux's Arch. Dev. Biol.* **204**, 46–53.

Hyman, A. A. (1989). Centrosome movement in the early divisions of *Caenorhabditis elegans*: A cortical site determining centrosome position. *J. Cell Biol.* **109**, 1185–1193.

Hyman, A. A., and White, J. G. (1987). Determination of cell division axes in the early embryogenesis of *Caernorhabditis elegens*. *J. Cell Biol.* **105**, 2123–2135.

Izumi, T., Walker, D. H., and Maller, J. L. (1992). Periodic changes in phosphorylation of the *Xenopus* cdc25 phosphatase regulate its activity. *Mol. Biol. Cell* **3**, 927–939.

Kawamura, K. (1977). Microdissection studies on the dividing neuroblast of the grasshopper, with special reference to the mechanism of unequal cytokinesis. *Exp. Cell Res.* **106**, 127–137.

King, R. W., Desiales, R. J., Peters, J. M., and Kirschner, M. W. (1996). How proteolysis drives the cell cycle. *Science* **274**, 1652–1659.

Knoblich, J. A., Jan, L. Y., and Jan, Y. N. (1995). Asymmetric segregation of Numb and Prospero during cell division. *Nature (London)* **377**, 624–627.

Kraut, R., Chia, W., Jan, L. Y., Jan, Y. N., and Knoblich, J. A. (1996). Role of *inscuteable* in orienting asymmetric cell divisions in *Drosophila*. *Nature (London)* **383**, 50–55.

Kumagai, A., and Dunphy, W. G. (1992). Regulation of the cdc25 protein during the cell cycle in *Xenopus* extracts. *Cell (Cambridge, Mass.)* **70**, 139–151.

Li, P., Yang, X., Wasser, M., Cai, Y., and Chia, W. (1997). Inscuteable and Staufen mediate asymmetric localization and segregation of *prospero* RNA during *Drosophila* neuroblast cell divisions. *Cell (Cambridge, Mass.)* **90**, 437–447.

Lutz, D. A., Hamaguchi, Y., and Inoue, S. (1988). Miocromanipulation studies of the asymmetric positioning of the maturation spindle in *Chaetopterus* sp. oocytes: I. Anchorage of the spindle to the cortex and migration of a displaced spindle. *Cell Motil. Cytoskeleton* **11**, 83–96.

Morgan, D. O. (1995). Principles of CDK regulation. *Nature (London)* **374**, 131–134.

Murray, A. W., and Kirschner, M. W. (1989). Cyclin synthesis drives the early embryonic cell cycle. *Nature (London)* **339**, 275–280.

Murray, A. W., Solomon, M. J., Kirshner, M. W. (1989). The role of cyclin synthesis and degradation in the control of maturation promoting factor activity. *Nature (London)* **399**, 280–286.

Nardelli-Haefliger, D., and Shankland, M. (1993). *Lox10*, a member of the *NK-2* homeobox gene class, is expressed in a segmental pattern in the endodem and in the cephalic nervous system of the leech *Helobdella*. *Development (Cambridge, UK)* **118**, 877–892.

Nelson, B. H., and Weisblat, D. A. (1992). Cytoplasmic and cortical determinants interact to specify ectodem and mesoderm in the leech embryo. *Development (Cambridge, UK)* **115**, 103–115.

Nigg, E. A. (1995). Cyclin-dependent protein kinases: Key regulators of the eukaryotic cell cycle. *BioEssays* **17**, 471–480.

Rameríz, F., Wedeen, C. J., Stuart, D. K., Lans, D., and Weisblat, D. A. (1995). Identification of a neurogenic sublineage required for CNS segmentation in an annelid. *Development (Cambridge, UK)* **121**, 2091–2097.

Sandig, M., and Dohle, W. (1988). The cleavage pattern in the leech *Theromyzon tessulatum* (Hirudinea, Glossiphoniidae). *J. Morphol.* **196,** 217–252.

Shankland, M. (1987a). Differentiation of the O and P cell lines in the embryo of the leech: I. Sequential commitment of blast cell sublineages. *Dev. Biol.* **123,** 85–96.

Shankland, M. (1987b). Differentiation of the O and P cell lines in the embryo of the leech: II. Genealogical relationship of descendant pattern elements in alternative developmental pathways. *Dev. Biol.* **123,** 97–107.

Smith, C. M., and Weisblat, D. A. (1994). Micromere fate maps in leech embryos: Lineage-specific differences in rates of cell proliferation. *Development (Cambridge, UK)* **120,** 3427–3438.

Spana, E. P., and Doe, C. Q. (1995). The prospero transcription factor is asymmetrically localized to the cell cortex during neuroblast mitosis in *Drosophila. Development (Cambridge, UK)* **121,** 3187–3195.

Spana, E. P., Kopczynski, C., Goodman, C. S., and Doe, C. Q. (1995). Asymmetric localization of numb autonomously determines sibling neuron identity in the *Drosophila* CNS. *Development (Cambridge, UK)* **121,** 3489–3494.

Strome, S., and Wood, W. B. (1983). Generation of asymmetry and segregation of germ-line granules in early *C. elegans* embryos. *Cell (Cambridge, Mass.)* **35,** 15–25.

van den Biggelaar, J. A. M. (1971). Timing of the phases of the cell cycle during the period of asynchronous division up to the 49-cell stage in *Lymnaea. J. Embryol. Exp. Morphol.* **26,** 367–391.

Waddle, J. A., Cooper, J., and Waterston, R. H. (1994). Transient localized accumulation of actin in *Caenorhabditis elegans* blastomeres with oriented asymmetric divisions. *Development (Cambridge, UK)* **120,** 2317–2328.

Weisblat, D. A., and Shankland, M. (1985). Cell lineage and segmentation in the leech. *Philos. Trans. R. Soc. London* **313,** 39–56.

Weisblat, D. A., Kim, S. Y., and Stent, G. S. (1984). Embryonic origins of cells in the leech *Helobdella triserialis. Dev. Biol.* **104,** 65–85.

Whitman, C. O. (1887). A contribution to the history of germ layers in *Clepsine. J. Morphol.* **1,** 105–182.

Zackson, S. L. (1984). Cell lineage, cell–cell interaction, and segment formation in the ectoderm of a glossiphoniid leech embryo. *Dev. Biol.* **104,** 143–160.

14

Anteroposterior Pattern Formation in the Leech Embryo

Marty Shankland
Department of Zoology and
Institute of Cellular and Molecular Biology
University of Texas, Austin
Austin, Texas 78712

I. Introduction

A long-standing problem in developmental biology is the means by which developing embryos first establish the axial organization of the mature body plan, and then deploy detailed features of the body plan along those axes. For one major group of animals, the Bilateria, the body plan is distinguished by a single, median plane of bilateral symmetry, and differentiation within that plane is defined with respect to anteroposterior (AP) and dorsoventral (DV) axes. Differentiation along the AP axis has been a major topic of research in a wide variety of experimental systems, and we now understand many features of this process—some of which are obviously conserved and some of which are outwardly dissimilar—in a few phyletically disparate animal groups (McGinnis and Krumlauf, 1992; Wang et al., 1993).

This chapter reviews what is currently known about AP pattern formation in the embryo of the leech. As outlined in Chapter 11 by Stent, leeches are annelid worms and have a body plan that is bilaterally symmetric and segmentally organized along most of the length of the AP axis. The leech embryo develops via a stereotyped sequence of cell divisions, and thus it is feasible to uniquely identify cells of the developing embryo by their lineage history, and to know the normal developmental fate of these identified cells with a high degree of precision. In glossiphoniid leeches such as *Helobdella triserialis* and *Theromyzon rude*, the early embryo is large and readily amenable to experimental manipulation, and considerable progress has been made in understanding the cellular events, and to some extent also the molecular events, that underlie the initial patterning of the AP axis. In hirudinid leeches such as *Hirudo medicinalis* the early embryology is not as accessible, but the large size of the nervous system in late embryos has permitted elegant analysis of AP patterning events associated with terminal cell differentiation.

The body plan of the adult leech (see Fig. 1 in Chapter 11 by Stent) is composed in large part of 32 segments tandemly arrayed along the AP axis, and at the extreme anterior end there is another unsegmented and embryologically distinct tissue domain known as the prostomium (Shankland and Savage, 1997). The leech body plan does not have an obvious anatomical separation into head and trunk, but for the sake of simplicity and comparison with other animals we will here use these terms to designate the unsegmented prostomium and the segmented portion of the body plan respectively. It should be noted that many annelid embryos also have a discrete anal domain (pygidium) and hindgut that lie posterior to the last trunk segment, but in the leech these latter structures are vestigial and only make a minor contribution to the adult (Anderson, 1973).

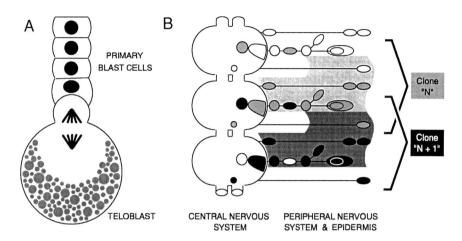

Figure 1 The segmental tissues of the leech embryo arise via a stereotyped cell lineage from a set of embryonic stem cells called teloblasts. (A) Each teloblast undergoes a sequence of highly asymmetric cell divisions to produce a linear chain or bandlet of primary blast cells whose position in the bandlet reflects the rank order of their birth. (B) Schematic diagram showing the distribution of two primary p blast cell clones on the right side of the body near the end of embryonic development. Three segmental ganglia of the central nervous system are shown in outline for reference. There is 1 p blast cell clone/segment, but note that clones "N" (light gray) and "N+1" (black and dark gray) show a stereotyped pattern of intermixing at their border. Neurons and epidermal specializations are represented as ellipses, while patches of squamous epidermis are represented as broader stretches of lighter labeling. Adapted from Shankland and Weisblat (1984).

II. Segmentation

A central theme in many studies of leech embryology has been the elucidation of developmental mechanisms that bring about the segmentation of the body trunk. Leeches are not amenable to genetic analysis of the sort that has been employed in the fruit fly *Drosophila*, but the leech embryo has been used to describe the cell lineages that generate the segmental body plan, to perform experimental manipulations that reveal how and when embryonic cells become committed to particular segmental identities, and to characterize the degree to which developmental mechanisms are conserved between the leech and other segmented animals.

A. Segmentation of the Mesoderm and Ectoderm

As with other aspects of embryogenesis, the segmental body plan of the leech is established by an essentially invariant cell lineage (Weisblat and Shankland, 1985). This is in stark contrast to those model systems, (i.e., insects and vertebrates) that have been used most extensively for studies of segmentation. In those animals, the segmental organization of the mature body plan does not bear a fixed lineal relationship to the cells or nuclei of the early embryo. However, the segmental organization of the body plan of the leech is implicit in the spatiotemporal pattern of its embryonic cell divisions, and the segmental identity of cells in the adult leech is reliably predicted by their lineage histories. This lineally stereotyped mode of segmentation is widespread among the clitellate annelids (Storey, 1989), and there are also

certain arthropods (e.g., the malacostracan crustaceans; Dohle and Scholtz, 1988) that utilize an outwardly similar lineage-based mechanism of segmentation in the rear portion of their body trunk.

A key step in the segmentation of the leech embryo is the formation during early cleavage of a set of large, uniquely identifiable stem cells, known as teloblasts (Fig. 1A), which define the prospective posterior end of the embryo. Unlike the progenitors of the prostomial head domain (see Section III,B), these teloblasts arise exclusively from only one quadrant (D) of the 4-cell embryo. The segmented mesoderm will arise from a bilaterally symmetric pair of M teloblasts, each of which contributes to only the right or left side of the embryo. The segmented ectoderm arises from a bilaterally paired set of four teloblasts designated as N, O, P, and Q. Each of these ectodermal teloblasts gives rise to a distinct subset of tissues in the nervous system and epidermis, and as with the mesoderm, the ectodermal blast cells generate clones that are limited to only one side of the midline. The endoderm of the leech also exhibits a segmental pattern of differentiation (Nardelli-Haefliger and Shankland, 1993; Wedeen and Shankland, 1997), but this germ layer does not arise from teloblastic stem cells, and its development will be discussed separately.

Blastomere ablation studies suggest that the M, N, and Q teloblasts are autonomously committed to produce different sets of descendant tissue from the time of their formative cleavage. In contrast the O and P teloblasts appear to be equipotent, and their descendant clones become committed to distinct developmental pathways only as a result of inductive cell interactions (Shankland and Weisblat, 1984; Huang and Weisblat,

1996). Because the generative O and P teloblasts are distinguished by fate rather than developmental potential, they are cumulatively referred to as O/P teloblasts excepting situations in which cell fate can be verified experimentally (Weisblat and Blair, 1984).

It is the cell division pattern of the teloblast that accounts for the segmental organization of its descendant clone. All five teloblasts undertake a rapid sequence of highly asymmetric cell divisions, and thereby produce a linear column or bandlet of much smaller daughters known as primary blast cells (Fig. 1A). The blast cells are designated by a lowercase letter corresponding to their parent teloblast and, when known, a subscript number corresponding to their birth rank in the stem cell lineage of that teloblast. All five teloblasts begin generating blast cells at roughly the same time[1] (Sandig and Dohle, 1988), and the individual bandlets are tied together at their leading ends by close association with the micromere clones that generate the head tissues. Thus, the firstborn blast cells will contribute to the anteriormost segment of the body trunk, and subsequent blast cells will contribute to progressively more posterior segments in the order of their birth.

In the M, O, and P lineages, each primary blast cell represents a single segmental repeat of the descendant clone of the parent teloblast (Fig. 1B). For example, every primary blast cell in the o bandlet undergoes the same stereotyped sequence of cell divisions, and by the end of embryogenesis each of these blast cells has given rise to a very similar clone of approximately 70 descendant cells including specific central and peripheral neurons, epidermal cells, and a portion of the excretory nephridium (Shankland, 1987a). However, the individual o blast cells contribute clones to different body segments in accordance with their differing birth ranks, and these outwardly similar clones do express some uniquely segment-specific features in the development of certain identified postmitotic cells (Martindale and Shankland, 1988; Nardelli-Haefliger *et al.*, 1994; Berezovskii and Shankland, 1996). Thus, both the segmental periodicity and the segmental specificity of the descendants of the O teloblast appear to be implicit in the spatiotemporal organization of its cell lineage.

Is this correlation coincidental, or does the lineage history of a blast cell play a causal role in defining its segmental fate? As noted above, the individual blast cell clones within an m bandlet, an o bandlet, or a p bandlet represent a tandem array of segmentally homologous repeats. One can entertain the possibility that the primary blast cells are restricted to their "one blast cell per seg-

ment" periodicity by regulative cell interactions, but experimental data argue that such interactions are not necessary for the manifestation of segmental periodicity. For instance, the germinal band of the leech will develop segmentally iterated patterns of cell division (Zackson, 1984) and, in many cases, cell differentiation (Stuart *et al.*, 1989; Torrence *et al.*, 1989; Martindale and Shankland, 1990a) even though one or more bandlets are missing, suggesting that inductive interactions between bandlets are not required for periodicity. And if one or more primary blast cells are ablated so as to create a gap in the bandlet, the blast cell clones bordering that gap anteriorly and posteriorly do not regulate to produce more than one segmental complement (Gleizer and Stent, 1993; Ramirez *et al.*, 1995; E. Seaver and M. Shankland, unpublished). These experiments argue that primary blast cells are autonomously restricted to produce a descendant clone of a certain size, which suggests in turn that the iterative stem cell lineage of the parent teloblast is in fact a causal determinant of the segmental organization of the bandlet. As discussed in the Section II,D, cell lineage also appears to be playing a causal role in determining the segmental specificity of blast cell clones (Martindale and Shankland, 1990b; Gleizer and Stent, 1993; Nardelli-Haefliger *et al.*, 1994).

It should be noted that two of the ectodermal teloblasts of the leech, N and Q, utilize an alternative mode of cell division in which two successive blast cell clones serve together to form a single segmental repeat (Fig. 2). Thus, one n or q blast cell will produce a clone

---

[1]There is one exception to this generality. The OP proteloblast divides asymmetrically to produce four op primary blast cells before cleaving symmetrically to form the definitive O and P teloblasts. These two blastomeres then resume blast cell production, and both of their firstborn blast cells (o_1 and p_1) remain closely associated with the last op blast cell (op_4) so as to create a fork in the bandlet.

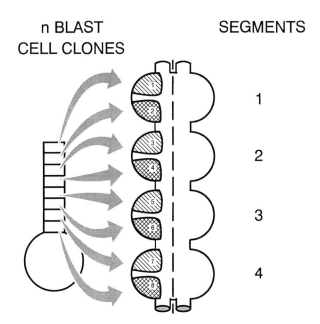

Figure 2 The N teloblast generates two primary blast cells per segment. Note how alternating blast cell daughters contribute descendant clones to anterior and posterior regions of the segmental ganglia respectively.

that extends roughly half the length of the segment, and the next blast cell within that bandlet produces a distinct set of clonal descendants situated in the next posterior half segment. Thus, N and Q differ from the other teloblasts in that they must generate twice as many primary blast cells (64) to occupy the same number of segments (32), and each n or q blast cell is homologous to every second blast cell within its particular bandlet (Fig. 2). But these minor differences do not alter the conclusions drawn above, since experimental studies indicate that individual n blast cells have an intrinsic segment identity (Martindale and Shankland, 1990b; Stuart *et al.*, 1989; Nardelli-Haefliger *et al.*, 1994), and are committed from an early stage to contribute either the anterior or posterior half of a segment (Bissen and Weisblat, 1987).

While there is a precise numerical relationship between primary blast cell clones and segments of the adult leech, it is important to realize that the blast cell clones do not function as developmental compartments (Martinez-Arias and Lawrence, 1985). For example, Fig. 1B shows the spatial relationship of three successive p blast cell clones after cell differentiation. First, it should be noted that the primary blast cell clone straddles the anatomically defined segment boundary, a phenomenon which is also true of m and o blast cell clones (Weisblat and Shankland, 1985). And while there is an obvious 1:1 relationship between blast cell clones and segments, each clone intermixes with the next anterior and next posterior clones such that a single p blast cell clone spans a length of roughly 1⅔ segments. Similar intermixing has been observed in all five teloblast lineages (Weisblat and Shankland, 1985; Ramirez *et al.*, 1995).

This observation may seem surprising given that clonal restrictions on cell mixing are a central part of the segmentation process observed in some other systems such as the *Drosophila* blastoderm (Martinez-Arias and Lawrence, 1985) and the vertebrate hindbrain (Fraser *et al.*, 1990). But the fact that the segmental periodicity of the leech is implicit in the iterated organization of its blast cell clones insures in turn that clonal intermixing occurs identically in every segment, and hence does not disrupt the preexisting periodicity (Shankland, 1991). Looking at this disparity in an evolutionary context, it is easy to understand how genetic mutations that permit cell mixing between segments would be more likely to disrupt segmental patterning—and hence be under strong negative selection—in an animal such as the fruit fly that uses boundaries of lineage restriction to establish and define its segments than in an animal such as the leech whose segmental periodicity is already implicit in the sequence of cell divisions by which the segmental founder cells are produced.

B. Segment Polarity

In order to function as a segmental unit, a primary blast cell must generate a descendant clone that is properly differentiated along its AP axis. All primary blast cells undergo a stereotyped sequence of divisions that is specific to their particular teloblast lineage, or, in the case of the N and Q lineages, to every second blast cell within that lineage (Zackson, 1984; Bissen and Weisblat, 1989). Many of these early blast cell divisions are asymmetric to a measurable degree (Shankland, 1987b), and thus it is possible to identify the various lineal descendants of the primary blast cell by their differing size and position within the clone (Fig. 3A). This morphological asymmetry is of some mechanistic consequence because it reveals that primary blast cells must acquire an intrinsic AP polarity no later than the onset of their first cytokinesis, a conclusion that is supported by the positioning of the microtubule-organizing centers during interphase (Weisblat *et al.*, 1987).

In the O and P teloblast lineages, the segregation of cell fates at the first few divisions of the primary blast cells has been characterized with intracellularly injected

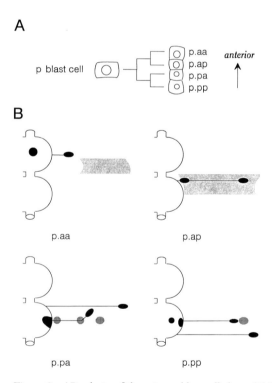

Figure 3 AP polarity of the primary blast cell clone. (A) The first two rounds of division in the p blast cell clone generate four progeny cells that align in a column along the AP axis. (B) Each of these four progeny cells gives rise to a stereotyped set of descendants, including neurons and epidermal specializations (black) and patches of squamous epidermis (gray). Each clone is shown in relationship to the right side profile of the same two segmental ganglia, and note that the AP order of the four founder cells (A) is reflected in the AP order of their descendant clones. However, the borders between clones are not sharp, and there is some intermixing.

lineage tracers (Shankland, 1987a,c). Although o and p blast cells manifest distinct patterns of cell division, in both lineages the first few cell divisions produce anterior and posterior daughter cells (Zackson, 1984). These early blast cell divisions subdivide the descendant fate of the parent cell in a stereotyped fashion, and the anterior and posterior location of the daughter cells predicts the relative positioning of their descendants along this same axis (Fig. 3B). However, the boundaries between subclones, like those between primary blast cell clones, do exhibit a limited degree of overlap and mixing.

How is it then that the various progeny of a primary blast cell become specified to produce phenotypically distinct AP subsets of its descendant clone? Because the early cell divisions segregate anterior and posterior positional values within a segment, one might anticipate that the underlying mechanism would employ some of the same "Segment Polarity" genes that are responsible for the establishment of subsegmental values in the *Drosophila* embryo (DiNardo *et al.*, 1988). To get at this point, Wedeen and co-workers cloned a leech orthologue of the *Drosophila* segment polarity gene *engrailed* (*en*) from the leech *H. triserialis*, and characterized its pattern of expression (Wedeen and Weisblat, 1991; Lans *et al.*, 1993). This gene, named *ht-en*, is not expressed in primary blast cells or their first generation of daughter cells, but its gene products are expressed during later cell divisions in all five teloblast lineages.

In the O, P, and Q lineages, expression of ht-en protein occurs sufficiently early that it can be ascribed to particular lineally identified cells (Lans *et al.*, 1993). For example, the primary o and p blast cell clones begin to express ht-en protein after their first five and four cell divisions respectively, and in each of these clones the protein is expressed only in one particular sublineage that located slightly behind the anterior boundary of the clone (Fig. 4). The situation is somewhat different in the Q lineage, where distinct q_f and q_s primary blast cells generate a single segmental repeat. The q_s blast cell clone does not express *ht-en* during segmentation, whereas the q_f cell divides to produce four viable progeny cells that all express ht-en protein (Fig. 4). Segmental expression of ht-en protein also has been observed in the N and M teloblast lineages, but not until a later stage when it is no longer feasible to trace blast cell divisions (Lans *et al.*, 1993).

These data indicate that *ht-en* is expressed at precise AP locations within the segmentally repeated cell lineages of the various teloblasts. In *Drosophila*, blastoderm cells that express the en transcription factor give rise to a posterior compartment of the mature segment, and this compartment is structurally distinct from the anterior compartment produced by nonexpressing cells (Lawrence, 1992). If one envisions that the segmentation of annelids such as the leech is homologous to the

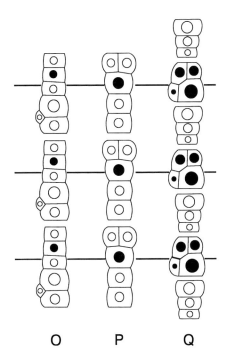

Figure 4 Expression of a leech *engrailed* orthologue (*ht-en*) during the early blast cell divisions of the O, P, and Q lineages. Outlined blocks represent primary blast cell clones at the earliest stage of *ht-en* expression in each lineage. Consecutive clones within a bandlet have been artificially separated for the purpose of representation, and are here aligned with the boundaries of the adult segments to which they will contribute. The individual nuclei within each blast cell clone are shown either as solid circles (*ht-en* expression) or hollow circles (no *ht-en* expression). In all three lineages, the expression of *ht-en* roughly correlates with the position of the segment boundary. Note that both O and P lineages consist of a single type of blast cell clone, and that only one cell within each clone expresses *ht-en* at this stage. The Q lineage consists of two alternating types of blast cell clone, one of which expresses *ht-en* and one of which does not. Expression data are taken from Lans *et al.* (1993); position of blast cell clones relative to the segment boundary is taken from Weisblat and Shankland (1985).

segmentation of arthropods such as *Drosophila*, then a conservation of mechanism would predict that expression of ht-en protein in the o and p blast cell clones might specify one of the progeny cells as distinct from the other cells within the same clone, and in doing so help to establish and diversify AP positional values within the segment.

During the early development of the *Drosophila* segment primordium, short-range cell interactions along the AP axis (i.e., within a segment, or between immediately adjacent segments) play an important role in establishing and maintaining subsegmental AP values (DiNardo *et al.*, 1988). This idea has not yet been tested explicitly in the leech embryo, but it should be feasible to do so given that the stereotyped cell lineage allows one to uniquely identify and manipulate the progeny of individual blast cells in living embryos (Shankland, 1987a,c). Such experiments will prove to be an important avenue for future research since very little is currently known about the cellular basis of segment polarity determination outside of *Drosophila.*

There are also some interesting distinctions between the expression of *ht-en* in the leech and *en* expression in arthropods. In both insects and crustaceans the cells that express *en* remain physically associated as a developmental compartment, and the posterior border of the *en*-expressing compartment maps to the boundary between adjacent segments [precisely in insects (Lawrence, 1992) and at least roughly in crustaceans—compare Scholtz *et al.* (1993) and Dohle and Scholtz (1988)]. In contrast, the the blast cell sublineages that express *ht-en* in the leech embryo do not remain together as an immiscible compartment for the remainder of development, and in the P and Q lineages the descendants of the early *ht-en*-expressing cells actually come to straddle the segment boundary (Fig. 4).

One way of reconciling this discrepancy between phyla is to imagine that *ht-en* expression in the leech embryo defines a morphogenetic unit that is comparable to the posterior compartment of arthropods in some functional context, but this compartment exists only during a transient phase of leech morphogenesis and has therefore been overlooked (Wedeen, 1995). It is interesting to note that in both insects (Vincent and O'Farrell, 1992) and crustaceans (Scholtz *et al.,* 1993) some of the cells at the posterior edge of the compartment express *en* transiently, and at least in *Drosophila* such cells cease to behave as part of the morphogenetic compartment. In the leech *Helobdella,* the early pattern of segmental *ht-en* expression is entirely transient. Hence, one cannot at this time rule out the possibility that *ht-en* has a transient effect on cell affinities within the leech germinal band, and that the intermingling of blast cell clones seen at later stages take place after its influence is past.

In *Drosophila* it has been shown that the prolonged maintenance of *en* expression during embryonic segmentation requires an intercellular feedback loop involving wingless (wg) protein expressed by neighboring cells (DiNardo *et al.,* 1988; Chapter 17 by Siegfried). The transient segmental expression of *ht-en* in leech suggests that this particular feedback loop may be missing. In keeping with this idea, the only leech homologue of *wg* that has been characterized to date does not show segmental expression (Kostriken and Weisblat, 1992).

C. Segment Alignment

One of the more unusual features of leech segmentation is the fact that segmental founder cells whose descendant clones are fated to occupy the same segment must travel a great distance along the AP axis to become juxtaposed in a transverse plane (Weisblat and Shankland, 1985; Lans *et al.,* 1993). In addition, this alignment is not complete until a relatively late stage in de-

velopment when these clones are already committed to many of their segmental characteristics (Nardelli-Haefliger *et al.,* 1994), indicating that the correct morphogenetic assembly of already specified segmental units is a critical part of the segmentation process. Comparable phenomena have not yet been reported in other segmented animals, and may not occur in some systems due to peculiar features of their development. For instance, the ingressing mesoderm of insects undergoes predominately dorsoventral rearrangements, and thus remains in more or less the same segmental register as the ectoderm throughout gastrulation (Lawrence, 1992).

In the leech *Helobdella,* the teloblasts generate blast cells at disparate points within the embryo, and a primary blast cell does not come into proximity with the other bandlets until it enters the germinal band 10–15 hours after its birth. The descendant clone of the blast cell must thereafter align itself with other blast cell clones that share the same segmental identity before they can undergo transverse mixing and organogenesis (Weisblat and Shankland, 1985). Among the M, O, and P lineages the necessary alignment is relatively minor since these teloblasts produce 1 blast cell per segment and have the same rate of mitosis (Wordeman, 1983). Thus, blast cells with the same birth rank, and hence the same segment identity, are thought to enter the germinal band in close proximity.

In contrast, the N and Q teloblasts produce 2 primary blast cells per segment, but their rate of mitosis is only 10–20% faster than the other teloblasts (Wordeman, 1983). As a result, the later born n and q blast cells initially enter the germinal band next to m, o, and p blast cells that are fated for more posterior segments, and the n and q bandlets must undergo an extensive anterior displacement (Fig. 5A) in order that their constitutent blast cell clones come into segmental register with the m, o, and p bandlets (Weisblat and Shankland, 1985). The final alignment is not achieved until more than a day after the primary blast cells are born (Weisblat and Shankland, 1985), and the process of alignment spans a period of time during which the primary blast cell is undergoing many of its subsidiary divisions.

As shown in Fig. 5B, the anterior displacement of n and q blast cell clones within the germinal band can be accounted for in large part by bandlet-specific changes in the shape of the primary blast cell clones. The early divisions of the o and p blast cells produce clones that are considerably elongated in their AP dimension. The n and q blast cells clones do not elongate to the same degree, and as a consequence the n and q bandlets slide anteriorly on either side of the o and p bandlets once within the germinal band (Fig. 5A). Much of this differential shape change occurs during the period from 20 to 40 hours after the primary blast

A **B**

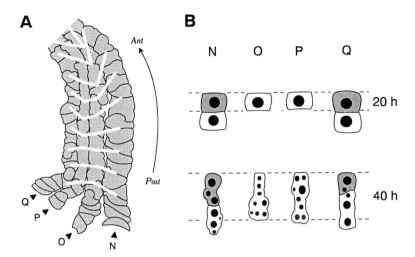

Figure 5 Different teloblast lineages must slide past one another to achieve their final segmental register. (A) Camera lucida tracing of the four ectodermal cell lineages (N, O, P, and Q) in the right germinal band of a stage 7 *Helobdella* embryo. The bandlets first come together at the posterior end of the germinal band, and shortly thereafter the primary blast cells begin their subsidiary divisions to generate descendant clones. Primary blast cell clones are outlined; individual cells within the clones are not. White lines connect sequential segmental primordia in each of the four bandlets, taking into account the fact that the O and P lineages have one blast cell clone per segment and the N and Q lineages have two blast cell clones per segment. Note that the N and Q lineages slide anteriorly past the O and P lineages as they progress further into the germinal band. Also note some anterior sliding of the P lineage relative to the O lineage, presumably as a consequence of its lying toward the inner side of the curvature of the band. (B) Shape changes of the primary blast cell clones account for the differential sliding of the bandlets within the germinal band. At 20 hours the primary blast cells are undivided in all four ectodermal bandlets, and roughly similar in shape. The N and Q lineages have two blast cells per segment, with the n_s and q_s blast cells shown in gray and the n_f and q_f blast cells shown in white. By 40 hours the primary blast cells have undergone several rounds of cell division (only the nuclei are shown). The o and p clones have elongated much more than the n and q clones, and as a result the paired segmental founder cells in the N and Q lineages are only slightly longer than the singular segmental founder cells of the O and P lineages. Clonal morphologies are redrawn from Bissen and Weisblat (1989), © Company of Biologists, Ltd.

cell is born (Fig. 5B). However, the blast cell divisions have not yet been characterized at later times, and so it is not known when segmental register is finally achieved.

If these bandlet-specific changes in clonal shape were determined solely by factors intrinsic to the clones, then the segmental alignment process might be explained simply by the close packing of clones whose size, shape, and order along the AP axis are already determined. However, experimental studies indicate that this is not a complete explanation. If the right and left mesoderm are artificially slipped one segment out of register, there is a very strong tendency for the m blast cell clones to realign themselves at the midline according to their already determined segment identities (Gleizer and Stent, 1993). This finding suggests that blast cell clones with the same segment identity can recognize one another across the midline, and are able to actively pull themselves into register accordingly. But the contribution of active recognition to morphogenesis as a whole would appear to be slight, since (i) these same embryos continue to tolerate segmental mismatches in the mesoderm over stretches of several segments before they realign, and (ii) the realignment of right and left mesoderm causes in turn an uncorrected misalignment of segment identity between the mesoderm and overlying ectoderm (Gleizer and Stent, 1993). It is also to be noted that when ectodermal blast cells

are shifted many segments out of segmental register, their descendant clones retain their original segment identity but nonetheless integrate seamlessly with tissues of the ectopic host segment (Shankland, 1984; Martindale and Shankland, 1988, 1990b).

The final steps of the alignment process are also intriguing, although they have not been a subject of active research to date. Examination of clonal boundaries within the germinal band indicates that even adjacent o and p blast cell clones, which should be roughly aligned with respect to segment identity as they enter the germinal band, do in fact shift relative to one another subsequent to their initial contact (Fig. 5A). And blast cell clones that have forced many segments out of alignment with the other bandlets show a nearly unerring ability to position themselves correctly within the boundaries of the ectopic host segment (Shankland, 1984; Martindale and Shankland, 1988). These findings indicate that blast cell clones in different bandlets very likely express subsegmental AP values, and that they use these subsegmental markers to fine tune the later stages of the alignment process. Genes such as *ht-en* that are expressed at a particular AP position within the primary blast cell clone are obvious candidates to control this subsegmental alignment (Shankland, 1994; Wedeen, 1995), but as yet there are no experimental data on this point.

D. Segmental Differences

Despite the many similarities between segments, the leech body plan also exhibits a large number of predictable segmental differences both in the presence or absence of particular organs, and in the detailed differentiation of individual cell types that are repeated in successive body segments (Shankland and Martindale, 1992). No segmental differences have been reported in the early blast cell divisions (Zackson, 1984; Bissen and Weisblat, 1989), and thus it would appear that most if not all of the segmental differences within the mesodermal and ectodermal tissues of the leech arise because lineally homologous blast cell progeny undertake distinct developmental pathways in different body segments.

Experimental analysis of segmental differentiation in a variety of leech species has revealed two major steps. The primary segmental differences are determined early in development, and appear to be an intrinsic property of the individual blast cell clones. Primary blast cells that are forced to develop in body segments inappropriate for their lineal identity will generate descendant clones containing segment-specific cells or or-

gans appropriate for their normal fates, not for the segments in which they develop (Fig. 6). This sort of blast cell autonomy has been demonstrated for the mesodermal M lineage (Gleizer and Stent, 1993) and three of the four ectodermal lineages (Martindale and Shankland, 1990b; Shankland and Martindale, 1992; Nardelli-Haefliger *et al.*, 1994; M. Q. Martindale and M. Shankland, unpublished), but has not yet been tested in the Q lineage due to a paucity of early-differentiating segmental markers. These results suggest that the primary blast cells acquire segment identity around the time of their birth, although it is still unclear whether those identities are specified by molecular events associated with the division of the parent teloblast, or by cell interactions occurring shortly after the division. But regardless of the way by which segment identity is first established, transplantation experiments clearly suggest this identity is passed along in a lineally restricted fashion from the blast cell to its descendants, much like the clonal inheritance of segment identity observed following cellularization of the *Drosophila* embryo (Simcox and Sang, 1983; Prokop and Technau, 1994). Some of the segmental phenotypes that are determined by the identity of the ancestral blast

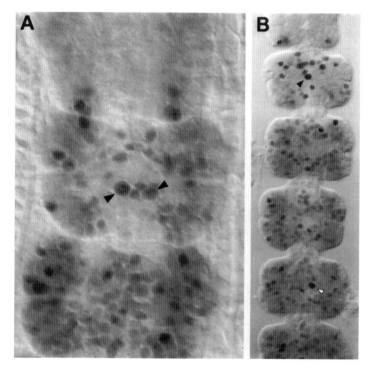

Figure 6   Primary blast cells of the leech embryo are committed with respect to segment identity from an early developmental stage. (A) Midbody ganglia M6-M8 stained with an antibody to the leech Hox gene product LOX2 using a horseradish peroxidase (HRP)-labeled secondary antibody. The dark HRP reaction product can be seen in a number of neuronal nuclei, including a large bilateral pair of OM7 neurons (arrowheads) located only at the ventral midline of ganglion M7. Dissected tissue is viewed with DIC optics, and aligned muscle fibers can be seen to either side of the neural ganglia. Cell lineage analysis reveals that each OM7 neuron is descended from the ipsilateral O teloblast (Nardelli-Haefliger *et al.*, 1994). (B) Nerve cord dissected from an embryo in which blast cells of the right O cell lineage were shifted posteriorly by three segments with respect to the other segmented tissues. Note the presence of a normally situated OM7 neuron on the left side (solid arrowhead), and a second OM7 neuron located three segments more posteriorly on the right (hollow arrowhead). This result demonstrates that the primary blast cell which would normally give rise to the OM7 neuron does so even if forced to differentiate ectopically in another segment.

cell involve segment-specific differences in the terminal differentiation of identified neurons (Fig. 6; see also Martindale and Shankland, 1990b; Nardelli-Haefliger *et al.*, 1994). This latter finding suggests that segment identity can be passed along over the roughly six rounds of cell division separating the primary blast cell from its postmitotic descendants before acting to direct otherwise homologous cells into divergent developmental pathways.

The intrinsic identity of blast cell clones does not account for all aspects of the segmental diversification of the leech, as there are also a number of secondary segmental differences that are determined at a later developmental stage by segment-specific cell interactions. For example, the o blast cell clone generates the distal tubule cell of the nephridium, but in those segments where the mesodermal component of the nephridium is lacking, the distal tubule cell undergoes a programmed cell death shortly after it is formed (Martindale and Shankland, 1988). If o blast cell clones are transplanted into segments inappropriate for their cell lineage, it is the presence or absence of the mesodermal nephridium in the host segment, not the identity of the ancestral blast cell, that determines whether the distal tubule cell will survive (Martindale and Shankland, 1988). Another example is the Retzius neuron, which is found in all 32 body segments but manifests a distinctive pattern of morphogenesis and synaptogenesis in the two reproductive segments. Ablation and transplantation experiments suggest that these particular Retzius neurons are induced to follow a segment-specific developmental pathway because they alone innervate the male genitalia (Loer *et al.*, 1987; Loer and Kristan, 1989a,b). Prior to that innervation the Retzius cells in all segments are morphologically similar (Jellies *et al.*, 1987), and may be equipotent in terms of their competence for this induction (Shankland and Martindale, 1992).

In each of these two examples of secondary segmental differentiation, the development of a segmentally iterated cell type depends on segment-specific interactions with another organ, and it is the preexisting distribution of that organ which provides the segmental specificity. The organs that provide this segmental template are, at least in these examples, derived from the mesoderm, and their segmental distribution was in fact determined at the level of the ancestral m blast cell (Gleizer and Stent, 1993). Thus, secondary segmental differences can serve to elaborate on primary segmental differences established at an earlier developmental stage. Cell interactions have also been shown to play an important role in shaping segment-specific patterns of axonal outgrowth (Gao and Macagno, 1987a,b), and in a secondary wave of neurogenesis that is also restricted to the two reproductive segments (Baptista *et al.*, 1990).

In most animals that have been studied, a central component of the molecular mechanism for diversifying ectodermal and mesodermal tissues along the AP axis is the Hox gene cluster. These dozen or so homeobox genes are typically found in tandem array along a single chromosome, and are expressed in overlapping domains along the AP body axis in an order that mirrors their chromosomal arrangement (McGinnis and Krumlauf, 1992; see also Chapter 26 by Ho *et al.*). In *Drosophila*, genetic analysis has shown that the Hox (or homeotic) genes play little or no role in the development of segment number or periodicity, but rather are essential for the development of differences between segments. This dichotomy is not so clear-cut in some other organisms, but still it is clear that the differential expression of Hox genes along the AP axis is a widespread mechanism for committing otherwise homologous cells at different AP positions (e.g., different segments) to follow distinct developmental pathways.

Sequence and expression data have to date been published on seven Hox genes from two different genera of leeches (Aisemberg *et al.*, 1993; Kourakis *et al.*, 1997), and a number of additional Hox-like homeobox fragments have been found by sequence homology (E. R. Macagno, M. Q. Martindale, and M. Shankland, personal communication). Those Hox genes that have been characterized are expressed in segmentally restricted domains along the AP axis (Fig. 7), and with one possible exception (Aisemberg and Macagno, 1994) their expression domains exhibit the same AP order as that seen for the orthologous genes from other animals (Kourakis *et al.*, 1997). This similarity suggests that the Hox genes do play a role in the segmental diversification of the leech embryo; however, it has not yet been possible to test this hypothesis by modifications of gene expression. In one series of experiments a leech Hox protein, LOX2, was misexpressed in ectopic segments by mRNA injection of teloblasts (Nardelli-Haefliger *et al.*, 1994), but the heterologous protein did not persist to the late stage in development when the endogenous gene is normally expressed, and no phenotypic changes in segment identity were observed (Bruce, 1997).

Alternative support for this hypothesis comes from two examples in which segmental patterns of Hox gene expression have been shown to correlate quite precisely with segmental differences in the phenotype of identified neurons. In the medicinal leech *Hirudo*, the RPE homologue neurons display three distinct temporal profiles of expression of the Hox gene *Lox4*—no expression, transient expression, and maintained expression—in different midbody segments, and each profile is correlated with a distinct morphological pattern of axonal pathfinding and branching (Wong *et al.*, 1995). In *Helobdella*, the MPS homologue neurons show an even more complicated pattern of segmental diversification, with four distinct segment-specific phenotypes that are distinguished by cell body size and expression of the Small Cardioactive Peptide (Berezovskii and Shankland,

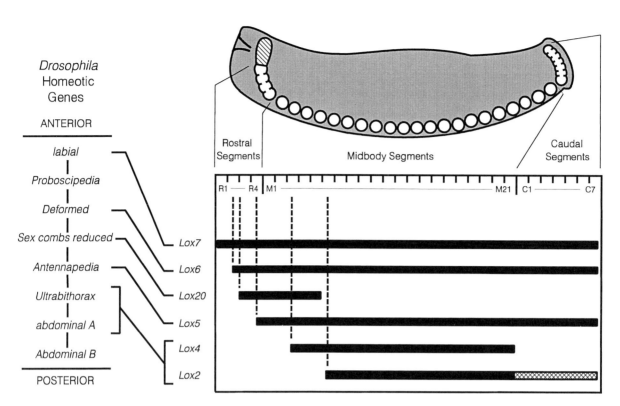

Figure 7 Expression domains of leech Hox genes. A stage 10 *Helobdella* embryo is schematized at the top, with the 32 segmental neuromeres of the central nervous system shown in white. The unsegmented head ganglion is hatched. For each gene, a solid horizontal bar marks all segments in which it shows detectable RNA or protein expression. Caudal expression of the *Lox2* gene is cross-hatched to note that it is expressed there in *Hirudo* (Wysocka-Diller *et al.*, 1989), but not in *Helobdella* (Nardelli-Haefliger *et al.*, 1994). Note that the anterior expression boundaries show the same AP order as the orthologous *Drosophila* homeotic genes, while the posterior boundaries vary considerably. Expression data and orthologies are taken from the following: *Lox7, Lox6, Lox20*, and *Lox5*, Kourakis *et al.*, 1997; *Lox4*, Wong *et al.*, 1995.

1996). The MPS homologue neurons express the Hox gene *Lox2* in a stretch of nine contiguous segments, and the anterior and posterior boundaries of that expression domain correlate precisely with two of the four phenotypic boundaries. The other two phenotypic boundaries of MPS differentiation do not have any obvious relationship to *Lox2* expression, suggesting the segmental diversification of this neuron may depend in a combinatorial fashion on some second, as yet undiscovered Hox gene that is expressed in an overlapping set of segments.

In insects such as *Drosophila* the Hox genes are first expressed several cell divisions prior to terminal differentiation (Lawrence, 1992). But most of the leech Hox genes that have been characterized to date are only expressed at detectable levels during late stages of development in terminally differentiating neurons and muscle cells (Aisemberg and Macagno, 1994; Nardelli-Haefliger *et al.*, 1994; Wong *et al.*, 1995; Kourakis *et al.*, 1997), and none are expressed uniformly throughout a blast cell clone. Thus, while the leech Hox genes are almost certainly involved in specifying differences between segments, both the temporal and spatial characteristics of their expression patterns suggest that they are not responsible for providing the repeating units of the segmental body plan of the leech (the blast cell

clones) with an integral segment identity. This raises the very interesting and at present completely unanswered question as to what molecular mechanism might be responsible for the initial establishment of segment identity at the level of the primary blast cell. To date there is no clear understanding on this point, but it seems likely that some form of segmental specification occurs at or shortly after the time when the primary blast cell is born from the teloblast, and that this initial specification acts upstream of the Hox genes to coordinate their segment-specific expression during the later stages of development.

E. Segmentation of the Endoderm

Annelids are unique among the major groups of segmented animals in that the segmental organization of the adult body plan extends to the endodermal tissues of the midgut (cf. Chapter 40 by Gannon and Wright). Not only does the digestive tract exhibit a segmental morphology associated with the periodic structure of the overlying mesoderm, but molecular studies indicate that the leech endoderm already exhibits segmentally organized patterns of homeobox gene expression prior to morphogenesis (Nardelli-Haefliger and Shankland,

1993; Wysocka-Diller *et al.,* 1995; Wedeen and Shankland, 1997).

The endoderm of the leech embryo develops in a manner quite distinct from the mesoderm and ectoderm, and does not involve teloblastic stem cells with segmentally iterated patterns of cell division. Rather, the definitive endoderm arises from three large yolk-filled blastomeres (A‴, B‴, and C‴ macromeres) that are generated during the early embryonic cleavages. In the glossiphoniid leeches it has been shown that these macromeres fuse to form a multinucleate syncytium (Liu *et al.,* 1998), and at a much later stage in development this midgut syncytium cellularizes to form a definitive layer of mononucleate endodermal cells located immediately beneath the visceral layer of the segmental mesoderm (Whitman, 1878; Nardelli-Haefliger and Shankland, 1993). The fate of the three original macromere nuclei has not been followed carefully, and they may undergo repeated karyokinesis to generate the hundreds of nuclei in the definitive endodermal cell layer. However, at the end of their stem cell lineages the mesodermal and ectodermal teloblasts generate numerous extra blast cells that do not contribute to the formation of body segments (Zackson, 1982; Shankland, 1984), and it appears that these supernumerary blast cells also fuse with and contribute their nuclei to the midgut syncytium (Fig. 8). Whether these supernumerary blast cell nuclei then make a contribution to the definitive cellular endoderm remains to be seen.

The mononucleate endodermal cells produced by the midgut syncytium differentiate to form the inner layer of the mature gut wall. It is not yet clear when the nascent endodermal cells lose cytoplasmic contact with the syncytium, but around the time that the endoderm first becomes apparent as a discrete cell layer it begins to express the homeobox genes *Lox3* and *Lox10*. The *Lox3* gene is a leech homologue of the vertebrate *pdx-1* gene (Wysocka-Diller *et al.,* 1995; Wedeen and Shankland, 1997), which is required for normal development of pancreas and duodenum (Offield *et al.,* 1996; see also Chapter 40 by Gannon and Wright). *Lox3* gene products are initially expressed in a diffuse pattern throughout the nascent endoderm, and this expression subsequently resolves into transverse segmental stripes as the endoderm matures (Fig. 9). *Lox10* is a member of the NK-2 gene family (Nardelli-Haefliger and Shankland, 1993), and it is expressed in transverse segmental stripes or spots from the onset. In *Helobdella* these stripes precede the segmental morphogenesis of the midgut, and mark future sites of intersegmental constriction. For both genes the size and shape of the stripes differs in a predictable pattern between segments, and the three AP regions defined by differing patterns of *Lox3* and *Lox10* expression subsequently develop into the crop, intestine, and rectum (Nardelli-Haefliger and Shankland, 1993; Wedeen and Shankland, 1997). The developmental function of these two leech genes is unknown, but the correlation of the ex-

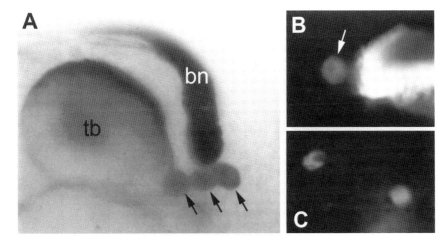

Figure 8 The supernumerary blast cells and spent teloblasts appear to fuse with the midgut syncytium, and may contribute nuclei to the definitive endoderm. (A) Photomicrograph of an early stage 8 *Helobdella* embryo in which the left O teloblast had been previously injected with HRP lineage tracer. Note the intense labeling of that portion of the o bandlet (bn) that has been incorporated into the segmental germinal band in contrast to the much fainter labeling of the injected teloblast (tb) and three supernumerary blast cells (arrows) that are still connected to it. The latter cells originally showed the same intense labeling, and the HRP has presumably leaked out of them as a result of fusion with the surrounding (unlabeled) syncytium. Note that the supernumerary blast cell nuclei are more intensely labeled than their cytoplasm. (B) Confocal optical section of the posterior end of the germinal band (anterior to the right) in an embryo whose right P teloblast had been previously injected with rhodamine-dextran lineage tracer. The last blast cell in the labeled bandlet has almost completely lost its cytoplasmic tracer, presumably by diffusion into the surrounding syncytium. Fluorescent dextran is still readily visible in the nucleus of this cell (arrow), which shows a normal morphology including a nucleolus made visible by dextran exclusion. (C) Optical section revealing two more labeled supernumerary blast cell nuclei located deep within the embryo. These cells show little or no cytoplasmic dextran, which has presumably been lost by diffusion into the syncytium.

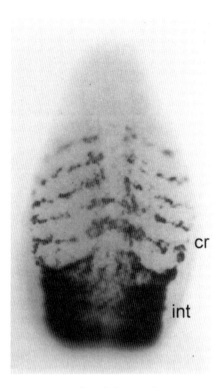

Figure 9 Leech endoderm undergoes a segmentally periodic pattern of molecular differentiation. This stage 10 *Helobdella* embryo was stained by *in situ* hybridization for expression of the homeobox gene *Lox3*, and is shown from the ventral surface with anterior towards the top. In the crop (cr), *Lox3* is expressed in a segmentally organized pattern of narrow transverse stripes. In the intestine (int), the stripes of gene expression are much wider, and the narrow interstripes are obscured by the curvature of the embryo. This curvature completely obscures the rectum, which is located on the far side of the embryo and shows a third, nonsegmental pattern of *Lox3* expression.

tion is imprinted onto the endoderm by means of inductive cell interactions.

Initial attempts to address this problem have involved localized ablations of the segmental mesoderm prior to the normal appearance of the definitive cellular endoderm. Such ablations have a catastrophic effect on the endoderm, which fails to form as a discrete cell layer in either small or large areas where the visceral layer of mesoderm is missing (Wedeen and Shankland, 1997). The spatial correlation between the experimentally imposed mesodermal deficit and the resulting endodermal deficit is quite precise, suggesting that the mesoderm provides the endoderm with some sort of short-range signal necessary for its initial formation and/or survival. The portion of the endodermal cell layer that surrounds the deficit shows a pattern of *Lox3* expression that is normal in terms of periodicity and segmental specificity (Wedeen and Shankland, 1997), as would be expected if the segmental pattern were imprinted by vertical signals from the mesoderm rather than arising from lateral cell signaling within the plane of the endoderm. However, given the complete absence of differentiated endoderm in regions where the mesoderm is missing, it is not possible to draw any strong conclusions from these experiments as to what sort of inductive role the mesoderm may be playing after the endodermal cell layer has formed.

III. Development of the Head Domain

A. Cell Lineage

While the trunk teloblasts arise from only one quadrant of the 4-cell leech embryo, the prostomial head tissues arise from a distinct set of much smaller blastomeres called micromeres that are generated by all four quadrants (Shankland and Savage, 1997). The micromeres arise at the animal pole of the embryo as a result of asymmetric cleavage, and there proliferate to form a cap that will spread out to envelop the entire embryo during gastrulation.

Lineage analyses indicate that micromere derivatives can be separated into two distinct cell populations. The first is a squamous epithelium that spreads away from the animal pole as the outer layer of a provisional yolk sac (Ho and Weisblat, 1987; Smith and Weisblat, 1994). These cells make no known contribution to the body plan of the adult leech. The second population of micromere derivatives remains at the animal pole during gastrulation, and there differentiates to form prostomial tissues, that is, the most anterior structures of the mature body plan. The mature prostomium includes the circumoral epidermis, the supraesophageal ganglion of the central nervous system (CNS), and the foregut. In

pression stripes with sites of morphogenetic constriction suggests a possible role in regulating changes in cell shape that cause the endodermal sheet to buckle at those sites.

It is not yet clear how the leech endoderm acquires its segmental pattern of molecular differentiation. It is possible that the endoderm has intrinsic mechanisms for generating periodicity, although it clearly does not do so by the sort of iterated cell lineages seen in the mesoderm and ectoderm. Alternatively, segmentation may be imposed onto an intrinsically unsegmented endoderm by inductive interactions with the other germ layers. In *Helobdella* the mesoderm has already developed into segmentally repeating somites (Zackson, 1982) and has a finely resolved pattern of segment identities (Gleizer and Stent, 1993) by the time the endoderm begins to express *Lox3* and *Lox10*, and once expression begins the two germ layers are already in segmental register. Thus it seems likely that the mesoderm plays some role in determining the location of segmental repeats within the endoderm, and may in fact serve as a template whose own periodic organiza-

the glossiphoniid leech *Helobdella,* the foregut becomes subdivided into the oral cavity, an eversible proboscis and its enveloping proboscis sheath, and probably also gives rise to the esophagus, although this latter point has not yet been clarified using modern lineage-tracing techniques.

In contrast to the trunk segments, there has been only limited experimental embryology performed on the leech micromeres, and that has focused on the development of the nonprostomial micromeres during gastrulation (Smith *et al.,* 1996). Nonetheless, a number of interesting observations have been made regarding patterns of cell lineage and gene expression.

B. Symmetry Properties

A major component of the prostomial head domain comes from clonal derivatives of the primary micromeres (cells a′, b′, c′, and d′) which are the first set of micromeres produced after the embryo has divided into its four quadrants (Weisblat *et al.,* 1984; Nardelli-Haefliger and Shankland, 1993). Once formed these micromeres lie together at the apex of the animal pole,

and are arranged in clockwise order as viewed from above. The primary micromeres divide repeatedly (Sandig and Dohle, 1988) to produce clones of descendants, which then break apart as a portion of each clone spreads or migrates away from the prostomium to join the provisional yolk sac (Smith and Weisblat, 1994). The remainder of the primary micromeres clones gives rise to the ectodermal tissues of the prostomium.

The fate map of the primary micromeres has been determined in *Helobdella* by injection of cell lineage tracers (Weisblat *et al.,* 1984; Nardelli-Haefliger and Shankland, 1993). On the right side of the prostomium, the b′ micromere gives rise to an anterior quadrant of the circumoral head ganglion and an overlying patch of epidermis shaped like a pie wedge, and the c′ micromere gives rise to a contiguous posterior clone in each of these same two tissues (Fig. 10A,B). The a′ and d′ clones are situated on the left side of the embryo, and both the position and gene expression (Nardelli-Haefliger and Shankland, 1993) of their clones suggest that a′ and d′ are bilaterally homologous to b′ and c′, respectively. The a′–d′ micromeres also make a minor contribution to the foregut (M. Shankland, unpub-

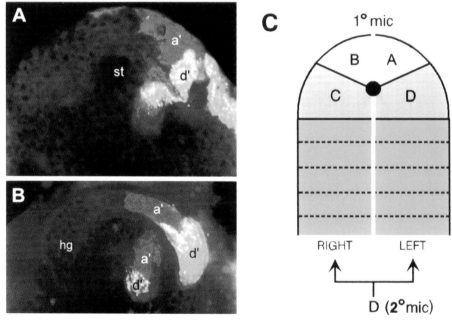

Figure 10 Symmetry properties of primary micromere clones in the prostomial head domain of the leech *Helobdella.* (A,B) Confocal optical sections of the prostomium of a mid-stage 8 embryo in which the a′ micromere had been previously injected with rhodamine-dextran (red) and the d′ micromere injected with fluorescein-dextran (green) at stage 4a. Specimen is viewed from the ventral surface, with the midline at the center and anterior towards the top. Both a′ and d′ clones are located on the left side of the prostomium, with the a′ clone more anteriorly. A superficial section (A) reveals the distribution of these two clones within the epidermis of the prostomium. The hollow stomadeum (st) marks the center of the prostomial head domain. A deeper section (B) reveals the distribution of these same two clones in the annular head ganglion and more centrally in the foregut. The unlabeled right side of the head ganglion can be distinguished by background fluorescence, and is marked hg. (C) Schematic diagram showing the symmetry properties of ectodermal contributions from the four embryonic quadrants in a stage 8 germinal plate viewed from the ventral surface. Clones produced by the primary micromeres (yellow and orange) are situated in the prostomial head domain, with the A and D quadrants on the left side of the midline and the B and C quadrants symmetrically on the right. In contrast, the ectoderm of the trunk segments (turquoise) is entirely generated by the D quadrant, and is bisected by the midline. Thus, the relationship of the D quadrant to the median plane is rotated by 45° in the head and the trunk. Note that the blastomere that generates the trunk ectoderm of the leech (DNOPQ) is homologous to secondary micromere 2d of other spiralian embryos (Shankland and Savage, 1997).

220Marty Shankland

lished), which is mostly derived from secondary and tertiary micromeres that are produced following the primary micromeres (Anderson, 1973; Smith and Weisblat,
1994).

The symmetry properties of the primary micromere clones in the head domain differ dramatically
from those of the trunk ectoderm and mesoderm (Fig.
10C). The latter structures derive with bilateral symmetry from the D quadrant teloblasts, and the median
plane thus bisects the D quadrant derivatives in the
trunk. But in the head ectoderm, the median plane separates A and D quadrant derivatives on the left from B
and C quadrant derivatives on the right.

This apparent discrepancy can be understood by
examining the development of a more basal annelid
group, the polychaetes, from which leeches are thought
to have evolved (McHugh, 1997). In polychaetes the
even- and odd-numbered micromeres exhibit distinct
patterns of symmetry that alternate by a 45° rotation
about the animal pole of the embryo (Fig. 11), and the
embryo first develops into a trochophore larva whose
body plan is derived from micromere tissues of all four
quadrants (Anderson, 1975). In polychaetes a segmented body trunk is added to this larval body plan during postembryonic development, and the ectodermal
and mesodermal components of that trunk arise respectively from the secondary and quaternary D quadrant
micromeres. Thus, the sole progenitors of the trunk ectoderm and mesoderm are even-numbered micromeres

whose symmetry properties should be rotated by 45°
with respect to the primary micromere clones in the
head, in complete accordance with the fate map that
has been obtained for the leech.

Annelids utilize a spiral pattern of cleavage that
is widespread among protostomatous invertebrates
(Brusca and Brusca, 1990), and it seems likely that the
alternating symmetry of odd- and even-numbered micromeres reflects the clockwise or counterclockwise orientation of their formative cell divisions. It is worth noting that a similar alternation of symmetry properties has
been found for successive odd- and even-numbered
micromere quartets in the nemertean *Cerebratulus*
(M. Q. Martindale and J. Henry, personal communication), and also for the second and third micromere quartets of the mollusc *Ilyanassa* (Render, 1997). However,
these symmetry properties are not universal among
spiralian phyla, since the primary micromere clones
of *Ilyanassa* adopt a pattern of symmetry that is characteristic of the even-numbered micromeres in the
other groups (Render, 1991). But despite such coun-
terexamples, the rotational alternation of successive
micromere quartets in the larvae and/or head domain
of spiralian animals is a point worthy of some note
since most reviews on spiral cleavage focus primarily
on the fate of trunk precursor cells, and as a result the
distinct symmetry properties of the adult head have
been routinely overlooked in that literature for many
decades.

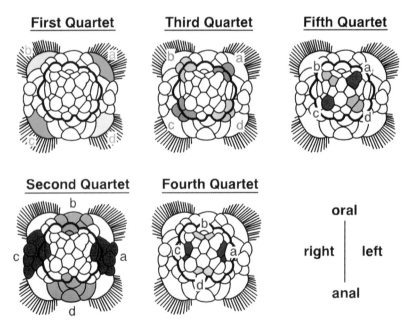

Figure 11 Odd- and even-numbered micromere quartets alternate by 45° with respect to the median plane in an embryo of the polychaete annelid *Polygordius*. Gastrulae are shown from the vegetal surface, and the edge of the blastopore is marked as a heavy line. (Blastomeres inside
the blastopore will be internalized.) For each micromere quartet, the clones derived from the A and C quadrants are shown in bold color, and
clones derived from B and D quadrants are shown in a lighter hue. Note that for the odd-numbered micromeres, the B and C quadrants on the
right of the embryo are bilaterally symmetrical to the A and D quadrant micromeres on the left of the embryo. For even-numbered micromeres,
the median plane bisects clones from the B (oral) and D (anal) quadrants, with the A and C quadrants being bilaterally symmetrical. Adapted
from Woltereck (1904).

C. Gene Expression

Not only are the head and trunk tissues segregated at a very early stage in the cell lineage of the leech embryo, they also seem to depend on distinct genetic mechanisms to bring about patterned cell differentiation. As already discussed, the body trunk of the leech is similar to that of other bilaterian organisms in that it appears to use the Hox genes as a mechanism for segmental diversification, and an orthologue of the *en* gene to help establish the early patterning of positional values within each segment. However, none of these genes is expressed in the prostomial head tissues (Wedeen and Weisblat, 1991; Kourakis *et al.*, 1997). A similar situation is found in at least some other bilaterian organisms, since in both insects and vertebrates the extreme anterior region of the AP axis is largely devoid of Hox gene expression, and seems to depend instead on a set of head genes known as Otx (*orthodenticle*-like) and Emx (*empty spiracles*-like) for its regional specification (Finkelstein and Boncinelli, 1994).

The homeobox gene *Lox22-Otx* is a *Helobdella* orthologue of the Otx genes described in other animals (Bruce and Shankland, 1998), and it is expressed almost exclusively in the prostomial head region of the leech (Fig. 12). *Lox22-Otx* is expressed in defined regions of the surface ectoderm and the foregut, and also in a few disparate cells of the supraesophageal head ganglion. Analysis of *Lox22-Otx* expression has not yet been combined with lineage tracers, but the expression pattern would appear to include derivatives of all four primary micromeres, and some of the later micromeres as well (Bruce and Shankland, 1998). Onset of *Lox22-Otx* expression begins at a time when the prostomium consists

of hundreds or thousands of cells, and hence this gene is unlikely, at least in terms of zygotic expression, to play a role in the initial specification of the micromeres. Rather, it seems likely that the *Lox22-Otx* gene functions in the regionalization of the prostomium at the onset of organogenesis.

Lox22-Otx is expressed in concentric domains of cells in both the surface ectoderm and foregut, and each of these two circular domains is composed of a radially symmetric array of cells that will experience similar fates during the complicated morphogenesis of the *Helobdella* foregut and mouth (Bruce and Shankland, 1998). The radially symmetric character of these gene expression patterns is particularly interesting given that the primary micromere clones are also radially deployed about the mouth (Fig. 10C). As described in more detail elsewhere, this radial organization of the head domain may have its origin in the symmetry properties of a radially organized prebilaterian ancestor (Bruce and Shankland, 1998).

IV. Conclusions

Leeches have proven to be an excellent experimental system for characterizing various aspects of AP patterning. The AP axis of the adult leech body plan is subdivided into a small, unsegmented head region and an extensive body trunk comprised of 32 outwardly similar body segments. These two regions of the body plan become separated early in the embryonic cell lineage, and seem to utilize largely distinct cellular and molecular patterning mechanisms during subsequent development. The prostomial head domain has a generally radial pattern of organization about the mouth, and has been shown to be the predominant site of expression of a gene that is also known to play a head-patterning role in the development of arthropods and vertebrates.

The ectoderm and mesoderm of the segmented body trunk are generated by a distinct set of embryonic stem cells, the teloblasts, that are situated at the posterior end of the developing body plan. These teloblasts undergo iterative patterns of cell division that first establish the segmental periodicity of the body trunk. The blast cell progeny of these stem cells contribute descendant clones to particular segments in accordance with their lineage history, and some blast cell clones must undergo extensive morphogenetic displacement along the AP axis before coming into register with other segmental tissues. Experiments in which blast cells are repositioned along the AP axis indicate that they possess an intrinsic segment identity from an early developmental stage, and the primary segmental differences established during early development serve as a template for the elaboration of secondary segmental differences that depend on later cell interactions. Blast cell reposition-

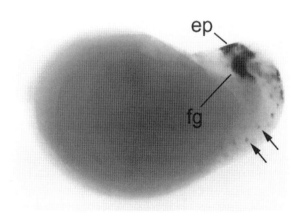

Figure 12 The homeobox gene *Lox22-Otx* is predominantly expressed within the unsegmented head domain. This stage 9 *Helobdella* embryo was stained by *in situ* hybridization, revealing multicellular domains of *Lox22-Otx* expression in the head epidermis (ep) and foregut (fg), as well as a small number of neurons in the anteriormost body segments (arrows). Anterior is to the right, and dorsal toward the top.

ing experiments further indicate that these cells have only a limited ability to actively align themselves with respect to other segmented cell lineages during morphogenesis. Thus, a major component of segmentation in the leech embryo is the morphogenetic assembly of already specified segmental founder cells into the final pattern of segmental primordia.

As seen with the head domain, the trunk ectoderm and mesoderm of the leech embryo utilize some of the same regulatory genes known to be involved in the segmental patterning of a wide variety of other bilaterian animals. The leech Hox genes have been extensively characterized, and very likely play a role in segment diversification, although the timing of their expression suggests that they function downstream of some other mechanism for the initial establishment of segment identities. A leech *en* gene has also been characterized and shown to be expressed around the time when segment primordia first become defined as spatially organized units, although its exact role in the establishment of segment polarity and/or morphogenesis remains unclear. These broad stroke similarities in the regional utilization of widely conserved patterning genes suggest that annelids such as the leech are derived along with other bilaterian animals from a common ancestor whose body plan exhibited a genetically defined head versus trunk distinction (Bruce and Shankland, 1998). This being the case, mechanistic differences in the pattern formation of extant bilaterian animals very likely reflect distinct evolutionary modifications of some common set of ancestral patterning mechanisms.

One such modification must involve the degree to which segmentation or other aspects of AP patterning depend on an invariant pattern of cell lineage. Many features of leech segmentation are closely tied to a segmentally repetitive sequence of cell lineages, and in this regard the leech embryo provides an interesting counterpoint to the other segmented animals currently in use as model systems.

References

Aisemberg, G. O., Wysocka-Diller, J., Wong, V. Y., and Macagno, E. R. (1993). *Antennapedia*-class homeobox genes define diverse neuronal sets in the embryonic CNS of the leech. *J. Neurobiol.* **24**, 1423–1432.

Aisemberg, G. O., and Macagno, E. R. (1994). *Lox1,* an *Antennapedia*-class homeobox gene, is expressed during leech gangliogenesis in both transient and stable central neurons. *Dev. Biol.* **161**, 445–455.

Anderson, D. T. (1973). "Embryology and Phylogeny in Annelids and Arthropods." Pergamon, New York.

Baptista, C. A., Gershon, T. R., and Macagno, E. R. (1990). Peripheral organs control central neurogenesis in the leech. *Nature (London)* **346**, 855–858.

Berezovskii, V. K., and Shankland, M. (1996). Segmental diversification of an identified leech neuron correlates with the segmental domain in which it expresses *Lox2,* a member of the Hox gene family. *J. Neurobiol.* **29**, 319–329.

Bissen, S. T., and Weisblat, D. A. (1987). Early differences between alternate n blast cells in leech embryo. *J. Neurobiol.* **18**, 251–269.

Bissen, S. T., and Weisblat, D. A. (1989). The durations and compositions of cell cycles in embryos of the leech, *Helobdella triserialis. Development (Cambridge, UK)* **106**, 105–118.

Bruce, A. E. E. (1997). Characterization of an *orthodenticle* homolog in the leech and investigation of ectopic gene expression techniques. Ph.D Thesis, Program in Neuroscience, Harvard Medical School, Cambridge, Massachusetts.

Bruce, A. E. E., and Shankland, M. (1998). Expression of the head gene *Lox22-Otx* in the leech *Helobdella* and the origin of the bilaterial body plan. *Dev. Biol.* **200.** In press.

Brusca, R. C., and Brusca, G. J. (1990). "Invertebrates." Sinauer, Sunderland, Massachusetts.

DiNardo, S., Sher, E., Heemskerk-Jongens, J., Kassis, J. A., and O'Farrell, P. H. (1988). Two-tiered regulation of spatially patterned *engrailed* gene expression during *Drosophila* embryogenesis. *Nature (London)* **332**, 604–609.

Dohle, W., and Scholtz, G. (1988). Clonal analysis of the crustacean segment: the discordance between genealogical and segmental borders. *Development (Cambridge, UK) Suppl.* **104**, 147–160.

Finkelstein, R., and Boncinelli, E. (1994). From fly head to mammalian forebrain: The story of *otd* and *Otx. Trends Genet.* **10**, 310–315.

Fraser, S. E., Keynes, R., and Lumsden, A. (1990). Segmentation in the chick embryo hindbrain is defined by cell lineage restrictions. *Nature (London)* **344**, 431–435.

Gao, W.-Q., and Macagno, E. R. (1987a). Extension and retraction of axonal projections by some developing neurons in the leech depends upon the existence of neighboring homologues. I. The HA cells. *J. Neurobiol.* **18**, 43–59.

Gao, W.-Q., and Macagno, E. R. (1987b). Extension and retraction of axonal projections by some developing neurons in the leech depends upon the existence of neighboring homologues. II. The AP and AE neurons. *J. Neurosci.* **18**, 295–313.

Gleizer, L., and Stent, G. S. (1993). Developmental origin of segment identity in the leech mesoderm. *Development (Cambridge UK)* **117**, 177–189.

Ho, R. K., and Weisblat, D. A. (1987). A provisional epithelium in leech embryo: Cellular origins and influence on a developmental equivalence group. *Dev. Biol.* **120**, 520–534.

Huang, F. Z., and Weisblat, D. A. (1996). Cell fate determination in an annelid equivalence group. *Development (Cambridge, UK)* **122**, 1839–1847.

Jellies, J., Loer, C. M., and Kristan, W. B. (1987). Morphological changes in leech Retzius neurons after target contact during embryogenesis. *J. Neurosci.* **7**, 2618–2629.

Kostriken, R., and Weisblat, D. A. (1992). Expression of a *Wnt* gene in embryonic epithelium of the leech. *Dev. Biol.* **151**, 225–241.

Kourakis, M. J., Master, V. A., Lokhorst, D. K., Nardelli-Haefliger, D., Wedeen, C .J., Martindale, M. Q., and Shankland, M. (1997). Conserved anterior boundaries of Hox gene expression in the central nervous system of the leech *Helobdella. Dev. Biol.* **190**, 284–300.

Lans, D., Wedeen, C. J., and Weisblat, D. A. (1993). Cell lineage analysis of the expression of an *engrailed* homolog in leech embryos. *Development (Cambridge, UK)* **117**, 857–871.

Lawrence, P. A. (1992). "The Making of a Fly." Blackwell Scientific, Oxford.

Liu, N.-J. L., Isaksen, D. E., Smith, C. M., and Weisblat, D. A. (1998). Movements and stepwise fusion of endodermal precursor cells in leech. *Dev. Genes & Evol.* **208**, 117–127.

Loer, C. M., and Kristan, W. B. (1989a). Central synaptic inputs to identified leech neurons are determined by peripheral targets. *Science* **244**, 64–66.

Loer, C. M., and Kristan, W. B. (1989b). Peripheral target choice by homologous neurons during embryogenesis of the medicinal leech. II. Innervation of ectopic reproductive tissue by nonreproductive Retzius cells. *J. Neurosci.* **9**, 528–538.

Loer, C. M., Jellies, J., and Kristan, W. B. (1987). Segment-specific morphogenesis of leech Retzius neurons requires particular peripheral targets. *J. Neurosci.* **7**, 2630–2638.

Martindale, M. Q., and Shankland, M. (1988). Developmental origin of segmental differences in the leech ectoderm: Survival and differentiation of the distal tubule cell is determined by the host segment. *Dev. Biol.* **125**, 290–300.

Martindale, M. Q., and Shankland, M. (1990a). Neuronal competition determines the spatial pattern of neuropeptide expression by identified neurons of the leech. *Dev. Biol.* **139**, 210–226.

Martindale, M. Q., and Shankland, M. (1990b). Segmental founder cells of the leech embryo have intrinsic segmental identity. *Nature (London)* **347**, 672–674.

Martinez-Arias, A., and Lawrence, P. A. (1985). Parasegments and compartments in the *Drosophila* embryo. *Nature (London)* **313**, 639–642.

McGinnis, W., and Krumlauf, R. (1992). Homeobox genes and axial patterning. *Cell (Cambridge, Mass.)* **68**, 283–302.

McHugh, D. (1997). Molecular evidence that echiurans and pogonophorans are derived annelids. *Proc. Natl. Acad. Sci. U.S.A.* **94**, 8006–8009.

Nardelli-Haefliger, D., and Shankland, M. (1993). *Lox10,* a member of the *NK-2* homeobox gene class, is expressed in a segmental pattern in the endoderm and in the cephalic nervous system of the leech *Helobdella. Development (Cambridge, UK)* **118**, 877–892.

Nardelli-Haefliger, D., Bruce, A. E. E., and Shankland, M. (1994). An axial domain of HOM/Hox gene expression is formed by the morphogenetic alignment of independently specified cell lineages in the leech *Helobdella. Development (Cambridge, UK)* **120**, 1839–1849.

Offield, M. F., Jetton, T. L., Labosky, P. A., Ray, M., Stein, R., Magnuson, M. A., Hogan, B. L. M., and Wright, C. V. E. (1996). PDX-1 is required for pancreatic outgrowth and differentiation of the rostral duodenum. *Development (Cambridge, UK)* **122**, 983–995.

Prokop, A., and Technau, G. M. (1994). Early tagma-specific commitment of *Drosophila* CNS progenitor NB1-1. *Development (Cambridge, UK)* **120**, 2567–2578.

Ramirez, F.-A., Wedeen, C. J., Stuart, D. K., Lans, D., and Weisblat, D. A. (1995). Identification of a neurogenic sublineage required for CNS segmentation in an Annelid. *Development (Cambridge, UK)* **121**, 2091–2097.

Render, J. (1991). Fate maps of the first quartet micromeres in the gastropod *Ilyanassa obsoleta. Development (Cambridge, UK)* **113**, 495–501.

Render, J. (1997). Cell fate maps in the *Ilyanassa obsoleta* embryo beyond the third division. *Dev. Biol.* **189**, 301–310.

Sandig, M., and Dohle, W. (1988). The cleavage pattern in the leech *Theromyzon tessulatum* (Hirudinea, Glossiphoniidae). *J. Morphol.* **196**, 217–252.

Scholtz, G., Dohle, W., Sandeman, R. E., and Richter, S. (1993). Expression of engrailed can be lost and regained in cells of one clone in crustacean embryos. *Intl J. Dev. Biol.* **37**, 299–304.

Shankland, M. (1984). Positional determination of supernumerary blast cell death in the leech embryo. *Nature (London)* **307**, 541–543.

Shankland, M. (1987a). Differentiation of the O and P cell lines in the embryo of the leech. I. Sequential commitment of blast cell sublineages. *Dev. Biol.* **123**, 85–96.

Shankland, M. (1987b). Determination of cleavage pattern in embryonic blast cells of the leech. *Dev. Biol.* **120**, 494–498.

Shankland, M. (1987c). Differentiation of the O and P cell lines in the embryo of the leech. II. Genealogical relationship of descendant pattern elements in alternative developmental pathways. *Dev. Biol.* **123**, 97–107.

Shankland, M. (1991). Leech segmentation: Cell lineage and the formation of complex body patterns. *Dev. Biol.* **144**, 221–231.

Shankland, M. (1994). Leech segmentation: A molecular perspective. *BioEssays* **16**, 801–808.

Shankland, M., and Martindale, M. Q. (1992). Segmental differentiation of lineally homologous neurons in the central nervous system of the leech. *In* "Determinants of Neuronal Identity" (M. Shankland and E. R. Macagno, eds.), pp. 45–77. Academic Press, New York.

Shankland, M., and Savage, R. M. (1997). Annelids, the segmented worms. *In* "Embryology: Constructing the Organism" (S. F. Gilbert and A. M. Raunio, eds.), pp. 219–235. Sinauer, Sunderland, Massachusetts.

Shankland, M., and Weisblat, D. A. (1984). Stepwise commitment of blast cell fates during the positional specification of the O and P cell lines in the leech embryo. *Dev. Biol.* **106**, 326–342.

Simcox, A., and Sang, J. (1983). When does determination occur in *Drosophila* embryos? *Dev. Biol.* **97**, 212–221.

Smith, C. M., and Weisblat, D. A. (1994). Micromere fate maps in leech embryos: Lineage-specific differences in rates of cell proliferation. *Development (Cambridge, UK)* **120**, 3427–3438.

Smith, C. M., Lans, D., and Weisblat, D. A. (1996). Cellular mechanisms of epiboly in leech embryos. *Development (Cambridge, UK)* **122**, 1885–1894.

Storey, K. G. (1989). Cell lineage and pattern formation in the earthworm embryo. *Development (Cambridge, UK)* **107**, 519–532.

Stuart, D. K., Torrence, S. A., and Law, M. I. (1989). Leech neurogenesis. I. Positional commitment of neural precursor cells. *Dev. Biol.* **136**, 17–39.

Torrence, S. A., Law, M. I., and Stuart, D. K. (1989). Leech neurogenesis. II. Mesodermal control of neuronal patterns. *Dev. Biol.* **136**, 40–60.

Vincent, J.-P., and O'Farrell, P. H. (1992). The state of *engrailed* expression is not clonally transmitted during early *Drosophila* development *Cell (Cambridge, Mass.)* **68**, 923–931.

Wang, B., Müller-Immergluck, M. M., Austin, J., Robinson, N. T., Chisholm, A., and Kenyon, C. (1993). A homeotic gene cluster patterns the anteroposterior body axis of C. *elegans. Cell (Cambridge, Mass.)* 74, 29–42.

Wedeen, C. J. (1995). Regionalization and segmentation of the leech. *J. Neurobiol.* **27**, 277–293.

Wedeen, C. J., and Shankland, M. (1997). Mesoderm is required for the formation of a segmented endodermal cell layer in the leech *Helobdella. Dev. Biol.* **191**, 202–214.

Wedeen, C. J., and Weisblat, D. A. (1991). Segmental expression of an *engrailed*-class gene during early development and neurogenesis in an annelid. *Development (Cambridge, UK)* **113**, 805–814.

Weisblat, D. A., and Blair, S. S. (1984). Developmental indeterminancy in embryos of the leech *Helebodella triserialis. Dev. Biol.* **101**, 326–335.

Weisblat, D. A., and Shankland, M. (1985). Cell lineage and segmentation in the leech. *Philos. Trans. R. Soc. London Ser. B* **312**, 39–56.

Weisblat, D. A., Kim, S. Y., and Stent, G. S. (1984). Embryonic origins of cells in the leech *Helobdella triserialis. Dev. Biol.* **104**, 65–85.

Weisblat, D. A., Astrow, S. H., Bissen, S. T., Ho, R. K., Holton, B., and Settle, S. A. (1987). Early events associated with determination of cell fate in leech embryos. *In* "Genetic Regulation of Development" (W. F. Loomis, ed.), pp. 265–285. A. R. Liss, New York.

Whitman, C. O. (1878). The embryology of *Clepsine. Q. J. Microsc. Sci.* **18**, 215–315.

Woltereck, R. (1904). Beiträge zur praktischen Analyse der *Polygordius* Entwicklung nach dem "Nordsee"- und "Mittelmeer"-Typus. I. Die für beide Typen gleichverlaufende entwicklungsabschnitt: Vorm Ei bis zum jungsten Trochophora-Stadium. *Arch. Entwick lungs, mech. Org.* **18**, 377–403.

Wong, V. Y., Aisemberg, G. O., Gan, W.-B., and Macagno, E. R. (1995). The leech homeobox gene *Lox4* may determine segmental differentiation of identified neurons. *J. Neurosci.* **15**, 5551–5559.

Wordeman, L. (1983). Kinetics of primary blast cell production in the embryo of the leech *Helobdella triserialis.* Honors Thesis, Department of Molecular Biology, University of California, Berkeley.

Wysocka-Diller, J., Aisemberg, G. O., and Macagno, E. R. (1995). A novel homeobox cluster expressed in repeated structures of the midgut. *Dev. Biol.* **171**, 439–447.

Zackson, S. L. (1982). Cell clones and segmentation in leech development. *Cell (Cambridge, Mass.)* **31**, 761–770.

Zackson, S. L. (1984). Cell lineage, cell–cell interaction and segment formation in the ectoderm of a glossiphoniid leech embryo. *Dev. Biol.* **104**, 43–60.

IV

Drosophila

15

Studies on Cell Lineage and Cell Fate Determination in Drosophila

JOSÉ A. CAMPOS-ORTEGA
Institut für Entwicklungsbiologie
Universität zu Köln
50923 Köln, Germany

I. Introduction
II. Embryonic Development
III. The Imaginal Discs and the Histoblasts
IV. Studies on Cell Lineage
    A. Genetic Markers

V. Studies on Cell Fate Determination
VI. Summary
    References

I. Introduction

The fruit fly has proved to be an excellent organism in which to approach a large variety of developmental problems—the study of cell lineage and cell fate determination is a prime example. On the one hand, the biological characteristics of *Drosophila,* as a holometabolous insect, offer the opportunity to analyze questions related to cell lineage and determination in either the larva, by means of experiments carried out on the developing embryo, or in the imaginal discs (which will give rise to the imaginal epidermis), by means of experiments carried out on the larva. On the other hand, a rich repertoire of experimental techniques is available with which to approach these questions. In spite of its small size, the embryo of *Drosophila* is amenable to manipulation with relative ease; techniques of experimental embryology have attained a considerable degree of sophistication: intracellular injection of dyes, DiI application, cell transplantations, etc., are routinely used for different purposes.

In addition to embryology, developmental genetics, with a profusion of tools, has become established as a firm pillar in the field of ontogenic investigations in *Drosophila.* First of all, a large number of easily scorable marker mutations, which have been collected over the years, segregate with each of the chromosomes, and are expressed in all epidermal derivatives, can be used to

ask questions related to epidermal development. Second, a large number of reagents have been generated in studies on the many genes already cloned and analyzed at the molecular level: cDNA and genomic clones as probes for *in situ* hybridization, antibodies directed against either naturally occurring or artificially engineered proteins, and "enhancer traps," expressed in particular cells and tissues. This material can be used as cell specific markers to elucidate cytological aspects of questions related to cell lineage and determination in internal organs, as the central nervous system (CNS), the musculature, or the alimentary canal. Third, the analysis of the phenotypic effects of a large number of mutations has immensely contributed to our understanding of specific aspects of development, thus increasing the conceptual richness of the problems treated. As a whole, *Drosophila* is one of the most convenient organisms in which to approach the questions that concern us here.

The following is not intended to be an exhaustive treatment of cell lineage and determination in *Drosophila,* but just to present a brief overview of the major techniques used in the analysis of cellular lineages and the assignment of a particular developmental fate. Because most investigations related to the subject of this chapter have been carried out either in the embryo or in the imaginal discs, a short summary of the main features of embryonic and imaginal disc development is given below.

II. Embryonic Development

The *Drosophila* embryo is a dorsoventrally slightly flattened ovoid, with an average length of 500 μm and a diameter of 180 μm. Developing structures can be unambiguously located within the embryo by reference to the percentage egg length (% EL, 0% at the posterior pole) and the percentage of ventrodorsal circumference (% VD, 0% at the ventral midline). After fertilization, the zygotic nuclei divide 13 times before cellularization occurs at the blastoderm stage. Formation of the somatic cells involves extension of membrane furrows between the syncytial blastoderm nuclei. Embryogenesis lasts for about 22 hours at 25°C and the entire process has been subdivided into 17 stages, each defined by specific embryogenetic features (Bownes, 1975; Campos-Ortega and Hartenstein, 1985, 1997; Wieschaus and Nüsslein-Volhard, 1986).

This scheme provides a very useful frame of reference for morphogenetic processes and has been widely used during the past several years (refer to Fig. 1). The first five embryogenetic stages comprise the events preceding cell formation, which takes place in stages 5–6. Gastrulation, namely, the formation of germ layers, takes place in stages 6–7. Germ band elongation starts at stage 7 and ends in stage 11, causing the posterior half of the germ band to fold over on top of the anterior half. The germ band comprises the primordia of the three gnathal segments (mandible, maxilla, and labium), three thoracic segments (pro-, meso-, and metathorax), and nine abdominal segments. In stage 10 a group of cells invaginates at the anteroventral tip of the egg to form the stomodeum, which will give rise to the foregut. Germ band shortening takes place in stage 12 and the normal anatomical relationships of the larva are restored.

Definitive segmentation in the territory of the germ band is completed by the end of germ band shortening, when deep intersegmental furrows develop at ventral and dorsal levels in the epidermis. However, in *Drosophila* the process of segmentation is initiated by the formation of transitory structures called parasegmental furrows, which are later replaced by the definitive segmental furrows (Martinez-Arias and Lawrence, 1985).

Neurogenesis is the process of development of the nervous system. Neuroblast segregation begins at the end of stage 8 and lasts for 2.5 hours, until stage 11; it takes place in pulses, in which different subpopulations of neuroblasts arise (Hartenstein and Campos-Ortega, 1984; Doe, 1992; Broadus *et al.*, 1995). After segregation, the neuroblasts become arranged according to a characteristic, constant pattern, and soon afterwards neuroblasts start dividing to produce ganglion mother cells; the progeny of these latter cells will differentiate as either neurons or glial cells.

The period of postblastoderm mitotic activity corresponds roughly to the extended germ band stage (stages 8–12). All embryonic cells, except those of the amnioserosa, the neuroblasts, the progenitors of the epidermal sensilla, and the pole cells, divide on average only three times during the entire postblastoderm period according to a characteristic spatiotemporal pattern (Hartenstein and Campos-Ortega, 1985; Foe, 1989; Campos-Ortega and Hartenstein, 1997). Each neuroblast divides according to a specific pattern and produces a stereotypic lineage. Amnioserosa cells do not divide at all.

Dorsal closure of the embryo occurs by extension of the dorsal epidermal primordium on either side, and fusion at the dorsal midline. At the time of germ band shortening, the complex morphogenetic movements that lead to head involution begin. After dorsal closure and head involution, morphogenesis of the larva is essentially complete.

III. The Imaginal Discs and the Histoblasts

The epidermis of the adult fly develops from the imaginal discs and the histoblasts, groups of epidermal cells that become obvious toward the end of embryogenesis (refer to Bate and Martinez-Arias, 1991; Cohen *et al.*, 1991, 1993; Hartenstein and Jan, 1992; Cohen, 1993; Campos-Ortega and Hartenstein, 1997, for more detailed descriptions). The primordia of the imaginal discs invaginate from the epidermis during embryonic stages 15–17 to form small pouches connected with the outside, whereas the histoblasts remain as specialized cell nests integrated in the larval abdominal epidermis. There are three pairs of ventral thoracic, or leg, discs and three pairs of dorsal thoracic, that is, humerus, wing, and haltere, discs. Wing and haltere disc cells originate in association with the meso- and metathoracic leg discs during stage 12, but become displaced dorsally prior to invagination in stage 15 (Cohen *et al.*, 1993; Cohen, 1993; Meise and Janning, 1993). The genital disc is an unpaired cluster of 12–15 cells in the posterior of the embryo. The imaginal head develops from three pairs of imaginal discs, labial, clypeo-labral, and eye-antennal. The size of the imaginal disc primordia in the embryo varies between about 10 and 60 cells.

The cells of all imaginal discs proliferate during the larval period, the cell population increasing exponentially until approximately 10 hours after puparium formation, when mitotic activity ceases (see Nöthiger, 1972). Some aspects of cell differentiation, particularly those related to the sensory neurons, occur during the third larval instar. However, the bulk of cytodifferentiation takes place during metamorphosis.

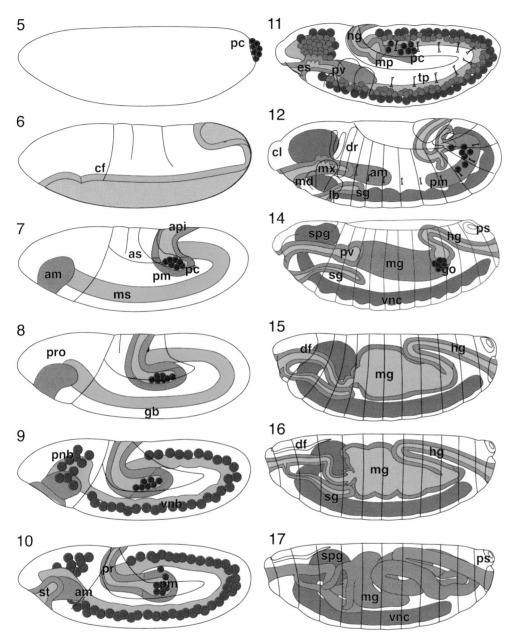

Figure 1 Twelve drawings of embryos of increasing age illustrating the major events of *Drosophila* embryonic development. The number at the top left of each drawing indicates the embryonic stage. Stage 5 shows the blastoderm with the pole cells (pc, red) at the posterior pole. Stage 6 illustrates morphogenetic movements during gastrulation. Light shading indicates the larval anlagen that will invaginate during gastrulation, that is, anlagen of the endodermal anterior midgut primordium (am, pink), mesoderm (ms, green), proctodeum (blue), and posterior midgut primordium (pm/pr, pink). The cephalic furrow (cf) separates the procephalon from the prospective metameric germ band. as denotes amnioserosa. Mesoderm (green) and endoderm (am, pm, pink) have completed invagination. Invaginated regions are shaded in lateral projection (hollow spaces, or lumina, in this drawing are hatched), and pole cells are included in the posterior midgut primordium. Stage 7 illustrates the beginning of germ band elongation. During stage 8 germ band (gb) elongation proceeds further; the primordia of the posterior midgut and the proctodeum become individualized. In stage 9 germ band elongation enters its slow phase; the mesoderm transiently exhibits segmental bulges; segregation of neuroblasts (purple: vnb, ventral neuroblasts; pnb, procephalic neuroblasts) begins. In stage 10 stomodeal invagination (st) takes place; segregation of the neuroblasts continues, and first neuroblast divisions occur to give rise to ganglion mother cells (light purple). In stage 11 parasegmental furrows become evident (tp, tracheal pits; es, oesophagus); pole cells leave the posterior midgut pocket (mp) and become arranged laterally. The germ band retracts during stage 12, and the drawing shows the final stages of this retraction. During this process fusion of the anterior and posterior midgut primordia takes place and the definitive segmental boundaries appear. The salivary glands (blue, sg) and the dorsal ridge (dr) appear. cl denotes clypeolabrum; hg, hindgut (blue); md, mandibulary bud; mx, maxillary bud; lb, labial bud; pm, posterior midgut. In stage 14 dorsal closure of the midgut (mg) and epidermis, and head involution begin. go denotes gonads (red); ps, posterior spiracles; pv, proventriculus; spg, supraesophageal ganglion; vnc: ventral cord (purple). In stage 15 head involution is well advanced. Notice growth of the hindgut. df denotes dorsal fold; sg, salivary glands. In stage 16 head involution is almost complete, midgut (mg) constrictions appear, and condensation of the ventral cord begins. Stage 17 corresponds to the fully developed embryo. [Modified from Campos-Ortega, J. A. and Hartenstein, V. (1997) "The Embryonic Development of *Drosophila melanogaster*," 2nd Ed., with permission of Springer-Verlag. Copyright 1997 Springer-Verlag.]

The histoblasts give rise to the abdominal epidermis. Three histoblast nests can already be distinguished in each segment in stage 14: two are located dorsally, one anterior to the other at the level of the longitudinal tracheal trunk, and the third is ventral. The cells of the histoblasts are mitotically quiescent during larval stages, remaining in the G2 phase of the cycle and starting their mitotic divisions after pupariation.

IV. Studies on Cell Lineage

In *Drosophila*, the direct observation of developing cells is of limited value for cell lineage studies due to the high cell density and the refringency of yolk granules. Thus, a prerequisite for such studies is to label the cell whose lineage is to be analyzed. Two main types of label have been used, genetic markers and dyes.

A. Genetic Markers

1. Gynandromorphs

Cell lineage studies in Drosophilidae using genetically marked cells have a long history. They were initiated in 1929, when Alfred Sturtevant constructed the first fate map derived from the analysis of a number of gynandromorphs using the *claret* mutation of *Drosophila simulans*. For many years, cell lineage studies in *Drosophila* concentrated on fate map construction. More elaborate blastoderm fate maps were established by Garcia-Bellido and Merriam (1969), using Sturtevant's data on 379 gynandromorphs of *D. simulans*, as well as by Hotta and Benzer (1972), and many others (see Janning, 1978) using gynandromorphs of *Drosophila melanogaster*. These initial fate maps were restricted to imaginal structures, since the markers used permitted one to score derivatives of the imaginal discs only. Janning (1974, 1976) constructed the first fate map of larval structures using aldehyde oxidase expression as a marker for internal organs and the *maroon-like* mutation (Janning, 1972). Kankel and Hall (1976) used the acid phosphatase locus, translocated to the X-chromosome, for gynandromorph studies on the development of the CNS.

2. Mitotic Recombination

A new dimension in the cell lineage studies in *Drosophila* was inaugurated with the discovery of spontaneous mitotic recombination by Curt Stern (1936, 1968). X-ray induced mitotic recombination was soon introduced (Friesen, 1936) and it has been extensively used for studies related to various aspects of clonal analysis (Becker, 1957, 1978; Bryant, 1970; Garcia-Bellido and Merriam, 1971; Garcia-Bellido and Ripoll, 1978; Wieschaus, 1978). The analysis of cell clones ini-

tiated by inducing mitotic recombination with X-rays at defined developmental stages has permitted conclusions chiefly related to the parameters of growth of the different imaginal discs in the wild type. It has allowed inferences concerning the size of the embryonic anlagen, the number of mitoses carried out by imaginal disc cells, the orientation of the mitotic spindle, developmental compartments, etc. As with gynandromorphs, enzymatic markers can be used to analyze the development of the nonepidermal, internal organs; thus, a temperature-sensitive allele of succinate dehydrogenase (*sdh*) has been used for clonal analysis of muscle development (Lawrence, 1982). Transplantations of nuclei homozygous for the *sdh*[ts] allele in embryos that are wild type for the *sdh* gene have been used to generate mosaics and thus study some aspects of development of internal organs (Fischbach and Technau, 1984; Lawrence and Johnston, 1986). Xu and Rubin (1993) have engineered chimeric proteins that are expressed either in the nucleus or in the cell membrane of developing and adult cells. Antibodies against these proteins can be used for clonal analysis of internal structures in genetic mosaics generated by mitotic recombination.

Clonal analysis also can be very useful for morphogenetic studies. The most important use of mitotic recombination in this context is in the analysis of the morphogenetic effects of mutant genes which when homozygous have lethal consequences for the whole animal. Development of a cell clone that is homozygous for a lethal mutation can be easily initiated in an animal that is heterozygous for this mutation; the cell clone will manifest the morphogenetic effects of the mutation without affecting the remainder of the organism (see Garcia-Bellido and Dapena, 1974; Ferrús and Garcia-Bellido, 1976; Xu and Rubin, 1993). This technique thus makes most of the genome of the fruit fly available for developmental genetic studies. Among other aspects, studies using this technique have defined the morphogenetic effects of particular mutations, both on their own and when intermingled with wild-type cells, as well as answering questions concerning cell autonomy of phenotypic expression (see a discussion in Garcia-Bellido and Ripoll, 1978).

In spite of all these advantages, mitotic recombination has two important disadvantages: the frequency of spontaneous mitotic recombination is very low; this frequency can be enhanced considerably by ionizing irradiation, but irradiation is associated with cell death. Both these problems can be overcome using the technique developed by Golic (1991), which makes use of the yeast recombinase FLP, triggered by heat-shock, to delete sequences flanked by its target sequence FRT (Golic and Lindquist, 1989). This technique is considerably less traumatic than X-ray treatment and induces mitotic recombination at a much higher frequency. Xu and Rubin (1993) have generated a large number of strains carrying FRT insertions in the neighborhood of the centromeres of each chromosome arm.

3. *Labeling Cells with Dyes*

Injection techniques for cell lineage studies in *Drosophila*, that is, labeling by intracellular injection of dyes into cells of the blastoderm and studying the progeny of the injected cells in the fully differentiated embryo, were introduced in the 1980s. However, injections of individual cells are difficult and, consequently, injections of several cells were mainly directed towards fate-mapping embryonic organs and tissues, such as the nervous system, the alimentary canal, the somatic and visceral musculature, etc. (Technau and Campos-Ortega, 1985; Hartenstein *et al.*, 1985; Schmidt-Ott and Technau, 1994). However, more recently, DiI techniques were developed to label single cells. In this manner, the composition of the lineages of all types of *Drosophila* neuroblasts was determined (Bossing and Technau, 1994; Udolph *et al.*, 1993; Bossing *et al.*, 1996; Schmidt *et al.*, 1997).

A variation of cell lineage studies by dye injection consists in the transplantation of single cells previously labeled with horseradish peroxidase (HRP), FITC, or a mixture of both (Technau, 1986). The protocol for this kind of experiment is simple: the label is injected into an embryo prior to cell formation. Before cell membranes form, the injected label diffuses throughout the egg and is later incorporated into all blastoderm cells, which are thus labeled. A few cells are then removed from specific positions in the donor embryos, and individually transplanted at specific positions into unlabeled host embryos. Since the number of times the transplanted cell divides is low, its progeny carry the label at high concentration and, thus, can be visualized in the fully developed embryo. Still another variation of this technique, consisting of transplanting single cells from genetically labeled donor embryos expressing *lacZ* in larval and imaginal cells (Prokop and Technau, 1991; Meise and Janning, 1993, 1994; Holz *et al.*, 1997), has been used for cell lineage and fate mapping studies of imaginal cells. Since the *lacZ* gene is expressed throughout development, and the β-galactosidase protein is very stable, the label is not diluted significantly even by many mitotic divisions. Prokop and Technau (1991) demonstrated that larval and imaginal nerve cells in the CNS share lineages. Meise and Janning (1993, 1994; Holz *et al.*, 1997) succeeded in mapping the location of the thoracic discs in the late blastoderm–early gastrula stage embryo, showing in addition that progenitor cells for larval and imaginal tissues are not yet segregated at this stage (see below).

The data on cell lineage collected using either genetic markers or dyes are numerous and have contributed a great deal to our understanding of embryonic and larval development in the fruit fly. However, these data have failed to further dissect individual clones in terms of genealogy and temporal relationships, two essential aspects of lineage development, that is, which cell is related to which other cell within a given clone, and when do the individual cells become postmitotic and differentiate within a given branch of the genealogical tree. However, thanks to recent progress in microscopy and electronics, it is now possible in *Drosophila* to approach the genealogical and temporal aspects of at least some cell lineages, following by time-lapse methods the behavior of cells labeled by any of various means.

V. Studies on Cell Fate Determination

Mitotic recombination has allowed inferences as to the time at which cells become specified as components of a particular imaginal disc (Steiner, 1976; Wieschaus and Gehring, 1976), or part of a disc (Becker, 1957; Garcia-Bellido *et al.*, 1973). However, it is not possible to deduce principles of cell determination from the analysis of somatic mosaics, since the concept of cell fate determination is a purely operational one. This means that, strictly speaking, the claim that a cell or a group of cells is determined for a given developmental fate requires an experiment, generally a transplantation experiment, to show that the cell behavior remains unchanged (see a discussion in Stent, 1985) in an altered environment. In *Drosophila*, cell transplantation experiments directed toward answering questions related to cell fate determination have been carried out in various types of embryonic cells and in the imaginal discs of the larva.

As to the embryo, experiments have been carried out with individual embryonic cells labeled either with genetic marker mutations (Illmensee, 1978), or with a mixture of HRP and FITC (Technau, 1986). In the latter case, a few cells are removed from specific positions of any of the three germ layers of, generally, a stage 7 embryo and transplanted individually iso- or heterotopically, iso- or heterochronically into unlabeled host embryos (Technau, 1986; Technau and Campos-Ortega, 1986a,b; Beer *et al.*, 1987; Technau *et al.*, 1988; Ludolph *et al.*, 1995; Holz *et al.*, 1997). A few xenoplastic transplantations also have been carried out between different *Drosophila* species (Becker and Technau, 1990). Experiments on isotopic transplantation of pole cells by Illmensee (see 1978 for review) showed that these cells are firmly committed to their fate as progenitors of the germ line. Single cell transplantation of somatic cells has shown that individual cells develop according to the germ layer from which they derive, but that they also possess a broad capability to regulate. Thus, single cells from all three germ layers in the early gastrula stage (stage 7–8) behave as though firmly committed to the fate of a given germ layer; however, specification as a particular derivative is largely dependent on the relationships that the individual cell establishes with its neighbors. By transplanting single cells that express *lacZ* in all imaginal and larval cells into unlabeled hosts, Holz *et al.* (1997) have shown that abdominal cells from gastrula stage donors cannot participate in the formation of thoracic imaginal discs. That is to say, abdominal cells have attained a determination state along the anteroposterior axis earlier than along the dorsoventral axis. With respect to the developing CNS, Technau and co-workers have collected a number of interesting observations on determination and specification of neuroblasts, recently compiled by Urban and Technau (1997).

A large number of studies on cell determination have been carried out by transplanting fragments of imaginal discs into third-instar larvae that are ready to pupate, a technique developed by Ephrussi and Beadle (1936) to study eye-color mutants. The host pupates, the transplanted blasteme differentiates with the host, and the differentiated tissue can be removed from the fly and identified. This technique has been used to establish fate maps of imaginal discs, and, in addition to determination, studies related to problems of transdetermination (Hadorn, 1965) relied on the same technique (see an excellent discussion of the older literature in Nöthiger, 1972; Gehring, 1972).

VI. Summary

The contribution of work on *Drosophila* has been of paramount importance to our current understanding of the mechanisms of cell type specification and cell determination. This is explained by the fact that, in approaching particular ontogenetic processes, work from several disciplines, namely, embryology, cell biology, classic genetics, and molecular genetics, could be combined. In the following three chapters, specific aspects of the development of the fruit fly are discussed. Chapter 16 by Carthew *et al.* deals with the mechanisms and elements controlling cell type specification in the compound eye. Chapter 17 by Siefried deals with the Wingless signaling pathway and its role in cell type specification and pattern formation. Chapter 18 by Broadus and Spana presents the current results of the analysis of embryonic lineages in the developing CNS. The development of the compound eye and pattern formation in limbs and metameres are discussed chiefly from the point of view of molecular genetics. A large number of mutations interfering with these processes, along with a painstaking dissection of mutant phenotypes and the affected genes, have allowed investigators to achieve a high degree of resolution and a deep level of under-

standing in their studies. In contrast, neural lineages are still analyzed mainly at the cellular level, whereas genetic and molecular genetic studies are just beginning. The technical difficulties associated with the elucidation of these lineages and the definition of the various cell types in the developing CNS are formidable, but the reader will agree that a very promising start has been made.

References

Bate, C. M., and Martinez-Arias, A. (1991). The embryonic origin of imaginal discs in *Drosophila*. *Development* (*Cambridge, UK*) **112,** 755–761.

Becker, H. J. (1957). Über Röntgenmosaikflecken und Defektmutationen am Auge von *Drosophila* und die Entwicklungsphysiologie des Auges. *Z. Indukt. Abstamm. Vererbungsl.* **88,** 333–373.

Becker, H. J. (1978). Mitotic recombination and position effect variegation. *In* "Genetic Mosaics and Cell Differentiation" (W. J. Gehring, ed.), pp. 29–49. Springer-Verlag, Berlin, Heidelberg, and New York.

Becker, T., and Technau, G. M. (1990). Single cell transplantation reveals interspecific cell communication in *Drosophila* chimeras. *Development* (*Cambridge, UK*) **109,** 821–832.

Beer, J., Technau, G. M., and Campos-Ortega, J. A. (1987). Lineage analysis of transplanted individual cells in embryos of *Drosophila melanogaster*. IV. Commitment and proliferative capabilities of mesodermal cells. *Roux's Arch. Dev. Biol.* **196,** 220–230.

Bossing, T., and Technau, G.M. (1994). The fate of the CNS midline progenitors of *Drosophila* as revealed by a new method for single cell labelling. *Development* (*Cambridge, UK*) **120,** 1895–1906.

Bossing, T., Udolph, G., Doe, C. Q., and Technau, G. M. (1996). The embryonic central nervous system lineages of *Drosophila melanogaster*. I. Neuroblast lineages derived from the ventral half of the truncal neuroectoderm. *Dev. Biol.* **179,** 41–64.

Bownes, M. (1975). A photographic study of development in the living embryo of *Drosophila melanogaster*. *J. Embryol. Exp. Morphol.* **33,** 789–801.

Broadus, J., Skeath, J. B., Spana, E. P., Bossing, T., Technau, G. M., and Doe, C. Q. (1995). New neuroblast markers and the origin of the aCC/pCC neurons in the *Drosophila* central nervous system. *Mech. Devl.* **53,** 393–402.

Bryant, P. J. (1970). Cell lineage relationships in the imaginal wing disc of *Drosophila melanogaster*. *Dev. Biol.* **22,** 389–411.

Campos-Ortega, J. A., and Hartenstein, V. (1997). "The Embryonic Development of *Drosophila melanogaster*," 2nd Ed. Springer-Verlag, Berlin, Heidelberg, New York, and Tokyo.

Cohen, B., Wimmer, E. A., and Cohen, S. M. (1991). Early development of leg and wing primordia in the *Drosophila* embryo. *Mech. Dev.* **33,** 229–240.

Cohen, B., Simcox, A. A., and Cohen, S. M. (1993). Allocation of the imaginal primordia in the *Drosophila* embryo. *Development (Cambridge, UK)* **117**, 597–608.

Cohen, S. M. (1993). Imaginal disc development. *In* "Development of *Drosophila melanogaster*" (C. M. Bate and A. Martinez-Arias, eds.), pp. 609–685. Cold Spring Harbor Laboratory, Cold Spring Harbor, New York.

Doe, C. Q. (1992). Molecular markers for identified neuroblasts and ganglion mother cells in the *Drosophila* nervous system. *Development (Cambridge, UK)* **116**, 855–863.

Ephrussi, B., and Beadle, G. W. (1936). A technique of transplantation for *Drosophila*. *Am. Nat.* **70**, 218–225.

Ferrus, A., and Garcia-Bellido, A. (1976). Morphogenetic mutants detected in mitotic recombination clones. *Nature (London)* **260**, 425–426.

Fischbach, K. F., and Technau, G. M. (1984). Cell degeneration in the developing optic lobes of the sine oculis and small optic lobes mutants of *Drosophila melanogaster*. *Dev. Biol.* **104**, 219–239.

Foe, V. A. (1989). Mitotic domains reveal early commitment of cells in *Drosophila* embryos. *Development (Cambridge, UK)* **107**, 1–22.

Friesen, H. (1936). Spermatogoniales Crossing-over bei *Drosophila*. *Z. Indukt. Abst. Vererbungsl.* **71**, 501–526.

Garcia-Bellido, A., and Dapena, J. (1974). Induction, detection and characterization of cell differentiation mutants in *Drosophila*. *Mol. Gen. Genet.* **128**, 117–130 .

Garcia-Bellido, A., and Merriam, J. R. (1969). Cell lineage of the imaginal disks in *Drosophila* gynandromorphs. *J. Exp. Zool.* **170**, 61–76.

Garcia-Bellido, A., and Merriam, J. R. (1971). Parameters of the wing imaginal disk development of *Drosophila melanogaster*. *Dev. Biol.* **24**, 61–87.

Garcia-Bellido, A., and Ripoll, P. (1978). Cell lineage and differentiation in *Drosophila*. *In* "Genetic Mosaics and Cell Differentiation" (W. J. Gehring, ed.), pp. 119–156. Springer-Verlag, Berlin, Heidelberg, and New York.

Garcia-Bellido, A., Ripoll, P., and Morata, G. (1973). Developmental compartmentalization of the wing disc of *Drosophila*. *Nature (London)* **245**, 251–253.

Gehring, W. J. (1972). The stability of the determined state in cultures of imaginal disks in *Drosophila*. *In* "The Biology of Imaginal Disks" (H. Ursprung and R. Nöthiger, eds.), pp. 35–58, Springer-Verlag, Berlin, Heidelberg, and New York.

Golic, K. G. (1991). Site-specific recombination between homologous chormosomes in *Drosophila*. *Science* **252**, 958–961.

Golic, K. G., and Lindquist, S. (1989). The FLP recombinase of yeast catalyzes site-specific recombination in the *Drosophila* genome. *Cell (Cambridge, Mass.)* **59**, 499–509.

Hadorn, E. (1965). Problems of determination and transdetermination. *Brookhaven Symp. Biol.* **18**, 148–161.

Hartenstein, V., and Campos-Ortega, J. A. (1984). Early neurogenesis in wild-type. *Drosophila melanogaster*. *Roux's Arch. Dev. Biol.* **193**, 308–325.

Hartenstein, V., and Campos-Ortega, J. A. (1985). Fate-mapping in wild-type *Drosophila melanogaster*. I. The spatio-temporal pattern of embryonic cell divisions. *Roux's Arch. Dev. Biol.* **194**, 181–195.

Hartenstein, V., and Jan, Y. N. (1992). Studying *Drosophila* embryogenesis with P-*lac* Z enhancer trap lines. *Roux's Arch. Dev. Biol.* **201**, 194–220.

Hartenstein, V., Technau, G. M., and Campos-Ortega, J. A. (1985). Fate-mapping in wild-type *Drosophila melanogaster*. III. A fate map of the blastoderm. *Roux's Arch. Dev. Biol.* **194**, 213–216.

Holz, A., Meise, M., and Janning, W. (1997). Adepithelial cells in *Drosophila melanogaster*: Origin and cell lineage. *Mech. Dev.* **62**, 93–101.

Hotta, Y., and Benzer, S. (1972). Mapping of behaviour in *Drosophila* mosaics. *Nature (London)* **240**, 527–535.

Illmensee, K. (1978). Drosophila chimeras and the problem of determination. *In* "Genetic Mosaics and Cell Differentiation" (W. J. Gehring, ed.), pp. 51–69. Springer-Verlag, Berlin, Heidelberg, and New York.

Janning, W. (1972). Aldehyde oxidase as a cell marker for internal organs in *Drosophila melanogaster*. *Naturwissenschaften.* **59**, 516–517.

Janning, W. (1974). Entwicklungsgenetische Untersuchungen an Gynandern von *Drosophila melanogaster*. I. Die inneren Organe der Imago. *Roux's Arch. Dev. Biol.* **174**, 313–332.

Janning, W. (1976). Entwicklungsgenetische Untersuchungen an Gynandern von *Drosophila melanogaster*. IV. Vergleich der morphologischen Anlagepläne larvaler und imaginaler Strukturen. *Roux's Arch. Dev. Biol.* **179**, 349–372.

Janning, W. (1978). Gynandromorph fate maps in *Drosophila*. *In* "Genetic Mosaics and Cell Differentiation" (W. J. Gehring, ed.), pp. 1–28. Springer-Verlag, Berlin, Heidelberg, and New York.

Kankel, D. R., and Hall, J. C. (1976). Fate mapping of nervous system and other internal tissues in genetic mosaics of *Drosophila melanogaster*. *Dev. Biol.* **48**, 1–24.

Lawrence, P. A. (1981). A general cell marker for clonal analysis of *Drosophila* development. *J. Embryol. Exp. Morphol.* **64**, 321–332.

Lawrence, P. A. (1982). Cell lineage of the thoracic muscles of *Drosophila*. *Cell (Cambridge, Mass.)* **29**, 493–503.

Lawrence, P. A., and Johnston, P. (1986). Observations on cell lineage of internal organs of *Drosophila*. *J. Embryol. Exp. Morphol.* **91**, 251–266.

Martinez-Arias, A., and Lawrence, P. A. (1985). Parasegments and compartments in the *Drosophila* embryo. *Nature (London)* **313**, 639–642.

Meise, M., and Janning, W. (1993). Cell lineage of larval and imaginal thoracic anlagen cells of *Drosophila melanogaster*, as revealed by single cell transplantations. *Development (Cambridge, UK)* **118**, 1107–1121.

Meise, M., and Janning, W. (1994). Localization of thoracic imaginal precursor cells in the early embryo of *Drosophila melanogaster*. *Mech. Dev.* **48**, 109–117.

Nöthiger, R. (1972). The larval development of imaginal disks. *In* "The Biology of Imaginal Disks" (H. Ursprung and R. Nöthiger, eds.), pp. 1–34. Springer-Verlag, Berlin, Heidelberg, and New York.

Prokop, A., and Technau, G. M. (1991). The origin of postembryonic neuroblasts in the ventral nerve cord of *Drosophila melanogaster*. *Development (Cambridge, UK)* **111**, 79–88.

Schmidt, H., Rickert, C., Bossing, T., Vef, O., Urban, J., and Technau, G.M. (1997). The embryonic central nervous system lineages of *Drosophila melanogaster*. II. Neuroblast lineages derived from the dorsal part of the neuroectoderm. *Dev. Biol.* **189,** 186–204.

Schmidt-Ott, U., and Technau, G.M. (1994). Fate-mapping in the procephalic region of the embryonic *Drosophila* head. *Roux's Arch. Dev. Biol.* **203,** 367–373.

Steiner, E. (1976). Establishment of compartments in the developing leg imaginal disks of *Drosophila melanogaster*. *Wilhelm Roux's Arch. Entwicklungsmech. Org.* **180,** 9–30.

Stent, G. (1985). The role of cell lineage in development. *Philos. Trans. R. Soc. London Ser. B* **312,** 3–19.

Stern, C. (1936). Somatic crossing over and segregation in *Drosophila melanogaster*. *Genetics* **21,** 625–730.

Stern, C. (1968). Genetic mosaics in animal and man. *In* "Genetic Mosaics and Other Essays," pp. 27–129. Harvard Univ. Press, Cambridge, Massachusetts.

Sturtevant, A. H. (1929). The *claret* mutant type of *Drosophila simulans*: A study of chromosome elimination and cell lineage. *Z. Wiss. Zool. Abt. A* **135,** 323–356.

Technau, G. M. (1986). Lineage analysis of transplanted individual cells in embryos of *Drosophila melanogaster*. I. The method. *Roux's Arch. Dev. Biol.* **195,** 389–398.

Technau, G. M., and Campos-Ortega, J. A. (1985). Fate-mapping in wild-type *Drosophila melanogaster*. II. Injections of horseradish peroxidase in cells of the early gastrula stage. *Roux's Arch. Dev. Biol.* **194,** 196–212.

Technau, G. M., and Campos-Ortega, J. A. (1986a). Lineage analysis of transplanted individual cells in embryos of *Drosophila melanogaster*. II. Commitment and proliferative capabilities of neural and epidermal cell progenitors. *Roux's Arch. Dev. Biol.* **195,** 445–454.

Technau, G. M., and Campos-Ortega, J. A. (1986b). Lineage analysis of transplanted individual cells in embryos of *Drosophila melanogaster*. III. Commitment and proliferative capabilities of pole cells and midgut progenitors. *Roux's Arch. Dev. Biol.* **195,** 489–498.

Technau, G. M., Becker, T., and Campos-Ortega, J.A. (1988). Reversible commitment of neural and epidermal progenitor cells during embryogenesis of *Drosophila melanogaster*. *Roux's Arch. Dev. Biol.* **197,** 413–418.

Udolph, G., Prokop, A., Bossing, T., and Technau, G. M. (1993). A common precursor for glia and neurons in the embryonic CNS of *Drosophila* gives rise to segment specific lineage variants. *Development (Cambridge, UK)* **118,** 765–775.

Udolph, G., Lüer, K., Bossing, T., and Technau, G. M. (1995). Commitment of CNS progenitors along the dorsoventral axis of *Drosophila* neuroectoderm. *Science* **269,** 1278–1281.

Urban, J., and Technau, G. M. (1997). Cell lineage and cell fate specification in the embryonic CNS of *Drosophila*. *Semin. Cell Dev. Biol.* **8,** 391–400.

Wieschaus, E. (1978). Cell lineage relationships in the *Drosophila* embryo. *In* "Genetic Mosaics and Cell Differentiation" (W. J. Gehring, ed.), pp. 97–118. Springer-Verlag, Berlin, Heidelberg, and New York.

Wieschaus, E., and Gehring, W. (1976). Clonal analysis of primordial disc cells in the early embryo of *Drosophila melanogaster*. *Dev. Biol.* **50,** 249–263.

Wieschaus, E., and Nüsslein-Volhard, C. (1986). Looking at embryos. *In* "Drosophila, a Practical Approach" (D. B. Roberts, ed.), pp. 199–226. IRL Press, Oxford and Washington, D.C.

Xu, T., and Rubin, G. M. (1993). Analysis of genetic mosaics in developing and adult *Drosophila* tissues. *Development (Cambridge, UK)* **117,** 1223–1237.

Cell Determination in the Drosophila Eye

Richard W. Carthew
Rachele C. Kauffmann
Susan Kladny
Songhui Li
Jianjun Zhang
Department of Biological Sciences
University of Pittsburgh
Pittsburgh, Pennsylvania 15260

I. Introduction

Cells are prevented from spontaneously differentiating by their dependence on inductive signals from their environment. Cells progress down a pathway of differentiation, with each step often relying on a separate inductive signal. Historically, it has been thought that these signals are proactive, and responding cells are naive and easily influenced. However, many inductive factors, such as members of the growth factor family, communicate to cells through a common signal transduction pathway once they bind to their cognate receptors. If this is true, then three questions come to mind. First, how does a cell get specific information from an inductive factor? Second, how does a cell resist the influences of many other inductive factors that it may contact, especially if it migrates within an embryo? Third, how does a cell resist passing through several steps of differentiation at once if most factors trigger the same signal transduction pathway? These questions suggest that cells have inhibitory mechanisms that prevent them from overstepping their prescribed responses. In a sense, the inhibitors can be thought of as gatekeepers. This chapter is an attempt to explore the role of molecular gatekeepers in cell differentiation and their relationship to induction.

II. The *Drosophila* Compound Eye

The experimental system we use to study molecular gatekeepers is the compound eye of *Drosophila*. The compound eye is a collection of 800 subunits called ommatidia (Fig. 1A). Each ommatidium consists of 21 cells that are differentiated into 14 different cell types (Fig. 1B). There are eight different photoreceptor cells (labeled R1–R8) arranged in a trapezoid pattern that is at the core of each ommatidium. There are four cells called cone cells that together are responsible for constructing the lens and cone (Ready *et al.*, 1976). Pigment cells form a pigmented sheath to optically insulate each ommatidium. This complex pattern develops from a monolayer epithelium known as the eye imaginal disc (Ready *et al.*, 1976). Differentiation begins in the mid-third instar stage and at the point where the dorsoventral midline of the eye disc intersects the posterior margin of the disc. Differentiation progresses from posterior to anterior following a moving furrow in the eye epithelium called the morphogenetic furrow (Fig. 1C,D). The furrow moves across the eye disc as a straight line aligned along the dorsal/ventral (D/V) axis, taking 2 days to travel from posterior to anterior edge (Campos-Ortega and Hofbauer, 1977). Progenitor cells begin to differentiate once the furrow has passed. Differentia-

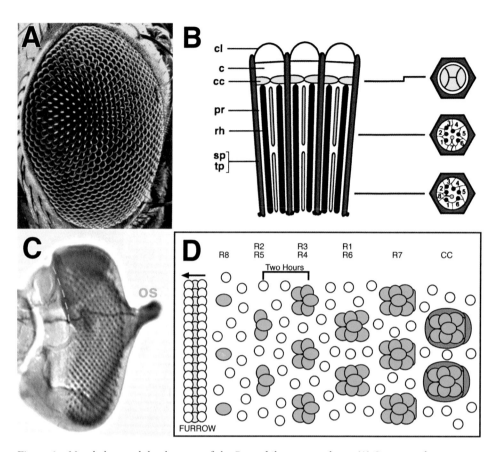

Figure 1 Morphology and development of the *Drosophila* compound eye. (A) Scanning electron micrograph of a compound eye. A hexagonal array of facets or ommatidia is evident. The surface of each facet has a simple lens and a short mechanosensory bristle which protrudes outwards. (B) Schematic ommatidium. On the left is a side view of three ommatidia. cl, Cuticular lens; c, pseudocone; cc, cone cell; pr, photoreceptor; rh, rhabdomere; sp, secondary pigment cell; tp, tertiary pigment cell. On the right are cross-sectional views of one ommatidium at different depths. Numbers refer to photoreceptor identities. In each ommatidium, the six outer rhabdomeres (black circular structures) of photoreceptors R1–R6 are patterned into a trapezoid shape. These surround the centrally projecting R7 (blue) and R8 (yellow) rhabdomeres. (C) An eye imaginal disc stained with an antibody against Neuroglian which is a neural-specific protein. Anterior is to the left. The morphogenetic furrow is marked by a dashed line and is proceeding from posterior to anterior (right to left). Behind the furrow neuronal differentiation is evident in evenly spaced clusters of staining. The underlying larval Bolwig's nerve is shown with an arrow to proceed from anterior to the posterior end of the eye disc where it goes down the optic stalk (os) to the brain. (D) Schematic representation of ommatidia formation depicting the cellular organization within the developing eye disc along the anterior–posterior axis (anterior left). The furrow is moving from posterior to anterior. Immediately posterior to the furrow, preclusters containing R8, R2, R5, R3, and R4 are found. Neural differentiation (in blue) occurs sequentially with R8 differentiating first. A second wave of mitosis gives rise to R1, R6, and R7. Once all photoreceptors have begun to differentiate, four cone cells begin their differentiation (in orange). Visualization of the eye disc permits analysis of sequentially more mature ommatidia as one proceeds toward the posterior margin of the eye disc.

tion occurs in a prescribed sequence beginning with differentiation of the R8 photoreceptors at uniformly spaced positions to form incipient ommatidia (Tomlinson, 1985). Undetermined progenitor cells surround each developing ommatidium. Sequential differentiation of the other seven photoreceptors occurs in four steps that are spaced approximately 4 hours apart. These seven cells cluster around each founder R8 cell to form the photoreceptor core. The four cone cells then differentiate and their cell bodies assemble over the photoreceptor core. Bristle cell and pigment cell differentiation follows beginning several hours later (Cagan and Ready, 1989a).

All retinal progenitors appear to have the same de-

velopmental potential to differentiate into any retinal cell type. A clone of related cells can be composed of any combination of cell types in the eye (Ready *et al.*, 1976; Lawrence and Green, 1979; Wolff and Ready, 1991). Moreover, the number of mitotic cycles a cell undergoes plays no role in fate determination. Cells that do not form the first five photoreceptors undergo one more round of mitosis than the first five photoreceptors, but it was found that there is no causal relationship between cell division and fate choice (de Nooij and Hariharan, 1995). Instead, positional information is exchanged between cells to inform them of their fates. Ommatidia that are isolated genetically or physically are still able to develop normally, suggesting that the cell–cell interactions are short-range (Lebovitz and Ready, 1986; Baker and Rubin, 1989). These features suggested to Tomlinson and Ready (1987a) a model in which the fate of a cell is controlled by the contacts it makes with previously determined cells. Thus, a differentiating cell would send signals to its neighbors and trigger their differentiation. A great deal of molecular evidence now supports this model for eye development.

The first mutation found to perturb cell–cell interactions in the eye was in the *sevenless* gene. The R7 cell is not specified as a neuron but instead becomes a cone cell in a *sevenless* mutant (Tomlinson and Ready, 1987b). The *sevenless* gene encodes a receptor tyrosine kinase (RTK) that is only required in the presumptive R7 cell (Hafen *et al.*, 1987). The ligand for Sevenless was identified (Bride of Sevenless or Boss) as a membrane-bound protein found on the surface of the differentiating R8 cell (Van Vactor *et al.*, 1991). Although Sevenless is expressed in most ommatidial cells, it is only activated in the presumptive R7 cell (Tomlinson *et al.*, 1987; Banerjee *et al.*, 1987). This is due to two features of the ligand: only those cells that directly contact R8 can interact with the ligand, and cells that have already begun to differentiate are unable to respond to the ligand (Van Vactor *et al.*, 1991). Thus, the Boss/Sevenless signaling mechanism directly confirms the predictions of the Tomlinson and Ready model.

Further support for a sequential induction model comes from molecular genetic studies of another RTK, the *Drosophila* homolog of the epidermal growth factor receptor (EGFR). Clones of cells missing the *Egfr* gene fail to differentiate into photoreceptors (Xu and Rubin, 1993). Expression of a dominant-negative form of EGFR in the eye blocks not only differentiation of all photoreceptors (except R8 which was not tested), but also cone cell and pigment cell differentiation (Freeman, 1996). The universal need for EGFR suggests that all ommatidial cells are determined similarly, and implies that different cell types do not use distinct inductive mechanisms. This interpretation is supported by characterization of the EGFR ligand in the eye. The *spitz* gene encodes a tumor growth factor α (TGFα)-like secreted factor that is a major activating ligand of EGFR in the embryo (Rutledge *et al.*, 1992). Spitz also is required for the formation of all photoreceptors except R8 (Freeman, 1994; Tio *et al.*, 1994; Tio and Moses, 1997). Mature Spitz protein is generated from differentiating R8, R2, R5, R3, and R4 cells in an order that follows their onset of differentiation (Tio *et al.*, 1994; Tio and Moses, 1997). Overexpression of secreted Spitz recruits ectopic photoreceptors, cone cells, and pigment cells, suggesting that Spitz induces determination of neuronal and nonneuronal cell types (Freeman, 1996). Altogether, it suggests a model in which differentiating cells secrete Spitz and induce the next set of progenitor cells to undergo differentiation by activating EGFR signal

transduction. An exception is R7 cell induction which requires both EGFR and Sevenless activation by their respective ligands.

Since Spitz and its receptor are used to generate almost all cell types in the eye, the progenitor cells must undergo some change that allows them to differentially respond to the same inducer. This change is not in the mechanism of signal transduction within the cells. A series of genetic experiments have demonstrated that the Ras/Raf/ERK pathway is the primary effector of Sevenless signaling (Zipursky and Rubin, 1994). Moreover, EGFR signaling utilizes the same pathway both in the eye and in other tissues (Rogge *et al.*, 1991; Doyle and Bishop, 1993; Diaz-Benjumea and Hafen, 1994; Brunner *et al.*, 1994a). Inductive determination of all photoreceptor cells requires activation of the Ras/Raf/ERK pathway (Rogge *et al.*, 1991; Brunner *et al.*, 1994a; Simon *et al.*, 1991). The way in which cells then differentially respond to induction must reflect a change in how each cell interprets activation of its own Ras/Raf/ERK pathway. Recently, it has been suggested that transcription factors expressed in defined subsets of cells could modulate Ras/Raf/ERK activation to produce specific cell types (Kumar and Moses, 1997). Indeed, two such transcription factors play key roles as molecular gatekeepers of differentiation.

III. Gatekeepers in Eye Development

Yan (also called Pokkuri or Aop) is a negative regulator of differentiation. The polypeptide sequence of Yan contains a central ETS domain that is a sequence-specific DNA-binding domain shared among a family of transcription activators and repressors (Lai and Rubin, 1992; Tei *et al.*, 1992). Hypomorphic *yan* alleles produce spontaneous ectopic differentiation of photoreceptors. Yan protein is detected in all progenitor cells behind the morphogenetic furrow (Lai and Rubin, 1992). As progenitor cells assume defined positions that indicate their readiness to respond to inductive signals, their nuclei migrate in a basal to apical direction. During this time, Yan protein levels are dramatically reduced. When Yan is overexpressed in differentiating retinal cells as an activated protein, it blocks differentiation of photoreceptors and cone cells, and results in their undergoing programmed cell death (Rebay and Rubin, 1995). Based on these data, it has been suggested that Yan normally inhibits differentiation of neuronal and nonneuronal cell types in the eye, and that Yan protein levels must be reduced before cells can undergo differentiation.

The *tramtrack* (*ttk*) gene encodes two protein isoforms (p69 and p88) that contain an amino-terminal

BTB/POZ domain and a sequence specific DNA-binding domain composed of two zinc fingers (Harrison and Travers, 1990; Brown *et al.*, 1991; Read and Manley, 1992). The BTB/POZ domain is found in a diverse family of proteins with transcriptional repressor or chromatin remodeling activities, most of which have zinc finger DNA-binding domains (Zollman *et al.*, 1994; Bardwell and Treisman, 1994). Indeed, Ttk69 represses transcription of genes containing p69-binding sites in their regulatory regions (Harrison and Travers, 1990; Brown *et al.*, 1991; Read and Manley, 1992). When the BTB/POZ domain of the oncogene BCL-6 was fused to the GAL4 DNA-binding domain, the fusion protein repressed transcription of a reporter gene containing GAL4 binding sites in its promoter (Seyfert *et al.*, 1996). This suggests that the BTB/POZ domain acts as a transrepression domain for transcription. It also mediates protein dimerization and has a curious ability to cis-inhibit the DNA-binding activity of a zinc finger domain (Bardwell and Treisman, 1994). The two isoforms of Ttk are produced by alternative splicing. Both share a common amino terminal region including the BTB/POZ domain, but diverge for the remainder of the polypeptide sequence such that each isoform contains a distinct zinc finger domain with unrelated sequence specificity.

Tramtrack is a negative regulator of differentiation. Xiong and Montell (1993) discovered that the *ttk*[1] mutation produces ommatidia with multiple R7 cells because cone cell progenitors were transformed to a neural fate. The *ttk*[1] allele selectively disrupts synthesis of the p88 isoform but not the p69 isoform. They concluded that Ttk88 normally prevents progenitor cells from becoming photoreceptors. Using a p88-specific antibody, Lai *et al.* (1996) showed that Ttk88 is found in eye progenitor cells and in cone cells but is not detected in photoreceptor cells. This suggests that Ttk88 does not block cone cell differentiation. If Ttk88 is able to block photoreceptor differentiation, then forced expression of Ttk88 in photoreceptor cells should inhibit their differentiation. To test this, we misdirected expression of Ttk88 by genetically combining a *UAS-Ttk88* transgene with a *GMR-Gal4* transgene (Li *et al.*, 1997). *GMR-Gal4* drives transcription of *UAS-Ttk88* in all photoreceptor and cone cells. Ttk88 expressed in all photoreceptors completely inhibits their differentiation (Fig. 2). However, cone cells form normally and produce normal cones and lenses. This block in photoreceptor differentiation occurs early, with affected cells failing to express early markers of neural differentiation such as Elav.

Our conclusion is that Ttk88 normally inhibits differentiation of neuronal cell types but not nonneuronal cell types such as cone cells. Ttk88 protein levels must be reduced before cells can undergo neuronal differentiation. If Ttk88 is missing, then a cell is free to become

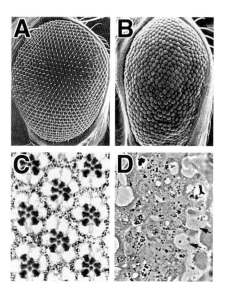

Figure 2 Overexpression of Ttk88 inhibits photoreceptor differentiation. Adult eyes are shown by scanning electron microscopy (A and B) and tangential sections through eyes (C and D). Wild-type eyes (A, C) have a uniform array of ommatidia containing photoreceptors. In each ommatidium, the rhabdomeres of photoreceptors are patterned into a trapezoid shape. The R8 photoreceptor lies below this plane of section. In eyes from *GMR-Gal4/UAS-Ttk88* flies (B, D), the external array is disorganized and all bristles are missing. Photoreceptors are not seen in sections of the retina, though cone cells (arrows) can be found.

a neuron and actually may do this as a default. These qualities and the described qualities of Yan are consistent with Yan and Ttk88 acting as molecular gatekeepers to block differentiation.

IV. How Do Gatekeepers Arrest Differentiation?

Both Yan and Ttk88 are repressors of transcription. Transfection of Yan into *Drosophila* tissue culture cells represses transcription of a cotransfected gene bearing canonical ETS-binding sites in its promoter (O'Neill *et al.*, 1994). The canonical binding site of Ttk88 is not as well established precluding similar analysis although Ttk88 appears to repress expression of *engrailed* (Xiong and Montell, 1993; Read *et al.*, 1992). These molecular activities suggest a simple mechanism for gatekeepers to inhibit differentiation. They could block differentiation by repressing transcription of genes that are required for differentiation. In the absence of repressor, the target genes could be transcribed and consequently direct the cell to differentiate. If this model is correct, then it should be possible to find target genes important for differentiation that are regulated in this manner. The *Drosophila* gene *prospero* (*pros*) is one such target gene.

Pros is required in R7 photoreceptors to ensure their correct differentiation as functional sensory neurons (Kauffmann *et al.*, 1996). The pattern of synaptic connections made by *pros* mutant R7 cells is abnormal. Terminals are nonuniformly spaced in the medullar optic ganglion and they lie at several different ganglion layers. This is in contrast to wild-type R7 cells which terminate in a uniform pattern lying at one layer of the medulla. The Pros protein is detected in R7 and cone cells for the first 50 to 60 hours of their differentiation program (Kauffmann *et al.*, 1996). The onset of Pros protein synthesis actually precedes the synthesis of early neuronal marker gene products such as Elav by 2 or more hours, suggesting that its activation is an early event in differentiation of R7 cells. The cell-specific protein pattern of Pros appears to be entirely dependent on transcriptional regulation, as determined by comparing the patterns of Pros protein with an enhancer-trap insertion within the *pros* gene. Enhancer-trap P element transposons carry the *lacZ* coding region coupled to a minimal TATA promoter. If a transposon inserts into a gene, then nearby enhancer elements regulate not only the endogenous gene but also the *lacZ* gene, generating a *lacZ* pattern of expression that mimics the transcriptional pattern of the endogenous gene. Enhancer traps inserted in the *pros* gene give a *lacZ* expression pattern identical to Pros.

Yan is partially responsible for this expression pattern. Many progenitor cells ectopically express Pros in a *yan* loss-of-function mutant (Fig. 3). Therefore, Yan normally represses Pros expression in progenitor cells, consistent with our gatekeeper model. If Yan directly represses Pros transcription, then it should be possible to find the silencer elements to which Yan binds. We have taken an approach to finding such elements by making chimeric genes in which 5′ genomic flanking DNA from the *pros* gene is fused to *lacZ* coding DNA. Flies are transformed with the chimeric genes by P element mediated transformation, and transformant eyes are examined to see if *lacZ* is expressed in the same pattern as

Pros. If it is possible to find the regulatory DNA sufficient to generate a Pros pattern of expression, then the DNA can be tested for direct interaction with Yan. Approximately 12.2 kb of *pros* flanking DNA including the promoter is sufficient to drive *lacZ* in a Pros expression pattern (Fig. 4). In contrast, 6.8 kb of flanking DNA is unable to generate a pattern. A combination of 5′ and 3′ deletions within −6.8 to −12.2 kb have narrowed down a region from −8.0 to −9.2 kb that is necessary for generating a Pros expression pattern. To test if this region is sufficient to direct expression, tandem copies of the −8.0 to −9.2 region were fused to a heterologous promoter (HSP70) linked to *lacZ* and were tested *in vivo*. Indeed, this region directs *lacZ* expression in differentiating R7 and cone cells.

The sequence of DNA within the 1200-bp region contains a total of 11 canonical ETS-binding sites (C/GA/CGGAA/TA/GC/T). To test if any of these sites or other sequences bind specifically to Yan protein, recombinant Yan protein was assayed for DNA binding. The carboxyl terminal half of Yan (including the complete ETS domain) was fused in frame to glutathione S-transferase (GST), and the fusion protein, GST-YanC, was purified to homogeneity from *Escherichia coli*. GST-YanC was assayed for DNA binding by the electrophoretic mobility shift assay. An example of such an assay is shown in Fig. 5 where a radioactive DNA fragment containing one potential ETS site has been incubated with GST-YanC. A slower migrating complex is observed that is effectively competed by addition of an excess of unlabeled fragment but not by addition of an excess of unlabeled fragment carrying a point mutation in the ETS site. These data are consistent with GST-YanC specifically binding to the ETS sequence within the DNA fragment. A total of four ETS sites in the −8.0 to −9.2 kb region are specifically bound by GST-YanC as determined by these assays. The coordinate effects of Yan action on the Pros enhancer *in vivo* and the

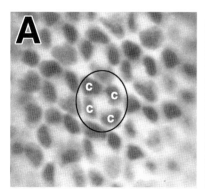
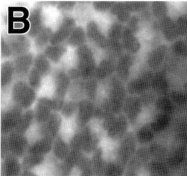

Figure 3   Regulation of Pros expression by Yan. Eye discs from wild-type (A) *and yan¹/yan^{e0166}* (B) larvae were labeled with anti-Pros antibody. Anterior is to the left. The focal plane in each photo shows staining of cone cell nuclei. A single ommatidium from the wild-type disc is circled with each cone cell nucleus labeled c. The mutant disc contains many more Pros-positive cells in each ommatidium rather than the normal four cells in wild-type ommatidia.

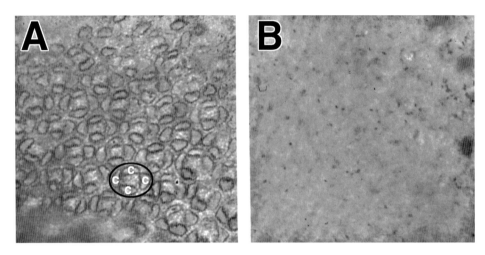

Figure 4 *LacZ* expression from Pros promoter *LacZ* fusion transgenes. Expression was detected by Xgal activity assay of β-galactosidase. (A) Expression of *LacZ* directed by 12.2 kb of the Pros promoter is virtually indistinguishable from that of the endogenous Pros gene. Activity is detected in the perinuclear region of the four cone cells and the R7 cell (not seen in this plane of focus). Although expression is restricted to cone and R7 cells like the endogenous gene, there is no elevated expression in the R7 cells unlike the endogenous gene. A single ommatidium is circled with each cone cell nucleus labeled c. (B) Expression of *LacZ* directed by 6.8 kb of the Pros promoter. No significant expression is observed in the eye disc.

sequence-specific binding of Yan to the enhancer *in vitro* are evidence for Yan playing a direct role in regulating Pros transcription.

An important point about Yan repressing Pros transcription is that it implies that there are one or more activators of Pros transcription. Indeed, this is borne

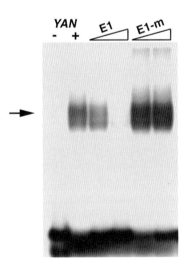

E1:   CAGGAAGTG

E1-m: CACTAAGTG

Figure 5 Electrophoretic mobility shift assays with Pros enhancer sequences and GST-YAN protein. ^{32}P-labeled DNA fragment (20 fmol) from the -8.0 to -9.2 kb enhancer region was incubated with 120 ng GST-YAN purified from bacteria. Arrow shows position of DNA–protein complex in gel. Some reactions also received varying amounts (0.4 or 2 pmol) of unlabeled DNA fragment with wild-type sequence (E1) or with two base substitutions in the canonical ETS sequence contained within the fragment (E1-m). The unlabeled wild-type DNA effectively competes with the labeled DNA for GST-YAN binding whereas the unlabeled mutant DNA does not compete with labeled DNA for binding to GST-YAN.

out by our observation that mutations in the *lozenge* (*lz*) gene completely abolish Pros expression in the eye. Lozenge plays a key role in promoting differentiation of photoreceptors R1, R6, R7, and cone cells (Daga *et al.*, 1996). The *lz* gene encodes an α subunit of the *Drosophila* Core Binding Factor (CBF) transcription factor (Daga *et al.*, 1996). The α subunits of CBF factors, which contain a DNA-binding domain known as the RUNT domain, interact with a β subunit to form a heterodimer that potently activates transcription of target genes (Golling *et al.*, 1996). In mice, CBFA1 is a transcriptional activator of osteoblast differentiation during embryonic development (Rodan and Harada, 1997). This suggested association between CBF molecular activity and cell differentiation is consistent with Lz activating Pros transcription for proper R7 cell differentiation. Moreover, a single canonical binding site (Pu/T A C C Pu C A) for RUNT domain proteins is located within the Pros enhancer, and this sequence is specifically bound by recombinant Lz protein *in vitro*. Interestingly, the Lz binding site is located between two of the four Yan-binding sites in the enhancer. Altogether, these data indicate that Lz and Yan work antagonistically to control Pros transcription in the eye by interacting with a common enhancer.

A simple model then might explain how Pros is regulated to give its pattern of expression (Fig. 6). If a cell does not contain Lz, then the upstream enhancer is not activated. If a cell contains Lz, then the upstream enhancer is activated. However, if a cell contains both Lz and Yan, the upstream enhancer is prevented from being activated. Yan and Lz are themselves expressed in overlapping patterns; Yan is found in progenitor cells only, and Lz is found in R1, R6, R7, cone cells, and progenitor cells. Following the model, R2, R3, R4, R5

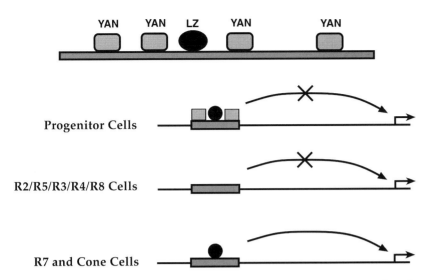

Figure 6 Summary of combinatorial interactions underlying Pros transcriptional regulation. The top hatched line represents the Pros eye enhancer from −9.2 to −8.0 kb upstream of the start site. Shown are the relative positions of specific binding sites for Yan and Lozenge (Lz) proteins as determined by mobility shift experiments. Below is illustrated the combinatorial interactions guiding regulation of promoter activity. In progenitor cells, Yan blocks Lz from activating transcription. The nature of the repression is unknown though it is not due to steric hindrance in binding since the binding sites are nonoverlapping. In precluster photoreceptor cells, the absence of Lz renders the enhancer inactive. However, in R7 and cone cells, Lz is able to activate transcription since Yan is neutralized following induction.

and R8 cells would not express Pros because they lack Lz; progenitor cells would not express Pros because they contain both Yan and Lz; R1, R6, R7, and cone cells would express Pros because they contain only Lz. The model does not explain why Pros is not expressed in R1 and R6. Presumably other repressors inhibit Lz in these two cells. Nevertheless, it suggests that an important function of gatekeepers such as Yan is to prevent the expression of differentiation-specific genes in progenitor cells containing transcription factors that would otherwise activate gene expression.

V. The Role of Yan in Induction

Available genetic data suggest that Yan functions downstream of the Sevenless signal transduction cascade. The evidence for this conclusion comes from examining the effect of *sevenless* mutations on the ability of hypomorphic *yan* alleles to generate ectopic R7 cells. In each case, ectopic R7 cells are produced regardless of the presence or absence of Sevenless (Lai and Rubin, 1992; Tei *et al.*, 1992). Moreover, *yan* mutations are required cell-autonomously to generate supernumerary R7 cells (Lai and Rubin, 1992), and they show genetic interactions with mutations in *Ras1, DRaf,* and the MEK gene *Dsor* (Lai *et al.*, 1996; Yamamoto *et al.*, 1996). If Yan acts in the Sevenless signaling pathway, it would suggest that inductive signaling is not simply a positive force but that it functions to neutralize gatekeepers such as Yan.

Further evidence suggests that Yan functions downstream of the EGFR signal transduction cascade as well. Four sets of experiments have addressed the is-

sue of the role of EGFR activation in regulating Yan. First, we have found that constitutive activation of the EGFR in progenitor cells results in ectopic expression of Pros in these cells. Freeman (1996) produced a constitutively activated EGFR by fusing the intracellular domain of EGFR with the transmembrane and extracellular domains of the Torso RTK. When we expressed such a Torso/EGFR protein under heat-shock control in the eye, supernumerary Pros-positive cells were detected. We observed a similar result when a constitutively activated form of Ras1 (Ras1^{v12}) was expressed in the eye. These data suggest that EGFR signaling and Yan play antagonistic roles in regulating Pros transcription. Second, we and other labs showed that Yan is phosphorylated by activated ERK *in vitro* (Brunner *et al.*, 1994b). Indeed, Yan has eight canonical ERK phosphorylation sites as deduced from its amino acid sequence. ERK is the final enzymatic component acting in the Ras signal transduction pathway downstream of EGFR. Third, O'Neill *et al.* (1994) used a *Drosophila* cell culture assay to show how Yan might be regulated by ERK activation. Yan represses transcription of a reporter gene containing multiple ETS-1 binding sites when the two constructs are cotransfected. The ability of Yan to repress reporter expression is compromised when it is cotransfected with constitutively activated Ras1 or ERK. Fourth, Rebay and Rubin (1995) altered all eight ERK phosphorylation sites by site-directed mutagenesis to prevent ERK phosphorylation of Yan. This Yan^{ACT} was transformed into *Drosophila* and specifically expressed in the eye, resulting in complete inhibition of photoreceptor and cone cell differentiation. Thus, ERK phosphorylation of Yan normally derepresses differentiation, likely as a result of relieving inhibition of differentiation-specific gene transcription. Indeed, when we expressed Yan^ACT in R7 and cone cells using the *sev* enhancer, we observed a constitutive repression of Pros transcription (Kauffmann *et al.*, 1996). When Yan^ACT was transfected into tissue culture cells, it localized into the nucleus regardless of whether the cells also contained activated Ras1 (Rebay and Rubin, 1995). On the other hand, cotransfecting wild-type Yan with activated Ras1 shifts its subcellular localization from the nucleus to the cytoplasm where it is rapidly degraded. This offers an explanation for how inductive signals overcome Yan. Activation of the Ras pathway leads to Yan phosphorylation by ERK and its rapid degradation, thereby relieving repression of target genes such as *pros*.

VI. The Role of Tramtrack in Induction

Genetic experiments suggest that Ttk also functions downstream of the Sevenless signal transduction pathway. Hypomorphic *ttk* alleles generate ectopic R7 cells regardless of the presence or absence of Sevenless (Lai *et al.*, 1996; Yamamoto *et al.*, 1996). Moreover, *ttk* mutations are required cell-autonomously to generate supernumerary R7 cells, and they show genetic interactions with mutations in *Ras1*, *yan*, and the ERK gene *rolled*. Epistasis analysis places Ttk downstream of Yan, separated from Yan by the action of two other genes: *phyllopod* (*phyl*) and *seven in absentia* (*sina*). Phyl plays an important role in initiating R7 differentiation. Mutation of *phyl* leads to the tranformation of R1, R6, and R7 photoreceptors into cone cells (Chang *et al.*, 1995; Dickson *et al.*, 1995). Directed expression of Phyl in the cone-cell precursors leads to their differentiation into R7-like cells. It is generally thought that Phyl expression is regulated as an immediate-early gene in inductive signal transduction. Two pieces of evidence support this view. First, Phyl is transcribed in photoreceptors R1, R6, and R7 but not the cone-cell precursors (Chang *et al.*, 1995; Dickson *et al.*, 1995). However, Phyl expression in the cone cells is observed when either Ras1 is hyperactivated or Yan is removed from these cells. Second, the loss-of-function phenotype of *phyl* is essentially unchanged in a *yan* mutant background, indicating that Phyl acts at some point downstream of Yan (Dickson *et al.*, 1995).

Sina, like Phyl, is required for R7 determination at some point downstream of Yan. R7 cells are transformed into cone cells in a *sina* mutant (Carthew and Rubin, 1990), and this phenotype is unchanged in a *yan* mutant background (Lai and Rubin, 1992; Tei *et al.*, 1992). Sina also acts downstream or in parallel to Phyl since Phyl-directed transformation of cone cells into R7 cells requires Sina activity (Chang *et al.*, 1995; Dickson *et*

al., 1995). Sina differs from Phyl in other respects. Sina is widely expressed in most cells in the eye disc and its expression is independent of Sevenless (Carthew and Rubin, 1990). Overexpression of Sina in cone cells does not transform them into R7 cells.

The structures of Phyl and Sina proteins offer few clues as to their biochemical activities. The amino acid sequence of Phyl is novel (Chang *et al.*, 1995; Dickson *et al.*, 1995). Sina contains a RING finger domain located near the amino terminus (Carthew and Rubin, 1990). Although both proteins are found in nuclei of cells, Sina is also located in the cytoplasmic compartment. Highly conserved paralogs of Sina have been found in the plant kingdom (*Arabadopsis*) and throughout the animal kingdom (fly, mouse, human). Indeed, we have found that one of the mouse genes, *Siah-2,* can functionally replace the *Drosophila sina* gene, and can be considered a homolog.

Epistasis analysis suggests that Ttk acts downstream of Sina and presumably Phyl also. Formation of supernumerary R7 cells by a *ttk* mutation is unaltered in a *sina* mutant background (Lai *et al.*, 1996; Yamamoto *et al.*, 1996). This result suggests that Sina, and possibly Phyl, regulates Ttk. To test this hypothesis, we examined *sina* and *phyl* mutant eye discs for the presence of Ttk88 protein. In wild-type discs, Ttk88 is detected in progenitor cells and cone cells but not photoreceptors. Strikingly, in a *phyl* mutant disc Ttk88 also is detected in one to three presumptive photoreceptor cells (Li *et al.*, 1997). These cells likely correspond to R1, R6, and R7. Moreover, in a *sina* mutant eye disc Ttk88 also is detected in one or two presumptive photoreceptor cells. Since Phyl is normally expressed exclusively in R1, R6, and R7, and Sina is also found in these cells, it suggests that Phyl and Sina normally repress Ttk88 in R1, R6, and R7 cells. If this is correct then misexpression of Phyl in cone cells should repress Ttk88 levels in cone cells. We observed that no Ttk88 is detected in cone cells of mutants carrying a *Sev-phyl* transgene which directs Phyl synthesis in R1, R3, R4, R6, R7, and cone cells (Li *et al.*, 1997). However, Ttk88 is detected in cone cells of a fly carrying the *Sev-phyl* transgene in a loss-of-function *sina* mutant background. Therefore, Phyl in combination with Sina, is both necessary and sufficient to repress Ttk88.

Sina and Phyl repress Ttk88 posttranscriptionally. An enhancer trap inserted near the promoter of the *ttk* gene expresses *lacZ* in all photoreceptors and cone cells (Lai *et al.*, 1996). This discrepancy between the Ttk88 protein pattern and the pattern of Ttk transcription indicates that posttranscriptional regulation excludes p88 from photoreceptor cells. Moreover, the expression pattern of the enhancer trap *lacZ* gene is unaltered in either a *Sev-phyl* or *Sev-Ras^{v12}* background (Li *et al.*, 1997). A similar result has been obtained looking at Ttk mRNA levels in eye discs by *in situ* hybridization (Tang *et al.*, 1997). Sina and Phyl could be regulating Ttk88 at the level of protein translation or stability. To investigate this issue, we examined the effect of coexpressing Phyl and Sina with Ttk88 in *Drosophila* tissue culture cells. An expression construct containing the Ttk88 cDNA was transiently transfected into S2 cells and Ttk88 protein was analyzed by Western blot. Two polypeptides (hereafter called p88α and p88β) were observed from transfected cells (Li *et al.*, 1997). To examine a potential effect of Sina and Phyl on Ttk88 protein levels, we carried out the same experiment in the presence of vectors directing the expression of Sina and Phyl. Cotransfection of Sina with the Ttk88 construct had no effect on p88 protein levels whereas cotransfection of Phyl with Ttk88 caused a decrease in the level of p88α. Cotransfection of Sina and Phyl had no greater effect on p88α than Phyl alone. The absence of an effect of Sina on Ttk88 is likely because endogenous Sina protein is constitutively produced in S2 cells. The reduction of p88α levels in the presence of Phyl is not due to conversion of that form to the p88β form since no increase in p88β was observed at the expense of p88α.

To investigate the effect of Phyl on p88 protein stability, we performed a pulse-chase experiment (Li *et al.*, 1997). Ttk88 was transiently transfected in S2 cells and protein was metabolically labeled with [^{35}S]methionine. The decay of p88 was monitored after the radioactive methionine in the culture medium was replaced with nonradioactive methionine. Both p88α and p88β have similar half-lives of 8 hours in the absence of cotransfected Phyl. However, when cotransfected with Phyl, the half-life of p88α is reduced to 40 minutes. Cotransfection of Phyl and Sina have an additional effect on p88α, reducing the half-life of p88α to 25 minutes. These results indicate that Phyl and Sina regulate the half-life of Ttk88 protein.

Many short-lived proteins are targeted for degradation by the ubiquitin/proteasome pathway. We performed experiments aimed at identifying the proteolytic pathway responsible for Ttk88 degradation (Li *et al.*, 1997). S2 cells which had been cotransfected with Phyl and Ttk88 plasmids were treated with inhibitors of different proteolytic pathways, and cells were subsequently analyzed for Ttk88 protein. Several proteasome-specific inhibitors increased the steady-state levels of p88α whereas other inhibitors had no effect. To show that the proteasome-specific inhibitors stabilized p88, we performed a pulse-chase experiment in which cells cotransfected with Ttk88, Sina, and Phyl were treated with inhibitor during the radioactive-labeling phase and the nonradioactive chase phase. The half-life of p88α is markedly increased in cells treated with inhibitor.

A key step in proteasome-dependent proteolysis involves the covalent attachment of ubiquitin polypep-tide chains to the substrate, which targets the substrate for rapid degradation by the proteasome. These ubiquitin–substrate conjugates are highly unstable and are often visualized only by treatment of cells with proteasome inhibitors. Cells cotransfected with Phyl, Sina, and Ttk88 were treated with proteasome inhibitors, and lysates were probed for p88 protein (Li *et al.*, 1997). An immunoreactive species of greater molecular weight than p88α was observed. In a parallel blot this species was found to react with an anti-ubiquitin antibody. Thus, Ttk88 is ubiquitinated in the presence of Sina and Phyl and efficiently degraded by the proteasome pathway.

The proteins Sina and Phyl physically associate to form a complex (Kauffmann *et al.*, 1996). Phyl synthesized by *in vitro* translation specifically binds to Sina-GST fusion protein immobilized to glutathione beads. By deletion analysis, we have localized the region of Phyl necessary for this interaction to between amino acids 110 and 127. This region is sufficient to confer specific Sina binding to a chimeric protein with the Phyl region fused to the carboxyl terminus of β-galactosidase. We also find that Sina interacts with itself (Li *et al.*, 1997). Sina synthesized by *in vitro* translation specifically binds to Sina-GST protein. Using a quantitative yeast-interaction trap assay, we find additional evidence for a direct physical interaction between Sina polypeptides. Thus, Sina not only physically interacts with Phyl but also with itself.

Ttk88 is negatively regulated by Sina and Phyl through a protein destabilization mechanism. We tested whether Sina and Phyl might interact directly with Ttk88 to potentiate its destabilization (Li *et al.*, 1997). Using a GST pull-down assay, we detected a specific interaction between Ttk88 and Sina-GST. A truncated polypeptide containing amino acids 1–116 of Ttk88 results in a strong interaction with Sina-GST. Interestingly, this region corresponds precisely to the BTB/POZ domain. We next considered the possibility that Ttk88 also physically interacts with Phyl. An immunoprecipitation assay was used in which Phyl and Ttk88 were produced by cotranslation, and Phyl was immunoprecipitated. We observed efficient coprecipitation of Ttk88 with Phyl. When Sina, Phyl, and Ttk88 were together cotranslated and immunoprecipitated with anti-Phyl, both Ttk88 and Sina coprecipitated with Phyl.

Altogether these results suggest a model (Fig. 7A) in which Sina and Phyl form a complex with Ttk88 and target it for ubiquitination and degradation by the proteasome. The proteasome pathway is complex and can be divided into two phases (Ciechanover, 1994). First, a target protein is conjugated to ubiquitin. Conjugation involves three enzymes which act in series. E1 conjugating enzyme transfers ubiquitin via a thioester linkage to an E2 conjugating enzyme which donates its ubiquitin to an E3 ligase. The E3 enzyme specifically binds to the substrate protein and donates its ubiquitin to the substrate. Once the substrate is fully ubiquitinated, it interacts with the 26S proteasome where it is degraded and the conjugated ubiquitin is released. Our model tentatively places Sina and Phyl acting together as an E3 ligase with two lines of evidence offered for support. One, E3 enzymes are the specificity factors which target selected proteins for degradation and they do this by binding to the proteins (Ciechanover, 1994). Indeed, Phyl and Sina together specifically bind to Ttk88. Two, E3 en-

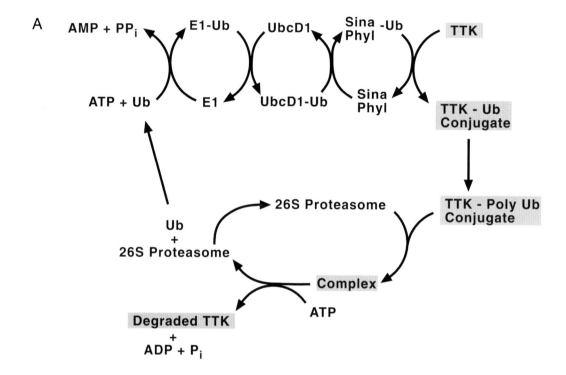

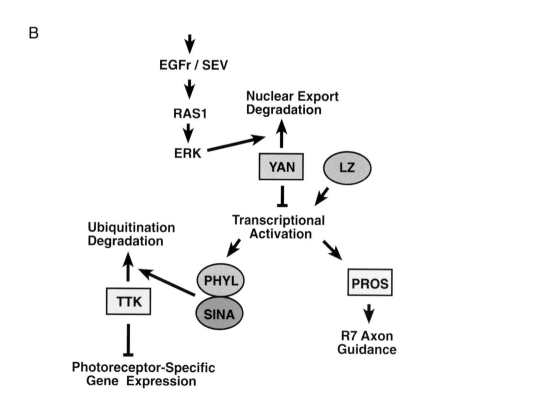

zymes intimately function in combination with E2 enzymes (Ciechanover, 1994). We find that the *phyl* gene interacts with an E2-coding gene, *UbcD1*, in the eye. Ectopic transformation of cone cells into R7 cells by *Sev-phyl* is suppressed by a 2-fold reduction in the dosage of the *UbcD1* gene. Moreover, Sina interacts with UbcD1 in a yeast two-hybrid assay, suggesting a physical interaction between these two proteins (Tang *et al.*, 1997). If a Sina and Phyl complex constitutes an E3 ligase, then it represents a new member of the E3 family since no other identified E3 enzyme subunits structurally resemble either Sina or Phyl. Because Sina is a member of a conserved protein family, it is likely to serve as the prototype for a new class of E3 ligases.

Our results argue that photoreceptor differentiation is regulated by an inductive signal causing targeted proteolysis of Ttk88 (Fig. 7B). In progenitor cells, Ttk88 protein is stable because Phyl is not expressed in these cells. Consequently, Ttk88 is able to repress transcription of neural-specific genes. When the Ras/ERK pathway is activated in certain cells receiving an inductive signal (Spitz or Boss), it triggers the expression of Phyl as an immediate-early gene. The Phyl protein product then forms a complex with Sina, and this complex binds to Ttk88 protein. Ubiquitin is conjugated with Ttk88, and Ttk88 is consequently degraded. Rapid departure of Ttk88 relieves the repression of neural-specific gene transcription, and cells initiate differentiation as neurons.

VII. Perspectives

Gatekeepers exist to keep cells arrested at different points along a differentiation pathway. We have identified two transcription repressors whose function is to repress expression of certain genes critical for retinal differentiation. They might repress these genes to prevent transcription activators from otherwise activating their expression. For example, in progenitor cells the transcription activator Lz is prevented from activating Pros transcription because of Yan. Gatekeepers might also prevent cells from erroneously expressing genes when a multipurpose signal transduction pathway is activated. For example, the Ras pathway is switched on in both cone and photoreceptor cells but is prevented from activating neural-specific gene expression in cone cells by Ttk88. Why are gatekeepers transcription repressors? Repressors with sequence-specific DNA-binding activity provide exquisite specificity to inhibit transcription of a defined subset of genes. Repressors also generally exhibit greater individual potency in regulating transcription than activators.

Ubiquitin-mediated proteolysis of regulatory inhibitors is a common theme in biology. Two well-known examples are p53 degradation by Mdm2 or the E6 product of papillomavirus, and IκB degradation by extracellular signals inducing its phosphorylation (Scheffner *et al.*, 1993; Chen *et al.*, 1995; Haupt *et al.*, 1997). Three features of ubiquitin-mediated proteolysis make it an attractive regulatory mechanism. First, proteolysis is a rapid means of removing a protein, much faster than simply shutting off its synthesis. This may be critical for a cell to make rapid developmental decisions in a constantly changing environment. Second, ubiquitination is highly specific, depending on direct interaction between the protein target and E3 ligase. There are a growing number of E3 ligase classes being discovered, each with different substrate specificity (Ciechanover, 1994). Indeed, the F-Box family of proteins has been recently discovered to recruit phosphorylated substrates to the SCF E3 ligase (Skowyra *et al.*, 1997; Feldman *et al.*, 1997). Each F-Box member appears to have a binding specificity for different phosphorylated proteins. Third, proteolysis can occur at different points along the same regulatory pathway. In induced photoreceptor cells, Yan is degraded on phosphorylation by activated ERK, and immediate early genes such as Phyl are expressed. This leads to ubiquitination of Ttk and its degradation. Sequential proteolysis of gatekeepers allows each gatekeeper to be regulated at a discrete time within a cell as it follows a differentiation program even if the gatekeepers are inhibited by a common regulatory pathway. This timing of proteolysis may be a critical component

Figure 7 How proteolysis controls cell differentiation in the *Drosophila* eye. (A) A speculative model for the regulated proteolysis of Tramtrack by Sina and Phyl in R1, R6, and R7 cells. Synthesis of Phyl enables ubiquitination of Tramtrack by components of the E1–E3 machinery. Ubiquitin (Ub) is activated by the conjugating enzyme E1 and the *Drosophila* E2 enzyme UbcD1. Ttk88 protein is bound to Sina and Phyl perhaps acting as substrate-specific components of the destruction pathway. E2 then transfers ubiquitin to Ttk88 with Sina and Phyl acting as a bridge. Ttk88 becomes polyubiquitinated, and is consequently degraded by the 26S proteasome. Ubiquitin is released from proteolytic products by C-terminal hydrolases and is free to be reactivated. (B) A model depicting relationship between inductive signals and gatekeeper proteolysis. Gatekeepers maintain a cell in a particular regulatory state. The regulatory states are interconnected by a series of dependencies. Each regulatory state has two functions: to trigger a differentiation event such as regulating expression of a new set of genes, and to enable the transition to a subsequent regulatory state. For example, ERK-induced destruction of Yan triggers retinal differentiation by activating expression of certain genes such as *Prospero*. Destruction of Yan also enables the destruction of Tramtrack by allowing Phyl to be synthesized. The destruction of Tramtrack triggers the expression of neural-specific genes. Although the switch in regulatory states is initiated by a cell receiving an inductive signal, each switch might be potentially modified by other factors. For example, induction does not enable the destruction of Tramtrack in all retinal cells but only a subset of cells. Other factors restrict inductive signals from activating Phyl expression in all retinal cells, and consequently Tramtrack is not destroyed in cells lacking Phyl.

In programming appropriate cellular responses to inductive signals.

Gatekeepers may also be regulated by inhibitory signals sent between cells. For example, Notch signaling positively regulates Ttk expression in sensory organ cells of the embryonic peripheral nervous system (Guo et al., 1996). Since Notch signaling is present in many developing tissues including the compound eye (Cagan and Ready, 1989b), Ttk may represent a point of intersection between negative signals and inductive signals.

In the cell cycle, three checkpoints arrest progression until either DNA synthesis, DNA repair, or mitotic spindle assembly are complete (Hartwell and Kastan, 1994). Autonomous regulatory mechanisms release a cell from each checkpoint when these processes are complete. In the linear pathway of differentiation, checkpoints arrest progression through the action of proteins such as Yan and Ttk. Unlike the cell cycle, the regulatory mechanisms that release a cell from each checkpoint appear to be nonautonomous. Depending on a cell's environment, inductive signals that it receives triggers passage through a particular checkpoint, and the cell progresses until arresting at the next checkpoint. We envision that there may be a diverse array of checkpoints guarding differentiation of most cell types. The recent discovery of a link between protein kinase signaling and protein abundance (Skowyra et al., 1997; Feldman et al., 1997) suggests that one of the primary functions of inductive signals is to neutralize checkpoint inhibitors. It will be of great interest to determine the validity of this idea.

References

Baker, N. E., and Rubin, G. M. (1989). Effect on eye development of dominant mutations in Drosophila homologue of the EGF receptor. Nature (London) 340, 150–153.

Banerjee, U., Renfranz, P. J., Hinton, D. R., Rabin, B. A., and Benzer, S. (1987). The sevenless protein is expressed apically in cell membranes of developing Drosophila retina; it is not restricted to cell R7. Cell (Cambridge, Mass.) 51, 151–158.

Bardwell, V. J., and Treisman, R. (1994). The POZ domain: A conserved protein–protein interaction motif. Genes Dev. 8, 1664–1677.

Brown, J. L., Sonoda, S., Ueda, H., Scott, M. P., and Wu, C. (1991). Repression of the Drosophila fushi tarazu (ftz) segmentation gene. EMBO J. 10, 665–674.

Brunner, D., Oellers, N., Szabad, J., Biggs III, W. H., Zipursky, S. L., and Hafen, E. (1994a). A gain-of-function mutation in Drosophila MAP kinase activates multiple receptor tyrosine kinase signaling pathways. Cell (Cambridge, Mass.) 76, 875–888.

Brunner, D., Ducker, K., Oellers, N., Hafen, E., Scholz, H., and Klambt, C. (1994b). The ETS domain protein Pointed-P2 is a target of MAP kinase in the Sevenless signal transduction pathway. Nature (London) 370, 386–389.

Cagan, R. L., and Ready, D. F. (1989a). The emergence of order in the Drosophila pupal retina. Dev. Biol. 136, 346–362.

Cagan, R. L., and Ready, D. F. (1989b). Notch is required for successive cell decisions in the developing Drosophila retina. Genes Dev. 3, 1099–1112.

Campos-Ortega, J. A., and Hofbauer, A. (1977). Cell clones and pattern formation: On the lineage of photoreceptor cells in the compound eye of Drosophila. Roux's Arch. Dev. Biol. 181, 227–245.

Carthew, R. W., and Rubin, G. M. (1990). seven in abstentia, a gene required for specification of R7 cell fate in the Drosophila eye. Cell (Cambridge, Mass.) 63, 561–577.

Chang, H. C., Solomon, N. M., Wassarman, D. A., Karim, F. D., Therrien, M., Rubin, G. M., and Wolff, T. (1995). phyllopod functions in the fate determination of a subset of photoreceptors in Drosophila. Cell (Cambridge, Mass.) 80, 463–472.

Chen, Z., Hagler, J., Palombella, V. J., Melandri, F., Scherer, D., Ballard, D., and Maniatis, T. (1995). Signal-induced site-specific phosphorylation targets IκBα to the ubiquitin–proteasome pathway. Genes Dev. 9, 1586–1597.

Ciechanover, A. (1994). The ubiquitin–proteasome proteolytic pathway. Cell (Cambridge, Mass.) 79, 13–21.

Daga, A., Karlovich, C. A., Dumstrei, K., and Banerjee, U. (1996). Patterning of cells in the Drosophila eye by lozenge, which shares homologous domains with AML1. Genes Dev. 10, 1194–1205.

de Nooij, J. C., and Hariharan, I. K. (1995). Uncoupling cell fate determination from patterned cell division in the Drosophila eye. Science 270, 983–985.

Dickson, B. J., Dominquez, M., Straten, A. V. D., and Hafen, E. (1995). Control of Drosophila photoreceptor cell fates by Phyllopod, a novel nuclear protein acting downstream of the Raf kinase. Cell (Cambridge, Mass.) 80, 453–462.

Diaz-Benjumea, F. J., and Hafen, E. (1994). The Sevenless signaling cassette mediates Drosophila EGF receptor function during epidermal development. Development (Cambridge, UK) 120, 569–578.

Doyle, H., and Bishop, J. M. (1993). Torso, a receptor tyrosine kinase required for embryonic pattern formation, shares substrates with the sevenless and EGF–R pathways in Drosophila. Genes Dev. 7, 633–646.

Feldman, R. M. R., Correll, C. C., Kaplan, K. B., and Deshaies, R. J. (1997). A complex of Cdc4p, Skp1p, and Cdc53p/Cullin catalyzes ubiquitination of the phosphorylated CDK inhibitor Sic1p. Cell (Cambridge, Mass.) 91, 221–230.

Freeman, M. (1994). The spitz gene is required for photoreceptor determination in the Drosophila eye where it interacts with the EGF receptor. Mech. Dev. 48, 25–33.

Freeman, M. (1996). Reiterative use of the EGF receptor triggers differentiation of all cell types in the Drosophila eye. Cell (Cambridge, Mass.) 87, 651–660.

Golling, G., Li, L. H., Pepling, M., Stebbins, M., and Gergen, J. P. (1996). Drosophila homologs of the proto-oncogene product PEBP2/CBFβ regulate the DNA-binding properties of Runt. Mol. Cell. Biol. 16, 932–942.

Guo, M., Jan, L. Y., and Jan, Y. N. (1996). Control of daughter cell fates during asymmetric division: Interaction of Numb and Notch. *Neuron* **17,** 27–41.

Hafen, E., Basler, K., Edstroem, J. E., and Rubin, G. M. (1987). Sevenless, a cell-specific homeotic gene of *Drosophila,* encodes a putative transmembrane receptor with a tyrosine kinase domain. *Science* **236,** 55–63.

Harrison, S. D., and Travers, A. A. (1990). The *tramtrack* gene encodes a *Drosophila* finger protein that interacts with the *ftz* transcriptional regulatory region and shows a novel embryonic expression pattern. *EMBO J.* **9,** 207–216.

Hartwell, L. H., and Kastan, M. B. (1994). Cell cycle control and cancer. *Science* **266,** 1821–1828.

Haupt, Y., Maya, R., Kazaz, A., and Oren, M. (1997). Mdm2 promotes the rapid degradation of p53. *Nature (London)* **387,** 296–303.

Kauffmann, R. C., Li, S., Gallagher, P., Zhang, J., and Carthew, R. W. (1996). Ras1 signaling and transcriptional competence in the R7 cell of *Drosophila. Genes Dev.* **10,** 2167–2178.

Kumar, J., and Moses, K. (1997). Transcription factors in eye development: A gorgeous mosaic? *Genes Dev.* **11,** 2023–2028.

Lai, Z. C., and Rubin, G. M. (1992). Negative control of photoreceptor development in *Drosophila* by the product of the *yan* gene, an ETS domain protein. *Cell (Cambridge, Mass.)* **70,** 609–620.

Lai, Z. C., Harrison, S. D., Karim, F., Li, Y., and Rubin, G. M. (1996). Loss of *tramtrack* gene activity results in ectopic R7 cell formation, even in a *sina* mutant background. *Proc. Natl. Acad. Sci. U.S.A.* **93,** 5025–5030.

Lawrence, P. A., and Green, S. M. (1979). Cell lineage in the developing retina of *Drosophila. Dev. Biol.* **71,** 142–152.

Lebovitz, R., and Ready, D. F. (1986). Ommatidial development in *Drosophila* eye disc fragments. *Dev. Biol.* **117,** 663–671.

Li, S., Li, Y., Carthew, R. W., and Lai, Z. C. (1997). Photoreceptor cell differentiation requires regulated proteolysis of the transcriptional repressor Tramtrack. *Cell (Cambridge, Mass.)* **90,** 469–478.

O'Neill, E. M., Rebay, I., Tjian, R., and Rubin, G. M. (1994). The activities of two Ets-related transcription factors required for *Drosophila* eye development are modulated by the Ras/MAPK pathway. *Cell (Cambridge, Mass.)* **78,** 137–147.

Read, D., and Manley, J. L. (1992). Alternatively spliced transcripts of the *Drosophila tramtrack* gene encode zinc finger proteins with distinct DNA binding specificities. *EMBO J.* **11,** 1035–1044.

Read, D., Levine, M., and Manley, J. L. (1992). Ectopic expression of the *Drosophila tramtrack* gene results in multiple embryonic defects, including repression of *even-skipped* and *fushi tarazu. Mech. Dev.* **38,** 183–196.

Ready, D. F., Hanson, T. E., and Benzer, S. (1976). Development of the *Drosophila* retina, a neurocrystalline lattice. *Dev. Biol.* **53,** 217–240.

Rebay, I., and Rubin, G. M. (1995). Yan functions as a general inhibitor of differentiation and is negatively regulated by activation of the Ras1/MAPK pathway. *Cell (Cambridge, Mass.)* **81,** 857–866.

Rodan, G. A., and Harada, S.-I. (1997). The missing bone. *Cell (Cambridge, Mass.)* **89,** 677–680.

Rogge, R. D., Karlovich, C. A., and Banerjee, U. (1991). Genetic dissection of a neurodevelopmental pathway: *Son of sevenless* functions downstream of the sevenless and EGF receptor tyrosine kinases. *Cell (Cambridge, Mass.)* **64,** 39–48.

Rutledge, B. J., Zhang, K., Bier, E., Jan, Y. N., and Perrimon, N. (1992). The *Drosophila spitz* gene encodes a putative EGF-like growth factor involved in dorsal–ventral axis formation and neurogenesis. *Genes Dev.* **6,** 1503–1517.

Scheffner, M., Huibregtse, J. M., Vierstra, R., and Howley, P. M. (1993). The HPV-16 E6 and E6-AP complex function as a ubiquitin-protein ligase in the ubiquitination of p53. *Cell (Cambridge, Mass.)* **75,** 495–505.

Seyfert, V. L., Allman, D., He, Y., and Staudt, L. (1996). Transcriptional repression by the proto-oncogene BCL-6. *Oncogene* **12,** 2331–2342.

Simon, M. A., Bowtell, D. D., Dodson, G. S., Laverty, T. R., and Rubin, G. M. (1991). Ras1 and a putative guanine nucleotide exchange factor perform crucial steps in signaling by the sevenless protein tyrosine kinase. *Cell (Cambridge, Mass.)* **67,** 701–716.

Skowyra, D., Craig, K. L., Tyers, M., Elledge, S. J., and Harper, J. W. (1997). F-box proteins are receptors that recruit phosphorylated substrates to the SCF ubiquitin–ligase complex. *Cell (Cambridge, Mass.)* **91,** 209–219.

Tang, A. H., Neufeld, T. P., Kwan, E., and Rubin, G. M. (1997). PHYL acts to downregulate TTK88, a transcriptional repressor of neuronal cell fates, by a SINA-dependent mechanism. *Cell (Cambridge, Mass.)* **90,** 459–467.

Tei, H., Nihonmatsu, K., Yokokura, I., Ueda, T., Sano, R., Okuda, Y., Sato, T., Hirata, K., Fujita, S. C., and Yamamoto, D. (1992). *pokkuri,* a *Drosophila* gene encoding an E-26-specific (Ets) domain protein, prevents overproduction of the R7 photoreceptor. *Proc. Natl. Acad. Sci. U.S.A.* **89,** 6856–6860.

Tio, M., and Moses, K. M. (1997). The *Drosophila* TGFα homolog Spitz acts in photoreceptor recruitment in the developing retina. *Development (Cambridge, UK)* **124,** 343–351.

Tio, M., Ma, C., and Moses, K. M. (1994). *spitz,* a *Drosophila* homolog of transforming growth factor-alpha, is required in the founding photoreceptor cells of the compound eye facets. *Mech. Dev.* **48,** 13–23.

Tomlinson, A. (1985). The cellular dynamics of pattern formation in the eye of *Drosophila. J. Embryol. Exp. Morphol.* **89,** 313–331.

Tomlinson, A., and Ready, D. F. (1987a). Neuronal differentiation in the *Drosophila* ommatidium. *Dev. Biol.* **120,** 366–376.

Tomlinson, A., and Ready, D. F. (1987b). Cell fate in the *Drosophila* ommatidium. *Dev. Biol.* **123,** 264–275.

Tomlinson, A., Bowtell, D. D., Hafen, E., and Rubin, G. M. (1987). Localization of the sevenless protein, a putative receptor for positional information, in the eye imaginal disc of *Drosophila. Cell (Cambridge, Mass.)* **51,** 143–150.

Van Vactor, J. L., Jr., Cagan, R. L., Kramer, H., and Zipursky, S. L. (1991). Induction in the developing compound eye of *Drosophila*: Multiple mechanisms restrict R7 induction to a single retinal precursor cell. *Cell (Cambridge, Mass.)* **67**, 1145–1155.

Wolff, T., and Ready, D. F. (1991). The beginning of pattern formation in the *Drosophila* compound eye: The morphogenic furrow and the second mitotic wave. *Development (Cambridge, UK)* **113**, 841–850.

Xiong, W. C., and Montell, C. (1993). *tramtrack* is a transcriptional repressor required for cell fate determination in the *Drosophila* eye. *Genes Dev.* **7**, 1085–1096.

Xu, T., and Rubin, G. M. (1993). Analysis of genetic mosaics in developing and adult *Drosophila* tissues. *Development (Cambridge, UK)* **117**, 1223–1237.

Yamamoto, D., Nihonmatsu, I., Matsuo, T., Miyamoto, H., Kondo, S., Hirata, K., and Ikegami, Y. (1996). Genetic interactions of *pokkuri* with *seven in absentia, tramtrack* and downstream components of the *sevenless* pathway in R7 photoreceptor induction in *Drosophila melanogaster. Roux's Arch. Dev. Biol.* **205**, 215–224.

Zipursky, S. L., and Rubin, G. M. (1994). Determination of neuronal cell fate: Lessons from the R7 neuron of *Drosophila. Annu. Rev. Neurosci.* **17**, 373–397.

Zollman, S., Godt, D., Prive, G. G., Couderc, J. L., and Laski, F. (1994). The BTB domain, found primarily in zinc finger proteins, defines an evolutionarily conserved family that includes several developmentally regulated genes in *Drosophila. Proc. Natl. Acad. Sci. U.S.A.* **91**, 10717–10721.

17

Role of Drosophila Wingless Signaling in Cell Fate Determination

ESTHER SIEGFRIED
Departments of Biology and Biochemistry and Molecular Biology
Pennsylvania State University
University Park, Pennsylvania 16802

I. Wingless Is a Member of the Wnt Family of Cell Signaling Proteins

A hallmark of all multicellular organisms is the ability of cells to communicate and coordinate their actions. Cell communication is used to regulate growth, and to determine both the position and specialized role of individual cells. In the 1990s, tremendous advances have been made in identifying the extracellular signals which mediate the cell signaling essential for the development of invertebrate and vertebrate embryos. Perhaps one of the most intensely studied class of cell signaling molecules is the Wnt family of secreted growth factors. The Wnts comprise a large family of proteins that are expressed in many different species (McMahon, 1992; Nusse and Varmus, 1992). Int-1, the first Wnt encoding gene to be described, was identified as a protooncogene activated by the insertion of a mouse mammary tumor

virus which caused ectopic expression of Int-1, resulting in mammary tumors (Nusse and Varmus, 1982). Subsequently, it was recognized that Int-1 is the mouse homologue of the *Drosophila* segment polarity gene *wingless* and that these genes are just two members of a large family of related genes (Rijsewik *et al.*, 1987). The gene family was renamed Wnt in recognition of shared evolutionary history of its members (Nusse *et al.*, 1991). Each *Wnt* gene is predicted to encode a secreted protein with 22–24 conserved cysteine residues and the overall sequence identity among the different Wnts is between 30–60%.

The oncogenic potential of Wnts is just part of the story; work in a number of model systems has demonstrated that Wnts and Wnt signaling are crucial for normal development as well. In vertebrates, individual Wnts have been implicated in such diverse developmental processes as patterning of the dorsal–ventral axis, embryonic brain, spinal cord, limb bud, and kidney (Moon *et al.*, 1997). In *Drosophila*, there are four *Wnt* genes: *wingless* (*wg*), *DWnt-2*, *Dwnt-3/5A*, and *DWnt-4* (Baker, 1987; Cabrera *et al.*, 1987; Eisenberg *et al.*, 1992; Graba *et al.*, 1995; Russell *et al.*, 1992). However, *wingless* is the only one characterized in detail and for which loss-of-function mutations have been described. The Wingless protein (Wg) is an extremely versatile molecule, and it is used to direct cell fate specification and patterning in a variety of tissues in the embryo including the ectoderm, nervous system, and mesoderm. In the imaginal discs, the precursors of the adult appendages, Wg is involved in long-range patterning, establishing the dorsal–ventral axis in wing discs, and both the dorsal–ventral and the proximal–distal axes in the leg disc.

Genetic studies in *Drosophila*, augmented by biochemical and cell biological approaches, have defined a conserved pathway for transduction of the Wg/Wnt signal from the plasma membrane to the nucleus. The availability of loss-of-function and conditional mutations of *wg*, combined with the powerful genetic techniques of clonal analyses and tissue specific ectopic expression, have facilitated investigations into the role, mechanisms, and range of action of Wg protein. Genetic screens have lead to the discovery of intracellular components that are necessary to transduce the Wg signal and hold out the promise of uncovering novel components of this pathway. Given the extensive conservation between Wg and Wnt signaling, the investigation of Wg signaling in *Drosophila* will continue to contribute to our understanding of Wnt signaling in vertebrates. This chapter will highlight some of the functions of Wg in cell fate determination and patterning and review genetic evidence for the proposed mechanism of Wg signaling.

II. Wg Is Required for Correct Differentiation and Patterning of Embryonic Tissues

Wg has a role in patterning a variety of embryonic tissues; however, the best studied developmental role for Wg is in embryonic segmental patterning. *Drosophila* segmentation is regulated by both maternal and zygotic genes that pattern the embryo in a stepwise fashion. Initially, the anterior–posterior axis is established and then the embryo is subdivided into segmental units. This occurs in the syncytial blastoderm, a unique feature of *Drosophila* embryos, where the first 13 nuclear divisions occur in the absence of cellular divisions. After the cells have formed, the embryo begins to gastrulate and patterning within each segmental unit begins. At this point, the embryonic epidermis is an epithelial monolayer that

must establish and maintain polarity and specific cell fates within each segmental unit, as well as the borders between them. Prior to gastrulation, patterning of the blastoderm is directed by maternally and zygotically expressed genes, most of which encode transcription factors. However, once cellularization is complete and segments have been established, cell fate is determined by intercellular communication. The two signaling pathways that determine polarity and cell fate within each segmental unit are the Wg signaling pathway, the subject of this review, and the Hedgehog signaling pathway (DiNardo *et al.*, 1994; Perrimon, 1994).

The first morphological sign of segmentation in the developing embryo are the transient parasegmental borders (PSB), which occur at the juxtaposition of Wg-expressing cells and cells expressing the homeodomain protein Engrailed (En; Fig. 1). Anterior of the PSB are the Wg-expressing cells and posterior are the En-expressing cells (Martinez Arias, 1993; Martinez Arias and Lawrence, 1985). The PSB act as organizing centers for future patterning of the embryo and in *wg* mutant embryos these borders do not form. Subsequent to the PSB formation, the permanent segmental borders (SB) form at the posterior edge of the En-expressing cells. The posterior half of the segment is demarcated by *en* expression, which is required for the commitment of cells to posterior segmental fates (Fjose *et al.*, 1985; Kornberg *et al.*, 1985; Poole *et al.*, 1985). Wg establishes segmental borders and the polarity within each segmental unit by signaling to adjacent cells and stabilizing *en* expression and posterior cell fate determination. The pattern of segmentation that occurs in embryogenesis is reflected in the ventral cuticle pattern of the mature first instar larva. In addition to being able to identify the number of segments, the polarity and specific cell fates within each segment also are visible. Each segmental unit is demarcated by a region of naked cuticle and a region covered with small cuticle projections called denticles. The posterior half of the segment is mostly naked and the anterior half is decorated with denticles; each denticle type indicates a unique cell specification (Fig. 1). Wg also is required for the diversity of cell differentiation in each segment.

A. Wg Signaling Determines Cell Fate and the Patterning of the Embryonic Ectoderm

The initial of *wg* and *en* expression are determined by early acting maternal and zygotic genes; however, the maintenance of their expression is mutually dependent (Martinez Arias, 1993). In a *wg* mutant embryo, epidermal *en* expression decays prematurely and the posterior half of the segment, including the naked portion of the cuticle, is lost (Fig. 2) (DiNardo *et al.*, 1988; Heemskerk *et al.*, 1991; Martinez Arias *et al.*, 1988).

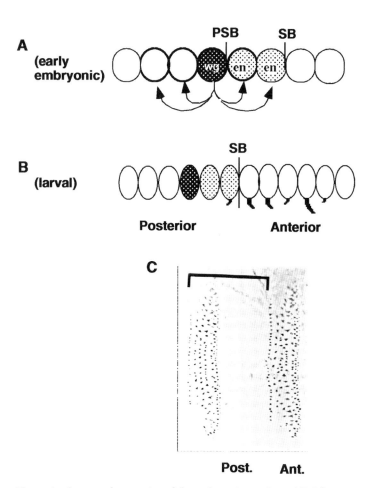

Figure 1 Segmental patterning of the embryonic ectoderm. (A) Schematic representation of the early embryonic ectoderm illustrating two adjacent segmental units along the anterior–posterior axis of the embryo. The parasegmental border (PSB) forms between the Wg-expressing (dark stippled) and En-expressing (light stippled) cells and the segmental border (SB) forms at the posterior edge of the En-expressing cells. Wg signals to neighboring cells, stabilizing the expression of *en* and the accumulation of Arm protein (indicated as cells with bold outline). Anterior is to the right. (B) Schematic representation of the larval ventral ectoderm showing the posterior half of one segment and the anterior half of the adjacent posterior segment. Wg and En-expressing cells are indicated with dark and light stippling, respectively. Individual denticles are illustrated. (C) Phase-contrast photomicrograph of the larval ventral cuticle demonstrating denticular (dark dots) and smooth cuticle. The bracket indicates a single segment.

Conversely, in *en* mutant embryos epidermal *wg* expression fades. Maintenance of *en* expression is mediated by Wg signaling; whereas, the reciprocal signaling from the En-expressing cells to Wg-expressing cells is mediated by the secreted protein Hedgehog (Hh). Mosaic analyses in embryos have demonstrated that stable *en* expression is only maintained in cells adjacent to the Wg-expressing cells, suggesting that cell fate is directly determined by Wg signaling (Vincent and Lawrence, 1994; Vincent and O'Farrell, 1992). In addition to maintenance of *en* expression, Wg appears to have a role in the maintenance of it own expression (Li and Noll, 1993). This occurs indirectly through Hh signaling from the adjacent En-expressing cells, as well as independently of Hh signaling through an autocrine pathway (Bejsovec and Wieschaus, 1993; Hooper, 1994; Yoffe *et al.*, 1995). Wg signaling also results in the stabilization and accumulation of the Armadillo protein in

adjacent cells in the embryonic ectoderm (Peifer *et al.*, 1994a; Riggleman *et al.*, 1989). Armadillo (Arm) is a multifunctional protein, required both for Wg signaling and for cadherin mediated cell adhesion (Orsulic and Peifer, 1996; Sanson *et al.*, 1996). Although stabilization of Arm is necessary to properly transduce the Wg signal into the nucleus (discussed in greater detail below), it also can be used as a measure of Wg signaling.

Wg also has a role in the differentiation of the ventral cuticle pattern (Bejsovec and Martinez Arias, 1991; Bejsovec and Wieschaus, 1993; Dougan and DiNardo, 1992). In *wg* mutant embryos, the "naked" posterior half of the cuticle of each segment is absent, leaving the embryos completely covered in a "lawn" of a single type of denticle (Fig. 2). Inactivation of Wg activity through the use of a temperature sensitive allele of *wg*, has shown that Wg has distinct functions in early and late embryonic development. Early signaling is necessary for

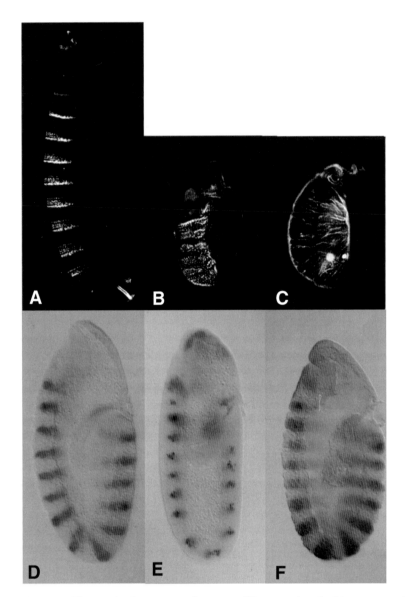

Figure 2 The cuticle phenotypes and patterns of En expression of wild-type, *wg*, and *zw3* mutant embryos. (A–C) Dark-field photomicrographs of cuticle preparations of wild-type (A), *wg* mutant (B), and *zw3* mutant (C) embryos. Bright staining indicates denticular cuticle. (D–F) Anti-En immunostaining of stage 10 wild-type (D), *wg* mutant (E), and *zw3* mutant embryos (F).

the stabilization of En expression but later signaling is required for naked cuticle secretion. Posterior to the Wg-expressing cell are two En-expressing cells that give rise to different cell fates, one secretes naked cuticle (the anterior cell closest to the Wg-expressing cell) and the other a denticle (Dougan and DiNardo, 1992). The distinction between these two cells is regulated by signaling through two distinct and opposing pathways: the Wg signaling pathway and the Epidermal Growth Factor Receptor (EGFR) pathway activated by the secreted ligand Spitz (O'Keefe *et al.*, 1997, Szats *et al.*, 1997). Wg and Spitz define domains anterior and posterior of the En-expressing cells respectively. Both En cells can respond to either signal and the final pattern of cell fates is determined by the relative levels of activation of

these two pathways. Specification of naked cuticle and the differentiation of denticle types are two properties of Wg which are separable within the Wg molecule (Bejsovec and Wieschaus, 1995; Hays *et al.*, 1997). There are distinct mutations of *wg* which disrupt either specification of naked cuticle or specification of denticle diversity. Furthermore, is has been suggested that these may occur through distinct pathways.

B. Wg Signaling in the Cell Fate Determination of the Nervous System

Wg is required in a nonautonomous fashion for neuroblast determination in the embryonic central nervous system (Chu-LaGraff and Doe, 1993). Neuroblasts normally delaminate from the neuroectoderm in a precise and stereotypical fashion, giving rise to approximately 30 identifiable neurons per hemisegment. *wg* is expressed in a row of cells in the neuroectoderm and in the emerging neuroblasts. Transient inactivation of Wg function, by shifting embryos that are homozygous for the temperature sensitive allele of *wg* to nonpermissive temperatures, results in embryos that are defective in neuroblast specification. Under these conditions, inactivation of *wg* does not affect ectodermal patterning, but it does result in changes in gene expression in the neuroectoderm prior to neuroblast delamination and in a transformation of the fate of one type of neuroblast which gives rise to the RP2 neuron. Thus, Wg functions nonautonomously in the neuroectoderm as an instructive signal to determine the fate of developing neuroblasts; in the absence of Wg the RP2 neuron is missing.

An additional role for Wg in nervous system developmental has been found in the patterning of the stomatogastric nervous sytem (SNS) which innervates the gut (Gonzalez-Gaitan and Jackle, 1995). The SNS is composed of four ganglia derived from an epithelial primordia, the dorsal surface of the developing foregut. This epithelium undergoes three invaginations which give rise to three vesicular structures containing the neural precursors of the SNS. The number and position of the invagination centers in the epithelium are controlled by lateral inhibition; in *Notch* mutant embryos the entire SNS primordia invaginates. Wg is expressed in the SNS primordia and serves to restrict the range of Notch mediated lateral inhibition. Thus, in *wg* mutant embryos, there is only a single invagination of the SNS primordia and only a single neural precursor, presumably due to Notch signaling throughout the primordia. In this case, Wg appears to be a permissive signal which modulates the range of lateral inhibition rather than an instructive signal determining cell fate.

C. Wg Signaling in Tissue Induction: Embryonic Mesoderm Specification

The best studied case of Wg signaling is in local cell–cell signaling to establish cell fate within adjacent cells (described above); however, another important function of Wg is its role as a signal in germ layer interactions. In embryonic development, there are two instances of Wg signaling from one tissue affecting the patterning and differentiation of a second juxtaposed tissue; mesoderm specification (somatic muscles and the heart) and midgut differentiation. At cellular blastoderm, the presumptive mesoderm is derived from a strip of ventral cells (Bate and Rushton, 1993). During gastrulation these cells invaginate along the ventral furrow and become juxtaposed against the ectoderm. The mesoderm then segregates into three distinct muscle progenitors: precursors to the somatic musculature, the

visceral musculature, and the cardiac. The formation of the visceral and cardiac mesoderm is specified by the expression of two homeodomain genes, *bagpipe* and *tinman* (Azpiazu and Frasch, 1993; Bodmer, 1993). *tinman* expression and the specification of the visceral and cardiac mesoderm is regulated in part by signaling from the ectoderm through Decapentaplegic (Dpp), a member of the transforming growth factor β (TGF-β) family of growth factors (Frasch, 1995; Staehling-Hampton *et al.*, 1994).

One aspect of mesoderm patterning that is still unclear is how segmental patterning along the anterior–posterior axis is established. Clearly the mesoderm must come into correct register with the ectoderm because the body wall muscles must attach to the correct ectodermal segments to allow coordinated movement of the larvae. There is some evidence that mesoderm segmentation is autonomous; the pattern of expression of homeotic genes in the mesoderm is out of register with their pattern in the ectoderm and ectopic expression of *Ubx* specifically in the mesoderm can alter the segmental identities within this germ layer (Greig and Akam, 1993; Michelson, 1994). It has been suggested that tissue induction between the ectoderm and mesoderm is responsible for the specification of distinct cell types within the developing mesoderm (Baker and Schubinger, 1995). The correct juxtaposition of ectoderm and mesoderm is therefore essential for correct mesoderm differentiation, and it appears that extracellular signals, such as Dpp and Wg, are mediating these interactions. However, there is some uncertainty if Wg signaling from the ectoderm is patterning the mesoderm or if low levels of expression in the mesoderm is sufficient for correct specification (Baylies *et al.*, 1995; Lawrence *et al.*, 1995; Lawrence *et al.*, 1994).

1. Cardiac Development

Wg regulates the specification of the cardiac precursors in the developing mesoderm (Lawrence *et al.*, 1995; Park *et al.*, 1996; Wu *et al.*, 1995). Loss of Wg activity results in a selective loss of cardiac precursors and the absence of the heart. Conversely, overexpression of *wg* results in an increased number of cardiac precursor cells in the mesoderm. The use of mosaic embryos has demonstrated that *wg* expression in the mesoderm alone is sufficient for specification of cardiac precursors. Interestingly, uniform expression of *wg* does not result in ectopic cardiac precursors throughout the mesoderm. The ectopic precursors cells are in the normal position along the dorsal–ventral axis and are still found in a segmentally repeated pattern along the anterior-posterior axis. This suggests that Wg activity is permissive rather than instructive for cardiac determination.

2. Somatic Muscle Specification

The specification of somatic muscle fibers also is regulated by ectoderm–mesoderm interactions mediated by Wg. Each hemisegment contains 30 unique body wall muscles; each fiber is a syncytium formed by the fusion of anywhere from 5 to 25 cells, depending on the individual muscle fiber (Bate, 1993). The specification and precise patterning of these muscle fibers arise from so-called muscle founder cells (Bate and Rushton, 1993; Rushton *et al.*, 1995). Individual muscle founders are set aside in the developing mesoderm and specify the precise muscle to be made; these cells fuse with surrounding unspecified fusion-competent cells, imparting on them distinct muscle identity. The specification of founder cells is likely to be controlled by the expression of combinations of regulatory genes encoding transcription factors such as *nautilus, S59, apterous,* and *vestigial,* all of which are expressed in subsets of muscle founder cells (Abmayr *et al.*, 1995).

The precise mechanism specifying the founder cell identity is unknown but signaling from the ectoderm through Wg may be involved. In *wg* mutant embryos, the mature somatic muscle pattern is severely disrupted, but not all muscles are affected. For example, the muscle precursors marked by S59 expression and a subset of those marked by *nautilus* expression are absent in *wg* mutant embryos; however, other muscle precursors are still present indicating that this is not a general effect (Baylies *et al.*, 1995; Ranganayakulu *et al.*, 1996). Although there is low-level expression of *wg* in the mesoderm, it is likely that *wg* expressed in the ecto-

derm can directly signal to the mesoderm. The loss of muscle precursors in *wg* mutant embryos can be rescued by either expression of *wg* solely in the ectoderm or the mesoderm (Baylies *et al.*, 1995; Lawrence *et al.*, 1994; Ranganayakulu *et al.*, 1996).

D. Wg Signaling in Tissue Induction: Embryonic Midgut Patterning

The segmental patterning of the embryonic midgut epithelium occurs by tissue induction that is mediated in part by Wg and by Dpp (Bienz, 1994). The *Drosophila* embryonic midgut is composed of two tissues, the visceral mesoderm and an underlying endodermal epithelium (see Fig. 3). At the time of midgut formation the visceral mesoderm is segmented; this is revealed by the nonoverlapping domains of homeotic gene expression along the anterior–posterior axis. This contrasts with the unsegmented endodermal cell layer which comes to lie below the visceral mesoderm and requires induction by the visceral mesoderm for correct patterning. Ultimately, these two tissues will coordinate to form the midgut, undergoing three specific morphological constrictions to form a four chambered structure and to give rise to distinct cells in the larval midgut. The extracellular signals in this induction are encoded by *wg* and *dpp* which are both expressed in spatially restricted domains within the visceral mesoderm. *dpp* and *wg* ex-

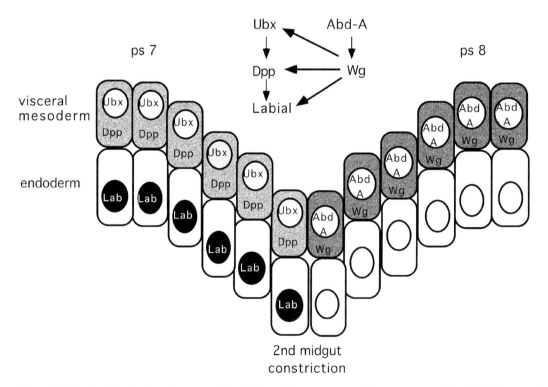

Figure 3 Wg signaling in the embryonic midgut. *Ubx* and *dpp* are expressed in parasegment 7 (ps 7) of the visceral mesoderm. *abd-A* and *wg* are expressed in ps 8 of the visceral mesoderm. Wg and Dpp signal to the underlying endoderm and establish a domain of *labial* expression. Wg also signals in the mesoderm to maintain *Ubx* and *dpp* expression.

pression is regulated by two homeotic genes, *Ultra-bithorax* (*Ubx*) and *abdominal-A* (*abd-A*), respectively (Reuter *et al.*, 1990; Tremml and Bienz, 1989). Wg and Dpp then signal to the underlying endoderm which in turn specifies the expression of the homeotic gene *labial* in the midgut epithelium and the formation of the central midgut constriction (Immergluck *et al.*, 1990; Panganiban *et al.*, 1990). In addition to Wg signaling across germ layers to pattern the endoderm, there is Wg signaling within a germ layer where it is required to maintain expression of *Ubx* and *dpp* in the adjacent parasegment of the visceral mesoderm (Thuringer and Bienz, 1993; Thuringer *et al.*, 1993; Yu *et al.*, 1996).

Correct *labial* expression is necessary for the specification and differentiation of a unique cell type in the larval midgut, the copper cells (Hoppler and Bienz, 1994). Specification and differentiation of the larval copper cells is regulated directly by different levels of Wg (Hoppler and Bienz, 1995). Cells in the endoderm which are immediately juxtaposed to the Wg-expressing cells experience the highest level of Wg and are prevented from undergoing copper cell differentiation giving rise instead to the large flat cells of the larval midgut. Cells experiencing intermediate levels of Wg differentiate into copper cells and cells that receive no Wg signal will not give rise to either copper cells or flat cells. The fate of larval midgut cells can be altered by ectopically expressing high or intermediate levels of Wg. Here Wg functions as an instructive signal, regulating cell fate in a concentration dependent fashion.

III. Wg Is Required for the Growth and Patterning of the Adult Appendages

As the name implies, Wg has a role in the differentiation and patterning of the adult wing. *wg* was discovered by the isolation of a viable allele that exhibited a transformation of wing to notum (Sharma and Chopra, 1976). Other pupal lethal alleles of *wg* exhibit defects in other imaginal tissue including the legs (Baker, 1988a). The appendages of the adult fly, like much of the adult tissues, are derived from specific imaginal discs; these are epithelial sacs which give rise to the ectodermal tissue of each appendage. Initially, the imaginal disc are composed of 10–40 cells derived from the embryonic ectoderm (Blair, 1995; Cohen, 1993). During larval stages, these clusters of cells proliferate and give rise to thousands of cells in each disc. These cells rearrange and the disc elongates during the pupal stages to give rise to the final adult forms. The emerging model for limb patterning in *Drosophila* proposes that positional information and patterning are acquired through both

short- and long-range interactions (Blair, 1995; Lawrence and Struhl, 1996). Imaginal discs are initially subdivided into compartments which are established by the stable expression of selector genes, such as *en* which marks the posterior compartment. This positional information is derived from the embryo; embryonic En-expressing cells are incorporated into the imaginal disc precursor. Interactions between neighboring cells of the two compartments induces a source of long-range graded signals, such as Wg and Dpp, along the compartment boundary. These long-range signals pattern the anterior–posterior and dorsal–ventral axis, as well as functioning in establishing the proximal–distal axis.

A. Wg Is Required for Growth, Patterning, and Cell Specification of the Wing

The developing wing imaginal disc, which gives rise to the notum and wing, is subdivided into four distinct compartments by two orthogonal boundaries, the anterior–posterior (A–P) compartment boundary and dorsal–ventral (D–V) compartment boundary. The posterior compartment is inherited from the En-expressing embryonic ectodermal cells which comprise the early disc, and the dorsal compartment is acquired in the second larval instar through the localized expression of *apterous* (Blair, 1995). Interaction between the anterior and posterior compartments is mediated through the secreted protein Hh, which is expressed in the posterior compartment in response to En. Hh signaling to adjacent cells in the anterior compartment induces *dpp* expression along the A–P boundary (Basler and Struhl, 1994; Tabata and Kornberg, 1994). Dpp is a long range signal which is required for growth and patterning of the A-P axis and specifies distinct patterns of gene expression (Lecuit *et al.*, 1996; Nellen *et al.*, 1996). Ectopic Dpp can alter patterns of gene expression in both anterior and posterior compartments, and appears to do so in a concentration dependent fashion. These and other results taken together have suggested that Dpp is a long-range morphogen, the source of which is established at the A–P compartment boundary and which directly determine positional information and cell fate along the anterior–posterior axis.

Wg expression is dynamic in the developing wing discs and appears to have three distinct roles: specification of the wing, organizing pattern from the D–V compartment boundary, and specification of the wing margin. *wg* expression is first detected in the anterior compartment of the disc, but by the second instar is predominately localized to a ventral quadrant in both the anterior and posterior compartment (Baker, 1988a; Couso *et al.*, 1993; Ng *et al.*, 1996). By the third instar stage, the pattern of expression has changed again and fills most of the presumptive wing blade. Finally, the *wg* pattern resolves to a stripe straddling the length of the dorsal–ventral compartment boundary, two bands that encircle the wing presumptive wing blade, and a broad band covering the notal region of the disc. Wg has been implicated in the specification of the domain of the disc which will give rise to the wing rather than notum (Ng *et al.*, 1996). Inactivation of *wg*, through the use of the temperature sensitive allele, has demonstrated that the early anterior expression of *wg* is required for the specification of the wing primoridia; in its absence there is duplication of the notum primordia. Furthermore, ectopic *wg* expression can reprogram the fate of cells from notum to wing identity. Expression of *wg* may not be sufficient for specification of the wing primordia; there is evidence that expression of *vestigial* also is required for correct specification of the wing (Kim *et al.*, 1996).

Once the wing primordia has been correctly specified, local interaction between the dorsal and ventral compartments of the disc trigger a long-range signal necessary for correct growth and patterning of the wing (Diaz-Benjumea and Cohen, 1993; Williams *et al.*, 1994). Localized *apterous* expression, which arises in mid-second larval instar stage, defines the dorsal compartment (Blair, 1995). Signaling between the dorsal and ventral compartments occurs through Fringe and Serrate, expressed in the dorsal compartment, and Delta, which is expressed in the ventral compartment (Panin *et al.*, 1997). Delta and Serrate signaling is mediated through Notch which activates Wg expression at the D–V border (Diaz-Benjumea and Cohen, 1995; Kim *et al.*, 1996; Neumann and Cohen, 1996). It has been suggested that Wg expression at the D–V compartment boundary functions as a long-range signal which regulates proliferation and patterning of the wing. Examination of clones of *wg* mutant tissue has demonstrated that *wg* expression is required in at least one side of the D–V compartment boundary for normal wing development (Diaz-Benjumea and Cohen, 1995). Conversely, clones ectopically expressing *wg* induce ectopic wing margin and outgrowth in a nonautonomous fashion.

Wg expression at the D–V border may provide long-range signaling activity; however, later in development it also has a role in specification of the cells at this location, as it gives rise to the margin of the wing (Couso *et al.*, 1994; Phillips and Whittle, 1993). The margin is decorated by distinct chemosensory bristles and the expression of the genes of the *achaete–scute* complex (ASC) have been correlated with cell specification in this region. High levels of expression of ASC determine the sensory mother cells (SMCs), the precursors of the marginal chemosensory bristles. Wingless signaling is required for regulating the expression of these genes, as well as for refining its own expression (Rulifson *et al.*, 1996). Wg expression straddles the D–V border in a 3- to 6-cell stripe. High levels of ASC expression and subsequent specification of SMC, occurs a further 3–6 cells away on either side of the margin. Inactivation of *wg* during pupal stages, results in a selective loss of the margin and SMCs (Phillips and Whittle, 1993). Furthermore, ectopic *wg* expression results in the generation of an ectopic margin in the midst of the wing blade, ectopic expression of the ASC, and ectopic SMCs (Diaz-Benjurnea and Cohen, 1995; Zecca *et al.*, 1997).

B. Wg Is Required for Growth and Patterning of the Leg

Axis formation in the leg imaginal disc is regulated in a similar fashion to the wing, but with one important distinction; there is no apparent selector gene for the dorsal fate. Dpp and Wg signaling appear to regulate the specification of dorsal and ventral fates, respectively. The posterior compartment is determined by the expression of *en* and the short-range signaling between the anterior and posterior compartments is mediated by Hh, in a similar fashion to wing imaginal disc (Basler and Struhl, 1994; Diaz-Benjumea *et al.*, 1994). Hh induces expression of *dpp* along the A–P compartment border but its expression is greater in the dorsal half than in the ventral half. Hh signaling also induces the expression of *wg* in the ventral anterior quadrant. The expression of *wg* is correlated with the induction of ventral cell fates and loss-of function of *wg* results in the deletion of these cell fates (Baker, 1988a; Couso *et al.*, 1993; Peifer *et al.*, 1991). Conversely, ectopic expression of *wg* on the dorsal side of the disc has a ventralizing effect on neighboring cells and results in the expansion of the ventral cell fate or a duplication of the leg (Maves and Schubinger, 1995; Struhl and Basler, 1993; Wilder and Perrimon, 1995). Work from a number of laboratories has demonstrated that the complementary patterns of expression of *wg* and *dpp* are important for specifying the D–V axis in the leg imaginal disc, and that the patterns of expression are maintained by mutually antagonistic interactions. This was demonstrated by expressing either *wg* or *dpp* through the GAL4/UAS binary expression system, which allows ectopic gene expression through the use of heterologous promoters (Brand and Perrimon, 1993). Ectopic expression of *wg* in the Dpp domain repress endogenous *dpp* expression and overexpression of *dpp* in its endogenous pattern results in a repression of *wg* expression (Brook and Cohen, 1996; Jiang and Struhl, 1996; Johnston and Schubiger, 1996; Theisen *et al.*, 1996). In addition, clones that are mutant for the Dpp receptors (*thickvein* and *punt*), and so are unable to respond to the Dpp signal, exhibit ectopic *wg* expression due to derepression (Penton and Hoffmann, 1996). It has also been demonstrated that Dpp blocks the ability of Wg to activate one of its target genes in the leg disc (Brook and Cohen, 1996).

At center of the leg disc is the point that endogenous *dpp* and *wg* expression overlap, and it is here that Wg and Dpp act synergistically to establish the proximal–distal (P–D) axis. The center of the disc gives rise to the most distal structures of the leg and is also the site of expression of *distalless* and *aristaless,* which are required for differentiation of distal structure. Inactivation of *wg* and *dpp* results in loss of *distalless* and *aristaless* expression and a failure to form the distal structures of the leg (Campbell *et al.,* 1993; Diaz-Benjumea *et al.,* 1994; Lecuit and Cohen, 1997). Ectopic expression of *wg* and *dpp* results in ectopic expression of *distalless* and *aristaless* and the induction of ectopic P–D axes, resulting in supernumerary legs. It has been demonstrated that the activation of *distalless* expression by Wg and Dpp is likely to be direct; clones which have constitutively active Dpp and Wg signaling express *distalless* in a cell autonomous manner (Lecuit and Cohen, 1997). These authors have shown that Wg and Dpp pattern the P–D axis in a concentration dependent fashion and that two genes, *distalless* and *dachshund,* which are expressed in distinct domains along the P–D axis of the leg, respond to different thresholds of Dpp and Wg activity. Patterning along the P–D axis is further elaborated by the proliferation of these domains and the creation of a domain where both *dachshund* and *distalless* are expressed. In the leg imaginal disc, Wg, in conjunction with Dpp, functions as an instructive signal to establish D–V and P–D axes.

IV. The Mechanism of Wg Signal Transduction

A. Wg Is the Extracellular Signal

What is the evidence that Wg protein itself is the extracellular signal that triggers cell fate determination and patterning? There are several observations that support this hypothesis. First, early mosaic analysis of clones of *wg* mutant cells surrounded by wild-type cells showed that the Wg deficient patches are rescued by the surrounding wild-type tissue, establishing that Wg functions nonautonomously (Baker, 1988b). Second, Wg protein has been detected 2–3 cell diameters away from the source of Wg-expression by immunoelectron microscopy (Gonzales *et al.,* 1991; van den Heuvel *et al.,* 1989). Third, experiments reconstructing Wg signaling in heterologous cultured cells have demonstrated that the Wg protein is required to achieve the same effects of signaling observed *in vivo* (Cumberledge and Krasnow, 1993). Cultures of embryonic En-expressing cells rapidly loss *en* expression; however, when these cells are mixed with Wg-expressing cells, *en* expression persists. Furthermore, media from Wg-expressing cultured Schneider cells can promote Arm stabilization in the Wg responsive imaginal cell line clone-8 and this effect can be specifically blocked by the application of anti-Wg antibodies (van Leeuwen *et al.,* 1994).

B. A Model for Wg Signaling

A variety of genetic and biochemical studies support the current model for the Wg/Wnt signaling pathway (Fig. 4). Furthermore the tremendous conservation between the Wg signaling pathway in *Drosophila* and the vertebrate Wnt signaling pathway is demonstrated by the results that homologous components, where tested, have analogous function (Moon *et al.,* 1997). Some components were first identified by mutations in *Drosophila,* and then shown to have vertebrate homologues. The ability of these vertebrate homologues to transduce Wnt signals is determined in the *Xenopus* embryo, where Wnt signaling regulates the patterning of the dorsal–ventral axis. Other proteins in the putative Wg/Wnt pathway were identified by biochemical and cell biological means and have been tested in both *Xenopus* and *Drosophila.* The following model is based on an amalgam of these genetic, cell biological, and biochemical data (Nusse, 1997; Perrimon, 1996). Wg is secreted and this secretion is likely to be regulated by the *porcupine* gene product (Kadowaki *et al.,* 1996). On the surface of the Wg responsive cell, Wg is bound to the putative receptor, Dfrizzled2 (Dfz2), a seven pass transmembrane protein

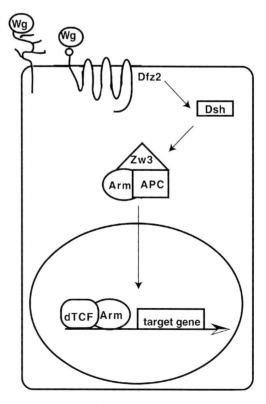

Figure 4 A model for Wg signaling.

which is a member of a family of related proteins found in both vertebrates and invertebrates (Bhanot et al., 1996). Wg also binds to proteoglycans on the cell surface and to the extracellular matrix, and these interactions appear to play a role in Wg signaling (Binari et al., 1997; Hacker et al., 1997; Haerry et al., 1997; Reichsman et al., 1996). The precise interactions among Wg, its receptor, and cell surface proteoglycans is unclear; however, binding of Wg to a proteoglycan may regulate stability, diffusion, and local concentration of Wg and thereby altering its interaction with Dfz2 and the efficiency of Wg signaling. Downstream of Dfz2 is the novel protein Dishevelled (Dsh) which also is conserved in evolution (Klingensmith et al., 1994; Theisen et al., 1994). The function of this protein is unknown but it is phosphorylated in response to Wg signaling (Yanagawa et al., 1995). Dsh indirectly regulates the activity of Zeste white 3 kinase (Zw3); inactivation of this kinase is likely to be the consequence of Wg signaling although the precise mechanism for this is unclear (Siegfried et al., 1992; Siegfried and Perrimon, 1994). Zw3 is the fly homologue of the vertebrate serine/threonine protein kinase glycogen synthase kinase 3β (Bourouis et al., 1990; Siegfried et al., 1990; Woodgett, 1990). Zw3 in turn regulates the stability of Armadillo (Arm) protein, the Drosophila homologue of β-catenin, possibly by direct phosphorylation (Peifer et al., 1994a,b; Siegfried and Perrimon, 1994). Arm protein has two distinct roles, the first in Wg signaling and the second in cadherin mediated cell adhesion (Orsulic and Peifer, 1996; Sanson et al., 1996). Zw3 regulates the levels of Arm; when Zw3 is active, cytoplasmic Arm is rapidly degraded while the cadherin associated Arm is relatively stable. Arm also may interact with an additional protein, APC, the product of the tumor suppresser gene adenomatous polyposis coli, which regulates its stability in combination with Zw3 (Hayashi et al., 1997; Rubinfeld et al., 1996). Accumulation of cytoplasmic Arm results in the association of Arm with an HMG box transcription factor, dTCF, and the translocation of the complex into the nucleus (Brunner et al., 1997; van de Wetering et al., 1997). There the Arm–dTCF complex binds to DNA and can directly activate transcription of Wg responsive genes (Fig. 4).

C. Genetic Dissection of the Wg Signaling Pathway

Although Wnts were first described in the 1980s, the signal transduction mechanism mediated by these growth factors has only come to light in the last several years. One important line of investigation that has advanced our understanding of Wg/Wnt signaling has been the identification of signaling components by mutational analysis. Genetic screens uncovered several maternal effect mutations which caused embryonic pheno-

types similar to wg, and were therefore considered good candidates for components of the signaling pathway (Perrimon et al., 1989). These genes are maternally expressed and the mRNAs are uniformly distributed throughout the early embryo; followed by uniform zygotic expression. It is the loss of both maternal and zygotic expression which results in the mutant phenotype. Mutations in three maternally expressed genes, dishevelled (dsh), armadillo (arm), and porcupine (porc) produce phenotypes very similar to wg mutant embryos; en expression fades prematurely and the ventral cuticle is a lawn of one single denticle type (Kadowaki et al., 1996; Klingensmith et al., 1994; Peifer et al., 1991; Riggleman et al., 1989, 1990; Siegfried et al., 1994; Theisen et al., 1994).

A fourth gene, zeste white 3 (zw3, also referred to as shaggy), appears to act antagonistically to Wg and the phenotype produced by zw3 mutant embryos is reciprocal to wg mutant embryos; the cuticle is entirely naked, en expression persists in the epidermis and is broader than wild type (see Fig. 2) (Siegfried et al., 1992). This suggests that the role of Zw3 is to downregulate en expression and removing Zw3 activity results in derepression of en. This phenotype is similar to that documented for the ubiquitous expression of wg in transgenic embryos carrying a copy of the wg gene driven by the heat shock protein 70 promoter (Noordermeer et al., 1992). These observations suggested that Zw3 functions in the Wg signaling pathway and we tested this hypothesis by determining the epistatic relationship of zw3 and wg. In embryos which are mutant for both wg and zw3, the pattern of en expression and the cuticle phenotype is identical to embryos mutant for zw3 alone: en expression is expanded and the cuticle is naked (Siegfried et al., 1992). This demonstrated that in the absence of Zw3 activity, Wg is no longer required for stable en expression. Furthermore, we proposed that Wg signaling is mediated by the inactivation or repression of Zw3 as follows. Normally, Wg signaling alleviates Zw3 repression of en expression in a small number of cells which are adjacent to the Wg-expressing cells; but, in the case of uniform wg expression Zw3 activity is repressed throughout the segment. This is similar to a loss of activity in zw3 mutant embryos and results in a derepression of en throughout the En competent domain.

The mutant phenotypes of dsh, arm, and porc suggest that these gene products also act in the Wg signaling pathway. From the phenotypes alone it is difficult to determine if these gene products function upstream or downstream of Wg in the signaling pathway and genetic epistasis with wg mutants is out of the question because the phenotypes are identical. However, the order of these gene products in a linear pathway was determined in two different ways, relative to zw3 and relative to ectopic wg expression. By examining embryos which are double mutants for zw3 and porc, dsh, and arm, respectively, we were able to determine if any one mutant is upstream or downstream of *zw3* in a linear pathway (Peifer *et al.*, 1994a; Siegfried *et al.*, 1994). Embryos that are mutant for both *zw3* and *porc*, or *zw3* and *dsh* are identical to *zw3* mutant embryos: *en* expression is broader than wild type and the ventral cuticle is devoid of denticles. This indicates that both Porc and Dsh function upstream of Zw3. However, embryos that are double mutant for *zw3* and *arm* are similar to *arm* mutant embryos, *en* expression fades prematurely and the ventral cuticle is a lawn of denticles. This indicates that Arm functions downstream of Zw3 and that Zw3 activity is mediated by regulation of Arm.

The necessity for *porc, dsh,* and *arm* to transduce the Wg signal also was tested by the ability of mutations in each of these genes to suppress the dominant phenotype generated by ectopic *wg* expression (Noordermeer *et al.*, 1994). Both *dsh* and *arm* were epistatic to the *hs-wg* phenotypes which is consistent with these gene products functioning downstream of Wg. Porc on the other hand, is not essential for the ectopic *wg* phenotypes suggesting that it functions upstream of Wg. Finally, additional evidence that Arm is downstream of, and regulated by, Zw3 came from examination of levels of cytoplasmic Arm in *zw3* mutant embryos. In *zw3* mutant embryos, cytoplasmic Arm protein accumulates to high levels; consistent with the epistasis experiments which predicted that Zw3 activity is mediated through Arm (Peifer *et al.*, 1994a; Siegfried and Perrimon, 1994). These epistasis experiments lead to the initial proposal of a pathway for Wg signaling: the Wg signal is transduced through Dsh to inactivate Zw3 kinase, which in turn regulates the stability and activity of Arm (Fig. 4).

D. Wg Signal Is Mediated through the Same Pathway in Different Tissues

The genetic dissection of the Wg signaling pathway is largely derived from examining the effects of mutants on Wg signaling in the embryonic ectoderm. However, as described earlier, Wg signaling is required in other tissues in the embryo and in the imaginal discs. We will consider the evidence that indicates that Wg signaling is conserved in different embryonic and imaginal tissue. The role of Zw3, Dsh, and Arm have been considered in the Wg dependent patterning of the stomatogastric system and the pathway appears to be conserved (Gonzalez-Gaitan and Jackle, 1995). Loss of *wg* results in a single epithelial invagination and the differentiation of a single neural precursor instead of the normal three; this phenotype is identical in *dsh* and *arm* mutant embryos. In contrast, there are three invaginations and the subsequent differentiation of three neural precursors in *zw3* mutant embryos. The role of the Wg signaling pathway in the determination of neuroblast identity in the

central nervous system has not been fully examined; however, it has been demonstrated that overexpression of *dsh* can rescue the loss of the RP2 neuron in *wg* mutant embryos (Park *et al.*, 1996).

Wg is required for the specification of somatic muscle precursors and cardiac precursors in the mesoderm. It has been demonstrated that Wg signaling in cardiac specification is mediated by Dsh and Arm, but that Zw3 is not required (Park *et al.*, 1996). In *dsh* and *arm* mutant embryos the precursors of the heart are absent. If the Wg signal is mediated through Zw3 the number of heart precursors should be expanded in *zw3* mutant embryos; however, this is not observed and the number of heart precursors appear normal in *zw3* mutant embryos. We have begun to examine the role of Wg signaling in the specification of somatic muscles by examining the fate of muscle precursors expressing S59 in *dsh, arm,* and *zw3* mutant embryos. S59 is expressed in a subset of muscle precursors in a Wg dependent fashion; in *dsh* and *arm* mutant embryos mesodermal S59 expression is absent (data not shown). In *zw3* mutant embryos, there are a greater number of S59-expressing cells in the mesoderm than wild type (Fig. 5). These ectopic cells arise in a segmentally repeated fashion. These observations are consistent with the hypothesis that Wg signaling is mediated through Dsh, Arm, and Zw3 in the embryonic mesoderm. Wg signaling in the patterning of the embryonic midgut occurs through the pathway identified in the embryonic ectoderm; embryos which are mutant for *dsh* or *arm* exhibit phenotypes similar to *wg* mutant embryos (Yu *et al.*, 1996). Direct targets of Wg signaling, such as the second midgut constriction, *dpp,* and *Ubx* expression, are absent or reduced, respectively. This is in contrast to *zw3* mutant embryos, where the second midgut constriction has multiple invaginations and *dpp* and *Ubx* expression are expanded. Mutants in *dTCF* also exhibit a failure to form the second midgut constriction and reduced *Ubx* expression (van de Wetering *et al.*, 1997).

The Wg signaling pathway also is conserved in a number of Wg dependent processes in the imaginal discs. Specification of the wing primordia is mediated through Dsh; in *dsh* mutant adults and pharate pupae the wing is absent and is replaced with a duplication of the notum similar to *wg* mutants (Klingensmith *et al.*, 1994; Theisen *et al.*, 1994). The ability of Wg to organize the D–V axis is also mediated through Dsh and Arm. This has been tested by examining the expression of *vestigial* and *distalless*, which respond to Wg activity originating from the D–V compartment boundary, in clones mutant for *dsh* or *arm* (Neumann and Cohen, 1997; Zecca *et al.*, 1997). In these mutant clones there is nonautonomous reduction of expression of Wg target genes. Mutations in *dTCF* result in small wings which are missing the margin, indicating a role in Wg signaling in the wing disc (Brunner *et al.*, 1997). Arm and Dsh

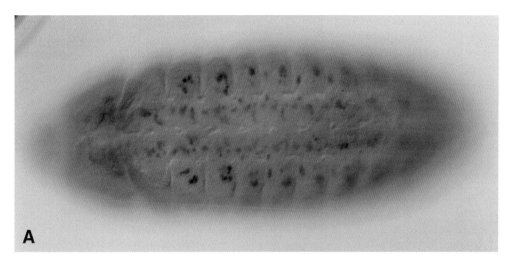

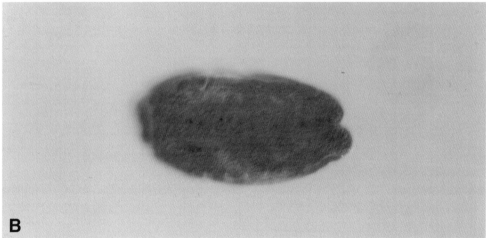

Figure 5 Specification of muscle precursors expressing S59 in wild-type, *wg*, and *zw3* mutant embryos. Anti-S59 immunostaining of stage 12 wild-type (A), *wg* mutant (B), and *zw3* mutant (C) embryos. Embryos are shown anterior to the right and dorsal up.

are necessary for the specification of the wing margin and clones of *zw3* mutant tissue give rise to ectopic margin, similar to ectopic expression of *wg*, suggesting that the Wg pathway is conserved in this instance as well (Blair, 1994; Couso *et al.*, 1994).

In the leg imaginal disc, clones of *dsh* or *arm* mu-

tant tissue in the ventral region of the discs result in a loss of ventral structures and duplication of dorsal structures, sometimes resulting in leg duplications similar to a loss of *wg* (Klingensmith *et al.*, 1994; Peifer *et al.*, 1991; Theisen *et al.*, 1994). Mutations in *dTCF* also result in dorsalization of leg structures (Brunner *et al.*,

1997). Conversely, the expansion of ventral cell fate is observed in *zw3* mutant clones (Diaz-Benjumea and Cohen, 1994; Wilder and Perrimon, 1995). Recently, it has been demonstrated that Zw3 and Dsh are required for Wg repression of *dpp* in the leg disc (Heslip *et al.,* 1997; Jiang and Struhl, 1996). Clones of cells mutant for *dsh* in the ventral region of the disc lose the ability to repress *dpp* resulting in ectopic expression of *dpp*. In contrast, clones of cells mutant for *zw3* in the dorsal region of the disc have lost *dpp* expression through Wg independent repression of *dpp*. The Wg dependent pattern of gene expression along the P–D axis of the leg also is mediated through Zw3 (Diaz-Benjumea and Cohen, 1994; Wilder and Perrimon, 1995; Zecca *et al.,* 1997). Therefore, it appears that in the imaginal disc the Wg dependent processes are mediated through the same pathway identified in the embryonic ectoderm.

E. Molecular Characterization of the Components of Wg/Wnt Signaling

1. *Porc and Wg Secretion*

The models of Wnt/Wg signaling propose that Wg is secreted and received by neighboring cells (Figs. 1 and 4). The distance that Wg protein may travel *in vivo* is an interesting one and will be addressed below; however, here we will consider what is known about the secretion of Wg protein. Three distinct forms of Wg protein have been isolated from the media of cultured cells that secrete active Wg (Reichsman *et al.,* 1996). Of these electrophoretic isoforms, one is unglycosylated and two are distinct N-glycosylated proteins. All three forms are found associated with the cell surface and the extracellular matrix, and a small amount of each is found free in the media. The relative abundance of the three forms appears unchanged in the different extracellular locations. Porc is likely to function in the secretion of Wg protein since in *porc* mutant embryos Wg protein is not diffuse and appears to be confined to those cells which express it (Siegfried and Perrimon, 1994). *porc* encodes a novel transmembrane protein that is found in the endoplasmic reticulium and expression of *porc* in cultured cells affects the processing of Wg protein (Kadowaki *et al.,* 1996).

2. *The Role of Proteoglycans*

Wg is a heparin binding protein and Wg binding to the cell surface and the extracellular matrix can be competed with sulfonated glycosaminoglycans (SO$_4$-GAGs) such as heparin, heperan sulfate or chondroitin sulfate (Reichsman *et al.,* 1996). The binding of different isoforms of Wg are sensitive to the different SO$_4$-GAGs, suggesting that Wg binds to the cell surface and extracellular matrix through specific interactions. Moreover, the enzymatic cleavage of SO$_4$-GAGs or the inhibition of sulfanation of proteoglycans of Wg-responsive clone-8 cells results in a reduction of Wg signaling when media from Wg-expressing cells is applied. In contrast, application of exogenous SO$_4$-GAGs results in the stimulation of Wg signaling. These results suggest that Wg signaling is mediated by interaction with cell surface proteoglycans.

Interestingly, in 1997, three groups conducting genetic screens for additional components of the Wg signaling pathway have simultaneously identified mutations in the same gene encoding UDP-glucose dehydrogenase, an essential enzyme in the synthesis of heparan sulfate and chondroitin sulfate (Binari *et al.,* 1997; Hacker *et al.,* 1997; Haerry *et al.,* 1997). These mutations, named independently by each group as *kiwi, sugarless,* and *suppenkasper,* were identified in genetic screens for maternal effect mutations which resulted in phenotypes similar to *wg* mutant embryos (*kiwi* and *sugarless*) (Binari *et al.,* 1997; Hacker *et al.,* 1997) or screens for mutations which suppressed or enhanced a viable adult *dsh* mutant phenotype (*suppenkasper*) (Haerry *et al.,* 1997). Loss of maternal and zygotic gene products for these genes results in embryos which resemble *wg* mutant embryos: the cuticles are a lawn of denticles and *en* expression fades from the ectoderm. In addition, *sugarless* mutant embryos have similar phenotypes in the SNS and embryonic midgut to *wg* mutant embryos, although not as severe. Furhermore, it has been demonstrated that ectopic *wg* can bypass the need for these genes in the secretion of naked cuticle and the expansion of *en* expression (Hacker *et al.,* 1997). These results indicate that the gene encoding UDP-glucose dehydrogenase is necessary for ensuring the correct presentation of Wg to the Wg responsive cell. This may occur by the ability of a proteoglycan to bind Wg, increasing the local concentration and/or altering the diffusion of Wg and increasing the stability of Wg/Dfz2 interaction. Of great interest is the identity of a specific proteoglycan that might serve this role in Wg signaling. One possible candidate is the glypican encoded by the gene *dally,* which has been demonstrated to interact with and affect Dpp signaling (Jackson *et al.,* 1997).

3. *Dfz2, the Wg Receptor*

The putative receptor for Wg is the transmembrane protein Dfz2 (Fig. 4), although it has not been demonstrated *in vivo* that Dfz2 is required for Wg signaling. Dfz2 is a member of a family of related proteins that do not have great sequence similarity but exhibit similar overall hydrophobicity and structure (Bhanot *et al.,* 1996). Expression of Dfz2 in Schneider cells, which are normally not responsive to Wg, causes the accumulation of Wg protein on the cell surface and the accumulation of cytoplasmic Arm protein.

4. *Dsh, a Transducer of the Wg Signal*

The gene *dsh* encodes a novel protein of unknown function; however, it is critical for Wg signaling. Ectopic expression of *wg* in the embryo results in the secretion of naked cuticle. This dominant effect of Wg can be completely suppressed by inactivation of *dsh*, indicating that Dsh is necessary for transducing the Wg signal (Noordermeer *et al.*, 1994). The only domain within Dsh with homology to a known protein is a stretch of 60 amino acids (DHR domain) that is similar to a repeated sequence found in the *Drosophila* tumor suppresser gene, *discs-large* (Klingensmith *et al.*, 1994; Theisen *et al.*, 1994). This domain has been found in a number of proteins that are thought to be associated with the cell junctions. The cellular distribution of Dsh in embryos has been determined by immunostaining and it appears to be cytoplasmic and shows no apparent redistribution of the protein in response to Wg signaling. Dsh is phosphorylated in response to Wg signaling; this has been detected in cultured clone-8 cells as well as in embryos (Yanagawa *et al.*, 1995). There is no direct evidence that this phosphorylation is important to the activity of Dsh or that Dsh interacts with Dfz2 directly. Finally, overexpression of *dsh* results in ectopic activation of the Wg signaling pathway in cultured cells and *in vivo*. Overexpression of *dsh* in clone-8 cells results in phosphorylation of Dsh and accumulation of Arm protein in a similar manner to the application of secreted Wg to these cells (Yanagawa *et al.*, 1995). In the wing imaginal disc, overexpression of *dsh* results in ectopic bristles near the wing margin (Axelrod *et al.*, 1996). The specification of these marginal bristles is a Wg dependent process and high levels of Dsh appears to enhance Wg signaling.

Dsh is a fascinating molecule because it appears to be involved in more than one signal transduction pathway. In the specification of the wing margin, Dsh is required to transduce the Wg signal. Dsh also has a role in inhibition of Notch function at the wing margin and has been demonstrated to interact directly with the cytoplasmic domain of Notch (Axelrod *et al.*, 1996). Dsh has an additional role in the pathway that coordinates planar cell polarity in epithelium. The planar cell polarity can be visualized in the bristles and hairs that cover much of the adult fly; these are cytoskeletal projections from individual cells and the direction in which they point is a reflection of the overall polarity of the tissue. *dsh* has been placed genetically downstream of *frizzled*, the putative receptor in the signal transduction cascade which determines tissue polarity (Krasnow *et al.*, 1995). Interestingly, Frizzled is related to the putative Wg receptor Dfz2; although these two proteins share only 33% identity at the amino acid level, they share a number of structural motifs. Dsh appears to be the only member of the Wg signaling pathway which is shared with the Frizzled signaling pathway. The precise function of Dsh in either of these pathways is still obscure but given the similarity of the putative receptors of these divergent pathway it is tempting to speculate that Dsh has a role in receptor function, stabilization, or localization.

5. *Wg Signaling Is Mediated through Zw3 Kinase*

In the model for Wg signaling, Dsh transduces the Wg signal which results in downregulation of Zw3 activity, leading to the correct cell fate determination. This model was proposed based on the genetic epistasis of *wg* and *zw3* which demonstrated that in the absence of *zw3*, Wg signaling is not required for maintenance of *en* expression or differentiation of naked cuticle (Siegfried *et al.*, 1992). The loss of Zw3 is thus analogous to hyperactivation of the Wg signaling pathway and this is consistent with the observation that uniform expression of *wg* in the embryo results in phenotypes that are similar to a loss of maternal *zw3* activity.

Zw3 is a serine/threonine protein kinase and is a member of a family of related kinases found in numerous plant and animal species including yeast, *Dictyostelium*, *Arabadopsis*, *Xenopus*, rat, and human. Interestingly, the homologue in *Dictyostelium* is required for cell fate determination in response to cAMP signaling (Harwood *et al.*, 1995). The vertebrate homologue is glycogen synthase kinase 3β (GSK-3β) and has a role in glycogen metabolism in response to insulin stimulation as well as a role in Wnt signaling (Woodgett, 1990). Zw3 is 75% identical at the amino acid level to GSK-3β, one of the two isoforms found in mammals which are encoded by distinct genes. The functional homology of these proteins has also been tested; heterologous expression of a rat cDNA encoding GSK-3β can rescue the segmentation phenotype of *zw3* mutant embryos which lack endogenous *zw3* (Ruel *et al.*, 1993; Siegfried *et al.*, 1992). In *Xenopus*, the role of GSK-3β in Wnt signaling has been addressed by injection of mRNA encoding a kinase-dead version of GSK-3β (Dominguez *et al.*, 1995; He *et al.*, 1995; Pierce and Kimelman, 1995). This protein has a dominant negative effect and results in an axis duplication similar to ectopic Wnt mRNA injection. The mechanism for Wg modulation of Zw3 kinase activity is unclear but studies in heterologous cultured cells have demonstrated that in mouse 10T1/2 fibroblast cells Wg inactivates the activity of GSK-3β through a protein kinase C (Cook *et al.*, 1996).

One prediction of the hypothesis that Wg signaling is mediated by Zw3 inactivation is that high levels of Zw3 will saturate the mechanism for inactivation and be resistant to Wg modulation, resulting in a block of Wg signaling. We have tested this prediction by overexpression of *zw3* in a variety of embryonic tissue using the GAL4/UAS binary expression system (Steitz *et al.*, 1998). We have used a GAL4 line which is expressed in the visceral mesoderm (GAL4-24B) to drive high levels of *zw3* expression in that tissue (Brand and Perrimon, 1993), and have examined the effects on Wg mediated patterning of the embryonic midgut. As described above, Wg signaling is necessary for the formation of the second midgut constriction. In addition, Wg signaling is required in the mesoderm to maintain the levels of expression of *dpp* and *Ubx* in the adjacent cells in the mesoderm. Correct levels of Wg and Dpp in the visceral mesoderm determines the pattern of *labial* expression in the endoderm and ultimately the specification of larval copper cell fate. Overexpression of *zw3* in the visceral mesoderm causes in a failure of the second midgut constriction and a reduction of *Ubx* and *dpp* expression in the mesoderm (Fig. 6). This is turn results in reduced *labial* expression in the endoderm and a loss of larval copper cell fates. Zw3 behaves as we have predicted, and its overexpression in the mesoderm blocks Wg signaling in that tissue, presumably by saturating some component of the signaling pathway. We have observed similar effects by overexpression of *zw3* in other embry-onic and imaginal tissues where Wg signaling is necessary for patterning and cell fate determination.

6. Regulation of Arm Accumulation

Genetic data suggest that Zw3 regulates the levels of Arm phosphorylation and stability, and further bio-chemical experiments indicate that this may be a direct interaction between Zw3 and Arm. In response to Wg signaling and inactivation of Zw3, intracellular levels of Arm protein are stabilized (Peifer *et al.*, 1994a,b). Fur-thermore it has been observed that in *zw3* mutant em-bryos, Arm protein is hypophosphorylated and that the levels of cytoplasmic Arm are elevated. Finally, a genetic interaction between *zw3* and *arm* has been noted in embryos which lack both *zw3* and *arm* maternally but ex-press one copy of each of these genes zygotically; these

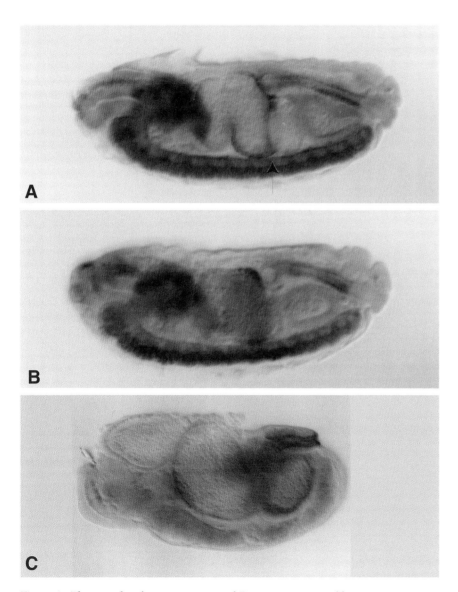

Figure 6 The second midgut constriction and Dpp expression in wild-type, *zw3* overexpressing, and *wg* mutant embryos. Anti-Dpp immuno-staining of stage 16 wild-type (A), *zw3* overexpressing embryos (B), and *wg* mutant embryos (C). The arrow indicates the position of the second midgut constriction. Embryos are shown anterior to the right and dorsal up.

embryos are completely rescued and appear wild type (Peifer *et al.*, 1994a; Siegfried *et al.*, 1994). This is in contrast to *zw3* mutant embryos which lack the maternal product but can not be rescued by one wild-type zygotic gene (Siegfried *et al.*, 1992). These results suggest that Zw3 regulates the level of Arm protein and that this is critical for correct Wg signaling. *Xenopus* GSK-3β can phosphorylate β-catenin *in vitro* and mutant forms of β-catenin which lack the consensus phosphorylation sites are more stable and more active *in vivo* than wild-type β-catenin (Yost *et al.*, 1996). In *Drosophila* a direct phosphorylation of Arm by Zw3 has not been detected; however, a N-terminal truncation of Arm which removes potential Zw3 sites is more stable *in vivo* and behaves as a constitutively active form of the protein, independent of Wg signaling and Zw3 regulation (Pai *et al.*, 1997).

7. APC, a Partner in Arm Regulation

The regulation of Arm stability requires Zw3, but there is evidence that another protein may interact with and regulate the Zw3 Arm interaction (Fig. 4). Protein complexes containing β-catenin, GSK-3β, and the product of the tumor suppresser gene *adenomatous polyposis coli* (APC) have been isolated (Rubinfeld *et al.*, 1996). Mutations in APC have been observed in most human colon cancers and more recently it has been demonstrated that β-catenin levels are elevated in these and other carcinomas (Korinek *et al.*, 1997; Morin *et al.*, 1997; Rubinfeld *et al.*, 1997). These results have lead to the conclusion that APC in conjunction with GSK-3β negatively regulates β-catenin levels. The role of APC in Wg signaling has been tested by the identification of the *Drosophila* homologue of APC (Hayashi *et al.*, 1997). The *Drosophila* APC (D-APC) gene shows 46% similarity to the human gene and within the identified β-catenin binding domains the identity is ~60%. D-APC binds Arm *in vitro* and *in vivo* has similar activity to mammalian APC. Furthermore, expression of *D-APC* in cancer cells which lack endogenous APC downregulates the levels of cytoplasmic β-catenin. However, despite these indications pointing to a function for D-APC in Arm regulation, embryos which lack any zygotic D-APC product appear completely normal for Wg signaling, suggesting that zygotic D-APC does not play a role in Wg signaling and Arm regulation *in vivo*.

8. Stabilization of Arm and Association with dTCF Transduces the Wg Signal into the Nucleus

As we have seen, Arm protein levels and intracellular distribution are critical for Wg signaling. Arm can be detected in a number of locations in the cell; the plasma membrane associated with adherens junctions, the cytoplasm, and the nucleus. Mutations throughout the Arm protein are associated with different mutant phenotypes and have identified domains of Arm that

function in cell adhesion, which are separable from domains involved in Wg signaling (Orsulic and Peifer, 1996). Cytoplasmic Arm is rapidly degraded, either directly or indirectly through interactions with Zw3 and APC (Pai *et al.*, 1997). An amino-truncated form of Arm, missing the putative Zw3 phosphorylation sites, is more stable than wild type and accumulates in the cytoplasm and nucleus independently of Wg signaling. Expression of this mutant Arm results in mutant phenotypes that are similar to hyperactivation of the Wg signaling pathway, by ectopic *wg* expression or loss of maternal *zw3*.

It has been demonstrated that the level of stable Arm is the link between Wg signal transduction in the cytoplasm and changes in gene expression in the nucleus. The first evidence came from yeast two hybrid screens which identified the TCF/Lef-1 HMG domain DNA binding proteins, as partners which could bind β-catenin; furthermore, β-catenin–TCF complexes bind to DNA and activate transcription (Behrens *et al.*, 1996; Molenaar *et al.*, 1996). In *Xenopus*, injection of mRNA encoding either TCF or Lef-1 demonstrated that they function in axis formation, in a similar fashion as Wnt and other components of the signaling pathway. Ectopic expression of mouse Lef-1 in *Drosophila* results in phenotypes that resemble ectopic activation of Wg signaling, suggesting a Lef-1 homologue function in *Drosophila* (Riese *et al.*, 1997). The *Drosophila* homologue of TCF/Lef-1 was cloned and mutations were identified in two independent studies. In the first study *Drosophila* TCF, *dTCF*, was cloned and mapped to the fourth chromosome, close to previously isolated mutations (van de Wetering *et al.*, 1997). Subsequently, these mutations were shown to be alleles of *dTCF*. In an independent genetic screen for modifiers of a dominant phenotype due to ectopic *wg* expression, the fourth chromosome mutation *pangolin* was recovered (Brunner *et al.*, 1997). The authors took an educated guess that this locus might encode dTCF. They isolated the *dTCF* gene and demonstrated it was mutant in two different *pangolin* alleles. Both groups demonstrated that dTCF binds Arm and moreover that it is genetically required downstream of Arm to mediate the effects of a constitutively active Arm (Fig. 4). A loss of zygotic *dTCF* results in segmental patterning defects in the ectoderm, and the absence of the second midgut constriction and *Ubx* expression in the embryonic midgut that are reminiscent of weak *wg* mutant phenotypes. A likely target of Arm–dTCF complex *in vivo* is the *Ubx* promoter which contains a consensus site for TCF binding and whose expression in the midgut is activated by heterologous expression of TCF (Riese *et al.*, 1997). In addition to the embryonic phenotype, mutations in *dTCF* result in patterning defects of the adult wing and leg, again similar to a loss of Wg activity (Brunner *et al.*, 1997).

The current model for Arm function in transcriptional activation suggests that once Arm is stabilized by Wg signaling, there is accumulation of Arm in the cytoplasm and nucleus, where it binds to dTCF. The ability of Arm to translocate into the nucleus is likely to be independent of dTCF binding, since mutant forms of Arm which do not bind dTCF are still localized to the nucleus by an unknown mechanism (van de Wetering *et al.*, 1997). Once in the nucleus the Arm–dTCF complex binds DNA via dTCF and activates transcription with Arm contributing an activation domain (Fig. 4). How Arm–dTCF interacts with other transcriptional activators and the basal transcriptional complexes remains to be investigated. Recently, the crystal structure of the repeated motif of β-catenin was determined and revealed an elongated structure composed of α-helices, wrapped in a superhelical structure (Huber *et al.*, 1997). This unique structure forms a positively charged groove which is likely to serve as the binding surface for interaction with cadherins, APC, and TCF.

V. Is Wg a Morphogen?

The existence of morphogenetic gradients of substances which organize pattern and cell fate has been debated for many years (Slack, 1987; Vincent, 1994). Wg is a good candidate for a morphogenetic substance; it is a secreted protein and, as described above, does direct the fate of adjacent cells. In the embryonic ectoderm, Wg acting coordinately with Spitz directs the fate of the two adjacent En-expressing cells (Fig. 1), and the proximity of these cells from the source of Wg secretion determines the differentiation of these cells (Dougan and DiNardo, 1992; Vincent and Lawrence, 1994; Vincent and O'Farrell, 1992; O'Keefe *et al.*, 1997; Szüts *et al.*, 1997). In addition, Wg appears to direct the differentiation of the cuticle of the entire embryonic segment (Bejsovec and Martinez Arias, 1991). In the development of the embryonic midgut, the levels of Wg determine the differentiation of different cell types in the endoderm (Hoppler and Bienz, 1995). In both of these examples Wg signaling is short-range, affecting cells that are relatively close to the source of Wg signal. However, when considering the role of Wg in patterning the D–V and P–D axes of the imaginal discs signaling must occur across a great number of cells and also specify a greater complexity of patterns of gene expression and cell types. There had been some suggestion that Wg acts directly to pattern the D–V axis in the leg and wing imaginal disc, and some indication that the levels of Wg specify different elements of the pattern along the axis (Diaz-Benjumea and Cohen, 1995; Struhl and Basler, 1993).

Wg signaling can be acting directly to pattern the entire disc or alternatively, Wg can be acting as a short-range signal inducing a cascade of signals which pattern the imaginal discs sequentially. These two possibilities are difficult to distinguish, but recent work has demonstrated that Wg acts directly, and at a distance, to regulate specific cell fate and pattern in a concentration dependent fashion. The ability of Wg to act at a distance was tested directly by examining the effects of ectopic expression of wild-type secreted Wg protein, a membrane tethered form of the protein and a constitutively active form of Arm (Zecca *et al.*, 1997). Ectopic expression of wild-type Wg results in upregulation of Wg target genes, such as *distalless* and *vestigial*, in surrounding cells (as much as 10 cell diameters away) and the boundary of expression is not sharp but graded. In contrast, the ectopic expression of the membrane tethered form of Wg upregulates expression in adjacent cells only and the boundary of this expression is sharp. Ectopic expression of activated Arm activates Wg signaling in a cell autonomous fashion. To determine if Wg is acting directly on cells which respond to Wg signaling, clones which are mutant for *dsh* or *arm* were examined for expression of *distalless* and *vestigial* (Neumann and Cohen, 1997; Zecca *et al.*, 1997). In both *dsh* and *arm* mutant clones, the expression of Wg target genes is reduced in a cell autonomous fashion. These results argue that Wg signals directly to cells in the discs to specify domains of gene expression, rather than as a short-range inducer of additional signals. Finally, ectopic expression of Wg reveals that there are distinct thresholds for Wg activity for defining domains of gene expression (Lecuit and Cohen, 1997; Neumann and Cohen, 1997; Zecca *et al.*, 1997). Taken together these results are compelling evidence for the role of Wg as a morphogen in imaginal disc development.

VI. Future Perspectives

The past decade has seen great advances in the investigation of the role and mechanism of Wg signaling in *Drosophila* development. As I have described here, there is a great deal of knowledge on Wg signaling in specification of cell fate and pattern in the embryo and imaginal disc. In addition, a model for Wg signaling has been proposed, although the precise molecular mechanism is unknown and there are likely to be components which have not been identified to date. Completing our understanding of the mechanism of Wg signaling is clearly a goal for future investigations in this field. Further genetic screens, as well as other approaches, are likely to uncover novel components of this pathway; whereas, additional biochemical studies are necessary to determine the precise protein–protein interactions. Another interesting issue which remains to be addressed is the interaction of the Wg signaling pathway with other signaling pathways. It is striking that in a number of instances Wg signaling and Dpp signaling acts either synergistically (embryonic midgut patterning and establishment of the P–D axis in the leg imaginal disc) or antagonistically (establishment of the D–V axis in the leg imaginal disc) to control cell fate and pattern. In these examples the two signaling pathways are affecting the same target genes; however, it would be interesting to determine if Wg signaling components can respond to any other signals. In at least one case, Dsh, that does appear to be the case; although it appears independent of Wg signaling (Krasnow *et al.,* 1995). Finally, of great interest is the specificity of the Wg signaling pathway for Wg rather than the other Wnts expressed in *Drosophila*. It is not clear if the other Wnts are transduced through the same pathway as Wg or through some independent pathways.

Acknowledgments

I thank H. Rizkalla and M. Steitz for comments, discussion, and assistance with the figures. I also thank Dr. G. Thomas for comments on the manuscript and Dr. E. Wilder for helpful and enjoyable discussion. Work in my laboratory is supported by National Science Foundation and March of Dimes.

References

Abmayr, S. M., Erickson, M. S., and Bour, B. A. (1995). Embryonic development of the larval body wall musculature of *Drosophila melanogaster. Trends Genet.* **11,** 153–159.

Axelrod, J. D., Matsuno, K., Artavanis-Tsakonas, S., and Perrimon, N. (1996). Interaction between wingless and Notch signaling pathways mediated by dishevelled. *Nature (London)* **271,** 1826–1832.

Azpiazu, N., and Frasch, M. (1993). *tinman* and *bagpipe:* Two homeo box genes that determine cell fates in the dorsal mesoderm of *Drosophila. Genes Dev.* **7,** 1325–1340.

Baker, N. E. (1987). Molecular cloning of sequences from *wingless,* a segment polarity gene in *Drosophila:* The spatial distribution of a transcript in embryos. *EMBO J.* **6,** 1765–1773.

Baker, N. E. (1988a). Transcription of the segment-polarity gene *wingless* in the imaginal discs of *Drosophila,* and the phenotype of a pupal-lethal *wg* mutation. *Development (Cambridge, UK)* **102,** 489–497.

Baker, N. E. (1988b). Embryonic and imaginal requirements for *wingless,* a segment-polarity gene in *Drosophila. Dev. Biol.* **125,** 96–108.

Baker, R., and Schubinger, G. (1995). Ectoderm induces muscle-specific gene expression in *Drosophila* embryos. *Development (Cambridge, UK)* **121,** 1387–1398.

Basler, K., and Struhl, G. (1994). Compartment boundaries and the control of *Drosophila* limb pattern by Hedgehog protein. *Nature (London)* **368,** 208–214.

Bate, M. (1993). The mesoderm and its derivatives. *In* "The Development of *Drosophila melanogaster*" (M. Bate and

A. Martinez Arias, eds.), Vol 2, pp. 1013–1090. Cold Spring Harbor Laboratory, Cold Spring Harbor, New York.

Bate, M., and Rushton, E. (1993). Myogenesis and muscle patterning in *Drosophila, C. R. Acad. Sci. Paris, Ser. 3* **316,** 1055–1061.

Baylies, M. K., Arias, M. A., and Bate, M. (1995). *wingless* is required for the formation of a subset of muscle founder cells during *Drosophila* embryogenesis. *Development (Cambridge, UK)* **121,** 3829–3837.

Behrens, J., von Kries, J. P., Kuhl, M., Bruhn, L., Wedlich, D., Grosschedl, R., and Birchmeier, W. (1996). Functional interaction of β-catenin with the transcription factor LEF-1. *Nature (London)* **382,** 638–642.

Bejsovec, A., and Martinez Arias, A. (1991). Roles of *wingless* in patterning the larval epidermis of *Drosophila. Development (Cambridge, UK)* **113,** 471–485.

Bejsovec, A., and Wieschaus, E. (1993). Segment polarity gene interactions modulate epidermal patterning in *Drosophila* embryos. *Development (Cambridge, UK)* **119,** 501–517.

Bejsovec, A., and Wieschaus, E. (1995). Signaling activities of the *Drosophila wingless* gene are seperately mutable and appear to be transduced at the cell surface. *Genetics* **139,** 309–320.

Bhanot, P., Brink, M., Samos, C. H., Hsieh, J., Wang, Y., Macke, J. P., Andrew, D., Nathans, J., and Nusse, R. (1996). A new member of the *frizzled* family from *Drosophila* functions as a Wingless receptor. *Nature (London)* **382,** 225–230.

Bienz, M. (1994). Homeotic genes and positional signalling in the *Drosophila* viscera. *Trends Genet.* **10,** 22–26.

Binari, R. C., Staveley, B. E., Johnson, W. A., Godavarti, R., Sasisekharan, R., and Manoukian, A. S. (1997). Genetic evidence that heparin-like glycosaminoglycans are involved in *wingless* signaling. *Development (Cambridge, UK)* **124,** 2623–2632.

Blair, S. S. (1994). A role for the segment polarity gene *shaggy-zeste white 3* in the specification of regional identity in the developing wing of *Drosophila. Dev. Biol.* **162,** 229–244.

Blair, S. S. (1995). Compartments and appendage development in *Drosophila. BioEssays* **17,** 299–309.

Bodmer, R. (1993). The gene *tinman* is required for specification of the heart and visceral muscles in *Drosophila. Development (Cambridge, UK)* **118,** 719–729.

Bourouis, M., Moore, P., Ruel, L., Grau, Y., Heitzler, P., and Simpson, P. (1990). An early embryonic product of the gene *shaggy* encodes a serine/threonine protein kinase related to the CDC28/cdc2 subfamily. *EMBO J.* **9,** 2877–2884.

Brand, A., and Perrimon, N. (1993). Targeted gene expression as a means of altering cell fates and generating dominant phenotypes. *Development (Cambridge, UK)* **118,** 401–415.

Brook, W. J., and Cohen, S. M. (1996). Antagonistic interactions between Wingless and Decapentaplegic responsible for dorsal–ventral pattern in the *Drosophila* leg. *Science* **273,** 1373–1377.

Brunner, E., Peter, O., Schweizer, L., and Basler, K. (1997). *pangolin* encodes a Lef-1 homologue that acts downstream of Armadillo to transduce the Wingless signal in *Drosophila*. *Nature (London)* **385**, 829–833.

Cabrera, C. V., Alonso, M. C., Johnston, P., Phillips, R. G., and Lawrence, P. A. (1987). Phenocopies induced with antisense RNA identify the *wingless* gene. *Cell (Cambridge, Mass.)* **50**, 659–663.

Campbell, G., Weaver, T., and Tomlinson, A. (1993). Axis specification in the developing *Drosophila* appendage: The role of *wingless, decapentaplegic*, and the homeobox gene *aristaless*. *Cell (Cambridge, Mass.)* **74**, 1113–1123.

Chu-LaGraff, Q., and Doe, C. Q. (1993). Neuroblast specification and formation regulated by *wingless* in the *Drosophila* CNS. *Science* **261**, 1594–1597.

Cohen, S. M. (1993). Imaginal disc development. *In* "The Development of *Drosophila melanogaster*" (M. Bate and A. Martinez Arias, eds.), Vol. 2, pp. 747–841. Cold Spring Harbor Laboratory, Cold Spring Harbor, New York.

Cook, D., Fry, M. J., Hughes, K., Sumathipala, R., Woodgett, J. R., and Dale, T. C. (1996). Wingless inactivates glycogen synthase kinase-3 via an intracellular signalling pathway which involves a protein kinase C. *EMBO J.* **15**, 4526–4536.

Couso, J. P., Bate, M., and Martinez, A. A. (1993). A *wingless*-dependent polar coordinate system in *Drosophila* imaginal discs. *Science* **259**, 484–489.

Couso, P. J., Bishop, S. A., and Martinez, Arias, A. (1994). The wingless signaling pathway and the patterning of the wing margin in *Drosophila*. *Development (Cambridge, UK)* **120**, 621–636.

Cumberledge, S., and Krasnow, M. A. (1993). Intercellular signalling in *Drosophila* segment formation reconstructed *in vitro*. *Nature (London)* **363**, 549–552.

Diaz-Benjumea, F. J., and Cohen, S. M. (1993). Interaction between dorsal and ventral cells in the imaginal discs directs wing development in *Drosophila*. *Cell (Cambridge, Mass.)* **75**, 741–752.

Diaz-Benjumea, F., and Cohen, S. M. (1994). *wingless* acts through the *shaggy/zeste-white 3* kinase to direct dorsal–ventral axis formation in the *Drosophila* leg. *Development (Cambridge, UK)* **120**, 1661–1670.

Diaz-Benjumea, F. J., and Cohen, S. M. (1995). Serrate signals through Notch to establish a Wingless-dependent boundary of the *Drosophila* wing. *Development (Cambridge, UK)* **121**, 4215–4225.

Diaz-Benjumea, F. J., Cohen, B., and Cohen, S. M. (1994). Cell interaction between compartments establishes the proximal–distal axis of *Drosophila* legs. *Nature (London)* **372**, 175–179.

DiNardo, S., Heemsterk, J., Dougan, S., and O'Farrell, P. H. (1994). The making of a maggot: Patterning the *Drosophila* embryonic epidermis. *Curr. Opin. Genet. Dev.* **4**, 529–534.

DiNardo, S., Sher, E., Heemskerk, J. J., Kassis, J. A., and O'Farrell, P. H. (1988). Two-tiered regulation of spatially patterned *engrailed* gene expression during *Drosophila* embryogenesis. *Nature (London)* **332**, 604–609.

Dominguez, I., Itoh, K., and Sokol, S. Y. (1995). Role of glycogen synthase kinase 3β as a negative regulator of dorsoventral axis formation in *Xenopus* embryos. *Proc. Natl. Acad. Sci. U.S.A.* **92**, 8498–8502.

Dougan, S., and DiNardo, S. (1992). *Drosophila wingless* generates cell type diversity among engrailed expressing cells. *Nature (London)* **360**, 347–350.

Eisenberg, L. M., Ingham, P. W., and Brown, A. M. (1992). Cloning and characterization of a novel *Drosophila* Wnt gene, *Dwnt-5*, a putative downstream target of the homeobox gene *distal-less*. *Dev. Biol.* **154**, 73–83.

Fjose, A., McGinnis, W. J., and Gehring, W. J. (1985). Isolation of a homeobox containing gene from the *engrailed* region of *Drosophila* and the spatial distribution of its transcript. *Nature (London)* **313**, 284–289.

Frasch, M. (1995). Induction of visceral and cardiac mesoderm by ectodermal Dpp in the early *Drosophila* embryo. *Nature (London)* **374**, 464–467.

Gonzalez, F., Swales, L., Bejsovec, A., Skaer, H., and Martinez, A. A. (1991). Secretion and movement of Wingless protein in the epidermis of the *Drosophila* embryo. *Mech. Dev.* **35**, 43–54.

Gonzalez-Gaitan, M., and Jackle, H. (1995). Invagination centers within the *Drosophila* stomatogastric nervous system anlage are positioned by *Notch*-mediated signaling which is spatially controlled through *wingless*. *Development (Cambridge, UK)* **121**, 2313–2325.

Graba, Y., Geiseler, K., Aragnol, D., Laurenti, P., Mariol, M. C., Berenger, H., Sagnier, T., and Pradel, J. (1995). DWnt-4, a novel *Drosophila* Wnt gene acts downstream of homeotic complex genes in the visceral mesoderm. *Development (Cambridge, UK)* **121**, 209–218.

Greig, S., and Akam, M. (1993). Homeotic genes autonomously specify one aspect of pattern in the *Drosophila* mesoderm. *Nature (London)* **362**, 630–632.

Hacker, U., Lin, X., and Perrimon, N. (1997). The *Drosophila sugarless* gene modulates Wingless signaling and encodes an enzyme involved in polysaccharide biosynthesis. *Development (Cambridge, UK)* **124**, 3565–3573.

Haerry, T. E., Heslip, T. R., Marsh, J. L., and O'Connor, M. B. (1997). Defects in glucuronate biosynthesis disrupt Wingless signaling in *Drosophila*. *Development (Cambridge, UK)* **124**, 3055–3064.

Harwood, A. J., Plyte, S. E., Woodgett, J., Strutt, H., and Kay, R. R. (1995). Glycogen synthase kinase 3 regulates cell fate in *Dictyostelium*. *Cell (Cambridge, Mass.)* **80**, 139–148.

Hayashi, S., Rubinfeld, B., Souza, B., Polakis, P., Wieschaus, E., and Levine, A. (1997). A *Drosophila* homolog of the tumor suppressor gene *adenomatous polyposis coli* downregulates β-catenin but its zygotic expression is not essential for the regulation of Armadillo. *Proc. Natl. Acad. Sci. U.S.A.* **94**, 242–247.

Hays, R., Gibori, G. B., and Bejsovec, A. (1997). Wingless signaling generates epidermal pattern through two distinct mechanisms. *Development (Cambridge, UK)* **124**, 3727–3736.

He, X., Saint-Jeannet, J.-P., Woodgett, J. R., Varmus, H. E., and Dawid, I. B. (1995). Glycogen synthase kinase-3 and dorsoventral patterning in *Xenopus* embryos. *Nature (London)* **374**, 617–622.

Heemskerk, J., DiNardo, S., Kostriken, R., and O'Farrell, P. H. (1991). Multiple modes of *engrailed* regulation in the progression towards cell fate determination. *Nature (London)* **352**, 404–410.

Heslip, T. R., Theisen, H., Walker, H., and Marsh, J. L. (1997). SHAGGY and DISHEVELLED exert opposite effects on *wingless* and *decapentaplegic* expression and on positional identity in imaginal discs. *Development (Cambridge, UK)* **124**, 1069–1078.

Hooper, J. E. (1994). Distinct pathways for autocrine and paracrine Wingless signalling in *Drosophila* embryos. *Nature (London)* **372**, 461–464.

Hoppler, S., and Bienz, M. (1994). Specification of a single cell type by a *Drosophila* homeotic gene. *Cell (Cambridge, Mass.)* **76**, 689–702.

Hoppler, S., and Bienz, M. (1995). Two different thresholds of *wingless* signalling with distinct developmental consequences in the *Drosophila* midgut. *EMBO J.* **14**, 5016–5026.

Huber, A. H., Nelson, W. J., and Weis, W. I. (1997). Three-dimensional structure of the Armadillo repeat region of β-catenin. *Cell (Cambridge, Mass.)* **90**, 871–882.

Immergluck, K., Lawrence, P. A., and Bienz, M. (1990). Induction across germ layers in *Drosophila* mediated by genetic cascade. *Cell (Cambridge, Mass.)* **62**, 261–268.

Jackson, S. M., Nakato, H., Sugiura, M., Jannuzi, A., Oakes, R., Kaluza, V., Golden, C., and Selleck, S. B. (1997). *dally*, a *Drosophila* glypican, controls cellular responses to the TGF-β-related morphogen, Dpp. *Development (Cambridge, UK)* **124**, 4113–4120.

Jiang, J., and Struhl, G. (1996). Complementary and mutually exclusive activities of Decapentaplegic and Wingless organize axial patterning during *Drosophila* leg development. *Cell (Cambridge, Mass.)* **86**, 401–409.

Johnston, L. A., and Schubiger, G. (1996). Ectopic expression of *wingless* in imaginal discs interferes with *decapentaplegic* expression and alters cell determination. *Development (Cambridge, UK)* **122**, 3519–3529.

Kadowaki, T., Wilder, E., Kligensmith, J., Zachary, K., and Perrimon, N. (1996). The segment polarity gene *porcupine* encodes a putative multitransmembrane protein involved in Wingless processing. *Genes Dev.* **10**, 3116–3128.

Kim, J., Sebring, A., Esch, J. J., Kraus, M. E., Vorweck, K., Magee, J., and Carroll, S. B. (1996). Integration of positional signals and regulation of wing formation and identity by *Drosophila vestigal* gene. *Nature* **382**, 133–138.

Klingensmith, J., Nusse, R., and Perrimon, N. (1994). The *Drosophila* segment polarity gene *dishevelled* encodes a novel protein required for response to the *wingless* signal. *Genes Dev.* **8**, 118–130.

Korinek, V., Barker, N., Morin, J., van Wichen, D., de Weger, R., Kinzler, K.W., Vogelstein, B., and Clevers, H. (1997). Constitutive transcriptional activation by a β-catenin-Tcf complex in APC-/-colon carcinoma. *Science* **275**, 1784–1787.

Kornberg, T., Siden, I., O'Farrell, P., and Simon, A. M. (1985). The *engrailed* locus of *Drosophila*: *In situ* localization of transcripts reveals compartment-specific expression. *Cell (Cambridge, Mass.)* **40**, 45–53.

Krasnow, R. E., Wong, L. L., and Adler, P. N. (1995). Dishevelled is a component of the Frizzled signalling pathway in *Drosophila*. *Development (Cambridge, UK)* **121**, 4095–4102.

Lawrence, P. A., and Struhl, G. (1996). Morphogens, compartments, and pattern: Lessons from *Drosophila*? *Cell (Cambridge, Mass.)* **85**, 951–961.

Lawrence, P. A., Johnston, P., and Vincent, J. (1994). Wingless can bring about a mesoderm-to-ectoderm induction in *Drosophila* embryos. *Development (Cambridge, UK)* **120**, 3355–3359.

Lawrence, P. A., Bodmer, R., and Vincent, J. (1995). Segmental patterning of heart precursors in *Drosophila*. *Development (Cambridge, UK)* **121**, 4303–4308.

Lecuit, T., and Cohen, S. (1997). Proximal–distal axis formation in the *Drosophila* leg. *Nature (London)* **388**, 139–145.

Lecuit, T., Brook, W. J., Ng, M., Callega, M., Sun, H., and Cohen, S. M. (1996). Two distinct mechanisms for long-range patterning by Decapentaplegic in the *Drosophila* wing. *Nature (London)* **381**, 387–393.

Li, X., and Noll, M. (1993). Role of the *gooseberry* gene in *Drosophila* embryos: Maintenance of *wingless* expression by a *wingeless-gooseberry* autoregulatory loop. *EMBO J.* **12**, 4499–4509.

McMahon, A. P. (1992). The Wnt family of developmental regulators. *Trends Genet.* **8**, 236–242.

Martinez, Arias, A. (1993). Larval epidermis of *Drosophila*. In "The Development of *Drosophila melanogaster*" (M. Bate and A. Martinez Arias, eds.), Vol. 1, pp. 517–608. Cold Spring Harbor Laboratory, Cold Spring Harbor, New York.

Martinez Arias, A., and Lawrence, P. A. (1985). Parasegments and compartments in the *Drosophila* embryo. *Nature (London)* **313**, 639–642.

Martinez Arias, A., Baker, N. E., and Ingham, P. W. (1988). Role of segment polarity genes in the definition and maintenance of cell states in the *Drosophila* embryo. *Development (Cambridge, UK)* **103**, 157–170.

Maves, L., and Schubinger, G. (1995). *wingless* induces transdetermination in developing *Drosophila* imaginal discs. *Development (Cambridge, UK)* **121**, 1263–1272.

Michelson, A. M. (1994). Muscle pattern diversification is determined by the autonomous function of homeotic genes in the embryonic mesoderm. *Development (Cambridge, UK)* **120**, 755–768.

Molenaar, M., van de Wetering, M., Oosterwegel, M., Peterson-Maduro, J., Godsave, S., Korinek, V., Roose, J., Destree, O., and Clevers, H. (1996). XTcf-3 transcription factor mediates β-catenin-induced axis formation in *Xenopus* embryos. *Cell (Cambridge, Mass.)* **86**, 391–399.

Moon, R. T., Brown, J. D., and Torres, M. (1997). Wnts modulate cell fate and behavior during vertebrate development. *Trends Genet.* **13**, 157–162.

Morin, P. J., Sparks, A. B., Korinek, V., Barker, N., Clevers, H., Vogelstein, B., and Kinzler, K. W. (1997). Activation of β-catenin-Tcf signaling in colon cancer by mutations in β-catenin or APC. *Science* **275**, 1787–1790.

Nellen, D., Burke, R., Struhl, G., and Basler, K. (1996). Direct and long-range action of a Dpp morphogen gradient. *Cell (Cambridge, Mass.)* **85**, 357–368.

Neumann, C. J., and Cohen, S. M. (1996). A hierarchy of cross-regulation involving *Notch, wingless, vestigal* and *cut* organizes the dorsal–ventral axis of the *Drosophila* wing. *Development (Cambridge, UK)* **122**, 3477–3485.

Neumann, C. J., and Cohen, S. M. (1997). Long-range action of Wingless organizes the dorsal–ventral axis of the *Drosophila* wing. *Development (Cambridge, UK)* **124**, 871–880.

Ng, M., Diaz-Benjumea, F. J., Vincent, J.-P., Wu, J., and Cohen, S. M. (1996). Specification of the wing by localized expression of *wingless* protein. *Nature (London)* **381**, 316–318.

Noordermeer, J., Johnston, P., Rijsewijk, F., Nusse, R., and Lawrence, P. A. (1992). The consequences of ubiquitous expression of the *wingless* gene in the *Drosophila* embryo. *Development (Cambridge, UK)* **116**, 711–719.

Noordermeer, J., Klingensmith, J., Perrimon, N., and Nusse, R. (1994). *dishevelled* and *armadillo* act in the Wingless signaling pathway in *Drosophila*. *Nature (London)* **367**, 80–83.

Nusse, R. (1997). A versatile transcriptional effector of Wingless signaling. *Cell (Cambridge, Mass.)* **89**, 321–323.

Nusse, R., and Varmus, H. E. (1982). Many tumors induced by the mouse mammary tumor virus contain a provirus integrated in the same region of the host genome. *Cell (Cambridge, Mass.)* **31**, 99–109.

Nusse, R., and Varmus, H. E. (1992). Wnt genes. *Cell (Cambridge, Mass.)* **69**, 1073–1087.

Nusse, R., Brown, A., Papkoff, J., Scambler, P., Schackleford, G., McMahon, A., Moon, R., and Varmus, H. (1991). A new nomenclature for int-1 and related genes: The Wnt gene family. *Cell (Cambridge, Mass.)* **64**, 231.

O'Keefe, L., Dougan, S. T., Gabay, L., Raz, E., Shilo, B., and DiNardo, S. (1997). Spitz and Wingless, emanating from distinct borders, cooperate to establish cell fate across the Engrailed domain in the *Drosophila* epidermis. *Development (Cambridge, UK)* **124**, 4837–4845.

Orsulic, S., and Peifer, M. (1996). An *in vivo* structure–function study of Armadillo, the β-catenin homologue, reveals both separate and overlapping regions of the protein required for cell adhesion and for Wingless signaling. *J. Cell Biol.* **134**, 1283–1300.

Pai, L.-M., Orsulic, S., Bejsovec, A., and Peifer, M. (1997). Negative regulation of Armadillo, a Wingless effector in *Drosophila*. *Development (Cambridge, UK)* **124**, 2255–2266.

Panganiban, G. E. F., Reuter, R., Scott, M. P., and Hoffmann, F. M. (1990). A *Drosophila* growth factor homolog, *decapentaplegic*, regulates homeotic gene expression within and across germ layers during midgut morphogenesis. *Development (Cambridge, UK)* **110**, 1041–1050.

Panin, V. M., Papayannopoulos, V., Wilson, R., and Irvine, K. D. (1997). Fringe modulates Notch-ligand interactions. *Nature (London)* **387**, 908–912.

Park, M., Wu, X., Golden, K., Axelrod, J. D., and Bodmer, R. (1996). The Wingless signaling pathway is directly involved in *Drosophila* heart development. *Dev. Biol.* **177**, 104–116.

Peifer, M., Rauskolb, C., Williams, M., Riggleman, B., and Wieschaus, E. (1991). The segment polarity gene *armadillo* interacts with the *wingless* signaling pathway in both embryonic and adult pattern formation. *Development (Cambridge, UK)* **111**, 1029–1043.

Peifer, M., Sweeton, D., Casey, M., and Wieschaus, E. (1994a). Wingless signal and Zeste-white 3 kinase trigger opposing changes in the intracellular distribution of Armadillo. *Development (Cambridge, UK)* **120**, 369–380.

Peifer, M., Pai, L.-M., and Casey, M. (1994b). Phosphorylation of the *Drosophila* adherens junction protein Armadillo: Roles for Wingless signal and Zeste-white 3 kinase. *Dev. Biol.* **166**, 543–556.

Penton, A., and Hoffman, F. M. (1996). Decapentaplegic restricts the domain of *wingless* during *Drosophila* limb patterning. *Nature (London)* **382**, 162–165.

Perrimon, N. (1994). The genetic basis of patterned baldness in *Drosophila*. *Cell (Cambridge, Mass.)* **76**, 781–784.

Perrimon, N. (1996). Serpentine proteins slither into the wingless and hedgehog fields. *Cell (Cambridge, Mass.)* **86**, 513–516.

Perrimon, N., Engstrom, L., and Mahowald, A. P. (1989). Zygotic lethals with specific maternal effect phenotypes in *Drosophila melanogaster*. I. Loci on the X-chromosome. *Genetics* **121**, 333–352.

Phillips, R. G., and Whittle, J. R. S. (1993). *wingless* expression mediates determination of peripheral nervous system elements in late stages of *Drosophila* wing disc development. *Development (Cambridge, UK)* **118**, 427–438.

Pierce, S. B., and Kimelman, D. (1995). Regulation of Spemann organizer formation by the intracellular kinase Xgsk-3. *Development (Cambridge, UK)* **121**, 755–765.

Poole, S. J., Kauvar, L. M., Drees, B., and Kornberg, T. (1985). The *engrailed* locus of *Drosophila*: Structural analysis of an embryonic transcript. *Cell (Cambridge, Mass.)* **40**, 37–43.

Ranganayakulu, G., Schulz, R. A., and Olson, E. N. (1996). Wingless signaling indices *nautilus* expression in the ventral mesoderm of the *Drosophila* embryo. *Dev. Biol.* **176**, 143–148.

Reichsman, F., Smith, L., and Cumberiedge, S. (1996). Glycosaminoglycans can modulate extracellular localization of the *wingless* protein and promote signal transduction. *J. Cell Biol.* **135**, 819–827.

Reuter, R., Panganiban, G. E. F., Hoffmann, F. M., and Scott, M. P. (1990). Homeotic genes regulate the spatial expression of putative growth factors in the visceral mesoderm of *Drosophila* embryos. *Development (Cambridge, UK)* **110**, 1031–1040.

Riese, J., Yu, X., Munnerlyn, A., Eresh, S., Hsu, S.-C., Grosschedl, R., and Bienz, M. (1997). LEF-1, a nuclear factor coordinating signaling inputs from *wingless* and *decapentaplegic*. *Cell (Cambridge, Mass.)* **88**, 777–787.

Riggleman, B., Wieschaus, E., and Schedl, P. (1989). Molecular analysis of the *armadillo* locus: Uniformly distributed transcripts and a protein with novel internal repeats are associated with a *Drosophila* segment polarity gene. *Genes Dev.* **3**, 96–113.

Riggleman, B., Schedl, P., and Wieschaus, E. (1990). Spatial expression of the *Drosophila* segment polarity gene *armadillo* is posttranscriptionally regulated by *wingless*. *Cell (Cambridge, Mass.)* **63**, 549–560.

Rijsewijk, F., Schuermann, M., Wagenaar, E., Parren, P., Weigel, D., and Nusse, R. (1987). The *Drosophila* homolog of the mouse mammary oncogene int-1 is identical to the segment polarity gene *wingless*. *Cell (Cambridge, Mass.)* **50**, 649–657.

Rubinfeld, B., Albert, I., Porfiri, E., Fiol, C., Munemitsu, S., and Polakis, P. (1996). Binding of GSK3β to the APC-β-catenin complex and regulation of complex assembly. *Science* **272**, 1023–1026.

Rubinfeld, B., Robbins, P., El-Gamil, M., Albert, I., Porfiri, E., and Polakis, P. (1997). Stabilization of β-catenin by genetic defects in melanoma cell lines. *Science* **275**, 1790–1792.

Ruel, L., Bourouis, M., Heitzler, P., Pantesco, V., and Simpson, P. (1993). *Drosophila* shaggy kinase and rat glycogen synthase kinase-3 have conserved activities and act downstream of *Notch*. *Nature (London)* **362**, 557–560.

Rulifson, E. J., Michhelli, C. A., Axelrod, J. D., Perrimon, N., and Blair, S. S. (1996). *wingless* refines its own expression domain on the *Drosophila* wing margin. *Nature (London)* **384**, 72–74.

Rushton, E., Drysdale, R., Abmayr, S. M., Michelson, A. M., and Bate, M. (1995). Mutations in a novel gene, *myoblast city*, provide evidence in support of the founder cell hypothesis for *Drosophila* muscle development. *Development (Cambridge, UK)* **121**, 1979–1988.

Russell, J., Gennissen, A., and Nusse, R. (1992). Isolation and expression of two novel *Wnt/wingless* gene homologues in *Drosophila*. *Development (Cambridge, UK)* **115**, 475–485.

Sanson, B., White, P., and Vincent, J.-P. (1996). Uncoupling cadherin-based adhesion from *wingless* signaling in *Drosophila*. *Nature (London)* **383**, 627–630.

Sharma, R. P., and Chopra, V. I. (1976). Effect of the *wingless* (*wg¹*) mutation on wing and haltere development in *Drosophila melanogaster*. *Dev. Biol.* **48**, 461–465.

Siegfried, E., and Perrimon, N. (1994). *Drosophila* Wingless: A paradigm for the function and mechanism of Wnt signaling. *BioEssays* **16**, 395–404.

Siegfried, E., Perkins, L. A., Capaci, T. M., and Perrimon, N. (1990). Putative protein kinase product of the *Drosophila* segment-polarity gene *zeste-white 3*. *Nature (London)* **345**, 825–829.

Siegfried, E., Chou, T. B., and Perrimon, N. (1992). *wingless* Signaling Acts through *zeste-white 3*, the *Drosophila* homolog of glycogen synthase kinase-3, to regulate *engrailed* and establish cell fate. *Cell (Cambridge, Mass.)* **71**, 1167–1179.

Siegfried, E., Wilder, E. L., and Perrimon, N. (1994). Components of Wingless signaling in *Drosophila*. *Nature (London)* **367**, 76–80.

Slack, J. M. W. (1987). Morphogenetic gradients-past and present. *Trends Biochem.* **12**, 200–204.

Staehling-Hampton, K., Hoffman, F. M., Baylies, M. K., Rushton, E., and Bate, M. (1994). *dpp* induces mesodermal gene expression in *Drosophila*. *Nature (London)* **372**, 783–786.

Steitz, M. C., Wickenheisser, J. K., and Siegfried, E. (1998). Over-expression of Zeste white 3 blocks Wingless signaling in the *Drosophila* embryonic midgut. *Dev. Biol.* **197**, 218–233.

Struhl, G., and Basler, K. (1993). Organizing activity of wingless protein in *Drosophila*. *Cell (Cambridge, Mass.)* **72**, 527–540.

Szüts, D., Freeman, M., and Bienz, M. (1997). Antagonism between EGFR and Wingless signaling in the larval cuticle of *Drosophila*. *Development (Cambridge, UK)* **124**, 3209–3219.

Tabata, T., and Kornberg, T. B. (1994). Hedgehog is a signaling protein with a key role in patterning *Drosophila* imaginal discs. *Cell (Cambridge, Mass.)* **76**, 89–102.

Theisen, H., Purcell, J., Bennett, M., Kansagara, D., Syed, A., and Marsh, J. L. (1994). *dishevelled* is required during *wingless* signaling to establish both cell polarity and cell identity. *Development (Cambridge, UK)* **120**, 347–360.

Theisen, H., Haerry, T., O'Conner, M. B., and Marsh, J. L. (1996). Development territories created by mutual antagonism between Wingless and Decapentaplegic. *Development (Cambridge, UK)* **122**, 3939–3948.

Thuringer, F., and Bienz, M. (1993). Indirect autoregulation of a homeotic *Drosophila* gene mediated by extracellular signaling. *Proc. Natl. Acad. Sci. U.S.A.* **90**, 3899–3903.

Thuringer, F., Cohen, S. M., and Bienz, M. (1993). Dissection of an indirect autoregulatory response of a homeotic *Drosophila* gene. *EMBO J.* **12**, 2419–2430.

Tremml, G., and Bienz, M. (1989). Homeotic gene expression in the visceral mesoderm of *Drosophila* embryos. *EMBO J.* **8**, 2677–2685.

van den Heuvel, M., Nusse, R., Johnston, P., and Lawrence, P. A. (1989). Distribution of the *wingless* gene product in *Drosophila* embryos: A protein involved in cell–cell communication. *Cell (Cambridge, Mass.)* **59**, 739–749.

van de Wetering, M., Cavallo, R., Dooijes, D., van Beest, M., van Es, J., Loureiro, J., Ypma, A., Hursh, D., Jones, T., Bejsovec, A., Peifer, M., Mortin, M., and Clevers, H. (1997). Armadillo coactivates transcription driven by the product of the *Drosophila* segment polarity gene *dTCF*. *Cell (Cambridge, Mass.)* **88**, 789–799.

van Leeuwen, F., Harryman Samos, C., and Nusse, R. (1994). Biological activity of soluble *wingless* protein in cultured *Drosophila* imaginal disc cells. *Nature (London)* **368**, 342–344.

Vincent, J. (1994). Morphogens dropping like flies. *Trends Genet.* **10**, 383–385.

Vincent, J.-P., and Lawrence, P. A. (1994). *Drosophila wingless* sustains *engrailed* expression only in adjoining cells: Evidence from mosaic embryos. *Cell (Cambridge, Mass.)* **77**, 909–915.

Vincent, J. P., and O'Farrell, P. H. (1992). The state of *engrailed* expression is not clonally transmitted during early *Drosophila* development. *Cell (Cambridge, Mass.)* **68**, 923–931.

Wilder, E. L., and Perrimon, N. (1995). Dual functions of *wingless* in the *Drosophila* leg imaginal disc. *Development (Cambridge, UK)* **121**, 477–488.

Williams, J. A., Paddock, S. W., Vorwerk, K., and Carroll, S. B. (1994). Organization of wing formation and induction of a wing-patterning gene at the dorsal/ventral compartment boundary. *Nature (London)* **368**, 299–305.

Woodgett, J. R. (1990). Molecular cloning and expression of glycogen synthase kinase-3/factor A. *EMBO J.* **9**, 2431–2438.

Wu, X., Golden, K., and Bodmer, R. (1995). Heart development in *Drosophila* requires the segment polarity gene *wingless*. *Dev. Biol.* **169**, 619–628.

Yanagawa, S.-I., van Leeuwen, F., Wodarz, A., Klingensmith, J., and Nusse, R. (1995). The Dishevelled protein is modified by Wingless signaling in *Drosophila. Genes Dev.* **9**, 1087–1097.

Yoffe, K. B., Manoukian, A. S., Wilder, E. L., Brand, A. H., and Perrimon, N. (1995). Evidence for *engrailed*-independent *wingless* autoregulation in *Drosophila. Dev. Biol.* **170**, 636–650.

Yost, C., Torres, M., Miller, J., Huang, E., Kimelman, D., and Moon, R. T. (1996). The axis-inducing activity, stability, and subcellular distribution of β-catenin is regulated in *Xenopus* embryos by glycogen synthase kinase 3. *Genes Dev.* **10**, 1443–1454.

Yu, X., Hoppler, S., Eresh, S., and Bienz, M. (1996). *decapentaplegic*, a target gene of the wingless signaling pathway in the *Drosophila* midgut. *Development (Cambridge, UK)* **122**, 849–858.

Zecca, M., Basler, K., and Struhl, G. (1997). Direct and long-range action of a Wingless morphogen gradient. *Cell (Cambridge, Mass.)* **87**, 833–844.

18

Asymmetric Cell Division and Fate Specification in the Drosophila Central Nervous System

Julie Broadus
Howard Hughes Medical Institute
Department of Human Genetics
University of Utah
Salt Lake City, Utah 84112

Eric P. Spana
Department of Genetics
Harvard Medical School
Boston, Massachusetts 02115

I. Introduction

The generation of cell diversity during development requires asymmetric cell divisions wherein cell division and fate specification are coordinated. In this way, one precursor cell can divide to produce two differently fated sibling cells. The *Drosophila* central nervous system (CNS) arises from asymmetric cell divisions of neural precursors, called neuroblasts. The sibling cells produced at each neuroblast division are the regenerated neuroblast and its daughter cell, called a ganglion mother cell (GMC). Protein and RNA determinants are unequally inherited by the neuroblast and GMC as the neuroblast divides (Rhyu et al., 1994; Knoblich et al., 1995; Hirata et al., 1995; Spana and Doe 1995; Spana et al., 1995; Kraut and Campos-Ortega, 1996; Kraut et al., 1996; Ikeshima-Kataoka et al., 1997; Li et al., 1997; Shen et al., 1997; Broadus et al., 1998). This differential partitioning of gene products regulates neuroblast and GMC cell fates (Doe et al., 1991; Vaessin et al., 1991; Spana et al., 1995; Ikeshima-Kataoka et al., 1997; Shen et al., 1997; Broadus et al., 1998). Asymmetric localization of several of these proteins also has been observed in precursors of other *Drosophila* tissues (Rhyu et al., 1994; Spana and Doe, 1995; Kraut and Campos-Ortega, 1996; Ikeshima-Kataoka et al., 1997; Shen et al., 1997; Broadus et al., 1998), and at least one homologous protein in vertebrates shows asymmetric localization in dividing nervous system precursors (Zhong et al., 1996). The utilization of similar asymmetrically localized proteins in diverse tissues and species suggests the existence of a fundamental and evolutionary conserved pathway for asymmetric divisions.

This chapter reviews recent work on neuroblast polarity and asymmetric cell division in the *Drosophila* CNS. We begin with a brief introduction of neuroblast formation and specification. Second, we describe morphological and molecular features of neuroblast polarity and cell division. Third, we present what is known about the functions of asymmetrically localized proteins and RNAs in neuroblasts. Fourth, we address cell cycle and cytoskeletal regulatory mechanisms which are required for asymmetric protein and RNA localization in neuroblasts. Last, we consider asymmetric divisions of neuroblast daughter cells (GMCs).

II. Neuroblast Formation and Fate

In the *Drosophila* CNS, cell diversification occurs in a stepwise manner. First, individual fates are assigned within the neuroblast population. Second, each neuroblast undergoes stem cell-like divisions to generate a characteristic set of progeny.

Immediately following gastrulation, neuroblasts delaminate from the ventral ectoderm of the embryo, enlarging and moving dorsally into the embryo to lie between the ectoderm and mesoderm. A stereotyped array of 30 neuroblasts forms in each hemisegment. Within this group, each neuroblast has a unique identity that is reflected by its position within the hemisegment, time of formation, gene expression profile, and the set of progeny that it produces (Doe, 1992; Udolph *et al.,* 1993; Broadus *et al.,* 1995; Bossing *et al.,* 1996; Schmidt *et al.,* 1997). At the level of individual neuroblasts, the heterogeneity of the precursor population is fundamental to the diversity of neurons and glia found in the mature CNS. If a neuroblast is eliminated, the progeny normally produced by that neuroblast are lost; other neuroblasts do not compensate for the lost neurons by altering their lineages (Chu-LaGraff and Doe, 1993; Zhang *et al.,* 1994; Skeath *et al.,* 1995; Bhat, 1996; Parras *et al.,* 1996; Skeath and Doe, 1996; Duman-Scheel *et al.,* 1997). Conversely, mutations or ectopic expression that result in neuroblast duplication show a duplication of neuroblast progeny (Zhang *et al.,* 1994; Skeath *et al.,* 1995; Bhat, 1996; Duman-Scheel *et al.,* 1997). Therefore, each neuroblast contributes a unique set of neurons and glia to the final complement of cells found in the mature embryonic CNS.

What are the molecular mechanisms that create differences among neuroblasts? The specification of unique neuroblast fates is directed by the same genes that establish the anterior–posterior (A–P) and dorsal–ventral (D–V) axes of the embryo. The intersection of pair-rule and segment polarity gene expression along the A–P axis and D–V patterning gene expression along the D–V axis defines clusters of 5–7 ectodermal cells from which a single neuroblast delaminates (Skeath and Carroll,

1994). In response to these positional cues, gene expression is activated in neuroblasts and thereby establishes neuroblast fates. Mutations which expand or eliminate regions of patterned ectoderm duplicate or delete the neuroblasts that normally delaminate at these positions. (Skeath and Carroll, 1992; Chu-LaGraff and Doe, 1993; Zhang *et al.,* 1994; Skeath *et al.,* 1995; Bhat, 1996; Matsuzaki and Saigo, 1996; Duman-Scheel *et al.,* 1997; McDonald and Doe, 1997).

The second process of cell diversification in the CNS occurs as each neuroblast undergoes its lineage. Neuroblasts divide repeatedly by budding off GMC daughter cells from their basal side, and each GMC divides once to produce a pair of postmitotic neurons or glia. The family of neurons and glia derived from each neuroblast is nearly invariant (Udolph *et al.,* 1993; Bossing *et al.,* 1996; Schmidt *et al.,* 1997).

While the lineage generated by each neuroblast is unique, most neuroblasts complete similar asymmetric divisions. The neuroblast and GMC are sibling cells produced by division of the neuroblast, and they are distinguished by several criteria: the neuroblast is larger, is positioned apically, has greater mitotic potential, and expresses neuroblast-specific genes such as *deadpan* and *asense*; the GMC is smaller, is positioned basally, completes a single division, and expresses GMC-specific markers such as *even-skipped, fushi-tarazu,* and nuclear Prospero (Doe *et al.,* 1988a,b; Vaessin *et al.,* 1991; Bier *et al.,* 1992; Matsuzaki *et al.,* 1992; Brand *et al.,* 1993; Spana and Doe, 1995). These differences are established in part by the differential partitioning of cytoplasmic determinants during neuroblast mitosis (see below).

One exception to this division profile is the MP2 neuroblast. MP2 divides only once to produce two neurons, vMP2 and dMP2 (Bate and Grunewald, 1981). The MP2 division is oriented like that of a neuroblast, producing one apical cell (vMP2) and one basal cell (dMP2), and these two neurons have unique fates: dMP2 is an interneuron that projects posteriorly while vMP2 is an interneuron that projects anteriorly (Lin *et al.,* 1994; Spana *et al.,* 1995). Like the neuroblast and GMC, the vMP2 and dMP2 siblings are specified by the unequal inheritance of cytoplasmic determinants as MP2 divides (Spana *et al.,* 1995).

III. Asymmetrically Localized Proteins and RNAs in Neuroblasts

Currently, five proteins and two RNAs are known to be asymmetrically localized in neuroblasts: Inscuteable (Insc), Prospero (Pros), Staufen (Stau), Numb, Miranda (Mira), *pros* RNA, and *insc* RNA (Table I). During in-

TABLE I
Asymmetrically Localized Proteins and RNAs in Neuroblasts

Asymmetrically localized protein and RNAs	Protein type	Localization in mitotic neuroblasts	CNS function	References
Numb	Phosphotyrosine-binding protein	Basal	Specifies dMP2 fate in MP2 lineage	1–4[a]
Prospero	Homeodomain transcription factor	Basal	Enables GMCs to develop differently than neuroblasts	5–11
prospero RNA	—	Basal	Regulates GMC fate specification in neuroblast lineages	12, 13
Staufen	RNA-binding protein	Basal	Required for *prospero* RNA localization	12–16
Miranda	Novel	Basal	Required for Prospero and Staufen/*prospero* RNA localization	17, 18
Inscuteable	Putative cytoskeletal adaptor protein	Apical	Coordinates basal protein/RNA localization and mitotic spindle orientation	12, 18–20
inscuteable RNA	—	Not localized	Unknown	12

[a]Key to references: (1) Bork and Margolis, 1995; (2) Spana *et al.*, 1995; (3) Spana and Doe, 1996; (4) Uemura *et al.*, 1989; (5) Spana and Doe, 1995; (6) Knoblich *et al.*, 1995; (7) Hirata *et al.*, 1997; (8) Vaessin *et al.*, 1991; (9) Matsuzaki *et al.*, 1992; (10) Hassan *et al.*, 1997; (11) Doe *et al.*, 1991; (12) Li *et al.*, 1997; (13) Broadus *et al.*, 1998; (14) Ferrandon *et al.*, 1994; (15) St. Johnston *et al.*, 1991; (16) St. Johnston *et al.*, 1992; (17) Ikeshima-Kataoka *et al.*, 1997; (18) Shen *et al.*, 1997; (19) Kraut and Campos-Ortega, 1996; (20) Kraut *et al.*, 1996.

terphase, Insc, Pros, Stau, *pros* RNA, and *insc* RNA show apical localization in neuroblasts, either in the apical cytoplasm or at the apical cortex (Spana and Doe, 1995; Kraut and Campos-Ortega, 1996; Kraut *et al.*, 1996; Li *et al.*, 1997; Broadus *et al.*, 1998); Numb distribution in interphase neuroblasts has not been assayed, and Mira is uniformly distributed at the cell cortex (Ikeshima-Kataoka *et al.*, 1997). As neuroblasts enter mitosis, Insc remains localized apically as a cortical crescent, but is delocalized at telophase and inherited equally by the neuroblast and GMC at cell division (Kraut *et al.*, 1996). *insc* RNA is not localized during mitosis (Li *et al.*, 1997). In contrast, Pros, Stau, and *pros* RNA are translocated to the basal cell cortex at the beginning of mitosis, where they colocalize with Numb and Mira. All are localized as tight crescents and are inherited exclusively by the GMC at neuroblast division (Knoblich *et al.*, 1995; Spana and Doe, 1995; Spana *et al.*, 1995; Ikeshima-Kataoka *et al.*, 1997; Li *et al.*, 1997; Shen *et al.*, 1997; Broadus *et al.*, 1998). In the MP2 precursor, Numb is asymmetrically localized to the basal cortex, just as in neuroblasts, and is inherited solely by the basal daughter cell (dMP2) as MP2 divides (Spana *et al.*, 1995). Pros is nuclear in MP2 and is equally inherited by dMP2 and vMP2 (Broadus *et al.*, 1995; Skeath and Doe, 1996). The distribution of Insc, Mira, Stau, *pros* RNA, and *insc* RNA in MP2 is unknown.

Asymmetric protein and RNA localization in neuroblasts can be reproduced in primary cell culture with two notable exceptions. First, while *in vivo* observations of Insc suggest that it becomes delocalized late in the mitotic cycle and is inherited equally by the neuroblast and GMC (Kraut *et al.*, 1996), in cultured neuroblasts Insc remains apical through telophase and is inherited solely by the neuroblast (Broadus and Doe, 1997). The

persistent apical localization of Insc in cultured neuroblasts may be due to the increased sensitivity of immunolocalization *in vitro*, or it may reflect an artificial situation that occurs *in vitro* but not *in vivo*. Second, apical protein localization during interphase is not observed in culture, suggesting that extrinsic cues are required for apical protein localization or anchoring (Broadus and Doe, 1997).

IV. Functions of Asymmetrically Localized Proteins and RNA in Neuroblasts

What are the functions of asymmetrically localized proteins and RNA in neuroblasts? So far, the roles of localized proteins and RNA fall into two classes: those that influence developmental fate (Pros, *pros* RNA, and Numb) and those that operate to localize the fate determinants (Stau, Mira, and Insc). In neuroblast lineages, Pros and *pros* RNA help specify GMC fate (Doe *et al.*, 1991; Vaessin *et al.*, 1991; Broadus *et al.*, 1998). Stau is responsible for *pros* RNA localization, and Mira is required for localization of Pros, Stau, and *pros* RNA (Ikeshima-Kataoka *et al.*, 1997; Li *et al.*, 1997; Shen *et al.*, 1997; Broadus *et al.*, 1998; S. Fuerstenberg, P. Alvarez, and C. Q. Doe, 1998). In the MP2 lineage, Numb functions to specify the fate of dMP2 (Spana *et al.*, 1995; Spana and Doe, 1996). Finally, Insc coordinates the localization of all other known localized proteins and RNA with the mitotic spindle such that asymmetrically localized proteins and RNAs are distributed into a single daughter cell at cell division (Kraut *et al.*, 1996; Li *et al.*, 1997; Shen *et al.*, 1997).

A. Prospero, *prospero* RNA, Staufen, and Miranda Enable Ganglion Mother Cells to Develop Differently Than Neuroblasts

Asymmetry between the neuroblast and GMC cell fates is partly established by the differential inheritance of Pros (Doe *et al.*, 1991; Vaessin *et al.*, 1991; Spana and Doe, 1995). *pros* was cloned independently by three groups utilizing a P[*lacZ*] insertion in the *pros* locus that directs *lacZ* expression in all neuroblasts and results in nervous system defects and embryonic lethality (Doe *et al.*, 1991; Vaessin *et al.*, 1991; Matsuzaki *et al.*, 1992). *pros* encodes a divergent homeodomain transcription factor that is required in the GMC for activation of GMC-specific genes and repression of neuroblast-specific genes (Fig. 1) (Chu-LaGraff *et al.*, 1991; Doe *et al.*,

1991; Vaessin *et al.*, 1991; Matsuzaki *et al.*, 1992). This combination of target genes results in the GMC adopting a fate distinct from that of the neuroblast. In embryos lacking *pros* function, GMCs are misspecified: they fail to initiate GMC-specific expression of *even-skipped*, show persistent expression of the neuroblast-specific genes *deadpan* and *asense*, and ultimately divide to produce neurons that show defects in gene expression and axon trajectory (Doe *et al.*, 1991; Vaessin *et al.*, 1991).

The DNA-binding and transactivation functions of Pros require both the homeodomain and the C-terminal region of the protein. In addition, Pros can regulate the DNA-binding activity of other homeodomain proteins and thereby impart specificity to its transregulatory functions (Hassan *et al.*, 1997). This mechanism is postulated to be responsible for *pros*-mediated gene activation in subsets of GMCs, despite Pros localization in the nucleus of all GMCs. Interestingly, this regulatory function can occur in the absence of Pros binding to DNA and does not require the C-terminal region but may require helices 1 and 2 of the homeodomain (Hassan *et al.*, 1997). The Pros homeodomain and C-terminal region have been highly conserved during evolution from flies through humans, suggesting that Pros may interact with similar homeodomain cofactors and regulate similar target genes in diverse organisms (Oliver *et al.*, 1993; Burglin, 1994; Tomarev *et al.*, 1996; Zinovieva *et al.*, 1996).

Pros function in the GMC is dependent on its inheritance from the neuroblast during neuroblast division. In neuroblasts, Pros is translated at relatively low levels (*pros* transcripts are abundant yet Pros protein is barely detectable in interphase neuroblasts), localizes briefly to the apical cell cortex at late interphase, and then localizes as a tight basal cortical crescent during mitosis. As the neuroblast divides, Pros is sequestered into the budding GMC, and ultimately translocates to the GMC nucleus (Fig. 2A–C,E) (Hirata *et al.*, 1995; Knoblich *et al.*, 1995; Spana and Doe, 1995). By examining the localization of truncated Pros proteins and *pros-lacZ* fusion proteins, a 234-amino acid domain, the Pros asymmetric localization domain, is defined as both necessary and sufficient for Pros localization during neuroblast mitosis (Hirata *et al.*, 1995).

Localization of *pros* RNA is similar to Pros protein localization: *pros* transcript is enriched in the apical cytoplasm of interphase neuroblasts, forms a basal cortical crescent at metaphase, and is sequestered into the GMC at cell division (Fig. 2D,E) (Li *et al.*, 1997; Broadus *et al.*, 1998). The Stau RNA-binding protein, which colocalizes with and is required for the localization of *oskar* and *bicoid* RNAs during oogenesis (St. Johnston *et al.*, 1991, 1992; Ferrandon *et al.*, 1994), is also asymmetrically localized in neuroblasts, and precisely matches *pros* RNA localization (Fig. 2A–C,E) (Li *et al.*,

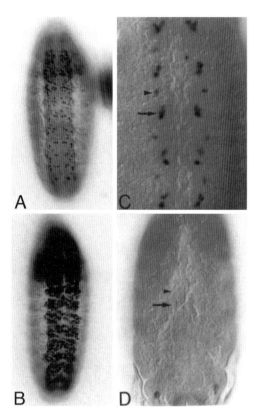

Figure 1 *prospero* is required in GMCs to repress neuroblast-specific genes and activate GMC-specific genes. (A, B) Dpn staining in wild-type (A) and *pros* (B) embryos. Ventral view, stage 16. (C, D) Eve staining in wild-type (C) and *pros* (D) embryos. Ventral view, stage 11. (A) In wild-type embryos, Dpn is detected in all neural precursors and shows progressively less staining as neuroblasts compete their lineages. At stage 16, a small subset of neuroblasts show Dpn staining. (B) In *pros* embryos, Dpn is detected in neuroblasts and neuroblast progeny. (C) In wild-type embryos, Eve is detected in a subset of GMCs and their progeny neurons, including RP2 (arrowhead) and aCC/pCC (arrow). (D) In *pros* embryos, Eve is lost in medial GMCs and neurons; the approximate positions of the unstained RP2 (arrowhead) and aCC/pCC (arrow) are indicated. (A, B reprinted with permission from Vaessin *et al.*, 1991, © Cell Press; C, D reprinted with permission from Doe *et al.*, 1991, © Cell Press.)

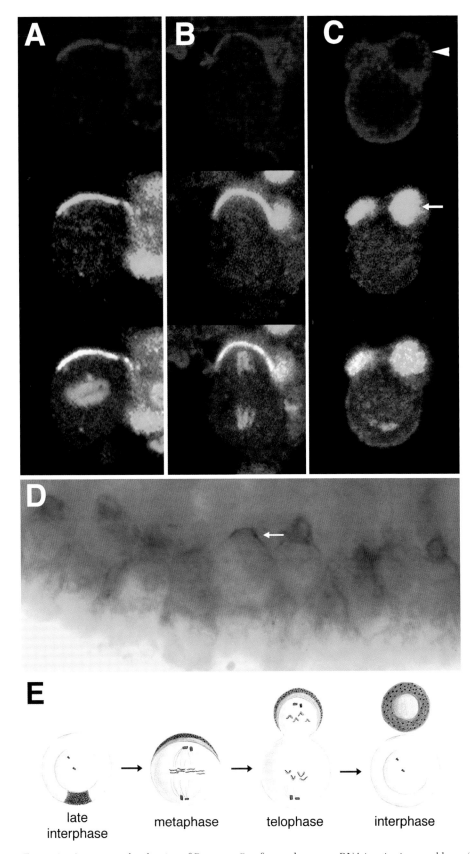

Figure 2 Asymmetric localization of Prospero, Staufen, and *prospero* RNA in mitotic neuroblasts. (A–C) Cultured neuroblasts triple-labeled for Stau (red), Pros (green), and DNA (blue); triple labels are shown in the bottom row. (A) In metaphase neuroblasts, Pros and Stau are tightly colocalized as a cortical crescent. (B) In anaphase neuroblasts, Pros and Stau segregate into the budding GMC. (C) During interphase Pros is undetectable in neuroblasts and Stau reaccumulates in the neuroblast cytoplasm. In cultured neuroblasts, Stau is delocalized during interphase (C), while *in vivo* Stau is apically localized during interphase (E). In GMCs, Pros is translocated to the nucleus (arrow) and Stau fills the GMC cytoplasm (arrowhead). (D) Lateral view of a stage 9 embryo hybridized with a *pros* probe shows basal localization of *pros* RNA in mitotic neuroblasts (arrow). (E) Summary of Pros, Stau, and *pros* RNA localization and *pros* transcription in neuroblasts and GMCs. Large cell, neuroblast; small cell, GMC; Light blue, Pros; red, Stau; dark blue, *pros* RNA; dots in neuroblast nucleus, sites of *pros* transcription; green, DNA; pink, centrosomes. In all panels, apical is down and basal is up.

1997; Broadus *et al.*, 1998). During interphase, Stau is localized apically in neuroblasts (Li *et al.*, 1997; Broadus *et al.*, 1998), and during mitosis Stau is localized as a basal cortical crescent (Broadus *et al.*, 1998). In embryos lacking Stau, *pros* RNA is delocalized in most neuroblasts, demonstrating that Stau mediates *pros* RNA localization (Li *et al.*, 1997; Broadus *et al.*, 1998). Stau regulation of *pros* RNA is likely direct as Stau can bind *pros* RNA through a domain in the *pros* 3' UTR (Li *et al.*, 1997).

Loss-of-function *stau* embryos have been used to address the functional significance of *pros* RNA localization. By all measures assayed, CNS development is unaltered in *stau* embryos: neuroblast formation and asymmetric cell division appear normal, Pros protein is localized normally (Li *et al.*, 1997; Broadus *et al.*, 1998), and expression of the GMC-specific markers Even-skipped and Fushi-Tarazu and the neuroblast marker Asense is unchanged (Broadus *et al.*, 1998). These results suggest that Stau and *pros* RNA localization are not essential for normal CNS development.

Although both neuroblasts and GMCs contain high levels of cytoplasmic *pros* RNA (Chu-LaGraff *et al.*, 1991; Doe *et al.*, 1991; Vaessin *et al.*, 1991), *pros* is not transcribed in GMCs (Broadus *et al.*, 1998), suggesting that asymmetric inheritance of *pros* RNA and Pros protein may be functionally redundant. *stau;pros* double mutants show that loss of *stau* enhances a *pros* hypomorphic GMC phenotype. Embryos homozygous for any of three hypomorphic *pros* alleles show partial loss of Eve-positive GMCs and neurons. In combination with a null allele of *stau*, the number of Eve-positive GMCs and neurons is reduced further (Broadus *et al.*, 1998). These results suggest a role for Stau/*pros* RNA in regulating the neuroblast/GMC fate decision.

How are Pros and Stau/*pros* RNA localized? Using the yeast two-hybrid system to detect proteins that interact with the Pros asymmetric localization domain, two groups have identified Mira, a protein that is asymmetrically localized identically to Pros in mitotic neuroblasts and is required for Pros localization (Ikeshima-Kataoka *et al.*, 1997; Shen *et al.*, 1997). More recently, it has been determined that Stau has a domain that resembles the Pros asymmetric localization domain and that Stau also requires *mira* for its localization (S. Fuerstenberg, P. Alvarez, and C. Q. Doe, 1998). Therefore, one pathway that affects GMC specification operates by Mira-dependent localization of Pros and Stau/*pros* RNA. Mira also interacts directly with Numb in a yeast two-hybrid assay, but it is not required for Numb localization in neuroblasts (Shen *et al.*, 1997).

The gene *mira* encodes a unique protein that contains multiple coiled-coil structures implicated in mediating protein–protein interactions, two leucine zipper motifs, and eight repeats of consensus protein kinase C

(PKC) phosphorylation sites (Ikeshima-Kataoka *et al.*, 1997; Shen *et al.*, 1997). To examine Mira function, six *mira* alleles were isolated during a screen to identify mutations that alter cell fates in the embryonic CNS (Ikeshima-Kataoka *et al.*, 1997). Pros is mislocalized in all six alleles. Five alleles show Pros delocalization in mitotic neuroblasts, resulting in equal partitioning of Pros to the neuroblast and GMC, and subsequently, ectopic Pros in the neuroblast nucleus (Ikeshima-Kataoka *et al.*, 1997). This phenotype also is observed when the entire gene is deleted (Shen *et al.*, 1997). In the sixth allele, asymmetric localization of Pros to the basal neuroblast cortex at mitosis occurs normally, Pros segregates to the GMC normally, but then fails to be released from the GMC cortex such that it may be transported to the GMC nucleus. This interesting allele identifies an additional role for Mira in regulating release of Pros from the cortex into the GMC nucleus (Ikeshima-Kataoka *et al.*, 1997).

B. Numb Antagonizes Notch Signaling to Specify the dMP2 Neuronal Cell Fate

The mutation *numb* was identified as causing a near complete loss of peripheral nervous system (PNS) neurons in the *Drosophila* embryo. During normal development, *numb* is responsible for distinguishing sibling fates found within the external sense organ lineage of the PNS. In embryos lacking *numb*, external cell fates are duplicated at the expense of the neuronal fate, and overexpression of *numb* produces the opposite cell fate transformation (Uemura *et al.*, 1989; Rhyu *et al.*, 1994). Immunolocalization of the Numb protein shows that Numb is associated with the cell cortex and is asymmetrically distributed in mitotic sense organ precursors (SOPs). During SOP division, Numb is segregated into the pIIb cell where it is autonomously required for neuron formation (Rhyu *et al.*, 1994).

In the CNS, Numb is asymmetrically localized in mitotic neuroblasts, and it is inherited solely by the GMC as the neuroblast divides (Rhyu *et al.*, 1994; Knoblich *et al.*, 1995; Spana *et al.*, 1995). However, a role for Numb in GMC specification has not been demonstrated (Uemura *et al.*, 1989; Spana *et al.*, 1995). In contrast, Numb does function to specify cell fate in the simple MP2 lineage. During mitosis, Numb is asymmetrically localized to the basal side of MP2 and is segregated into the basal daughter cell, dMP2. Two molecular markers exist to assay the different fates of the MP2 neurons. Their axonal projections can be labeled with the monoclonal antibody 22C10, and the nuclear transcription factor Odd-skipped (Odd) has a differential expression pattern in the two MP2 neurons. While both neurons inherit Odd from MP2, vMP2 quickly downregulates Odd, whereas dMP2 maintains Odd expression until the end of embryonic development. These markers show that in *numb* embryos, the dMP2 fate is lost and the vMP2 fate is duplicated (Fig. 3A,B) (Spana et al., 1995).

How does Numb function to specify cell fate? In both the CNS and PNS, *numb* mutants have phenotypes that are opposite to that of mutants in the Notch pathway (Hartenstein and Posakony, 1990; Parks and Muskavitch, 1993; Guo *et al.*, 1996; Spana and Doe, 1996). Embryos doubly mutant for *numb* and *Notch* pathway components exhibit the *Notch* phenotype (Guo *et al.*, 1996; Spana and Doe, 1996). In the MP2 lineage, *Notch* pathway mutants show a transformation of vMP2 into dMP2 (Fig. 3c) (Spana and Doe, 1996). These results suggest that *numb* functions to block the Notch-induced signal. Thus, during normal development, dMP2 inherits Numb as MP2 divides and can thereby block Notch signaling that induces the vMP2 fate (Fig. 4) (Spana and Doe, 1996).

The gene *numb* encodes a protein with an SHC phosphotyrosine-binding domain, which is implicated in tyrosine-kinase signaling (Uemura *et al.*, 1989; Bork and Margolis, 1995). Several studies highlight the importance of this domain for Numb function. First, the Numb phosphotyrosine-binding domain interacts with the intracellular domain of Notch in a yeast two-hybrid assay, supporting a model wherein repression of Notch by Numb is direct (Guo *et al.*, 1996). Second, in cultured *Drosophila* cells, Numb can block Notch-induced nuclear translocation of the Suppressor of Hairless (Su(H)) transcription factor, and this block requires the Numb phosphotyrosine binding domain (Frise *et al.*, 1996). Third, overexpression of Numb transgenes which code for a truncated Numb protein that lacks the phosphotyrosine domain are unable to generate a Numb gain-of-function phenotype (Frise *et al.*, 1996).

Recent work places another gene, *sanpodo* (*spdo*), in the *numb/Notch* pathway for sibling cell fate specification. Loss of *spdo* does not induce general neural hypertrophy characteristic of neurogenic mutants such as *Notch*, but does result in similar cell fate transformations in both the CNS and PNS (Skeath and Doe, 1998; Dye *et al.*, 1998). For example, in the CNS of *spdo* embryos, vMP2 is transformed to dMP2. Genetic epistasis of embryos doubly mutant for *spdo* and *numb* show that *numb* functions upstream of *spdo* (Skeath and Doe, 1998; Dye *et al.*, 1998). *spdo* encodes an actin tropomyosin binding protein, whose molecular action in Notch signaling is unknown (Dye *et al.*, 1998).

Future work will undoubtedly reveal the nuclear

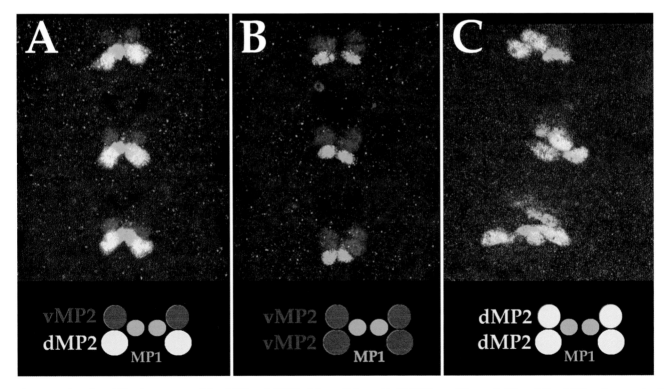

Figure 3 Numb and Notch are required for vMP2 and dMP2 sibling cell fates. (A–C) Wild-type (A), *numb* (B), and *Notch* (C) embryos stained for Odd (green) and the AJ96 enhancer trap line (red). Both dMP2 and vMP2 express AJ96; dMP2 and the lineally unrelated MP1 neurons express Odd. In all panels three segments of a stage 15 CNS are shown; anterior up. The schematic below each panel summarizes the cell fates. (A) In wild-type embryos, there is one vMP2 neuron, one dMP2 neuron, and one MP1 neuron per hemisegment. (B) In *numb* mutants there are two vMP2 neurons per hemisegment; dMP2 has been transformed into vMP2. (C) In *Notch* mutant embryos, there is no vMP2 neuron, two dMP2 neurons, and one MP1 neuron per hemisegment. (Reprinted with permission from Spana and Doe, 1996, © Cell Press).

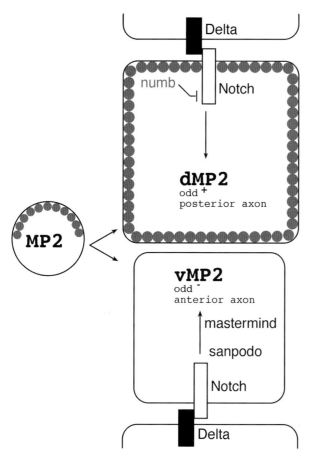

Figure 4 Asymmetric localization of Numb and Numb inhibition of Notch signaling in the MP2 lineage. Numb (gray dots) is asymmetrically localized in MP2 and is segregated into dMP2 at mitosis. The ligand Delta is expressed by cells adjacent to the MP2 neurons, but not in them, whereas the transmembrane receptor Notch is expressed by both dMP2 and vMP2. In vMP2, Delta–Notch signaling requires *sanpodo* and *mastermind* to downregulate Odd and promote an anterior axon. In dMP2, the Delta–Notch interaction can occur, but signal transduction is blocked by the cortical Numb protein. Numb may directly interact with the intracellular domain of Notch. See text for details. (Modified from Spana and Doe, 1996, © Cell Press).

components that are induced or repressed by the Notch signaling cascade and function to specify each sibling cell's final fate. So far, there are no known target genes (i.e., *Enhancer of split, tramtrack*) whose expression is induced by Notch signaling in the MP2 lineage. *odd* is downregulated by Notch signaling in vMP2, but it is unknown whether Odd is required for, or only diagnostic of, the dMP2 fate.

C. Inscuteable Coordinates Basal Protein Localization and Spindle Orientation during Mitosis

To ensure that asymmetrically localized determinants are differentially inherited by sibling cells, the timing of localization must be precisely coordinated with the cell cycle and the localized position must be coordinated

with mitotic spindle orientation. Mechanisms to achieve this synchrony are clearly in operation: basal protein localization occurs only during mitosis (although the proteins are present during interphase) and the crescents of localized protein are always centered over the centrosomes. In neuroblasts that acquire two basal centrosomes by cytokinesis failure, two Pros and Numb crescents can form and are positioned over the normal basal centrosome and the ectopic basal centrosome (Knoblich *et al.*, 1995).

The earliest molecular indication of neuroblast polarity is Insc protein in the apical endfoot of delaminating neuroblasts. In neuroblasts that have delaminated fully, Insc is asymmetrically localized to the apical side during interphase and persists at this site at least through metaphase (Kraut and Campos-Ortega, 1996; Kraut *et al.*, 1996). Insc is therefore a good candidate for coordinating the later asymmetric localization of other proteins during neuroblast mitosis.

Similar to *pros, insc* was identified by virtue of a P[*lacZ*] insertion in the *insc* locus that showed *lacZ* expression in all neuroblasts and sensory organ precursors and perturbed nervous system development (Kraut and Campos-Ortega, 1996). To investigate the possibility that Insc helps establish neuroblast polarity, two features of neuroblast polarity—Pros and Numb localization during mitosis and mitotic spindle orientation—were assayed in *insc* mutants (Kraut *et al.*, 1996). In embryos lacking Insc, Pros and Numb colocalize as a crescent and the mitotic spindle is assembled, but these events are randomized with respect to the apical–basal axis. Moreover, protein localization and spindle orientation are uncoupled from each other, producing neuroblasts that can bisect Pros and Numb crescents at cell division. More recently, it has been determined that Insc also regulates Stau/*pros* RNA and Mira localization. Similar to Pros and Numb, Stau/*pros* RNA and Mira crescents are incorrectly positioned in neuroblasts of *insc* embryos (Li *et al.*, 1997; Shen *et al.*, 1997). Ectopic expression of Insc in the procephalic ectoderm is sufficient to alter the mitotic spindle orientation of ectodermal cells, which normally divide within the ectodermal layer, to resemble that of the procephalic neuroblasts, which divide perpendicular to the ectoderm and along the apical–basal axis (Kraut *et al.*, 1996). Collectively, these studies demonstrate that Insc provides essential information about the apical–basal axis in dividing neuroblasts.

How does Insc function at a molecular level to coordinate basal protein localization and spindle orientation? *insc* encodes a protein with polyproline-rich regions that define target SH3 binding sites common in many cytoskeleton-associated proteins and shows limited similarity to several of these proteins (Kraut and Campos-Ortega, 1996). Insc interacts with Stau in a yeast two-hybrid assay, suggesting that Insc regulation of protein localization may be direct (Li *et al.*, 1997). In neuroblasts, Stau and Insc are colocalized apically for a brief period at the end of interphase (Li *et al.*, 1997; Broadus *et al.*, 1998). It has been suggested that apically localized Insc nucleates a complex that includes Stau and that this interaction is required for the subsequent basal localization of Stau (Li *et al.*, 1997). However, it is unknown whether Insc and Stau interact directly *in vivo* and more specifically in neuroblasts. Moreover, colocalization of Insc and Stau is rarely observed in neuroblasts grown in culture, yet these proteins localize normally to opposite sides of the neuroblast during mitosis (Broadus and Doe, 1997). These results suggest that apical colocalization of Insc and Stau may not be a prerequisite for basal localization of Stau during mitosis. Insc also regulates the localization of Pros, Numb, and Mira, but it is unknown whether Insc interacts directly with these proteins. Discerning how Insc coordinates mitotic spindle orientation and asymmetric protein localization during mitosis will require further biochemical characterization of Insc–protein interactions that occur *in vivo*.

D. Summary of Asymmetric Protein and RNA Localization in Neuroblasts

Recent work from several labs has established a hierarchy of asymmetrically localized proteins and RNA that contribute to neuroblast polarity and asymmetric division (Fig. 5). There are at least two pathways for asym-

metric protein localization in mitotic neuroblasts. One pathway is comprised of Mira, Pros, Stau, and *pros* RNA and regulates GMC cell fate specification. The other pathway utilizes Numb and functions to specify dMP2 fate in the MP2 lineage. Insc regulates both pathways and coordinates protein localization with mitotic spindle orientation.

V. Coordinating Asymmetric Cell Division and the Cell Cycle

Does cell cycle progression control the timing of basal protein localization in neuroblasts? To address this question, Spana and Doe (1995) examined protein localization in embryos deficient for the cell cycle regulator *string*. In *string* embryos, neuroblasts are arrested at the G2/M checkpoint (Edgar and O'Farrell, 1989) and fail to localize Pros to the basal cortex (Spana and Doe, 1995). These experiments demonstrate that entry into mitosis is required for Pros basal localization. Since Mira, Pros, and Stau/*pros* RNA are basally localized as a complex, it is likely that entry into mitosis is a requirement for all of these factors. In addition, neuroblasts arrested at mitosis using colcemid depolymerization of microtubules accumulate Insc, Pros, Stau, and *pros* RNA crescents, suggesting that exit from mitosis is required for release of proteins and RNA from their localized cortical sites (Broadus and Doe, 1997; Broadus *et al.*, 1998). The molecular regulatory events that coordinate

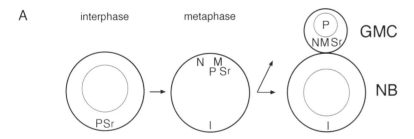

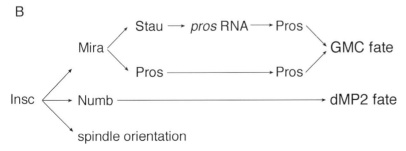

Figure 5 Summary of protein and RNA localization in mitotic neuroblasts. (A) Protein and RNA localization in wild-type neuroblasts. Large cell, neuroblast; small cell, GMC; P, Pros; S, Stau; r, *pros* RNA; M, Mira; I, Insc; N, Numb; apical, down; basal, up. (B) The Mira/Pros/Stau/*pros* RNA pathway is independent of the Numb pathway and is required for GMC fate specification. The Numb pathway functions to specify dMP2 fate. Both protein/RNA localization pathways operate independently of mitotic spindle orientation. Asymmetric protein/RNA localization and mitotic spindle orientation require Insc. See text for details.

asymmetric protein localization and the mitotic cell cycle are unknown.

VI. Cytoskeletal Mechanisms for Asymmetric Localization in Neuroblasts

In order for proteins and RNAs to be localized, the cytoskeleton must be deployed for active transport to and/or anchoring at the localized site. Experiments using drugs which specifically disrupt cytoskeletal components show that actin microfilaments, but not microtubules, play a central role in protein and RNA localization in mitotic neuroblasts (Knoblich *et al.*, 1995; Broadus and Doe, 1997; Broadus *et al.*, 1998). Asymmetrically localized proteins are tightly associated with and dependent on the actin cortex, but actin itself is not asymmetrically distributed in neuroblasts (Spana and Doe 1995; Broadus and Doe, 1997).

Both *in vivo* and *in vitro*, colcemid depolymerization of microtubules arrests neuroblasts in metaphase, and protein crescents accumulate normally. Therefore, microtubules are dispensable for asymmetric protein localization in neuroblasts (Knoblich *et al.*, 1995; Broadus and Doe, 1997). In contrast, cultured neuroblasts treated with cytochalasin show partial depletion of the actin cortex and frequent protein delocalization at mitosis. Neuroblasts treated with latrunculin, which nearly abolishes the actin cortex after brief exposure, shows a more severe delocalization phenotype: Pros and Stau are almost always delocalized in the cytoplasm and Insc is delocalized uniformly at the cortex. These experiments demonstrate that actin is essential for asymmetric protein localization or anchoring (Broadus and Doe, 1997).

Double drug experiments show that actin functions to anchor asymmetrically localized proteins at the cortex. In these studies, neuroblasts are arrested at metaphase using colcemid, protein crescents accumulate in this condition, then cytochalasin or latrunculin is added. Following addition of the microfilament inhibitors, proteins fall off the cortex. This effect is reversible: if cytochalasin is washed out, protein crescents reform. These experiments demonstrate that proteins are anchored at their localized site by an actin-dependent mechanism (Broadus and Doe, 1997). Actin also may actively transport proteins to their asymmetric positions. Alternatively, proteins may diffuse to the localized site at which they become anchored.

VII. Asymmetric Ganglion Mother Cell Divisions

Although GMC divisions appear physically symmetric, they often produce sibling cells that have distinct devel-opmental fates. For example, the firstborn GMC of neuroblast 4-2 divides to make the RP2 motor-neuron and the RP2 sibling neuron that degenerates shortly after its formation (Doe, 1992). These asymmetric divisions utilize some of the same components that polarize neuroblast divisions. For example, as seen in the MP2 lineage, differential specification of some sibling neuron fates requires Numb to block Notch-mediated signal transduction (Skeath and Doe, 1998). Thus, Numb inhibition of Notch signaling may represent a widespread mechanism for generating binary sibling fates.

VIII. Perspectives

Recent advances in our understanding of asymmetrically localized proteins and RNAs in mitotic neuroblasts of *Drosophila* have unveiled many new questions. Among the most challenging is: what are the origins of neuroblast polarity? Neuroblasts may inherit polar qualities from the polarized epithelium from which they delaminate. The molecular nature of apical and basal polarity cues, however, is completely unknown. With the recent discovery of several proteins that define asymmetric localization pathways in *Drosophila* neuroblasts, there exist new opportunities to investigate fundamental aspects of cell polarity and asymmetric cell division.

Another major direction for future studies will be to investigate how stem cells generate a unique progeny at each cell division. For example, how do neuroblasts generate differently fated GMCs? GMCs produced by the same neuroblast are produced in an invariant sequence and are clearly different from one another based on their patterns of gene expression and the neurons/glia which they produce. For example, the first GMC produced by neuroblast 1-1 is Eve-positive and divides to produce the aCC and pCC neurons; the second GMC of neuroblast 1-1 is Eve-negative and probably produces a pair of glia (Udolph *et al.*, 1993; Broadus *et al.*, 1995). Several mechanistic models have been proposed to explain how GMCs derived from the same neuroblast acquire different identities (Doe and Smouse, 1990). One possibility is that each GMC inherits a unique determinant from the neuroblast. Alternatively, extrinsic factors—such as signaling events between the neuroblast and GMC—may be regulated at each neuroblast cell cycle to generate different GMCs at each neuroblast mitosis. Several genes are known to be expressed in neuroblasts during only a subset of a neuroblast's lineage (Cui and Doe, 1992; Doe, 1992; Broadus *et al.*, 1995). Cell cycle progression and neuroblast cytokinesis are important regulators of these gene expression transitions within neuroblast lineages (Cui and Doe, 1995; Weigmann and Lehner, 1995). Experiments that address how these transcriptional changes are initiated will uncover additional mechanisms for cell fate specification in stem cell lineages.

Acknowledgments

We thank Laurina Manning for drawing Fig. 2E; Harald Vaessin for permission to reprint Fig. 1A,B; Chris Doe for permission to reprint Fig. 1C,D, for communicating unpublished results, and for comments on the manuscript; Sal Fuerstenberg and Pedro Alvarez for communicating unpublished results; and Jim Skeath and Hugo Bellen for communicating unpublished results and for comments on the manuscript. J. B. is a Research Associate of the Howard Hughes Medical Institute, and E. P. S. is a Fellow of the Helen Hay Whitney Foundation. We also thank Carl Thummel and Norbert Perrimon for their gracious support.

References

Bate, C. M., and Grunewald, E. B. (1981). Embryogenesis of an insect nervous system II. A second class of neuron precursor cells and the origin of the intersegmental connectives. *J. Embryol. Exp. Morphol.* **61**, 317–330.

Bhat, K. M. (1996). The *patched* signaling pathway mediates repression of *gooseberry* allowing neuroblast specification by *wingless* during Drosophila neurogenesis. *Development (Cambridge, UK)* **122**, 2921–2932.

Bier, E., Vaessin, H., Younger-Shepard, S., Jan, L. Y., and Jan, Y. N. (1992). *deadpan,* an essential pan-neural gene in *Drosophila,* encodes a helix-loop-helix protein with a structure similar to the *hairy* product. *Genes Dev.* **6**, 2137–2151.

Bork, P., and Margolis, B. (1995). A phosphotyrosine interaction domain. *Cell (Cambridge, Mass.)* **80**, 693–694.

Bossing, T., Udolph, G., Doe, C. Q., and Technau, G. T. (1996). The embryonic central nervous system lineages of Drosophila melanogaster I. Neuroblast lineages derived from the ventral half of the neuroectoderm. *Dev. Biol.* **179**, 41–64.

Brand, M., Jarman, A. P., Jan, L. Y., and Jan, Y. N. (1993). *asense* is a Drosophila neural precursor gene and is capable of initiating sense organ formation. *Development (Cambridge, UK)* **119**, 1–17.

Broadus, J., and Doe, C. Q. (1997). Extrinsic cues, intrinsic cues, and microfilaments regulate asymmetric protein localization in Drosophila neuroblasts. *Curr. Biol.* **7**, 827–835.

Broadus, J., Skeath, J. B., Spana, E. P., Bossing, T., Technau, G., and Doe, C. Q. (1995). New neuroblast markers and the origin of the aCC/pCC neurons in the Drosophila central nervous system. *Mech. Dev.* **54**, 1–10.

Broadus, J., Fuerstenburg, S., and Doe, C. Q., (1998). Staufen-dependent asymmetric localization of *prospero* RNA contributes to neuroblast daughter cell fate. *Nature (London)* **391**, 792–795.

Burglin, T. R. (1994). A Caenorhabditis elegans prospero homologue defines a novel domain. *Trends Biochem. Sci.* **19**, 70–71.

Chu-LaGraff, Q., and Doe, C. Q. (1993). Neuroblast specification and formation is regulated by *wingless* in the Drosophila CNS. *Science* **261**, 1594–1597.

Chu-LaGraff, Q., Wright, D. M., McNeil, L. K., and Doe, C. Q. (1991). The *prospero* gene encodes a divergent

homeodomain protein that controls neuronal identity in Drosophila. *Development (Cambridge, UK) Suppl.* **2**, 79–85.

Cui, X., and Doe, C. Q. (1992). *ming* is expressed in neuroblast sublineages and regulates gene expression in the Drosophila central nervous system. *Development (Cambridge, UK)* **116**, 943–952.

Cui, X., and Doe, C. Q. (1995). The role of the cell cycle and cytokinesis in regulating gene expression in the Drosophila CNS. *Development (Cambridge, UK)* **121**, 3233–3243.

Doe, C. Q. (1992). Molecular markers for identified neuroblasts and ganglion mother cells in the Drosophila central nervous system. *Development (Cambridge, UK)* **116**, 855–863.

Doe, C. Q., Smouse, and D. (1990). The origins of cell diversity in the insect central nervous system. *Semin. Cell Biol.* **1**, 211–218.

Doe, C. Q., Hiromi, Y., Gehring, W. J., and Goodman, C. S. (1988a). Expression and function of the segmentation gene *fushi tarazu* during Drosophila neurogenesis. *Science* **239**, 170–175.

Doe, C. Q., Smouse, D., and Goodman, C. S. (1988b). Control of neuronal fate by the Drosophila segmentation gene *even-skipped. Nature (London)* **333**, 376–378.

Doe, C. Q., Chu-LaGraff, Q., Wright, D. M., and Scott, M. P. (1991). The *prospero* gene specifies cell fates in the Drosophila central nervous system. *Cell (Cambridge, Mass.)* **65**, 451–464.

Duman-Scheel, M., Li, X. Orlov, I., Noll, M., and Patel, N. H. (1997). Genetic separation of the neural and cuticular patterning functions of *gooseberry. Development (Cambridge, UK)* **124**, 2855–2865.

Dye, C. A., Lee, J.-K., Atkinson, R. C., Brewster, R., Han, P.-L., and Bellen, H. J. (1998). The Drosophila sanpodo gene controls sibling cell fate and encodes a tropomodulin homolog, an actin/tropomyosin-associated protein. *Development (Cambridge, UK)* **125**, 1845–1856.

Edgar, B. A., and O'Farrell, P. H. (1989). Genetic control of cell division patterns in the Drosophila embryo. *Cell (Cambridge, Mass.)* **57**, 177–187.

Ferrandon, D., Elphick, L., Nusslein-Volhard, C., and St. Johnston, D. (1994). Staufen protein associates with the 3′ UTR of *bicoid* mRNA to form particles that move in a microtubule-dependent manner. *Cell (Cambridge, Mass.)* **79**, 1221–1232.

Frise, E., Knoblich, J., Younger-Shepherd, S., L. Jan, Y., and Jan, Y. N. (1996). The Drosophila Numb protein inhibits signaling of the Notch receptor during cell–cell interaction in sensory organ lineage. *Proc. Natl. Acad. Sci. U.S.A.* **93**, 11925–11932.

Guo, M., Jan, L. Y., and Jan, Y. N. (1996). Control of daughter cell fates during asymmetric division: Interaction of Numb and Notch. *Neuron* **17**, 27–41.

Hartenstein, V., and Posakony, J. W. (1990). A dual function of the *Notch* gene in sensillum development. *Dev. Biol.* **142**, 13–30.

Hassan, B., Li, L., Bremer, K. A., Chang, W., Pinsonneault, J., and Vaessin, H. (1997). Prospero is a panneural transcription factor that modulates homeodomain protein activity. *Proc. Natl. Acad. Sci. U.S.A.* **94**, 10091–10096.

Hirata, J., Nakagoshi, H., Nabeshima, Y., and Matsuzaki, F. 1995. Asymmetric segregation of a homeoprotein, *prospero*, during cell divisions in neural and endodermal development. *Nature (London)* **377**, 627–630.

Ikeshima-Kataoka, H., Skeath, J. B., Nabeshima, Y., Doe, C. Q., and Matsuzaki, F. (1997). Miranda directs Prospero to a daughter cell during *Drosophila* asymmetric divisions. *Nature (London)* **390**, 625–629.

Knoblich, J. A., Jan, L. Y., and Jan, Y. N., (1995). Asymmetric segregation of Numb and Prospero during cell division. *Nature (London)* **377**, 624–627.

Kraut, R., and Campos-Ortega, J. A. (1996). *inscuteable*, a neural precursor gene of *Drosophila,* encodes a candidate for a cytoskeleton adaptor protein. *Dev. Biol.* **174**, 65–81.

Kraut, R., Chia, W., Jan, L. Y., Jan, Y. N., and Knoblich, J. A. (1996). Role of inscuteable in orienting asymmetric cell divisions in *Drosophila*. *Nature (London)* **383**, 50–55.

Li, P., Yang, X., Wasser, M., Cai, Y., and Chia, W. (1997). Inscuteable and Staufen mediate asymmetric localization and segregation of *prospero* RNA during *Drosophila* neuroblast cell divisions. *Cell (Cambridge, Mass.)* **90**, 437–447.

Lin, D. M., Fetter, R. D., Kopczynski, C., Greeningloh, G., and Goodman, C. S. (1994). Genetic analysis of *fasciclin II* in *Drosophila*: Defasciculation, refasciculation and altered fasciculation. *Neuron* **13**, 1055–1069.

McDonald, J. A., and Doe, C. Q. (1997). Establishing neuroblast-specific gene expression in the *Drosophila* CNS: *huckebein* is activated by Wingless and Hedgehog and repressed by Engrailed and Gooseberry. *Development (Cambridge, UK)* **124**, 1079–1087.

Matsuzaki, F., Koisumi, K., Hama, C., Yoshioka, T., and Nabeshima, Y. (1992). Cloning of the *Drosophila prospero* gene and its expression in ganglion mother cells. *Biochem. Biophys. Res. Commun.* **182**, 1326–1332.

Matsuzaki, M., and K. Saigo, K. (1996). *hedgehog* signaling independent of *engrailed* and *wingless* required for post-S1 neuroblast formation in *Drosophila* CNS. *Development (Cambridge, UK)* **122**, 3567–3575.

Oliver, G., Sosa-Pineds, B., Geisendorf, S., Spana, E. P., Doe, C. Q., and Gruss, P. (1993). Proxl, a prospero-related homeobox gene expressed during mouse development. *Mech. Dev.* **44**, 3–16.

Parks, A. L., and Muskavitch, M. A. (1993). *Delta* function is required for bristle organ determination and morphogenesis in *Drosophila*. *Dev. Biol.* **157**, 484–496.

Parras, C., Garcia-Alonso, L. A., Rodriguez, I., and Jimenez, F. (1996). Control of neural precursor specification by proneural proteins in the CNS of *Drosophila*. *EMBO J.* **15**, 6394–6399.

Rhyu, M. S., Jan, L. Y., and Jan, Y. N. (1994). Asymmetric distribution of numb protein during division of the sensory organ precursor cell confers distinct fates to daughter cells. *Cell (Cambridge, Mass.)* **76**, 477–491.

Schmidt, H., Rickert, C., Bossing, T., Vef, O. U., Jr., and Technau, G. M. (1997). The embryonic central nervous system lineages of *Drosophila melanogaster*. II. Neuroblast lineages derived from the dorsal part of the neuroectoderm. *Dev. Biol.* **189**, 186–204.

Shen, C.-P., Jan, L. Y., and Jan, Y. N. (1997). Miranda is re-

quired for the asymmetric localization of prospero during mitosis in *Drosophila*. *Cell (Cambridge, Mass.)* **90**, 449–458.

Skeath, J. B., and Carroll, S. B. (1992). Regulation of proneural gene expression and cell fate during neuroblast segregation in the *Drosophila* embryo. *Development (Cambridge, UK)* **114**, 939–946.

Skeath, J. B., and Carroll, S. B. (1994). The *achaete–scute* complex: Generation of cellular pattern and fate within the *Drosophila* nervous system. *FASEB J.* **8**, 714–721.

Skeath, J. B., and Doe, C. Q. (1996). The achaete–scute complex proneural genes contribute to neural precursor specification in the *Drosophila* CNS. *Curr. Biol.* **6**, 1146–1152.

Skeath, J. B., and Doe, C. Q. (1998). Sanpodo and Notch act in opposition to Numb to distinguish sibling neuron fates in the *Drosophila* CNS. *Development (Cambridge, UK)* **125**, 1857–1865.

Skeath, J. B., Zhang, Y., Holmgren, R., Carroll, S. B., and Doe, C. Q. (1995). Specification of neuroblast identity in the *Drosophila* embryonic central nervous system by *gooseberry-distal*. *Nature (London)* **376**, 427–430.

Spana, E., and Doe, C. Q. (1995). The prospero transcription factor is asymmetrically localized to the cell cortex during neuroblast mitosis in *Drosophila*. *Development (Cambridge, UK)* **121**, 3187–3195.

Spana, E. P., and Doe, C. Q. (1996). Numb antagonizes Notch signaling to specify sibling neuron cell fates. *Neuron* **17**, 21–26.

Spana, E. P., Kopczynski, C., Goodman, C. S., and Doe, C. Q. (1995). Asymmetric localization of numb autonomously determines sibling neuron identity in the *Drosophila* CNS. *Development (Cambridge, UK)* **121**, 3489–3494.

St. Johnston, D., Beuchle, D., and Nusslein-Volhard, C. (1991). *staufen*, a gene required to localize maternal RNAs in the *Drosophila* eggs. *Cell (Cambridge, Mass.)* **66**, 51–63.

St. Johnston, D., Brown, N. H., Gall, J. G., and Janntsch, M. (1992). A conserved double-stranded RNA-binding domain. *Proc. Natl. Acad. Sci. U.S.A.* **89**, 10979–10983.

Tomarev, S. I., Sundin, O., Banerjee-Basu, S., Duncan, M. K., Yang, J. M., and Piatigorsky, J. (1996). Chicken homeobox gene *Prox 1* related to *Drosophila prospero* is expressed in the developing lens and retina. *Dev. Dyn.* **206**, 354–367.

Udolph, G., Prokop, A., Bossing, T., and Technau, G. M. (1993). A common precursor for glia and neurons in the embryonic CNS of *Drosophila* gives rise to segment-specific lineage variants. *Development (Cambridge, UK)* **118**, 765–775.

Uemura, T., Sheperd, S., Ackerman, L., Jan, L. Y., and Jan, Y. N. (1989). *numb*, a gene required in determination of cell fate during sensory organ formation in *Drosophila* embryos. *Cell (Cambridge, Mass.)* **58**, 349–360.

Vaessin, H., Grell, E., Wolff, E., Bier, E., Jan, L. Y., and Jan, Y. N. (1991). *prospero* is expressed in neuronal precursors and encodes a nuclear protein that is involved in the control of axonal outgrowth in *Drosophila*. *Cell (Cambridge, Mass.)* **67**, 941–953.

Weigmann, K., and Lehner, C. F. (1995). Cell fate specification by *even-skipped* expression in the *Drosophila* nervous system is coupled to cell cycle progression. *Development (Cambridge, UK)* **121,** 3713–3721.

Zhang, Y., Ungar, A., Fresquez, C., and Holmgren, R. (1994). Ectopic expression of either the *Drosophila gooseberry-distal* or *proximal* gene causes alterations of cell fate in the epidermis and central nervous system. *Development (Cambridge, UK)* **120,** 1151–1161.

Zhong, W., Feder, J. N., Jiang, M.-M., Jan, L. Y., and Jan, Y. N. (1996). Asymmetric localization of a mammalian Numb homologue during mouse cortical neurogenesis. *Neuron* **17,** 43–53.

Zinovieva, R. D., Duncan, M. K., Johnson, T. R., Torres, R., Polymeropoulos, M. H., and Tomarev, S. I. (1996). Structure and chromosomal localization of the human homeobox gene *Prox 1. Genomics* **35,** 517–522.

V

Frog

19

Introduction to the Frog:
Your Origin Is Your Fate—Or Is It?

IGOR B. DAWID
Laboratory of Molecular Genetics
National Institute of Child Health and Human Development
National Institutes of Health
Bethesda, Maryland 20892

I. Introduction

The somewhat facetious title of this chapter is meant to convey a dichotomy inherent in lineage studies in the past and continuing at present. The importance of lineage relationships in embryogenesis is so intuitively obvious, such a powerful source of understanding how a tissue or organ arises that, in various ways at various times, workers in the field have felt that lineage explains everything and that without knowing the lineage derivation of a structure we cannot understand anything. Of course, time has passed and much has been learned since Driesch concluded that the regulative properties of blastomeres make it impossible (not just difficult, futile) to search for mechanisms underlying development (as reviewed in Wilson, 1925). And yet, even today some comments imply that knowing the lineage relationships in a process explains its mechanism or that failure to know these relationships dooms any attempt at understanding. Such assumptions need to be raised to explicit status; then one can examine rationally what contribution to understanding does or does not derive from a knowledge of the lineage relationships.

Having written what might be considered an assault on the significance of lineage studies, let me hasten to clarify. Without doubt, knowing lineage relationships is a critical aspect of attempts to elucidate the mechanism of development in any multicellular organism, at every level of organization from the entire embryo to the formation of organs and tissues. Defining lineage is often the basis for further work, a baseline without which many other types of experimentation are difficult to interpret. Yet, it is important to remember what one can and cannot learn from a known lineage. If lineage is indeterminate, as for example, in the eight-cell mouse embryo, it is a necessary conclusion that developmental processes and the emergence of pattern depend on cell interactions. The converse, however, is not true. If cell lineage is strictly determinate and the fate map invariant, as it is in the nematode *Caenorhabditis elegans*, it does not follow that differentiation is necessarily cell autonomous. In fact, while some cell autonomous differentiation does occur in this animal, numerous cell interactions take place and are required for embryogenesis of *C. elegans*, which has consequently become one of the most powerful systems for studying signaling mechanisms in development (see Section III in this volume). And it is precisely the knowledge of the lineage of all cells in this animal that, together with other factors, makes it such an effective object for studying cell interaction mechanisms.

Xenopus laevis is a useful study object for several reasons, not the least of which is that different regions of the egg and the early embryo are fated to contribute to specific structures; in other words, one can generate a fate map for the early embryo. At the same time, it is clear that cell interactions play a major role in amphibian embryogenesis, as has been widely appreciated ever since the time of the celebrated studies of Spemann and Mangold (1924). By representing a situation in which aspects of both regulative and lineage-dependent development are apparent, amphibians and specifically *Xenopus* have long been and continue to be favored as a system for studying many aspects of embryogenesis. Thus, a substantial fraction of our insight into events during early stages of vertebrate development derives from work with *Xenopus,* as discussed in some detail in two chapters in this section (Chapter 20 by Sullivan *et al.* and Chapter 21 by Kessler). Furthermore, elucidation of the cellular interactions and regulatory genes involved in organogenesis have been facilitated by the ease of manipulating tissues and organ anlagen, as will be discussed in two other chapters in this section (Chapter 22 by Newman and Krieg and Chapter 23 by Perron and Harris; see also chapters in Section IX).

The *Xenopus* life cycle is illustrated briefly in Fig. 1. During the long process of oogenesis, large amounts of reserve substances are accumulated that allow rapid cell division and differentiation in the early embryo. In particular, a large store of mRNA, preformed organelles like mitochondria, and ribosomes are stored so that components for the basic machinery of cellular metabolism need not be produced during early embryogenesis, thus allowing rapid development. After fertilization, cleavage proceeds without RNA synthesis to the mid blastula transition at about 6 hours of development (Newport and Kirschner, 1982), using maternal mRNA whose expression and stability are regulated by several mechanisms. The basic body pattern is laid down during gastrulation, and is further elaborated during a period of organogenesis that leads to metamorphosis, a dramatic reorganization of the body plan from which the adult, albeit immature, frog emerges. After further growth and maturation, the life cycle is closed with the attainment of fertility at the age of about 1 year.

II. The Emergence of Body Pattern in the *Xenopus* Embryo

By the end of gastrulation the future body pattern of the embryo is well established. This pattern is based on an initial state defined by the constituents and organization of the unfertilized egg; this initial state is elaborated by multiple cell interactions, ultimately leading to the emergence of embryonic form. The unfertilized egg exhibits a rotational axis of symmetry, the animal–vegetal axis, which is transformed in a complex manner into the anterior–posterior axis of the larva. A rearrangement of the cytoplasm, the cortical-cytoplasmic rotation, is initiated after fertilization and leads, again through several steps, to the establishment of the dorsal–ventral axis (Gerhart *et al.,* 1989). With the setting of the two major axes of the body the embryo aquires a fate map, that is, the progeny of individual blastomeres in the cleavage stage embryo give rise to distinct tissues and organs of the tadpole. Fate maps for amphibian embryos have been available for a long time (Vogt, 1929), but it is the work on *Xenopus* that has been of critical importance in the recent history of the field. Keller (1975, 1976) mapped the fates of different regions of the blastula and early gastrula embryo, while Moody (1987a,b) and Dale and Slack (1987) generated fate maps for the 16- and 32-cell stage blastomeres. The latter maps are particularly useful in the design of overexpression experiments, as discussed below. While fate maps earlier than the 32-cell stage necessarily have more limited resolution (see Chapter 20 by Sullivan *et al.*), they suggest that the late one-cell embryo already carries the projection of the 32-cell stage map in it.

An important feature of the Moody and the Dale

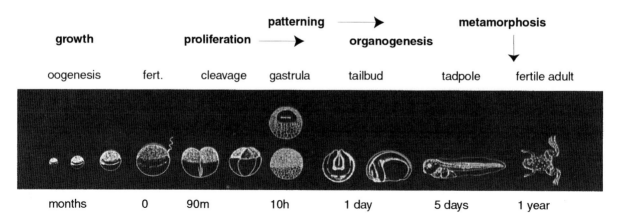

Figure 1 A general outline of the life cycle of *Xenopus laevis.* Time of development and stages are listed, and major developmental events are summarized at the top. See Nieuwkoop and Faber (1967), and Hausen and Riebesell (1991) for detailed descriptions of *Xenopus* development.

and Slack fate maps is their statistical, approximate nature. Fates are not precisely defined—most blastomeres give rise to certain predominant fates, but can at low frequency contribute to almost any tissue in the body. Fate can properly be expressed as the probability of forming a certain tissue or cell type. Some blastomeres have rather restricted fates, for example, vegetal tier cells form mostly gut and do not contribute to the central nervous system (CNS), while animal tier cells form ectodermal derivatives but not gut. In general, though, blastomeres give rise to a wide array of progeny while, at the same time, exhibiting predominant fates that allow predictions that are of great value in a variety of experimental settings.

III. Axis Specification and Mesoderm Induction

While the broad fate map of the embryo is likely to be established by the time of first cleavage, the way in which these fates emerge is complex and involves multiple steps of cell interaction. Much has been learned in recent years about the underlying mechanisms through which the state of specification in the early embryo is transformed into the pattern in the gastrula. The events that take place are usually considered under the headings of axis specification and mesoderm induction. This field has been reviewed frequently (Nieuwkoop, 1973; Smith *et al.*, 1985; Gerhart *et al.*, 1989; Kimelman *et al.*, 1992; Kessler and Melton, 1994; Dawid, 1994; Miller and Moon, 1996; Sasai and De Robertis, 1997; Moon *et al.*, 1997; Hemmati-Brivanlou and Melton,

1997; Heasman, 1997; see also Chapter 20 by Sullivan *et al.*, and Chapter 21 by Kessler); here I provide a brief overview by way of introduction.

A. Formation of the Nieuwkoop Center

The cortical–cytoplasmic rotation that is necessary and sufficient for dorsal axis formation in *Xenopus* rearranges components in the egg, but the nature of the critical translocated material is not entirely clear. It is generally accepted that, as a consequence of rotation, a dorsal signaling center is generated that has been named the Nieuwkoop center (Fig. 2A). While there are different views of its properties, an operational definition exists and is useful: the Nieuwkoop center is a region of the embryo which can induce a dorsal axis but whose progeny does not contribute to axial structures like notochord and CNS, but rather populates the endoderm. The Nieuwkoop center has been associated with signaling through the Wnt pathway since certain Wnt factors and downstream components of its signal transduction pathway can create a Nieuwkoop center in ventral vegetal or marginal regions (reviewed in Miller and Moon, 1996; Heasman, 1997). However, it appears that no Wnt factor is involved in dorsal axis initiation in the normal embryo, but rather that the pathway is activated somewhere halfway down its signal transduction path. Whatever initiates Nieuwkoop center formation, it appears to be associated with the movement of β-catenin, the downstream component of the Wnt cascade, into the nucleus where it associates with other factors to regulate target gene expression (see Miller and Moon, 1996; Heasman, 1997).

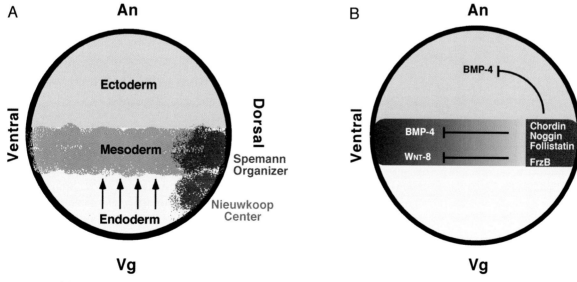

Figure 2 (A) Signaling in the early frog embryo. Mesoderm induction is symbolized by arrows, and the Nieuwkoop center and Spemann organizer are indicated. See text for further description. (B) Signaling interactions in the *Xenopus* gastrula. Ventralizing and epidermal inducers are counteracted by organizer factors. See text for further description. An, Animal; Vg, vegetal.

B. Mesoderm Induction and Formation of the Spemann Organizer

Dorsoventral polarity is first established in the mesoderm, and axis formation and mesoderm induction are intimately linked. Mesoderm derives from cells in the equatorial, or marginal, region of the embryo as a result of inductive signaling (Dawid, 1994; Kimelman *et al.,* 1992; Heasman, 1997) although cell-autonomous cues contribute as well (Lemaire and Gurdon, 1994). Members of the transforming growth factor-β (TGF-β) and fibroblast growth factor (FGF) families have been implicated in mesoderm induction by methods that will be summarized below. In spite of much effort and considerable progress it remains unclear exactly which factor(s) are responsible, and in which way mesoderm inducers interact with dorsalizing (Nieuwkoop) signals. While the schematic representation in Fig. 2A suggests that the inducing effects point uniformly in a vegetal-to-marginal direction, mesoderm inducers of the TGF-β class also modulate the dorsoventral character of the mesoderm they induce. Nevertheless it is clear that the establishment of a normal pattern at the initiation of gastrulation requires, at a minimum, Nieuwkoop signaling through β-catenin and the induction of mesoderm.

The events described above have led, by the beginning of gastrulation, to the specification of the equatorial region of the embryo as mesoderm which exhibits dorsoventral polarity; the dorsal region of the mesoderm is synonymous with the organizer. The Spemann organizer has held a prominent place in developmental biology for over 70 years because of its ability to self-organize and to induce an embryonic axis in ventral tissue (Lemaire and Kodjabachian, 1996). It is clear that the Nieuwkoop signal is required for generating an organizer; it is not clear how this is achieved. While the two centers can be distinguished operationally as dorsalizing regions whose progeny does not (Nieuwkoop) or does (Spemann) contribute to axial tissues, the distinction is often difficult to make. The expression of marker genes widely accepted as suitable markers, for example, *siamois* for the Nieuwkoop center (Lemaire *et al.,* 1995) and *goosecoid* for the organizer (Cho *et al.,* 1991), overlaps temporally and spatially, and both the temporal and spatial extent of either region are difficult to define precisely. In particular, it is not clear whether an intercellular signal is provided by Nieuwkoop cells that is required for organizer formation; if such a signal exists, its nature is unknown.

C. Organizer Function

While its formation is still not fully understood, much has been learned about organizer function by using organizer-specific gene expression as an entry to its study. Among early isolates were genes encoding transcription factors, for example, *goosecoid* (Cho *et al.,* 1991) or *Xlim-1* (Taira *et al.,* 1992), and genes encoding secreted factors notably noggin (Smith and Harland, 1992), chordin (Sasai *et al.,* 1994), and follistatin (Hemmati-Brivanlou *et al.,* 1994). These factors have at least partial axis-inducing properties (see Lemaire and Kodjabachian, 1996), implying that they are functional components of the organizer. At the same time, other studies focused on the previously neglected ventral side of the gastrula embryo. It emerged that the ventral state is actively induced by BMP-4 and Wnt-8, and BMP-4 also specifies epidermal differentiation in the ectoderm; these activities are counteracted by organizer factors, and suppression of ventralizing activity is sufficient for neural induction and axis specification (Fig. 2B; reviewed in Lemaire and Kodjabachian, 1996; Hogan, 1996; Hemmati-Brivanlou and Melton, 1997; Sasai and De Robertis, 1997). Organizer factors thus are negative regulators of axis-inhibiting signals.

D. Head Induction—A Role for Anterior Endoderm

Organizer function is required for the formation of the entire axis including the head, yet recent evidence has pointed to endoderm as a region of critical importance in head induction (Bouwmeester and Leyns, 1997). Indications that this is so come from several directions, but the key piece of evidence is based on the properties of a novel secreted factor, cerberus. Cerberus is not expressed in prechordal mesoderm but rather in anterior endoderm, a region that gives rise to liver and foregut, and overexpression of cerberus in *Xenopus* embryos elicits the formation of heads including cyclopic eye and cement gland (Bouwmeester *et al.,* 1996). It remains to be clarified in which way anterior endoderm and the organizer interact in head formation.

IV. Basic Techniques for the Study of *Xenopus* Embryogenesis

How have we learned all of the facts and generated all of the hypotheses that I summarized in the previous section? Many approaches have contributed, but much of the progress comes from relatively few techniques. Explant culture and transplantation experiments have a long history in the field, and the suitability of amphibians for these techniques is a major reason for their popularity as study objects. Lineage studies, whose significance for the understanding of development has been recognized since the nineteenth century, have had a major influence in making *Xenopus* a useful experimental model. When these embryological methods were combined with molecular biology, starting about 1960 and accelerating greatly with the introduction of recombinant DNA techniques, the stage was set for rapid progress. Many of the experimental stratagems used are common to work in *Xenopus* and other animals, but a few techniques are, if not uniquely frog-specific, particularly effective in this system.

The ability of *Xenopus* oocytes to translate injected mRNA (Gurdon *et al.,* 1971) has been immensely useful in numerous ways, making the oocyte one of the most widely used "living test tubes." From this application it was a short, though critical, step to inject mRNA into embryos and watch for phenotypic consequences of the expression of the exogenously generated protein. When this approach was combined with the ability to synthesize functional mRNA from cloned genes *in vitro* (Krieg and Melton, 1984), a method of broad applicability ensued. Such RNA injection experiments are particularly powerful in *Xenopus* because of our knowledge of the lineage map, as illustrated in Fig. 3A. Because we can target the injected RNA to the future dorsal or ventral side (in this example), results become more readily

interpretable. In addition, since the fate map has some statistical scatter, a lineage label like β-galactosidase RNA is often coinjected with the test RNA so that the progeny of the injected cell can be determined in each case. The outcome of the experiment can be determined by visual inspection when the effect is strong as in the example in Fig. 3A, but may also be determined by histology and by testing for expression of marker genes by *in situ* hybridization or immunocytochemistry.

A second powerful approach available to frog researchers is based on the properties of the animal region of the blastula embryo. This region is fated to form ectoderm and, when explanted and cultured in salt solution, forms epidermis. However, the animal cap, as it is called, can easily be diverted to different fates. Nieuwkoop and colleagues showed that the animal ectoderm is induced to form mesoderm when cultured together with the vegetal region of the blastula embryo (Boterenbrood and Nieuwkoop, 1973), and an assay based this observation demonstrated that soluble growth factors likewise could convert animal caps into

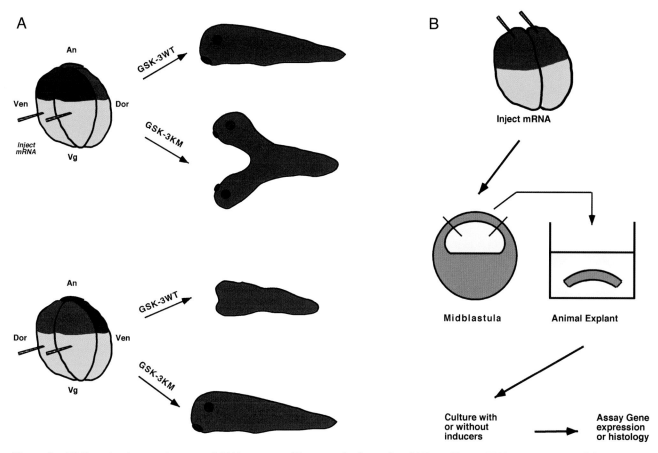

Figure 3 (A) Functional assays by targeted RNA injection. This example shows that GSK-3 wild type (WT), a component of the Wnt signal transduction pathway, inhibits axis formation when injected dorsally while a dominant-negative form, GSK-3KM, induces a secondary axis when injected ventrally. Based on results of He *et al.* (1995). (B) The animal cap assay. The animal region dissected from a blastula embryo forms epidermis in culture but can be induced to change into mesoderm or neural tissue by the addition of growth factors. Alternatively, injection of various RNAs into the early embryo leads to the production of the encoded proteins within the cells of the animal cap, allowing the assay of phenotypic consequences. An, Animal; Vg, vegetal; Ven, ventral; Dor, dorsal.

mesoderm (Smith, 1987; Slack *et al.*, 1987). This assay is illustrated in Fig. 3B, together with a variation in which RNA encoding a protein to be tested is injected into the animal region of a 2- to 4-cell embryo, followed by animal cap explant culture. A great deal has been learned by this technique about the function of various factors in the embryo. Most of the conclusions summarized in the previous section are based on the application of such RNA injection and explant experiments and their variations.

V. Conclusion and Outlook

Xenopus laevis is among the most useful systems for the study of vertebrate embryogenesis. The combination of embryological, specifically lineage information with the ability to carry out explant and transplant experiments, and the facility of expressing proteins from cloned genes in targeted areas of the embryos, provide a powerful set of tools for the elucidation of the molecular and cellular basis of development. The present is an exciting time: just in a few years we have learned much about such fundamental processes as the specification of the dorsal axis and the functions of the Spemann organizer. We can confidently expect progress in the near future to rapidly expand our understanding of the frog embryo and through it, of vertebrate development in general.

References

Boterenbrood, E. C., and Nieuwkoop, P. D. (1973). The formation of the mesoderm in urodelan amphibians. V. Its regional induction by the endoderm. *Roux's Arch. Dev. Biol.* **173**, 319–332.

Bouwmeester, T., and Leyns, L. (1997). Vertebrate head induction by anterior primitive ectoderm. *Bioessays* **19**, 855–863.

Bouwmeester, T., Kim, S., Sasai, Y., Lu, B., and De Robertis, E. M. (1996). Cerberus is a head-inducing secreted factor expressed in the anterior endoderm of Spemann's organizer. *Nature (London)* **382**, 595–601.

Cho, K. W., Blumberg, B., Steinbeisser, H., and De Robertis, E. M. (1991). Molecular nature of Spemann's organizer: The role of the *Xenopus* homeobox gene goosecoid. *Cell (Cambridge, Mass.)* **67**, 1111–1120.

Dale, L., and Slack, J. M. (1987). Fate map for the 32-cell stage of *Xenopus laevis. Development (Cambridge, UK)* **99**, 527–551.

Dawid, I. B. (1994). Intercellular signaling and gene regulation during early embryogenesis of *Xenopus laevis. J. Biol. Chem.* **269**, 6259–6262.

Gerhart, J., Danilchik, M., Doniach, T., Roberts, S., Rowning, B., and Stewart, R. (1989). Cortical rotation of the *Xenopus* egg: Consequences for the anteroposterior pattern of embryonic dorsal development. *Development (Cambridge, UK)* **107** (Suppl.), 37–51.

Gurdon, J. B., Lane, C. D., Woodland, H. R., and Marbaix, G. (1971). Use of frog eggs and oocytes for the study of messenger RNA and its translation in living cells. *Nature (London)* **233**, 177–182.

Hausen, P., and Riebesell, M. (1991). "The Early Development of *Xenopus laevis.*" Springer-Verlag, Berlin.

He, X., Saint-Jeannet, J. P., Woodgett, J. R., Varmus, H. E., and Dawid, I. B. (1995). Glycogen synthase kinase-3 and dorsoventral patterning in *Xenopus* embryos. *Nature (London)* **374**, 617–622.

Heasman, J. (1997). Patterning the *Xenopus* blastula. *Development (Cambridge, UK)* **124**, 4179–4191.

Hemmati-Brivanlou, A., and Melton, D. (1997). Vertebrate embryonic cells will become nerve cells unless told otherwise. *Cell (Cambridge, Mass.)* **88**, 13–17.

Hemmati-Brivanlou, A., Kelly, O. G., and Melton, D. A. (1994). Follistatin, an antagonist of activin, is expressed in the Spemann organizer and displays direct neuralizing activity. *Cell (Cambridge, Mass.)* **77**, 283–295.

Hogan, B. L. (1996). Bone morphogenetic proteins: Multifunctional regulators of vertebrate development. *Genes Dev.* **10**, 1580–1594.

Keller, R. (1975). Vital dye mapping of the gastrula and neurula of *Xenopus laevis.* I. Prospective areas and morphogenetic movements of the superficial layer. *Dev. Biol.* **42**, 222–241.

Keller, R. (1976). Vital dye mapping of the gastrula and neurula of *Xenopus laevis.* II. Prospective areas and morphogenetic movements of the deep layer. *Dev. Biol.* **51**, 118–137.

Kessler, D. S., and Melton, D. A. (1994). Vertebrate embryonic induction: Mesodermal and neural patterning. *Science* **266**, 596–604.

Kimelman, D., Christian, J. L., and Moon, R. T. (1992). Synergistic principles of development: Overlapping patterning systems in *Xenopus* mesoderm induction. *Development (Cambridge, UK)* **116**, 1–9.

Krieg, P. A., and Melton, D. A. (1984). Functional messenger RNAs are produced by SP6 *in vitro* transcription of cloned cDNAs. *Nucleic Acids Res.* **12**, 7057–7070.

Lemaire, P., and Gurdon, J. B. (1994). A role for cytoplasmic determinants in mesoderm patterning: Cell-autonomous activation of the *goosecoid* and *Xwnt-8* genes along the dorsoventral axis of early *Xenopus* embryos. *Development (Cambridge, UK)* **120**, 1191–1199.

Lemaire, P., and Kodjabachian, L. (1996). The vertebrate organizer: Structure and molecules. *Trends Genet.* **12**, 525–531.

Lemaire, P., Garrett, N., and Gurdon, J. B. (1995). Expression cloning of *Siamois*, a *Xenopus* homeobox gene expressed in dorsal-vegetal cells of blastulae and able to induce a complete secondary axis. *Cell (Cambridge, Mass.)* **81**, 85–94.

Miller, J. R., and Moon, R. T. (1996). Signal transduction through β-catenin and specification of cell fate during embryogenesis. *Genes Dev.* **10**, 2527–2539.

Moody, S. A. (1987a). Fates of the blastomeres of the 16-cell stage *Xenopus* embryo. *Dev. Biol.* **119**, 560–578.

Moody, S. A. (1987b). Fates of the blastomeres of the 32-cell-stage *Xenopus* embryo. *Dev. Biol.* **122**, 300–319.

Moon, R. T., Brown, J. D., and Torres, M. (1997). WNTs modulate cell fate and behavior during vertebrate development. *Trends Genet.* **13**, 157–162.

Newport, J., and Kirschner, M. (1982). A major developmental transition in early *Xenopus* embryos: I. Characterization and timing of cellular changes at the midblastula stage. *Cell (Cambridge, Mass.)* **30**, 675–686.

Nieuwkoop, P. D. (1973). The "organization center" of the amphibian embryo: Its origin, spatial organization, and morphogenetic action. *Adv. Morphogenet.* **10**, 1–39.

Nieuwkoop, P. D., and Faber, J. (1967). "Normal Table of *Xenopus laevis* (Daudin)." North-Holland Publ., Amsterdam.

Sasai, Y., and De Robertis, E. M. (1997). Ectodermal patterning in vertebrate embryos. *Dev. Biol.* **182**, 5–20.

Sasai, Y., Lu, B., Steinbeisser, H., Geissert, D., Gont, L. K., and De Robertis, E. M. (1994). *Xenopus* chordin: A novel dorsalizing factor activated by organizer-specific homeobox genes. *Cell (Cambridge, Mass.)* **79**, 779–790.

Slack, J. M., Darlington, B. G., Heath, J. K., and Godsave, S. F. (1987). Mesoderm induction in early *Xenopus* embryos by heparin-binding growth factors. *Nature (London)* **326**, 197–200.

Smith, J. C. (1987). A mesoderm inducing factor is produced by a *Xenopus* cell line. *Development (Cambridge, UK)* **99**, 3–14.

Smith, J. C., Dale, L., and Slack, J. M. (1985). Cell lineage labels and region-specific markers in the analysis of inductive interactions. *J. Embryol. Exp. Morphol.* **89** (Suppl.), 317–331.

Smith, W. C., and Harland, R. M. (1992). Expression cloning of noggin, a new dorsalizing factor localized to the Spemann organizer in *Xenopus* embryos. *Cell (Cambridge, Mass.)* **70**, 829–840.

Spemann, H., and Mangold, H. (1924). Über Induktion von Embryonalanlagen durch Implantation artfremder Organisatoren. *Wilhelm Roux's Arch. Entwicklungsmech. Org.* **100**, 599–638.

Taira, M., Jamrich, M., Good, P. J., and Dawid, I. B. (1992). The LIM domain-containing homeo box gene *Xlim-1* is expressed specifically in the organizer region of *Xenopus* gastrula embryos. *Genes Dev.* **6**, 356–366.

Vogt, W. (1929). Gestaltungsanalyse am Amphibienkeim mit örtlicher Vitalfärbung. II. Teil. Gastrulation und Mesodermbildung bei Anuren und Urodelen. *Wilhelm Roux's Arch Entwicklungsmech Org.* **120**, 384–706.

Wilson, E. B. (1925). "The Cell in Development and Heredity," 3rd Ed. Macmillan, New York.

20

Early Events in Frog Blastomere Fate Determination

Steven A. Sullivan
Kathryn B. Moore
Sally A. Moody
Department of Anatomy and Cell Biology
Institute for Biomedical Sciences
The George Washington University
Washington, D.C. 20037

I. Introduction

The process by which embryonic cells acquire differentiated phenotypes initially begins during oogenesis. In many invertebrates it is during this period that RNAs and proteins are synthesized which after fertilization will establish the cardinal axes, the gametic cell line, germ layers, and tissues. Much less is known about this process in vertebrates, but where it has been studied, there also is a period after fertilization during which the zygotic genome is silent, and therefore all activities depend on molecules synthesized during oogenesis. This period of maternal orchestration of development can last a few to many hours in vertebrates, and depending on the length of the cell cycle, can encompass just the first few cleavages (as in mouse), or nearly the entire pregastrulation period (as in frog). In order for maternal molecules to establish differences in fate between cells that essentially are parcels of the oocyte cytoplasm, they

must be differentially localized, either in the oocyte or fertilized egg, or they must be differentially activated in certain lineages.

A first step in deciphering the maternal contribution to early cell fate decisions is to identify the fate differences between blastomeres. Marine invertebrates were the first to be fate mapped because unequal distributions in pigment granules could be followed visually through particular cell lineages. However, in most animals it is very difficult to optically monitor a molecular difference between individual cleavage cells. The cell divisions of some small, transparent embryos have been documented visually using differential interference contrast optics (Sulston *et al.*, 1983; Chapter 34 by Selwood and Hickford), but complete cell lineages for many species were not known until intracellular tracer molecules were developed to mark single cells (Weisblat *et al.*, 1978; see Chapter 11 by Stent).

Xenopus was the first vertebrate to which these

techniques were applied (Hirose and Jacobson, 1979; Jacobson and Hirose, 1981). Because *Xenopus* (and other amphibian) embryos are amenable to experimental perturbation throughout all stages of development (see Chapter 19 by Dawid; Moody, 1998), many experimental tests of fate specification, determination, and commitment are possible. In *Xenopus* there is the added advantage that naturally fertilized eggs often cleave in predictable, stereotypic patterns (Hirose and Jacobson, 1979; Jacobson and Hirose, 1981) and the three cardinal axes can be recognized by the first cleavage (Fig. 1). The animal-vegetal axis, which will be transformed into the anterior-posterior axis at gastrulation, can be identified in unfertilized eggs by the asymmetric distribution of pigment granules to the animal hemisphere. The dorsal-ventral axis can be identified shortly after fertilization by a change in the animal hemisphere pigmentation caused by cytoplasmic reorganizations in response to the entry of the sperm. Finally, the first cleavage furrow defines the midsagittal plane in naturally fertilized eggs, identifying left and right sides (Klein, 1987; Masho, 1990). These characteristics make it possible to identify specific lineages (Fig. 2), just as has been done in invertebrates with invariant cleavage patterns.

Several classic studies implied that fate commitment is not manifested in amphibians until just before gastrulation, suggesting that cleavage stage cells were naive and totipotent. Much later, lineage tracing techniques were used to identify whether there are regional differences or tissue restrictions in the ultimate fates of cleavage blastomeres (Hirose and Jacobson, 1979; Jacobson and Hirose, 1981; Jacobson, 1983; Masho and Kubota 1986; Dale and Slack, 1987a; Moody, 1987a,b; Takasaki, 1987; Masho, 1988; Moody and Kline, 1990). Putting these maps together into a lineage diagram (Fig. 2; Moody and Kline, 1990) demonstrated several important points about the progressive segregation of fate during the maternal phase of frog development. All cleavage stage blastomeres give rise to endoderm, mesoderm, and ectoderm (even neural) derivatives, albeit in widely varying amounts, demonstrating that there is no early fate restriction to germ layers in the intact embryo. Although not every blastomere contributed to every organ, no organ descended from a single progenitor even as late as the 32-cell embryo (Moody, 1987b). Quantitative mapping of specific cell phenotypes also has not revealed a monoclonal origin for several different kinds of neurons (Moody and Kersey, 1989; Moody, 1989; Huang and Moody, 1992, 1995, 1997). Thus, unlike some invertebrates, all cleavage stage blastomeres give rise to multiple germ layer derivatives and defined organs arise from more than one progenitor. However, blastomeres are not equivalent in fate. There is a regional organization to the parceling of fate, based on

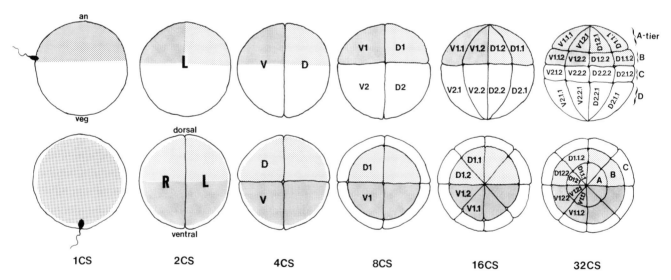

Figure 1 Diagrams of the stereotypic cleavages of first five cleavages in *Xenopus* from fertilization of the 1-cell egg (1CS) to the 32-cell stage (32CS). Top row is a left side (L) view with the animal pole (an) to the top, vegetal pole (veg) to the bottom and dorsal to the right. Bottom row is an animal pole view with dorsal to the top; the animal is facing the reader so right (R) and left (L) are reversed. Shading indicates the pigmented animal hemisphere, which becomes asymmetrically distributed after fertilization in response to entry of the sperm aster. Pigment condenses on the ventral side and becomes lighter on the dorsal side (cf. 1CS and 2CS). Letters within the blastomeres indicate the nomenclature designed by Jacobson and Hirose (1981; Hirose and Jacobson 1979). Basically, the first character of the blastomere name indicates whether the cell is dorsal (D) or ventral (V); this separation happens at the 4CS. The second character indicates whether the cell is in the animal (1) or vegetal (2) hemisphere; this separation happens at the 8CS. The third character indicates whether the daughters at the 16CS lie on the midline (1) or on the lateral meridian (2). The fourth character indicates whether the daughter at the 32CS lies at its respective pole (1) or toward the equator (2). Thus, D1.2.2 is a dorsal, animal, lateral, equatorial blastomere. Another common nomenclature used for the 32CS embryo is indicated by A, B, C, and D tiers (Nakamura and Kishiyama, 1971). The correspondence between these two systems is shown in Fig. 2.

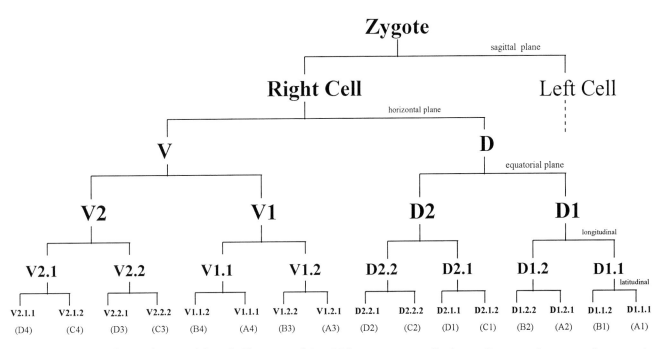

Figure 2 A complete lineage diagram of the right blastomere of the 2CS from a stereotypically cleaving *Xenopus* embryo, using the nomenclature of Jacobson and Hirose (1979, 1981). The same pattern would be repeated for the left cell. Nomenclature in parentheses at the 32CS (bottom row) is that of Nakamura and Kishiyama (1971). Planes of cleavage, with respect to the embryo body are indicated on the far right for each cell division.

blastomere position. Anterior epidermal structures descend from animal branches (D1, V1) and *not* from vegetal branches (D2, V2) of the 8-cell lineages (Figs. 1 and 2). The rostral central nervous system (CNS), including retina, descends almost exclusively from the D lineage at the 4-cell stage, and is further restricted to the D1 daughter at the next cleavage. Even within tissues there is a spatial segregation of the contributing blastomere lineages. For example, the central regions of somites derive from dorsal midline blastomeres, whereas dorsal and ventral regions of somites derive from lateral blastomeres (Moody and Kline, 1990; Moody *et al.*, 1996).

These detailed fate maps clearly demonstrate that at cleavage stages germ layer, organ, and tissue fates are not restricted to a single blast cell, as occurs in some invertebrates. But neither is cell fate random, as might be expected for uncommitted, totipotent progenitors, as described in zebrafish (Kimmel and Law, 1985) and mouse (Pedersen *et al.*, 1986). Spatial fate is segregated with each cell division. Furthermore, the fate of each identified blastomere is consistent, not random, when compared across a large population of embryos. Although the exact repertoire of progeny within a lineage is not identical between animals, as in several invertebrates, it is highly reproducible. Even the number of particular neurons produced by each blastomere is statistically distinct (Moody, 1989; Huang and Moody, 1992, 1995, 1997). This suggests that the general regionalization of the body plan of *Xenopus* emerges during cleavage stages by fate restrictions, but the determinant instructions for making specific organs occurs at later stages.

The consistency and predictability of the *Xenopus* fate map make it possible to examine the relative roles of maternal versus zygotic gene products, and intrinsic versus cell interactive mechanisms of fate determination. Lineage tracing alone, however, only documents the range of phenotypes that descend from a blastomere and can not provide any information as to how and when fate restrictions occur. Amphibians eggs are large enough that single cells can be marked and experimentally manipulated to ask these questions. The first experimental study of this type in frogs tested whether one blastomere of the 2-cell embryo can recreate the entire embryo or is restricted already to produce only a half embryo. When one blastomere was killed with a hot needle, the remaining blastomere made an organized half embryo (Roux, 1888), at least by superficial inspection. But, if instead the cells were completely separated by a ligature, that same blastomere could reconstitute the total body plan (McClendon, 1910). These pioneering experiments illustrate that exactly how the manipulation is done in amphibians can effect the outcome. Although not specifically addressed in these studies, they also remind us, that in amphibians pattern often will seem to be reconstituted or unperturbed if analyzed only at the gross morphological level. Lineage tracing has provided an increased level of resolution for assessing fate changes of single cells, and enables detection of subtle lineage changes after manipulations that heretofore had seemed to have no effect on fate. The frog embryo, which had been considered the archetype of "regulative" development, that is of having virtually no maternal influences on fate specification, now can be subjected to experimental tests for the potential influences of asymmetric distributions of maternal gene products and signaling pathways.

In this chapter we review two research projects that have sought to determine whether two different aspects of the fate expressed by a dorsal animal lineage are specified during the cleavage stages by maternal molecules. We define a specified fate as one that is still subject to alteration by experimental means such as cell culture or transplantation; this is also called a biased fate. In contrast, a determined fate is unalterable. The D1 lineage (Figs. 1 and 2) gives rise to the majority of the dorsal axial tissues, including the nervous system, notochord, and central third of the somites. First, we discuss the establishment of the dorsal axis in the frog, and the evidence that the dorsal axial fate of the D1 blastomere is specified by the activation of maternal molecules. This same lineage also gives rise to most of the retina, providing the opportunity to test the commitment to organ fate. In the second part of this chapter we demonstrate which signaling pathways (both maternal and zygotic) are necessary for descendants of animal and vegetal blastomeres to become progenitors of the retina. Together, these studies illustrate that the fate of frog blastomeres is specified progressively, first regionally by maternal molecules that set up the cardinal axes, then within organs and tissues by intercellular signaling that utilizes both maternal and zygotic molecules. Furthermore, they demonstrate that beginning at cleavage stages, *Xenopus* blastomeres are not totipotent, but exhibit different states of competence and bias in fate.

II. Blastomeres Are Specified to a Dorsal Fate Shortly after Fertilization

Amphibians have been the subjects of numerous studies to discover the molecular mechanisms by which dorsal-ventral (D/V) axial identities are achieved. Descriptions of the molecular and cellular asymmetries of the egg, of the initial cellular changes in response to fertilization and of the experimental potencies of individual blastomeres were initiated many years prior to the use of intracellular lineage tracers. As described below, asymmetries were detected, but until more recently remained unidentified at the molecular level. In this section we review approximately a decade's worth of studies that show that regionalization of the embryo at the earliest cleavage stages is essential for establishing dorsal fates in *Xenopus*.

A. Precleavage Regionalization of the Egg

Oogenesis endows the spawned *Xenopus* egg with animal-vegetal polarity. The animal hemisphere has a pigmented cortex and contains small yolk platelets, the maternal pronucleus, and a central region of yolk-poor cytoplasm, whereas the vegetal hemisphere is not pigmented and harbors medium and large yolk platelets. Around the virtual axis connecting animal and vegetal poles the egg is radially symmetric. The first important developmental hurdle facing the embryo is to superimpose bilateral symmetry on radial symmetry by establishing regional differences along the second, D/V axis, a process initiated by fertilization (Fig. 1).

A sperm can fertilize the egg at any point in the animal hemisphere; the sperm entry point (SEP), visible as a transient concentration of pigment, marks the future ventral side of the embryo (Fig. 1; Palacek *et al.*, 1978; Vincent and Gerhart, 1986). Sperm entry triggers the elaboration of two microtubular arrays in the cytoskeleton (Fig. 3). An extensive "sperm aster" radiates from the migrating male pronucleus in the animal hemisphere. Associated with it is an expanding front of thinner tubulin fibrils which may be involved in shifting the yolk-poor, RNA- and mitochondria-rich central cytoplasm to the prospective dorsal marginal region (Denegre and Danilchik, 1993; Brown *et al.*, 1993; Yost *et al.*, 1995). The second array consists of parallel tracks of microtubules located 3–4 µm beneath the surface of the vegetal hemisphere, at the interface between the cortical cytoplasm and the yolk mass (Elinson and Rowning, 1988). The subcortical microtubules align themselves parallel to the meridian defined by the SEP and the animal-vegetal axis, an orientation likely influenced by the sperm aster (Manes and Barbieri, 1977; Houliston and Elinson, 1991a,b). The subcortical array is required for the 30° rotation of the thin, yolk-free cortical layer with respect to the yolk mass, which occurs during the first cell cycle (Fig. 3).

Cortical rotation normally moves the vegetal cortex away from the SEP, creating a presumptive dorsal center (the more lightly pigmented grey crescent of classic embryology) on the opposite side (Figs. 1 and 3; Vincent and Gerhart, 1987; Gerhart *et al.*, 1989). Although this motion reliably predicts D/V axis orientation (and thus the location of the future dorsal side), the natural requirements for fertilization and microtubular array formation can be experimentally bypassed. A needle prick or electrical shock substituting for fertilization induces cortical rotation and its sequelae; the direction of rotation, although bearing no relation to the site of the activating stimulus, still reliably predicts the orientation of the D/V axis. Treatments that disrupt the subcortical

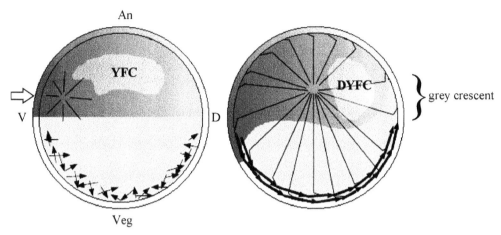

Figure 3 Cytoplasmic rearrangements during the first cell cycle. Cardinal axes: An, animal pole; Veg, vegetal pole; D, presumptive dorsal midline; V, presumptive ventral midline. Left, an embryo approximately a third of the way through the first cell cycle, prior to cortical rotation. The thickness of the cortex (outer concentric circle) has been exaggerated for illustrative purposes. The animal hemisphere is outwardly distinguishable from the yolky vegetal hemisphere by the presence of pigment granules (brown) in the cortex and underlying cytoplasm. Soon after fertilization these granules collect on the presumptive ventral-animal side around the sperm entry point (large arrowhead). The sperm pronucleus has migrated partly into the underlying cytoplasm and induced the formation of a microtubular sperm aster (starburst). The animal hemisphere is also notable for containing a central region of RNA- and mitochondria-rich yolk-free cytoplasm (YFC), itself a remnant of the oocyte nucleus (germinal vesicle). In the vegetal hemisphere, subcortical microtubules are not yet organized (jumbled arrows). Right, The same embryo at the end of the first cell cycle. Sperm aster-associated microtubules have become continuous with the cortical microtubules in the vegetal hemisphere, and contribute to aligned parallel microtubular bundles (aligned arrows) whose plus ends point in the direction of cortical rotation (Schroeder and Yost, 1996). Putative "motor" molecules associated with the parallel array are thought to drive the apparent 30° rotation of the cortex relative to the deep cytoplasm (Larabell *et al.*, 1996). This rotation also produces some mixing of subcortical animal and vegetal cytoplasm at the presumptive dorsal and ventral poles (Danilchik and Denegre, 1991). The juxtaposition of nonpigmented vegetal cortex and subcortical cytoplasm with underlying pigmented animal cytoplasm yields the "gray crescent" at the dorsal equator, long a focus of investigations into dorsal axis formation (e.g., Gerhart *et al.*, 1981). Deep cytoplasmic rearrangements in the animal hemisphere, perhaps resulting from the expansion of the sperm aster, shift the YCF towards the presumptive dorsal side, where it is referred to as the dorsal yolk-free cytoplasm (DYFC) (Herkovits and Ubbels, 1979; Danilchik and Denegre, 1991; Yost *et al.*, 1995).

array affect the expression of D/V polarity but can be overridden by gravity. For example, a high dose of UV irradiation at the vegetal half of fertilized eggs blocks the formation of the microtubular array and results in the complete suppression of dorsal development (Zisckind and Elinson, 1990). But, tilting these embryos during the first cleavage stage reorients their cytoplasm in response to gravity, and rescues the D/V axis (Chung and Malacinski, 1980; Scharf and Gerhart, 1980; Zisckind and Elinson, 1990). Conversely, under microgravity conditions of space flight, sperm entry is sufficient to cue cortical rotation and D/V axis formation (Ubbels, 1997). Taken together, these observations suggest that although sperm entry and gravity serve as natural orientation cues for the D/V axis, the truly crucial requirement for its establishment is that maternal cytoplasmic constituents be redistributed during the first cell cycle.

After cortical rotation and deep cytoplasmic shifts break radial symmetry during the first cell cycle, cell cleavages establish new regional differences along a prospective D/V axis. As shown in Fig. 1, the first two cell cleavages are stereotypically meridional and perpendicular to each other. In the majority of naturally fertilized embryos (Klein, 1987; Masho, 1990), the first cleavage furrow separates presumptive right and left sides along the mid-sagittal plane of the animal, and the second furrow delineates presumptive dorsal and ventral regions along the mid-horizontal plane. The third cleavage is equatorial (transverse) and separates the animal and vegetal masses. Maternal molecules drive these and subsequent events until the midblastula transition (MBT) of the 12th cell cycle, which marks the onset of zygotic transcription (Newport and Kirschner, 1982a,b).

B. Evidence for Early Specification of Dorsal Fate of Blastomeres

Many lines of evidence converge to support the idea that maternal determinants of dorsal fate are unequally distributed during the cytoplasmic rearrangements of the first cell cycle. Lineage tracing demonstrates that the dorsal axial tissues derive predominantly from the D lineage (Figs. 1 and 2; Moody and Kline, 1990). Studies of the regulative response of embryos subjected to blastomere ablation, of the differentiative capacities of transplanted and explanted blastomeres, and of the effects of transfer of cytoplasmic material all support a role for localized cytoplasmic factors in establishing the D/V axis.

When a stereotypically cleaving embryo (Fig. 1) is surgically halved at early cleavage stages and development examined at a later stage, the fates of the halves depend on the plane of surgery. Embryos divided into left and right halves by separation along the first cleavage plane develop into two small but otherwise normal tadpoles. Halves separated along the second cleavage plane develop into embryos with structural defects reflecting a D/V prepattern, for example, the dorsal half-embryo is missing the normal derivatives of the ventral half and vice versa (Kageura and Yamana, 1983; Cooke and Webber, 1985a,b). Blastomere ablations at subsequent stages define localized dorsal identities with increasing precision. Systematic deletion of blastomeres between the 8- and 32-cell stages (CS) shows that information (i.e., maternal determinants) essential for complete dorsal development is restricted largely to cells along the entire presumptive dorsal midline, that is, on either side of the first cleavage plane, opposite the SEP, from the animal to vegetal poles. The most profound dorsal axial defects arise from deletion of the dorsal equatorial blastomeres [D1.1.2 (B1) and D2.1.2 (C1)]; Figs. 1 and 2] (Kageura and Yamana, 1984; Gimlich, 1986; Takasaki, 1987; Gallagher *et al.*, 1991; Kageura, 1995).

Blastomere transplantation complements deletion studies by revealing states of commitment of donor and host regions. When transplanted to ventral sites, the various dorsal midline blastomeres (Fig. 1) often produce secondary dorsal axes, and lineage studies show that their descendants both contribute to mesodermal derivatives (e.g., prechordal mesoderm, notochord, somites) and induce dorsal respecification of ventral host cells (Kageura and Yamana, 1986; Gimlich, 1986; Takasaki and Konishi, 1989; Kageura, 1990; Gallagher *et al.*, 1991). In fact, even in embryos in which no secondary axis was formed, a significant proportion of the descendants of the transplanted blastomere were part of the host primary dorsal axial tissues (Gallagher *et al.*, 1991). Transplanted dorsal equatorial 32CS blastomeres (D1.1.2 and D2.1.2) produced the highest incidence of secondary axes (Kageura, 1990; Gallagher *et al.*, 1991), and transplantation of 32-64CS dorsal midline blastomeres into UV-ventralized embryos significantly rescues the D/V axis (Gimlich and Gerhart, 1984; Gimlich, 1986). Ventral-to-dorsal transplants exhibit stage-dependent effects. At the 8CS ventral animal blastomeres (V1) can substitute for dorsal animal (D1) blastomeres (Kageura and Yamana, 1986), but by the 32CS, the ventral equatorial granddaughters can neither substitute for their dorsal counterparts nor rescue UV-ventralized embryos (Gimlich and Gerhart, 1984; Gimlich, 1986; Takasaki and Konishi, 1989). These studies suggest a fate-defining event at the end of the 8CS.

Fine temporal resolution of early D/V commitment was achieved by Cardellini (1988), who rotated the animal halves of 8CS embryos 180° relative to their vegetal halves. He found that this manipulation often inverts the normal D/V polarity or induces twins if performed up to 2 minutes before the beginning of the fourth cleavage. Cardellini posited a prepatterning signal that passes from dorsal animal to subjacent vegetal regions prior to the end of the 8CS, inducing a vegetal dorsal signaling center (i.e., the Nieuwkoop center), which thereafter is dominant in vegetal-animal combinations. Interestingly, prior to these experiments, Kageura and Yamana (1986) reported qualitatively similar results for 8CS animal hemispheres isochronically transplanted in inverted orientations. In addition, dorsal animal blastomeres are competent to respond to activin by the end of the 8CS (Kinoshita *et al.*, 1993). Together these studies suggest that dorsal midline blastomeres begin to be biased to express dorsal fates near the end of the 8CS, and that a dorsal center is localized to the dorsal equatorial cells [both animal (D1.1.2) and vegetal (D2.1.2) to the equatorial cleavage plane] shortly thereafter.

Culturing isolated blastomeres in the absence of growth factors or influential neighbors is another way to assess their developmental commitment. Because frog blastomeres have their own internal yolk supply, they can develop for a few days in a simple, defined, salt medium. The elongation of the explant (Symes and Smith, 1987), which mimics the convergent-extension movements of dorsal axial tissues during gastrulation (Keller and Danilchik, 1988), and the differentiation of dorsal structures, using tissue-specific markers, can be assayed. These criteria demonstrated that the 16CS dorsal midline, animal blastomeres (D1.1) can autonomously express dorsal fates in the absence of exogenous growth factors or their vegetal neighbors. But, neither the midline ventral animal blastomeres (V1.1) nor the dorsal vegetal blastomeres (D2.1) have this ability (Gallagher *et al.*, 1991). However, the mother of D1.1 (D1, Figs. 1 and 2) does not differentiate into dorsal structures under the same culture conditions (Hainski and Moody, 1996), suggesting that the ability to autonomously express dorsal axial fate is acquired by dorsal animal blastomeres beginning at the 16CS. These data confirm observations by Cardellini (1988) that in these blastomeres important determinative events occur during a narrow window between the 8- and 16CS.

Vegetal signaling also has an important role in the formation of the dorsal axis (see Chapter 21 by Kessler). The 16CS dorsal vegetal blastomeres can induce dorsal mesoderm in animal cap assays (Dale and Slack, 1987b), and 32-64CS dorsal vegetal blastomeres act as a Nieuwkoop center to induce secondary axes and rescue UV-irradiated embryos (Gimlich and Gerhart, 1984; Gimlich, 1986). It is likely that these vegetal cells signal the overlying animal blastomeres, and reinforce their commitment to a dorsal fate. To test this, we cultured different animal blastomeres together with D2.1, the progenitor of the Nieuwkoop center (Moody, 1987a; Bauer *et al.*, 1994). Coculture of D1 blastomeres with D2.1 produced dorsal axis tissue and co-culture of D1.1 with D2.1 enhanced the frequency of explant elongation (Hainski and Moody, 1996). Lineage tracing demonstrated that the notochord in the D1.1/D2.1 explants derived predominantly from the animal blastomere, not from the signaling vegetal blastomere. The ventral animal blastomere (V1.1) also responded to D2.1 signals, albeit to a lesser degree. These results suggest that vegetal-to-animal signaling on the dorsal side circa the 16CS strengthens the commitment of the D1 lineage to a dorsal program. Furthermore, they are consistent with the deletion and transplantation data in demonstrating the requirement for both animal and vegetal components in establishing the dorsal axis. In fact, we chanced on embryos in which the third cleavage furrow is oblique instead of horizontal, and dips into the vegetal equatorial region. This creates D1 blastomeres with an inclusion of D2 equatorial cytoplasm. These blastomeres elongated and formed dorsal tissues at high frequency when cultured in isolation (Hainski and Moody, 1996), further confirming that the combination of animal and vegetal components (or signals) is essential.

C. Molecular Asymmetry in the *Xenopus* Egg and Embryo

What is the molecular basis for these numerous instances of cell-cell interactions and early steps in dorsal fate specification? A number of asymmetries in the early *Xenopus* embryo have been described, including gradients of yolk platelets (Neff *et al.*, 1984), metabolic activity (Black, 1989), and animal-vegetal or D/V differences in protein (Smith and Knowland, 1984; Shiokawa *et al.*, 1984; Smith, 1986; Miyata *et al.*, 1987; Klein and King, 1988; Capco and Mecca, 1988) and RNA (Wakahara, 1981; Phillips, 1982, 1985; King and Barklis, 1985) profiles. Gross cytoplasmic differences also have been extensively described, especially a dorsal yolk-free, mitochondria-, and RNA-rich region that contains the oocyte nucleoplasm (Fig. 3; Herkovits and Ubbels, 1979; Ubbels *et al.*, 1983; Imoh, 1984; Danilchik and Denegre, 1991; Denegre and Danilchik, 1993; Brown *et al.*, 1993; Yost *et al.*, 1995). However, further characterization is frustrated by the enormous difficulty in obtaining enough tissue to generate the microgram amounts of protein needed for sequencing.

In more recent years these studies have been supplemented by functional tests of the abilities of embryonic constituents to specify axial tissue, with the ultimate goal being to elucidate the molecular components of axis specification. This goal has been approached from the top down by testing cellular components first and then their fractions, and from the bottom up by assessing known molecules for roles in axis formation. D/V axis specification is likely to involve molecules localized both physically (e.g., factors distributed along the animal-vegetal axis during oogenesis or translocated during the cytoplasmic reorganizations of the first cell cycle) and functionally (e.g., more or less ubiquitous factors differentially activated in different regions) within the early embryo. Although the number of maternal *Xenopus* RNAs and proteins identified as being localized to different embryonic regions has increased rapidly in the 1990s (Table 1), the relevance of differential processing to early *Xenopus* development *in vivo* is still largely inferential (however, see Chapter 21 by Kessler). Moreover, many gene products have both maternal and zygotic transcripts whose respective roles in dorsal axis formation are far from clear (Table I).

Observations of animal-vegetal differences in the egg prompted the first molecular screens for differentially localized mRNAs along that axis (King and Barklis, 1985; Rebagliati *et al.*, 1985). The most extensively studied vegetally localized molecule is Vg1, whose mRNA is tightly localized to the vegetal pole cortex during oogenesis and is slowly released into the vegetal hemisphere during oocyte maturation and the first cell cycle (Rebagliati *et al.*, 1985; Weeks and Melton, 1987). The third cleavage plane effectively segregates the Vg1 protein to the vegetal half of the embryo. Vg1, which encodes a protein of the transforming growth factor β (TGFβ) superfamily (Weeks and Melton, 1987; Dale *et al.*, 1993), is an example of a maternal factor that may be localized both literally and functionally by region-specific posttranslational processing within the vegetal hemisphere (Chapter 21 by Kessler). In addition, there are several other maternal mRNAs localized to the vegetal hemisphere that appear to have a role in some aspect of dorsal axial tissue formation (Table I). A family of noncoding mRNAs called Xlsirts is required to anchor Vg1 RNA to the vegetal cortex during oogenesis (Kloc and Etkin, 1994). The mRNA of Xwnt-11, a member of the wingless-int (Wnt) family of secreted proteins (reviewed in Moon *et al.*, 1997), can rescue partial posterior dorsal axes in UV-treated embryos (Ku and Melton, 1993). Maternal members of the TBox family of transcription factors—*Xombi* (Lustig *et al.*, 1996), *Antipodean* (Stennard *et al.*, 1996), *VegT* (Zhang and King, 1996), and *Brat* (Horb and Thomsen, 1997)—bear highly similar sequences and expression patterns, suggesting that they represent the same gene. Overexpression of TBox mRNAs leads to ectopic mesoderm and endoderm formation, whereas blocking their function inhibits mesoderm formation. Other vegetally localized maternal transcripts either do not appear to be involved in D/V axis specification or have not yet been rigorously tested for such a role (Table I).

Table I
Maternal Molecules in Cleavage-Stage *Xenopus* Embryos

	Putative product[b]	Notes	References
Animal → vegetal			
An1a,b	NP w/ubiquitin-like sequences		Linnen *et al.* (1993)
An2	Mitochondrial ATPase		Weeks and Melton (1987)
An3	RNA helicase		Gururajan *et al.* (1991)
βTrCP 2.5	NP w/b-transducin repeats		Hudson *et al.* (1996)
bFGF, eFGF, FGF-9	Fibroblast growth factor-like	eFGF and FGF-9 are probably secreted	Reviewed in Isaacs (1997)
FGFr1	Type-1 FGF receptor tyrosine kinase		Reviewed in Isaacs (1997)
HNF4α[a]	Zn-finger transcription factor	Enters nucleus by 64CS; gene is activin-inducible	Holewa *et al.* (1996)
HNF4β	Zn-finger transcription factor	Protein enters nucleus by 64CS	Holewa *et al.* (1997)
An4a,b	NP		Hudson *et al.* (1996)
PABP	Cytoplasmic RNA-binding protein	Protein binds to poly(A) tracts	Schroeder and Yost, (1996)
x121	Transcription factor?		Kloc *et al.* (1991)
XFLIP	FKhd/winged helix transcription factor		Sullivan *et al.*, in preparation
xlan4	NP		Reddy *et al.* (1992)
XGβ1	G-protein subunit		Devic *et al.* (1996)
XLPOU-60	POU-domain transcription factor		Whitfield *et al.* (1993)
Xrel	Transcription factor	Protein enters nucleus prior to MBT	Bearer (1994)
Xtcf-3	Transcription factor	Protein translocated to nucleus with β-catenin	Molenaar *et al.* (1996)
Xwnt-8b	Secreted Wnt signal	Induces complete axes	Cui *et al.* (1995)
Vegetal → animal			
Antipodean/Brat/etc.	T-box transcription factor(s)	Meso- and endoderm inducer	Reviewed in Stennard *et al.* (1997)
βTrCP 3.5, 4.9	NP w/b-transducin repeats		Hudson *et al.* (1996)
Vg1	Secreted TGFb signal	Modified protein induces complete axes	Weeks and Melton (1987); Thomsen and Melton (1993); Forristall *et al.* (1995)
Xcat-2	Nanos-like RNA binding protein	Localized to germ plasm	Forristall *et al.* (1995)
Xlsirts	Noncoding RNA repeats	Required for cortical Vg1 RNA localization	Kloc *et al.* (1993); Kloc and Etkin (1994)
Xwnt-11	Secreted Wnt signal	Partial (posterior) axis inducer	Ku and Melton (1993)
Dorsal → ventral			
β-catenin[a]	Wnt signal transducer	Required maternally for dorsal axis formation Protein dorsally enriched at 2 CS and nuclear at 16CS	Wylie *et al.* (1996); Larabell *et al.* (1997); Rowning *et al.* (1997)
Xa5B6[a]	Unidentified antibody target		Suzuki *et al.* (1991)
Ventral → dorsal			
Xwnt-8b	Secreted Wnt signal	Includes complete axes	Sullivan *et al.*, in preparation
Selected axis-influencing maternal molecules with ubiquitous or uncharacterized maternal distribution			
activin[a]	Secreted TGFβ-like signal	Partial axis inducer; maternal mRNA undetectable	Asashima *et al.* (1991); Smith and Harland (1991)
BMP-2,−4,−7	Secreted TGFβ-like signals	Ventralizing signals	Reviewed in Heasman (1997)
Follistatin	Secreted	Activin and BMP-4 antagonist	Hemmati-Brivanlou *et al.* (1994b)
Goosecoid	Homeobox transcription factor	Axis inducer	Cho *et al.* (1991)
Noggin	Secreted	Induces complete axes; can also directly neuralize	Smith and Harland (1992)
TGFβ receptors	Activin/BMP receptor serine kinases	Dominant negative forms can suppress or induce axes	Harland (1994); New *et al.* (1997)
Xdsh	Cytoplasmic Xgsk3 antagonist	Partial axis inducer	Soko *et al.* (1995)
Xgsk3	Cytoplasmic β-catenin antagonist	Dominant negative form is axis-inducing	Dominguez *et al.* (1995)
Xwnt-5A	Secreted Wnt signal antagonist?	Modulates morphogenetic movements; not an axis inducer; can act via G-protein pathway	Moon *et al.* (1993); Torres *et al.* (1996); Slusarski *et al.* (1997)

[a] Protein; all other data are for mRNA.

[b] NP, novel protein.

An1, 2, and *3*, the first animal hemisphere-specific mRNAs to be cloned (Rebagliati *et al.*, 1985), encode a ubiquitin-like protein, a mitochondrial ATPase subunit, and an RNA helicase, respectively (Dagle *et al.*, 1990; Linnen *et al.*, 1993). Of these, An3 may have a role in regional specification of the embryo, since in *Drosophila* the localized *vasa* RNA helicase contributes to establishment of posterior identity by regulating translation (Hay *et al.*, 1988). A maternal member of the Wnt family, Xwnt8b, also is localized to the animal hemisphere and induces complete dorsal axes if supplied before the onset of zygotic transcription (Cui *et al.*, 1995). Since other Xwnts act as modifiers of mesodermal patterning rather than as mesoderm inducers themselves (Ku and Melton, 1993; Sokol *et al.*, 1995; Moon *et al.*, 1997), Xwnt8b may modify the response of the animal hemisphere along the future D/V axis to a Vg1-like vegetally derived mesoderm-inducing signal (Cui *et al.*, 1996). In addition, a diverse group of molecules localized to the animal hemisphere or exhibiting an animal-to-vegetal RNA gradient have not been tested for dorsalizing activity (Table I). One which may prove interesting on further investigation is a *Xenopus* homologue of the *Drosophila* gene *dorsal*, a transcription factor required for D/V axis formation. The *Xenopus* protein, Xrel-1, is localized mainly in the animal blastomeres (both dorsal and ventral) by the 8CS (Bearer, 1994). Given the regional changes in dorsal fate commitment that occurs around the 16CS, it is intriguing that the Xrel-1 protein moves into the nucleus beginning at the 32CS. Another potentially relevant animal hemisphere mRNA encodes a poly(A) binding protein (PABP) (Schroeder and Yost, 1996) that is thought to enhance the translatability of messages to which it binds (reviewed in Jacobson, 1996). Lastly, we have determined recently that maternal mRNA encoding XFLIP, a forkhead/winged helix transcription factor that induces elongation in animal caps, is localized to the animal hemisphere in cleavage-stage embryos (Sullivan *et al.*, 1998).

In contrast to the numerous animal-vegetal molecular asymmetries cited above, to date few maternal molecules have displayed asymmetries along the presumptive D/V axis (Table 1). Suzuki *et al.* (1991) raised a monoclonal antibody to a protein that accumulates in the equatorial zone of the dorsal half at the 2CS, and in the tier-3 dorsal blastomeres by the 32CS, but no further characterization of this protein has been reported. Much recent work has focused on maternal β-catenin, a cytoplasmic protein initially noted for its role in cell adhesion (McCrea *et al.*, 1991, 1993; Heasman, 1997). Although present throughout the embryo, by the end of cortical rotation it accumulates preferentially within a 60°-90° swath on the future dorsal side, in association with subcortical microtubules (Rowning *et al.*, 1997). By the 16CS, β-catenin begins to move into cell nuclei of both animal and vegetal dorsal blastomeres in associ-

ation with the Xtcf-3 transcription factor. One of the zygotic gene targets of this complex is *siamois* (Molenaar *et al.*, 1996; Carnac *et al.*, 1996; Larabell *et al.*, 1997; Brannon *et al.*, 1997; see Chapter 21 by Kessler), which in turn positively regulates the transcription of organizer-specific genes. Dorsal accumulation and nuclear translocation of β-catenin can be increased or inhibited by ectopic effectors (e.g., Xwnt-8) or inhibitors (e.g., Xgsk) of the Wnt pathway (Larabell *et al.*, 1997), and depletion of maternal β-catenin completely blocks dorsal axis formation (Wylie *et al.*, 1996). Again, considering the acquisition of autonomous dorsal fate by midline dorsal blastomeres at the 16CS (Gallagher *et al.*, 1991; Hainski and Moody, 1992, 1996), the regulation of β-catenin function at these stages is important to investigate.

Several other maternal molecules exhibit strong dorsalizing ability when overexpressed in cleavage stage blastomeres or in animal cap assays (Table I). These include maternal activin protein, a potent mesoderm inducer (Thomsen *et al.*, 1990); noggin mRNA, which encodes a novel secreted protein that can directly induce neural tissue (Smith and Harland, 1992); goosecoid mRNA, a partial axis inducer whose zygotic transcript is expressed in Spemann's organizer (Cho *et al.*, 1991); follistatin mRNA, an activin antagonist with direct neuralizing ability (Hemmati-Brivanlou *et al.*, 1994b); and several receptor mRNAs that are members of the TGFβ and Xwnt signal transduction pathways (reviewed in Heasman, 1997). The mRNAs all appear to play important roles in the zygotic aspects of dorsal axis formation, but whether they have a maternal role is uncertain. All of these mRNAs and the activin protein (Asashima *et al.*, 1991; Rebagliati and Dawid, 1993) are present at cleavage stages, but either are not localized or exist at such low concentrations that localization data by whole mount *in situ* hybridization techniques have not been obtainable. We tested whether several of these maternal transcripts are distributed asymmetrically along the animal D/V axis by dissecting D1.1 and V1.1 blastomeres, extracting their RNA and performing reverse transcriptase-polymerase chain reaction (RT-PCR) analyses. We cannot detect significant D/V blastomere differences in goosecoid, follistatin, an activin receptor (ALK4; Hudson *et al.*, 1996), noggin, or β-catenin mRNAs. Of course, significant differences in protein localization or activity would not be revealed by this assay (e.g., for β-catenin). But, these results suggest that the localization per se of these mRNAs plays little, if any, role in early dorsal fate specification. However, we do find that maternal Xwnt-8b mRNA is localized to ventral animal blastomeres, suggesting the possibility that this quadrant may have unique, intrinsic signaling properties of its own (P. Pandur, S. Sullivan, and S. A. Moody, in preparation). Elsewhere it has been shown that ventral animal blastomeres are required during cleavage stages to "program" the attenuation of dorsal zygotic expression of *siamois* and *Xnr3* that occurs after stage 8.5 (Ding *et al.* 1998). We are currently investigating the possibility that Xwnt-8b signaling mediates this phenomenon.

D. Axis-Inducing Cytoplasmic Asymmetries

Is there evidence for regional differences in axis-inducing activities? The idea that vegetally localized cytoplasmic factors are shifted towards the future dorsal side after fertilization inspired direct tests of the dorsalizing ability of early embryonic cytoplasm (reviewed in Elinson and Kao, 1989). More recently, ectopic injection (Yuge *et al.*. 1990; Fujisue *et al.*, 1993), transplantation (Holowacz and Elinson, 1993; Kageura, 1997), and deletion (Sakai, 1996; Kikkawa *et al.*, 1996) of cortical cytoplasm indicates that a UV-insensitive factor(s) is required for D/V axis formation and is restricted initially to the vegetal pole of the uncleaved egg. By the first cleavage (2CS) this cortical factor is distributed in a narrow swath along the entire animal-vegetal meridian opposite the SEP (Fig. 4). Although maximum axis-

inducing ability resides in the dorsal-vegetal cortex through the 32CS, significant activity also is detected in the dorsal-animal cortex. In addition, the competence of an 8CS host to respond to transplanted dorsal-vegetal cortex has been mapped (Kageura, 1997; Darras *et al.*, 1997). Competence extends along the entire animal-vegetal axis but is strongest at the equatorial zone (Fig. 4). These data suggest that a deep cytoplasmic cofactor is distributed in a vegetal-to-animal gradient, which is required to interact with the cortical cofactor to form the "actual" dorsal determinant (Kageura, 1997; Fig. 4). These studies indicate that there are at least two maternal factors required for establishing the dorsal axis, and both are found in varying concentrations or activities along the animal-vegetal axis after cortical rotation.

Independently, our laboratory investigated the molecular basis of the dorsal fate prepatterning of animal blastomeres, as revealed by single cell deletion, transplantation, and culture experiments (Section IIB; Gallagher *et al.*, 1991). Direct transfer of cytoplasm from 16CS dorsal animal (D1.1) blastomeres to ventral blastomeres had a minimal dorsalizing effect, but mi-

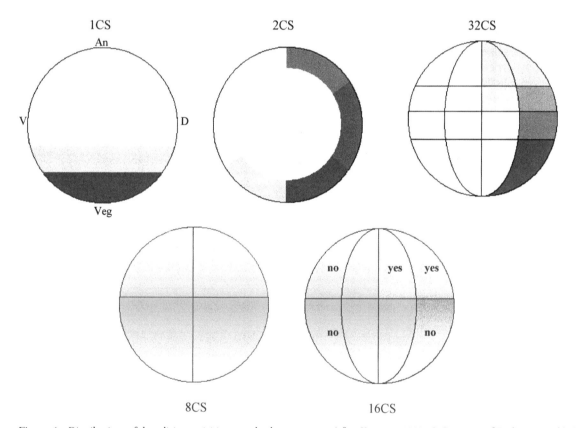

Figure 4 Distributions of dorsalizing activities at early cleavage stages (after Kageura, 1997, © Company of Biologists, Ltd.). (Top) Red indicates the distribution of cortical dorsalizing activity, derived from cortical transplantation at either the 1CS, 2CS , or 32CS into 8CS ventral vegetal hosts. Density of stippling is proportional to the degree of activity. (Bottom left) Green indicates the distribution in an 8CS host of competence to form secondary axes in response to transplants of 2CS vegetal pole cortex. This competence gradient has been posited to represent the distribution of a deep cytoplasmic "core cofactor" with which the cortical factor (red, top row) interacts to establish actual dorsal determinants. Competence extends to both animal and vegetal poles, but is most active at the equator. (Bottom right) Brown indicates the theoretical distribution of actual dorsal determinants resulting from the interaction of cortical and core factors, extrapolated for the 16CS. Superimposed is the ability of total RNA from 16CS blastomeres to induce secondary axes (yes and no) when injected into ventral vegetal hosts (Hainski and Moody, 1992, 1996).

croinjection of total RNAs purified from these blastomeres induced ectopic D/V axes and rescued UV-ventralized embryos (Hainski and Moody, 1992). An axis-inducing activity of total RNAs from the dorsal animal sibling (D1.2) blastomere also was detected (Hainski and Moody, 1996), but there was no activity from total RNA extracted from either ventral animal (V1.1), ventral vegetal (V2.1) or dorsal vegetal blastomeres (D2.1) (Fig. 4; Hainski and Moody, 1992). These studies were the first to identify a dorsal-*animal* dorsalizing activity. Although the majority of dorsal determinant studies have focused on axis-inducing activities in the dorsal-vegetal quadrant, our findings are consistent with the model that the commingling of both vegetal-cortical and deep cytoplasmic factors are necessary for axis-inducing activity (Fig. 4). It seems paradoxical that the total RNA fraction from the D2.1 blastomere did not exhibit dorsalizing activities in these studies, since both putative cofactors are at high activity levels in this blastomere (Fig. 4). Either our methods of RNA extraction inactivated or lost the required molecule(s), or in this blastomere one or both factors already is in protein, not mRNA, form.

These results suggested that the cytoplasmic reorganizations resulting from cortical rotation simply place maternal RNAs that have a role in D/V axis formation in the dorsal animal quadrant. If true, then RNA from earlier stages also should exhibit this activity. Surprisingly, total RNA from D1 blastomeres had no dorsalizing activity (Hainski and Moody, 1996), indicating that the RNAs are masked prior to the 16CS (since there is virtually no zygotic transcription; Newport and Kirschner, 1982a,b). It is notable that the transition between the 8CS and 16CS is the same window during which vegetal-to-animal signaling dominance commences (Cardellini, 1988; Jones and Woodland, 1987), and about the time when β-catenin (Larabell *et al.*, 1997) and Xrel-1 (Bearer, 1994) translocate into the nuclei of dorsal animal blastomeres. In an attempt to unmask the activity, we cultured D1 blastomeres in a few growth factors involved in mesoderm induction (Hainski and Moody, 1996). Only the addition of activin (PIF medium from S. Sokol; Harvard Medical School) to the culture for the first 2 hours (between the 8- and 256CS) caused D1 explants to subsequently elongate and express differentiated dorsal axial tissues. In addition, total RNA extracted from these blastomere explants at the end of the 2 hour incubation acquired dorsal-axis-inducing activity. In contrast, although ventral blastomeres also elongated when exposed to activin during the first 2 hours of culture, total RNA extracted from them did not have axis-inducing activity. These results imply that an activin-like signal around the 8-16CS period unmasks maternal RNAs in the dorsal animal hemisphere that then function in dorsal axis formation. There are two obvious sources of this signal. Maternal activin protein is present in the cleavage embryo

(Asashima *et al.*, 1991; Rebagliati and Dawid, 1993), but is present at such a low concentration that it is not known if it is locally concentrated. Maternal Vg1 RNA and its inactive proprotein are both present throughout the vegetal hemisphere, but the active form of the protein may only be present in the dorsal vegetal blastomeres (Chapter 21 by Kessler), and thus may be a region-specific signal to dorsal animal neighbors.

E. Mechanisms of Dorsal Determinant Activation

Since the early embryo is transcriptionally quiescent (Newport and Kirschner, 1982a,b), the regional activation of dorsal axis-inducing ability likely results from a combination of initial localization of maternal molecules during oogenesis and cortical rotation, and subsequent localization of mRNA and/or protein processing (under which rubric we include all manner of splicing, augmentation, proteolysis, dimerization, nuclear translocation, etc.). The latter would result in site-specific enhancement of translation or protein function, for example, as posited for Vg1 protein (Chapter 21 by Kessler).

Translational regulation is a well-established mode of developmental control of cell fate in several invertebrates, where it complements early polarized transport of maternal molecules (reviewed in Wickens *et al.*, 1996; Hake and Richter, 1997). Two of the most common forms of translational regulation are mRNA masking by bound proteins, which negatively regulates translation (reviewed in Hake and Richter, 1997), and cytoplasmic poly(A) tail elongation, which is mediated by cis- and trans-acting factors that generally result in translational enhancement (reviewed in Colgan and Manley, 1997). It is unlikely that differences seen in the dorsal axis-inducing ability of D1.1 and V1.1 RNA are due to adherent masking proteins, since blastomere RNA isolated for injection assays is deproteinated during preparation. On the other hand, if localized cytoplasmic polyadenylation between the 8 and 16CS is responsible for the onset of axis-inducing ability, then *in vitro* adenylation of D1 RNA should allow it to prematurely acquire dorsal-axis inducing activity. Indeed, this is the case (Sullivan and Moody, 1996). We have characterized RNAs from D1 further by separating them into poly(A)+ (translatable mRNAs) and poly(A)− (ribosomal, transfer, and untranslatable mRNAs) fractions and assaying the axis-inducing abilities (Sullivan and Moody, 1996; P. Pandur, S. Sullivan, and S. A. Moody, in preparation). The polyA+ (translatable mRNAs) fraction of D1 has axis-inducing ability, suggesting that the mRNA species responsible for this activity are present but represent too small a percentage of the total RNA at the 8CS to function efficiently in the ectopic injection assays used (Hainski and Moody, 1996). Perhaps total RNA isolated from D1.1 and D1.2 daughters can induce dorsal axes because the responsible mRNA(s) represent a higher proportion of the translatable pool by the 16CS. Consis-tent with this explanation, *in vitro* adenylation of the poly(A)− fraction from the D1 blastomere imparts axis-inducing activity to that fraction. The simplest explanation is that maternal RNA inherited by the D1 lineage is subjected to differential cytoplasmic adenylation between the 8 and 16CS, which moves one or more mRNA species from the inefficiently translated, poly(A)− pool into the efficiently translated poly(A)+ pool. If true, this would indicate that regional translational regulation of a maternal RNA is critical for dorsal fate specification.

Another approach is to start with known mRNAs that might play a role in dorsal axis specification (e.g., Table I) and examine their adenylation status during development. One drawback to this approach is the large number of maternal molecules available for investigation. However, several *Xenopus* maternal mRNAs that are developmentally regulated contain a polyuridine cytoplasmic polyadenylation element (CPE) in their 3' untranslated regions, which is necessary for the cleavage-stage extension of their poly(A) tails (Simon *et al.*, 1992; Simon and Richter, 1994). We and others have been investigating the adenylation status of developmentally relevant sequences containing such CPEs. For example, Richter's group has shown that one of the CPE-containing activin receptor mRNAs is polyadenylated during cleavage and blastula stages (Simon *et al.*, 1996). We have sequenced the 3' untranslated region of β-catenin and found that it, too, contains a potential CPE.

Moreover, we find that endogenous β-catenin mRNA is polyadenylated between the 2CS and 64CS (P. Pandur, S. Sullivan, and S. A. Moody, in preparation). We are currently trying to determine if this adenylation occurs selectively on the dorsal side. Continued studies using this approach may prove fruitful in elucidating the means by which mRNAs that are not physically localized may function in the regional fate specification of the early embryo.

F. Summary

In this section we have reviewed cytoplasmic rearrangments and positioning of maternal factors that may be involved in specifying the dorsal axis during the early cleavage stages. The symmetry-breaking events of the first cell cycle are analogous to superimposing one half-filled sphere on another at a 90° angle, effectively creating four different quadrants. *In vivo*, animal-vegetal differences, for example, opposing animal-vegetal gradients of Xwnt-8b or XFLIP and Vg1, are supplemented by dorsal-ventral gradients generated during the first cell cycle by cortical rotation (Fig. 5). Conceptually, at least, this is sufficient to generate presumptive dorsal-animal, dorsal-vegetal, ventral-animal, and ventral-vegetal quadrants with distinct molecular profiles as early as the 2CS. This early molecular loading of quadrants, in itself and by way of intercellular signaling arising from it, could be the basis for differences in dorsal axis-generating ability observed in blastomeres and their RNAs only a few cell cycles later. Specifically, interaction of cortical factors, translocated to the animal hemisphere during the first cell cycle, with cytoplasmic factors localized

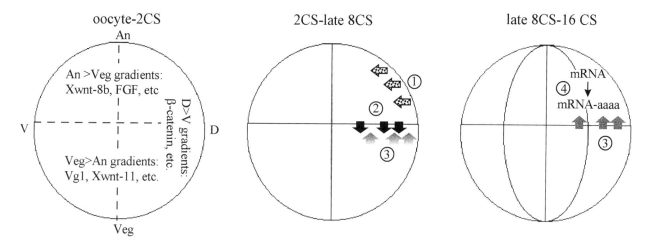

Figure 5 A model of early D/V axis-establishing events in *Xenopus*. (*Left*) Molecular gradients along the animal–vegetal and presumptive D/V axes are established during oogenesis and cortical rotation, creating four qualitatively different quadrants (see also Table I). (*Middle*) (1) The interaction of factors uniquely combined in the dorsal-animal quadrant as a result of these early events leads to (2) signaling in the animal-to-vegetal direction during the period between cortical rotation and the late 8CS, thereby establishing functional dorsal-ventral polarity (3) in the vegetal half (Cardellini, 1988). (*Right*) During the late 8CS and 16CS a vegetal-to-animal signal (3), presumably resulting from the previous reciprocal signal (2), imparts functional dorsal identity to the D1 lineage. This is manifested in autonomous dorsal development *in vitro* (Gallagher *et al.*, 1991; Hainski and Moody 1996) and in the ability of D1.1 RNA to induce ectopic dorsal axes (Hainski and Moody, 1992, 1996). The latter property may in turn result from region-specific cytoplasmic polyadenylation (4) in response to the vegetal signal.

to the animal half (or confined mainly to the dorsal-animal quadrant, e.g., the dorsal yolk-free cytoplasm), could lead to animal-to-vegetal signaling during the first few cell cycles (Fig. 5). Reception of these signals by vegetally localized factors, in turn, could lead to reciprocal vegetal-to-animal signaling resulting in specification of dorsal-animal fates, as manifested in the activation (via cytoplasmic polyadenylation) of dorsalizing mRNAs in the D1 lineage circa the 16 CS (Fig. 5).

Although much progress has been made in recent years, numerous aspects of early axis specification remain to be addressed. What is the dorsalizing cortical factor(s), and how does it come to be distributed further toward the animal pole than one would expect from cortical rotation alone? When and how is the gradient of deep cytoplasmic cofactor(s) established, and what is its molecular identity? How much axis-specifying cross talk actually goes on between blastomeres during the early cleavage stages, and by what signaling pathways? What early roles do differentially processed RNAs play *in vivo* in establishing the D/V axis, and how is processing regulated? Which of the burgeoning number of identified maternal molecules are relevant to specification of the D/V axis? As Heasman (1997) cogently reminds us, drastic yet common experimental interventions such as cell transplantation or injection of supraphysiological amounts of RNA run the risk of: (a) activating molecular pathways that never function in undisturbed embryos; (b) merely mimicking the actual, unidentified, endogenous positive and negative regulators of dorsal development; and (c) acting at any time during the long developmental interval between their application and assessment of their effects. Such considerations thwart the easy identification of physiologically important dorsalizing factors by typical experimental means, and part of the challenge of answering the questions posed above will be to tease out the truly relevant threads of data from the web of interesting artifacts.

III. The Initial Steps in Blastomere Specification to a Retinal Fate

The dorsal animal lineage (D1) comprises the major progenitors of dorsal axial tissues, including the CNS (Hirose and Jacobson, 1979; Jacobson and Hirose, 1981; Moody, 1989; Huang and Moody, 1992). In the previous section we demonstrated that this blastomere receives early maternal instructions regarding its axial fate. Is this lineage also specified early to produce its neuronal descendants? The retina has been used to investigate this question because quantitative fate maps are available for both the entire retina (Huang and Moody, 1993) and for several specific, differentiated cell types within it (Huang and Moody, 1995, 1997). This

allows us to precisely monitor retinal fate changes after experimental manipulations. As discussed in detail elsewhere (Chapter 16 by Carthew *et al.*; Chapter 23 by Perron and Harris; Chapter 31 by Adler and Belecky-Adams), specific retinal cell fate appears to be established close to the terminal mitosis of a cell. In the *Xenopus* optic cup mitotic cells give rise to numerous neuronal and glial cell types (Holt *et al.*, 1988; Wetts and Fraser, 1988). However, there also is evidence for some degree of lineage restriction to the retina early in development (Williams and Goldwitz, 1992; Huang and Moody, 1993), including the fact that several neurotransmitter subtypes of amacrine cells descend from specific blastomeres in a lineage-biased pattern (Huang and Moody, 1995, 1997).

A. Blastomere Origin of the Retina and Specification of This Lineage

As a first step to determine if blastomere lineage plays a role in determining retinal fate, the retina progenitors of the 32CS embryo were mapped (Huang and Moody, 1993). An individual retina descends from a restricted and invariant group of blastomeres, which are a subset of those blastomeres that produce the forebrain (Moody, 1987b; Huang and Moody, 1992). Each retina is composed of the descendants of nine animal blastomeres, one ventral and eight dorsal, four of which are contralateral to the retina being studied (Fig. 6A, B). Consistent with lineage studies of optic vesicle clones (Holt *et al.*, 1988; Wetts and Fraser 1988), the descendants of a single blastomere are not restricted to a single cell type. Each blastomere contributes progeny to all three cellular laminae (photoreceptor layer, inner nuclear layer, and ganglion cell layer) in identical proportion to the overall composition of the retina (Holt *et al.*, 1988). However, the clones are neither identical in size, as each produces a characteristic proportion of the retina (Fig. 6A), nor identical in cellular composition, when fates are examined by specific neurotransmitter types (Huang and Moody, 1995, 1997). Therefore, some aspect of the blastomere lineage appears to bias both quantitative and qualitative aspects of its retinal fate.

To test whether blastomeres are committed to their retinal fates during cleavage stages the major retinal progenitors (left and right D1.1.1; Figs. 1 and 6A) were deleted, and the retinal contribution of each remaining blastomere evaluated (Fig. 6C; Huang and Moody, 1993). Even though the ancestors for greater than one-half of the retina were removed, a complete retina, with the normal complement and number of cells, was reconstituted. Therefore, the 32CS blastomeres in the animal hemisphere are not determined to make retina, as the remaining blastomeres regulate their lineages to produce a full-sized organ. In fact, each of the remaining retina-producing blastomeres signifi-

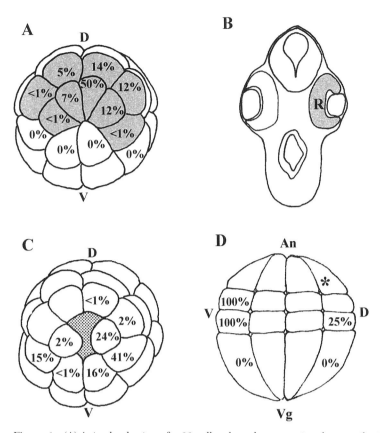

Figure 6 (A) Animal pole view of a 32-cell embryo demonstrating the contribution of each blastomere to one retina. Five ipsilateral and four contralateral blastomeres contribute descendants to each retina (shaded). The numbers within each blastomere indicate the percentages of cells in one retina (B) that descend from each blastomere. (B) Transverse section of a tail bud embryo showing the right retina (R) that descends from the blastomeres shaded in A. (C) Animal pole view of a 32-cell embryo demonstrating the percentage of the retina that descends from each blastomere following deletion of both D1.1.1 blastomeres (stippling). Both ventral and dorsal blastomeres change their contributions, their lineages regulating to produce a normal-sized eye. Dorsal blastomeres make fewer retinal cells and ventral blastomeres make more retinal cells. Some ventral blastomeres that normally never make retina become significant retinal progenitors (cf. A). (D) A lateral view of a 32-cell embryo demonstrating the competence of blastomeres, which normally do not produce retina, to contribute to the retina after transplantation to the D1.1.1 position (*). Numbers represent the percentage of cases in which the progeny from transplanted blastomeres populated the retina. D, Dorsal; V, ventral; An, animal; Vg, Vegetal. Data are from Huang and Moody (1993).

cantly changed their numerical contributions to retina (Fig. 6C). Dorsal blastomeres that normally contribute large numbers of cells to the retina contributed significantly smaller numbers, and ventral blastomeres that normally contribute very small numbers of cells contributed significantly larger numbers. Furthermore, some ventral blastomeres that normally never contribute to the retina became significant progenitors. A series of deletion experiments demonstrated that there are specific shifts in the position of the remaining clones (ventral to dorsal, animal to vegetal), consistent with the directions that normal midline clones move during epiboly and gastrulation (Hainski and Moody, 1992; Bauer *et al.*, 1994), that affect whether clones will inhabit the anterior neural plate after wound healing (Huang and Moody, 1993). Thus, initial blastomere position serves to ensure that clones are in the correct positions during and after gastrulation to contribute progeny to the anterior neural plate, and thus to the retina (see also Chapter 23 by Perron and Harris).

The importance of blastomere position was confirmed by transplantation of other blastomeres to the D1.1.1 position. Ventral equatorial blastomeres (V1.1.2 and V2.1.2) are competent to form retina if they are moved to the proper location (Fig. 6D). Presumably cells at the D1.1.1 position receive signals that allow them to contribute to the retinal fields of the neural plate. To discover the molecular nature of these signals, we analyzed whether overexpression or inhibition of signaling pathways involved in mesodermal, neural, or anteroposterior patterning alters a blastomere's contribution to the retina.

B. What Molecular Signals Initially Specify the Retinal Cell Fate of Animal Blastomeres?

A number of signaling pathways are important for the many different cell fate decisions and early inductive events in early *Xenopus* development (see Chapter 19 by

Dawid). Are these molecules also involved in regulating whether blastomeres and their descendants respond to the positional cues that direct them into retinal lineages? We tested whether perturbation of three different growth factor pathways: the activin/Vg1, fibroblast growth factor (FGF), and bone morphogenetic protein (BMP) pathways, affect animal blastomere contributions to the retina. These factors function through signal transduction pathways involving receptors and cascades of kinase and phosphatase regulatory components (Amaya et al., 1991; LaBonne et al., 1995; He et al., 1995; Northrop et al., 1995). This allows the functional ablation of a pathway in a specific lineage by microinjection of mRNAs encoding dominant-negative receptors. These mutant receptors are abundantly expressed, form inactive dimers with the endogenous receptors, and essentially eliminate the intracellular signaling cascade. Conversely, a signaling pathway can be provided or enhanced by the ectopic expression of wild-type or constitutively active receptors. In all cases, mRNA encoding green fluorescent protein (GFP) was coinjected as a tracer to assess whether descendants of the mutated lineage contributed to the retina. Many previous studies have shown that activin/Vg1, FGFs, and BMPs each play specific roles in mesoderm induction during blastula stages by microinjection of these mRNA constructs prior to the 4CS. However, because each pathway has multiple functions encompassing many developmental stages, we targeted identified blastomeres of the 32CS embryo to avoid disrupting mesoderm induction and potentially masking any later, specific effects on the establishment of the retina.

Activin, a member of the TGFβ superfamily, has been implicated in mesoderm induction in *Xenopus* (Symes et al., 1994; Dyson and Gurdon, 1997), and its overexpression can elicit formation of neural structures including brain and eyes (Sokol et al., 1990; Thomsen et al., 1990). However, inhibition of this pathway by overexpression of the dominant-negative activin/Vg1 receptor (tAR, truncated type II activin receptor; Hemmati-Brivanlou and Melton, 1992; Hemmati-Brivanlou et al., 1994a) in the D1.1.1 lineage did not prevent progeny from expressing retinal fates (Table II). In fact, the D1.1.1 clone was indistinguishable in size from control clones, and its members extended through all retinal layers (Moore and Moody, 1997). The volumes of the resulting retinas were normal in size, indicating that neighboring blastomeres also were not inhibited from expressing their full retinal fate. The FGF pathway also plays a role in mesoderm induction (Cornell and Kimelman, 1994; Doniach, 1995), and disruption of this pathway in the animal hemisphere prior to the 16CS leads to developmental deficiencies in the dorsal axis (Amaya et al., 1991; Slack, 1994; Doniach, 1995), including the reduction or absence of eyes (Launay et al., 1996). At later stages mRNA and protein for FGF-1,

TABLE II
Effects of Overexpression of Signaling Components in Retina-Producing Blastomeres

RNA	Site of injection	Clone in retina?	Size of clone
tAR	D1.1.1	Yes	Wild type
XFD	D1.1.1	Yes	Wild type
BMPR	D1.1.1	Yes	Wild type
	D1.2.1	Yes	Wild type
	D1.2.2	No	—
	V1.2.1	No	—
BMP4	D1.1.1	No	—
	D1.2.1	No	—
Noggin	V1.2.1	Yes	Larger
FGFR1	D1.1.1	Yes	Wild type
FGFR2	D1.1.1	Yes	Smaller
cFGFR1	D1.1.1	No	—
cFGFR2	D1.1.1	No	—
cFGFR2 + pax6	D1.1.1	Yes	Small to wild type

FGF-2 and their receptors are detected in the developing CNS, including the developing eye (Friesel and Brown, 1992; Tannahill et al., 1992; Tcheng et al., 1994). We found, however, that inhibition of this pathway in the D1.1.1 lineage by overexpression of a dominant-negative FGF receptor (XFD, Amaya et al., 1991) had no effect on the D1.1.1 clone or the size of the retina (Table II). Thus, neither the activin/Vg1 nor FGF signaling pathway determines the retinal fate of the D1.1.1 blastomere after their involvement in dorsal mesoderm specification.

BMPs, also members of the TGFβ superfamily, have been implicated in instructing both mesoderm to be ventral in character and ectoderm to be epidermis (Jones et al., 1992, 1996; Dale et al., 1992; Graff et al., 1994; Schmitt et al., 1995; Wilson and Hemmati-Brivanlou, 1995). BMP signaling both induces epidermis formation and inhibits neural induction (Schmidt et al., 1995; Wilson and Hemmati-Brivanlou, 1995; Kroll and Amaya, 1996; Weinstein and Hemmati-Brivanlou, 1997). Studies indicate that neural induction occurs by the direct binding of two secreted molecules, noggin and chordin, to BMPs (Sasai et al., 1995; Xu et al., 1995; Piccolo et al., 1996; Zimmerman et al., 1996). This prevents the ligand from activating its receptor. Thus, where there are low levels of BMP signaling in the embryo, the CNS forms (Sasai et al., 1995; Wilson and Hemmati-Brivanlou, 1995; see also Chapter 29 by Streit and Stern). To assess whether regulating BMP signaling, and thus neural induction, is sufficient to instruct animal blastomeres to contribute to the retina, mRNA encoding a receptor for BMP 4/2 (BMPR; Sasai et al., 1995) was injected into four different dorsal animal blastomeres that normally give rise to the retina (Fig. 6A). The effects were position dependent: the descendants of D1.1.1 and D1.2.1 remained in the retina in lineage-appropriate numbers, but clones derived from D1.2.2 and V1.2.1 no longer contributed to the retina (Table II). Conversely, injection of noggin mRNA into V1.2.1 increases dramatically the number of progeny that contribute to the retina. These studies indicate that the opposing gradients of noggin and BMP acting within the extracellular environment of these blastomeres (Sasai and De Robertis, 1997), which modulate the extent of neural plate induction, dictate whether lateral blastomeres become retina progenitors.

Why were the clones of D1.1.1 and D1.2.1 unaffected by this manipulation of the BMP receptor? Evidence suggests that the BMP ligand is at such a low level in the region of the descendants of these blastomeres that the mutant receptor has no effect. BMP proteins are expressed widely in the *Xenopus* gastrula except in the dorsal marginal ectoderm [occupied by D1.1.1 progeny (Bauer *et al.*, 1994)] that will become the neural plate (Dale *et al.*, 1992; Hemmati-Brivanlou and Thomsen, 1995). Therefore, overexpression of their receptors probably is not sufficient to activate this pathway in the D1.1.1 and D1.2.1 lineages. This hypothesis was confirmed by demonstrating that overexpression of the BMP4 ligand drives both the D1.1.1 and D1.2.1 clones out of the retina (Table II). Taken together, these studies show that the initial step in specifying dorsal animal blastomeres to contribute to retina relies on blastomere position within a field of signaling gradients that regulate the extent of neural plate formation.

As mentioned above, deletion of D1.1.1 can cause ventral animal blastomeres that normally are not retinal precursors to contribute to retina (Huang and Moody, 1993). In particular, the V1.1.1 lineage produces 16% of the retina (Fig. 6C). Since the descendants of this blastomere normally are found in the animal pole region during gastrulation (Bauer *et al.*, 1994) where BMP4 expression is strongest (Fainsod *et al.*, 1994; Schmidt *et al.*, 1995), we hypothesized that high levels of endogenous BMP signaling normally prevent this blastomere from entering the retinal lineage. Blocking the BMP pathway in V1.1.1 by overexpression of either a dominant-negative truncated BMP receptor (tBMPR), noggin, or tBMPR in combination with noggin (tBMPR/noggin) allowed the progeny of this blastomere to contribute to the retina (Table III). The tBMPR and tBMPR/noggin-expressing cells contributed progeny to approximately 10–25% of the retina (the same amount as after D1.1.1-deletion), and noggin alone-expressing cells contributed even higher numbers of descendants. These results indicate that when the D1.1.1 blastomere is deleted, the subsequent dorsal-ward movements of the V1.1.1 clone at the gastrula stage must place its descendants in a region of high noggin (and/or chordin) expression, which in turn significantly reduces the surrounding levels of endogenous BMP ligand (Smith and Harland, 1992). Thus, one reason that initial position of retinal precursors in the an-

TABLE III
Effects of Overexpression of Signaling Components in Blastomeres That Do Not Produce Retina

RNA	Site of injection	Clone in retina?	Size of clone	Transplanted clone in retina?
tBMPR	V1.1.1	Yes	Small	—
Noggin	V1.1.1	Yes	Large	—
tBMPR/ noggin	V1.1.1	Yes	Small to medium	—
AR	V2.1.1	No	—	No
FGFR1	V2.1.1	No	—	No
FGFR2	V2.1.1	No	—	—
c-ras	V2.1.1	No	—	—
c-raf	V2.1.1	No	—	—
tAR	V2.1.1	No	—	No
XFD	V2.1.1	No	—	No
tBMPR	V2.1.1	No	—	No
Noggin	V2.1.1	No	—	No
tBMPR/ noggin	V2.1.1	No	—	No
Cerberus	V2.1.1	Yes	Small	—
Xwnt8	V2.1.1	No	—	—
dnXwnt8	V2.1.1	No	—	—
dnXwnt8 + tBMPR	V2.1.1	No	—	—

imal hemisphere of the cleavage stage embryo is important is to ensure that the BMP pathway remains inactive, allowing induction of the anterior neural plate, and subsequent expression of a retinal fate.

This finding is consistent with studies that demonstrate that the ability to become retina is not determined until the neural plate is established (see Chapter 23 by Perron and Harris). During neural plate stages not only are the eye fields established, but the CNS becomes regionalized and the major subdivisions and dorsal-ventral identities begin to become evident. Many growth factors and transcription factors are involved in this process, including FGFs. FGFs are implicated in determining posterior fates during early neural plate formation (Doniach, 1995; Cox and Hemmati-Brivanlou, 1995; Godsave and Durston, 1997; Lamb and Harland, 1995; Lamb *et al.*, 1993; Isaacs, 1997; see Chapter 29 by Streit and Stern). Later they also may be necessary for neural retina differentiation (Pittack *et al.*, 1997) and the specification of some retinal cell types (Guillemot and Cepko, 1992; Pittack *et al.*, 1991, 1997). To test if FGF signaling in the neural plate affects whether blastomere clones enter the retina, we overexpressed two wild-type FGF receptors in the D1.1.1 lineage. FGFR1 is the major embryonic receptor present prior to gastrulation (Friesel and Dawid, 1991), and FGFR2 is the major receptor expressed in the embryonic CNS (Friesel and Brown, 1992). Overexpression of FGFR1 had no effect, but FGFR2 overexpression caused the D1.1.1 clone in the retina to be 25-75% smaller (Table II). Since limiting amounts of ligand may be the reason that FGFR2 overexpression did not completely block the retinal fate of D1.1.1, we overexpressed a constitutively active construct of either the FGFR1 or FGFR2 receptor (cFGFRs; Neilson and Freisel, 1995, 1996). Under these conditions D1.1.1 blastomeres did not contribute any progeny to the retina, indicating that high levels of FGF signaling after mesoderm induction can prevent cells from entering the retinal lineage (Table II). However, unlike the effects seen in embryos with high BMP signaling, the D1.1.1 clone remained neural, populating the forebrain as well as large masses of neural tissue extending from the brain. Lineage tracing of these FGF-activated clones demonstrated that early convergent-extension movements of gastrulation were unaffected, but lateral dispersion of the anterior members of the clone into the eye fields of the neural plate was inhibited. Thus, high levels of FGF signaling does not interfere with the induction of the D1.1.1 progeny to become neural, but it prevents them from assuming the specific regional fate of becoming part of the retina.

In other systems activated FGF enhances or maintains sonic hedgehog (Shh) signaling, for example, during limb development (Vogel *et al.*, 1996). Shh also is a candidate signal for retinal field resolution (Macdonald *et al.*, 1995; Li *et al*, 1997; Chapter 23 by Perron and Harris), since it represses the expression of *pax6*, a gene that is critical for retinal development (Hirsch and Harris, 1997; Macdonald and Wilson, 1997). Since co-expression of *pax6* mRNA with the activated FGF mRNA is sufficient to restore the D.1.1.1 progeny to the retinal lineage (Table II), we predict that the activated FGF signaling in the D1.1.1 clones suppressed the retinal fate via a similar mechanism. These studies demonstrate that maintaining low levels of FGF signaling in the anterior neural plate is a critical step in specifying the retinal fate of the descendants of animal hemisphere blastomeres.

C. Why Are Vegetal Blastomeres Not Competent to Express a Retinal Fate?

Transplantation, deletion, and gene misexpression studies demonstrate that animal hemisphere blastomeres all are competent to contribute to the retina, provided they are located in the correct position in the blastula field of neural inducers, and in the neural plate field of low FGF signaling. However, D-tier vegetal cells are not competent to contribute to the retinal lineage, even if transplanted to the D1.1.1 position (Fig. 6D; Huang and Moody, 1993). These blastomeres either lack a maternal molecular component(s) of a critical signaling event (mesoderm induction, neural induction, regionalization, or retinal fate determination), or they contain an inhibitory molecule(s) that prevents their descendants from responding to retinogenic signals. We tested the first possibility by overexpressing mRNA for the wild-type activin/Vg1 receptor (AR), the wild-type FGFR1 and FGFR2 receptors and their downstream

signaling molecules, *c-ras* and *c-raf* (Freisel and Dawid, 1991; Freisel and Brown, 1992; LaBonne and Whitman, 1994) in blastomere V2.1.1. None of these molecules enabled vegetal blastomere progeny to contribute to the retina (Table III), nor were other V2.1.1 fates significantly altered. Thus, these cells are not simply missing elements of the mesoderm induction pathway. Since the normal high levels of FGF signaling in the vegetal region imparts a posterior fate to the CNS (Doniach, 1995; Kengaku and Okamoto, 1995; Lamb and Harland, 1995; Godsave and Durston, 1997), we next tested whether this pathway commits V2.1.1 progeny to posterior fates and prevents expression of anterior fates. Injecting the mRNA encoding the dominant-negative FGF receptor (XFD) into V2.1.1, however, also could not render this lineage competent to produce retina (Table III).

It could be argued, of course, that in spite of providing signaling molecules, if the clone remains in the vegetal pole, it has little chance to access the anterior neural plate, in which the retinal fields will be committed. To ensure that access to the correct position is not a confounding variable, vegetal blastomeres were injected with each of these mRNA constructs and transplanted into the D1.1.1 position of unlabeled hosts. None of the progeny of these transplanted vegetal clones ever entered the retina (Table III), but remained predominantly endodermal.

Can components of neural induction cause V2.1.1 progeny to enter the retinal lineage? Overexpression of the truncated BMP receptor (tBMPR), which blocks the BMP pathway and thereby promotes neural fates (Sasai *et al.*, 1995; Xu *et al.*, 1995), in V2.1.1 resulted in the formation of secondary axes, but the V2.1.1 clone remained in the gut and lateral plate mesoderm of the primary axis. Overexpression of the truncated activin/Vg1 receptor (tAR) in V2.1.1, which induces neural structures including eyes (Sokol *et al.*, 1990; Thomsen *et al*, 1990; Hemmati-Brivanlou and Melton, 1992) presumably by also blocking BMP signaling (Chang *et al.*, 1997), did cause V2.1.1 descendants to adopt more rostral fates, including CNS. However, it was not sufficient to drive the V2.1.1 clone into the retina (Table III). Finally, overexpression of noggin, alone or in combination with tBMPR, in V2.1.1 blastomeres induced secondary axes that contained heads and eyes, but the descendants of V2.1.1 only populated the gut and lateral plate mesoderm of the primary axis (Table III). Therefore, in contrast to animal blastomeres, inhibiting BMP signaling is not sufficient to direct vegetal clones to express a retinal fate. The ectopic expression of some of these molecules does induce extra brains and eyes, but the vegetal clone does not populate the retina. Even if V2.1.1 is injected with these constructs and then transplanted to the

D1.1.1 position, they do not contribute to the retina (Table III).

These results demonstrate that vegetal cells require some molecule downstream from neural induction in order to express a retinal fate. They are intrinsically different from animal blastomeres. This implies that they lack a maternal molecule, other than what we have provided experimentally, which is necessary to activate a zygotic program leading to production of the retina. There are several potential zygotic candidates involved in head and eye formation to investigate. We first studied *cerberus*, a putative secreted factor expressed in the deep layers (mesoendoderm) of the Organizer, because its overexpression promotes the differentiation of anterior head structures including eyes (Bouwmeester *et al.*, 1996). Injection of *cerberus* mRNA into V2.1.1 caused its progeny to populate both the retinas of the induced secondary axes, as well as the retinas of the primary axis in those embryos in which ectopic eyes did not form (Table 3). Recently it has been proposed that head induction requires the simultaneous repression of both BMP and Wnt signaling (Glink *et al.*, 1997). Since *cerberus* can downregulate BMP4 signaling (Boumeester *et al.*, 1996) and inhibit Wnt signaling (Glinka *et al.*, 1997), our results demonstrating that *cerberus* is sufficient to drive the V2.1.1 clone into the retinal lineage are consistent with this model. However, co-repression of Wnt and BMP signaling using mRNAs encoding a dominant negative Xwnt8 and tBMPR is not sufficient to drive V2.1.1 progeny into the retinal lineage (Table III). Thus, *cerberus* is evidently doing more than simply repressing Wnt and BMP signaling in this vegetal cell. To date *cerberus* is the only molecule we have tested that not only overrides the endogenous BMP signaling in this most ventral vegetal blastomere, but also substitutes for the retinogenic signal normally present in the animal hemisphere to which V2.1.1 progeny are refractory. These results demonstrate that D-tier blastomeres are not competent to contribute to retina because they lack or are blocked at a signaling component downstream of the noggin/BMP interaction that results in neural plate formation, and upstream of *cerberus*-mediated signaling.

Possible candidates for this signaling component include *siamois* and *Lim1*, both of which are proposed to be upstream of *cerberus* action. *Siamois* is a target of the Wnt dorsalization pathway and it activates Organizer-specific genes (LeMaire *et al.*, 1995; Carnac *et al.*, 1996; Fan and Sokol, 1997). *Lim1*-deficient mice are characterized by a lack of head structures (Shawlot and Behringer, 1995), and an activated form of Lim1 generates partial secondary axes in *Xenopus* when overexpressed in the ventral vegetal cells (Taira *et al.*, 1994). However, we need to test whether *siamois* or *Lim1* can overcome the inhibitory mechanisms that normally render vegetal pole blastomeres refractory to a retinal fate.

D. Summary

In these studies we have demonstrated that the position of a blastomere is critical for establishing whether its descendants will contribute to the retina, and the initial steps in the pathway of retinal cell specification. Since not all blastomeres can contribute progeny to the retina, both intrinsic blastomere properties (competence) and extrinsic cues (position-based signaling pathways) establish which blastomeres will give rise to retinal progeny. Blastomeres in the animal hemisphere all are competent to make retina, and the degree to which they do in normal or experimentally perturbed embryos depends inversely on the degree to which the BMP signaling pathway is activated. It is likely that the degree of BMP inhibition (e.g., via noggin and/or chordin expression) in these blastomeres results from earlier steps in dorsal axial fate specification, as discussed in Section II, so that under normal developmental conditions, only dorsal animal blastomeres make retina. However, a direct link between the molecular pathway of dorsal axis specification and that of retinal fate specification in the D1 lineage has not yet been made. Blastomeres that reside at the vegetal pole, however, are refractory to the overexpression of components of mesoderm and neural inductive or regionalizing signaling pathways, or to a dorsal animal position. However, overexpression of a gene well downstream of the onset of dorsal-axis-specific zygotic transcription, *cerberus*, can overcome this lack of retinal fate competence. This implies that animal-vegetal asymmetries in maternal molecules profoundly affect the later zygotic patterns of expression leading to a retinal fate. Further functional studies will build a more complete picture of the hierarchy of genes involved in specification of both animal and vegetal blastomere clones to a retinal fate.

Acknowledgments

This work has been supported by NIH Grants HD08055 (S.A.S.), EY06649 (K.B.M.), NS23158 (S.A.M.) and EY10096 (S.A.M.). We thank Petra Pandur and Kristy Kenyon for assistance and critical reading of the manuscript.

References

Amaya, E., Musci, T. J., and Kirschner, M. W. (1991). Expression of a dominant negative mutant of the FGF receptor disrupts mesoderm function in *Xenopus* embryos. Cell (*Cambridge, Mass.*) **66**, 257–270.

Asashima, M., Nakano, H., Sugino, H., Nakamura, T., Eto, Y., Ejima, D., Nishimatsu, S., Ueno, N., and Kinoshita, K. (1991). Presence of activin (eryhoid differentiation factor) in unfertilized eggs and blastulae of *Xenopus laevis*. *Proc. Natl. Acad. Sci. U.S.A.* **88**, 6511–6514.

Bauer, D. V., Huang, S., and Moody, S. A. (1994). The cleavage stage origin of Spemann's Organizer: Analysis of the movements of blastomere clones before and after gastrulation in *Xenopus*. *Development (Cambridge, UK)* **120**, 1179–1189.

Bearer, E. L. (1994). Distribution of *Xrel* in the early *Xenopus* embryo: A cytoplasmic and nuclear gradient. *Eur. J. Cell Biol.* **63**, 255–268.

Black, S. D. (1989). Experimental reversal of the normal dorsal-ventral timing of blastopore formation does not reverse axis polarity in *Xenopus* laevis embryos. *Dev. Biol.* **134**, 376–381.

Bouwmeester, T., Kim, S., Sasai, Y., and DeRobertis, E. M. (1996). Cerberus is a head-inducing secreted factor expressed in the anterior endoderm of Spemann's organizer. *Nature (London)* **383**, 595–601.

Brannon, M., Gomperts, M., Sumoy, L., Moon, R. T., and Kimelman, D. (1997). A β-catenin/XTCf-3 complex binds to the *siamois* promoter to regulate dorsal axis specification in *Xenopus*. *Genes Dev.* **11**, 2359–2370.

Brown, E. E., Denegre, J. M., and Danilchik, M. V. (1993). Deep cytoplasmic rearrangements in ventralized *Xenopus* embryos. *Dev. Biol.* **160**, 148–156.

Capco, D. G., and Mecca, M. D. (1988). Analysis of proteins in the peripheral and central regions of amphibian oocytes and eggs. *Cell. Differ.* **23**, 155–164.

Cardellini, P. (1988). Reversal of dorsoventral polarity of *Xenopus laevis* embryos by 180° rotation of the animal micromeres at the eight-cell stage. *Dev. Biol.* **128**, 428–434.

Carnac, G., Kodjabachian, L., Gurdon, J. B., and Lemaire, P. (1996). The homeobox gene *Siamois* is a target of the Wnt dorsalisation pathway and triggers organiser activity in the absence of mesoderm. *Development (Cambridge, UK)* **122**, 3055–3065.

Chang, C., Wislon, P. A., Mathews, L. S., and Hemmati-Brivanlou, A. (1997). A *Xenopus* type I activin receptor mediates mesodermal but not neural specification during embryogenesis. *Development (Cambridge, UK)* **124**, 827–837.

Cho, K. W., Blumberg, B., Steinbeisser, H., and De Robertis, E. M. (1991). Molecular nature of Spemann's organizer: the role of the *Xenopus* homeobox gene *goosecoid*. *Cell (Cambridge, Mass.)* **67**, 1111–1120.

Chung, H. M., and Malacinski, G. M. (1980). Establishment of the dorsal/ventral polarity of the amphibian embryo: Use of ultraviolet irradiation and egg rotation as probes. *Dev. Biol.* **80**, 120–133.

Colgan, D. F., and Manley, J. L. (1997). Mechanism and regulation of mRNA polyadenylation. *Genes Dev.* **11**, 2755–2766.

Cooke, J., and Webber, J. A. (1985a). Dynamics of the control of body pattern in the development of *Xenopus laevis*. I. Timing and pattern in the development of dorsoanterior and posterior blastomere pairs, isolated at the 4-cell stage. *J. Embryol. Exp. Morphol.* **88**, 85–112.

Cooke, J. and Webber, J. A. (1985b). Dynamics of the control of body pattern in the development of *Xenopus laevis*. II. Timing and pattern in the development of single blastomeres (presumptive lateral halves) isolated at the 2-cell stage. *J. Embryol. Exp. Morphol.* **88**, 113–133.

Cornell, R. A., and Kimelman, D. (1994). Combinatorial signaling in development. *BioEssays* **16**, 577–581.

Cox, W. G., and Hemmati-Brivanlou, A. (1995). Caudalization of neural fate by tissue recombination and bFGF. *Development (Cambridge, UK)* **121**, 4349–4358.

Cui, Y., Brown, J. D., Moon, R. T., and Christian, J. L. (1995). *Xwnt-8b*: A maternally expressed *Xenopus* Wnt gene with a potential role in establishing the dorsoventral axis. *Development (Cambridge, UK)* **121**, 2177–2186.

Cui, Y., Tian, Q., and Christian, J. L. (1996). Synergistic effects of Vg1 and Wnt signals in the specification of dorsal mesoderm and endoderm. *Dev. Biol.* **180**, 22–34.

Dagle, J. M., Walder, J. A., and Weeks, D. L. (1990). Targeted degradation of mRNA in *Xenopus* oocytes and embryos directed by modified oligonucleotides: studies of An2 and cyclin in embryogenesis. *Nucleic Acids Res.* **18**, 4751–4757.

Dale, L., and Slack, J. M. W. (1987a). Fate map of the 32-cell stage of *Xenopus laevis*. *Development (Cambridge, UK)* **99**, 527–551.

Dale, L., and Slack, J. M. W. (1987b). Regional specification within the mesoderm of early embryos of *Xenopus laevis*. *Development (Cambridge, UK)* **100**, 279–295.

Dale, L., Howes, G., Price, B. M. J., and Smith, J. C. (1992). Bone morphogenetic protein 4: A ventralizing factor in early *Xenopus* development. *Development (Cambridge, UK)* **115**, 573–583.

Dale, L., Matthews, G., and Colman, A. (1993). Secretion and mesoderm-inducing activity of the TGF-β-related domain of *Xenopus* Vg1. *EMBO J.* **12**, 4471–4480.

Danilchik, M. V., and Denegre, J. M. (1991). Deep cytoplasmic rearrangements during early development in *Xenopus laevis*. *Development (Cambridge, UK)* **111**, 845–856.

Darras, S., Marikawa, Y., Elinson, R. P., and Lemaire, P. (1997). Animal and vegetal pole cells of early *Xenopus* embryos respond differently to maternal dorsal determinants: Implications for the patterning of the organiser. *Development (Cambridge, UK)* **124**, 4275–4286.

Denegre, J. M., and Danilchik, M. V. (1993). Deep cytoplasmic rearrangements in axis-respecified *Xenopus* embryos. *Dev. Biol.* **160**, 157–164.

Devic, E., Paquereau, L., Rizzoti, K., Monier, A., Knibiehler, B., and Audigier, Y. (1996). The mRNA encoding a β subunit of heterotrimeric GTP-binding proteins is localized to the animal pole of *Xenopus laevis* oocyte and embryos. *Mech. Dev.* **59**, 141–151.

Ding, X., Hausen, P., and Steinbesser, H. (1988). Pre-MBT patterning of early gene regulation in *Xenopus*: The role of the cortical rotation and mesoderm induction. *Mech. Devel.* **70**, 15–24

Dominguez, I., Itoh, K., and Sokol, S. Y. (1995). Role of glycogen synthase kinase 3 β as a negative regulator of dorsoventral axis formation in *Xenopus* embryos. *Proc. Natl. Acad. Sci. U.S.A.* **92**, 8498–8502.

Doniach, T. (1995). Basic FGF as an inducer to anteroposterior neural pattern. *Cell (Cambridge, Mass.)* **83**, 1067–1070.

Dyson, S., and Gurdon, J. B. (1997). Activin signalling has a necessary function in *Xenopus* early development. *Curr. Biol.* **7**, 81–84.

Elinson, R. P., and Kao, K. R. (1989). The location of dorsal information in frog early development. *Dev.* Growth Differ. **31**, 423–430.

Elinson, R. P., and Rowning, B. (1988). A transient array of parallel microtubules in frog eggs: Potential tracks for a cytoplasmic rotation that specifies the dorso-ventral axis. *Dev. Biol.* **128**, 185–197.

Fainsod, A., Steinbeisser, H., and DeRobertis, E. M. (1994). On the function of BMP-4 in patterning the marginal zone of the *Xenopus* embryo. *EMBO J.* **13**, 5015–5025.

Fan, M. J. and Sokol, S. Y. (1997). A role for Siamois in Spemann organizer formation. *Development (Cambridge, UK)* **124**, 2581–2589.

Forristall, C., Pondel, M., Chen, L., and King, M. L. (1995). Patterns of localization and cytoskeletal association of two vegetally localized RNAs, Vg1 and Xcat-2. *Development (Cambridge, UK)* **121**, 201–208.

Friesel, R., and Brown, S. (1992). Spatially restricted expression of fibroblast growth factor receptor 2 during *Xenopus* development. *Development (Cambridge, UK)* **116**, 1051–1058.

Friesel, R., and Dawid, I. (1991). cDNA cloning and developmental expression of fibroblast growth factor receptors from *Xenopus laevis*. *Mol. Cell. Biol.* **11**, 2481–2488.

Fujisue, M., Kobayakawa, Y., and Yamana, K. (1993). Occurrence of dorsal axis-inducing activity around the vegetal pole of an uncleaved *Xenopus* egg and displacement to the equatorial region by cortical rotation. *Development (Cambridge, UK)* **118**, 163–170.

Gallagher, B. C., Hainski, A. M., and Moody, S. A. (1991). Autonomous differentiation of dorsal axial structures from an animal cap cleavage stage blastomere in *Xenopus*. *Development (Cambridge, UK)* **112**, 1103–1114.

Gerhart, J., Ubbels, G., Black, S., and Kirschner, M. (1981). A reinvestigation of the role of the gray crescent in axis formation in *Xenopus laevis*. *Nature (London)* **292**, 511–516.

Gerhart, J., Danilchik, M., Doniach, T., Roberts, S. Rowning, B., and Stewart, R. (1989). Cortical rotation of the *Xenopus* egg: Consequences for the anteroposterior pattern of embryonic dorsal development. *Development (Cambridge, UK)* **107**, (Suppl.) 37–52.

Gimlich, R. L. (1986). Acquisition of developmental autonomy in the equatorial region of the *Xenopus* embryo. *Dev. Biol.* **115**, 340–352.

Gimlich, R. L., and Gerhart, J. C. (1984). Early cellular interactions promote embryonic axis formation in *Xenopus laevis*. *Dev. Biol.* **104**, 117–130.

Glinka, A., Wu, W., Onichtchouk, D., Blumenstock, C., and Niehrs, C. (1997). Head induction by simultaneous repression of BMP and Wnt signalling in *Xenopus*. *Nature (London)* **389**, 517–519.

Godsave, S. F., and Durston, A. J. (1997). Neural induction and patterning in embryos deficient in FGF signaling. *Int. J. Dev. Biol.* **41**, 57–65.

Graff, J. M., Thies, R. S., Song, J. J., and Gerhart, J. C. (1994). Studies with a *Xenopus* BMP receptor suggests that ventral mesoderm-inducing signals override dorsal signals *in vivo*. *Cell (Cambridge, Mass.)* **79**, 168–179.

Guillemot, F., and Cepko, C. L. (1992). Retinal fate and ganglion cell differentiation are potentiated by acidic FGF in an *in vitro* assay of early retinal development. *Development (Cambridge, UK)* **114**, 743–754.

Gururajan, R., Perry, O. K., Melton, D.A., and Weeks, D. L. (1991). The *Xenopus* localized messenger RNA An3 may encode an ATP-dependent RNA helicase. *Nature (London)* **349**, 717–719.

Hainski, A. M., and Moody, S. A. (1992). *Xenopus* maternal RNAs from a dorsal animal blastomere induce a secondary axis in host embryos. *Development (Cambridge, UK)* **116**, 347–355.

Hainski, A. M., and Moody, S. A. (1996). Activin-like signal activates dorsal-specific maternal RNA between 8- and 16-cell stages of *Xenopus*. *Dev. Genet.* **19**, 210–221.

Hake, L. E., and Richter, J. D. (1997). Translational regulation of maternal mRNA. *Biochim. Biophys. Acta* **1332**, M31–M38.

Harland, R. M. (1994). The transforming growth factor β family and induction of the vertebrate mesoderm: Bone morphogenetic proteins are ventral inducers. *Proc. Natl. Acad. Sci. U.S.A.* **91**, 10243–10246.

Hay, B., Jan, L. Y., and Jan, Y. N. (1988). A protein component of *Drosophila* polar granules is encoded by vasa and has extensive sequence similarity to ATP-dependent helicases. *Cell (Cambridge, Mass.)* **55**, 577–587.

He, X., Saint-Jeannet, J. P., Woodgett, J. R., Varmus, H. E., and Dawid, I. (1995). Glycogen synthase kinase-3 and dorsoventral patterning in *Xenopus* embryos. *Nature (London)* **374**, 617–622.

Heasman, J. (1997). Patterning the *Xenopus* blastula. *Development (Cambridge, UK)* **124**, 4179–4191.

Hemmati-Brivanlou, A., and Melton, D. A. (1992). A truncated activin receptor inhibits mesoderm induction and formation of axial structures in *Xenopus* embryos. *Nature (London)* **359**, 609–614.

Hemmati-Brivanlou, A., and Thomsen, G. (1995). Ventral mesoderm patterning in *Xenopus* embryos: Expression pattern and activity of BMP-2 and BMP-4. *Dev. Genet.* **17**, 8–89.

Hemmati-Brivanlou, A., Kelly, O. G., and Melton, D. A. (1994a). Inhibition of activin receptor signaling promotes neuralization in *Xenopus*. *Cell (Cambridge, Mass.)* **77**, 283–295.

Hemmati-Brivanlou, A., Kelly, O. G., and Melton, D. A. (1994b). Follistatin, an antagonist of activin is expressed in the Spemann organizer and displays direct neuralizing activity. *Cell (Cambridge, Mass.)* **77**, 283–295.

Herkovits, J., and Ubbels, G. A. (1979). The ultrastructure of the dorsal yolk-free cytoplasm and the immediately surrounding cytoplasm in the symmetrized egg of *Xenopus laevis*. *J. Embryol. Exp. Morpho.* **51**, 155–164.

Hirose, G., and Jacobson, M. (1979). Clonal organization of the central nervous system of the frog. I. Clones stemming from the individual blastomeres of the 16-cell and earlier stages. *Dev. Biol.* **71**, 191–202.

Hirsch, N., and Harris, W. A. (1997). *Xenopus* Pax6 and retinal development. *J. Neurobiol.* **32**, 45–61.

Holewa, B., Strandmann, E. P., Zapp, D., Lorenz, P., and Ryffel, G. U. (1996). Transcriptional hierarchy in *Xenopus* embryogenesis: HNF4 a maternal factor involved in the developmental activation of the gene encoding the tissue specific transcription factor HNF1 α (LFB1). *Mech. Dev.* **54**, 45–57.

Holewa, B., Zapp, D., Drewes, T., Senkel, S., and Ryffel, G. U. (1997). HNF4β, a new gene of the HNF4 family with distinct activation and expression profiles in oogenesis and embryogenesis of *Xenopus laevis. Mol. Cell. Biol.* **17**, 687–694.

Holowacz, T., and Elinson, R. P. (1993). Cortical cytoplasm, which induces dorsal axis formation in *Xenopus*, is inactivated by UV irradiation of the oocyte. *Development* (*Cambridge, UK*) **119**, 277–285.

Holt, C.E., Bertsch, T. W., Ellis, H. M., and Harris, W. A. (1988). Cellular determination in the *Xenopus* retina is independent of lineage and birth date. *Neuron* **1**, 15–26.

Horb, M. E., and Thomsen, G. H. (1997). A vegetally localized T-box transcription factor in *Xenopus* eggs specifies mesoderm and endoderm and is essential for embryonic mesoderm formation. *Development* (*Cambridge, UK*) **124**, 1689–1698.

Houliston, E., and Elinson, R. P. (1991a). Patterns of microtubule polymerization relating to cortical rotation in *Xenopus laevis* eggs. *Development* (*Cambridge, UK*) **112**, 107–117.

Houliston, E., and Elinson, R. P. (1991b). Evidence for the involvement of microtubules, ER, and kinesin in the cortical rotation of fertilized frog eggs. *J. Cell Biol.* **114**, 1017–1028.

Huang, S., and Moody, S. A. (1992). Does lineage determine the dopamine phenotype in the tadpole hypothalamus?: A quantitative analysis. *J. Neurosci.* **12**, 1351–1362.

Huang, S., and Moody, S. A. (1993). The retinal fate of *Xenopus* cleavage stage progenitors is dependent upon blastomere position and competence: Studies of normal and regulated clones. *J. Neurosci.* **13**, 3183–3210.

Huang, S., and Moody, S. A. (1995). Asymmetrical blastomere origin and spatial domains of dopamine and Neuropeptide Y amacrine subtypes in *Xenopus* tadpole retina. *J. Comp. Neurol.* **360**, 2–13.

Huang, S., and Moody, S. A. (1997). Three types of serotonin-containing amacrine cells in tadpole retina have distinct clonal origins. *J. Comp. Neurol.* **387**, 42–52.

Hudson, J. W., Alarcon, V. B., and Elinson, R. P. (1996). Identification of new localized RNAs in the *Xenopus* oocyte by differential display PCR. *Dev. Genet.* **19**, 190–198.

Imoh, H. (1984). Appearance and distribution of RNA-rich cytoplasms in the embryo of *Xenopus laevis* during early development. *Dev. Growth Differ.* **26**, 167–176.

Isaacs, H. V. (1997). New perspectives on the role of the fibroblast growth factor family in amphibian development. *Cell. Mol. Life Sci.* **53**, 350–361.

Jacobson, A. (1983). Clonal organization of the central nervous system of the frog. III. Clones stemming from individual blastomeres of the 128-, 256-, and 512-cell stages. *Dev. Biol.* **71**, 191–202.

Jacobson, M. (1996). Poly (A) metabolism and translation: The closed-loop model. *In* "Translational Control" (J. W. B. Hershey, M. B. Mathews, and N. Sonenberg, eds.), pp. 451–480. Cold Spring Harbor Laboratory, Plainview, New York.

Jacobson, M., and Hirose, G. (1981). Clonal organization of the central nervous system of the frog. II. Clones stemming from individual blastomeres of the 32- and 64-cell stages. *J. Neurosci.* **1**, 271–284.

Jones, C. M., Lyons, K. M., Lapan, P. M., Wright, C. V. E., and Hogan, B. J. (1992). DVR-4 (Bone morphogenetic protein-4) as a postero-ventralizing factor in *Xenopus* mesoderm induction. *Development* **115**, 639–647.

Jones, C. M., Armes, N., and Smith, J. C. (1996). Signalling by TGF-β members: Short range effects of Xnr-2 and BMP-4 contrast with the long-range effects of activin. *Curr. Biol.* **6**, 1468–1475.

Jones, E. A., and Woodland, H. R. (1987). The development of animal cap cells in *Xenopus*: A measure of the start of animal cap competence to form mesoderm. *Development* (*Cambridge, UK*) **101**, 557–563.

Kageura, H. (1990). Spatial distribution of the capacity to initiate a secondary embryo in the 32-cell embryo of *Xenopus laevis. Dev. Biol.* **142**, 432–438.

Kageura, H. (1995). Three regions of the 32-cell embryo of *Xenopus laevis* essential for formation of a complete tadpole. *Dev. Biol.* **170**, 376–386.

Kageura, H. (1997). Activation of dorsal development by contact between the cortical dorsal determinant and the equatorial core cytoplasm in eggs of *Xenopus laevis. Development* (*Cambridge, UK*) **124**, 1543–1551.

Kageura, H and Yamana, K. (1983). Pattern regulation in isolated halves and blastomeres of early *Xenopus laevis. J. Embryol. Exp. Morph.* **74**, 221–234.

Kageura, H., and Yamana, K. (1984). Pattern regulation in defect embryos of *Xenopus laevis. Dev. Biol.* **101**, 410–415.

Kageura, H., and Yamana, K. (1986). Pattern formation in 8-cell composite embryos of *Xenopus laevis. J. Embryol. Exp. Morphol.* **91**, 79–100.

Keller, R. E., and Danilchik, M. (1988). Regional expression, patterning, and timing of convergence and extension during gastrulation of *Xenopus laevis. Development* (*Cambridge, UK*) **103**, 193–209.

Kengaku, M., and Okamoto, H. (1995). bFGF as a possible morphogen for the anteroposterior axis of the central nervous system in *Xenopus. Development* **121**, 3121–3130.

Kikkawa, M., Takano, K., and Shinagawa, A. (1996). Location and behavior of dorsal determinants during first cell cycle in *Xenopus* eggs. *Development* (*Cambridge, UK*) **122**, 3687–3696.

Kimmel, C. B., and Law, R. D. (1985). Cell lineage of zebrafish blastomeres. II. Clonal analyses of the blastula and gastrula stages. *Dev. Biol.* **108**, 94–101.

King, M. L., and Barklis, E. (1985). Regional distribution of maternal messenger RNA in the amphibian oocyte. *Dev. Biol.* **112**, 203–212.

Kinoshita, K., Bessho, T., and Asashima, M. (1993). Competence prepattern in the animal hemisphere of the 8-cell-stage *Xenopus* embryo. *Dev. Biol.* **160**, 276–284.

Klein, S. L. (1987). The first cleavage furrow demarcates the dorsal-ventral axis in *Xenopus* eggs. *Dev. Biol.* **120**, 299–304.

Klein, S. L., and King, M. L. (1988). Correlations between cell fate and the distribution of proteins that are synthesized before the midblastula transition in *Xenopus*. *Dev. Biol.* **197**, 275–281.

Kloc, M., and Etkin, L. D. (1994). Delocalization of Vg1 mRNA from the vegetal cortex in *Xenopus* oocytes after destruction of Xlsirt RNA. *Science* **265**, 1101–1103.

Kloc, M., Reddy, B. A., Miller, M., Eastman, E., and Etkin, L. D. (1991). x121: A localized maternal transcript in *Xenopus laevis*. *Mol. Reprod. Dev.* **28**, 341–345.

Kloc, M., Spohr, G., and Etkin, L. D. (1993). Translocation of repetitive RNA sequences with the germ plasm in *Xenopus* oocytes. *Science* **262**, 1712–1714.

Kroll, K. L., and Amaya, E. (1996). Transgenic *Xenopus* embryos from sperm nuclear transplantations reveal FGF signaling requirements during gastrulation. *Development* (*Cambridge, UK*) **122**, 3173–3183.

Ku, M., and Melton, D. A. (1993). *Xwnt-11*: A maternally expressed *Xenopus* Wnt gene. *Development* (*Cambridge, UK*) **119**, 1161–1173.

LaBonne, C., and Whitman, M. (1994). Mesoderm induction by activin requires FGF-mediated intracellular signals. *Development* (*Cambridge, UK*) **120**, 463–472.

LaBonne, C., Burke, B., and Whitman, M. (1995). Role of MAP kinase in mesoderm induction and axial patterning during *Xenopus* development. *Development* (*Cambridge, UK*) **121**, 1475–1486.

Lamb, T. M., and Harland, R. M. (1995). Fibroblast growth factor is a direct neural inducer, which combined with noggin generates anterior-posterior neural pattern. *Development* (*Cambridge, UK*) **121**, 3627–3636.

Lamb, T. M., Knecht, A. K., Smith, W. C., Stachel, S. E., Economides, A. N., Stahl, N., Yancopolous, G. D., and Harland, R. M. (1993). Neural induction by the secreted polypeptide noggin. *Science* **262**, 713–718.

Larabell, C. A., Rowning, B. A., Wells, J., Wu, M., and Gerhart, J. C. (1996). Confocal microscopy analysis of living *Xenopus* eggs and the mechanism of cortical rotation. *Development* **122**, 1281–1289.

Larabell, C. A., Torres, M., Rowning, B. A., Yost, C., Miller, J. R., Wu, M., Kimelman, D., and Moon, R. T. (1997). Establishment of the dorso-ventral axis in *Xenopus* embryos is presaged by early asymmetries in β-catenin that are modulated by the Wnt signaling pathway. *J. Cell Biol.* **136**, 1123–1136.

Launay, C., Fromentoux, V., Shi, D. L., and Boucaut, J. C. (1996). A truncated FGF receptor blocks neural induction by endogenous *Xenopus* inducers. *Development* (*Cambridge, UK*) **122**, 869–880.

Lemaire, P., Garrett, N., and Gurdon, J. B. (1995). Expression cloning of Siamois, a *Xenopus* homeobox gene expressed in dorsal-vegetal cells of blastulae and able to induce a complete secondary axis. *Cell* (*Cambridge, Mass.*) **81**, 85–94.

Li, H., Tierney, C., Wen, L., Wu, J. Y., and Rao, Y. (1997). A single morphogenetic field gives rise to two retina primordia under the influence of the prechordal plate. *Development* (*Cambridge, UK*) **124**, 603–615.

Linnen, J. M., Bailey, C. P., and Weeks, D. L. (1993). Two related localized mRNAs from *Xenopus laevis* encode ubiquitin-like fusion proteins. *Gene* **128**, 181–188.

Lustig, K. D., Kroll, K. L., Sun, E. E., and Kirschner, M. W. (1996). Expression cloning of a *Xenopus* T-related gene (*Xombi*) involved in mesodermal patterning and blastopore lip formation. *Development* (*Cambridge, UK*) **122**, 4001–4012.

McClendon, J. F. (1910). The development of isolated blastomeres of the frog's egg. *Am. J. Anat.* **10**, 425–430.

McCrea, P. D., Turck, C. W., and Gumbiner, B. (1991). A homolog of the armadillo protein in *Drosophila* (plakoglobin) associated with E-cadherin. *Science* **254**, 1359–1361.

McCrea, P. D., Brieher, W. M., and Gumbiner, B. M. (1993). Induction of a secondary body axis in *Xenopus* by antibodies to β-catenin. *J. Cell Biol.* **123**, 477–484.

Macdonald, R., Barth, K., Xu, Q., Holder, N., Mikkola, I., and Wilson, S. (1995). Midline signaling is required for *Pax* gene regulation and patterning of the eyes. *Development* (*Cambridge, UK*) 121, (Suppl). 3267–3278.

Macdonald, R., and Wilson, S. W. (1997). Distribution of Pax6 protein during eye development suggests discrete roles in proliferative and differentiated visual cells. *Dev. Genes Evol.* **206**, 363–369.

Manes, M. E., and Barbieri, F. D. (1977). On the possibility of sperm aster involvement in dorso-ventral polarization and pronuclear migration in the amphibian egg. *J. Embryol. Exp. Morphol.* **40**, 187–197.

Masho, R. (1988). Fates of animal-dorsal blastomeres of eight-cell stage *Xenopus* embryos vary according to the specific patterns of the third cleavage plane. *Dev. Growth Differ.* **30**, 347–359.

Masho, R. (1990). Close correlation between the first cleavage plane and the body axis in early *Xenopus* embryos. *Dev. Growth Differ.* **32**, 57–64.

Masho, R., and Kubota, H. Y. (1986). Developmental fates of blastomeres of the eight-cell stage *Xenopus* embryos. *Dev. Growth Differ.* **28**, 113–123.

Miyata, S., Kageura, H., and Kihara, H. K. (1987). Regional differences of proteins in isolated cells of early embryos of *Xenopus laevis*. *Cell. Differ.* **21**, 47–52.

Molenaar, M., Van, D. W., Oosterwegel, M., Peterson-Maduro, J., Godsave, S., Korinek, V., Roose, J., Destree, O., and Clevers, H. (1996). XTcf-3 transcription factor mediates β-catenin-induced axis formation in *Xenopus* embryos. *Cell* (*Cambridge, Mass.*) **86**, 391–399.

Moody, S. A. (1987a). Fates of the blastomeres of the 16-cell *Xenopus* embryo. *Dev. Biol.* **119**, 560–578.

Moody, S. A. (1987b). Fates of the blastomeres of the 32-cell *Xenopus* embryo. *Dev. Biol.* **122**, 300–319.

Moody, S. A. (1989). Quantitative lineage analysis of the origin of frog primary motor and sensory neurons from cleavage stage blastomeres. *J. Neurosci.* **9**, 2919–2930.

Moody, S. A. (1998). Testing the cell fate commitment of single blastomeres in *Xenopus laevis*. *In* "Advances in Molecular Biology: A Comparative Approach to the Study of Oocytes and Embryo" (J. Richter, ed.). Oxford Univ. Press, New York.

Moody, S. A, and Kersey, K. S. (1989). Clonal analysis of GABAergic neurons in frog embryo spinal cord. *Soc. Neurosci. Abstr.* **15**, 885.

Moody, S. A., and Kline, M. J. (1990). Segregation of fate during cleavage of frog (*Xenopus laevis*) blastomeres. *Anat. Embryol.* **182**, 347–362.

Moody, S. A., Bauer, D. V., Hainski, A. M., and Huang, S.

(1996). Determination of *Xenopus* cell lineage by maternal factors and cell interactions. *In* "Current Topics in Developmental Biology" (R. A. Pedersen and G. P. Schatten, eds.), pp. 103–138. Academic Press, New York.

Moon, R. T., Campbell, R. M., Christian, J. L., McGrew, L. L., Shih, J., and Fraser, S. (1993). Xwnt-5A: A maternal Wnt that affects morphogenetic movements after overexpression in embryos of *Xenopus laevis*. *Development* (*Cambridge, UK*) **119**, 97–111.

Moon, R.T., Brown, J. D., and Torres, M. (1997). WNTs modulate cell fate and behavior during vertebrate development. *Trends Genet.* **13**, 157–162.

Moore, K. B., and Moody, S. A. (1997). Perturbation of activin, FGF, and BMP signaling affect retinal cell fates in *Xenopus laevis*. *Mol. Biol. Cell Supp.* **8**, 439.

Nakamura, O., and Kishiyama, K. (1971). Prospective fates of blastomeres at the 32-cell stage of *Xenopus laevis* embryos. *Proc. Jpn Acad.* **47**, 406–412.

Neff, A. W., Wakahara, M., Jurand, A., and Malacinski, G. M. (1984). Experimental analyses of cytoplasmic rearrangements which follow fertilization and accompany symmetrization of inverted *Xenopus* eggs. *J. Embryol. Exp. Morphol.* **80**, 197–224.

Neilson, K. M., and Friesel, R. E. (1995). Constitutive activation of fibroblast growth factor receptor-2 by a point mutation associated with Crouzon syndrome. *J. Biol. Chem.* **270**, 26037–26040.

Neilson, K. M., and Friesel, R. (1996). Ligand-independent activation of fibroblast growth factor receptors by point mutations in the extracellular, transmembrane, and kinase domains. *J. Biol. Chem.* **271**, 25049–25057.

New, H. V., Kavka, A. I., Smith, J. C., and Green, J. B. (1997). Differential effects on *Xenopus* development of interference with type IIA and type IIB activin receptors. *Mech. Dev.* **61**, 175–186.

Newport, J., and Kirschner, M. (1982a). A major developmental transition in early *Xenopus* embryos. I. Characterization and time of cellular changes at the midblastula stage. *Cell* (*Cambridge, Mass.*) **30**, 675–686.

Newport, J., and Kirschner, M. (1982b). A major developmental transition in early *Xenopus* embryos: II. Control of the onset of transcription. *Cell* (*Cambridge, Mass.*) **30**, 687–696.

Nieuwkoop, P. D., and Faber, J. (1994). "Normal Table of *Xenopus* (Daudin)", Garland, New York.

Northrop, J., Woods, A., Seger, R., Suzuki, A., Ueno, N., Krebs, E., and Kimelman, D. (1995). BMP-4 regulates the dorsal-ventral differences in FGF/MAPKK-mediated mesoderm induction in *Xenopus*. *Dev. Biol.* **172**, 242–252.

Palacek, J., Ubbels, G., and Rzehak, K. (1978). Changes of the external and internal pigment pattern upon fertilization in the egg of *Xenopus laevis*. *J. Embryol. Exp. Morphol.* **45**, 203–214.

Pedersen, R.A., Wu, K., and Balakier, H. (1986). Origin of the inner cell mass in mouse embryos: Cell lineage analysis by microinjection. *Dev. Biol.* **117**, 581–595.

Phillips, C. R. (1982). The regional distribution of poly (A) and total RNA concentrations during early *Xenopus* development. *J. Exp. Zool.* **223**, 265–275.

Phillips, C. R. (1985). Spatial changes in poly (A) concentra-

tions during early embryogenesis in *Xenopus laevis*: Analysis by *in situ* hybridization. *Dev. Biol.* **109**, 299–310.

Piccolo, B., Sasai, Y., Lu, B., and DeRobertis, E. M. (1996). Dorso-ventral patterning in *Xenopus*: Inhibition of ventral signals by direct binding of Chordin to BMP-4. *Cell* (*Cambridge, Mass.*) **86**, 589–598.

Pittack, C., Jones, M., and Reh, T. A. (1991). Basic fibroblast growth factor induces retinal pigment epithelium to generate neural retina *in vitro*. *Development* (*Cambridge, UK*) **113**, 577–588.

Pittack, C., Grunwald, G. B., and Reh, T. A. (1997). Basic fibroblast growth factors are necessary for neural retina but not pigmented epithelium differentiation in chick embryos. *Development* (*Cambridge, UK*) **124**, 805–816.

Rebagliati, M. R., and Dawid, I. B. (1993). Expression of activin transcripts in follicle cells and oocytes of *Xenopus laevis*. *Dev. Biol.* **159**, 574–580.

Rebagliati, M. R., Weeks, D. L., Harvey, R. P., and Melton, D. A. (1985). Identification and cloning of localized maternal RNAs from *Xenopus* eggs. *Cell* (*Cambridge, Mass.*) **42**, 769–777.

Reddy, B. A., Kloc, M., and Etkin, L. D. (1992). The cloning and characterization of a localized maternal transcript in *Xenopus laevis* whose zygotic counterpart is detected in the CNS. *Mech. Dev* **39**, 143–150.

Roux, W. (1888). Beitrage zur Entwicklungsmechnik des embryo. *Virchows Arch. Pathol. Anat. Physiol. Klin. Med.* **114**, 113–153. Translated to English (1974). *In* "Foundations of Experimental Embryology" (B. H. Willer, and J. M. Oppenheimer, eds H. Laufer, transl.), pp. 2–37. Hafner, New York.

Rowning, B. A., Wells, J., Wu, M., Gerhart, J. C., Moon, R. T., and Larabell, C. A. (1997). Microtubule-mediated transport of organelles and localization of β-catenin to the future dorsal side of *Xenopus* eggs. *Proc. Natl. Acad. Sci. U.S.A.* **94**, 1224–1229.

Sakai, M. (1996). The vegetal determinants required for the Spemann organizer move equatorially during the first cell cycle. *Development* (*Cambridge, UK*) **122**, 2207–2214.

Sasai, Y., and De Robertis, E. M. (1997). Ectodermal patterning in vertebrate embryos. *Dev. Biol.* **182**, 5–20.

Sasai, Y., Lu, B., Steinbeisser, H., Geissert, D., Gont, L. K., and DeRobertis, E. M. (1995). Regulation of induction by the chd and BMP-4 antagonistic patterning signals in *Xenopus*. *Nature* (*London*) **376**, 333–336.

Scharf, S. R., and Gerhart, J. C. (1980). Determination of the dorsal-ventral axis in eggs of *Xenopus laevis*: Complete rescue of uv-impaired eggs by oblique orientation before first cleavage. *Dev. Biol.* **79**, 181–198.

Schmidt, J. E., Suzuki, A., Ueno, N., and Kimelman, D. (1995). Localized BMP-4 mediates dorso/ventral patterning in the early *Xenopus* embryo. *Dev. Biol.* **169**, 37–50.

Schroeder, K. E., and Yost, H. J. (1996). Xenopus poly (A) binding protein maternal RNA is localized during oogenesis and associated with large complexes in blastula. *Dev. Genet.* **19**, 268–276.

Shawlot, W., and Behringer, R. R. (1995). Requirement for *Lim-1* in head-organizer function. *Nature* (*London*) **374**, 4225–4230.

Shiokawa, K., Saito, A., Kageura, H., Higuchi, K., Koga, K., and Yamana, K. (1984). Protein synthesis in dorsal, ventral, animal and vegetal half- embryos of *Xenopus laevis* isolated at the 8-cell stage. *Cell Struct. Funct.* **9**, 369–380.

Simon, R., and Richter, J. D. (1994). Further analysis of cytoplasmic polyadenylation in *Xenopus* embryos and identification of embryonic cytoplasmic polyadenylation element-binding proteins. *Mol. Cell. Biol.* **14**, 7867–7875.

Simon, R., Tassan, J. P., and Richter, J. D. (1992). Translational control by poly (A) elongation during *Xenopus* development: Differential repression and enhancement by a novel cytoplasmic polyadenylation element. *Genes Dev.* **6**, 2580–2591.

Simon, R., Wu, L., and Richter, J. D. (1996). Cytoplasmic polyadenylation of activin receptor mRNA and the control of pattern formation in *Xenopus* development. *Dev. Biol.* **179**, 239–250.

Slack, J. M. W. (1994). Inducing factors in *Xenopus* early embryos. *Curr. Biol.* **4**, 116–126.

Slusarski, D. C., Corces, V. G., and Moon, R. T. (1997). Interaction of Wnt and a Frizzled homologue triggers G-protein-linked phosphatidylinositol signalling. *Nature (London)* **390**, 410–413.

Smith, R. C. (1986). Protein synthesis and messenger RNA levels along the animal- vegetal axis during early *Xenopus* development. *J. Embryol. Exp. Morphol.* **95**, 15–35.

Smith, R. C., and Knowland, J. (1984). Protein synthesis in dorsal and ventral regions of *Xenopus laevis* embryos in relation to dorsal and ventral differentiation. *Dev. Biol.* **103**, 355–368.

Smith, W. C. and Harland, R. M. (1991). Injected *Xwnt-8* RNA acts early in *Xenopus* embryos to promote formation of a vegetal dorsalizing center. *Cell (Cambridge, Mass.)* **67**, 753–765.

Smith, W. C., and Harland, R. M. (1992). Expression cloning of *noggin*, a new dorsalizing factor localized to the spemann organizer in *Xenopus* embryos. *Cell (Cambridge, Mass.)* **70**, 829–840.

Sokol, S., Wong, G., and Melton, D. (1990). A mouse macrophage factor induces head structures and organizes a body axis in *Xenopus*. *Science* **249**, 561–564.

Sokol, S., Klingensmith, J. Perrimon, N., and Itoh, K. (1995). Dorsalizing and neuralizing properties of *Xdsh*, a maternally expressed *Xenopus* homolog of *dishevelled*. *Development (Cambridge, UK)* **121**, 1637–1647.

Stennard, F., Carnac, G., and Gurdon, J. B. (1996). The *Xenopus* T-box gene, *Antipodean*, encodes a vegetally localised maternal mRNA and can trigger mesoderm formation. *Development (Cambridge, UK)* **122**, 4179–4188.

Stennard, F., Ryan, K., and Gurdon, J. B. (1997). Markers of vertebrate mesoderm induction. *Curr. Opin. Genet. Dev.* **7**, 620–627.

Sullivan, S. A., and Moody, S. A. (1996). Polyadenylation of maternal RNAs activates dorsal axis-inducing activity at the *Xenopus* 8-cell stage. *Mol. Biol. Cell Suppl.* **7**, 640a.

Sullivan, S. A., Jamrich, M., and Moody, S. A. (1998). Molecular and functional characterization of XFLIP, a maternal member of the *fork head/winged helix* family with mesoderm-inducing ability. *Dev. Biol.* **198**, 174.

Sulston, J. E., Schierenberg, E., White, J. G., and Thomsen, J. N. (1983). The embryonic cell lineage of the nematode *Cnaenorhabditis elegans*. *Dev. Biol.* **100**, 64–119.

Suzuki, A. S., Manabe, J., and Hirakawa, A. (1991). Dynamic distribution of region-specific maternal protein during oogenesis and early embryogenesis of *Xenopus laevis*. *Roux's Arch. Dev. Biol.* **200**, 213–222.

Symes, K., and Smith, J. C. (1987). Gastrulation movements provide an early marker of mesoderm induction in *Xenopus laevis*. *Development (Cambridge, UK)* **101**, 339–349.

Symes, K., Yordan, C., and Mercola, M. (1994). Morphological differences in *Xenopus* embryonic mesodermal cells are specified as an early response to distinct threshold concentrations of activin. *Development (Cambridge, UK)* **120**, 2339–2346.

Taira, M., Otani, H., Saint-Jeannet, J. P., and Dawid, I. B. (1994). Roles of the LIM class homeodomain protein *Xlim-1* in neural and muscle induction by the Spemann organizer in *Xenopus*. *Nature (London)* **372**, 677–679.

Takasaki, H. (1987). Fates and roles of the presumptive organizer region in the 32-cell embryo in normal development of *Xenopus laevis*. *Dev. Growth Differ.* **29**, 141–152.

Takasaki, H., and Konishi, H. (1989). Dorsal blastomeres in the equatorial region of the 32-cell *Xenopus* embryo autonomously produce progeny committed to the organizer. *Dev. Growth Differ.* **31**, 147–156.

Tannahill, D., Isaacs, H. V., Close, M. J., Peters, G., and Slack, J. M. W. (1992). Developmental expression of the *Xenopus int-2* (FGF-3) gene: Activation by mesodermal and neural induction. *Development (Cambridge, UK)* **115**, 695–670.

Tcheng, M., Fuhrmann, G., Hartmann, M. P., Courtois, Y., and Jeanny, J. C. (1994). Spatial and temporal expression patterns of FGF receptor genes type I and type 2 in the developing chick retina. *Exp. Eye Res.* **58**, 351–358.

Thomsen, G. H. and Melton, D. A. (1993). Processed Vg1 protein is an axial mesoderm inducer in *Xenopus*. *Cell (Cambridge, Mass.)* **74**, 433–441.

Thomsen, G., Woolf, T., Whitman, M., Sokol, S., Vaughan, J., Vale, W., and Melton, D. (1990). Activins are expressed early in *Xenopus* embryogenesis and can induce axial mesoderm and anterior structures. *Cell (Cambridge, Mass.)* **63**, 485–493.

Torres, M. A., Yang-Snyder, J. A., Purcell, S. M., DeMarais, A. A., McGrew, L. L., and Moon, R. T. (1996). Activities of the Wnt-1 class of secreted signaling factors are antagonized by the Wnt-5A class and by a dominant negative cadherin in early *Xenopus* development. *J. Cell Biol.* **133**, 1123–1137.

Ubbels, G. A. (1997). Establishment of polarities in the oocyte of *Xenopus laevis*: The provisional axial symmetry of the full-grown oocyte of *Xenopus laevis*. *Cell. Mol. Life Sci.* **53**, 382–409.

Ubbels, G. A., Hara, K., Koster, C. H., and Kirschner, M. W. (1983). Evidence for a functional role of the cytoskeleton in determination of the dorsoventral axis in *Xenopus laevis* eggs. *J. Embryol. Exp. Morphol.* **77**, 15–37.

Vincent, J. P., and Gerhart, J. C. (1986). A reinvestigation of the process of gray crescent formation in *Xenopus* eggs. *Prog. Clin. Biol. Res.* **217B**, 349–352.

Vincent, J., and Gerhart, J. C. (1987). Subcortical rotation in *Xenopus* eggs: An early step in embryonic axis specification. *Dev. Biol.* **123**, 526–539.

Vogel, A., Rodriguez, C., and Izpisua-Belmonte, J. C. (1996). Involvement of FGF-8 in initiation, outgrowth, and patterning of the vertebrate limb. *Development (Cambridge, UK)* **122**, 1737–1750.

Wakahara, M. (1981). Accumulation, spatial distribution and partial characterization of poly(A)+RNA in the developing oocytes of *Xenopus laevis. J. Embryol. Exp. Morphol.* **66**, 127–140.

Weeks, D. L. ,and Melton, D. A. (1987). A maternal mRNA localized to the vegetal hemisphere in *Xenopus* eggs codes for a growth factor related to TGF-β. *Cell (Cambridge, Mass.)* **51**, 861–867.

Weinstein, D. C., and Hemmati-Brivanlou, A. (1997). Neural induction in *Xenopus laevis*: evidence for the default model. *Curr. Opin. Neurobiol.* **7**, 7–12.

Weisblat, D. A., Sawyer, R. T., and Stent, G. S. (1978). Cell lineage analysis by intracellular injection of a tracer enzyme. *Science* **202**, 1295–1298.

Wetts, R., and Fraser, S. E. (1988). Multipotent precursors can give rise to all major cell types of the frog retina. *Science* **239**, 1142–1145.

Whitfield, T., Heasman, J., and Wylie, C. (1993). XLPOU-60, a *Xenopus* POU-domain mRNA, is oocyte-specific from very early stages of oogenesis, and localised to presumptive mesoderm and ectoderm in the blastula. *Dev. Biol.* **155**, 361–370.

Wickens, M., Kimble, J., and Strickland, S. (1996). Translational control of developmental decisions. *In* "Translational Control" (J. W. B. Hershey, M. B. Mathews, and N. Sonenberg, Eds.), pp. 411–450. Cold Spring Harbor Laboratory, Plainview, New York.

Williams, R. W., and Goldowitz, D. (1992). Structure of clonal and polyclonal cell arrays in chimeric mouse retina. *Proc. Natl. Acad. Sci. U.S.A.* **89**, 1184–1188.

Wilson, P. A., and Hemmati-Brivanlou, A. (1995). Induction of epidermis and inhibition of neural fate by BMP-4. *Nature (London)* **376**, 331–333.

Wylie, C., Kofron, M., Payne, C., Anderson, R., Hosobuchi, M., and Heasman, E. (1996). Maternal β-catenin establishes a 'dorsal signal' in early *Xenopus* embryos. *Development (Cambridge, UK)* **122**, 2987–2996.

Xu, R. H., Kim, J., Taira, M., Zhan, S., Sredni, D., and Kung, H. F. (1995). A dominant negative bone morphogenetic protein 4 receptor causes neuralization in *Xenopus* ectoderm. *Biophys. Biochem. Res. Commun.* **212**, 212–219.

Yost, H. J., Phillips, C. R., Boore, J. L., Bertman, J., Whalon, B., and Danilchik, M. V. (1995). Relocation of mitochondria to the prospective dorsal marginal zone during *Xenopus* embryogenesis. *Dev. Biol.* **170**, 83–90.

Yuge, M., Kobayakawa, Y., Fujisue, M., and Yamana, K. (1990). A cytoplasmic determinant for dorsal axis formation in an early embryo of *Xenopus laevis. Development (Cambridge, UK)* **110**, 1051–1056.

Zhang, J., and King, M. L. (1996). *Xenopus* VegT RNA is localized to the vegetal cortex during oogenesis and encodes a novel T-box transcription factor involved in mesodermal patterning. *Development (Cambridge, UK)* **122**, 4119–4129.

Zimmerman, L. B., De Jesus-Escobar, J. M., and Harland, R. M. (1996). The Spemann organizer signal noggin binds and inactivates bone morphogenetic protein-4. *Cell (Cambridge, Mass.)* **86**, 599–606.

Zisckind, N., and Elinson, R. P. (1990). Gravity and microtubules in dorsoventral polarization of the *Xenopus* egg. *Dev. Growth Differ.* **32**, 575–581.

21

Maternal Signaling Pathways and the Regulation of Cell Fate

Daniel S. Kessler

Department of Cell and Developmental Biology
University of Pennsylvania School of Medicine
Philadelphia, Pennsylvania 19104

I. Regulation of Mesoderm and Endoderm Formation by Vg1

A. Germ Layer Formation in *Xenopus*

1. Mesodermal Development

The importance of vegetal blastomeres in the induction of mesoderm in the frog, *Xenopus laevis*, was established by Nieuwkoop and colleagues. In isolation, explanted blastula animal and vegetal pole cells form only ectoderm and endoderm, respectively, but ectoderm can be induced to form mesodermal structures in recombinants containing both presumptive ectoderm and endoderm (Nieuwkoop, 1969a; Sudarwati and Nieuwkoop, 1971; Nieuwkoop, 1973). In addition, although explants of the marginal zone (presumptive mesoderm) from a 32-cell stage embryo fail to form mesoderm, blastula stage explants will form mesoderm, implicating a progressive interaction between endoderm and ectoderm to form mesoderm (Nakamura and Takasaki, 1970; Dale and Slack, 1987a). These observations suggest that vegetal blastomeres (prospective endoderm) produce a mesoderm-inducing signal during cleavage stages.

In addition to inducing mesoderm, vegetal blas-

tomeres can confer a dorsal–ventral pattern on mesoderm. Dorsal vegetal cells induce dorsal mesoderm (notochord and muscle), whereas lateral and ventral vegetal cells induce ventrolateral mesoderm (mesenchyme, blood, and small amounts of muscle) (Nieuwkoop, 1969b; Boterenbrood and Nieuwkoop, 1973; Dale and Slack, 1987a). In addition, transplanted dorsal vegetal blastomeres can induce ectopic dorsal axial structures, an activity that has led to these cells being designated the Nieuwkoop center (Gimlich and Gerhart, 1984; Gerhart *et al.*, 1989, 1991; Kageura, 1990).

In *Xenopus*, the dorsal–ventral axis is established at fertilization with sperm entry stimulating a reorganization of egg contents by cortical rotation, leading to demarcation of future dorsal tissues opposite the site of sperm entry. Cortical rotation, a displacement of the surface, or cortex, of the egg relative to inner cytoplasm, is thought to result in the formation of a "dorsal determinant" in the presumptive endoderm (Vincent *et al.*, 1986). Disruption of cortical rotation by UV-irradiation results in loss of dorsal axial structures (Vincent *et al.*, 1986; Elinson and Rowning, 1988; Elinson and Pasceri, 1989; Gerhart *et al.*, 1989). Both cortical rotation and subsequent axis formation can be rescued by manual tipping of the egg, which causes gravity-induced re-arrangements (Gerhart *et al.*, 1981; Cooke, 1986). Mesodermal patterning continues during gastrulation, as evidenced by the fact that as late as the gastrula stage explanted lateral marginal zone tissue forms ventral mesoderm rather than the intermediate mesodermal tissues (muscle and kidney) predicted from the fate map. Organizer tissue can induce ventral and lateral marginal zone to form intermediate mesoderm, suggesting that the gastrula stage organizer "dorsalizes" neighboring lateral mesoderm (Yamada, 1950; Slack and Foreman, 1980; Dale and Slack, 1987a).

These studies illuminate the cellular basis of mesoderm induction. Cortical rotation generates a dorsal determinant that resides in dorsal vegetal blastomeres, the Nieuwkoop center, and that subsequently induces formation of Spemann's organizer. Although promising candidates for endogenous mesoderm inducers have been identified, it has not yet been possible to assign them to specific inducing functions in vivo.

2. Endodermal Development

The initiation of vertebrate endodermal development is poorly understood in any experimental system. Induction and patterning of premorphogenesis endoderm, forming epithelia of the gut and associated organs, results in the precise organization of the differentiated digestive tract. In contrast to the detailed analysis of neural and mesodermal lineages, molecular studies of early endodermal development have only more recently utilized amphibian systems. Morphogenesis of the vertebrate gut begins at gastrulation with the formation of a gut tube of endodermal epithelia surrounded by adjacent mesoderm (Nieuwkoop and Faber, 1967). While embryological studies of epithelial–mesenchymal interactions in the gut suggested an instructive role for associated mesenchyme (Gumpel-Pinot *et al.*, 1978; Fukumachi and Takayama, 1980; Haffen *et al.*, 1987), the source of endodermal tissue is the primary determinant of the type of endodermal tissue that forms in coculture experiments (Okada, 1954, 1957; Takata, 1960; Kedinger *et al.*, 1990). For example, midgut endoderm gives rise to intestinal tissues even when recombined with anterior mesenchyme. Therefore, endodermal lineages are specified prior to morphogenesis and differentiation, and their pattern underlies the anteroposterior and dorsoventral organization of the digestive tract. The processes that underlie segregation of endodermal lineages from the other germ layers are yet to be defined. Moreover, the gene products responsible for endodermal fate decisions and the pathways that generate regional pattern in primitive gut endoderm, so crucial to the future organization of the functional digestive tract, have only more recently received scrutiny.

Fate mapping of *Xenopus* embryos at the 32-cell stage indicated that vegetal blastomeres are the primary source of endodermal tissues in tadpoles (Dale and

Slack, 1987b; Moody, 1987; Bauer *et al.*, 1994). Transplantation of single vegetal blastomeres demonstrated that commitment to endodermal fate occurs by the gastrula stage, long before overt differentiation of endodermal tissue (Heasman *et al.*, 1984). Recent studies, using molecular markers of gut endoderm, indicate that blastula vegetal pole explants can initiate a program of endodermal development in the absence of mesoderm (Jones *et al.*, 1993; Gamer and Wright, 1995; Henry *et al.*, 1996). As suggested in earlier work (Wylie *et al.*, 1987), these results indicate that commitment to endodermal fate is autonomous (independent of mesodermal or neural interactions), and may be a consequence of unidentified maternal determinants present in vegetal blastomeres. In addition to founding endodermal lineages, cleavage-stage vegetal blastomeres produce inducing signals regulating both mesodermal and dorsal development (Ogi, 1967; Nieuwkoop, 1969a,b; Boterenbrood and Nieuwkoop, 1973; Gimlich and Gerhart, 1984; Gimlich, 1986; Dale and Slack, 1987a). Thus, analysis of early endodermal development is central to an understanding of germ layer induction and patterning, as well as later development of the digestive tract.

A vegetal pole explant assay has been utilized to examine the autonomous development of gut endoderm in the absence of mesodermal or neural tissues. Vegetal pole tissue is dissected from blastula stage embryos, cultured in isolation, and examined for the expression of endoderm-specific genes by reverse transcriptase–polymerase chain reaction (RT–PCR) or *in situ* hybridization. Vegetal explants lacking any detectable mesoderm express gut-specific markers when cultured in isolation. The intestinal markers 4G6 (Jones *et al.*, 1993) and IFABP (intestinal fatty acid binding protein) (Shi and Par Hayes, 1994), and the pancreatic marker Xlhbox8 (Wright *et al.*, 1988) are expressed in vegetal explants. Preparation of dorsal and ventral vegetal explants indicated that Xlhbox8 expression is restricted to dorsal vegetal tissue while IFABP is expressed throughout the vegetal pole (Fig. 1) (Henry *et al.*, 1996). Consistent with previous work, these results suggest that commitment to endodermal fate is autonomous, and may be a consequence of maternal determinants present in vegetal blastomeres (Wylie *et al.*, 1987). Moreover, regionalization of gut endoderm occurs long before cytodifferentiation and in the absence of mesodermal influences. Therefore, the analysis of the earliest stages of endodermal development should provide an understanding of primary induction events underlying germ layer segregation and axial organization, as well as later organogenesis in the digestive tract.

B. Localization and Posttranslational Regulation of Vg1

The studies of Nieuwkoop demonstrated the existence of a vegetally localized mesoderm inducing factor. Vg1, a maternal mRNA encoding a member of the transform-

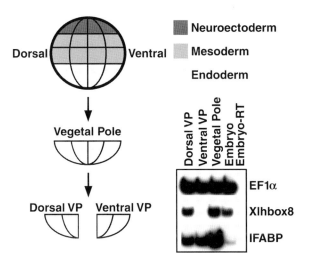

Figure 1 Autonomous expression of endodermal markers in vegetal explants. The fate map of a 32-cell embryo is represented by four tiers of cells: top tier cells (animal) form neuroectoderm, middle tier cells (marginal) form mesoderm, and bottom tier cells (vegetal) form endoderm. RT–PCR analysis of cultured vegetal explants reveal autonomous expression of the pancreas marker Xlhbox8 and the intestinal marker IFABP. Analysis of dorsal or ventral vegetal explants indicates that Xlhbox8 expression is restricted to dorsal regions. Vegetal explants are devoid of mesoderm (data not shown), and therefore, patterned expression of endodermal markers is likely due to localized determinants. For all RT–PCR experiments presented ubiquitous EF1α is a control for RNA recovery and loading. Sibling embryos served as a positive control (Embryo) and an identical reaction without reverse transcriptase controlled for PCR contamination (Embryo-RT). This figure is adapted from Henry et al., 1996.

ing growth factor β (TGFβ) superfamily, is localized to the vegetal pole of *Xenopus* eggs and early embryos (Rebagliati *et al.*, 1985; Weeks and Melton, 1987; reviewed by Vize and Thomsen, 1994). Vg1 mRNA is synthesized early in oogenesis and becomes tightly localized to the vegetal cortex by the end of oogenesis (Fig. 2A) (Melton, 1987; Yisraeli and Melton, 1988; Mowry and Melton, 1992), and therefore, Vg1 mRNA and protein become partitioned within vegetal pole blastomeres (inducing tissue) in the early embryo (Dale *et al.*, 1989; Tannahill and Melton, 1989). TGFβ-like proteins are synthesized as inactive precursors, which form disulfide-linked dimers and are proteolytically cleaved, releasing a mature C-terminal bioactive dimer (Massague *et al.*, 1994). Although abundantly expressed, endogenous Vg1 protein accumulates as an unprocessed precursor (46 kDa) and little or no mature Vg1 (18 kDa) has been detected (Dale *et al.*, 1989; Tannahill and Melton, 1989; Thomsen and Melton, 1993), suggesting that processing of Vg1 is tightly regulated during development. Consistent with this observation, injection of embryos with Vg1 mRNA produces high levels of Vg1 precursor, but no processed protein and, consequently, neither mesoderm induction nor developmental effects are observed (Dale *et al.*, 1989; Tannahill and Melton, 1989; Thomsen and Melton, 1993). In contrast, injection of activin mRNA directs efficient production of mature protein and ani-

mal pole explants are induced to form mesoderm (Thomsen *et al.*, 1990). Therefore, although Vg1 localization suggests that it is an *in vivo* mesoderm-inducing signal, posttranslational regulation of Vg1 processing may negatively regulate Vg1 activity.

C. Mesoderm Induction by Processed Vg1

1. Injection of Chimeric Vg1 mRNA

To stimulate production of mature Vg1, hybrid Vg1 molecules have been prepared, consisting of the N-terminal proregion and dibasic cleavage site of a bone morphogenetic protein (BMP) fused to the C-terminal mature region of Vg1 (Fig. 2B). Microinjection of BMP-Vg1 mRNA directed synthesis and processing of this hybrid molecule, resulting in efficient production of mature Vg1 protein. Expression of processed Vg1 in animal pole explants strongly induced dorsal mesoderm. However, blood, a ventral mesodermal tissue, is not induced, suggesting that additional factors are required during normal development. Injection of BMP-Vg1 mRNA into UV-ventralized embryos directed formation of a complete dorsal axis (Dale *et al.*, 1993; Thomsen and Melton, 1993). These observations suggest that a transient or localized production of mature Vg1 ligand may be sufficient for induction of dorsal mesoderm. It has been proposed that cortical rotation may stimulate Vg1 processing in dorsal vegetal cells, perhaps by localized translation or activation of posttranslational processing (Thomsen and Melton, 1993; Klein and Melton, 1994). Detection of endogenous, mature Vg1 and a description of its temporal and spatial regulation is now needed to substantiate the role of Vg1 *in vivo*.

2. Production of Bioactive, Mature Vg1 Protein

In contrast to the BMP-Vg1 hybrids, functional activin βB ligand is efficiently secreted in both COS cells and oocytes (Thomsen *et al.*, 1990). Therefore, to obtain soluble mature Vg1, an additional hybrid molecule was prepared containing the activin βB signal sequence, proregion, tetrabasic cleavage site, and four amino acids of the activin βB mature region fused to the mature region of Vg1 (Fig. 2B). Defolliculated oocytes were injected with 50 ng of *in vitro* transcribed RNA and cultured for 3 days. Injection of activin βB mRNA resulted in secretion of processed protein, detected as a reduced monomer of ~14 kDa. Injection of activin-Vg1 mRNA resulted in secretion of abundant mature Vg1, detected as a series of bands of ~18 kDa (Fig. 2C). The different mature species presumably reflect varying glycosylation states, as described for the endogenous precursor (Dale *et al.*, 1989, 1993; Tannahill and Melton, 1989). The ability of activin-Vg1 to direct secretion of mature protein was comparable in efficiency to activin βB, and in both cases secreted precursor (~46 kDa) is also detected. No secreted mature Vg1 was detected fol-

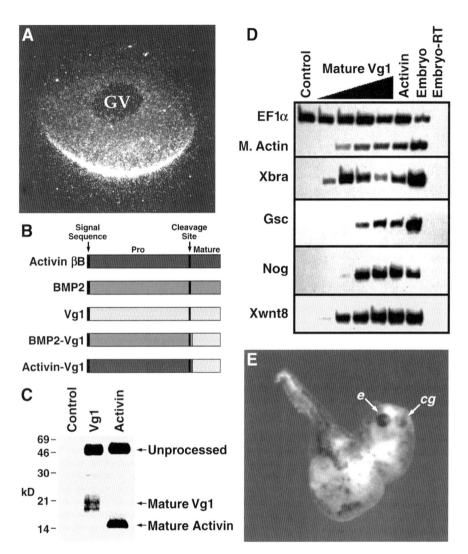

Figure 2 Mesoderm induction by processed Vg1. (A) *In situ* hybridization analysis of a sectioned *Xenopus* oocyte reveals a tight localization of Vg1 mRNA (white autoradiographic grains) to the cortex of the vegetal pole. The centrally located structure is the germinal vesicle (GV). (B) Schematic representation of activin βB, BMP2, Vg1, and chimeric BMP2-Vg1 and activin-Vg1 proteins. Members of the TGFβ superfamily, these gene products contain a signal sequence, proregion, tetrabasic cleavage site, and mature region. Activin βB and BMP2, but not Vg1, form disulfide-linked dimers that are subsequently cleaved, releasing the mature C-terminal peptide as a secreted bioactive dimer. The chimeric constructs, fused four amino acids downstream of the cleavage site, are designed to facilitate processing and secretion of mature Vg1. (C) Defolliculated oocytes were injected with *in vitro* transcribed mRNA and incubated with ^{35}S-methionine. Following 3 days of incubation, supernatants were collected and analyzed by 15% SDS–PAGE and autoradiography. No labeled proteins were detected in the supernatants of uninjected oocytes (Control), while proteins of the appropriate size for mature Vg1 and activin were detected in activin-Vg1 (Vg1) or activin-injected oocytes. (D) Blastula stage animal pole explants were treated with increasing doses of mature Vg1 supernatant, or control or activin βB supernatant, and analyzed at the neurula stage by RT–PCR. At low doses the general marker Xbra is induced; at intermediate doses the ventrolateral marker Xwnt8 and the dorsolateral marker muscle actin (M. Actin) are induced; and at high doses the dorsoanterior markers goosecoid (Gsc) and noggin (Nog) are induced. Activin βB treatment results in a similar response and control supernatants have no effect. (E) Blastula animal pole explants treated with a high dose of mature Vg1 and cultured to the late tadpole stage differentiate into embryoids displaying a rudimentary axial organization, with anterior–posterior pattern and head structures. The embryoid has a clear head to tail pattern with pigmented eye (e) and cement gland (cg). Treatment with supernatant of uninjected oocytes has no effect and explants form atypical epidermis (data not shown). (Part of this figure is adapted from Kessler and Melton, 1995.)

lowing injection of native Vg1 mRNA, and a small amount of mature Vg1 secretion was directed by BMP2-Vg1 (data not shown).

3. Induction of Dorsal Mesoderm by Mature Vg1 Protein

Oocyte supernatants were tested for mesoderm inducing activity on animal pole explants. Blastula-stage animal poles (prospective ectoderm) were explanted, incubated with supernatant, cultured to the neurula stage, and scored for the formation of mesodermal tissues and expression of mesodermal markers. Supernatants of uninjected, Vg1 or BMP2-Vg1 injected oocytes all failed to induce morphogenetic movements or expression of mesodermal markers (data not shown). Supernatants of activin-Vg1 or activin βB injected oocytes strongly induced morphogenetic movements indicative of mesoderm induction, expression of mesodermal markers, and differentiation of the mesodermal tissues, muscle and notochord (Fig. 2D and data not shown).

Low doses of mature Vg1 induced expression of the general mesodermal marker Xbra (Smith *et al.*, 1991). Intermediate doses induced the ventrolateral marker Xwnt-8 (Christian *et al.*, 1991) and the dorsal marker muscle actin (Stutz and Sphor, 1986). High doses of mature Vg1 induced expression of the dorsal mesodermal markers goosecoid (Blumberg *et al.*, 1991) and noggin (Smith and Harland, 1992). These markers were also induced by activin βB, which was a positive control, but not by supernatant of uninjected oocytes (Fig. 2D). Globin, a definitive marker of ventral mesoderm was not induced by Vg1 or activin (data not shown). Examination of mature Vg1-treated explants cultured to the late neurula stage by immunohistochemistry and histology revealed the differentiation of the mesodermal tissues, notochord, and somitic muscle, as well as neural tissue (data not shown). Explants cultured to the late tadpole stage, following treatment with mature Vg1, often developed a high degree of axial organization. These "embryoids" displayed a rudimentary anterior–posterior pattern with organized head structures, including eyes, and a functional neuromuscular system (Fig. 2E).

4. Inhibition of Dorsal Mesoderm Formation by Inhibitors of Vg1

To address the relation between mature Vg1 and endogenous mesoderm-inducing signals the ability of dominant inhibitory receptors to block the activity of mature Vg1 was tested. A truncated form of an activin type II receptor (tAR), lacking the intracellular kinase domain, blocked formation of detectable mesoderm and expression of mesodermal markers in embryos (Hemmati-Brivanlou and Melton, 1992). Animal pole explants of embryos injected with 4 ng of tAR mRNA were treated with mature Vg1 or activin βB. tAR expression fully inhibited induction of morphogenetic movements and expression of brachyury and cardiac actin by mature Vg1. As expected, tAR also blocked activin βB activity and resulted in NCAM expression regardless of treatment with supernatants (Fig. 3A) (Schulte-Merker *et al.*, 1994). A dominant inhibitory FGF receptor (tFGFR), lacking the kinase domain, blocked formation of trunk and posterior mesoderm, resulting in loss of axial mesoderm and tail structures (Amaya *et al.*, 1991). In addition to blocking FGF activity, tFGFR also blocks mesoderm induction by activin βB in animal pole explants (Cornell and Kimelman, 1994; LaBonne and Whitman, 1994). Injection of 4 ng of tFGFR mRNA resulted in a complete block of mesoderm induction by both mature Vg1 and activin βB (Fig. 3A) (Schulte-Merker *et al.*, 1994). These observations suggest that the ability of the truncated activin and FGF receptors to

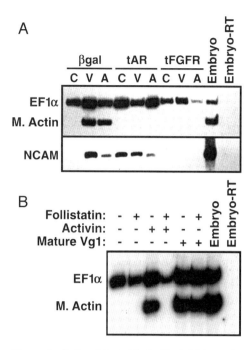

Figure 3 Vg1 activity is blocked by inhibitors of mesodermal development. (A) Truncated activin and FGF receptors block muscle actin induction by mature Vg1. At the 2-cell stage embryos were injected with 4 ng of mRNA encoding the truncated activin type II receptor (tAR), the truncated FGF receptor (tFGFR) or β-galactosidase (βgal). Blastula stage animal pole explants were prepared and treated with mature Vg1 (V), activin βB (A), or control (C) supernatants. Following incubation to the neurula stage samples were examined by RT–PCR. The truncated receptors fully block muscle actin (M. Actin) induction by both mature Vg1 and activin, and tAR stimulates NCAM expression regardless of treatment. (B) Follistatin fails to inhibit muscle actin (M. Actin) induction by mature Vg1. Blastula animal pole explants were treated with mature Vg1 or activin only, or with these supernatants preincubated with follistatin. RT–PCR analysis at the neurula stage indicates that while follistatin fully inhibits muscle actin induction by activin, no effect on mature Vg1 activity is detected. (Part of this figure is adapted from Kessler and Melton, 1995.)

perturb mesoderm induction in the embryo may, in fact, be due to an inhibition of endogenous Vg1 signaling. In contrast, animal cap expression of a truncated BMP2/4 receptor (tBR), capable of converting ventral mesoderm to dorsal mesoderm in embryos, (Graff *et al.*, 1994; Suzuki *et al.*, 1994) failed to inhibit the mesoderm inducing activity of mature Vg1 (data not shown).

A natural inhibitor of activin function is the activin-binding protein follistatin (Nakamura *et al.*, 1990; Kogawa *et al.*, 1991). *Xenopus* follistatin is maternally expressed and can block activin induction of animal pole explants (Tashiro *et al.*, 1991; Fukui *et al.*, 1993, 1994; Hemmati-Brivanlou *et al.*, 1994). The ability of follistatin to inhibit mature Vg1 activity was examined. Mature Vg1 or activin supernatants were combined with an equal volume of *Xenopus* follistatin (XFS-319, Hemmati-Brivanlou *et al.*, 1994) supernatant prior to addition to blastula animal pole explants. At these doses XFS fully blocked activin-induced morphogenetic movements and expression of muscle actin, whereas mature

Vg1 activity was unaffected (Fig. 3B). Consistent with the inability of follistatin to inhibit Vg1 activity, injection of whole embryos with XFS mRNA did not perturb development of dorsal mesoderm (data not shown). In fact, follistatin resulted in a consistent loss of posterior structures and an apparent enhancement of dorsoanterior structures, an effect likely due to the inhibition of BMP4, a TGFβ-related factor implicated in ventral development.

The precise correspondence between the inhibition profile of mature Vg1 and that of endogenous dorsal mesoderm induction supports a role for Vg1 in dorsal mesoderm induction. In conjunction with the potent activity of mature Vg1 and the vegetal localization of Vg1 mRNA and protein (the endogenous source of mesoderm inducing signals), the observations establish Vg1 as a strong candidate for the natural inducer of dorsal mesoderm during *Xenopus* development.

D. Endodermal Specification by Processed Vg1

Numerous protein factors can induce formation of mesodermal and neural tissue during *Xenopus* embryonic development (reviewed in Kessler and Melton, 1994). More recently, it has been determined that some of these factors also can induce endodermal markers in animal pole explants. For example, the mesoderm-inducing molecules FGF, activin, and chimeric BMP2-Vg1 (BVg1), induce endodermal markers (Jones et al., 1993; Gamer and Wright, 1995; Henry et al., 1996). Using inhibitors of signaling, the signal transduction pathways responsible for expression of endodermal markers also have been examined. Dominant inhibitory mutants of ras or an activin receptor (Hemmati-Brivanlou and Melton, 1992; Whitman and Melton, 1992), both inhibitors of activin and Vg1, block endoderm formation in vegetal explants (Gamer and Wright, 1995; Henry et al., 1996). The potential contributions of activin and Vg1 to endogenous endoderm induction are distinguished by two observations. First, follistatin, an inhibitor of activin (Nakamura et al., 1990; Asashima et al., 1991; Kogawa et al., 1991) that does not inhibit Vg1 activity (Schulte-Merker et al., 1994; Kessler and Melton, 1995), fails to inhibit the expression of endodermal markers in vegetal explants (Henry et al., 1996). Second, the localization of endogenous Vg1, unlike other inducers, is consistent with a role in endodermal development. Therefore, in addition to its described mesoderm inducing activity, Vg1 may be an endogenous regulator of endodermal development.

1. Induction of Endodermal Gene Expression by Vg1

To examine the role of Vg1 in endodermal development, the ability of mature Vg1 protein to induce en-

dodermal markers in ectodermal explants was tested. The induction of a pancreatic gene (Xlhbox8) and an intestinal gene (IFABP) was assessed using an RT–PCR assay. Blastula-stage ectodermal explants were treated with mature Vg1 protein prepared as described (Kessler and Melton, 1995), and following culture to the tadpole stage, the expression of tissue-specific markers was scored. Significantly, in the picomolar range mature Vg1 induced both Xlhbox8 and IFABP (Fig. 4A). In addition, endodermin, a marker of the primitive gut (Sasai et al., 1996), and 4G6, an endodermal antigen restricted to intestinal epithelium (Jones et al., 1993), were induced in response to mature Vg1 (data not shown). The expression of both anterior (Xlhbox8, pancreas) and posterior (IFABP and 4G6, small intestine) endodermal markers indicates that Vg1 induces endodermal lineages present throughout the digestive tract.

The dose of mature Vg1 required for endodermal marker induction was similar to that resulting in muscle actin induction (data not shown), and no dose tested resulted in the exclusive expression of endodermal markers or muscle actin. The results demonstrate that mature Vg1 protein is a potent inducer of endodermal markers. However, the induction of endoderm and dorsal mesoderm cannot be disjoined in these experiments, and therefore the primary inducing activity of Vg1, whether endodermal or mesodermal or both, cannot be discriminated.

2. Specification of Endodermal Fate by Mature Vg1

The ability of Vg1 to direct cells to an endodermal fate was examined by lineage-mapping of injected embryos. At the 32-cell stage, a prospective neural blastomere (dorsal-animal) was injected with a combination of BVg1 and β-galactosidase (βgal) mRNAs or βgal mRNA alone. Embryos were cultured to the late tadpole stage, βgal visualized, and samples analyzed histologically. Since morphogenesis and terminal differentiation of the digestive tract occurs in the late tadpole, a definitive identification of endodermal tissues was possible. While control injections of βgal mRNA resulted in labeling of neural tissues, but not endodermal lineages, processed Vg1 directed a complete change in cell fate. Rather than forming neural derivatives, labeled cells populated endodermal lineages of the gut, including the epithelial lining of the pharynx, pharyngeal pouches, esophagus, liver, stomach, and small intestine (Fig. 4B–E). The morphology and location of the labeled tissues suggested that Vg1-expressing cells differentiated as endodermal epithelium. The results demonstrated the endodermal fate of Vg1-expressing cells, and support a role for Vg1 in endodermal specification during normal development.

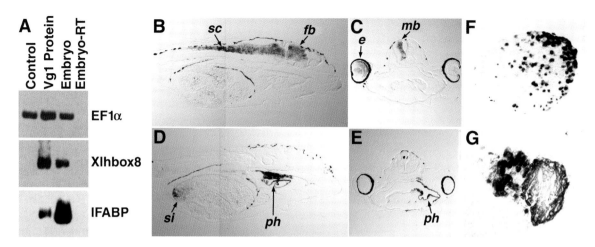

Figure 4 Specification of endodermal fate by mature Vg1. (A) Blastula-stage animal pole explants were incubated in control (Control) or mature Vg1-containing (Vg1 Protein) supernatant, and at the tailbud stage gene expression was analyzed by RT–PCR. Mature Vg1 induced the endoderm-specific genes Xlhbox8 (pancreas) and IFABP (intestine). (B–E) A single dorsal-animal blastomere was injected at the 32-cell stage with β-galactosidase (βgal) lineage marker (B, C) or βgal plus BVg1 (D, E) mRNAs. Following culture to the point of morphogenesis and differentiation of the digestive tract (stage 46), βgal-positive cells were visualized and embryos analyzed by sagital (B, D) and transverse sectioning (C, E). βgal mRNA injection resulted in dorsal midline staining of the central nervous system, including forebrain (fb), midbrain (mb), eye (e), and spinal cord (sc), but not tissues of the digestive tract. BVg1 mRNA injection resulted in a complete change in position of βgal-positive cells with staining present in the epithelial lining of the pharynx and the pharyngeal pouches (ph) and small intestine (si), as well as liver (data not shown). The tissues populated by the βgal-positive progeny of BVg1-injected cells are definitively endodermal by histological criteria. (F, G) At the 32-cell stage a single animal blastomere was injected with βgal (F) or βgal plus BVg1 (G) mRNAs, animal pole explants were prepared at the blastula stage and cultured to the tadpole stage. Explants were stained in whole-mount for βgal activity (blue), using chromogenic substrate, for somitic mesoderm (brown) with a muscle-specific antibody (12/101) and were subsequently sectioned. (F) Control explants contained βgal-positive cells interspersed with nonstaining cells, but no muscle. (G) Cells expressing mature Vg1 (βgal-positive) formed a coherent group that did not stain for muscle, while a group of adjacent cells not expressing Vg1 (βgal-negative) differentiated as muscle. (Adapted from Kessler and Melton, 1998.)

3. Cells Expressing Mature Vg1 Adopt Endodermal Fate

Given the previously described mesoderm-inducing activity of mature Vg1, an apparent discrepancy may be perceived in these results. However, during normal development, mesoderm-inducing activity is initially displayed by vegetal blastomeres that are founders of endodermal lineages. Therefore, endodermal tissue specified by Vg1 may subsequently induce mesoderm. Alternatively, processed Vg1 may induce a mesendodermal precursor that can contribute to both lineages. To distinguish these possibilities, expression of processed Vg1 and βgal in ectodermal explants was used to examine the fate of Vg1-expressing cells and the fate of non-expressing neighbors.

At the 32-cell stage a single animal blastomere was injected with either βgal mRNA or a combination of βgal and BVg1 mRNAs, and blastula ectodermal explants were prepared. Because only one blastomere was injected with mRNA, explants contained a group of Vg1-expressing cells surrounded by nonexpressing neighbors. Following culture to the tadpole stage, βgal-positive cells were visualized and explants were analyzed by immunocytochemistry. Explants expressing mature Vg1 contained a coherent group of βgal-positive cells adjacent to differentiated muscle that was βgal-negative (Fig. 4F, G). Using histological criteria and additional

tissue-specific molecular markers, βgal-positive tissue appeared to be neither mesodermal nor neural. These observations suggest that cells expressing processed Vg1 become organized as a spatially discrete group and induce mesodermal differentiation in neighboring cells. At a later stage, the fate of Vg1-expressing cells was examined using 4G6, an antigen marker of intestinal epithelium (Jones *et al.*, 1993). 4G6 was expressed in the majority of βgal-positive cells (data not shown), consistent with an endodermal fate for Vg1-expressing cells. Moreover, few 4G6-positive cells were present outside of βgal-positive areas. In control explants, βgal-positive cells were scattered throughout the explant and only atypical epidermis was observed. These observations suggest that cells expressing mature Vg1 are organized into a discrete group of cells that adopt endodermal fate and induce mesodermal differentiation in neighboring cells.

The formation of both dorsal mesoderm and endoderm in induced animal pole explants raises questions about the lineage relation and interaction of these tissues. In explants, mesoderm and endoderm may be derived from an induced mesendodermal precursor, or alternatively, one tissue may be initially induced and, by subsequent interactions, result in formation of the second tissue. Lineage mapping of induced explants indicated that a coherent group of Vg1-expressing cells adopted an endodermal fate while adjacent, nonexpressing cells, formed dorsal mesoderm. Consistent with these results, Minuth and Grunz (1980) observed that formation of endoderm in explants preceeds mesoderm, and subsequent secondary interactions with noninduced ectoderm results in mesoderm formation. Furthermore, previous studies showed that disaggregation and reaggregation of a mixed cell population, derived from distinct germ layers, resulted in sorting into two homogeneous groups of cells, one internal to the other (Townes and Holtfreter, 1955; Turner *et al.*, 1989). The segregation of Vg1-expressing endodermal cells from neighboring dorsal mesoderm is consistent with the sorting of germ layers, and this suggests that cells secreting mature Vg1 adopt endodermal fates, perhaps by an autocrine mechanism, and induce dorsal mesoderm in adjacent cells. Furthermore, the induction of both anterior and posterior endodermal markers suggested that mature Vg1 acts early in endodermal fate decisions without restriction to particular organs or anteroposterior position, resulting in specification of endodermal lineages present throughout the digestive tract.

E. Is Vg1 a Functional Inducer during Vertebrate Embryogenesis?

Chimeric molecules, facilitating the production of mature Vg1 protein, are used to analyze Vg1 function. Although endogenous Vg1 precursor is abundant, posttranslational processing is tightly regulated and the endogenous levels of mature Vg1 are below the sensitivity of standard detection methods (Dale *et al.*, 1989; Tannahill and Melton, 1989). In fact, since mature Vg1 is active in the picomolar range (Kessler and Melton, 1995), expression of an inducing quantity of chimeric Vg1 does not result in levels of mature protein sufficient for immunodetection. These observations suggest that endogenous mature Vg1, present at levels far below detectability, would be a potent inducer.

Analysis of the zebrafish (zVg1) and chick (cVg1) homologs of Vg1 (Helde and Grunwald, 1993; Seleiro *et al.*, 1996; Shah *et al.*, 1997) revealed a conservation of Vg1 regulation and function in other vertebrate systems (Dohrmann *et al.*, 1996; Seleiro *et al.*, 1996; Shah *et al.*, 1997). Like *Xenopus* Vg1, zVg1 is present maternally as an inactive precursor, suggesting a similar degree of posttranslational regulation. Interestingly, zVg1 expressed in *Xenopus* was properly cleaved, forming a mature dimer with potent mesoderm- and endoderm-inducing activity (Dohrmann *et al.*, 1996) (D.S. Kessler, 1997, unpublished). Thus, under appropriate conditions, a native Vg1 precursor can direct formation of bioactive protein. In chick, cVg1 is expressed in the posterior marginal zone, a region with demonstrated axis-inducing activity. Similar to Vg1 and zVg1, processing of cVg1 is tightly regulated,

but when processed, potent inducing activity is observed (Seleiro *et al.*, 1996; Shah *et al.*, 1997). These results indicate that regulation of processing is not unique to *Xenopus* and may be a conserved mechanism for restricting Vg1 activity in vertebrates. Furthermore, the specific activity and induced tissues were similar for each homolog, demonstrating conservation of function as well.

A truncated activin type II receptor, lacking the cytoplasmic kinase domain, fully inhibited mesodermal and axial development (Hemmati-Brivanlou and Melton, 1992). Due to the broad inhibitory effects of this receptor (block of activin, BMP4, Vg1, and Xnr2 activity), it was not possible to determine which inducer(s) was responsible for the observed defects. In contrast, an activin type II receptor lacking both the transmembrane and cytoplasmic domains inhibited activin, but not Vg1, BMP4, or Xnr2 (Dyson and Gurdon, 1997), consistent with the ligand binding properties of this receptor (Kessler and Melton, 1995). In embryos expressing this activin-specific inhibitor, mesoderm develops, despite a delay, and although anterior development is abnormal, axial structures are clearly present (Dyson and Gurdon, 1997). Given that BMP4 inhibition fails to block dorsal mesoderm or axis induction in embryos (Graff *et al.*, 1994; Suzuki *et al.*, 1994; Hawley *et al.*, 1995), and that the described nodal-related factors are not maternally expressed, the combined results implicate Vg1, or a Vg1-like factor, in early induction.

Expression of putative dominant negative mutants of Vg1 has provided direct evidence of a requirement for functional Vg1 during early development. Due to their homodimeric structure, mutation of conserved residues in the mature region of TGFβ-related molecules permits specific inhibition of endogenous inducer function (Wittbrodt and Rosa, 1994). Preliminary results of mutant Vg1 expression reveal defects in axial development, mesoderm induction, and, importantly, loss of endodermal marker expression (E. Josephs and D. Melton, 1998, unpublished). These observations, suggesting a requirement for endogenous Vg1 function in endodermal development and subsequent processes, support the present conclusions.

II. Dorsal Specification by Maternal Wnt Signaling and Zygotic Siamois

A. Dorsal Development in *Xenopus*

In *Xenopus*, maternal factors establish dorsoventral pattern in the cleavage embryo, resulting in formation of Spemann's organizer at the gastrula stage (Kessler and Melton, 1994). Maternal dorsal determinants, localized to the vegetal pole at fertilization, are displaced by cortical rotation to the future dorsal domain of the cleavage embryo (reviewed in Elinson and Kao, 1989; Fujisue *et al.*, 1993; Holowacz and Elinson, 1993; Kikkawa *et al.*, 1996; Sakai, 1996), a region defined functionally as the Nieuwkoop center. An early response to these determinants is the nuclear accumulation, in dorsal blastomeres, of β-catenin, a component of the wnt pathway required for dorsal development (Heasman *et al.*, 1994; reviewed in Miller and Moon, 1996; Schneider *et al.*, 1996; Larabell *et al.*, 1997). These observations suggest that stimulation of a maternal wnt pathway upstream of, or at β-catenin results in dorsal development. While the identified components of the wnt pathway are maternally expressed (reviewed in Moon *et al.*, 1997), transcriptional targets that respond to maternal signals and are zygotic effectors of dorsal development are not yet defined. A strong candidate for a zygotic effector of maternal dorsal signals is the wnt-inducible factor siamois (Lemaire *et al.*, 1995).

B. Induction of Siamois by the Wnt Pathway

The homeobox gene siamois was isolated in a functional screen for factors with axis-inducing activity (Lemaire *et al.*, 1995). Siamois is expressed in dorsal blastomeres at the mid-blastula transition and ventral injection of siamois mRNA results in complete axial duplications. In contrast to other zygotic axis inducers, siamois expression is stimulated by components of the wnt signaling pathway (Xwnt8, frizzled, dishevelled, dominant negative GSK3β, APC, and β-catenin), but not by other factors regulating axial or mesodermal development (Fig. 5A, B) (Brannon and Kimelman, 1996; Carnac *et al.*, 1996; Yang-Snyder *et al.*, 1996; Fagotto *et al.*, 1997; Vleminckx *et al.*, 1997). The induction of siamois by wnt signaling is unaffected by cycloheximide (Fig. 5C), consistent with a direct transcriptional activation mediated by a nuclear complex of β-catenin and LEF-1/XTcf-3 (Behrens *et al.*, 1996; Huber *et al.*, 1996; Molenaar *et al.*, 1996; Brannon *et al.*, 1997). In animal explants, siamois activates expression of organizer-specific genes, in the absence of mesodermal gene expression or differentiation (Fig. 5D) (Carnac *et al.*, 1996; Fagotto *et al.*, 1997). The results suggest a role for siamois in organizer formation, while other zygotic factors such as noggin, chordin, and goosecoid, are likely to play a role in organizer function.

C. Regulation of Organizer Formation by Siamois

1. Siamois Functions as a Transcriptional Activator

Siamois is composed of a C-terminal paired-type homeodomain and N-terminal sequences unrelated to previously described transcriptional regulatory domains,

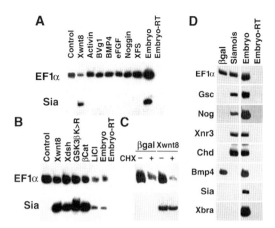

Figure 5 Direct activation of siamois by the wnt pathway and regulation of organizer gene expression by siamois. (A) At the 2-cell stage, animal pole blastomeres were injected with mRNAs of the indicated inducers, explants prepared at the blastula stage and analyzed by RT–PCR at the early gastrula stage. Siamois (Sia) expression is strongly induced by Xwnt8 and not by other inducers. The activity of the other inducers was verified using additional markers (data not shown). (B) Components of the wnt signaling pathway were tested, as in (A), for the ability to induce siamois. Siamois was induced by each component including Xwnt8, dishevelled (Xdsh), dominant negative GSK3β (GSK3β K>R), β-catenin (βCat), and the GSK3β inhibitor LiCl. (C) At the 1-cell stage, embryos were injected with β-galactosidase (βgal) or Xwnt8 mRNA and at the early blastula stage (prior to the onset of zygotic transcription) animal explants were prepared. Explants were cultured in the presence or absence of the protein synthesis inhibitor cycloheximide (CHX) and were harvested at the gastrula stage for RT–PCR analysis. Siamois induction in response to Xwnt8 was unaffected by cycloheximide, indicating that siamois is a direct transcriptional response to the wnt pathway. (D) Siamois or βgal mRNA was injected into the animal pole of a 1-cell stage embryo and blastula-stage animal explants were harvested for RT–PCR analysis at the gastrula stage. Siamois induced expression of several organizer-specific genes, including goosecoid (Gsc), noggin (Nog), nodal-related 3(Xnr3), and chordin (Chd). BMP4, a marker of ventral mesoderm and ectoderm, was repressed. The panmesodermal marker brachyury (Xbra) was not induced, as was the case for endogenous siamois.

so it is unclear whether siamois functions as a transcriptional activator or repressor. To define the transcriptional activity of siamois that results in axis induction, well-characterized regulatory domains, the HSV VP16 activator (Sadowski *et al.*, 1988; Triezenberg *et al.*, 1988) or the *Drosophila* engrailed repressor (Jaynes and O'Farrell, 1991; Han and Manley, 1993; Badiani *et al.*, 1994), were fused to the siamois homeodomain, and the axis inducing activity of the fusion proteins was determined (Fig. 6A). At the 4-cell stage, a single ventral blastomere was injected with mRNA encoding siamois, the VP16-siamois fusion (VP16-Sia), or the engrailed-siamois fusion (Eng-Sia), and axial development was assessed at the tailbud stage. Ventral injection of VP16-Sia induced complete axial duplication at a frequency similar to siamois, and Eng-Sia did not induce axis formation (Fig. 6B–E). This indicates that siamois functions as a transcriptional activator in inducing axial development.

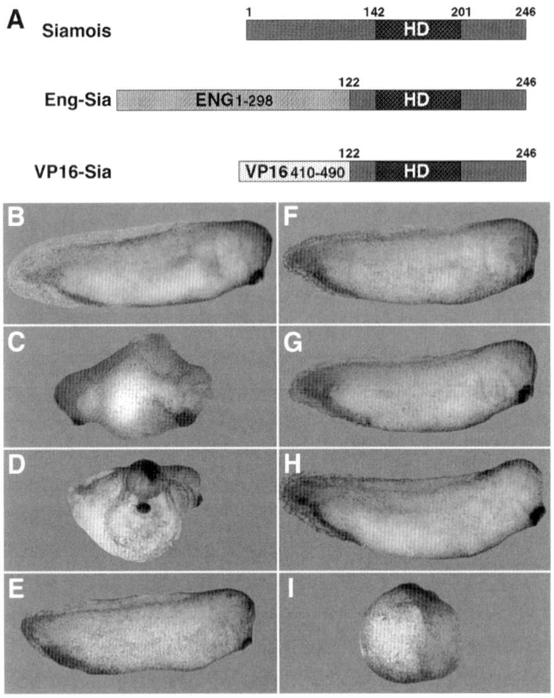

Figure 6 Axis formation is affected by siamois fusions. (A) Schematic of the siamois fusion constructs. A C-terminal region of siamois (residues 122–246), containing the homeodomain, was fused to the engrailed repressor (residues 1–298) (Eng-Sia) or the VP16 activator (residues 410–490) (VP16-Sia). At the 4-cell stage, one ventral (B–E) or two dorsal (F–I) blastomeres were injected with 30 pg of β-galactosidase (B, F), siamois (C, G), VP16-Sia (D, H), or Eng-Sia (E, I) mRNA. Like native siamois, ventral injection of VP16-Sia induced complete axial duplication, while Eng-Sia had no effect. In contrast, dorsal injection of Eng-Sia fully blocked axial development, while siamois and VP16-Sia had no effect. (Adapted from Kessler, 1997.)

2. *An Engrailed Repressor-Siamois Homeodomain Fusion Blocks Organizer Formation*

The demonstration that siamois functions as a transcriptional activator suggested the possibility that Eng-Sia may antagonize endogenous siamois by targeting an active repressor to siamois-binding sites. To assess the potential dominant inhibitory activity of Eng-Sia, both dorsal blastomeres of 4-cell stage embryos were injected with mRNAs encoding siamois, VP16-Sia, or Eng-Sia (Fig. 6F–I). Dorsal injection of Eng-Sia resulted in a severe or complete loss of axial development. Greater than 90% of injected embryos displayed a reduction or loss of anterior and axial structures, resulting in the absence of somitic muscle, notochord, and neural tube. In contrast, dorsal injection of siamois or VP16-Sia had no effect on axis formation. The reciprocal response to VP16-Sia and Eng-Sia suggests that the activator function of siamois is required for axial development and that Eng-Sia inhibits axis formation by repressing transcriptional targets of endogenous siamois. In control experiments, embryos were injected with the individual protein domains used to construct the fusions, including the VP16 activator, the engrailed repressor, and the siamois homeodomain, as well as siamois lacking the homeodomain. In no case did these individual domains alter axial development (data not shown).

The inhibitory action of Eng-Sia is predicted to result from a specific block of endogenous siamois function. However, it is possible that the siamois homeodomain, as part of an overexpressed fusion protein, may bind targets not normally regulated by siamois and inhibit additional transcriptional regulators, particularly other homeodomain proteins. To assess the specificity of axis inhibition by Eng-Sia, the ability of Xwnt8 (Smith and Harland, 1991; Sokol *et al.*, 1991), β-catenin (Funayama *et al.*, 1995), and siamois (Lemaire *et al.*, 1995) to rescue axis formation in Eng-Sia-injected embryos was examined. At the 4-cell stage, both dorsal blastomeres were coinjected with mRNAs encoding Eng-Sia and β-galactosidase, Xwnt8, β-catenin, or siamois. No rescue of axial development was observed with coinjection of Eng-Sia and Xwnt8 or β-catenin, consistent with a role for siamois downstream of the wnt pathway. In contrast, a 3-fold excess of siamois resulted in complete rescue of axial development, indicating that the effects of Eng-Sia are due to a block of endogenous siamois (Fig. 7A–E).

The inducing activity of siamois, the induction of siamois by the wnt pathway, and the inability of Xwnt8 and β-catenin to rescue axis inhibition by Eng-Sia suggest that siamois acts as a transcriptional mediator of wnt signaling. To test this idea, the ability of Eng-Sia to inhibit axis induction by the "upstream" factors, Xwnt8 and β-catenin, and the "downstream" factors, noggin and chordin, as well as siamois itself, was examined. At the 4-cell stage, a single ventral blastomere was injected with Xwnt8, β-catenin, siamois, noggin, or chordin mRNAs in combination with β-galactosidase or Eng-Sia mRNA. Whereas axial duplication in response to noggin and chordin was unaffected by Eng-Sia, axis induction by Xwnt8, β-catenin, and siamois (at a dose equal to Eng-Sia) was fully blocked (data not shown). The results indicate that siamois mediates transcriptional responses to the wnt pathway that are required for axis induction. Furthermore, axis induction by noggin and chordin in the presence of Eng-Sia suggests that these factors act downstream of siamois, consistent with the ability of siamois to induce organizer-specific genes, including noggin and chordin.

Eng-Sia is likely to function by targeting the repressor domain to specific sequences bound by the siamois homeodomain; therefore, altering the DNA-binding properties of Eng-Sia should lessen or abolish activity. The requirement for specific DNA binding by siamois and Eng-Sia was examined by generating a mutation in residue 50 (amino acid 9 of helix 3) of the

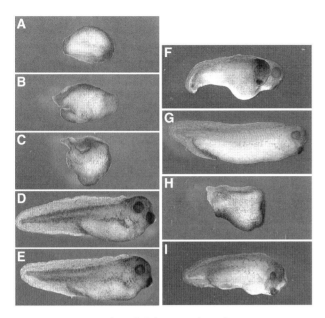

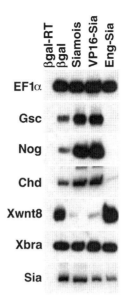

Figure 7 Rescue of axial defects resulting from Eng-Sia injection and a requirement for specific DNA-binding. At the 4-cell stage, both dorsal blastomeres were injected with 30 pg of Eng-Sia (A–D) in combination with 1 ng of βgal (A), 5 pg of Xwnt8 (B), 1 ng of β-catenin (C), or 100 pg of siamois (D) mRNA. (E) As a control, both dorsal blastomeres were injected with 1 ng of βgal mRNA. Site-directed mutagenesis was used to target a glutamine at position 50 (Q191) of the siamois homeodomain, an amino acid crucial for DNA-binding activity and specificity, transforming it into a glutamate residue (Q191E) in both siamois and Eng-Sia. At the 4-cell stage, a single ventral blastomere was injected with wild-type (F) or mutant (G) siamois, or both dorsal blastomeres were injected with wild-type (H) or mutant (I) Eng-Sia. The Q191E mutation resulted in a loss of axis inducing and axis inhibition activity for siamois and Eng-Sia, respectively, indicating a requirement for sequence-specific DNA-binding. (Part of this figure is adapted from Kessler, 1997.)

Figure 8 Organizer formation is affected by siamois fusions. At the 2-cell stage, both blastomeres were injected with 30 pg of β-galactosidase, siamois, VP16-Sia, or Eng-Sia mRNA. Embryos were harvested at the early gastrula stage and processed for RT–PCR analysis of the organizer genes goosecoid (Gsc), noggin (Nog), and chordin (Chd), the ventrolateral gene Xwnt8, the pan-mesodermal gene brachyury (Xbra), and endogenous siamois (Sia). Injection of siamois or VP16-Sia inhibited Xwnt8 expression and enhanced expression of goosecoid, noggin, and chordin, without affecting siamois, or brachyury expression. Eng-Sia inhibited expression of goosecoid, noggin, and chordin, enhanced Xwnt8 expression, and had no effect on siamois or brachyury. (Adapted from Kessler, 1997.)

homeodomain, a key amino acid in determining DNA-binding specificity (Hanes and Brent, 1989; Treisman *et al.*, 1989; reviewed in Mann, 1995). The glutamine at position 50 of the siamois homeodomain was mutated to glutamate, which was predicted to diminish DNA binding. The glutamate mutation resulted in a loss of axis induction by siamois and axis inhibition by Eng-Sia (Fig. 7F–I). Therefore, both the dorsalizing activity of siamois and the ventralizing activity of Eng-Sia are dependent on appropriate sequence-specific binding.

Dorsal injection of Eng-Sia resulted in ventralized embryos with morphological features indistinguishable from UV-irradiated embryos (DAI 0), suggesting a complete loss of Spemann's organizer. To assess organizer formation, both blastomeres of 2-cell stage embryos were injected with βgalactosidase, siamois, VP16-Sia, or Eng-Sia mRNA, and organizer-specific genes were examined in gastrulae (Fig. 8). A series of gastrula markers were examined by RT–PCR. The organizer markers goosecoid (Cho *et al.*, 1991), noggin (Smith and Harland, 1992), and chordin (Sasai *et al.*, 1994) were expressed

at elevated or normal levels in response to siamois and VP16-Sia, and were greatly reduced or undetectable in response to Eng-Sia. In contrast, the ventrolateral marker Xwnt8 (Christian *et al.*, 1991) was repressed by siamois and VP16-Sia, and was elevated in response to Eng-Sia. Expression of pan-mesodermal brachyury (Smith *et al.*, 1991), as well as endogenous siamois, was unaffected by siamois, VP16-Sia, or Eng-Sia expression. The results indicate that siamois and VP16-Sia induce an expansion of organizer, resulting in dorsalization, while Eng-Sia blocked organizer formation without interfering with general mesoderm induction.

D. Direct Activation of Goosecoid Transcription by Siamois

The potential role for siamois as a transcriptional mediator of wnt signaling was directly tested using a defined, 50 bp wnt-responsive goosecoid promoter element (−155 to −105). A luciferase reporter containing 155 bp of goosecoid promoter sequence responded strongly to Xwnt8, while a reporter containing 104 bp of promoter sequence was unresponsive (Watabe *et al.*, 1995). Induction of the reporter constructs (−155 or −104) by Xwnt8, siamois, and Eng-Sia, as well as mixtures of the mRNAs, was tested by injecting a single ventral blastomere at the 4-cell stage and assaying luciferase activity at the gastrula stage (Fig. 9A). The −155 reporter was induced 8- to 10-fold by siamois and 6- to 8-fold by Xwnt8. In contrast, Eng-Sia repressed the −155 reporter, resulting in a 6- to 7-fold decrease in basal activity. Coinjection of Eng-Sia with Xwnt8 or siamois repressed induction of the −155 reporter, resulting in activity below basal levels. The −104 reporter, lacking the wnt-response element, was unresponsive to Xwnt8, siamois, or Eng-Sia. Activation of the wnt-response element by siamois and the ability of Eng-Sia to block activation by Xwnt8 suggest that siamois directly mediates transcriptional responses to wnt signaling. This is confirmed by the demonstration that siamois directly binds the wnt-response element (Fig. 9B). Furthermore, the results support the conclusion that siamois functions as a transcriptional activator and that Eng-Sia can repress transcriptional targets of siamois.

The results demonstrate that siamois is required for development of Spemann's organizer and subse-

quent axis formation. Using a similar approach, Fan and Sokol (1997) have analyzed siamois function and have come to identical conclusions.

The ability of siamois to rescue axis inhibition by Eng-Sia indicates a specific block of endogenous siamois. However, other transcriptional activators with similar DNA-binding specificity may also be inhibited by Eng-Sia. In addition, the cooperative binding of paired-type homeodomain proteins suggests that Eng-Sia may indirectly influence factors that interact with siamois at target promoters (Wilson *et al.*, 1993; reviewed in Mann, 1995). This latter possibility is supported by the *in vitro* interaction of siamois with Mix.1, a paired-type homeodomain protein implicated in ventral development (Mead *et al.*, 1996). At the late blastula stage, expression of siamois and Mix.1 overlaps and interactions may regulate dorsoventral pattern, a process potentially influenced by Eng-Sia.

Siamois is induced by all components of the wnt pathway, including β-catenin, and this induction occurs

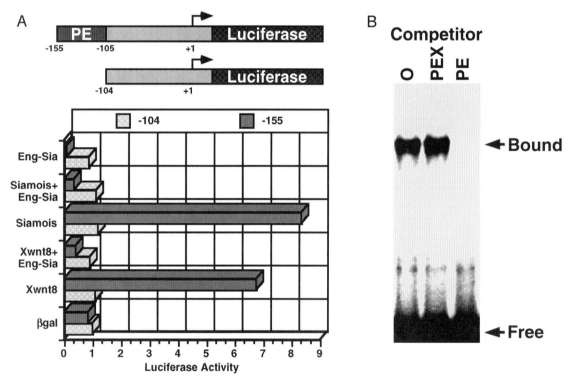

Figure 9   Siamois activates the goosecoid promoter by binding a wnt-responsive regulatory element. (A) A 50-bp wnt-responsive proximal element (PE) is located between bases −155 and −105 of the goosecoid promoter (Watabe *et al.*, 1995). Luciferase reporter constructs containing 155 bp of promoter sequence (including PE) or the 104 bp minimal promoter were tested for responsiveness to Xwnt8, siamois, Eng-Sia, or a mixture of mRNAs. At the 4-cell stage, one ventral blastomere was injected with the −155 or −104 reporter plasmid in combination with βgal, Xwnt8, siamois, or Eng-Sia mRNA, or mixtures of Xwnt8 or siamois with Eng-Sia mRNA. At the gastrula stage, extract was prepared and luciferase activity measured. Basal activity of uninjected embryos was subtracted for all values, averages were determined for duplicate samples, and values were normalized to the activity of the −104 reporter coinjected with β-galactosidase. (B) Within the wnt-responsive element are sequences that match a consensus binding site for paired-type homeodomains (Wilson *et al.*, 1993) that are sufficient for transcriptional activation by siamois (data not shown). Using purified, bacterially expressed siamois protein in a gel shift assay, direct binding to the responsive sequence is observed (Bound). Specificity of binding was established by competition with an excess of unlabeled DNA (PE), while a mutant element (PEX), insufficient for transcriptional activation by siamois, did not compete for complex formation. The results demonstrate that siamois binds directly to conserved consensus sites within the goosecoid promoter, and these sites are both necessary and sufficient for transcriptional activation by siamois and the wnt pathway. (Part of this figure is adapted from Kessler, 1997.)

in the presence of cycloheximide, indicating that preexisting maternal components directly activate siamois transcription. This suggests that siamois may mediate transcriptional responses to wnt signaling, and two observations support this proposal. First, the effects of Eng-Sia injection and antisense ablation of β-catenin on axial development are indistinguishable (Heasman *et al.*, 1994). Second, axis inhibition by Eng-Sia is not rescued by Xwnt8 or β-catenin, consistent with a dependence of wnt dorsalizing activity on siamois function. The ability of β-catenin to enter the nucleus as a complex with LEF-1 (Behrens *et al.*, 1996; Huber *et al.*, 1996; Molenaar *et al.*, 1996), a maternal transcription factor, points to the potential role of a β-catenin–LEF-1 complex in directly activating siamois at the mid-blastula transition (Brannon *et al.*, 1997). In agreement with this idea, siamois transcripts are present in dorsal cells containing nuclear β-catenin and siamois can rescue axis formation following antisense depletion of β-catenin (Wylie *et al.*, 1996). Alternatively, undescribed maternal components of the wnt pathway, acting either downstream of, or in a complex with β-catenin, may regulate siamois transcription. It should be noted that dorsal cells containing nuclear β-catenin are present in a broad domain along the animal–vegetal axis, and only a subset of these cells express siamois, suggesting that additional signals may play a role in regulating siamois expression.

In *Xenopus*, dorsal determinants are displaced from the vegetal pole to future dorsal regions by cortical rotation, establishing dorsoventral polarity that results in formation of Spemann's organizer (reviewed in Elinson and Kao, 1989; Fujisue *et al.*, 1993; Holowacz and Elinson, 1993; Kikkawa *et al.*, 1996; Sakai, 1996). Although the identity of the dorsal determinants is undefined, their position corresponds to the site of nuclear β-catenin and siamois transcription. In UV-irradiated embryos, dorsal determinants remain at the vegetal pole, resulting in vegetal cells containing nuclear β-catenin and siamois transcripts (Brannon and Kimelman, 1996). This suggests that in mediating the transcriptional response to wnt signaling, siamois is regulating zygotic events that have their origin in maternal dorsal determinants. However, vegetal expression of siamois is not sufficient for axis formation, suggesting that vegetal cells are not competent to respond to siamois, or that signals not present in vegetal cells act in conjunction with siamois to regulate organizer formation.

The requirement for siamois function in the development of Spemann's organizer in *Xenopus* suggests that siamois homologs may play a similar role in the development of other vertebrate organizers. Support for this proposal awaits the isolation and analysis of siamois homologs from the fish, mouse, and chick. Furthermore, the interacting components of the wnt pathway comprise a conserved signaling system that functions in diverse biological processes in both vertebrates and in-

vertebrates (Moon *et al.*, 1997). A conserved role for siamois, or siamois-like factors in other wnt signaling events, such as neural patterning or limb development, is an intriguing possibility to pursue.

III. Conclusions

The observations presented here lead to two apparently distinct conclusions: (i) Vg1, or Vg1-like factors localized to vegetal blastomeres regulate endodermal fate as a primary effect, and mesodermal fate, perhaps as a secondary effect; (ii) Siamois, as a mediator of maternal dorsal determinants, regulates dorsal specification, and organizer formation. These conclusions are based on a variety of overexpression experiments and, as discussed here, I propose a model by which the endogenous signaling pathways, each active at low levels, act in concert to specify dorsal fates and control organizer gene expression.

Vg1 induction of markers of both anterior (pancreas) and posterior (intestine) endodermal cell types suggests that Vg1 acts early in the endodermal fate decision, resulting in endodermal lineages present throughout the digestive tract. In the case of mesoderm induction, Vg1 induces dorsal markers and results in differentiation of notochord and somitic muscle. However, Vg1 also induces ventral–lateral markers such as Xwnt8, and therefore, Vg1 can induce gene expression representing all domains of the gastrula marginal zone, with the exception of definitive ventral fates that form blood. The results point to a role for Vg1 in establishing endodermal and mesodermal fates, without a specific role in the dorsal-ventral pattern of the induced tissues.

In contrast to the germ layer induction observed for Vg1, the wnt pathway confers dorsal identity on tissues independent of germ layer identity. Whether activated in neuroectoderm, mesoderm, or endoderm, the maternal wnt pathway shifts cells toward dorsal fates without altering germ layer identity. A number of approaches, including dominant-negative siamois and antisense ablation of β-catenin mRNA, indicate that maternal wnt signaling is essential for organizer development and subsequent axis formation. In addition, acting through siamois, the wnt pathway induces organizer gene expression, suggesting a role in both organizer formation and function.

Despite the apparent differences between Vg1 and wnt activity, both factors can induce ectopic axis formation and organizer gene expression. It may be concluded that axis induction by Vg1 may simply be an overexpression artifact, since the studies implicating the wnt pathway in axis formation are compelling (reviewed in Miller and Moon, 1996). However, a recent analysis of goosecoid, an organizer-specific gene, suggests a requirement for the combined action of Vg1 and wnt signaling in organizer formation and/or function (see Cui *et al.*, 1996). Present in the promoters of both the *Xenopus* and mouse goosecoid genes are two conserved transcriptional regulatory elements that are targets of Vg1/activin signals (distal element) and wnt/siamois signals (proximal element) (Watabe *et al.*, 1995). Overexpression of Vg1/activin or wnt/siamois can induce a strong transcriptional response of the goosecoid promoter, so at high levels each pathway can stimulate a response independent of the other, consistent with axis induction by each pathway. At lower levels, perhaps similar to endogenous signals, neither pathway alone is sufficient for activation, but the combined action of both pathways at low levels results in synergistic, strong activation of the goosecoid promoter. This synergy is observed for the goosecoid promoter, as well as the endogenous goosecoid gene (D. S. Kessler, 1997, unpublished). In the cleavage embryo, Vg1-like signals are present at equal levels throughout the vegetal pole, and wnt-like signals are restricted to dorsal blastomeres (Watabe *et al.*, 1995). Therefore, synergistic activation of goosecoid transcription may take place in an overlap region corresponding to the normal expression domain of endogenous goosecoid.

Thus, the organization and regulation of the goosecoid promoter points to a requirement for both Vg1-like and wnt signals in the control of this organizer-specific gene. The analysis of additional organizer genes will be necessary to determine whether synergistic activation of transcription is a general mechanism for coordinate regulation of organizer genes, a process that may result in the formation and function of Spemann's organizer.

Acknowledgments

I am grateful to Doug Melton, in whose laboratory much of this work was initiated, for constant support and encouragement; Lee Henry and Doug Melton for providing figures; Olivia Kelly for assistance with follistatin experiments; and Geoff Moorer and Bridget Munson for excellent technical assistance. This work was supported in part by grants from the NIH (HD 35159), the American Digestive Health Foundation/American Gastroenterological Association/Schering-Plough Corporation Research Scholar Award, and the Pew Scholars Program in Biomedical Sciences.

References

Amaya, E., Musci, T. J., and Kirschner, M. W. (1991). Expression of a dominant negative mutant of the FGF receptor disrupts mesoderm formation in *Xenopus* embryos. *Cell (Cambridge, Mass.)* **66**, 257–270.

Asashima, M., Nakano, H., Uchiyama, H., Sugino, H., Nakamura, T., Eto, Y., Ejima, D., Nishimatsu, S. I., Ueno, N., and Kinoshita, K. (1991). Presence of activin (erythroid differentiation factor) in unfertilized eggs and blastulae of *Xenopus laevis*. *Proc. Natl. Acad. Sci. U.S.A.* **88**, 6511–6514.

Badiani, P., Corbella, P., Kioussis, D., Marvel, J., and Weston, K. (1994). Dominant interfering alleles define a role for c-Myb in T-cell development. *Genes Dev.* **8**, 770–782.

Bauer, D. V., Huang, S., and Moody, S. A. (1994). The cleavage stage origin of Spemann's Organizer: Analysis of the movements of blastomere clones before and during gastrulation in *Xenopus*. *Development (Cambridge, UK)* **120**, 1179–1189.

Behrens, J., von Kries, J. P., Kuhl, M., Bruhn, L., Wedlich, D., Grosschedl, R., and Birchmeier, W. (1996). Functional interaction of beta-catenin with the transcription factor LEF-1. *Nature (London)* **382**, 638–642.

Blumberg, B., Wright, C. V., De Robertis, E. M., and Cho, K. W. (1991). Organizer-specific homeobox genes in *Xenopus laevis* embryos. *Science* **253**, 194–196.

Boterenbrood, E. C., and Nieuwkoop, P. D. (1973). The formation of the mesoderm in urodelean amphibians. V. Its regional induction by the endoderm. *Roux's Arch. Entwicklungsmech Org.* **173**, 319–332.

Brannon, M., and Kimelman, D. (1996). Activation of Siamois by the Wnt pathway. *Dev. Biol.* **180**, 344–347.

Brannon, M., Gomperts, M., Sumoy, L., Moon, R. T., and Kimelman, D. (1997). A β-catenin/XTcf-3 complex binds to the *siamois* promoter to regulate axis specification in *Xenopus*. *Genes Dev.* **11**, 2359–2370.

Carnac, G., Kodjabachian, L., Gurdon, J. B., and Lemaire, P. (1996). The homeobox gene Siamois is a target of the Wnt dorsalisation pathway and triggers organiser activity in the absence of mesoderm. *Development (Cambridge, UK)* **122**, 3055–3065.

Cho, K. W. Y., Morita, E. A., Wright, C. V. E., and De Robertis, E. M. (1991). Overexpression of a homeodomain protein confers axis-forming activity to uncommitted *Xenopus* embryonic cells. *Cell (Cambridge, Mass.)* **65**, 55–64.

Christian, J. L., Gavin, B. J., McMahon, A. P., and Moon, R. T. (1991). Isolation of cDNAs partially encoding four *Xenopus* wnt-1/int-1 related proteins and characterization of their transient expression during embryonic development. *Dev. Biol.* **143**, 230–234.

Cooke, J. (1986). Permanent distortion of the positional system of the *Xenopus* embryo by early perturbation in gravity. *Nature (London)* **319**, 60–63.

Cornell, R., and Kimelman, D. (1994). Activin-mediated mesoderm induction requires FGF. *Development (Cambridge, UK)* **120**, 453–462.

Cui, Y., Tian, Q., and Christian, J. L. (1996). Synergistic effects of Vg1 and Wnt signals in the specification of dorsal mesoderm and endoderm. *Dev. Biol.* **180**, 22–34.

Dale, L., and Slack, J. M. W. (1987a). Regional specification within the mesoderm of early embryos of *Xenopus laevis*. *Development (Cambridge, UK)* **100**, 279–295.

Dale, L., and Slack, J. M. W. (1987b). Fate map of the 32 cell stage of *Xenopus laevis*. *Development (Cambridge, UK)* **99**, 527–551.

Dale, L., Matthews, G., Tabe, L., and Colman, A. (1989). Developmental expression of the protein product of Vg1, a localized maternal mRNA in the frog *Xenopus laevis*. *EMBO J.* **8**, 1057–1065.

Dale, L., Matthews, G., and Colman, A. (1993). Secretion and mesoderm-inducing activity of the TGF-β-related domain of *Xenopus* Vg1. *EMBO J.* **12**, 4471–4480.

Dohrmann, C. D., Kessler, D. S., and Melton, D. A. (1996). Induction of axial mesoderm by zDVR-1, the zebrafish orthologue of *Xenopus* Vg1. *Dev. Biol.* **175**, 108–117.

Dyson, S., and Gurdon, J. B. (1997). Activin signaling has a necessary function in *Xenopus* early development. *Curr. Biol.* **7**, 81–84.

Elinson, R. P., and Kao, K. R. (1989). The location of dorsal information in Frog early development. *Dev. Growth Differ.* **31**, 423–430.

Elinson, R. P., and Pasceri, P. (1989). Two UV-sensitive targets in dorsoanterior specification of frog embryos. *Development* (*Cambridge, UK*) **106**, 511–518.

Elinson, R. P., and Rowning, B. (1988). A transient array of parallel microtubules in frog eggs: Potential tracks for a cytoplasmic rotation that specifies the dorso-ventral axis. *Dev. Biol.* **128**, 185–197.

Fagotto, F., Guger, K., and Gumbiner, B. M. (1997). Induction of the primary dorsalizing center in *Xenopus* by the Wnt/GSK/beta-catenin signaling pathway, but not by Vg1, Activin or Noggin. *Development* (*Cambridge, UK*) **124**, 453–460.

Fan, M. J., and Sokol, S. Y. (1997). A role for siamois in Spemann organizer formation. *Development* (*Cambridge, UK*) **124**, 2581–2589.

Fujisue, M., Kobayakawa, Y., and Yamana, K. (1993). Occurence of dorsal axis-inducing activity around the vegetal pole of an uncleaved *Xenopus* egg and displacement to the equatorial region by cortical rotation. *Development* (*Cambridge, UK*) **118**, 163–170.

Fukui, A., Nakamura, T., Sugino, K., Takio, K., Uchiyama, H., Asashima, M., and Sugino, H. (1993). Isolation and characterization of *Xenopus* follistatin and activin. *Dev. Biol.* **159**, 131–139.

Fukui, A., Nakamura, T., Uchiyama, H., Sugino, K., Sugino, H., and Asashima, M. (1994). Identification of activins A, AB, and B and follistatin proteins in *Xenopus* embryos. *Dev. Biol.* **163**, 279–281.

Fukumachi, H., and Takayama, S. (1980). Epithelial–mesenchymal interaction in differentiation of duodenal epithelium of fetal rats in organ culture. *Experientia* **36**, 335–336.

Funayama, N., Fagotto, F., McCrea, P., and Gumbiner, B. M. (1995). Embryonic axis induction by the armadillo repeat domain of beta-catenin: Evidence for intracellular signaling. *J. Cell Biol.* **128**, 959–968.

Gamer, L. W., and Wright, C. V. (1995). Autonomous endodermal determination in *Xenopus*: Regulation of expression of the pancreatic gene XlHbox 8. *Dev. Biol.* **171**, 240–251.

Gerhart, J., Ubbels, G., Black, S., Hara, K., and Kirschner, M. (1981). A reinvestigation of the role of the gray crescent in axis formation in *Xenopus laevis*. *Nature* (*London*) **292**, 511–516.

Gerhart, J., Danilchik, M., Doniach, T., Roberts, S., Rowning, B., and Stewart, R. (1989). Cortical rotation of the *Xenopus* egg: Consequences for the anteroposterior pattern of embryonic dorsal development. *Development* (*Cambridge, UK*) **107** (Suppl.), 37–51.

Gerhart, J., Doniach, T., and Stewart, R. (1991). Organizing the *Xenopus* Organizer. *In* "Gastrulation: Movements, Patterns, and Molecules" (R. Keller, W. H. Clark, Jr., and F. Griffin, eds.), pp. 57–77. Plenum, New York.

Gimlich, R. L. (1986). Acquisition of developmental autonomy in the equatorial region of the *Xenopus* embryo. *Dev. Biol.* **115**, 340–352.

Gimlich, R. L., and Gerhart, J. C. (1984). Early cellular interactions promote embryonic axis formation in *Xenopus laevis*. *Dev. Biol.* **104**, 117–130.

Graff, J. M., Thies, R. S., Song, J. J., Celeste, A. J., and Melton, D. A. (1994). Studies with a *Xenopus* BMP receptor suggest that ventral mesoderm-inducing signals override dorsal signals *in vivo*. *Cell* (*Cambridge, Mass.*) **79**, 169–179.

Gumpel-Pinot, M., Yasugi, S., and Mizuno, T. (1978). Differentiation of the endodermal epithelium associated with the splanchnic mesoderm. *C. R. Acad. Sci. Hebd. Seances Acad. Sci. D* **286**, 117–120.

Haffen, K., Kedinger, M., and Simon-Assmann, P. (1987). Mesenchyme-dependent differentiation of epithelial progenitor cells in the gut. *J. Pediatr. Gastroenterol. Nutr.* **6**, 14–23.

Han, K., and Manley, J. L. (1993). Functional domains of the *Drosophila* Engrailed protein. *EMBO J.* **12**, 2723–2733.

Hanes, S. D., and Brent, R. (1989). DNA specificity of the bicoid activator protein is determined by homeodomain recognition helix residue 9. *Cell* (*Cambridge, Mass.*) **57**, 1275–1283.

Hawley, S. H., Wunnenberg-Stapleton, K., Hashimoto, C., Laurent, M. N., Watabe, T., Blumberg, B. W., and Cho, K. W. (1995). Disruption of BMP signals in embryonic *Xenopus* ectoderm leads to direct neural induction. *Genes Dev.* **9**, 2923–2935.

Heasman, J., Wylie, C. C., Hausen, P., and Smith, J. C. (1984). Fates and states of determination of single vegetal pole blastomeres of *X. laevis*. *Cell* (*Cambridge, Mass.*) **37**, 185–194.

Heasman, J., Crawford, A., Goldstone, K., Garner-Hamrick, P., Gumbiner, B., McCrea, P., Kintner, C., Noro, C. Y., and Wylie, C. (1994). Overexpression of cadherins and underexpression of beta-catenin inhibit dorsal mesoderm induction in early *Xenopus* embryos. *Cell* (*Cambridge, Mass.*) **79**, 791–803.

Helde, K. A., and Grunwald, D. J. (1993). The DVR-1 (Vg1) transcript of zebrafish is maternally supplied and distributed throughout the embryo. *Dev. Biol.* **159**, 418–426.

Hemmati-Brivanlou, A., and Melton, D. A. (1992). A truncated activin receptor dominantly inhibits mesoderm induction and formation of axial structures in *Xenopus* embryos. *Nature* (*London*) **359**, 609–614.

Hemmati-Brivanlou, A., Kelly, O. G., and Melton, D. A. (1994). Follistatin, an antagonist of activin, is present in the Spemann organizer and displays direct neuralizing activity. *Cell* (*Cambridge, Mass.*) **77**, 283–295.

Henry, G. L., Brivanlou, I. H., Kessler, D. S., Hemmati-Brivanlou, A., and Melton, D. A. (1996). TGF-β signals and a prepattern in *Xenopus laevis* endodermal development. *Development* (*Cambridge, UK*) **122**, 1007–1015.

Holowacz, T., and Elinson, R. P. (1993). Cortical cytoplasm, which induces dorsal axis formation in *Xenopus*, is inactivated by UV irradiation of the oocyte. *Development (Cambridge, UK)* **119,** 277–285.

Huber, O., Korn, R., McLaughlin, J., Ohsugi, M., Herrmann, B. G., and Kemler, R. (1996). Nuclear localization of beta-catenin by interaction with transcription factor LEF-1. *Mech. Dev.* **59,** 3–10.

Jaynes, J. B., and O'Farrell, P. H. (1991). Active repression of transcription by the engrailed homeodomain protein. *EMBO J.* **10,** 1427–1433.

Jones, E. A., Abel, M. H., and Woodland, H. R. (1993). The possible role of mesodermal growth factors in the formation of endoderm in *Xenopus laevis. Roux's Arch. Dev. Biol.* **202,** 233–239.

Kageura, H. (1990). Spatial distribution of the capacity to initiate a secondary embryo in the 32-cell embryo of *Xenopus laevis. Dev. Biol.* **142,** 432–438.

Kedinger, M., Simon-Assmann, P., Bouziges, F., Arnold, C., Alexandre, E., and Haffen, K. (1990). Smooth muscle actin expression during rat gut development and induction in fetal skin fibroblastic cells associated with intestinal embryonic epithelium. *Differentiation* **43,** 87–97.

Kessler, D. S. (1997). Siamois is required for formation of Spemann's organizer. *Proc. Natl. Acad. Sci. U.S.A.,* **94,** 13017–13022.

Kessler, D. S., and Melton, D. A. (1994). Vertebrate embryonic induction: Mesodermal and neural patterning. *Science* **266,** 596–604.

Kessler, D. S., and Melton, D. A. (1995). Induction of dorsal mesoderm by soluble, mature Vg1 protein. *Development (Cambridge, UK)* **121,** 2155–2164.

Kessler, D. S., and Melton, D. A. (1998). Development of endodermal lineages: Specification by mature Vg1. *Dev. Biol.* in press.

Kikkawa, M., Takano, K., and Shinagawa, A. (1996). Location and behavior of dorsal determinants during first cell cycle in *Xenopus* eggs. *Development (Cambridge, UK)* **122,** 3687–3696.

Klein, P. S., and Melton, D. A. (1994). Hormonal regulation of embryogenesis: The formation of mesoderm in *Xenopus laevis. Endocr. Rev.* **15,** 326–340.

Kogawa, K., Nakamura, T., Sugino, K., Takio, K., Titani, K., and Sugino, H. (1991). Activin-binding protein is present in pituitary. *Endocrinology (Baltimore)* **128,** 1434–1440.

LaBonne, C., and Whitman, M. (1994). Mesoderm induction by activin requires FGF-mediated intracellular signals. *Development (Cambridge, UK)* **120,** 463–472.

Larabell, C. A., Torres, M., Rowning, B. A., Yost, C., Miller, J. R., Wu, M., Kimelman, D., and Moon, R. T. (1997). Establishment of the dorso-ventral axis in *Xenopus* embryos is presaged by early asymmetries in beta-catenin that are modulated by the Wnt signaling pathway. *J. Cell Biol.* **136,** 1123–1136.

Lemaire, P., Garrett, N., and Gurdon, J. B. (1995). Expression cloning of Siamois, a *Xenopus* homeobox gene expressed in dorsal–vegetal cells of blastulae and able to induce a complete secondary axis. *Cell (Cambridge, Mass)* **81,** 85–94.

Mann, R. S. (1995). The specificity of homeotic gene function. *BioEssays* **17,** 855–863.

Massague, J., Attisano, L., and Wrana, J. L. (1994). The TGF-β family and its composite receptors. *Trends Cell Biol.* **4,** 172–178.

Mead, P. E., Brivanlou, I. H., Kelley, C. M., and Zon, L. I. (1996). BMP-4-responsive regulation of dorsal–ventral patterning by the homeobox protein Mix.1. *Nature (London)* **382,** 357–360.

Melton, D. A. (1987). Translocation of a localized maternal mRNA to the vegetal pole of *Xenopus* oocytes. *Nature (London)* **328,** 80–82.

Miller, J. R., and Moon, R. T. (1996). Signal transduction through β-catenin and specification of cell fate during embryogenesis. *Genes Dev.* **10,** 2527–2539.

Minuth, M., and Grunz, H. (1980). The formation of mesodermal derivatives after induction with vegetalizing factor depends on secondary cell interactions. *Cell Differ.* **9,** 229–238.

Molenaar, M., van de Wetering, M., Oosterwegel, M., Peterson-Maduro, J., Godsave, S., Korinek, V., Roose, J., Destree, O., and Clevers, H. (1996). XTcf-3 transcription factor mediates beta-catenin-induced axis formation in *Xenopus* embryos. *Cell (Cambridge, Mass.)* **86,** 391–399.

Moody, S. A. (1987). Fates of the blastomeres of the 16-cell stage *Xenopus* embryo. *Dev. Biol.* **119,** 560–578.

Moon, R. T., Brown, J. D., and Torres, M. (1997). WNTs modulate cell fate and behavior during vertebrate development. *Trends Genet.* **13,** 157–162.

Mowry, K., and Melton, D. (1992). Vegetal messenger RNA localization directed by a 340-nt sequence element in *Xenopus* oocytes. *Science* **255,** 991–994.

Nakamura, O., and Takasaki, H. (1970). Further studies on the differentiation capacity of the dorsal marginal zone in the morula of *Triturus pyrrhogaster. Proc. Jpn. Acad.* **46,** 700–705.

Nakamura, T., Takio, K., Eto, Y., Shibai, H., Titani, K., and Sugino, H. (1990). Activin-binding protein from rat ovary is follistatin. *Science* **247,** 836–838.

Nieuwkoop, P. D. (1969a). The formation of mesoderm in urodelean amphibians. I. Induction by the endoderm. *Roux's Arch. Entwicklungsmech. Org.* **162,** 341–373.

Nieuwkoop, P. D. (1969b). The formation of the mesoderm in urodelean Amphibians II. The origin of the dorso-ventral polarity of the mesoderm. *Roux's Arch. Entwicklungsmech. Org.* **163,** 298–315.

Nieuwkoop, P. D. (1973). The "organisation center" of the amphibian embryo: Its origin, spatial organisation and morphogenetic action. *Adv. Morphogen.* **10,** 1–39.

Nieuwkoop, P. D., and Faber, J. (1967). "Normal Table of *Xenopus laevis* (Daudin)." North Holland Publ, Amsterdam.

Ogi, K.-I. (1967). Determination in the development of the amphibian embryo. *Sci. Rep. Tohoku Univ. Ser. IV (Biol.)* **33,** 239–247.

Okada, T. S. (1954). Experimental studies on the differentiation of the endodermal organs in amphibia. II. Differentiating potencies of the presumptive endoderm in the presence of mesodermal tissues. *Mem. Coll. Sci. Univ. Kyoto* **21,** 7–14.

Okada, T. S. (1957). The pluripotency of the pharyngeal primordium in Urodelan neurulae. *J. Embryol. Exp. Morphol.* **5,** 438–448.

Rebagliati, M. R., Weeks, D. L., Harvey, R. P., and Melton, D. A. (1985). Identification and cloning of localized maternal mRNAs from *Xenopus* eggs. *Cell (Cambridge, Mass.)* **42**, 769–777.

Sadowski, I., Ma, J., Triezenberg, S., and Ptashne, M. (1988). GAL4-VP16 is an unusually potent transcriptional activator. *Nature (London)* **335**, 563–564.

Sakai, M. (1996). The vegetal determinants required for the Spemann organizer move equatorially during the first cell cycle. *Development (Cambridge, UK)* **122**, 2207–2214.

Sasai, Y., Lu, B., Steinbeisser, H., Geissert, D., Gont, L. K., and De Robertis, E. M. (1994). *Xenopus* chordin: A novel dorsalizing factor activated by organizer-specific homeobox genes. *Cell (Cambridge, Mass.)* **79**, 779–790.

Sasai, Y., Lu, B., Piccolo, S., and De Robertis, E. M. (1996). Endoderm induction by the organizer-secreted factors chordin and noggin in *Xenopus* animal caps. *EMBO J.* **15**, 4547–4555.

Schneider, S., Steinbeisser, H., Warga, R. M., and Hausen, P. (1996). Beta-catenin translocation into nuclei demarcates the dorsalizing centers in frog and fish embryos. *Mech. Dev.* **57**, 191–198.

Schulte-Merker, S., Smith, J. C., and Dale, L. (1994). Effects of truncated activin and FGF receptors and of follistatin on the inducing activities of BVg1 and activin: Does activin play a role in mesoderm induction? *EMBO J.* **13**, 3533–3541.

Seleiro, E. A. P., Connolly, D. J., and Cooke, J. (1996). Early development and experimental axis determination by the chicken Vg1 gene. *Curr. Biol.* **6**, 1476–1486.

Shah, S. B., Skromne, I., C. R., H., Kessler, D. S., Lee, K. J., Stern, C. D., and Dodd, J. (1997). Misexpression of Chick Vg1 in the marginal zone induces primitive streak formation. *Development (Cambridge, UK)*, **124**, 5127–5138.

Shi, Y., and Par Hayes, W. (1994). Thyroid hormone-dependent regulation of the intestinal fatty acid-binding protein gene during amphibian metamorphosis. *Dev. Biol.* **161**, 48–58.

Slack, J. M. W., and Foreman, D. (1980). An interaction between dorsal and ventral regions of the marginal zone in early amphibian embryos. *J. Embryol. Exp. Morphol.* **56**, 283–299.

Smith, J. C., Price, B. M. J., Green, J. B. A., Weigel, D., and Herrmann, B. G. (1991). Expression of a *Xenopus* homolog of *Brachyury* (*T*) in an immediate-early response to mesoderm induction. *Cell (Cambridge, Mass.)* **67**, 79–87.

Smith, W. C., and Harland, R. M. (1991). Injected Xwnt-8 RNA acts early in *Xenopus* embryos to promote formation of a vegetal dorsalizing center. *Cell (Cambridge, Mass.)* **67**, 753–765.

Smith, W. C., and Harland, R. M. (1992). Expression cloning of noggin, a new dorsalizing factor localized to the Spemann organizer in *Xenopus* embryos. *Cell (Cambridge, Mass.)* **70**, 829–840.

Smith, W. C., Knecht, A. K., Wu, M., and Harland, R. M. (1993). Secreted noggin protein mimics the Spemann organizer in dorsalizing *Xenopus* mesoderm. *Nature (London)* **361**, 547–549.

Sokol, S., Christian, J. L., Moon, R. T., and Melton, D. A. (1991). Injected wnt RNA induces a complete body axis in *Xenopus* embryos. *Cell (Cambridge, Mass.)* **67**, 741–752.

Stutz, F., and Sphor, G. (1986). Isolation and characterization of sarcomeric actin genes expressed in *Xenopus laevis* embryos. *J. Mol. Biol.* **187**,, 349–361.

Sudarwati, S., and Nieuwkoop, P. D. (1971). Mesoderm formation in the anuran *Xenopus laevis* (Daudin). *Roux's Arch. Entwicklungsmech. Org.* **166**, 189–204.

Suzuki, A., Thies, R. S., Yamaji, N., Song, J. J., Wozney, J. M., Murakami, K., and Ueno, N. (1994). A truncated BMP receptor affects dorsal–ventral patterning in the early *Xenopus* embryo. *Proc. Natl. Acad. Sci. U.S.A.* **91**, 10255–10259.

Takata, C. (1960). The differentiation *in vivo* of the isolated endoderm under the influence of the mesoderm in *Triturus Pyrrhogaster. Embryologica* **5**, 38–70.

Tannahill, D., and Melton, D. A. (1989). Localized synthesis of the Vg1 protein during early *Xenopus* development. *Development (Cambridge, UK)* **106**, 775–785.

Tashiro, K., Yamada, R., Asano, M., Hasimoto, M., Muramatsu, M., and Shiokawa, K. (1991). Expression of mRNA for activin-binding protein (follistatin) during early embryonic development of *Xenopus laevis. Biochem. Biophy. Res. Commun.* **174**, 1022–1027.

Thomsen, G. H., and Melton, D. A. (1993). Processed Vg1 protein is an axial mesoderm inducer in *Xenopus. Cell (Cambridge, Mass.)* **74**, 433–441.

Thomsen, G., Woolf, T., Whitman, M., Sokol, S., Vaughan, J., Vale, W., and Melton, D. A. (1990). Activins are expressed early in Xenopus embryogenesis and can induce axial mesoderm and anterior structures. *Cell (Cambridge, Mass.)* **63**, 485–493.

Townes, P. L., and Holtfreter, J. (1955). Directed movements and selective adhesion of embryonic amphibian cells. *J. Exp. Zool.* **128**, 53–120.

Treisman, J., Gonczy, P., Vashishtha, M., Harris, E., and Desplan, C. (1989). A single amino acid can determine the DNA-binding specificity of homeodomain proteins. *Cell (Cambridge, Mass.)* **59**, 553–562.

Triezenberg, S. J., Kingsbury, R. C., and McKnight, S. L. (1988). Functional dissection of VP16, the trans-activator of herpes simplex virus immediate early gene expression. *Genes Dev.* **2**, 718–729.

Turner, A., Snape, A. M., Wylie, C. C., and Heasman, J. (1989). Regional identity is established before gastrulation in the *Xenopus* embryo. *J. Exp. Zool.* **251**, 245–252.

Vincent, J. P., Oster, G. F., and Gerhart, J. C. (1986). Kinematics of gray crescent formation in *Xenopus* eggs. The displacement of cortical cytoplasm relative to the egg surface. *Dev. Biol.* **113**, 484–500.

Vize, P. D., and Thomsen, G. H. (1994). Vg1 and regional specification in vertebrates: A new role for an old molecule. *Trends Genet.* **10**, 371–376.

Vleminckx, K., Wong, E., Guger, K., Rubinfeld, B., Polakis, P., and Gumbiner, B. M. (1997). Adenomatous polyposis coli tumor suppressor protein has signaling activity in *Xenopus laevis* embryos resulting in the induction of an ectopic dorsoanterior axis. *J. Cell Biol.* **136**, 411–420.

Watabe, T., Kim, S., Candia, A., Rothbacher, U., Hashimoto, C., Inoue, K., and Cho, K. W. (1995). Molecular mechanisms of Spemann's organizer formation: Conserved growth factor synergy between *Xenopus* and mouse. *Genes Dev.* **9**, 3038–3050.

Weeks, D. L., and Melton, D. A. (1987). A maternal mRNA localized to the vegetal hemisphere in *Xenopus* eggs codes for a growth factor related to TGF-β. *Cell (Cambridge, Mass.)* **51**, 861–867.

Whitman, M., and Melton, D. (1992). Involvement of p21ras in *Xenopus* mesoderm induction. *Nature (London)* **357**, 252–254.

Wilson, D., Sheng, G., Lecuit, T., Dostatni, N., and Desplan, C. (1993). Cooperative dimerization of paired class homeo domains on DNA. *Genes Dev.* **7**, 2120–2134.

Wittbrodt, W., and Rosa, F. R. (1994). Disruption of mesoderm and axis formation in fish by ectopic expression of activin variants: The role of maternal activin. *Genes Dev.* **8**, 1448–1462.

Wright, C. V., Schnegelsberg, P., and De Robertis, E. M. (1988). XlHbox 8: A novel *Xenopus* homeoprotein restricted to a narrow band of endoderm. *Development (Cambridge, UK)* **105**, 787–794.

Wylie, C. C., Snape, A., Heasman, J., and Smith, J. C. (1987). Vegetal pole cells and commitment to form endoderm in *Xenopus laevis*. *Dev. Biol.* **119**, 496–502.

Wylie, C., Kofron, M., Payne, C., Anderson, R., Hosobuchi, M., Joseph, E., and Heasman, J. (1996). Maternal beta-catenin establishes a 'dorsal signal' in early *Xenopus* embryos. *Development (Cambridge, UK)* **122**, 2987–2996.

Yamada, T. (1950). Regional differentiation of the isolated ectoderm of the *Triturus* gastrula induced through a protein extract. *Embryologia* **1**, 1–20.

Yang-Snyder, J., Miller, J. R., Brown, J. D., Lai, C. J., and Moon, R. T. (1996). A frizzled homolog functions in a vertebrate Wnt signaling pathway. *Curr. Biol.* **6**, 1302–1306.

Yisraeli, J., and Melton, D. A. (1988). The maternal mRNA Vg1 is correctly localized following injection into *Xenopus* oocytes. *Nature (London)* **336**, 592–595.

22

Specification and Differentiation
of the Heart in Amphibia

CRAIG S. NEWMAN
PAUL A. KRIEG
Institute for Cellular and Molecular Biology and
Department of Zoology
University of Texas, Austin
Austin, Texas 78712

I. Historical Introduction

The cardiovascular system is among the first functional organ systems to develop in the vertebrate embryo. This fact, along with the relatively simple cellular composition of the cardiovascular system makes it an ideal model for the study of the mechanisms underlying organogenesis. Historically, the amphibian has been an attractive model system for the investigation of cardiac development. Whereas much of the early embryological research has centered on urodele salamanders, the advent of molecular biology has resulted in a refocusing of interest toward the development of the anuran embryo, in particular that of the frog *Xenopus laevis*. Before discussing the more recent data regarding the molecular mechanisms of cardiovascular specification and differentiation we will first examine the large body of embryological data concerning this topic. Primarily completed in the early and mid parts of the twentieth century, this work has laid the foundations for the current molecular research in the field.

II. Heart Development in Urodeles

In contrast to the position of the mature heart on the ventral side of the vertebrate body, fate mapping experiments have demonstrated that the cells destined to form cardiac tissue originate in the amphibian as two patches of cells within the dorsal mesoderm on either side of the embryonic organizer (Jacobson, 1961, and references within; Dale and Slack, 1987). This precardiac tissue lies adjacent to the prechordal mesoderm, overlying the anterior endoderm. During gastrulation and neurulation, the patches of precardiac tissue undergo a series of migrations, until, in the postneurula embryo, they fuse to form a single cardiac region on the extreme ventral side of the organism. More details of these cellular movements will be presented when discussing heart development in *Xenopus*. As with many other structures within the embryo, a series of specific tissue interactions must occur to bring about the specification of cells to the cardiac lineage. Most studies dealing with these interactions fall into one of two types—either extirpation or explant analysis. In the former type of experiment, specific tissues are excised from an embryo which is then allowed to heal and continue development, whereas in the latter, precardiac tissue is removed from the embryo and cultured separately. Often in explant studies, specific tissues are recombined with the precardiac cells in order to determine their effect on cardiogenesis. In both types of experiments, the presence of beating tissue has historically been used as the assay for successful cardiac differentiation.

The use of extirpation and explant studies in several urodele species demonstrated that, while the process of cardiac specification occurs in the gastrula stage embryo, additional interactions between the cardiac anlage and surrounding tissues in the neurula are also important for the formation of the differentiated organ. Experiments in which the entire endoderm was removed at a variety of developmental stages demonstrate that the endoderm provides essential instructive signals to cardiac precursor cells. For example, removal of the endoderm at the early neurula stage results in an almost complete lack of heart development (Jacobson, 1960). As development progresses, the reliance on endodermal signals weakens, until by the late neurula or tail bud stage, removal of the endoderm has no affect on cardiac differentiation. Complementary experiments involving explantation and subsequent *in vitro* culturing of prospective cardiac mesoderm with a variety of other tissue types confirms that the presence of endoderm greatly enhances cardiac development (Jacobson, 1960; Jacobson and Duncan, 1968). More specifically, it has been shown that it is the anterior endoderm that supports cardiac differentiation, and that when anterior endoderm is excised and replaced with posterior endoderm, cardiogenesis does not occur (Jacobson, 1961). Explant studies also have confirmed that heart inducing activity is concentrated within the anterior endoderm (Jacobson and Duncan, 1968; Fullilove, 1970). It also seems likely that ectodermal tissue plays a role in cardiac development. Removal of ectoderm from the embryo adversely affects migration of the cardiac progenitors whereas inclusion of ectodermal tissue in explant experiments increases the frequency with which beating tissue develops (Jacobson, 1960).

Studies on urodele embryos also demonstrated that additional inhibitory tissue interactions are required for normal cardiac development. For example, not all of the cells capable of forming cardiac tissue, as assayed in explant studies, contribute to the mature heart in the embryo. Rather, the heart field—the total region of mesodermal tissue with the potential to contribute to the myocardium—is trimmed by inhibitory signals, such that only a subset ultimately contribute to the heart. In urodeles, inclusion of neural tissue in explants has an inhibitory effect on cardiac development suggesting that neural tissue may be one source of the signals that limit the heart field (Jacobson, 1960).

Although these experiments clearly demonstrate that various tissue interactions play an active role in the development of cardiac tissue, they have not addressed the identity of the molecules mediating the processes. Jacobson and Duncan (1968) showed that an extract of pharyngeal endoderm was capable of inducing cardiac differentiation in mesodermal explants, but technical limitations of that time prevented any further characterization. Similar experiments demonstrate that the inhibitory signal emanating from the neural folds is also present in tissue extracts, but once again, further characterization of the molecular nature of this signal was not possible (Jacobson and Duncan, 1968). More re-

cently, Muslim and Williams (1991) have shown that the application of several purified protein growth factors to mesodermal explants can induce the formation of cardiac tissue. Specifically, the application of either transforming growth factor β-1 (TGF-β1) or platelet derived growth factor (PDGF) results in an increase in the tendency of early neurula stage cardiac explants to form beating tissue. In contrast, the application of basic fibroblast growth factor (bFGF) decreases the frequency of heart formation. Interestingly, work in *Drosophila* has concluded that *decapentaplegic (DPP)*, a member of the TGF superfamily of growth factors, also plays a critical role in activating cardiac markers (Frasch, 1995).

III. Heart Development in Anurans

Many of these embryological experiments have been replicated in the anuran embryo, using *Xenopus* as the model system. Fate mapping has demonstrated that cardiac tissue in *Xenopus* is derived from the dorsal vegetal blastomeres, in the equatorial region (Fig. 1A), very similar to the location of cardiac precursors in other amphibia (Dale and Slack, 1987; Moody, 1987). By the time of gastrulation, the heart progenitors lie between 30° and 45° from the dorsal midline, immediately adjacent to the dorsal blastopore lip (Fig. 1B). Explant studies by Sater and Jacobson (1989) demonstrate that specification of the cardiac cells in *Xenopus* is essentially complete by the time of gastrulation, much earlier than in urodele embryos in which specification of cardiac tissues occurs during neurulation. In the *Xenopus* studies, only explants taken early in gastrulation, prior to stage 10.5, show significantly less than 100% appearance of differentiated cardiac tissue. Explants of cardiac tissue taken at the late gastrula or early neurula stages form beating tissue in virtually all cases.

How do these mesodermal cells adjacent to the dorsal midline become specified to develop into heart tissue? It seems clear from the results of Sater and Jacobson (1990a) that signals arising from the dorsal blastopore lip, the embryonic organizer, are required for the specification of the cardiac lineage (Fig. 1B). Removal of the dorsal blastopore region early in gastrulation completely eliminates subsequent heart formation in the embryo. On the other hand, incubation of the blastopore lip tissue with ventral mesoderm (tissue that would never normally contribute to the heart) results in the formation of differentiated cardiac tissue. The dorsal lip region therefore acts as an inducer of cardiac development, and it appears that all mesodermal tissue has the ability to respond (to a greater or lesser extent), to the instructive signals arising from the inducing tissue. In addition, consistent with results obtained in salamanders and other vertebrates, dorsal endodermal tissue effects the efficiency with which

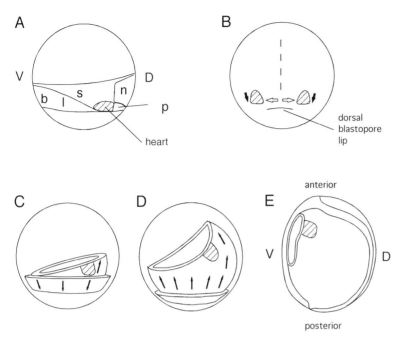

Figure 1 Fate map of the heart in the frog embryo and movements of heart progenitors during embryonic development. (A) A fate map of the mesodermal tissues of the embryo, showing the location of the heart primordium (crosshatched) at a dorsal position, adjacent to the prechordal plate (p). Dorsal (D) is to the right and ventral (V) to the left in all figures. The location of tissues fated to become blood (b), lateral plate mesoderm (l), somitic mesoderm (s), and notochord (n) are indicated. (B) View from the dorsal side of the embryo at the early gastrula stage. The dorsal midline is indicated by the dashed line. During gastrulation, the two regions of precardiac cells will move to the inside of the embryo through the blastopore lip (movement indicated by black arrows). At about the same time, inductive signals from the dorsal lip region (open arrows) instruct the precardiac cells to adopt a cardiac fate. (C, D) Movement of cardiac precursor cells during gastrulation. The cell movements of gastrulation bring the mesodermal tissues deeper into the embryo. The precardiac mesoderm (indicated by the crosshatching) is at the leading edge of the involuting mesoderm. Note that the dorsal tissues involute further than the ventral tissues and the precardiac tissue is on the dorsal side. (E) Neurula stage embryo. Once gastrulation is complete, the heart primordia are now at the anterior end of the embryo, closer to the ventral side. During subsequent development, the heart primordia will move further ventrally and fuse at the ventral midline, where the mature heart will differentiate.

cardiac cells are specified in *Xenopus* (Nascone and Mercola, 1995), although it is not absolutely required for heart development. After the two patches of precardiac tissue have been specified in the gastrula embryo, they are carried by the involution movements of gastrulation to a more anterior position in the embryo (Fig. 1C,D). During neurulation, the precardiac tissues migrate ventrally to their final position at the ventral midline of the embryo (Fig. 1E). As in urodeles, the early cardiac field in the *Xenopus* embryo becomes smaller as development proceeds (Sater and Jacobson, 1990b). However, in contrast to urodeles, in which neural tissues appear to be the source of inhibitory signals, tissues in the region of the third pharyngeal arch are inhibitory in *Xenopus* (Sater and Jacobson, 1990b). Thus, although the basic program of cardiac specification appears to be conserved across amphibians, the details of tissue interactions may differ between urodeles and anurans.

Long after the initial specification of cardiac tissues in the frog embryo, the early morphological events of heart development become visible. During early tail bud stages, at approximately stage 30, a sheet of cells comprising the future myocardium delaminates from the overlying pharyngeal endoderm. The folding of this sheet into the linear heart tube begins first in the middle of the heart, and then proceeds both rostrally and caudally. Soon afterwards, at the linear heart tube stage, the organ consists of several cell types arranged in concentric tubes. The innermost layer of the heart, the endocardium, lines the lumen of the organ and is continuous with the endothelial layer of the great blood vessels entering and exiting the heart. Surrounding the endocardium is the muscular myocardial layer. Although the precise origins of the endocardial and myocardial cells are not clearly understood, both originate from the dorsal mesoderm and evidence from other species suggests that both cell types may be derived from a common precursor (Linask and Lash, 1993; Lee *et al.*, 1994, Chapter 30 by Mikawa). Soon after formation of the heart tube is complete, the heart begins to beat and then undergoes a series of stereotyped movements resulting in a characteristic S shape. The folding is complete by about stage 35/36 at which time the chambers of the heart are distinct, with the atrium positioned dorsally relative to the ventricle. At the tadpole stage (stage

42), the valves of the heart form, separating the chambers, and shortly thereafter septation of the atrium results in the mature three-chambered heart found in adult amphibians.

To summarize, the following conclusions can be drawn concerning specification of the amphibian heart. (1) The progenitors of the cardiac tissue arise in the dorsal mesoderm, in two patches located immediately adjacent to the embryonic organizer. (2) These cardiac progenitor tissues migrate laterally and ventrally during neurulation until they occupy a position at the ventral midline. (3) Important inductive signals emanate from surrounding tissues, most notably the anterior endoderm and in the case of *Xenopus* the dorsal blastopore lip. (4) Inhibitory interactions between tissues are also important, and these serve to restrict the number of cells that will ultimately differentiate into myocardial tissue. In the sections below we will discuss the application of molecular biology to the questions of amphibian cardiovascular development, noting the complementarity of the embryological and molecular approaches.

IV. Molecular Markers of Cardiac Differentiation

Although the presence of beating tissue is a clear and easily assayed feature of differentiated myocardial tissue, it relies on completion of the entire developmental program and is thus inappropriate for the study of the intermediate developmental steps linking cardiac specification to the formation of the mature organ. Furthermore, using morphological criteria alone, it is not possible to identify cells of the myocardial and endocardial lineages prior to the assembly and folding of the heart tube. In order to facilitate studies of early heart development, it is important to have molecular markers which can act as sensitive substitutes for the beating heart assay and which serve to unambiguously identify the differentiated myocardial cell type. Although a number of sarcomeric structural proteins, including α-cardiac actin (Mohun *et al.*, 1984; Hemmati-Brivanlou *et al.*, 1990) and the α-myosin heavy chain gene (MHCα) (Logan and Mohun, 1993; Cox and Neff, 1995), have been characterized for some time and are known to be expressed within the myocardium, these genes also are transcribed in skeletal or facial muscles and therefore are not appropriate for the unambiguous identification of myocardial tissue (Fig. 2B,C).

More recently however, two differentiation markers that are completely cardiac specific have been described in *Xenopus*. The first is the *cardiac troponin I* gene, *XTnIc* (Drysdale *et al.*, 1994) which encodes a sarcomeric protein involved in regulating the actin/myosin interaction in muscular contractions. By *in situ*

hybridization analysis, *XTnIc* transcripts are first detected in the early tail bud embryo at stage 26 in two patches of cells on either side of the ventral midline, immediately posterior to the cement gland, at the position of the cardiac precursor cells (Fig. 2A). As development proceeds, the two patches of *XTnIc* positive cells fuse and then delaminate as a sheet from the overlying pharyngeal endoderm. The sheet of cardiac precursor cells then fold into the myocardium of the linear heart tube. Expression of *XTnIc* is maintained specifically in the myocardium throughout the remainder of cardiac differentiation (Fig. 2D) and transcripts remain abundant in the adult heart, accounting for approximately 1% of all mRNAs. A second myocardial specific marker gene, *myosin light chain 2* (*MLC2*) (Chambers *et al.*, 1994) shows an expression profile effectively identical to that of *XTnIc* (Fig. 2E), with transcription first detected in precardiac tissues of the stage 26 embryo. It is interesting to observe that expression of all described myocardial differentiation genes occurs simultaneously during *Xenopus* heart development. This includes the genes which are not strictly restricted to the heart, like α-*cardiac actin* and *MHCα*. This is quite different from the situation in mouse, in which expression of different cardiac markers commences over the course of several days of embryonic development (Sassoon *et al.*, 1988; Lyons *et al.*, 1990; Lyons, 1996).

The isolation of cardiac specific molecular markers has allowed a more thorough characterization of the effects of agents that perturb normal embryonic heart development. For example, previous experiments showed that treatment of embryos with lithium at the 32-cell stage results in abnormal embryos displaying an overabundance of dorsal structures at the expense of more ventral ones (Kao and Elinson, 1988). Whole-mount *in situ* hybridization analysis of lithium treated embryos (Drysdale *et al.*, 1994) reveals an enlarged region of cardiac tissue organized in a near radial band of *XTnIc* positive tissue which is excluded from only the dorsal most region (Fig. 3A). These experiments once again demonstrate the dorsal origins of the cardiac tissue. With time, this atypical cardiac tissue localizes to a more central position in the embryo and ultimately forms beating tissue. In contrast to the expansion of dorsal tissues in lithium treated embryos, exposure to low levels of retinoic acid (RA) can result in a loss of anterior embryonic structures including the heart (Durston *et al.*, 1989; Sive *et al.*, 1990). Treatment of *Xenopus* embryos with increasing doses of RA at the gastrula stage results in a progressive reduction in the eyes, olfactory pits, cement gland, hatching gland, and heart of the resultant embryos. However, the results of this early RA treatment are somewhat difficult to interpret since the loss of cardiac tissue in these experiments may not be directly attributable to the effects of RA, but rather may be an indirect consequence of the loss of the more

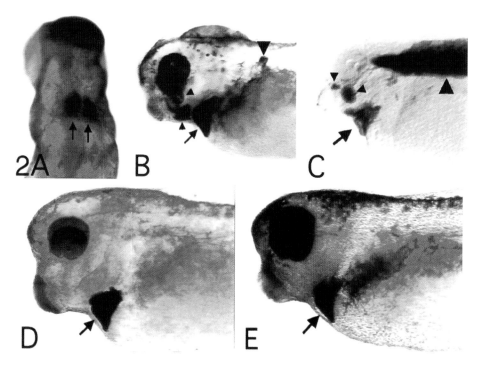

Figure 2 Expression patterns of cardiac differentiation markers revealed by whole-mount *in situ* hybridization. (A) Ventral view of a tail bud stage embryo (stage 28) assayed for myosin heavy chain shortly after the onset of terminal differentiation. The paired cardiac rudiments are clearly visible (arrows). (B) Expression of the myosin heavy chain gene in the heart (arrow), lymph heart (large arrowhead), and facial muscles (small arrowheads). (C) Lateral view of a late tail bud stage embryo stained for α-cardiac actin transcripts. In addition to expression in the folded heart (arrow), transcripts are also apparent in the somites (large arrowhead) and facial muscles (small arrowheads). (D, E). Cardiac specific expression of cardiac troponin I (D) and myosin light chain (E) in the late tail bud embryo.

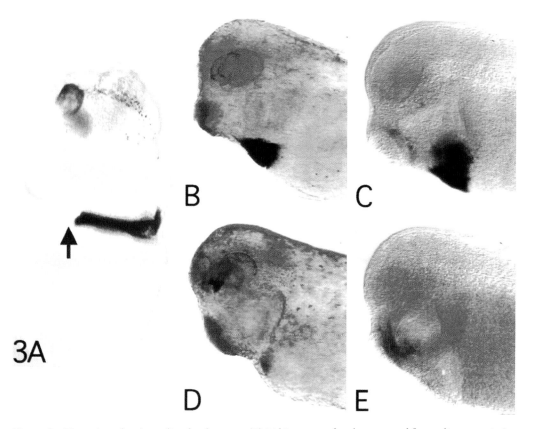

Figure 3 Disruption of early cardiac development. (A) Lithium treated embryo assayed for cardiac troponin I transcripts. Lithium treatment at the 32-cell stage results in hyperdorsalized embryos which display cardiac tissue arranged in a near radial band, excluded from only the dorsal-most region (arrow). (B–E) Control (B, C) and retinoic acid treated embryos (D, E) assayed for either *cardiac troponin I* (B, D) or *XNkx2-5* (C, E). RA exposure at the late neurula stage results in morphologically normal embryos displaying severely downregulated expression of cardiac markers.

dorsal cardiac inducing tissue and subsequent failure of specification of the heart fate. More interesting and interpretable results are obtained when RA exposure is commenced near the end of neurulation, well after cardiac specification. Drysdale and colleagues (1997) have shown that this treatment results in embryos with complete anterior–posterior axes and virtually normal overall morphology, but which lack expression of my-ocardial markers including *XTnIc* and *XNkx2-5* (Fig. 3D and E). Use of a general muscle marker, *α-cardiac actin*, reveals that the effects of RA are specific to cardiac muscle because expression in the skeletal muscle in the somite is unaffected, whereas expression in myocardial tissue is clearly inhibited (Drysdale *et al.*, 1997). Similar results are obtained when RA treatment is initiated at any stage prior to the onset of myocardial differentiation. Once myocardial gene expression has commenced however, the cardiac tissue becomes refractory to RA treatment. In contrast to the severe inhibitory effects on myocardial gene expression, endocardial gene expression appears undisturbed. Histological examination of experimental embryos reveals that the myocardial precursor cells delaminate as normal and form a structure resembling the linear heart tube. This heart tube encloses apparently normal endocardial cells that are expressing endocardial marker genes. During subsequent development, however, looping of the heart is abnormal and beating tissue is never observed. Thus the pathway leading to myocardial sarcomeric differentiation, including the formation of beating tissue is experimentally separable from the events of morphological differentiation. It is interesting to note that genetic ablation of a number of different RA receptor proteins in mouse results in atypical cardiac development, suggesting a conserved role for RA signaling throughout the vertebrates (reviewed in Mark *et al.*, 1997).

The isolation of molecular markers for the differentiated cardiac phenotype also has allowed the assessment of the heart inducing capabilities of various growth factors in *Xenopus*. Previous studies have clearly demonstrated the ability of both activin A, a TGF-β family member, and basic fibroblast growth factor (bFGF) to induce mesodermal cell types in *Xenopus* animal caps (Chapter 19 by Dawid). Importantly, whereas activin A produces mesoderm of the dorsoanterior phenotype, bFGF induces mesoderm with more ventral characteristics. Consistent with both the dorsal origins of the cardiac precursors cells as well as previous results obtained using axolotl explants (Muslin and Williams, 1991), *Xenopus* animal caps can be induced to express *MHCα* on exposure to high concentrations of activin A (Logan and Mohun, 1993). No cardiac marker expression is detected following exposure to bFGF. Thus, it appears that signaling molecules capable of patterning the dorsal mesoderm prior to gastrulation also may be important in the specification of the cardiac precursors.

V. The *tinman* Genes

A major step forward in the study of the molecular mechanisms of cardiogenesis occurred with the isolation and characterization of *tinman*, a homeobox containing gene in *Drosophila* (Kim and Nirenberg, 1989; Bodmer *et al.*, 1990; Azpiazu and Frasch, 1993; Bodmer, 1993). Initially widely expressed in the mesoderm prior to gastrulation, *tinman* transcripts later become localized to cardiac precursors located on the dorsal side of the embryo. Loss of *tinman* function leads to a disorganization of both somatic and splanchnic embryonic musculature, but more strikingly, results in the complete absence of the dorsal vessel, a structure generally regarded to be the insect equivalent of the heart (Azpiazu and Frasch, 1993; Bodmer, 1993). Since this initial observation, a flurry of activity in the field has resulted in the description of a family of *tinman*-related homeobox genes, the NK-2 family, in a variety of organisms ranging from nematodes to humans.

The first vertebrate *tinman*-related gene, *Nkx2-5/csx* was cloned from the mouse and found to be expressed in a pattern consistent with that of *tinman* in the fly (Komuro and Izumo, 1993; Lints *et al.*, 1993). In mouse, *Nkx2-5* transcripts are first detectable at the embryonic headfold stage in a set of cells destined to form the heart. These transcripts are first observed near the time of cardiac specification and several hours prior to the appearance of any differentiation products. Expression is maintained in the myocardium throughout development. In view of this conserved expression pattern in evolutionarily distant organisms, the results of the gene ablation studies are surprising. Although elimination of *Nkx2-5* function in the mouse results in embryonic lethality due to impaired cardiovascular function, an almost normal heart tube containing beating tissue is formed (Lyons *et al.*, 1995). The major defects in $Nkx2-5^{-/-}$ mice are a failure of correct cardiac looping, impaired endocardial cushion formation, and abnormal development of the ventricular walls. At the molecular level, expression of most cardiac differentiation markers is unaffected in the knock-out mice. The presence of beating tissue in the knock-out mice has lead to much speculation regarding the role of *Nkx2-5* in cardiac specification. For example, it has been suggested that whereas *Nkx2-5* shares similarities in both sequence and expression pattern to *tinman*, it may not play an equivalent role in the specification of cardiac tissue, but rather may be involved primarily in the final stages of cardiac morphogenesis (Biben and Harvey, 1997). Alternatively, the mouse genome may contain additional genes functionally redundant to *Nkx2-5* and expression of these other related genes may be capable of rescuing early specification defects in the knock-out embryos. This latter situation is certainly plausible, since similar redundant pathways have previously been described in vertebrates, including the MyoD/myogenin pathway involved in somitic muscle differentiation (reviewed in Rudnicki and Jaenisch, 1995; see also Chapter 41 by Buckingham and Tajbakhsh).

In support of the functional redundancy hypothesis is the fact that multiple *tinman* family members have been cloned in a variety of vertebrates and in many cases these genes also are expressed in the developing cardiac tissue (Patterson *et al.*, 1998; Newman and Krieg, 1998). Thus far, three *tinman*-related genes have been isolated from *Xenopus* and each is found to be expressed in cardiac tissue. The first to be described, *XNkx2-5*, is the *Xenopus* ortholog of the murine *Nkx2-5* gene and displays a very similar expression profile (Tonissen *et al.*, 1994). When analyzed by RNase protection, *XNkx2-5* is first expressed in the dorsal region of the embryo shortly after the onset of gastrulation (K. D. Patterson and P. A. Krieg, unpublished observation). This temporal and spatial pattern of expression is consistent with the location of the cardiac precursors in *Xenopus*. At present, however, it is not clear if the initial activation of *XNkx2-5* expression is a direct consequence of signaling from the dorsal blastopore lip or whether it lies further downstream in the signaling cascade. By whole-mount *in situ* hybridization, *XNkx2-5* transcripts can first be seen in the early neurula in a crescent of tissue just ventral to the anterior neural plate (Fig. 4A). This region of the embryo is comprised not only of the paired cardiac anlage (Fig. 4B) but also the underlying pharyngeal endoderm known to be a potent inducer of cardiac tissue. As the heart begins to differentiate at the early tail bud stage, *XNkx2-5* expression is maintained in the myocardium of the heart as well as the ventral pharyngeal endoderm (Fig. 4C), but is clearly absent from the endocardial layer of the heart. As development proceeds, transcripts are lost in the endoderm, whereas expression in the myocardium remains robust (Fig. 4D). In experimental embryos treated with RA prior to cardiac differentiation, *XNkx2-5* transcript levels decline rapidly (Fig. 3E) becoming almost undetectable within 3 hours of the initiation of treatment (Drysdale *et al.*, 1997). This rapid downregulation suggests that the *XNkx2-5* promoter may be directly regulated by RA levels in the embryo, although at this time, there is no evidence that the *XNkx2-5* promoter contains RA responsive elements.

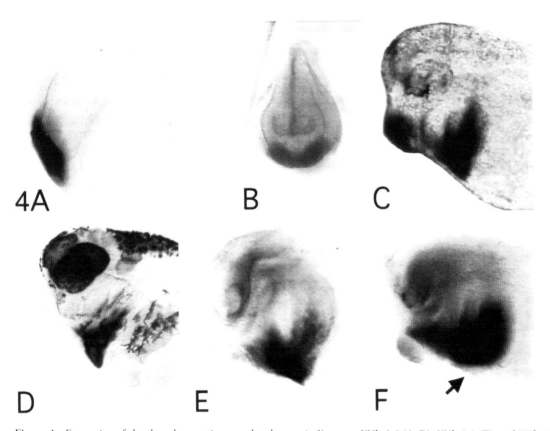

Figure 4 Expression of the three known *tinman*-related genes in *Xenopus*, *XNkx2-5* (A–D), *XNkx2-3* (E), and *XNkx2-9* (F). (A) Expression of *XNkx2-5* at the late neurula stage is restricted to the precardiac tissue as well as the overlying pharyngeal endoderm. (B) Anterior view of an early tail bud stage embryo assayed for *XNkx2-5*. Paired cardiac precursors are visible immediately below the neural plate. (C, D) *XNkx2-5* transcripts are retained in cardiac and pharyngeal tissue at the mid (C) and late (D) tail bud stages as the heart folds and loops. (E) Expression of *XNkx2-3* at the mid tail bud stage. Note that the pattern of expression is similar to that of *XNkx2-5*. (F) At the mid tail bud stage, *XNkx2-9* is strongly expressed in the pharyngeal endoderm but in contrast to the other *tinman*-related genes in *Xenopus*, is absent from the folding heart (arrow).

In addition to *XNkx2*-5, a very closely related gene, *XNkx2*-3, also is expressed in the developing cardiac tissues in the *Xenopus* embryo (Evans *et al.*, 1995; Cleaver *et al.*, 1996). At the amino acid level, XNkx2-3 is 92% identical to XNkx2-5 within the homeodomain and the *XNkx2*-3 gene is expressed with a virtually identical expression pattern (Fig. 4E).

A third gene, *XNkx2*-9, is the most recent addition to the *tinman* family in frogs and shows an early expression pattern similar to the other two genes (Newman and Krieg, 1998). In the neurula, *XNkx2*-9 can be seen in the precardiac region in a crescent of cells just posterior to the cement gland. Expression is maintained in these cells until the time of cardiac differentiation at which point *XNkx2*-9 is downregulated in the presumptive heart tissue. In contrast to *XNkx2*-3 and *Nkx2*-5, little or no *XNkx2*-9 expression is observed in cardiac tissues by the time that the linear heart tube is formed (Fig. 4F). Despite this downregulation in the cardiac tissues, very broad, intense expression is maintained in the pharyngeal endoderm in the epithelial layer of cells as well as in the branchial clefts. At present, the relative importance of these different *tinman*-related genes for embryonic heart development remains to be determined.

Since Nkx2-5 is a homeodomain transcription factor, it must play a role in the regulation of transcription of downstream genes. Studies in mouse have indicated that *Nkx2*-5 is capable of regulating the expression of several diverse cardiac genes including a structural protein (MLC2V), a nuclear ankyrin-like repeat protein (CARP), the atrial natriuretic factor (ANF) gene, and a basic helix-loop-helix transcription factor (eHand) (Lyons *et al.*, 1995; Zou *et al.*, 1997; Biben and Harvey, 1997; Durocher *et al.*, 1997). Despite this, the presence of beating heart tissue in the *Nkx2*-5 knock-out mice has led to speculation that, unlike *Drosophila tinman*, vertebrate *tinman*-related genes may not be playing a major regulatory role in cardiac development but may be important for regulating heart looping (Biben and Harvey, 1997). The problem is exacerbated by the fact that additional *tinman* family genes that may be capable of rescuing heart forming ability in *Nkx2*-5 mutant mice have not been reported. Unlike the frog, in which *Nkx2*-5 and *Nkx2*-3 have very similar expression profiles, the mouse orthologue of *Nkx2*-3 is expressed in the developing gut and is never detected in cardiac tissues (Pabst *et al.*, 1997). Overall therefore, the importance of the *tinman* genes for heart development in the mouse remains an open question.

Recent experiments in *Xenopus* have used a different approach to investigate the possible importance of *tinman* genes for heart development. The results of these experiments suggest that *tinman* family genes may indeed be critical for the process of cardiac specification and development. The overexpression or ectopic expression of specific gene products in the early *Xenopus* embryo provides a simple yet powerful mechanism for investigating gene function. This is generally achieved by injection of synthetic mRNA encoding the sequence of interest into the early blastula stage embryo. By injecting a specific blastomere, expression of the injected sequences can be directed to a limited subset of tissues in the developing embryo (Dale and Slack, 1987; Moody, 1987). Using this strategy, Cleaver *et al.* (1996) have shown that overexpression of either *XNkx2*-3 or *XNkx2*-5 in the heart-forming region of the embryo results in a significant increase in the amount of cardiac tissue in the early embryo. Analysis of injected embryos by whole-mount *in situ* hybridization for *XTnIc* reveals an increased myocardial layer, an effect most noticeable shortly after the onset of cardiac differentiation (Fig. 5A, B). Quantitation of cardiac cell nuclei shows an almost 2-fold increase in the number of cells making up the myocardium. Related experiments have been carried out by Fu and Izumo (1995) who reported precocious activation of the myosin heavy chain differentiation marker in *XNkx2*-5 overexpressing embryos. In contrast, to these results however, Cleaver and colleagues found no evidence of temporal misregulation of the cardiac troponin gene. Significantly, the endocardial layer is apparently unaffected in these *Nkx2*-5 overexpression experiments suggesting that if the myocardium and the endocardium do share a common precursor cell, the *XNkx* genes are acting after the separation of these two cell lineages. At present, it is not clear from these overexpression experiments if the *XNkx* genes are (a) acting to recruit additional cells to the heart field, (b) limiting the reduction of the heart field by inhibitory signals, or (c) prompting an overproliferation of the already specified cardiac precursors. Interestingly, the overexpression of another predifferentiation myocardial transcription factor, GATA-6 has yielded a slightly different result (Gove *et al.*, 1997). This zinc-finger transcription factor is expressed in myocardial precursor cells but is rapidly downregulated at the time of differentiation. This observation suggests that GATA-6 may play a role in the maintenance of the precardiac cells in an undifferentiated state. Consistent with this hypothesis, the injection of *GATA*-6 synthetic mRNA results in a delay in the appearance of myocardial differentiation markers but ultimately in a larger heart. The authors speculate that the increase in the number of cardiac cells may be due to an extra round of cell division prior to the delayed onset of differentiation.

Previous experiments have demonstrated that Tinman-family proteins are capable of homodimerization and heterodimerization with other family members as well as being able to physically interact with other transcription factors including GATA-4 and MADS domain proteins (Chen and Schwartz, 1996; Durocher *et al.*, 1997). The observation that Tinman-family proteins

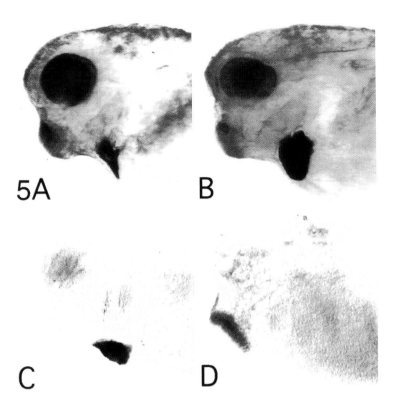

Figure 5 *In situ* hybridization analysis of XTnIc expression in control (A, C) and injected embryos (B, D). (A) A mid tail bud stage embryo after injection with water showing a normal heart. (B) Injection of synthetic *XNkx2-5* mRNA into one cell at the 16-cell stage results in an enlarged myocardium at the mid tail bud stage. (C) A control embryo stained for XTnIc. (D) An embryo resulting from the double-sided injection of mRNA coding for a dominant negative *XNkx2-5* protein. Expression of cardiac differentiation markers is greatly reduced in these experimental embryos.

interact with other proteins to form transcription complexes, allows the possibility of carrying out overexpression experiments using a dominant negative form of the XNkx2-5 protein. Previous studies have demonstrated that a single amino acid change from a nonpolar residue to a proline between helices two and three of the homeodomain is sufficient to disrupt DNA binding (Mead *et al.,* 1996). It is believed however that the mutant protein remains capable of forming stable complexes with the complement of molecules normally associated with the wild-type protein. In this situation therefore, overexpression of the mutant protein may sequester cofactors necessary for wild-type gene activity (Mead *et al.,* 1996). In such experiments, not only is the wild-type gene activity ablated, but it seems likely that the activity of any redundant protein capable of interacting with the same set of cofactors also will be compromised. Thus, in contrast to mouse gene ablation experiments, closely related protein activities will not be able to rescue the loss of function.

When synthetic mRNA encoding the dominant negative XNkx2-5 protein is directed toward the heart-forming region of the *Xenopus* embryo, a reduction in the expression of cardiac differentiation markers is observed on the manipulated side. Double-sided injections, in which both blastomeres fated to give rise to

cardiac tissue are targeted, can result in a complete lack of detectable myocardial marker gene expression in an otherwise phenotypically normal embryo (Fig. 5C, D). Furthermore, sectioning of doubly injected embryos reveals that no morphological cardiac differentiation is apparent, suggesting that a defect in the specification of the cardiac precursors may be responsible for the observed phenotype. Significantly, injections of an XNkx2-3 dominant negative construct result in an identical embryonic phenotype, suggesting that interference with any member of the *tinman* gene family is sufficient to interrupt cardiac development. Although these studies are preliminary, use of dominant negative versions of these homeobox regulatory genes suggests a crucial role for the *tinman* gene family in vertebrate cardiac development. Further experiments however, will be required to determine the specific mechanisms of *tinman* gene function.

VI. Conclusions

The ability to combine experimental embryological manipulations with the modern tools of molecular biology makes the *Xenopus* embryo an extremely valuable model system for investigating the mechanisms underlying embryonic development. In recent years, a number of laboratories have focused on the question of organ development in vertebrates. However, the molecular and cellular events leading to organ development are extremely complex, and it is increasingly apparent that it will be necessary to study a number of different model systems in order to fully understand the rules governing the growth and differentiation of organs in the embryo. In the case of embryonic heart development, for example, genetic studies in *Drosophila* have pointed to a critical role for the *tinman* homeobox gene in cardiac development. Based on these observations, gene ablation studies in the mouse have confirmed that the *tinman* family of genes are required for development of the vertebrate heart, although in these studies, mice lacking Nkx2-5 function only exhibit abnormalities during the later stages of heart development. Examination of tinman genes in *Xenopus*, chick, and zebrafish has demonstrated the presence of multiple *tinman*-related genes that are expressed during embryonic heart development. These genes potentially play redundant roles in regulating embryonic cardiogenesis. Finally, overexpression experiments and dominant negative expression studies in the frog embryo have provided evidence that the *tinman*-related genes are essential for heart development in vertebrates. Ultimately therefore, a combination of approaches, each exploiting the specific advantages of the different model systems, will contribute to a clear understanding of the molecular and cellular mechanisms underlying embryonic heart development.

Acknowledgments

We thank Matthew Grow for providing information prior to publication. This work was supported by Grant HL52476 to PAK.

References

Azpiazu, N., and Frasch, M. (1993). *tinman* and *bagpipe:* Two homeobox genes that determine cell fates in the dorsal mesoderm of *Drosophila*. *Genes Dev.* 7, 1325–1340.

Biben, C., and Harvey, R. P. (1997). The homeodomain factor Nkx2-5 controls of left–right asymmetric expression of the bHLH gene *eHand* during murine heart development. *Genes Dev.* 11, 1357–1369.

Bodmer, R. (1993). The gene *tinman* is required for specification of the heart and visceral muscles in *Drosophila*. *Development (Cambridge, UK)* 118, 719–729.

Bodmer, R., Jan, L. Y., and Jan, Y. N. (1990). A new homeobox-containing gene, *msh-2*, is transiently expressed early during mesoderm formation in *Drosophila*. *Development (Cambridge, UK)* 110, 661–669.

Chambers, A. E., Logan, M., Kotcha, S., Towers, N., Sparrow,

D., and Mohun, T. J. (1994). The RSRF/MEF2 protein SL1 regulates cardiac muscle-specific transcription of a myosin light-chain gene in *Xenopus* embryos. *Genes Dev.* 8, 1324–1334.

Chen, C. Y., and Schwartz, R. J. (1996). Recruitment of the tinman homolog Nkx2-5 by serum response factor activates cardiac α-actin gene transcription. *Mol. Cell. Biol.* 16, 6372–6384.

Cleaver, O. B., Patterson, K. D., and Krieg, P. A. (1996). Overexpression of the *tinman*-related genes *XNkx2-5* and *XNkx2-3* in *Xenopus* embryos results in myocardial hyperplasia. *Development (Cambridge, UK)* 122, 3549–3556.

Cox, W. G., and Neff, A. W. (1995). Cardiac myosin heavy chain expression during heart development in *Xenopus laevis*. *Differentiation* 58, 269–280.

Dale, L., and Slack, J. M. W. (1987). Fate map for the 32-cell stage of *Xenopus laevis*. *Development (Cambridge, UK)* 99, 527–551.

Drysdale, T. A., Tonissen, K. F., Patterson, K. D., Crawford, M. J., and Krieg, P. A. (1994). Cardiac troponin I is a heart-specific marker in the *Xenopus* embryo: Expression during abnormal heart morphogenesis. *Dev. Biol.* 165, 432–441.

Drysdale, T. A., Patterson, K. D., Saha, M., and Krieg, P. A. (1997). Retinoic acid can block differentiation of the myocardium after heart specification. *Dev. Biol.* 188, 205–215.

Durocher, D., Charron, F., Warren, R., Schwartz, R. J., and Nemer, M. (1997). The cardiac transcription factors Nkx2-5 and GATA-4 are mutual cofactors. *EMBO J.* 16, 5687–5696.

Durston, A. J., Timmermans, J. P. M., Hage, W. J., Hendriks, H. F. J., de Vries, N. J., Heideveld, M., and Nieuwkoop, P. D. (1989). Retinoic acid causes an anteroposterior transformation in the developing central nervous system. *Nature (London)* 340, 140–144.

Evans, S. M., Yan, W., Murillo, M. P., Ponce, J., and Papalopulu, N. (1995). *tinman*, a *Drosophila* homeobox gene required for heart and visceral mesoderm specification may be represented by a family of genes in vertebrates: *XNkx2-3*, a second vertebrate homologue of *tinman*. *Development (Cambridge, UK)* 121, 3889–3899.

Frasch, M. (1995). Induction of visceral and cardiac mesoderm by ectodermal Dpp in the early *Drosophila* embryo. *Nature (London)* 374, 464–467.

Fu, Y., and Izumo, S. (1995). Cardiac myogenesis: Overexpression of XCsx2 or XMEF2A in whole *Xenopus* embryos induces the precocious expression of *XMHCα* gene. *Roux's Arch. Dev. Biol.* 205, 198–202.

Fullilove, S. L. (1970). Heart induction: Distribution of active factors in newt endoderm. *J. Exp. Zool.* 175, 323–326.

Gove, C., Walmsley, M., Nijar, S., Bertwistle, D., Guille, M., Partington, G., Bomford, A., and Patient, R. (1997). Over-expression of GATA-6 in *Xenopus* embryos blocks differentiation of heart precursors. *EMBO J.* 16, 355–368.

Hemmati-Brivanlou, A., Frank, D., Bolce, M. E., Brown, B. D., Sive, H. L., and Harland, R. M. (1990). Localization of specific mRNAs in *Xenopus* embryos by whole-mount *in situ* hybridization. *Development (Cambridge, UK)* **110**, 325–330.

Jacobson, A. G. (1960). Influences of ectoderm and endoderm on heart differentiation in the newt. *Dev. Biol.* **2**, 138–154.

Jacobson, A. G. (1961). Heart determination in the newt. *J. Exp. Zool.* **146**, 139–151.

Jacobson, A. G., and Duncan, J. T. (1968). Heart induction in salamanders. *Am. J. Exp. Zool.* **167**, 79–103.

Kao, K. R., and Elinson, R. P. (1988). The entire mesodermal mantle behaves as Spemann's organizer in dorsoanterior enhanced *Xenopus laevis* embryos. *Dev. Biol.* **127**, 64–77.

Kim, Y., and Nirenberg, M. (1989). *Drosophila* NK homeobox genes. *Proc. Natl. Acad. Sci. U.S.A.* **86**, 7716–7720.

Komuro, I., and Izumo, S. (1993). Csx: a murine homeobox-containing gene specifically expressed in the developing heart. *Proc. Natl. Acad. Sci. U.S.A.* **90**, 8145–8149.

Lee, R. K. K., Stainier, D. Y. R., Weinstein, B. M., and Fishman, M. C. (1994). Cardiovascular development in the zebrafish. ii. Endocardial progenitors are sequestered within the heart field. *Development (Cambridge, UK)* **120**, 3361–3366.

Linask, K. K., and Lash, J. W. (1993). Early heart development: Dynamics of endocardial cell sorting suggests a common origin with cardiomyocytes. *Dev. Dyn.* **195**, 62–69.

Lints, T., Parsons, L., Hartley, L., Lyons, I., and Harvey, R. (1993). *Nkx-2.5*: A novel murine homeobox gene expressed in early heart progenitor cells and their myogenic descendants. *Development (Cambridge, UK)* **119**, 419–431.

Logan, M., and Mohun, T. (1993). Induction of cardiac muscle differentiation in isolated animal pole explants of *Xenopus laevis* embryos. *Development (Cambridge, UK)* **118**, 865–875.

Lyons, I., Parsons, L. M., Hartley, L., Li, R., Andrews, J. E., Robb, L., and Harvey, R. P. (1995). Myogenic and morphogenetic defects in the heart tubes of murine embryos lacking the homeobox gene *Nkx-2.5*. *Genes Dev.* **9**, 1654–1666.

Lyons, G. E. (1996). Vertebrate heart development. *Curr. Opin. Genet. Dev.* **6**, 454–460.

Lyons, G. E., Schiaffino, S., Sassoon, D., Parton, P., and Buckingham, M. (1990). Developmental regulation of myosin gene expression in mouse cardiac cells. *J. Cell Biol.* **111**, 2427–2436.

Mark, M., Kastner, P., Ghyselinck, N. B., Krezel, W., Dupe, V., and Chambon, P. (1997). Genetic control of the development by retinoic acid. *C. R. Acad. Sci. Paris* **191**, 77–90.

Mead, P. E., Hemmati-Brivanlou, I., Kelley, C. M., and Zon, L. I. (1996). BMP-4-responsive regulation of dorsal–ventral patterning by the homeobox protein Mix.1. *Nature (London)* **382**, 357–360.

Mohun, T. J., Brennan, S., Dathan, N., Fairman, S., and Gurdon, J. B. (1984). Cell type-specific activation of actin genes in the early amphibian embryo. *Nature (London)* **311**, 716–721.

Moody, S. A. (1987). Fates of the blastomeres of the 32-cell-stage *Xenopus* embryo. *Dev. Biol.* **122**, 300–319.

Muslin, A. J., and Williams, L. T. (1991). Well-defined growth factors promote cardiac development in axolotl mesodermal explants. *Development (Cambridge, UK)* **112**, 1095–1101.

Nascone, N., and Mercola, M. (1995). An inductive role for the endoderm in *Xenopus* cardiogenesis. *Development (Cambridge, UK)* **121**, 515–523.

Newman, C. S., and Krieg, P. A. (1998). *Tinman*-related genes expressed during heart development in *Xenopus*. *Dev. Gen.* **22**(3), 230–238.

Pabst, O., Schneider, A., Brand, R., and Arnold, H.-H. (1997). The mouse *Nkx2-3* homeodomain gene is expressed in gut mesenchyme during pre- and postnatal mouse development. *Dev. Dyn.* **209**, 29–35.

Patterson, K. D., Cleaver, O. B., Gerber, W. V., Grow, M. W., Newman, C. S., and Krieg, P. A. (1998). The role of homeobox genes in the development of the cardiovascular system. *Curr. Topics Dev. Biol.* **40**, 1–44.

Rudnicki, M. A., and Jaenisch, R. (1995). The MyoD family of transcription factors and skeletal myogenesis. *BioEssays* **17**, 203–209.

Sassoon, D. A., Garner, I., and Buckingham, M. (1988). Transcripts of α-cardiac and α-skeletal actin are early markers for myogenesis in the mouse embryo. *Development (Cambridge, UK)* **105**, 155–164.

Sater, A. K., and Jacobson, A. G. (1989). The specification of heart mesoderm occurs during gastrulation in *Xenopus laevis*. *Development (Cambridge, UK)* **105**, 821–830.

Sater, A. K., and Jacobson, A. G. (1990a). The role of the dorsal lip in the induction of heart mesoderm in *Xenopus laevis*. *Development (Cambridge, UK)* **108**, 461–470.

Sater, A. K., and Jacobson, A. G. (1990b). The restriction of the heart morphogenetic field in *Xenopus laevis*. *Dev. Biol.* **140**, 328–336.

Sive, H. L., Draper, B. W., Harland, R. M., and Weintraub, H. (1990). Identification of a retinoic acid-sensitive period during axis formation in *Xenopus laevis*. *Genes Dev.* **4**, 932–942.

Tonissen, K. F., Drysdale, T. A., Lints, T. J., Harvey, R. P., and Krieg, P. A. (1994). *XNkx-2.5*, a *Xenopus* gene related to *Nkx-2.5* and *tinman*: Evidence for a conserved role in cardiac development. *Dev. Biol.* **162**, 325–328.

Zou, Y., Evans, S., Chen, J., Kuo, H.-C., Harvey, R. P., and Chien, K. R. (1997). CARP, a cardiac ankyrin repeat protein, is downstream in the *Nkx2-5* homeobox pathway. *Development (Cambridge, UK)* **124**, 793–804.

Cellular Determination in Amphibian Retina

Muriel Perron
William A. Harris
Department of Anatomy
University of Cambridge
Cambridge CB2 3DY, United Kingdom

I. Introduction

In terms of cellular diversity, the nervous system is exceptionally complex. One of the goals of developmental neurobiology is to understand how this cellular complexity is generated. The retina, as "an approachable part of the brain" (Dowling, 1987) and where a fairly limited array of cell types are organized in a laminated structure, is an excellent model for such studies. Because amphibians present many advantages for embryological and molecular experiments, they have been extensively used for developmental studies of the visual system. We first present the different steps of amphibian eye development. Then, we present cell lineage, experimental embryology, and cell culture experiments that provide some insights into the mechanisms of cell fate determination in the amphibian retina. Finally, we present gene expression and gene manipulation studies and in particular, we have focused on the involvement of the neurogenic/

proneural pathway in the control of cell fate determination in the *Xenopus* retina.

II. *Xenopus* Eye Development

A. Origins of the *Xenopus* Eye

The region of the *Xenopus* embryo fated to form the eye has been mapped in blastula embryos using dye injection in a single blastomere (Fig. 1A; Huang and Moody, 1993). As this part is largely described in Chapter 25 by Fraser *et al.*, we will not give a detailed account of it here. The retina descends from nine blastomeres of the 32-cell stage embryos. Each of these nine blastomeres give rise to various proportions of the retina, one of the blastomeres, for example, produces about half of the retina. Hoechst and DiI or DiO have been used to label the descendants of a group of cells in the neural plate

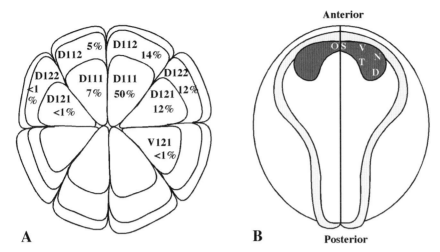

Figure 1 Fate map of the presumptive retina in *Xenopus*. (A) Animal pole view of a 32-cell stage embryo showing the nine blastomeres that will give rise to the right retina. The name of these blastomeres is indicated together with the percentage of the retina these blastomeres produce (Huang and Moody, 1993). (B) Dorsal view of a neural plate (stage 15). The areas that will give rise to the eyes are in dark gray. The future dorsoventral (D-V) and nasotemporal (N-T) axis of the retina are indicated. The medial region will give rise to the optic stalk (OS).

(Fig. 1B; Eagleson and Harris, 1990; Eagleson *et al.*, 1995). The area that forms the eye has been mapped to the anterior neural plate. The study of cell movements that drive morphogenesis of the eye shows that the ventral to dorsal axis of the retina maps to the anteromedial to posterolateral axis of the neural plate (Fig. 1B; Eagleson *et al.*, 1995).

B. Establishment of the Eye Field during *Xenopus* Development

Transplantation experiments have demonstrated that the anterior neural plate containing the presumptive eye originates under the influence of the invaginating mesoderm during gastrulation (Adelmann, 1937; Spemann, 1938). In order to know precisely when the eye field is specified, transplantations have been used (Spemann, 1938), and more recently the presence of different molecular markers has been assayed on explants from different developmental stages (Saha and Grainger, 1992). At midgastrula stage the entire presumptive neural plate is capable of producing eyes. The specification of the eye region in the anterior neural plate seems to be firmly fixed only at neural plate stages. Several factors, such as Noggin, Chordin, or Follistatin have been shown to be involved in neural induction (review in Hemmati-Brivanlou and Melton, 1997), but factors involved specifically in the induction of the eye field are still unknown. The expression of *Pax6* in the eye field (Hirsch and Harris, 1997a) suggested that it might define this field as competent to form an eye. *Pax6* is a vertebrate homologue of the *Drosophila* gene *eyeless* (*ey*) that is able to give ectopic eye structures when ectopically expressed (Halder *et al.*, 1995). However, optic vesi-

cle formation does occur in *Pax6* loss-of-function mutations in mammals, either the mutation *Small eye* in mice (Hogan *et al.*, 1988; Grindley *et al.*, 1995) or the aniridia syndrome in human (Ton *et al.*, 1991). Moreover, in *Xenopus*, *Pax6* ectopic expression induces only lens marker expression (Altmann *et al.*, 1997). Therefore, *Pax6* does not seem to be sufficient to determine the eye field to form an eye. *Xrx1* is another transcription factor expressed in the eye field (Casarosa *et al.*, 1997). Optic cups do not form in mice carrying *Xrx1* loss-of-function mutation (Mathers *et al.*, 1997). Moreover, ectopic expression of *Xrx1* in *Xenopus* embryos leads to ectopic retinal tissue formation (Mathers *et al.*, 1997), suggesting that *Xrx1* is involved in the specification of the eye field. *Six3* is the murine homologue (Oliver *et al.*, 1995) of the homeobox gene *sine oculis* implicated in *Drosophila* eye development (Serikaku and O'Tousa, 1994; Cheyette *et al.*, 1994). It has been shown that the murine *Six3* is able to induce an ectopic lens in fish (Oliver *et al.*, 1996). Recently, an homologue of *Six3* has been isolated in *Xenopus*, *XSix3* (Zuber *et al.*, 1998). *XSix3* also is expressed in the eye field of *Xenopus* embryos but its role in eye field determination and its relationships with *Pax6* or *Xrx1* remain to be discovered.

During neurulation, the presumptive eye tissue in the anterior neural plate separates into two primordia (Keating and Kennard, 1976). If the prechordal plate is removed from an early neurula, this process of separation does not occur and a single eye forms in the median position (Adelmann, 1937; Lopashov and Stroeva, 1964). Two mechanistic explanations can be proposed. Cyclopia either results from the fusion of two originally separate eyes, or from the failure in separating a single vesicle (reviewed in Adelmann, 1936). The availability of molecular markers makes possible the visualization of the morphogenetic eye field and its resolution in two primordia. For example, the expression patterns of *ET, Pax6, Xrx1* and *XSix3* genes in the anterior neural plate provide direct support for the existence of a single retina field in *Xenopus* embryos. The early expressions of these genes consist of a continuous band in the anterior neural plate which subsequently splits into two fields that give rise to the retinas (Fig. 2; Li *et al.*, 1997; Hirsch and Harris, 1997a; Mathers *et al.*, 1997; Zuber *et al.*, 1998). These expression patterns argue in favor of the hypothesis that both eyes originate from only one field. What happens to the median cells remains to be explained. Results from fate mapping experiments, using DiI injection coupled to *in situ* hybridizations, rule out the possibility that the median region of the retina field contains retinal precursor cells which migrate into the lateral retina primordia (Li *et al.*, 1997). Indeed, cells in the midline of the retina field were found in the ventral hypothalamus and optic stalk (Eagleson *et al.*, 1995; Li *et al.*, 1997). These experiments in *Xenopus* embryos suggest that retina field resolution results at least partially from the suppression of retina formation in the median region.

Only one retina forms from neural plate explants in which the prechordal mesoderm has been removed, showing that the prechordal mesoderm is important for the formation of two retina (Li *et al.*, 1997). Therefore, the function of the prechordal mesoderm on retina field resolution was examined using specific markers of the eye field, on neural plate explants (Li *et al.*, 1997). In the presence of the prechordal mesoderm, the retina field resolved into two primordia as expected. In contrast, in the absence of the prechordal mesoderm, the retina field did not resolve, suggesting that the prechordal mesoderm is essential for this process. The signaling molecules from the prechordal mesoderm that instruct the resolution of the retina field remain to be

discovered. *Sonic hedgehog* (*shh*) is a good candidate since mutant mice lacking *shh* display a cyclopean phenotype (Chiang *et al.*, 1996) and in zebrafish *cyclops* mutants, which exhibit fusion of the eyes (Hatta *et al.*, 1991), *shh* expression fails to be induced within the neurectoderm during early stage of development (Krauss *et al.*, 1993). These data suggest that *shh* is required for the formation of two eyes.

C. Development and Compartmentalization of the Optic Vesicle

After the rolling up of the neural plate into the neural tube, cells of the anterior ventral region of the neural tube evaginate to form the optic vesicle (Fig. 3; Lopashov and Stroeva, 1964). The optic vesicles reach the epidermis which later forms the lens. The lens formation is dependent on this interaction with the eye vesicle since ablation of the vesicle blocks lens formation (Spemann, 1938). Optic vesicles moved under non-lens epidermis are able to induce lens formation (Lewis, 1904, 1907). Similar experiments done more recently using lineage tracer to distiguish between host and donor tissues (Grainger *et al.*, 1988), however, suggest that the optic vesicle by itself is not sufficient to induce lens formation in any piece of epidermis. These results suggested that lens formation is the result of a series of inductive interactions which initiate preceding contact between the ectoderm and the optic vesicle. During gastrula stages, the presumptive anterior neural plate tissue induces, through planar signaling, the presumptive lens ectoderm and surrounding ectoderm to become competent to respond to the optic vesicle (Henry and Grainger, 1990; Grainger, 1992).

Cells located in different positions within the optic vesicle give rise to three morphologically and functionally different compartments: the optic stalk, the retinal pigmented epithelium (RPE), and the neural retina (Fig. 3). While the optic vesicles grow out later-

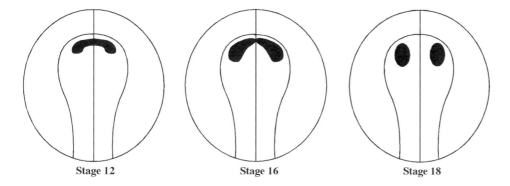

Stage 12   Stage 16   Stage 18

Figure 2 Schematic representation of the resolution of one eye field into two. (A) At stage 12, the eye field, revealed by the expression of *Pax6, ET, Xrx1,* or *XSix3*, consists of a continuous band in the anterior neural plate. (B, C) Progressively, under the influence of the underlying prechordal mesoderm, the eye field splits into two by stage 18.

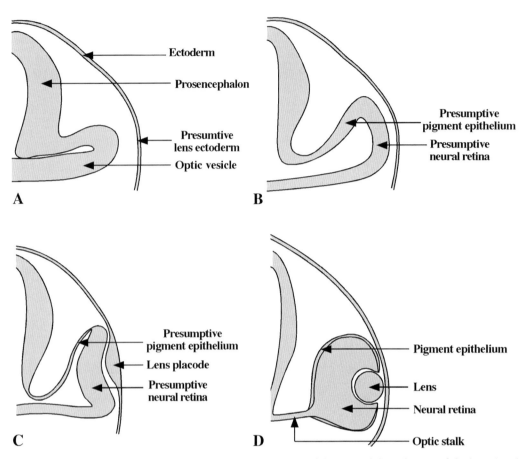

Figure 3 Schematic representation of the compartmentalization of the eye and the induction of the lens. (A, B) At stage 22, the optic vesicles evaginate from the brain until they reach the surface of the ectoderm. (C, D) While the optic vesicles invaginate, they induce the lens placode in the overlying ectoderm, which later invaginates to form the lens. The invagination of the optic vesicle leads to the formation of the optic cup in which the outer layer gives rise to the retinal pigment epithelium and the inner layer forms the neural retina. The optic cup remains in contact with the brain through the optic stalk.

ally, they become constricted from the brain, remaining connected with it only by the optic stalks. The anterior part of the vesicle forms the neural retina whereas the posterior part forms the RPE. The invagination of the optic vesicles from the outside creates the two-layered cuplike structure, the optic cup. The outer thin layer develops into the RPE, and the inner thick layer develops into the neural retina. Transcription factors expressed in the optic vesicle, such as *Otx2, Pax6,* or *Pax2,* might be involved in its compartmentalization. *Otx2* is one of the two homologues of *orthodenticle,* a regulatory gene controlling the determination of specific head segments in *Drosophila* (Finkelstein *et al.,* 1990). A mouse heterozygous mutant for *Otx2* shows severe eye defects (Matsuo *et al.,* 1995), supporting the idea that *Otx2* might be a homeobox gene involved in the control of eye development. In chick embryos, *Otx2* is first expressed throughout the optic vesicles but becomes restricted to the outer layer of the optic cup which gives rise to the retinal pigment epithelium, suggesting that it might be necessary for the specification of the RPE (Bovolenta *et al.,*

1997). However, even though the expression of *Otx2* *Xenopus* homologue has not yet been investigated in the RPE in *Xenopus* retina, it has been showed that it is expressed in some cells of the neural retina (M. Perron and W. A. Harris, 1998). Whether there is another member of the *Otx* family in *Xenopus* that is retricted to the RPE or whether it is a true difference between chick and *Xenopus,* remains to be elucidated. *Pax6* also is expressed first in the entire vesicle but is progressively downregulated in the cells that compose the RPE and the optic stalk and becomes restricted to the neural retina (Hirsch and Harris, 1997a; Macdonald and Wilson, 1997). *Pax2* is initially expressed in the ventral portion of the vesicle becoming restricted to the optic stalks and to cells around the choroid fissure, which establishes continuity between the inner retina and the optic stalks (Krauss et al., 1991; Macdonald *et al.,* 1997). Using zebrafish cyclops mutants and shh misexpression in zebrafish embryos, it has been suggested that shh might regulate the spatial distribution of *Pax2* and *Pax6,* which in turn might regulate the subdivision of the optic vesicle into optic stalk and retina (Macdonald *et al.*, 1995; Ekker *et al.*, 1995). Such a role for *shh* has not yet been investigated in *Xenopus*.

III. Generation of Neurons, Lamination, and Mosaic Patterning of Embryonic Retina

A. Cell Lineage

The determination of blastomeres to give a particular type of retinal cells has been investigated through lineage tracing technique (Huang and Moody, 1993). Each of the blastomeres that participate in the eye formation give clones composed of about 25% of photoreceptor cells, 50% of inner nuclear layer cells, and 25% of ganglion cells, which corresponds to the proportions of these cells in the retina (Holt *et al.*, 1988). This result suggests that the different retina blastomere progenitors are not determined to give a particular combination of retinal cell types. However, when the origins of different subtypes of amacrine cells is traced using a combination between lineage tracing and immunofluorescence techniques, unexpected results were obtained (Huang and Moody, 1995). Indeed, the production of two amacrine cell subtypes by the retina blastomere progenitors did not correspond to the proportion of these cells in the retina, and the contribution of each progenitor to these two subtypes was different. Therefore, it has been suggested that early lineage has an influence on specific neuronal type determination.

Throughout early *Xenopus* development, and well into the growth of the optic vesicle, there is a logarithmic increase in cell number and a logarithmic decrease in clone size, indicative of symmetric divisions in retinal precursor cells (Holt *et al.*, 1988; Harris and Hartenstein, 1991). As cells begin to leave the cell cycle, it is likely that there is an increasing switch to an asymmetric mode of cell division, in which one daughter is postmitotic while the other daughter remains in the cell cycle. To estimate cell lineage importance in retinal cell fate determination at this stage of development, single neuroepithelial cells of the optic vesicle have been injected with a lineage tracer (Holt *et al.*, 1988; Wetts and Fraser, 1988). Clonaly related cells in the central retina are arranged in columns running perpendicularly to the retinal layering (Holt *et al.*, 1988). This suggests that dividing cells in the neuroepithelium and their postmitotic progeny do not migrate laterally in the epithelium. Rather, neuroepithelium cells remain at their locations, and postmitotic cells migrate vertically giving rise to a full array of cell types. Indeed, labeled descendant of all major types were observed in all three retinal layers even in small clones,

indicating that retinal stem cells are not committed to particular classes of progeny. At the extreme, when clones consisted of only two or three cells, indicative of one or two divisions after labeling and before differentiation, the composition of cell types is mixed and unpredictable, showing that even the penultimate retinoblasts are not committed to particular retinal fates, and suggesting that retinal progenitors make their final decisions about cell fate postmitotically. However, since blastomeres seem biased to give rise to particular combinations of different amacrine subtypes, molecular markers might be required to identify different neuron subtypes among such clones.

B. Histogenesis and Lamination

The process of retinal cell differentiation leads to the subdivision of the retina into different layers. Like all vertebrate retinas, *Xenopus* retina contains two synaptic layers (outer and inner plexiform layers), which are intercalated between three nuclear layers: the outer nuclear layer, the inner nuclear layer, and the ganglion cell layer (Fig. 4). The outer nuclear consists of two major classes of photoreceptors, the rods and the cones. Rods mediate dim light vision, whereas cones function in bright light and are responsible for color vision. Two populations of rod photoreceptors have been identified, one major green-sensitive rod type and one minor blue-sensitive rod type (Hollyfield *et al.*, 1984). The *Xenopus* retina appears to contain three types of cones, a red-sensitive one, a blue-or-green sensitive one and a violet-sensitive one (Chang and Harris, 1998). The inner nuclear layer contains Müller cells, the only intrinsic glial cell in the retina, as well as horizontal, bipolar, and amacrine neurons. There are several subtypes of each class of neurons, and there are estimates, based on neurochemical differences, of 20 or more different types of amacrine cells. The ganglion cell layer also consists of a variety of subtypes of retinal ganglion cells (RGC) that send their axons through the optic nerve to various targets in the brain, including the dorsal thalamus, the ventral hypothalamus, the basal optic nucleus, and the tectum. There are at least three distinct types of RGC, based on differences in the size of the dendritic arbors alone. These differences arise during early larval development after central connections have been established, but they are independent of central connectivity because eye buds transplanted to the gut region develop these distinct classes of ganglion cell in the absence of central nervous system (CNS) innervation. Based on physiological, ultrastructural, and neurochemical criteria, there are clearly a much larger variety of RGCs.

Cell-type gradients of retinal histogenesis are largely conserved in vertebrate retinas. Ganglion cells are first to be born, followed by cones and horizontal

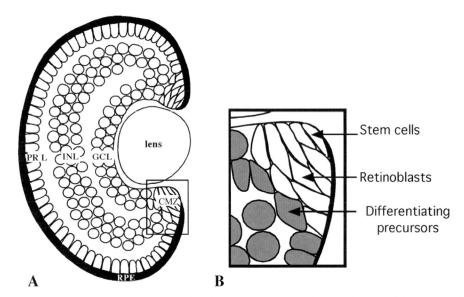

Figure 4 Schematic representation of the laminated retina. (A) The central region of the laminated retina contains three nuclear layers, the photoreceptor layer (PRL), the inner nuclear layer (INL), and the ganglion cell layer (GCL), separated by two plexiform layers. In the peripheral region of the laminated retina, called ciliary marginal zone (CMZ), cells continue to divide and to add new cells to the retina. The black box indicates the area that is enlarged in (B). (B) Schematic representation of the CMZ. The CMZ contains stem cells in the most peripheral region, nonrenewing retinoblasts in the middle, and postmitotic precursors in the process of differentiation in the central edge of the CMZ. Postmitotic cells are grey.

cells, while rods and Müller cells are the last (Sidman, 1961; Morest, 1970; Kahn, 1974; Holt *et al.*, 1988). Retinal histogenesis takes weeks to months in mammals, and therefore ganglion cells may all be postmitotic before the first rod cells are born. Within a specific class of cells, there may be further structure to the histogenesis, with some subtypes of ganglion cells, for example, being born before others. In contrast to mammals, there is a temporal compression in *Xenopus* histogenesis (Jacobson, 1968; Holt *et al.*, 1988). The central sector of embryonic retina completes histogenesis within 24 hours, so cells of all retinal layers are born nearly simultaneously. Using injections of tritiated thymide at closely spaced stages to study retinal cell birth date in *Xenopus*, it was shown that even though a late born cell can enter any of the layers, the general vertebrate plan of retinal histogenesis is maintained (Fig. 5). Thus ganglion cells tend to be born first and Müller cells tend to be born last. This minimal correlation between birth date and histogenesis suggests that the birth date may not be crucial for determination of retinal cell type.

In order to assess the role of cell division in retinal cell determination experimentally, cell division was blocked using a cocktail of hydroxyurea and amphidicolin (HUA) (Harris and Hartenstein, 1991). The results of this experiment showed that animals treated with HUA from early gastrula stages have far fewer cells than normal, yet the gross morphology of the eye is remarkably normal. Examination of the different cell types in the treated retina accomplished by immunolabeling on retinal sections (Harris and Hartenstein, 1991) showed that each cell type assayed for was present in the HUA

treated retinas. These results prove that the fate of a retinal stem cell is not determined either by counting the number of divisions that a retinoblast undergoes nor is the birth date of a cell a determining factor since in these animals all the cells were forced to be born at the same time, yet a full array of cell fates developed.

In spite of these results, the conserved general vertebrate pattern of histogenesis seen in *Xenopus* (Holt *et al.*, 1988) might provide a clue to the process regulating cellular diversity. For example, time of birth may bias the sequential determination of cells in which cell fate is a product of cellular interactions with other differentiating cells and the emerging environment.

C. Importance of Cell–Cell Interactions

Among the different retinal cell types, photoreceptor cells provide an excellent model for a molecular biological approach since many different markers are available to identify them *in vivo* and *in vitro*. In particular, opsins, the protein components of visual photopigments allow us to identify different types and subtypes of photoreceptors. To assess the role of cell–cell interactions in the determination of photoreceptors cells in *Xenopus* retina, an *in vitro* assay with different conditions controlling cell–cell interactions, was used (Harris and Messersmith, 1992). Isolated uncommitted cells from *Xenopus* retinas do not become photoreceptors, whereas those cells grown as clumps produce photoreceptor cells. These observations suggest that photoreceptor differentiation is promoted by cell–cell interactions *in vitro*. Moreover, using a marker for all the photoreceptors,

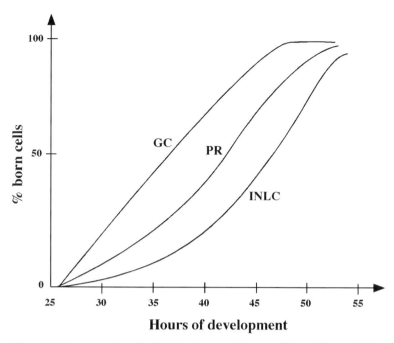

Figure 5 Representation of cell birth date according to cell layer. This graph shows that neurogenesis in the different layers is largely overlapping. However, ganglion cells (GC) tend to be born before photoreceptors (PR), themselves tending to be born before inner nuclear layer cells (INLC).

XAP-1, and a specific rod marker, XAP-2, it was shown that there are two cellular inductions involved in photoreceptor determination (Harris and Messersmith, 1992). The first induction promotes the expression of XAP-1 and later in development, another one promotes the expression of XAP-2. This suggests that it is only after a cell has chosen a photoreceptor fate that its rod fate is decided. To position cone determination in this cascade of inductions, several specific cone markers have been used (Chang and Harris, 1998). This led to the observation that the earliest cones are born before the first rods and in isolated culture, most of the cells that express XAP-1 at early stages also go on to express cone opsins. Thus, it is possible that cones are a default state of general photoreceptor determination, and that the same cells may become rods through a second induction. The determination of rods after cones, or the possibility that cones are a default state of a generalized photoreceptor fate, emphasizes the importance of a temporal sequence of events in postmitotic retinal neurons that correlates only weakly with birth date and provides a good illustration of a succession of cell–cell interactions that progressively restrict the fate of a cell. In other systems, other interactions have been studied. For instance, in the developing chick retina, early differentiating retinal ganglion cells appear to produce a factor that inhibits newly postmitotic cells from assuming a ganglion cell fate (Waid and McLoon, 1995). Thus, we might imagine a host of positive and negative interactions between cells in the differentiating neuroepithelium that work together to influence the proper succession of cell types.

D. Control of Proliferation and Determination by Growth Factors

To identify signaling molecules that control cell fate decision, dissociated cell culture is often used to test *in vitro* the effects of potential candidates. Molecules that have been tested in a variety of systems include basic fibroblast growth factor (FGF-2), transforming growth factor α (TGFα), epidermal growth factor (EGF), insulinlike growth factor (IGF), ciliary neurotrophic factor (CNTF), retinoic acid, and thyroxin, among others. However, effects seen in these *in vitro* experiments raise the question whether the observed effects reveal *in vivo* mechanisms. To get closer to an *in vivo* experiment, an eye bud technique has been elaborated (Stiemke and Hollyfield, 1995; S. Arora and W. A. Harris, unpublished data) in which eye buds from stage 23 *Xenopus* embryos are dissected and grown in a simple saline solution without any serum. These eye buds develop into eyes with lens, pigment epithelium, and normal neural retinal layers. To investigate growth factor effects on retinal cell determination, the eye buds can be cultured in various conditions. For example, the effects of a number of factors such as retinoic acid, CNTF, hedgehog, and FGF-2 on retinal cell determination were assayed for several antibodies specific to different retinal cell-types (S. Arora and W. A. Harris, unpublished data). In the case of FGF-2, it was shown that the number of both photoreceptors and retinal ganglion cells decreased in response to exogenous FGF-2. These findings suggested that FGF-2 was causing these changes in differentiated retinal cells either by killing some cells or by delaying the transition from undifferentiated progenitors to the differentiated cell types. To test these hypotheses, dividing cells in the treated eye buds were counted after BrdU labeling. The CMZ of the treated eye buds contained more BrdU positive cells than the untreated eye buds, suggesting that FGF-2 delays the transition of progenitors to differentiated cells. This increases the number of dividing cells in the CMZ, and decreases the number of differentiated cells in the central retina. There also is some evidence that FGF signaling is involved in cell fate decision since blocking the FGF pathway with a dominant negative form of FGF receptor changes the proportion of the different retinal cell types (McFarlane *et al.*, 1998).

Results with other signaling molecules and their receptors in *Xenopus* and other systems, suggest that there are many factors that are potentially involved in the determination of cell type identity. For instance retinoic acid in fish and rat promotes early rod determination (Kelley *et al.*, 1994; Hyatt *et al.*, 1996), whereas thyroid hormone favors cone determination (Kelley *et al.*, 1995), and CNTF favors rod determination in some systems and inhibits rod determination in others (Kirsch *et al.*, 1996). Combinations of factors may have different effects than the factors applied alone, suggesting a complex and possibly combinatorial scheme of signaling. Expression studies on these factors show that there are dynamic and laminar specific changes in the ligands and receptors (reviewed in Harris, 1997). Together, these data indicate that much work needs to be done to sort out the important external influences on cell determination in the developing retina.

E. Mosaic Patterning

In many vertebrates cell subtypes are arranged in equidistant mosaic arrays (Engstrom, 1963; Wassle *et al.*, 1981; Wikler and Rakic, 1991). This spacing mechanism assures that the appropriate cell types are available to cover the local neuronal processing in each bit of the retina. Cones, particular types of amacrine cells, or indeed almost any cell type in the retina, exhibit geometrical precision in their spacing. For example, the cone array in some teleosts rivals the spacing of cell types seen in the compound eye of insects. In contrast, where this mosaic patterning has been looked at in the *Xenopus* retina, although there is some degree of regularity in the patterning, the distribution of cone subtypes is more imprecise (Chang and Harris, 1998), suggesting that the patterning is somewhat rudimentary or degenerate in *Xenopus*. In any case, the mechanism responsible for geometrically arraying cells, causing them to spread out from each other, is suggestive of a lateral inhibitory mechanism of some sort.

IV. Postembryonic Retinogenesis

A. The Ciliary Marginal Zone: A Postembryonic Retinal Growing Zone

The retina of amphibians, like that of fish, continues to grow postembryonically and throughout life, keeping pace with the increasing size of the larval and adult animals by adding new cells of all types from the *ora serrata*, or ciliary margin (Johns, 1977; Beach and Jacobson, 1979; Hollyfield, 1968; Straznicky and Gaze, 1971; Reh, 1989; Wetts *et al.*, 1989). This region, situated at the peripheral edge of the retina, is composed of self renewing stem cells, nonrenewing retinoblasts, and differentiating neuronal precursors (Fig. 4). The ciliary marginal zone (CMZ) has the exceptional advantage of being spatially ordered with respect to cellular development and differentiation. The youngest and least determined stem cells are the closest to the periphery, the pluripotent retinoblasts are in the middle, whereas the cells at the central edge have stopped dividing and are in the process of differentiating (Dorsky *et al.*, 1995, see below). The relative peripheral to central position in the CMZ of the retina, therefore, reflects development along a spatial dimension, and a spatial analysis of cells in the CMZ provides a view of the developmental sequence that takes place during retinal neuron differentiation.

The influence of cell lineage on cell fate determination of CMZ cells has been investigated using lineage tracing techniques (Wetts *et al.*, 1989). Clones of cells generated from the ciliary margin give rise to cells of all retinal types, including glial cells. It also has been reported that one-quarter of the clones have descendants in both the neural retina and the RPE. These data demonstrated that CMZ stem cells are multipotent for all retinal cell types and that the regulation of cell fate must occur late in the cell lineage of the CMZ cells. Again, however, a molecular study of different subtypes of neurons remains to be undertaken. The broad range of size of the clones, compared to clones derived from optic vesicle cells, and their distribution, suggested that there may be two types of CMZ cells with different proliferative fates: self-renewing stem cells and cells that undergo a limited number of divisions.

Transplantation and ablation experiments have provided evidence that CMZ cell fate is regulated by cell–cell interactions (Hunt *et al.*, 1987; Reh and Tully, 1986; Reh, 1987). Small groups of eye cells from pigmented donor embryos have been transplanted into the eyes of albino hosts, transposing cells from the central retina (where cells are mitotically quiescent) to the CMZ (germinal zone) and vice versa (Hunt *et al.*, 1987). Only implants into the host germinal zone behaved like germinal cells. Conversely, when donor germinal cells were implanted in the central retina, they ceased dividing. One can conclude from these experiments that local environmental cues, rather than intrinsic cellular determinants, specify retinal cell fate. In an ablation experiment, the distribution of a differentiated cell in the central retina has been experimentally altered and the effects on the CMZ cell production monitored. Treatment with the neurotoxin 6-hydroxydopamine, which kills the central retinal dopaminergic amacrine cells, leads temporally to an increase in production of this cell type from the CMZ (Reh and Tully, 1986). Similarly, killing the central glutaminergically stimulated cells with kainic acid leads to an increased production from the CMZ of those cells (Reh, 1987). These results also suggest that the CMZ precursors are not committed to a particular cell type and produce daughter cells that differentiate as particular cell types in response to signals from more central retina.

B. Transdifferentiation during Retina Regeneration

Another aspect of postembryonic retinogenesis in amphibians concerns retinal regeneration. Retinal regeneration in mammals can occur in fetal or embryonic stages but not in adults. In contrast, regeneration of an eye is possible after complete retinectomy in adult urodeles (Stone, 1950). As a mechanistic explanation, it has been suggested that the RPE might be able to transdifferentiate to give rise to a new retinal neural epithelium (Stone, 1950). RPE cells have been followed step by step during the regeneration of neural retina of adult salamander eye, after excision of the entire neural retina (Stone, 1950; review in Okada, 1980). It was observed that the pigment cells change both their morphology and function, and undergo mitosis when they lose contact with the neural retina tissue. These cells migrate, lose their pigment and by proliferation give rise to the new neural retina. In contrast with urodeles, following complete retinectomy, retinal regeneration has not been observed in several anuran species (Lopashov, 1963; Sologub, 1968; Sakaguchi *et al.*, 1997). However, some studies have found that transplantation of RPE sheets into the posterior eye chamber of *Rana* or *Xenopus* leads to transdifferentiation of the RPE and production of a new retina (Lopashov and Sologub, 1972; Sologub, 1977; Reh and Nagy, 1987), suggesting that anurans and urodeles exhibit similar mechanisms for retinal regeneration. Regeneration of retina has been described in several newt species and in *Rana catesbienna* after extensive degeneration of the neural retina induced by devascularization (Keefe, 1973; Reyer, 1977; Reh and Nagy, 1987). These studies confirmed that RPE transdifferentiation is a source of new retinal cells. However,

it also has been shown that an increase in cell proliferation of the remaining CMZ is an additional source of the new retina.

Since RPE also can transdifferentiate to either neuron or lens in culture, an *in vitro* approach has been used to elucidate the factors that regulate this phenomenon (Reh and Nagy, 1987). Transdifferentiation is profoundly influenced by the substrate in which cells are cultured (Reh *et al.*, 1987). When RPE cells are cultured on plastic, polylysine, collagen type I, or fibronectin, no cells resembling neurons are observed. However, when cultured on laminin, the RPE cells change their morphology, extend fine processes resembling neurites with growth cones at their tips, and express neuronal specific antigens. These results thus suggest that laminin induces the transdifferentiation of RPE cells into neurons *in vitro*. Moreover, it was reported that retinal regeneration can be blocked *in vivo* by an antibody that inhibits the interaction of cells with the laminin-heparan sulfate proteoglycan complex (Nagy and Reh, 1994). Therefore, the process of retinal regeneration in amphibians might be regulated by the extracellular matrix.

Retinal pigmented epithelium from *Xenopus* tadpoles cultured in the presence of FGF-2 also transdifferentiates, generating a new retina containing both neurons and glial cells (Sakaguchi *et al.*, 1997). These results suggest that FGF-2 plays an important role in promoting retina regeneration *in vitro*. Continuing research to identify molecules involved in RPE transdifferentiation is needed to understand the retinal regeneration phenomenon and also to better understand retinal cell determination and differentiation during normal development.

V. Genetic Cascade of Retinogenesis

To identify the genes involved in the determination of retinal cells, researchers in vertebrate eye development have often taken clues from the *Drosophila* retina. The vertebrate retina, besides sharing genes that are involved in the early determination of the eye field suggesting evolutionary conservation in the early development of the eye, also share cellular mechanisms of retinal determination. For example, in both flies and vertebrates, it is clear that cell–cell interactions mediated by particular signals and receptors are critical to generate cellular diversity in the retina. In particular, the neurogenic genes mediating an inhibition of differentiation are involved in cell fate determination during *Drosophila* and vertebrate retinal development (Cagan and Ready, 1989; Fortini *et al.*, 1993; Dorsky *et al.*, 1995, 1997). The neurogenic genes interact with the proneural genes which promote differentiation. These too are used in both fly and frog retinal cell determination (Jarman *et al.*, 1993, 1994, 1995; Kanekar *et al.*, 1997). Thus, although *Drosophila* eyes do not look much like a vertebrate eye, they both use very similar mechanisms during development.

A. Isolation and Characterization of Neurogenic and Proneural Genes in *Xenopus*

In *Xenopus*, homologues of the *Drosophila* neurogenic genes *Notch*, *Delta*, *E(spl)* and *Su(H)* have been identified (Coffman *et al.*, 1990; Chitnis *et al.*, 1995; Turner, unpublished data; Wettstein *et al.*, 1997). The homologues of *Notch* and *Delta* in *Xenopus*, as in *Drosophila*, seem to control the number of neurons that comprise the primary nervous system (Coffman *et al.*, 1993; Chitnis *et al.*, 1995). Several homologues of proneural genes also have been identified in *Xenopus*. *Xash1* and *Xash3* are two members of the *achaete scute* complex (Ferreiro *et al.*, 1992, 1994; Zimmerman *et al.*, 1993), whereas *neuroD* (Lee *et al.*, 1995), *Xngnr-1* (Ma *et al.*, 1996), *ATH3*, (Takebayashi *et al.*, 1997) and *Xath-5* (Kanekar *et al.*, 1997) are related to *Drosophila atonal* gene (Jarman *et al.*, 1993). Both *achaete scute* and *atonal* related genes encode transcription factors belonging to the basic-helix-loop-helix (bHLH) family. *XMyT-1*, encoding a C2HC-type zinc finger transcription factor, in cooperation with *Xash3* or *Xngnr-1*, also has a proneural activity (Bellefroid *et al.*, 1996).

B. Expression of Neurogenic and Proneural Genes in *Xenopus* Retina

In order to know whether neurogenic and proneural genes are involved in vertebrate retinogenesis, their expression patterns have been studied during Xenopus development. During early development, both *X-Notch-1* and *X-Delta-1* are expressed in the prospective primary neurons in three longitudinal stripes of the neural plate. Later, they maintain expression in the germinal zone of the neural tube, and in the developing retina (Dorsky *et al.*, 1997). This expression is restricted to dividing cells. When the retina is laminated for example, they are expressed in the ciliary marginal zone but not in the central retina where cells are postmitotic (Dorsky *et al.*, 1997). However neither *X-Notch-1* nor *X-Delta-1* are expressed in the most peripheral part of the CMZ where stem cells are present (Dorsky *et al.*, 1995, 1997). Therefore, these expression patterns, similar to the expression of *Notch* in *Drosophila* eye disc (Kidd *et al.*, 1989), suggest that *X-Notch-1* and *X-Delta-1* are involved in retinoblast development. Two *E(spl)* homologues (*ESR1* and *ESR3*) are expressed in a similar pattern. Since ESR expression is dependent on Delta Notch signaling, it appears that the signaling pathway is active in these cells (Perron *et al.*, 1998).

All the proneural genes *Xash1*, *Xash3*, *ATH-3*, *Xath5*, *neuroD*, and *X-MyT1* are expressed in the developing optic vesicle (Ferreiro *et al.*, 1992; Zimmerman *et al.*, 1993; Takebayashi *et al.*, 1997; Kanekar *et al.*, 1997; Lee *et al.*, 1995; Bellefroid *et al.*, 1996). *Xash1*, *Xash3*, and *Xath-5* are, as the neurogenic genes, restricted to the CMZ in the laminated retina (Ferreiro *et al.*, 1992; Turner and Weintraub, 1994; Kanekar *et al.*, 1997). The proneural genes *ATH3*, *neuroD*, and *X-MyT1* also are expressed in the CMZ but in addition they are expressed in some cells of the central retina (Takebayashi *et al.*, 1997; Kanekar *et al.*, 1997; Perron *et al.*, 1998; Bellefroid *et al.*, 1996). Together, these expression patterns suggest that the neurogenic/proneural pathway is involved during *Xenopus* retinogenesis and that some proneural genes are still involved in differentiated retinal cells. The expression of *Xngnr-1* also appears to be restricted to the CMZ but the expression level is very weak (Perron *et al.*, 1998). Because another *neurogenin* related gene in mouse, *ngn2*, is expressed in the retina, other *neurogenin* genes might be involved in retinogenesis in *Xenopus*.

To position neurogenic and proneural genes in a developmental hierarchy during retinogenesis, the *Xenopus laevis* retinal CMZ has been used to compare gene expression patterns by double *in situ* hybridizations (Perron *et al.*, 1998). Such double *in situ* experiments allow us to compare on the same sections the expression patterns of two genes. In the CMZ, if one gene is expressed more peripherally than the other one, that would suggest that it is expressed earlier during retinogenesis, since spatial cellular organization reflects cellular development in the CMZ. To position neurogenic and proneural genes in a more general genetic hierarchy, their expressions in the retina have also been compared to the expression patterns of other genes involved at different steps in eye development, such as *Xrx1* (Casarosa *et al.*, 1997; Mathers *et al.*, 1997), *XSix3* (Oliver *et al.*, 1995, 1996; Zuber *et al.*, 1998), *Xotx2* (Matsuo *et al.*, 1995; Pannese *et al.*, 1995; Blitz and Cho, 1995), *Pax6* (Hirsch and Harris, 1997a; Altmann *et al.*, 1997), or *Brn-3.0* (Hirsch and Harris, 1997b; Gan *et al.*, 1996). For more information on *Xrx1*, *XSix3*, *Pax6*, and *Otx2* genes, see Sections II,C and IV,A. *Brn-3.0* is a member of the *Brn-3* family of POU domain transcription factors. It is expressed in retina ganglion cells in *Xenopus* (Hirsch and Harris, 1997b). The disruption of murine *Brn-3* family genes leads to a loss of retinal ganglion cells (Gan *et al.*, 1996; Erkman *et al.*, 1996). These genes thus represent genes involved in either a very early or rather a late step of eye development.

The cell mitotic state and the relative expression patterns of neurogenic, proneural genes and those other genes involved in eye development suggest four phases of retinal development in the CMZ (Fig. 6). The first zone contains stem cells in the most peripheral part of

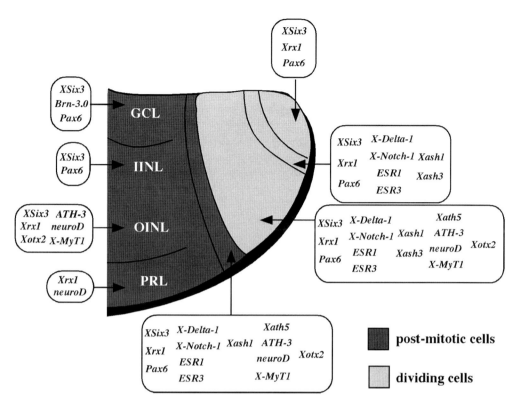

Figure 6 Genetic hierarchy during retinogenesis. According to gene expression, the CMZ region of the retina has been subdivided into four zones. Since spatial cellular organization reflects cellular development in the CMZ, if one gene is expressed more peripherally than the other one, it suggests that it is expressed earlier during retinogenesis. Genes expressed in the different zones of the CMZ and in the different layers of the retina are depicted. The region where cells are dividing, according to BrdU staining, is in light gray, and the region where cells are postmitotic is in dark gray. The first genes to be expressed, in the most peripheral region of the retina probably corresponding to stem cells, are *XSix3*, *Xrx1*, and *Pax6*. Cells in the next stage in retinogenesis still express these genes but express in addition *X-Delta-1*, *X-Notch-1*, *ESR1*, *ESR3*, *Xash1*, and *Xash3*. These cells correspond to proliferating retinoblasts. Cells in the third stage still express all the previous genes but express in addition *ATH-3*, *Xath5*, *X-MyT1*, *neuroD*, and *Xotx2*. In this stage *Xash3* is progressively downregulated. Cells in the next stage express exactly the same combination of genes (except *Xash3*) but are postmitotic. Cells in the ganglion cell layer turn off all the genes expressed in the previous stage except *XSix3* and *Pax6*. In addition, those cells turn on *Brn-3.0*. Cells in the inner part of the inner nuclear layer turn off all the genes except *XSix3*, *Pax6*, and *X-MyT1*. Cells in the outer part of the inner nuclear layer turn off all the genes except *XSix3*, *X-MyT1*, *ATH-3*, *neuroD*, and *Xotx2*. In addition, some cells in this layer still express *Xrx1*. Cells in the outer nuclear layer turn off all the genes except *neuroD*, and some also still express *Xrx1*. GCL, ganglion cell layer; IINL, inner part of the inner nuclear layer; OINL, outer part of the inner nuclear layer; ONL, outer nuclear layer.

the CMZ that do not express any of the neurogenic nor proneural genes but do express *Xrx1*, *XSix3*, and a low level of *Pax6*. The neurogenic genes *X-Notch-1*, *X-Delta-1*, *ESR1*, and *ESR3* are turned on in the next phase, zone 2, together with two *achaete scute* related genes, *Xash1 and Xash3*. In zone 3, *Otx2*, *X-MyT1*, and the *atonal*-like proneural genes, *ATH-3*, *Xath5*, and *neuroD* are turned on. Finally, in zone 4, the cells in the CMZ express the same genes as that in zone 3 (except for *Xash3* which is turned off) but have stopped dividing. In the central retina, neurogenic genes are turned off, together with some other proneural genes depending on the neural layer where cells will differentiate (Fig. 6). Indeed, each cell layer maintains a particular combinatorial of the above genes. These results suggest that retinoblasts, when they reach the central edge of the CMZ, may receive different signals from postmitotic differentiated cells. Such signals could induce or maintain the expression of genes specific to the different retinal

layers and lead to the differentiation of each type of retinal neuron.

The developmental positioning presented above leads us to make some suggestions about gene interactions during retinogenesis, which can be subsequently functionally tested. For instance, the staggered expressions of *achaete-scute*-like and *atonal*-like proneural genes suggest that *Xash1* and *Xash3* are upstream of *neuroD*, *Xath5*, and *ATH-3*, and may induce their expressions. Indeed, overexpression of *Xash3* is able to induce ectopic *neuroD* expression in neurula embryos, but not vice versa (Kanekar *et al.*, 1997). The coexpressions of *neuroD* and *Xath5* in the CMZ suggested that these genes may interact. It has indeed been shown that overexpression of *Xath5* is able to induce ectopic *neuroD* expression and that, conversely, overexpression of *neuroD* is able to induce ectopic *Xath5* expression (Kaneker *et al.*, 1997). These results are consistent with the expression patterns of both genes in the CMZ, and suggest that such a cross-regulation occurs in the retina. More functional studies, according to this global developmental cascade, remain to be done to better understand the complex relationships between neurogenic and proneural genes during retinogenesis.

C. Perturbation of Delta/Notch Signaling in *Xenopus* Retina

To gain further insights into the function of the neurogenic pathway during *Xenopus* retinogenesis, gene misexpression experiments in the retina have been undertaken (Fig. 7; Dorsky *et al.*, 1995, 1997; Kanekar *et al.*, 1997). An *in vivo* lipofection experiment has been elaborated in order to study transfected cells during eye development (Holt *et al.*, 1990). This technique has been used to transfect retinal precursors with an activated form of *X-Notch-1*, Notch∆E (Dorsky *et al.*, 1995). The eye phenotype was analyzed on laminated retinas. In contrast to control transfected cells, which exhibit characteristics of mature neurons, only few Notch∆E transfected cells had a morphology of mature neurons. Moreover, Notch∆E transfected cells with a neuroepithelial morphology did not express molecular markers of differentiated retinal neurons. These results suggest that an activated form of *X-Notch-1* in retinal cells causes them to retain a neuroepithelial morphology instead of differentiating into a particular type of retinal neuron. Therefore, one can predict from these results that the normal function of *X-Notch-1* during eye development, by delaying or inhibiting cell differentiation, might be to control the time when a cell is ready to differentiate.

A retina progenitor blastomere has been injected with *X-Delta-1* and the blastomere descendants in the retina have been analyzed in order to study *X-Delta-1* function during *Xenopus* retinogenesis (Dorsky *et al.*, 1997). Cells surrounding cells expressing exogenous *X-Delta-1* express a low level of endogenous *X-Delta-1* compared to nonneighbor cells. This suggests that a cell expressing *X-Delta-1* inhibits expression of *X-Delta-1* in neighbor cells. Those *X-Delta-1* positive cells, which have wild-type neighbors, showed a bias towards early differentiation (Fig. 7), probably because they no longer receive *X-Delta-1* signaling from their wild-type neighbors. Therefore, consistent with the role of *Delta* in *Drosophila* (Parks *et al.*, 1995; Parody and Muskavitch, 1993), this suggests that X-Delta-1 signaling in *Xenopus* retinal cells prevents them from differentiating. In accordance with this hypothesis, cells within a large clone of *X-Delta-1* positive cells remain neuroepithelial. They do not exhibit differentiation, probably because they receive the inhibitory *X-Delta-1* signaling from their *X-Delta-1* positive neighbor cells. A dominant negative form of X-Delta-1, Delta^{STU}, has been used to block endogenous X-Delta-1 signaling. In all sizes of clones, Delta^{STU} positive cells ex-

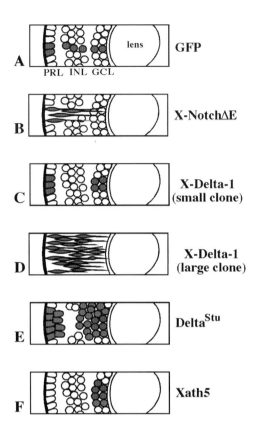

Figure 7 Effects of misexpression of proneural and neurogenic genes in *Xenopus* retina. Using *in vivo* lipofection or blastomere injection, the effects of misexpressing neurogenic or proneural genes have been investigated. The descendant cells in stage 41 retina are in gray. (A) Control GFP positive cells include cells of all the different layers. (B) X-Notch∆E positive cells have a neuroepithelial morphology. (C) X-Delta-1 positive cells, in small clones, show a bias toward ganglion cell and cone differentiation. (D) X-Delta-1 positive cells, in large clones, have a neuroepithelial morphology. (E) DeltaSTU (a dominant negative form of X-Delta-1) positive cells exhibit a bias towards ganglion cell and cone differentiation. (F) Xath5 positive cells show a bias toward ganglion cells differentiation. PRL, photoreceptor layer; INL, inner nuclear layer; GCL, ganglion cell layer.

hibit a bias toward early differentiation, consistent with the idea that releasing cells from X-Delta-1 inhibitory signaling pushes them to a premature differentiation. Moreover, this experiment has been done both using the blastomere injection technique and the *in vivo* lipofection experiment (Dorsky *et al.*, 1997). This last experiment results in blocking X-Delta-1 signaling later in development than when injection is done in blastomeres. The consequence is that, even though in both cases a bias toward early differentiation is observed, the fate of the Delta^{STU} positive cells is different in both techniques. This suggests that depending of the developmental stage when the cell is released from X-Delta-1 inhibitory signaling, the differentiating signals received by the cell also are different. Together, these results suggest that X-Delta-1 signaling controls neuronal diversity in the retina by regulating the competence of precursors to respond to signals that promote different cell types.

The effects on retinal cell determination of proneural genes, such as *Xath5*, have been analyzed using the *in vivo* lipofection technique (Kanekar *et al.,* 1997). This work showed that transfected cells with *Xath5* present a bias toward retinal ganglion cell fate at the expense of amacrine, bipolar, and Müller cells (Fig 7). Since ganglion cells are the first born retinal cells, this suggest that transfection of *Xath5* promotes early differentiation of progenitor cells. Therefore, *Xath5* might be involved in the regulation of cell differentiation during normal retinogenesis.

VI. Conclusions and Perspectives

What emerges from these data is that neither cell birth date, nor the numbers of divisions that a cell undergoes, seem to be crucial for retinal cell fate determination. Studies of the influence of lineage relationships on retinal cell fate determination present a conundrum. When cells are identified according to the layer they are in, cell lineage seems to have no influence on their fate. However, when neuronal cell subtypes in the same layer are identified using molecular marker, cell lineage seems to have an influence on their fate. Nevertheless, both embryological and molecular approaches converge to suggest that cell fate is largely influenced by cell–cell interactions. Following results obtained from the genetic system of *Drosophila* eye development, it was suggested that the proneural and neurogenic pathways might be a good candidate to play a role in such interactions in vertebrates. Results of expression pattern analyses of neurogenic and proneural genes and misexpression experiments during *Xenopus* retinogenesis support the hypothesis that this pathway regulates retinal cell fate determination in vertebrates. However, more functional studies are needed to better understand the role played by all these genes and their relationships during retinogenesis. Moreover, the implication of other signaling pathways in eye development remains to be investigated. Hedgehog signaling pathway, for example, is another genetic cascade involved in eye development in *Drosophila*. It has been proposed that it is implicated in the progression of the morphogenetic furrow of the eye disc (Chapter 16 by Carthew *et al.*). Since many signaling pathways are conserved between insects and vertebrates, the investigation of the role played by the EGF pathway, the hedgehog pathway, and other pathways, during *Xenopus* retinogenesis will, in the future, better our understanding of retinal cell fate determination in vertebrates. In this context, *Xenopus* retina is and will remain an excellent model. But clearly the conjunction of all the data obtained through the study of different vertebrate species, each of them having their own particular advantages, is necessary to provide a comprehensive understanding of vertebrate retinal development.

References

Adelmann, H. B. (1936). The problem of cyclopia. *Q. Rev. Biol.* **11**, 161 182 and 284–304.
Adelmann, H. B. (1937). Experimental studies on the development of the eye. IV. The effect of partial and complete excision of the prechordal substrata on the development of the eyes of *Ambystoma punctatum*. *J. Exp. Zool.* **75**, 199–227.
Altmann, C. R., Chow, R. L., Lang, R. A., and Hemmati-Brivanlou, A. (1997). Lens induction by *Pax-6* in *Xenopus laevis*. *Dev. Biol.* **185**, 119–123.
Beach, D. H., and Jacobson, M. (1979). Patterns of cell proliferation in the retina of the clawed frog during development. *Comp. Neurol.* **183**, 603–613.
Bellefroid, E. J., Bourguignon, C., Hollemann, T., Ma, Q., Anderson, D. J., Kintner, C., and Pieler, T. (1996). X-MyT1, a *Xenopus* C2HC-type zinc finger protein with a regulatory function in neuronal differentiation. *Cell (Cambridge, Mass.)* **87**, 1191–1202.
Blitz, I. L., and Cho, K. W. (1995). Anterior neurectoderm is progressively induced during gastrulation: The role of the *Xenopus homeobox gene* orthodenticle. *Development (Cambridge, UK)* **121**, 993–1004.
Bovolenta, P., Mallamaci, A., Briata, P., Corte, G., and Boncinelli, E. (1997). Implication of OTX2 in pigment epithelium determination and neural retina differentiation. *J. Neurosci.* **17**, 4243–4252.
Cagan, R. L., and Ready D. F. (1989). *Notch* is required for successive cell decisions in the developing *Drosophila* retina. *Genes Dev.* **3**, 1099–1112.
Casarosa, S., Andreazzoli, M., Simeone, A., and Barsacchi, G. (1997). Xrx1, a novel *Xenopus* homeobox gene expressed during eye and pineal gland development. *Mech. Dev.* **61**, 187–198.
Chang, W. S., and Harris, W. A. (1998). Sequential genesis and induction of cone and rod photoreceptors in *Xenopus*. *J. Neurobiol.* **35**, 227–244.
Cheyette, B. N., Green, P. J., Martin, K., Garren, H., Hartenstein, V., and Zipursky, S. L. (1994). The *Drosophila sine oculis* locus encodes a homeodomain-containing protein required for the development of the entire visual system. *Neuron* **12**, 977–996.
Chiang, C., Litingtung, Y., Lee, E., Young, K. E., Corden, J. L., Westphal, H., and Beachy P. A. (1996). Cyclopia and defective axial patterning in mice lacking *sonic hedgehog* gene function. *Nature (London)* **383**, 407–413.
Chitnis, A., Henrique, D., Lewis, J., Ish-Horowicz, D., and Kintner, C. (1995). Primary neurogenesis in *Xenopus* embryos regulated by a homologue of the *Drosophila* neurogenic gene *Delta*. *Nature (London)* **375**, 761–766.
Coffman, C., Harris, W., and Kintner, C. (1990). Xotch, the Xenopus homolog of *Drosophila Notch*. *Science* **249**, 1438–1441.
Coffman, C., Skoglund, P., Harris, W., and Kintner, C. (1993). Expression of an extracellular deletion of Xotch alters cell fate in *Xenopus* embryos. *Cell (Cambridge, Mass.)* **73**, 659–667.
Dorsky, R. I., Rapaport, D. H., and Harris, W. A. (1995). Xotch inhibits cell differentiation in the *Xenopus* retina. *Neuron* **14**, 487–496.

Dorsky, R. I., Chang, W. S., Rapaport, D. H., and Harris, W. A. (1997). Regulation of neuronal diversity in the *Xenopus* retina by Delta signaling. *Nature (London)* **385,** 67–70.

Dowling, J. E. (1987). "The Retina. An Approachable Part of the Brain." Belknap Press, Cambridge, Massachusetts.

Eagleson, G. W., and Harris, W. A. (1990). Mapping of the presumptive brain regions in the neural plate of *Xenopus laevis. J. Neurobiol.* **21,** 427–440.

Eagleson, G. W., Ferreiro, B., and Harris, W. A. (1995). Fate of the anterior ridge and the morphogenesis of the *Xenopus* forebrain. *J. Neurobiol.* **28,** 146–158.

Ekker, S. C., Ungar, A. R., Greenstein, P., von Kessler, D. P. Porter, J. A., Moon, R. T., and Beachy, P. A. (1995). Patterning activities of vertebrate hedgehog proteins in the developing eye and brain. *Curr. Biol.* **5,** 944–955.

Engstrom, K. (1963). Cone types and cone arrangements in the teleost retina. *Acta Zool.* **44,** 179–243.

Erkman, L., McEvilly, R. J., Luo L., Ryan, A. K., Hooshmand, F., O'Connell, S. M., Keithley, E. M., Rapaport, D. H., Ryan, A. F., and, Rosenfeld, M. G. (1996). Role of transcription factors *Brn-3.1* and *Brn-3.2* in auditory and visual system development. *Nature (London)* **381,** 603–606.

Ferreiro, B., Skoglund, P., Bailey, A., Dorsky, R., and Harris, W. A. (1992). Xash1, a Xenopus *homolog of achaete-scute:* A proneural gene in anterior regions of the vertebrate CNS. *Mech. Dev.* **40,** 25–36.

Ferreiro, B., Kintner, C., Zimmerman, K., Anderson, D., and Harris, W. A. (1994). XASH genes promote neurogenesis in *Xenopus* embryos. *Development (Cambridge, UK)* **120,** 3649–3655.

Finkelstein, R., Smouse, D., Capaci, T. M., Spradling, A. C., and Perrimon, N. (1990). The *orthodenticle* gene encodes a novel homeo domain protein involved in the development of the *Drosophila* nervous system and ocellar visual structures. *Genes Dev.* **4,** 1516–1527.

Fortini, M. E., Rebay, I., Caron, L. A., and Artavanis-Tsakonas, S. (1993). An activated Notch receptor blocks cell-fate commitment in the developing *Drosophila* eye. *Nature (London)* **365,** 555–557.

Gan, L., Xiang, M., Zhou, L., Wagner, D. S., Klein, W. H., and Nathans, J. (1996). POU domain factor Brn-3b is required for the development of a large set of retinal ganglion cells. *Proc. Natl. Acad. Sci. U.S.A.* **93,** 3920–3925.

Grainger, R. M. (1992). Embryonic lens induction: Shedding light on vertebrate tissue determination. *Trends Genet.* **8,** 349–355.

Grainger, R. M., Herry, J. J., and Henderson, R. A. (1988). Reinvestigation of the role of the optic vesicle in embryonic lens induction. *Development (Cambridge, UK)* **102,** 517–526.

Grindley, J. C., Davidson, D. R., and Hill, R. E. (1995). The role of Pax-6 in eye and nasal development. *Development* **121,** 1433–1444.

Halder, G., Callaerts, P., and Gehring, W. J. (1995). Induction of ectopic eyes by targeted expression of the eyeless gene in *Drosophila. Science* **267,** 1788–1792.

Harris, W. A. (1997). Cellular diversification in the vertebrate retina. *Curr. Opin. Genet. Dev.* **7,** 651–658.

Harris, W. A., and Hartenstein, V. (1991). Neuronal determi-nation without cell division in *Xenopus* embryos. *Neuron* **6,** 499–515.

Harris, W. A., and Messersmith, S. L. (1992). Two cellular inductions involved in photoreceptor determination in the *Xenopus* retina. *Neuron* **9,** 357–72.

Hatta, K., Kimmel, C. B., Ho, R. K., and Walker, C. (1991). The *cyclops* mutation blocks specification of the floor plate of the zebrafish central nervous system. *Nature (London)* **350,** 339–341.

Hemmati-Brivanlou, A., and Melton, D. (1997). Vertebrate neural induction. *Annu. Rev. Neurosci.* **20,** 43–60.

Henry, J. J., and Grainger, R. M. (1990). Early tissue interactions leading to embryonic lens formation in *Xenopus laevis. Dev. Biol.* **141,** 149–163.

Hirsch, N., and Harris, W. A. (1997a). *Xenopus Pax6* and retinal development. *J. Neurobiol.* **32,** 45–61.

Hirsch, N., and Harris, W. A. (1997b). *Xenopus Brn-3.0,* a POU-domain gene expressed in the developing retina and tectum. Not regulated by innervation. *Invest. Ophthalmol. Visual Sci.* **38,** 960–969.

Hogan, B. L., Hirst, E. M., Horsburgh, G., and Hetherington, C. M. (1988). *Small eye (Sey):* A mouse model for the genetic analysis of craniofacial abnormalities. *Development (Cambridge, UK)* **103** (Suppl.), 115–119.

Hollyfield, J.G. (1968). Differential addition of cells to the retina in *Rana pipiens* tadpoles. *Dev. Biol.* **18,** 163–179.

Hollyfield, J. G., Rayborn, M. E., and Rosenthal, J. (1984). Two populations of rod photoreceptors in the retina of *Xenopus laevis* identified with 3H-fucose autoradiography. *Vision Res.* **24,** 777–782.

Holt, C. E., Bertsch, T.W., Ellis, H. M., and Harris, W. A. (1988). Cellular determination in the *Xenopus* retina is independent of lineage and birth date. *Neuron* **1,** 15–26.

Holt, C. E., Garlick, N., and Cornel, E. (1990). Lipofection of cDNAs in the embryonic vertebrate central nervous system. *Neuron* **4,** 203–214.

Huang, S., and Moody S. A. (1993). The retinal fate of *Xenopus* cleavage stage progenitors is dependent upon blastomere position and competence: Studies of normal and regulated clones. *J. Neurosci.* **13,** 3193–3210.

Huang, S., and Moody, S. A. (1995). Asymmetrical blastomere origin and spatial domains of dopamine and neuropeptide Y amacrine subtypes in *Xenopus* tadpole retina. *J. Comp. Neurol.* **360,** 442–453.

Hunt, R. K., Cohen, J. S., and Mason, B. J. (1987). Cell patterning in pigment-chimeric eyes of *Xenopus:* Local cues control the decision to become germinal cells. *Proc. Natl. Acad. Sci. U.S.A.* **84,** 5292–5296.

Hyatt, G. A., Schmitt, E. A., Fadool, J. M., and Dowling, J. E. (1996). Retinoic acid alters photoreceptor development in vivo. *Proc. Natl. Acad. Sci. U.S.A.* **93,** 13298–13303.

Jacobson, M. (1968). Development of neuronal specificity in retinal ganglion cells of *Xenopus. Dev. Biol.* **17,** 202–218.

Jarman, A. P., Grau, Y., Jan, L. Y., and Jan, Y. N. (1993). *Atonal* is a proneural gene that directs chordotonal organ formation in the *Drosophila* peripheral nervous system. *Cell (Cambridge, Mass.)* **73,** 1307–1321.

Jarman, A. P., Grell, E. H., Ackerman, L., Jan, L. Y., and Jan, Y. N. (1994). *Atonal* is the proneural gene for Drosophila photoreceptors. *Nature (London)* **369,** 398–400.

Jarman, A. P., Sun, Y., Jan, L. Y., and Jan, Y. N. (1995). Role of the proneural gene, *atonal,* in formation of *Drosophila* chordotonal organs and photoreceptors. *Development (Cambridge, UK)* **121**, 2019–2030.

Johns, P. R. (1977). Growth of the adult goldfish eye. III. Source of the new retinal cells. *J. Comp. Neurol.* **176**, 343–357.

Kahn, A. J. (1974). An autoradiographic analysis of the time of appearance of neurons in the developing chick neural retina. *Dev. Biol.* **38**, 30–40.

Kanekar, S., Perron, M., Dorsky, R., Harris, W. A., Jan, L. Y., Jan, Y. N., and Vetter, M. (1997). *Xath5* participates in a network of bHLH genes in the developing *Xenopus* retina. *Neuron* **19**, 981–994.

Keating, M. J., and Kennard, C. (1976). The amphibian visual system as a model for development neurobiology. *In* "The Amphibian Visual System. A Mutidisciplinary Approach" (K. V. Fite, ed.), pp. 267–315. Academic Press, New York.

Keefe, J. R. (1973). An analysis of urodelian retinal regeneration. I. Studies of the cellular source of retinal regeneration in *Notophthalmus viridescens* utilizing 3 H-thymidine and colchicine. *J. Exp. Zool.* **184**, 185–206.

Kelley, M. W., Turner, J. K., and Reh, T. A. (1994). Retinoic acid promotes differentiation of photoreceptors *in vitro.* *Development (Cambridge, UK)* **120**, 2091–2102.

Kelley, M. W., Turner, J. K., and Reh, T. A. (1995). Ligands of steroid/thyroid receptors induce cone photoreceptors in vertebrate retina. *Development (Cambridge, UK)* **121**, 3777–3785.

Kidd, S., Baylies, M. K., Gasic, G. P., and Young, M. W. (1989). Structure and distribution of the Notch protein in developing *Drosophila. Genes Dev.* **3**, 1113–1129.

Kirsch, M., Fuhrmann, S., Wiese, A., and Hofmann, H. D. (1996). CNTF exerts opposite effects on *in vitro* development of rat and chick photoreceptors. *Neuroreport* **7**, 697–700.

Krauss, S., Johansen, T., Korzh, V., and Fjose, A. (1991). Expression pattern of zebrafish *pax* genes suggests a role in early brain regionalization. *Nature (London)* **353**, 267–270.

Krauss, S., Concordet, J. P., Ingham, P. W. (1993). A functionally conserved homolog of the *Drosophila* segment polarity gene hh is expressed in tissues with polarizing activity in zebrafish embryos. *Cell (Cambridge, Mass.)* **75**, 1431–1444.

Lee, J. E., Hollenberg, S. M., Snider, L., Turner, D. L., Lipnick, N., and Weintraub, H. (1995). Conversion of *Xenopus* ectoderm into neurons by NeuroD, a basic helix-loop-helix protein. *Science* **268**, 836–844.

Lewis, W. H. (1904). Experimental studies on the development of the eye in amphibia I. On the origin of the lens. *Rana palustris. Am. J. Anim.* **3**, 505–536.

Lewis, W. H. (1907). Experimental studies on the development of the eye in amphibia III. On the origin and differentiation of the lens. *Am. J. Anim.* **6**, 473–509.

Li, H., Tierney, C., Wen, L., Wu, J. Y., and Rao, Y. (1997). A single morphogenetic field gives rise to two retina primordia under the influence of the prechordal plate. *Development (Cambridge, UK)* **124**, 603–615.

Lopashov, G. V. (1963). "Developmental Mechanisms of Vertebrate Eye Rudiments" Pergamon, Oxford.

Lopashov, G.V., and Sologub, A. A. (1972). Artificial metaplasia of pigmented epithelium into retina in tadpoles and adult frogs. *J. Embryol. Exp. Morphol.* **28**, 521–546.

Lopashov, G. V., and Stroeva, O. G. (1964). "Development of the Eye. Experimental Studies." S. Monson, Jerusalem.

Ma, Q., Kintner, C., and Anderson, D. J. (1996). Identification of neurogenin, a vertebrate neuronal determination gene. *Cell (Cambridge, Mass.)* **87**, 43–52.

Mathers, P. H., Grinberg, A., Mahon, K. A., and Jamrich, M. (1997). The *Rx* homeobox gene is essential for vertebrate eye development. *Nature (London)* **387**, 603–607.

Matsuo, I., Kuratani, S., Kimura, C., Takeda, N., and Aizawa, S. (1995). Mouse *Otx2* functions in the formation and patterning of rostral head. *Genes Dev.* **9**, 2646–2658.

Macdonald, R., and Wilson, S. W. (1997). Distribution of Pax6 protein during eye development suggests discrete roles in proliferative and differentiated visual cells. *Dev. Genes Evol.* **206**, 363–369.

Macdonald, R., Barth, K., Xu, Q., Holder, N., Mikkola, I., and Wilson, S. (1995). Midline signaling is required for *Pax* gene regulation and patterning of the eyes. *Development (Cambridge, UK)* **121**, (Suppl.) 3267–3278.

Macdonald, R., Scholes, J., Strähle, U., Brennan, C., Holder, N., Brand, M., and Wilson, S. (1997). The Pax protein Noi is required for commissural axon pathway formation in the rostral forebrain. *Development (Cambridge, UK)* **124**, (Suppl.) 2397–2408.

McFarlane, S., Zuber, M. E., and Holt, C. E. (1998). A role for the fibroblast growth factor receptor in cell fate decisions in the developing vertebrate retina. *Development,* In Press.

Morest, D. K. (1970). The pattern of neurogenesis in the retina of the rat. *Z. Anat. Entwicklungsgesch.* **131**, 45–67.

Nagy, T., and Reh, T. A. (1994). Inhibition of retinal regeneration in larval *Rana* by an antibody directed against a laminin-heparan sulfate proteoglycan. *Brain Res. Dev. Brain Res.* **81**, 131–134.

Okada, T. S. (1980). Cellular metaplasia or transdifferentiation as a model for retinal cell differentiation. *Curr. Top. Dev. Biol.* **16**, 349–380.

Oliver, G., Mailhos, A., Wehr, R., Copeland, N. G., Jenkins, N. A., and Gruss, P. (1995). *Six3,* a murine homologue of the *sine oculis* gene, demarcates the most anterior border of the developing neural plate and is expressed during eye development. *Development (Cambridge, UK)* **121**, 4045–4055.

Oliver, G., Loosli, F., Koster, R., Wittbrodt, J., and Gruss, P. (1996). Ectopic lens induction in fish in response to the murine homeobox gene *Six3. Mech. Dev.* **60**, 233–239

Pannese, M., Polo, C., Andreazzoli, M., Vignali, R., Kablar, B., Barsacchi, G., and Boncinelli, E. (1995). The *Xenopus* homologue of *Otx2* is a maternal homeobox gene that demarcates and specifies anterior body regions. *Development (Cambridge, UK)* **121**, 707–720.

Parks, A. L., Turner, F. R., and Muskavitch, M. A. (1995). Relationships between complex *Delta* expression and the specification of retinal cell fates during *Drosophila* eye development. *Mech. Dev.* **50**, 201–216.

Parody, T. R., and Muskavitch, M. A. (1993). The pleiotropic function of Delta during postembryonic development of *Drosophila melanogaster. Genetics* **135**, 527–539.

Perron, M., Kanekar, S., Vetter, M. L., and Harris, W. A. (1998). The genetic sequence of retinal development in the ciliary margin of the *Xenopus* eye. *Dev. Biol.* In press.

Reh, T. A. (1987). Cell-specific regulation of neuronal production in the larval frog retina. *J. Neurosci.* **7**, 3317–3324.

Reh, T. A. (1989). The regulation of neuronal production during retinal neurogenesis. *In* "Development of the Vertebrate Retina". (Finlay, B. L. and D. R. Sengelaub, eds.), pp. 43–67. Plenum, New York.

Reh. T. A., and Nagy, T. (1987). A possible role for the vascular membrane in retinal regeneration in *Rana catesbienna* tadpoles. *Dev. Biol.* **122**, 471–482.

Reh, T. A., and Tully, T. (1986). Regulation of tyrosine hydroxylase-containing amacrine cell number in larval frog retina. *Dev. Biol.* **114**, 463–469.

Reh, T. A., Nagy, T., and Gretton, H. (1987). Retinal pigmented epithelial cells induced to transdifferentiate to neurons by laminin. *Nature* (*London*) **330**, 68–71.

Reyer, R.W. (1977). The amphibian eye: Development and regeneration. *In* "The Visual System in Vertebrate" (E. F. Crescitelli, ed.), Vol. 8, pp. 309–390. Springer-Verlag, Berlin.

Saha, M. S., and Grainger, R. M. (1992). A labile period in the determination of the anterior-posterior axis during early neural development in *Xenopus. Neuron* **8**, 1003–1014.

Sakaguchi, D. S., Janick, L. M., and Reh, T. A. (1997). Basic fibroblast growth factor (FGF-2) induced transdifferentiation of retinal pigment epithelium: Generation of retinal neurons and glia. *Dev. Dyn.* **209**, 387–398.

Serikaku, M. A., and O'Tousa, J. E. (1994). *sine oculis* is a homeobox gene required for *Drosophila* visual system development. *Genetics* **138**, 1137–1150.

Sidman, R. L. (1961). Histogenesis of mouse retina studied with [^{3}H] thymidine. *In* "The Structure of the Eye" (G. Smeller, ed.), pp. 487–506. Academic Press, New York.

Sologub, A. A. (1968). On the capacity of eye pigmented epithelium for transformation into retina in anuran amphibian tadpoles. *Tsitologiya* **10**, 1526–1532.

Sologub, A. A. (1977). Mechanisms of replacement derepression of artificial transformation of pigmented epithelium into retina in *Xenopus laevis. Wilhelm Roux's Arch. Dev. Biol.* **182**, 277–292.

Spemann, H. (1938). "Embryonic Development and Induction." Yale Univ. Press, New Haven, Connecticut.

Stiemke, M. M., and Hollyfield, J. G. (1995). Cell birthdays in *Xenopus laevis* retina. *Differentiation,.* **58**, 189–193.

Stone, L. S. (1950). The role of retina pigment cells in regenerating neural retinae of adult salamander eyes. *J. Exp. Zool.* **113**, 9–31.

Straznicky, K., and Gaze, R. M. (1971). The growth of the retina in *Xenopus laevis,* an autoradiographic study. *J. Embryol. Exp. Morphol.* **26**, 67–79.

Takebayashi, K., Takahashi, S., Yokota, C., Tsuda, H., Nakanishi, S., Asashima, M., and Kageyama, R. (1997). Conversion of ectoderm into a neural fate by ATH-3, a vertebrate basic helix-loop-helix gene homologous to *Drosophila* proneural gene atonal. *EMBO J.* **16**, 384–395.

Ton, C. C., Hirvonen, H., Miwa, H., Weil, M. M., Monaghan, P., Jordan, T., van Heyningen, V., Hastie, N. D., Meijers-Heijboer, H., Drechsler, M., Royer-Pokora, B., Collins, F., Swaroop, A., Strong, L. C., and Saunders, G. F. (1991). Positional cloning and characterization of a paired box-and homeobox-containing gene from the aniridia region. *Cell* (*Cambridge, Mass.*) **67**, 1059–1074.

Turner, D. L., and Weintraub, H. (1994). Expression of *achaete-scute* homolog 3 in *Xenopus* embryos converts ectodermal cells to a neural fate. *Genes Dev.* **8**, 1434–1447.

Waid, D. K., and McLoon, S. C. (1995). Factors produced by ganglion cells influence cell commitment in the developing retina. *Neurosc. Abstr.* **21**, 529.

Wassle, H., Boycott, B. B., and Illing, R. B. (1981). Morphology and mosaic of on- and off-beta cells in the cat retina and some functional considerations. *Proc. R. Soc. London Ser. B Biol. Sci.* **212**, 177–195.

Wetts, R., and Fraser, S. E. (1988). Multipotent precursors can give rise to all major cell types of the frog retina. *Science* **239**, 1142–1145.

Wetts, R., Serbedzija, G. N., and Fraser, S. E. (1989). Cell lineage analysis reveals multipotent precursors in the ciliary margin of the frog retina. *Dev. Biol.* **136**, 254–263.

Wettstein, D. A., Turner, D. L., and Kintner, C. (1997). The *Xenopus* homolog of *Drosophila Suppressor of Hairless* mediates Notch signaling during primary neurogenesis. *Development* (*Cambridge, UK*) **124**, 693–702.

Wikler, K. C., and Rakic, P. (1991). Relation of an array of early-differentiating cones to the photoreceptor mosaic in the primate retina. *Nature* (*London*) **351**, 397–400.

Zimmerman, K., Shih, J., Bars, J., Collazo, A., and Anderson, D. J. (1993). XASH-3, a novel *Xenopus* achaete-scute homolog, provides an early marker of planar neural induction and position along the mediolateral axis of the neural plate. *Development* (*Cambridge, UK*) **119**, 221–232.

Zuber, M. E., Perron, M., Bang, A., Holt, C. E., and Harris, W. A. (1997). Molecular cloning and expression analysis of the homeobox gene *Six3* in *Xenopus laevis. Dev. Biol.* **186**, 314.

VI

Zebrafish

24

Introduction to the Zebrafish

WOLFGANG DRIEVER
Department of Developmental Biology
University of Freiburg
D-79104 Freiburg, Germany

I. Zebrafish—A Recent Addition to the Palette of Model Systems for Studying Vertebrate Development

Hardly any developmental biology textbook written before the early 1990s even mentions zebrafish, and few of us have learned about them during biology classes at universities. Why is a whole section of this book dedicated to this creature?

Two independent types of experimental approaches have introduced zebrafish into the world of science. First, in the 1930s, Jane Oppenheimer (1936) and others realized that zebrafish are a good system for experimental embryology, since eggs can be readily obtained all year long and embryos are transparent—albeit, at about 700 μm egg diameter, they are a bit small for experimental manipulations and require skillful hands and/or good micromanipulators as elegantly demonstrated in the following chapters. Then, in the 1970s, George Streisinger started to exploit the short generation time and high fecundity and developed zebrafish into a genetic model to study vertebrate development and behavior (Streisinger *et al.*, 1981). It is exactly the combination of two features, the accessibility of the transparent embryo at any stage and the potential to perform genetic experiments similar to those that led to the breakthroughs in our understanding of *Drosophila* and *Caenorhabditis* development, that over the last decade made zebrafish the animal system of choice for hundreds of scientists worldwide, and one of the best understood vertebrates in terms of its development.

Zebrafish are small tropical freshwater teleosts that are at home in the shallow rivers of India and southeast Asia (Hamilton and Buchanan, 1822). Mature fish are about 4–5 cm long but weigh not much more than 2–3 g. Generation time is about 3 months, and fish live 2 to 3 years. Females may shed 50 to 100 eggs about every other week, but in a laboratory setting, 1000 and more embryos have been obtained in one clutch from a single female. Eggs are usually shed following short but intense courtship within 1 hour after the dawn of light, such that all eggs are fertilized within 10 to 15 minutes and develop pretty much synchronously. Eggs are fertilized externally while the male chases the female, and in natural settings fall between gravel, where they are protected, not the least by their transparency, from predators. In a laboratory setting, one has to have fertilized eggs fall through a "false bottom" type mesh, or between marbles, inside the breeding aquarium, otherwise parents immediately will eat their fry (Westerfield, 1994).

II. Early Development of Zebrafish

Zebrafish develop in a manner typical of teleosts. Kimmel *et al.* (1995) have established a normal table of developmental stages and describe morphological aspects of embryonic and early larval development in detail (Table I and Figs. 1 and 2). In freshly laid eggs, yolk and cytoplasm are intermixed, and the egg is surrounded by a transparent chorion, which swells and lifts away from

TABLE I
Zebrafish Developmental Stages[a]

Period	Stages	Duration (hours)	Development
Zygote	1-cell	0–0:45	Eggs are fertilized within a minute after the female sheds them. Segregation of polar body/completion of the second meiotic division within 5 minutes. Pronuclear fusion. Cytoplasm streams toward animal pole to establish animal (blastodisc) and vegetal (yolk mass) asymmetry.
Cleavage	2-cell through 64-cell	0:45–2:15	Rapid progression through cell cycles 2–7 with cleavage occurring every 15 minutes in a synchronous fashion. Meroblastic, partial early cleavages leave connection to yolk cell. At 64-cell stage, 3 regular tiers of blastomers located on top of yolk cell.
Blastula	128-cell through 1000-cell	2:15–3:20	Rapid (15 minutes) metasynchronous cell cycles 8 and 9; cycle 10 slightly lengthened; midblastula transition; cell cycles 11 and following more and more asynchronous.
	High	Till 3:40	Blastoderm sits high on yolk cell, constriction where marginal cells meet YSL; cell cycle 12.
	Oblong	Till 4	Animal–vegetal axis of blastoderm flattens, marginal constriction disappears, and blastula has smoothly outlined ellipsoidal shape.
	Sphere	Till 4:20	Spherical shape of blastula, caused by continued animal–vegetal flattening of blastoderm. Interface between blastodisc and yolk cell is flat.
	Dome	Till 4:40	In the center of the embryo, the yolk syncytial cell bulges (domes) up toward animal pole. Onset of epiboly.
	30% epiboly	Till 5	Margin of blastoderm at about 30% of distance between animal and vegetal poles.
Gastrula	50% epiboly	5:15–5:40	Margin of blastoderm at about 50% of distance between animal and vegetal poles.
	Germ ring	5:40–6	Onset of hypoblast formation; germ ring visible all around margin.
	Shield	6–8	Convergence movements form a dorsal thickening the embryonic shield; still at 50% epiboly.
	60 to 100% epiboly	8–10	Epiblast and hypoblast can be clearly distinguished; embryonic axis extends on the dorsal side; at the anterior end, prechordal plate extends just past animal pole.
	Tail bud	10–10:20	The tail bud is prominent; notochord primordium distinct from neural keel.
Segmentation	1-somite through 26-somite	10:20–22	The first somitic furrow will be the boundary between somite one and two. Somites develop at a regular pace during the second day of development, about one every 30 minutes (with the first six forming a bit faster, within 2 hours.), and their count is conveniently used for staging. During this period, neurulation is completed, organogenesis is initiated, and the embryonic axis extends and straightens off the yolk. Development of the remaining somites (between 30 and 34 pairs of somites form) is slower and less regular.
Pharyngula (second day of development)	Prim-5 through Prim-25	24–42	During the second day of development, the most precise staging is achieved by determining the position, expressed as somite number, of the posterior (leading) tip of the migrating lateral line organ primordium (moving at about 100 μm/hour). This requires a compound microscope with DIC optics. Stages: prim-5, early pigmentation, heartbeat; prim-15, early touch/escape reflex, first aortic arch; prim-25, strong circulation through caudal artery and vein extends into tail.
	High-pec	42–48	Pectoral fin bud's height is about equal to the width at its base. Xanthophores and iridophores appear; mandibular and hyoid arch visible.
Hatching (third day of development)	—	48–72	Hatching is somewhat independent of progress in development and happens during the third and into the fourth day of development. Pectoral fins develop. Second aortic arch and branchial arches 1–4 form. Completion of morphogenesis of primary organ systems. Cartilage development.
Early larva (fourth and fifth day of development)	—	72–120	Embryogenesis is usually considered to end with hatching. On day 4, the first bone (cleithrum) forms, the swim bladder starts to inflate and the digestive tract fully differentiates. The embryo begins to swim. Toward the end of day 4 and from day 5 on, the larvae is fully functional with food seeking and active avoidance behavior.

[a] Progress of development at 28.5°C; time indicated in hours post fertilization; modified from Kimmel *et al.* (1995).

the egg on contact with water. The animal–vegetal axis is already preset during oogenesis, and sperm can enter the egg only at the future animal pole through the micropyle, a specialization in the otherwise sperm-impermeable chorion (Wolenski and Hart, 1987). After fertilization, cytoplasm streams to the animal pole as it segregates from the yolk. About 30 minutes after fertil-

ization, the cytoplasm forms the blastodisc at the animal pole, and surrounds the vegetal yolk mass as a thin yolk cytoplasmic layer (YCL) (Fig. 1A; Fig. 2).

At 40 minutes postfertilization, the first meroblastic cleavage occurs (Fig. 1B). Four more cleavages occur in stereotypic orientation at 15-minute intervals, followed by five synchronous but not oriented cleavages

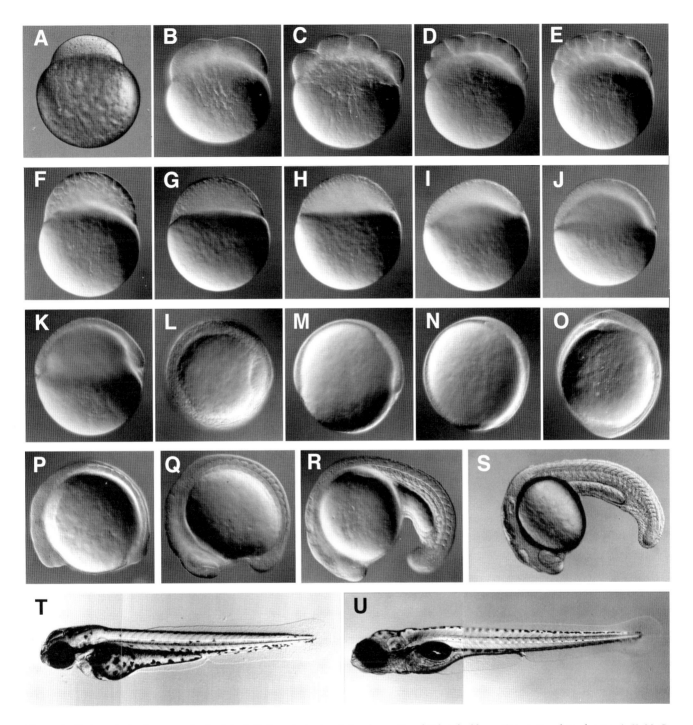

Figure 1 Embryonic development of zebrafish. A–S show dechorionated embryos, T and U hatched larva. Orientation: lateral views: A–K, M–O, animal to the top and dorsal to the right; P–U anterior to the left, dorsal top; L is a view of the animal pole, dorsal at right. Diameter of embryos in A–Q is about 700 µm; length of embryo in T is about 3 mm, and length of larvae in U is about 3.5 mm. The age of the embryos is indicated below in hours after fertilization at 28.5°C. Stages are as follows: A, 1-cell (0.2 hours); B, 2-cell (0.75 hours); C, 8-cell (1.25 hours); D, 32-cell (1.75 hours); E, 64-cell (2 hours); F, 256-cell (2.5 hours); G, 1000-cell (3 hours); H, sphere (4.0 hours.); I, dome (4.3 hours); J, 40% epiboly (4.75 hours); K and L, shield (6 hours.); M, 75% epiboly (8 hours); N, yolk sac closure (9.6 hours.); O, 1-somite (10.3 hours); P, 5 somites (11.7 hours); Q, 14 somites (16 hours.); R, 20 somites (19 hours); S, 26 somites (22 hours.); T, end of pharyngula stage (48 hours); U, feeding larvae (5 days old).

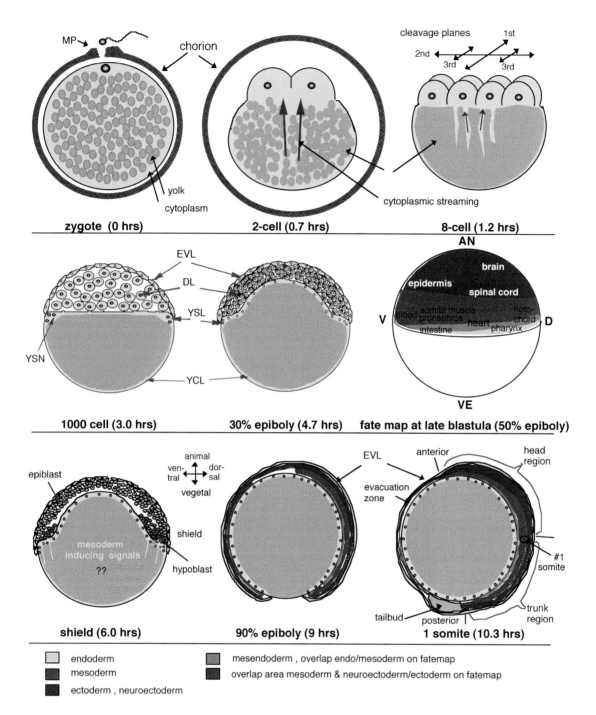

Figure 2 Events during early development of zebrafish. The development of zebrafish embryos from fertilization to the end of gastrulation is represented by schematic drawings of mid-sagittal sections of the embryos. Developmental times are in hours post-fertilization at 28°C . The chorion is shown only for the zygote and the 2-cell embryo. Fertilization occurs through the micropyle (MP). In the zygote, yolk (gray) and cytoplasm (light blue) are mixed, but separate during the first 2 hours of development by cytoplasmic streaming to the animal pole (blue arrows). The stereotypical cleavage planes are indicated above the 8-cell embryo. The 1000-cell embryo represents the mid-blastula. The different embryonic and extraembryonic lineages can be clearly distinguished (DL= deep layer, embryo proper; EVL = enveloping layer; yolk cell with the YSL = yolk syncytial layer with the YSN = yolk syncytial nuclei, the yolk depicted gray, and the YCL = yolk cytoplasmic layer in light blue). Fate map at 50% epiboly, just before the onset of gastrulation V = ventral; D = dorsal; AN = animal pole; VE = vegetal pole). During fate mapping experiments, significant regions of overlap have been found. Orange represents areas with cells that give rise to endo- or mesodermal fates. Purple represents areas where both cells with mesodermal as well as ectodermal or neuroectodermal fate were found. At the shield stage, gastrulation movements start to create the hypoblast. The yellow arrows represent hypothetical mesoderm inducing signals, the green arrow a dorsal inducing signal. At 90% epiboly during late gastrula, the hypoblast has almost reached the animal pole. Convergence of cells to the dorsal side results in a thickening of the embryonic axis. On the ventral side of the yolk, only very thin layers remain, and the hypoblast disappears from the evacuation zone. At the one somite stage, gastrulation is complete, and the major regions along the anterior–posterior axis (head, trunk, and tail) can be distinguished. The organization of germ layers in the tail bud (green) is not known for zebrafish (but see Chapter 26 by Ho *et al.*) (Modified from Driever, 1995, with permission.)

(Fig. 1C–G). During these cleavages, the marginal, vegetal blastomeres maintain large cytoplasmic bridges with the yolk cell. These ten cleavages generate a mound of blastomeres on top of the vegetal yolk cell. Subsequently, during mid-blastula transition, activation of zygotic transcription coincides with the generation of the first three separate lineages of the embryo (Kane *et al.*, 1992, Kane and Kimmel, 1993) (Figs. 2 and 3). Two of these lineages constitute extraembryonic lineages: the yolk syncytial layer (YSL) derived from the collapse of vegetal marginal blastomeres into the yolk cell, and the enveloping layer (EVL) forming the outer surface of the blastoderm. The third lineage, termed the deep cell layer (DL), will form the embryo proper. At the end of the mid-blastula transition, the first coordinated cell movements occur in the embryo. The cells of the blastoderm spread vegetalward over the yolk cell during epiboly (Strähle and Jesuthasan, 1993; Solnica-Krezel and Driever, 1994; Solnica-Krezel *et al.*, 1995; Jesuthasan and Strähle, 1996) (Fig. 1I–J). No morphological dorsal–ventral asymmetry has been detected during these stages. However, the first molecular asymmetries include dorsal nuclear localization of β-catenin protein (Kelly *et al.*, 1995) and the onset of expression of dorsal specific genes like *goosecoid* (Schulte-Merker *et al.*, 1994; Thisse *et al.*, 1994).

Gastrulation movements result in the formation of the germ layers. The formation of the hypoblast (the mesendodermal germ layer) is initiated on the dorsal side of the embryo, and soon continues all around the margin. Formation of the hypoblast leads to the appearance of the germ ring, a thickened marginal region, all around the blastoderm rim. On the dorsal side, a pronounced thickening, the embryonic shield, forms (Warga and Kimmel, 1990) (Fig. 1K, 1L, 2). Gastrulation in zebrafish is characterized by movement of single cells rather than coherent cell layers. In clear distinction to the involution of layers of cells in amphibia, ingression might be a more precise term to describe the movements of cells during hypoblast formation (Ballard, 1973; Collazo *et al.*, 1994; Shih and Fraser, 1995; Trinkaus, 1996; J. Shih, personal communications). These movements are detailed in Chapter 25 by Fraser. As gastrulation continues, the hypoblast extends toward the animal pole, while epiboly expands both layers toward the vegetal pole (Fig. 1M,N). At the same time, both hypoblast and epiblast cells converge from ventral and lateral positions toward the dorsal side. Mediolateral intercalation of cells converging to the dorsal side leads to elongation of the embryonic shield along the anteroposterior axis of the embryo (Warga and Kimmel, 1990; Kane and Warga, 1994). Convergence results in a ventro-animal region free of hypoblastic cells, the so-called evacuation zone (Fig. 2). At the end of epiboly, the embryo extends along the dorsal side of the yolk sphere, with the head positioned at the former animal pole and

the tail bud developing at the former vegetal pole of the egg (Fig. 1O). This also marks completion of gastrulation after about 10 hours of development.

The segmentation period extends through the second half of the first day of development, and is characterized by formation of somites at regular time intervals as well as the completion of neurulation (Fig. 1O–S) and the formation of the organ rudiments. Neurulation, in contrast to higher vertebrates, proceeds by a "sinking in" of the neural keel, or better, epithelial infolding at the midline, rather than closure of the neural fold (Papan and Campos-Ortega, 1994). The cavitation of the neural rod proceeds only secondarily to form the neurocoel in zebrafish. Since segmentation and neurulation coincide in time, zebrafish do not have a distinct neurula period, such as amphibians do. The roles of *Hox* genes in the events of this period are detailed in Chapter 26 by Ho *et al.* The second day of development is termed the pharyngula period, since embryos develop the characteristic phylotypic features, with arch primordia forming and organogenesis proceeding (Fig. 1T shows larvae at end of pharyngula period; see also Table I). Some time during the third day, zebrafish hatch (hatching period), but are still very "incomplete"—head and pharyngeal development, organ differentiation, and formation of a functional nervous system proceed during this period, including the establishment of neural crest lineages, as detailed in Chapter 27 by Raible and Eisen. Only on the fourth to fifth day of development is the alimentary canal functional, and larvae swim in a coordinated fashion, start to chase food, and avoid attacks.

III. Some Special Features of Teleost Early Development: Control of Epiboly

Slightly oversimplified, one could transform the mode of early amphibian development into teleostean development by taking the yolk out of the vegetal blastomeres and placing it into a single large cell at the vegetal pole. This would make epiboly, which is an inherent feature of amphibian gastrulation too, appear more prominent, as it does in teleosts. During epiboly, the deep cells, the EVL, and the YSL cell lineages expand toward the vegetal pole, eventually to cover the yolk sphere completely. However, different mechanisms seem to be involved in epibolic movements of the three cell types (Fig. 3). For the deep cells, it appears that epiboly is largely dependent on epiboly of the YSL—these findings are mainly based on work with another teleost, *Fundulus heteroclitus*. Epiboly of the blastoderm will not take place without concurrent epiboly of the underlying YSL. In contrast, epiboly of the syncytial layer can be completed even after the entire blastoderm has been removed (Trinkaus, 1951). This suggests that deep cells might

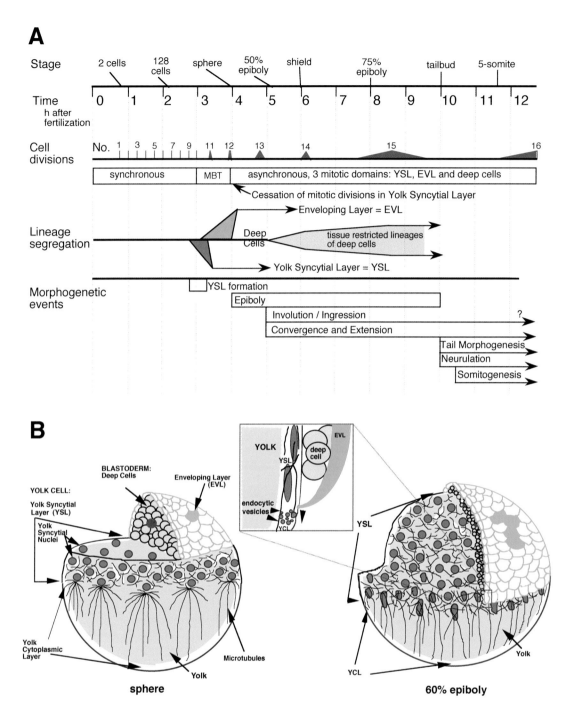

Figure 3 Segregation of the embryonic lineages and epiboly in zebrafish. (A): Temporal correlation between cell cycles, formation of the early embryonic lineages and morphogenesis. Developmental times are given in hours after fertilization at 28.5°C. The timing of the first 12 cell divisions is indicated as an average for all cells. For divisions 13 through 16, the timing including variation are for deep cells exclusively (Kimmel *et al.*, 1994). See text for details. (B.) The yolk syncytial cell during epiboly: sphere stage (left) and 60% epiboly (right). At sphere stage, the blastoderm is composed of the internal mass of deep cells and the superficial enveloping layer (EVL), and is positioned on top of the syncytial yolk cell. Two compartments can be distinguished in the cortical layer of the yolk cell. A cortical yolk syncytial layer (YSL) is populated by the syncytial nuclei and covers the animal portion of the yolk sphere, underlying the blastoderm. A thinner anuclear yolk cytoplasmic layer surrounds the bulk of the yolk mass with the vegetal pole. The yolk cell is furnished with two types of microtubular arrays. A network of microtubules is present in the YSL whereas animal–vegetal-oriented microtubules that originate in the YSL span the yolk cytoplasmic layer. Epiboly starts with the yolk cell bulging toward the animal pole. This so-called doming is correlated with and might drive radial intercalation of deep cells that scatter clonally related cells (illustrated by scattering of a red-marked clone of deep cells). Simultaneously the deep cell layer becomes thinner and spreads toward the vegetal pole. The YSL with the nuclei and intercrossed microtubule network spread vegetally. The surface of the yolk cytoplasmic layer (purple line) decreases due to membrane internalization (see inset). This is correlated with expansion of the YSL surface (black line). Epiboly of the EVL involves only limited rearrangements of clonally related cells (marked as solid green) that occurs exclusively in the plane of the monolayer. The EVL margin is tightly linked to the surface of the YSL (see inset). [Parts A and B modified from Solnica-Krezel, L, Stemple, D. L., and Driever, W. (1995). *BioEssays* **17,** 931–939. Copyright © 1995. Reprinted by permission of Wiley-Liss, Inc., a subsidiary of John Wiley & Sons, Inc.]

follow passively into the space generated by epiboly of YSL and EVL, which is not true, since genetic studies have made it clear that specific genetic functions, such as the *half baked* (*hab*) locus (Kane *et al.*, 1996), are required. *hab* mutant embryos complete epiboly of the YSL and EVL, but the deep cells arrest at 70–80% epiboly.

At the beginning of epiboly, the embryo is spherical with a mound of blastoderm cells atop a flattened yolk cell (Fig. 3B). The cortex of the yolk cell comprises two distinct regions. The YSL, a thickened cytoplasmic layer populated by the yolk nuclei, is located in the animal part of the yolk cell and is partially covered by the blastoderm. The remainder of the yolk cortex is a thin anuclear layer of cytoplasm called the yolk cytoplasmic layer. The beginning of epiboly is marked by a major morphological change in the yolk cell, which bulges toward the animal pole in a process called doming (Warga and Kimmel, 1990). Subsequently, the YSL expands toward the vegetal pole. In contrast, the thin anuclear cytoplasmic layer of the yolk cell progressively disappears (Betchaku and Trinkaus, 1978; Solnica-Krezel and Driever, 1994). The surface of the yolk cytoplasmic layer decreases most likely due to a process of endocytosis occurring just vegetal to the YSL (Fig. 3B, insert) (Betchaku and Trinkaus, 1986; Solnica-Krezel and Driever, 1994). Concomitant with the decrease in the surface of the yolk cytoplasmic layer is an expansion of the surface of the YSL, which is most likely mediated by the release of the excess of membrane stored in a form of membranous microvilli (Erickson and Trinkhaus, 1976; Betchaku and Trinkaus, 1978).

Epiboly of the YSL is at least partially driven by microtubule-dependent forces (Strähle and Jesuthasan, 1993; Solnica-Krezel and Driever, 1994; Jesuthasan and Strähle, 1996). In zebrafish, the yolk cell is equipped with two distinct microtubule arrays that change their organization during epiboly (Fig. 3B). One array, an extensive network of intercrossing microtubules, is part of the YSL and expands as the epiboly of the YSL proceeds. By contrast, another array of microtubules oriented along the animal–vegetal axis within the yolk cytoplasmic layer becomes shorter as this layer diminishes. If microtubules are completely depolymerized, by treatment with high doses of nocodazole, the movements of the syncytial nuclei are blocked. However, endocytosis of the yolk cell and epiboly of the blastoderm are only partially inhibited. Thus, the yolk cell microtubules may be involved in the vegetal movements of the YSN and may also contribute to epibolic expansion of the EVL and deep cells.

As the yolk cell begins to dome, the mound of deep cells becomes thinner while simultaneously expanding vegetalward. This is accomplished by the radial intercalation of cells from deeper to more superficial positions (Fig. 3) (Warga and Kimmel, 1990). It has been proposed that such radial intercalation movements are

driven by the doming of the yolk cell (Wilson *et al.*, 1995). These movements thoroughly scatter deep blastomeres, dispersing clonally related cells (Kimmel and Law, 1985; Warga and Kimmel, 1990). There are regional differences in the degree of cell mixing. Cells in the center of the blastula undergo more extensive scattering and mixing than cells located near the blastoderm margin (Wilson *et al.*, 1993; Helde *et al.*, 1994). It has been suggested that these reproducible patterns of cell mixing during epiboly result from the passive response of deep cells to the doming of the underlying yolk cell.

In contrast to deep layer cells, when the EVL undergoes epiboly, cell rearrangements are observed to occur only within the plane of the monolayer (Fig. 3) (Keller, 1987, Warga and Kimmel, 1990). The expansion of the EVL in *Fundulus* appears to be driven by the epiboly of the surface of the YSL to which the monolayer is tightly linked by its margin (Fig. 3). Epiboly of the EVL is accompanied by an extension of the intercellular tight junctions and the increase in the tension within the monolayer (Lentz and Trinkaus, 1987).

IV. Fate Maps and Cell Determination

As in amphibians (Chapter 19 by Dawid), zebrafish embryonic cells are readily marked by cell lineage tracers. At the onset of gastrulation, analysis of cellular fates using this technique reveals that the endoderm will derive from the vegetalmost, marginal blastomeres (Fig. 2). Mesoderm forms from the vegetal third of the blastoderm, whereas ectoderm originates from the animal half of the blastoderm. Neuroectoderm in particular derives from the dorsal section of the animal half (Kimmel *et al.*, 1990). Notochord derives from the dorsal side, where the shield forms, whereas somitic mesoderm, heart and blood develop from lateral and ventral positions, respectively (see also Lee *et al.*, 1994). The organization of the zebrafish fate map, therefore, is similar to the one of *Xenopus*.

Attempts have been made to establish a fate map based on the first cleavages, which are of stereotype orientation and, at the 8-cell stage, produce an asymmetric array of two by four cells on top of the yolk cell, but, a correlation could not be obtained (Abdelilah *et al.*, 1994; Helde *et al.*, 1994). A mechanism by which localized (dorsal?) determinants are distributed in a stereotype fashion among the early blastomeres in correlation to the cleavage planes can be excluded to be involved in axis formation in zebrafish.

When do cells become irreversibly committed to a specific fate? Heterotopic and heterochronic transplantation experiments (Ho and Kimmel, 1993) reveal that cells are not committed to a certain fate at late blastula stages (5 hours postfertilization), when tissue restricted lineages arise. However, when hypoblast cells are transplanted heterotopically to the epiblast at mid-gastrula (8 hours postfertilization), they will predominantly give rise to hypoblast fates. Thus cells become committed to a specific germ layer at mid-gastrula stages. Earlier tissue specific lineage restrictions do not necessarily reflect the state of commitment of the cell. These issues are addressed in detail in the following chapters with regard to neural fate (Chapter 25 by Fraser), neural crest commitment (Chapter 27 by Raible and Eisen), and tailbud lineages (Chapter 26 by Ho *et al.*)

V. Experimental Analysis of Axis Formation

Establishment of the dorsoventral axis is best understood in *Xenopus* (reviewed in Gerhart *et al.*, 1989; Kessler and Melton, 1994; Chapter 19 by Dawid; Chapter 20 by Sullivan *et al.*). Cortical rotation during the first cell cycle is microtubule dependent, UV sensitive, and results in the formation of the Nieuwkoop center in dorsal vegetal blastomeres (reviewed in Chapter 20 by Sullivan *et al.*). The Nieuwkoop center appears to induce the Spemann organizer in the dorsal blastopore lip (see also Chapter 21 by Kessler), which again is the source of signals patterning the dorsoventral and anteroposterior axis. Are there similarities to zebrafish development?

Ultraviolet irradiation of zebrafish eggs between 10 and 25 minutes after fertilization depletes dorsal structures, and results in radially symmetric embryos (Strähle and Jesuthasan, 1993). Within the zebrafish YCL at that stage, there is a prominent parallel array of microtubules, oriented in the animal–vegetal direction. However, the target of UV action is unknown—so far there is no evidence that cortical rotation takes place in zebrafish, and the target of UV action could be maternal RNAs as well as the microtubules. Indeed, strong evidence points at an involvement of microtubules in the transport of substances required for organizer formation from the vegetal hemisphere to the blastomeres (Jesuthasan and Strähle, 1996). Nocodazole treatment before the 32-cell stage disrupts microtubules and prevents axis formation, as evident from the absence of goosecoid expression in treated embryos. These cortical microtubule arrays have been demonstrated to be able to account for the transport of particles from vegetal cortex to the blastomeres, as judged from transport of vegetally injected fluorescent beads to marginal blastomeres.

Similarities to amphibia are evident in the sensitivity to LiCl exposure during cleavage stages, which results in dorsalized zebrafish embryos (Stachel *et al.*, 1993). On lithium treatment, dorsal genes like *goosecoid* are expressed all around the margin of the gastrula, whereas the expression of ventral and posterior genes like *eve1* are suppressed (Joly *et al.*, 1993). In lithium treated, radialized embryos, which appear to have no

dorsoventral axis, the anterior–posterior order of gene expression is maintained between the animal pole and the vegetal margin at late gastrulation stages (Xu *et al.*, 1994). The lithium sensitive period is between the 16- and 1024-cell stages (Stachel *et al.*, 1993), before midblastula transition. Therefore, the activities of maternal factors appear to be responsible for the initiation of dorsoventral axis formation.

Experimental evidence points at an involvement of the yolk syncytial cell in mesoderm induction and in the specification of dorsal fates (Mizuno *et al.*, 1996). Transplantation of a "bald" yolk cell, from which all blastomeres have been removed, onto the animal pole of a recipient embryo blastula induces a second ring of *ntl/Brachyury* expression at the margin between recipient animal cells and the transplanted yolk cell (see yellow arrows in Fig. 2, shield stage). Further, it appears as if the yolk cell also induced, at a position random with respect to the recipient organizer, a second organizer, as judged from the appearance of a second, ectopic domain of *goosecoid* expression in experimental sandwiched blastoderms (see green arrows in Fig. 2, shield stage). What kind of maternal gene products are localized in the zebrafish yolk syncytial cell, and what contribution to patterning is provided by transcription from the yolk syncytial nuclei, needs to be determined. One gene product which is likely involved in axis formation in zebrafish is β-catenin, providing further evidence for similarities to amphibian development. In zebrafish, the nuclear translocation of β-catenin protein occurs first in a restricted number of nuclei in the yolk syncytial layer at the late blastula stage (Schneider *et al.*, 1996). Slightly later, β-catenin also is translocated into a small number of deep cells close to the dorsal margin. Overexpression of β-catenin mRNA in zebrafish results in the induction of a second axis (Kelly *et al.*, 1995).

In summary, it appears that the yolk cell has similar functions and hosts similar activities when compared to vegetal blastomeres in *Xenopus*, including Nieuwkoop center-like activities.

VI. The Fish Organizer and Dorsoventral Patterning

Classic experiments demonstrate that the zebrafish shield has organizer activities similar to the dorsal blastopore lip of amphibia. Shield mesoderm, when ectopically transplanted to the ventral side of a host embryo, induces secondary axis formation (Ho, 1992; Shih and Fraser, 1996). The molecular nature of the signals from the organizer appears to be conserved, since zebrafish shield can induce axis duplications in amphibia (Oppenheimer, 1936). The high degree of conservation of dorsoventral patterning mechanisms among vertebrates has been demonstrated in zebrafish both by analysis of protein activities as well as through the analysis of mutations affecting dorsoventral patterning. Systematic screens (see below) have revealed a number of genetic loci involved in specification of dorsal or ventral fates (Hammerschmidt *et al.*, 1996a; Mullins *et al.*, 1996). These mutations reveal that the basic molecular mechanism of dorsoventral patterning, the antagonism between ventral bone morphogenetic protines (BMPs) and organizer derived inhibitors of BMP activity, also applies to zebrafish. The locus *swirl* encodes BMP2, and mutations severely affect development of ventral structures (Kishimoto *et al.*, 1997; Nguyen *et al.*, 1998). The *dino* locus codes for the zebrafish chordin homolog, and mutations affect specification of dorsal fate (Hammerschmidt *et al.*, 1996b; Miller-Bertoglio *et al.*, 1997; Schulte-Merker *et al.*, 1997). Also, modulation of these signals is highly conserved as revealed from the study of the zebrafish *tolloid* homolog (Blader *et al.*, 1997). Zebrafish have opened an exciting opportunity for studying dorsoventral patterning using a combination of genetic and experimental tools to investigate the role of antagonizing morphogenetically active signaling gradients in early vertebrate development.

VII. Genetic Analysis of Zebrafish Development

The previous paragraph highlighted one of the major advantages of working with zebrafish. Zebrafish are highly amenable to genetic analysis. Genetic screens have already generated more than 2000 published mutations, probably representing more than a thousand genetic loci important for embryogenesis, organogenesis, and behavior. Further mutagenesis screens are underway in many laboratories, such that it might be possible within a few years to understand zebrafish development at mechanistic detail that was previously achieved for invertebrates only.

Predominantly two aspects of zebrafish genetics have contributed to its progress so far, and have been reviewed in detail (Kimmel, 1989; Driever *et al.*, 1994). First, zebrafish are easy to breed, and the short generation time helps to drive a newly induced mutation to homozygosity through two generations of breeding within about half a year. The large number of progeny in an individual egg clutch has been instrumental in actually identifying new mutations: given that one obtains on average 100 or so eggs at a time from a female, about 25 are expected to be mutant if the phenotype segregates as a Mendelian trait. This sounds trivial, but is actually very helpful during a genetic screen to distinguish nongenetic abnormalities from true mutant alleles.

Second, a number of genetic "tricks" are available to investigate phenotypes caused by a recessive mutation without having to drive the mutation to homozygosity in a time-consuming two-generation breeding scheme. From individual founder females, which are heterozygous for a new mutation, embryos expressing the mutant phenotype can be obtained directly in one of the following ways: (i) Eggs are fertilized *in vitro* with UV-inactivated sperm. There is no contribution of male genetic material to the development of the zygote; the resulting embryos are haploid, and 50% of an egg clutch express the mutant phenotype of a given mutant allele. Haploid zebrafish embryos per se develop slightly abnormally, and die after 3 to 4 days. However, most of the organs form, and the presence or absence of a structure can be efficiently screened for. (ii) Eggs are fertilized with UV-inactivated sperm *in vitro* and the second meiotic division is suppressed by a pulse of high pressure at a defined time point after fertilization ("early pressure" technique). Embryos develop as diploid, representing meiotic half tetrads. Thus, due to nonsegregation of recombination products, depending on gene–centromere distance, between 50% (centromere close loci) and only a few percent (loci close to telomeres) are homozygous and express the mutant phenotype. (iii) Eggs are fertilized with UV-inactivated sperm *in vitro* and the first mitotic cleavage is inhibited by a heat shock. Such embryos develop as homozygous diploid, and 50% express the mutant phenotype if the mother was heterozygous. The disadvantage of the heat shock technique is that a significant portion of the embryos can be damaged physically by the heat shock, resulting in substantially abnormal development. Using these techniques, a vast array of mutations has been characterized after gamma ray or chemical mutagenesis (Ho and Kane, 1990; Hatta *et al.*, 1991; Driever *et al.*, 1996; Mullins *et al.*, 1996).

Alternative techniques for mutagenesis have been developed that give more rapid access to cloning of the mutated genes. The simplest of such techniques has been insertional mutagenesis by injection of plasmid DNA into early zygotes (Bayer and Campos-Ortega, 1992), but it has not been very efficient in generating mutants. The use of pantropic retroviruses, generated by VSV-G pseudotyping of MoMLV-based viral backbones (Burns *et al.*, 1993), has made generation of many hundred retroviral insertions possible (Lin *et al.*, 1994; Gaiano *et al.*, 1996), and the first molecular characterizations of mutated genes have been published (Allende *et al.*, 1996). However, these retroviral vectors do not even come close to the efficiency of P-element mediated transpose mutagenesis in *Drosophila* because expression of genes from the viral construct, for example, as gene trap reporters, is hampered by inactivation after germ line transmission of the proviral DNA. Recently, a breakthrough with respect to future insertional mutagenesis screens has been the demonstration of mobilization of heterologous transposable elements in zebrafish (Ivics *et al.*, 1997; Raz *et al.*, 1998). A mariner family transposable element from *Caenorhabditis elegans*, Tc3, can integrate into and excise from the zebrafish genome in a transposase dependent fashion. More importantly, the transposon elements can be modified to express green fluorescent protein (GFP) in zebrafish, and expression is stable even after several generations. Thus, gene trap and enhancer trap mutagenesis screens can now be performed, similar to what is possible in *Drosophila* (Cooley *et al.*, 1988). It is still a daunting task to saturate a vertebrate genome for such insertional mutants, since several hundred thousands of insertion events will have to be investigated. However, since such screens can be performed as F₁ screens, they are not impossible. For comparison, the previous large chemical mutagenesis screens have each analyzed between 500,000 (Driever *et al.*, 1996) and 1 million embryos (Haffter *et al.*, 1996).

While such insertional mutagenesis screens are currently in preparation in several laboratories, one is aware that most transposable elements have some site specificity, and therefore current efforts in many laboratories are focusing on positional cloning of chemically induced mutations, rather than waiting a long time for new insertion alleles to be recovered at a specific locus. Molecular characterization of chemically induced point mutations has been made possible by advances in the characterization of the zebrafish genome. High resolution genetic maps are available, with markers spaced at near centimorgan distance (1 cM is equivalent to 500–800 kb in zebrafish), and many genes already are placed on the map (Postlethwait *et al.*, 1994, 1998, Knapik *et al.*, 1996, 1998). These genomic resources have enabled researchers to identify candidate genes for zebrafish mutations based on their map position and segregation pattern (Talbot *et al.*, 1995) as well as by chromosomal walks starting from genetic markers closely linked to a mutation (Zhang *et al.*, 1997). The effort of creating a genetic map and placing onto it many genes with homologs in other species, which also have been mapped onto the respective genomes, revealed exciting insights into chromosome evolution (Postlethwait *et al.*, 1998). An example is the discovery of extra families of *Hox* genes (Chapter 27 by Ho *et al.*) The combination of such genomic analysis with molecular characterization of genes, phenotypic analysis of mutations and genetic pathways, overexpression studies, and other experimental embryological approaches make zebrafish a unique system to study cell fate during early development. This multidisciplinary approach is elegantly demonstrated in the following three chapters. With fish being several hundred million years apart from mammals in evolution, and at least some morphological aspects of early development being quite divergent, it is amazing to see how well conserved the fundamental molecular control and signaling mechanisms are.

References

Abdelilah, S., Solnica-Krezel L., Stainier, D. Y. and Driever, W. (1994). Implications for dorsoventral axis determination from the zebrafish mutation janus. *Nature (London)* **370**, 468–471.

Allende, M. L., Amsterdam, A., Becker, T., Kawkami, K., Gaiano, N., and Hopkins, N. (1996). Insertional mutagenesis in zebrafish identifies two novel genes, pescadillo and dead eye, essential for embryonic development. *Genes Dev.* 10, 3141–3155.

Ballard, W. W. (1973). A new fate map for *Salmo gairdneri*. *J. Exp. Zool.* **184**, 49–74.

Bayer, T. A., and Campos-Ortega, J. A. (1992). A transgene containing lacZ is expressed in primary sensory neurons in zebrafish. *Development (Cambridge, UK)* **115**, 421–426.

Betchaku, T., and Trinkaus, J. P. (1978). Contact relations, surface activity, and cortical microfilaments of marginal cells of the enveloping layer and of the yolk syncytial and yolk cytoplasmic layers of fundulus before and during epiboly. *Exp. Zool.* **206**, 381–426.

Betchaku, T., and Trinkaus, J. P. (1986). Programmed endocytosis during epiboly of *Fundulus heteroclitus*. *Am. Zool.* **26**, 193–199.

Blader, P., Rastegar, S., Fischer, N., and Strähle, U. (1997). Cleavage of the BMP-4 antagonist chordin by zebrafish tolloid. *Science* **278**, 1937–1940.

Burns, J. C., Friedmann, T., Driever, W., Burrascano, M., and Yee, J. K. (1993). Vesicular stomatitis virus G glycoprotein pseudotyped retroviral vectors: Concentration to very high titer and efficient gene transfer into mammalian and nonmammalian cells. *Proc. Natl. Acad. Sci. U.S.A.* **90**, 8033–8037.

Collazo, A., Bolker, J. A., and Keller, R. (1994). A phylogenetic perspective on teleost gastrulation. *Am. Nat.* **144**, 133–152.

Cooley, L., Kelley, R., and Spradling, A. (1988). Insertional mutagenesis of the *Drosophila* genome with single P elements. *Science* **239**, 1121–1128.

Driever, W. (1995). Axis formation in zebrafish. *Curr. Opin. in Genet. Dev.* **5**, 610–618.

Driever, W., Stemple, D., Schier, A., and Solnica-Krezel, L. (1994). Zebrafish: Genetic tools for studying vertebrate development. *Trends in Genetics* **10**, 152–159.

Driever, W., Solnica-Krezel. L., Schier, A. F., Neuhauss, S. C. F., Malicki, J., Stemple, D. L., Stainier, D. Y. R., Zwartkruis, F., Abdelilah, S., Rangini, Z., Belak, J., and Boggs, C. (1996). A genetic screen for mutations affecting embryogenesis in zebrafish. *Development (Cambridge, UK)* **123**, 37–46.

Erickson, C. A., and Trinkhaus, J. P. (1976). Microvilli and blebs as source of reserve surface membrane during cell spreading. *Exp. Cell Res.* **99**, 375–384.

Gaiano, N., Allende, M., Amsterdam, A., Kawakami, K., and Hopkins, N. (1996). Highly efficient germ-line transmission of proviral insertions in zebrafish. *Proc. Natl. Acad. Sci. U.S.A.* **93**, 7777–7782.

Gerhart, J., Danilchik, M., Doniach, T., Roberts, S., Rowning, B., and Stewart, R. (1989). Cortical rotation of the *Xenopus* egg: Consequences for the anteroposterior pattern of embryonic dorsal development. *Development (Cambridge, UK)* (Suppl.), 37–51.

Haffter, P., Granato, M., Brand, M., Mullins, M.C., Hammerschmidt, M., Kane, D. A., Odenthal, J., van-Eeden, F. J. M., Jiang, Y.-J., Heisenberg, C.-P., Kelsh, R. N., Furutani-Seiki, M., Warga, R. M., Vogelsang, E., Beuchle, D., Schach, U., Fabian, C., and Nüsslein-Volhard, C. (1996). The identification of genes with unique and essential functions in the development of the zebrafish, *Danio rerio*. *Development (Cambridge, UK)* **123**, 1–36.

Hamilton, F., and Buchanan, F. (1822). "An Account of the Fishes Found in the River Ganges and its Branches." Archibald Constable and Company. Edinburgh and London.

Hammerschmidt, M., Pelegri, F., Kane. D. A., Mullins, M. C., Serbedzija, G., van-Eeden, F. J. M., Granato, M., Brand, M., Furutani-Seiki, M., Haffter, P., Heisenberg, C. P., Jiang, Y.-J., Kelsh, R. N., Odenthal, J., Warga, R. M., and Nüsslein-Volhard, C. (1996a). *dino and mercedes*, two genes regulating dorsal development in the zebrafish embryo. *Development (Cambridge, UK)* **123**, 95–102.

Hammerschmidt, M., Serbedzija, G. N., and McMahon, A. P.. (1996b). Genetic analysis of dorsoventral pattern formation in the zebrafish: Requirement of a BMP-like ventralizing activity and its dorsal repressor. *Genes Dev.* **10**, 2452–2461.

Hatta, K., Kimmel, C. B., Ho, R. K., and Walker, C. (1991). The cyclops mutation blocks specification of the floor plate of the zebrafish central nervous system. *Nature (London)* **350**, 339–341.

Helde, K. A., Wilson, E. T., Cretekos, C. J., and Grunwald, D. J. (1994). Contribution of early cells to the fate map of the zebrafish gastrula. *Science* **265**, 517–520.

Ho, R. (1992). Axis formation in the embryo of the zebrafish *Brachydanio rerio. Semin. Dev. Biol.* **3**, 53–64.

Ho, R. K., and Kane, D. A. (1990). Cell-autonomous action of zebrafish spt-1 mutation in specific mesodermal precursors. *Nature (London)* **348**, 728–30.

Ho, R. K., and Kimmel, C. B. (1993). Commitment of cell fate in the early zebrafish embryo. *Science* **261**, 109–111.

Ivics, Z., Hackett, P. B., Plasterk, R. H., and Izsvák, Z.. (1997). Molecular reconstruction of sleeping beauty, a Tc1-like transposon from fish, and its transposition in human cells. *Cell (Cambridge, Mass.)* **81**, 501–510.

Jesuthasan, S., and Strähle, U. (1996). Dynamic microtubules and specification of the zebrafish embryonic axis. *Curr. Biol.* **7**, 31–42.

Joly, J. S., Joly, C., Schulte-Merker, S., Boulekbache, H., and Condamine, H. (1993). The ventral and posterior expression of the zebrafish homeobox gene eve1 is perturbed in dorsalized and mutant embryos. *Development (Cambridge, UK)* **119**, 1261–1275.

Kane, D. A., Hammerschmidt, M., Mullins, M. C., Maischein, H.-M., Brand, M., van Eeden, F. J. M., Furutani-Seiki, M., Granato, M., Haffter, P., Heisenberg, C.-P., Jiang, Y.-J., Kelsh, R.N., Odenthal, J., Warga, R. M., and Nüsslein-Volhard, C. (1996). The zebrafish epiboly mutants. *Development (Cambridge, UK)* **123**, 37–55.

Kane, D. A., and Kimmel, C. B. (1993). The zebrafish midblastula transition. *Development (Cambridge, UK)* **119**, 447–456.

Kane, D. A., and Warga, R. M. (1994). Domains of movement in the zebrafish gastrula. *Semin. Dev. Biol.* **5**, 101–109.

Kane, D. A., Warga, R. M., and Kimmel, C. B. (1992). Mitotic domains in the early embryo of the zebrafish. *Nature (London)* **360**, 735–737.

Keller, R. E. (1987). Rearrangement of enveloping layer cells without disruption of the epithelial permeability barrier as a factor in *Fundulus* epiboly. *Dev. Biol.* **120**, 12–24.

Kelly, G. M., Erezyilmaz, D. F., and Moon, R. T. (1995). Induction of secondary embryonic axis in zebrafish occurs following the overexpression of β-catenin. *Mech. Dev.* **53**, 261–273.

Kessler, D. S., and Melton, D.A. (1994). Vertebrate embryonic induction: Mesodermal and neural patterning. *Science* **266**, 596–604.

Kimmel, C. B. (1989). Genetics and early development of zebrafish. [Review]. *Trends Genet.* **5**, 283–288.

Kimmel, C. B., Warga, R. M., and Schilling, T. F. (1990). Origin and organization of the zebrafish fate map. *Development (Cambridge, UK)* **108**, 581–594.

Kimmel, C. B., and Law, R. D. (1985). Cell lineage of zebrafish blastomeres. III. Clonal analyses of the blastula and gastrula stages. *Dev. Biol.* **108**, 94–101.

Kimmel, C. B., Warga, R. M., and Kane, D. A. (1994). Cell cycles and clonal strings during formation of the zebrafish central nervous system. *Development (Cambridge, UK)* **120**, 265–76.

Kimmel, C. B., Ballard, W. W., Kimmel, S. R., Ullmann, B., and Schilling, T. F. (1995). Stages of embryonic development of the zebrafish. *Dev. Dyn.* **203**, 253–310.

Kishimoto, Y., Lee, K., Zon, L., Hammerschmidt, M., and Schulte-Merker, S. (1997). The molecular nature of zebrafish *swirl*: BMP2 function is essential during early dorsoventral patterning. *Development (Cambridge, UK)* **124**, 4457–4466.

Knapik, E. W., Goodman, A., Atkinson, S., Roberts, C. T., Shiozawa, N., Sim, C. U., Weksler-Zangen, S., Trolliet, M., Futrell, C., Innes, B. A., Koike, G., McLaughlin, M. G., Pierre, L., Simson, J. S., Vilallonga, E., Roy, M., Chiang, P., Fishman, M. C., Driever, W., and Jacob, H. J. (1996). A reference cross for zebrafish (*Danio rerio*). *Development (Cambridge, UK)* **123**, 451–460.

Knapik, E., Goodman, A., Ekker, M., Chevrette, M., Delgado, J., Neuhauss, S., Shimoda, N., Driever, W., Fishman, M. C., and Jacob, H. J. (1998). A microsatellite genetic linkage map for zebrafish (*Danio rerio*). *Nat. genet.* **18**, 338–343.

Lee, R. K., Stainier, D. Y., Weinstein, B. M., and Fishman, M. C. (1994). Cardiovascular development in the zebrafish. II. Endocardial progenitors are sequestered within the heart field. *Development (Cambridge, UK)* **120**, 3361–3366.

Lentz, T., and Trinkaus, J. P. (1987). Differentiation and junctional complex of surface cells in the developing *Funulus* blastoderm. *J. Cell Biol.* **48**, 455–472.

Lin, S., Gaiano, N., Culp, P., Burns, J. C., Friedmann, T., Yee, J. K., and Hopkins, N. (1994). Integration and germline transmission of a pseudotyped retroviral vector in zebrafish. *Science* **265**, 666–669.

Miller-Bertoglio, V. E., Fisher, S., Sánchez, A., Mullins, M. C., and Halpern, M. E. (1997). Differential regulation of *chordin* expression domains in mutant zebrafish. *Dev. Biol.* **192**, 537–550.

Mizuno, T., Yamaha, E., Wakahara, M., Kuroiwa, A., and Takeda, H. (1996). Mesoderm induction in zebrafish. *Nature (London)* **383**, 131–132.

Mullins, M. C., Hammerschmidt, M., Kane, D. A., Odenthal, J.,

Brand, M., van Eeden, F. J. M., Furutani-Seiki, M., Granato, M., Haffter, P., Heisenberg, C.-P., Jiang, Y.-J., Kelsh, R. N., Warga, R. M., and Nüsslein-Volhard, C. (1996). Genes establishing dorsal–ventral pattern formation in the zebrafish embryo: The ventral specifying genes. *Development (Cambridge, UK)* **123,** 81–93.

Nguyen, V. H., Schmid, B., Trout, J., Connors, S. A., Ekker, M., and Mullins, M. C. (1998). Ventral and lateral regions of the zebrafish gastrula, including the neural crest progenitors, are established by a bmp2b/swirl pathway of genes. *Dev. Biol.* **199,** 93–110.

Oppenheimer, J. M. (1936). Structures developed in amphibians by implantation of living fish organizers. *Proc. Soc. Exp. Biol. Med.* **34,** 461–463.

Papan, C., and Campos-Ortega, J. A. (1994). On the formation of the neural keel and neural tube in the zebrafish *Danio (Brachydanio) rerio. Roux Arch Dev. Biol.* **203,** 178–186.

Postlethwait, J. H., Johnson, S. L., Midson, C. N., Talbot, W. S., Gates, M., Ballinger, E. W., Africa, D., Andrews, R., Carl, T., Eisen, J. S., Horne, S., Kimmel, C. B., Hutchinson, M., Johnson, M., and Rodriguez, A. (1994). A genetic linkage map for the zebrafish. *Science* **264,** 699–703.

Postlethwait, J. H., Yan Y.-L., Gates, M. A., Horne, S., Amores, A., Brownlie, A., Donovan, A., Egan, E.S., Force, A., Gong, Z., Goutel, C., Fritz, A., Kelsh, R., Knapik, E., Liao, E., Paw, B., Ransom, D., Singer, A., Thomson, M., Abduljabbar, T. S., Yelick, P., Beier, D., Joly, J.-S., Larhammar, D., Rosa, F., Westerfield, M., Zon, L. I., Johnson, S. L., and Talbot, W. S. (1998). Vertebrate genome evolution and the zebrafish gene map. *Nat. Genet.* **18,** 345–349.

Raz, E., van Luenen, H. G., Schaerringer, B., Plasterk, R. H. A., and Driever, W. (1998). Transposition of the nematode *Caenorhabditis elegans* Tc3 element in the zebrafish *Danio rerio. Curr. Biol.* **8,** 82–88.

Schneider, S., Steinbeisser, H., Warga, R. M., and Hausen, P. (1996). beta-Catenin translocation into nuclei demarcates the dorsalizing centers in frog and fish embryos. *Mech. Deve.* **57,** 191–198.

Schulte-Merker, S., Hammerschmidt, M., Beuchle, D., Cho K. W., Derobertis, E. M., and Nusslein-Volhard, C. (1994). Expression of zebrafish goosecoid and no tail gene products in wild-type and mutant no tail embryos. *Development (Cambridge, UK)* **120,** 843–852.

Schulte-Merker, S., Lee, K. J., McMahon, A. P., and Hammerschmidt, M. (1997). The zebrafish organizer requires *chordino. Nature (London)* **387,** 862–863.

Shih, J., and Fraser, S. E. (1995). Distribution of tissue progenitors within the shield region of the zebrafish gastrula. *Development (Cambridge, UK)* **121,** 2755–2765.

Shih, J., and Fraser, S. E. (1996). Characterizing the zebrafish organizer: Microsurgical analysis at the early-shield stage. *Development (Cambridge, UK)* **122,** 1313–1322.

Solnica-Krezel, L., and Driever, W. (1994). Microtubule arrays of the zebrafish yolk cell: organization and function dur-

ing epiboly. *Development (Cambridge, UK)* **120,** 2443–2455.

Solnica-Krezel, L., Stemple, D. L., and Driever, W. (1995). Transparent things—cell fates and cell movements during early embryogenesis of zebrafish. *BioEssays* **17,** 931–939.

Stachel, S. E., Grunwald, D. J., and Myers, P. Z. (1993). Lithium perturbation and goosecoid expression identify a dorsal specification pathway in the pregastrula zebrafish. *Development (Cambridge, UK)* **117,** 1261–1274.

Strähle, U., and Jesuthasan, S. (1993). Ultraviolet irradiation impairs epiboly in zebrafish embryos: Evidence for a microtubule-dependent mechanism of epiboly. *Development (Cambridge, UK)* **119,** 909–919.

Streisinger, G., Walker, C., Dower, N., Knauber, D., and Singer, F. (1981). Production of clones of homozygous diploid zebra fish (*Brachydanio rerio*). *Nature (London)* **291,** 293–296.

Talbot, W. S., Trevarrow, B., Halpern, M. E., Melby, A. E., Farr, G., Postlethwait, J. H., Jowett, T., Kimmel, C. B., and Kimelman, D. (1995). A homeobox gene essential for zebrafish notochord development. *Nature (London)* **378,** 150–157.

Thisse, C., Thisse, B., Halpern, M. E., and Postlethwait, J. H. (1994). Goosecoid expression in neurectoderm and mesendoderm is disrupted in zebrafish cyclops gastrulas. *Dev. Biol.* **164,** 420–429.

Trinkaus, J. P. (1951). A study of the mechanism of epiboly in the egg of *Fundulus heteroclitus. J. Exp. Zool.* **118,** 269–320.

Trinkaus, J. P. (1996). Ingression during early gastrulation of *Fundulus. Dev. Biol.* **177,** 356–370.

Warga, R. M., and Kimmel, C. B. (1990). Cell movements during epiboly and gastrulation in zebrafish. *Development* **108,** 569–580.

Westerfield, M. (1994). "The Zebrafish Book." Univ. of Oregon Press. Eugene.

Wilson, E. T., Helde, K. A., and Grunwald, D. J. (1993). Something's fishy here—rethinking cell movements and cell fate in the zebrafish embryo. [Review]. *Trends Genet.* **9,** 348–352.

Wilson, E. T., Cretekos, C. J., and Helde, K. A. (1995). Cell mixing during early epiboly in the zebrafish embryo. *Dev. Genet.* **17,** 6–15.

Wolenski, J. S., and Hart, N. H. (1987). Scanning electron microscope studies of sperm incorporation into the zebrafish (*Brachydanio*) egg. *J. Exp. Zool.* **243,** 259–273.

Xu, Q., Holder, N., Patient, R., and Wilson, S. W. (1994). Spatially regulated expression of three receptor tyrosine kinase genes during gastrulation in the zebrafish. *Development (Cambridge, UK)* **120,** 287–299.

Zhang, J., Talbot, W. S., and Schier, A. F. (1997). Positional cloning identifies zebrafish *one-eyed pinhead* as a permissive EGF-related ligand required during gastrulation. *Cell (Cambridge, Mass.)* **92,** 241–251.

25

Cell Interactions and Morphogenetic Motions Pattern the Zebrafish Nervous System

Scott E. Fraser

Division of Biology
Biological Imaging Center
Beckman Institute
California Institute of Technology
Pasadena, California 91125

I. Introduction

The combined attributes of the zebrafish (*Danio rerio*; formerly known as *Brachydanio rerio*) has made it one of the most rapidly growing experimental systems for studies of embryonic patterning. In addition to its advantages for genetic studies, its size, clarity, and rapid development makes it an ideal system for the analysis of cell lineages and morphogenetic movements. By combining cell labeling techniques and intravital microscopy, cell movement and differentiation can be assessed directly, allowing the construction of fate maps of the early embryo. Fate maps are depictions of what cells in various regions of an embryo will become during normal development. Although they do not by themselves tell us whether cells are committed in either the dynamic or material sense (Slack, 1991), they provide critical insights into the mechanisms of cell fate determination, embryonic induction, and tissue morphogenesis. Fate maps of different vertebrates have been invaluable in aiding the design of grafting and explant experiments, as well as the interpretation of molecular markers. Furthermore, they have helped to highlight both similarities and differences between different vertebrate embryos

Some characteristic features appear to be conserved, such as a notochord anlage centered at the dorsal midline, somitic mesoderm on either side of the notochord primordium, and a neural primordium located closer to the presumptive animal pole (or anterior end) of the embryo than is the mesoderm.

Despite the clear similarities in the fate maps of different vertebrate embryos, care must be taken in extrapolating results from one species to another. There are clear differences in the early cellular dynamics between species. For example, there is limited intermixing of blastomeres during cleavage and blastula stages in *Xenopus* (Wetts and Fraser, 1989; see Chapter 20 by Sullivan *et al.*); in contrast, cell mixing begins earlier and is more dramatic in zebrafish (Kimmel and Law, 1985; Warga and Kimmel, 1990; Helde *et al.*, 1994). Such differences may permit and/or reflect different cellular interactions, involving distinct signaling mechanisms to pattern the embryo. Therefore, similarities in some features of embryonic fate maps should not be taken as proof of the presence of "universal" molecular mechanisms. Similarly, systematic changes in fate maps between systems cannot be taken as clear evidence of either the conservation or systematic changes in molecular mechanism. Instead, fate maps are best viewed as providing the means to construct hypotheses and to execute more definitive test experiments.

Construction of a fate map requires a means of labeling a cell (or distinct group of cells) in a defined region of the embryo, cataloging the labeled region, identifying the progeny of the labeled cell(s) over time, and scoring the final phenotypes and positions of the progeny. The zebrafish embryo offers advantages over most other vertebrate systems for each of these steps: It can be cultured in a simple medium of pond water at near room temperature, making it much easier to culture than avian or mouse embryos; its size and transparency is ideal for light microscopy, avoiding the limitations imposed by the highly scattering yolk inclusions in amphibian embryos. Because of these obvious advantages, my colleagues and I have been performing fate mapping and grafting experiments that build on the gastrula fate map generated by Kimmel and colleagues (1990). In the following sections, I will review some of the aspects of the various zebrafish fate maps, and touch on some of the test experiments they have motivated.

II. The Appearance of Embryonic Order: The Gastrula Fate Map

The gastrula fate map (Kimmel *et al.*, 1990) has motivated many other fate mapping efforts, provided a major resource for cell transplantation experiments, and served as an important tool in the experimental analysis of mutant phenotypes. By labeling a single cell during cleavage stages, the normal scattering of descendants by the late blastula stage resulted in isolated labeled cells that could be followed by video microscopy. The time-lapse recordings showed that cell fates could be predicted from their positions at 50% epiboly (Kimmel and Warga, 1986; Kimmel *et al.*, 1990). Furthermore, the appearance of the embryonic shield by 6 hours of development provided a reliable landmark for the dorsal midline, a key landmark for any fate map. The results suggest strongly that a fate map of clear predictive value could be made for the zebrafish at the onset of gastrulation.

Detailed topological fate maps during blastula stages are as yet unavailable, due mainly to the lack of reliable landmarks necessary for their construction. Clonal analyses have shown that three distinct mitotic domains can be identified during the zebrafish mid-blastula transition (cycle 10–12), corresponding to the YSL domain (yolk syncytial layer), EVL domain (enveloping layer), and deep cell domain, which give rise to all embryonic lineages (Kane *et al.*, 1992). There is a broad correspondence between fate (though not necessarily commitment) and mitotic domain for the YSL and EVL; however, within the deep cell domain, no correlations have been found between cell cycle characteristics and eventual cell fate. Schmitz and Campos-Ortega

(1994) have suggested the possibility of identifying "dorsal" in certain batches of embryos prior to shield formation, which may facilitate fate map construction at these stages; however, this asymmetry may prove too difficult for most workers to detect reliably. Furthermore, Cooper and D'Amico (1996) have shown a group of cells with endocytotic activity at the future dorsal margin of the blastoderm; this may provide a useful landmark for moving the analysis of fate maps back to earlier stages of development.

Claims of a restricted fate map established in early cleavage (Strehlow and Gilbert, 1993; Strehlow *et al.*, 1994) based on injections of very large fluoresceinated dextrans (2000 kDa) into single blastomeres at the 8- and 16-cell stages are generally regarded as incorrect. The authors were able to label single blastomeres because such large dextran dyes do not pass through the cytoplasmic bridges that are present until the fifth cleavage. Contrary to expectations based on the gastrula fate map, they reported that the blastomeres were not identical in their prospective fates. For example, blastomeres contributing to the notochord lineage almost never contributed to blood, ventral epidermis, and pronephros. When the somitic progeny derived from a labeled cell, they were almost always exclusively on the left- or right-hand side of the embryo; labeled progeny in the nervous system were typically either anterior to or posterior to the mesencephalon. The authors (Strehlow and Gilbert, 1993; Strehlow *et al.*, 1994) proposed that the first cleavage separated a dorsal blastomere from a ventral blastomere, the second division separated a left from a right blastomere, and the third separated more anterior blastomeres from more posterior blastomeres. These claims were immediately challenged by experiments designed to determine whether the second cleavage plane did in fact correspond to the plane of bilateral symmetry (Helde *et al.*, 1994). If the relationship between the plane of cleavage and the body axis were fixed, then injections of neighboring end blastomeres at the 8-cell stage with different color lineage tracers should result in different colored cells on the left and right sides of either the dorsal midline or the ventral midline. The results showed that the plane of bilateral symmetry did not correspond to the second cleavage plane and that the relative position of the two labeled groups were not strictly related to the axis. The conclusion of no strict fate map was independently arrived at by Abdelilah *et al.*, (1994) while exploring a patterning mutant, and is in agreement with classic studies in other teleosts.

Although it appears that the proposal of Strehlow and colleagues must be incorrect, the question of whether there is any dorsoventral pattern within the 8-cell stage zebrafish embryo remains unanswered. Since the dorsoventral axis in *Xenopus* is defined by a cortical rotation prior to the first cleavage (reviewed in Chapter 20 by Sullivan *et al.*), it seems reasonable to suspect the presence of some degree of patterning in the 8- to 16-

cell stage zebrafish embryo. For such patterns to persist to later stages, cell movement cannot be completely random. In agreement with this, cell mixing during blastula stages is not uniform; instead, the progeny of marginal blastomeres appear to disperse less than animal blastomeres (Helde *et al.*, 1994). Of course, if the patterning information were contained within the yolk cell, then there would be no need for the individual blastomeres to be restricted in their motions in the early embryo; instead, the yolk cell could convey the needed information to blastomeres that arrived in a given location at random. Interestingly, such a possibility is favored by more recent experiments on teleosts (Mizuno *et al.*, 1996; 1997) showing that important determinants within the yolk cell can be experimentally manipulated. These experiments and those that build on them promise exciting new insights into the nature and action of polarizing signals in the early fish embryo.

III. The Elaboration of Pattern: The Fate Map of the Embryonic Shield

The embryonic shield, which appears near the onset of germ ring formation as a slight thickening along the blastoderm margin, offers the first clear morphological criterion for distinguishing the dorsal from the ventral blastoderm. By analogy with the dorsal lip of the blastopore in amphibians, the embryonic shield has been proposed to function as a teleostean organizer. Grafting experiments, ranging from those in other teleosts (cf. Oppenheimer, 1936) to more recent approaches in zebrafish (cf. Ho, 1992; Shih and Fraser, 1996), show that the embryonic shield can induce ectopic axes. The most recent experiments employed lineage markers, and demonstrated that both the donor and the host tissues collaborate in the formation of an ectopic axis (Fig. 1).

When looked at in detail, there were some unexpected aspects. First, when the grafts were performed to the ventral germ ring, the induced axes were truncated at hindbrain levels. While at the time we took this to suggest the presence of distinct head and trunk organizing activities (Shih and Fraser, 1996), more recent experiments show that it likely results from signals emanating from the germ ring that caudalize nearby neuronal tissue (Woo and Fraser, 1997; see neural fate map section below). Such a transforming activity can signal even forebrain tissue to develop into hindbrain. Thus any axis near the germ ring should terminate at the hindbrain independent of the axial level originally induced. Second, the relative contribution of the host and donor tissues is somewhat different than might have been expected. The labeled donor piece gave rise to the notochord and occasional somites (as expected), as well as unexpected descendants such as the floor plate and neurons within the neural tube. This motivated a more careful examination of the fate of the early shield.

Homotopic grafting experiments of the early embryonic shield from labeled to unlabeled animals confirmed that the grafts contained mesodermal, endodermal, and neural tissues (Shih and Fraser, 1996). One easy way to explain this appearance of a neural contribution where none was expected is that the donor piece was too large, extending out of the organizer and into the region of the prospective neural plate. To address this, small grafts were performed, of as few as a half dozen cells. Even these grafts, which were well within the boundaries of the early embryonic shield, contributed to both mesodermal and neural derivatives (Fig. 2). This suggests that the unexpected contribution of the grafted tissues to the neuraxis is the product of the normal fate of the embryonic shield. Either the early embryonic shield contains cells of indeterminant fate or it consists of an intermixed set of determined precursor cells.

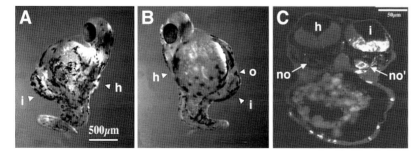

Figure 1 Secondary axis induced by grafting the embryonic shield to the ventral blastoderm margin. The graft was performed from a fluorescent dextran labeled donor to an unlabeled host at the early shield stage of development. As in previous experiments, in several species such grafts induce host tissues to participate in the formation of dorsal ectodermal and mesodermal tissues. Because the graft was placed very near the blastoderm margin, the secondary axis is truncated at the level of the hindbrain. (A, B) Dorsal and ventral views of the experimental animal showing the complete host axis (h) and incomplete induced axis (i) that terminates at the level of the otic placode (o). (C) Cross section through an experimental animal, viewed in combined bright field and fluorescence, in which the fluorescently labeled donor cells appear bright white. Descendants of the grafted embryonic shield are visible in the secondary notochord (no'), a small amount of underlying endoderm, and a significant amount of the induced neuraxis (i). Labeled cells are not found in the host axis (h) or notochord (no). The scattered cells among the ventral ectoderm are hatching gland cells.

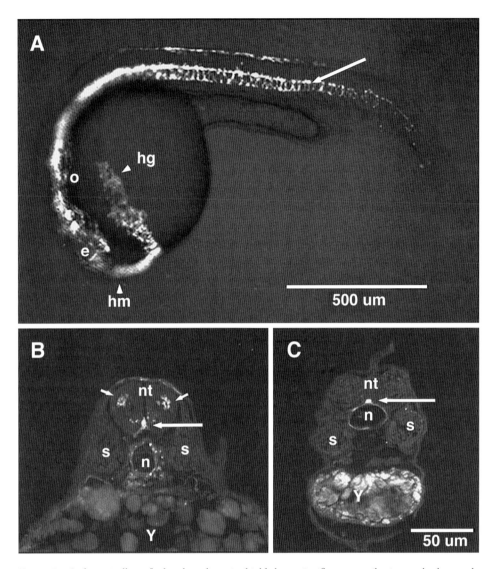

Figure 2 Orthotopically grafted early embryonic shield shows significant contributions to both mesodermal and ectodermal derivatives. In these combined fluorescence and bright field images, the descendants of the grafted cells appear white because the donor was labeled with fluorescent dextran. (A) Side view of the experimental animal, showing the expected contribution of grafted cells to the hatching gland (hg), head mesenchyme (hm), and notochord. In addition, a bright rank of cells can be seen just dorsal to the notochord within the floorplate of the neural tube (long arrow). (B, C) Cross sections show that descendants of the grafted embryonic shield contributed to the notochord (n), as well as to the floorplate (long arrow) and to cells (short arrows) within the more dorsal neural tube (nt). e, eye; o, otic placode; s, somite.

Because even limited intermixing between axial mesodermal and neural precursors would permit different classes of cell-to-cell interactions between neural progenitors and their inducers, this is an important issue to resolve. Any overlap should be revealed in the fate map of the early shield, but at the time we were first doing the grafts, there was little information at the relevant stage and position. The data in the gastrula fate map (Kimmel *et al.*, 1990) suggest that the distributions of axial mesodermal and neural progenitors might overlap in the region of the embryonic shield. Of course, this apparent overlap could be the product of slight misalignments of the data from different animals. In addition, it could be that the side view of the map gives a false impression of overlap, because in such a side view, the shield region is nearly orthogonal to the page, and foreshortening might make regions next to the shield appear to superimpose on the shield.

To resolve this issue, we assessed the fates of single cells within the embryonic shield region (the marginal region of the dorsal blastoderm at the onset of gastrulation; Shih and Fraser, 1995). Single cells were labeled by iontophoretic injection of fluorescent dextrans at the early-shield stage (just as the shield becomes visible as a slight thickening). Because no fate map can be better than the accuracy with which the position of the labeled cell is recorded, we carefully cataloged the position of the labeled cell with respect to both the dorsal midline and the blastoderm margin. Low light level video microscopy was used to document at least three views of the embryo (side view; view from the animal pole; view from the embryonic shield). This offered two independent measures of the position of the labeled cell in each of three dimensions (Fig. 3). The fates of the cells were assessed at multiple time points over the next 1.5 days, again using low light level video microscopy.

As expected, descendants with the characteristic morphology of notochord cells (disklike cells ventral to the neurectoderm) were present within the embryonic shield region. Somite and endodermal cells also descended from precursors within the embryonic shield region. The somite cells were easy to recognize by their elongated cell bodies that spanned the rostrocaudal extent of a somite. The endoderm progeny were typically thin, spread-out cells that, over time, take on a more-rounded appearance, positioned beneath the notochord. The fate mapping confirmed the surprising result of the tissue grafting experiment: neural progenitors were found within the embryonic shield region. The neural descendants were consistently found in the ventral portion of the neural tube, either as cells in the floor plate or more dorsally in the neural tube. These cells often had the large cell bodies and axons characteristic of neurons.

The early-shield fate map shows that there is a considerable intermixing of progenitors near the beginning of gastrulation, distinct from the results reported for amphibian fate maps (Wetts and Fraser, 1989; Bauer et al., 1994). The 143 injections used to construct the fate map show that this intermixing is not subtle: nearly half of the injections labeled neurectodermal progenitors, 10% labeled endodermal progenitors, and the remaining injections labeled notochord and/or somitic precursors (Fig. 3). There was no strict organization to the position of the progeny and the position of the progenitors within the early shield, but this is not surprising given the small size of the early shield. Radial and lateral convergence movements would be expected to intermix the cells dramatically. Although the injections show that there were multiple fates within the embryonic shield, they do not suggest that individual cells are multipotent; single cells labeled at the early-shield stage gave rise to descendants that were restricted to single tissues.

In contrast to these results, a more recent fate map of the embryonic shield performed at slightly later stages has been interpreted to show that there is little or no overlap between the progenitors (Melby et al., 1996). The difference in staging between the two studies may suggest that the fate map refines relatively quickly into unifated domains during the process of shield elaboration. In fact, the cell movements between layers needed to accomplish this have been observed (see below). It is also possible that differences in the guidelines for which injections to include in the fate map result in the major difference in conclusions. Given the dramatic intermixing we find at the beginning of embryonic shield formation, it seems impossible that our conclusions are the result of a few mistakenly identified progenitor positions.

To examine this potential in more detail, we studied cell injections made exactly at the blastoderm margin. These are excellent test cases, as there is no mistaking the animal–vegetal position of a cell at the margin. Of the 22 injections made into positions at the margin, 9 were neural progenitors (41%) and 8 were notochord progenitors (36%). As both notochordal and neural progenitor types were present at the blastoderm margin, the overlap we have observed between these two tissue progenitors could not be rooted solely in an error in our assessment of the position of the labeled cell.

There are implications to the intermixing of fates in the embryonic shield, given the common belief that the topology of vertebrate gastrulae are conserved (Kimmel et al., 1990; Lawson et al., 1991). We and others have found that most of the notochord progenitors in the zebrafish gastrula are located close to the dorsal midline and that somitic progenitors are located lateral to the notochord (Shih and Fraser, 1995; Kimmel et al., 1990; Melby et al., 1996) in agreement with this belief. However, the intermingled relationship between mesoderm and neurectoderm progenitors in the zebrafish gastrula argues against a strict conservation of topological fate relationships in vertebrate embryos. Except for limited statistical overlap, intermingling of neural and mesodermal progenitors has not been reported in amphibian fate maps (Pasteels, 1942; Keller, 1975, 1976). In contrast, a detailed fate map of Hensen's node in the chick suggests that neural and mesodermal progenitors may intermingle during the gastrulation of avian embryos as single labeled cells could give rise to both neural and mesodermal progeny (Selleck and Stern, 1991). Whether the zebrafish progenitor distribution map we have presented is representative of a second distinct class of vertebrate fate maps, of which chick may also be a member, remains to be determined.

The presence of neural progenitors in the shield region of the dorsal blastoderm poses three challenges. First, it may alter some of the current interpretations of *in situ* or antibody staining patterns in the zebrafish gastrula. For example, at 50% epiboly, whole-mount antibody staining of the Brachyury gene product, Zf-T, labels cells described as endodermal and mesodermal progenitors, as well as cells of the enveloping layer within the blastoderm margin (Schulte-Merker et al., 1992). Since all the nuclei within at least the first five cell diameters of the blastoderm margin stain for Zf-T, it follows then that some neural progenitors must also express Zf-T at one time. Could these Zf-T-expressing neural progenitors represent a special population of neural precursors? This is possible since a sizable portion (30–40%) of the neural progenitors we labeled in the embryonic shield region gave rise to floor plate. Whether Zf-T expression is involved in floor plate formation remains to be shown. However, it is clear that the zebrafish is not alone in having Brachyury expression in neural precursors, as the T gene is also briefly expressed in prospective neuro-

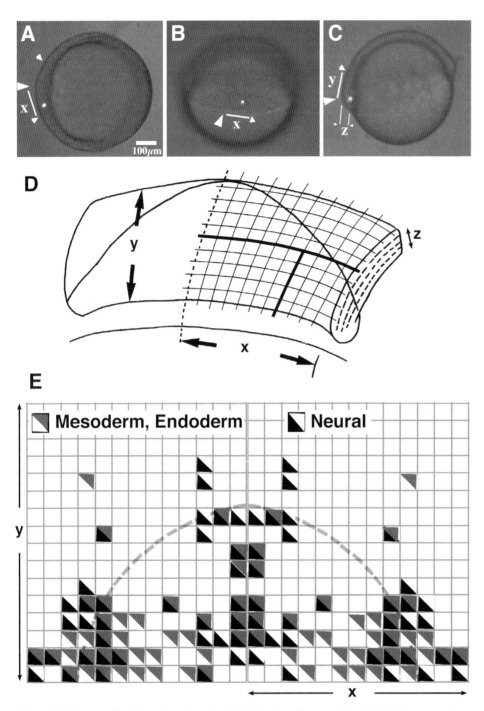

Figure 3 Fate-mapping the early embryonic shield. Single cells were injected with fluorescent dextran by iontophoresis, which appears bright white in these combined fluorescence and bright field images. The initial position of each labeled cell was assessed in three orientations: animal pole view (A); frontal view (B); profile view (C). In each view, the position of the cell was recorded with respect to the middle (large pointer) and margins (small pointer) of the forming embryonic shield as well as the depth below the blastoderm surface. (D) This allowed each cell to be assigned an unambiguous x, y, z address with respect to the embryonic shield. (E) The fates of the labeled cells, scored in living animals as well as in tissue sections, were recorded with respect to each cell's position. In this rendering, the z axis is collapsed so they are displayed in x and y with respect to the margin of the early embryonic shield (dashed line). Mesendodermal and neural descendants are recorded with the symbols shown. A given location has more than one symbol type in many cases indicating that some of the injections into that position gave one fate, and other injections gave a different fate.

ectoderm during mouse development (Kispert and Herrmann, 1994). Second, it raises questions about the state of commitment of the cells in the embryonic shield. Our lineage analysis describes only what the labeled cells differentiated into during development and not their states of commitment at the time of labeling; therefore, it is unclear whether our distribution map describes an intermingling of differently committed cells or one of wholly uncommitted cells. This is an important distinction, because, if a number of these cells are committed, then the timing and manner of cell–tissue interactions underlying neural induction and mesoderm dorsalization in the zebrafish may be more involved than is presently believed. Consistent with the notion of some degree of early committment, cells within the embryonic shield region display a considerable degree of molecular specialization (cf. Strahle *et al.*, 1993; Thisse *et al.*, 1994; Hammerschmidt and Nüsslein-Volhard, 1993; Xu *et al.*, 1994; Krauss *et al.*, 1993; Schulte-Merker *et al.*, 1992). Third, any intermixing requires some sorting mechanism or cell movement pattern other than that demonstrated during the gastrulation of the frog. It is this third issue that has motivated our studies of cell movements in fish gastrulation.

IV. Involution, Ingression and Egression in Fish Gastrulation

During vertebrate gastrulation, mesoderm and endoderm progenitors are internalized from the surface of the embryo. As the mechanics of gastrulation link together the primary axes of the embryo with later patterning events such as the induction of the nervous system, it has quite correctly been viewed as central to development. During gastrulation in most vertebrates, the blastoderm becomes divided into two layers, an outer and an inner lamina, separated by a cleft that is often visible in histological sections (Brachet, 1935). The outer lamina is the primary ectoderm (PE, see Dettlaff, 1993) and the inner lamina has been named the primitive hypoblast (PH, Wilson, 1889). Following gastrulation, the cells of the primitive hypoblast give rise to mesoderm and endoderm derivatives; primary ectoderm cells not internalized during gastrulation give rise to ectodermal derivatives. The division between the PE and PH in zebrafish is the cleft (sometimes referred to as Brachet's cleft; Brachet, 1935), visible as an optical discontinuity that runs parallel to the surface near the margin of the blastoderm.

In amphibians, the gastrulation begins with bottle cell formation at the blastopore producing an invagination; the primitive hypoblast then arises from the sheet-like motion of a group of cells (chordamesoderm and endoderm progenitors), flowing around the dorsal lip of the blastopore, leaving behind and eventually laying beneath cells of the primary ectoderm. The movement of the cells as a sheet around an edge defines this class of motion as involution (Trinkaus, 1984). In contrast, the primitive hypoblast could form by individual cells delaminating into the interior, losing their contacts with near neighbors many of which remain in the primary ectoderm, and then reassociating with cells to form the primitive hypoblast. This appears to occur during gastrulation in avian embryos (cf. Stern and Canning, 1990; see review by Schoenwolf, 1991) and perhaps in mouse embryos (Parameswaran and Tam, 1995; Hashimoto and Nakatsuji, 1989). The individual nature of the cell movements and the changes in neighbor relationships define this class of motion as ingression (see Trinkaus, 1984). Involution and ingression movements are not mutually exclusive; for example, during sea urchin gastrulation (cf. Fink and McClay, 1985), the archenteron forms by involution but the primary and the secondary mesoderm both ingress.

The fate mapping studies in the previous section raised questions concerning whether involution or ingression drive gastrulation in the zebrafish. The early embryonic shield region, at the dorsal margin of the blastoderm, contains progenitors for neurons as well as the expected progenitors of the mesoderm and endoderm (Shih and Fraser, 1995). The primary data of other fate maps are not inconsistent with this intermingling of neural and mesoderm progenitors in the forming embryonic shield (Kimmel *et al.*, 1990; Melby *et al.*, 1996), although the number of neural progenitors within the shield must decrease with development (Kimmel *et al.*, 1990; Shih and Fraser, 1995; Melby *et al.*, 1996). Any intermingling of progenitor populations at the dorsal midline requires a rearrangement of neighbor relationships (i.e., ingression) as the germ layers organize. A recent time-lapse study of gastrulation in a teleost fish, *Fundulus heteroclitus*, showed that many of the cells moved deepward at or near the blastoderm margin as individuals or in small groups, not as part of a sheet (Trinkaus, 1996). Because of these time-lapse results and our previous fate-mapping studies we have used cell labeling and confocal microscopy to determine whether chordamesoderm and endoderm progenitors in the zebrafish embryo move by involution, by ingression, or both.

Three different cell-labeling methods were used to assure that the results were not the product of the experimental technique itself. The first of the techniques, "scatter-labeling," involved injections of a single cell with a fluorescent lineage tracer at cleavage stages resulting in labeled progeny dispersed in the embryo that could be used for time-lapse recording (cf. Kimmel *et al.*, 1990). In the second approach, microsurgical transplantation of 5–10 cells was used to label small contiguous groups of cells within the shield region from a donor embryo labeled with a fluorescent lineage tracer to an unlabeled sibling recipient (cf. Shih and Fraser, 1996). In the third approach, BODIPY-ceramide was used to label the membranes of the cells so their outlines could be followed over time (Cooper and D'Amico, 1996).

All three labeling techniques showed the presence of two basic categories of cell trajectory. As shown in previous work from others (cf. Warga and Kimmel, 1990; Wood and Timmermans, 1988), a number of labeled cells traveled a "circum-marginal" trajectory: the cells first moved toward the blastoderm margin; near the margin, they moved away from the surface; after this deepward movement, the cells moved away from the margin. A second category of cell movement, involving zebrafish dorsal deep cells that changed laminar position before reaching the blastoderm margin, was observed in a number of cases with each of the three labeling techniques (Fig. 4). Ingression is somewhat surprising, as it re-

quires that cells traverse the cleft, often viewed as a major division or boundary between germ layers. Even more surprising, many cells egressed, moving from the primitive hypoblast to the primary ectoderm. Because the cells that egress must have earlier reached the primitive hypoblast either by ingression or by circummarginal movements, the multiple pathways may be best viewed as different facets of a common mechanism for gastrulation in embryos in which the cells are not constrained to sheetlike tissue movements.

The distinctions between involution and ingression are more than semantic, as they imply distinct mechanisms of morphogenesis and offer distinct opportunities for patterning interactions. For example, because the zebrafish deep cells are not constrained to move together,

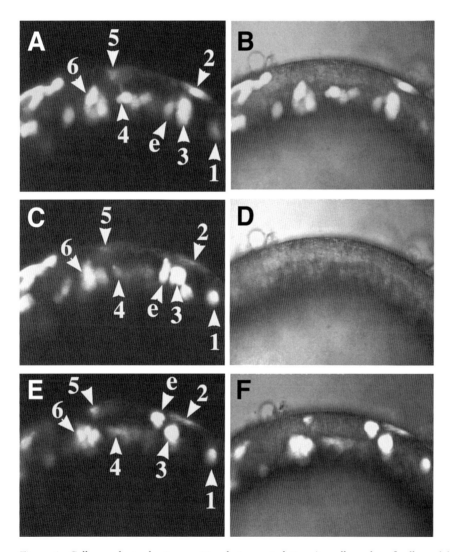

Figure 4 Cells can change laminar position during gastrulation. A small number of cells are labeled in this confocal time-lapse series by a previous iontophoretic injection of fluorescent dextran. The series covers a 30-minute period, beginning at 85–90% epiboly (A, B) and ending at ~90% epiboly (E, F). In this profile view of the dorsal axis of the embryo, rostral is to the right (cells are labeled 1–3, "e," and 4–6 in rostrocaudal order). (A, B) Fluorescence and combined bright field/fluorescence images showing that the cell labeled "e" is clearly beneath the cleft separating the primary ectoderm from the primitive hypoblast at the beginning of the series. (C, D) Fluorescence and bright field images, a few minutes later, showing that cell "e" is now spanning the cleft. (E, F) Fluorescence and combined bright field/fluorescence images showing that cell "e" has completed its egression from the primitive hypoblast into the primary ectoderm. The stable positioning of the other cells in the field shows the individual nature of the movements that cross the cleft.

some cells can reside near or within the blastoderm margin during gastrulation, such as the recently identified noninvoluting endocytic marginal cells (Cooper and D'Amico, 1996); such cell behaviors are inconsistent with the sheetlike motions of involution. Moreover, the prolonged and dispersed nature of ingression/egression shown here permits different progenitor populations to remain intermingled during zebrafish gastrulation, such as the intermixing of the putative neural ectoderm and the putative mesoderm within the shield region fate map (Shih and Fraser, 1995; see also data in Melby *et al.*, 1996 and Kimmel *et al.*, 1990). This may have implications for neural induction, as the separation of inducing and responding cells by the cleft into two distinct cell laminae would normally require the exchange of longer-range signals (cf. Jessell and Melton, 1992). Because the cells in zebrafish are not constrained by sheetlike cell movements, neural specification could result from direct neighbor interactions between interspersed cells within the same lamina; this represents an extreme version of the planar component of neural induction proposed in *Xenopus* (cf. Doniach, 1992a,b; 1993).

Given the traffic of cells between laminae, the sharp boundaries of gene expression observed in the embryo cannot be maintained solely by a restriction of cell movements across the cleft; instead, cells crossing the cleft must rapidly turn on and off the genes characteristic of the laminae they enter and leave, respectively. A number of genes thought to be important to vertebrate axis development display a restricted expression pattern within either the primary ectoderm or the primitive hypoblast during zebrafish gastrulation, including *goosecoid* (Thisse *et al.*, 1994; Schulte-Merker *et al.*, 1994a), *Brachyury* (*zf-T*; Schulte-Merker *et al.*, 1992, 1994b), *shh* (Krauss *et al.*, 1993), *not* (Talbot *et al.*, 1995), *otx* (Mori *et al.*, 1994; Li *et al.*, 1994), and *axial* (Strahle *et al.*, 1993). Some of these genes are known to exhibit dynamic patterns of expression. For example, expression of *zf-T* changes from a broad expression by all cells within the germ ring to expression only in the deep cells of the shield region (Schulte-Merker *et al.*, 1992, 1994b). Similarly, *axial,* a member of the *fork head* family of transcription regulators, is expressed initially by cells in the dorsal primitive hypoblast (Strahle *et al.*, 1993), but changes to include dorsal midline cells of the primary ectoderm just before yolk plug closure, which has been interpreted as a response to neural induction. The movement of cells across the cleft suggests that environmental cues may play a major role in determining the ongoing expression of even some of the genes that have been taken as markers of future cell fate.

V. The Neuronal Fate Map of the Zebrafish

The cell movements at the dorsal midline presented above bring to the fore the issue of neuronal patterning

because they offer different mechanisms for cells to interact during embryonic patterning. We have studied the regional neural fate map of the zebrafish at the beginning and end of gastrulation (6 hours and 10 hours of development, respectively). By 6 hours, the embryonic shield first becomes morphologically pronounced, enabling unambiguous identification of "dorsal" (Westerfield, 1994; Kimmel *et al.,* 1995). At 10 hours, the epiblast of the zebrafish embryo completely covers the yolk, and the distinct dorsal axis makes alignment of the data straightforward. As before, the first steps in the construction of the fate map was the labeling of a cell by iontophoretic injection of fluorescent dextran and the careful recording of the position of the injected cells. As we were exploring the fate map and not the presence of lineage restrictions, we included embryos with two or more labeled cells when the cells were in close proximity (48% of the 6-hour injections, and 31% of the 10-hour injections labeled a single cell). At least two different aligned views were taken of each embryo to assure an accurate alignment. The descendants of labeled cells were assigned to one or more of six categories: Telencephalon, including the olfactory placode; Diencephalon, excluding the neural retina; Retina; Midbrain; Hindbrain; and Other, including spinal cord neurons, somite, neural crest, and epidermal (including placodal) derivatives. We included the olfactory placode in the telencephalon category since it is too closely associated with the brain proper at the time of scoring to be distinguished reliably.

The fate maps showed a significant amount of order even at 6 hours (Fig. 5). The majority of the embryos had labeled cells in only one brain region, even in those cases in which multiple cells were injected. Although the fates of the individual progenitors might suggest the presence of some fate restrictions, no fate map can provide evidence for the presence of restrictions. The predictable order to the fate map is not fully refined, and different brain regions appear somewhat intermixed. Coherence is shown by the regions of the fate map that give rise to only one brain region. For example, between $-20°$ and $+20°$ longitude, only retinal precursors (green) are found at $65°–80°$ latitude; whereas, at $40°–50°$ latitude, only diencephalic precursors (black) are seen. Imprecision is shown by the presence of progenitors for telencephalon (red), diencephalon (black), retina (green), and midbrain (orange) all within the area at $55°–65°$ latitude and $30°–70°$ longitude.

At 10 hours, the number of injections yielding a single regional fate increased across all brain regions, especially the retina. This homogeneity in fate and the reduced overlap between domains in the 10-hour fate map suggest that regional patterning has become significantly more pronounced. In the 4 hours from the 6-hour fate map, the entire neuroectoderm has narrowed from 6 hours, placing the neural–epidermal border at $40°–70°$ from the midline, consistent with previous observations in the anterior spinal region at the two-somite stage

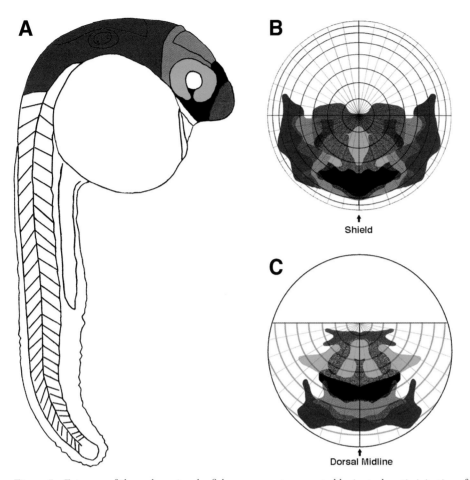

Figure 5 Fate map of the embryonic zebrafish nervous system, created by iontophoretic injection of the neurectoderm. Using techniques similar to those employed in Fig. 3, single cells or pairs of cells were iontophoretically labeled with fluorescent dextran. After allowing the embryo to develop, the positions of the labeled cells were judged with respect to their brain region (A). The fate maps depicted at 6 hours (B) and 10 hours (C) of development show the regions of the ectoderm that later form the color-coded regions of the nervous system. Most labeled clones gave rise to only a single brain area. The regions of overlap shown with two different colors are mostly the result of different clones at what was judged to be identical positions giving rise to different fates in different animals.

(Schmitz *et al.*, 1993). The amount of convergence from 6 to 10 hours varies with axial level. For example, the lateral limit of retinal domain changes from 90° to 60° (a 30° change) during this time interval, while the hindbrain domain narrows from 110° to 50° (a 60° change). These movements have brought the central nervous system (CNS) regions into roughly the same order found later in the neural tube. Beginning at the front of the axis and moving caudally, one sees the progenitors for the telencephalon, retina, diencephalon, midbrain, and hindbrain (Fig. 5). The shapes of three domains have changed significantly between 6 and 10 hours. The telencephalic domain has adopted a more mature shape with most of the cells lateral to the midline. The midbrain and hindbrain domains, which were both initially split across the midline at 6 hours, have coalesced into single, coherent masses directly posterior to the forebrain.

An interesting aspect of the fate map is that both neural retinas map to a single coherent region at the anteromedial portion of the forebrain (Fig. 5). Single cells injected on one side can give rise to progeny in both retinae, indicating that both retinas originate from this

singular domain. In more recent fate mapping experiments (Woo and Fraser, 1998) we have imaged labeled retinal precursors as the retinal field is bisected by the tissue more caudally located in the fate map (diencephalon progenitors). Thus, it appears that the retinal field is separated by physical interactions with ventral diencephalic progenitors that move cells from a central to a more lateral position, not by a signal that recruits cells away from a retinal fate as has been proposed for *Xenopus* (Li *et al.*, 1997; but see also Chapter 20 by Sullivan *et al.*; and Chapter 23 by Perron and Harris). This may have implications for the interpretation of mutant phenotypes. For example, cyclopia, a notable defect in the cyclops and other mutants, is believed to result from a massive deletion of the ventral forebrain (Hatta *et al.*, 1991), which may stem from impaired cell proliferation (Hatta *et al.*, 1991; Strahle *et al.*, 1993). Our results suggest that some mutations leading to cyclopia might instead impair motility of the ventral diencephalic progenitors during gastrulation; cyclopia in such a case would result from a failure of the retinal anlage to be physically bifurcated. This scenario does not preclude a change in the specification state of diencephalic progenitors due to alteration in cell signaling or cell fate in some mutants (Hatta *et al.*, 1991), but suggests another hypothesis, focused on cell motility, that should be ruled out in each case before concluding otherwise.

VI. Neuronal Commitment and the Acquisition of Regional Fates

The localized tissue domains observed in the 6-hour and 10-hour fate maps (Woo and Fraser, 1995) suggest that cell fate may be assigned to the neurectoderm around these stages. However, fate maps cannot be used to pinpoint the time of cell fate decisions. Determining the timing of cell fate assignments requires confronting cells with a foreign set of cell interactions; if the cells remain true to their original fates, they are said to be committed. Fate maps cannot define the stage of commitment because it is not clear if the cells are being challenged by different environments; for example, the orderly motions of the cells in the neural plate keep neighboring cells together in coherent groups and do not challenge the state of specification of the individual cells. Similarly, the timing of cell fate restrictions within the zebrafish neurectoderm can only be loosely inferred from analyses of gene expression patterns. *Otx1* expression (and, slightly later, *otx2*) is first observed at 65% epiboly in the presumptive forebrain region (Li *et al.*, 1994). This is followed by the appearance of *wnt1* and *pax2* expression in the presumptive mid-hindbrain region (Kelly and Moon, 1995), and *goosecoid* in the midline axial region (Thisse *et al.*, 1994). By the end of gastrulation, the expression of a number of genes including *pax6* (Krauss *et al.*, 1991; Puschel *et al.*, 1992a), *pax2* (Kelly and Moon, 1995; Puschel *et al.*, 1992b), and *Krox20* (Oxtoby and Jowett, 1993) appears restricted within defined domains. Although gene expression patterns are suggestive, they are no more conclusive than the fate maps in establishing the commitment state of cell groups.

To test when regional commitment occurs in the neural plate, we transplanted prospective hindbrain neural tissue at different stages into the ventral-most ectoderm of a shield-stage host. The fate of the transplanted tissue was assessed using neural-specific antibody markers, hindbrain-specific *in situ* probes, and the ability of the grafted tissue to induce ectopic otic vesicles and neural crest cells. We concentrated on hindbrain progenitors because their fate map location was clearly defined and regional-specific markers were available. More importantly, it was possible to isolate these cells free of axial-mesoderm contamination during the relevant stages. We chose the ventral-most region of the zebrafish epiblast at 6 hours as the "neutral" recipient site for the transplants. We reasoned that this region, being the furthest away from the organizer, would be least affected by its neuralizing influence. Homotopic replacement transplants showed that this ventral-most epiblast normally

gives rise to the ectoderm of the yolk and fin structures, in agreement with the fate map (Woo and Fraser, 1995). When grafted at 6 hours, hindbrain progenitors integrated seamlessly into the host. It was impossible to distinguish the boundary between host and graft tissue without the aid of the fluorescence marking of the donor tissue. When scored at 24–36 hours, the labeled descendants of the grafts had differentiated into ectodermal cells of the trunk and tail, some somites in the tail, and scattered cells in the fin, identical to that documented in the homotopic grafts of the ventral epiblast. Thus, at 6 hours the predictable order of the neural fate map is not indicative of commitment of the cells to either neural or hindbrain fates.

In contrast, when transplanted to the ventral midline of the 6-hour embryo, hindbrain progenitors from 80% epiboly embryos retained both their neural and regional fates. The grafted tissues appeared to integrate into the host immediately after surgery, only to later segregate from the ectoderm of the host. In histological sections, the descendants of the grafted tissue often appeared somewhat like the neural keel. Analysis with zns-1 and zns-4 antibodies showed that grafted cells gave rise to neural tissue in all cases. In addition, the transplants maintained their regional identity, as they maintained their expression of Krox20, a zinc-finger transcription factor normally expressed in presumptive rhombomeres 3 and 5 of the hindbrain starting at 10 hours (Oxtoby and Jowett, 1993). Interestingly, Krox20-positive cells always appeared as coherent groups organized in no more than two bands. The domains of ectopic Krox20 expression often aligned with ectopic otic vesicles formation, suggesting that the positive cells represented rhombomere 5, which normally juxtaposes the otic vesicle. The retention of the neural fate was not due to the grafting from an 80% epiboly donor to the ventral epiblast of 6-hour hosts. Control grafts between these stages gave rise to epidermis and occasional tail somite cells, as expected. The hindbrain progenitors did not maintain their fate because we had accidentally included axial mesoderm in the transplants; the transplanted pieces were negative for the axial mesoderm markers (zf-T, Schulte-Merker *et al.*, 1992; and MyoD, Weinberg *et al.*, 1996). Thus, we conclude that the commitment to neural fate observed in the presumptive hindbrain transplants was not the result of experimental limitations.

Interestingly, the grafts of committed hindbrain tissue often recruited neighboring host ectoderm to express neural markers (zns-1 and zns-4) and Krox20. This may offer evidence for the action of "homeogenetic induction," first defined by the ability of neural plate tissue to induce nearby blastocoel ectoderm to become neural in *Triturus* (Mangold and Spemann, 1927; Spemann, 1938). The role of homeogenetic induction in normal development remains to be understood. Homeogenetic induction has been proposed to mediate the propagation of organizer-derived patterning information received by midline-neural tissues outward to the lateral limits of the neural plate (Albers, 1987; Nieuwkoop and Albers, 1990). In chick, ectopically placed floor plate can induce floor plate from neural tube cells that normally would not give rise to floor plate (Placzek *et al.*, 1993; see also Chapter 29 by Streit and Stern). Similar phenomena have been suggested to explain the development of floor plate cells in the zebrafish mutant cyclops after the grafting of wild-type floor plate precursors (Hatta *et al.*, 1991; Hatta, 1992). However, these zebra-fish genetic mosaic experiments do not rigorously demonstrate homeogenetic induction (Gurdon *et al.*, 1993), as the responding cells may have been already specified for the floor plate fate but arrested along the floor plate developmental program due to the mutation. Here, our experiments using ectopic transplantation between wide-type embryos clearly indicate homeogenetic induction of neural tissue. Further dissection of this phenomenon may advance our insights into the role of homeogenetic induction in the normal mechanisms of neural patterning.

Taken together, these results show that the significant order and coherence of the 6-hour fate map does not reflect any cell fate restrictions. Presumptive hindbrain cells at shield stage (6 hours) gave rise to epidermis when transplanted to the ventral side of the embryo, demonstrating that the cells have not undergone either neural or regional commitment. In contrast, at 80% epiboly (8–8.5 hours) the grafted cells maintained both their neural and regional fates in both neutral (ventral ectoderm) and nonneutral (forebrain) environments. Because neither surgical manipulation nor heterochronic transplant controls produced ectopic hindbrain development, the timing determined in our experiments most likely reflects the normal timing of developmental events. These results show that conclusions that the progenitors of different epiblast territories may be defined to some degree by shield stage, based on the patterning of regionally expressed genes, are incorrect. At present, it is not clear whether the hindbrain neural fate is achieved by a single step, or by a "neural" restriction followed by a "hindbrain" restriction. It seems that there might be such a stepwise progression, and that not all regional fate decisions become fixed immediately. Several lines of evidence suggest that early anteroposterior (AP) specification within the neurectoderm is labile for some period. For example, the anteroposterior regional characteristic of the ectoderm in *Xenopus* is not firmly established until neural plate stages (Saha and Grainger, 1992); similarly, in chick, prospective ectodermal cells can adopt the neural fate as late as the onset of neurulation (Schoenwolf and Alvarez, 1991). Further studies are needed to define the interactions and signals that take place during the 2-hour period (6 hours, 80% epiboly) during which hindbrain progenitors become committed to their neuronal and regional fates. Because this commitment takes place before the hindbrain progenitors contact the axial mesoderm, alternative sources of patterning information must be examined.

VII. Nonaxial Signals Provide Rostrocaudal Patterning

The gastrula organizer ("dorsal lip" in amphibians; "node" in amniotes; "shield" in fish) is thought to be responsible for providing both anterior–posterior and dorsal–ventral patterning information to the overlying neurectoderm. Classic embryological analyses using the amphibian embryo have suggested that two temporally distinct signals emanate from the Spemann organizer to pattern the neural axis in the anteroposterior (AP) direction (cf. Doniach, 1993): An activator signal, originating from the anterior axial mesoderm, broadly neuralizes ectodermal tissue with anterior characteristics; whereas a transformer signal, emanating from the chordamesoderm, repatterns nearby neural tissue into more-posterior types in a dose-dependent manner. The results presented in the previous section show that the fates of hindbrain progenitors become fixed at 80% epiboly, while they are located dorsolaterally and have yet to establish vertical contact with the axial mesoderm. Thus, the cells are becoming patterned before they contact the cells within or derived from the shield that are thought to convey regional patterning information. Because vertical contact with axial mesoderm is not required for cells to adopt a hindbrain regional identity, it seems clear that positional signals can be transmitted by a planar mechanism. The source of this patterning signal resides, at least in part, outside the shield region, as microsurgical deletion of the shield at 6 hours does not prevent hindbrain formation (Shih and Fraser, 1996).

Based on the predictable order present in the 6-hour neural fate map (Woo and Fraser, 1995) we hypothesized that cells in the lateral germ ring were ideally positioned for conveying rostrocaudal patterning information. Forebrain progenitors span the dorsal midline near the animal pole and are far from the germ ring; in contrast, hindbrain progenitors lay close to the germ ring, and remain juxtaposed for the early to mid-stages of epiboly (Fig. 5). The precursors of the midbrain are in contact with the germ ring more briefly, until the normal motions of early epiboly and neurulation move them more rostral. This direct correlation between the length of time that the cells remain near the germ ring and their degree of posterior transformation suggests a role for planar or radial signaling from the nonaxial mesoderm in neuronal patterning. Cells that will give rise to the posterior neural axis remain close to the germ ring throughout gastrulation and would receive the most transforming signal. Midbrain progenitors would receive only a moderate amount of the signal during their brief time near the germ ring, whereas forebrain progenitors would have little, if any, exposure to the transforming signal.

Consistent with the germ ring providing a transforming signal that promotes more-posterior neural fates, labeled forebrain progenitors adopt the hindbrain fate when grafted to the position of the presumptive

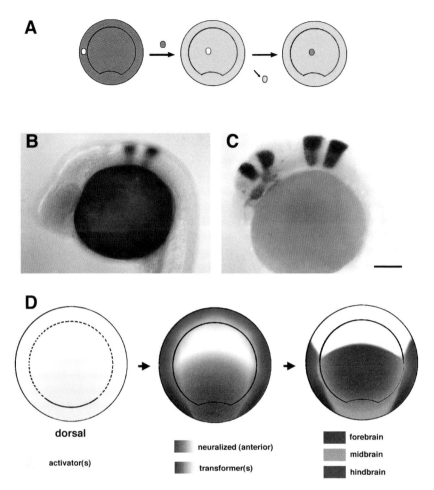

Figure 6 Nonaxial mesoderm can provide axial information to the neurectoderm. (A) Schematic of the grafting operation to test the proposed role of nonaxial mesoderm in assigning posterior fates to the neurectoderm. Labeled cells (fluorescent dextran and biotin dextran) were grafted from the germ ring to the region of the putative forebrain (compare to Fig. 5). (B) Krox20 expression in a control embryo in which the histochemical stain for the Krox20 (purple) marks rhombomeres 3 and 5 in the hindbrain. (C) Krox20 expression in an embryo with germ ring graft (orange-brown). Note that the forebrain above the grafted cells now displays bands of ectopic Krox20 indicating a transformation to hindbrain. (D) A model for patterning the zebrafish neural axis. An activator signal (yellow) originating from the dorsal midline acts to induce neural tissue with a broad anterior character (red). A transformer signal from the germ ring (blue) is generated that modulates the axial character of the nearby tissue that has previously been neutralized. This results in distinct forebrain (deep red), midbrain (orange), and hindbrain (blue) territories.

hindbrain (Woo and Fraser, 1997). Such a conversion of forebrain to hindbrain suggests that the signals that normally instruct cells to adopt the hindbrain fate are still active *in vivo* at 6 hours. This result is consistent with the notion that the position of cells in the gastrula is a reliable indicator of their eventual fate, and with the fact that presumptive hindbrain cells are not committed to their positional fate at this stage (see Section VI).

A more definitive test of the possible role of nonaxial germ ring tissue in patterning the neural axis would be to transform cells that would normally not make hindbrain. To do so, we transplanted different regions of the germ ring into the animal pole of shield-stage zebrafish (Fig. 6), a region fated to become forebrain (Fig. 5). If the germ ring were the source of a transforming signal, such grafts should cause the presumptive forebrain cells to adopt more-posterior fates. The embryos that received a fluorescently labeled germ ring graft to the animal pole exhibited the predicted dis-

ruption of normal patterning. Hindbrainlike structures appeared in the forebrain, and a large fraction developed ectopic otic vesicles derived from host tissue (Woo and Fraser, 1997). Using Krox20 expression as a marker for rhombomeres 3 and 5 of the hindbrain confirmed that the region close to the germ ring transplant had adopted a hindbrainlike fate (Fig. 6). The ectopic Krox20 was expressed on a normal time course, appearing as either a single patch or as two discrete bands. All regions of the lateral and ventral germ ring possessed similar transforming activities; the normal fates of the grafted cells include somitic mesoderm, posterior mesoderm, and endoderm, in varying ratios. In contrast, grafting the embryonic shield to the animal pole did not lead to hindbrainlike transformation.

These grafting results demonstrate that tissues from the nonaxial germ ring of the zebrafish gastrula can transform the presumptive forebrain of the same developmental stage *in vivo*. The few direct tests of defined factors have not mimicked the activity of the germ ring. For example, bFGF beads at the animal pole of a shield-stage gastrula did not induce ectopic Krox20 (Woo and Fraser, 1997), making it unlikely to act directly on the neurectoderm in this patterning process. While the nature of the signal remains unclear, the clear separation of inducing and transforming activity in the zebrafish suggests a revision of the model described for *Xenopus* (Doniach, 1995; see also Chapter 29 by Streit and Stern). We propose an "activating" influence that originates from the embryonic shield to neuralize the presumptive neurectoderm at or before the onset of shield formation (Fig. 6). A second "transforming" signal emanates radially from the germ ring of the gastrula to posteriorize the activated tissue. Little, if any, of the transforming signal is present in the zebrafish embryonic shield. This simple reworking of the model is sufficient to fit our grafting results and predicts the normal patterning of the zebrafish neural fate map. In a normal gastrula, the presumptive forebrain progenitors are displaced toward the animal pole (Woo and Fraser, 1995), away from the transforming signal(s). Hindbrain progenitors, in contrast, remain close to the germ ring mesendoderm as they converge to the dorsal side, permitting their transformation to more-posterior neural fates. In addition, it explains a puzzling aspect of our previous experiments on grafting the embryonic shield (Shih and Fraser, 1996). In those studies, the second axes that were formed after an embryonic shield was grafted to the ventral midline lacked defined midbrain and forebrain structures. When viewed from the zebrafish two-signal model (Woo and Fraser, 1997), this is exactly the expected result, as the induced tissue would have been well within the influence of the transforming signal from the ventral germ ring. Consistent with this, grafts that are placed closer to the animal pole are more rostrally complete.

VIII. Conclusion

The results and deductions that are briefly reviewed above show the power of the zebrafish for investigations of embryonic patterning. Its clarity and rapid development make it nearly ideal for studies of fate maps, the interactions that organize them and the morphogenetic events that are driven by them. For example, the combination of fate mapping, grafting, and observation permitted a direct assay of the events in forebrain development, and resulted in a revised model for neural activation/transformation (Woo and Fraser, 1997). Such an integrated approach would have been much more difficult in other experimental models. The rapid advance in large scale mutagenesis screens in the zebrafish offers a tremendous resource for developmental biologists. It seems clear that the new families of mutants, together

with the next generation of high-resolution fate maps will continue to contribute to and to challenge our present understanding of vertebrate development.

References

Abdelilah, S., Solnica-Krezel, L., Stainier, D. Y. R., and Driever, W. (1994). Implications for dorsoventral axis determination from the zebrafish mutation Janus. *Nature (London)* **370**, 468–471.

Albers, B. (1987). Competence as the main factor determining the size of the neural plate. *Dev. Growth Differ.* **29**, 535–545.

Bauer, D. V., Huang, S., and Moody, S. A. (1994). The cleavage stage origin of Spemann's Organizer: Analysis of the movements of blastomere clones before and during gastrulation in *Xenopus*. *Development (Cambridge, UK)* **120**, 1179–1189.

Brachet, A. (1935). Traite d'Embryologie des Vertebres, 2nd Ed. Masson et cie, Paris.

Cooper, M. S., and D'Amico, L. A. (1996). A cluster of noninvoluting endocytic cells at the margin of the zebrafish blastoderm marks the site of embryonic shield formation. *Dev. Biol.* **180**, 184–198.

Dettlaff, T. A. (1993). Evolution of the historical and functional structure of the ectoderm, chordamesoderm, and their derivatives in Anamnia. *Roux's Arch. Dev. Bio.* **203**, 3–9.

Doniach, T. (1992a). Induction of anteroposterior neural pattern in *Xenopus* by planar signals. *Development (Cambridge, UK)* Suppl., 183–193.

Doniach, T. (1992b). Planar induction of anteroposterior pattern in the central nervous system of *Xenopus laevis*. *Science* **257**, 542–545.

Doniach, T. (1993). Planar and vertical induction of anteroposterior pattern during the development of the amphibian central nervous system. *J. Neurobiol.* **24**, 1256–1275.

Doniach T. (1995). *Cell (Cambridge, Mass.)* **83**, 1067–1070.

Fink, R. D., and McClay, D. R. (1985). Three cell recognition changes accompany the ingression of sea urchin primary mesenchyme cells. *Dev. Biol.* **107**, 66–74.

Gurdon, J. B., Lemaire, P., and Kato, K. (1993). Community effects and related phenomena in development. *Cell (Cambridge, Mass.)* **75**, 831–834.

Hammerschmidt, M., and Nüsslein-Volhard, C. (1993). The expression of a zebrafish gene homologous to *Drosophila* snail suggests a conserved function in invertebrate and vertebrate gastrulation. *Development (Cambridge, UK)* **119**, 1107–1118.

Hashimoto, K., and Nakatsuji, N. (1989). Formation of the primitive streak and mesoderm cells in mouse embryos–detailed scanning electron microscopical study. *Dev. Growth Diff.* **31**, 209–218.

Hatta, K. (1992). Role of the floor plate in axonal patterning in the zebrafish CNS. *Neuron* **9**, 629–642.

Hatta, K., Kimmel, C. B., Ho, R. K., and Walker, C. (1991). The cyclops mutation blocks specification of the floor plate of the zebrafish CNS. *Nature (London)* **350**, 339–341.

Helde, K. A., Wilson, E. T., Cretekos, C. J., and Grunwald, D. J. (1994). Contributions of early cells to the fate map of the zebrafish gastrula. *Science* **265**, 517–520.

Ho, R. K. (1992). Axis formation in the embryo of the zebrafish, *Brachydanio rerio*. *Semin. Dev. Biol.* **3**, 53–64.

Jessell, T. M., and Melton, D. A. (1992). Diffusible factors in vertebrate embryonic induction. *Cell (Cambridge, Mass.)* **68**, 257–270.

Keller, R. E. (1975). Vital dye mapping of the gastrula and neurula of *Xenopus laevis*. I. Prospective areas and morphogenetic movements of the superficial layer. *Dev. Biol.* **42**, 222–241.

Keller, R. E. (1976). Vital dye mapping of the gastrula and neurula of *Xenopus laevis*. II. Prospective areas and morphogenetic movements of the deep layers. *Dev. Biol.* **51**, 118–137.

Kelly, G. M., and Moon, R. T. (1995). Involvement of Wnt1 and Pax2 in the formation of the midbrain–hindbrain boundary in the zebrafish gastrula. *Dev. Genet.* **17**, 129–140.

Kimmel, C. B., and Warga, R. M. (1986). Tissue specific cell lineages originate in the gastrula of the zebrafish. *Science* **231**, 365–368.

Kimmel, C. B., and Law, R. D. (1985). Cell lineage of zebrafish blastomeres III. Clonal analysis of the blastula and gastrula stages. *Dev. Biol.* **108**, 94–101.

Kimmel, C. B., Warga, R. M., and Schilling, T. F. (1990). Origin and organization of the zebrafish fate map. *Development (Cambridge, UK)* **108**, 581–594.

Kimmel, C. B., Ballard, W. W., Kimmel, S. R., Ullmann, B., and Schilling, T. F. (1995). Stages of embryonic development of the zebrafish. *Dev. Dyn.* **203**, 253–310.

Kispert, A., and Herrmann, B. G. (1994). Immunohistochemical analysis of the brachyury protein in wild-type and mutant mouse embryos. *Dev. Biol.* **161**, 179–193.

Krauss, S., Johansen, T., Korzh, V., and Fjose, A. (1991). Expression patterning of zebrafish pax genes suggests a role in early brain regionalization. *Nature (London)* **353**, 267–270.

Krauss, S., Concordet, J-P., and Ingham, P. W. (1993). A functionally conserved homolog of the *Drosophila* segment polarity gene hh is expressed in tissues of polarizing activity in Zebrafish embryos. *Cell* **75**, 1431–1444.

Lawson, K. A., Meneses, J. J., and Pedersen, R. A. (1991). Clonal analysis of epiblast fate during germ layer formation in the mouse embryo. *Development (Cambridge, UK)* **113**, 891–911.

Li, Y., Allende, M. L., Finkelstein, R., and Weinberg, E. S. (1994). Expression of two zebrafish orthodenticle-related genes in the embryonic brain. *Mech. Dev.* **48**, 229–244.

Li, H.-S., Tierney, C., Wen, L., Wu, J. Y. and Rao, Y. (1997). A single morphogenetic field gives rise to two retina primordia under the influence of the prechordal plate. *Development* **124**, 603–615.

Mangold, O., and Spemann, H. (1927). Über Induktion von Medullarplatte durch Medullarplatte in jüngeren Keim, ein Beispeil homoeogenetischer oder assimilatorischer Induktion. *Wilhelm Roux Arch. Entwicklungsmech. Org.* **111**, 341–342.

Melby, A. E., Warga, R. M., and Kimmel, C. B. (1996). Specification of cell fates at the dorsal margin of the zebrafish gastrula. *Development (Cambridge, UK)* **122**, 2225–2237.

Mizuno, T., Yamaha, E., Wakahara, M., Kuroiwa, A., and Takeda, H. (1996). Mesoderm induction in zebrafish. *Nature (London)* **383**, 131–132.

Mizuno, T., Yamaha, E., and Yamazaki, F. (1997). Localized axis determinant in the early cleavage embryo of the goldfish, *Carassius auratus*. *Dev. Genes Evol.* **206**, 389–396.

Mori, H., Miyazaki, Y., Morita, T., Nitta, H., and Mishina, M. (1994). Different spatio-temporal expressions of three otx homeoprotein transcripts during zebrafish embryogenesis. *Mol. Brain Res.* **27**, 221–231.

Nieuwkoop, P. D., and Albers, B. (1990). The role of competence in the cranio-caudal segregation of the central nervous system. *Dev. Growth Differ.* **32**, 23–31.

Oppenheimer, J. M. (1936). Transplantation experiments on developing teleosts (Fundulus and Perca). *J. Exp. Zool.* **72**, 409–437.

Oxtoby, E., and Jowett, T. (1993). Cloning of the zebrafish *Krox-20* gene (krx-20) and its expression during hindbrain development. *Nucleic Acids Res.* **21**, 1087–1095.

Parameswaran, M., and Tam, P. P. (1995). Regionalization of cell fate and morphogenesis movement of the mesoderm during mouse gastrulation. *Dev. Genet.* **17**, 16–28.

Pasteels, J. (1942). New observations concerning the maps of presumptive areas of the young amphibian gastrula (Ambystoma and Discoglossus). *J. Exp. Zool.* **89**, 255–281.

Placzek, M., Jessell, T. M., and Dodd, J. (1993). Induction of floor plate differentiation by contact dependent, homeogenetic signals. *Development (Cambridge, UK)* **117**, 205–218.

Puschel, A. W., Gruss, P., and Westerfield, M. (1992a). Sequence and expression pattern of pax-6 are highly conserved between zebrafish and mice. *Development (Cambridge, UK)* **114**, 643–651.

Puschel, A. W., Westerfield, M., and Dressler, G. R. (1992b). Comparative analysis of Pax-2 protein distributions during neurulation in mice and zebrafish. *Mech. Dev.* **38**, 197–208.

Saha, M. S., and Grainger, R. M. (1992). A labile period in the determination of the anterior–posterior axis during early neural development in *Xenopus*. *Neuron* **8**, 1003–1014.

Schmitz, B., and Campos-Ortega, J. A. (1994) Dorso-ventral polarity of the zebrafish embryo is distinguishable prior to the onset of gastrulation. *Roux's Arch. Dev. Biol.* **203**, 374–380.

Schmitz, B., Papan, C., and Campos-Ortega, J. A. (1993). Neurulation in the anterior trunk region of the zebrafish brachydanio-rerio. *Roux-A-DB* **202**, 250–259.

Schoenwolf, G. C. (1991). Cell movements in the epiblast during gastrulation and neurulation in avian embryos. *In* "Gastrulation: Movements, Patterns, and Molecules" (R. E. Keller, ed.), pp. 1–28. Plenum, New York.

Schoenwolf, G. C., and Alvarez, I. S. (1991). Specification of neurepithelium and surface epithelium in avian transplantation chimeras. *Development (Cambridge, UK)* **112**, 713–772.

Schulte-Merker, S., Ho, R. K., Herrmann, B. G., and Nüsslein-Volhard, C. (1992). The protein product of the zebrafish homologue of the mouse T gene is expressed in nuclei of the germ ring and the notochord of the early embryo. *Development (Cambridge, UK)* **116**, 1021–1032.

Schulte-Merker, S., Hammerschmidt, M., Beuchle, D., Cho, K. W., De Robertis, E. M., and Nüsslein-Volhard, C. (1994a). Expression of zebrafish goosecoid and no tail gene products in wild-type and mutant no tail embryos. *Development (Cambridge, UK)* **120**, 843–852.

Schulte-Merker, S., van Eeden, F., Halpern, M. E., Kimmel, C. B., and Nüsslein-Volhard, C. (1994b). no tail (ntl) is the zebrafish homolog of the mouse T (brachyury) gene. *Development (Cambridge, UK)* **120**, 1009–1015.

Selleck, M. A. J., and Stern, C. D. (1991). Fate mapping and cell lineage analysis of Hensen's node in the chick embryo. *Development (Cambridge, UK)* **112**, 615–626.

Shih, J., and Fraser, S. E. (1995). The distribution of tissue progenitors within the shield region of the zebrafish gastrula. *Development (Cambridge, UK)* **121**, 2755–2765.

Shih, J., and Fraser, S. E. (1996). Characterizing the zebrafish organizer—microsurgical analysis at the early-shield stage. *Development (Cambridge, UK)* **122**, 1313–1322.

Slack, J. M. W. (1991). "From Egg to Embryo: Regional Specification in Early Development." (P. Barlow, D. Bray, P. Green, and J. Slack, eds.), 2nd Ed. Cambridge Univ. Press, Cambridge.

Spemann, H. (1938). "Embryonic Development and Induction." Yale Univ. Press, New Haven, Connecticut.

Stern, C. D., and Canning, D. R. (1990). Origin of cells giving rise to mesoderm and endoderm in chick-embryo. *Nature (London)* **343**, 273–275.

Strahle, U., Blader, P., Henrique, D., and Ingham, P. W. (1993). Axial, a zebrafish gene expressed along the developing body axis, shows altered expression in cyclops mutant embryos. *Genes Dev.* **7**, 1436–1446.

Strehlow, D., and Gilbert, W. (1993). A fate map for the first cleavages of the zebrafish. *Nature (London)* **361**, 451–453.

Strehlow, D., Heinrich, G., and Gilbert, W. (1994). The fate of the blastomeres of the 16-cell zebrafish embryo. *Development (Cambridge, UK)* **120**, 1791–1798.

Talbot, W. S., Trevarrow, B., Halpern, M. E., Melby, A. E., Farr, G., Postlethwait, J. H., Jowett, T., Kimmel, C. B.,

and Kimelman, D. (1995). A homeobox essential for zebrafish notochord development. *Nature (London)* **378**, 150–157.

Thisse, C., Thisse, B., Halpern, M. E., and Postlethwait, J. H. (1994). Goosecoid expression in neurectoderm and mesendoderm is disrupted in zebrafish cyclops gastrulas. *Dev. Biol.* **164**, 420–429.

Trinkaus, J. P. (1984). "Cells into Organs: The Forces that Shape the Embryo," 2nd Ed. Prentice-Hall, Englewood Cliffs, New Jersey.

Trinkaus, J. P. (1996). Ingression during early gastrulation of Fundulus. *Dev. Biol.* **177**, 356–370.

Warga, R. M., and Kimmel, C. B. (1990). Cell movements during epiboly and gastrulation in zebrafish. *Development (Cambridge, UK)* **108**, 569–580.

Weinberg, E. S., Allende, M. L., Kelly, C. S., Abdelhamid, A., Murakami, T., Andermann, P., Doerre, O. G., Grunwald, D. J., and Riggleman, B. (1996). Developmental regulation of zebrafish MyoD in wild-type, no tail and spadetail embryos. *Development (Cambridge, UK)* **122**, 271–280.

Westerfield, M. F. (1994). "The Zebrafish Book: A Guide for the Laboratory Use of Zebrafish (*Brachydanio rerio*)," 2nd Ed. Univ. of Oregon Press, Eugene.

Wetts, R., and Fraser, S. E. (1989). Slow intermixing of cells during *Xenopus* embryogenesis contributes to the consistency of the blastomere fate map. *Development (Cambridge, UK)* **108**, 9–15.

Wilson, H. V. (1889). The embryology of the sea bass (*Serranus atrarius*). *Bull. U.S. Fish. Comm.* **9**, 209–278.

Woo, K., and Fraser, S. E. (1995). Order and coherence in the fate map of the zebrafish nervous system. *Development (Cambridge, UK)* **121**, 2595–2609.

Woo, K., and Fraser, S. E. (1997). Specification of the zebrafish nervous system by nonaxial signals. *Science* **277**, 254–257.

Woo, K., and Fraser, S. E. (1998). Characterizing the ventral diencephalon in the zebrafish: Morphogenesis and inductive capacity. Submitted for publication.

Wood, A., and Timmermans, L. P. (1988). Teleost epiboly: A reassessment of deep cell movement in the germ ring. *Development (Cambridge, UK)* **102**, 575–585.

Xu, Q., Holder, N., Patient, R., and Wilson, S. W. (1994). Spatially regulated expression of three receptor tyrosine kinase genes during gastrulation in the zebrafish. *Development (Cambridge, UK)* **120**, 2287–2299.

26

Patterning of the Zebrafish Embryo along the Anteroposterior Axis

Robert K. Ho
John P. Kanki
Victoria E. Prince*
Department of Molecular Biology
Princeton University
Princeton, New Jersey 08544

Lucille Joly
Marc Ekker
Loeb Institute for Medical Research
Ottawa Civic Hospital
Anatomy and Neurobiology
University of Ottawa
Ottawa, Ontario, Canada K1N 6NS

Andreas Fritz
Institute of Neuroscience
University of Oregon
Eugene, Oregon 97403

I. Introduction

II. Hox Cluster Genes
  A. Why Study the Hox Clusters in Zebrafish?
  B. Genomic Organization of the Zebrafish Hox Clusters Is Significantly Altered Relative to Tetrapods
  C. Expression Patterns of the Hox Genes in the Zebrafish Head Region Appear Similar to Tetrapods
  D. Modified Colinear Expression Patterns of the Hox Genes in the Zebrafish Trunk

III. Posterior Body Development
  A. Different Patterning Mechanisms in the Vertebrate Posterior Body
  B. Fate Map of the Zebrafish Tail Bud
  C. Unique Cell Movements within the Developing Tail Bud Region

IV. Summary
  References

I. Introduction

Understanding how development proceeds requires knowledge gleaned from many disciplines, including experimental embryology, molecular biology, genetics,

and evolutionary biology. The past decade has seen a growing emphasis placed on genetic analyses which have allowed new insights into various mechanisms of development, including axis formation, cell cycle control, organogenesis, and cell death. Much of the excitement in the scientific community surrounding the use of zebrafish as a research preparation has been generated by its potential as a genetic system, especially with regards to the possibility of performing successful large-

*Present address: Department of Organismal Biology and Anatomy, University of Chicago, Chicago, IL 60615

scale mutagenesis screens. However, despite many decades of study, a great deal remains to be learned about normal teleost development and anatomy. Modern techniques such as the use of molecular markers and new cell labeling methods will afford us further findings into zebrafish development and allow researchers in this field to contribute to our expanding knowledge of vertebrate development.

In this chapter, we summarize work undertaken in our laboratory to better describe how different regional identities may be assigned along the anteroposterior (AP) body axis of the developing zebrafish embryo, namely in the head, trunk, and tail regions. We begin with a description of our work on the isolation of the hox cluster genes in the zebrafish. Although the characterization of this family of genes in the frog, chick, mouse, and human has often been dogmatically reported in the literature and textbooks as applicable to all vertebrates, we describe important differences in the genomic organization and expression patterns of the zebrafish hox cluster genes relative to the Hox cluster genes in tetrapods. We find that the zebrafish hox cluster genes are useful markers for regions of the zebrafish hindbrain and trunk, however, as in tetrapods, the genes which we have studied to date do not appear to have unique expression patterns within the tail region. To show how different regional identities may be specified in the vertebrate tail region, we describe cell marking and cell movement analyses of the developing tail bud in the living zebrafish embryo. We have found that although the tail bud region of the zebrafish, like most vertebrates, appears initially as a homogeneous blastema without easily recognizable germ layer distinctions, it is actually a very ordered tissue with distinct fate-map regions of cell fate identity. In addition, we describe a set of cell movements that appear to be unique to the tail bud region.

II. Hox Cluster Genes

The genes described in this section encode a family of proteins containing a 60-amino acid domain called the "homeobox." Homeobox-containing proteins act as transcriptional regulators to affect the expression of downstream factors. Although there are many homeobox-containing genes, a subset of these genes are clustered together in a precise order within the genome and termed the *Hox* Genes. The clustered hox genes were first described for *Drosophila melanogaster* where two groups of genes form the homeotic complex, Hom-C (Lewis, 1978; McGinnis and Krumlauf, 1992). In *Drosophila*, the Hom-C cluster is split into two major regions, the Bithorax complex and the Antennapedia complex of genes. It has been shown that these clustered genes exhibit a quality termed colinearity in that the position of the gene within the Hom-C cluster accurately predicts the expression domains of the gene within the

developing embryo. That is to say, a more 3′ located gene within the cluster has a more anterior expression limit within the embryo.

In *Drosophila*, the Hom-C genes have been shown to play an essential role in the assignment of different regional identities along the anteroposterior axis of the body. Gene knockouts of members of this complex lead to segmental transformations where, in general, a posterior region of the fly body takes on the identity of a more anterior region of the body (Lewis, 1978). For instance, deletions in the Bithorax and Utrabithorax genes of fly will cause the haltere-bearing third thoracic segment to transform into a wing-bearing second thoracic segment.

In arguably one of the most startling discoveries of modern biology, it has been shown that not only do homologues of these Hom-C group of genes exist in higher vertebrates such as chickens, mice, and humans, but also that these genes appear to play a completely analogous role in the assignment of regional identity along the vertebrate body axis. (McGinnis and Krumlauf 1992; Burke *et al.*, 1995). In the vertebrates described to date, the six homeotic genes of *Drosophila* have undergone both lateral duplications of single genes along the cluster and whole-scale duplications of the entire cluster to yield the organization shown in Fig. 1. The tetrapod Hox complex consists of 38–39 homeodomain-containing genes distributed into four separate clusters termed clusters A to D; each cluster is located on a different chromosome. Within each of the four tetrapod clusters, individual genes can be assigned according to sequence homology to one of 13 possible cognate or paralogy groups, where paralogy group 1 genes lie most 3′ in the cluster and paralogy group 13 genes lie most 5′; genes from different clusters which fall into the same cognate group are termed paralogs. As in flies, the vertebrate Hox cluster genes reported to date in frogs, chickens, mice, and humans also exhibit colinear expression patterns within the embryo. Therefore a more 3′ gene in a cluster has a more anterior expression limit in the embryo relative to a more 5′ gene. The notable exception to this rule concerns the expression domains of the group 2 paralogs; both *Hoxa-2* and *Hoxb-2* are expressed more anteriorly than their respective group 1 counterparts (Prince and Lumsden, 1994).

Hox cluster genes have more recently been isolated from the cephalochordate *Amphioxus*, a representative of a sister taxon to the vertebrates. In *Amphioxus* only one hox cluster appears to be present, comprised of at least 10 hox genes with similar genomic organization to the vertebrate paralogy groups (Garcia-Fernandez and Holland, 1994). The discovery of this single hox cluster suggests that *Amphioxus* may represent an example of a more ancestral condition in the evolution of the organization of the *hox* genes relative to vertebrates. The *Amphioxus* data have reemphasized the importance of several related issues from evolutionary, anatomical, and developmental biology viewpoints. The most in-

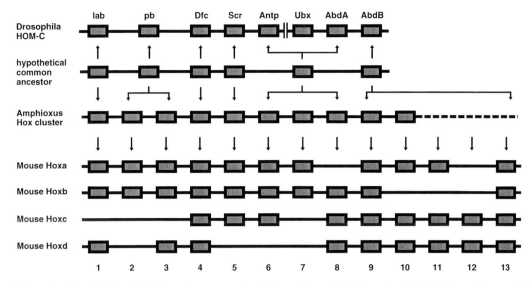

Figure 1 Organization of the Hox cluster genes in the insect *Drosophila*, the cephalochordate *Amphioxus*, and the vertebrate tetrapod mouse. Arrows and brackets define proposed lateral duplications.

triguing hypothesis to be drawn from the *Amphioxus* data concerns the time when duplications of the ancestral single hox cluster may have occurred to give rise to the four clusters described for vertebrates. The existence of only one hox cluster in *Amphioxus* suggests that this duplication event may have occurred very close to the origins of the vertebrate lineage, perhaps right at the time of divergence between the cephalochordate and vertebrate lines.

A. Why Study the Hox Clusters in Zebrafish?

Given that the Hox cluster genes have been so extensively studied especially in mice and chickens, what new insights could we hope to uncover through characterization of these genes in zebrafish? As discussed above, it is likely that the ancestral hox cluster has gone through multiple rounds of large-scale duplications in the lines leading to tetrapod vertebrates. However, the textbook generalization that all vertebrates possess four Hox clusters is actually based on a very small sampling of vertebrate species, and, until relatively recently, almost no fish species have been surveyed (Misof *et al.*, 1996; Aparicio *et al.*, 1997). Thus our knowledge of genomic organization and the evolution of this conserved gene family is incomplete. Fish are the oldest true vertebrate line; the organization of the hox cluster genes in fish could represent an organization more similar to the ancestral condition. On the other hand, fish are also the most diverse vertebrate line in terms of number of species. Therefore, different fish could possess novel arrangements of hox cluster genes reflecting different evolutionary derivations and/or selection pressures. Also, by virtue of being nontetrapods, fish form the natural outgroup to all of the vertebrates whose Hox gene clusters have been extensively studied to date, namely frog, chicken, mouse, and human, that is, tetrapods.

Our main objective for isolating the zebrafish hox genes was to provide molecular markers for different anterior–posterior regions of the body. As already mentioned, the zebrafish embryo is being increasingly used as a research preparation and having markers for different anterior–posterior identities would be instructive and helpful in a number of embryological and genetic studies. Possible differences in the expression and function of the hox cluster genes in fish versus tetrapods also could be expected based on the simplified anatomy of the fish relative to tetrapods. Therefore, one question is whether the very simplified body plan of fish is reflected in a simplified expression of the hox cluster genes along the anteroposterior body axis. To address these and related questions, we have more recently isolated a large number of the zebrafish hox cluster genes.

B. Genomic Organization of the Zebrafish Hox Clusters Is Significantly Altered Relative to Tetrapods

Before our survey, very few zebrafish hox cluster genes had been isolated and used for developmental studies. In the most ambitious study prior to our work, Misof *et al.* (1996) characterized the homeobox regions of a number of zebrafish hox cluster genes to compare homeobox sequences across species. We have extended these initial studies by utilizing a 3'RACE-PCR strategy (Frohman, 1993). This method allowed us to obtain large cDNAs suitable for production of riboprobes consisting of most of the homeobox, the entire 3' coding region, and the 3' untranslated region to the polyA tail. For many of these genes, assignment to a specific paralogy group was straightforward, being based on sequence homologies to diagnostic amino acid residues both within and outside of the homeodomain (Sharkey *et al.*,

1997). Assignment to specific clusters was determined through linkage group analyses using both mouse:zebrafish hybrid cell lines (Ekker *et al.*, 1996) and zebrafish PAC libraries.

Figure 1 shows the organization of the Hox complex in chicken, mouse, and human (tetrapods) consisting of 38–39 homeodomain-containing genes distributed into the four clusters A to D. In Fig. 2, this canonical tetrapod organization is shown with the zebrafish hox cluster genes superimposed. The placement of a small circle inside of a box represents those hox cluster genes that we have found to be present in both zebrafish and tetrapods. An empty box represents those instances where a corresponding zebrafish gene has not been found, however, an important caveat is that our searches were not necessarily exhaustive and the absence of a corresponding gene should not be taken to mean that it does not exist in zebrafish. However, it is obvious that zebrafish possess several hox cluster genes which are not present in tetrapods, both within and beyond the canonical four tetrapod clusters. These "extra" zebrafish genes are represented as large hatched circles in Fig. 2. For example, within the four tetrapod clusters A to D, zebrafish have a *hoxa8* gene, a *hoxb10* gene, and a *hoxc3* gene which are not present in chicken, mice, or humans.

Even more surprising is the existence of zebrafish hox genes which are unlinked to any of the PAC clones or mouse:zebrafish hybrid lines which were previously identified as containing members of the A, B, C, or D hox clusters. Sequence homologies have identified these genes as members of the hox cluster family, that is, they contain diagnostic residues within and outside of the homeodomain regions which have allowed paralogy as-

signments (Sharkey *et al.*, 1997). These are expressed genes, not pseudogenes, as they can be found in cDNA isolated from embryonic stages. We have identified at least two linkage groups containing extra hox cluster genes; these linkage groups represent extra hox clusters within the zebrafish genome which are separate from the canonical four hox clusters described in tetrapods. Following current terminology, we have tentatively named these extra clusters the hoxe and hoxf clusters. This terminology may need modification in the future as we collect more data on the possible derivations of the extra clusters; for instance, the hoxe appears most similar to the hoxa cluster, based on sequence and expression data. This suggests that the hoxe cluster may be a duplication of the hoxa cluster and likewise, based on the same criteria, the hoxfb gene appears to be a duplication of the hoxcb gen.

C. Expression Patterns of the Hox Genes in the Zebrafish Head Region Appear Similar to Tetrapods

Some expression patterns for the zebrafish hox cluster paralogs from groups 1 to 4 are shown in Fig. 3. In tetrapods such as chickens and mice, the hox cluster genes generally are first expressed in the caudal region of the body. Expression then extends rostrally until reaching its definitive anterior limit in the body (Gaunt and Strachan, 1994). In zebrafish, however, the onset of expression does not involve an anterior creeping of expression boundaries. Rather, the zebrafish genes are first expressed with their definitive anterior expression limit already set (Prince *et al.*, 1997a). Another general difference is that many of the zebrafish hox genes ap-

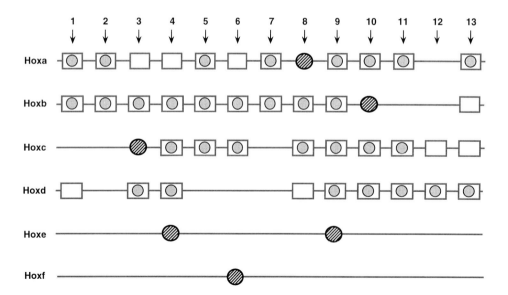

Figure 2 Organization of the zebrafish hox cluster genes. Gray boxes represent the Hox cluster genes described for the mouse. Small gray circles represent those genes found in zebrafish which have orthologues in the chicken, mouse, and human. Large hatched circles represent hox cluster genes that are unique to the zebrafish relative to tetrapod vertebrates. Note that mapping and linkage data have identified three "extra" clusters of hox genes in the zebrafish; these clusters have been tentatively named the hoxe and hoxf clusters.

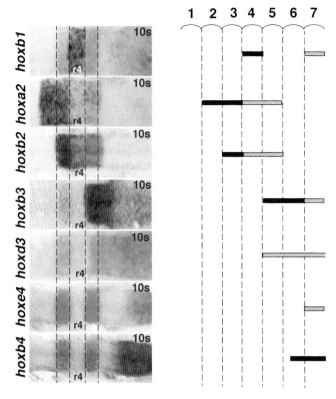

Figure 3 Expression patterns of the group 1 to 4 zebrafish hox genes within the hindbrain regions of embryos at the 10 somite stage. Embryos have been flat-mounted between coverslips after yolk removal; orientation is dorsal toward the viewer, anterior to the left. The embryos show expression of the particular gene listed on the left; in addition the embryos have been double-labeled for Krox20 which is expressed in rhombomeres 3 and 5. The Krox20 positive regions have been aligned as shown by the two sets of dashed lines and rhombomere 4 (r4) which lies between the two expression stripes of Krox20 has been designated. Alignment of the Krox20 positive regions allows a comparison of the anterior expression boundaries of the individual hox cluster genes. Anterior limits are shown also diagramatically in the right part of the figure with stronger and lighter regions of expression designated by shading. Numbers at the top of the diagram represent particular rhombomeres. As is common for other vertebrates, the group 2 genes of zebrafish are expressed with a more anterior boundary than the group 1 genes. The remaining group 3 and 4 genes have expression boundaries more posterior. Approximate width of a single rhombomere is 50 μm.

pear to possess a sharp posterior boundary of expression as well as the expected anterior boundary. Most tetrapod Hox cluster genes maintain a high level of expression anteriorly with a gradual fading of expression posteriorly to the tail region. Many of the zebrafish genes lack the lower level posterior expression domains exhibited in tetrapods, however, functional data from other organisms have shown that the Hox genes exert their influence probably only within the most anterior domains of their expression. Therefore the lack of associated posterior expression domains in zebrafish does not necessarily represent differences in function between the Hox genes of tetrapods and zebrafish.

In general, the expression patterns of the group 1 to 4 genes in the hindbrain are similar to those re-

ported from tetrapods (most of our analyses have concentrated on those genes within the canonical four clusters although at this writing we have also begun to characterize several of the extra hox cluster genes). Colinear expression limits are maintained for the group 1 to 4 genes, for instance, the paralog group 1 genes, *hoxb1*, and *hoxa1*, have anterior expression domains in r4 (where r stands for rhombomere); the group 3 genes, *hoxb3* and *hoxd3*, have expression limits further posterior at the r4/r5 boundary and the r6/r7 boundary, respectively. The paralog group 4 genes, *hoxb4* and *hoxe4*, have expression limits at the r6/r7 boundary. These expression patterns are remarkably similar to those reported to occur in the hindbrains of tetrapods with the homologous genes. It is interesting to note that even though *hoxe4* is a gene not present in tetrapods, its expression limits in the zebrafish embryo are not significantly different from those limits described for group 4 genes in other vertebrates and are therefore compatible with its assignment as a bona fide group 4 gene.

One noteworthy difference in the expression patterns of homologous genes between zebrafish and tetrapods had been previously predicted from comparative anatomical data (Gilland and Baker, 1993). In the mouse, the abducens nucleus (cranial nerve VI motoneurons) is derived exclusively from r5 cells, whereas in zebrafish the abducens nucleus is derived from both r5 and r6 cells. The *Hoxb3* gene of mouse has an anterior limit at the r4/r5 boundary, with a high level expression domain confined to r5. The zebrafish *hoxb3* gene, likewise has an anterior expression boundary at the r4/r5 boundary but zebrafish *hoxb3* has an elevated expression domain in *both* r5 and r6. Thus the high level expression domains of both mouse and zebrafish *hoxb3* genes correspond to the origins of the abducens nuclei in both species even though the actual rhombomeric derivations of the abducens differs between the two animals. This is a particularly unambiguous example where differences in neuroanatomy between two organisms are accurately reflected in differences in Hox gene expression. If one believes that changes in regional identity need to be ultimately patterned at the level of gene expression, then corresponding differences in gene expression patterns should be expected, as shown in this example. However, it still very often remains the dream of a hopeful investigator that these differences would always be logical and understandable. But to paraphrase a much used line, "Alas, such is not always the case."

D. Modified Colinear Expression Patterns of the Hox Genes in the Zebrafish Trunk

1. Expression in the Central Nervous System

Some of the patterns of the zebrafish hox cluster genes that are expressed in the trunk region are shown in Fig. 4. By the 20s stage (20s, where s stands for

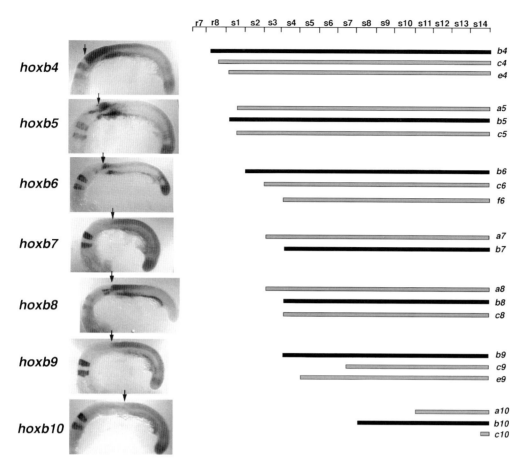

Figure 4 Expression of hoxb cluster genes in the CNS of zebrafish. Lateral views of 20s stage embryos, dorsal toward top of page, anterior to left; underlying yolk cells have been dissected away. Embryos have been colabeled with Krox20 which is visible as two stripes in the hindbrain. Arrows point to anterior expression limits of individual hox genes within the CNS. Note that most expression patterns are colinear, however hoxb-7, hoxb-8, and hoxb-9 share similar expression limits in the CNS. Diagrams to the right show the expression limits of a number of hox genes in zebrafish arranged by paralogy groups; expression limits of hoxb cluster genes are shown in darker shading. Note that hoxa-7 and hoxa-8 also share identical expression limits in the CNS. r, Rhombomere number; s, Somite number.

somite number present; zebrafish typically form 30–34 somites) the anterior limits of hox gene expression have been reached in the central nervous system (CNS) for all members of the different paralog groups. All of the embryos in Fig. 4 have been double-labeled with krox-20, a marker of rhombomeres 3 and 5 (Oxtoby and Jowett, 1993) and with the anti-myosin antibody F59 (Crow and Stockdale, 1986) to provide fixed reference points within the embryo. As with the hox genes expressed in the hindbrain, the trunk expression patterns have an obvious anterior limit with highest levels of expression in the anterior which gradually fade posteriorly. A quick listing of some representative genes shows that colinearity is generally maintained along the anteroposterior axis of the zebrafish nervous system, however, with an interesting modification displayed by those genes from the 7, 8, and 9 groups. For instance, within the b cluster group of genes: the anterior expression limit of hoxb4 is at the r6/r7 boundary, *hoxb*5 is at the anterior of s1, *hoxb*6 is at the s1/s2 boundary, *hoxb*7 is at s3/s4 boundary, and *hoxb10* is at the s7/s8 boundary (Prince *et al.*, 1997b). Thus, as in other organisms, the

more 3′ genes are expressed more anteriorly in the zebrafish embryo in an apparently conserved manifestation of colinearity for the majority of the zebrafish hox cluster genes.

A modification to the general rule of colinear expression patterns displayed by the majority of the zebrafish hox cluster genes is seen in the case of genes from paralogs 7, 8, and 9. As mentioned above, *hoxb*7 has an anterior expression limit at the s3/s4 boundary. However, somewhat surprisingly, this limit at the s3/s4 boundary is also shared by the *hoxb*8 and *hoxb*9 genes. This is surprising because the three homologous tetrapod genes have different, nonidentical expression limits which are nested along the AP axis. Therefore, in zebrafish, the shared expression limits displayed by the *hoxb*7, *hoxb*8, and *hoxb*9 genes in the CNS are a modification of the colinear expression patterns displayed by the majority of the hox cluster genes. However, these modified expression patterns are not confined to only the B cluster genes, two group A genes of zebrafish also exhibit shared expression limits. As shown in Fig. 5, *hoxa*7 and *hoxa*8 both have an anterior expression limit

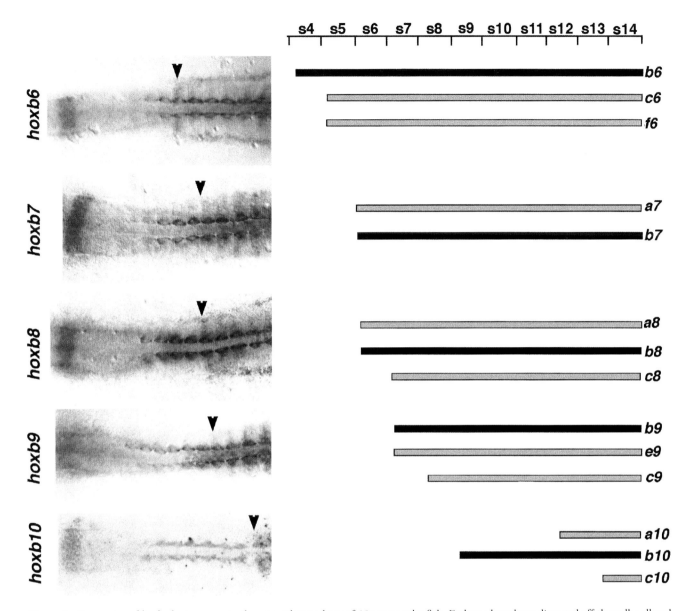

Figure 5 Expression of hoxb cluster genes in the paraxial mesoderm of 10s stage zebrafish. Embryos have been dissected off the yolk cell and flat-mounted between coverslips. Orientation is dorsal toward the viewer, anterior to the left. The Krox 20 expressing rhombomere 5 is just visible in each figure at the left side. In addition, embryos have been labeled with the f59 antibody which recognizes the darkly stained somites. Arrows designate the anterior expression limit of individual hox cluster genes in the lateral region of the mesodermal plate. Note that the expression limits of the *hoxb-7* and *hoxb-8* are identical, however *hoxb-9* is further posterior. Diagram to the right shows expression limits for a number of hox genes in the trunk region. Expression limits of hoxb cluster genes are shaded darker. Note that expression limits for *hoxa-7* and *hoxa-8* are also identical in the mesoderm. s, Somite number.

within the CNS adjacent to the s2/s3 boundary. Thus, similar to the situation for the *hoxb7*, *hoxb8*, and *hoxb9* genes, the *hoxa7* and *hoxa8* genes share identical anterior expression limits in the CNS.

2. Expression in the Paraxial Mesoderm

Many of the zebrafish genes from paralog groups 4–13 also are expressed in the paraxial mesoderm. In general, the mesodermal expression domains are expressed early and reach their anterior expression limits by the 10s stage (Fig. 5). Many of these genes have sharp expression boundaries in the somitic mesoderm

and the expression patterns are generally colinear as expected among different paralog groups. For instance, again using the zebrafish B cluster genes as an example: *hoxb6* has an anterior limit at s4, *hoxb7* at s6, *hoxb8* at s6, *hoxb9* at s7, and *hoxb10* at s9 (Prince *et al.*, 1997b). Once again, note that in the zebrafish different paralog members appear to share identical anterior limits. As mentioned above, this is apparently a different situation to what has been described in tetrapods where different paralog members of the hox cluster genes have nested, nonidentical expression limits. Thus, as in the CNS, the anterior expression limits for zebrafish *hoxb7* and *hoxb8* are shared, but unlike the situation in the CNS, *hoxb9* does not share this limit. For the A cluster genes, *hoxa*7 is expressed in the paraxial mesoderm with an anterior limit at s6 and *hoxa*8 is also expressed with a limit at s6. Therefore, similar to their expression in the CNS, *hoxa*7 and *hoxa*8 share an identical anterior expression limit in the mesoderm.

Why would the expression patterns of the hox cluster genes be modified in the zebrafish? In tetrapod vertebrates like frog, chicken, and mouse, the Hox cluster genes exhibit strict spatial colinearity whereby genes lying more 3' in a Hox cluster have more anterior expression limits than those genes lying more 5' within the cluster. However, in the zebrafish, it is clear that members of the paralog group 7 and 8 genes share equivalent anterior expression limits; in addition, the limits of the group 6 paralogs have dispersed only a very short distance along the AP axis. One simple explanation may be that this lack of dispersed expression represents a remnant of an ancestral condition. The group 6, 7, and 8 genes are hypothesized to derive from lateral duplications of a common ancestral gene (Fig. 1). One proposal consistent with our expression results in zebrafish would be that in an ancestral organism, the newly duplicated group 6, 7, and 8 genes may originally have had completely overlapping, identical expression limits. In fish, the oldest vertebrate group, these expression domains have only shifted slightly with respect to the group 6 genes and not at all with respect to the group 7 and 8 genes. In more derived vertebrates such as tetrapod mammals, these expression limits have diverged to a much greater extent and the 6, 7, and 8 paralogs have nonidentical, yet colinear, expression limits. It would be informative to compare the known patterns in vertebrates with expression data from the group 6, 7, and 8 genes of *Amphioxus* which has a single hox cluster consisting of at least 10 paralogs (including 6, 7, and 8) and is the closest living representative of a sister taxon to vertebrates. In this scheme, the group 6, 7, and 8 genes of *Amphioxus* may share identical limits pointing to a less derived nature from the common ancestral gene or genes. Unfortunately, expression data for these genes in *Amphioxus* has not yet been reported.

Another possible reason for the apparent lack of greater dispersal of the 6, 7, 8, and 9 expression patterns of the genes could be due to the simplicity of the zebrafish body plan relative to tetrapods. The first obvious simplification is that zebrafish possess fewer pre-anal somites (~ 17) relative to many tetrapods (~ 30 for mammals). It has been suggested that the posterior limit of hox cluster gene function is fixed at the level of the anus (van der Hoeven *et al.*, 1996), and in fact, no reported hox cluster genes have anterior expression limits in the post-anal body region. Therefore, the compaction or modified colinearity of expression patterns exhibited by the zebrafish may in part be due to the fact that zebrafish has approximately half the number of

segmental units over which to disperse Hox gene expression patterns relative to amniotes. A second obvious simplification of the zebrafish body plan is dramatically manifested by morphological transformations in vertebral structure. The two most studied areas of the body with respect to hox gene function have been the CNS and the sclerotome-derived vertebrae. In tetrapods the axial skeletal elements are characterized as belonging to one of five distinct subclasses: cervical, thoracic, lumbar, sacral, or coccygeal. In fish, the axial vertebrae are classified into only two subdivisions, namely trunk vertebrae consisting of a centrum articulating with a neural arch dorsally and ribs ventrally, and tail vertebrae, in which the ribs are replaced by ventral hemal arches (Kent, 1992; van Eeden *et al.*, 1996). Therefore, at least for the case of skeletal vertebrae, the anatomy of the fish has undergone fewer diversifications and modifications along the AP axis. However, this becomes a classic chicken or egg problem: is this a case where the functions of the zebrafish hox genes have not been required to diverge because the body plan of the fish has remained simple, or, alternatively, is this a case where the body plan of fish has remained simply due to the fact that the zebrafish hox genes have not diverged in their functions? Of course, this is a gross simplification of how actual changes take place in the evolution of body structure, however, it is a helpful framework in which to place our thinking about the functions of the Hox cluster genes in AP patterning of different body regions.

This brings up a seeming paradox when discussing the zebrafish hox cluster genes. To whit, the argument could be made that the body plan of the fish is simpler relative to tetrapods. In the paragraph above, this simpler body plan of fish appears to be mirrored in the less diversified expression limits of some of the hox cluster genes. The Hox cluster genes are postulated to be important in the assignment of different regional identities along the AP axis; however, it would appear strange that zebrafish, which arguably might require fewer different regional AP identities, actually possess many more hox cluster genes than tetrapods. Could it be that the hox cluster genes of the zebrafish have gone through extra duplications and the fish has simply been able to fix this extra level of redundancy into its developmental repertoire or could the "extra" hox genes be assuming new divergent functions? The assumption of new functions is not without precedent. Cephalochordates (*Amphioxus*) are the chordate sister group to vertebrates which diverged about 450–500 million years ago. *Amphioxus* has a single hox cluster consisting of at least 10 hox gene paralogs (Garcia-Fernadez and Holland, 1994). *Amphioxus* with its single cluster could specify at most probably 13 different compartments along the AP axis. Tetrapods, with their duplications of clusters, possess 38–39 hox cluster genes. Though most of these genes still roughly recognize the original boundaries of the original 13 compartments, that is, the 13 paralog groups roughly adhere to 13 roughly defined expression domains, each of the separate 38–39 genes display subtle yet important and unique modifications of their expression patterns. For instance, though two paralogous genes may share the same anterior boundary they may maintain expression within different dorsoventral regions. Hence, the modern tetrapod actually has the potential to form 38–39 unique "compartments" of possible functional differences. In fact, mouse knock-outs of single Hox cluster genes in some cases display only very subtle defects or changes which very likely represent the loss of only the unique functions of that particular gene; the redundant functions of the other members of that paralog group may prevent more deleterious effects from being expressed (Fromental-Ramain *et al.*, 1996; Peichel *et al.*, 1997). The zebrafish, with its extra duplications of whole hox clusters, possesses a greater number of hox cluster genes than tetrapods. These extra zebrafish genes could theoretically form the basis for a larger number of unique compartments; this number is currently represented by the 46 zebrafish genes isolated to date. Among those zebrafish genes which share identical anterior boundaries, differences can easily be seen in the dorsoventral or mediolateral extent of expression.

Given the possibility of a very large number of expressed zebrafish hox cluster genes, future studies will be performed with an eye toward careful analyses of the subtle differences in displayed expression patterns especially among related paralogous genes. Characterization of these differences may allow us to read the course whereby duplicated and initially redundant genes, perhaps released from some of the selective pressure to maintain an ancestral function, may diverge into new and unique functions. One hypothesis which we hope to test is whether the paralog genes at either end of the zebrafish clusters, the paralog 1 and 13 genes, may be assuming a broader range of boundaries. In tetrapods the range of hox gene anterior limits are bounded by the hindbrain anteriorly and the anus posteriorly. In the zebrafish future work will be directed to determining if the extra group 1 genes have extended their expression limits anterior to the hindbrain or if group 13 genes have expression limits that are located post-anally.

III. Posterior Body Development

A. Different Patterning Mechanisms in the Vertebrate Posterior Body

To date, the patterns reported for the tetrapod Hox cluster genes have expression limits which reside at diagnostic positions within the hindbrain and trunk regions. It has been argued, possibly due to their shared role in patterning the genitalia, that the Hox genes may be constrained to have expression limits only pre-anally (van der Hoeven *et al.*, 1996). To date, we have not found any of the zebrafish Hox cluster genes to have anterior limits of expression within the tail region, consistent with data obtained from other vertebrates. Thus, while the Hox cluster genes are useful markers for different AP identities in the head and trunk, they do not recognize specific boundaries in the tail region and are therefore not useful as expression markers of different identities within the body regions posterior to the anus. It is possible that a different gene family may be responsible for setting up regional differences in the tail, perhaps even a gene family whose functional expression is confined to the tail region only. The expression patterns of at least two genes, that of the caudal and the even skipped genes, have been shown in zebrafish to be present only within the tail bud region (Joly *et al.*, 1992; Joly *et al.* 1993).

The posterior body or tail region of the vertebrate embryo has, in fact, presented a formal controversy to developmental biologists for almost 80 years. In the 1920s, Holmdahl proposed that the vertebrate body is formed by two separate and distinct programs of development. The first program, which he termed primary body development, leads to the formation of the head and the trunk while a separate process, termed secondary body development, is responsible for the formation of the tail. According to Holmdahl, the tissues of the head and trunk derive "indirectly" through the formation of the intermediary germ layers of ectoderm, mesoderm, and endoderm, but the anatomically similar tissues of the tail form "directly" from a pluripotent blastema he called the tail bud (Holmdahl, 1925). Other investigators, especially Pasteels, disagreed with the view that tail formation was a separate process (Pasteels, 1943). They maintained that tail formation was an extension of the processes initiated during gastrulation and was therefore developmentally related to the formation of the trunk region. Several workers, most notably Schoenwolf (1977, 1981), have made outstanding contributions toward this issue by comparing serial sections of different-staged embryos, however, knowledge of the cell movements accompanying tail development have been lacking. More recently, the availability of molecular markers has allowed a reexamination of the issue of whether the tail region formed out of a homogeneous blastema as Holmdahl proposed or if tail formation was a continuation of gastrulation as Pasteels proposed. For instance, Gont *et al.* (1993) investigated the expression domains of two genes by *in situ* analysis, *Xbra* and *Xnot2*, in the *Xenopus* embryo and were led to conclude that the frog tail bud could indeed be histologically defined into separate regions, thereby weakening the argument that the vertebrate tail bud exists as a homogeneous blastema.

B. Fate Map of the Zebrafish Tail Bud

We have used a variety of techniques in an attempt to determine how anteroposterior, as well as dorsoventral, patterning occurs within the zebrafish tail region. The advent of new vital cell marking and microscopy techniques allows the detailed examination and recording of individual cell movements within a living embryo. The zebrafish, because of its optical transparency, is ideally suited to cell labeling and time-lapse recording techniques. We have taken advantage of these methods to carry out fate map and cell movement analyses of the developing tail bud region in the zebrafish (Kanki and Ho, 1997).

Briefly, by labeling single cells within the tail bud at the end of gastrulation, we have determined that cells in specific regions of the tail bud will undergo highly stereotyped patterns of cell movements. The zebrafish

tail bud is derived from marginal cells which meet at the yolk plug closure at the end of the classically defined gastrulation stage. Although there are cells all around the margin region at this time, two main groupings of cells at the respective dorsal and ventral aspects of the embryo meet at the yolk plug. The larger dorsal group of cells is contiguous with the anterior body axis and forms the anterior half of the tail bud whereas the ventrally derived group of cells forms the posterior half of the tail bud (Fig. 6). Our analyses show that these two groups of cells do not intermingle extensively but rather maintain two distinct domains which are adjoined roughly at the location of Kupffer's vesicle. Kupffer's vesicle is a teleost-specific structure whose function has been long debated, but it serves as an easily recognizable landmark for morphological studies (Laale, 1985).

A fate map for the tail region is shown in Fig. 7. What is immediately obvious is the striking difference in

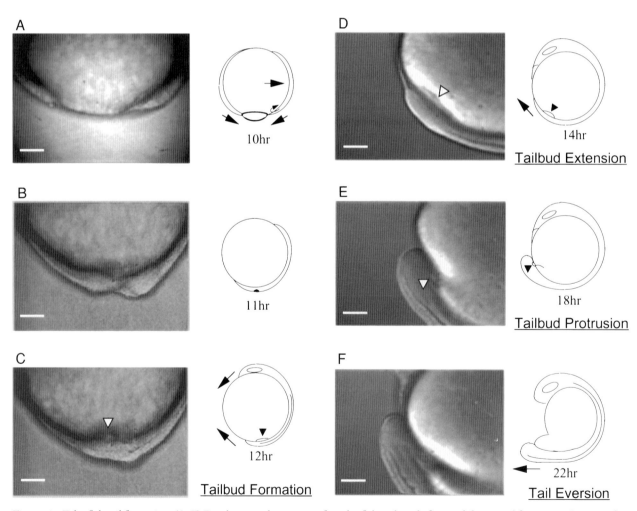

Figure 6   Zebrafish tail formation. (A–F) Developmental sequence of a zebrafish embryo before and during tail formation. Arrows indicate the direction of cell movements, whereas arrowheads indicate the position of Kupffer's vesicle. The right side of each panel shows a drawing of the entire embryo at the indicated developmental stages while the left side shows a closer view of the corresponding tail forming regions in developing embryos (bar: 100 μm). The cell movements at the end of gastrulation are shown in A, and the fusion of the blastoderm margin is shown in B. In C, cells to the left of the arrowhead constitute the posterior tail bud, whereas cells immediately to the right of the arrowhead constitute the anterior tail bud region. Specific stages of zebrafish tail development are indicated (C–F).

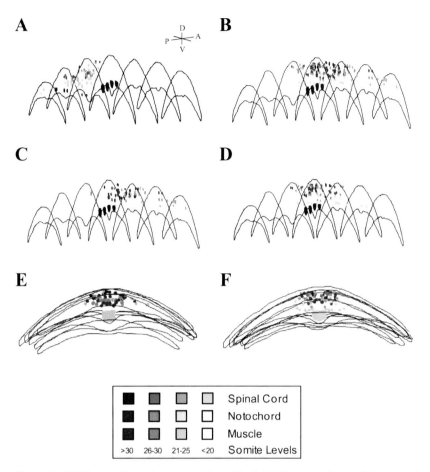

Figure 7 Cell fates within subregions of the tail bud. (A) Positions of ventral-derived cells that were labeled within the posterior half of the tail bud. (B) Positions of dorsal-derived cells that were labeled within the anterior half of the tail bud. (C–F) Subgroups of cells that were labeled in the anterior tail bud. A–D show slightly oblique lateral views of the tail bud model with posterior to the left and dorsal upward, as indicated by the axes in the upper right corner of A. Kupffer's vesicle is represented by four dark spots in the middle section. (E and F show posterior views of the tail bud with dorsal upward and Kupffer's vesicle shown in light pink. Cells C and E gave rise to spinal cord (blue), or notochord (purple), whereas cells in D and F gave rise to muscle tissue. Final tissue types were classified as either neural (blue), axial mesoderm (purple), or paraxial mesoderm (yellow/orange/red). The intensity of the color indicates the final anteroposterior (A–P) level such that darker shades of colors indicate more posterior positions along the A–P axis. The A–P axis was divided into four regions according to somite levels. From anterior to posterior these regions represent (1) A–P levels anterior to somite 20 (<20), (2) somite level 21–25, (3) somite level 26–30, and (4) somite level 31–35 (>30).

the tissue types generated by the distinct anterior and posterior halves of the tail bud region. The posterior half of the tail bud gives rise exclusively to paraxial mesodermal derivatives, namely, somitic muscle cells (Fig. 7A). In comparison, the anterior half of the tail bud gives rise to various cell fates including somitic muscle but also including all of the spinal cord and notochord derivatives eventually contained within the posterior body region (Fig. 7B).

In Figs. 7C–7F, we show that the anterior tail bud can be further subdivided into distinct tissue-restricted regions. The cells fated to become spinal neurons were located both medially and laterally within the most dorsal, superficial layers of the tail bud. The notochord precursors were located more ventrally and confined to the medial aspect of the tail bud region. The anlages for the spinal cord and notochord of the tail bud were both

contiguous with the already forming spinal and notochord regions, respectively, of the more developed trunk region. These results strongly suggest that the formation of CNS and notochord in the tail region may be an extension of the morphological processes responsible for formation of these tissues in the trunk region. This finding, taken by itself would no doubt make Pasteels (1943) very happy as he held to the extreme position that tail development was no different than trunk development. However, as described below, there are, in fact, significant differences between development of the trunk and tail regions.

Cells of the anterior tail bud which give rise to somitic mesodermal derivatives were located laterally in both dorsal and ventral tissue layers (Figs. 7C, F). Interestingly, many of the muscle cell precursors are intermingled with spinal neuron precursors in the dorsal, lateral regions of the tail bud. Despite their colocalization, cells of this region were never observed to give rise to mixed fates; that is, labeled cells formed either all muscle cell derivatives or neuronal derivatives but never both. This colocalization of mesodermal and ectodermal fates is somewhat unusual, (although it has been reported that a similar intercalated mixture of germ layer fates might possibly exist within the embryonic shield region of the zebrafish at gastrulation, Shih and Fraser, 1995; see Chapter 25 by Fraser). However, an intercalation of different fates within the tail bud would constitute a region with the appearance of a blastema without separated germ layer distinctions. This finding would lend support to the views of Holmdahl (1925) who believed that tail formation was through a "direct" type of development without the prior formation of germ layers and that the different tissues of the posterior body derived from a blastema-like region of the tail bud.

Thus, in agreement with the finding that the vertebrate tail bud has defined boundaries of molecular expression patterns (Gont *et al.*, 1993), we have found that the zebrafish tail bud does indeed have regions of defined cell fate distinctions. However, in the early tail bud, there are also regions where different germ layer fates are initially colocalized and interspersed among each other. It seems that the tail bud of the zebrafish forms not according to either of the two extreme hypotheses suggested by Holmdahl (1925) or Pasteels (1943). We have found that there are regions of the tail bud that are contiguous with already formed tissues of the more anterior trunk (such as the notochord and spinal cord anlage) as well as regions of mixed fates not dissimilar to a blastema (such as the dorsolateral region which gives rise to both muscle and neuronal derivatives). We have decided that one very important way to further study this very interesting region of the vertebrate body is to carefully follow the movements of individual cells within the living embryo. The section below outlines some of the cell movements that are responsible for formation of the posterior body of the zebrafish embryo.

C. Unique Cell Movements within the Developing Tail Bud Region

A time-lapse analysis of labeled cells within a live zebrafish tail bud is shown in Fig. 8. In the descriptions below, the terms anterior and posterior refer to specific *regions* of the tail bud as defined in the section above and the terms rostral and caudal refer to *directions* along the body axis. Three intracellularly labeled cells and their progeny have been colored in order to show the dynamic movements associated with tail bud outgrowth. The blue cells were initially positioned in the posterior, lateral tail bud region. The green cells were positioned further anterior in the tail bud and the yel-

low cells were located in the most anterior portion of the tail bud as well as the most dorsomedial. In timelapse, it is easy to follow the movements and mitotic divisions of the cells. The yellow and green cells could be seen to move caudally past the blue cells during tail bud extension. At the stage when the tail bud begins to protrude off the yolk mass, the green cells moved from their more dorsomedial position into a more lateral position and stopped advancing further caudal. The yellow cells, which initially were located most rostrally as well as most medially, continued their movements out into the most caudal tip of the extending tail region. Thus by comparing panels A and D, it can be seen that the three groups of cells have reversed their relative anteroposterior positions during tail bud outgrowth.

Therefore, the extent of movements of individual cells into the tail region appears correlated with mediolateral as well as initial anteroposterior position of the cell in the tail bud. Although the yellow cells were initially located most anteriorly in the tail bud, they eventually moved furthest posteriorly into the extending tail. However, the yellow cells were also the most medially (as well as dorsally) located cells as shown in the bottom panel of Fig. 8. The green cells were located more laterally than the yellow cells and they also moved into the extending tail region, but to a lesser extent. This example shows that there are different domains of movements within the tail bud region. Cells within the dorsomedial region of the tail bud will move furthest into the extending tail; however, cells leave this strong domain of movement by "falling out" laterally to either side, as the green cells appear to do.

Whereas the yellow and green labeled cells were initially located in the anterior tail bud, the blue cells were initially located in the posterior tail bud region. As described above, the posterior tail bud is derived exclusively from a particular group of cells located on the most ventral side of gastrula-staged embryo. These posterior tail bud cells do not intermingle with the anterior tail bud cells. The posterior tail bud cells undergo a very unique set of movements which place them into more rostral locations even as the majority of tail bud cells are extending in the caudal direction. Fate map and time-lapse data clearly show that the medial tissues of the notochord and spinal cord extend into the tail region concomitant to, and perhaps driving, tail bud extension in the caudal direction. During this extension, the cells of the posterior tail bud are pushed under and to either side of the extending medial tissues, not unlike the physical movements of water being displaced by the bow of a boat. This movement has two consequences: one, the cells of the posterior tail bud are placed into a less superficial position within the future hypoblast region which gives rise to mesoderm and two, the posterior tail bud cells are split to either side of the medial tissues and migrate in a relatively rostral direction along

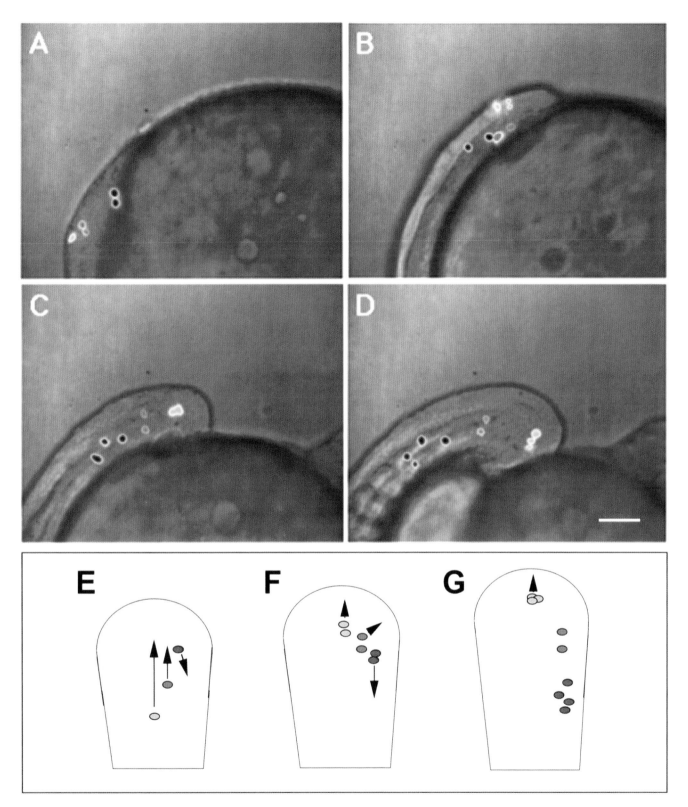

Figure 8 Movements of single cells in the developing tail. (A–D) Images from a time-lapse sequence in which the movements of three labeled cells were recorded as the tail developed. The tail is viewed laterally as it extends posteriorly along the yolk cell toward the upper right. Dorsal is toward the upper left. Clones of these cells were colored (yellow, green, or blue), in order to distinguish them from each other. These images show tail development from approximately 15.5 hours postfertilization (h.p.f.) until 20 h.p.f. with about 1.5 hours between each image (bar: 50 μm). (E–G) Dorsal views of the tail bud summarizing the relative positions and movements of these clones. The colors of these cells are the same as in A–D. The direction of cell movements are indicated by the arrows.

the most lateral aspect of the body axis. We have termed this movement "subduction" in recognition of this unusual form of mass ingression of the posterior tail bud cells (Kanki and Ho, 1997). As already mentioned, the cells of the posterior tail bud give rise exclusively to lateral somitic mesoderm; this shared fate is understandable given that the subduction movement places all of these cells into deeper and more lateral positions within the body axis.

We have used the term subduction to distinguish between the tissue-wide movement of the entire posterior tail bud and the ingression of single cells which occurs in the anterior tail bud region (Fig. 9). Both of these movements, subduction and ingression, are correlated with the specification of cells to form the paraxial mesoderm of the tail region. Subduction movements as described above place the entire group of posterior tail bud cells into the mesodermal layer of the tail. However, in the anterior tail bud region where ectodermal and

mesodermal precursors are initially intermingled together, the future muscle derivatives are separated from the future neuronal derivatives by the ingression movements of individual cells into deeper layers of the tail. Thus, whereas tissue-wide involution movements around the marginal regions of the gastrula-stage embryo appear responsible for the major formation of the mesendodermal germ layer of the head and trunk regions, the mesendodermal germ layer of the tail region is not formed by classically defined involution movements but by the two very different mechanisms of subduction in the posterior tail bud and ingression in the anterior tail bud (Fig. 9).

IV. Summary

In this chapter we have outlined some of our work to elucidate how anteroposterior patterning information is

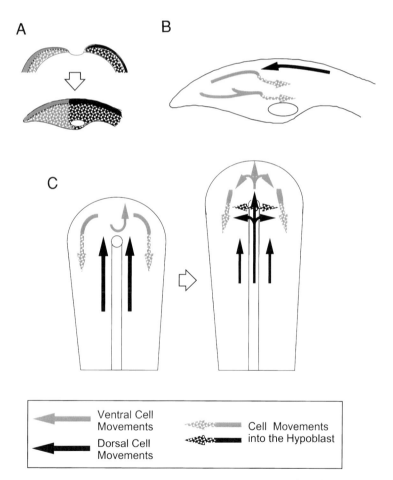

Figure 9 Summary of tail bud cell movements. (A) Fusion of the blastoderm margin over the yolk plug during tail bud formation. The black color indicates cells from the dorsal side of the embryo (right), and the gray color indicates the ventrally derived cells (left). The solid color represents superficial cell regions, and the stippled pattern represents the hypoblast. The top drawing shows a lateral view of the late gastrula embryo with the vegetal pole upward. The lower drawing shows the tail bud after blastopore closure and the respective contribution of dorsal and ventral-derived cells to the anterior (right) and posterior (left) halves of the tail bud. (B) The ensuing subductive movements of cells within the newly formed tail bud. The gray and black arrows indicate the direction of cell movements from their respective halves of the tail bud. The half-stippled arrows indicate the transition of cells from superficial regions (solid color) into the deeper hypoblast (stippled). (C) The general movements of cells in dorsal views of the tail bud during its formation (left) and extension (right). Posterior is upward. Movements of cells are represented by arrows as described above.

imparted along the embryonic body axis of the zebrafish. Toward this goal, we have isolated and characterized a number of the hox cluster genes from the zebrafish. Unexpectedly, the zebrafish hox cluster genes exhibit several unusual features relative to their tetrapod counterparts. First, mapping and linkage studies have shown that the zebrafish genome contains extra hox cluster genes relative to tetrapods. Some of the extra zebrafish hox genes reside within the canonical four tetrapod clusters, however, other zebrafish hox genes appear to be organized into separate new clusters; to date we have identified two novel zebrafish clusters and named them the hox<u>e</u> and the hox<u>f</u> clusters. Second, some of the hox cluster genes' expression patterns appear to be modified relative to tetrapods. The expression patterns of some zebrafish hox cluster genes appear to be less dispersed relative to tetrapods such that some genes within a cluster appear to share common anterior boundary limits. Therefore, in the case of zebrafish, an organism with a much simpler body plan relative to most tetrapods, there appears to be less dispersal of hox gene expression boundaries along the AP axis. Finally, this situation presents something of a paradox: if the zebrafish embryo exhibits fewer AP distinctions relative to the tetrapod body plan, then it would logically need fewer hox genes rather than more. So what are the extra zebrafish hox cluster genes doing? We are currently investigating the possibility that the hox cluster genes of the zebrafish may have taken on novel functions to pattern different tissues of the zebrafish embryo, possibly the anterior head regions or midline structures such as the notochord.

In order to characterize axis formation in the posterior body of the zebrafish, we have used cell lineage and fate-mapping techniques to follow the development of cells within the tail bud. We have found that different regions of the tail bud are separated into specific cell fate and cell movement domains. These movements are much more complex than we had first imagined. For instance, extensive ingression and subduction movements are responsible for placing cells within the mesendodermal layers of the tail region, substituting for the involution movements which in large part characterize gastrulation movements in the head and trunk regions. In addition, time-lapse data of individually labeled cells within the tail bud region show that an unusual reversal of anteroposterior position occurs during the tail bud extension phase: cells of the anterior tail bud will migrate further caudally into the tail region whereas cells of the posterior tail bud will migrate less caudally and eventually become located in relatively more rostral positions.

Characterization of hox gene expression patterns in the head and trunk region and also the description of a fate map for the tail region have given us a foundation for studying how anteroposterior patterning information may be imparted in the zebrafish embryo. With this knowledge of wild-type expression patterns and cell movements, we may now begin a survey of available zebrafish mutants which show patterning defects along the anteroposterior aspect of the head, trunk, or tail regions. In particular, we are interested in the cellular interactions that may play a role in the proper expression of the hox cluster genes or may be responsible for setting up an organizer region in the tail bud. The versatility of the zebrafish embryo as a biological preparation will allow us to study these problems using a variety of techniques including cell lineage analyses, cell transplantation methods, the characterization of molecular markers for anteroposterior patterning information, and the analyses of developmental mutants that lack specific regions of the body plan.

Acknowledgments

We thank Tracy Roskoph for technical assistance, fish care, and snippets of information from the outside world. Work from our laboratories was supported by a donation from the Rathmann Families Foundation to the Department of Molecular Biology at Princeton University, also by the Lucille P. Markey Charitable Trust Foundation, by the European Molecular Biology Organization, by the International Human Frontiers Science Program Organization, by the March of Dimes Birth Defects Foundation, by the Rita Allen Foundation, by the National Science Foundation, by the National Institutes of Health, and by the Canadian Genome Analysis and Technology Program.

References

Aparicio, S., Hawker, K., Cottage, A., Mikawa, Y., Zuo, L., Venkatesh, B., Chen, E., Krumlauf, R., and Brenner, S. (1997). Organization of the *Fugu rubripes Hox* clusters, evidence for continuing evolution of vertebrate *Hox* complexes. *Nat. Genet.* **16**, 79–83.

Burke, A. C., Nelson, C. E., Morgan, B. A., and Tabin, C. (1995). *Hox* genes and the evolution of vertebrate axial morphology. *Development* (*Cambridge, UK*) **121**, 333–346.

Crow, M. T., and Stockdale, F. E. (1986). Myosin expression and specialization among the earliest muscle fibers of the developing avian limb. *Dev. Biol.* **113**, 238–254.

Ekker, M., Speevak, M. D., Martin, C. C., Joly, L., Giroux, G., and Chevrette, M. (1996). Stable transfer of zebrafish chromosome segments into mouse cells. *Genomics* **33**, 65–74.

Frohman, M. A. (1993). Rapid amplification of complementary DNA ends for generation of full-length complementary DNAs: Thermal RACE. *In* "Methods in Enzymology" (R. Wu, ed.), Vol. 218, pp. 340–356. Academic Press, San Diego.

Fromental-Ramain, C., Warot, X., Lakkaraku, S., Favier, B.,

Haack, H., Birling, C., Dierich, A., Dolle, P., and Chambon, P. (1996). Specific and redundant functions of the paralogous *Hoxa-9* and *Hoxd-9* genes in forelimb and axial skeletal patterning. *Development (Cambridge, UK)* **122**, 461–472.

Garcia-Fernandez, J., and Holland, P. W. H. (1994). Archetypal organization of the amphioxus *Hox* gene cluster. *Nature (London)* **370**, 563–566.

Gaunt, S. J., and Strachan, L. (1994). Forward spreading in the establishment of a vertebrate Hox expression boundary: The expression domain separates into anterior and posterior zones, and the spread occurs across implanted glass barriers. *Dev. Dyn.* **199**, 299–240.

Gilland, E., and Baker, R. (1993). Conservation of neuroepithelial and mesodermal segments in the embryonic vertebrate head. *Acta Anat.* **148**, 110–123.

Gont, L. K., Steinbeisser H., Blumberg B., and De Robertis E. M. (1993). Tail formation as a continuation of gastrulation: The multiple cell populations of the *Xenopus* tail bud derived from the late blastopore lip. *Development (Cambridge, UK)* **119**, 991–1004.

Holmdahl, D. E. (1925). Experimentalle Untersuchgen über die Lage der Grenze zwischen primärer und sekundärer körperentwicklung beim Huhn. *Anat. Anz.* **59**, 393–396.

Joly, J.-S., Muary, M., Joly, C., Duprey, P., Boulebache, H., and Condamine, H. (1992). Expression of a zebrafish *caudal* homeobox gene correlates with the establishment of posterior cell lineages at gastrulation. *Differentiation* **50**, 75–87.

Joly, J. S., Joly, C., Schulte-Merker, S., Boulebache, H., and Condamine, H. (1993). The ventral and posterior expression of the zebrafish homeobox gene eve1 is perterbed in dorsalized and mutant embryos. *Development (Cambridge, UK)* **119**, 1261–1275.

Kanki, J. P. and Ho, R. K. (1997). The development of the posterior body in zebrafish. *Development (Cambridge, UK)* **124**, 881–893.

Kent, G. C. (1992). In "Comparative Anatomy of the Vertebrates," Chapter 8, 7th Ed. Mosby-Year Book, Louisiana, Missouri.

Laale, H. W. (1985). Kupffer's vesicle in *Brachydanio rerio*: Multivesicular origin and proposed function *in vitro*. *Can. J. Zool.* **63**, 2408–2415.

Lewis, E. B. (1978). A gene complex controlling segmentation in *Drosophila*. *Nature (London)* **276**, 565–570.

McGinnis, W., and Krumlauf, R. (1992). Homeobox genes and axial patterning. *Cell (Cambridge, Mass.)* **68**, 283–302.

Misof, B. Y., Blanco, M. J., and Wagner, G. P. (1996). PCR-survey of Hox-genes of the zebrafish: New sequence information and evolutionary implications. *J. Exp. Zool.* **274**, 193–206.

Oxtoby, E., and Jowett, T. (1993). Cloning of the zebrafish *Krox-20* gene (*krx-20*) and its expression during hindbrain development. *Nucleic Acids Res.* **21**, 1087–1095.

Pasteels, J. (1943). Prolifèrations at croissance dans la gastrulation at la formation de la queue des Vertebrés. *Arch. Biol.* **54**, 1–51.

Peichel, C. L., Prabhakaran, B., and Vogt, T. F. (1997). The mouse *Ulnaless* mutation deregulates posterior *HoxD* gene expression and alters appendicular patterning. *Development (Cambridge, UK)* **124**, 3481–3492.

Prince, V., and Lumsden, A. (1994). *Hoxa-2* expression in normal and transposed rhombomeres: Independent regulation in the neural tube and neural crest. *Development (Cambridge, UK)* **120**, 911–923.

Prince, V. E., Moens, C. B., Kimmel, C. B., and Ho, R. K. (1998a). Zebrafish *hox* genes: Expression in the hindbrain region of wild-type and mutants of the segmentation gene, *valentino*. *Development (Cambridge, UK)* **125**, 393–406.

Prince, V. E., Joly, L., Ekker, M., and Ho, R. K. (1998b). Zebrafish *hox* genes: Genomic organization and modified colinear expression patterns in the trunk. *Development (Cambridge, UK)* **125**, 407–420.

Schoenwolf, G. C. (1977). Tail (end) bud contributions to the posterior region of the chick embryo. *J. Exp. Zool.* **201**, 227–246.

Schoenwolf, G. C. (1981). Morphogenetic processes involved in the remodeling of the tail region of the chick embryo. *Anat. Embryol.* **162**, 183–197.

Sharkey, M., Graba, Y., and Scott, M. P. (1997). *Hox* genes in evolution: Protein surfaces and paralog groups. *Trends Genet.* **13**, 145–151.

Shih, J., and Fraser, S. F. (1995). Distribution of tissue pro-genitors within the shield region of the zebrafish gastrula. *Development (Cambridge, UK)* **121**, 2755–2765.

van der Hoeven, F., Sordino, P., Fraudeau, N., Izpisua-Belmonte, J.-C. and Duboule, D. (1996). Teleost *HoxD* and *HoxA* genes: Comparison with tetrapods and functional evolution of the *HOXD* complex. *Mech. Dev.* **54**, 9–21.

van Eeden, F. J. M., Granato, M., Schach, U., Brand, M., Furutani-Seiki, M., Haffter, P., Hammerschmidt, M., Heisenberg, C.-P., Jiang, Y.-J., Kane, D. A., Kelsh, R. N., Mullins, M. C., Odenthal, J., Warga, R. M., Allende, M. L., Weinberg, E. S., and Nüsslein-Volhard, C. (1996). Mutations affecting somite formation and patterning in the zebrafish, *Danio rerio*. *Development (Cambridge, UK)* **123**, 153–164.

27

Specification of Neural Crest Cell Fate in the Embryonic Zebrafish

DAVID W. RAIBLE
Department of Biological Structure
University of Washington
Seattle, Washington 98195

JUDITH S. EISEN
Institute of Neuroscience
University of Oregon
Eugene, Oregon 97403

I. Introduction

Embryonic development can be thought of as a process during which cells acquire specific cell fates and become progressively different from one another. Cell fate defines what a cell will do in its usual environment as the normal outcome of development. Cell fate differs from cell potential, which encompasses all the possible fates a cell may undertake if it is exposed to a variety of environmental conditions in addition to those it normally experiences. Restrictions in fate occur as development proceeds, so that the progeny of a cell express only a subset of possible fates. Cell fate restrictions are experimentally defined by following individual cells and identifying times after which their progeny give rise to a more limited set of derivatives. Cell fate restriction could result from many different factors. For example, cells may no longer have access to specific signals because morphogenetic movements restrict their local environment. Alternatively, cell fate restrictions may be the result of intrinsic changes in cell potential. Restrictions in cell potential are experimentally revealed by challenging cells with new environments and deter-

mining whether they change their developmental programs.

Cell fate restrictions are the result of specification, a process in which cells acquire specific characteristics such as differential gene expression and distinct cell movements (Davidson, 1990; Kimmel et al., 1991). Specification may be conditional, since specified cells can alter their fates when transplanted to new locations (Ho, 1992; Ho and Kimmel, 1993), and are thus different from commitment events, which represent irreversible changes in cell fate (Kimmel et al., 1991). Specification results in cell fate restrictions, but not necessarily cell potential restrictions, while commitment results in cell potential restrictions and thus necessarily restrictions in cell fate.

We are interested in how cell fates are determined within the neural crest, which begins as a seemingly homogeneous population of cells on the dorsolateral aspect of the neural tube that migrates away to specific locations and makes distinct derivatives. These include neurons and glia of the peripheral nervous system, craniofacial mesenchyme that forms cartilage and bone of the face and neck, and pigment cells. In this chapter, we describe our work examining when zebrafish trunk neural crest cells undergo fate restrictions and whether these restrictions are the result of restrictions in cell potential. We have asked when cell fate restrictions occur by performing cell lineage analysis: labeling individual cells within the trunk neural crest, following them during development, recording cell divisions, and keeping track of all progeny. We have asked whether cell fate restrictions result from restrictions in cell potential by comparing the fates of individual cells from different crest subpopulations when transplanted to the same environmental conditions. These experiments have led us to propose a model of neural crest development in which regulative interactions among neural crest cells influence their fates.

II. Development of Zebrafish Neural Crest

Zebrafish neural crest is amenable to study both as a population and at the level of the single cell. Zebrafish neural crest has many fewer cells relative to other vertebrates; in the trunk there are only 10–12 neural crest cells per body segment compared to hundreds in other species (Raible *et al.*, 1992). Despite the relative simplicity of the zebrafish neural crest, almost all aspects of its development are similar to those of other vertebrates including the migration pathways it takes and types of derivatives it makes. Trunk neural crest cells generate neurons of the dorsal root ganglia (DRG) and sympathetic ganglia, glial cells and pigment cells (Raible *et al.*, 1992). Zebrafish cranial neural crest also forms cartilage and connective tissue (Schilling and Kimmel,

1994), as in other animals. In zebrafish, crest-derived cartilage precursors interact with mesodermally derived muscle precursors to generate an early craniofacial pattern (Schilling and Kimmel, 1997), much as they do in other vertebrates (Noden, 1991).

Zebrafish trunk neural crest cells segregate from the neural tube and migrate ventrally in a rostrocaudal sequence (Raible *et al.*, 1992). Neural crest cells first migrate on a medial pathway between the somite and neural tube (Fig. 1). Neural crest cells that migrate ventrally along the medial path encounter and intermingle with sclerotome cells migrating dorsally from the ventral-most somite (Morin-Kensicki and Eisen, 1997). Later, neural crest cells migrate on a lateral pathway between the somite and overlying ectoderm (Raible *et al.*, 1992), after an inhibitory signal that prevents their entry onto this pathway is removed (Jesuthasan, 1996). The timing and directions of migration of zebrafish crest correspond to trunk crest cell migration in other vertebrates. In avian embryos, neural crest cells first migrate along the medial aspect of the myotome as well as intermingle with mesenchymal sclerotome cells (Tosney *et al.*, 1994). Cells later enter a lateral path between dermamyotome and ectoderm after an inhibitory signal abates (Erickson *et al.*, 1992).

Neural crest cells in the zebrafish trunk that migrate at different times generate different derivatives (Fig. 1; Raible and Eisen, 1994). Crest cells that migrate early on the medial path produce DRG neurons as well as glia and pigment cells, while late-migrating cells on the medial path produce only nonneuronal derivatives. Cells that migrate on the lateral path appear to produce only pigment cells. In *Xenopus laevis*, cells migrating medially form similar derivatives as in zebrafish (Collazo *et al.*, 1993), while in avian embryos, cells mi-

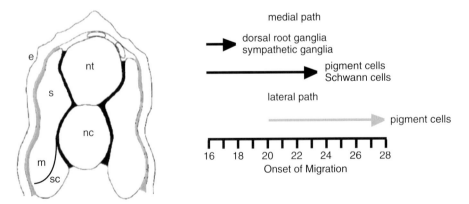

Figure 1 Neural crest cell migration in the zebrafish trunk. The diagram on the left shows a schematic cross section through the trunk of a 20-hour zebrafish embryo. Neural crest cells first migrate on a medial pathway, shown in black, between the somite (s) and neural tube (nt). Neural crest cells migrate along the medial surface of the myotome (m) and encounter sclerotome (sc) cells that migrate dorsally along the same pathway. Later, neural crest cells migrate on a lateral pathway, shown in gray, between somite and overlying ectoderm (e). The graph on the right depicts the relationship between the time of neural crest cell migration on the medial (black arrows) and lateral (gray arrow) paths and the derivative types they generate. Cells migrating early on the medial path generate all derivative types, both neuronal and nonneuronal, while cells migrating later only produce nonneuronal cell types. Cells that migrate on the lateral path generate pigment cells.

grating medially do not generate pigment cells (Le Douarin, 1982).

Study of neural crest cell fate determination has several advantages in zebrafish embryos. Zebrafish embryos are optically clear and develop rapidly outside the mother, and individual cells can be labeled and followed live through development. Individual neural crest cells can be transplanted between embryos to test how changing environment alters cell fate. These characteristics have allowed us to begin to understand how individual cells develop in the context of the neural crest population as a whole.

III. Cell Fate Restriction within the Neural Crest

Do neural crest cells undergo restrictions in cell fate? Clonal analysis *in vitro* (Sieber-Blum *et al.*, 1993; Le Douarin and Dupin, 1993) and *in vivo* (Bronner-Fraser and Fraser, 1989, 1991) has shown that individual neural crest cells can generate more than one type of derivative. In the same experiments, many neural crest cells can also give rise to clones of progeny that are all the same, suggesting that some cells have restricted fates. However, these studies could not resolve when and where these fate restrictions occurred. Moreover, tissue culture may not accurately reproduce the normal fates of crest cells or reflect when different fates first become apparent. Clonal analysis done *in vivo* typically has been analyzed in fixed and sectioned tissue, which cannot adequately describe dynamic temporal processes. These difficulties are overcome by studying cells in zebrafish embryos, in which individual cells can be followed over time and lineages constructed showing temporal sequences of divisions and phenotypes of all progeny (Kimmel and Warga, 1986).

Clonal analysis in zebrafish demonstrated that most neural crest cells are cell fate restricted before they begin to migrate. Indeed, within the zebrafish cranial neural crest all cells are fate restricted as soon as they are recognizable as distinct from the neuroepithelium (Schilling and Kimmel, 1994). When individual cells were injected with fluorescent lineage tracer, they generated progeny of all the same cell type, either neurons, glia, cartilage or pigment cells. These cell-type specific progenitors were spatially organized, with cartilage, connective tissue and pigment progenitors situated more medially than neurogenic crest cells, and glial progenitors more evenly distributed. Zebrafish cranial neural crest cells also displayed segmental restrictions, so that progeny of individual cells were confined to a single pharyngeal arch primordium.

In contrast to zebrafish cranial crest cells, some trunk neural crest cells are not fate restricted before they begin to migrate (Raible and Eisen, 1994), although the majority of cells generated only a single type of progeny. Individual cells were labeled with fluorescent lineage tracer while they were situated on the dorsolateral aspect of the neural tube before they began to migrate. Cells that gave rise to multiple types of derivatives were among the earliest crest cells to migrate from the neural tube. All later-migrating cells were fate restricted.

To resolve whether crest cells that are initially unrestricted generate fate-restricted progenitors, we carried out lineage analysis in which we labeled cells with vital dye and followed them in living embryos, keeping track of all cell divisions and the fates of all progeny (Raible and Eisen, 1994). In these experiments, we were able to construct lineages of progeny derived from individual premigratory cells (Fig. 2). Most lineages were comprised of progeny of only a single cell type, consistent with our clonal analysis. About a third of the cells generated more than one type of progeny. In most cases, after the first division the cells were fate restricted and subsequently gave rise to a single type of derivative. This lineage analysis, together with clonal analysis of zebrafish crest, suggests that neural crest cells generate different derivative types by first generating fate-restricted precursors. Clonal analysis of cultured avian neural crest cells corroborates our findings in zebrafish that neural crest cells generate fate-restricted precursors (Henion and Weston, 1997).

Neural crest cells have been proposed to go through progressive fate restrictions to generate different derivative types (Weston, 1991; Le Douarin *et al.*, 1991). In these models, as neural crest cells divide, daughter cells generate sequentially more limited subsets of crest derivatives. Pigment/Schwann cell progenitors or neurogenic progenitors have been proposed to be such intermediates. We examined the combinations of cell types within two-derivative clones to see whether such intermediate sublineages exist. We observed almost all combinations of cell types (Table I), suggesting individual zebrafish cells do not generate specific restricted intermediates.

In contrast to our results, two-derivative clones from analysis of cultured avian crest contained neuron–glia and glia–melanocyte but not neuron–melanocyte combinations (Henion and Weston, 1997). However, in zebrafish, pigment cells are derived from early-migrating and late-migrating crest cells, while in avians pigment cells are derived only from late-migrating crest cells. In both zebrafish and avians, neurons are derived from early-migrating cells. Thus, other factors controlling pigment cell differentiation in avians may account for the generation of restricted sublineages.

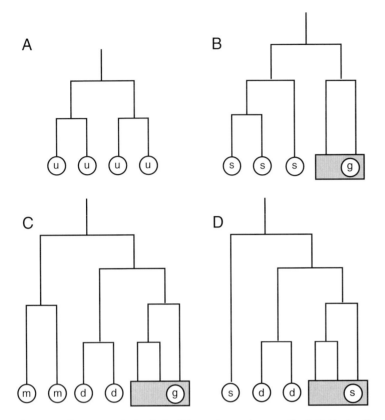

Figure 2 Lineage analysis of zebrafish trunk neural crest cells. Individual cells were labeled with fluorescent vital dye before they began to migrate from the neural tube and were followed in living embryos. (A–D) Lineages of individual cells where horizontal lines represent cell divisions and circles represent all descendants. Where progeny are boxed, individual cells could not be distinguished because divisions occurred in close proximity to one another. Most neural crest cells generated progeny of all the same phenotype (A). However, some lineages gave rise to progeny of two (B) or three (C) cell types. In these lineages, fate-restricted precursors were initially generated at the first divisions. In some lineages, progeny of the same type were not formed from the same branch of the lineage (D). However, these lineages resemble those where three restricted precursors were first formed (C), except that two precursors generated progeny of the same type. U, unmelanized pigment cell; m, melanized pigment cell; g, glial cell; d, dorsal root ganglion neuron; s, sympathetic neuron.

IV. Genetic Analysis of Zebrafish Neural Crest Development

A powerful way to examine cell fate decisions in neural crest development is to take a genetic approach. We and others have screened for mutations that disrupt neural crest derivatives in zebrafish (Henion *et al.*, 1996; Kelsh *et al.*, 1996; Schilling *et al.*, 1996a,b; Piotrowski *et al.*, 1996; Neuhauss *et al.*, 1996). Mutations were isolated that affected development of neurons, cartilage, and all three classes of zebrafish pigment cells. Most of these mutations affect a single crest-derived cell type without obviously altering other neural crest lineages. Embryos mutant for *nosedive* (Henion *et al.*, 1996) or *spike* (K. Latham and D. W. Raible, 1997, Fig. 3) are missing DRG neurons, yet have normal pigment patterning and initiate cartilage development properly. However, these mutations also affect central nervous system (CNS) development, so may specifically affect neuronal development irrespective of whether neurons are derived from neuroepithelium or from neural crest.

In embryos homozygous for *chinless*, crest-derived cartilages fail to develop while neurons, glia, and pigment cells develop normally (Schilling *et al.*, 1996a). Several mutations affect individual pigment cell types, which include the black melanophores, iridescent iridophores, and yellow xanthophores, without interfering with the

TABLE I
Distribution of Cell Types within Two-Derivative Clones[a]

	SYMP	GLIA	MEL	UNPIG
DRG	7	2	2	2
SYMP		2	0	1
GLIA			1	1
MEL				0

[a]Cell types in two-derivative clones were observed in almost all combinations, suggesting that neural crest cells do not first generate partially restricted intermediates. The table shows the number of times each fate combination (indicated by row and column headings) was observed. Clonal analysis of 154 labeled cells yielded 18 that generated exactly two types of derivatives. For details, see Raible and Eisen (1994).

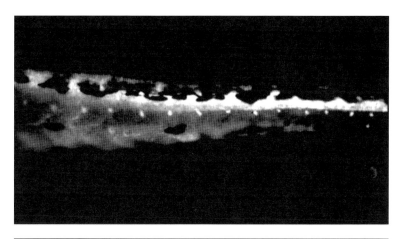

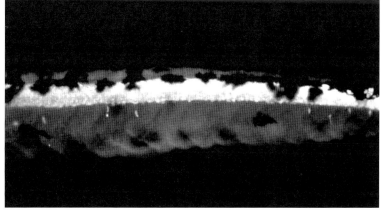

Figure 3 *spike* mutant embryos are missing dorsal root ganglia. Embryos were fixed and stained with anti-Hu antibodies, which identifies neurons. In wild-type embryos (top), the dorsal root ganglia are segmentally reiterated, and contain 2–3 neurons at this stage. Ganglia are missing variably in spike mutants (bottom). ×100

development of other crest derivatives. These include *salz* and *pfeffer*, which reduce the number of xanthophores (Kelsh *et al.*, 1996; Odenthal *et al.*, 1996), *shady*, which reduces the number of iridophores (Kelsh *et al.*, 1996), and *nacre*, which eliminates melanophores (Lister *et al.*, 1998; Fig. 4). All of these pigment mutations are viable as homozygous adults.

If neural crest cells produce intermediate sublin-

eages that give rise to specific subsets of derivative types, one might expect to find mutations that specifically affect the cell types generated from such a sublineage. For example, if there were a precursor cell that gave rise to all pigment cell types, it might be possible to isolate mutations in a gene that affects all pigment lineages. The mutation *colourless* appears to be such a gene; in *colourless* mutant embryos, melanophores, xan-

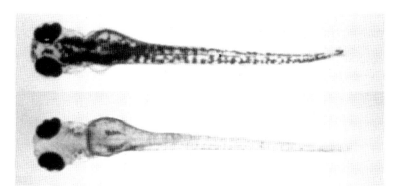

Figure 4 *nacre* mutants are missing melanophores. A dorsal view shows wild-type (top) and *nacre* (bottom) mutant embryos. Eye pigmentation, which is not generated by neural crest, is normal in *nacre* mutants. ×50

thophores, and iridophores are all greatly reduced (Kelsh et al., 1996). However, neuronal derivatives are also affected in mutants (R. N. Kelsh, 1998, personal communication), suggesting that *colourless* might be a more general regulator of neural crest development. Analyses of mutations isolated so far suggest that most identify regulators of individual crest-derived cell types but these mutations have not, as yet, revealed specified intermediates. However, none of the screens performed to date has been saturating for crest phenotypes, meaning that all possible mutations were not identified under the given screen paradigm (Haffter et al., 1996; Henion et al., 1996), and leaving open the possibility that such genes may be identified in future screens.

Further analysis of mutations may reveal when and where regulatory genes function, thus helping to identify when fate choices are being made. Mosaic analysis, transplanting cells between wild-type and mutant embryos, will identify whether genes act cell-autonomously within the neural crest. The eventual molecular identification of these genes will reveal when they are initially expressed. The genetic approach thus yields promise for greater understanding of mechanisms behind cell fate decisions.

V. Cell Potential Restriction within the Neural Crest

Do neural crest cells undergo restrictions in cell potential? Transplantation studies have been interpreted to support both heterogeneity and homogeneity of cell potential within the neural crest. Neural crest populations from embryos of different stages behave differently when placed under the same environmental conditions *in vivo* (Artinger and Bronner-Fraser, 1992; Erickson and Goins, 1995). Similarly, differences in fates of neural crest cells along the body axis have been proposed to be cell-intrinsic. Although neural crest from all axial levels produces several cell types in common, including neurons, glia, and pigment cells, crest-derived ectomesenchyme is unique for each axial level (Hall and Horstadius, 1988). Cranial crest ectomesenchyme forms skeletal components, cardiac crest generates heart ectomesenchyme, and in amphibians and fish, trunk crest forms dorsal fin mesenchyme. When neural crest from one axial level is replaced with crest from another level, ectomesenchymal structures do not develop properly, and in some cases complete structures appropriate for the level of the donor crest are formed, suggesting that crest-derived ectomesenchyme is committed to generate specific structures (Le Douarin, 1982; Noden, 1983; Kirby, 1989; Graveson et al., 1995). In contrast, cranial crest is able to contribute to trunk ganglia and trunk crest is able to contribute to cranial ganglia (Le Douarin, 1982). Some transplanted cells are also plastic with respect to regionalized homeobox gene

expression (Birgbauer et al., 1995; Saldivar et al., 1996). Heterochronic transplantation of cranial neural crest cells suggests that timing of migration controls final embryonic position (Baker et al., 1997).

It is important to note that experiments testing neural crest potential by transplantation have been performed on populations of cells, not individuals (e.g., Baker et al., 1997). If cell populations are heterogeneous, different individuals among the population may act differently under test conditions, yet, taken as a whole, the outcomes for the population may be the same. For example, if a population contains two types of cells, neuronal precursors and pigment cell precursors, when placed in conditions favoring neurons the neuronal precursors may proliferate and when placed under conditions favoring pigment cells the pigment precursors may proliferate. If analyzed as a whole, the population would develop neurons under one condition and pigment cells under another, and if assumed to be homogeneous the population would be described as multipotent. However, if analyzed as individual cells, the population would be revealed to be heterogeneous, and a subset of cells would respond under each condition. It is important to realize that each cell fate decision is made by an individual cell, not by the cell population. Furthermore, a growing body of work demonstrates that cells behave differently when transplanted singly than they do when transplanted in groups (Ho, 1992), a phenomenon commonly referred to as a community effect (Gurdon et al., 1993). Thus transplanting populations of cells carries not only individual cells but also some of their neighbors, and thus some of their original environment, to the new position.

We therefore sought to test neural crest cell potential by transplanting individual cells instead of cell populations. We took advantage of our observations that zebrafish trunk neural crest cells that migrate at different times on the medial pathway have different fates (Fig. 1; Raible and Eisen, 1994). Early-migrating crest (EMC) cells, which migrate before 18 hours postfertilization (h), generate derivatives of all types, including DRG neurons, glia, and pigment cells. In contrast, late-migrating crest (LMC) cells generate only glia and pigment cells, but not neurons. Thus cells that migrate at different times on the same pathway differ in their production of neurons: some EMC cells generate neurons while LMC cells do not.

To test whether differences between EMC and LMC cells were intrinsic, we transplanted individual cells of both types so that they now encountered the same environmental conditions, and asked whether they still differed in their production of neurons (Raible and Eisen, 1996). When individual EMC cells were transplanted so that they migrated 1–3 hours later than normal, and with host LMC cells, they still produced DRG neurons (Fig. 5). LMC cells transplanted under identical conditions were never able to produce neurons.

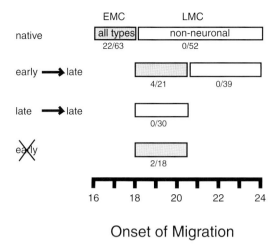

Onset of Migration

Figure 5 Zebrafish trunk neural crest cells have similar potential but different intrinsic biases. The figure summarizes transplantation and ablation experiments. Numbers below the bars show the proportion of cells that formed DRG neurons for each condition. Fate mapping of neural crest cells showed EMC cells generate all derivative types while LMC cells form nonneuronal cell types (native, top row). EMC cells transplanted so they migrated 1–3 hours later than normal still generated neurons (second row, gray box) whereas LMC cells transplanted under the same conditions (third row) never formed neurons. These results demonstrate that different cells under identical conditions have different intrinsic biases. EMC cells transplanted so they migrate 3–6 hours later than normal are no longer able to generate neurons (second row, white box). When EMC cells are ablated, LMC cells are now able to form neurons (fourth row).

These results suggest that there are intrinsic differences between the two cell populations. Interestingly, when EMC cells were transplanted so that they migrated 3–6 hours later than normal, they no longer produced neurons, suggesting that specific, labile environmental conditions are necessary for crest cells to generate neurons. Alternatively, prolonged exposure to signals from the neural tube may inhibit crest cells from producing neurons.

Ablation of neuron-producing crest cells allows cells that would normally produce other derivative types to generate neurons (Raible and Eisen, 1996). When EMC cells were removed just before they began to migrate, presumptive LMC cells quickly filled in so that by 18 hours, the time LMC cells normally begin to migrate, ablated segments were indistinguishable from controls. When embryos were examined at 3 days, the DRG appeared normal in segments where EMC cell ablation had been performed. When EMC cells were ablated by laser irradiation after they began migration, DRGs still formed normally even though LMC cells migrated at their normal times (Fig. 5). LMC cells were shown directly to contribute to DRGs by labeling individual LMC cells in the ablated segment with fluorescent vital dye. These studies confirmed that LMC cells produce neurons after EMC ablation, even when they migrated at their normal times. These results suggest that EMC cells influence the fates of

later migrating cells by preventing them from forming neurons.

What do transplantation and ablation studies tell us about cell potential within the neural crest? As initially defined, cell potential encompasses all possible fates a cell may undertake if it is exposed to a variety of environmental conditions. LMC cells, which are fate restricted and normally generate only nonneuronal derivatives, clearly have the potential to generate DRG neurons when EMC cells are removed. Therefore, fate restrictions displayed by LMC cells do not result from loss of neurogenic potential. Although both EMC and LMC cells have the potential to generate DRG neurons, this does not mean the two cell types are intrinsically the same, since when both cell types are transplanted to similar conditions they behave differently. Thus zebrafish crest cells appear to have intrinsic biases in which fates they adopt, but these biases are not the result of irreversible commitment events.

VI. Regulative Interactions among Neural Crest Cells

Several scenarios could explain the fate differences between EMC cells and LMC cells observed in the experiments described above. LMC cells may be the same as EMC cells but normally not generate neurons because environmental cues that direct neuron formation are absent by the time LMC cells migrate. However, after ablation of EMC cells some LMC cells are able to generate neurons even when they migrate at their normal times, suggesting that environmental factors that direct neurogenesis are still present. Alternatively, EMC cells may preferentially compete for differentiation/survival factors so that not enough factor is left for the LMC cells that follow. However, transplanted EMC cells that migrate with host LMC cells are still able to generate neurons even after the putative factor would have been removed by host EMC cells that migrated before them. We favor a model in which interactions between EMC and LMC cells maintain intrinsic biases in neurogenic ability.

These results have led us to propose a model of neural crest development in which regulative interactions among neural crest cells influence their fates (Raible and Eisen, 1994; 1995; 1996). We propose that there are two types of neural crest cells, a population initially biased to generate neurons, and a second population that begins unbiased but is directed through interactions with other neural crest cells to select nonneural fates. Our transplant data demonstrate that there are neural crest cells with intrinsically different biases, and our ablation experiments suggest that interactions maintain these biases. An early segregating neurogenic lineage within the neural crest has been proposed by others (Sieber-Blum, 1989; Weston, 1991; Henion and Weston, 1997). It is possible that, within the neural crest, there are several different subpopulations that are each biased to produce a different derivative type.

Anderson and colleagues have demonstrated that mammalian neural crest cells in culture have the properties of stem cells that undergo self-renewing divisions and give rise to a variety of derivatives (Stemple and Anderson, 1992). In this system, progeny of individual founder cells can be analyzed by examining clonal progeny. Differentiation of cultured stem cells can be regulated by a variety of environmental cues (Stemple and Anderson, 1992), including addition of specific growth factors (Shah *et al.*, 1994, 1996; Shah and Anderson, 1997). Two models were proposed to account for the effects of growth factors on the clonal progeny derived from individual neural crest stem cells. In one model, stem cells first give rise to cell-type specific progenitors with restricted potential, and different growth factor combinations select a subset of these progeny by affecting survival or proliferation, thus acting permissively. Alternatively, stem cells give rise to progeny with unrestricted potential, and different growth factors may directly influence cell fate, thus acting instructively. Addition of BMP2 (bone morphogenetic protein 2) or TGFβ1 (transforming growth factor-β1) result in differentiation of neurons or smooth muscle, respectively, without differentially affecting survival (Shah *et al.*, 1996), suggesting that these factors act instructively. Consistent with this idea, in 52–67% of clones all the cells differentiated into a single derivative type under each condition.

Although tissue culture has the distinct advantage of precise control of environmental conditions for the study of cell fate, *in vitro* experiments probably do not accurately represent all cells found *in vivo*. The isolation methods used to establish cultures are inherently selective of specific cell types that initially survive or proliferate *in vitro* before test conditions are applied. For example, neural crest stem cells were selected through several conditions: initial outgrowth on fibronectin, immunoselection of cells expressing 25% highest levels of p75, and then culture in a complex defined medium (Stemple and Anderson, 1992). If the neural crest was initially heterogeneous, it seems likely that a subset of cells would be differentially selected by such procedures. Indeed, different populations of neural crest cells have different culture requirements. Avian neural crest cells require specific trophic factors to retain the ability to generate neurons (Henion *et al.*, 1995; Henion and Weston, 1994). Similarly, murine melanogenic precursors require steel factor to survive and differentiate *in vitro* and *in vivo* (Morrison-Graham and Weston, 1993; Lahav *et al.*, 1994; Wehrle-Haller and Weston, 1995; Reid *et al.*, 1995).

The results of these culture studies are compatible with our experiments with zebrafish trunk neural crest cells *in vivo*. Zebrafish LMC cells have the potential to generate all neural crest derivative types, and thus possibly be unbiased and have stem cell properties. Interactions with other neural crest cells therefore may be one of several influences on stem cell fate. More recent data suggest that cultured neural crest stem cells are biased toward generating neurons (Shah and Anderson, 1997). Thus zebrafish EMC cells, some of which have biases toward generating neurons, may also have stem cell properties. Whether zebrafish neural crest cells act as stem cells *in vitro* remains to be investigated.

Regulative interactions have been proposed to occur later in neural crest development, in cell fate choices between neurons and glia within developing peripheral ganglia (Shah *et al.*, 1994). Differentiated neurons produce glial growth factor (GGF), which promotes gliogenesis and inhibits neurogenesis by neural crest cells *in vitro* (Shah *et al.*, 1994; Shah and Anderson, 1997). Thus, a feedback loop would be established in which neural crest cells differentiate as neurons, inhibit other undifferentiated crest cells from forming neurons, and promote their development as glial cells. In zebrafish, cells do not begin to differentiate as DRGs until almost a day after the interactions we have described. Thus interactions among crest cells may occur at several stages of neural crest development.

Regulative interactions are also the hallmark of lateral specification within invertebrate equivalence groups (Greenwald and Rubin, 1992). Cells of an equivalence group make hierarchical fate choices between two fates: a primary or default fate and a secondary or alternative fate. Initially, cells have the same potential, but this equivalence is broken by lateral interactions in which a cell inhibits its neighbors from assuming the fate it has already assumed. These lateral inhibitory signals are mediated by members of the Notch family of transmembrane receptors (Greenwald, 1994; Artavanis-Tsakonis *et al.*, 1995; Simpson, 1997).

Although invertebrate equivalence groups have some similarities with our observations of zebrafish neural crest cells, the analogy is not exactly parallel. We have observed that neural crest cells have different biases towards neurogenesis, and are thus not exactly equivalent. Therefore interactions may be occurring in the neural crest between nonequivalent cells. Notch family members also mediate interactions between nonequivalent cells in invertebrates (reviewed in Simpson, 1997), and correspondingly, could also be involved in neural crest development. Genes encoding proteins involved in all aspects of Notch signaling are found in neural crest cells and their derivatives (Johnston *et al.*, 1990; Reume *et al.*, 1992; Weinmaster *et al.*, 1992; Lardelli *et al.*, 1994; Bettenhausen *et al.*, 1995;

Williams *et al.*, 1995; Myat *et al.*, 1996; Uyttendaele *et al.*, 1996; de la Pompa *et al.*, 1997; Dunwoodie *et al.*, 1997; Johnston *et al.*, 1997; Ma *et al.*, 1997; Zhong *et al.*, 1997). However, to date there is no direct evidence for Notch signaling influencing neural crest cell fate decisions.

We envision regulative interactions among neural crest cells to be one of several mechanisms influencing their development. Neural crest cells may be biased by signals from the neural tube or interact with growth factors encountered after migration, such as steel factor (Wehrle-Haller and Weston, 1995) or BMPs (Shah *et al.*, 1996). Indeed, regulative interactions may influence how neural crest cells respond to these environmental cues, as they do in the vertebrate retina (Dorsky *et al.*, 1997) and spinal cord (Appel and Eisen, 1998). A combination of factors, both intrinsic and extrinsic, are thus likely to determine neural crest cell fate.

Acknowledgments

We thank Kirsten Stoesser, and the staff of the University of Oregon zebrafish facility, for their help in the work described. Work was supported by grants from the Dysautonomia Foundation, the American Heart Association, and the National Institutes of Health (DWR) and by NIH HD22486 (JSE).

References

Appel and Eisen (1998). Regulation of neuronal specification in the zebrafish spinal cord by Delta function. *Development* **125**, 371–380.

Artavanis-Tsakonas, S., Matsuno, K., and Fortini, M. E. (1995). Notch signaling. *Science* **268**, 225–232.

Artinger, K. B., and Bronner-Fraser, M. (1992). Partial restriction in the developmental potential of late emigrating avian neural crest cells. *Dev. Biol.* **149**, 149–157.

Baker, C. V., Bronner-Fraser, M., Le Douarin, N. M., and Teillet, M. A. (1997). Early- and late-migrating cranial neural crest cell populations have equivalent developmental potential *in vivo*. *Development (Cambridge, UK)* **124**, 3077–3087.

Bettenhausen, B., Hrabe de Angelis, M., Simon, D., Gu'enet, J. L., and Gossler, A. (1995). Transient and restricted expression during mouse embryogenesis of Dll1, a murine gene closely related to *Drosophila* Delta. *Development (Cambridge, UK)* **121**, 2407–2418.

Birgbauer, E., Sechrist, J., Bronner-Fraser, M., and Fraser, S. (1995). Rhombomeric origin and rostrocaudal reassortment of neural crest cells revealed by intravital microscopy. *Development (Cambridge, UK)* **121**, 935–945.

Bronner-Fraser, M., and Fraser, S. (1989). Developmental potential of avian trunk neural crest cells *in situ*. *Neuron* **3**, 755–766.

Bronner-Fraser, M., and Fraser, S. E. (1991). Cell lineage analysis of the avian neural crest. *Development (Cambridge, UK)* **2**, 17–22.

Collazo, A., Bronner-Fraser, M., and Fraser, S. E. (1993). Vital dye labeling of *Xenopus laevis* trunk neural crest reveals multipotency and novel pathways of migration. *Development (Cambridge, UK)* **118**, 363–376.

Davidson, E. H. (1990). How embryos work: A comparative view of diverse modes of cell fate specification. *Development (Cambridge, UK)* **108**, 365–389.

de la Pompa, J. L., Wakeham, A., Correia, K. M., Samper, E., Brown, S., Aguilera, R. J., Nakano, T., Honjo, T., Mak, T. W., Rossant, J., and Conlon, R. A. (1997). Conservation of the Notch signaling pathway in mammalian neurogenesis. *Development (Cambridge, UK)* **124**, 1139–1148.

Dorsky, R. I., Chang, W. S., Rapaport, D. H. and Harris, W. A. (1997). Regulation of neuronal diversity in the *Xenopus* retina by Delta signaling. *Nature (London)* **385**, 67–70.

Dunwoodie, S. L., Henrique, D., Harrison, S. M., and Beddington, R. S. (1997). Mouse Dll3: A novel divergent Delta gene which may complement the function of other Delta homologues during early pattern formation in the mouse embryo. *Development (Cambridge, UK)* **124**, 3065–3076.

Erickson, C. A., and Goins, T. L. (1995). Avian neural crest cells can migrate in the dorsolateral path only if they are specified as melanocytes. *Development (Cambridge, UK)* **121**, 915–924.

Erickson, C. A., Duong, T. D., and Tosney, K. W. (1992). Descriptive and experimental analysis of the dispersion of neural crest cells along the dorsolateral path and their entry into ectoderm in the chick embryo. *Dev. Biol.* **151**, 251–272.

Graveson, A. C., Hall, B. K., and Armstrong, J. B. (1995). The relationship between migration and chondrogenic potential of trunk neural crest cells in *Ambystoma mexicanum*. *Roux's Arch. Dev. Biol.* **204**, 477–483.

Greenwald, I. (1994). Structure/function studies of lin-12/Notch proteins. *Curr. Opin. Genet. Dev.* **4**, 556–562.

Greenwald, I., and Rubin, G. M. (1992). Making a difference: The role of cell–cell interactions in establishing separate identities for equivalent cells. *Cell (Cambridge, Mass.)* **68**, 271–281.

Gurdon, J. B., Lemaire, P., and Kato, K. (1993). Community effects and related phenomena in development. *Cell (Cambridge, Mass)* **75**, 831–834.

Haffter, P., Granato, M., Brand, M., Mullins, M. C., Hammerschmidt, M., Kane, D. A., Odenthal, J., van Eeden, F. J., Jiang, Y. J., Heisenberg, C. P., Kelsh, R. N., Furutani-Seiki, M., Vogelsang, E., Beuchle, D., Schach, U., Fabian, C., and Nusslein-Volhard, C. (1996). The identification of genes with unique and essential functions in the development of the zebrafish, *Danio rerio*. *Development (Cambridge, UK)* **123**, 1–36.

Hall, B. K. and Horstadius, S. (1988). "The Neural Crest." Oxford Univ. Press, London.

Henion, P. D., and Weston, J. A. (1994). Retinoic acid selectively promotes the survival and proliferation of neurogenic precursors in cultured neural crest cell populations. *Dev. Biol.* **161**, 243–250.

Henion, P. D., and Weston, J. A. (1997). Timing and pattern of cell fate restrictions in the neural crest lineage. *Development (Cambridge, UK)* **124**, 4351–4359.

Henion, P. D., Garner, A. S., Large, T. H., and Weston, J. A. (1995). trkC-mediated NT-3 signaling is required for the early development of a subpopulation of neurogenic neural crest cells. *Dev. Biol.* **172**, 602–613.

Henion, P. D., Raible, D. W., Beattie, C. E., Stoesser, K. L., Weston, J. A., and Eisen, J. S. (1996). Screen for mutations affecting development of zebrafish neural crest. *Dev Genet.* **18**, 11–17.

Ho, R. K. (1992). Cell movements and cell fate during zebrafish gastrulation. *Development (Cambridge, UK)* (Suppl), 65–73.

Ho, R. K., and Kimmel, C. B. (1993). Commitment of cell fate in the early zebrafish embryo. *Science* **261**, 109–111.

Jesuthasan, S. (1996). Contact inhibition/collapse and pathfinding of neural crest cells in the zebrafish trunk. *Development (Cambridge, UK)* **122**, 381–389.

Johnson, J. E., Birren, S. J., and Anderson, D. J. (1990). Two rat homologues of *Drosophila* achaete-scute specifically expressed in neuronal precursors. *Nature (London)* **346**, 858–861.

Johnston, S. H., Rauskolb, C., Wilson, R., Prabhakaran, B., Irvine, K. D., and Vogt, T. F. (1997). A family of mammalian Fringe genes implicated in boundary determination and the Notch pathway. *Development (Cambridge, UK)* **124**, 2245–2254.

Kelsh, R. N., Brand, M., Jiang, Y. J., Heisenberg, C. P., Lin, S., Haffter, P., Odenthal, J., Mullins, M. C., van Eeden, F. J., Furutani-Seiki, M., Granato, M., Hammerschmidt, M., Kane, D. A., Warga, R. M., Beuchle, D., Vogelsang, L., and Nusslein-Volhard, C. (1996). Zebrafish pigmentation mutations and the processes of neural crest development. *Development (Cambridge, UK)* **123**, 369–389.

Kimmel, C. B., and Warga, R. M. (1986). Tissue-specific cell lineages originate in the gastrula of the zebrafish. *Science* **231**, 365–368.

Kimmel, C. B., Kane, D. A., and Ho, R. K. (1991). Lineage specification during early embryonic development of the zebrafish. *In* "Cell–Cell Interactions in Early Development" (J. Gerhart, ed.), pp. 203–225. Wiley-Liss, New York.

Kirby, M. L. (1989). Plasticity and predetermination of mesencephalic and trunk neural crest transplanted into the region of the cardiac neural crest. *Dev. Biol.* **134**, 402–412.

Lahav, R., Lecoin, L., Ziller, C., Nataf, V., Carnahan, J. F., Martin, F. H., and Le Douarin, N. M. (1994). Effect of the Steel gene product on melanogenesis in avian neural crest cell cultures. *Differentiation* **58**, 133–139.

Lardelli, M., Dahlstrand, J., and Lendahl, U. (1994). The novel Notch homologue mouse Notch 3 lacks specific epidermal growth factor-repeats and is expressed in proliferating neuroepithelium. *Mech. Dev.* **46**, 123–136.

Le Douarin, N. M. (1982). "The Neural Crest." Cambridge Univ. Press, Cambridge.

Le Douarin, N. M., and Dupin, E. (1993). Cell lineage analysis in neural crest ontogeny. *J. Neurobiol.* **24**, 146–161.

Le Douarin, N. M., Dulac, C., Dupin, E., and Cameron-Curry, P. (1991). Glial cell lineages in the neural crest. *Glia* **4**, 175–184.

Lister, J. A., Robertson, C., Le Page, T., Ungos, J., and Raible, D. W. (1998). *Nacre,* a gene required for melanophore development in zebrafish. *Dev. Biol.* **198**, 121.

Ma, Q., Sommer, L., Cserjesi, P., and Anderson, D. J. (1997). Mash1 and neurogenin1 expression patterns define complementary domains of neuroepithelium in the developing CNS and are correlated with regions expressing notch ligands. *J. Neurosci.* **17**, 3644–3652.

Morin-Kensicki, E. M., and Eisen, J. S. (1997). Sclerotome development and peripheral nervous system segmentation in embryonic zebrafish. *Development (Cambridge, UK)* **124**, 159–167.

Morrison-Graham, K., and Weston, J. A. (1993). Transient steel factor dependence by neural crest-derived melanocyte precursors. *Dev. Biol.* **159**, 346–352.

Myat, A., Henrique, D., Ish Horowicz, D., and Lewis, J. (1996). A chick homologue of Serrate and its relationship with Notch and Delta homologues during central neurogenesis. *Dev. Biol.* **174**, 233–247.

Neuhauss, S. C., Solnica-Krezel, L., Schier, A. F., Zwartkruis, F., Stemple, D. L., Malicki, J., Abdelilah, S., Stainier, D. Y., and Driever, W. (1996). Mutations affecting craniofacial development in zebrafish. *Development (Cambridge, UK)* **123**, 357–367.

Noden, D. M. (1983). The role of the neural crest in patterning of avian cranial skeletal, connective, and muscle tissues. *Dev. Biol.* **96**, 144–165.

Noden, D. M. (1991). Vertebrate craniofacial development: The relation between ontogenetic process and morphological outcome. *Brain Behav. Evol.* **38**, 190–225.

Odenthal, J., Rossnagel, K., Haffter, P., Kelsh, R. N., Vogelsang, E., Brand, M., van Eeden, F. J., Furutani-Seiki, M., Granato, M., Hammerschmidt, M., Heisenberg, C. P., Jiang, Y. J., Kane, D. A., Mullins, M. C., and Nusslein-Volhard, C. (1996). Mutations affecting xanthophore pigmentation in the zebrafish, *Danio rerio. Development (Cambridge, UK)* **123**, 391–398.

Piotrowski, T., Schilling, T. F., Brand, M., Jiang, Y. J., Heisenberg, C. P., Beuchle, D., Grandel, H., van Eeden, F. J., Furutani-Seiki, M., Granato, M., Haffter, P., Hammerschmidt, M., Kane, D. A., Kelsh, R. N., Mullins, M. C., Odenthal, J., Warga, R. M., and Nusslein-Volhard, C. (1996). Jaw and branchial arch mutants in zebrafish II: Anterior arches and cartilage differentiation. *Development (Cambridge, UK)* **123**, 345–356.

Raible, D. W., and Eisen, J. S. (1994). Restriction of neural crest cell fate in the trunk of the embryonic zebrafish. *Development (Cambridge, UK)* **120**, 495–503.

Raible, D. W., and Eisen, J. S. (1995). Lateral specification of cell fate during vertebrate development. *Curr. Opin. Genet. Dev.* **5**, 444–449.

Raible, D. W., and Eisen, J. S. (1996). Regulative interactions in zebrafish neural crest. *Development (Cambridge, UK)* **122**, 501–507.

Raible, D. W., Wood, A., Hodsdon, W., Henion, P. D., Weston, J. A., and Eisen, J. S. (1992). Segregation and early dispersal of neural crest cells in the embryonic zebrafish. *Dev. Dyn.* **195**, 29–42.

Reaume, A. G., Conlon, R. A., Zirngibl, R., Yamaguchi, T. P., and Rossant, J. (1992). Expression analysis of a Notch homologue in the mouse embryo. *Dev. Biol.* **154**, 377–387.

Reid, K., Nishikawa, S., Bartlett, P. F., and Murphy, M. (1995). Steel factor directs melanocyte development *in vitro* through selective regulation of the number of c-kit⁺ progenitors. *Dev. Biol.* **169**, 568–579.

Saldivar, J. R., Krull, C. E., Krumlauf, R., Ariza-McNaughton, L., and Bronner-Fraser, M. (1996). Rhombomere of origin determines autonomous versus environmentally regulated expression of Hoxa-3 in the avian embryo. *Development (Cambridge, UK)* **122**, 895–904.

Schilling, T. F., and Kimmel, C. B. (1994). Segment and cell type lineage restrictions during pharyngeal arch development in the zebrafish embryo. *Development (Cambridge, UK)* **120**, 483–494.

Schilling, T. F., and Kimmel, C. B. (1997). Musculoskeletal patterning in the pharyngeal segments of the zebrafish embryo. *Development (Cambridge, UK)* **124**, 2945–2960.

Schilling, T. F., Piotrowski, T., Grandel, H., Brand, M., Heisenberg, C. P., Jiang, Y. J., Beuchle, D., Hammerschmidt, M., Kane, D. A., Mullins, M. C., van Eeden, F. J., Kelsh, R. N., Furutani-Seiki, M., Granato, M., Haffter, P., Odenthal, J., Warga, R. M., Trowe, T., and Nusslein-Volhard, C. (1996a). Jaw and branchial arch mutants in zebrafish I: Branchial arches. *Development (Cambridge, UK)* **123**, 329–344.

Schilling, T. F., Walker, C., and Kimmel, C. B. (1996b). The chinless mutation and neural crest cell interactions in zebrafish jaw development. *Development (Cambridge, UK)* **122**, 1417–1426.

Shah, N. M., and Anderson, D. J. (1997). Integration of multiple instructive cues by neural crest stem cells reveals cell-intrinsic biases in relative growth factor responsiveness. *Proc. Natl. Acad. Sci. U.S.A.* **94**, 11369–11374.

Shah, N. M., Marchionni, M. A., Isaacs, I., Stroobant, P., and Anderson, D. J. (1994). Glial growth factor restricts mammalian neural crest stem cells to a glial fate. *Cell (Cambridge, Mass.)* **77**, 349–360.

Shah, N. M., Groves, A. K. and Anderson, D. J. (1996). Alternative neural crest cell fates are instructively promoted by TGFbeta superfamily members. *Cell (Cambridge, Mass.)* **85**, 331–343.

Sieber-Blum, M. (1989). Commitment of neural crest cells to the sensory neuron lineage. *Science* **243**, 1608–1611.

Sieber-Blum, M., Ito, K., Richardson, M. K., Langtimm, C. J., and Duff, R. S. (1993). Distribution of pluripotent neural crest cells in the embryo and the role of brain-derived neurotrophic factor in the commitment to the primary sensory neuron lineage. *J. Neurobiol.* **24**, 173–184.

Simpson, P. (1997). Notch signaling in development. *Perspect. Dev. Neurobiol.* **4**, 297–304.

Stemple, D. L., and Anderson, D. J. (1992). Isolation of a stem cell for neurons and glia from the mammalian neural crest. *Cell (Cambridge, Mass.)* **71**, 973–985.

Tosney K. W., Dehnbostel, D. B., and Erickson, C. A. (1994). Neural crest cells prefer the myotome's basal lamina over the sclerotome as a substratum. *Dev. Biol.* **163**, 389–406.

Uyttendaele, H., Marazzi, G., Wu, G., Yan, Q., Sassoon, D. and Kitajewski, J. (1996). Notch4/int-3, a mammary proto-oncogene, is an endothelial cell-specific mammalian Notch gene. *Development (Cambridge, UK)* **122**, 2251–2259.

Wehrle-Haller, B., and Weston, J. A. (1995). Soluble and cell-bound forms of steel factor activity play distinct roles in melanocyte precursor dispersal and survival on the lateral neural crest migration pathway. *Development (Cambridge, UK)* **121**, 731–742.

Weinmaster, G., Roberts, V. J., and Lemke, G. (1992). Notch2: A second mammalian Notch gene. *Development (Cambridge, UK)* **116**, 931–941.

Weston, J. A. (1991). Sequential segregation and fate of developmentally restricted intermediate cell populations in the neural crest lineage. *Curr. Top. Dev. Biol.* **25**, 133–153.

Williams, R., Lendahl, U., and Lardelli, M. (1995). Complementary and combinatorial patterns of Notch gene family expression during early mouse development. *Mech. Dev.* **53**, 357–368.

Zhong, W., Jiang, M. M., Weinmaster, G., Jan, L. Y., and Jan, Y. N. (1997). Differential expression of mammalian Numb, Numblike and Notch1 suggests distinct roles during mouse cortical neurogenesis. *Development (Cambridge, UK)* **124**, 1887–1897.

VII

Chick

The Avian Embryo: A Model for Descriptive and Experimental Embryology

Gary C. Schoenwolf

Department of Neurobiology and Anatomy
University of Utah School of Medicine
Salt Lake City, Utah 84132

"I have invested years of pondering in order to track down the possibilities of how a structure as complex and typically shaped as the chick could arise from a relatively or seemingly simple state without external formative influences."

Roux, 1885, p. 419; translated and cited by Sander (1997)

I. Introduction

The avian embryo, and especially the chick embryo, has provided yeoman's service to the field of embryology/developmental biology. Since the time of Aristotle, who advocated that one only needed to crack open a hen's egg each day to see development unfolding before one's very eyes (unequivocally supporting the theory of epigenesis and refuting the theory of preformation), the avian embryo has served as a popular model system for both descriptive and experimental studies of animal embryonic development. In this introductory chapter, I will answer the question of why the chick embryo is such an excellent model.

Birds are classified with reptiles and mammals as amniotes; that is, during embryogenesis they develop within an extraembryonic membrane called the amnion. Birds and mammals are further classified as warmblooded higher vertebrates or vertebrates in which

embryonic development is accompanied by extensive growth with exponential increase in volume. In lower vertebrates, such as fish, amphibians, and reptiles, the embryo undergoes little or no growth during its morphogenesis. Rather the original single-cell egg becomes partitioned through mitotic divisions into multiple daughter cells, each of which is approximately one-half the volume of its parental cell (see Chapter 19 by Dawid; Chapter 24 by Driever). The growth that occurs during morphogenesis in higher vertebrates not only accompanies morphogenesis but actually drives it. Thus, understanding differential growth is a key to understanding the morphogenesis of birds and mammals. Moreover, differential growth gone awry leads to dysmorphogenesis, which in turn results in the formation of birth defects. Thus, understanding differential growth provides keys to understanding both normal and abnormal development of higher vertebrates, including humans.

Before addressing the question of why the chick is such an excellent model, I will provide a brief overview of avian embryogenesis (Fig. 1). For additional details, see Schoenwolf (1995, 1997a).

II. Overview of Avian Embryogenesis

"The egg is, to us, a familiar and hence seemingly prosaic structure for the breakfast table. It is in reality the

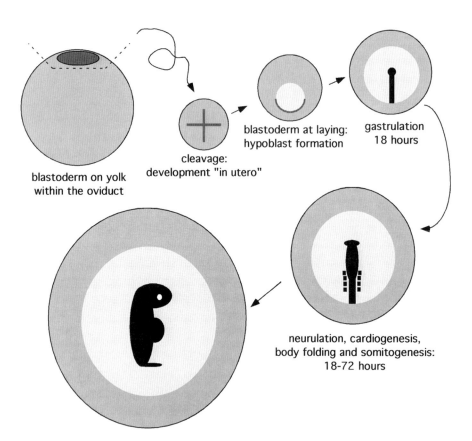

Figure 1 Overview of avian embryogenesis. Events up through blastoderm formation occur within the reproductive track of the hen. Later events occur after the egg is laid and are therefore accessible to experimental manipulations. Light gray denotes area pellucida, dark gray denotes area opaca, and black denotes primitive streak and embryo proper.

most marvelous single 'invention' in the whole history of vertebrate life."

Romer (1968), p. 183.

A. Formation of the Egg: Prenatal Development within the Hen

Several important events occur during avian embryogenesis before the hen presents her egg to the embryologist (i.e., lays it). The ovum, freshly ovulated from the ovary, is enclosed by acellular membranes called the vitelline membranes. The ovum and its vitelline membranes quickly enter the oviduct after ovulation. As the ovum and surrounding vitelline membranes traverse the oviduct, they are wrapped in layers of albumen, shell membranes, and calcified shell. However, if development of the ovum is going to occur, it must be fertilized by sperm within the oviduct before these layers are added (sperm can penetrate the vitelline membranes but not the albumen). Also within the oviduct and following fertilization, two important developmental events occur (Fig. 1): the blastodisc, the cytoplasmic cap that floats on the yolk at the animal pole of the ovum, undergoes cleavage, subdividing into thousands of cells; and the future rostrocaudal axis of the blastoderm (i.e., the blastodisc after cleavage is initiated and multiple cells or

blastomeres have formed) becomes established owing to the action of gravity and rotational (centrifugal) forces, which presumably cause a redistribution of cytoplasmic components within the perimeter of the forming blastoderm. The ovum (or developing blastoderm if fertilization occurs) spends slightly more than 24 hours passing through the oviduct. Shortly after the egg is laid, another ovum is ovulated from the ovary of the hen and the process of egg manufacturing begins again.

B. Formation of the Embryo: Prenatal Development within the Eggshell during Incubation

After laying, further development of the blastoderm requires incubation at about 38°C. This temperature can be easily maintained by the hen who sits on her eggs within the nest, or by the embryologist who uses a simple, inexpensive incubator. If all goes well, from the deceptively simple hen's egg, a fully formed chick will emerge after some 3 weeks of incubation (Fig. 1).

1. Gastrulation: Formation of the Layers of the Blastoderm

The egg is laid during the phase of development called gastrulation. More specifically, the egg is laid during the stage of gastrulation during which the hypoblast is forming. The hypoblast is the first layer of the blastoderm to form. It is a deep (i.e., close to the yolk of the egg) layer of cells that is ultimately replaced by endoderm during subsequent gastrulation. As the endoderm forms (see below), the hypoblast is displaced to the extraembryonic region where it is associated with the primordial germ cells. During subsequent development after laying, formation of the hypoblast is completed, after which two layers can be distinguished: a superficial epiblast (just beneath the vitelline membranes) and the deep hypoblast. Thus, with formation of the hypoblast, the blastoderm becomes bilaminar. Three zones can also be distinguished within the blastoderm at this stage: a central area pellucida, a peripheral area opaca, and a ring of cells, the marginal zone, at the interface between the area opaca and the area pellucida (Fig. 1).

After formation of the hypoblast is completed, a linear thickening, the primitive streak, forms at the caudal midline of the blastoderm within the area pellucida and adjacent to the caudal part of the marginal zone (Figs. 2, 3, and 4a,b). During the next few hours of incubation, the primitive streak elongates rostrocaudally, and epiblast cells move toward, within, and through the primitive streak (i.e., ingress) to form the endodermal and mesodermal layers (Fig. 3); after gastrulation is completed, the cells remaining within the epiblast form the ectoderm. With gastrulation, the blastoderm becomes trilaminar, consisting of an inner endoderm, middle mesoderm, and outer ectoderm.

2. Neurulation: Formation of the Rudiments of an Organ System

Neurulation, the formation of the neural tube—the rudiment of the entire adult central nervous system, occurs concomitantly with gastrulation. The ectoderm located rostral to the primitive streak thickens, as cells undergo apicobasal elongation, forming the neural plate (Fig. 4c,d). During subsequent development, the neural plate narrows transversely, extends longitudinally, and rolls up into a gutterlike structure called the neural groove (Fig. 4c,e). The neural groove is flanked laterally by the ectodermal neural folds, which ultimately meet in the dorsal midline, closing the neural groove and establishing the neural tube (Fig. 5a). For additional details, see Smith and Schoenwolf (1997). Eventually, cells derived from the roof of the newly formed neural tube migrate bilaterally to form the neural crest, the precursor cells of much of the peripheral nervous system as well as the progenitors of many of the cell types that populate the future head and neck regions.

3. Embryonic Staging

At this point, it is useful to digress briefly to explain embryonic staging. Two systems are used to define precisely the stages of the developing chick embryo. The development of the blastoderm prior to and shortly after laying of the egg is characterized using Roman numerals I–XIV and the system of Eyal-Giladi and Kochav (1976). Cleavage occurs during stages I–VI, formation of the area pellucida (and area opaca) occurs during

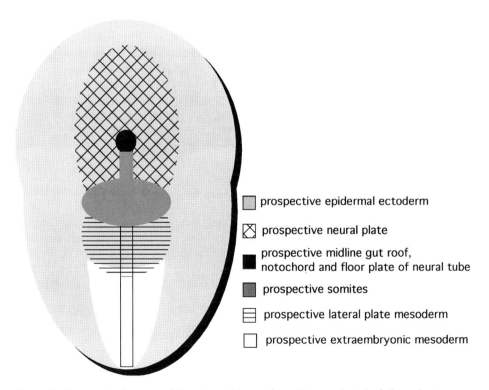

□ prospective epidermal ectoderm

⊠ prospective neural plate

■ prospective midline gut roof, notochord and floor plate of neural tube

▨ prospective somites

▤ prospective lateral plate mesoderm

□ prospective extraembryonic mesoderm

Figure 2 Prospective fate map of the avian epiblast and primitive streak at the full-streak stage.

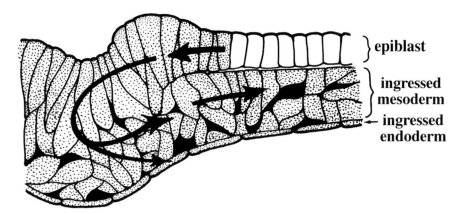

Figure 3 Diagram of a cross section through the avian primitive streak illustrating cell ingression.

stages VII–IX, and formation of the hypoblast occurs during stages X–XIV. The egg is usually laid at stage X (i.e., at the initial stage of hypoblast formation). After laying and with incubation at 38°C, the development of the blastoderm and, subsequently, of the embryo is characterized using Hamburger and Hamilton's (1951) stages 1–46. Eyal-Giladi and Kochav's stages X–XIV are substages (more precisely defined) of Hamburger and Hamilton's stage 1. Hamburger and Hamilton stages 2–4 define stages of formation and elongation (also called progression) of the primitive streak. During stages 5–11, the rostral end of the primitive streak moves caudally (i.e., regresses) and the embryonic body forms in its wake. During these stages, the straight-heart tube, the neural tube, the notochord, the somites, the foregut, the head fold of the body, and the lateral body folds all form, and they do so in only about 24 hours (total "gestation" time of 21 days for chick and 19 days for quail). Somites form at the rate of about one pair every 90 minutes, and Hamburger and Hamilton stages 7–14+ are defined based solely on the number of somite pairs present. For example, a stage 7 embryo has one somite, a stage 7+ embryo has two somites, a stage 8− embryo

has three somites and a stage 8 embryo has four somites. A stage 14 embryo has 22 somites and a stage 14+ embryo has 23 somites. Thus, during stages 7–14+, the presence of each pair of somites is indicated by a stage number and a + or − suffix.

Embryos older than Hamburger and Hamilton stage 14+ have widely variable numbers of somite pairs for a fixed period of incubation, so other criteria are used for staging. These include the amount of development of the extraembryonic membranes, the presence of flexures and areas of body rotation (torsion), the size and shape of the limb buds, and the number or state of development of the branchial arches and grooves. During these stages, organogenesis and growth of the embryonic body predominate.

4. Formation of the Vertebrate Body Plan

Returning to the development of the chick embryo during early postlaying stages, by the end of 2 days of incubation, the body plan characteristic of vertebrates has formed in developing chick embryos. This body plan, called the vertebrate body plan, is a tube-within-a-tube

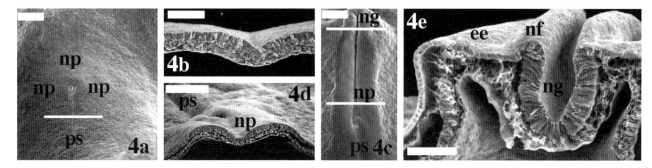

Figure 4 Scanning electron micrographs of early chick embryos. (a) Dorsal view showing locations of the neural plate (np) and primitive streak (ps). White line indicates the level of the transverse section shown in b. (b) Transverse section through the primitive streak at the level shown by the white line in a. (c) Dorsal view showing locations of the neural groove (ng), neural plate (np), and primitive streak (ps). Lower white line indicates the level of the transverse section shown in d. Upper white line indicates the level of the transverse section shown in e, from a slightly older embryo. (d) Transverse section through the neural plate (np) at the level shown by the lower white line in c. ps, Primitive streak. (e) Transverse section through the neural groove (ng) at the level shown by the upper white line in c. ee, Epidermal ectoderm; nf, neural fold. Bars: 100 μm (a–d); 50 μm (e).

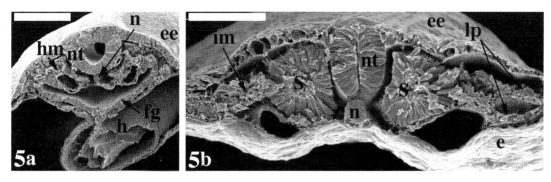

Figure 5 Scanning electron micrographs of early chick embryos. (a) Transverse section through the incipient neural tube (nt) at the future brain level. ee, Epidermal ectoderm; fg, foregut; h, heart; hm, head mesenchyme; n, notochord. (b) Transverse section through the incipient neural tube (nt) at the future spinal cord level. e, Endoderm; ee, epidermal ectoderm; im, intermediate mesoderm; lp, lateral plate mesoderm; n, notochord; s, somite. Bar: 50 μm.

body plan (Fig. 5a,b). The vertebrate body as seen in transverse section consists of an outer tube derived from ectoderm, which later forms the adult skin (epidermis) and associated appendages (feathers, scales), and an inner tube derived from endoderm, consisting of the gut and later forming the lining of the respiratory and gastrointestinal systems. These two tubes are separated from the underlying yolk of the egg of the action of the body folds. Between the ectoderm and endoderm is a loose, mesodermally derived connective tissue, which quickly becomes regionally specialized into a series of rudiments. In the future trunk region (Fig. 5b), these consist of the following types of structures listed in medial-to-lateral order: a single midline rod called the notochord (forms a skeletal support for the early embryo and functions as a signaling center in dorsoventral polarization of the neural tube and somites; see below), paired segmental blocks called somites (source of the dermis, skeletal muscle, and bony elements of the trunk), paired bands called intermediate mesoderm (source of the urogenital system), and paired plates called lateral plate mesoderm (each is split frontally into a dorsal somatic mesoderm, which lies subjacent to the epidermal ectoderm, and a ventral splanchnic mesoderm, which lies subjacent to the endoderm; the somatic mesoderm and ectoderm together constitute the somatopleure, the source of the body wall and part of the coelomic epithelium, and the splanchnic mesoderm and endoderm together constitute the splanchnopleure, the source of the gut wall and remainder of the coelomic epithelium). Within the developing head (Fig. 5a), the somites and more lateral mesodermal subdivisions are not present; instead the mesoderm of the head remains as loosely organized cells constituting part of the head mesenchyme (the remainder of the head mesenchyme is derived from the ectodermal neural crest, which arise from the roof of the neural tube). In addition, a localized mesodermal subdivision, initially formed within the head but later displaced to the trunk, forms a pair of heart tubes (one on either side of the developing foregut), which through

the action of the lateral body folds, forms a single midline heart tube composed of two layers separated by an extracellular matrix called cardiac jelly.

The endoderm forming the gut also becomes regionally subdivided. The rostral endoderm forms the foregut, the trunk endoderm forms the midgut, and the caudal endoderm (near the forming tail) forms the hindgut. Each of the three regions of the early gut contributes to distinct regions of the adult gastrointestinal system; the adult respiratory system derives entirely from the foregut.

The neural tube also undergoes regional specialization. Four rostrocaudal subdivisions form early in neurulation—the forebrain, midbrain, hindbrain, and spinal cord—and each of these forms specific areas of the adult central nervous system. For example, the forebrain becomes subdivided into a rostromost telencephalon and a more caudal diencephalon. Optic vesicles, the initial rudiments of the retinae of the eyes, evaginate bilaterally from the diencephalon (Fig. 6) and contact the adjacent ectoderm, which forms a lens on each side. Thus, the eyes are derived from two sources of ectoderm: neuroectoderm of the neural tube and the overlying epidermal ectoderm.

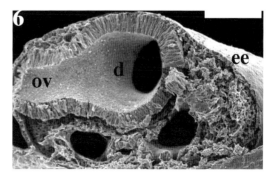

Figure 6 Scanning electron micrograph of an early chick embryo. Transverse section through the level of the future diencephalon (d). ee, Epidermal ectoderm; ov, optic vesicle. Bar: 50 μm.

5. Secondary Body Development

Formation of the vertebrate body plan as described above can be referred to as primary body development, or the formation of rudiments from the three primary germ layers: the ectoderm, the mesoderm, and the endoderm. In addition to primary body development, another event occurs called secondary body development, in which the caudal part of the body arises from an area called the tail bud (Fig. 7). During secondary body development, a tail bud forms at the caudal end of the embryo from persisting remnants of the primitive streak. During secondary body development, the tail bud forms the caudal (mainly tail) neural tube (caudomost spinal cord) and somites. In contrast, the notochord of the tail derives from an ingrowth from more rostral levels.

6. Gene Expression during Formation of the Body Plan

Many genes are expressed during formation of the body plan and can be detected within early embryonic rudiments using *in situ* hybridization. Three types of molecules tend to be expressed: secreted signaling molecules (e.g., Sonic hedgehog), signaling transduction intermediaries (usually kinases), and transcription factors (e.g., paraxis). Although many details still remain to be discerned, these genes likely function in the induction and early patterning of the vertebrate body plan. For example, signaling molecules secreted by the rostral end of the primitive streak, an area called Hensen's node, induce neural plate from the ectoderm (in large part by preventing the formation of epidermal ectoderm by suppressing growth factors required for epidermis formation). Subsequently, cells derived from the primitive streak and Hensen' node secrete factors that regionally pattern the developing neural tube and adjacent structures. The notochord, which is derived from Hensen's node during gastrulation, patterns the neural tube and somites dorsoventrally through two actions: suppression of the activity of growth factors required for dorsalization, and secretion of ventralizing signals such as Sonic Hedgehog. For additional details, see Placzek and

Furley (1996); Sasai and De Robertis (1997); Chapter 29 by Streit and Stern).

III. Why Is the Avian Embryo Such an Excellent Model System?

The avian embryo, and especially the chick embryo, is an excellent model system for studying vertebrate development. There are five key reasons for this. The first reason is a practical one. Fertile chick and quail eggs are inexpensive, animal care costs do not accrue to the scientist, any number of eggs can be obtained (a few to tens of dozens), suppliers are available throughout the world, seasonal changes have little impact on obtaining and using eggs, eggs can be stored simply until needed without having expensive storage facilities, and egg incubators are inexpensive and readily available. The avian egg is a self-contained system (i.e., the avian egg is a cleidoic egg) and further development with incubation occurs outside of the mother (hen) providing outstanding accessibility. Such an egg was used by dinosaurs and has continued with birds today some 45–75 million years later. Chick and quail eggs develop rapidly (3 weeks or less from laying to hatching).

Second, a huge data base exists on avian development. This data base includes a wealth of descriptive and experimental embryology as well as a plethora of toxicological and teratological studies on avian embryos. Descriptive data have been obtained with conventional morphological techniques including light microscopy, histology, electron microscopy (scanning, transmission and freeze fracture), histochemistry, immunocytochemistry, and *in situ* hybridization. Experimental techniques have been devised for observation of embryo behavior, recording physiological parameters (e.g., heart or respiration rate), microsurgery, drug treatment, analyses of cell cycle, analyses of cell death, assays of protein synthesis and accumulation, assessment of RNA expression, etc. Experimental techniques can be applied *in ovo* or *in vitro,* the latter utilizing many methods of whole-embryo culture, organ culture, tissue culture, and cell culture. As part of the descriptive data base, complete stage series exist as do detailed atlases of various stages of development. Accurate and detailed fate maps have been constructed to show the origins and displacements of cells during gastrulation, neurulation, and segmentation (e.g., Fig. 2). Chimeras can be constructed by transplanting quail cells into chick embryos or vice versa, and highly specific markers have been developed to label and track transplanted cells. Cell markers such as fluorescent dyes or replication-incompetent retroviruses containing reporter genes can be microinjected into embryos, and groups of labeled cells can be followed over time (e.g., Chapter 30 by Mikawa). Various inhibitors can be applied to embryos to inactivate com-

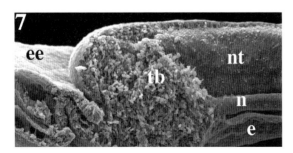

Figure 7 Scanning electron micrograph of an early chick embryo. Sagittal section through the level of the incipient tail (caudal is to the left). e, Endoderm; ee, epidermal ectoderm; n, notochord; nt, neural tube; tb, tail bud. Bar: 50 μm.

ponents of the cytoskeleton (e.g., cytochalasin D to depolymerize microfilaments; colchicine or nocodazole to depolymerize microtubules) or extracellular matrix (e.g., hyaluronidase), and DNA analogs can be injected to clock cell cycles (e.g., tritiated thymidine or BrdU) (e.g., Chapter 31 by Adler and Belecky-Adams). A strain of chickens has been developed that is deficient in vitamin A, essentially providing chick embryos with retinoic acid "knockouts" (Maden *et al.*, 1996).

Third, cell fate is determined progressively over time by cell–cell interactions mediated by the secretion of factors (ligands) that bind to receptors on responding cells. Thus, which tissues interact to irrevocably commit cells to their prospective fates can be assessed with currently available strategies and methodologies. Such approaches involve the addition of tissue (i.e., transplantation), to test whether the added tissue is sufficient to determine the fate of the responding tissue (e.g., Chapter 29 by Streit and Stern); and the substraction (i.e., ablation) of tissue to test whether the removed tissue is necessary to determine the fate of the responding tissue. In addition, culture of avian cells has been a standard technique for decades, and cell–cell interactions that influence fate decisions can be precisely defined in these assays (e.g., Chapter 31 by Adler and Belecky-Adams).

Fourth, molecular techniques to identify genes controlling developmental events have been developed for avian embryos. As a result, several developmental control genes that direct early developmental events have been identified. Such techniques include construction of libraries, differential display, and polymerase chain reaction (PCR). Riboprobes isolated through such techniques can then be used with *in situ* hybridization to provide cell-, tissue-, or region-specific markers.

Fifth, gain-of-function and loss-of-function mutations can now be generated in chick embryos. For example, for gain-of-function studies, COS cells or QT6 cells can be transfected with cDNAs for secreted proteins, and expressing cells (which can also be cotransfected with reporter genes) can be transplanted into avian embryos, either in whole-embryo culture or *in ovo* (e.g., Chapter 29 by Streit and Stern). Growth factors can be used to coat small chromatography beads, which can be implanted into embryos to provide a targeted source of protein. Retroviruses can be engineered to express sense constructs of transcription factors or growth-factor receptors, as well as reporter genes, and embryos can be infected with either replication-competent or replication-incompetent retroviruses, depending on the experimental goal. For loss-of-function studies, retroviruses containing antisense or dominant-negative receptor constructs can be injected into embryos, as can antisense oligonucleotides and function-blocking antibodies. Although both gain-of-function and loss-of-function mutations can be generated with such techniques, only mo-

saic patches of mutant cells can produced, not entire mutant embryos.

Several studies have been published that provide detailed methodologies/descriptions for maximizing the use of avian embryos as model systems. For details, see Hamilton (1952), Waddington (1952), Hamburger (1960), Romanoff (1960), Patten (1971), Stern (1993, 1994), Darnell and Schoenwolf (1996, 1998a–c), Selleck (1996), Schoenwolf (1997b), Stern and Bachvarova (1997), Darnell *et al.* (1988), and Mathews and Schoenwolf (1998).

IV. Limitations in the Use of the Avian Embryo as a Model System

Although the avian embryo is an excellent model for embryology/developmental biology, it is not a perfect model. It has four significant limitations. The first limitation concerns the structure of the embryo. The blastoderm at the time the egg is laid is a small structure (about 2 mm in diameter), with cells having different prospective fates lying in close proximity to one another, making isolation of cell populations difficult. Moreover, cell size is small (a few microns) making single-cell injections difficult and the process of single-cell targeting unreliable (however, see Fraser *et al.*, 1990). Although translucent, cells contain yolk and embryos are not transparent as are, for example, zebrafish embryos. In addition, considerable development occurs before the egg is laid, when blastoderms are difficult to access, culture, and study.

The second limitation concerns the sensitivity of the early blastoderm and nascent embryo to undergo abnormal development. Windowing eggs to view and access embryos during neurulation stages invariably induces neural tube defects unless special precautions are taken. Fortunately, the use of modifications in the windowing procedure and whole-embryo culture obviates this problem.

The third limitation concerns the ability to culture blastoderms and embryos. Embryos can be readily cultured from the time of laying but conventional whole-embryo culture (i.e., New culture) is limited to a 24 to 48 hour period. Additional whole-embryo culture methods can be used to extend the culture period, but they introduce other disadvantages such as less access to embryos and more elaborate preparation procedures.

The final limitation concerns the lack of ability to genetically manipulate avian embryos and to induce mutations. Transgenic procedures are being attempted but no current method exists. Additionally, few spontaneous mutations have been obtained and preserved for additional studies. However, many of those that have been maintained have become important models, especially for studies of limb development (e.g., Niswander, 1997). Finally, few immortalized cell lines and no equivalent of mouse embryonic stem cells exist.

V. Conclusions

For a number of reasons, the avian embryo has served as an excellent model system for well over the last 100 years. The application of modern cell biological, marking and molecular techniques have further extended its usefulness greatly. Development of genetic procedures in avian embryos, including the promising ways of altering gene expression on an embryo-by-embryo basis in a temporally and spatially restricted manner provides further promise that the avian embryo will be a favored experimental model for well into twenty-first century.

Acknowledgments

Original work described herein from the Schoenwolf laboratory was supported by Grants NS 18112 and HD 28845 from the National Institutes of Health.

References

Darnell, D. K., and Schoenwolf, G. C. (1996). Modern techniques for cell labeling in avian and murine embryos. *In* "Molecular and Cellular Methods in Developmental Toxicology" (G. P. Daston, ed.), pp. 231–273. CRC Press, Boca Raton, Florida.

Darnell, D. K., and Schoenwolf, G. C. (1998a). The chick embryo as a model system for analyzing mechanisms of development. *In* "Methods in Molecular Biology. Developmental Biology Protocols" (R. S. Tuan and C. W. Lo, eds.), Vol. 1. Humana, Totowa, New Jersey.

Darnell, D. K., and Schoenwolf, G. C. (1998b). Culture of avian embryos. *In* "Methods in Molecular Biology. Developmental Biology Protocols" (R. S. Tuan and C. W. Lo, eds.), Vol. 1. Humana, Totowa, New Jersey.

Darnell, D. K., and Schoenwolf, G. C. (1998c). Transplantation chimeras: Use in analyzing mechanisms of avian development. *In* "Methods in Molecular Biology. Developmental Biology Protocols" (R. S. Tuan and C. W. Lo, eds.), Vol. 1. Humana, Totowa, New Jersey.

Darnell, D. K., Garcia-Martinez, V., Lopez-Sanchez, C., Yuan, S., and Schoenwolf, G. C. (1998). Dynamic labeling techniques for fate mapping, testing cell commitment and following living cells in avian embryos. *In* "Methods in Molecular Biology. Developmental Biology Protocols" (R. S. Tuan and C. W. Lo, eds.), Vol. 1. Humana, Totowa, New Jersey.

Eyal-Giladi, H., and Kochav, S. (1976). From cleavage to primitive streak formation: A complementary normal table and a new look at the first stages of the development of the chick. I. General morphology. *Dev. Biol.* **49**, 321–337.

Fraser, S., Keynes, R., and Lumsden, A. (1990). Segmentation in the chick embryo hindbrain is defined by cell lineage restrictions. *Nature (London)* **344**, 431–435.

Hamburger, V. (1960). "A Manual of Experimental Embryology," Revised Ed. Univ. of Chicago Press, Chicago.

Hamburger, V., and Hamilton, H. L. (1951). A series of normal stages in the development of the chick embryo. *J. Morphol.* **88**, 49–92.

Hamilton, H. L. (1952). "Lillie's Development of the Chick," 3rd Ed. Holt, Rinehart, and Winston, New York.

Maden, M., Gale, E., Kostetskii, I., and Zile, M. (1996). Vitamin A-deficient quail embryos have half a hindbrain and other neural defects. *Curr. Biol.* **6**, 417–426.

Mathews, W. W., and Schoenwolf, G. C. (1998). "Atlas of Descriptive Embryology," 5th Ed. Prentice-Hall, Upper Saddle River, New Jersey.

Niswander, L. (1997). Limb mutants: What can they tell us about normal limb development? *Curr. Opin. Genet. Dev.* **7**, 530–536.

Patten, B. M. (1971). "Early Embryology of the Chick," 5th Ed. McGraw-Hill, New York.

Placzek, M., and Furley, A. (1996). Neural development: Patterning cascades in the neural tube. *Curr. Biol.* **6**, 526–529.

Romanoff, A. L. (1960). "The Avian Embryo." Macmillan, New York.

Romer, A. S. (1968). "The Procession of Life." The World Publishing Company, Cleveland, Ohio.

Sander, K. (1997). "Landmarks in Developmental Biology. 1883–1924." Springer-Verlag, Heidelberg.

Sasai, Y., and De Robertis, E. M. (1997). Ectodermal patterning in vertebrate embryos. *Dev. Biol.* **182**, 5–20.

Schoenwolf, G. C. (1995). "Laboratory Studies of Vertebrate and Invertebrate Embryos. Guide and Atlas of Descriptive and Experimental Embryology," 7th Ed. Prentice-Hall, Englewood Cliffs, New Jersey.

Schoenwolf, G. C. (1997a). Reptiles and birds. *In* "Embryology. Constructing the Organism" (S. F. Gilbert and A. M. Raunio, eds.), pp. 437–458. Sinauer, Sunderland, Massachusetts.

Schoenwolf, G. C. (1997b). "Embryo: CD-Color Atlas for Developmental Biology." Prentice-Hall, Englewood Cliffs, New Jersey.

Selleck, M. A. J. (1996). Culture and microsurgical manipulation of the early avian embryo. *In* "Methods in Cell Biology" (M. A. Bronner-Fraser, ed.), Vol. 51. pp. 1–21. Academic Press, San Diego.

Smith, J. L., and Schoenwolf, G. C. (1997). Neurulation: Coming to closure. *Trends Neurosci.* **20**, 510–517.

Stern, C. D. (1993). Avian embryos. *In* "Essential Developmental Biology. A Practical Approach" (C. D. Stern and P. W. H. Holland, eds.), pp. 45–54. Oxford Univ. Press, Oxford.

Stern, C. D. (1994). The avian embryo: A powerful model system for studying neural induction. *FASEB J.* **8**, 687–691.

Stern, C. D., and Bachvarova, R. (1997). Early chick embryos *in vitro*. *Int. J. Dev. Biol.* **41**, 379–387.

Waddington, C. H. (1952). "The Epigenetics of Birds." Cambridge Univ. Press, Cambridge.

29

More to Neural Induction Than Inhibition of BMPs

Andrea Streit
Claudio D. Stern
Department of Genetics and Development
Columbia University
New York, New York 10032

I. Introduction

The development of the nervous system is initiated by the induction of the neural plate from naive ectodermal cells. Spemann and Mangold (1924) showed that when the dorsal lip of the blastopore (Spemann's organizer) is grafted to a new position, the overlying ectoderm gives rise to an ectopic neural plate: cells change their fate from epidermal to neural. It has been known for a long time that Hensen's node (the tip of the primitive streak) is the amniote equivalent of the amphibian organizer (Waddington 1932a,b; Gallera, 1971; Dias and Schoenwolf, 1990; Storey et al., 1992; Beddington, 1994). Transplantation of the node to an ectopic site results in the induction of a secondary neural axis in host tissue, with correct anteroposterior and dorsoventral patterning.

Between the 1920s and 1980s, an enormous amount was learned about the cellular events associated with neural induction and patterning in both amphibians and the chick (reviewed by Nakamura and Toivonen, 1978; Lumsden and Krumlauf, 1996; Tanabe and

Jessell, 1996). For example, it was established that the organizer itself as well as some of its descendants emit neuralizing and regionalizing signals (Spemann and Mangold, 1924; Waddington, 1932b; Mangold, 1933; Nakamura and Toivonen, 1978; Hamburger, 1988), that both planar (within the ectoderm) and vertical (between mesendoderm and ectoderm) signals may play a role in neural induction (Doniach, 1993; Ruiz i Altaba, 1993), that the neural plate itself is capable of inducing more neural plate from competent ectoderm ("homoiogenetic induction") (Nakamura and Toivonen, 1978; Waddington, 1932b), that competence is an important factor in determining the position and extent of the neural plate (Albers, 1987), and that signals from the notochord and overlying epidermis set up the dorsoventral character of the neural tube (Tanabe and Jessell, 1996). Interspecific grafts of the organizer between different classes of vertebrates (fish, amphibian, avian, and mammalian in almost every combination) lead to neural induction, suggesting that at least some of the signaling mechanisms are highly conserved (Waddington, 1934; Waddington and Waterman, 1934; Oppenheimer, 1936; Kintner and

Dodd, 1991; Blum et al., 1992; Hatta and Takahashi, 1996). However, until very recently, the nature of these signals had remained elusive.

Within the last few years, studies in Xenopus have revealed that neural induction depends on a balance between antagonizing signals: the ventralizing bone morphogenetic protein-4 (BMP-4), which acts as an epidermal inducer and neural antagonist, and the dorsalizing and neuralizing molecules chordin, noggin, follistatin, and Xnr3 (for review see Sasai and DeRobertis, 1997; Wilson and Hemmati-Brivanlou, 1997; Weinstein and Hemmati-Brivanlou, 1997; Fig. 1). In Xenopus, dissociation of early gastrula ectoderm leads to the formation of neural tissue (Godsave and Slack, 1989; Grunz and Tacke, 1989; Sato and Sargent, 1989), which has been explained by the idea that release from inhibitory signals can cause neural development. Indeed, inhibition of the BMP-4 signaling pathway by dominant negative receptors (Hemmati-Brivanlou and Melton, 1992, 1994; Suzuki et al., 1995; Xu et al., 1995), dominant negative BMP-4 or -7 ligand (Hawley et al., 1995), or antisense BMP-4 RNA (Sasai et al., 1995) causes ectodermal cells to adopt a neural fate instead of becoming epidermis. In addition, BMP-4 can prevent the differentiation of neural cells in dissociated gastrula ectoderm (Wilson and Hemmati-Brivanlou, 1995). The expression pattern of BMP-4 in Xenopus is in accordance with its proposed antineuralizing function: in the early gastrula, BMP-4 transcripts are widely expressed in the entire ectoderm and then are cleared from the presumptive neural plate at the time when the organizer appears (Fainsod et al., 1994; Hemmati-Brivanlou and Thomsen, 1995; Schmidt et al., 1995).

The candidate endogenous neural inducers noggin (Lamb et al., 1993), chordin (Sasai et al., 1995), follistatin (Hemmati-Brivanlou et al., 1994), and Xnr3 (Hansen et al., 1997) are all expressed in the organizer

and its descendants. Ectopic expression of any of them results in the development of neural tissue from gastrula ectoderm (Lamb et al., 1993; Hemmati-Brivanlou et al., 1994; Sasai et al., 1995; Hansen et al., 1997). They all seem to counteract BMP4 signaling, and chordin, noggin, and follistatin do so by direct binding to BMPs (Zimmermann et al., 1996; Piccolo et al., 1996; Fainsod et al., 1997). All four molecules promote the development of anterior neural structures (see also Chapter 20 by Sullivan et al.), whereas proteins of the fibroblast growth factor family (FGF; Isaacs et al., 1992; Cox and Hemmati-Brivanlou, 1995; Kengaku and Okamoto, 1995; Lamb and Harland, 1995; Launay et al., 1996), retinoids (Durston et al., 1989; Sive et al., 1990; Ruiz i Altaba and Jessell, 1991), and Wnts (McGrew et al., 1995) are thought to be important for the development of posterior neural tissue (Fig. 1).

Nieuwkoop and colleagues (Nieuwkoop et al., 1952; Nieuwkoop and Nigtevecht, 1954) proposed that initial signals from the amphibian organizer generate anterior neural tissue, a process they termed "activation," whereas later interactions with other tissues generate posterior neural structures ("transformation"). Thus, the more recent molecular evidence in Xenopus appears to support Nieuwkoop's proposal: BMP antagonists first generate anterior nervous system, while FGF, retinoids, and Wnts subsequently posteriorize it.

Despite the attractive simplicity of the model shown in Fig. 1, several important features of neural induction and patterning are left unresolved. For example, homoiogenetic induction cannot be explained since none of the neural inducing molecules identified to date are expressed in newly induced neural plate. This implies that there are still some molecules capable of initiating neural induction remaining to be discovered. This conclusion is further supported by the finding that, in mouse, disruption of the genes encoding BMP-2 (Zhang and Bradley, 1996), BMP-4 (Winnier et al., 1995), BMP-7 (Dudley et al., 1995; Jena et al., 1997), BMP-receptor type I (Mishina et al., 1995), and follistatin (Matzuk et al., 1995) do not produce specific defects in the mouse central nervous system (CNS).

II. Neural Induction in Amniotes

From the earliest days of studies on neural induction, experiments also have been conducted in amniote embryos, following from the pioneering work of Waddington (1930, 1932a,b). In the experiments of Waddington and his followers, Hensen's node was generally transplanted to the edge of the area pellucida of a host embryo (sometimes of a different species so that host and donor cells can be distinguished) (Fig. 2). This position lies either very close to, or even within, the prospective neural plate of the host, and it is therefore impossible to rule out the possibility that the organizer

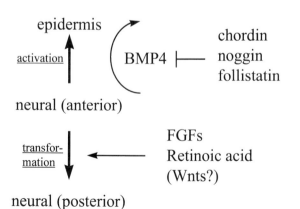

Figure 1 Simplified diagram of the model proposed in amphibians to explain neural induction and patterning. Neural ectoderm represents a "default" state, which requires inhibition of BMP signaling to manifest itself as the development of anterior neural structures. Other signals, such as FGF, Retinoids, and Wnts can transform some of this tissue into posterior nervous system.

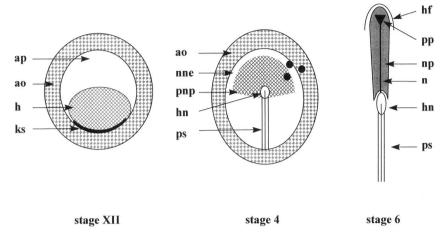

stage XII   **stage 4**   **stage 6**

Figure 2   Diagrams of three stages of chick development. (a) Stage XII, showing Koller's sickle (ks), the hypoblast (h), and the positions of embryonic (area pellucida; ap) and extraembryonic (area opaca; ao) regions; (b) stage 4 (definitive primitive streak) embryo, showing the primitive streak (ps) and Hensen's node (hn), the extent of the prospective neural plate (pnp), the nonneural ectoderm (nne), and the positions for the organizer grafts (black circles) discussed in the text; (c) stage 6 (early neural plate), showing the extent of the neural plate (np), the notochord (n), prechordal plate (pp), the head fold (hf), and the remnants of the primitive streak (ps) and Hensen's node (hn).

merely recruits already induced cells into a second neural plate. However, even Waddington himself (1934) showed that a more peripheral region (the area opaca), which normally gives rise only to extraembryonic tissue, also can respond to an organizer graft by producing a patterned nervous system. For this reason, the most recent and cautious assays on neural induction have been done within the area opaca.

Early studies using only morphological criteria (Waddington, 1936, 1940; Woodside, 1937; Gallera and Ivanov, 1964; Gallera and Nicolet, 1969; Gallera, 1970, 1971), and more recent studies using molecular markers (Dias and Schoenwolf, 1990; Storey *et al.*, 1992, 1995; Streit *et al.*, 1997) have established the time-course of changes in both the inducing ability of the organizer (Hensen's node) and the responsiveness of the ectoderm (competence) in the chick embryo. They showed that neural inducing ability is lost gradually, starting at the full primitive streak stage (stage 4). As the head process emerges (stage 5), the inducing capacity of the node diminishes to about half, and only posterior nervous system can be generated. By stage 7, the node can no longer induce neural structures from the host, and can only self-differentiate (Storey *et al.*, 1992).

Although these studies have provided complementary information on neural induction to that obtained in amphibians, work on the two vertebrate groups has sometimes led to opposing views. The remainder of this chapter deals with these differences, which concern three aspects of early neural development: (i) how competence is regulated; (ii) whether or not regionalization can be dissociated from the initial inductive events, and (iii) whether inhibition of BMP signaling is sufficient for neural induction.

III. Regulation of Neural Competence

Inductive interactions during development are governed not only by the properties of the inducing cells, but also largely by the responsive capacity, or competence, of the tissue receiving the inducing signals (Holtfreter, 1933, 1938; Waddington, 1940; for review see Gurdon, 1987). In the chick, the embryonic region that is competent to respond to neural inducing signals is much broader than the future neural plate, reaching out into the extraembryonic epiblast. As development proceeds, competence is gradually lost from more lateral regions. The $L5^{220}$ glycoprotein has been suggested to be a marker for competence, since its expression pattern closely mirrors the changes in competence of the epiblast in time and in space (Streit *et al.*, 1995, 1997). Functional experiments indicate that $L5^{220}$ is directly involved in the response of cells to inducing signals: in the presence of L5 antibody, the response to neural inducing signals from Hensen's node is completely inhibited (Roberts *et al.*, 1991).

Studies in *Xenopus* proposed that competence to respond to successive inductive stimuli (e.g., mesodermal, neural, eye, and ear inductions) is a cell autonomous property of the responding tissue, which does not require external control (Gurdon, 1987; Grainger and Gurdon, 1989; Servetnik and Grainger, 1991; Steinbach *et al.*, 1997). Recent experiments in the chick, however, suggest that neural competence can be regulated by external signals such as the secreted factor hepatocyte growth factor/scatter factor (HGF/SF). HGF/SF is expressed in a subpopulation of cells in Hensen's node at the time when neural induction occurs (Streit *et al.*, 1995). When exposed to HGF/SF *in vitro* or *in vivo*, responsive epiblast continues to express and upregulates $L5^{220}$ beyond the normal time, whereas the expression of neural specific markers ($L5^{450}$, *Sox-2*, *Sox-3*) is not affected (Streit *et al.*, 1995, 1997). Moreover, embryos in which $L5^{220}$ has been prolonged with a source of HGF/SF also retain their competence to respond to a graft of Hensen's node beyond the normal time. This suggests that HGF/SF may be able to maintain competence, but cannot induce neural tissue in naive ectoderm, and that Hensen's node itself may release factors that prolong the responsiveness of the surrounding epiblast to neural inducing signals (Streit *et al.*, 1997). Taken together, these experiments suggest that $L5^{220}$ is causally related to neural competence, and that competence can be regulated by signaling molecules from outside the prospective neural plate.

IV. Neural Induction Can Be Separated from Regionalization

As in amphibians, experiments in amniotes have led to the view that neural induction consists of at least two distinct steps (Waddington and Needham, 1936; Waddington, 1940). The first, which Waddington termed "evocation" consists of a diversion of the fates of the ectodermal cells from epidermal to neural, but the induced cells lack all regional character. The second step, called "individuation," is the process by which the early neural primordium acquires stable anteroposterior characteristics. There are two main differences between the proposals of Waddington (working on amniotes) and those of Nieuwkoop *et al.* (1952; working on amphibians; see above). First, whether or not initial inductive interactions also lead to regionalization, and second, whether it is possible to anteriorize the neural plate (which would be compatible with Waddington's model but not with Nieuwkoop's).

Two unrelated observations suggest that regionalization of the neuroepithelium can at least continue to stages of development long after the node has lost its neural inducing activity. First, when metencephalic neuroepithelium is transplanted to the forebrain at stage 10, cells surrounding the graft acquire the expression of the midbrain marker *En-2* and a cerebellar phenotype (Itasaki *et al.*, 1991; Martínez *et al.*, 1991; Bally-Cuif *et al.*, 1992; Nakamura *et al.*, 1994; Crossley *et al.*, 1996). Second, transplantation of somitic mesoderm can respecify the Hox-code of neighboring hindbrain neuroepithelium (Itasaki *et al.*, 1996), as can relocations of the hindbrain neuroepithelium itself (Grapin-Botton *et al.*, 1995; Itasaki *et al.*, 1996). Most of these experiments have demonstrated that neuroepithelium can be posteriorized, which is compatible with both Waddington's and Nieuwkoop's models. However,

more recent work from several laboratories has shown that one embryonic structure, the prechordal mesendoderm, can also anteriorize neuroepithelium fairly late in its development (Ang and Rossant, 1993; Ang *et al.*, 1994; Foley *et al.*, 1997; Shimamura and Rubenstein, 1997; Pera and Kessel, 1997; Dale *et al.*, 1997).

Finally, another recent observation seems to fit with Waddington's proposal rather well. The neural plate that develops in the "competence rescue" experiment described above expresses the panneural marker *Sox-2*, but does not express any of the early expressed regional markers tested, which cover the forebrain (*Tailless*), diencephalon (3A10), hindbrain (3A10 and *Krox-20*), and posterior spinal cord (*Hoxb-9*). Thus, neuralization can be separated experimentally from regionalization: these neural plates are neuralized, but fail to become regionalized along the anteroposterior axis. These data lend strong support to Waddington's hypothesis that initial neural induction gives gives rise to nonregionalized neural tissue, which acquires regional identity in subsequent steps, but are difficult to reconcile with Nieuwkoop's views.

V. The Roles of Chordin and BMP Inhibition in Neural Induction

Whereas in *Xenopus* the release from the inhibitory, epidermal inducing signal BMP-4 by antagonizing secreted factors such as chordin, noggin, and follistatin, or by simple cell dissociation, seems to be sufficient to make ectodermal cells develop into neural tissue (see also Chapter 20 by Sullivan *et al.*), several lines of evidence in amniotes suggest a much higher degree of complexity: in the chick, inhibition of the BMP signaling pathway does not lead to neural differentiation, nor can BMPs prevent initial induction. However, antagonism between BMP-4 and chordin seems to play a role in controlling the extent of the neural plate once it is formed. We will discuss these points in detail below.

A. Inhibition of BMP Signaling Does Not Appear to Be Sufficient for Neural Induction

In contrast with *Xenopus*, where dissociation of animal caps leads to formation of neural ectoderm (see above), dissociation of early chick epiblast from a variety of stages, does not elicit neural differentiation, but rather causes an increase in the proportion of cells that differentiate into skeletal muscle (George-Weinstein *et al.*, 1996; A. Streit and C. D. Stern, unpublished observations, 1997).

Unlike in *Xenopus* where overexpression of the BMP-4 antagonist chordin results in the formation of anterior neural tissue, misexpression of chordin in the nonneural epiblast of the area pellucida or the inner margin of the area opaca of the chick embryo does not elicit the formation of a neural plate or expression of neural markers. Therefore, chordin on its own cannot mimic a graft of the organizer in either region. Furthermore, chordin (Streit *et al.*, 1998) as well as other BMP antagonists such as noggin (Connolly *et al.*, 1997) and follistatin (Connolly *et al.*, 1995) remain expressed in Hensen's node well beyond the period when this structure has lost its neural inducing ability.

In addition, neither *BMP-4* nor *BMP-7* is expressed in the prospective neural plate at the stages when neural induction is thought to start (Schultheiss *et al.*, 1997; Watanabe and Le Douarin, 1996; Streit *et al.*, 1998) as is the case in *Xenopus* (Fainsod *et al.*, 1994; Hemmati-Brivanlou and Thomsen, 1995; Schmidt *et al.*, 1995), and misexpression of BMP-4 or -7 within the prospective neural plate region at any stage of development (even before primitive streak formation) does not prevent initial neural induction (Streit *et al.*, 1998).

Together, these observations suggest that neural induction by the organizer in the chick is not explicable merely as the release from negative, epidermal-inducing signals mediated by BMPs.

B. Chordin and BMPs at the Edges of the Neural Plate

A graft of Hensen's node to the area opaca induces expression of BMP-4 as a ring at a distance from the transplanted organizer (A. Streit and C. D. Stern, unpublished, 1997; Fig. 3), which raises the possibility that BMP-4 plays a role at the borders of the developing neural plate. This is consistent with the reported role of many members of the transforming growth factor β (TGFβ) superfamily, including BMP-4 and -7, in dorsoventral patterning of the neural tube, where they induce neural crest and specific, dorsally located neuronal cell types (Liem *et al.*, 1995, 1997). The expression of chordin (Streit *et al.*, 1998), as well as that of noggin and follistatin (Connolly *et al.*, 1995, 1997) in the notochord and node of these later stage embryos is consistent with this, as they might serve to restrict the range of action of BMPs to the dorsal portion of the neural tube. Also consistent with this, we have found that grafts of BMP-4 or -7 secreting cells at the border of the fairly mature neural plate expand the domain of expression of the dorsal marker *msx-1* (A. Streit and C. D. Stern, unpublished observations, 1997).

In addition, when BMP-4 is misexpressed at the border of the *early* neural plate, this appears slightly reduced in size, while chordin slightly expands the neural plate (Streit *et al.*, 1998, and A. Streit and C. D. Stern, unpublished observations, 1997; Fig. 3). This suggests that one role of chordin and BMPs is to adjust the po-

sition of the lateral edges of the neural plate, fine-tuning the results of initial induction and perhaps playing a role in determining the shape of the neural plate. A similar conclusion can be reached by examination of the phenotype of zebrafish mutants for the homologs of these molecules (Hammerschmidt *et al.*, 1996; Kishimoto *et al.*, 1997). In *chordino* (chordin mutation), the neural plate is reduced but not absent, while in *swirl* (BMP-2 mutation) the neural plate is expanded but does not completely replace the epidermis. *chordino* × *swirl* double-mutants display a phenotype identical to that of *swirl*, suggesting that no additional BMPs are present that overlap functionally with *swirl* (Hammerschmidt *et al.*, 1996). Taken together, the results summarized above indicate that the chordin/BMP system is probably involved in controlling the width of the early neural plate and later its dorsoventral pattern, but is not solely responsible for initiating neural induction.

VI. Multiple Steps in Neural Induction Revealed by Molecular Markers

In *Xenopus*, most assays for neural induction are generally based only on the expression of late, neural-specific genes. Typical markers include NCAM (which is neural plate-specific in *Xenopus* but not in chick), an acetylated form of tubulin expressed in differentiating neurons, neurofilament proteins, markers for the cement gland (an amphibian-specific anterior structure), as well as several regional markers. In the chick, equivalent markers for cells that have acquired their neural and/or regional identities are also available. In addition, many markers are expressed in a clear temporal sequence both during normal development of the neural plate and in response to a graft of Hensen's node (Fig. 4).

Since chordin alone is not sufficient to elicit neural induction in the competent region of the area opaca epiblast, but can regulate the lateral extent of the induced neural plate (see above), the possibility was tested that signals from the organizer may be required in conjunction with, or upstream of chordin. When area opaca epiblast is exposed to Hensen's node for a short time (insufficient by itself to induce a neural plate) and the graft then replaced with chordin-expressing cells, a neural plate forms in host tissue, which expresses the general neural marker *Sox-3* (Streit *et al.*, 1998). To test whether HGF/SF is the sole signal produced by the node that is required upstream of chordin to generate a neural plate, the epiblast was exposed to both HGF/SF and chordin: no expression of *Sox-3* was seen (Streit *et al.*, 1998). This experiment shows that the node emits signal(s) other than HGF/SF, which are required upstream of, or in parallel with chordin to

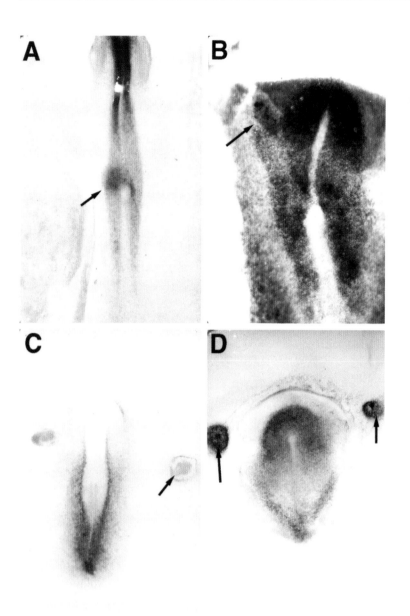

Figure 3 (A) When BMP-4 is misexpressed at the edge of the future neural plate at stage 4 using a graft of transfected COS cells (arrow), the width of the neural plate on the side of the graft is reduced. The neural plate is visualized here by *in situ* hybridization for Sox-2. (B) When the BMP-4 inhibitor chordin is misexpressed at the edge of the neural plate, it prevents the downregulation of Sox-3 and expands the neural plate laterally (arrow, position of transfected COS cells). (C) A graft of quail Hensen's node (arrow) into the extraembryonic area opaca at stage 4 induces BMP-4 expression (blue) as a ring surrounding the graft. This expression is reminiscent of the endogenous pattern of BMP-4 at the border of the neural plate, as seen in the host embryo. Graft-derived quail cells are visualized with the QCPN antibody (brown). (D) Expression of *R1* in the normal stage 4 embryo and its induction by grafts of quail Hensen's node in the area opaca (arrows), 4 hours after grafting. Quail cells are labeled with the QCPN antibody (brown).

produce a neural plate in competent epiblast of the area opaca.

To gain insight into the nature of such upstream signals and the responses to them, we have more recently performed a screen for genes expressed in area opaca epiblast in response to a short exposure to a grafted node. An initial differential screen identified several novel genes, one which was investigated further and temporarily named *R1*. It contains a putative open reading frame of about 1.3 kb, without obvious homol-

ogy to any known gene, and flanked by long 3' and 5' untranslated regions. *In situ* hybridization analysis in the normal embryo showed that *R1* expression is restricted to the area pellucida epiblast from prior to primitive streak formation, that it starts to be downregulated after stage 4, and that it is induced after just 2 hours following a graft of Hensen's node to the area opaca (Fig. 3). These results identify *R1* as a very early response gene to inducing signals from the node, upstream of Sox-3 and Sox-2 (Berliner *et al.*, 1998; Figs. 4 and 5).

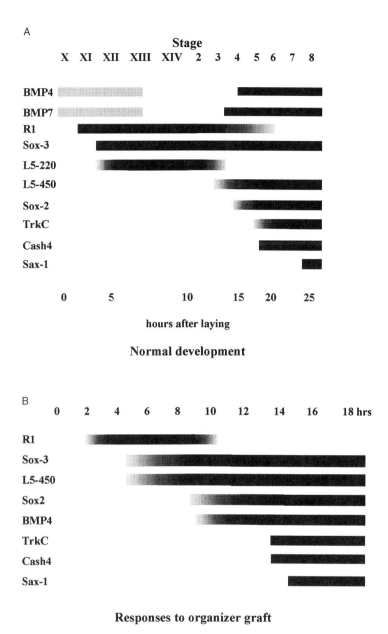

A

Normal development

B

Responses to organizer graft

Figure 4 Time course of expression of several genes during the course of neural induction. (A) Normal expression of the markers at different stages of development. Roman numbers indicate Eyal-Giladi and Kochav (1976) pre-primitive streak stages and arabic numerals are Hamburger and Hamilton (1951) stages after formation of the primitive streak. (B) Time course (in hours) of expression of the same markers following a graft of Hensen's node to the area opaca. Levels of gray density represent relative levels of gene expression.

VII. A Model for Neural Induction

The observations discussed above argue that BMP inhibition by chordin is not sufficient to elicit neural induction in competent epiblast of the chick embryo, and that other steps upstream of this are required. The newly discovered gene *R1* appears to mark one of these steps, and can be induced by signals from the organizer. BMP and its inhibition by chordin appear to control the lateral extent of the neural plate. Taken together, these arguments suggest that the normal events of neural induction constitute a hierarchical chain, and that regions

progressively more remote from the prospective neural plate lack progressively more upstream components of this pathway (see Fig. 5).

Initially, the marker for competence, L5^{220}, is expressed throughout the prospective neural plate as well as in prospective epidermis and extraembryonic epiblast. Competence and expression of L5^{220} is lost rapidly when no maintenance signal, such as HGF/SF is supplied. The next step in the chain results in induction of *R1* which is not normally expressed in extraembryonic regions, but which can be upregulated by signals from the organizer. The region at the border of the fu-

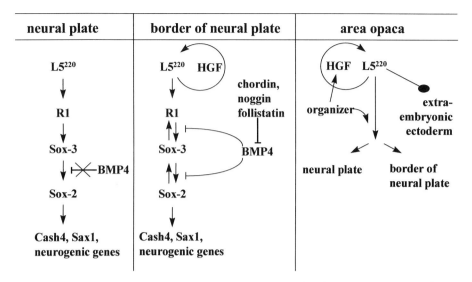

Figure 5 A model for neural induction in the chick. A hierarchy of gene expression precedes neural differentiation in the neural plate and at its border in the normal embryo. BMP-4 can destabilize the expression of *Sox-3* and *Sox-2* at the border of the neural plate but has no effect within the neural plate itself. The extraembryonic (area opaca) epiblast normally expresses the marker for competence, L5^{220}, but no further markers from the pathway unless an organizer is grafted. Organizer grafts result in the formation of an ectopic neural plate and a region that behaves as its border (including expression of BMP-4; Fig. 3).

ture neural plate expresses more downstream genes including *Sox-2* during normal development, but this appears to be an unstable state. If chordin is supplied in this region, the expression of *Sox* genes is maintained and the width of the neural plate is increased; if BMP-4 is expressed in this region, the width of the neural plate is reduced and expression of *Sox* genes is lost.

Ectoderm cells responding to organizer signals may themselves emit a signal that upregulates *BMP-4* expression, while chordin and other BMP inhibitors secreted by the organizer antagonize this effect. This results in *BMP-4* expression being restricted to a ring delimiting the borders of the neural plate. Finally, cells that have received neural inducing signals and whose neural status has been maintained by BMP inhibition may now express other genes (such as *Cash-4*, *Sax-1*, *Notch*, *delta*, and *serrate*; see Henrique *et al.*, 1995, 1997; Myat *et al.*, 1996), some of which may be involved in neuro- or gliogenesis. Patterning of the neural plate along the rostrocaudal and dorsoventral axes may take place simultaneously with the later events of neural induction, and continue after neural induction is complete, under the influence of neighboring axial (notochord, prechordal mesendoderm) and paraxial (somitic mesoderm) structures.

According to this model, therefore, neural induction can be viewed as a series of consecutive steps, at least some of which may be reversible. Each subsequent step contributes to the stability of previous states. In addition, the model provides a means of generating neural plate lacking regional character, as proposed by Waddington (Waddington and Needham, 1936; Waddington, 1940).

VIII. The Origin of the Organizer and the Onset of Neural Induction

While the above model attempts to explain the events occurring during neural induction, and how responsiveness to signals from the organizer is lost, it leaves open the question of when neural induction truly begins. Recent fate mapping studies have shown that cells that contribute to Hensen's node arise from two separate sites. One population forms a sparse middle layer of cells at stage XII, adjacent to Koller's sickle (Fig. 2)— these cells already express the organizer markers *goosecoid*, *HNF3 β*, *Otx2*, and *chordin* (Izpisúa-Belmonte *et al.*, 1993; Bally-Cuif *et al.*, 1995; Ruiz i Altaba *et al.*, 1995; Streit *et al.*, 1998). As the primitive streak forms, these cells move along with the extending streak towards the center of the blastoderm and become located in the node, from where they emerge mostly as embryonic endoderm by about stage 4^{+}–5 (Izpisúa-Belmonte *et al.*, 1993).

The second population of cells that contributes to the organizer is situated in the epiblast immediately anterior to Koller's sickle at stage X, but by stages XI–XII it undergoes migration within the epiblast sheet, reaching the middle of the blastoderm by stage XII–XIII, where it remains until the node forms. This region contains cells fated to contribute to prechordal mesendo-derm, notochord, and some somite precursors. Although no molecular markers have been found that characterize this cell population at this early stage, it has been located by detailed, stage-by-stage fate mapping analysis (Hatada and Stern, 1994).

The two cell populations come together between stages 3 and 3^+, and a morphologically recognizable node forms immediately thereafter at stages 3^+–4^-. By this time, the cells that make up the node cannot be distinguished according to their origin, and most node cells express the same markers as each other.

Some of the genes that are induced quickly (within a few hours) in response to a graft of Hensen's node in the area opaca also are expressed in the normal embryo at very early stages of development (including *R1* and *Sox3*; Fig. 4). This raises the question of whether some of the earliest steps in neural induction may begin before gastrulation, adjacent to the location of these organizer precursors. If so, do different signals in the hierarchy outlined in Fig. 5 emanate from different cell populations? This is one of the many questions that remain to be addressed.

IX. Comparison with the Amphibian Model

In the chick, BMPs and chordin play a role in primitive streak formation (Streit *et al.*, 1998), consistent with the finding in *Xenopus* that inhibitors of BMP signaling act as dorsalizing agents (Dale *et al.*, 1992; Sasai *et al.*, 1994; Jones *et al.*, 1996). However, unlike *Xenopus* (Sasai *et al.*, 1995; Piccolo *et al.*, 1996), chordin is not sufficient to neuralize nonneural epiblast and neither BMP-4 nor -7 inhibit neural induction. Rather, the chordin/BMP system appears to act either downstream of, or in conjunction with, other factors produced by the organizer. Similar conclusions are reached in *Drosophila*, where the *sog* mutant phenotype in the nervous system is detectable only during midgastrulation (François *et al.*, 1994; Biehs *et al.*, 1996). Thus, *sog* mutants have ventral neural progenitors despite some reduction in the size of the domains of *rhomboid*, *lethal of scute*, and *thick veins*, which has led to the proposal that the major role of *sog* in the nervous system is to stabilize or maintain a subdivision of the primary ectoderm into neural and nonneural territories, established previously by other signals (Biehs *et al.*, 1996).

The chordin/BMP (dpp/sog) antagonistic system of signaling molecules has been tightly conserved in evolution (Holley *et al.*, 1995, 1996; Sasai *et al.*, 1995; for review see, De Robertis and Sasai, 1996; Ferguson, 1996; Holley and Ferguson, 1997). In *Drosophila*, like

in *Xenopus* and chick, it is used many times during development and one function is to control dorsoventral polarity of the embryo. It seems unlikely therefore that *Xenopus* has recruited this system for a new function (neural induction). We believe that a more likely explanation for these differences concerns the assays used for neural induction in *Xenopus* and chick. For example, animal caps excised from chordin-injected (and therefore dorsalized) embryos may contain part of an expanded prospective neural plate which arises from an increase in the size of the organizer. It is also possible that in *Xenopus* the region equivalent to what we have called the "border of the neural plate" (Fig. 5) actually extends much further, covering most of the nonneural ectoderm of the animal cap. In addition, it is interesting to note that tissue dissociation is only effective as a means of eliciting neural differentiation in *Xenopus* during a brief time period (Knecht and Harland, 1997), indicating that mechanisms other than release from negative signals might be involved.

In the chick, a large area of extraembryonic epiblast (the area opaca) is competent to respond to signals from the organizer by generating a complete, patterned nervous system. *Xenopus* lacks such a region. While it might be argued that signals in addition to neural inducing molecules may be required in this extraembryonic region first to convert it into "embryonic" epiblast and then to neuralize it, chordin fails to produce ectopic neural structures even in the embryonic nonneural epiblast (overlying the germinal crescent) at any stage in development unless an ectopic primitive streak is generated first (Streit *et al.*, 1998). These results suggest that chordin can only elicit the formation of an ectopic neural plate through prior induction of tissue with organizer properties, or in conjunction with other neural inducing signals from the organizer.

References

Albers, B. (1987). Competence as the main factor determining the size of the neural plate. *Dev. Growth Differ.* **29**, 535–545.

Ang, A. L., and Rossant, J. (1993). Anterior mesendoderm induces mouse engrailed genes in explant cultures. *Development (Cambridge, UK)* **118**, 139–149.

Ang, S. L., Conlon, R. A., Jin, O., and Rossant, J. (1994). Positive and negative signals from mesoderm regulate the expression of mouse Ots 2 in ectoderm explants. *Development (Cambridge, UK)* **120**, 2979–2989.

Bally-Cuif, L., Alvarado-Mallart, R. M., Darnell, D. K., and Wasseff, M. (1992). Relationship between Wnt-1 and En-2 expression domains during early development of normal and ectopic met-mesencephalon. *Development (Cambridge, UK)* **115**, 999–1099.

Bally-Cuif, L., Gulisano, M., Broccoli, V., and Boncinelli, E.

(1995). c-otx2 is expressed in two different phases of gastrulation and is sensitive to retinoic acid treatment in chick embryo. *Mech. Dev.* **49,** 49–63.

Beddington, R. S. P. (1994). Induction of a second neural axis by the mouse node. *Development (Cambridge, UK)* **120,** 613–620.

Berliner, A., Sirulnik, A., Stern, C. D., and Streit, A. (1998). In preparation.

Biehs, B., François, V., and Bier, E. (1996). The *Drosophila* short gastrulation gene prevents dpp from autoactivating and suppressing neurogenesis in the neuroectoderm. *Genes Dev.* **10,** 2922–2934.

Blum, M., Gaunt, S. J., Cho, K. W. Y., Steinbeisser, H., Blumberg, B., Bittner, D., and De Robertis, E. M. (1992). Gastrulation in the mouse: The role of the homeobox gene *goosecoid. Cell (Cambridge, Mass.)* **69,** 1097–1106.

Connolly, D. J., Patel, K., Seleiro, E. A., Wilkinson, D. G., and Cooke, J. (1995). Cloning, sequencing and expressional analysis of the chick homologue of follistatin. *Dev. Genet.* **17,** 65–77.

Connolly, D. J., Patel, K., and Cooke, J. (1997). Chick noggin is expressed in the organizer and neural plate during axial development, but offers no evidence of involvement in primary axis formation. *Int. J. Dev. Biol.* **41,** 389–396.

Cox, W. G., and Hemmati-Brivanlou, A. (1995). Caudalization of neural fate by tissue recombination and bFGF. *Development (Cambridge, UK)* **121,** 4349–4358.

Crossley, P. H., Martínez, S., and Martin, G. R. (1996). Midbrain development induced by FGF8 in the chick embryo. *Nature (London)* **380,** 66–68.

Dale, J. K. M., Vesque, C., Lints, T. J., Sampath, T. K., Furley, A., Dodd, J., and Placzek, M. (1997). Cooperation of BMP7 and SHH in the induction of forebrain ventral midline cells by prechordal mesoderm. *Cell (Cambridge, Mass.)* **90,** 257–269.

Dale, L., Howes, G., Price, B. M., and Smith, J. C. (1992). Bone morphogenetic protein 4: A ventralizing factor in early *Xenopus* development. *Development (Cambridge, UK)* **115,** 573–585.

De Robertis, E. M., and Sasai, Y. (1996). A common plan for dorsoventral patterning in *Bilateria. Nature (London)* **380,** 37–40.

Dias, M. S., and Schoenwolf, G. C. (1990). Formation of ectopic neuroepithelium in chick blastoderms: Age related capacities for induction and self-differentiation following transplantation of quail Hensen's nodes. *Anat. Rec.* **229,** 437–448.

Doniach, T. (1993). Planar and vertical induction of anteroposterior pattern during the development of the amphibian central nervous system. *J. Neurobiol.* **24,** 1256–1275.

Dudley, A. T., Lyons, K. M., and Robertson, E. J. (1995). A requirement for bone morphogenetic protein-7 during development of the mammalian kidney and eye. *Genes Dev.* **9,** 2795–2807.

Durston, A. J., Timmermans, J. P., Hage, W. J., Hendriks, H. F., de Vries, N. J., Heideveld, M., and Nieuwkoop, P. D. (1989). Retinoic acid causes an anteroposterior

transformation in the developing central nervous system. *Nature (London)* **340,** 140–145.

Eyal-Giladi, H., and Kochav, S. (1976). From cleavage to primitive streak formation: A complementary normal table and a new look at the first stages of the development of the chick. *Dev. Biol.* **49,** 321–337.

Fainsod, A., Steinbeisser, H., and De Robertis, E. M. (1994). On the function of BMP-4 in patterning the marginal zone of the *Xenopus* embryo. *EMBO J.* **13,** 5015–5025.

Fainsod, A., Deissler, K., Yelin, R., Marom, K., Epstein, M., Pillemer, G., Steinbeisser, H., and Blum, M. (1997). The dorsalizing and neural inducing gene follistatin is an antagonist of BMP-4. *Mech. Dev.* **63,** 39–50.

Ferguson, E. L. (1996). Conservation of dorsal–ventral patterning in arthropods and chordates. *Curr. Opin. Genet. Dev.* **6,** 424–431.

Foley, A. C., Storey, K. G., and Stern, C. D. (1997). The prechordal region lacks neural inducing ability, but can confer anterior character to more posterior neuroepithelium. *Development (Cambridge, UK)* **124,** 2983–2996.

François, V., Soloway, M., O'Neill, J. W., Emery, J., and Bier, E. (1994). Dorsal–ventral patterning of the *Drosophila* embryo depends on a putative negative growth factor encoded by the short gastrulation gene. *Genes Dev.* **8,** 2602–2616.

Gallera, J. (1970). Différence de la réactivité à l'inducteur neurogène entre l'ectoblaste de l'aire opaque et celui de l'aire pellucide chez le poulet. *Experientia* **26,** 1353–1354.

Gallera, J. (1971). Primary induction in birds. *Adv. Morphol.* **9,** 149–180.

Gallera, J., and Ivanov, I. (1964). La compétence neurogène du feuillet externe du blastoderme de Poulet en fonction du facteur 'temps.' *J. Embryol. Exp. Morphol.* **12,** 693–711.

Gallera, J., and Nicolet, G. (1969). Le pouvoir inducteur de l'endoblaste présomptif contenu dans la ligne primitive jeune de poulet. *J. Embryol. Exp. Morphol.* **21,** 105–118.

George-Weinstein, M., Gerhart, J., Reed, R., Flynn, J., Callihan, B., Battiacci, M., Miehle, C., Foti, G., Lash, J. W., and Weintraub, H. (1996). Skeletal myogenesis: The preferred pathway of chick embryo epiblast cells *in vitro. Dev. Biol.* **173,** 279–291.

Godsave, S. F., and Slack, J. M. W. (1991). Single cell analysis of mesoderm formation in the *Xenopus* embryo. *Development* **111,** 523–530.

Grainger, R. M., and Gurdon, J. B. (1989). Loss of competence in amphibian induction can take place in single nondividing cells. *Proc. Natl. Acad. Sci. U.S.A.* **86,** 1900–1904.

Grapin-Botton, A., Bonnin, M. A., McNaughton, L. A., Krumlauf, R., and Le Douarin, N. (1995). Plasticity of transposed rhombomeres: Hox gene induction is correlated with phenotypic modifications. *Development (Cambridge, UK)* **121,** 2707–2721.

Grunz, H., and Tacke, L. (1989). Neural differentiation of *Xenopus laevis* ectoderm takes place after disaggregation and delayed reaggregation without inducer. *Cell. Differ. Dev.* **28**, 211–218.

Gurdon, J. B. (1987). Embryonic induction—molecular prospects. *Development (Cambridge, UK)* **99**, 285–306.

Hamburger, V. (1988). "The Heritage of Experimental Embryology: Hans Spemann and the Organizer." Oxford Univ. Press, Oxford.

Hamburger, V., and Hamilton, H. L. (1951). A series of normal stages in the development of the chick embryo. *J. Morphol.* **88**, 49–92.

Hammerschmidt, M., Serbedzija, G. N., and McMahon, A. P. (1996). Genetic analysis of dorsoventral pattern formation in the zebrafish: Requirement of a BMP-like ventralizing activity and its dorsal repressor. *Genes Dev.* **10**, 2452–2461.

Hansen, C. S., Marion, C. D., Steele, K., George, S., and Smith, W. C. (1997). Direct neural induction and selective inhibition of mesoderm and epidermis inducers by Xnr3. *Development (Cambridge, UK)* **124**, 483–492.

Hatada, Y., and Stern, C. D. (1994). A fate map of the epiblast of the early chick embryo. *Development (Cambridge, UK)* **120**, 2879–2890.

Hatta, K., and Takahashi, Y. (1996). Secondary axis induction by heterospecific organizers in zebrafish. *Dev. Dyn.* **205**, 183–195.

Hawley, S. H. B., Wünnenberg-Stapleton, K., Hashimoto, C., Laurent, M. N., Watabe, T., Blumberg, B. W., and Cho, K. W. Y. (1995). Disruption of BMP signals in embryonic *Xenopus* ectoderm leads to direct neural induction. *Genes Dev.* **9**, 2923–2935.

Hemmati-Brivanlou, A., and Melton, D. A. (1992). A truncated activin receptor inhibits mesoderm induction and formation of axial structures in *Xenopus* embryos. *Nature (London)* **359**, 609–614.

Hemmati-Brivanlou, A., and Melton, D. A. (1994). Inhibition of activin receptor signaling promotes neuralization in *Xenopus*. *Cell (Cambridge, Mass.)* **77**, 273–281.

Hemmati-Brivanlou, A., and Thomsen, G. H. (1995). Ventral mesodermal patterning in *Xenopus* embryos: Expression patterns and activities of BMP-2 and BMP-4. *Dev. Genet.* **17**, 78–89.

Hemmati-Brivanlou, A., Kelly, O. G., and Melton, D. A. (1994). Follistatin, an antagonist of activin, is expressed in the Spemann organizer and displays direct neuralizing activity. *Cell (Cambridge, Mass.)* **77**, 238–295.

Henrique, D., Adam, J., Myat, A., Chitnis, A., Lewis, J., and Ish-Horowicz, D. (1995). Expression of a delta homologue in prospective neurons in the chick. *Nature (London)* **375**, 787–790.

Henrique, D., Tyler, D., Kintner, C., Heath, J. K., Lewis, J. H., Ish-Horowicz, D., and Storey, K. G. (1997). Cash4, a novel achaete-scute homolog induced by Hensen's node during generation of the posterior nervous system. *Genes Dev.* **11**, 603–615.

Holley, S. A., and Ferguson, E. L. (1997). Fish are like flies are like frogs: Conservation of dorsal–ventral patterning mechanisms. *BioEssays* **19**, 281–284.

Holley, S. A., Jackson, P. D., Sasai, Y., Lu, B., De Robertis, E. M., Hoffmann, F. M., and Ferguson, E. L. (1995). A conserved system for dorsal–ventral patterning in insects and vertebrates involving sog and chordin. *Nature (London)* **376**, 249–253.

Holley, S. A., Neul, J. L., Attisano, L., Wrana, J. L., Sasai, Y., O'Connor, M. B., De Robertis, E. M., and Ferguson, E. L. (1996). The *Xenopus* dorsalizing factor noggin ventralizes *Drosophila* embryos by preventing DPP from activating its receptor. *Cell (Cambridge, Mass.)* **86**, 607–617.

Holfreter, J. (1933). Der Einfluss von Wirtsalter und verschiedenen Organbezirken auf die Differenzierung von angelagertem Gastrulaektoderm. *Arch. Entwicklungsmech. Org.* **127**, 619–775.

Holtfreter, J. (1938). Veränderung der Reaktionsweise im alternden isolierten Gastrulaektoderm. *Arch. Entwicklungsmech. Org.* **138**, 163–196.

Isaacs, H. V., Tannahill, D., and Slack, J. M. (1992). Expression of a novel FGF in the *Xenopus* embryo. A new candidate inducing factor for mesoderm formation and anteroposterior specification. *Development (Cambridge, UK)* **114**, 711–720.

Itasaki, N., Ichijo, H., Hama, L., Matsuno, T., and Nakamura, H. (1991). Establishment of rostrocaudal polarity in rectal primordium: Engrailed expression and subsequent rectal polarity. *Development (Cambridge, UK)* **113**, 1133–1144.

Itasaki, N., Sharpe, J., Morrison, A., and Krumlauf, R. (1996). Reprogramming Hox expression in the vertebrate hindbrain: Influence of paraxial mesoderm and rhombomere transposition. *Neuron* **16**, 487–500.

Izpisúa-Belmonte, J. C., De Robertis, E. M., Storey, K. G., and Stern, C. D. (1993). The homeobox gene *goosecoid* and the origin of the organizer cells in the early chick blastoderm. *Cell (Cambridge, Mass.)* **74**, 645–659.

Jena, N., Martin-Seisdedos, C., McCue, P., and Croce, C. M. (1997). BMP7 null mutation in mice: Developmental defects in skeleton, kidney, and eye. *Exp. Cell Res.* **230**, 28–37.

Jones, C. M., Dale, L., Hogan, B. L., Wright, C. V., and Smith, J. C. (1996). Bone morphogenetic protein-4 (BMP-4) acts during gastrula stages to cause ventralization of *Xenopus* embryos. *Development (Cambridge, UK)* **122**, 1545–1554.

Kengaku, M., and Okamoto, H. (1995). bFGF as a possible morphogen for the anteroposterior axis of the central nervous system in *Xenopus*. *Development (Cambridge, UK)* **121**, 3121–3130.

Kintner, C. R., and Dodd, J. (1991). Hensen's node induces neural tissue in *Xenopus* ectoderm. Implications for the action of the organizer in neural induction. *Development (Cambridge, UK)* **113**, 1495–1505.

Kishimoto, Y., Lee, K., Zon, L., Hammerschmidt, M., and Schulte-Merker, S. (1997). The molecular nature of zebrafish *swirl*: BMP2 function is essential during early dorsoventral patterning. *Development (Cambridge, UK)* **124**, 4457–4466.

Knecht, A. K., and Harland, R. M. (1997). Mechanisms of dorsal–ventral patterning in noggin-induced neural tissue. *Development (Cambridge, UK)* **124**, 2477–2488.

Lamb, T. M., and Harland, R. M. (1995). Fibroblast growth

factor is a direct neural inducer, which combined with noggin generates anterior–posterior neural pattern. *Development (Cambridge, UK)* **121**, 3627–3636.

Lamb, T. M., Knecht, A. K., Smith, W. C., Stachel, S. E., Economides, A. N., Stahl, N., Yancopoulos, G. D., and Harland, R. M. (1993). Neural induction by the secreted polypeptide noggin. *Science* **262**, 713–718.

Launay, C., Fromentoux, V., Shi, D. L., and Boucaut, J. C. (1996). A truncated FGF receptor blocks neural induction by endogenous *Xenopus* inducers. *Development (Cambridge, UK)* **122**, 869–880.

Liem, K. F., Tremmel, G., Roelink, H., and Jessell, T. M. (1995). Dorsal differentiation of neural plate cells induced by BMP-mediated signals from epidermal ectoderm. *Cell (Cambridge, Mass.)* **82**, 969–979.

Liem, K. F., Tremml, G., and Jessell, T. M. (1997). A role for the roof plate and its resident TGFβ-related proteins in neuronal patterning in the dorsal spinal cord. *Cell (Cambridge, Mass.)* **91**, 127–138.

Lumsden, A., and Krumlauf, R. (1996). Patterning the vertebrate neuraxis. *Science* **274**, 1109–1114.

McGrew, L. L., Lai, C. J., and Moon, R. T. (1995). Specification of the anteroposterior neural axis through synergistic interaction of the Wnt signaling cascade with noggin and follistatin. *Dev. Biol.* **172**, 337–342.

Mangold, O. (1933). Über die Induktionsfähigkeit der verschiedenen Bezirke der Neurula von Urodelen. *Naturwissenschaften* **21**, 716–766.

Martínez, S., Wassef, M., and Alvarado-Mallart, R. M. (1991). Induction of a mesencephalic phenotype in the 2-day-old chick prosencephalon is preceded by the eary expression of the homeobox gene en. *Neuron* **6**, 971–981.

Matzuk, M. M., Lu, N., Vogel, H., Sellheyer, K., Roop, D. R., and Bradley, A. (1995). Multiple defects and perinatal death in mice deficient in follistatin. *Nature (London)* **374**, 360–363.

Mishina, Y., Suzuki, A., Ueno, N., and Behringer, R. R. (1995). Bmpr encodes a type I bone morphogenetic protein receptor that is essential for gastrulation during mouse embryogenesis. *Genes Dev.* **9**, 3027–3037.

Myat, A., Henrique, D., Ish-Horowicz, D., and Lewis, J. (1996). A chick homologue of serrate and its relationship with Notch and delta homologues during central neurogenesis. *Dev. Biol.* **174**, 233–247.

Nakamura, H., Itasaki, N., and Matsuno, T. (1994). Rostrocaudal polarity formation of chick optic rectum. *Int. J. Dev. Biol.* **38**, 281–286.

Nakamura, O., and Toivonen, S. (1978). "Organizer: A Milestone of a Half Century from Spemann." Elsevier/North-Holland Biomedical, Amsterdam.

Nieuwkoop, P. D., and Nigtevecht, G. V. (1954). Neural activation and transformation in explants of competent ectoderm under the influence of fragments of anterior notochord in urodeles. *J. Embryol. Exp. Morphol.* **2**, 175–193.

Nieuwkoop, P. D., Boterenbrood, E. C., Kremer, A., Bloesma, F. F. S. N., Hoessels, E. L. M. J., Meyer, G., and Verheyen, F. J. (1952). Activation and organization of the central nervous system in amphibians. *J. Exp. Zool.* **120**, 1–108.

Oppenheimer, J. M. (1936). Structures developed in amphibians by implantation of living fish organizer. *Proc. Soc. Exp. Biol. Med.* **34**, 461–463.

Pera, E. M., and Kessel, M. (1997). Patterning of the chick forebrain anlage by the prechordal plate. *Development (Cambridge, UK)* **124**, 4153–4162.

Piccolo, S., Sasai, Y., Lu, B., and De Robertis, E. M. (1996). Dorsoventral patterning in *Xenopus*: inhibition of ventral signals by direct binding of chordin to BMP-4. *Cell (Cambridge, Mass.)* **86**, 589–598.

Roberts, C., Platt, M., Streit, A., Schachner, M., and Stern, C. D. (1991). The L5 epitope: An early marker for neural induction in the chick embryo and its involvement in inductive interactions. *Development (Cambridge, UK)* **112**, 959–970.

Ruiz i Altaba, A. (1993). Induction and axial patterning of the neural plate: Planar and vertical signals. *J. Neurobiol.* **24**, 1276–1304.

Ruiz i Altaba, A., Placzek, M., Baldassare, M., Dodd, J., and Jessell, T. M. (1995). Early stages of notochord and floor plate development in the chick embryo defined by normal and induced expression of HNF-3β. *Dev. Biol.* **170**, 299–313.

Ruiz i Altaba, A., and Jessell, T. M. (1991). Retinoic acid modifies the pattern of cell differentiation in the central nervous system of neurula stage *Xenopus* embryos. *Development (Cambridge, UK)* **112**, 945–958.

Sasai, Y., and De Robertis, E. M. (1997). Ectodermal patterning in vertebrate embryos. *Dev. Biol.* **182**, 5–20.

Sasai, Y., Lu, B., Steinbeisser, H., Geissert, D., Gont, L. K., and De Robertis, E. M. (1994). *Xenopus* chordin: A novel dorsalizing factor activated by organizer-specific homeobox genes. *Cell (Cambridge, Mass.)* **79**, 779–790.

Sasai, Y., Lu, B., Steinbeisser, H., and De Robertis, E. M. (1995). Regulation of neural induction by the Chd and BMP-4 antagonistic patterning signals in *Xenopus*. *Nature (London)* **376**, 333–336.

Sato, S. M., and Sargent, T. D. (1989). Development of neural inducing capacity in dissociated *Xenopus* embryos. *Dev. Biol.* **134**, 263–266.

Schmidt, J. E., Suzuki, A., Ueno, N., and Kimelman, D. (1995). Localized BMP-4 mediates dorso/ventral patterning in the early *Xenopus* embryo. *Dev. Biol.* **169**, 37–50.

Schultheiss, T. M., Burch, J. B., and Lassar, A. B. (1997). A role for bone morphogenetic proteins in the induction of cardiac myogenesis. *Genes Dev.* **11**, 451–462.

Servetnik, M., and Grainger, R. M. (1991). Changes in neural and lens competence in *Xenopus* ectoderm: Evidence for an autonomous developmental timer. *Development (Cambridge, UK)* **112**, 177–188.

Shimamura, K., and Rubenstein, J. L. R. (1997). Inductive interactions direct early regionalization of the mouse forebrain. *Development (Cambridge, UK)* **124**, 2709–2718.

Sive, H. L., Draper, B. W., Harland, R. M., and Weintraub, H. (1990). Identification of a retinoic acid-sensitive period during primary axis formation in *Xenopus laevis*. *Genes Dev.* **4**, 932–942.

Spemann, H., and Mangold, H. (1924). Über Induction von

Embryonalanlagen durch Implantation artfremder Organisatoren. *Wilhelm Roux's Arch. Entwicklungsmech Org.* **100**, 599–638.

Steinbach, O. C., Wolffe, A. P., and Rupp, A. W. (1997). Somatic linker histones cause loss of mesodermal competence in *Xenopus. Nature (London)* **389**, 395–399.

Storey, K. G., Crossley, J. M., De Robertis, E. M., Norris, W. E., and Stern, C. D. (1992). Neural induction and regionalisation in the chick embryo. *Development (Cambridge, UK)* **114**, 729–741.

Storey, K. G., Selleck, M., and Stern, C. D. (1995). Neural induction and regionalization by different subpopulations of cells in Hensen's node. *Development (Cambridge, UK)* **121**, 417–420.

Streit, A., Stern, C. D., Théry, C., Ireland, G. W., Aparicio, S., Sharpe, M. J., and Gherardi, E. (1995). A role for HGF/SF in neural induction and its expression in Hensen's node during gastrulation. *Development (Cambridge, UK)* **121**, 813–824.

Streit, A., Sockanathan, S., Pérez, L., Rex, M., Scotting, P. J., Sharpe, P. T., Lovell-Badge, R., and Stern, C. D. (1997). Preventing the loss of competence for neural induction: HGF/SF, L5 and Sox-2. *Development (Cambridge, UK)* **124**, 1191–1202.

Streit, A., Lee, K. J., Woo, I., Roberts, C., Jessell, T. M., and Stern, C. D. (1998). Chordin regulates primitive streak development and the stability of induced neural cells, but is not sufficient for neural induction in the chick embryo. *Development (Cambridge, UK)* **125**, 507–519.

Suzuki, A., Thies, R. S., Yamaji, N., Song, J. J., Wozney, J. M., Murakami, K., and Ueno, N. (1995). A truncated bone morphogenetic protein receptor affects dorsal–ventral patterning in the early *Xenopus* embryo. *Proc. Natl. Acad. Sci. U.S.A.* **91**, 10255–10259.

Tanabe, Y., and Jessell, T. M. (1996). Diversity and pattern in the developing spinal cord. *Science* **274**, 1115–1123.

Waddington, C. H. (1930). Developmental mechanics of chicken and duck embryos. *Nature (London)* **125**, 924–925.

Waddington, C. H. (1932a). Experiments on the development of chick and duck embryos. *Philos. Trans. R. Soc. London Ser. B* **221**, 179–230.

Waddington, C. H. (1932b). Induction by the primitive streak and its derivatives in the chick. *J. Exp. Embryol.* **10**, 38–46.

Waddington, C. H. (1934). Experiments on embryonic induction. *J. Exp. Biol.* **11**, 211–227.

Waddington, C. H. (1936). Organizers in mammalian development. *Nature (London)* **138**, 125.

Waddington, C. H. (1940). "Organizers and Genes" (C. H. Waddington, ed.) Cambridge Univ. Press, London.

Waddington, C. H., and Waterman, A. J. (1934). The development *in vitro* of young rabbit embryos. *J. Anat.* **67**, 356–370.

Waddington, C. H., and Needham, J. (1936). Evocation, individuation and competence in amphibian organizer action. *Proc. Kon. Akad. Wetensch. Amsterdam* **39**, 887–891.

Watanabe, Y., and Le Douarin, N. M. (1996). A role for BMP-4 in the development of subcutaneous cartilage. *Mech. Dev.* **57**, 69–78.

Weinstein, D. C., and Hemmati-Brivanlou, A. (1997). Neural induction in *Xenopus laevis:* Evidence for the default model. *Curr. Opin. Neurobiol.* **7**, 7–12.

Wilson, P. A., and Hemmati-Brivanlou, A. (1995). Induction of epidermis and inhibition of neural fate by BMP-4. *Nature (London)* **376**, 331–333.

Wilson, P. A., and Hemmati-Brivanlou, A. (1997). Vertebrate neural induction: Inducers, inhibitors and a new synthesis. *Neuron* **18**, 699–710.

Winnier, G., Blessing, M., Labosky, P. A., and Hogan, B. L. (1995). Bone morphogenetic protein-4 is required for mesoderm formation and patterning in the mouse. *Genes Dev.* **9**, 2105–2116.

Woodside, G. L. (1937). The influence of host age on induction in the chick blastoderm. *J. Exp. Biol.* **75**, 259–281.

Xu, R. H., Kim, J., Taira, M., Zhan, S., Sredni, D., and Kung, H. F. (1995). A dominant negative bone morphogenetic protein 4 receptor causes neuralization in *Xenopus* ectoderm. *Biochem. Biophys. Res. Commun.* **212**, 212–219.

Zhang, H., and Bradley, A. (1996). Mice deficient for BMP2 are nonviable and have defects in amnion/chorion and cardiac development. *Development (Cambridge, UK)* **122**, 2977–2986.

Zimmermann, L. B., De Jesús-Escobar, J. M., and Harland, R. M. (1996). The Spemann organizer signal noggin binds and inactivates bone morphogenetic protein 4. *Cell (Cambridge, UK)* **86**, 599–606.

30

Determination of Heart Cell Lineages

Takashi Mikawa
Department of Cell Biology
Cornell University Medical College
New York, New York 10021

I. Introduction
II. Origin of the Cardiomyocyte and Endocardial
    Cell Lineages
    A. The Cardiogenic Mesoderm
    B. Myocyte Lineage Commitment
    C. The Role of Hypoblast and Endoderm in Myocyte
       Commitment and Differentiation
    D. Common versus Separate Origins of Endocardial
       and Myocyte Cell Lineages

III. Diversification within the Myocyte Lineage
     A. Origin of the Atrial and Ventricular Myocyte
        Lineages
     B. Myocardial Wall Morphogenesis
     C. Origin of the Cardiac Conduction System
     D. Potential Mechanisms Inducing Purkinje Fibers
        within the Myocyte Lineage
IV. Concluding Remarks
    References

I. Introduction

The heart of amniotes such as birds and mammals is established through the integrated and sequential processes of cell commitment, cell proliferation, and cell movements of multiple cell types derived from different embryonic sites (Fig. 1). These cardiac cell types include atrial and ventricular myocytes, cells of the conduction system, the endocardium and valves of the heart, the coronary vessels, and neural elements. During gastrulation, the amniote embryo establishes the heart field in the rostrolateral areas of the lateral mesoderm, from which the cardiomyocyte and endocardial lineages arise. Fusion of the bilateral heart field forms the primitive heart, a single tube consisting of two epithelial layers: the inner endocardium and the outer myocardium (Fig. 2). The primitive tubular heart partitions further into two chambers, atrial and ventricular, separated by the atrioventricular septa. Myocardial contractions begin during this double-walled stage of heart formation. Initial pulsations are generated at the right myocardium and spread posteriorly to anteriorly over the whole myocardium. The two-chambered heart further undergoes a series of morphogenetic steps: looping, septation, trabeculation, and thickening of the ventricular walls; and the cranial shift of atrial chambers.

Associated with these morphogenetic events, fate diversification takes place within both the myocyte and endocardial endothelial lineages (Fig. 1). Diversification

in the myocyte lineage generates three distinct cardiac cell types, the atrial myocytes, the ventricular myocytes, and the cells of the cardiac conduction system that coordinates the rhythmic heart beat. A subpopulation of the endocardial cell lineage further differentiates into valves of the heart. These fate diversification events of cardiogenic mesoderm-derived lineages are regulated by local cell to cell communications of neighboring endoderm, ectoderm, migratory neural crest, and coronary vascular cells derived from outside of the heart field. Development and integration of these subcardiac components are sequentially programmed during heart formation in higher vertebrates. This chapter reviews the current understanding of the timing and mechanism that establish the myocyte and endocardial cell lineage and their derivatives in the avian heart. The origin of coronary vascular cell lineage and the fate of the cardiac neural crest are reviewed elsewhere (Mikawa and Fischman, 1992; Kirby, 1993; Noden *et al.*, 1995; Mikawa and Gourdie, 1996; Mikawa, 1998).

II. Origin of the Cardiomyocyte and Endocardial Cell Lineages

A. The Cardiogenic Mesoderm

In the chicken embryo, gastrulation begins with primitive streak formation. Once the primitive streak forma-

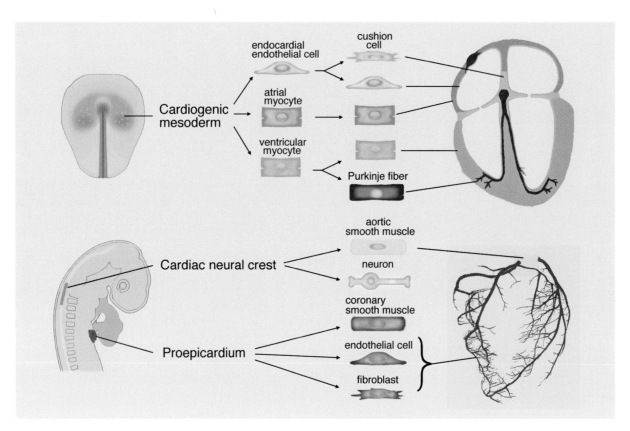

Figure 1   The origin and lineage relationship of cells forming the amniote heart. (Modified from Mikawa, 1998.)

tion is completed at Hamburger and Hamilton (HH) stage 3 (Hamburger and Hamilton, 1951), mesodermal cell migration from the primitive streak becomes evident at HH-stage 3+. By HH-stage 4, Hensen's node forms at the anterior tip of the primitive streak, and the mesoderm layer extends rostrolaterally (Fig. 2). Fate map studies have established that the mesoderm rostrolaterally flanking Hensen's node contains precursor cells of all three early cardiac lineages (Rawles, 1943; Rosenquist and DeHaan, 1966; Mikawa *et al.,* 1992a,b, 1996). Therefore, this mesodermal region is called the cardiogenic mesoderm (Fig. 1). Cells of the cardiogenic mesoderm arise from the rostral half of the primitive streak (Garcia-Martinez and Schoenwolf, 1993), and continue their migration rostrolaterally during HH-stage 5–6, forming a horseshoe-shaped cardiac primordia (Rosenquist and DeHaan, 1966). During the lateral body fold of the embryo, the bilateral cardiac primordia fuse and establish a single tubular heart on the ventral side of the foregut (Fig. 2). At this stage, only two cell types constitute the tubular heart, endocardial endothelia and myocardial myocytes (Manasek, 1968).

During the fusion to form the tubular heart, all cells of the bilateral cardiac primordia begin to express N-cadherin, a calcium-dependent cell adhesion molecule (Takeichi, 1991), and become epithelialized by HH-stages 6–7 (Linask, 1992). The majority of cells of the epithelialized cardiogenic mesoderm maintain N-cadherin expression, remain as epithelia and differentiate into presumptive myocytes (Fig. 2). A minor population downregulates N-cadherin expression and segregates from the original epithelial layer to form the presumptive endocardial endothelia (Manasek, 1968; Linask and Lash, 1993). Endocardial progenitor cells begin expression of endothelial markers, such as flk and QH1, as they segregate from the epithelialized cardiogenic mesoderm (Linask and Lash, 1993; Sugi and Markwald, 1996). Expression of muscle-specific proteins begins in the epithelioid myocardial cells substantially before initiation of the heart beat (Han *et al.,* 1992; Yutzey *et al.,* 1994). Thus, both myocyte and endocardial endothelial cell lineages arise from the cardiogenic mesoderm and begin to function before any other organ system is established.

B. Myocyte Lineage Commitment

Cells isolated and cultured from the cardiogenic mesoderm at HH-stage 4 are able to initiate cardiomyocyte-specific gene expression (Gonzales-Sanchez and Bader, 1990), indicating that a cell population in the cardiogenic mesoderm is already committed to the myocyte lineage. Consistent with these results, explants from the posterior region of prestreak blastodisc differentiate into heart muscle at a high frequency (Yatskievych *et al.,* 1997). To date, there are no ideal molecular mark-

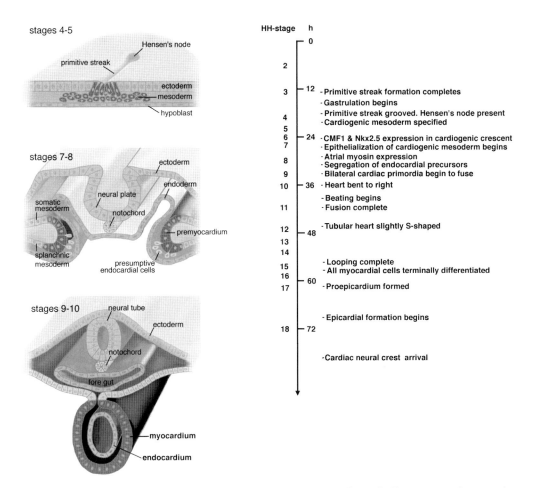

stages 4-5

Hensen's node
primitive streak
ectoderm
mesoderm
hypoblast

stages 7-8

ectoderm
endoderm
neural plate
somatic mesoderm
notochord
premyocardium
splanchnic mesoderm
presumptive endocardial cells

stages 9-10

neural tube
ectoderm
notochord
fore gut
myocardium
endocardium

HH-stage	h	
	0	
2		
3	12	- Primitive streak formation completes
		- Gastrulation begins
4		- Primitive streak grooved. Hensen's node present
5		- Cardiogenic mesoderm specified
6	24	- CMF1 & Nkx2.5 expression in cardiogenic crescent
7		- Epithelialization of cardiogenic mesoderm begins
8		- Atrial myosin expression
		- Segregation of endocardial precursors
9		- Bilateral cardiac primordia begin to fuse
10	36	- Heart bent to right
		- Beating begins
11		- Fusion complete
12	48	- Tubular heart slightly S-shaped
13		
14		
15		- Looping complete
16		- All myocardial cells terminally differentiated
17	60	- Proepicardium formed
18	72	- Epicardial formation begins
		- Cardiac neural crest arrival

Figure 2 Morphogenetic processes in the early heart development. Left panels illustrate gastrulating cardiac mesodermal cells at HH-stage 4, segregation of endocardial precursor cells from epithelioid presumptive myocardium during lateral body fold at the neurula stage, and subsequent tubular heart formation by fusion of bilateral heart primordia, respectively. (Modified from Mikawa, 1998.) The right diagram summarizes the time table of main morphogenetic events during heart development, according to Hamburger and Hamilton (1951).

ers that distinguish the cardiogenic mesoderm at HH-stage 4, the time when the myocyte lineage commitment is completed. However, by HH-stage 6–8 when myofibrillar protein expression begins, the cardiogenic mesodermal cells can be distinguished from other mesodermal cells by probing for specific marker gene expression, such as CMF1 (Wei *et al.*, 1996) and Nkx2.5 (Schultheiss *et al.*, 1995; Harvey, 1996). CMF1, a cloned factor from the embryonic heart, is expressed specifically in the myocyte lineage only between HH-stage 7–15. It encodes zinc finger motifs, a nuclear localization signal, and an E-box binding site (Wei *et al.*, 1996). Nkx2.5 is a homeodomain homolog of *Drosophila* tinman (Bodmer, 1993; Azpiazu and Frasch, 1993) and is expressed in both myocytes and endothelial cells in addition to pharyngeal endoderm of the tubular heart stage embryo. Nkx2.5 expression becomes restricted to the myocyte population later in development.

Several other transcription factors have also been identified in myocyte lineages, including members of Tbx, GATA, HAND, and MEF2 gene families (reviewed in Olson and Srivastava, 1996; Fishman and Chien, 1997). Although expression of these factors is not re-

stricted to the myocyte lineage, deletion of each of these factors in mice results in a loss of survival and/or proliferation of already differentiated myocytes either directly or indirectly. However, unlike the role of the MyoD gene family in the skeletal muscle lineage, each of these factors alone appear to be insufficient for specification of the cardiomyocyte lineage (Fishman and Chien, 1997).

C. The Role of Hypoblast and Endoderm in Myocyte Commitment and Differentiation

Endoderm underlying the cardiogenic mesoderm has been proposed as an inducer of the heart field (Jacobson and Sater, 1988). This is evidenced in amphibians, since its removal results in either a non beating heart or noncontractile myocytes (Jacobson, 1961). In the avian embryo however, it is still unclear which step of myocyte commitment and differentiation is regulated by underlying endoderm. Removal of rostral endoderm at HH-stage 4 gives rise to a heart tube that expresses myocyte marker genes but lacks contractile function (Gannon and Bader, 1995). In the absence of endoderm, explants of cardiogenic mesoderm can develop myofibrils, if fibronectin, which is expressed by endoderm, is supplied (Gannon and Bader, 1995). The model drawn from these studies suggests that the rostral endoderm does not induce or maintain cardiac gene expression, nor is it required for terminal differentiation. Rather, it is necessary at the onset of contraction (Gannon and Bader, 1995). Similar results were obtained in quail embryos. Explants of cardiogenic mesoderm at stage 4+ differentiate into cardiac myocytes in the absence of endoderm, although the endoderm enhances the rate at which myocytes differentiate and also the onset of beating (Antin *et al.*, 1994).

Another model proposes that paracrine factors from hypoblast and/or endoderm underlying cardiogenic precursor cells may play a role in myocyte specification and terminal differentiation (Sugi and Lough, 1994; Antin *et al.*, 1994; Schultheiss *et al.*, 1995; Yatskievych *et al.*, 1997). Proliferation, survival, and/or muscle-specific gene expression of epiblast and mesodermal cell explants can be regulated by fibroblast growth factor 1 (FGF1) (Zhu *et al.*, 1996), FGF2 (Sugi *et al.*, 1993), FGF4 (Zhu *et al.*, 1996; Lough *et al.*, 1996), bone morphogenetic protein-2 (BMP-2) (Lough *et al.*, 1996; Schultheiss *et al.*, 1997), BMP-4 (Schultheiss *et al.*, 1997), insulinlike growth factor (IGF) (Antin *et al.*, 1996), and activin (Sugi and Lough, 1995; Yatskievych *et al.*, 1997). Interestingly, expression of Nkx2.5 and cardiac muscle genes can be induced in the noncardiac paraxial mesoderm in the presence of rostrolateral endoderm (Schultheiss *et al.*, 1995). BMP-2 and BMP-4 can mimic the instructive activity of rostrolateral endoderm (Schultheiss *et al.*, 1997). Furthermore, differentiation of cardiogenic mesoderm is completely blocked in whole embryo culture by the secreted protein nogin, which binds to BMPs and antagonizes BMP activity. These studies on mesoderm explants or whole embryo culture will stimulate researchers to take *in vivo* approaches to clarifying the exact roles of secreted factors in induction, maintenance, proliferation, and/or terminal differentiation of the myocyte lineage.

D. Common Versus Separate Origins of Endocardial and Myocyte Cell Lineages

Although both endocardial and myocyte cell lineages arise from the cardiogenic mesoderm (Garcia-Martinez and Schoenwolf, 1993; Cohen-Gould and Mikawa, 1996), the primitive heart tube, as in the adult heart, consists of a greater proportion of myocardial cells than endocardial cells (Manasek, 1968). It remains unclear how more myocardial cells than endocardial cells segregate from the cardiogenic mesoderm. There is a paradox in the understanding of the genesis of endocardial and myocardial cell lineages. Morphological heterogeneity within the cardiogenic mesoderm suggested that the

two lineages already were separated when their progenitors migrate to the heart field (Pardanaud *et al.*, 1987a,b; Coffin and Poole, 1988; DeRuiter *et al.*, 1992). In contrast, expression patterns of myocardial and endocardial cell markers in cardiogenic mesoderm (Linask and Lash, 1993) and in an immortalized myogenic cell line (Eisenberg and Bader, 1995) led to the hypothesis that cells of the cardiogenic mesoderm commonly produce both cell lineages.

The debate was recently settled by fate mapping analysis of individual progenitor cells within the heart field. Retroviral cell lineage studies in the chicken embryo revealed that individual cells in the heart field give rise to a clone consisting only of one cell type, either endocardial or myocardial (Cohen-Gould and Mikawa, 1996). No mesodermal cells generate clones containing both of these two cell types. Importantly, approximately 95% of the mesoderm-derived clones are localized in the myocardium only, while about 5% of them are found in endocardium only. Thus, the heart field mesoderm consists of at least two distinct subpopulations with substantially more premyocardial cells than preendocardial cells (Fig. 3). Consistent with this model, it was shown that endothelial cell commitment occurs before and independent of gastrulation, whereas myocyte commitment has not yet occurred at the presteak stage (von Kirschhoffer *et al.*, 1994). Single cell marking and tracing studies in zebrafish (Lee *et al.*, 1994) have identified a blastomere population which generates only endocardial or myocardial cells, indicating that the separation of these two lineages can occur at blastula stage prior to formation of mesoderm.

Differentiation of endocardial endothelia appears to be regulated by foregut endoderm (Sugi and Markwald, 1996). Vascular endothelial growth factor (VEGF), necessary for induction of the endothelial lineage (Shalaby *et al.*, 1995; Ferrara *et al.*, 1996; Fong *et al.*, 1995), is expressed in the endoderm (Flamme *et al.*, 1994), and flk, a cognate receptor for VEGF, is expressed in endocardial endothelial precursors. It remains to be seen whether mechanisms inducing the endocardial endothelial lineage are distinct from those regulating commitment and differentiation of other vascular endothelial cells (Fig. 3).

III. Diversification within the Myocyte Lineage

A. Origin of the Atrial and Ventricular Myocyte Lineages

During the tubular heart formation cells of the caudal region of the presumptive myocardium begin to express atrial myocyte-specific contractile protein genes (O'Brien *et al.*, 1993; Yutzey *et al.*, 1994). Before the on-

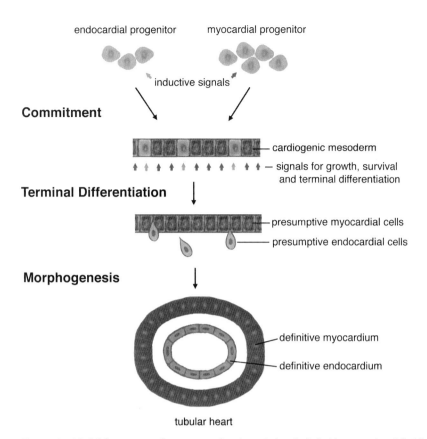

Figure 3 Model for genesis of myocyte and endocardial endothelial lineages (modified from Cohen-Gould and Mikawa, 1996; Mikawa, 1998). Prior to or at the initiation of gastrulation, each cardiac lineage in the epiblast or primitive streak cells is established by receiving distinct inductive signals. During gastrulation, precursors of both lineages comigrate to the cardiogenic mesoderm area. Responding to instructive signals from endoderm and/or ectoderm, the precursors undergo terminal differentiation and morphogenetic processes.

set of contraction, the future atrial population exhibits electrophysiological properties distinct from those differentiating into ventricular myocytes (Kamino *et al.*, 1981). Fate mapping studies have shown that more rostral cells of the primitive streak tend to generate a cell population forming the more rostral, ventricular region of the tubular heart, whereas the more caudal cells form

the caudal, atrial regions (Fig. 4; Garcia-Martinez and Schoenwolf, 1993). Retroviral-mediated genetic marking and subsequent fate analyses of individual cells present in the cardiogenic mesoderm have proven that individual cell in the rostral cardiogenic mesoderm only enter the ventricular myocyte lineage, whereas cells in the caudal region differentiate into the atrial myocyte

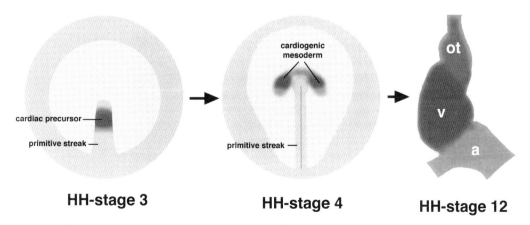

Figure 4 Fate maps of myocyte and endocardial endothelial precursors during gastrulation and tubular heart formation of the avian embryo. ot, Outflow trunk; v, ventricle; a, aorta.

lineage (Fig. 4; Mikawa *et al.*, 1992a,b, 1996). In agreement with these results, explants from the caudal region of cardiogenic mesoderm can differentiate into atrial but not ventricular myocytes in culture (Yutzey *et al.*, 1995).

Although these two myocyte lineages are not derived from common parental cells, myogenic progenitors within the cardiogenic mesoderm seem to remain bipotential until terminal differentiation into either atrial or ventricular phenotypes becomes evident. Implantation of the caudal region of cardiogenic mesoderm into the rostral heart-forming region changes its beat rate from atrial type to ventricular type (Satin *et al.*, 1988). An atrial myosin heavy chain can be induced in presumptive ventricular myocytes ectopically by treatment of cardiogenic mesoderm with retinoic acid (Yutzey *et al.*, 1994). Such plasticity completely disappears once atrial and ventricular myocyte phenotypes become apparent. These observations suggest that the terminal differentiation of either atrial or ventricular lineages is defined by positionally delineated extracellular signal(s). Limited migratory activities of myocyte progenitors within the epithelioid presumptive myocardium (Mikawa *et al.*, 1992a,b) may play a role in stabilizing the terminal differentiation process. Molecular signals inducing the two myocyte lineages still remain to be identified.

B. Myocardial Wall Morphogenesis

All myocardial cells, both atrial and ventricular, complete their terminal-differentiation and become contractile by HH-stage 15, leaving no stem cell population (Manasek, 1968). As illustrated in Fig. 5, the beating myocytes divide, delaminate, and migrate more vertically than horizontally, creating ridge-like protrusions (trabeculae) (Manasek, 1968). The trabeculation process is more active in the ventricule than in the atrium. The trabeculated patterning increases surface area and facilitates diffusion of oxygen and nutrient into the avascular ventricular myocardium, prior to the coronary vessel system development. Subsequent coalescence of the trabeculae develops the thickened myocardium. To date little is known about the mechanism underlying asymmetrical muscle growth between the atrial-ventricular chambers, left-right ventricles, muscular septa and papillary muscles.

Retroviral cell fate analysis of individual myocytes during trabeculation, subsequent thickening and multilayering of the ventricular myocardial wall has revealed the basic plan of myocardial wall morphogenesis (Fig. 5; Mikawa *et al.*, 1992a,b; Mikawa, 1995). Individual precursors of ventricular myocytes generate a series of daughter cells which proliferate and migrate toward endocardium as a tight cluster, thereby generating a clone that forms one or at most two trabeculae. During fusion of trabeculae, myocyte proliferation becomes greater at

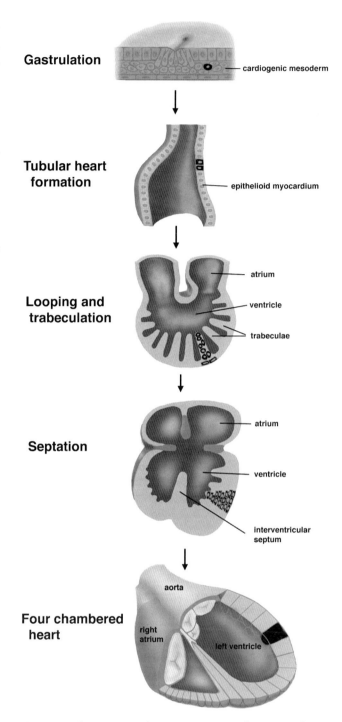

Figure 5 The main morphogenetic events in the myocyte lineage during formation of four-chambered heart. (Modified from Mikawa, 1995, 1998.) A clonal myocyte population is highlighted by filled cytoplasm. The endoderm is omitted from the diagram.

the periphery than the deeper layers, resulting in clonal domains of cone- or wedge-shaped sectors which span the entire thickness of the myocardium. The clonally related sectors serve as the fundamental growth units of the myocardial wall. Two dimensional arrays of these cone-shaped growth units give rise to three-dimensional ovoid structures of the ventricular walls. The interventricular septum forms by generating a myocyte clone with more axially elongated dimensions than those in the lateral walls. Thus, the thickened and multi-layered ventricular myocardium is defined by the locally regulated migration and proliferation of progeny derived from individual epithelioid parental myocytes (Fig. 5).

Neuregulin, a peptide secreted by endocardial endothelia, plays a critical role in initiating trabeculation via its cognate tyrosine kinase receptors (erbB2 and erbB4) expressed by epithelioid myocytes (Meyer and Birchmeier, 1995; Lee *et al.*, 1995; Gassmann *et al.*, 1995). Signaling attributed to fibroblast growth factor (FGF) through its tyrosine kinase receptor, serves as a potent mitogen during myocardial wall thickening (Mima *et al.*, 1995; Mikawa, 1995). In addition to its known roles for myocyte epithelialization (Radice *et al.*, 1997) and intercalated disc formation (Goncharova *et al.*, 1992; Solar and Knudsen, 1994; Herting *et al.*, 1996), N-cadherin appears to contribute to homotypic interactions between nonepithelial migratory myocytes during trabecular formation of the embryonic heart (Ong *et al.*, 1998). N-cadherin mediated cell adhesion is one of the molecular components that allows clonally related myocytes to develop as a tight cluster during myocardial wall morphogenesis (Ong *et al.*, 1998). Analysis of the cell diversification process in an individual myocyte clonal population has led to the indentification of the cardiac conduction system origin described below.

C. Origin of the Cardiac Conduction System

In contrast to the open circulation system of invertebrates, in which hearts stochastically reverse the direction of contractile wave, the vertebrate heart unidirectionally pumps blood by the precisely timed contractions of atrial and ventricular chambers. During the tubular heart formation, cells of the presumptive atrium become electrically active even before initiation of contraction. The action potentials propagate to the rostral end of the tubular heart through gap junctions between the epithelioid myocytes (Kamino *et al.*, 1981) and produce a caudal-to-rostral contractile wave in the tubular heart (Fig. 6). During topological transition of the heart chambers from two to four, the impulse conduction pathway is dramatically remodeled (Fig. 6). The rhythmic beat of the four-chambered heart is coordinated by electrical impulses from a specialized tissue called the cardiac conduction system (Tawara, 1906). The pacemaking site in the mature heart is the sinuatrial (SA) node (Davies, 1930; Lamers *et al.*, 1991). From this site, impulses are spread through atrial myocardium and then focused into the atrioventricular (AV) node. The action potentials, leaving from the AV-node, propagate along the AV-bundle, and are finally spread into ventricular muscle via a distributive network of Purkinje fiber cells, thereby synchronizing contraction of the

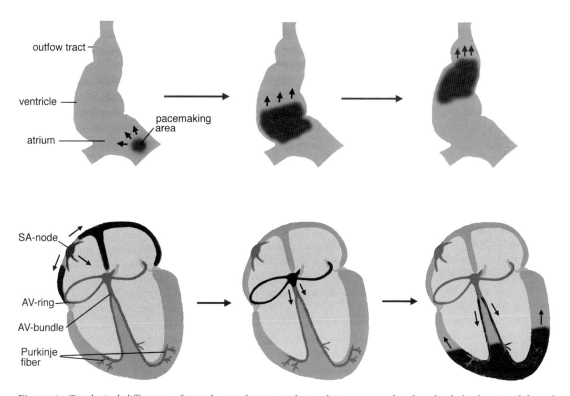

Figure 6 Topological difference of impulse-conducting pathways between two-chambered tubular heart and four-chambered heart. Arrows mark the direction of propagation of action potentials and contractile wave; black, action potential-positive cells; dark gray, cells of the cardiac conduction system.

ventricular chambers of the heart (reviewed in Viragh and Challice, 1973; Lamers *et al.*, 1991).

Conduction cells coexpress neural and muscle genes (Sartore *et al.*, 1978; Gonzalez-Sanchez and Bader, 1985; reviewed in Gorza *et al.*, 1994). This unique gene expression pattern has led to the suggestion of two possible origins, myogenic (Pattern and Kramer, 1933; Pattern, 1956) and neural crest (Gorza *et al.*, 1988, 1994; Vitadello *et al.*, 1990). It also was unknown whether the branched network of the entire conduction system is established by outgrowth from a common progenitor (Wessels *et al.*, 1992; Lamers *et al.*, 1991; Chan-Thomas *et al.*, 1993; Fishman and Chien, 1997) or by *in situ* linkage of subcomponents with independent origins (reviewed in Mikawa and Fischman, 1996). A clear resolution of these lineage relationships has been determined by retroviral cell lineage studies on the Purkinje fiber network of the chicken embryonic heart (Gourdie *et al.*, 1995).

In the chicken heart, the Purkinje fiber network is localized in a precise association with the coronary arterial bed (Fig. 7; Davies, 1930; Vassal-Adams, 1982). In the developing heart, differentiating Purkinje fibers can first be detected along the growing coronary arteries on embryonic day 10 as a group of cells highly expressing Cx-42, a member of the connexin family (Gourdie *et al.*, 1993). Retroviral-mediated genetic marking of individual differentiated and contractile myocytes in the tubular heart and subsequent examination of their clonal populations have revealed that a subset of clonally related myocytes differentiates into conducting Purkinje fibers, invariably in close spatial association with forming coronary arterial blood vessels (Fig. 7). Myocyte clones containing a Purkinje fiber population never generate cells of more proximal components of the conduction system, such as the atrioventricular node and bundles, or sinoatrial components (Gourdie *et al.*, 1995). These data indicate that impulse generating and conduction cells are derived by localized recruitment of differentiated, beating myocytes specifically along the developing coronary arterial bed (Fig. 7) and the Purkinje fibers have a different parental lineage than that of the proximal conduction system (Gourdie *et al.*, 1995). The definitive mapping of Purkinje fiber progenitor cells to a myocyte lineage, and not to neural crest, now allows us to study the mechanism of Purkinje fiber differentiation by analyzing the process by which heart cells are converted from a contractile to a conductive lineage.

D. Potential Mechanisms Inducing Purkinje Fibers within the Myocyte Lineage

The coronary vasculature does not arise by outgrowth from the root of the aorta, but by migration into the tubular heart from extracardiac mesenchyme (Mikawa and Fischman, 1992; Poelmann *et al.*, 1993; Mikawa and Gourdie, 1996) along with a proepicardial sheet (Hiruma and Hirakow, 1989; Ho and Shimada, 1988). In the avian system, entry of these coronary precursors into the heart begins at HH-stage 17–18 (Mikawa and Fischman, 1992; Poelmann *et al.*, 1993). Following inward migration, vasculogenic cells first form discontinuous endothelial channels. Subsequent fusion between the endothelial channels by embryonic day 6 and then connection to the aorta, establishes the closed coronary vessel network by embryonic day 14 (Bogers *et al.*, 1989; Waldo *et al.*, 1990; Mikawa and Fischman, 1992; Mikawa and Gourdie, 1996). Coincident with this early vasculogenic process, recruitment of Purkinje fibers begins exclusively in myocyte subpopulations juxtaposed to developing coronary arteries but not veins (Gourdie *et al.*, 1995). The close spatiotemporal relationship between Purkinje fiber differentiation and coronary blood vessel development suggests that an inductive role of coronary vasculature, namely, coronary arteries, may recruit contractile myocytes to form Purkinje fibers (Fig. 7). If coronary arteries play a role in the recruitment of Purkinje fibers from contractile myocytes, the vessel network may be a key factor in defining the branching pattern of the peripheral conduction system.

Ablation of the cardiac neural crest alters the pattern of the coronary arterial tree (Hood and Rosenquist, 1992). Additionally, neural crest derivatives are necessary for the survival of branches of the coronary artery system (Waldo *et al.*, 1994). Thus, neural crest-derived cells affect the development of coronary arteries. Although cell lineage studies prove a myogenic origin for the Purkinje fibers, neural crest-derived cells may indirectly contribute to their differentiation and network patterning through regulation of coronary arterial development. It still remains to be determined if the induction and patterning of Purkinje fibers are linked with the development of coronary arteries and the differentiation of cardiac neural crest cells.

IV. Concluding Remarks

Recent genetic approaches in fly, zebrafish, and mouse have been highly successful in identifying genes or gene networks involved in the regulation of heart development. Chicken embryos have been a powerful system for molecular and cell biological approaches to address the potential or plasticity of embryonic cells in their lineage commitment and terminal differentiation. Because of its accessibility and amenability to clone-based cell lineage and fate mapping, the chicken embryos allow us to define the timing and location at which embryos accurately induce the cardiac lineages and their differentiation and patterning.

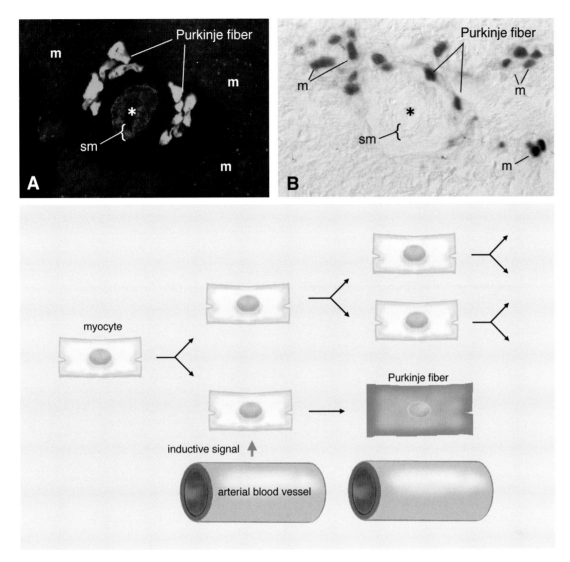

Figure 7 Purkinje fiber differentiation within the myocyte lineage. (Modified from Mikawa, 1998.) (A) Periatrial localization of Purkinje fibers in the avian heart, showing Purkinje fibers (green) and the atrial bed (red). (B) A subpopulation of clonally related myocytes expressing nuclear-directed β-galactosidase (blue-stained nuclei) differentiate into periarterial Purkinje fibers (arrows). (C) Proposed model of induction of Purkinje fibers within the myocyte lineage. m, myocytes; sm, smooth muscle.

Acknowledgment

I thank members of my laboratory for their comments on this chapter, and the NIH and AHA for their support.

References

Antin, P. B., Taylor, R. G., and Yatskievych, T. (1994). Precardiac mesoderm is specified during gastrulation in quail. *Dev. Dyn.* **200,** 144–154.

Antin, P. B., Yatskievych, T., Dominguez, J. L., and Chieffi, P. (1996). Regulation of avian precardiac mesoderm development by insulin and insulin-like growth factors. *J. Cell. Physiol.* **168,** 42–50.

Azpiazu, N., and Frasch, M. (1993). *tinman* and *bagpipe:* Two homeobox genes that determine cell fates in the dorsal mesoderm of *Drosophila. Genes Dev.* **7,** 1325–1340.

Bodmer, R. (1993). The gene *tinman* is required for specification of the heart and visceral muscle in *Drosophila. Development (Cambridge, UK)* **118,** 719–729.

Bogers, A. J. J. C., de Groot, A. C., Poelmann, R. E., and Huysmans, H. A. (1989). Development of the origin of the coronary arteries, a matter of ingrowth or outgrowth? *Anat. Embryol.* **180,** 437–441.

Chan-Thomas, P. S., Thompson, R. P., Robert, B. Y., Yacoub, M. H., and Barton, P. J. R. (1993). Expression of homeobox genes Msx-1 (Hox-7) and Msx-2 (Hox-8) during cardiac development in the chick. *Dev. Dyn.* **197,** 203–216.

Coffin, J. D., and Poole, T. J. (1988). Embryonic vascular development: Immunohistochemical identification of the origin and subsequent morphogenesis of the major vessel primordia in quail embryos. *Development (Cambridge, UK)* **102**, 735–748.

Cohen-Gould, L., and Mikawa, T. (1996). The fate diversity of mesodermal cells within the heart field during chicken early embryogenesis. *Dev. Biol.* **177**, 265–273.

Davies, F. (1930). The conducting system of the bird's heart. *J. Anat.* **64**, 129–146.

DeRuiter, M. C., Poelmann, R. E., Vander Plas-de Vries, I., Mentink, M. M. T., and Gittenberger-de Groot, A. C. (1992). The development of the myocardium and endocardium in mouse embryos. *Anat. Embryol.* **185**, 461–473.

Eisenberg, C. A., and Bader, D. (1995). QCE-6: A clonal cell line with cardiac myogenic and endothelial cell potentials. *Dev. Biol.* **167**, 469–481.

Ferrara, N., Carver-Moore, K., Chen, H., Dowd, M., Lu, L., O'Shea, K. S., Powell-Braxton, L., Hillan, K. J., and Moore, M. W. (1996). Heterozygous embryonic lethality induced by targeted inactivation of the VEGF gene. *Nature (London)* **380**, 439–442.

Fishman, M. C., and Chien, K. R. (1997). Fashioning the vertebrate heart: Earliest embryonic decisions. *Development (Cambridge, UK)* **124**, 2099–2117.

Flamme, I., Breier, G., and Risau, W. (1994). Vascular endothelial growth factor (VEGF) and VEGF receptor 2 (flk-1) are expressed during vasculogenesis and vascular differentiation in the quail embryo. *Dev. Biol.* **169**, 699–712.

Fong, G. H., Rossant, J., Gertsenstein, M., and Breitman, M. L. (1995). Role of the Flt-1 receptor tyrosine kinase in regulating the assembly of vascular endothelium. *Nature (London)* **376**, 66–70.

Gannon, M., and Bader, D. (1995). Initiation of cardiac differentiation occurs in the absence of anterior endoderm. *Development (Cambridge, UK)* **121**, 2439–2450.

Garcia-Martinez, V., and Schoenwolf, G. C. (1993). Primitive-streak origin of the cardiovascular system in avian embryos. *Dev. Biol.* **159**, 706–719.

Gassmann, M., Casagranda, F., Orioli, D., Simon, H., Lai, C., Klein, R., and Lemke, G. (1995). Aberrant neural and cardiac development in mice lacking the ErbB4 neuregulin receptor. *Nature (London)* **378**, 390–394.

Goncharova, E. J., Kam, Z., and Geiger, B. (1992). The involvement of adherens junction components in myofibrillogenesis in cultured cardiac myocytes. *Development (Cambridge, UK)* **114**, 173–183.

Gonzalez-Sanchez, A., and Bader, D. (1985). Characterization of a myosin heavy chain in the conductive system of the adult and developing chicken heart. *J. Cell Biol.* **100**, 270–275.

Gonzalez-Sanchez, A., and Bader, D. (1990). *In vitro* analysis of cardiac progenitor cell differentiation. *Dev. Biol.* **139**, 197–209.

Gorza, L., Schiaffino, S., and Vitadello, M. (1988). Heart conduction system: A neural crest deivative. *Brain Res.* **457**, 360–366.

Gorza, L., Vettore, S., and Vitadello, M. (1994). Molecular and cellular diversity of heart conduction system myocytes. *Trends Card. Med.* **4**, 153–159.

Gourdie, R. G., Green, C. R., Severs, N. J., Anderson, R. H., and Thompson, R. P. (1993). Evidence for a distinct gap-junctional phenotype in ventricular conduction tissues of the developing and mature avian heart. *Circ. Res.* **72**, 278–289.

Gourdie, R. G., Mima, T., Thompson, R. P., and Mikawa, T. (1995). Terminal diversification of the myocyte lineage generates Purkinje fibers of the cardiac conduction system. *Development (Cambridge, UK)* **121**, 1423–1431.

Hamburger, V., and Hamilton, H. L. (1951). A series of normal stages in the development of the chick embryo. *J. Morphol.* **88**, 49–92.

Han, Y., Dennis, J. E., Cohen-Gould, L., Bader, D. M., and Fischman, D. A. (1992). Expression of sarcomeric myosin in the presumptive myocardium of chicken embryos occurs within six hours of myocyte commitment. *Dev. Dyn.* **193**, 257–265.

Harvey, R. (1996). Nk-2 homeobox genes and heart development. *Dev. Biol.* **178**, 203–216.

Herting, C. M., Eppenberger-Eberhardt, M., Koch, S., and Eppenberger, H. M. (1996). N-cadherin in adult rat cardiomyocytes in culture. I. Functional role of N-cadherin and impairment of cell–cell contact by a truncated N-cadherin mutant. *J. Cell Sci.* **109**, 1–10.

Hiruma, T., and Hirakow, R. (1989). Epicardial formation in embryonic chick heart: Computer-aided reconstruction, scanning, and transmission electron microscopic studies. *Am. J. Anat.* **184**, 129–138.

Ho, E., and Shimada, Y. (1988). Formation of the epicardium studied with the scanning electron microscope. *Dev. Biol.* **66**, 579–585.

Hood, L. A., and Rosenquist, T. H. (1992). Coronary artery development in the chick: Origin and development of smooth muscle cells, and effects of neural crest ablation. *Anat. Rec.* **234**, 291–300.

Jacobson, A. G. (1961). Heart determination in the newt. *J. Exp. Zool.* **146**, 139–151.

Jacobson, A. G., and Stater, A. K. (1988). Features of embryonic induction. *Development (Cambridge, UK)* **104**, 341–359.

Kamino, K., Hirota, A., and Fujii, S. (1981). Localization of pacemaking activity in early embryonic heart monitored using voltage-sensitive dye. *Nature (London)* **290**, 595–597.

Kirby, M. L. (1993). Cellular and molecular contributions of the cardiac neural crest to cardiovascular development. *Trends Cardiovasc. Med.* **3**, 18–23.

Lamers, W. H., De Jong, F., De Groot, I. J. M., and Moorman, A. F. M. (1991). The Development of the avian conduction system, a review. *Eur. J. Morphol.* **29**, 233–253.

Lee, K., Simon, H., Chen, H., Bates, B., Hung, M., and Hauser, C. (1995). Requirement for neuregulin receptor erbB2 in neural and cardiac development. *Nature (London)* **378**, 394–398.

Lee, R. R. K., Stainier, D. Y. R., Weinstein, B. M., and Fishman, M. C. (1994). Cardiovascular development in the zebrafish. II. Endocardial progenitors are sequestered within the heart field. *Development (Cambridge, UK)* **120**, 3361–3366.

Linask, K. K. (1992). N-cadherin localization in early heart development and polar expression of Na$^+$, K$^+$-ATPase, and integrin during pericardial coelom formation and epithelialization of the differentiating myocardium. *Dev. Biol.* **151,** 213–224.

Linask, K. K., and Lash, J. W. (1993). Early heart development: Dynamics of endocardial cell sorting suggests a common origin with cardiomyocytes. *Dev. Dyn.* **195,** 62–66.

Lough, J., Barron, M., Brogley, M., Sugi, Y., Bolender, D. L., and Zhu, X. (1996). Combined BMP-2 and FGF-4, but neither factor alone, induces cardiogenesis in nonprecardiac embryonic mesoderm. *Dev. Biol.* **178,** 198–202.

Manasek, F. J. (1968). Embryonic development of the heart: A light and electron microscopic study of myocardial development in the early chick embryo. *J. Morphol.* **125,** 329–366.

Meyer, D., and Birchmeier, C. (1995). Multiple essential functions of neuregulin in development. *Nature (London)* **378,** 386–390.

Mikawa, T. (1995). Retroviral targeting of FGF and FGFR in cardiomyocytes and coronary vascular cells during heart development. *Ann. N. Y. Acad. Sci.* **752,** 506–516.

Mikawa, T. (1998). Cardiac lineages. *In* "Heart Development" (R. P. Harvey and N. Rosenthal, eds.), pp. 19–33. Academic Press, San Diego.

Mikawa, T., and Fischman, D. A. (1992). Retroviral analysis of cardiac morphogenesis: Discontinuous formation of coronary vessels. *Proc. Natl. Acad. Sci. U.S.A.* **89,** 9504–9508.

Mikawa, T., and Fischman, D. A. (1996). The polyclonal origin of myocyte lineages. *Annu. Rev. Physiol.* **58,** 509–521.

Mikawa, T., and Gourdie, R. G. (1996). Pericardial mesoderm generates a population of coronary smooth muscle cells migrating into the heart along with ingrowth of the epicardial organ. *Dev. Biol.* **173,** 221–232.

Mikawa, T., Borisov, A., Brown, A. M. C., and Fischman, D. A. (1992a). Clonal analysis of cardiac morphogenesis in the chicken embryo using a replication-defective retrovirus: I. Formation of the ventricular myocardium. *Dev. Dyn.* **193,** 11–23.

Mikawa, T., Cohen-Gould, L., and Fischman, D. A. (1992b). Clonal analysis of cardiac morphogenesis in the chicken embryo using a replication-defective retrovirus. III: Polyclonal origin of adjacent ventricular myocytes. *Dev. Dyn.* **195,** 133–141.

Mikawa, T., Hyer, J., Itoh, N., and Wei, Y. (1996). Retroviral vectors to study cardiovascular development. *Trends Cardiovasc. Med.* **6,** 79–86.

Mima, T., Ueno, H., Fischman, D. A., Williams, L. T., and Mikawa, T. (1995). FGF-receptor is required for *in vivo* cardiac myocyte proliferation at early embryonic stages of heart development. *Proc. Natl. Acad. Sci. U.S.A.* **92,** 467–471.

Noden, D. M., Poelmann, R. E., and Gittenberger-de Groot, A. C. (1995). Cell origins and tissue boundaries during outflow tract development. *Trends Cardiovasc. Med.* **5,** 69–75.

O'Brien, T. X., Lee, K. J., and Chien, K. R. (1993). Positional specification of ventricular myosin light chain 2 expression in the primitive murine heart tube. *Proc. Natl. Acad. Sci. U.S.A.* **90,** 5157–5161.

Olson, E. N., and Srivastava, D. (1996). Molecular pathways controlling heart development. *Science* **272,** 671–676.

Ong, L. L., Kim, N., Mima, T., Cohen-Gould, L., and Mikawa, T. (1998). Trabecular myocytes of the embryonic heart require N-cadherin for migratory unit identity. *Dev. Biol.* **193,** 1–9.

Pardanaud, L., Altmann, C., Kitos, P., Dieterlen-Lievre, F., and Buck, C. A. (1987a). Vasculogenesis in the early quail blastodisc as studied with a monoclonal antibody recognizing endothelial cells. *Development (Cambridge, UK)* **100,** 339–349.

Pardanaud, L., Buck, C., and Dieterlen-Lievre, D. (1987b). Early germ cell segregation and distribution in the quail blastodisc. *Cell Differ.* **22,** 47–60.

Pattern, B. M. (1956). The development of the sinoventricular conduction system. *Univ. Mich. Med. Bull.* **22,** 1–21.

Pattern, B. M., and Kramer, T. C. (1993). The initiation of contraction in the embryonic chick heart. *Am. J. Anat.* **53,** 349–375.

Poelmann, R. E., Gittenberger-de Groot, A. C., Mentink, M. T., Bokenkamp, R., and Hogers, B. (1993). Development of the cardiac coronary vascular endothelium, studied with anti-endothelial antibodies, in chicken-quail chimeras. *Circ. Res.* **73,** 559–568.

Radice, G. L., Rayburn, H., Matsunami, H., Knundsen, K. A., Takeichi, M., and Hynes, R. O. (1997). Developmental defects in mouse embryos lacking N-cadherin. *Dev. Biol.* **181,** 64–78.

Rawles, M. E. (1943). The heart forming regions of the early chick blastoderm. *Physiol. Zool.* **16,** 22–42.

Rosenquist, G. C., and DeHaan, R. L. (1996). Migration of precardiac cells in the chick embryo: A radioautographic study. *Carnegie Inst. Wash. Publ. 625, Contrib. Embryol.* **38,** 111–121.

Sartore, S., Pierobon-Bormioli, S., and Schiaffino, S. (1978). Immuno-histochemical evidence for myosin polymorphism in the chicken heart. *Nature (London)* **274,** 82–83.

Satin, J., Fujii, S., and DeHaan, R. L. (1988). Development of cardiac beat rate in early chick embryos is regulated by regional cues. *Dev. Biol.* **129,** 103–113.

Schultheiss, T. M., Xydas, S., and Lassar, A. B. (1995). Induction of avian cardiac myogenesis by anterior endoderm. *Development (Cambridge, UK)* **121,** 4203–4214.

Schultheiss, T. M., Burch, J. B., and Lassar, A. B. (1997). A role for bone morphogenetic proteins in the induction of cardiac myogenesis. *Genes Dev.* **11,** 451–462.

Shalaby, F., Rossant, J., Yamaguchi, T. P., Gertsentein, M., Wu, X. F., Breitman, M. L., and Schuh, A. C. (1995). Failure of blood-island formation and vasculogenesis in Flk-1-deficient mice. *Nature (London)* **376,** 62–66.

Soler, A. P., and Knudsen, K. A. (1994). N-cadherin involvement in cardiac myocyte interaction and myofibrillogenesis. *Dev. Biol.* **162,** 9–17.

Sugi, Y., and Lough, J. (1994). Anterior endoderm is a specific effector of terminal cardiac myocyte differentiation of cells from the embryonic heart forming region. *Dev. Dyn.* **200,** 155–162.

Sugi, Y., and Lough, J. (1995). Activin-A and FGF-2 mimic the inductive effects of anterior endoderm on terminal cardiac myogenesis *in vitro*. *Dev. Biol.* **168**, 567–574.

Sugi, Y., and Markwald, R. R. (1996). Formation and early morphogenesis of endocardial endothelial precursor cells and the role of endoderm. *Dev. Biol.* **175**, 66–83.

Sugi, Y., Sasse, J., and Lough, J. (1993). Inhibition of precardiac medoderm cell proliferation by antisense oligodeoxynucleotide complementary to fibroblast growth factor-2 (FGF-2). *Dev. Biol.* **157**, 28–37.

Takeichi, M. (1991). Cadherin cell adhesion receptors as a morphogenetic regulator. *Science* **251**, 1451–1455.

Tawara, S. (1906). Das reizleitungs system des Säugetierherzens. Gustv Fischer, Jena.

Vassal-Adams, P. R. (1982). The development of the atrioventricular bundle and its branches in the avian heart. *J. Anat.* **134**, 169–183.

Viragh, S., and Challice, C. E. (1973). The development of the conduction system in the mouse embryo heart. IV. Differentiation of the atrioventricular conduction system. *Dev. Biol.* **89**, 25–40.

Vitadello, M., Matteoli, M., and Gorza, L. (1990). Neurofilament proteins are co-expressed with desmin in heart conduction system myocytes. *J. Cell Sci.* **97**, 11–21.

von Kirschhofer, K., Grim, M., Christ, B., and Wachtler, F. (1994). Emergence of myogenic and endothelial cell lineages in avian embryos. *Dev. Biol.* **163**, 270–278.

Waldo, K. L., Willner, W., and Kirby, M. L. (1990). Origin of the proximal coronary artery stems and a review of ventricular vascularization in the chick embryo. *Am. J. Anat.* **188**, 109–120.

Waldo, K. L., Kumiski, D. H., and Kirby, M. L. (1994). Association of the cardiac neural crest with development of the coronary arteries in the chick embryo. *Anat. Rec.* **239**, 315–331.

Wei, Y., Bader, D., and Litvin, J. (1996). Identification of a novel cardiac-specific transcript essential for cardiac myocyte differentiation. *Development (Cambridge, UK)* **22**, 2779–2789.

Wessels, A., Vermeulen, J. L. M., Verbeek, F. J., Viragh, S., Kalman, F., Lamers, W. H., and Moorman, A. F. M. (1992). Spatial distribution of "tissue-specific" antigens in the developing human heart. *Anat. Rec.* **232**, 97–111.

Yatskievych, T. A., Ladd, A. N., and Antin, P. B. (1997). Induction of cardiac myogenesis in avian pregastrula epiblast: The role of the hypoblast and activein. *Development (Cambridge, UK)* **124**, 2561–2570.

Yutzey, K. E., Rhee, J. T., and Bader, D. (1994). Expression of the atrial-specific myosin heavy chain AMHC1 and the establishment of anteroposterior polarity in the developing chicken heart. *Development (Cambridge, UK)* **120**, 871–883.

Yutzey, K., Gannon, M., and Bader, D. (1995). Diversification of cardiomyogenic cell lineages *in vitro*. *Dev. Biol.* **170**, 531–541.

Zhu, X., Sasse, J., McAllister, D., and Lough, J. (1996). Evidence that fibroblast growth factors 1 and 4 participate in regulation of cardiogenesis. *Dev. Dyn.* **207**, 429–438.

31

Cell Fate Determination in the Chick
Embryo Retina

Ruben Adler
Teri Belecky-Adams
The Wilmer Ophthalmological Institute
The Johns Hopkins University
School of Medicine
Baltimore, Maryland 21287

I. The Issues

The adult neural retina consists of an assortment of differentiated cell types that are postmitotic and occupy precise positions in one of the characteristic retinal layers. Both the diversity of cell types and their stereotyped and orderly laminar distribution are generated during embryonic development by a series of modifications of a morphologically homogeneous pseudostratified neuroepithelium; all cells in this neuroepithelium are mitotically active at the onset of retinal embryogenesis, as first shown autoradiographically by Fujita and Horii (1963). The various retinal cell types are generated (i.e., become postmitotic) in a nonrandom, predictable sequence (reviewed in Altshuler et al., 1991), but lineage tracing studies have shown that in chick embryos, as well as in other species, the proliferating neuroepithelial cells are multipotential, that is to say, can give rise to progenies comprising two or more types of cells (Turner and Cepko, 1987; Holt et al., 1988; Wetts

and Fraser, 1988; Turner et al., 1990; Fekete et al., 1994). Retinal histogenesis, therefore, does not seem to involve a deterministic lineage mechanism (see also Chapter 16 by Carthew et al.; Chapter 23 by Perron and Harris).

The question as to whether cell determination occurs at the time of, or some time after, terminal mitosis could not be elucidated by lineage-tracing studies such as those mentioned above, and its investigation has been difficult due to the temporal and spatial complexity of the developing retina. A good example of the problem is the existence of a developmental gradient according to which many developmental events occur earlier in the fundal than in the peripheral region of the retina. The duration of the period of generation (terminal mitosis) of the neuronal elements of the chick embryo retina takes approximately 4–5 days, but this period begins and ends earlier in the fundal region than in the periphery (Dutting et al., 1983). The complexity derived from this asynchrony is accompanied by the existence of considerable overlap in the time of birth of different cell types within each region of the retina. During the second half of the first embryonic week, therefore, the retina will contain at any point in time a mixture of cells that are dividing, cells that are undergoing terminal mitosis, and postmitotic cells that are migrating to or have reached their definitive laminar positions. An additional complicating factor is the rather variable interval between the time at which cells undergo terminal mitosis, and the time at which they express overt differentiated properties. This period is relatively short for retinal ganglion cells, which are already identifiable as such in the fundal region of the retina before embryonic day (ED) 8. On the other hand, photoreceptor cells (which also are generated before ED 8) will not express visual pigments or begin to form outer segments before ED 14 (Govardovsky and Kharkeevich, 1965; Bruhn and Cepko, 1996), that is to say, over a week after the cells are born. However, other photoreceptor-specific properties are expressed much earlier, as exemplified by the photoreceptor-specific molecule visinin, which by ED 6 can already be detected in, and restricted to photoreceptor cells, by _in situ_ hybridization and immunocytochemistry (Bruhn and Cepko, 1996; T. Belecky-Adams and R. Adler, 1998, unpublished observations). Such a protracted appearance of differentiated properties could suggest that each one of the properties is induced by a separate signal (or group of signals), acting around the time when it first becomes detectable; however, the possibility that many or even most of those differentiated properties are coordinated by a "developmental master program" cannot be dismissed _a priori_ based exclusively on the descriptive analysis of the timing of their expression (see below).

In summary, then, the development of the retina can conveniently (but also arbitrarily) be subdivided into several different stages. The first is the neuroepithelial phase, lasting only until ED 3–4, during which all cells are morphologically homogeneous, mitotically active, and not yet committed to specific phenotypic fates. Between ED 4 and 7, the retina contains a mixture of proliferating neuroepithelial cells, cells that are undergoing terminal mitosis, postmitotic cells at various stages of their migration to their laminar positions, and even some cells (ganglion cells) which are undergoing some degree of overt differentiation. In the fundal region, cell proliferation is largely complete by ED 8 (with the exception of some cells that will give rise to Müller glia), but most retinal cells still appear undifferentiated even in this region. After ED 8, retinal development seems to be dominated by the progressive differentiation of various cell types, with each cell type having a characteristic timetable for the expression of various phenotypic properties. The complexity of this developmental sequence is further increased by the existence of fundus-to-periphery gradients in retinal maturation,

with the fundus generally being in advance of the periphery in its development.

II. The Approaches

We have investigated the issues summarized in the preceding section using three complementary methodological approaches, which address some (but, unfortunately, not all) of the experimental challenges derived from the above mentioned spatial and temporal complexity of the embryonic retina. These methodological approaches will be summarized in this section, before describing the insights into mechanisms of retinal cell differentiation that were derived from our studies.

A. Low Density Cell Cultures

Isolation and transplantation experiments are the two most commonly used approaches to test precursor cell "commitment" (i.e., stable restriction of the developmental potential of a precursor cell to a specific differentiated fate). For our studies we have used the first of these approaches, isolating retinal precursor cells from the chick embryo retina at different stages of development, and growing them at low densities, on highly adhesive substrata to which the cells attach as individual units; this minimizes the opportunities for contact-mediated intercellular interactions. Unlike other systems in which dissociated cells are grown at higher density, these cultures are devoid of "flat cells," such as glial cells (which fail to develop when the formation of multicellular clumps is prevented), fibroblasts and endothelial cells (not normally present in the chick embryo retina), or pigment epithelial cells (from which the retina is separated by mechanical dissection). The absence of flat cells is experimentally advantageous because it eliminates a putative source of signals that could influence cell differentiation, and allows direct observation of individual precursor cells without interference from other cells. An additional advantage is that growth factors and other putative regulatory molecules produced by physiologically relevant sources, such as glia and the retinal pigment epithelium, can be tested for effects on retinal neurons and photoreceptors under defined conditions.

By challenging the cells with conditions that are completely different from those present in the embryo, this experimental system makes it possible to investigate three interrelated questions: (1) Can retinal precursor cells differentiate _in vitro_ when they are isolated before the onset of overt differentiation, and grown in the absence of contact-mediated intercellular interactions? (2) If they can differentiate, do all retinal precursor cells undergo the same pattern of differentiation when grown under uniform culture conditions, or can different precursors follow divergent pathways of differentiation despite the homogeneity of their microenvironment?

(3) How complex is the differentiation pattern achieved by isolated cells *in vitro*, and to what extent does their phenotype resemble that of their *in vivo* counterparts?

B. Use of a Battery of Complementary Analytical Techniques to Characterize the Differentiated Phenotype of Cultured Cells

To avoid possible problems of interpretation associated with the utilization of isolated cell markers for cell identification, in our studies the differentiated phenotypes expressed by retinal precursor cells *in vitro* have been characterized with a battery of complementary techniques, including not only the detection of cell-specific molecules, but also the characterization of the structural organization of the cells using light and electron microscopy, and the investigation of physiological activities such as the responses of photoreceptor cells to light (see below). Molecular markers have some advantages for studies of this type because they can be objectively determined by immunocytochemistry or *in situ* hybridization; they are not devoid of drawbacks, however, particularly when the cells are isolated from their normal microenvironment to study their differentiation under experimental conditions. One of the fundamental problems that emerges when cell differentiation is evaluated through the detection of individual markers is that it is frequently difficult to distinguish bona fide changes in the differentiated fate of precursor cells from changes in the expression of one particular gene that do not necessarily involve changes in cell commitment. In addition, many immunocytochemical markers used to study the development of the retina are cell-specific for only short periods of time, and are expressed by increasing numbers of cell types as the retina differentiates (Reh and Kljavin, 1989; Watanabe and Raff, 1990; 1992; Altshuler and Cepko, 1992; Guillemot and Cepko, 1992; Reh, 1992; Kelley *et al.*, 1994; Snow and Robson, 1994; Austin *et al.*, 1995; Waid and McLoon, 1995; Ezzeddine *et al.*, 1997) This may lead to confusion between qualitative changes in the differentiated fate of precursor cells, on the one hand, and increases or decreases in the rate of cell differentiation which might occur in response to changes in microenvironmental conditions *in vitro*.

C. High Resolution Analysis of the Time of Terminal Mitosis

As already indicated, the time at which neurons undergo their last mitotic division (i.e., are "born") represents a fundamental landmark in the life history of differentiating neuronal precursor cells. Given the temporal and spatial heterogeneity of the embryonic retina (see above), precise knowledge of the stage at which each cell is exposed to putative regulatory signal(s), relative to its terminal mitosis, is crucial for the analysis and interpretation of cell differentiation studies. Information available in the literature about the time of cell birth in the retina of chick embryos and other vertebrates was obtained by radioactive thymidine (^{3}HT) autoradiography or (less frequently) bromodeoxyuridine (BrDU) immunocytochemistry. These precursors, which are taken up and incorporated into the DNA of dividing cells, were used in either a "cumulative" labeling paradigm (in which the DNA precursor molecule is kept constantly available after its initial administration), or in "pulse-labeling" methods (in which the DNA precursor molecule is only made available to the cells for a limited period of time).

Although both cumulative and pulse-labeling methods have provided useful information about the kinetics of cell generation *in vivo* (see below), they lack the high degree of temporal resolution necessary for experimental analysis of correlations between time of cell birth and cell differentiation. To overcome these limitations, many of our studies employ a "window-labeling" technique developed in this laboratory which allows determining the time of terminal mitosis with a resolution of hours (Repka and Adler, 1992a; Belecky-Adams *et al.*, 1996). The method is based on an initial injection of ^{3}HT, followed at a predetermined interval (usually 3–5 hours) by administration of BrDU. Additional administration of BrDU is subsequently used as necessary to maintain its constant availability to the cells (see Fig. 1A). When the tissue labeled in this manner is analyzed by ^{3}HT autoradiography combined with BrDU immunocytochemistry, three populations of cells can be distinguished (Fig. 1B): the cells born before ^{3}HT and BrDU administration appear unlabeled, the cells born after the onset of BrDU administration are BrDU (+) [and frequently also ^{3}HT(+)], and the cells undergoing their last round of DNA duplication after ^{3}HT but before BrDU administration are the only ones that appear ^{3}HT(+)/BrDU(−). Therefore, the technique provides higher temporal resolution than what can be achieved with pulse- or cumulative labeling paradigms using individual DNA precursor molecules, and also makes it possible to identify cells born at particular time points not only within their normal microenvironment, but also when they are grown *in vitro* outside their normal milieu.

III. Evidence for a Developmental Master Plan Regulating Retinal Cell Differentiation

A. *In Vitro* Studies

The results to be summarized below have shown that retinal precursor cells, isolated before the onset of overt differentiation, not only can differentiate *in vitro* when

A

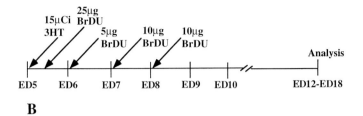

B

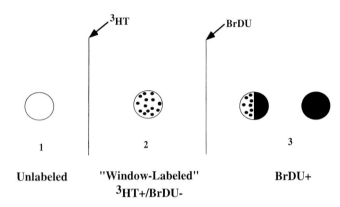

Figure 1 Diagrammatic representation of the window-label paradigm. (A) In the example shown, ³HT was added on ED 5, followed 5 hours later by BrDU. BrDU administration was repeated until ED 8, to keep it available to dividing cells. Embryos were then allowed to develop further, fixed, and processed for BrDU immunocytochemistry and autoradiography. (B) The diagram illustrates the three populations that can be observed with the window-labeling technique (adapted from Repka and Adler, 1992b). Cells born before the addition of ³HT and BrDU appear unlabeled (1). The only cells that appear ³HT(+)/BrDU(−) are those born after the addition of ³HT, but before BrDU administration (window-labeled cells), whereas cells born after the addition of ³HT and BrDU are BrDU(+) and, in many cases, also ³HT(+) (3). ³HT, Tritiated thymidine; BrDU, bromodeoxyuridine; ED, embryonic day. From Belecky-Adams *et al.*, 1996, with permission from Academic Press.

grown in the absence of contact-mediated cell interactions, but also can follow divergent pathways of differentiation, with some cells becoming photoreceptors and others differentiating as nonphotoreceptor neurons. The differentiation of these isolated cells involves the expression of many molecular, structural, and functional properties, which closely resemble those of their *in vivo* counterparts. Since the divergent differentiation of these cells takes place as they develop within the same microenvironment, the data are strongly consistent with the possible existence of "developmental master programs" coordinating the expression of networks of phenotypic properties characteristic of one or another of the cell types present in the retina.

Cultures of dissociated ED 6–ED 8 retinal cells appear morphologically homogeneous at the onset of the experiments, with the cells showing a process-free, undifferentiated round morphology. Their morphological differentiation begins within a few hours, and already at those early stages of *in vitro* development it is evident that neighboring cells frequently follow divergent pathways of differentiation (Adler *et al.*, 1984; Madreperla and Adler, 1989). As the cultures continue to develop over the next 3–6 days *in vitro*, some of the

cells express many of the characteristic features of photoreceptors, whereas others differentiate as nonphotoreceptor, multipolar neurons. The formation of neuritic processes that takes place during this process of differentiation leads to the establishment of contacts between many of the cells, but many others remain devoid of intercellular contacts. The cells that differentiate as nonphotoreceptor, multipolar neurons give rise to several long, branched neurites, whereas their cell body is larger than that of photoreceptors, and has a circular or polyhedrical appearance. Many of these nonphotoreceptor neurons display amacrine cell markers, such as immunoreactivity with the HPC-1, Pax-6, and GABA antibodies, and show a high affinity uptake mechanism for the neurotransmitter GABA (Pessin and Adler, 1985; Politi and Adler, 1986, 1987; and R. Adler and T. Belecky-Adams, 1997, unpublished observations). The cultured precursor cells that develop as photoreceptors, on the other hand, follow a completely different pattern of differentiation. Their transformation from round precursor cells into highly elongated and polarized cells has been documented by sequential photography, which shows that the process involves a stereotyped series of changes (Madreperla and Adler, 1989). At the end of this process the cultured photoreceptors are highly elongated and compartmentalized, with a cell body occupied almost exclusively by the nucleus, and a single short neurite. The nuclear compartment is in direct continuity with the inner segment, which contains the metabolic machinery of the cell (endoplastic reticulum, Golgi apparatus, mitochondria, etc.), and has an apical cilium that in many photoreceptors expands into a visual pigment-rich, outer segment-like structure in which membranous disks can be demonstrated by electron microscopy (Adler *et al.*, 1984; Adler, 1986; Madreperla and Adler, 1989; Saga *et al.*, 1996). The finding that the structural transformations through which round photoreceptor precursors acquire this complex elongated and polarized phenotype occur in cells devoid of intercellular contacts, suggests that intracellular forces play a protagonistic role in determining photoreceptor organization; this hypothesis was supported by experiments using cytoskeletal inhibitors, such as nocodazole (which depolimerizes microtubes) and cytochalasin D (which disrupts actin filaments) (Madreperla and Adler, 1989). These experiments showed that the development and maintenance of the structural polarity of photoreceptors results from the balance between microtubule-dependent forces, which tend to elongate the cells, and actin-dependent forces, which tend to shorten them. It is noteworthy that the disorganization triggered by the cytoskeletal inhibitors is reversible, with the cells recovering their normal pattern of organization when the drugs are removed (Madreperla and Adler, 1989).

The cells that differentiate as photoreceptors acquire not only structural polarity, but molecular polarity as well. Visual pigments, for example, are not yet present in the cells when they are isolated from the retina; when their expression begins as the cells differentiate in culture, opsin immunoreactivity becomes concentrated in the small outer segmentlike processes of the photoreceptors (Adler, 1986; Saga *et al.*, 1996). On the other hand, immunoreactivity for the enzyme Na^+-K^+-ATPase (which plays a critical role in photoreceptor response to light) is already detectable in morphologically undifferentiated (circular) photoreceptor precursors, in which it is diffusely distributed. As the cells elongate, however, ATPase immunoreactivity becomes localized to the plasma membrane of the inner segment region of the cultured photoreceptors (Madreperla *et al.*, 1989), which is also the site where it is localized *in vivo*. This polarized distribution of Na^+-K^+-ATPase appears to depend on its interactions with the subcortical cytoskeleton, as indicated by its immunocytochemical colocalization with spectrin, its resistance to detergent extraction, and its lack of mobility in the plane of the membrane as determined by fluorescent recovery after photobleaching (Madreperla *et al.*, 1989). As significant as the positivity of photoreceptors for these immunocytochemical markers is their negativity for markers characteristic of

nonphotoreceptor neurons, such as GABA, HPC-1, Pax-6, and Prox 1 (Politi and Adler, 1986, 1987; T. Belecky-Adams and R. Adler, 1997, unpublished observations).

Photoreceptor cell differentiation *in vitro* also involves the acquisition of the capacity to perform cell-specific functional activities (Stenkamp and Adler, 1993, 1994; Stenkamp *et al.*, 1994; Argamaso-Hernan and Adler, 1997). As their *in vivo* counterparts, cultured photoreceptors respond to light with photomechanical movements, with most of the responsive photoreceptors elongating in light and contracting in darkness. Cultured photoreceptors also develop a diurnal rhythm in iodopsin mRNA levels, indicating that photoreceptors that develop in the absence of cellular interactions are capable of translating light into transcriptional control of key genes (Pierce *et al.*, 1993; Argamaso-Hernan and Adler, 1997). Photomechanical responses in culture also resemble the *in vivo* situation in that they are influenced by the neuromodulators dopamine and melatonin (Stenkamp *et al.*, 1994); the data indicate that the cultures contain a complex network of neuroregulatory mechanisms involving not only the photoreceptors (which produce melatonin) but also nonphotoreceptor neurons, which produce dopamine.

B. Differential Expression of Homeobox Genes in Retinal Precursor Cells before the Onset of Overt Differentiation

The concept of a "master regulatory program" controlling cell differentiation suggests that, even before their overt differentiation, the precursors of different cell types may express different transcriptional regulators; such a hypothesis appears reasonable in light of extensive demonstration of the role of DNA-binding transcription factors in the control of the coordinated expression of networks of cell-specific genes both in invertebrates (Banerjee and Zipursky, 1990) and in vertebrates (Kageyama *et al.*, 1995; Ludolph and Konieczny, 1995; Groves and Anderson, 1996; Boucher and Pedersen, 1996; Quinn *et al.*, 1996; Bally-Cuif and Boncinelli, 1997).

Our laboratory is investigating three candidate genes belonging to the homeobox/paired domain families, namely, Pax-6, Prox 1, and Chx10. These genes have been shown to participate in various aspects of eye development in other species (Hogan *et al.*, 1986; Hill *et al.*, 1991; Glaser *et al.*, 1994; Li *et al.*, 1994; Quiring *et al.*, 1994; Del Rio-Tsonis *et al.*, 1995; Grindley *et al.*, 1995; Halder *et al.*, 1995; Schedl *et al.*, 1996; Quinn *et al.*, 1996; Altmann *et al.*, 1997; Hirsch and Harris, 1997; Tomarev *et al.*, 1997; see Chapter 23 by Perron and Harris); the respective chicken homologs have been cloned by Tomarev *et al.*, 1997 (for Pax-6), Li *et al.*, 1994 (for Pax-6) and Belecky-Adams *et al.*, 1997 (for Chx10). We used three complementary techniques

(Northern blot analysis, *in situ* hybridization, and immunocytochemistry) to investigate the expression of these genes during retinal development *in vivo* (Belecky-Adams *et al.*, 1997).

Northern blot analysis showed that all three genes were expressed at all the developmental stages studied, including ED 5, when the future retina is still predominantly a proliferating neuroepithelium. Quantitative changes were particularly noticeable for Prox 1, which was only detectable in overexposed blots on ED 5 but became very abundant thereafter. Pax-6 also showed some increases between ED 5 and ED 8 but remained unchanged thereafter, whereas Chx10 levels were fairly constant throughout development.

By *in situ* hybridization, the three genes appeared diffusely distributed by ED 4–ED 5, before the onset of retinal histogenesis. However, there were dramatic changes between these stages and ED 8, with the initially diffuse pattern of distribution being replaced by a topographically restricted, laminar pattern of expression that correlates with the positions that, as cell differentiation advances, are occupied by particular retinal cell types. The only region of the developing retina that is completely devoid of signals for all three gene products is the scleral-most area, occupied by putative photoreceptors; all the remaining layers, on the other hand, are enriched for at least one of the mRNAs (see Fig. 2). Prox 1, for example, shows a stronger signal in putative horizontal cells, decreasing toward the vitreal aspect of the inner nuclear layer and becoming undetectable in the ganglion cell layer. The latter shows very high levels of expression of Pax-6, which is also abundant in the amacrine sublayer of the inner nuclear layer, and detectable in a subpopulation of putative horizontal cells. Chx10 mRNA is concentrated in the putative Müller cell/bipolar region of the inner nuclear layer, decreasing toward both the scleral and vitreal sides of this layer. At later developmental stages (between ED 15 and PD 1), the intensity of the *in situ* hybridization signal decreases, but the overall pattern of distribution remains unchanged. Immunocytochemical analysis of the distribution of the corresponding proteins showed similar patterns of laminar distribution, and documented also their nuclear localization, which is consistent with their putative role as transcriptional regulators.

Taken together, these findings, summarized in Table I, show that each retinal cell type has a characteristic assortments of transcriptional regulators before the onset of their overt differentiation, which is consistent with data obtained by others (Xiang *et al.*, 1993, 1995, 1996; Carriere *et al.*, 1993; Levine *et al.*, 1994; Liu *et al.*, 1994; Gan *et al.*, 1996; Chapter 23 by Perron and Harris). The functional significance of these transcriptional regulators in the control of cell differentiation has not been established, although gene deletion experiments by homologous recombination have demonstrated the importance of Brn-3b (Xiang *et al.*, 1993, 1995, 1996) for ganglion cell differentiation (Gan *et al.*, 1996). In the particular case of chick embryo retinal cells, we have observed by *in situ* hybridization and immunocytochemistry that Pax-6, Prox 1, and Chx10 are expressed in subpopulations of neurons, but are never detectable in photoreceptor cells *in vitro* (T. Belecky-Adams and R. Adler, 1997, unpublished observations); such a differential pattern of expression in the cells that differentiate *in vitro* open experimental possibilities to investigate the functional role of these transcriptional regulators in cell differentiation by antisense oligonucleotide treatments and gene overexpression experiments. Such studies are currently underway in the laboratory.

IV. Developmental Plasticity of Postmitotic Retinal Precursor Cells

The studies summarized in the preceding sections suggest that the coordinated expression of many of the phenotypic properties that characterize each differentiated retinal cell type is governed by a master regulatory program. Those programs appear to be already in place before the onset of overt cell differentiation, but it is not

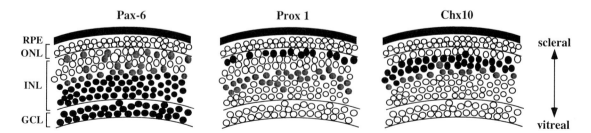

Figure 2 Topographically specific distribution of mRNA for the homeobox genes Pax-6, Prox 1, and Chx10 in ED8 chick retinas. See text for details. The darkness of circles is intended as an approximate representation of the relative intensity of the *in situ* hybridization signals, with unfilled circles representing cells that appear completely negative. RPE, Retinal pigment epithelium; ONL, outer nuclear layer; INL, inner nuclear layer, GCL, ganglion cell layer. From Belecky-Adams *et al.* 1996, with permission from Academic Press.

TABLE I

Comparison of Pax-6, Prox 1, and Chx10 Signals at ED 4–5 and ED 8[a]

Cell type	Pax-6	Prox-1	Chx10
ED 4–5			
Neural epithelium	+	+	+
ED 8			
Photoreceptors	−	−	−
Horizontal	+/−	++++	−
Bipolar/Müller	−	+/−	++++
Amacrine	++++	−	+
Ganglion	++++	−	−

[a]At ED 4–5, a stage at which most cells in the retinal neuroepithelium are mitotically active, the *in situ* hybridization signals for all three homeobox genes were faintly positive and diffusely distributed. At ED 8, when most cells are already postmitotic but generally not yet overtly differentiated, the expression pattern of these homeobox genes already has become compartmentalized with a laminar pattern that denotes the future differentiated fate of the cells. ++++, Strong signals; +, faint signals; +/−, signal present only in some cells of a particular layer; −, no signal.

known when cells become committed to a particular developmental program. That such commitment does not occur while the cells are proliferating was suggested by lineage tracing studies (Turner and Cepko, 1987; Holt *et al.*, 1988; Wetts and Fraser, 1988; Turner *et al.*, 1990, Fekete *et al.*, 1994). These studies, however, did not elucidate whether cells become determined at the time of, or some time after terminal mitosis. We have addressed this issue using the "window-labeling" (WL) technique, which allows identifying cohorts of contemporary precursor cells born during narrow (5 hour) time periods. The studies summarized below strongly suggest that many retinal precursor cells remain plastic for some time after terminal mitosis.

A. *In Vivo* Studies

As described in Section II,C, the window-labeling (WL) technique is based on an initial injection of [3]HT, followed 5 hours later by the administration of BrDU; the cells born during this 5-hour period can thus be identified by being labeled with thymidine but not with BrDU (Fig. 1; Belecky-Adams *et al.*, 1996). For the studies summarized below, separate groups of embryos were labeled during 5-hour intervals on ED 4, 5, 6, 7, or 8, and the fundal regions of their retinas were quantitatively analyzed on ED 18 by BrDU immunocytochemistry and [3]HT autoradiography, in order to determine the number of WL cells present in various retinal layers. It was found instructive to plot the data to indicate either the fate of cells born during each of the window-labeling periods (Fig. 3A), or the time of generation of cells in each retinal layer (Fig. 3B). Several conclusions can be

drawn from a comparison of both graphs. First, the generation of photoreceptors and nonphotoreceptor neurons is largely completed by embryonic day 8 (at least in the fundal region of the retina). Second, the time of terminal mitosis does not seem to determine the fate of precursor cells, because the cell cohorts born during each of the 5-hour periods studied on ED 4-6 were found to contribute to all retinal layers (Fig. 3). It is noteworthy, however, that the range of developmental fates followed by dividing precursor cells becomes more restricted at later stages, since the generation of ganglion cells and photoreceptors is practically completed by ED 6 in the fundal region, with cells that continue proliferating until ED 8 giving rise to inner nuclear layer neurons. A related conclusion is that there is no obvious inside-out or outside-in pattern of cell generation, since cells generated at early embryonic stages contribute to all retinal layers, and addition of new cells ceases more or less simultaneously in the inner most and outer most layers of the retina (ganglion cells and photoreceptor cells, respectively). The overall sequence of cell generation and the extensive overlap in the generation of different cell types observed in our studies are in general agreement with those obtained in the chick embryo retina using either pulse- or cumulative thymidine labeling (Fujita and Horii, 1963; Morris, 1973; Kahn, 1974; Dutting *et al.*, 1983; Spence and Robson, 1989; Prada *et al.*, 1991; Snow and Robson, 1994). There are, however, some discrepancies among studies regarding the onset of ganglion cell birth and the overall duration of the period of cell generation in the chick retina, which may be due to differences in strains of chickens, incubation conditions, etc.

B. *In Vitro* Studies

One of the useful features of the window-labeling technique is that it allows the identification of cells from one defined population under different experimental conditions (see Section II,C). This made it possible to challenge precursor cells at different times after terminal mitosis, to determine when they switch from being plastic to being rigidly committed to particular differentiated fates. The experiments were carried out with cells window-labeled for 5 hours on ED 5 (WL$_5$), which were either allowed to develop until ED 18 (for histological analysis of the fate of window-labeled cells *in vivo*), or isolated for dissociated retinal cell cultures on ED 6 or on ED 8. *In vivo*, approximately 80% of the WL$_5$ cells gave rise to nonphotoreceptor neurons, with the remaining cells developing as photoreceptors. The results were remarkably similar in cultures of retinas from ED 8 embryos, in which approximately 80% of the window-labeled cells gave rise to nonphotoreceptor, multipolar neurons. The behavior of the WL$_5$ cells, however, was completely different in cultures of retinas dissociated on

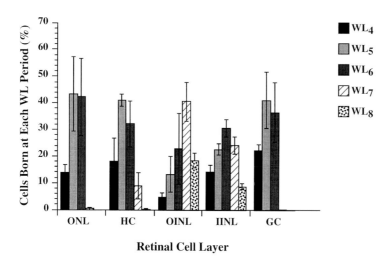

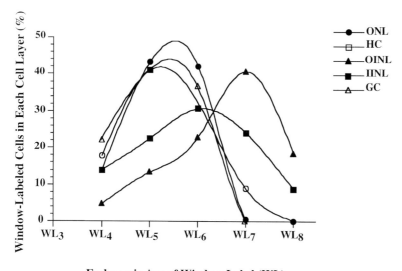

Figure 3 Laminar fate of cells born between ED 4 and ED 8, as determined using the window-label technique. Chick embryos were window-labeled for 5 hours on ED 4, 5, 6, 7, or 8 (WL$_{4-8}$), and allowed to develop until ED 18, when their retinas were fixed, sectioned, and processed for BrDU immunocytochemistry and autoradiography. The number of ^{3}HT/BrDU($-$) (window-labeled) cells was analyzed in the fundal region of the retinas, and plotted either as the percentage of cells from each window-label period that migrated to specific cell layers within the retina (A) or as curves representing the percentage of cells in each layer that are born at specific ages (B). ONL, Outer nuclear layer; HC, horizontal cells; OINL, outer half of the inner nuclear layer; IINL, inner half of the inner nuclear layer; GC, ganglion cell layer; WL, window-label period. From Belecky-Adams *et al.* 1996, with permission from Academic Press.

ED 6, since the window-labeled cells gave rise predominantly to photoreceptors, with only some 20% of the WL$_5$ cells differentiating as nonphotoreceptor neurons (Belecky-Adams *et al.*, 1996). A series of controls demonstrated that the differences in the behavior of the WL$_5$ cells could not be explained by differential cell death. Therefore, the experiments indicate that the behavior of the WL$_5$ cells changes as a function of the length of time that they are exposed to the retinal microenvironment prior to their isolation for culture, and

are consistent with the hypothesis that many retinal precursor cells remain plastic (i.e., uncommitted to particular differentiated fates) for some time after terminal mitosis. Similar conclusions had been reached by Adler and Hatlee (1989) in experiments using a cumulative thymidine labeling paradigm to compare the fate of cells born before ED 5 and isolated for culture either on ED 6 or on ED 8.

Both sets of experiments also are consistent with the notion that photoreceptor differentiation may be the "default pathway" followed by many retinal precursor cells when they are prevented from interacting, while they are still uncommitted, with inductive signals from the retinal microenvironment. This hypothesis also was supported by the results from Repka and Adler (1992b), who observed that practically 100% of dissociated retinal cells that undergo their terminal mitosis in culture (i.e., in isolation from the retinal microenvironment) give rise to photoreceptor cells. It must be noted, however, that our conclusions may not apply to all retinal precursor cells. For example, Waid and McLoon (1995) observed that at least some retinal precursor cells express ganglion cell markers immediately after their terminal mitosis. Moreover, Austin et al. (1995) have reported predominant expression of ganglion cell markers when retinal cells were isolated from younger chick embryos and grown in vitro. The discrepancy between these results and those in our experiments could be due to differences in culture media and substrata, which could influence cell differentiation (Belecky-Adams et al., 1996). It is also conceivable that the immunocytochemical markers used by Austin et al. (1995) to identify their cells as putative ganglion cells could be transiently expressed by cells that are not yet committed to specific cell fates, a phenomenon that has previously been discussed for other cell types (Linser and Moscona, 1981; Lemmon and Rieser, 1983; Hansson et al., 1989; Chabot and Vincent, 1990; Larison and BreMiller, 1990; Bodenant et al., 1991; Hutchins, 1994; Lake, 1994; Pow et al., 1994; Chien and Liem, 1995; Hutchins et al., 1995). Moreover, one of the "ganglion cell markers" used by Austin et al. (1995) (e.g., immunoreactivity with antibodies against the Islet-1 protein) is only restricted to ganglion cells at early stages of development, and becomes broadly distributed among other types of neurons as development progresses (T. Belecky-Adams and R. Adler, 1997, unpublished observations; Henrique et al., 1997); increases in numbers of islet-1 positive cells observed with some experimental treatments in those studies, therefore, do not necessarily reflect increased ganglion cell differentiation.

In vivo studies, described in Section IV,A, showed that retinal precursor cells that undergo terminal mitosis before ED 6 contributed to all the differentiated populations in the fundus of the retina, whereas no photoreceptors or ganglion cells are generated after ED 7 in this region. Two different (but not necessarily exclusive) scenarios could be proposed to explain these changes. One of these models would propose that the developmental potential of retinal precursor cells becomes progressively restricted as the cells undergo increasing numbers of divisions before their terminal mitosis; the alternative hypothesis would suggest that the intrinsic developmental potential of the precursor cells remains constant, but that photoreceptor- and ganglion cell-inducing signals are present at early, but not at later de-

velopmental stages. While a possible contribution by microenvironmental signals cannot be excluded, some experimental results suggest very strongly that retinal precursor cells do indeed become restricted in their developmental potential as they undergo terminal mitosis at progressively later stages. We have observed, for example, that the tendency of precursor cells to give rise to photoreceptor cells in vitro, when they are isolated from the embryo and placed in culture immediately after terminal mitosis, is observed in over 80% of the cells born on ED 6, but is not seen with cells born on ED 7 or 8. The WL$_7$ cells, moreover, fail to give rise to photoreceptors even when cocultured with ED 6 cells (Belecky-Adams et al., 1996, and T. Belecky-Adams, 1997, unpublished observations). A restriction in the capacity of older cells to give rise to photoreceptors was also observed by Repka and Adler (1992b) in their investigation of the fate of isolated cells undergoing terminal mitosis in low density cultures.

V. A Working Hypothesis

Our understanding of the mechanisms regulating cell differentiation in the vertebrate retina is far from complete, and it is in fact likely that significant differences may exist between different vertebrate species. On the other hand, significant progress has been made in this field since the early 1990s, which allows the formulation of working models that provide useful conceptual frameworks for further experimentation. In the case of the chick embryo, the studies from this and other laboratories, summarized in the preceding sections, have led us to propose the following working model (1) retinal precursor cells remain uncommitted to specific phenotypic fates while they are mitotically active, but their developmental potential becomes progressively restricted with time; (2) the precursor cells also remain uncommitted (i.e., plastic) for some time after terminal mitosis; (3) regulatory signals capable of inducing cells to follow particular differentiated fates are differentially distributed across the thickness of the retina, such that postmitotic precursor cells migrating to different laminar positions would be exposed to different microenvironmental influences; (4) the precursor cells that remain at the ventricular surface of the retina after terminal mitosis are prevented from interacting with neuron-inducing signals located more vitreally, and follow a photoreceptor "default pathway" of development; and (5) even before the onset of overt differentiation, the commitment of precursor cells to specific differentiated fates involves the expression of specific sets of transcriptional regulators, which set in motion a "master plan" of development that the cells can express to a considerable (but not absolute) degree in a cell autonomous manner, if they encounter a permissive environment. Obviously, only limited components of this model have received experimental support so far and, even for those that have, the evidence is incomplete. Tentative as the model is, however, it has nonetheless the advantage to generate testable predictions, some of which are currently under investigation.

Acknowledgments

This work was supported by Grants EYO4859, EYO5404 (R.A.), EYO6642 (T.B-A.), NIH Core Grant EYO1765, and an unrestricted grant from Research to Prevent Blindness, Inc. R. A. is a Senior Investigator of Research to Prevent Blindness, Inc. The authors are grateful to Ms. Elizabeth M. Bandell for secretarial assistance.

References

Adler, R. (1986). Developmental predetermination of the structural and molecular polarization of photoreceptor cells. *Dev. Biol.* **117**, 520–527.

Adler, R., and Hatlee, M. (1989). Plasticity and differentiation of embryonic retinal cells after terminal mitosis. *Science* **243**, 391–393.

Adler, R., Lindsey, J. D., and Elsner, C. L. (1984). Expression of cone-like properties by chick embryo neural retina cells in glial-free monolayer cultures. *J. Cell Biol.* **99**, 1173–1178.

Altmann, C. R., Chow, R. L., Lang, R. A., and Hemmati-Brivanlou A. (1997). Lens induction by Pax-6 in *Xenopus laevis*. *Dev. Biol.* **185**, 119–123.

Altshuler, D., and Cepko, C. (1992). A temporally regulated, diffusible activity is required for rod photoreceptor development *in vitro*. *Development (Cambridge, UK)* **114**, 947–957.

Altshuler, D. M., Turner, D. L., and Cepko, C. L. (1991). Specification of cell type in vertebrate retina. In "Development of the Visual System" (D. M.-K. Lam and C. J. Shatz, eds.), pp. 37–58. MIT Press, Cambridge, Massachusetts and London.

Argamaso-Hernan, S. M., and Adler, R. (1997). Diurnal rhythm in visual pigment mRNA expression in isolated chick embryo photoreceptor cells. *Invest. Ophthalmol. Visual Sci.* **38**, S53.

Austin, C. P., Feldman, D. E., Ida, J. A., and Cepko, C. L. (1995). Vertebrate retinal ganglion cells are selected from competent progenitors by the action of Notch. *Development (Cambridge, UK)* **121**, 3637–3650.

Bally-Cuif, L., and Boncinelli, E. (1997). Transcription factors and head formation in vertebrates. *BioEssays* **19**, 127–135.

Banerjee, U., and Zipursky, S. L. (1990). The role of cell–cell interaction in the development of the *Drosphila* visual system. *Neuron* **4**, 177–187.

Belecky-Adams, T., Cook, B., and Adler, R. (1996). Correlations between terminal mitosis and differentiated fate of retinal precursor cells *in vivo* and *in vitro*: Analysis with the "window-label" technique. *Dev. Biol.* **178**, 304–315.

Belecky-Adams, T., Tomarev, S., Li, H.-S., Ploder, L.,

McInnes, R. R., Sundin, O., and Adler, R. (1997). Prox 1, Pax-6 and Chx10 homeobox gene expression correlate with phenotypic fate of retinal precursor cells. *Invest. Ophthalmol. Visual. Sci.* **38**, 1293–1303.

Bodenant, C., Leroux, P., Gonzalez, B. J., and Vaudry, H. (1991). Transient expression of somatostatin receptors in the rat visual system during development. *Neuroscience (Oxford)* **41**, 595–606.

Boucher, D. M., and Pedersen, R. A. (1996). Induction and differentiation of extra-embryonic mesoderm in the mouse. *Reprod. Fertil. Dev.* **8**, 765–777.

Bruhn, S. L., and Cepko, C. L. (1996). Development of the pattern of photoreceptors in the chick retina. *J. Neurosci.* **16**, 1430–1439.

Carriere, C., Plaza, S., Martin, P., Quantannens, B., Bailly, M., Stehelin, D., and Saule, S. (1993). Characterization of quail Pax-6 (Pax-QNR) proteins expressed in the neuroretina. *Mol. Cell. Biol.* **13**, 7157–7166.

Chabot, P., and Vincent, M. (1990). Transient expression of an intermediate filament-associated protein (IFAPa-400) during *in vivo* and *in vitro* differentiation of chick embryonic cells derived from neuroectoderm. *Brain Res. Dev. Brain Res.* **54**, 195–204.

Chien, C. L., and Liem, R. K. (1995). The neuronal intermediate filament, alpha-internexin is transiently expressed in amacrine cells in the developing mouse retina. *Exp. Eye Res.* **61**, 749–756.

Del Rio-Tsonis, K., Washabaugh, C. H., and Tsonis, P. A. (1995). Expression of Pax-6 during urodele eye development and lens regeneration. *Proc. Natl. Acad. Sci. U.S.A.* **92**, 5092–5096.

Dutting, D., Gierer, A., and Hansmann, G. (1983). Self renewal of stem cells and differentiation of nerve cells in the developing chick retina. *Dev. Brain Res.* **10**, 21–32.

Ezzeddine, Z. D., Yang, X., DeChiara, T., Yancopoulos, G., and Cepko, C. L. (1997). Postmitotic cells fated to become rod photoreceptors can be respecified by CNTF treatment of the retina. *Development (Cambridge, UK)* **124**, 1055–1067.

Fekete, D. M., Perez-Miguelsanz, J., Ryder, E. F., and Cepko, C. L. (1994). Clonal analysis in the chicken retina reveals tangential dispersion of clonally related cells. *Dev. Biol.* **166**, 666–682.

Fujita, S. and Horii, S. (1963). Analysis of cytogenesis in the chick retina by tritiated thymidine autoradiography. *Archumhistol. Jpn.* **23**, 295–366.

Gan, L., Xiang, M. Q., Zhou, L. J., Wagner, D. S., Klein, W. H., and Nathans, J. (1996). POU domain factor Brn-3B is required for the development of a large set of retinal ganglion cells. *Proc. Natl. Acad. Sci. USA* **93**, 3920–3925.

Glaser, T., Jepeal, L., Edwards, J. G., Young, S. R., Favor, J., and Maas, R. L. (1994). Pax-6 gene dosage effect in a family with congenital cataracts, aniridia, anophthalmia and central nervous system defects. *Nat. Genet.* **8**, 203.

Govardovsky, V. J., and Kharkeevich, T. A. (1965). Histochemical and electron microscopic study of photoreceptive cell development under conditions of tissue culture. *Arkh. Anat. Gistol. Embriol.* **49**, 50–56.

Grindley, J. C., Davidson, D. R., and Hill, R. E. (1995). The role of Pax-6 in eye and nasal development. *Development (Cambridge, UK)* **121**, 1433–1442.

31. Cell Fate Determination in the Chick Embryo Retina

Groves, A. K., and Anderson, D. J. (1996). Role of environmental signals and transcriptional regulators in neural crest development. *Dev. Genet.* **18**, 64–72.

Guillemot, F. and Cepko, C. L. (1992). Retinal fate and ganglion cell differentiation are potentiated by acidic FGF in an *in vitro* assay of early retinal development. *Development* (*Cambridge, UK*) **114**, 743–754.

Halder, G., Callaerts, P., and Gehring, W. J. (1995). Induction of ectopic eyes by targeted expression of the *eyeless* gene in *Drosophila*. *Science* **267**, 1788–1792.

Hansson, H. A., Holmgren, A., Norstedt, G., and Rozell, B. (1989). Changes in the distribution of insulin-like growth factor I, thioredoxin, thioredoxin reductase and ribonucleotide reductase during the development of the retina. *Exp. Eye Res.* **48**, 411–420.

Henrique, D., Hirsinger, E., Adam, J., Le Roux, I., Pouriquie, O., Ish-Horowicz, D., and Lewis J. (1997). Maintenance of neuroepithelial progenitor cells by Delta-Notch signalling in the embryonic chick retina. *Curr. Biol.* **7**, 661–670.

Hill, R. E., Favor, J., Hogan, B. L., Ton, C. C., Saunders, G. F., Hanson, I. M., Prosser, J., Jordan, T., Hastie, N. D., and van Heyningen, V. (1991). Mouse small eye results from mutations in a paired-like homeobox-containing gene. *Nature* (*London*) **354**, 522–525.

Hirsch, N., and Harris, W. A. (1997). *Xenopus* Brn-3.0, a POU-domain gene expressed in the developing retina and tectum. Not regulated by innervation. *Invest. Ophthalmol. Visual Sci.* **38**, 960–969.

Hogan, R. N., Bowman, K. A., Baringer, J. R., and Prusiner, S. B. (1986). Replication of scrapie prions in hamster eyes precedes retinal degeneration. *Ophthalmic Res.* **18**, 230–235.

Holt, C. E., Bertsch, T. W., Ellis, H. M., and Harris, W. A. (1988). Cellular determination in the *Xenopus* retina is independent of lineage and birth date. *Neuron* **1**, 15–26.

Hutchins, J. B. (1994). Development of muscarinic acetylcholine receptors in the ferret retina. *Brain Res. Dev. Brain Res.* **82**, 45–61.

Hutchins, J. B., Bernanke, J. M., and Jefferson, V. E. (1995). Acetylcholinesterase in the developing ferret retina. *Exp. Eye Res.* **60**, 113–125.

Kageyama, R., Sasai, Y., Akazawa, C., Ishibashi, M., Takebayashi, K., Shimizu, C., Tomita, K., and Nakanishi, S. (1995). Regulation of mammalian neural development by helix-loop-helix transcription factors. *Crit. Rev. Neurobiol.* **9**, 177–188.

Kahn, A. J. (1974). An autoradiographic analysis of the time of appearance of neurons in the developing chick neural retina. *Dev. Biol.* **38**, 30–40.

Kelley, M. W., Turner, J. K., and Reh, T. A. (1994). Retinoic acid promotes differentiation of photoreceptors *in vitro*. *Development* (*Cambridge, UK*) **120**, 2091–2102.

Lake, N. (1994). Taurine and GABA in the rat retina during postnatal development. *Visual Neurosci.* **11**, 253–260.

Larison, K. D., and BreMiller, R. (1990). Early onset of phenotype and cell patterning in the embryonic zebra fish retina. *Development* (*Cambridge, UK*) **109**, 567–576.

Lemmon, V. and Rieser, G. (1983). The development and distribution of vimentin in the chick retina. *Brain Res.* **313**, 191–197.

Levine, E. M., Hitchcock, P. F., Glasgow, E., and Schechter, N. (1994). Restricted expression of a new paired-class homeobox gene in normal and regenerating adult goldfish retina. *J. Comp. Neurol.* **348**, 596–606.

Li, H. S., Yang, J. M., Jacobson, R. D., Pasko, D., and Sundin, O. (1994). Pax-6 is first expressed in a region of ectoderm anterior to the early neural plate—implications for stepwise determination of the lens. *Dev. Biol.* **162**, 181–194.

Linser, P., and Moscona, A. A. (1981). Carbonic anhydrase C in the neural retina: Transition from generalized to glia-specific cell localization during embryonic development. *Proc. Natl. Acad. Sci. U.S.A.* **78**, 7190–7194.

Liu, I. S. C., Chen, J. D., Ploder, L., Vidgen, D., Vanderkooy, D., Kalnins, V. I., Mcinnes, R. R. (1994). Developmental expression of a novel murine homeobox gene (CHX10)—evidence for roles in determination of the neuroretina and inner nuclear layer. *Neuron* **13**, 377–393.

Ludolph, D. C., and Konieczny, S. F. (1995). Transcription factor families: Muscling in on the myogenic program. *FASEB J.* **9**, 1595–1604.

Madreperla, S. A., and Adler, R. (1989). Opposing microtubule-and actin-dependent forces in the development and maintenance of structural polarity in retinal photoreceptors. *Dev. Biol.* **131**, 149–160.

Madreperla, S. A., Edidin, M., and Adler, R. (1989). Na^+,K^+-Adenosine triphosphatase polarity in retinal photoreceptors: A role for cytoskeletal attachments. *J. Cell Biol.* **109**, 1483–1493.

Morris, V. B. (1973). Time differences in the formation of the receptor types in the developing chick retina. *J. Comp. Neurol.* **151**, 323–330.

Pessin, M., and Adler, R. (1985). Coexistence of high affinity uptake mechanisms for putative neurotransmitter molecules in chick embryo retinal neurons in purified culture. *J. Neurosci. Res.* **14**, 317–328.

Pierce, M. E., Sheshberadaran, H., Zhe, Z., Fox, L. E., Applebury, M. L., and Takahashi, J. S. (1993). Circadian regulation of iodopsin gene expression in embryonic photoreceptors in retinal cell culture. *Neuron* **10**, 579–584.

Politi, L. E., and Adler, R. (1986). Generation of enriched populations of cultured photoreceptor cells. *Invest. Ophthalmol. Vis. Sci.* **27**, 656–665.

Politi, L. E., and Adler, R. (1987). Selective destruction of photoreceptor cells by anti-opsin antibodies. *Invest. Ophthalmol. Visual Sci.* **28**, 118–125.

Pow, D. V., Crook, D. K., and Wong, R. O. (1994). Early appearance and transient expression of putative amino acid neurotransmitters and related molecules in the developing rabbit retina: An immunocytochemical study. *Visual Neurosci.* **11**, 1115–1134.

Prada, C., Puga, J., Perez-Mendez, L., Lopez, R., and Ramirez, G. (1991). Spatial and temporal patterns of neurogenesis in the chick retina. *Eur. J. Neurosci.* **3**, 559–569.

Quinn, J. C., West. J. D., and Hill, R. E. (1996). Multiple functions for Pax-6 in mouse eye and nasal development. *Genes Dev.* **10**, 435–446.

Quiring, R., Walldorf, U., Kloter, U., and Gehring, W. J. (1994). Homology of the *eyeless* gene of *Drosophila* to the S*mall* eye gene in mice and *Aniridia* in humans. *Science* **265**, 785–789.

Reh, T. A. (1992). Cellular interactions determine neuronal phenotypes in rodent retinal cultures. *J. Neurobiol.* **23,** 1067–1083.

Reh, T. A., and Kljavin, I. J. (1989). Age of differentiation determines rat retinal germinal cell phenotype: Induction of differentiation by dissociation. *J. Neurosci.* **9,** 4179–4189.

Repka, A. M., and Adler, R. (1992a). Accurate determination of the time of cell birth using a sequential labeling technique with ³H-thymidine and bromodeoxyuridine (window labeling). *J. Histochem. Cytochem.* **40,** 947–953.

Repka, A. M., and Adler, R. (1992b). Differentiation of retinal precursor cells born *in vitro. Dev. Biol.* **153,** 242–249.

Saga, T., Scheurer, D., and Adler, R. (1996). Development and maintenance of outer segments by isolated chick embryo photoreceptor cells in culture. *Invest. Ophthalmol. Visual Sci.* **37,** 561–573.

Schedl, A., Ross, A., Lee, M., Engelkamp, D., Rashbass, P., van Heyningen, V., and Hastie, N. D. (1996). Influence of Pax-6 gene dosage on development—overexpression causes severe eye abnormalities. *Cell (Cambridge, Mass.)* **86,** 71–82.

Snow, R. L., and Robson, J. A. (1994). Ganglion cell neurogenesis, migration and early differentiation in the chick retina. *Neuroscience (Oxford)* **58,** 399–409.

Spence, S. G., and Robson, J. A. (1989). An autoradiographic analysis of neurogenesis in the chick retina *in vivo* and *in vitro. Neuroscience (Oxford)* **32,** 801–812.

Stenkamp, D. L., and Adler, R. (1993). Photoreceptor differentiation of isolated retinal precursor cells includes the capacity for photomechanical responses. *Proc. Natl. Acad. Sci. U.S.A.* **90,** 1982–1986.

Stenkamp, D. L., and Adler, R. (1994). Cell-type- and developmental-stage-specific metabolism and storage of retinoids by embryonic chick retinal cells in culture. *Exp. Eye Res.* **58,** 675–687.

Stenkamp, D. L., Iuvone, P. M., and Adler, R. (1994). Photo-

mechanical movements of cultured embryonic photoreceptors: Regulation by exogenous neuromodulators and by a regulable source of endogenous dopamine. *J. Neurosci.* **14**(5 Part 2), 3083–3096.

Tomarev, S. I., Sundin, O., Banerjee-Basu, S., Duncan, M. K., Yang, J.-M., and Piatigorsky, J. (1997). A chicken homeobox gene Prox 1, related to *Drosophila prospero,* is expressed in the developing lens and retina. *Dev. Dyn.* **206,** 354–367.

Turner, D. L., and Cepko, C. L. (1987). A common progenitor for neurons and glia persists in rat retinas late in development. *Nature (London)* **328,** 131–136.

Turner, D. L., Snyder, E. Y. and Cepko, C. L. (1990). Lineage-independent determination of cell type in the embryonic mouse retina. *Neuron* **4,** 833–845.

Waid, D. K., and McLoon, S. C. (1995). Immediate differentiation of ganglion cells following mitosis in the developing retina. *Neuron* **14,** 117–124.

Watanabe, T., and Raff, M. C. (1990). Rod photoreceptor development *in vitro*: Intrinsic properties of proliferating neuroepithelial cells change as development proceeds in the rat retina. *Neuron* **2,** 461–467.

Wetts, R., and Fraser, S. E. (1988). Multipotent precursors can give rise to all major cell types of the frog retina. *Science* **239,** 1142–1145.

Xiang, M. Q., Zhou, L. J., Peng, Y. W., Eddy, R. L., Shows, T. B., and Nathans, J. (1993). Brn-3B —a POU domain gene expressed in a subset of retinal ganglion cells. *Neuron* **11,** 689–701.

Xiang, M. Q., Zhou, L. J., Macke, J. P., Yoshioka, T., Hendry, S. H. C., Eddy, R. L., Shows, T. B., and Nathans, J. (1995). The Brn-3 family of POU-domain factors—primary structure, binding specificity, and expression in subsets of retinal ganglion cells and somatosensory neurons. *J. Neurosci.* **15**(7 Part 1), 4762–4785.

Xiang, M. Q., Zhou, H., and Nathans J. (1996). Molecular biology of retinal ganglion cells. *Proc. Natl. Acad. Sci. U.S.A.* **93,** 596–601.

VIII

Mammals

32

Cell Lineage and Cell Fate Determination in Mammals

ROGER A. PEDERSEN
Reproductive Genetics Unit
Department of Obstetrics, Gynecology,
and Reproductive Sciences
University of California, San Francisco
San Francisco, California 94143

I. Introduction

Defining the embryonic cell lineage relationships for a particular species is one of the fundamental tasks that must be accomplished to establish a model system for developmental biology studies. In mammals, whose entire experimental history is approximately 100 years (see Biggers, 1991, for review), cell lineage studies have been a priority for more than three decades. The difficulties of working with mammalian systems have nonetheless required some unique approaches and novel adaptations of methods used to study cell lineage in other experimental systems. After 1961 when Tarkowski discovered that two entire cleavage stage mouse embryos could be aggregated together to form a chimera, extensive studies were performed to determine the fates of component embryos that were genetically marked to distinguish their descendant cells (reviewed by McLaren, 1976). In an important advance in chimera studies, Gardner (1968) developed approaches

for generating mosaicism at the blastocyst stage, then Beddington used similar approaches to make chimeras at postimplantation stages, thus enabling a characterization of the fate of cells that differentiate in mouse embryos during and after implantation into the uterus (reviewed by Rossant, 1986; Beddington, 1986). The introduction of intracellular injection of lineage tracers yielded additional insights about the fates of blastomeres and later cell types in cultured mouse embryos (Balakier and Pedersen, 1982; reviewed by Pedersen and Burdsal, 1994). In addition to these studies on early embryos, new approaches were developed for studying neural development using retroviral tracers (Walsh and Cepko, 1992). Very recently, techniques have been developed for visualizing mosaicism in living mouse embryos using a naturally fluorescent protein (Zernicka-Goetz et al., 1997). Some of these approaches for studying mouse cell lineages have been applied to other eutherian species, including other laboratory species (Papaioannou and Ebert, 1986), livestock species (Prather and

Robl, 1991), and even human embryos (Mottla *et al.*, 1995), and to some marsupial species (Cruz *et al.*, 1996; reviewed in Chapter 34 by Selwood and Hickford). Together, these studies have culminated in a substantial understanding of the fates of cells in early mouse development in particular, and have provided insight into the extent to which the mouse cell lineage and cell fate determination mechanisms apply to other mammals and are shared with other vertebrates.

One purpose of this introduction to the section on mammals is to review briefly the approaches that were used in the studies noted above, in order to distinguish the nature of the information that could be obtained in each approach. An additional topic is the advantages and limitations of the model species used, especially as these pertain to the chapters in the present volume. Finally, I identify several unresolved issues in mammalian cell lineage, focusing on the biological questions at stake, some of the studies that must be undertaken to resolve these questions, and what implications the findings might have for humans. This latter emphasis is based on the conviction that our strongest motivations for enduring the difficulties of studying mammalian embryos are curiosity about our own embryological origins and hope that such understanding might improve human health.

II. Methodological Considerations

A. Chimera Approaches

With the discovery of the chimera methodology for mouse embryos came a flood of information about the distribution of cell populations during embryogenesis and organogenesis. This novel technique provided mouse embryologists with a tool that *Drosophila* developmental geneticists had long exploited, the ability to generate mosaicism within an embryo. Initially, it appeared that a simple application of the binomial sampling thorem to the number of patches, or coherent clones, would reveal the time of determination and number of founder cells in each organ primordium (Mintz, 1967; Gearhart and Mintz, 1972; Moore and Mintz, 1972). Eventually, it was realized that the size and number of patches in a chimeric organ or tissue more closely reflected the relationship between the degree of coherence within a clone (i.e., the degree of mixing between cells of different clonal origin) and the total number of cells in the primordium at the time of allocation (i.e., the number of cells present in the organ rudiment when it ceased recruiting cells to its founder population). Because the degree of mixing within each organ rudiment and the time of allocation could not be readily determined for any organ rudiment, the observed patchiness of chimeras was less informative than originally hoped in terms of funda-

mental understanding of organogenesis (reviewed by McLaren, 1976; West, 1999).

Nevertheless, the very survival of chimeras revealed a momentous fact of mammalian embryogenesis, namely, that early embryos are highly regulative. Not only could whole cleavage stage embryos be aggregated together and develop normally, but also embryonic fragments, consisting of single blastomeres or groups of blastomeres, could be aggregated to form entire embryos consisting of parts of other embryos. Similarly, isolated two-cell blastomeres could be grown individually to generate small embryos, complete with both inner cell mass and trophectoderm populations (Tarkowski, 1959), although isolated four-cell and eight-cell blastomeres more frequently formed trophoblast vesicles devoid of an inner cell mass. On the basis of these observations, Tarkowski and Wroblewska (1967) concluded that trophoectoderm differentiation occurs whenever morula stage cells occupy an outer position, whereas inner cell mass differentiation occurs in cells that are enclosed by outer cells. This "inside–outside" hypothesis accounts for one of the few known "rules" for embryonic cell differentiation in eutherian mammals, namely that any cells that are exposed to the morula surface differentiate into trophectoderm (reviewed by Pedersen, 1986).

Given such a capacity for adaptation, preimplantation embryos from a variety of eutherian mammals can regulate for loss of blastomeres (e.g., from freezing, arrest, fragmentation, exclusion, or embryologists' curiosity) (reviewed by Pedersen, 1986; Papaioannou and Ebert, 1986; Prather and Robl, 1991). The demonstrated totipotency of early cleavage stage blastomeres in mouse and other eutherians has led to the current view that epigenetic cues, rather than oocyte cytoplasmic determinants, direct the differentiation of early lineages, and moreover, that epigenetic signals specify the embryonic axes that form during peri-implantation development (see Chapter 33 by Davidson *et al.*).

Creation of chimeras by blastocyst injection, more than any other chimera approach, has revealed the fate of the predominant cell types of peri-implantation mouse embryos. This body of knowledge provides the basis for understanding cell allocation at early stages to embryonic versus extraembryonic lineages. Briefly, inner mass cells isolated from expanded blastocysts (E3.5 = fourth day of gestation) can colonize both the primitive ectoderm (epiblast) and primitive endoderm (hypoblast) lineages. Cells isolated from E4.5 inner cell mass could be distinguished as either smooth (primitive ectoderm) or rough (primitive endoderm) and, when injected into E3.5 blastocysts, had as descendants only epiblast or hypoblast, respectively. Thus, the progenitors of these two lineages were distinct cell populations by E4.5. Moreover, these studies revealed that primitive endoderm contributed descendants only to endoderm of the yolk sac, and not to the gut (i.e., there was no contribution to the definitive endoderm of the embryo proper, or fetus). The fetus in its entirety, together with extraembryonic mesoderm and amnionic ectoderm, was descended from the primitive ectoderm. These observations established the similarity between mammalian and avian cell fates, where epiblast, not hypoblast, had also been demonstrated to be the source of all embryonic lineages, including the gut endoderm (Beddington, 1986).

Transplantation of cells from one embryo to another also has been carried out at postimplantation stages, generating chimeras in primitive ectoderm and embryonic mesoderm (see Chapter 33 by Davidson *et al.*). In this approach, the embryos were cultured after microsurgery, because returning manipulated postimplantation stage embryos to the uterus was not efficient enough for practical application (Beddington, 1985). Although technically different from the chimera studies at preimplantation stages, postimplantation stage chimeras have provided similar types of information as with blastocyst injection. The procedure has primarily been used to understand the potency and fate of transplanted epiblast cells. These studies show that most epiblast regions of E7.5 mouse embryos are capable of a wider range of developmental fates than they would express if left in their initial locations. That is, epiblast cells are not determined to become any particular tissue before they emerge from the primitive streak in the process of gastrulation (Chapter 33 by Davidson *et al.*). Nonetheless, transplantation of lineage-marked epiblast progenitor cells reveals a gastrula fate map of the mouse that is remarkably similar to that of the chick embryo (reviewed by Tam and Beddington, 1992; Tam and Behringer, 1997).

B. Intracellular Tracer Studies

In an effort to design experiments that distinguished between fate and potency, we adapted an approach from *Xenopus* and leech studies (Jacobson and Hirose, 1978; Weisblat *et al.*, 1978) for studying cells in minimally perturbed mouse embryos, using intracellular microinjection of lineage tracers to mark single or a few progenitor cells (Balakier and Pedersen, 1982). We used macromolecules (horseradish peroxidase and rhodamine-conjugated dextran) that were transferred only to lineal descendants to mark progenitor blastomeres at cleavage, morula, and blastocyst stages, as well as postimplantation stages (E6.5 and 7.5) (reviewed by Lawson and Pedersen, 1992a,b; Latimer and Pedersen, 1993; Pedersen and Burdsal, 1994; Tam and Behringer, 1997). This intracellular tracer approach revealed the fate of cells at either preimplantation or postimplantation stages in finer detail than previously possible, and also provided information about the cell cycle dynamics of the labeled progenitors and their descendants. In general, the find-

ings of cell lineage tracer injection confirmed the views obtained from previous studies of chimeras. First, cells of cleavage stage embryos underwent relatively little cell mixing during development to the blastocyst stage. Second, epiblast cells contributed to all three embryonic cell lineages (endoderm, mesoderm, and ectoderm) as well as to extraembryonic mesoderm and allantoic ectoderm. Third, the similarity of mouse and chick fate maps was evident, once corrections were made for their topographic differences. However, unpredicted observations also emerged from the intracellular studies, including the finding that the inner cell mass population was established in two cleavage divisions (fourth and fifth) (Pedersen *et al.*, 1986), and that there was massive cell death in the primitive endoderm cell population during gastrulation (Lawson and Pedersen, 1987).

Perhaps the most significant finding from the intracellular tracer approach was their confirmation of the important results of chimera studies using a minimally perturbing method. Similar results have been obtained using membrane-permeant dyes [such as 1,1'-dioctadecyl-3,3,3'3'-tetramethylindocarbocyanine perchlorate (DiI)] to label cells in intact embryos *in vitro* at postimplantation stages (e.g., Thomas *et al.*, 1998). Thus, the results of these three approaches could be used together to formulate a gastrula fate map for the mouse embryo that provides a basis for mechanistic studies on the molecular basis of axial development (reviewed by Tam and Behringer, 1997; Chapter 33 by Davidson *et al.*).

What has not emerged from either the intracellular tracer or the dye tracing studies, however, is an understanding of the cell fate relationships between preimplantation and postimplantation stages in minimally perturbed embryos. This failure to trace microinjected lineage markers across implantation may result from the expansive growth of the mouse embryo during this period, with corresponding passive dilution of the tracers. Cell counts of primitive ectoderm cells in implanting embryos show an 80-fold increase in epiblast cell number from E5.5 to the mid-gastrula stage (E7.5), with similar increases in endoderm/mesoderm cell number (reviewed by Snow, 1976). Thus, questions remain about whether the asymmetries in blastocysts connote axial relationships of the gastrula stage embryo, as subsequently discussed (reviewed by Gardner, 1998). Finally, the minimal invasiveness of the microinjected lineage tracer approach renders it unable to address the issue of potency. This is because the essence of a test of potency is its requirement for physical modification of the embryonic context of the cell whose potency is being tested.

C. Retroviral Labeling Studies

The appeal of a genetically based approach to lineage tracing is its indelibility. In contast to microinjected lineage tracers, which dilute or fade with time, cells marked with retroviral tracers offer the capacity to identify labeled descendants over a prolonged developmental period. Thus, retroviral tracers possess the advantages of both the tracer microinjection approach, in being minimally perturbing of the relationships between cells, as well as the genetic strength of the chimera approach. Since its introduction (Sanes *et al.*, 1986; Walsh and Cepko, 1992, 1993), the retroviral approach has been used effectively for the analysis of nervous system development. In a technically distinct approach that is similar in its outcome to retroviral infections, Tajbakhsh *et al.* (1996) used the *Myf*-5 promoter to activate a *lacZ* reporter gene in the population of stem cells of somites, thus learning about lineage relationships during muscle development (see also Chapter 41 by Buckingham and Tajbakhsh). Such studies reveal the lineage relationships between progenitors and their descendants, and are informative about cell migratory pathways (reviewed by Ware and Walsh in Chapter 36). In addition, the retroviral approach has been used to study cell fate in cleavage stage (Soriano and Jaenisch, 1986) and gastrulating mouse embryos (Carey *et al.*, 1995). The major limitation of retroviral tracer approaches for analyzing cell fate is the inherent inability to control the site (or precise time) of viral infection. Thus, the information derived from retroviral tracing studies is not a fate map (in the sense of a physical portrait of the progenitor locations and their subsequent descendant locations). Moreover, the retroviral approach, like the microinjected tracer approach, provides no information about cell potency.

D. Green Fluorescent Protein Labeling Studies

Is there a lineage tracing approach that would combine the features of a microinjected tracer with the persistence of genetics, yet still be capable of providing information about cell potency, as well as fate and lineage? Green fluorescent protein (GFP) may provide such a tool for mammalian cell lineage studies (reviewed in Chapter 35 by Zernicka-Goetz). The injection of mRNA encoding a thermostable form of GFP has been used for cell lineage tracing in preimplantation mouse embryos (Zernicka-Goetz *et al.*, 1997). This approach offers an intermediate between the rapidly diluting passive lineage tracers used previously for intracellular labeling (Beddington and Lawson, 1990) and the permanence of a genetic marker. Although the combined half-lives of injected GFP mRNA and its encoded protein may extend the time span of cell lineage tracing, both will eventually degrade to extinction, so injected mRNA is not equivalent to genetic marking. An intriguing possibility for achieving long-term marking using a minimally perturbing approach would be to use a Cre/lox or Flp/frt approach to mark progenitor cells in the embryo (O'Gor-

man *et al.*, 1991; Lewandoski *et al.*, 1997; Lewandoski and Martin, 1997). Accordingly, a reporter gene, such as *LacZ* or GFP, could be rendered inactive by placing a lox or frt-flanked sequence within the promoter to interrupt transcription. Then, injection of Cre or Flp protein or mRNA into a single blastomere at preimplantation stages would generate a permanent genetic mark for subsequent analysis at any stage of development. Such a labeled cell could be studied in its native state or by transplantation, thus alternatively providing information about either cell fate or capacity for regulation.

As a microinjected tracer in intact embryos, GFP mRNA should be minimally invasive, and thus would neither invoke nor address issues of cell potency. However, because GFP is a vital marker, GFP mRNA-labeled cells could be identified after disaggregation of labeled embryos, providing an opportunity to test the potency of cells from a similar embryonic stage and location as analyzed in intact embryos. An even more intriguing approach would be to combine the injected GFP mRNA with another mRNA selected to overexpress a gene product or to diminish a gene product using antisense mRNA (Zernicka-Goetz *et al.*, 1996). Moreover, the analysis of GFP-labeled cells in the intact embryo with sequential observations (time-lapse microscopy) would provide a more dynamic view of the behavior of embryonic cells at critical phases, such as during development of blastocyst and gastrula asymmetry. Using GFP as a transgene in embryos or ES cells may provide additional opportunities for long-term cell lineage analysis using chimera approaches. Although the latter approach would not differ in principle from previous chimera studies, the capacity for multiple observations could provide novel insights about the dynamic behavior of cells in cultured embryos.

E. Summary: A Postmodern View, or How Approaches Used to Study Cell Lineage Shape Our View of the Embryo

Our current views about the dominant role of epigenetic mechanisms in mammalian embryonic cell differentiation may be inherent in the methods used to answer the question, "What is the response of the mammalian embryo to experimental perturbation?" In both aggregation chimeras and blastocyst injection chimeras, incorporating the donor cells into the host embryo may evoke a regulative response from both donor and host, since neither would ordinarily have to accommodate the experimental procedure. Thus, in addition to providing information about cell fate, the combination of donor and host cells to create a chimera is a test of potency (i.e., the ability of the donor cell to regulate to its new environment, as well as the host cells to incorporate such a donor). By confounding the two parameters of potency and fate, chimeras could underestimate (or conceivably overestimate) the contribution of a particular cell type, as compared with its fate in the undisturbed embryo. Response to the experimental perturbation that establishes the mosaicism of chimeras is thus an inextricable part of any such experiment.

Cell lineage marking by intracellular injection, extracellular injection with membrane dyes, retroviral infection, or even GFP mRNA all can avoid confounding the assessment of fate and potency, thus serving purely as tests of fate. Yet, this very specificity prevents these procedures by themselves from delivering information about potency. Moreover, retroviral-type lineage tracing provides information only about lineal relationships of cells, not about their positional relationships. In order to learn about cell potency with such markers, they would need to be used in chimera or disaggregation studies, where the marker could provide additional specificity about the physical source of the donor cell than has previously been possible in such studies.

In sum, there is currently no single approach to cell lineage analysis in mammals that can provide a complete view of the potency, fate, and lineage of embryonic cells. Accordingly, the present state of our knowledge from studies of mammalian chimeras could conceal from view developmental mechanisms that operate in the unperturbed embryo, or under certain conditions of environmental or developmental stress. This may explain the disagreement, evident in this volume (e.g., Chapter 33 by Davidson *et al.*, Chapter 35 by Zernicka-Goetz) and elsewhere (Gardner, 1998), on the probable roles of epigenetic information and egg cytoplasm as determinants of early cell types and of the embryonic axes. The difference may be more apparent than real, if we recall the proverbial story of how three blindfolded observers variously describe an elephant as being a tree, a snake, or a rope, depending on whether the observer is holding its leg, trunk, or tail. Clearly, further cell lineage work is needed to clarify the nature of the mammalian embryo.

III. Model Systems for Cell Lineage: Advantages and Limitations

Because the description of embryonic cell lineage is an essential component of any model system, extensive studies of cell lineage have been performed in sea urchins and ascidians, *Caenorhabditis elegans*, leech, *Drosophila, Xenopus*, zebrafish, and chick, as described elsewhere in this volume. A substantial body of knowledge has been acquired about mouse cell lineage in particular (reviewed by Pedersen, 1986; Tam and Beddington, 1992; Latimer and Pedersen, 1993; Pedersen and Burdsal, 1994; Tam and Behringer, 1997; Chapter 33

by Davidson *et al.*). In addition, less extensive studies have been carried out in other mammalian species, such as the rat and the marsupial, *Sminthopsis macroura*, because of their unique experimental or conceptual benefits (reviewed by Selwood, 1992; Chapter 34 by Selwood and Hickford). In addition to the apparent value of intensive studies on model organisms, nonmodel organisms provide unique resources to studies of cell lineage, as in other areas of biology. Numerous examples attest to the research and therapeutic opportunities inherent in biodiversity, and for exploiting evolutionary history for its innovations. For example, telomeres were discovered in *Tetrahymena* (protozoan), cyclins in *Spisula* (clam), and green fluorescent protein in *Aequoria* (jellyfish).

The paramount advantage of working on the mouse is its genetics, consisting of the data base of spontaneous and induced mutations. The relative ease of inducing novel mutations in known or unknown genes using gene targeting or enhancer trapping further augment this data base (reviewed by Copp, 1995; see also Townley *et al.*, 1997). In addition, there is a substantial body of conceptual knowledge about other aspects of mouse embryos, including their biochemistry, physiology, cell biology, and axial development (reviewed in Rossant and Pedersen, 1986; Tam and Beddington, 1992; Hogan *et al.*, 1994; Tam and Behringer, 1997; Chapter 33 by Davidson *et al.*). The rapid development and small size of the mouse provide logistic and economic advantages, at least as compared with other mammals. However, there are some conspicuous disadvantages of the mouse as an experimental system, including its small size, which precludes certain physiological studies. Accordingly, our current understanding of mouse physiology is relatively limited. Moreover, there are several apparent disparities between rodent embryogenesis and mammalian embryogenesis in general, such as the "inversion of germ layers," which complicates comparisons between species (Cruz and Pedersen, 1991). (This germ layer inversion is actually illusory, consisting only of an indentation of the epiblast, which everts after gastrulation, thus restoring the conventional appearance with ectoderm outside and endoderm inside.) Chapter 33 by Davidson *et al.* summarizes knowledge about cell lineage relationships during gastrulation in the mouse, with an emphasis on the insights from transplantation studies and on the determination of fate in mesodermal and neural tissues through epigenetic interactions. Chapter 35 by Zernicka-Goetz describes the development and use of GFP for cell lineage tracing in mouse embryos, particularly for the analysis of postimplantation cell fate in the living embryo.

The conspicuous advantage of the rat as a model system derives from the abundant knowledge about its physiology, and thus its value as a model for disease. In terms of size, the rat provides a compromise between surgical accessibility and economy of animal husbandry. However, the genetic data base for the rat is relatively poor, and there is relatively little information about its early embryology, in comparison with the mouse. Nonetheless, much of our understanding of neural development and function is derived from rat studies, as summarized in Chapter 36 by Ware and Walsh, who also integrate information from studies of mouse, nonhuman primates, and humans in their compelling synthesis of knowledge about precursor proliferation, and neuronal cell specification and migration during development of the cerebral cortex.

Marsupials provide unique models for comparisons between eutherians and other amniotes, the birds and reptiles. Most marsupials are seasonal breeders, and only a few species have been adapted as pedigreed colonies for laboratory studies, including the stripe-faced dunnart (*S. macroura*), the fat-tailed dunnart (*Sminthopsis crassicaudata*), and the grey, short-tailed opossum (*Monodelphis domestica*). Chapter 34 by Selwood and Hickford, summarizes current knowledge about cell lineage in *S. macroura*. Because of the special evolutionary significance of marsupials (and the inavailability of monotremes for research), further cell lineage studies of marsupials are essential, despite the difficulties inherent in working with such species.

IV. Controversies and Paradigm Shifts

In describing scientific progress, Thomas Kuhn recognized that most research refines existing paradigms, while relatively little work actually establishes completely new views of the natural world (Kuhn, 1970). At any particular time, conflicting descriptions of reality may exist, depending on the status of work on existing paradigms, and the relative acceptance or recognition of new paradigms. Until recently, our understanding of nuclear determination during cell differentiation was based largely on work in amphibian embryos, where somatic cell nuclear transfer was used to assess the potency of donor nuclei obtained at sequential stages of development. The paradigm that resulted from the amphibian work was one of decreasing accessibility of genetic information; that view was fundamentally altered by the birth of Dolly, cloned from the nucleus of an adult ewe (Wilmut *et al.*, 1997). This new insight may qualify as "revolutionary" in Kuhn's sense, because it provides a new paradigm for nuclear status in cell differentiation, as discussed below.

The analysis of potency, lineage, and fate in mouse embryogenesis is a relatively mature subject, having received major experimental attention since the 1960s. As with nonmammalian systems, contrasting paradigms

of segregating cytoplasmic determinants versus epigenetic factors have been invoked to explain embryonic cell differentiation in mammals (reviewed by Gardner, 1998). However, the extensive regulative ability of mouse and other eutherian mammalian embryos is commonly interpreted as a conclusive demonstration that cytoplasmic determinants have no role in mammals, as discussed above and reiterated in Chapter 33 by Davidson *et al.* Nonetheless, controversy has reemerged about whether an epigenetic paradigm can completely explain axial development in mammalian embryos, as discussed below and in Chapter 35 by Zernicka-Goetz.

A third area of mammalian cell lineage that is ripe for novel insights is determination of the germ cell lineage. Current knowledge about the germ cell origins in the mouse indicate the primordial germ cells (PGCs) arise from epiblast progenitors during gastrulation, as discussed in Chapter 33 by Davidson *et al.* Recent work by Tam and Zhou (1996) implies that germ cells are determined by an epigenetic mechanism. The intriguing question is whether mammals share the genetic basis of germ cell origin that appears to be conserved between invertebrates and amphibians, as discussed below.

A. Nuclear Totipotency and the Role of Oocyte Cytoplasm: Lessons from Dolly

Before Dolly, the prevailing view of nuclear potency during vertebrate development was one of a gradually declining capacity to recapitulate embryogenesis as development progressed beyond larval stages. This view was based largely on experiments in amphibians, in which donor larval nuclei would support development into adults, but adult, differentiated nuclei supported development only to larval stages (reviewed by McKinnell, 1979). In the first mammalian somatic cell nuclear transfer, McGrath and Solter (1983, 1984) showed that transfer of a pair of pronuclei (but not late cleavage stage nuclei) to an enucleated mouse zygote could result in blastocyst formation. It was subsequently shown that fusion of blastomeres from 8-cell embryos to enucleated 2-cell embryos could lead to the development of blastocysts (Robl *et al.*, 1986) and after embryo transfer to the uterus could lead to full-term development (Tsunoda *et al.*, 1987). In 1986, Willadsen showed that enucleated oocytes could support the development of sheep 8- and 16-cell blastomere nuclei to term. As a result he was the first to produce genetically identical mammalian offspring by nuclear transfer. Shortly thereafter, products of nuclear transfer were produced in cattle (Prather *et al.*, 1987), rabbits (Stice and Robl, 1988), and pigs (Prather *et al.*, 1989). In a major breakthrough, Campbell *et al.*, (1996) showed that fetal fibroblasts could support development to term when used for nuclear transfer into sheep oocytes. Their success in achieving full-term development with apparently differentiated nuclei opened the field of mammalian nuclear transfer to further innovation. A year later the same authors took the further step of showing that adult cells also can support development to term, thus bringing widespread attention to the field of mammalian developmental biology (Wilmut *et al.*, 1997).

Why is Dolly's existence a revolutionary observation? In the earlier paradigm, nuclear differentiation was regarded as unidirectional, or irrevocable, at least with onset of adult development. The inaccessibility of genetic information in the adult was regarded as a possible mechanism for stability of the differentiated state, thus restraining spontaneous dedifferentiation, which occurs during carcinogenesis. While the efficiency of full-term development with adult nuclei may be reduced as compared with embryonic or fetal donor nuclei, even a single occurrence of nuclear totipotency in an adult cell is compelling. The existence of Dolly suggests an alternate paradigm for maintenance of a differentiated state, namely, that continual interactions between nucleus and cytoplasm reinforce a metastable state in which the cytoplasm "instructs" the nucleus to continue its present pattern of gene expression. In this view, many, perhaps most adult nuclei (other than cells of the immune system) could be totipotent and would be capable of recapitulating embryogenesis or dramatically altering their differentiated state if they were placed in a different cytoplasmic environment.

An important prediction of this alternate paradigm is that cytoplasmic factors are present in each cell that are capable of inducing the pattern of gene expression appropriate to the differentiated state of the cell. For example, oocyte cytoplasm should induce embryo-appropriate gene expression, whereas muscle cytoplasm should induce muscle genes, as actually observed by Blau and colleagues (1985). A major task will be to discover the cytoplasmic composition of oocytes and other cells that have these powerful roles.

The clinical implication of metastable nuclear differentiation governed by continual nuclear–cytoplasmic interactions is that customized cells might be generated as therapeutic materials for transplantation. If cells could be made that were genetically identical with an existing individual, it would circumvent problems of graft rejection. In principle, such customized cells could be generated by several routes: (1) Somatic cell nuclei might be transferred to oocytes, then the product could be used for deriving embryonic stem cell-like lines whose subsequent differentiation might be controlled *in vitro* by altering the epigenetic environment of the cultured cells. (2) Somatic cell nuclei might be altered by directly transferring them to the cytoplasm of differentiated cells, such as muscle or pancreatic β cells. (3) So-

matic cells might be altered by transfecting expression vector(s) for factor(s) necessary or sufficient for cell differentiation. Clearly, the identification of cytoplasmic factors responsible for tissue-specific differentiation would be a major milestone in such an undertaking.

B. Oocyte Organization and Specification of the Embryonic Axes

Although the extensive regulative ability of the mouse embryo is consistent with an epigenetic mechanism for determination of cell fate, recent observations suggest an alternative paradigm in which oocyte organization plays a role in subsequent events in embryogenesis. As concluded by Davidson *et al.* in Chapter 33, the well-established capacity of embryos of eutherian mammals to adjust to loss or gain of blastomeres provides compelling evidence for the importance of epigenetic factors in differentiative decisions of inner cell mass versus trophectoderm, hypoblast versus epiblast, and in the specialization of epiblast descendants.

How could any element of cell fate determination by cytoplasmic factors be consistent with such evidence? Virtually the entire body of cell fate, potency, and lineage knowledge in mouse embryos was acquired without reference to the oocyte axis. In terms of experimental design, the parameter of oocyte axis was confounded by randomization in the potency and fate studies. If certain orientations or combinations were nonviable because they disrupted essential cytoplasmic organization or information, these outcomes would be lumped together with other experimental failures resulting from damage or inefficiency of the procedures. Few, if any, of the studies of mouse blastomere potency and epigenetic determination resulted in 100% survival, thus it is difficult to rule out the possibility that some reconstructed embryos died because of incomplete cytoplasmic information.

Until recently, there was no apparent experimental basis for analyzing the role of mammalian oocyte organization, owing to the lack of a stable landmark for any feature of the oocyte. Early studies of mouse embryos by time-lapse cinemicrography were interpreted as revealing a highly motile second polar body, because of its constant undulations and apparent movement within the perivitelline space (Borghese and Cassini, 1963; R. A. Pedersen, 1977, unpublished observations). However, when Gardner (1997) reexamined this issue in a strain of mouse whose second polar body survived to the blastocyst stage, he found that the polar body was generally associated with the junction between polar and mural trophectoderm, rather than being randomly arrayed. Moreover, Gardner (1997) found that the blastocyst was bilaterally symmetrical, rather than radially symmetrical. As viewed from the embryonal (inner cell mass-containing) pole, the blastocyst was oval, rather than circular, and the polar body was typically aligned with the axis of greater diameter. Consequently, the oocyte animal–vegetal axis correlates with the axis of blastocyst bilateral symmetry.

This important observation raises a number of compelling questions about the causes and consequences of the correlation between oocyte and blastocyst organization. First, do certain aspects of oocyte axial organization induce the blastocyst axis to form where it does? If so, what are these components and how do they regulate blastocyst development? For example, since the first sign of blastocyst polarity is the eccentric formation of the blastocyst cavity, perhaps some aspect of oocyte organization disposes toward precocious trophoblast differentiation. Second, do oocyte cytoplasmic constituents or organizational elements become partitioned in an orderly way during cleavage (e.g., Gulyas, 1975)? If the distribution of cytoplasmic constituents plays any role in subsequent organization of the embryo, there should be cytoplasmic mechanisms controlling cleavage planes and/or cell differentiation. For example, there could be a role for the Numb/Notch/Delta pathway in blastomere differentiation into inner cell mass and trophectoderm, as discussed for asymmetric cell fates in the central nervous system in Chapter 36 by Ware and Walsh. Third, is there a correlation between the axis of bilateral symmetry in blastocysts and either the anterior–posterior axis or dorso–ventral axis at the gastrula stage (Gardner *et al.*, 1992)? Gardner (1998) suggests that extraembryonic components convey axial information from the blastocyst to the gastrula stage, which would be consistent with localized gene expression in anterior visceral endoderm at early postimplantations stages (e.g., Chen *et al.*, 1994; Thomas *et al.*, 1998). This hypothesis predicts that blastocyst cells from different circumferential regions of the polar–mural junction (using the second polar body as a landmark) would have unique postimplantation fates. The prospects for analyzing the relationship between blastocyst and gastrula axes using GFP mRNA as a lineage tracer are considered by Zernicka-Goetz in Chapter 35.

These and other issues are reviewed in detail by Gardner (1998), who considers the implications of a determinative mechanism for cell fate involving cytoplasmic constituents in mammalian embryos. Edwards and Beard (1997) also have strongly advocated the view that early cleavage divisions of mammalian embryos segregate the oocyte contents in a determinative way. Unfortunately, the latter authors' model for determination of trophoblast and inner cell mass lineages during early cleavage divisions contradicts the known fate of mouse and human blastomeres at these stages. Previous studies have shown that all 2- to 4-cell, half of 8-cell, and some of 16-cell blastomeres contribute to both trophectoderm and inner cell mass (Pedersen *et al.*, 1986; Mottla *et al.*, 1995; reviewed by Pedersen, 1986). Nonetheless,

the observation of differentially distributed proteins, such as leptin and STAT3 in mouse and human oocytes and embryos provides at least circumstantial evidence for the capacity of mammalian embryos to segregate oocyte components during oogenesis and cleavage (Antczak and Van Blerkom, 1997). Whether the oocyte contains factors that actually determine cell fate in the intact mammalian embryo remains to be seen.

The clinical significance of a cytoplasmic mechanism for determination of cell fate in mammalian embryos will depend on the capacity of the human embryo for regulation. Human embryos could be susceptible to perturbation by the manipulation involved in preimplantation genetic diagnosis if cytoplasmic determinants exist. In this clinical procedure, one or two blastomeres are removed from the 8-cell embryo for genetic analysis of aneuploidy or specific genetic dysfunctions (Handyside and Delhanty, 1997). The remaining blastomeres are generally sufficient for normal development (Hardy *et al.*, 1990) but any excess in aberrant outcomes (e.g., implantation failure) might be attributable to sampling of blastomeres from a region of the embryo containing determinative constituents inherited from the oocyte, as suggested by Edwards and Beard (1997) and Antczak and Van Blerkom (1997). Clearly, human cleavage stage embryos are capable of accommodating for the loss of cytoplasmic mass and any specialized cytoplasmic contents when blastomeres are damaged by cryopreservation or fragmentation (Mohr and Trounson, 1985). However, in view of the generally poor developmental efficiencies to the blastocyst stage and after implantation into the uterus, it is difficult to know whether there is a detrimental effect of blastomere losses on pre- or postimplantation development because of the high "noise" level of developmental failures. From this hypothetical perspective, an essential aspect of evaluating the clinical safety of preimplantation genetic diagnosis is to clarify whether oocyte constituents play any role in subsequent axial development of mammals.

C. Mechanisms of Germ Line Formation in Mammals

In several model systems whose cell lineage has been studied extensively, germ cells arise from a region of the embryo that inherits localized oocyte cytoplasmic components known as germ plasm. In such systems, one or more components of the germ plasm have been shown to be essential for germ cell formation, and thus for both male and female fertility. Because several reviews describe the identity and function of known germ plasm components (Lehmann and Ephrussi, 1994; Micklem, 1995; Rongo and Lehmann, 1996), it is not necessary to reiterate the subject in detail here. Rather, my purpose is to consider the possibility that mammalian germ cell determination also involves the family of genes whose products are present in the germ plasm of other model systems.

Germ cell origins have been examined directly by cell lineage analysis at the onset of gastrulation in mouse embryos. The progenitors of mouse germ cells are first recognized at E6.0 in the region of proximal epiblast abutting the extraembryonic ectoderm (Lawson and Hage, 1994). Thence, these cells become committed to a germ cell pathway, migrate to the posterior of the embryo between the base of the allantois and the posterior of the primitive streak, and are recognizable as primordial germ cells (PGC), owing to their staining for alkaline phosphatase (Ginzburg *et al.*, 1990). The transplantation studies by Tam and Zhou (1996) show that mouse PGC differentiation is epigenetically inducible, because epiblast progenitors of neuroectoderm can become PGC when they are placed in proximal epiblast; conversely, progenitors of PGC become neuroectoderm when placed in a distal epiblast location. Combined with evidence for extensive mixing within the epiblast cell population before and during gastrulation (Lawson *et al.*, 1991), it is not easy to see how PGC formation could result from inheritance of germ plasm contents in mammals. Nonetheless, it is interesting to consider whether gene products localized as cytoplasmic determinants in some amphibians and invertebrate embryos could influence germ cell differentiation in mammals.

The gene products involved in the germ plasm of *Drosophila*, *C. elegans*, and *Xenopus* encodes a set of mRNAs, RNA-binding proteins, RNA helicases, cytoskeletal components, and mitochondrial rRNA. Included in the mRNAs are OSKAR, NANOS, GERM CELL-LESS, and others, all of which contain 3' UTR regions capable of forming extensive secondary structures (Rongo and Lehmann, 1996). Germ plasm proteins capable of binding to and modifying mRNAs include BRUNO, which binds to bruno response elements in the 3' UTR of oskar mRNA, and NANOS and PUMILIO, which bind to nanos response elements in the 3' UTRs of hunchback mRNA, repressing its translation in the posterior part of the embryo (Rongo and Lehmann, 1996). There are also RNA helicases, such as VASA, a DEAD-box protein localized in the germ plasm. Although most genetic studies of germ plasm function have been carried out in *Drosophila*, *Xenopus* also has a nanos homologue, *Xcat-2*, which is localized to the vegetal cortex at early stages of oogenesis, and remains in the vegetal blastomeres during cleavage (Zhou and King, 1996). A *vasa*-like gene is essential for germ cell formation in *Xenopus*, as indicated by function-perturbing antibodies to the protein (Ikenishi and Tanaka, 1997). The conservation of such germ plasm functions between invertebrate and vertebrate species is a compelling observation, given the known universality of homeobox and cell cycle regulatory genes. Moreover, the RNA binding protein, DAZ,

is a component of *Xenopus* germ plasm. When the respective *Daz* homologues are mutated in either *Drosophila*, mouse, or humans, it causes severe infertility (Reijo *et al.*, 1995; , Eberhart *et al.*, 1996; Ruggiu *et al.*, 1997). Therefore, at least some members of the germ plasm family of gene products appear to have functions essential for fertility in mammals as in other model systems.

Some additional observations on germ cell allocation in mouse embryos appear to indicate a separation of somatic and germ cell lineages at preimplantation stages, which could be consistent with localization of germ cell determinants in the oocyte or early embryo. Soriano and Jaenisch (1986) marked cleavage stage embryos using retroviral infection, which yielded distinct infectious events depending on their integration sites. By analyzing the distribution of descendant cells, they found no correlation between somatic tissues and the germ line in 39% of integration events. Similarly, Wilkie *et al.* (1986) found that pronuclear DNA injection frequently generated somatic–germ line mosaicism. These observations are difficult to explain unless germ cell allocation were to occur at preimplantation stages. Another estimate for the timing of germ cell allocation was obtained by McMahon *et al.* (1983) based on the proportion of somatic and germ cells inactivating polymorphic X chromosomes. Their findings were consistent with germ cell allocation before X chromosome inactivation, which was estimated to occur as early as E4.5 and 5.5. How is it possible to reconcile these observations with evidence from Lawson and Hage (1994) for allocation to a germ cell fate several days later, at the gastrula stage? Conceivably, a restriction in mixing between progenitor cell populations of somatic and germ line could occur at the late blastocyst stage, when inhomogeneities in cytoplasmic constituents inherited from oocyte cytoplasm could still exist. The predictions of such a phenomenon would be either: (1) a nonrandom distribution of epiblast descendants of the germ line progenitors to the proximal epiblast region adjacent to extraembryonic ectoderm; or (2) localization of trophoblast descendants to the distal region of extraembryonic ectoderm, which might function in the epigenetic induction of germ cells in adjacent epiblast. Fate mapping of blastocyst regions into the postimplantation period, as discussed by Gardner (1996, 1997) and Zernicka-Goetz in Chapter 35, would reveal either of these potential outcomes.

The clinical relevance of knowledge about germ cell origins echos the concern about perturbing cytoplasmic organization during human embryo micromanipulation, as discussed above, and extends to fundamental issues about the basis of human infertility. The hypothetical risk of biopsying the portion of a human embryo that contains a hypothetical germ plasm may be considered highly speculative, but the inevitable long-term delay in assessing the effects of this procedure on human fertility justifies a sense of urgency in understanding whether there are any mechanisms of germ cell allocation in preimplantation stages of mammalian embryos. Even if such mechanisms did exist, it might be possible to avoid losing information necessary to the embryo, provided that the location of such cytoplasmic determinants or germ cell progenitors were precisely known. Beyond the safety issue, there is still relatively little information about the genetic basis of human fertility. The discovery of the role of the Y-chromosome gene, *DAZ*, in human fertility and its autosomal homologue, *Dazl*, in mouse fertility, has provided a paradigm for investigating the basis of both male and female infertility (Reijo *et al.*, 1995). As a genetic problem, the study of human germ line development can benefit enormously from the frequent perturbations that manifest as self-referred fertility problems. Thus, the human population can be considered a repository for functional genomics of germ line development as it has for other human genetic diseases (Strachan and Read, 1996).

V. Concluding Remarks

The following chapters in this section summarize the past decades of work in each area. I have tried to provide a context and orientation for additional studies in mammalian cell lineage and cell fate determination. In this exercise, I have included speculations beyond what would be appropriate in another forum, in order to provoke a reexamination of what we know about mammalian embryo cell potency, fate, and lineage. If this approach leads to new experimental knowledge about mammalian embryos, and to improvements in clinical care of infertile patients, I will have succeeded.

Acknowledgments

I thank my mentors: Clement L. Markert for providing me with a classical education in developmental biology, which included an appreciation for the role of cytoplasmic organization; John D. Biggers for introducing me to the enduring fascination and beauty of mammalian embryos; and Anne McLaren for reminding me of the importance of the germ line. I also thank Magdalena Zernicka-Goetz for refreshing my interest in the unsolved problems of mammalian cell lineage, and I thank Martin Evans, Azim Surani, Anne McLaren, and John Gurdon for their hospitality in accommodating me at the Wellcome/CRC Institute in Cambridge, U.K., where this was written. Marie-Cecile Lavoir contributed to the section regarding mammalian nuclear transfer. My own work in this area is supported by NIH/NICHD Grant PO-1 HD 26732.

References

Antczak, M., and Van Blerkom, J. (1997). Oocyte influences on early development: The regulatory proteins leptin and STAT3 are polarized in mouse and human oocytes and differentially distributed within the cells of the preimplantation stage embryo. *Mol. Hum. Reprod.* **3,** 1067–1086.

Balakier, H., and Pedersen, R. A. (1982). Allocation of cells to inner cell mass and trophectoderm cell lineages in preimplantation mouse embryos. *Dev. Biol.* **90,** 352–362.

Beddington, R. S. P. (1985). The development of 12th to 14th day foetuses following reimplantation of pre- and early primitive streak stage mouse embryos. *J. Embryol. Exp. Morphol.* **88,** 281–291.

Beddington, R. (1986). Analysis of tissue fate and prospective potency in the egg cylinder. *In* "Experimental Approaches to Mammalian Embryonic Development," (J. Rossant and R. A. Pedersen, eds., pp. 121–147. Cambridge Univ. Press, Cambridge.

Beddington, R. S. P., and Lawson, K. A. (1990). Clonal analysis of cell lineages. *In* "Postimplantation Development in the Mouse," (A. J. Copp and D. L. Cockroft, eds.), Ciba Found. Symp. 185, pp. 61–71, Wiley, Chichester.

Biggers, J. D. (1991). Walter Heape, FRS: A pioneer in reproductive biology. Centenary of his embryo transfer experiments. *J. Reprod. Fertil.* **93,** 173–186.

Blau, H. M., Pavlath, G. K., Hardeman, E. C., Chiu, C. P., Silberstein, L., Webster, S. G., Miller, S. C., and Webster, C. (1985). Plasticity of the differentiated state. *Science* **230,** 758–766.

Borghese, E., and Cassini, A. (1963). Cleavage of the mouse egg. *In* "Cinemicrography in Cell Biology" G. G. Rose, ed., pp. 263–277. Academic Press, New York.

Campbell, K. H. S., McWhir, J., Ritchie, W. A., and Wilmut, I. (1996). Sheep cloned by nuclear transfer from a cultured cell line. *Nature (London)* **380,** 64–66.

Carey, F. J., Linney, E., and Pedersen, R. A. (1995). Allocation of epiblast cells to germ layer derivatives during mouse gastrulation as studied with a retroviral vector. *Dev. Genet.* **17,** 29–37.

Chen, W. S., Manova, K., Weinstein, D. C., Duncan, S. A., Plump, A. S., Prezioso, V. R., Bachvarova, R. F., and Darnell, J. E., Jr. (1994). Disruption of the HNF-4 gene, expressed in visceral endoderm, leads to cell death in embryonic ectoderm and impaired gastrulation of mouse embryos. *Genes Dev.* **8,** 2466–2477.

Copp, A. J. (1995). Death before birth: Clues from gene knockouts and mutations. *Trends Genet.* **11,** 87–93.

Cruz, Y. P., and Pedersen, R. A. (1991). Origin of embryonic and extraembryonic cell lineages in mammalian embryos. *In* "Animal Applications of Research in Mammalian Development," pp. 147–204. Cold Spring Harbor Laboratory, Cold Spring Harbor, New York.

Cruz, Y. P., Yousef, A., and Selwood, L. (1996). Fate-map of the epiblast of the dasyurid marsupial *Sminthopsis macroura* (Gould). *Reprod. Fertil. Dev.* **8,** 779–788.

Eberhart, C. G., Maines, J. Z. and Wasserman, S. A. (1996). Meiotic cell cycle requirement for a fly homologue of human Deleted in Azoospermia. *Nature (London)* **381,** 783–785.

Edwards, G. R., and Beard, H. K. (1997). Oocyte polarity and cell determination in early mammalian embryos. *Mol. Hum. Reprod.* **3,** 863–905.

Gardner, R. L. (1968). Mouse chimaeras obtained by the injection of cells into the blastocyst. *Nature (London)* **220,** 596–597.

Gardner, R. L. (1996). Clonal analysis of growth of the polar trophectoderm in the mouse. *Hum. Reprod.* **11,** 1979–1984.

Gardner, R. L. (1997). The early blastocyst is bilaterally symmetrical and its axis of symmetry is aligned with the animal–vegetal axis of the zygote in the mouse. *Development (Cambridge, UK)* **124,** 289–301.

Gardner, R. L. (1998). Axial relationships between egg and embryo in the mouse. *Curr. Top. Dev. Biol.* **39,** 35–71.

Gardner, R. L., Meredith, M. R., and Altman, D. G. (1992). Is the anterior–posterior axis of the fetus specified before implantation in the mouse? *J. Exp. Zool.* **264,** 437–443.

Gearhart, J. D., and Mintz, B. (1972). Clonal origins of somites and their muscle derivatives: Evidence from allophenic mice. *Dev. Biol.* **29,** 27–37.

Ginzburg, M., Snow, M. H. L., and McLaren, A. (1990). Primordial germ cells in the mouse embryo during gastrulation. *Development (Cambridge, UK)* **110,** 521–528.

Gulyas, B. J. (1975). A reexamination of cleavage patterns in eutherian mammalian eggs: Rotation of blastomere pairs during second cleavage in the rabbit. *J. Exp. Zool.* **193,** 235–248.

Handyside, A. H., and Delhanty, J. D. (1997). Preimplantation genetic diagnosis: Strategies and surprises. *Trends Genet.* **13,** 270–275.

Hardy, K., Marten, K. L., Leese, H. J., Winston, R. M., and Handyside, A. H. (1990). Human preimplantation development *in vitro* is not severely affected by biopsy at the 8 cell stage. *Hum. Reprod.* **5,** 708–714.

Hogan, B. L., Beddington, R. S. P., Lacey, E., and Costantini, F. (1994). "Manipulating the Mouse Embryo," Second Edition. Cold Spring Harbor Laboratory Press, Cold Spring Harbor, New York.

Ikenishi, K., and Tanaka, T. S. (1997). Involvement of the protein of *Xenopus* vasa homolog (*Xenopus vasa*-like gene 1, XVLG1) in the differentiation of primordial germ cells. *Dev. Growth Differ.* **39,** 625–633.

Jacobson, M., and Hirose, G. (1978). Origin of the retina from both sides of the embryonic brain: A contribution to the problem of crossing at the optic chiasma. *Science* **202,** 637–639.

Kuhn, T. S. (1970). "The Structure of Scientific Revolutions," 2nd Ed. Univ. of Chicago Press, Chicago.

Latimer, J. J., and Pedersen, R. A. (1993). Epigenetic interactions and gene expression in peri-implantation mouse embryo development. *In* Genes in Mammalian Reproduction," (R. B. L. Gwatkin, ed.), pp. 131–171. Wiley-Liss, New York.

Lawson, K. A., and Hage, W. J. (1994). Clonal analysis of the origin of primordial germ cells in the mouse. *In* "Germline Development," Ciba Found. Symp. **182,** pp. 68–84. Wiley, West Sussex, U.K.

Lawson, K. A., and Pedersen, R. A. (1987). Cell fate, morphogenetic movement and population kinetics of embryonic endoderm at the time of germ layer formation in the mouse. *Development (Cambridge, UK)* **101,** 627–652.

Lawson, K. A., and Pedersen, R. A. (1992a). Clonal analysis of cell fate during gastrulation and early neurulation in the mouse. *In* "Postimplantation Development in the Mouse" Ciba Found. Symp. **165,** pp. 3–26. Wiley, Chichester.

Lawson, K. A., and Pedersen, R. A. (1992b). Early mesoderm formation in the mouse embryo. *In* "Formation and Differentiation of Early Embryonic Mesoderm" (R. Bellairs *et al.*, eds.), pp. 33–46. Plenum, New York.

Lawson, K. A., Meneses, J. J., and Pedersen, R. A. (1991). Clonal analysis of epiblast fate during germ layer formation in the mouse embryo. *Development (Cambridge, UK)* **113,** 891–911.

Lehmann, R., and Ephrussi, A. (1994). Germ plasm formation and germ cell determination in *Drosophila. In* "Germline Development," Ciba Found. Symp., **182,** pp. 282–296. Wiley, Chichester.

Lewandoski, M., and Martin G. R. (1997). Cre-mediated chromosome loss in mice. *Nat. Genet.* **17,** 223–225.

Lewandoski, M., Wassarman, K. M., and Martin, G. R. (1997). Zp3-cre, a transgenic mouse line for the activation or inactivation of loxP-flanked target genes specifically in the female germ line. *Curr. Biol.* **7,** 148–151.

McGrath, J., and Solter, D. (1983). Nuclear transplantation in the mouse embryo by microsurgery and cell fusion. *Science* **220,** 1300.

McGrath, J., and Solter, D. (1984). Inability of mouse blastomere nuclei transferred to enucleated zygotes to support development *in vitro. Science* **226,** 1317.

McKinnell, R. G. (1979). "Cloning: A Biologist Reports." Univ. of Minnesota Press, Minneapolis.

McLaren, A. (1976). "Mammalian Chimaeras," Cambridge Univ. Press, Cambridge.

McMahon, A., Fosten, M., and Monk, M. (1983). X-chromosome inactivation mosaicism in the three germ layers and the germ line of the mouse embryo. *J. Embryol. Exp. Morphol.* **74,** 207–220.

Micklem, D. R. (1995). mRNA localisation during development. *Dev. Biol.* **172,** 377–395.

Mintz, B. (1967). Gene control of mammalian pigmentary differentiation. I. Clonal origin of melanocytes. *Proc. Natl. Acad. Sci. U.S.A.* **58,** 344–351.

Mohr, L. R., and Trounson, A. O. (1985). Cryopreservation of human embryos. *Ann. N.Y. Acad. Sci.* **442,** 536–543.

Moore, W. J., and Mintz, B. (1972). Clonal model of vertebral column and skull development derived from genetically mosaic skeletons in allophenic mice. *Dev. Biol.* **27,** 55–70.

Mottla, G. L., Adelman, M. R., Hall, J. L., Gindoff, P. R., Stillman, R. J., and Johnson, K. E. (1995). Lineage tracing demonstrates that blastomeres of early cleavage-stage human pre-embryos contribute to both trophectoderm and inner cell mass. *Hum. Reprod.* **10**, 384–391.

O'Gorman, S., Fox, D. T., and Wahl, G. M. (1991). Recombinase-mediated gene activation and site-specific integration in mammalian cells. *Science* **251**, 1351–1355.

Papaioannou, V. E., and Ebert, K. M. (1986). Comparative aspects of embryo manipulation in mammals. *In* "Experimental Approaches to Mammalian Embryonic Development," (J. Rossant and R. A. Pedersen, eds.), pp. 67–96. Cambridge Univ. Press, Cambridge.

Pedersen, R. A. (1986). Potency, lineage and allocation in preimplantation mouse embryos. *In* "Experimental Approaches to Mammalian Embryonic Development" (J. Rossant and R. A. Pedersen, eds.), pp. 3–33. Cambridge Univ. Press, Cambridge.

Pedersen, R. A., and Burdsal, C. A. (1994). Mammalian embryogenesis. *In* "The Physiology of Reproduction" (E. Knobil, J. D. Neill, and C. L. Markert, eds.), 2nd Ed. pp. 319–390. Raven Press, New York.

Pedersen, R. A., Wu, K., and Balakier, H. (1986). Origin of the inner cell mass in mouse embryos: Cell lineage analysis by microinjection. *Dev. Biol.* **117**, 581–595.

Prather, R. S., and Robl, J. M. (1991). Cloning by nuclear transfer and embryo splitting in laboratory and domestic animals. *In* "Animal Applications of Research in Mammalian Development," pp. 205–232. Cold Spring Harbor Laboratory, Cold Spring Harbor, New York.

Prather, R. S., Barnes, F. L., Sims, M. M., Robl, J. M., Eyestone, W. H., and First, N. L. (1987). Nuclear transplantation in the bovine embryo: Assessment of donor nuclei and recipient oocyte stage. *Biol. Reprod.* **37**, 859.

Prather, R., Sims, M. M., and First, N. L. (1989). Nuclear transplantation in pig embryos. *Biol. Reprod.* **41**, 414.

Reijo, R., Lee, T. Y., Salo, P., Alagappan, R., Brown, L. G., Rosenberg, M., Rozen, S., Jaffe, T., Straus, D., Hovatta, O., and Page, D. (1995). Diverse spermatogenic defects in humans caused by Y chromosome deletions encompassing a novel RNA-binding protein gene. *Nat. Genet.* **10**, 383–393.

Robl, J. M., Gilligan, B., Critser, E. S., and First, N. L. (1986). Nuclear transplantation in mouse embryos: Assessment of recipient cell stage. *Biol. Reprod.* **34**, 733–739.

Rongo, C., and Lehmann, R. (1996). Regulated synthesis, transport and assembly of the *Drosophila* germ plasm. *Trends Genet.* **12**, 102–109.

Rossant, J. (1986). Development of extraembryonic cell lineages in the mouse embryo. *In* "Experimental Approaches to Mammalian Embryonic Development" (J. Rossant and R. A. Pedersen, eds.), pp. 97–120. Cambridge Univ. Press, Cambridge.

Rossant, J., and Pedersen, R. A., eds. (1986). "Experimental Approaches to Mammalian Embryonic Development." Cambridge University Press, Cambridge, U.K.

Ruggiu, M., Speed, R., Taggart, M., McKay, S. J., Kilanowski, F., Saunders, P., Dorin, J., and Cooke, H. J. (1997). The mouse Dazla gene encodes a cytoplasmic protein essential for gametogenesis. *Nature* (*London*) **389**, 73–77.

Sanes, J. R., Rubenstein, J. L., and Nicolas, J. F. (1986). Use of a recombinant retrovirus to study post-implantation cell lineage in mouse embryos. *EMBO J.* **5**, 3133–3142.

Selwood, L. (1992). Mechanisms underlying the development of pattern in marsupial embryos. *Curr. Top. Dev. Biol.* **27**, 175–233.

Snow, M. H. L. (1976). Embryo growth during the immediate postimplantation period. *In* "Embryogenesis in Mammals," Ciba Found. Symp. 40, pp. 53–70. Elsevier, Amsterdam.

Soriano, P., and Jaenisch, R. (1986). Retroviruses as probes for mammalian development: Allocation of cells to the somatic and germ cell lineages. *Cell* (*Cambridge, Mass.*) **46**, 19–29.

Stice, S. L., and Robl, J. M. (1988). Nuclear reprogramming in nuclear transplant rabbit embryos. *Biol. Reprod.* **39**, 657.

Strachan, T., and Read, A. (1996). "Human Molecular Genetics." BIOS Scientific.

Tajbakhsh, S., Bober, E., Babinet, C., Pournin, S., Arnold, H., and Buckingham, M. (1996). Gene targeting the myf-5 locus with nlacZ reveals expression of this myogenic factor in mature skeletal muscle fibres as well as early embryonic muscle. *Dev. Dyn.* **206**, 291–300.

Tam, P. P. L., and Beddington, R. S. P. (1992). Establishment and organization of germ layers in the gastrulating mouse embryo. *In* "Postimplantation Development in the Mouse," Ciba Found. Symp. 165, pp. 27–49. Wiley, Chichester.

Tam, P. P. L., and Behringer, R. R. (1997). Mouse gastrulation: The formation of a mammalian body plan. *Mech. Dev.* **68**, 3–25.

Tam, P. P. L., and Zhou, S. X. (1996). The allocation of epiblast cells to ectodermal and germ-line lineages is influenced by the position of the cells in the gastrulating mouse embryo. *Dev. Biol.* **178**, 124–132.

Tarkowski, A. K. (1959). Experiments on the development of isolated blastomeres of mouse eggs. *Nature* (*London*) **184**, 1286–1287.

Tarkowski, A. K. (1961). Mouse chimaeras developed from fused eggs. *Nature* (*London*) **190**, 858–860.

Tarkowski, A. K., and Wroblewska, J. (1967). Development of blastomeres of mouse eggs isolated at the 4- and 8-cell stage. *J. Embryol. Exp. Morphol.* **18**, 155–180.

Thomas, Q. P., Brown, A., and Beddington, R. S. P. (1998). Hex: A homeobox gene revealing peri-implantation asymmetry in the mouse embryo and an early transient marker of endothelial cell precursors. *Development* (*Cambridge, UK*) **125**, 85–94.

Townley, D. J., Avery, B. J., Rosen, B., and Skarnes, W. C. (1997). Rapid sequence analysis of gene trap integrations to generate a resource of insertional mutations in mice. *Genome Res.* **7**, 293–298.

Tsunoda, Y., Yasui, T., Shioda, Y., Nakamura, K., Uchida, T., and Sugie, T. (1987). Full-term development of mouse blastomere nuclei transplanted into enucleated two-cell embryos. *J. Exp. Zool.* **242**, 147.

Walsh, C., and Cepko, C. L. (1992). Widespread dispersion of neuronal clones across functional regions of the cerebral cortex. *Science* **255**, 434–440.

Walsh, C., and Cepko, C. L. (1993). Clonal dispersion in proliferative layers of developing cerebral cortex. *Nature (London)* **362**, 632–635.

Weisblat, D. A., Sawyer, R. T., and Stent, G. S. (1978). Cell lineage analysis by intracellular injection of a tracer enzyme. *Science* **202**, 1295–1298.

West, J. D. (1999). Insights into development and genetics from mouse chimeras. *Curr. Top. Dev. Biol.* **44**, 21–66.

Wilkie, T. M., Brinster, R. L., and Palmiter, R. D. (1986). Germline and somatic mosaicism in transgenic mice. *Dev. Biol.* **118**, 9–18.

Willadsen, S. M. (1986). Nuclear transplantation in sheep embryos. *Nature (London)* **320**, 63.

Wilmut, I., Schnieke, A. E., McWhir, J., Kind, A. J., and Campbell, K. H. S. (1997). Viable offspring derived from fetal and adult mammalian cells. *Nature (London)* **385**, 810–813.

Zernicka-Goetz, M., Pines J., Ryan, K., Siemering K. R., Haseloff, J., Evans, M. J., and Gurdon, J. B. (1996). An indelible lineage marker for *Xenopus* using a mutated green fluorescent protein. *Development (Cambridge, UK)* **122**, 3719–3724.

Zernicka-Goetz, M., Pines, J., Hunter McLean, S., Dixon, P. C., Siemering, K. R., Haseloff, J., and Evans, M. J. (1997). Following cell fate in the living mouse embryo. *Development (Cambridge, UK)* **124**, 1133–1137.

Zhou, Y., and King, M. L. (1996). RNA transport to the vegetal cortex of *Xenopus* oocytes. *Dev. Biol.* **179**, 173–183.

33

Cell Fate and Lineage Specification in the Gastrulating Mouse Embryo

Bruce P. Davidson
Anne Camus
Patrick P. L. Tam
Embryology Unit
Children's Medical Research Institute
Wentworthville, New South Wales 2145, Australia

I. Introduction

Early embryonic development is characterized by the proliferation of cells and the acquisition of tissue-specific properties by these cells as they differentiate. The entire collection of cell types that is derived from a common progenitor, irrespective of the divergence of morphological and molecular properties, constitutes the tissue lineage.

Central to the concept of embryogenesis is the understanding of the mechanisms that establish the vast array of tissue lineages in the embryo. The progenitor cells for each tissue lineage are derived from a subset of pluripotent cells in the early embryo. The specification of tissue lineages begins with the formation of these specific progenitor cells. The allocation of progenitors for specific lineages may occur as early as the first cell division of the totipotent zygote that leads to the formation of daughter cells with different lineage potency. Alternatively, the segregation of cells with different potency may occur at later stages of embryogenesis.

The establishment of lineage progenitors may be accomplished by the segregation of cytoplasmic or cell membrane-associated lineage determinants. The presence of cytoplasmic determinants in the zygote or early embryo and their subsequent unequal distribution to daughter cells, have been shown to play a role in cell fate determination in a number of species. Examples of such cytoplasmic factors are the myogenic determinant localized to the yellow crescent of the ascidian embryo (Jeffrey *et al.*, 1986), the pole plasm or germ plasm of the germ cell lineage in *Caenorhabditis elegans* and *Drosophila* (Strome, 1986; Illmensee and Mahowald, 1974), and the nuage material of the primordial germ cells in *Xenopus* (Smith *et al.*, 1983). To ensure a consistent pattern of distribution of the lineage determinants the early embryo often divides in a determinate (or regular) pattern of cell cleavage that results in the maintenance of spatial position of the mitotic descendants of specific progenitors (reviewed by Davidson, 1990; see Chapter 7 by Bowerman; Chapter 20 by Sullivan *et al.*). In contrast, there is little evidence of a major role of early cytoplasmic determinants at the origin of tissue diversification in the mouse embryo. The mouse zygote apparently does not undergo regular cleavages that enable a reproducible segregation of cytoplasmic materials to the blastomeres. The allocation of progenitor cells seems to be largely influenced by the position of the blastomeres in the preimplantation embryo. In this case, cells acquire different lineage potency by way of the cell–cell interaction and the interpretation of positional information (Johnson *et al.*, 1986). Experimental evidence obtained by testing lineage potency of embryonic cells indicates that the positional influence is critical for the segregation of the trophectoderm and inner cell mass (ICM) lineages (see Section II). The ICM cells remain pluripotent until gastrulation when further segregation of embryonic tissue lineages occurs.

It has been demonstrated by examining the pathways of differentiation undertaken by cells under different experimental conditions that the progenitor cells pass through an intermediate stage of lineage differentiation. At this stage, the cells are becoming progressively more restricted in their ability to differentiate into divergent tissue types. Cells that show diminishing developmental potency are taken as displaying commitment to specific fates. The eventual acquisition of a particular fate is often accompanied by complete loss of the plasticity of cell differentiation. Cells at this stage are regarded to be fully determined for specific fates. These cells will proceed to terminal differentiation independent of any environmental influences. The processes beginning with the specification, through progressive commitment and the culmination in the determination of cell fates represent the sequence of events of lineage differentiation. In this chapter, we shall review our current understanding of these processes in the gastrulating mouse embryo.

II. Regulative Mode of Embryogenesis Suggests No Role for Inheritable Lineage Determinants

During preimplantation development, two morphologically distinctive cell types are first found in the blastocyst. The ICM mainly gives rise to the embryo proper and some extraembryonic tissues, and the trophectoderm contributes to placental tissues. In the mouse conceptus early cleavages do not follow a fixed pattern. Therefore, an orderly segregation of determinants to daughter cells is unlikely to be responsible for the establishment of these two first lineages. This contention is further supported by the outcome of several experimental manipulations. The scrambling of cytoplasm of the zygote does not impair subsequent embryonic development (Evsikov *et al.*, 1994), although it is possible that there may be an efficient reconstitution of cyto-

plasmic organization in the embryos. Even the addition or rearrangement of blastomeres of the preimplantation embryo that presumably will disrupt any preexisting intercellular relationship or the distribution of lineage-specific information apparently has no effect on embryonic development (reviewed in Pedersen, 1986). Last, single blastomeres from the 2- to 8-cell stage can contribute to many tissues in chimeric blastocysts and postimplantation embryos (Gardner and Rossant, 1976), suggesting that individual blastomeres are likely to be pluripotent. The highly regulative mode of development of the preimplantation embryo is inconsistent with a mechanism of lineage specification by the segregation of inheritable lineage determinants in the blastomeres.

The allocation of cells to the ICM and the trophectoderm lineages has been shown to be influenced by the position of the cells in the embryo. Morphological differences are first seen in the blastomeres of the 8-cell embryo. Each of the eight blastomeres has part of the cell surface on the outside and the rest of the surface facing neighboring cells or the intercellular space in the embryo. The morphology of the cell surface and the distribution of organelles and cytoskeleton are found to be polarized along this outer to inner axis of the blastomeres. Subsequent cell division will generate blastomeres that either retain the polarized characteristics or show no polarization. Cells that are found on the outside will retain the polarized characteristics and contribute to the trophectoderm while those on the inside of the embryo lose the polar morphology and give rise to the ICM. Despite the phenotypic difference the polar and apolar blastomeres are still capable of switching between the two tissue fates and this is always accompanied by changes in the polarized morphology. It seems that the fate of early embryonic mouse cells is influenced more by the position of the cell in the embryo rather than by molecular and/or structural factors (reviewed by Johnson *et al.*, 1986; Gueth-Hallonet and Maro, 1992; and compare to marsupials, Chapter 34, Selwood and Hickford). This process probably provides the early embryo with enough developmental flexibility to compensate for the indeterminate pattern of cell cleavage which may lead to a variable distribution of cytoplasmic factors or specialized structures associated with the cell membrane.

The developmental potency of the ICM and its immediate derivative (the epiblast of the pregastrulation embryo at 4.5–5.5 days post coitum) has been tested by assessing the fate of these cells. Single ICM or epiblast cells introduced into a recipient blastocyst contributes extensively to every identifiable tissue type in the chimera. There is no appreciable difference in the range of tissues colonized by the descendants of ICM or epiblast cells in different embryos (Gardner and Beddington, 1988). Embryonic stem cells that are derived from the ICM *in vitro* retain their pluripotency and when introduced to a host embryo behave exactly like their parental ICM cells (Beddington and Robertson, 1989). The weight of experimental evidence therefore supports the concept that before the onset of gastrulation, the epiblast that descends from the ICM are pluripotent for all embryonic tissue lineages.

III. Scenarios of Lineage Specification during Gastrulation

Subsequent postimplantation development involves the process of gastrulation that leads to the formation of the three germ layers at the late-primitive-streak (Late-PS) stage (Fig. 1). The onset of gastrulation is heralded by the appearance of the primitive streak in the epiblast at the extraembryonic–embryonic junction on the posterior side of the embryo. The primitive streak is a transient embryonic structure where the epiblast cells undergo an epithelial to mesenchymal transition (Hashimoto and Nakatsuji, 1989). During gastrulation the primitive streak continues to expand anteriorly by progressive recruitment of the epiblast cells until it reaches the distal region of the egg cylinder. The mesoderm layer expands from both sides of the primitive streak between the epiblast and the primitive endoderm (Poelmann, 1981; see Figs. 1 and 2). Active recruitment of ingressing cells to the definitive endoderm occurs primarily in the anterior segment of the primitive streak (Lawson and Pedersen, 1987). The cells that are not involved with cellular ingression remain in the epithelial layer and constitute the ectoderm (Tam *et al.*, 1993).

At the cessation of gastrulation all three germ layers are formed (Fig. 1C,D) and subsequent patterning of the mouse embryo occurs by inductive interactions both between and within these germ layers. Thus the morphogenetic movements of gastrulation have resulted in a positioning of the germ layers and the polarization of the embryo which enable organogenesis to proceed.

The results of a series of fate mapping studies have revealed that there is a discernible difference in the developmental fate of cells in different regions of the epiblast of the early gastrula (reviewed by Tam and Behringer, 1997; Fig. 3). The construction of fate maps in the mouse by tracking the morphogenetic movements of labeled embryonic cells has revealed the types of tissues to which a particular cell population will contribute. However, fate maps reveal little about the changes in cell state and the processes involved in restricting the types of tissue that can be generated by a progenitor population. In other words, the demonstration of specific cell fate does not indicate the occurrence of lineage specification.

A direct test for lineage specification is to assess if

Early-primitive-streak stage

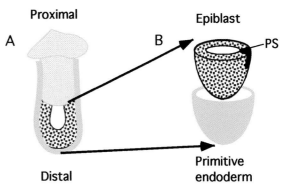

Late-primitive-streak stage

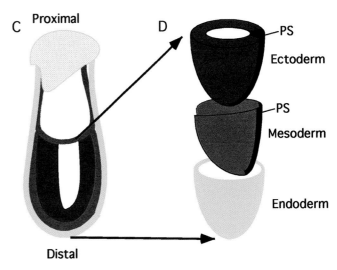

Figure 1 Schematic diagrams showing the germ layer composition of the gastrulating mouse embryo at the early- and late-primitive-streak (PS) stage. The early-PS embryo is composed of an inner epiblast and an outer primitive endoderm layer. At the completion of gastrulation, the late-PS embryo contains three germ layers, the ectoderm, the mesoderm, and the endoderm. (A, C) Longitudinal-section views of the embryos along the proximal–distal axis, proximal being the end attached to the uterus. (B, D) Exploded views of the germ layers of the embryos. The primitive streak (PS, black region) marks the posterior side of the embryo.

there is detectable restriction in the lineage potency of the cells. Lineage potency describes the variety of fates that a precursor cell can assume on experimental manipulation and is considered to be a measure of the degree of restriction or determination. Experimentally, the developmental potency of the epiblast cells can be studied by mapping the fate of subpopulations of epiblast cells and testing if they may acquire different fate by confronting them with different sets of environmental factors *in vitro* and *in vivo*. Transplantation of epiblast cells fated for a particular embryonic lineage into a novel environment, therefore, can reveal if a restriction of cell fate occurs. This manipulation is known as het-

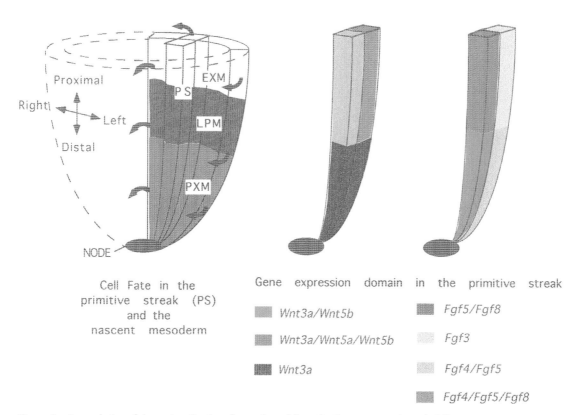

Figure 2   A correlation of the regionalization of mesodermal fate of cells ingressing through different segments of the primitive streak (PS) and the spatial pattern of the expression of the signaling molecules of the Wnt and FGF family. Cells passing through the distal segment of the primitive streak are predominantly fated for paraxial mesoderm (PXM), those in the middle segment are for lateral plate mesoderm (LPM), and those in the proximal segment are for extraembryonic mesoderm (EXM). The curved arrows show the direction of cell movement after ingression. The proximal–distal polarity is defined relative to the attachment of the egg cylinder to the uterus (see Fig. 1). The transcripts encoding the ligands are shown to be localized predominantly to specific segments of the primitive streak. The transcripts, however are often distributed in a gradient along the proximal–distal length of the primitive streak and are sometimes restricted to either the epithelial or the mesenchymal portion of the streak (these two portions are shown as sandwiched strips). The dark oval structure at the distal end of the primitive streak is the node, the organizer of the mouse gastrula. At late-PS the expression of *Wnt5a* and *Wnt5b* is posteriorly restricted, with *Wnt5a* detectable in the mesoderm whereas *Wnt5b* is expressed in both the ectoderm and mesoderm of the primitive streak. At this stage the expression of *Wnt3a* is apparent in both ectoderm and mesoderm of the entire primitive streak encompassing *Wnt5a* and *Wnt5b* but ceasing distal to the node region (Takada *et al.,* 1994). It is interesting to note that the combined expression of *Wnt3a, Wnt5a,* and *Wnt5b* mark the site of ingression of the extraembryonic mesoderm whereas *Wnt3a* is exclusively expressed in the distal region of the primitive streak where the embryonic mesoderm is formed. At late-PS *Fgf4* is restricted to the distal two-thirds of the primitive streak in both mesoderm and epithelial components (Niswander and Martin, 1992). *Fgf3,* is detectable in the proximal region (extraembryonic mesoderm) after it has ingressed, an expression pattern that is complementary to *Fgf4* (embryonic mesoderm) (Wilkinson *et al.,* 1988; Niswander and Martin, 1992). *Fgf5* is expressed in a proximal–distal gradient in the epithelial component of the streak and is down regulated proximally after cells have ingressed (Haub and Goldfarb, 1991; Hébert *et al.,* 1991). *Fgf8* also is expressed in a gradient and is reduced as cells pass through the streak (Crossley and Martin, 1995).

erotopic transplantation. If the cells are able to behave like the cells in their new surrounding they are deemed to be pluripotent or displaying developmental plasticity. If these transplanted cells are unable to acquire a novel fate in their new environment but differentiate in accordance to their presumed fate they are considered to have committed to a particular cell fate and their lineage potency is restricted. This committed state is passed onto daughter cells and may be the result of changes in genetic activity. However, it is important to note that irreversibility in genetic terms is difficult to define with the demonstration that nuclei from fully differentiated mammalian somatic cells are able to reca-

pitulate the developmental program if transplanted into an enucleated oocyte (Wilmut *et al.,* 1997).

The specification of tissue lineages may take place concomitantly with the formation of germ layers in the gastrulating mouse embryo. Four possible scenarios of lineage specification may be considered in relation to the formation of the ectoderm, mesoderm, and endoderm. In the first scenario, cells in the epiblast of the mouse gastrula are fully specified for the tissues of all three germ layers (Fig. 4, Model 1) and the morphogenetic activity of gastrulation serves primarily to distribute the specified cells to the appropriate germ layers. This would suggest that the primitive streak is effec-

Early-primitive-stage embryo

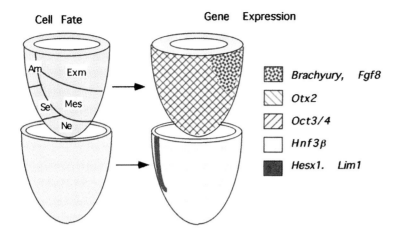

Cell Fate

Gene Expression

Brachyury, Fgf8

Otx2

Oct3/4

Hnf3β

Hesx1. Lim1

Late-primitive-streak stage embryo

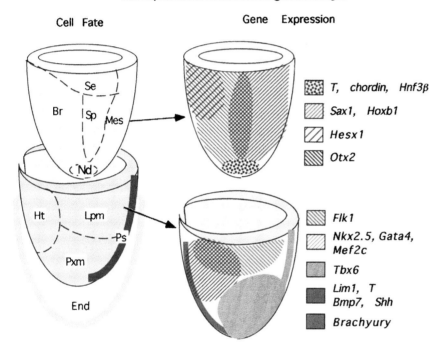

Cell Fate

Gene Expression

T, chordin, Hnf3β

Sax1, Hoxb1

Hesx1

Otx2

Flk1

Nkx2.5, Gata4, Mef2c

Tbx6

Lim1, T Bmp7, Shh

Brachyury

Figure 3 A comparison of cell fate and gene expression in the gastrulating mouse embryo. The orientation of the germ layers is shown in Fig. 1. At the early-primitive-streak stage, there is no obvious correlation of cell fate and gene expression. In the epiblast (blue), the regionalization of cell fate is revealed by the localization of the progenitors of major tissue types (Am, amnion; Exm, extraembryonic mesoderm; Mes, mesoderm; Ne, neurectoderm; Se, surface ectoderm). Concordant with the developmental plasticity of the epiblast cells, *Oct3/4* and *Otx2* are widely expressed in the epiblast. However, there is localized expression of the *Brachyury* (*T*) gene in the prospective extraembryonic mesoderm adjacent to the nascent primitive streak. The primitive endoderm (yellow) is fated for extraembryonic endoderm. There is widespread expression of genes such as *Hnf3β*, except for the localized expression of *Hesx1* in the anterior endoderm. For the late-primitive-streak stage embryo, detailed fate maps are available for the ectoderm (blue) and the mesoderm (red) but not yet for the endoderm (yellow). In the ectoderm, the localization of the progenitors of neurectoderm of the brain (Br) and spinal cord (Sp), mesoderm (Mes), and surface ectoderm (Se) is shown. There is a good correlation of the cell fate for the anterior and posterior neurectoderm with the expression of the *Hesx1* (forebrain), *Otx2* (fore- and midbrain), and *Sax1* genes (spinal cord). The node (Nd) expresses the organizer-specific genes such as *T*, *chordin* and *Hnf3β*. In the mesoderm, the localization of the heart (Ht), somitic (Pxm), and lateral mesoderm (Lpm) matches the expression of the tissue-specific genes. There is good correlation of cell fate with lineage-specific gene activity in the germ layers of the late-primitive streak stage embryo suggesting that, at the completion of gastrulation, lineage differentiation has advanced beyond the initial specification-phase. References: Fate maps: reviewed by Lawson *et al.*, 1991; Lawson and Pedersen, 1992; Tam and Behringer, 1997. Gene expression: *Bmp7* (Arkell and Beddington, 1997), *Brachyury* (*T*) (Wilkinson *et al.*, 1990; Hermann, 1991), *chordin* (T. Tsang, unpublished), *Flk1* (Yamaguchi *et al.*, 1993), *Gata4* (Heikenheimo *et al.*, 1994), *Hesx1* (Thomas and Beddington, 1996; Hermez *et al.*, 1996), *Hnf3β* (Sasaki and Hogan, 1993); *Hoxb1* (Frohman *et al.*, 1990), *Lim1* (Shawlot and Behringer, 1995), *Mef2c* (Edmonson *et al.*, 1994), *Nkx2.5* (Lints *et al.*, 1993), *Oct3/4* (Yeom *et al.*, 1996), *Otx2* (Acampora *et al.*, 1995; Matsuo *et al.*, 1995; Ang *et al.*, 1996), *Tbx6* (Chapman *et al.*, 1996), *Sax1* (Schubert *et al.*, 1995), and *Shh* (Echelard *et al.*, 1993).

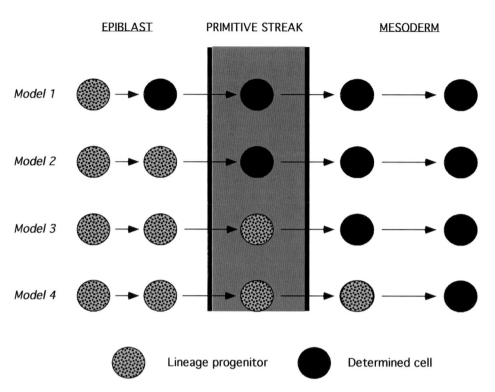

Figure 4 Models of the site of specification of the mesodermal lineage. Model 1: Presumptive mesodermal cells are fully determined in the epiblast. Model 2: Specification occurs during cellular ingression at the primitive streak. Model 3: Mesodermal lineages are specified after the cells are recently incorporated into the mesodermal layer. Model 4: Mesodermal fate is fully determined only after the cells have completed ingression and migration.

tively a channel for sorting an already specified cell population. In the second scenario, cells are still unspecified while they are within the epiblast, and the allocation of cells to specific mesodermal and endodermal lineages takes place as cells pass through the primitive streak (Fig. 4, Model 2). In this situation specification may be the result of an interaction between the ingressing cells with some localized specifying signals in the primitive streak. The source of such signals may emanate from cells within or adjacent to the primitive streak. It is also possible that lineage specification does not take place until after the embryonic cells have completed ingression. Therefore, in the third model, lineage specification takes place after the cells have been incorporated into the mesoderm or the endoderm (Fig. 4, model 3). This mode of specification requires the provision of specifying signals that are probably derived from cell–cell interactions within or between the germ layers. However, it is possible that lineage specification may be accomplished after the completion of morphogenesis of the germ layers (Fig. 4, Model 4). In this case, neither the ingression through the primitive streak nor the association with the newly formed germ layers has any permanent effect on cell differentiation. Determination of cell fate only occurs after the cells reach their final destination where position- or organ-specific specifying signals are available. It must be noted that models 2–4 are primarily concerned with the specifica-

tion of cells that are involved with cellular ingression and the formation of the mesoderm. The specification of the tissue lineages in the ectodermal cell population that do not ingress may be accomplished by other means.

The most likely mechanism is that specification is mediated by inductive interactions between germ layers. We shall examine if any of these scenarios of lineage specification fits the experimental evidence presently available for the specification of the ectoderm and mesoderm. We will not discuss the specification of the endoderm because although we have detailed information on its morphogenetic movement during gastrulation (see Lawson *et al.,* 1986; Lawson and Pedersen, 1987; Tam and Beddington, 1992; Thomas and Beddington, 1996), practically nothing is known of how this tissue is set aside in the gastrulating mouse embryo (see Chapter 40 by Gannon and Wright).

IV. The Epiblast Cells of the Early Gastrula Display Lineage Plasticity

An important issue of lineage specification is whether the cells of the epiblast in the early gastrula may have been irreversibly allocated to specific tissue lineages and are therefore restricted in the types of tissue to which they can contribute. It has been demonstrated that the clonal descendants of individual epiblast cells could contribute to a variety of embryonic and extraembryonic tissues that are derived from all three germ layers (Lawson *et al.*, 1991; Lawson and Pedersen, 1992). In other words, none of the clones tested is specified to differentiate into only one tissue type. A similar conclusion is reached when epiblast cells transplanted to heterotopic sites in the epiblast of another embryo are found to be capable of adopting the fate of the population in their new environment (Parameswaran and Tam, 1995). The finding of heterotopic transplantation, though suggestive of the plasticity of epiblast cell fate, must be interpreted with caution. In these experiments, groups of 5–20 cells instead of single cells are tested, because of the small size of the cells. It is not known if the apparent acquisition of novel cell fate may result from a differential selection of cells that are favored by the heterotopic sites. However, since in most cases the cells have taken cell fates that are totally different from their expected fates, this would argue that there is a change to novel cell fates. Clonal analysis, fate mapping, and transplantation studies therefore provide a consensus that epiblast cells are still pluripotent. It therefore seems paradoxical that on one hand epiblast cells are yet to be restricted in their lineage potency and on the other, a significant difference in the cell fate is found for cells localized to different parts of the embryo. Does this regionalization reflect a degree of cell specification?

Fate mapping studies have indicated that the ectodermal precursors (the neurectoderm and the surface ectoderm) are mostly localized to the anterior and distal regions of the early-PS embryo (45% of the total epiblast population) (Fig. 3). Not only is the entire precursor population of the neural tube present in the distal cap region but there is also a craniocaudal bias in the allocation of these precursor cells (Quinlan *et al.*, 1995). Whether this basic craniocaudal pattern of the neural precursors in the distal cap is an indication of a specified neurectodermal cell fate has not been resolved. Contrary to the ectodermal fate of distal epiblast cells, cells in the proximal region give rise to the extraembryonic mesoderm and primordial germ cells (PGCs) (Lawson and Hage, 1994). The developmental potency of these two epiblast populations with vastly divergent fates has been assessed by the reciprocal transplantation of cells between proximal and distal sites in the early-PS embryo (Tam and Zhou, 1996). This study has shown that distal epiblast cells transplanted to proximal sites colonize extraembryonic mesoderm and display a germ-line potency similar to that of the proximal epiblast (3.7% of the graft-derived cells exhibited alkaline phosphatase activity characteristic of PGCs). Furthermore, when proximal epiblast cells are transplanted to distal sites they behave like distal epiblast cells and contribute predominantly to the neurectoderm of the host early-somite-stage embryo (86% of graft-derived cells) and moderately to the surface ectoderm (10% of graft-derived cells). The findings of this study provide evidence that the epiblast cells of an early-PS embryo display remarkable plasticity in lineage potency. Moreover, the conversion of the grafted PGCs fate to another and vice versa by simple relocation to ectopic sites reinforce the idea that a prelocalization of germ-line determinants in the mouse oocyte is unlikely.

A more recent study has shown that the *Oct4* gene, encoding a POU transcription factor, may be involved in the establishment of the mammalian germline (Yeom *et al.*, 1996). *Oct4* is found in oocytes and morulae and is restricted to the ICM and the epiblast, and therefore it is a molecular signature for pluripotency (see Fig. 3). During gastrulation the expression in the epiblast is progressively downregulated such that *Oct4* activity is only maintained in the germ-line lineage. Two enhancer elements are found to be critical for this differential control of expression. It is postulated that the enhancer that specifically drives *Oct4* expression in the germ line is activated by factor(s) that are specific for totipotent cells. Whether the localization of germ cell precursors in the proximal region of the epiblast is necessary to maintain a state of germ-line totipotency has yet to be determined. However, the respecification of the fate of distal epiblast in the proximal epiblast may point to a local influence that activates the germ-line-specific *Oct4* enhancer. Proximal epiblast cells heterotypically grafted to distal sites may miss the local signal(s) that activate the germ-line enhancer and therefore retain the *Oct4* activity controlled by the non-germ-line enhancer.

In summary, although a regionalization of cell fate could be demonstrated by fate mapping, heterotopic transplantation studies have indicated that a lineage restriction or irreversible determination of cell fate seems not to have occurred in the epiblast at the onset of gastrulation. This suggests that cells in the epiblast are not committed to any particular cell fate. The scenario of lineage development (Fig. 4, Model 1) that requires complete determination of the tissue lineages before the epiblast cells engage with gastrulation is therefore not compatible with the experimental findings. Lineage specification that results in the acquisition of a more restricted cell fate is therefore likely to take place during gastrulation.

V. Determination of Neural Fate in the Ectoderm by Inductive Interactions

The dynamic tissue movements during gastrulation result in the juxtaposition of newly formed tissues. This also may allow novel tissue interactions via appositional or vertical signaling by inducing molecules that may be associated with a particular germ layer. Evidence that inductive interactions are important in cell fate determination in the mouse are starting to emerge from an analysis of expression patterns of genes encoding ligands and receptors of signaling systems and of the effect on cell differentiation in explants and recombinants of germ layers *in vitro.*

Recent evidence from the expression patterns of a number of genes in the pre- and early-PS embryo (Fig. 3) have suggested that the primitive endoderm may be involved in specifying neural fate in the mouse (reviewed by Bouwmeester and Leyns, 1997). A number of transcripts and molecules are expressed in a restricted region in the anterior part of the primitive endoderm either prior to, or during gastrulation in the mouse. They include the antigen marker Visceral Endoderm 1 (VE-1, Rosenquist and Martin, 1995), the homeobox gene *Hesx1/Rpx* (Thomas and Beddington, 1996; Hermez *et al.,* 1996), and the LIM-domain gene *Lim1* (Shawlot and Behringer, 1995; Fig. 3). During gastrulation, *Hesx1* activity is down-regulated in the endoderm but is later localized to the forebrain neuroectoderm (Fig. 3). Surgical ablation of the primitive endoderm that expresses the gene leads to diminished *Hesx1* activity in the embryonic forebrain. Although the effect of a loss-of-function mutation of the *Hesx1* gene is not yet known, the result of the endoderm ablation experiment strongly implicates an interaction of the two germ layers that successively express the *Hesx1* gene during gastrulation. It also raises the possibility that such interactions may be critical for the specification of neural fate of the epiblast. In the early-PS embryo transcripts of the *nodal* gene that encodes a transforming growth factor-β (TGFβ) molecule are detected at low levels in the primitive endoderm and a gradient of expression is apparent in the epiblast with the highest concentration in the proximal posterior region. Chimeric embryos comprised of a *nodal* deficient primitive endoderm fail to develop forebrain and possibly midbrain (Varlet *et al.,* 1997). Therefore, results of the *nodal* and *Hesx1* studies strongly suggest that there is a critical requirement of the endoderm for the specification of neural fate in the epiblast.

The combined inducing activity of the mesoderm and endoderm layers in the mouse gastrula has been demonstrated *in vitro* by the use of tissue recombinant explants (Ang and Rossant, 1993; Ang *et al.,* 1994). From these experiments it has been shown that ectodermal fragments isolated from the anterior region of mid- to late-PS embryos express brain-specific genes (e.g. *Otx2, En1,* and *En2*) in the absence of any other germ layer tissues. This implies that the ectoderm has all the necessary signals for neural differentiation by this stage. However, isolated epiblast from early-PS embryos is unable to express these genes unless they are cultured together with fragments of the mesendoderm isolated from the anterior region of the early headfold

stage embryos. Similar mesendoderm from the posterior part of the headfold stage embryo fails to induce the expression of the brain-specific markers. The posterior mesendoderm also seems to suppress the expression of *Otx2*, although it can still induce neural differentiation. These observations suggest that the mesendoderm is an important source of signal(s) for inducing neural differentiation in the epiblast and there may be region-specific morphogenetic factors associated with different mesendoderm populations.

Two issues remain unresolved regarding the inductive role of the germ layers in the specification of neural fate in the ectoderm. First, the studies on *Hesx1* and *Nodal* genes implicate a role of the primitive endoderm. However, this endodermal population will eventually give rise to extraembryonic endoderm and therefore has only a transient association with the prospective neural progenitors. If induction does occur, it is critical to elucidate the nature of the inductive signals and the timing required to achieve such an induction. Second, the tissue recombination experiment suggests that there is a regionalization of the interactive signals by mid to late gastrulation (Ang and Rossant, 1993; Ang *et al.,* 1994). It is not possible, however, to determine the source of these signals because the endoderm and the mesoderm have not been tested separately in these experiments. It is likely that signals for neural specification are different for the embryonic mesoderm and definitive endoderm. More recently, the midline anterior mesendoderm (prechordal plate and head process) of the chick gastrula has been shown to possess neural inductive activity (Pera and Kessel, 1997; Lemaire *et al.,* 1997). It is not clear if the mesendoderm isolated from the mid-PS mouse embryos for the induction experiment in the tissue recombination studies may contain cells that are destined for the midline mesendoderm, which could account for the neural inductive activity (Ang and Rossant, 1993; Ang *et al.,* 1994). Several gene mutations are known to be associated with deficiencies in cephalic neural tissues (e.g., *Otx2* and *Lim1:* Acampora *et al.,* 1995; Matsuo *et al.,* 1995; Ang *et al.,* 1996; Shawlot and Behringer, 1995). It is postulated that the phenotype is partially the result of the lack of neural induction by the midline mesendoderm (Bally-Cuif and Boncinelli, 1997). However, it is unclear whether this is caused by the loss of the axial mesendoderm tissues or a disruption of the inducing activity of the mutant cells.

VI. Restriction of Mesodermal Lineage Potency during Cellular Ingression in the Primitive Streak

As cells in the epiblast seem to be pluripotent before the onset of gastrulation, the allocation of cells to the mesodermal lineage may occur during the ingression of cells through the primitive streak (Fig. 4, Model 2). Fate mapping studies of the primitive streak have revealed that cells located in different regions along the streak display different mesodermal fates (Tam and Beddington, 1987; Lawson *et al.*, 1991; Smith *et al.*, 1994; Wilson and Beddington, 1996; Fig. 2). Hence cells present in the posterior segment of the streak contribute predominantly to extraembryonic mesoderm, cells emerging from the middle of the streak give rise almost exclusively to lateral mesoderm, whereas the most anterior positioned cells contribute to paraxial and axial mesoderm. However, it is not clear if the regionalization of mesodermal cells does reflect any real bias in lineage potency. Indeed, when different fragments of the primitive streak are allowed to differentiate as teratomas in ectopic sites, there is no significant variation in the tissue types that are produced (Chan, 1991). Heterotopic transplantation of cells between the distal and the proximal segments of the primitive streak of late-PS embryos also shows that the ingressing cells are developmentally plastic (Beddington, 1982). These findings suggest that although cells may adopt different fates as they emerge from different parts of the primitive streak, they may not have acquired any restriction in developmental potency as depicted in Model 2 (Fig. 4).

The impact of the act of cellular ingression on lineage potency has been tested by reciprocal transplantations of cells between the nascent mesoderm and the epiblast (Tam *et al.*, 1997). The posterior epiblast cells of the early-PS embryo are normally fated for the extraembryonic mesoderm and various embryonic mesoderm such as the heart, cranial mesenchyme, somites, and lateral plate mesoderm (Fig. 3). When these epiblast cells are transplanted to the nascent mesoderm of the mid-PS embryo, they are able to colonize a broader range of mesodermal tissues than that of the mesodermal population they have incorporated into (which mainly gives rise to yolk sac and heart mesoderm). This observation suggests that without ingression through the primitive streak the posterior epiblast cells have retained a broader tissue potency even when relocated to a mesodermal environment. Evidence for a restriction in lineage potency following ingression can be obtained by the reciprocal experiment, the transplantation of recently ingressed mesoderm cells back to the posterior region of an early-primitive-streak epiblast (Tam *et al.*, 1997). The graft-derived cells ingress for a second time and can contribute to most of the mesodermal tissues that are normally expected from the preingressed epiblast cells. The one exception is the contribution to lateral plate mesoderm: graft-derived cells are not found in the lateral mesoderm of the host embryo. However, such restriction in lineage potency may be found only for certain tissue types and the evidence has been based solely on the pattern of tissue colonization and not on the expression of tissue-specific characteristics (Fig. 3).

There is at present no evidence for any restriction in lineage potency of cells within the primitive streak (Fig. 4, Model 2). These experimental results are consistent with a restriction of mesodermal potency concomitant with the incorporation of the ingressed cells into the mesoderm (Fig. 4, Model 3).

VII. Mesoderm Induction in the Primitive Streak: Putative Roles of Inductive Molecules

Evidence for mesoderm induction by growth factors has primarily come from studies in *Xenopus* that show that secreted growth factors such as the fibroblast growth factor (FGF), Wnt, and TGFβ families may play a role in inducing mesodermal differentiation (reviewed by Isaacs, 1997; Chapter 19 by Dawid). Patterns of expression of genes encoding these factors in the primitive streak of the late-PS embryos are compatible with the idea that progenitors for different mesodermal lineages might be specified by the appropriate combination of factors during ingression (Sasaki and Hogan, 1993; Takada *et al.*, 1994). A number of the FGF molecules are expressed in restricted regions of the primitive streak of the gastrulating mouse embryo (Fig. 2). Targeted mutation of these genes have not been informative regarding a role in mesoderm specification during gastrulation due to the early lethal phenotype of *Fgf4* (Feldman *et al.*, 1995) and the later stage phenotypes in the case of *Fgf3* and *Fgf5* (Mansour *et al.*, 1993; Hébert *et al.*, 1994). In *Xenopus*, *Brachyury* expression and eFGF (closely related to mouse FGF-4 and FGF-6) may be regulated by an autocatalytic mechanism (Isaacs *et al.*, 1994; Isaacs, 1997). This potential autoregulatory relationship between a secreted peptide growth factor and a mesoderm expressed transcription factor may allow an amplification of an initial mesoderm inducing signal. However, an analysis of *Fgf4* expression in *Brachyury* mouse mutants has revealed that *Fgf4* expression is maintained in the primitive streak even in the absence of a functional *Brachyury* protein. Thus a regulatory interaction involving *Brachyury* and *Fgf4* appears not to be essential (Schmidt *et al.*, 1997). Furthermore, the expression of *Brachyury* in *Fgfr1* deficient embryos also casts doubt on this interaction (Yamaguchi *et al.*, 1994; Deng *et al.*, 1994).

Further evidence that FGF signaling may be involved in the specification of mesoderm comes from an analysis of the FGF receptor 1 (FGFR1) null mutant phenotype. *Fgfr1* is expressed initially throughout the epiblast and by mid-PS stages becomes concentrated in the posterior mesoderm lateral to the primitive streak (Orr-Urtreger *et al.*, 1991; Yamaguchi *et al.*, 1992). Homozygous null mutants show a disruption in mesoderm migration (Deng *et al.*, 1994; Yamaguchi *et al.*, 1994).

This deficiency in migration leads to an absence of lateral and paraxial mesoderm although an expansion of the axial mesoderm occurs. In chimeras that contain wild-type and mutant cells, the mutant cells are retained in the primitive streak and eventually form ectopic secondary neural tubes (Ciruna et al., 1997). Some mutant cells, however, can make variable contributions to the mesoderm of the chimera. Moreover, the ability of Fgfr1 deficient ES cells to differentiate into mesodermal cell types was found to be unaffected in teratoma assays (Deng et al., 1994). Collectively, the genetic evidence indicates that the FGF signaling system is critical mainly for the allocation of cells to the mesodermal tissues and may fulfill a competence or transforming role rather than mesodermal induction (Isaacs, 1997; Sasai and De Robertis, 1997; Lumsden and Krumlauf, 1996). Ingressed cells may be induced to form mesoderm by other factors but in the absence of a functional FGFR1 this specification process is interrupted and the cells may then differentiate atypically to form neurectoderm instead of mesoderm.

Several members of the Wnt family of secreted growth factors (Wnt3a, Wnt5a, and Wnt5b) also are present in the primitive streak (Fig. 2). Mutational analysis of Wnt3a has revealed that null mutants are deficient in the paraxial mesoderm caudal to the first 7–9 somites, but lateral plate mesoderm is present (Takada et al., 1994). The characterization of a hypomorphic allele of Wnt3a as a likely candidate for the vestigial tail mutation also has indicated that posterior paraxial mesoderm formation is sensitive to the level of Wnt3a activity (Greco et al., 1996; Yoshikawa et al., 1997). The disruption of caudal paraxial mesoderm formation in the null mutant and the expression of Wnt3a in the primitive streak reinforce the likelihood that Wnt3a is involved in mesodermal specification (Takada et al., 1994). A further analysis of the mutant phenotype has shown that epiblast cells normally fated to form the paraxial mesoderm contribute to an ectopic neural tube after ingression through the primitive streak. As previously indicated for the FGFR1 mutation, the null mutant phenotype of Wnt3a also suggests that ingression through the primitive streak has an impact on mesoderm specification. The deficiency in caudal paraxial mesoderm and the formation of ectopic neural tube after ingression indicate that the mesoderm is not correctly specified. This results in cells that may be responsive to the effects of neuralizing factors present in the embryo. If the problem was solely related to migration, an incorrect placement or accumulation of mesenchyme would be expected rather than an ectopic neural tube. The fact that ingressed cells can assume a neural fate after they have left the streak suggest that a specification event that involves Wnt3a is required during or soon after ingression. As chimeric or lineage potency studies have not been carried out for cells carry-

ing the Wnt3a null mutation it is unknown whether their mesodermal lineage potency is compromised.

VIII. Specification of Tissue Lineages on Completion of Gastrulation

A study was conducted on the cardiogenic mesoderm of the mouse gastrula to investigate whether the specification of myocardial cell fate occurs during gastrulation or after cells have completed morphogenetic movement. Progenitors of cardiac cells are found in the posterior region of the lateral epiblast. These cells ingress through the primitive streak and migrate anterior-laterally to the cardiogenic field by late gastrulation (Fig. 3). Transplantation experiments have shown that the epiblast cardiac progenitors can differentiate into myocardial and endocardial cells if they are transplanted directly into the cardiogenic field of the late-PS embryo. Even epiblast cells that are not normally fated to be cardiac mesoderm can also be induced to acquire a cardiac cell fate by similar transplantation to the late-PS embryo (Tam et al., 1997). These findings suggest strongly that the specification of a cardiac fate (particularly the myocardial fate) only requires signals provided by the tissues at the final cardiogenic site and is independent of cellular ingression and germ layer formation. They suggest that the main events associated with lineage restriction occur after the cells have reached their final position with commitment to a particular fate being the result of locally produced determinants (Fig. 4, Model 4). Therefore some tissue lineages can be correctly specified if the events associated with ingression are circumvented and the pluripotent epiblast is exposed to local influences.

Tbx6, a member of the T-box containing transcription factor family, is expressed in the primitive streak and newly recruited paraxial mesoderm during gastrulation (Chapman et al., 1996; Fig. 3). Similar to the Wnt3a mutant, the formation of trunk somites is disrupted in the Tbx6 null mutant. The prospective somitic mesoderm seems to have ingressed through the primitive streak and reached the appropriate paraxial sites. However, instead of forming somites, the tissue differentiates into ectopic neural tissue with proper dorsoventral organization of neural cell types. Lateral mesodermal tissues are spared from any effect of the mutation (Chapman and Papaioannou, 1998). The phenotype of this mutation suggests that the ability of the epiblast cells to migrate like the paraxial mesoderm during gastrulation does not necessarily confer a mesodermal characteristic on these cells. The choice of neural fate by the mutant cells is governed by the nature of the local specifying signals (Fig. 4 Model 4) and the competence of the cells to respond to such signals. The *Tbx6* gene product therefore is required for the acquisition of the somitic fate. The mutant phenotype, however, does not provide any clue as to when the specification of somite fate occurs during development.

Considered collectively, the three mutations *Fgfr1*, *Wnt3a*, and *Tbx6* all result in the incorrect specification of particular mesodermal lineages and mutant cells can adopt an ectopic neural fate. In two mutants, the ability of these cells to migrate also has been affected: *Fgfr1* mutant cells fail to exit the posterior primitive streak and *Wnt3a* mutant cells ingress and accumulate under the primitive streak whereas, in contrast, *Tbx6* mutant cells appear to migrate correctly to their final position. The *Tbx6* mutation suggests that even if cells can reach their final "mesodermal" destination the effect of local inductive interactions is not sufficient to specify a paraxial mesoderm fate but is sufficient for a neural fate.

IX. Summary

The cells of early mouse embryos exhibit a remarkable degree of plasticity in tissue fate. Lineage differentiation does not appear to involve the segregation of cytoplasmic determinants and the blastomeres of the preimplantation embryos are pluripotent. The ICM-derived epiblast of the early-gastrula also reveals little commitment to a particular fate. However, fate mapping experiments have revealed a consistent regionalization of the cells in the epiblast at this stage. Gastrulation results in the transformation of the pluripotent epiblast to three germ layers. The morphogenetic movements associated with this embryonic transition play a major role in partitioning these precursor populations to their appropriate location. The role that this process may play in the specification of an embryonic lineage has been considered from the perspective of four simple models. The four models place the event of lineage determination at different morphogenetic stages that may be experienced by the epiblast derived cells. Considering the experimental work summarized in this chapter a partitioning of an already specified epiblast population (Fig. 4, Model 1) appears unlikely. The remaining models primarily relate to the effect of primitive streak ingression on lineage potency and the potential inductive influences to which the germ layer tissue may be exposed.

Several regions of the mouse embryo have been identified that may have an effect on cell fate prior to or during gastrulation. These include the primitive endoderm, the axial and lateral mesendoderm, and the primitive streak. Gene expression patterns, null mutant phenotypes, and heterotopic transplantation experiments were assessed to determine whether an effect on cell fate may be occurring during ingression through the primitive streak. By considering this experimental evidence a probable mechanism for specifying cell fate in the gastrulating mouse embryo has emerged. Although simplistic, it appears that ingression is having an impact on cell fate and may result in a restriction of lineage potency of the pluripotent epiblast. However, any model that relates a function of the primitive streak to cell fate determination must also take into account of the heterogeneity of cell fate and the extent of determination that may be found within the dimension of the primitive streak. Future experiments to test the lineage potency of cells within the primitive streak and of those emerging from the primitive streak at different stages of gastrulation may help to resolve the relative degree of specification of different mesoderm and endodermal lineages during germ layer formation. Further specification of a mesodermal lineage appears to be dependent on other extrinsic cues that the cell is exposed to during its migration and once it is positioned. It is likely that the process of mesoderm specification is a continuum with cells being exposed to a variety of inducing influences which progressively commit a cell to a particular lineage.

Acknowledgment

We thank Peter Rowe and Devorah Goldman for comments on the manuscript, and Gabriel Quinlan for contribution of ideas on the models of lineage specification. Our work is supported by the Human Frontier of Science Program, the National Health and Medical Research Council of Australia, and Mr. James Fairfax.

References

Acampora, D., Mazan, S., Lallemand, Y., Avantaggiato, V., Maury, M., Simeone, A., and Brulet, P. (1995). Forebrain and midbrain regions are deleted in *Otx2* −/− mutants due to a defective anterior neurectoderm specification during gastrulation. *Development (Cambridge, UK)* **121**, 3279–3290.

Ang, S.-L., and Rossant, J. (1993). Anterior mesendoderm induces mouse *engrailed* genes in explant culture. *Development (Cambridge, UK)* **118**, 139–149.

Ang, S.-L., Conlon, R. A., Jin, O., and Rossant, J. (1994). Positive and negative signals from mesoderm regulate the expression of mouse *Otx2* in ectoderm explants. *Development (Cambridge, UK)* **120**, 2979–2989.

Ang, S.-L., Jin, O., Rhinn, M., Daigle, N., Stevenson, L., and Rossant, J. (1996). A targeted mouse *Otx2* mutation leads to severe defects in gastrulation and formation of axial mesendoderm and to deletion of rostral brain. *Development (Cambridge, UK)* **122**, 243–252.

Arkell, R., and Beddington, R. S. P. (1997). BMP-7 influences pattern and growth of the developing hindbrain of mouse embryos. *Development (Cambridge, UK)* **124**, 1–12.

Bally-Cuif, L., and Boncinelli, E. (1997). Transcription factors and head formation in vertebrates. *BioEssays* **19**, 127–135.

Beddington, R. S. P. (1982). An autoradiographic analysis of tissue potency in different regions of the embryonic egg cylinder. *J. Embryol. Exp. Morphol.* **69**, 265–285.

Beddington, R. S. P., and Robertson, E. J. (1989). An assessment of the developmental potential of embryonic stem cells in the midgestation mouse embryo. *Development (Cambridge, UK)* **105**, 733–737.

Bouwmeester, T., and Leyns, L. (1997). Vertebrate head induction by anterior primitive endoderm. *BioEssays* **19**, 855–862.

Chan, W. Y. (1991). Intraocular growth and differentiation of tissue fragments isolated from primitive-streak-stage mouse embryos. *J. Anat.* **175**, 41–50.

Chapman, D. L., and Papaioannou, V. E. (1998). Three neural tubes in mouse embryos with mutations in the T-box gene *Tbx6*. *Nature (London)* **391**, 695–697.

Chapman, D. L., Agulnik, I., Hancock, S., Silver, M. L., and Papaioannou, V. E. (1996). *Tbx6*, a mouse T-box gene is implicated in paraxial mesoderm formation at gastrulation. *Dev. Biol.* **180**, 534–542.

Ciruna, B. G., Schwartz, L., Harpal, K., Yamaguchi, T. P., and Rossant, J. (1997). Chimeric analysis of *fibroblast growth factor receptor-1* (*Fgfr1*) function: A role for FGFR1 in morphogenetic movement through the primitive streak. *Development (Cambridge, UK)* **124**, 2829–2841.

Crossley, P., and Martin, G. (1995). The mouse *Fgf-8* gene encodes a family of polypeptides that is expressed in regions that direct outgrowth and patterning in the developing embryo. *Development (Cambridge, UK)* **121**, 439–451.

Davidson, E. H. (1990). How embryos work: A comparative view of diverse modes of cell fate specification. *Development (Cambridge, UK)* **108**, 365–389.

Deng, C.-X., Wynshaw-Boris, A., Shen, M. M., Daugherty, D. M., Ornitz, D. M., and Leder, P. (1994). Murine FGFR-1 is required for early postimplantation growth and axial formation. *Genes Dev.* **8**, 3045–3057.

Echelard, Y., Epstein, D. J., St-Jaques, B., Shen, L., Mohlev, J., McMahon, J. A., and McMahon, A. P. (1993). Sonic hedgehog, a member of a family of putative signaling molecules, is implicated in the regulation of CNS polarity. *Cell* **75**, 1417–1430.

Edmonson, D. G., Lyons, G. E., Martin, J. F., and Olson, E. N. (1994). *Mef2* gene expression marks the cardiac and skeletal muscle lineages during mouse embryogenesis. *Development (Cambridge, UK)* **120**, 1251–1263.

Evsikov, S. V., Morozova, L. M., and Solomko, A. P. (1994). Role of ooplasmic segregation in mammalian development. *Roux's Arch. Dev. Biol.* **203**, 199–204.

Feldman, B., Poueymirou, W., Papaioannou, V. E., DeChiara, T. M., and Goldfarb, M. (1995). Requirement of FGF-4 for postimplantation mouse development. *Science* **267**, 246–249.

Flaub, O., and Goldfarb, M. (1991). Expression of the murine fibroblast growth factor 5 in the mouse embryo. *Development (Cambridge, UK)* **112**, 394–406.

Frohman, M. A., Boyle, M., and Martin, G. R. (1990). Isolation of the murine *Hox-2.9* gene: Analysis of embryonic expression suggests that positional information along the anterior–posterior axis is specified by mesoderm. *Development (Cambridge, UK)* **110**, 589–608.

Gardner, R. L., and Beddington, R. S. P. (1988). Multi-lineage 'stem' cells in the mammalian embryo. *J. Cell Sci.* **10**, (Suppl.) 11–27.

Gardner, R. L., and Rossant, J. (1976). Determination during embryogenesis. *In* "Embryogenesis in Mammals," Ciba Foundation Symp. 40, pp. 5–26. North-Holland: Elsevier, Amsterdam.

Greco, T. L., Takada, S., Newhouse, M. M., McMahon, J. A., McMahon, A. P., and Camper, S. A. (1996). Analysis of the *vestigial tail* mutation demonstrates that *Wnt3a* gene dosage regulates mouse axial development. *Genes Dev.* **10**, 313–324.

Gueth-Hallonet, C., and Maro, B. (1992). Cell polarity and cell diversification during mouse embryogenesis. *Trends Genet.* **8**, 274–279.

Hashimoto, K., and Nakatsuji, N. (1989). Formation of the primitive streak and mesoderm cells in mouse embryos: Detailed scanning electron microscopical study. *Dev. Growth Differ.* **31**, 209–218.

Haub, O., and Goldfarb, M. (1991).Expression of the murine fibroblast growth factor 5 in the mouse embryo. *Development (Cambridge, UK)* **112**, 394–406.

Hébert, J. M., Boyle, M., and Martin, G. R. (1991). mRNA localisation studies suggest that murine FGF-5 plays a role in gastrulation. *Development (Cambridge, UK)* **112**, 407–415.

Hébert, J. M., Rosenquist, T., Götz, J., and Martin, G. R. (1994). FGF5 as a regulator of the hair growth cycle: Evidence from targeted and spontaneous mutations. *Cell (Cambridge, Mass.)* **78**, 1017–1025.

Heikenheimo, M., Scandrett, J. M., and Wilson, D. B. (1994). Localization of transcription factor GATA-4 to regions of the mouse embryo involved in cardiac development. *Dev. Biol.* **164**, 361–373.

Hermez, E., Mackern, S., and Mahon, K. A. (1996). *Rpx*: A novel anterior-restricted homeobox gene progressively activated in the prechordal plate, anterior neural plate and Rathke's pouch of the mouse embryo. *Development (Cambridge, UK)* **122**, 41–52.

Herrmann, B. G. (1991). Expression pattern of the *Brachyury* gene in the whole mount *Twis/Twis* mutant embryos. *Development (Cambridge, UK)* **113**, 913–917.

Illmensee, K., and Mahowald, A. P. (1974). Transplantation of posterior polar plasm in *Drosophila*. Induction of germ cells at the anterior pole of the egg. *Proc. Natl. Acad. Sci. U.S.A.* **71**, 1016–1020.

Isaacs, H. V. (1997). New prespectives on the role of the fibroblast growth factor family in amphibian development. *Cell. Mol. Life Sci.* **53**, 350–361.

Isaacs, H. V., Pownall, M. E., and Slack, J. M. W. (1994). eFGF regulates *Xbra* expression during *Xenopus* gastrulation. *EMBO J.* **13**, 4469–4481.

Jeffery, W. R., Bates, W. R., Beach, R. L., and Tomlinson, C. R. (1986). Is maternal mRNA a determinant of tissue-specific proteins in ascidian embryos? *J. Embryol. Exp. Morphol.* **97**(Suppl.) 1–14.

Johnson, M. H., Chisholm, J. C., Fleming, T. P., and Houliston, E. (1986). A role for cytoplasmic determinants in the developing mouse embryo? *J. Embryol. Exp. Morphol.* **97**(Suppl.) 97–121.

Lawson, K. A., and Hage, W. J. (1994). Clonal analysis of the origin of primordial germ cells in the mouse. *In* "Germ Line Development," Ciba Foundation Symp. 182, pp. 68–92. Wiley, Chichester.

Lawson, K. A., and Pedersen, R. A. (1987). Cell fate, morphogenetic movement and population kinetics of embryonic endoderm at the time of germ layer formation in the mouse. *Development (Cambridge, UK)* **101**, 627–652.

Lawson, K. A., and Pedersen, R. A. (1992). Clonal analysis of cell fate during gastrulation and early neuralation in the mouse. *In* "Formation and Differentiation of Early Embryonic Mesoderm," NATO ASI Series, Vol. 231, pp. 33–46. Plenum, New-York.

Lawson, K. A., Meneses, J. J., and Pedersen, R. A. (1986). Cell fate and lineage in the endoderm of the presomite embryo, studied with an intracellular tracer. *Dev. Biol.* **115**, 325–339.

Lawson, K. A., Meneses, J. J., and Pedersen, R. A. (1991). Clonal analysis of epiblast fate during germ layer formation in the mouse embryo. *Development (Cambridge, UK)* **113**, 891–911.

Lemaire, L., Roeser, T., Izpisua-Belmonte, J. C., and Kessel, M. (1997). Segregating expression domains for two *goosecoid* genes during the transition from gastrulation to neurulation in chick embryos. *Development (Cambridge, UK)* **124**, 1443–1452.

Lints, T. J., Parson, L. M., Hartley, L., Lyons, I., and Harvey, R. P. (1993). Nkx-2.5, a novel murine homeobox gene expressed in early heart progenitor cells and their myogenic descendants. *Development (Cambridge, UK)* **119**, 419–431.

Lumsden, A., and Krumlauf, R. (1996). Patterning the vertebrate neuraxis. *Science* **274**, 1109–1115.

Mansour, S. I., Goddard, J. M., and Capecchi, M. R. (1993). Mice homozygous for a targeted disruption of the proto-oncogene *int-2* have developmental defects in the tail and inner ear. *Development (Cambridge, UK)* **117**, 13–28.

Matsuo, I., Kuratani, S., Kimura, C., Takeda, N., and Aizawa, S. (1995). Mouse Otx2 functions in the formation and patterning of rostral head. *Genes Dev.* **9**, 2646–2658.

Niswander, L., and Martin, G. R. (1992). Fgf-4 expression during gastrulation, myogenesis, limb and tooth development in the mouse. *Development (Cambridge UK)* **114**, 755–768.

Orr-Utreger, A., Givol, D., Yayon, A., Yarden, Y., and Lonai, P. (1991). Developmental expression of two murine fibroblast growth factor receptors, *flg* and *bek. Development (Cambridge, UK)* **113**, 1419–1434.

Parameswaran, M., and Tam, P. P. L. (1995). Regionalisation of cell fate and morphogenetic movement of the mesoderm during mouse gastrulation. *Dev. Genet.* **17**, 16–28.

Pedersen, R. A. (1986). Potency, lineage, and allocation in preimplantation mouse embryos. *In* "Experimental Ap-

proaches to Mammalian Embryonic Development" (R. A. Pedersen and J. Rossant, eds.), pp. 3–34. Cambridge Univ. Press, Cambridge.

Pera, E. M., and Kessel, M. (1997). Patterning of the chick forebrain analage by the prechordal plate. *Development (Cambridge, UK)* **124**, 4153–4162.

Poelmann, R. E. (1981). The formation of the embryonic mesoderm in the early postimplantation mouse embryo. *Anat. Embryol.* **162**, 29–40.

Quinlan, G. A., Williams, E. A., Tan, S.-S., and Tam, P. P. L. (1995). Neurectodermal fate of epiblast cells in the distal region of the mouse egg cylinder: Implication for body plan organization during early embryogenesis. *Development (Cambridge, UK)* **121**, 87–98.

Rosenquist, T. A., and Martin, G. R. (1995). Visceral endoderm-1 (VE-1): An antigen marker that distinguishes anterior from posterior embryonic visceral endoderm in the early post-implantation mouse. *Mech. Dev.* **49**, 117–122.

Sasai, Y., and De Robertis, E. M. (1997). Ectodermal patterning in vertebrate embryos. *Dev. Biol.* **182**, 5–20.

Sasaki, H., and Hogan, B. L. M. (1993). Differential expression of multiple forkhead related genes during gastrulation and axial pattern formation in the mouse embryo. *Development (Cambridge, UK)* **118**, 47–59.

Schubert, F. R., Fainsod, A., Gruenbaum, Y., and Gruss, P. (1995). Expression of a novel homeobox gene Sax1 in the developing nervous system. *Mech. Dev.* **51**, 99–114.

Schmidt, C., Wilson, V., Stott, D., and Beddington, R. (1997). T promoter activity in the absence of functional T protein during axis formation and elongation in the mouse. *Dev. Biol.* **189**, 161–173.

Shawlot, W., and Behringer, R. R. (1995). Requirement for *Lim1* in head-organizing function. *Nature (London)* **374**, 425–430.

Smith, J. D., Micheal, P., and Williams, M. A. (1983). Does a predetermined germ line exist in amphibian? *In* "Current Problems in Germ Cell Differentiation" (A. McLaren and C. C. Wylie, eds.), pp. 19–40. Cambridge Univ. Press, London.

Smith, J. L., Gesteland, K. M., and Schoenwolf, G. C. (1994). Prospective fate map of the mouse primitive streak at 7.5 days of gestation. *Dev. Dyn.* **201**, 279–289.

Strome, S. (1986). Asymmetric movement of cytoplasmic components in *Caenorhaditis elegans* zygotes. *J. Embryol. Exp. Morphol.* **97**, (Suppl.) 1–15.

Takada, S., Stark, K. L., Shea, M. J., Vassileva, G., McMahon, J. A., and McMahon, A. P. (1994). *Wnt-3a* regulates somite and tail bud development in the mouse embryo. *Genes Dev.* **8**, 174–189.

Tam, P. P. L., and Beddington, R. S. P. (1987). The formation of mesodermal tissues in the mouse embryo during gastrulation and early organogenesis. *Development (Cambridge, UK)* **99**, 109–126.

Tam, P. P. L., and Beddington, R. S. P. (1992). Establishment and organization of germ layers in the gastrulating mouse embryo. *In* "Postimplantation Development in the Mouse," Ciba Foundation Symp. Vol. 165, pp. 27–49. Wiley, Chichester.

Tam, P. P. L., and Behringer, R. R. (1997). Mouse gastrulation: The formation of the mammalian body plan. *Mech. Dev.* **68**, 3–25.

Tam, P. P. L., and Zhou, S. X. (1996). The allocation of epiblast cells to ectodermal and germ-line lineages is influenced by the position of the cells in the gastrulating mouse embryo. *Dev. Biol.* **178**, 124–132.

Tam, P. P. L., Williams, E., and Chan, W. Y. (1993). Gastrulation in the mouse embryo: Ultrastructural and molecular aspects of germ layer morphogenesis. *Microsc. Res. Tech.* **26**, 301–328.

Tam, P. P. L., Parmeswaran, M., Kinder, S. K., and Weinberger, R. P. (1997). The allocation of epiblast cells to the embryonic heart and other mesodermal lineages: The role of ingression and tissue movement during gastrulation. *Development (Cambridge, UK)* **124**, 1631–1642.

Thomas, P., and Beddington, R. (1996). Anterior primitive endoderm may be responsible for patterning the anterior neural plate in the mouse embryo. *Curr. Biol.* **6**, 1487–1496.

Varlet, I., Collignon, J., and Robertson, E. J. (1997). *nodal* expression in the primitive endoderm is required for specification of the anterior axis during mouse gastrulation. *Development (Cambridge, UK)* **124**, 1033–1044.

Wilkinson, D. G., Peters, G., Dickson, C., and McMahon, A. P. (1988). Expression of the FGF-related proto-oncogene *int-2* during gastrulation and neuralation in the mouse. *EMBO J.* **7**, 691–695.

Wilkinson, D. G., Bhatt, S., and Herrmann, B. (1990). Expression pattern of the mouse *T* gene and its role in mesoderm formation. *Nature (London)* **343**, 657–659.

Wilmut, I., Schneike, A. E., McWhir, J., Kind, A. J., and Campbell, K. H. S. (1997). Viable offspring derived from fetal and adult mammalian cells. *Nature (London)* **385**, 810–813.

Wilson, V., and Beddington, R. S. P. (1996). Cell fate and morphogenetic movement in the late mouse primitive streak. *Mech. Dev.* **55**, 79–90.

Yamaguchi, T. P., Conlon, R. A., and Rossant, J. (1992). Expression of the fibroblast growth factor receptor FGFR-1/*flg* during gastrulation and segmentation in the mouse embryo. *Dev. Biol.* **152**, 75–88.

Yamaguchi, T. P., Dumont, D. J., Conlon, R. A., Breitman, M. L., and Rossant, J. L. (1993). *flk-1*, and *flt*-related receptor tyrosine kinase is an early marker for endothelial cell precursors. *Development (Cambridge, UK)* **118**, 489–498.

Yamaguchi, T. P., Harpal, K., Hekenmeyer, M., and Rossant, J. (1994). *fgfr-1* is required for embryonic growth and mesodermal patterning during mouse gastrulation. *Genes Dev.* **8**, 3032–3044.

Yeom II, Y., Fuhrmann, G., Ovitt, C. E., Brehm, A., Ohbo, K., Gross, M., Hübner, K., and Schöler, H. R. (1996). Germline regulatory elements of Oct-4 specific for the totipotent cycle of embryonal cells. *Development (Cambridge, UK)* **122**, 881–894.

Yoshikawa, Y., Fujimori, T., McMahon, A., and Takada, S. (1997). Evidence that *Wnt-3a* signaling promotes neuralisation instead of paraxial mesoderm development in the mouse. *Dev. Biol.* **183**, 234–242.

34

Early Cell Lineages in Marsupial Embryos

LYNNE SELWOOD
DANIELLE HICKFORD
Department of Zoology
La Trobe University
Bundoora, Victoria 3083,
Australia

I. Introduction

In marsupials, allocation to the first three lineages occurs in a unilaminar structure, the unilaminar blastocyst. Also, axis formation and gastrulation occurs before the embryo is implanted. Marsupials therefore provide a unique opportunity to explore the evolution of early lineage allocation in mammals and to identify signals that are extrinsic or intrinsic to the embryo during axis formation. Since the early 1970s, embryonic development has been examined in a number of marsupials. Of these, three marsupials from the family Dasyuridae, *Sminthopsis macroura*, the stripe-faced dunnart, *Sminthopsis crassicaudata*, the fat-tailed dunnart, and *Antechinus stuartii*, the brown antechinus; *Macropus eugenii*, the tammar wallaby, from the Macropodidae; and *Monodelphis domestica*, the gray short-tailed opossum, from the Didelphidae, have received the most attention (for review, see Selwood, 1992, 1994). Studies on these animals have widened our knowledge of marsupial embryonic development, which until about 1980 was almost entirely confined to studies on *Dasyurus viverrinus*, the native cat (Hill, 1910), *Didelphis virginiana*, the Virginia opossum (Hartman, 1916, 1919; Krause and Cutts, 1985a,b), and the bandicoots *Perameles nasuta* and *Isoodon macrourus* (Lyne and Hollis, 1976, 1977; Hollis and Lyne, 1977). The more recent studies have provided some of the basic requirements for experimental analysis

of marsupial lineages, namely, a knowledge of the normal timetable of development, methods for examination of development *in vitro*, and development of manipulative and marking techniques. All these requirements have been met in the studies of the stripe-faced dunnart.

The stripe-faced dunnart, S. *macroura* (Marsupialia, Dasyuridae) (Fig. 1), is widespread throughout the arid and semiarid regions of Australia. It is small, 15–25 g at maturity, and insectivorous, and the breeding is seasonal in the wild (Morton, 1983) and in the laboratory. It is ideal for embryological studies because of its small size, previous studies on its reproduction (Godfrey, 1969; Woolley, 1990a,b; Selwood and Woolley, 1991), and knowledge of its embryological development (see below).

We have maintained a fully pedigreed laboratory colony of the stripe-faced dunnart since 1985. This would allow identification of genetically divergent strains should they appear. The original stock of the colony was 16 animals derived from a colony initiated by Dr. Meredith Smith at The Evolutionary Biology Unit, South Australian Museum in 1984 and comprised mainly animals from South Australia. Other animals were obtained from Dr. D. Evans from a colony started by Dr. Woolley at Latrobe University in 1980 and subsequently maintained by Dr. Evans, at the Arthur Rylah Research Institute in Victoria. The latter colony was reproductively moribund in 1985 and only two of 21 animals obtained from Dr. Evans bred and contributed

Figure 1 *Sminthopsis macroura,* the stripe-faced dunnart, ×1.

stock to the colony in the subsequent year. Their descendants did not survive past 1987. Few laboratory colonies of small marsupials survive more than 5 years because of decline in reproductive performance (Godfrey, 1975). One wild-caught animal from Alice Springs, Northern Territory, was incorporated into the colony in 1990, and since then, we have maintained two colony strains. One strain, the Laboratory Colony Strain, was not interbred with the Alice Springs animal or her descendants and hence has a colony age of 14 years from wild-caught stock. The other, the Alice Springs strain, was and hence has a colony age of 8 years since the introduction of wild-caught stock. Only three, fully pedigreed colonies of small marsupials have been established and maintained for more than 10 years, world-wide. Apart from the stripe-faced dunnart colony, these are the colony of the fat-tailed dunnart, set up by Professor Bennett at the Department of Genetics, Adelaide University (Bennett *et al.,* 1990) and the colony of the gray short-tailed opossum set up by Dr. J. VandeBerg at the Southwest Foundation for Biomedical Research, San Antonio, Texas (VandeBerg, 1990). Of these three colony animals, the most suitable animal for embryological research at present is the stripe-faced dunnart.

II. The Stripe-Faced Dunnart, a Model for Marsupial Embryological Studies

Several reasons contribute to making the stripe-faced dunnart eminently suitable as an embryological model.

It has the most detailed and comprehensive timetable of development (Selwood and Smith, 1990; Selwood and Woolley, 1991) for any small marsupial. Most is known about its development *in vitro* (Selwood, 1987; Selwood and Smith, 1990; Gardner *et al.,* 1996), and it can be cultured to within 18 hours of birth (Yousef and Selwood, 1993). It is polytocous (many offspring), and most experimental manipulations and analyses of marsupial embryonic development have been made using this animal (Selwood, 1989a,b; Selwood and Smith, 1990; Yousef and Selwood, 1996; Cruz *et al.,* 1996).

Its greatest advantage as a model is that known-age embryonic stages can be collected at intervals from time 0, defined as the time of minimum morning weight and marking the beginning of development (Selwood and Woolley, 1991). Time 0 is determined as shown in Fig. 2. Weight changes follow a fairly consistent profile during estrus and gestation (Woolley, 1990a). The day of parturition is associated with a fall in weight, red blood cells and yellow uterine secretions in the urine, clear secretion around the nipples, and the presence of pouch young (PY) in the pouch. Except for the appearance of PY, females that are not pregnant follow a similar pattern of weight changes, appearance of the pouch, and material in the urine.

In recent years the timetable of development has been refined so that specific stages can be collected at 12-hour intervals (Table I). Time 0 ± 12 hours is the time of fertilization and maturing oocytes in the ovary are collected on the day before this. Except for the 4-cell stage, which lasts for 34 hours, other cleavage cycles last for 8–10 hours (Selwood and Smith, 1990) so that cleavage is completed in 2.5 days. The duration of

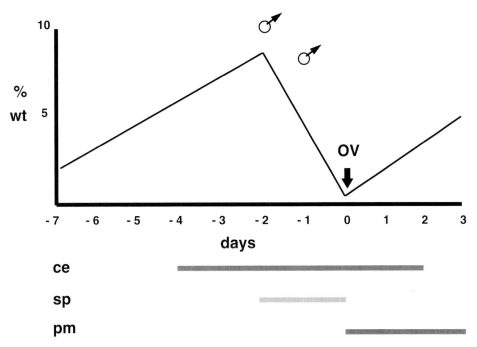

Figure 2 An idealized profile of changes in weight (wt) and cells in the urine during the estrous and early gestation period (days). Females are monitored during the breeding season to establish the initiation and length of the estrous cycle, whether females have mated, and the day of fertilization. Monitoring consists of weighing the females at the beginning of each day between 0800 and 1100 hours, collection of a urine sample, and examination of the pouch. The urine sample is examined microscopically at 100× to establish the presence and abundance of cornified epithelial cells (ce), polymorphonuclear leucocytes (pm), and spermatozoa (sp) (Selwood, 1980; Selwood and Woolley, 1991). The onset of estrous is associated over 4–5 days with a mean increase in weight of 9.5% and the appearance and increase in abundance of cornified epithelial cells. Weight falls over the next 2 days to the minimum weight at time 0. Time 0 also is associated with ovulation (ov), the appearance in the urine of polymorphonuclear leucocytes, and sometimes spermatozoa from the previous mating. Animals are paired and mating occurs around the time of peak weight and usually when epithelial cells are abundant in the urine (Selwood and Woolley, 1991). Weight increases during the first few days of gestation.

each stage *in vivo* and *in vitro* are shown in Table II. Implantation occurs on the 9th day when the embryo has 13–22 somites. The remainder of embryogenesis and fetal development takes 1.5 days. Because a suitable culture system has been developed for many of the developmental stages, stages that are earlier than the precise stage required can be cultured to that stage. The gestation period is 10.7 days ± 0.7 (S.D.) (Selwood and Woolley, 1991), which is the shortest for any mammal. Cleavage is uterine, asynchronous, and unequal at the determinate division. Early cleavage is associated with deutoplasmolysis (elimination of yolky cytoplasm) of either a large single "yolk mass" and other vesicular products or several smaller cytoplasmic bodies or yolk masses. Much of the yolk mass is extracellular matrix material. Because cell–zona adhesion precedes cell–cell adhesion, no morula is formed and cleavage results in a unilaminar blastocyst. Both pluriblast and trophoblast lie on the surface. The unilaminar blastocyst persists for several days (Tables I and II) and is transformed into a bilaminar blastocyst with the appearance of the hypoblast. Gastrulation and the trilaminar blastocyst stage is relatively rapid, as is embryogenic and fetal development (Tables I and II).

The class Mammalia has three subclasses: Monotremata, Metatheria (marsupials), and Eutheria (other mammals such as the mouse and humans). In general, embryonic manipulations are more difficult in marsupial embryos than in many eutherian embryos because of the presence of the mucoid and shell coats. The mucoid coat disappears during blastocyst expansion and the shell coat is lost just prior to implantation. The shell coat can be punctured by small bore needles of 1–2 μm outer diameter and the conceptuses will develop normally *in vitro* for several days at least (Selwood, 1986; Cruz *et al.*, 1996), but if a wider bore is used or the puncture is carelessly made, the shell is torn and the embryo collapses. This seriously limits the type of experiments that involve insertion or removal of cells, or complicated marking procedures requiring several puncture holes. The egg coats provide a framework for normal blastocyst construction and the separation of pluriblast and trophoblast cells during cleavage. In the stripe-faced dunnart and the brown antechinus, the cells of cleavage stages when removed from the coats undergo similar activities at each cleavage stage as in intact embryos (Selwood, 1989b). The nature of these activities, which include cell mobility and cell–substrate followed by cell–cell adhesion, mean that a normal blastocyst does not form once the framework is removed. In marsupials such as the tammar wallaby and the gray short-tailed opossum, which show a differ-

TABLE I

Normal Stages of Development of *Sminthopsis macroura*[a]

Day	Description of stage
−1.0	Preovulation; GV, M1, M2
−0.5	Preovulation, M1, M2; ovulation
Time 0	
0.5	Tubal oocytes; zygotes; uterine pronuclear zygotes (sticky with varying polarity)
1.0	Polarized zygote; 2-cell (early constriction of YM) (YM emission); 2-cell to 4-cell (shaggy, in division)
1.5	2- to 4-cell; 4-cell (rounded and in arrest)
2.0	4-cell (rounded); 4-cell (flattened on zona); 4- to 8-cell stage
2.5	4- to 8-cell stage; 8-cell; 16-cell (8 trophoblast, 8 pluriblast cells)
3.0	32-cell (complete 0.34 mm diameter UB); 32- to 64-cell (pluriblast and trophoblast visible)
3.5	Expanded spherical, 0.35 mm diameter UB (~64-cell, early preliminary expansion compressing mucoid layer, cells flattening, pluriblast and trophoblast visible)
4.0	Spherical, 0.4–0.5 mm diameter UB (definitive expansion, ~128 cells); spherical or egg-shaped UB (0.55 mm average diameter ~200 cells, pluriblast and trophoblast visible)
4.5	Egg-shaped UB, average diameter 0.6–0.7 mm (200–400 cells, no obvious pluriblast or trophoblast)
5.0	Egg-shaped UB, average diameter 0.7–0.9 mm (400–800 cells); spherical, 0.8 mm diameter UB (500–1000 cells)
5.5	Spherical, 0.8–1.2 mm diameter UB (1000–1500 cells, no obvious pluriblast and trophoblast)
6.0	Spherical, 1.0–1.4 mm diameter, early BB (1500–2500 cells, pluriblast and trophoblast variably distinct, YM visible and associated with pluriblast hemisphere)
6.5	Spherical, incomplete BB, 0.9–2.1 mm diameter (2500–4000 cells, epiblast, trophoblast, and YM distinct)
7.0	Incomplete to complete BB, 1.8–3.0 mm diameter (4500–7000 cells, round epiblast, YM loss in some blastocysts)
7.5	Complete BB, 2.2–3.0 mm diameter (oval epiblast, prestreak, YM lost, hypoblast epithelial in embryonic hemisphere, and a reticulum in abembryonic hemisphere)
8.0	Ovoid TB (2.5 by 2.7 mm to 4.0 by 4.6 mm diameter, pear-shaped epiblast, primitive streak later with node and notochord, hypoblast an epithelium in abembryonic hemisphere
8.5	Early somite (1–15), flat neural plate embryo stage (embryo GL 3.9–4.8 mm on 3.5–4.5 mm blastocyst; lateral heart tubes variably visible, head fold of proamnion)
9.0	Neurula flat embryo stage to late somite stage, CR 2.4 mm (bilaminar yolk sac loosely adherent to uterus; blood visible in heart)
9.5	Advanced embryos C-shaped, others partially flat, CR 4.3 mm (forelimb buds, complete amnion, anterior intestinal portal clear)
10.0	Embryo, C-shaped, lying on side, CR 2.0–3.0 mm, forelimbs digitiform, allantoic bud
10.5	Fetus, CR 3.4–4.9 mm, HL 1.3–1.9 mm.
11.0	Fetus, CR 5.2–5.4 mm, HL 2.0 mm, claws on forelimb, neonate PY

[a]This represents an expansion of the data found in Selwood and Woolley (1991) and also includes information in Yousef and Selwood (1993) and from L. Selwood and D. Hickford (1996, unpublished). To simplify the timetable, time 0 is indicated and the previous 24 hours is called Day −1.0. Further description of stages is added in parentheses. BB, Bilaminar blastocyst; CR, crown–rump measurement; GL, greater length measurement; HL, head length measurement; GV, germinal vesicle; M1, Meiosis I; M2, Meiosis 2; PY, pouch young; TB, trilaminar blastocyst; UB, unilaminar blastocyst; YM, yolk mass.

TABLE II

Duration of Each Stage *in Vivo* and *in Vitro* in the Stripe-Faced Dunnart[a]

Stage	Duration (hours)	
	In vivo	*In vitro*
Cl	60	71
UB	60–72	Incomplete culture (~6 hours)
BB	48	Incomplete culture (~20 hours)
TB	12	16
E	24	28
F	36	Culture to within 18 hours of birth

[a]Stages are defined as Cl (cleavage, uterine zygote to 32-cell stage), UB (unilaminar blastocyst, preliminary and definitive expansion), BB (bilaminar blastocyst, first hypoblast cells to complete BB), TB (trilaminar blastocyst, early streak to early neurula), E (embryo, flat early neurula to C-shaped embryo with fore- and hind-limb buds) and F (fetus).

ent cleavage pattern than the dasyurids and no evidence of cell mobility before cell–zona attachment, intact mucoid and shell coats are not essential for blastocyst formation (Moore and Taggart, 1993; Renfree and Lewis, 1996) but may be required for normal flattening of the blastocyst epithelium and blastocyst expansion.

Marsupial embryos can be used to great advantage in cell lineage studies in mammals because the conceptus lies free for most of development. In discussion of lineages the terminology suggested by Johnson and Selwood (1996), which is applicable to marsupial and eutherian mammals, will be used. The pluriblast, the cell population that will subsequently give rise to the embryo and extraembryonic membranes (Johnson and Selwood, 1996), and trophoblast, the first nutritive epithelium, are unilaminar epithelia. Because of this and the transparency of the egg coats, establishment of cell lineages during cleavage can be followed by time-lapse cinematography, without cell marking (Selwood and Smith, 1990). Formation of the hypoblast, gastrulation, and embryogenesis also can be readily studied because implantation does not occur until late in development so that the conceptus lies free in the uterus. The study of embryogenesis and early fetal development is facilitated by the relative ease of culture during these stages (Table II) and because implantation is superficial for most of this period. This makes the stripe-faced dunnart an excellent model for studies of organogenesis and of expression of *HOX* genes. Like all marsupials, the stripe-faced dunnart has accelerated and precocious development of anterior structures related to gaining access to the pouch, suckling, digestion, and respiration (Gemmell and Selwood, 1994).

III. Lineages

A. The Significance of Cleavage

Cleavage is the first of the many mesenchymal–epithelial transformations that characterize embryonic development. This is particularly obvious during cleavage in marsupials and a few primitive insectivores in which an unilaminar epithelium is built up from the very early cleavage stages (Selwood, 1992). It is less obvious but still true in many eutherian mammals, such as the mouse, that pass through a morula stage (Fleming and Johnson, 1988).

In eutherians, the morula represents the early stages of formation of a multilayered epithelium, which, during cavitation is transformed into the blastocyst. Extensive experimental evidence and the more recent transgenic embryo technology has shown that the embryos of eutherian mammals are regulative (see review by Pedersen, 1986). The evidence is strongest for the mouse but is also available for other mammals. Much of this research supports the "inside–outside hypothesis," that position in the mammalian embryo determines cell fate, instead of uneven distribution of maternal determinants (see review, Johnson *et al.*, 1986). A dependence on positional effects rather than uneven distribution of maternal determinants for cell determination is not so surprising in mammals, when one considers their amniote ancestry. The blastoderm and epiblast of the bird embryo is also highly regulative (Eyal-Giladi, 1984). The possibility of uneven distribution of determinants cannot be eliminated, but if they are present and play a role in determination, the embryo is capable of regulation for a disturbance of normal pattern when cells are experimentally added to or deleted from the conceptus.

Common themes emerge when the developing blastoderms of birds, reptiles, monotremes, and euthe-

rian mammals are compared with the developing blastocyst epithelium of marsupials (Selwood, 1994). In all of them, similar features are present that could generate diversity in the developing epithelium (Fig. 3). In an epithelium that is capable of regulation, but which later gives rise to a series of different lineages, such mechanisms are necessary. If the cells in an epithelium can regulate for loss or addition of cells, then all the cells are assumed to be pluripotent, at least. Differences that later appear between the cell types, such as when a new lineage appears, form as a result of diversity mechanisms within the epithelia or by localized application of some external influence, to change the potency of some cells. Features common to the developing epithelia and potentially capable of generating diversity are different topographic cell associations, such as inner and outer cells in the morula or central and peripheral cells in a unilaminar epithelium, asynchrony in cell division order, and oriented division planes. Additional external factors that could create diversity are conditions in the reproductive tract. Gravity and rotation of the egg in the oviduct have been implicated in hypoblast formation in birds (Eyal-Giladi, 1984; Olszanska *et al.*, 1984).

B. The First Dichotomy: Pluriblast–Trophoblast

1. Conceptus Polarity

The separation of the first two lineages in marsupials is preceded by variable degrees of polarity in the conceptus. Zygote polarity may be preceded by polarization of the maturing oocyte as it is in the oocytes of many dasyurids including the stripe-faced dunnart (see Selwood, 1992, for review). In the dunnart and brown antechinus second polar body emission and the sperm entry point occur in the hemisphere opposite the accumulating vesicles of the yolk mass. After fertilization, the sperm entry point is the site of formation of a large aster which further polarizes the zygote. In the gray short-tailed opossum oocyte, the position of the nucleus is polarized, but cytoplasmic polarity is conferred by sperm entry and the formation of a large aster (Breed *et al.*, 1994; see also Chapter 20 by Sullivan *et al.*). Polarization in these conceptuses is related to the early polarized emission of cellular material into the perivitelline space. Polarity in marsupial oocytes and conceptuses is manifested in a number of ways (Table III) including nuclear position, distribution of other organelles, features of the plasma membrane, cell–zona relationships, and possibly the point of sperm entry. The position of the polar bodies does not seem to be consistent after fertilization (Hill, 1910). Possibly the polar bodies move within the perivitelline space as they do in some mouse strains (Sutherland *et al.*, 1990), so that their

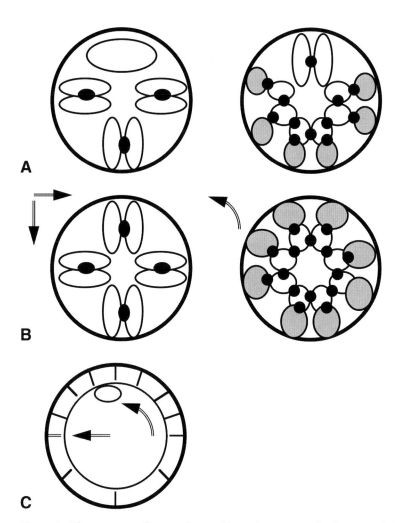

Figure 3 The various mechanisms that could give diversity to a developing epithelium during cleavage in dasyurids. (A) Asynchrony in cell division order at the third and fourth division in a dasyurid conceptus viewed from the upper pole. The cells that divide first have more cell–cell contacts than those that divide last at each division. This mechanism is capable of generating diversity in the pluriblast (inner tier of cells) and the trophoblast (shaded) (outer tier of cells). (B) Oriented division planes in a dasyurid conceptus viewed from the upper pole. Meridional division planes (straight arrows) produce eight similar cells at the third division. Latitudinal division planes (curved arrow) produce two cell types at the 16-cell stage, a tier of smaller pluriblast and one of larger trophoblast (shaded) cells. (C) Topographical relationship in an epithelium of a dasyurid blastocyst viewed here in transverse section. A radial division plane (straight arrow) lies parallel to a radius of the conceptus and contributes cells to the surface epithelium. A circumferential division plane (curved arrow) lies parallel to the circumference of the conceptus and contributes cells to inside the surface epithelium, shown here as an oval cell lying inside the epithelium. In addition, in the conceptuses at the 14-cell stage for A and 16-cell stage for B, the pluriblast cells at the center of the developing epithelium have earlier and more contacts than cells of the more peripheral trophoblast.

original location is impossible to determine and their position at particular cleavage stages varies from individual to individual.

In cleavage stages before allocation into pluriblast and trophoblast, polarity is expressed in each of the cells in a similar manner to that found in the zygote and the conceptus itself is frequently polarized (Table III). The extent of polarity and the features manifested vary from stage to stage and species to species.

2. Separation of Pluriblast and Trophoblast

The topographical relationship of cells in the developing epithelium is correlated with commitment to cell fate in marsupials (Selwood, 1994). The first dichotomy separates the totipotent cell population (polarblasts) into a nutritive pluripotent cell lineage (trophoblast) and a totipotent or pluripotent cell lineage (pluriblast) that will give rise to the embryo proper and the extraembryonic membranes (Fig. 4). The epithelium forms by spreading from a central focal area toward the periphery, so that it progressively lines the zona pellucida to enclose the cleavage cavity (Fig. 4). In the stripe-faced dunnart, the dichotomy into pluriblast and trophoblast occurs at the 16-cell stage, as it does in other dasyurid marsupials and the gray short-tailed opossum. In the Virginia opossum, the dichotomy occurs at the 2-cell stage. The difference in timing of the dichotomy is presum-

TABLE III

Polarized Features in Marsupial Zygotes and Conceptuses before the First Lineage Allocation[a]

Features	Oocyte	Zygote	Conceptus	Species
Polarized position of nucleus in cell(s)	+	+	+(2–8 c)	*A. stuartii*
	+	+	+(2–8 c)	*S. macroura*
	+	?	+(2–8 c)	*S. crassicaudata*
	+	+	+(2–8 c)	*D. viverrinus*
	+	?	?	*M. domestica*
	+	+	?	*D. virginiana*
	+	+	+(2–8 c)	*M. eugenii*
Cell–zona association	–	–	+(4 c)	*A. stuartii*
	–	–	+(2 c)	*S. macroura*
	–	–	+(4 c)	*S. crassicaudata*
	–	–	+(4 c)	*D. viverrinus*
	–	–	+(4 c)	*M. domestica*
	–	–	+(2 c)	*D. virginiana*
	–	–	+(4 c)	*M. eugenii*
Polarized location of organelles in cell(s)	+	+	+(2–8 c)	*A. stuartii*
	+	+	+(2–8 c)	*S. macroura*
	+	+	+(2–8 c)	*S. crassicaudata*
	+	+	+(2–8 c)	*D. viverrinus*
	–	+	?	*M. domestica*
	–	+	+(1 c)	*D. virginiana*
	–	–	+(4–8 c)	*M. eugenii*
Asymmetric mucoid coat	n.a.	–	–	*A. stuartii*
	n.a.	–	–	*S. macroura*
	n.a.	–	–	*S. crassicaudata*
	n.a.	–	–	*D. viverrinus*
	n.a.	+	+	*M. domestica*
	n.a.	–	+	*D. virginiana*
	n.a.	–	–	*M. eugenii*
Emission(s) into cleavage cavity	–	+(p)	+(2–4c, p)	*A. stuartii*
	–	+(p)	+(2–4c, p)	*S. macroura*
	–	+(p)	+(2c–?, p)	*S. crassicaudata*
	–	+(p)	+(2c–?, p)	*D. viverrinus*
	+	+	+(2–4c, p)	*M. domestica*
	+	+(p)	+(1c–?, p)	*D. virginiana*
	–	–	+	*M. eugenii*

[a]References as follows: *Antechinus stuartii* (Selwood and Smith, 1990; Selwood and Sathananthan, 1988; Sathananthan *et al.*, 1997), *Sminthopsis macroura* (Selwood and Smith, 1990; Merry *et al.*, 1995), *S. crassicaudata* (Selwood, 1987; Breed and Leigh, 1990; Breed, 1996), *Dasyurus viverrinus* (Hill, 1910), *Monodelphis domestica* (Baggott and Moore, 1990; Moore, 1996; Selwood *et al.*, 1997), *Didelphis virginiana* (McCrady, 1938), *Macropus eugenii* (Renfree and Lewis, 1996). The cell stages in which the features are present or first identified are shown in parentheses. p, Polarized; c, cell, n.a., not applicable.

ably because either determinants are distributed unevenly at two different stages or because mechanisms to separate the lineages operate at two different stages. Prior to separation into the two cell types in the dasyurids (Fig. 4, Table IV), the cells are similar to each other in size, appearance *in vivo* and *in vitro,* and in their markedly polarized distribution of organelles along the embryonic/abembryonic axis. Such cells have been described as polarblasts (Johnson and Selwood, 1996) because of their polarized character and because they subsequently undergo a determinate division into two cell types.

The separation of pluriblast and trophoblast is characterized by a number of common features outlined in Fig. 5. These features are polarization of the zygote or early cleavage stages; polarized secretion of extracellular matrix (ECM) that prevents cell–cell and cell–zona associations; polarized secretion of lysins that facilitate cell–zona and cell–cell associations; and expansion of the blastocyst epithelium from the site of the initial cell–zona adhesion to eventually completely line the zona. Cells in the center of the developing epithelium (future pluriblast) have cell–cell contact with more neighbors than cells at the periphery (future trophoblast). This epithelial sheet represents a two-dimensional equivalent of what happens in the more three-dimensional spherical compacted mouse conceptus.

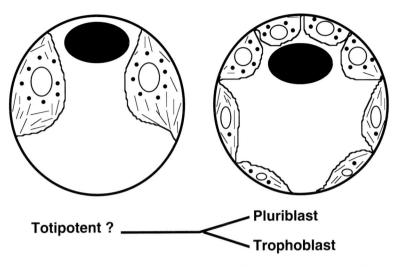

Figure 4 Sectional view of unilaminar blastocyst formation and separation of the pluriblast and trophoblast cells in *A. stuartii*. At the 8-cell stage on the left, the cells lie near the yolk mass (dark oval) in the upper region of the cleavage cavity. The cells (polarblasts) have a polarized distribution of organelles [nucleus (open circle), vesicular structures (closed dots), and fibrous arrays (lines)] along their long axis. The fourth division separates polarblasts into two cell types, the upper more rounded pluriblast cells which contain most of the vesicular material and lower, more flattened trophoblast cells that contain most of the fibrous arrays. The epithelium formed by these cells progressively spreads from the upper hemisphere (embryonic pole) to line the zona (thick circle). Bottom would represent the abembryonic pole.

3. Characteristics of Pluriblast and Trophoblast Lineages

The characteristics of pluriblast and trophoblast cells are shown in Table IV. It seems likely that the separation of pluriblast and trophoblast and many of the initial differences between the two cell types are attributable to maternal information. The polarized distribution of organelles is present as early as the zygote in some marsupials. Furthermore some studies suggest that the embryonic genome is not switched on until relatively late in marsupials. In the gray short-tailed opossum, an X-linked hypoxanthine phosphoribosyltransferase (HPRT) activity study (Johnston *et al.,* 1994) suggests the embryonic genome is not switched on until late cleavage. In the brown antechinus, nucleolar ultrastructural changes imply that the embryonic genome

is not switched on before the 8-cell stage (Sathananthan *et al.,* 1997). Light microscope observations suggest the stripe-faced dunnart is similar (L. Selwood, 1994, unpublished). Positional signals related to differences in cell–cell associations would be expected to operate during further differentiation of the two cell types. It would be interesting to determine whether disturbance of normal cell–cell associations interferes with the separation of pluriblast–trophoblast lineages.

Once separated, the pluriblast and trophoblast cells differentiate, their differences becoming more distinct over the next two cell divisions. The characteristics of the two populations have been most extensively investigated in the stripe-faced dunnart and the brown antechinus (Table IV). Because both cell types are epithelial in nature, they show a polarity relative to the zonal (apical) and basal surfaces of the epithelium. The relative differences between pluriblast and trophoblast are related to the relative differences in organelles along the embryonic–abembryonic axis of the 8-cell stage (Fig. 4). The changes in cell doubling time in the two cell types in the dunnart (Table IV) are very similar to what happens after pluriblast and trophoblast are first separated in the mouse. The initial retention of rapid divisions in trophoblast cells is presumably related to the importance of establishing the first nutritive lineage of the conceptus. It is not known whether pluriblast cells are capable of generating further trophoblast cells after the 16-cell stage.

Once the unilaminar blastocyst has formed and begun to expand, the pluriblast epithelium can regulate to replace blastomeres destroyed by a puncture (Selwood, 1986). Prior to this the epithelium does not

TABLE IV

Characteristics of Pluriblast and Trophoblast Cells in the Stripe-Faced Dunnart and Brown Antechinus

Cell characteristic	Cell type	
	Pluriblast	Trophoblast
Shape	Round	Long
Organelles	>Perinuclear vesicles, <fibers	<Perinuclear vesicles, >fibers
Zonal contact	Plugs, microvilli	Lamellipodia, plugs
Cell–cell contact	>16- to 32-cell stage	<16- to 32-cell stage
Cycling time (hours)		
<32-cell stage	8	8
>32-cell stage	24	8
In vitro	Stem cell-like	Granular

1. Oocyte or conceptus polarity:

2. Extra-cellular matrix secretion:

3. Lysins to facilitate cell-zona adhesion:

4. Cell-cell adhesion:

5. Cleavage ⟶ Unilaminar epithelium

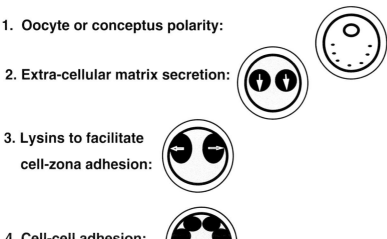

Figure 5 Steps leading to the separation of pluriblast and trophoblast lineages in marsupial conceptuses. 1. Polarity, reflected in nuclear position and distribution of organelles is found in the oocyte, zygote, or later cleaving conceptuses. 2. Extracellular matrix (ECM) secretion (white arrows), which may be polarized, separates cells from each other and from the zona pellucida (thick circle). 3. Polarized lysin secretion (dark arrows) facilitates cell–zona and later cell–cell adhesion (4). The order in which this occurs prevents a morula forming and results in a unilaminar epithelium. 5. Subsequent cleavages result in pluriblast and trophoblast formation (see Fig. 4) and a unilaminar epithelium which progressively lines the zona pellucida. Thin lined outer circle represents the mucoid and shell coats. Shading represents the cleavage cavity filled with ECM. Embryonic pole is to the top, abembryonic pole is to the bottom.

heal over when pluriblast or trophoblast are punctured. The latter result might be because insufficient time in culture was available for this healing to occur. If the pluriblast is similar to the germ of the bird, regulation to accommodate for loss or addition of blastomeres is likely to occur once the epithelium is established.

As the unilaminar blastocyst expands and the cells become extremely flattened, the morphological differences between trophoblast and pluriblast cells become less obvious (Selwood, 1994). This resulted in the mistaken assumption that the unilaminar blastocyst is a protoderm (primitive layer or undifferentiated primordial epithelium) (McCrady, 1938, 1944) in the Virginia opossum. Marked differences between the two cell types become obvious just prior to or coincident with the appearance of the hypoblast, the next major cell lineage. Cell counts made of blastocysts during this period of expansion suggest that different cell doubling times are maintained by pluriblast and trophoblast cells (Selwood, 1996). In the brown antechinus, cell doubling of pluriblast cells is arrested, while trophoblast cells divide extremely slowly over about 6 days. It is likely that variation in cell doubling time of pluriblast and trophoblast contribute to the different types of embryonic arrests found in the unilaminar blastocysts of marsupials. These arrests would vary from total arrest of pluriblast and trophoblast to arrest of pluriblast

and slowing of trophoblast or slowing down of both. When blastocyst cells are dissociated and examined *in vitro* pluriblast and trophoblast cells are distinctly different (Table IV). Trophoblast cells are larger, more granular, occasionally multinucleate, and without the stem cell-like appearance of the pluriblast cells (Yousef and Selwood, 1996).

C. The Second Major Dichotomy: Epiblast–Hypoblast

Little is known about this dichotomy in any marsupial. Hypoblast formation in marsupials occurs either in small blastocysts with few cells and a wide mucoid coat or large blastocysts with many cells and a narrow mucoid coat (Table V). It is not known why the hypoblast appears at different times in blastocyst development. Because the hypoblast lineage is involved in epiblast nutrition (Hollis and Lyne, 1977), its early appearance in blastocysts with an originally wide mucoid coat may be related to the nutritive requirements of these blastocysts. Its appearance in the smaller blastocysts is via the development of the enlarged endoderm mother cells, which are not found in the larger blastocysts.

The modes of hypoblast formation (Table V) can be classified as either direct or indirect (Selwood, 1992). In the direct formation found in larger blastocysts such as

TABLE V
Modes of Hypoblast Formation in Marsupials[a]

Species	Diameter (mm)	Cell number	Mode[b]	Initial mucoid width (μm)
D. virginiana	0.11	50–60	1b	140–230
M. rufogriseus	0.26	62	1b	60 (in diapause)
B. gaimardi	0.27	~80	2a	?
M. domestica	0.34	~100	1a	50
P. nasuta	0.80	~1000	2b	6.7
A. stuartii	1.3	~1000	2a	16.6
S. volans	1.5	>643	2a	5–28
S. macroura	1.3	2200	2a	7.9
D. viverrinus	4.5	2200	2a	15–22

[a]Features at the first appearance of hypoblast cells and the initial width of the mucoid coat are indicated.
[b]Hypoblast formation is indirect, via endoderm mother cells from either a distinct pluriblast (1a) or an indistinct pluriblast(1b) or direct from a distinct pluriblast (2a) or an indistinct pluriblast (2b).

in *S. macroura* (Fig. 6), the hypoblast cells develop directly from the pluriblast by either migration inwards or by oriented mitosis. The pluriblast may be distinct or indistinct at the time when hypoblast formation is initiated. Indirect formation occurs via large endoderm mother cells that subsequently generate hypoblast cells. Endoderm mother cells develop from indistinct and distinct pluriblasts in small blastocysts (Table V). The hypoblast is at first an open network but later consolidates to form a squamous epithelial inner lining of the blastocyst (Fig. 6), which is usually more flattened at the abembryonic pole of the blastocyst. Consolidation begins in the embryonic hemisphere and spreads to the abembryonic.

Where it has been studied in more detail, the pattern of appearance of the early stages of hypoblast cells can be correlated with events in cleavage. Such events include cell division order during the cleavage divisions and differences in cell–cell contacts in different parts of the pluriblast (Selwood, 1994). Thus in the Virginia opossum, in which cell–cell contacts initially are greatest around the edge of the developing pluriblast, endoderm mother cells first appear circumferentially. In the dunnart, in which cell–cell contacts initially are greatest on one side of the pluriblast, hypoblast cells first appear on one side of the pluriblast. Lineage tracing studies are needed to show whether this correlation is because of a causal relationship. Unlike in the mouse where the inner cell mass (pluriblast) is internal, all cells in the unilaminar external marsupial pluriblast have a similar exposure to both blastocoel and external conditions. Thus an inner position cannot account for hypoblast formation as it does in the mouse (Gardner, 1982). Some pre-pattern must be present in the marsupial pluriblast so that some cells are able to generate hypoblast cells either as a response to changes in the uterine milieu or as part of a developmental program. Studies *in vitro* of development of blastocysts or of dissociated blastocyst cells suggest that uterine signals may be required for hypoblast formation and proliferation (Yousef and Selwood, 1996).

D. Primary Germ Layers: Ectoderm, Endoderm, and Mesoderm and Primordial Germ Cell Lineages

The epiblast is initially round, without obvious signs of differentiation. It later becomes more opaque in an approximately triangular region at the circumference of the epiblast (Fig. 6) due to an increase in cell numbers in this region. The epiblast rapidly transforms to an oval and then pear shape with the long axis of the oval or the pear shape bisecting the apex of the triangular opacity (Fig. 6). The primitive streak is first visible near the opaque edge of the epiblast and extends forward from this region to the approximate center of the epiblast where the node develops. The appearance of the epiblast and streak formation are very similar to that in the chick.

Few studies have been made of gastrulation in marsupials. The only complete study of all stages of gastrulation is that by Kerr (1936). The field has been confounded for many years by the conclusions of McCrady (1938) for the Virginia opossum. He decided that the distinct embryonic area of the opossum blastocyst, which he called the "medullary plate," contained the presumptive neuroectoderm, mesoderm, and endoderm and that the embryonic ectoderm lay outside this region and was indistinguishable from the extraembryonic ectoderm. Despite the improbability of this lineage separation, in which neuroectoderm is allocated before mesoderm determination, and conclusions to the contrary for the Virginia opossum by Hartman (1919), his terminology was widely accepted (Selwood, 1992, for review). His hypothesis has been shown to be incorrect in the stripe-faced dunnart, by fate mapping studies of the epiblast (Cruz *et al.*, 1996). This study, which involved marking a small group of epiblast cells with 1,1′-dioctadecyl-3,3,3′,3′-tetramethylindocarbocyanine perchlorate (DiI) and following the fate of these cells in culture over 24 hours, clearly showed that the epiblast contained presumptive neuroectoderm and embryonic ectoderm. The latter was clearly demarcated from the trophoblast (extraembryonic ectoderm). A proposed lineage diagram of the dunnart is shown in Fig. 7. The three germ layers and extraembryonic mesoderm are derived from the epiblast. Ullmann *et al.* (1997) have shown that the primordial germ cell precursors are found in all three germ layers of the epiblast and in the extraembryonic endoderm of the bilaminar yolk sac. Neural crest cells appear and migrate at the neural plate stage at cranial and cervical levels prior to formation of the neural folds in both monotremes and marsupials

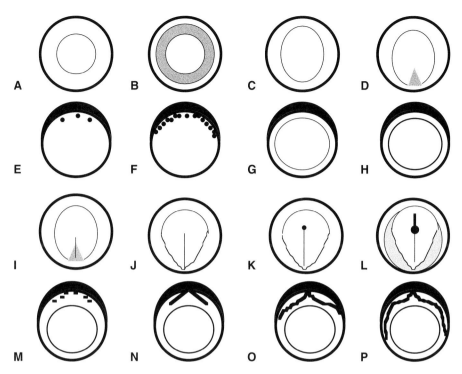

Figure 6 Diagrammatic representations of development of *Sminthopsis macroura* from the early bilaminar blastocyst stage to the late trilaminar blastocyst stage. Parts A–D and I–L represent surface views from the embryonic pole and E–H and M–P represent transverse sections in the posterior region of the blastocyst. Not drawn to scale. (A) A late unilaminar/early bilaminar blastocyst in early stages of hypoblast formation, in which the limit of the epiblast is represented by a thin inner circle. (B) A later incomplete bilaminar blastocyst stage, in which the hypoblast (shaded circle) has extended past the epiblast. (C) A complete bilaminar blastocyst in which the epiblast now has an oval shape. (D) A late complete bilaminar blastocyst, in which increasing cell densities has caused a triangular region (shaded triangle) to appear in the posterior of the epiblast. (E) Late unilaminar/early bilaminar blastocyst in which a few inward migrating hypoblast cells (small dark circles) lie under the epiblast (dark crescent) but not beneath the trophoblast (thick remainder of circle). (F) Incomplete bilaminar blastocyst in which hypoblast cells (dark circles) form a network beneath the epiblast. (G) Complete bilaminar blastocyst in which the hypoblast (fine inner circle) forms an attenuated inner lining of the blastocyst. (H) Complete bilaminar blastocyst in which the hypoblast has consolidated to a squamous epithelial layer (thick inner circle). (I) Early trilaminar blastocyst in which the first signs of the primitive streak (narrow line) bisects the triangular region. (J) Primitive streak (line) stage trilaminar blastocyst in which the epiblast has become pear-shaped. (K) Primitive streak stage trilaminar blastocyst with node (dark circle). (L) Late trilaminar blastocyst in which the notochord (thick line) is moving forward from the node and the mesodermal wings (stippled areas) extend either side of the epiblast. (M) Early trilaminar blastocyst in early stages of mesenchyme cell (thick bars) migration from the epiblast (dark crescent). (N) Primitive streak stage trilaminar blastocyst with mesoderm forming two wings (thick lines) under the epiblast. (O) Primitive streak stage trilaminar blastocyst in which the mesodermal wings have reached the limit of the epiblast. (P) Late trilaminar blastocyst in which the mesodermal wings have extended beyond the epiblast. Proximal to the epiblast, the trophoblast (outer thick circle), mesoderm, and hypoblast form the trilaminar yolk sac, and distal to epiblast, the trophoblast and hypoblast form the bilaminar yolk sac.

(Wilson and Hill, 1908; McCrady, 1938). Precocious neural crest lineages are an essential component of the precocious development of anterior structures associated with pouch life in marsupials. A small number of histological studies, mainly of isolated stages of gastrulation in marsupials (Selenka, 1892; Hartman, 1919; Kerr, 1936; McCrady, 1938; Lyne and Hollis, 1977; Krause and Cutts, 1985a) suggest that gastrulation is generally similar to the chick.

E. Extraembryonic Lineages and Membranes

1. Extraembryonic Mesoderm

Migration of the two wings of extraembryonic mesoderm from the streak to outside the epiblast and under the proximal trophoblast begins laterally (Fig. 6)

and posteriorly and spreads anteriorly. The mesodermal wings eventually fuse anterior to the epiblast. The extraembryonic mesoderm then extends to about the equator of the blastocyst where the sinus terminalis forms.

2. Trophoblast Lineages and the Yolk Sac

Migration of extraembryonic mesoderm converts abembryonic regions of the blastocyst into a trilaminar yolk sac proximal to the embryo and bilaminar yolk sac distal to the embryo. In dasyurid marsupials such as the dunnart, the trophectoderm of the yolk sac can be separated into three cell types; proximal trophectoderm of the trilaminar yolk sac, and two types of distal trophectoderm of the bilaminar yolk sac (Fig. 7). Of the latter, one type, of large cuboidal cells, lies adjacent to the sinus terminalis and is the site of adhesion to the uterus

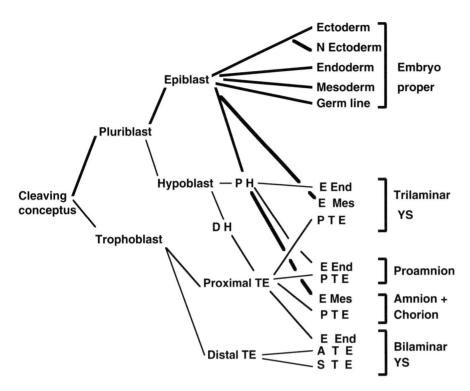

Figure 7 A proposed lineage diagram of *Sminthopsis macroura,* based on *in vitro,* fate mapping, and histological studies. The allantoic lineages are omitted because of insufficient information. ATE, Adhesive distal trophectoderm; E End, extraembryonic endoderm; E Mes, extraembryonic mesoderm; DH, distal hypoblast; N, neural; PH, proximal hypoblast; STE, squamous distal trophectoderm; TE, trophectoderm; YS, yolk sac. Lineages leading to epiblast and beyond are depicted by a wide line.

and the other comprises the squamous trophectoderm of the abembryonic pole (Hill, 1900; Roberts and Breed, 1996; Hughes, 1974). In the dasyurids, therefore, the bilaminar yolk sac contributes to the choriovitelline placenta, but in other marsupials the trilaminar yolk sac contributes to it (see Tyndale-Biscoe and Renfree, 1987, for review). Trophectoderm cells of the adhesive region of the bilaminar yolk sac form a syncytium (Hill, 1900). A low proportion of trophoblast cells dissociated from unilaminar and bilaminar blastocysts transform in culture to the multinucleate state (Yousef and Selwood, 1996). This and the histology suggests that this state is one of the features of at least one of the trophoblast lineages.

3. Hypoblast Lineages

The hypoblast does not differentiate further until relatively late in gastrulation when the extraembryonic mesoderm has extended to about the equator of the blastocyst. Its differentiation into the two lineages is presumably induced by its mesodermal covering. The fate of the hypoblast lineages is shown in Fig. 7. The proximal hypoblast forms the inner extraembryonic endoderm lining of what becomes the trilaminar yolk sac. The distal hypoblast forms the inner extraembryonic endoderm lining of the bilaminar yolk sac. The differentiation of these two cell types has been most extensively

described by Hollis and Lyne (1977) in the bandicoots and Krause and Cutts (1985a) in the Virginia opossum.

4. Amnion and Chorion

The amnion initially is formed of a proamnion (proximal trophectoderm and proximal hypoblast) anteriorly and a true amnion (proximal trophectoderm and extraembryonic mesoderm) posteriorly (Fig. 7). This is because the anterior margins of the mesodermal wings fuse anterior to the site of amnion formation but contribute to the lateral and posterior amniotic folds. In the stripe-faced dunnart, fate mapping studies have shown that the site of origin of the head fold of proamnion is located in the trophectoderm immediately anterior to the epiblast (Cruz *et al.,* 1996). The proamnion is transitory in a number of marsupials (Luckett, 1977). The lineages contributing to the chorion are shown in Fig. 7.

5. Allantois

The development of the allantois has only been studied in the opossum (McCrady, 1938) but isolated stages after its formation have been described in a number of marsupials (Sharman, 1961; Luckett, 1977). The allantois forms late in marsupials and its vascular state is short-lived in all except the bandicoots where the allantois contributes to a chorioallantoic placenta (Hill, 1895, 1900; Padykula and Taylor, 1977). McCrady

(1938) and Krause and Cutts (1985b) conclude that the allantois of the opossum is an evagination of the endoderm and mesoderm of the hindgut of the embryo proper. Because of the paucity of information available, especially for dasyurids, the lineages contributing to the allantois are not indicated in Fig. 7.

IV. Conclusions

Isolated from the constraints of implantation and occurring in simple unilaminar epithelia, the progressive separations of the early marsupial cell lineages show evidence of marsupial evolutionary history. The role of the trophoblast and hypoblast remains nutritional, but nutrient transfer is from the external uterine milieu to the pluriblast and then epiblast rather than from the internal yolk as in the lower amniotes. The yolk has been replaced by a cleavage cavity filled with ECM. The trophoblast and hypoblast probably play a role in signaling lineage separation, but the nature of this is unknown. Mechanisms to create diversity within developing epithelia also operate in these two lineages. A limited amount of evidence suggests that pluriblast and epiblast cells are pluripotential at least. They have a stem cell-like appearance ultrastructurally (Lyne and Hollis, 1977; Hollis and Lyne, 1977) and *in vitro*, but they do not need to be cultured over fibroblast feeder layers to maintain this state (Yousef and Selwood, 1996). A fate mapping study has confirmed that the neuroectoderm and embryonic ectoderm arise within the epiblast (Cruz *et al.*, 1996) and histological studies (Kerr, 1936) suggest that the epiblast gives rise to all cell lineages of the embryo and extraembryonic mesoderm. Further fate mapping studies and analysis of cells *in vitro* should confirm these earlier predictions.

Before progress could be made on analysis of the mechanisms of the separation of early cell lineages in marsupials, the misconceptions about the potency of the unilaminar blastocyst epithelium and of the epiblast had to be addressed. Studies to do this have identified possible common mechanisms in amniotes (higher vertebrates), which were not readily identified when eutherian models were examined. Further examination of these possible common mechanisms will give insights into the evolution of developmental mechanisms in all mammals during early lineage allocation. Marsupials also provide a unique model to separate intrinsic (embryonic) from extrinsic (uterine) signals specifying conceptus axis formation.

Acknowledgments

The support of the Australian Research Council, who funded studies leading to this review, and of Mr. M. Rudolf of Schumacher Pty. Ltd., Melbourne, who provided high quality shoe boxes as nest boxes for breeding females in the colony are gratefully acknowledged.

References

Baggott, L. M., and Moore, H. D. M. (1990). Early embryonic development of the grey short-tailed opossum *Monodelphis domestica, in vivo* and *in vitro. J. Zool.* **222**, 623–639.

Bennett, J. H., Breed, W. G., Hayman, D. L., and Hope, R. M. (1990). Reproductive and genetical studies with a laboratory colony of the dasyurid marsupial, *Sminthopsis crassicaudata. Aust. J. Zool.* **37**, 207–222.

Breed, W. G. (1996). Egg maturation and fertilization in marsupials. *Reprod. Fertil. Dev.* **8**, 617–643.

Breed, W. G., and Leigh, C. M. (1990). Morphological changes in the oocyte and its surrounding vestments during *in vivo* fertilization in the dasyurid marsupial *Sminthopsis crassicaudata. J. Morphol.* **204**, 177–196.

Breed, W. G., Simerly, C., Navara, C. S., VandeBerg, J. L., and Schatten, G. (1994). Microtubule configurations in oocytes, zygotes and early embryos of a marsupial, *Monodelphis domestica. Dev. Biol.* **164**, 230–240.

Cruz, Y. P., Yousef, A., and Selwood, L. (1996). Fate-map of the epiblast of the Dasyurid Marsupial *Sminthopsis macroura* (Gould). *Reprod. Fertil. Dev.* **8**, 779–788.

Eyal-Giladi, H. (1984). The gradual establishment of cell commitments during the early stages of chick development. *Cell Differ.* **14**, 245–255.

Fleming, T. P., and Johnson, M. H. (1988). From egg to epithelium. *Annu. Rev. Cell Biol.* **4**, 459–485.

Gardner, R. L. (1982). Investigation of cell lineage and differentiation in the extraembryonic endoderm of the mouse embryo. *J. Embryol. Exp. Morphol.* **68**, 175–198.

Gardner, D. K., Selwood, L., and Lane, M. (1996). Nutrient uptake and culture of *Sminthopsis macroura* (Stripe-faced Dunnart) embryos. *Reprod. Fertil. Dev.* **8**, 685–690.

Gemmell, R. T., and Selwood, L. (1994). Structural development in the newborn Marsupial, the stripe-faced Dunnart, *Sminthopsis macroura. Acta Anat.* **149**, 1–12.

Godfrey, G. K. (1969). Reproduction in a laboratory colony of the marsupial mouse *Sminthopsis larapinta* (Marsupialia: Dasyuridae). *Aust. J. Zool.* **17**, 637–654.

Godfrey, G. K. (1975). A study of oestrus and fecundity in a laboratory colony of mouse opossums (*Marmosa robinsoni*). *J. Zool.* **175**, 541–555.

Hartman, C. G. (1916). Studies in the development of the opossum *Didelphys virginiana*. I. History of the early cleavage. II. Formation of the blastocyst. *J. Morphol.* **27**, 1–83.

Hartman, C. G. (1919). Studies in the development of the opossum *Didelphys virginiana*. III. Description of new material on maturation, cleavage and endoderm formation. IV. The bilaminar blastocyst. *J. Morphol.* **32**, 1–139.

Hill, J. P. (1895). Preliminary note on the occurrence of a placental connection in *Perameles obesula*, and on the foetal membranes of certain macropods. *Proc. Linn. Soc. NSW* **10**, 578–581.

Hill, J. P. (1900). On the foetal membranes, placentation and parturition of the native cat (*Dasyurus viverrinus*). *Anat. Anz.* **18**, 364–373.

Hill, J. P. (1910). The early development of the marsupialia, with special reference to the native cat (*Dasyurus viverrinus*). Contributions to the embryology of the marsupialia, IV. *Q. J. Microsc. Sci.* **56**, 1–134.

Hollis, D. E., and Lyne, A. G. (1977). Endoderm formation in the blastocysts of the marsupial *Isoodon macrourus* and *Perameles nasuta*. *Aust. J. Zool.* **25**, 207–223.

Hughes, R. L. (1974). Morphological studies on implantation in marsupials. *J. Reprod. Fertil.* **39**, 173–186.

Johnson, M. H., and Selwood, L. (1996). Nomenclature of early development in mammals. *Reprod. Fertil. Dev.* **8**, 759–764.

Johnson, M. H., Chisholm, J. C., Fleming, T. P., and Houliston, E. (1986). A role for cytoplasmic determinants in the development of the mouse early embryo. *J. Embryol. Exp. Morphol. Suppl.* **97**, 97–121.

Johnston, P. G., Dean, D., VandeBerg, J. L., and Robinson, E. S. (1994). HPRT activity in embryos of a South American opossum *Monodelphis domestica*. *Reprod. Fertil. Dev.* **6**, 529–532.

Kerr, T. (1936). On the primitive streak and associated structures in the marsupial *Bettongia cuniculus*. *Q. J. Microsc. Sci.* **78**, 687–715.

Krause, W. J., and Cutts, J. H. (1985a). Morphological observations on the mesodermal cells of the 8 day opossum embryo. *Anat. Anz.* **158**, 273–278.

Krause, W. J., and Cutts, J. H. (1985b). Placentation in the opossum. *Didelphis virginiana*. *Acta Anat.* **123**, 156–171.

Luckett, W. P. (1977). Ontogeny of amniote fetal membranes and their application to phylogeny. *In* "Major Patterns in Vertebrate Evolution" (M. K. Hecht, P. C. Goody, and B. M. Hecht, eds.), pp. 439–516. Plenum, New York.

Lyne, A. G., and Hollis, D. E. (1976). Early embryology of the marsupials *Isoodon macrourus* and *Perameles nasuta*. *Aust. J. Zool.* **24**, 361–382.

Lyne, A. G., and Hollis, D. E. (1977). The early development of marsupials with special reference to bandicoots. *In* "Reproduction and Evolution" (J. H. Calaby and C. H. Tyndale-Biscoe, eds.), pp. 293–302. Australian Academy of Science, Canberra.

McCrady, E., Jr. (1938). The embryology of the opossum. *Am. Anat. Mem.* **16**, 1–233.

McCrady, E., Jr. (1944). The evolution and significance of the germ layers. *Tenn. Acad. Sci. J.* **19**, 240–251.

Merry, N. E., Johnson, M. H., Gehring, C. A., and Selwood, L. (1995). Cytoskeletal organization in the oocyte, zygote, and early cleaving embryo of the stripe-faced dunnart (*Sminthopsis macroura*). *Mol. Reprod. Dev.* **41**, 212–224.

Moore, H. D. M. (1996). Gamete biology of the New World marsupial, the grey short-tailed opossum, *Monodelphis domestica*. *Reprod. Fertil. Dev.* **8**, 605–616.

Moore, H. D. M., and Taggart, D. A. (1993). *In vitro* fertilization and embryo culture in the grey short-tailed opossum, *Monodelphis domestica*. *J. Reprod. Fertil.* **98**, 267–274.

Morton, S. R. (1983). Stripe-faced Dunnart. *In* "The Australian Museum Complete Book of Australian Mammals" (R. Strahan, ed.), p. 63. Angus & Robertson, Sydney.

Olszanska, B., Szolajska, E., and Lassota, Z. (1984). Effect of spatial position of uterine quail blastoderms cultured *in vitro* on bilateral symmetry formation. *Roux's Arch. Dev. Biol.* **193**, 108–110.

Padykula, H. A., and Taylor, J. M. (1977). Uniqueness of the bandicoot chorioallantoic placenta (Marsupialia: Peramelidae). Cytological and evolutionary interpretations. *In* "Reproduction and Evolution" (J. H. Calaby and C. H. Tyndale-Biscoe, eds.), pp. 303–323. Australian Academy of Science, Canberra.

Pedersen, R. A. (1986). Potency, lineage, and allocation in preimplantation mouse embryos. *In* "Experimental Approaches to Mammalian Embryonic Development" (J. Rossant and R. A. Pedersen, eds.), pp. 3–33. Cambridge Univ. Press, Cambridge.

Renfree, M. B., and Lewis, A. McD. (1996). Cleavage *in vivo* and *in vitro* in the marsupial *Macropus eugenii*. *Reprod. Fertil. Dev.* **8**, 725–742.

Roberts, C. T., and Breed, W. G. (1996). Variation in ultrastructure of mucoid coat and shell membrane secretion of a Dasyurid marsupial. *Reprod. Fertil. Dev.* **8**, 645–648.

Sathananthan, A. H., Selwood, L., Douglas, I., and Nanayakkara, K. (1997). Early cleavage to formation of the unilaminar blastocyst in the marsupial *Antechinus stuartii*: Ultrastructure. *Reprod. Fertil. Dev.* **9**, 201–212.

Selenka, E. (1892). Part 5. Beutelfuchs und kanguruhratte. *In* "Studien uber Entwickelungsgeschichte der Thiere." C. W. Kreidels, Wiesbaden, Germany.

Selwood, L. (1980). A timetable of embryonic development of the dasyurid marsupial *Antechinus stuartii* (Macleay). *Aust. J. Zool.* **28**, 649–668.

Selwood, L. (1986). Cleavage *in vitro* following destruction of some blastomeres in the marsupial *Antechinus stuartii* (Macleay). *J. Embryol. Exp. Morphol.* **92**, 71–84.

Selwood, L. (1987). Embryonic development in culture of two dasyurid marsupials, *Sminthopsis crassicaudata* (Gould) and *Sminthopsis macroura* (Spencer) during cleavage and blastocyst formation. *Gamete Res.* **16**, 355–370.

Selwood, L. (1989a). Marsupial pre-implantation embryos *in vivo* and *in vitro*. *In* "Development of Pre-implantation Embryos and Their Environment. Symposium of the 8th International Congress of Endocrinology, Kyoto, Japan, 1988" (K. Yoshinaga and T. Mori, eds.), pp. 225–236. Alan R. Liss, New York.

Selwood, L. (1989b). Development *in vitro* of investment-free marsupial embryos during cleavage and early blastocyst formation. *Gamete Res.* **23**, 399–413.

Selwood, L. (1992). Mechanisms underlying the development of pattern in marsupial embryos. *In* "Current Topics in Developmental Biology" (R. A. Pedersen, ed.). Vol. 27, pp. 175–233. Academic Press, New York.

Selwood, L. (1994). Development of early cell lineages in marsupials: An overview. *Reprod. Fertil. Dev.* **6**, 507–527.

Selwood, L. (1996). The blastocyst epithelium is not a protoderm in dasyurid marsupials: A review of the evidence. *Reprod. Fertil. Dev.* **8**, 711–723.

Selwood, L., and Sathananthan, A. H. (1988). Ultrastructure of early cleavage and yolk extrusion in the marsupial *Antechinus stuartii*. *J. Morphol.* **195**, 327–344.

Selwood, L., and Smith, D. (1990). Time-lapse analysis and normal stages of development of cleavage and blastocyst formation in the marsupials the brown antechinus and the stripe-faced dunnart. *Mol. Reprod. Dev.* **26**, 53–62.

Selwood, L., and Woolley, P. (1991). A timetable of embryonic development and ovarian and uterine changes during pregnancy, in the stripe-faced dunnart, *Sminthopsis macroura* (Marsupialia: Dasyuridae). *J. Reprod. Fertil.* **91**, 213–227.

Selwood, L., Robinson, E. S., Pedersen, R. A., and VandeBerg, J. L. (1997). Development *in vitro* of Marsupials: A comparative review of species and a timetable of cleavage and early blastocyst stages of development in *Monodelphis domestica*. *Int. J. Dev. Biol.* **41**, 397–410.

Sharman, G. B. (1961). The embryonic membranes and placentation in five genera of diprotodont marsupials. *Proc. Zool. Soc. London* **137**, 197–220.

Sutherland, A. E., Speed, T. P., and Calarco, P. G. (1990). Inner cell allocation in the mouse morula: The role of oriented division during fourth cleavage. *Dev. Biol.* **137**, 13–25.

Tyndale-Biscoe, C. H., and Renfree, M. B. (1987). "Reproductive Physiology of Marsupials." Cambridge Univ. Press, Cambridge.

Ullmann, S. L., Shaw, G., Alcorn, G. T., and Renfree, M. B. (1997). Migration of primordial germ cells to the developing gonadal ridges in the tammar wallaby *Macropus eugenii*. *J. Reprod. Fertil.* **110**, 135–143.

VandeBerg, J. L. (1990). The gray short-tailed opossum (*Monodelphis domestica*): A new laboratory animal. *ILAR News* **26**, 9–12.

Wilson, J. T., and Hill, J. P. (1908). Observations on the development of *Ornithorhynchus*. *Philos. Trans. R. Soc. London Ser. B* **199**, 31–168.

Woolley, P. A. (1990a). Reproduction in *Sminthopsis macroura*. (Marsupialia: Dasyuridae). 1. The female. *Aust. J. Zool.* **38**, 187–205.

Wooley, P. A. (1990b). Reproduction in *Sminthopsis macroura*: (Marsupialia: Dasyuridae). II. The male. *Aust. J. Zool.* **38**, 207–217.

Yousef, A., and Selwood, L. (1993). Embryonic development during gastrulation in *Antechinus stuartii* (MacLeay) and *Sminthopsis macroura* (Spencer) in culture. *Reprod. Fertil. Dev.* **5**, 445–458.

Yousef, A., and Selwood, L. (1996). The type and differentiation of cells *in vitro* from unilaminar and bilaminar blastocysts of two marsupials, *Antechinus stuartii* and *Sminthopsis macroura*. *Reprod. Fertil. Dev.* **8**, 743–752.

35

Green Fluorescent Protein: A New Approach to Understanding Spatial Patterning and Cell Fate in Early Mammalian Development

Magdalena Zernicka-Goetz
Wellcome/CRC Institute
and Department of Genetics
University of Cambridge
Cambridge CB2 1QR, United Kingdom

I. Introduction

The central problem in developmental biology is to understand how the single fertilized egg cell becomes organized into a three-dimensional organism, with its many differentiated cell types. Two distinct models have been proposed to explain the acquisition of spatial patterning and cell fate in animal development: determinative development, in which there is segregation of egg cytoplasmic determinants, and regulative development, in which epigenetic cues guide differentiation. In determinative development, cell lineage is the most important factor, and the fate of each cell is determined by its ancestry, whereas in epigenetic development the fate of each cell is established by its neighborhood. It previously had been assumed that the early development of various animal species occurs according to one or the other of these models. In fact, the situation is not so clear cut. Animals that show the determinative style of embryogenesis (such as leeches and nematodes) have some cell types whose fate is established through interactions with their neighbors. Similarly, in virtually all animals that develop

according to the epigenetic model (such as sea urchins and amphibians), the organization of the undivided egg helps to specify the patterning of the future embryo. Only mammals are thought to be exceptional in this respect (reviewed by Gardner, 1996). This is because the early embryos of eutherian mammals have an impressive ability to regulate their development after experimental manipulations such as removal, rearrangement, or addition of cells (reviewed by Pedersen, 1986). These experiments have led to the conclusion that the organization of the egg, and therefore cell lineage, cannot be important in establishing spatial patterning and cell fate during early stages of mammalian development. However, it is possible that the regulatory mechanisms that operate in response to experimental manipulation may differ from those that operate during the normal course of embryogenesis. Moreover, the observations by Gardner (1997) that the polarity of the mouse blastocyst bears relationship with the polarity of undivided egg raises the possibility that mammals use cytoplasmic determinants in some aspect of their development (see Chapter 32 by Pedersen for further discussion of this issue). It therefore remains critical to determine whether regions of the egg cortex or cytoplasm play a role in establishing specific cell lineages to determine the axis in the normal development of the early mammalian embryo.

By necessity, the approaches taken to address this issue must be carried out on intact, undisturbed embryos, as the regulative capacity of the embryos could obscure any effects of cell lineage. Previous manipulative studies of mouse embryos by forming chimeras between genetically marked embryonic cells have provided many insights into the origins of various cell types, but may not have addressed the possible role of cytoplasmic components (reviewed in McLaren, 1976). Intracellular approaches for lineage tracing have been used to study cell fate in minimally perturbed embryos cultured during their preimplantation development or postimplantation stages (e.g., Lawson *et al.*, 1991; Lawson and Pedersen, 1987; Pedersen *et al.*, 1986). However, attempts to follow postimplantation cell fate *in vivo*, in minimally-disturbed preimplantation embryos have proved difficult in the mouse, owing to the lack of an enduring way to mark the progeny of early cells. One aim of this chapter is to describe more recent approaches that have been taken to establish naturally fluorescent proteins as novel markers for studies of cell lineage in mammals.

Genes encoding naturally fluorescent proteins whose expression may be detected *in vivo*, also have the potential to provide reporters for following spatial and temporal patterns of gene expression controlled by cell specific promoters. When constitutively expressed, fluorescent proteins can enable us to determine cell lineages in transplants or chimeras. Until now the reporter gene that has been most useful in studies of mammalian development is *Escherichia coli lacZ*. Its β-galactosidase product is stable and its enzymatic activity can be readily detected at cellular resolution by histochemcal staining. Despite its great utility β-galactosidase has disadvantages as a reporter gene in early mammalian embryos in that the methods for staining are nonvital. Moreover, being a prokaryotic sequence, its message is relatively unstable in mammalian cells and *lacZ* transgenes are frequently inconsistent in their expression, being subjected to methylation and downregulation during development. Therefore the other aim of this chapter is to describe more recent attempts to establish naturally fluorescent proteins as *in vivo* markers to study patterns of gene expression and cell differentiation during early mammalian development.

II. The Ideal Cell Marker

For a marker to be suitable for cell lineage studies it should be stable for several cell generations, be cell au-

tonomous, and not be secreted or transferred to other cells. It cannot in any way perturb development. Ideally, the marker should be easily visualized and detectable in living tissues, to avoid possible artefacts that can be introduced during fixation. More importantly, this would enable cell fate to be followed directly over some period of time in living embryos. Exogenously applied dyes, such as 1,1: dioctadecyl-3,3,3'3'-tetramethylindocarbocyanine perchlora (DiI), carbohydrates, such as fluorescent dextrans, and proteins, such as horseradish peroxidase, are very useful but only for a restricted period, because as the marked cells divide, the markers become diluted and undetectable. Thus, an endogenously expressed marker is preferable. β-Galactosidase can be detected at cellular resolution, but only in fixed material. A fluorescent marker has the advantage that its localization within the living embryo can be studied by optical sectioning using confocal microscopy or deconvolution techniques. There exist fluorescent β-galactosidase substrates, but their use depends on preloading the cells (for instance by osmotic shock) and they produce a diffusible product, making the procedure difficult to apply to and analyze in organized tissues. The discovery of the green fluorescent protein (GFP) from the jellyfish *Aequorea victoria* provides a potential way to overcome these limitations. GFP is a naturally fluorescent 27-kDa protein which absorbs blue light and emits green light. It has the valuable property that the formation of its chromophore is an autocatalytic event that requires no cofactors (reviewed in (Cubitt *et al.*, 1995). Furthermore, the cDNA encoding GFP has been isolated (Prasher *et al.*, 1992), opening many interesting possibilities in developmental biology, because now GFP can be introduced, expressed, and become fluorescent in any organism or cell type.

III. Uses of Green Fluorescent Protein in Nonmammalian Systems

Green Fluorescent Protein has proved to be an ideal *in vivo* marker in many organisms. These include yeast (Doyle and Botstein, 1996; Nabeshima *et al.*, 1995), *Dictyostelium* (Gerisch *et al.*, 1995; Moores *et al.*, 1996), nematodes (Chalfie *et al.*, 1994), *Drosophila* (Davis *et al.*, 1995; Kerrebrock *et al.*, 1995; Wang and Hazelrigg, 1994; Yeh *et al.*, 1995), zebrafish (Amsterdam *et al.*, 1995), and *Xenopus* (Tannahill *et al.*, 1995; Zernicka-Goetz *et al.*, 1996). A consideration of these applications is beyond the scope of this chapter. However, I will consider approaches that we have developed to use GFP to follow cell lineages in the developing *Xenopus* embryo, as this has provided the foundation for some of our work with the mouse.

In studies on *Xenopus* development, GFP has proved to be very useful, not only to follow lineages of cells of early embryos, or the fate of GFP marked cells transplanted to an unlabeled host, but also to provide an enduring way to mark those cells in which an early regulatory gene has been experimentally overexpressed. In such studies, long-term identification of cells is essential. This permits us to relate a change in the fate of cells to the overexpression of an early regulatory gene in the ancestors of those cells, even when the gene product is ephemeral. Indeed, the injection of mRNA encoding GFP into early *Xenopus* blastomeres allows their progeny to be identified in much later stages, such as the feeding tadpole stage. Amaya (Kroll and Amaya, 1996) also has used GFP as a reporter in transgenic *Xenopus* embryos. Similarly, studies on sea urchin development have shown that GFP can be used as a reporter for the expression of transgenes, opening the way to several new experimental strategies (Arnone and Davidson, 1997).

IV. Properties of Green Fluorescent Protein

In attempts to extend the use of GFP to other systems, it became apparent that the wild-type protein is temperature sensitive and therefore only useful as a cell marker in a restricted number of organisms, notably those that develop at temperatures similar to that of jellyfish. This limitation originates from the fact that the protein only becomes fluorescent when correctly folded, and this process becomes much less efficient at higher temperatures (Siemering *et al.*, 1996). At above approximately 30°C, the majority of GFP molecules are misfolded and never become fluorescent. Therefore, to enable its use as an *in vivo* marker to follow lineages in other systems, a number of mutations have been introduced into GFP that enhance its ability to fold correctly at higher temperatures. For example the combination of Val163Ala and Ser175Gly improves the folding of GFP without altering its spectral characteristics (Siemering *et al.*, 1996). Other mutations enhance the thermostability of GFP, but also alter its spectral properties. These include the following combinations; Ser65Ala + Val68Leu + Ser72Ala, or Ser65Gly + Ser72Ala, or Phe64Leu + Ser65Thr (Cormack *et al.*, 1996). The consequence of these mutations is that GFP absorbs light at a longer wavelength (around 490 nm, with some variation between mutants) than the wild-type protein (which has a peak absorbance at 390 nm, and another peak at 470 nm). In fact these forms of GFP are preferable as markers in many instances, because living cells and animals better tolerate light of a longer wavelength due to the lower energies, which are less damaging.

V. MmGFP as an *in Vivo* Cell Lineage Marker in the Mouse

To enable tracing the fate of cells during mouse development, we have generated a novel form of GFP, named MmGFP (Zernicka-Goetz *et al.*, 1996, 1997). This provides an *in vivo* marker, either when introduced into individual blastomeres by the microinjection of its synthetic mRNA, or when expressed as a transgene (see below). MmGFP has a number of modifications that increase its solubility at higher temperatures. Consequently, a greater proportion of the molecules fold correctly and become fluorescent. The mutations include Val163Ala and Ser175Gly, and Phe64Leu + Ser65Thr, which also change the absorption peak to a longer wavelength (490 nm). In addition, MmGFP has a number of codon changes that bias the codon usage toward that of mammalian cells. These also remove a cryptic splice site that is recognized in plant cells (Haseloff *et al.*, 1997), but the importance of this in mammmalian cells is not clear. Because the absorption peak of MmGFP is 490 nm and emission peak is 510 nm, the protein can be visualized with a FITC filter set. However, a custom filter set, designed for its particular spectral properties, will enhance detectability.

VI. Introducing MmGFP into the Mouse

A. Microinjection of MmGFP RNA

A convenient way to follow cell lineages is to introduce MmGFP directly into the cell of interest. For this purpose we have explored the potential of marking the cell by injecting synthetic mRNA encoding MmGFP as this had proved to be an excellent method for tracing lineages in the *Xenopus* embryo (Zernicka-Goetz *et al.*, 1996). It appears that in comparison to other lineage markers so far used in this system, MmGFP synthetic mRNA has offered a number of important advantages: it is not only rapidly detectable and surprisingly stable, but, more importantly, emits very bright fluorescence in living embryos, and does not perturb their development.

For long term cell lineage studies it is essential that the synthetic mRNA is both stable and efficiently translated. Therefore the construct to produce synthetic MmGFP RNA is designed in such a way that the open reading frame is in an optimal context for translation, in a vector with a long 3′ untranslated region (UTR) that appears to protect the transcript from degradation. It is also important to ensure that the RNA is capped with m^7GpppG to protect against nucleases and to allow the RNA to be bound efficiently by the ribosome.

In studies on cell lineages during mouse development, synthetic mRNA encoding MmGFP provides a ready means to follow cells throughout the whole of preimplantation development. In the example shown in Fig. 1, one blastomere of a two-cell embryo was injected with MmGFP mRNA, and its progeny were traced until the blastocyst stage. Using synthetic mRNA encoding MmGFP to follow cells after implantation presents more of a challenge, because the mouse embryo begins to grow extensively at this time. As mentioned above, this has been a problem with the use of previously available lineage markers, because dyes and enzymes are rapidly diluted and soon become undetectable. However, mRNA should offer the advantage that its continued translation will amplify the signal, generating many copies of MmGFP in the cells that inherit it. This approach has the potential to provide an opportunity to follow the lineages of early blastomeres and to determine if cells derived from specific regions of egg cytoplasm contain positional information with respect to axial development of the mouse embryo during gastrulation.

B. Transgenic Mice

In the mouse, one attractive method to trace cell lineages is to introduce the reporter gene into embryonic stem (ES) cells. This is because ES cells are totipotent and can colonize a preimplantation embryo, contributing to the whole of embryogenesis (Evans and Kaufman, 1981; Martin, 1981). Therefore, one way in which to follow cell fate is to generate stable ES cell lines expressing MmGFP under specific promoters that are thought to be active throughout, or in a substantial part of, cell lineage. These can be used to produce chimeric and subsequently transgenic mice. Alternatively, mice expressing GFP under a ubiquitous promoter can be generated, and these animals can provide donor cells for transplantation experiments. We have generated transgenic lines of mice expressing MmGFP in two ways. First, we have established ES cell lines that stably express MmGFP, and introduced these into mouse morula or blastocyst stage embryos to obtain chimeric mice (Zernicka-Goetz et al., 1997). Second, we have inserted the MmGFP gene downstream of suitable promoters and injected these constructs into the

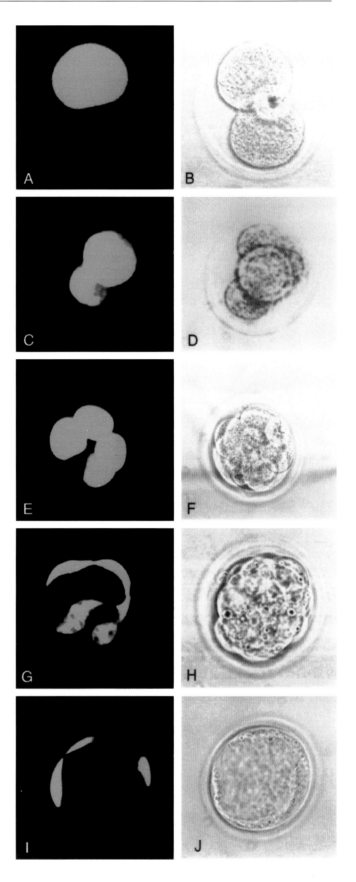

Figure 1 Blastomeres of the mouse embryo injected with MmGFP mRNA can be followed through preimplantation development. Fluorescence (A, C, E, G, I) and transmitted light (B, D, F, H, J) images of such a developing mouse embryo. (A, B) Two-cell stage; (C, D) 4-cell stage; (E, F) 8-cell stage; (G, H) morula; (I, J) blastocyst. The fluorescence images in panels G and I are single optical sections of the embryo. Magnification approx. 425×. (From Zernicka-Goetz et al., 1997, © Company of Biologists, Ltd.)

pronuclei of zygotes (M. Zernicka-Goetz, J. Pines, S. Barton, and M. Evans, 1998, unpublished).

MmGFP expression does not perturb normal embryogenesis: we have found that MmGFP-expressing cells can easily be identified throughout development, from the egg to the adult. An example of MmGFP expressing cells detected in the tissues of postimplantation embryos and those of newborn mice is shown in Fig. 2. This depicts the head region of the transgenic embryo a few days after its implantation (Fig. 2A) and the muscle tissue from a transgenic newborn mouse (Fig. 2B). In both of these cases MmGFP is expressed under the control of the EF1α promoter. Thus far, GFP has been reported to be expressed in mice under several different promoters such as the cell cycle specific cdc2 promoter (Zernicka-Goetz *et al.*, 1997), the elongation

factor 1α (EF1α) promoter (M. Zernicka-Goetz, 1998, unpublished results), the chicken β actin promoter (Ikawa *et al.*, 1995), and astrocyte-specific glial fibrillary acidic protein (GFAP) promoter (Zhuo *et al.*, 1997).

The EF1α and β actin promoters should give an ubiquitous pattern of expression. However, whether indeed the expression of a randomly integrated transgene is ubiquitous depends on the site of such integration. Nevertheless there are transgenic mice in which almost all the tissues were green when GFP was expressed from the β actin chicken promoter and CMV enhancer (Ikawa *et al.*, 1995; Okabe *et al.*, 1997). Those transgenic lines of mice in which GFP is expressed under the control of a cell specific promoter, such as GFAP, have been produced to enable *in vivo* studies on dynamic changes in the fate of astrocytes (Zhuo *et al.*, 1997).

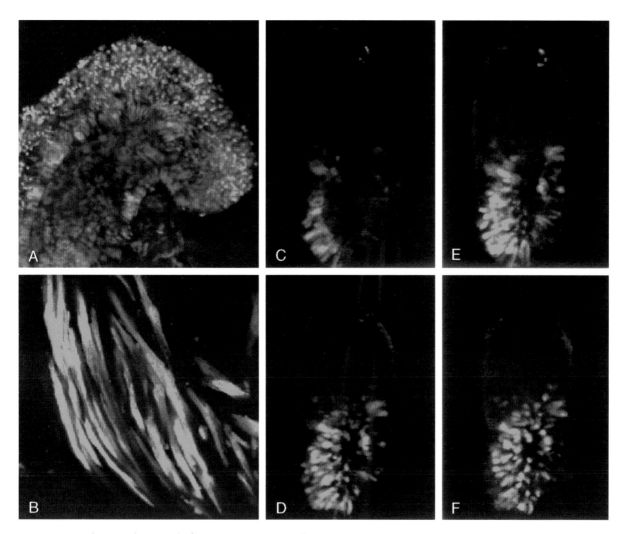

Figure 2 Developmental potential of MmGFP expressing cells in postimplantation mouse embryos. (A) Confocal fluorescence image of the head region of a 9-day-old heterozygous transgenic embryo expressing MmGFP from the EF1α promoter. (B) Confocal image of muscle cells from a newborn mouse expressing MmGFP from the EF1α promoter. (C–F) Samples from a Z-series of confocal sections of a 6.5-day-old chimeric embryo in which the progeny of ES cells are expressing MmGFP from the cdc2 promoter. Clones of fluorescent cells are clearly seen in the embryonic ectoderm. (From Zernicka-Goetz *et al.*, 1997, © Company of Biologists, Ltd.)

The cdc2 promoter was expected to provide a particularly good marker for proliferating tissues. Indeed, when cultured ES cells expressing MmGFP from this promoter are induced to differentiate, MmGFP expression is switched off (Zernicka-Goetz *et al.*, 1997). Therefore this construct is likely to find an application in studying lineages in early embryogenesis, where most cells are proliferating, as well as in specialized cell types later in development, or in tissues that undergo substantial regeneration.

Expression of MmGFP in ES cells enables us to follow the fate of ES cells themselves in living preimplantation embryos. An example of this can be seen in the embryo displayed in Fig. 2C-F, which shows a chimera formed by aggregating ES cells expressing MmGFP from the cdc2 promoter with the blastomeres from a wild-type embryo (Zernicka-Goetz *et al.*, 1997). The four confocal optical sections of the resulting 6.5-day-old chimeric embryo reveal the progeny of the MmGFP-expressing ES cells in the epiblast (Fig. 2C-F). Mouse embryos at this age (6.5 days old) begin gastrulation and this process has been shown to be associated with rapid cell proliferation. Therefore, as could be predicted, the embryonic ectoderm colonized by the progeny of transplanted ES cells should give very strong GFP expression. More importantly, however, analysis of patterns of distribution of the progeny of MmGFP expressing ES cells show us directly that there is extensive cell mingling in the epiblast even before gastrulation starts. This outcome supports the observations on interspecific chimeric mice that there is mixing of the clonal descendants of founder cells of the epiblast after implantation (reviewed in McLaren, 1976). However, these studies have not established when this takes place. Our observations suggest that the coherent growth is broken-down in the epiblast very early, even before gastrulation.

VII. Future Potential

Studies to explore the potential of GFP as a lineage marker in mice have only just begun. Apart from allowing us to follow lineages when GFP is introduced under the control of tissue specific promoters, the availability of totipotent ES cell lines expressing GFP will enable us to address many important issues about the fate of ES cells themselves. Previously it has been difficult to follow the colonization of a preimplantation embryo by ES cells, as it has been only possible to distinguish them from carrier embryo cells in fixed material. At its best, it has required the use of ES cells expressing *lacZ*, followed by the fixation and staining of embryos to reveal the ES cells lineages.

When expressed under the control of tissue specific promoters, GFP should be a useful reporter for

mapping the fates of specific cell types. Furthermore, GFP driven by the cdc2 promoter promises to allow the domains of cell proliferation to be traced throughout embryogenesis, as well as in postnatal development. Cells from mice ubiquitously expressing GFP will be invaluable as donors in transplantation experiments. The ability to culture embryos *in vitro* throughout preimplantation, and for restricted periods of time during posimplantation development, should permit the fate of transplanted cells to be followed directly in living embryos.

It would be invaluable to be able to follow the lineage of individual cells in tissues *in vivo* both under normal circumstances and following specific genetic change. It has been demonstrated that transgenic Cre-recombinase expression is compatibile with normal murine development and that this may be used to drive excision of tandem lox-p sites introduced into the genome. This has been demonstrated to be a particularly effective method of tissue specific gene ablation (Gu *et al.*, 1994). The approaches I have described to follow cell lineages can be combined with techniques to activate transcription of GFP at specific times during development using the Cre-recombinase. Such a GFP construct could be introduced as a transgene that will only be expressed, or, alternatively, inactivated, after recombination events mediated by Cre. Placing the Cre gene under a temporally regulated promoter would allow GFP to be expressed at particular times and in defined tissues during development either *in vivo* or *in vitro*.

GFP may also prove to be a superior reporter to *lacZ* in gene-targeting and in gene-trapping studies, because it can be visualized in living cells. The application of GFP as a vital reporter for tracing cell fate and gene expression will thus open the way to a number of new experimental strategies for studying unresolved issues of cell lineage and gene regulation in mammalian embryos.

Acknowledgments

I thank Roger Pedersen, David Glover, Martin Evans and Jonathon Pines for their comments and suggestions. I am grateful to the Lister Institute for Preventive Medicine for the award of a Senior Research Fellowship and to the Wellcome Trust for support of my work.

References

Amsterdam, A., Lin, S., and Hopkins, N. (1995). The Aequorea victoria green fluorescent protein can be used as a reporter in live zebrafish embryos. *Dev. Biol* **171**, 123–129.

Arnone, M. I. and Davidson, E. H. (1997). The hardwiring of development: Organization and function of genomic regulatory systems. *Development (Cambridge, UK)* **124**, 1851.

Chalfie, M., Tu, Y., Euskirchen, G., Ward, W. W., and Prasher, D. C. (1994). Green fluorescent protein as a marker for gene expression. *Science* **263**, 802–805.

Cormack, B. P., Valdivia, R. H., and Falkow, S. (1996). FACS-optimized mutants of the green fluorescent protein (GFP). *Gene* **173**, 33–38.

Cubitt, A. B., Heim, R., Adams, S. R., Boyd, A. E., Gross, L. A., and Tsien, R. Y. (1995). Understanding, improving and using green fluorescent proteins. *Trends Biochem. Sci.* **20**, 448–455.

Davis, I., Girdham, C. H., and O'Farrell, P. H. (1995). A nuclear GFP that marks nuclei in living *Drosohila* embryos; maternal supply overcomes a delay in the appearance of zygotic fluorescence. *Dev. Biol.* **170**, 726–729.

Doyle, T., and Botstein, D. (1996). Movement of yeast cortical actin cytoskeleton visualized *in vivo*. *Proc. Natl. Acad. Sci. U.S.A.* **93**, 3886–3891.

Evans, M. J., and Kaufman, M. H. (1981). Establishment in culture of pluripotential cells from mouse embryos. *Nature (London)* **292**, 154–156.

Gardner, R. L. (1996). Can developmentally significant spatial patterning of the egg be discounted in mammals? *Hum. Reprod.* **2**, 3–27.

Gardner, R. L. (1997). The early blastocyst is bilaterally symmetrical and its axis of symmetry is aligned with the animal-vegetal axis of the zygote in the mouse *Development (Cambridge, UK)* **124**, 289–301.

Gerisch, G., Albrecht, R., Heizer, C., Hodgkinson, S. and Maniak, M. (1995). Chemoattractant controlled acumulation of coronin at the leading edge of *Dictyostelium* cells monitored using green fluorescent protein-coronin fusion protein. *Curr. Biol.* **5**, 1280–1285.

Gu, H., Marth, J. D., Orban, P. C., Mossmann, H. and Rajewsky, K. (1994). Delation of a DNA polymerase beta gene segment in T cells using cell type-specific gene targeting. *Science* **265**, 103–106.

Haseloff, J., Siemering, K. R., Prasher, D. C., and Hodge, S. (1997). Removal of a cryptic intron and subcellular localization of green fluorescent protein are required to mark transgenic *Arabidopsis* plants brightly. *Proc. Natl. Acad. Sci. U.S.A.* **94**, 2122–2127.

Ikawa, M., Kominami, K., Yoshimura, Y., Tanaka, K., Nishimune, Y., and Okabe, M. (1995). A rapid and noninvasive selection of transgenic embryos before implantation using green fluorescent protein (GFP). *FEBS Lett.* **375**, 125–128.

Kerrebrock, A. W., Moore, D. P., Wu, J. S., and Orr Weaver, T. L. (1995). Mei-S332, a *Drosohila* protein required for sister-chromatid cohesion, can localize to meiotic centromere regions. *Cell (Cambridge, Mass.)* **83**, 247–256.

Kroll, K. L., and Amaya, E. (1996). Transgenic *Xenopus* embryos from sperm nuclear transplantations reveal FGF signaling requirements during gastrulation. *Development (Cambridge, UK)* **122**, 3173–3183.

Lawson, K. A. and Pedersen, R. A. (1987). Cell fate, morphogenetic movement and population kinetics of embryonic endoderm at the time of germ layer formation in the mouse. *Development (Cambridge, UK)* **101**, 627–652.

Lawson, K. A., Meneses, J. J., and Pedersen, R. A. (1991). Clonal analysis of epiblast fate during germ layer formation in the mouse embryo. *Development (Cambridge, UK)* **113**, 891–911.

McLaren, A. (1976). "Mammalian Chimaeras." Cambridge Univ. Press, Cambridge.

Martin, G. R. (1981). Isolation of a pluripotent cell line from early mouse embryos cultured in medium conditioned by teratocarcinoma stem cells. *Proc. Natl. Acad. Sci. U.S.A.* **78**, 7634–7368.

Moores, S. L., Sabry, J. H., and Spudich, J. A. (1996). Myosin dynamics in live *Dictyostelium* cells. *Proc. Natl. Acad. Sci. U.S.A.* **93**, 443–446.

Nabeshima, K., Kurooka, H., Takeuchi, M., Kinoshita, K., Nakaseko, Y. and Yanagida, M. (1995). p93dis1, which is required for sister chromatid separation, is a novel microtubule and spindle pole body-associating protein phosphorylated at the Cdc2 target sites. *Genes Dev.* **9**, 1572–1585.

Okabe, M., Ikawa, M., Kominami, K., Nakanishi, T., and Nishimune, Y. (1997). 'Green mice' as a source of ubiquitous green cells. *FEBS Lett.* **407**, 313–319.

Pedersen, R. A. (1986). Potency, lineage, and allocation in preimplantation mouse embryos. *In* "Experimental Approaches to Mammalian Embryonic Development" (J. Rosant and R. A. Pedersen, eds.), pp. 3–33. Cambridge Univ. Press, London.

Pedersen, R. A., Wu, K. and Balakier, H. (1986). Origin of the inner cell mass in mouse embryos: Cell lineage analysis by microinjection. *Dev. Biol.* **117**, 581–595.

Prasher, D. C., Eckenrode, V. K., Ward, W. W., Prendergast, F. G., and Cormier, M. J. (1992). Using GFP to see the light. *Gene* **111**, 229–233.

Siemering, K. R., Golbik, R., Sever, R., and Haseloff, J. (1996). Mutations that suppress the thermosensitivity of green fluorescent protein. *Curr. Biol.* **6**, 1653–1663.

Tannahill, D., Bray, S., and Harris, W. A. (1995). A *Drosophila* E(spl) gene is "neurogenic" in *Xenopus*: A green fluorescent protein study. *Dev. Biol.* **168**, 694–697.

Wang, S., and Hazelrigg, T. (1994). Implications for bcd mRNA localization from spatial distribution of exu protein in *Drosophila* oogenesis. *Nature (London)* **369**, 400–403.

Yeh, E., Gustafson, K., and Boulianne, G. L. (1995). Green fluorescent protein as a vital marker and reporter of gene expression in *Drosophila*. *Proc. Natl. Acad. Sci. U.S.A.* **92**, 7036–7040.

Zernicka-Goetz, M., Pines, J., Ryan, K., Siemering, K. R., Haseloff, J., and Gurdon, J. B. (1996). An indelible lineage marker for *Xenopus* using a mutated green fluorescent protein. *Development (Cambridge, UK)* **122**, 3719–3724.

Zernicka-Goetz, M., Pines, J., Dixon, J., Hunter, S., Siemering, K. R., Haseloff, J., and Evans, M. J. (1997). Following cell fate in the living mouse embryo. *Development (Cambridge, UK)* **124**, 1133–1137.

Zhuo, L., Sun, B., Zhang, C. L., Fine, A., Chiu, S. Y. and Messing, A. (1997). Live astrocytes visualized by green fluorescent protein in transgenic mice. *Dev. Biol.* **187**, 36–42.

Cell Fate and Cell Migration in the Developing Cerebral Cortex

Marcus L. Ware[1,2]
Christopher A. Walsh[1,3]

[1]Division of Neurogenetics, Department of Neurology, Beth Israel Deaconess Medical Center
and Program in Neuroscience, Harvard Medical School
[2]Harvard-MIT Division of Health Sciences and Technology,
and [3]Program in Biological and Biomedical Science, Harvard Medical School, Boston, Massachusetts 02115

I. Introduction

The mammalian cerebral cortex consists of numerous cell types whose diversity and connectivity underlie the integration of sensory information, cognition, and motor output. The formation of the neural networks that perform these tasks requires the creation, migration, and proper connections of neurons of different physiological types. In addition to the neurons of these networks, there are a number of other cell types required for the creation of a functional cerebral cortex. These include cells that support the function of neurons in the adult, such as astrocytes and oligodendrocytes, and cells that are required only during development, such as radial glia. Early studies of the cerebral cortex have described in some detail the sequence of events required for cortical development. More recent studies have implicated a number of proteins involved in the control

and coordination of these events. These proteins range from secreted growth factors which may act to control cortical proliferation and differentiation, such as basic fibroblast growth factor (bFGF) and neurotrophin-3 (NT-3), to transcription factors involved in neuronal induction, such as MASH-1 and HES-1. In this chapter, we review studies of the control of cortical progenitor proliferation, neuronal specification, and migration that have advanced our knowledge of the genetic pathways involved in development of the cerebral cortex.

II. Overview of Cortical Anatomy and Development

The adult cerebral cortex is organized horizontally into functional areas and vertically into cortical layers, or lamina. Vertically, each layer consists of different neuronal types that make characteristic connections to other neurons locally within the cortex, to distant sites within the cortex, and/or to sites outside the cerebral cortex. For example, layer 5 contains large pyramidal neurons that receive input locally and project to subcortical targets, whereas layer 4 contains stellate neurons that receive thalamic input and project locally within the cortex. Different functional areas of the cortex have the same basic laminar structure, but differ in the number of neurons within each layer and thickness of each layer. Also, neurons in different functional areas make connections that reflect their functional specification.

Birth-dating experiments have demonstrated that the neurons and macroglia of the adult cerebral cortex derive from the ventricular zone (VZ) and subventricular zone (SVZ) of the developing telencephalon. These studies show that the neurons of the cortex are formed by two distinct waves of cellular proliferation and migration (Fig. 1). During the first wave, postmitotic cells in the germinal zone migrate a short distance toward the pial surface to form the preplate, or primordial plexiform layer (Marin-Padilla, 1971, 1978). During the second wave, neurons from the germinal zone migrate into

the preplate to form the cortical plate within the preplate, thus splitting the preplate into the marginal zone and the subplate. Cogenerated neurons migrate past the previously generated cohort of neurons to form layers 6 through 2 of the cortical plate in an inside-out fashion. Thus, the neurons in layer 1 and in the white matter are remnants of the preplate and are the earliest born neurons (Allendoerfer and Shatz, 1994). Neurons in layer 6 are the earliest born neurons of the cortical plate and those in layer 2 are the latest born neurons of the cortex.

The majority of cortical neurons are formed from the pseudostratified neuroepithelium of the ventricular zone of the primordial cortex, making these cells the progenitor pool for the neurons of the cerebral cortex. These progenitors have an apically attached process that extends to the ventricle and a basally attached process that extends toward the pial surface. The position of the nuclei of these progenitors in the VZ is coordinated with their stage in the cell cycle (Fig. 2). The nuclei of mitotic progenitors begin G1 phase in a relatively apical position and ascend to a relatively basal position during DNA replication (S phase). The nuclei de-

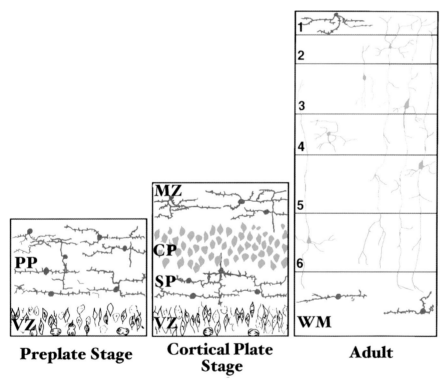

Preplate Stage   **Cortical Plate Stage**   **Adult**

Figure 1 Overview of cortical development. Cerebral cortical development is represented schematically by drawings of cortical cross sections at very early (left), intermediate (center), and late stages of development (right). The drawings correspond roughly to the E12 (left), E16 (middle), and adult rat cortex. The cerebral cortex arises from progenitor cells within the ventricular zone (VZ) lining the lateral ventricles. Early in neurogenesis these cells divide to form the preplate (PP) cells. These early formed neurons eventually differentiate to become the subplate neurons and the Cajal-Retzius neurons of layer 1, also known as the marginal zone (MZ). Over the course of development, other neurons arising from the germinal layer migrate to a position within the preplate, eventually splitting the preplate into layer 1 superficially and the subplate (SP) subjacently. The intervening neurons arrive to form the cortical plate (CP) in an inside-out fashion. Layer 6 neurons arrive first, then layer 5 and so on. wm, White matter. Reproduced with permission from Reid and Walsh (1996), Early development of the cerebral cortex. *Prog. Brain Res.* **108,** 17–30. © 1996 with kind permission of Elsevier Science, 1055 KV Amsterdam, The Netherlands.

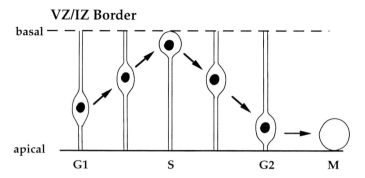

Figure 2 Interkinetic nuclear migration. Schematic representation shows interkinetic nuclear migration of proliferating progenitor cells in the ventricular zone. The nucleus begins G1 phase near the apical extent of the VZ. As the cell cycle progresses from G1 to S phase, the nucleus ascends basally to the border between the ventricular zone and intermediate zone (VZ/IZ border). As the cell cycle progresses from S phase to M phase, the nucleus descends apically toward the lateral ventricle. The cell divides with its nucleus at its most apical position.

scend through G2 phase and then divide in an apical position near the ventricle. This coupling of cell cycle stage and nuclear position is known as interkinetic nuclear migration (Sauer, 1935).

III. Control of Progenitor Proliferation

During the earlier stages of cortical development, there is extensive cell division of progenitor cells leading to expansion of the progenitor cell pool (Takahashi *et al.*, 1993). As development progresses, more progenitors give rise to postmitotic neurons. Recent studies in the mouse indicate that progenitor proliferation may be tightly regulated during cortical development. A study employing the S-phase maker [³H]thymidine has shown that different areas of the VZ may have marked differences in the rate of neuron production at the time when layers 6 and 5 are formed (Polleux *et al.*, 1997). Another study has shown that clonally related cells within the VZ remain in tight clusters and within approximate synchrony as they progress through the cell cycle (Cai *et al.*, 1997), suggesting that the history and/or immediate microenvironment of progenitors within the VZ may influence cycling kinetics. Together, these studies indicate that there may be local cues that control the proliferation of progenitor cells. The search for factors that may be involved in the control of progenitor proliferation has led to the implication of growth factors and transcription factors.

Recently, two growth factors have been implicated in control of neuronal precursor proliferation and differentiation. Basic fibroblast growth factor (bFGF) and neurotrophin-3 (NT-3) have been shown to regulate proliferation and differentiation of progenitor cells *in vitro*. In E14 cultures of rat cortical progenitors, the addition of bFGF promoted the proliferation of cortical precursors (Ghosh and Greenberg, 1995). The addition

of NT-3 to these cultures promoted the differentiation of neurons, and the addition of blocking antibodies to NT-3 decreased the number of cells differentiating in culture. Previous studies have shown that bFGF and its receptors (Gonzalez *et al.*, 1990; Powell *et al.*, 1991; Wanaka *et al.*, 1991), and NT-3 and its receptor (Lamballe *et al.*, 1994; Maisonpierre *et al.*, 1990; Tessarollo *et al.*, 1993), are present in the proliferative zone of the cerebral cortex. Taken together, these studies suggest that these secreted growth factors may be responsible for controlling pro-genitor proliferation and the differentiation of young neurons *in vivo*. However, there are no apparent cortical abnormalities in mice deficient in NT-3 (Ernfors *et al.*, 1994) or the NT-3 receptor, Trk C (Klein *et al.*, 1994), although the cortex was not examined in detail in either of these studies. These findings indicate that the role played by these growth factors in controlling progenitor proliferation *in vivo* may be more complicated than *in vitro* experiments suggest.

Two transcription factors have also been shown to be important in the control of progenitor proliferation. *Brain Factor 1* (*BF1*) encodes a winged helix transcription factor whose expression in the developing brain is restricted to the telencephalic neuroepithelium and the nasal half of the retina and optic stalk (Hatini *et al.*, 1994; Tao and Lai, 1992). Expression of *BF1* in mice is highest in the rapidly proliferating cells of the neuroepithelium, declining as the cells become postmitotic. Deletion of both copies of *BF-1* leads to a marked reduction in the size of the cerebral hemispheres by 95% by mass and death of animals only minutes after birth (Xuan *et al.*, 1995). By BrdU labeling, there was no difference in cell proliferation at E9.5, however, at E10.5 there was a 79% reduction in proliferation in the ventral telencephalon compared to heterozygous littermates. In addition, the ventricular zone of the cerebral cortex of mutants at E12.5 showed varying thickness that was probably due to premature neuronal differentiation. These results suggest that *BF-1* is necessary for the normal proliferation of progenitors in the VZ and the timing of neuronal differentiation.

Another transcription factor that has recently been shown to be important for normal cortical progenitor proliferation is the LIM homeobox gene, *Lhx2*. LIM homeobox genes are expressed widely in the mammalian central nervous system (CNS) and may be involved in neuronal specification and in the morphogenesis of other organ systems (Ahlgren *et al.*, 1997; Ericson *et al.*, 1992; Pfaff *et al.*, 1996). In rat embryos, *Lhx2* transcripts are first observed on E11 and by E12–E13 there is strong expression in the cells surrounding the telencephalic vesicles, in the diencephalon, and in the marginal layer of the myelencephalon. At the time of birth, *Lhx2* is intensely expressed in cortical layers 2–6, and CNS expression persists until adulthood (Xu *et al.*, 1993). Targeted disruption of *Lhx2* leads to anopthalmia, inefficient definitive erythropoiesis, agenesis of the hippocampal primordia, and hypoplasia of the basal ganglia and cortical plate (Porter *et al.*, 1997). Homozygous *Lhx2* mutants showed a decrease in mitotic figures in the VZ of the developing cerebral cortex with a marked reduction in the number of proliferating cells as measured by BrdU labeling; analysis of cell death by TUNEL staining showed no difference between *Lhx2* mutants and control brains (Porter *et al.*, 1997). The persistent expression of *Lhx2* until adulthood raises the possibility that this factor may also play additional roles in migration, differentiation, or cell maintenance that cannot be addressed by a standard knockout experiment. The role of *Lhx2* and *BF1* in the control of cortical progenitor proliferation suggests that these genes may possibly be within the same genetic pathway. However, studies have not been performed to confirm this possibility. Although there are large gaps in our knowledge of the control of progenitor cell proliferation, more detailed studies of the secreted growth factors and the transcription factors implicated in the control of progenitor proliferation should enrich our knowledge of the early events in cerebral cortical development.

IV. Genetic Control of Neuronal Cell Specification

Much of our knowledge of the genes involved in the control of neuronal specification comes from studies of the development of the peripheral nervous system (PNS) and the CNS of the fruit fly, *Drosophila melanogaster*. Neural development in the *Drosophila* PNS and CNS generally occurs with well-defined cell lineage patterns (Bossing *et al.*, 1996; Schmidt *et al.*, 1997). Knowledge of these lineages and mutagenesis leading to al-

terations of these lineage patterns have been used in a number of studies to identify genes important in neurogenesis.

A. Proneural Genes in *Drosophila*

The genes required early during neuronal specification in *Drosophila* development are the proneural genes (see also Chapter 18 by Broadus and Spana; Chapter 16 by Carthew). Proneural genes are expressed in undifferentiated ectodermal cells that give rise to proneural clusters (Cubas *et al.*, 1991; Skeath and Carroll, 1991), and the expression of proneural genes determines nervous system location. In the PNS, a single cell or a subset of cells expresses proneural genes at a higher level than the surrounding tissues and becomes a neural precursor, or SOP (sensory organ precursor) (Cabrera *et al.*, 1994; Ellis *et al.*, 1990). The proneural genes that have been studied most extensively are *achaete* and *scute*. These genes and the other genes of the *achaete-scute* complex encode helix–loop–helix (HLH) proteins that act as transcription factors by forming DNA-binding heterodimers with the ubiquitous HLH protein product of the *daughterless* (*da*) gene (Hoshijima *et al.*, 1995). Proneural genes have different functions during development as indicated by their loss-of-function phenotypes. The *extramacrochaetae* (*emc*) gene product is an HLH protein without the basic domain that is required to bind to DNA (Ellis *et al.*, 1990); the protein product of *emc* acts as a negative regulatory of proneural genes, presumably by forming heterodimers with and thereby sequestering the products of *achaete*, *scute*, or *da* (Cabrera *et al.*, 1994; Van Doren *et al.*, 1991). Thus, the relative levels of the products of *achaete*, *scute*, and *emc* determine the competence of neuronal precursors. Besides *emc* other negative regulators include the HLH gene, *hairy*, and the seven HLH genes in the *Enhancer of Split* complex of neurogenic genes.

B. Proneural Genes in the Developing Cerebral Cortex

A number of homologues of *Drosophila* proneural genes has been identified and shown to be important in the developing mammalian nervous system, including the murine *achaete-scute* homologue, *MASH-1*. *MASH-1* is first expressed on E8.5 in restricted domains in the neuroepithelium of the midbrain and ventral forebrain as well as in the spinal cord. Between E10.5 and E12.5, *MASH-1* transcripts are present in the ventricular zone in all regions of the brain. This pattern of expression suggests that *MASH-1* is involved in the early stages of neuronal development in the mouse (Guillemot and Joyner, 1993). Despite this expression pattern, null mutations in *MASH-1*, lead to extensive death in the olfactory epithelium with loss of neuronal progenitors at an early stage, but no cortical phenotype (Guillemot *et al.,* 1993). In sympathetic ganglia, the development of neuronal precursors arrests, preventing the generation of sympathetic neurons but not glial precursors. The expression pattern and loss-of-function phenotype suggest that *MASH-1*, like *achaete-scute*, promotes the development of neuronal progenitors in distinct lineages. These studies also indicate that *MASH-1* is not required for the development of all precursors in which it is normally expressed, suggesting that other proneural genes may promote neuronal development in these progenitors.

A proneural gene that is directly implicated in the development of the cerebral cortex is *HES-1*. *HES-1* is a murine homologue of *hairy*, the basic HLH factor that negatively regulates neurogenesis in *Drosophila*. In the developing mouse CNS, *HES-1* is normally expressed at high levels throughout the ventricular zone, but not in the outer layers where differentiated neurons and glia are present (Sasai *et al.,* 1992). Overexpression of *HES-1* using a retroviral vector severely perturbed the development of cortical progenitors (Ishibashi *et al.,* 1994). Retrovirally infected cells were found only in the ventricular and subventricular zones, were round in appearance, and had no mature processes. Infected progenitors never produced mature neurons. Conversely mice lacking *HES-1* had severe defects in the development of the CNS, coupled with precocious expression of *MASH-1* (Ishibashi *et al.,* 1995). These results suggest that *HES-1* normally inhibits the expression of *MASH -1*, just as *hairy* inhibits the expression of *achaete-scute* in *Droso-phila*. Although our knowledge of the genes involved in the development of the cerebral cortex is not nearly as extensive as our knowledge of *Drosophila* neuronal development, these studies indicate that neurogenesis in *Drosophila* and neurogenesis in the mammalian cerebral cortex may involve homologous genetic pathways.

C. Lateral Specification

After the expression of proneural genes in proneural clusters in *Drosophila*, individual cells are chosen to become neural precursors while other cells within proneural clusters are inhibited from choosing this fate. This process requires communication between cells and involves lateral specification in addition to a heterogeneous distribution of neural potential between different cells of the proneural group (Seugnet *et al.,* 1997). Through the interaction of these cells and subsequent cell divisions, nonequivalent daughter cells are produced. The genes of the Notch–Delta pathway are known to be required for lateral specification in the PNS and CNS of *Drosophila* and have now been impli-

cated in lateral specification in vertebrates (reviewed in Artavanis-Tsakonas *et al.,* 1995).

D. Notch Signaling in *Drosophila*

During the development of the PNS and CNS of *Drosophila*, the *Notch* gene is required for the normal selection of precursor cells from competent cells within a proneural cluster. In mutants lacking *Notch*, nearly all of the cells in the ventral ectoderm continue to express *achaete-scute* and nearly all precursors develop into neurons at the expense of other cell types (Cabrera, 1990). These mutants, termed neurogenic, have a lethal hypertrophy of the nervous system. Genetic mosaic analysis of Notch activity during the formation of adult sensory bristles has demonstrated that Notch acts cell autonomously (de Celis *et al.,* 1991), and that different dosages of Notch may bias cell fates during lateral specification (Heitzler and Simpson, 1991).

E. Notch Signaling in Vertebrates

There is a growing body of evidence that lateral specification in vertebrates may be mediated by a homologous Notch pathway. Highly conserved homologues of *Drosophila Notch* have been identified in zebrafish, frogs, mice, rats, and humans (Bierkamp and Campos-Ortega, 1993; del Amo *et al.,* 1993; Lardelli *et al.,* 1994; Reaume *et al.,* 1992; Uyttendaele *et al.,* 1996). In these organisms, Notch and Notchlike proteins are present throughout developing tissues at embryonic stages and in proliferative cell layers of mature tissues. In addition, homologues of other genes within the Notch signaling pathway in *Drosophila* also have been identified in vertebrates (Bettenhausen *et al.,* 1995; Dehni *et al.,* 1995; Helms *et al.,* 1994; Henrique *et al.,* 1995; Lindsell *et al.,* 1995; Myat *et al.,* 1996).

In *Xenopus*, expression of an activated form of the *Notch* homologue, *Xotch*, leads to loss of structures derived from the dorsal neural tube as well as to neural and mesodermal hypertrophy (Coffman *et al.,* 1993). Activated Xotch inhibits the early expression of epidermal and neural crest markers, and enhances and extends the normal time course over which ectoderm may be induced to form neural tissue. Ectopic expression of a homologue of *Delta*, *X-Delta-1* inhibits the production of primary neurons; interference of X-Delta-1 activity results in an overproduction of neurons (Chitnis *et al.,* 1995). Although these experiments provide evidence that Notch signaling may be central to lateral specification, the effect of extending the normal time course over which embryonic tissues may respond to inductive signals by activated Xotch suggests that Xotch may alter progenitor competency. Thus, Notch signaling may provide a permissive rather than an instructive role in determining cell fate, controlling the ability of progenitors to respond to other developmental cues.

Additional evidence that *Notch* may act by altering the ability of precursors to respond to inductive signals comes from studies in the vertebrate retina. In *Xenopus*, expression of an activated form of Xotch in retinal precursors led to retention of undifferentiated morphology and the absence of molecular markers of differentiated retinal neurons, suggesting the activation of Xotch maintains neuroepithelial cells in an undifferentiated state and inhibits their development into mature neurons (Dorsky *et al.*, 1995). Studies in the developing chick retina indicate that Notch may act to prevent progenitors from differentiating into specific neuron types. In the developing retina, ganglion cells are the first cells to differentiate. Overexpression of a constitutive active form of the *Notch* homologue, *Tan-1*, in the chick retina decreased the number of retinal ganglion cells formed (Austin *et al.*, 1995), presumably by keeping retinal precursors in an undifferentiated state or by restricting precursors to the fates of later born neurons. On the other hand, reduction of Notch expression using antisense oligonucleotides increased the number of retinal ganglion cells formed, presumably by removing the restrictive Notch signal. In addition, coculturing retinal progenitors with *Drosophila* Schneider 2 (S2) cells transfected with full-length *Drosophila Delta* inhibited retinal progenitors from differentiating as ganglion cells to the same degree as did TAN-1 (Austin *et al.*, 1995). Taken together, these results indicate that Notch signaling is involved in the specification of competent cells in the vertebrate retina for particular neuronal fates and that *Notch* may exert its influence by altering the ability of precursors respond to other developmental cues.

A role for Notch signaling has not been shown genetically in the developing cerebral cortex because mice deficient in *Notch1*, a *Notch* homologue, do not survive past midgestation (Swiatek *et al.*, 1994). However, overexpression of Tan-1 in cultures of P19 cells, a mouse embryonal carcinoma line that can be induced to differentiate into neuroectoderm or mesodermal derivatives, suppresses neurogenesis and myogenesis, but not gliogenesis (Nye *et al.*, 1994). These results suggest that *Notch1* may play a role in mammalian cells similar to that of *HES-1* in the cortex. Thus, there is a the possibility that within the cortex *Notch* homologues and *hairy* homologues may interact to inhibit neurogenesis.

F. Numb Signaling in *Drosophila*

In addition to *Notch* and *Delta*, other genes have been shown to be important in neuronal specification. One of these genes, *numb*, is particularly interesting because inheritance of the Numb protein into daughter cells during neural development is associated with differentiation into neuronal cells. The *numb* gene was identified in *Drosophila* in a screen for mutations that effect neuronal development (Uemura *et al.*, 1989). Loss of *numb* function causes an increased production of support cells in the SOP lineage with a concomitant decrease in the production of neurons, whereas overproduction of the Numb protein causes the production of neurons with a decrease in the production of support cells (Rhyu *et al.*, 1994). Observations in the *Drosophila* CNS indicate that Numb and Prospero, a homeodomain-containing protein involved in neural precursor development, are asymmetrically segregated into one daughter during cell division (Knoblich *et al.*, 1995) and that inheritance of Numb protein determines neural fate. Additional studies of the MP2 lineage in the *Drosophila* CNS have also shown that the inheritance of Numb can determine cell fate (Spana *et al.*, 1995). These studies are of particular importance because they show that asymmetric inheritance of Numb can determine cell fate in an autonomous fashion (see Chapter 18 by Broadus and Spana).

The role of the Numb pathway in lateral specification has been suggested by its interaction with Notch signaling. As discussed earlier, Notch overexpression and Numb loss-of-function have similar phenotypes. Epistatic studies in *Drosophila* indicate that loss of both Notch and Numb function produces a phenotype similar to loss of Notch signaling (overproduction of neurons) suggesting that *Notch* may be acting downstream of *numb* (Guo *et al.*, 1996). In a yeast two-hybrid interaction assay, there is a direct protein–protein interaction between the Notch protein, and the N-terminal portion of Numb (Guo *et al.*, 1996). In addition, there is evidence that Notch interferes with the downstream effectors in the Notch pathway. Activation of Notch causes the nuclear translocation of the suppressor of hairless [Su(H)] protein which is required for proper Notch signaling (Fortini and Artavanis-Tsakonas, 1994). Numb signaling inhibited the nuclear translocation of Su(H) caused by the activation of Notch when Notch, Numb, and Su(H) were expressed in S2 cells (Frise *et al.*, 1996). These studies indicate that not only do Notch and Numb genetically and physically interact, but that Numb determines daughter fates by inhibiting Notch signaling.

G. Vertebrate Numb Genes

With the similarities between Notch pathways in *Drosophila* and vertebrates and the role of Numb in *Drosophila*, it is not surprising that two mouse homologues of *Drosophila numb*, *m-numb* and *numblike* have been cloned and studied (Zhong *et al.*, 1996, 1997). When expressed in *Drosophila*, the m-Numb protein can rescue the *numb* mutant phenotype and is asymmetrically distributed in a pattern similar to that of

Drosophila Numb. When expressed in *Drosophila numb* mutants, the Numblike protein is inherited by both daughter cells and induces the fate of the daughter cell that normally inherits *Drosophila* Numb. Like their *Drosophila* homologue, both m-Numb and Numblike can directly bind Notch1 in the yeast two-hybrid assay (Zhong *et al.*, 1996, 1997). The predicted protein products of *m-numb* and *numblike* show strong sequence similarity of N-terminal amino acids, including the residues that are predictive of a phosphotyrosine binding (PTB) domain. PTB domains have been identified in a number of other proteins and bind to phosphotyrosine-containing proteins, such as the receptors for EGF, NGF, and insulin.

The expression pattern of the *m-numb* and *numblike* suggest that they may have different functions during development. At E12.5 in the mouse, *numblike* is highly expressed in the nervous system whereas *m-numb* is expressed at low levels in most embryonic structures (Zhong *et al.*, 1997). In the developing cerebral cortex, m-Numb is present in all layers from the ventricular zone to the cortical plate, whereas Numblike is present in the postmitotic neurons of the cortical plate that are undergoing active differentiation but not in the proliferating cells of the ventricular zone. Thus, *m-numb* may be involved in progenitor cell proliferation and neuronal specification and *numblike* may be involved in neuronal differentiation.

H. Asymmetric Distribution of Numb in Mammalian Cerebral Cortical Progenitors

In the developing mammalian cerebral cortex two types of cell divisions occur: symmetric and asymmetric. Symmetric divisions occur when a cell gives rise to two daughter cells of the same cell types, whereas asymmetric divisions occur when a cell gives rise to two daughter cells of different types. Thus, there are three possible types of cell divisions according to symmetry: asymmetric cell divisions that produce a progenitor cell and a more differentiated cell; symmetric divisions that produce two progenitor cells; and symmetric divisions that produce two differentiated cells. *In vitro* (Davis and Temple, 1994; Reynolds *et al.*, 1992; Reynolds and Weiss, 1992; Temple, 1989; Williams and Price, 1995) and retroviral (Price and Thurlow, 1988; Reid *et al.*, 1997; Reid and Walsh, 1996; Walsh and Cepko, 1992, 1993; Walsh, 1996) experiments suggest that there is a significant proportion of cortical progenitors that are multipotential and thus produce multiple cell types. The most conclusive evidence of symmetric and asymmetric divisions in the developing mammalian cerebral cortex was provided by Chenn and McConnell (1995) using an *in vitro* system that allowed them to directly

observe slices of E29 ferret cortex in real time. Cells that divided with vertical cleavage planes produced behaviorally similar daughters that resemble precursor cells. These divisions were considered the symmetrical divisions that expanded the progenitor pool. Cells that divided with horizontal cleavage planes produced basal daughter cells that behaved like young migratory neurons and apical daughters that remained within the ventricular zone. These divisions were considered to be the asymmetric divisions that led to the production of neurons.

As discussed earlier, the *Drosophila* Numb protein is localized in a crescent overlying one of the centrosomes during mitosis. This localization leads to asymmetric distribution of the Numb protein into primarily one of the daughter cells. In the mouse cortex, m-Numb crescents are always localized apically, or toward the ventricular surface, and Notch1 was evenly distributed around the membrane (Zhong *et al.*, 1996). In the ferret, Notch1 immunoreactivity is always localized basally, or toward the pial surface (Chenn and McConnell, 1995). Thus, when progenitors divide with a vertical cleavage plane Numb and Notch1 proteins are evenly distributed into both daughter cells (Fig. 3). On the other hand, when cells divide with a horizontal cleavage plane, Numb is selectively inherited by the apical daughter cell (presumed to remain as a progenitor), and Notch1 is selectively inherited by the basal daughter cell (presumed to be a young neuron) in the ferret and inherited by both daughters in the mouse.

The inheritance pattern of Notch1 and Numb in asymmetrical divisions in the cerebral cortex is seemingly paradoxical, since Notch1 apparently would be inherited by the cell interpreted as the postmitotic neuron. In experiments summarized earlier, Notch has been shown to generally inhibit neurogenesis, while Numb generally biases cells toward a neural fate. As discussed earlier, it has been suggested Notch plays a more important role in modulating the ability of a cell to respond to a variety of instructive signals. Artavanis-Tsakonas and colleagues (1995) have hypothesized that Notch activation freezes a cell in its current developmental state, deafening it to the reception of other developmental signals. If this is the case, Notch may influence cell fate decisions by inhibiting cells from following instructive cues from an array of developmental influences. Chenn and McConnell (1995) have suggested that newly generated neurons may require large amounts of Notch1 to delay differentiation long enough to escape the ventricular zone. However, our growing knowledge of the complexity of the patterns of cell migration in the VZ (see Section V) suggests that the identification of basal and apical daughter cells as progenitor and postmitotic neuron, respectively, might not be necessarily so simple.

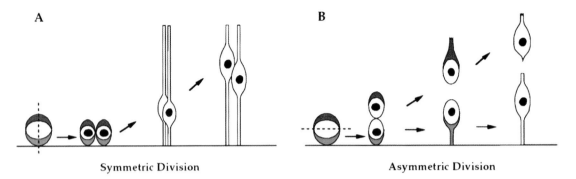

Figure 3 Asymmetric distribution of numb and notch1 in the ferret cortex. In the developing ferret cortex, Notch1 immunoreactivity is localized basally, or toward the pial surface. (A) When cells divide with a vertical cleavage plane, Numb (blue) and Notch1 (red) are evenly distributed into both daughter cells. (B) When cells divide with a horizontal cleavage plane, Numb is selectively inherited by the apical daughter cell (progenitor), and Notch1 is selectively inherited by the basal daughter cell (neuronal). Thus, horizontal cleavage planes would result in the asymmetric distribution of Numb during asymmetric divisions.

V. Radial and Nonradial Migration

Young neurons born in the ventricular zone find their way into the cortical plate by migrating in relation to a scaffolding of radial glial cells (Rakic, 1971, 1972). Radial glia are elongated nonneuronal cells that have their nuclei within the ventricular and subventricular zones and processes that stretch across the entire cerebral wall. During the cortical plate stage, young neurons become closely apposed to the fibers of radial glia and migrate along these fibers toward the pial surface.

The first firm evidence that neurons do not always follow strict radial pathways came from retroviral studies. These studies involved infecting progenitors with retroviruses containing unique DNA sequences that can be used to identify the multiple progeny of a single infected progenitor and thus establish clonal relationships. Using this technique, Walsh and Cepko (1992) first showed that clones derived from a singly infected progenitor may be dispersed tangentially over great distances in the cortex of the rat (Fig. 4). Evidence from experiments employing X-inactivation of transgenic *lacZ* in the mouse also provided strong evidence of nonradial as well as radial migration (Tan and Breen, 1993). X-inactivation provides a way to distinguish between clonally related populations. In the neocortex, stripes of *lacZ* positive cells showed mixing of neurons and glia, suggesting nonradial migration.

Time-lapse imaging showed a variety of potential nonradial routes in the intermediate zone (O' Rourke *et al.,* 1992). Subsequent studies suggested that there is rapid dispersion of cortical progenitors within the ventricular zone of cerebral cortex (Fishell *et al.,* 1993; Reid *et al.,* 1995; Walsh and Cepko, 1993). In addition, 1,1'-dioctadecyl-3,3,3,',3'-tetramethyl indocarbocyanine perchlorate (DiI) studies in living neocortex and in cortical slices have confirmed that there is significant tangential migration in the proliferative zones of the cortex

as well as in the intermediate zone (O' Rourke *et al.,* 1995, 1997).

Recent studies suggest that nonradial migration may be characteristic of particular stages of cortical development. In the ferret, retroviral injection studies showed that for injections performed at E33–E35, during the cortical plate stage, the neuronal clones labeled in the cortex consisted of single neurons (52%) or widespread clones (48%) of neurons and glia with an average of seven neurons per clones (Reid *et al.,* 1997). When the retroviral library was injected into developing ferrets at E25–E29, during the time when the cells of the preplate and earliest cells in the cortical plate are born, a similar pattern of single neurons and large clones were observed. In addition, however, a highly distinctive pattern was also seen only after early labeling in which very large clusters of labeled cells covered much smaller areas of the cortex (Ware *et al.,* 1997). The largest single cluster of this type consisted of more than 100 cells and had a diameter encompassing only 3% the rostral-caudal extent of the cortex. In addition, analysis of orientations of [³H]thymidine-labeled migrating cells in intact cortex showed that tangential migration occurs at all levels of the developing cerebral wall and that tangential migration increases during development (O' Rourke *et al.,* 1995). These experiments suggest that migratory behavior of cortical precursors may be markedly different depending on neuronal birth date.

The most dramatic evidence that the migratory paths of neurons may be complex comes from a more recent study (Anderson *et al.,* 1997), which finds that a significant population of predominantly interneurons arise in the lateral ganglionic eminence (LGE), the primordium of the striatum, then migrate into the cerebral cortex. In this study, DiI-labeled migrating cells from the LGE were detected in the neocortex in E12.5 in mouse brain slice cultures. Migrating cells from the LGE expressed GABA and calbindin, molecular markers

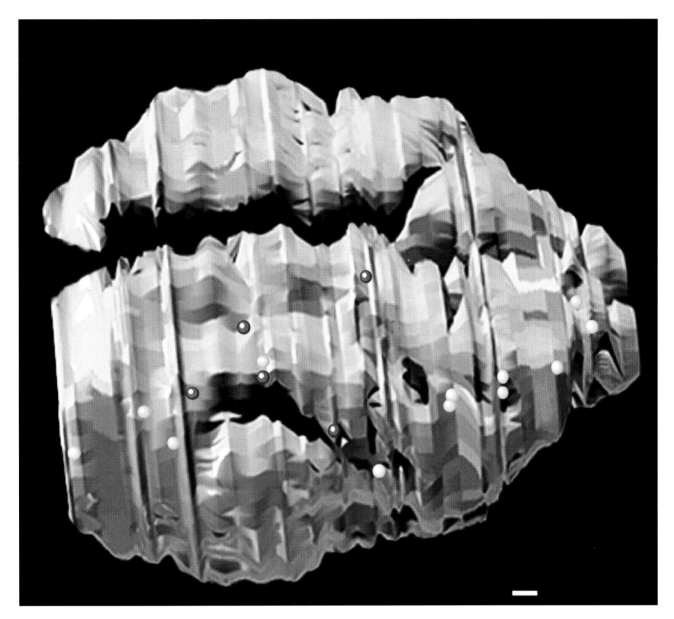

Figure 4 Three-dimensional reconstruction of two clones in ferret cortex labeled at E33 with a retroviral library. The computer-generated reconstruction of the ferret cerebral cortex is shown in gray. It is viewed from the right side and slightly from behind, and is oriented with the rostral end, including the olfactory bulbs, pointing toward the right. The more widespread clone (yellow) contributes cells to motor prefrontal, somatosensory, auditory, and visual areas. This clone is dispersed 16.7 mm in the A–P axis and 5.3 mm mediolaterally. Another clone (red) contributes cells to auditory and somatosensory cortices. The cells of this clone disperse 5.1 mm along the A–P axis and 4.4 mm mediolaterally. Scale bar: 1 mm. Reproduced with permission from Reid *et al.* (1997), © Company of Biologists, Ltd.

found in the interneurons of the cortex, as well as *Dlx-1* and *Dlx-2*, homeobox-genes encoding transcription factors. In the mouse, expression of *Dlx-1* and *Dlx-2* is initially restricted to the proliferative zones of the basal ganglia primordia, but by E13.5 they are expressed in the developing cortex as well. Unilateral transection of the cortical/subcortical angle, which ablates the pathway for migration from the LGE into the developing cortex, led to a 10-fold decrease in the number of calbindin and GABA expressing cells as well as a decrease

in the level of Dlx-1 expression in the cortex. In mice with mutations in both *Dlx-1* and *Dlx-2*, which are important for normal migration of cells out of the LGE (Anderson *et al.*, 1997), there was no detectable cell migration from the LGE into the cortex. These experiments show that there is a *Dlx-1/Dlx-2* dependent migration of a population of neurons from the LGE into the cortex, and that this population consists largely of interneurons. Taken together, these studies indicate that there are many possible migratory pathways into the developing cerebral cortex and that cortical precursors may derive from distant sites within the VZ of the cerebral cortex or from the LGE.

VI. Genetic Control of Cortical Migration

In contrast to cell specification, where genes discovered in *Drosophila* have been key to our understanding, the study of cortical migration has relied heavily on the study of spontaneous human and mouse mutations leading to aberrant cortical migration. Using these approaches, a number of genes required for normal cortical migration have been described. These genes are actively being studied to elucidate the genetic pathways that govern the migration and maturation of cortical neurons.

A. Lissencephaly

Lissencephaly (agyria–pachygyria) is a brain malformation occurring in humans that results from migrational arrest of virtually all cortical neurons short of their normal destination. Type 1 or classic lissencephaly is characterized by a markedly thickened cerebral cortex (Fig. 5) consisting of four coarse layers (Jellinger and Rett, 1976; Kuchelmeister *et al.*, 1993). Clinical manifestations include severe mental retardation, seizures, and other neurological abnormalities. These symptoms may appear in isolation or as Miller-Deiker syndrome (MDS), in which these symptoms are associated with craniofacial dysmorphisms (Dobyns *et al.*, 1991). About 15% of patients with isolated lissencephaly and more than

90% of patients with MDS have hemizygous microdeletions in a critical 350-kilobase region in chromosome 17p13.3 (Ledbetter *et al.*, 1992), implicating haploinsufficiency of a gene is this area. The *LIS-1* gene was later cloned and mutations in this gene were shown to be responsible for lissencephaly associated with MDS. The product of the *LIS-1* gene is identical to the human homologue of the 45-kDa regulatory subunit of bovine platelet activating factor (PAF) acetylhydrolase with a sequence identity of 99% (Hattori *et al.*, 1994).

The identification of *LIS-1* as a regulatory subunit of human platelet-activating factor acetylhydrolase (PAFAH) (Hattori *et al.*, 1994) and the identification of similar genes in other organisms have provided clues about possible functions of *LIS-1* during development, but the exact role of *LIS-1* is still unknown. PAFAH inactivates PAF by removing an acetyl group. PAF modulates the phosphoinositide second messenger systems (Asmis *et al.*, 1994; Catalan *et al.*, 1996; Lin and Rui, 1994; Yue *et al.*, 1992) and intracellular Ca^{2+} mobilization (Bito *et al.*, 1992; Rediske *et al.*, 1992) and is implicated in long-term synaptic potentiation (Bazan *et al.*, 1997; Kato and Zorumski, 1996; Kornecki *et al.*, 1996) and neuronal differentiation. A number of studies have shown the importance of Ca^{2+} to neuronal migration (Komuro and Rakic, 1993, 1996). Thus, *LIS-1* may mediate its effects by the regulation of PAF which in turn modulates intracellular Ca^{2+}.

Results from a recent study suggest that LIS-1 may also exert its effect by stabilizing cytoskeletal elements (Sapir *et al.*, 1997). The addition of LIS-1 to microtubule preparations in subphysiologic concentrations stabilized and resulted in an increased length of microtubules. Individual microtubules display dyna-

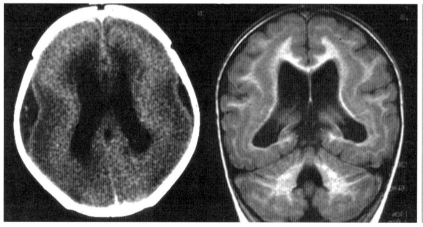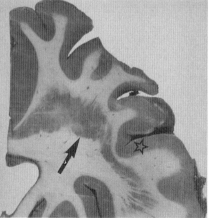

Figure 5 Lissencephaly and double cortex phenotype. MRI from a 4-year-old male patient with lissencephaly is shown on the left. The image is taken in the transverse plane. The MRI shown in the middle is from a female patient with double cortex syndrome. The image is taken in a coronal plane and shows the cortex and cerebellum. A postmortem specimen of a double cortex brain, in the same plane as in the MRI in the middle, is shown on the right. The outer cortex (star) is normal, but with a band of gray matter (arrow) embedded within the white matter beneath the cortex, representing the "double" cortex (also referred to as band heterotopia, or subcortical laminar heterotopia). Reproduced with permission from Gleeson *et al.* (1998), © Cell Press.

mic instability with transitions between growing and shortening phases (Horio and Hotani, 1986). The addition of LIS-1 to microtubule preparations led to a reduction in the transition from the growth to the shortening phase, also known as a catastrophe event. Thus, LIS-1 also may affect cortical migration by stabilizing the microtubules of the cytoskeleton during neuronal migration.

In addition, a homologue of *LIS-1* appears to be necessary for normal mitotic spindle function during cell division in *Saccharomyces cerevisiae* (Geiser *et al.*, 1997), and *nudF*, a gene similar to *LIS-1* (42% similar throughout the whole protein), in *Aspergillus nidulans* is required for normal nuclear migration during vegetative growth and development (Xiang *et al.*, 1995). Microtubules and cytoplasmic dynein, a microtubule-dependent motor, are known to be important in the nuclear migration of *A. nidulans*. In *Aspergillus*, mutations in the *nudA* gene, which encodes the cytoplasmic dynein heavy chain, suppresses mutations in *nudF* (Xiang *et al.*, 1995). Together, these studies suggest that the *nudF* protein product affects nuclear migration by acting on the dynein motor. Neuronal migration is believed to occur in two steps: the neuron first sends out a process toward its target position in the brain and the nucleus and accompanying perikaryon translocate through this process. Thus it is possible that LIS-1 may affect neuronal migration by the modulation of nuclear migration.

B. Double Cortex and X-Linked Lissencephaly

Linkage analysis from two studies have provided evidence that in addition to the *LIS-1* gene, an additional gene located on Xq21–24 may cause a very similar type of lissencephaly (des Portes *et al.*, 1997; Ross *et al.*, 1997). In affected pedigrees, males with X-linked lissencephaly (XLIS) have a phenotype that is indistinguishable from lissencephaly caused by *LIS-1* mutations (Berg *et al.*, 1998). Affected females heterozygous for the X-linked mutation display a milder phenotype. The brains of affected females show a population of neurons that behaves normally and a population of neurons that migrates approximately halfway to the cortex and then arrests in the subcortical white matter (Fig. 5), producing a band of neurons called subcortical band heterotopia or double cortex (DC) (Raymond *et al.*, 1995).

In a recent report, Gleeson *et al.* (1998) have mapped and identified a transcript, termed *doublecortin*, that is mutated in patients with double cortex and X-linked lissencephaly. Mutational analysis of DC/XLIS pedigrees and sporadic patients with DC has identified numerous deleterious mutations in *doublecortin* including nonconservative amino acid substitutions and frame shift mutations. Also in the patients analyzed, the se-

verity of the DC phenotype can be roughly correlated with the severity of the *doublecortin* mutation identified. Three sporadic patients with frame shift mutations show very severe mental retardation, while familial DC patients with single base substitutions display milder mental retardation.

The structure of *doublecortin* provides few clues to its role in neuronal migration. *doublecortin* encodes a novel 360-amino acid protein that at its amino terminus has a 75% identity to a predicted protein named KIAA0369 that in turn is 99% identical at its carboxyl terminus with a calcium calmodulin dependent (CAM) kinase found in rat. However, Doublecortin does not share significant homology to this kinase. Doublecortin has no apparent consensus signal peptide or transmembrane sequences suggesting that the protein is neither secreted nor membrane-bound. The protein does have a potential Abl phosphorylation site (Gleeson *et al.*, 1998). Thus it is possible that Doublecortin functions as part of an Abl-dependent signal transduction pathway or that it acts by modifying the activity of a CAM kinase. This is a particularly intriguing possibility because of the recent potential implication of Abl signaling in the mouse mutations that affect neuronal migration as well (see below).

C. *Reeler* and *Scrambler* Mutant Mice

The *reeler* mutation is an autosomal recessive mutation in the mouse that affects cortical migration. The *reeler* mouse has a marked malformation in the neocortex that is characterized by the inversion in the relative position of neurons such that the earlier born neurons and later born neurons are reversed (Caviness, 1982) and show a lack of cortical lamination (Fig. 6); however, appropriate neocortical neuronal connectivity occurs. In addition to the abnormalities of the cerebral cortex, *reeler* mutants also exhibit aberrant cell positioning and fasciculation of the pyriform cortex, hippocampus, and cerebellum (reviewed in Goffinet, 1984).

In the developing cortex of *reeler* mutants, neuronal formation and migration occur normally until the onset of the formation of the cortical plate (reviewed in Goffinet, 1984). When the neurons that normally form the deep layers of the cortical plate begin their ascent, they appear unable to enter the primordial plexiform layer. Subsequently born neurons are then unable to bypass the earlier born neurons resulting in the formation of a cortical plate below the primordial plexiform layer (Caviness, 1982) in an outside-in rather than the normal inside-out pattern.

The *scrambler* mutation (*scm*) is a more recently identified mutation that occurred spontaneously in mouse inbred strain that carried the Dancer mutation (Sweet *et al.*, 1996). Similar to *reeler* mice, *scrambler* mice show disorganized cerebral cortical lamination (Fig.

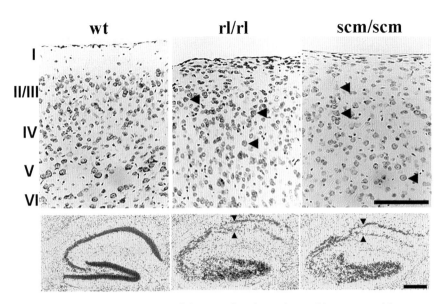

Figure 6 Histological appearance of the normal, reeler, and scrambler cortex and hippocampus. Shown are photomicrographs of cresyl violet-stained, sagittal sections of normal (wt), reeler (rl/rl), and scrambler (scm/scm) cortex (upper photomicrographs) and normal, reeler, and scrambler hippocampus (lower photomicrographs). In normal cortex there is the characteristic six-layer appearance that is not clearly distinct in reeler or scrambler cortex. Large pyramidal neurons (arrowheads) can be seen through all layers of the cortex and are somewhat more common at more superficial positions in reeler and scrambler cortex. Reeler and scrambler hippocampus differs from normal in that there is almost complete duplication of the pyramidal cell layer (arrowheads) and an overall level of disorganization. Reproduced with permission from Gonzalez *et al.* (1997), Figs. 2, 3. Scale bars = 400 μm

6) and cerebellar hypoplasia. Cerebral cortical neurons show the same dispersion and rough inversion of cortical birth dates, and detailed analysis of the cerebellar phenotype suggest close similarities to *reeler* (Gonzalez *et al.*, 1997). In addition Reelin protein expression is normal in *scm* mutants in its localization and timing, suggesting that the *scm* gene may act downstream of the Reelin protein.

The gene responsible for the *reeler* mutation, *reelin (rl)*, was cloned by two groups simultaneously, with complementary research coming from a third group. Using a strategy that employed the use of a *reeler* mutant produced by transgenic insertion, D'Arcangelo *et al.* (1995) employed polymerase chain reaction (PCR)-based technology to create probes to screen cDNA libraries and cloned the full length cDNA for *reelin*. In an independent effort, Hirotsune *et al.* (1995) identified *reelin* by constructing a genetic and physical map of the *reeler* locus. Ogawa *et al.* (1995), generated an antibody that was later shown to recognize Reelin protein by immunizing *reeler* mice with the brain extracts from wild-type mice.

Reelin is first expressed by E11.5 and expression increases up to birth and remains high from postnatal day 2 to 11 in both cerebellum and forebrain, declining thereafter (D'Arcangelo *et al.*, 1995). By *in situ* hybridization and immunohistochemistry, Reelin has been shown to be expressed at its highest levels in the Cajal-Retzius cells of the cerebral cortex and the hippocampus and in the external granule layer of the cerebellum

(Schiffmann *et al.*, 1997). The Cajal-Retzius cells are among the first neurons in the cerebral cortex and may serve as pioneer neurons for subsequent migrating cells. In addition, immunohistochemistry has shown that Reelin is an extracellular protein, suggesting that this protein modifies neuronal migration by altering the extracellular matrix. Furthermore, Ogawa *et al.* (1995), have shown that the addition of CR-50, the antibody that recognizes Reelin protein, to reaggregation assays of normal mouse brain, converts the normal histotypic organization to the abnormal pattern seen during reaggregation of *reeler* mouse brains. These results suggest that the Reelin protein may be important in regulating the cell–cell interactions of cortical neurons.

D. Cloning of mDab1 and Identification as the *scrambler* Gene

The gene responsible for the *scrambler* mutation has been identified (Sheldon *et al.*, 1997; Ware *et al.*, 1997) as the *mdab1* gene (Howell *et al.*, 1997a,b), which is a homologue of *Drosophila disabled*. We identified *mdab1* as a candidate gene for *scrambler* by genetic mapping and identifying a yeast artificial chromosome (YAC) containing an exon of the *mdab1* gene in the candidate region (Ware *et al.*, 1997). Northern analysis of RNA from *scm* mice showed that at least two different transcripts of *mdab1* migrated at aberrant rates. Analysis of abnormal transcripts showed that an intracisternal A particle (IAP) is inserted into all known transcripts in the antisense direction by a splicing event, introducing premature translational stop codons (Ware *et al.,* 1997). IAP elements are endogenous retroviruslike mobile elements that occur at ~1000 copies per haploid genome in the mouse (Kuff and Lueders, 1988; Lueders, 1987) and have been increasingly commonly recognized as causing insertional mutations in the mouse (Amariglio and Rechavi, 1993). In a similar study, Sheldon *et al.* (1997) also showed that *scm* is caused by an IAP insertion into *mdab1* transcripts. In support of the identification of *mdab1* as *scm*, the *mdab1* knockout has a phenotype that is almost indistinguishable from that of *scm* mutants (Howell *et al.,* 1997b). Simultaneously, an exon deletion of *mdab1* was found in the *yotari* mouse, which has a spontaneous mutation with a very similar phenotype to the *scm* mouse (Sheldon *et al.,* 1997).

Drosophila disabled (*dab*) was cloned in a screen for dominant second site mutations that exacerbate mutations in the *Drosophila Abl* (*abl*) gene, a nonreceptor kinase that is localized to the axons of the CNS (Gertler *et al.,* 1989). Mutations in *dab* and *abl* result in disruption of axonal bundles in the CNS. The *dab* gene encodes a 2412-residue protein that colocalizes with Abl to the cell bodies and axons of embryonic CNS neurons and is essential for CNS development (Gertler *et al.,* 1993). The predicted Dab protein contains 10 matches to a motif that is similar to the major autophosphorylation site of Abl. In *Drosophila* S2 cells, which endogenously express Abl and Dab, Dab is phosphorylated. Taken together, these findings suggest that *dab* is expressed within the CNS of *Drosophila,* and its protein product may be phosphorylated in axonal tracts by Abl, and this phosphorylation may be required for the proper function of Dab.

The mouse homologue of *dab*, *mdab1* was originally cloned based on the ability of its protein product to bind the Src protein in a two-hybrid screen (Howell *et al.,* 1997a). Whole-mount double-label immunostaining of E10.5 embryos revealed that mDab1 is localized in the head in neural tracts corresponding to the developing cranial nerves (Howell *et al.,* 1997a). In the body, *mdab1* is expressed in the spinal accessory nerve and dorsal root ganglia. At E13, all nerves identified by neurofilament antibody also expressed *mdab1*. Overexpression of mdab1 and activated Src in 293T fibroblast induced tyrosine phosphorylation of mDab1. Tyrosine phosphorylated mDab1, but not control mDab1, associated with the Src and Fyn SH2 domain *in vitro,* with that of Abl to a lesser degree, but not with the SH2 domain of Fyn (Howell *et al.,* 1997a). In addition, mDab1 was reimmunoprecipitated from Src immunoprecipitates made from lysates of differentiating S16 cells. Together, these experiments suggest that mDab1 is an axonal protein that may be a substrate for Src or a Src-activated tyrosine kinase.

Since mDab1 was cloned by virtue of its physical binding to Src, the identification of *mdab1* as *scm* possibly implicates nonreceptor tyrosine kinases such as Abl, Src, and Fyn in the control of neuronal migration. Assuming that mDab1 exerts its essential role by interacting with nonreceptor tyrosine kinases it is not clear which kinases are utilized. The homology of mDab1 to *Drosophila disabled,* which has been implicated in axon outgrowth, allows additional insight into its function during the development of the cerebral cortex. *Drosophila disabled* has been implicated in neuronal development through genetic interactions with *abl*. Another gene identified based on its interaction with *abl, enabled,* has a mouse homologue, MENA, that has recently been critically implicated in the migration of nonneuronal cells. The identification of *mdab1* as the *scm* gene has also raised the possibility that *scm* and *doublecortin,* which has a potential Abl phosphorylation site, are within the same genetic pathway. However, no studies have confirmed an interaction between these genes. Other genes that interact with *abl* include *fax* and *prospero*. Mutations in these genes cause abnormalities of neuronal process outgrowth as well (Gertler *et al.,* 1993). Kinases such as Src have been implicated for some time in axonal outgrowth. Since cell migration into the cortex occurs by the extension of a leading neuritelike process, there may be many analogies and conserved signaling pathways involved in both neuronal migration and axonal outgrowth.

E. *cdk5* and *p35*

Two additional genes have been shown to be important for normal neuronal migration in the cerebral cortex: *cdk5* and *p35*. *cdk5* was originally isolated on the basis of its close primary sequence homology to the human cdc2 serine/threonine kinase. Unlike other cell division cycle kinases, *cdk5* is expressed only in postmitotic cells in the developing cerebral cortex (Tsai *et al.,* 1993). Targeted disruption of *ckd5* in the mouse causes the lack of cortical lamination and cerebellar foliation similar to the *scm/reeler* phenotype (Ohshima *et al.,* 1996); however *cdk5* −/− mice die on the day of birth. Cortical birth-dating studies have not been performed to further analyze similarities with the *scm/reeler* phenotype. p35 is the regulatory subunit for cdk5 that has been shown to physically associate with cdk5 and to activate cdk5 activity. Targeted disruption of *p35* in the mouse causes a lack of cortical lamination and sporadic adult lethality and seizures (Chae *et al.,* 1997). Cortical birth-dating studies indicate that neurons of the cortical plate are born in the "outside in" pattern seen in the *scm/reeler* phenotype. In addition, cdk5/p35 has been implicated in axonal outgrowth. Expression of dominant negative mutants of cdk5 in cortical cultures led to an inhibition of neurite outgrowth (Nikolic *et al.,* 1996). In contrast, longer neurites were elaborated by neurons that coexpressed exogenous cdk5 and p35. Thus, *cdk5* and *p35* are similar to *mdab1* in that they have also been implicated in axonal outgrowth.

VII. Conclusions

The mammalian cerebral cortex is composed of the neural networks responsible for the behavioral diversity of mammals and intellectual life of humans. Studies describing cortical precursor proliferation, migration, and differentiation have revealed a highly ordered sequence of events by which cells from the relatively simple undifferentiated neuroepithelium of the developing telencephalon form the specialized neurons of the cerebral cortex in which cell position, morphology, and connectivity reflect a neuronal function.

Recently, studies in *Drosophila*, mice, and humans have identified a number of individual genes involved in the control of cortical development. However, our knowledge of the genetic pathways responsible for controlling cortical development is still in its infancy. For example, there have been a number of genes recently identified and shown to be required for the normal migration of cortical precursors. We know that mutations in *LIS-1* and *doublecortin* produce a near identical phenotype in humans, suggesting that they are within the same genetic pathway. However, LIS-1 is thought to act by controlling Ca^{2+} dynamics or nuclear migration, Doublecortin is a novel protein with no known function, and a relationship between the two has not been elucidated. Additionally, *scm/mdab1, reeler, cdk5,* and *p35* mutations also lead to similar phenotypes, suggesting that they are within the same genetic pathway. However, *reeler* encodes an extracellular protein, *scm/mdab1* encodes a cytoplasmic substrate for nonreceptor tyrosine kinases, and *Cdk5* and *p35* are involved in serine phosphorylation. If *scm* and *reeler* are, in fact, within the same genetic pathway, they are not likely to physically interact, as it was once thought, and the relationship between *Cdk5* and *p35* and *scm* and *reeler* is also a mystery. Moreover, *scm* and *doublecortin* may be within the same pathway as suggested by the possibility that they are both phosphorylated by Abl. However, their cortical phenotypes are only grossly similar and their cerebellar and behavioral phenotypes are very different, possibly partially due to differences between species.

Despite the seemingly daunting complexity of cortical development, recent discoveries have been very promising. Defining and understanding underlying genetic pathways will require the efforts of researchers studying different facets of development in different organisms. The integration of this fast-growing knowledge from many different sources will be a major challenge of developmental neurobiology and, indeed, of all of science.

References

Ahlgren, U., Pfaff, S. L., Jessell, T. M., Edlund, T., and Edlund, H. (1997). Independent requirement for ISL1 in formation of pancreatic mesenchyme and islet cells. *Nature (London)* **385**, 257–260.

Allendoerfer, K. L., and Shatz, C. J. (1994). The subplate, a transient neocortical structure: Its role in the development of connections between thalamus and cortex. *Annu. Rev. Neurosci.* **17**, 185–218.

Amariglio, N., and Rechavi, G. (1993). Insertional mutagenesis by transposable elements in the mammalian genome. *Environ. Mol. Mutagenesis* **21**, 212–218.

Anderson, S. A., Eisenstat, D. D., Shi, L., and Rubenstein, J. L. (1997). Interneuron migration from basal forebrain to neocortex: Dependence on Dlx Genes. *Science* **278**, 474–476.

Artavanis-Tsakonas, S., Matsuno, K., and Fortini, M. E. (1995). Notch signaling. *Science* **268**, 225–232.

Asmis, R., Randriamampita, C., Tsien, R. Y., and Dennis, E. A. (1994). Intracellular Ca^{2+}, inositol 1,4,5-trisphosphate and additional signalling in the stimulation by platelet-activating factor of prostaglandin E2 formation in P388D1 macrophage-like cells. *Biochem. J.* **298**, 543–551.

Austin, C. P., Feldman, D. E., Ida, J. A., Jr., and Cepko, C. L. (1995). Vertebrate retinal ganglion cells are selected from competent progenitors by the action of Notch. *Development (Cambridge, UK)* **121**, 3637–3650.

Bazan, N. G., Packard, M. G., Teather, L., and Allan, G. (1997). Bioactive lipids in excitatory neurotransmission and neuronal plasticity. *Neurochem. Int.* **30**, 225–231.

Berg, M. J., Schifitto, G., Powers, J., Maritinez-Capolino, C., Fong, C.-T., Myers, G., Epstein, L., and Walsh, C. (1998). X-linked female band heterotopia–male lissencephaly syndrome. *Neurology* **50**, 1143–1146.

Bettenhausen, B., Hrabe de Angelis, M., Simon, D., Guenet, J. L., and Gossler, A. (1995). Transient and restricted expression during mouse embryogenesis of Dl11, a murine gene closely related to *Drosophila Delta*. *Development (Cambridge, UK)* **121**, 2407–2418.

Bierkamp, C., and Campos-Ortega, J. A. (1993). A zebrafish homologue of the *Drosophila* neurogenic gene *Notch* and its pattern of transcription during early embryogenesis. *Mech. Dev.* **43**, 87–100.

Bito, H., Nakamura, M., Honda, Z., Izumi, T., Iwatsubo, T., Seyama, Y., Ogura, A., Kudo, Y., and Shimizu, T. (1992). Platelet-activating factor (PAF) receptor in rat brain: PAF mobilizes intracellular Ca^{2+} in hippocampal neurons. *Neuron* **9**, 285–294.

Bossing, T., Udolph, G., Doe, C. Q., and Technau, G. M. (1996). The embryonic central nervous system lineages of *Drosophila melanogaster*. I. Neuroblast lineages derived from the ventral half of the neuroectoderm. *Dev. Biol.* **179**, 41–64.

Cabrera, C. V. (1990). Lateral inhibition and cell fate during neurogenesis in *Drosophila*: The interactions between scute, Notch and Delta. *Development (Cambridge, UK)* **110**, 733–742.

Cabrera, C. V., Alonso, M. C., and Huikeshoven, H. (1994).

Regulation of scute function by extramacrochaete *in vitro* and *in vivo*. *Development (Cambridge, UK)* **120**, 3595–3603.

Cai, L., Hayes, N. L., and Nowakowski, R. S. (1997). Synchrony of clonal cell proliferation and contiguity of clonally related cells: Production of mosaicism in the ventricular zone of developing mouse neocortex. *J. Neurosci.* **17**, 2088–2100.

Catalan, R. E., Martinez, A. M., Aragones, M. D., Martinez, A., and Diaz, G. (1996). Endothelin stimulates phosphoinositide hydrolysis and PAF synthesis in brain microvessels. *J. Cerebral Blood Flow Metab.* **16**, 1325–1334.

Caviness, V. S., Jr. (1982). Neocortical histogenesis in normal and reeler mice: A developmental study based upon [³H]thymidine autoradiography. *Brain Res.* **256**, 293–302.

Chae, T., Kwon, Y. T., Bronson, R., Dikkes, P., Li, E., and Tsai, L. H. (1997). Mice lacking p35, a neuronal specific activator of Cdk5, display cortical lamination defects, seizures, and adult lethality. *Neuron* **18**, 29–42.

Chenn, A., and McConnell, S. K. (1995). Cleavage orientation and the asymmetric inheritance of Notch1 immunoreactivity in mammalian neurogenesis. *Cell (Cambridge, Mass.)* **82**, 631–641.

Chitnis, A., Henrique, D., Lewis, J., Ish-Horowicz, D., and Kintner, C. (1995). Primary neurogenesis in *Xenopus* embryos regulated by a homologue of the *Drosophila* neurogenic gene *Delta*. *Nature (London)* **375**, 761–766.

Coffman, C. R., Skoglund, P., Harris, W. A., and Kintner, C. R. (1993). Expression of an extracellular deletion of Xotch diverts cell fate in *Xenopus* embryos. *Cell (Cambridge, Mass.)* **73**, 659–671.

Cubas, P., de Celis, J. F., Campuzano, S., and Modolell, J. (1991). Proneural clusters of achaete-scute expression and the generation of sensory organs in the *Drosophila* imaginal wing disc. *Genes Dev.* **5**, 996–1008.

D'Arcangelo, G., Miao, G. G., Chen, S. C., Soares, H. D., Morgan, J. I., and Curran, T. (1995). A protein related to extracellular matrix proteins deleted in the mouse mutant reeler. *Nature (London)* **374**, 719–723.

Davis, A. A., and Temple, S. (1994). A self-renewing multipotential stem cell in embryonic rat cerebral cortex. *Nature (London)* **372**, 263–266.

de Celis, J. F., Mari-Beffa, M., and Garcia-Bellido, A. (1991). Cell-autonomous role of Notch, an epidermal growth factor homologue, in sensory organ differentiation in *Drosophila*. *Proc. Nat. Acad. Sci. U.S.A.* **88**, 632–636.

Dehni, G., Liu, Y., Husain, J., and Stifani, S. (1995). TLE expression correlates with mouse embryonic segmentation, neurogenesis, and epithelial determination. *Mech. Dev.* **53**, 369–381.

del Amo, F. F., Gendron-Maguire, M., Swiatek, P. J., Jenkins, N. A., Copeland, N. G., and Gridley, T. (1993). Cloning, analysis, and chromosomal localization of Notch-1, a mouse homolog of *Drosophila* Notch. *Genomics* **15**, 259–264.

des Portes, V., Pinard, J. M., Smadja, D., Motte, J., Boespflug-Tanguy, O., Moutard, M. L., Desguerre, I., Billuart, P., Carrie, A., Bienvenu, T., Vinet, M. C., Bachner, L., Beldjord, C., Dulac, O., Kahn, A., Ponsot, G., and

Chelly, J. (1997). Dominant X linked subcortical laminar heterotopia and lissencephaly syndrome (XSCLH/LIS): Evidence for the occurrence of mutation in males and mapping of a potential locus in Xq22. *J. Med. Genet.* **34**, 177–183.

Dobyns, W. B., Curry, C. J., Hoyme, H. E., Turlington, L., and Ledbetter, D. H. (1991). Clinical and molecular diagnosis of Miller-Dieker syndrome. *Am. J. Hum. Genet.* **48**, 584–594.

Dorsky, R. I., Rapaport, D. H., and Harris, W. A. (1995). Xotch inhibits cell differentiation in the *Xenopus* retina. *Neuron* **14**, 487–496.

Ellis, H. M., Spann, D. R., and Posakony, J. W. (1990). Extramacrochaetae, a negative regulator of sensory organ development in *Drosophila*, defines a new class of helix-loop-helix proteins. *Cell (Cambridge, Mass.)* **61**, 27–38.

Ericson, J., Thor, S., Edlund, T., Jessell, T. M., and Yamada, T. (1992). Early stages of motor neuron differentiation revealed by expression of homeobox gene Islet-1. *Science* **256**, 1555–1560.

Ernfors, P., Lee, K. F., Kucera, J., and Jaenisch, R. (1994). Lack of neurotrophin-3 leads to deficiencies in the peripheral nervous system and loss of limb proprioceptive afferents. *Cell (Cambridge, Mass.)* **77**, 503–512.

Fishell, G., Mason, C. A., and Hatten, M. E. (1993). Dispersion of neural progenitors within the germinal zones of the forebrain. *Nature (London)* **362**, 636–638.

Fortini, M. E., and Artavanis-Tsakonas, S. (1994). The suppressor of hairless protein participates in notch receptor signaling. *Cell (Cambridge, Mass.)* **79**, 273–282.

Frise, E., Knoblich, J. A., Younger-Shepherd, S., Jan, L. Y., and Jan, Y. N. (1996). The *Drosophila* Numb protein inhibits signaling of the Notch receptor during cell–cell interaction in sensory organ lineage. *Proc. Nat. Acad. Sci. U.S.A.* **93**, 11925–11932.

Geiser, J. R., Schott, E. J., Kingsbury, T. J., Cole, N. B., Totis, L. J., Bhattacharyya, G., He, L., and Hoyt, M. A. (1997). *Saccharomyces cerevisiae* genes required in the absence of the CIN8-encoded spindle motor act in functionally diverse mitotic pathways. *Mol. Biol. Cell* **8**, 1035–1050.

Gertler, F. B., Bennett, R. L., Clark, M. J., and Hoffmann, F. M. (1989). *Drosophila* abl tyrosine kinase in embryonic CNS axons: A role in axonogenesis is revealed through dosage-sensitive interactions with disabled. *Cell (Cambridge, Mass.)* **58**, 103–113.

Gertler, F. B., Hill, K. K., Clark, M. J., and Hoffmann, F. M. (1993). Dosage-sensitive modifiers of *Drosophila* abl tyrosine kinase function: Prospero, a regulator of axonal outgrowth, and disabled, a novel tyrosine kinase substrate. *Genes Dev.* **7**, 441–453.

Ghosh, A., and Greenberg, M. E. (1995). Distinct roles for bFGF and NT-3 in the regulation of cortical neurogenesis. *Neuron* **15**, 89–103.

Gleeson, J., Allen, K., Fox, J., Lamperti, E., Berkovic, S., Scheffer, I., Cooper, E., Dobyns, W., Minnerath, S., Ross, E., and Walsh, C. A. (1998). *doublecortin*, a brain-specific gene mutated in human X-linked lissencephaly and double cortex syndrome, encodes a putative signaling protein. *Cell (Cambridge, Mass.)* **92**, 63–72.

Goffinet, A. M. (1984). Events governing organization of post-migratory neurons: Studies on brain development in normal and reeler mice. *Brain Res.* **319**, 261–296.

Gonzalez, A. M., Buscaglia, M., Ong, M., and Baird, A. (1990). Distribution of basic fibroblast growth factor in the 18-day rat fetus: Localization in the basement membranes of diverse tissues. *J. Cell Biol.* **110**, 753–765.

Gonzalez, J. L., Russo, C., Goldowitz, D., Sweet, H. O., Davisson, M. T., and Walsh, C. A. (1997). Birthdate and cell marker analysis of scrambler: A novel mutation affecting cortical development with a reeler-like phenotype. *J. Neurosci.* **17**, 9204–9211.

Guillemot, F., and Joyner, A. L. (1993). Dynamic expression of the murine Achaete-Scute homologue Mash-1 in the developing nervous system. *Mech. Dev.* **42**, 171–185.

Guillemot, F., Lo, L. C., Johnson, J. E., Auerbach, A., Anderson, D. J., and Joyner, A. L. (1993). Mammalian achaete-scute homolog 1 is required for the early development of olfactory and autonomic neurons. *Cell (Cambridge, Mass.)* **75**, 463–476.

Guo, M., Jan, L. Y., and Jan, Y. N. (1996). Control of daughter cell fates during asymmetric division: Interaction of Numb and Notch. *Neuron* **17**, 27–41.

Hatini, V., Tao, W., and Lai, E. (1994). Expression of winged helix genes, BF-1 and BF-2, define adjacent domains within the developing forebrain and retina. *J. Neurobiol.* **25**, 1293–1309.

Hattori, M., Adachi, H., Tsujimoto, M., Arai, H., and Inoue, K. (1994). Miller-Dieker lissencephaly gene encodes a subunit of brain platelet-activating factor acetylhydrolase. *Nature (London)* **370**, 216–218.

Heitzler, P., and Simpson, P. (1991). The choice of cell fate in the epidermis of *Drosophila*. *Cell (Cambridge, Mass.)* **64**, 1083–1092.

Helms, J. A., Kuratani, S., and Maxwell, G. D. (1994). Cloning and analysis of a new developmentally regulated member of the basic helix-loop-helix family. *Mech. Dev.* **48**, 93–108.

Henrique, D., Adam, J., Myat, A., Chitnis, A., Lewis, J., and Ish-Horowicz, D. (1995). Expression of a Delta homologue in prospective neurons in the chick. *Nature (London)* **375**, 787–790.

Hirotsune, S., Takahara, T., Sasaki, N., Hirose, K., Yoshiki, A., Ohashi, T., Kusakabe, M., Murakami, Y., Muramatsu, M., Watanabe, S., *et al.* (1995). The reeler gene encodes a protein with an EGF-like motif expressed by pioneer neurons. *Nat. Genet.* **10**, 77–83.

Horio, T., and Hotani, H. (1986). Visualization of the dynamic instability of individual microtubules by dark-field microscopy. *Nature (London)* **321**, 605–607.

Hoshijima, K., Kohyama, A., Watakabe, I., Inoue, K., Sakamoto, H., and Shimura, Y. (1995). Transcriptional regulation of the Sex-lethal gene by helix-loop-helix proteins. *Nucleic Acids Res.* **23**, 3441–3448.

Howell, B. W., Gertler, F. B., and Cooper, J. A. (1997a). Mouse disabled (mDab1): A Src binding protein implicated in neuronal development. *EMBO J.* **16**, 121–132.

Howell, B. W., Hawkes, R., Soriano, P., and Cooper, J. A. (1997b). Neuronal position in the developing brain is regulated by mouse disabled-1. *Nature (London)* **389**, 733–736.

Ishibashi, M., Moriyoshi, K., Sasai, Y., Shiota, K., Nakanishi, S., and Kageyama, R. (1994). Persistent expression of helix-loop-helix factor HES-1 prevents mammalian neural differentiation in the central nervous system. *EMBO J.* **13**, 1799–1805.

Ishibashi, M., Ang, S. L., Shiota, K., Nakanishi, S., Kageyama, R., and Guillemot, F. (1995). Targeted disruption of mammalian hairy and Enhancer of split homolog-1 (HES-1) leads to upregulation of neural helix-loop-helix factors, premature neurogenesis, and severe neural tube defects. *Genes Dev.* **9**, 3136–3148.

Jellinger, K., and Rett, A. (1976). Agyria-pachygyria (lissencephaly syndrome). *Neuropadiatrie* **7**, 66–91.

Kato, K., and Zorumski, C. F. (1996). Platelet-activating factor as a potential retrograde messenger. *J. Lipid Mediators Cell Signalling* **14**, 341–348.

Klein, R., Silos-Santiago, I., Smeyne, R. J., Lira, S. A., Brambilla, R., Bryant, S., Zhang, L., Snider, W. D., and Barbacid, M. (1994). Disruption of the neurotrophin-3 receptor gene trkC eliminates Ia muscle afferents and results in abnormal movements. *Nature (London)* **368**, 249–251.

Knoblich, J. A., Jan, L. Y., and Jan, Y. N. (1995). Asymmetric segregation of Numb and Prospero during cell division. *Nature (London)* **377**, 624–627.

Komuro, H., and Rakic, P. (1993). Modulation of neuronal migration by NMDA receptors. *Science* **260**, 95–97.

Komuro, H., and Rakic, P. (1996). Intracellular Ca^{2+} fluctuations modulate the rate of neuronal migration. *Neuron* **17**, 275–285.

Kornecki, E., Wieraszko, A., Chan, J., and Ehrlich, Y. H. (1996). Platelet activating factor (PAF) in memory formation: Role as a retrograde messenger in long-term potentiation. *J. Lipid Mediators Cell Signalling* **14**, 115–126.

Kuchelmeister, K., Bergmann, M., and Gullotta, F. (1993). Neuropathology of lissencephalies. *Child's Nervous System* **9**, 394–399.

Kuff, E. L., and Lueders, K. K. (1988). The intracisternal A-particle gene family: Structure and functional aspects. *Adv. Cancer Res.* **51**, 183–276.

Lamballe, F., Smeyne, R. J., and Barbacid, M. (1994). Developmental expression of trkC, the neurotrophin-3 receptor, in the mammalian nervous system. *J. Neurosci.* **14**, 14–28.

Lardelli, M., Dahlstrand, J., and Lendahl, U. (1994). The novel Notch homologue mouse Notch 3 lacks specific epidermal growth factor-repeats and is expressed in proliferating neuroepithelium. *Mech. Dev.* **46**, 123–136.

Ledbetter, S. A., Kuwano, A., Dobyns, W. B., and Ledbetter, D. H. (1992). Microdeletions of chromosome 17p13 as a cause of isolated lissencephaly. *Am. J. Hum. Genet.* **50**, 182–189.

Lin, A. Y., and Rui, Y. C. (1994). Platelet-activating factor induced calcium mobilization and phosphoinositide metabolism in cultured bovine cerebral microvascular endothelial cells. *Biochim. Biophys. Acta* **1224**, 323–328.

Lindsell, C. E., Shawber, C. J., Boulter, J., and Weinmaster, G. (1995). Jagged: A mammalian ligand that activates Notch1. *Cell (Cambridge, Mass.)* **80**, 909–917.

Lueders, K. K. (1987). Specific association between type-II intracisternal A-particle elements and other repetitive sequences in the mouse genome. *Gene* **52**, 139–146.

Maisonpierre, P. C., Belluscio, L., Friedman, B., Alderson, R. F., Wiegand, S. J., Furth, M. E., Lindsay, R. M., and Yancopoulos, G. D. (1990). NT-3, BDNF, and NGF in the developing rat nervous system: Parallel as well as reciprocal patterns of expression. *Neuron* **5**, 501–509.

Marin-Padilla, M. (1971). Early prenatal ontogenesis of the cerebral cortex (neocortex) of the cat (*Felis domestica*). A Golgi study. I. The primordial neocortical organization. *Z. Anat. Entwicklungsgesch.* **134**, 117–145.

Marin-Padilla, M. (1978). Dual origin of the mammalian neocortex and evolution of the cortical plate. *Anat. Embryol.* **152**, 109–126.

Myat, A., Henrique, D., Ish-Horowicz, D., and Lewis, J. (1996). A chick homologue of Serrate and its relationship with Notch and Delta homologues during central neurogenesis. *Dev. Biol.* **174**, 233–247.

Nikolic, M., Dudek, H., Kwon, Y. T., Ramos, Y. F., and Tsai, L. H. (1996). The cdk5/p35 kinase is essential for neurite outgrowth during neuronal differentiation. *Genes Dev.* **10**, 816–825.

Nye, J. S., Kopan, R., and Axel, R. (1994). An activated Notch suppresses neurogenesis and myogenesis but not gliogenesis in mammalian cells. *Development (Cambridge, UK)* **120**, 2421–2430.

O'Rourke, N. A., Dailey, M. E., Smith, S. J., and McConnell, S. K. (1992). Diverse migratory pathways in the developing cerebral cortex. *Science* **258**, 299–302.

O'Rourke, N. A., Sullivan, D. P., Kaznowski, C. E., Jacobs, A. A., and McConnell, S. K. (1995). Tangential migration of neurons in the developing cerebral cortex. *Development (Cambridge, UK)* **121**, 2165–2176.

O'Rourke, N. A., Chenn, A., and McConnell, S. K. (1997). Postmitotic neurons migrate tangentially in the cortical ventricular zone. *Development (Cambridge, UK)* **124**, 997–1005.

Ogawa, M., Miyata, T., Nakajima, K., Yagyu, K., Seike, M., Ikenaka, K., Yamamoto, H., and Mikoshiba, K. (1995). The reeler gene-associated antigen on Cajal-Retzius neurons is a crucial molecule for laminar organization of cortical neurons. *Neuron* **14**, 899–912.

Ohshima, T., Ward, J. M., Huh, C. G., Longenecker, G., Veeranna, Pant, H. C., Brady, R. O., Martin, L. J., and Kulkarni, A. B. (1996). Targeted disruption of the cyclin-dependent kinase 5 gene results in abnormal corticogenesis, neuronal pathology and perinatal death. *Proc. Natl. Acad. Sci. U.S.A.* **93**, 11173–11178.

Pfaff, S. L., Mendelsohn, M., Stewart, C. L., Edlund, T., and Jessel, T. M. (1996). Requirement for LIM homeobox gene Is11 in motor neuron generation reveals a motor neuron-dependent step in interneuron differentiation. *Cell (Cambridge, Mass.)* **84**, 309–320.

Polleux, F., Dehay, C., Moraillon, B., and Kennedy, H. (1997). Regulation of neuroblast cell-cycle kinetics plays a crucial role in the generation of unique features of neocortical areas. *J. Neurosci.* **17**, 7763–7783.

Porter, F. D., Drago, J., Xu, Y., Cheema, S. S., Wassif, C.,

Huang, S. P., Lee, E., Grinberg, A., Massalas, J. S., Bodine, D., Alt, F., and Westphal, H. (1997). Lhx2, a LIM homeobox gene, is required for eye, forebrain, and definitive erythrocyte development. *Development (Cambridge, UK)* **124**, 2935–2944.

Powell, P. P., Finklestein, S. P., Dionne, C. A., Jaye, M., and Klagsbrun, M. (1991). Temporal, differential and regional expression of mRNA for basic fibroblast growth factor in the developing and adult rat brain. *Brain Res. Mol. Brain Res.* **11**, 71–77.

Price, J., and Thurlow, L. (1988). Cell lineage in the rat cerebral cortex: A study using retroviral-mediated gene transfer. *Development (Cambridge, UK)* **104**, 473–482.

Rakic, P. (1971). Guidance of neurons migrating to the fetal monkey neocortex. *Brain Res.* **33**, 471–476.

Rakic, P. (1972). Mode of cell migration to the superficial layers of fetal monkey neocortex. *J. Comp. Neurol.* **145**, 61–83.

Raymond, A. A., Fish, D. R., Boyd, S. G., Smith, S. J., Pitt, M. C., and Kendall, B. (1995). Cortical dysgenesis: Serial EEG findings in children and adults. *Electroencephalogr. Clin. Neurophysiol.* **94**, 389–397.

Reaume, A. G., Conlon, R. A., Zirngibl, R., Yamaguchi, T. P., and Rossant, J. (1992). Expression analysis of a Notch homologue in the mouse embryo. *Dev. Biol.* **154**, 377–387.

Rediske, J. J., Quintavalla, J. C., Haston, W. O., Morrissey, M. M., and Seligman, B. (1992). Platelet activating factor stimulates intracellular calcium transients in human neutrophils: Involvement of endogenous 5-lipoxygenase products. *J. Leukocyte Biol.* **51**, 484–489.

Reid, C. B., and Walsh, C. A. (1996). Early development of the cerebral cortex. *Prog. Brain Res.* **108**, 17–30.

Reid, C. B., Liang, I., and Walsh, C. (1995). Systematic widespread clonal organization in cerebral cortex. *Neuron* **15**, 299–310.

Reid, C. B., Tavazoie, S. F. and Walsh, C. A. (1997). Clonal dispersion and evidence for asymmetric cell division in ferret cortex. *Development (Cambridge, UK)* **124**, 2441–2450.

Reynolds, B. A., and Weiss, S. (1992). Generation of neurons and astrocytes from isolated cells of the adult mammalian central nervous system. *Science* **255**, 1707–1710.

Reynolds, B. A., Tetzlaff, W., and Weiss, S. (1992). A multipotent EGF-responsive striatal embryonic progenitor cell produces neurons and astrocytes. *J. Neurosci.* **12**, 4565–4574.

Rhyu, M. S., Jan, L. Y., and Jan, Y. N. (1994). Asymmetric distribution of numb protein during division of the sensory organ precursor cell confers distinct fates to daughter cells. *Cell (Cambridge, Mass.)* **76**, 477–491.

Ross, M. E., Allen, K. M., Srivastava, A. K., Featherstone, T., Gleeson, J. G., Hirsch, B., Harding, B. N., Andermann, E., Abdullah, R., Berg, M., Czapansky-Bielman, D., Flanders, D. J., Guerrini, R., Motte, J., Mira, A. P., Scheffer, I., Berkovic, S., Scaravilli, F., King, R. A., Ledbetter, D. H., Schlessinger, D., Dobyns, W. B., and Walsh, C. A. (1997). Linkage and physical mapping of X-linked lissencephaly/SBH (XLIS): A gene causing neuronal migration defects in human brain. *Hum. Mol. Genet.* **6**, 555–562.

Sapir, T., Elbaum, M., and Reiner, O. (1997). Reduction of microtubule catastrophe events by LIS-1, platelet-activating factor acetylhydrolase subunit. *EMBO J.* **16**, 6977–6984.

Sasai, Y., Kageyama, R., Tagawa, Y., Shigemoto, R., and Nakanishi, S. (1992). Two mammalian helix-loop-helix factors structurally related to *Drosophila* hairy and Enhancer of split. *Genes Dev.* **6**, 2620–2634.

Sauer, F. C. (1935). Mitosis in the neural tube. *J. Comp. Neurol.* **62**, 377–405.

Schiffmann, S. N., Bernier, B., and Goffinet, A. M. (1997). Reelin mRNA expression during mouse brain development. *Eur. J. Neurosci.* **9**, 1055–1071.

Schmidt, H., Rickert, C., Bossing, T., Vef, O., Urban, J., and Technau, G. M. (1997). The embryonic central nervous system lineages of *Drosophila melanogaster*. II. Neuroblast lineages derived from the dorsal part of the neuroectoderm. *Dev. Biol.* **189**, 186–204.

Seugnet, L., Simpson, P., and Haenlin, M. (1997). Transcriptional regulation of Notch and Delta: Requirement for neuroblast segregation in *Drosophila*. *Development (Cambridge, UK)* **124**, 2015–2025.

Sheldon, M., Rice, D. S., D'Arcangelo, G., Yoneshima, H., Nakajima, K., Mikoshiba, K., Howell, B. W., Cooper, J. A., Goldowitz, D., and Curran, T. (1997). *Scrambler* and *yotari* disrupt the *diabled* gene and produce a reeler-like phenotype in mice. *Nature (London)* **389**, 730–733.

Skeath, J. B., and Carroll, S. B. (1991). Regulation of *achaete-scute* gene expression and sensory organ pattern formation in the *Drosophila* wing. *Genes Dev.* **5**, 984–995.

Spana, E. P., Kopczynski, C., Goodman, C. S., and Doe, C. Q. (1995). Asymmetric localization of numb autonomously determines sibling neuron identity in the *Drosophila* CNS. *Development (Cambridge, UK)* **121**, 3489–3494.

Sweet, H. O., Bronson, R. T., Johnson, K. R., Cook, S. A., and Davidson, M. T. (1996). Scrambler, a new neurological mutation of the mouse with abnormalities of neuronal migration. *Mammal. Genome* **7**, 798–802.

Swiatek, P. J., Lindsell, C. E., del Amo, F. F., Weinmaster, G., and Gridley, T. (1994). Notch1 is essential for postimplantation development in mice. *Genes Dev.* **8**, 707–719.

Takahashi, T., Nowakowski, R. S., and Caviness, V. S., Jr. (1993). Cell cycle parameters and patterns of nuclear movement in the neocortical proliferative zone of the fetal mouse. *J. Neurosci.* **13**, 820–833.

Tan, S. S., and Breen, S. (1993). Radial mosaicism and tangential cell dispersion both contribute to mouse neocortical development. *Nature (London)* **362**, 638–640.

Tao, W., and Lai, E. (1992). Telencephalon-restricted expression of BF-1, a new member of the HNF-3/fork head gene family, in the developing rat brain. *Neuron* **8**, 957–966.

Temple, S. (1989). Division and differentiation of isolated CNS blast cells in microculture. *Nature (London)* **340**, 471–473.

Tessarollo, L., Tsoulfas, P., Martin-Zanca, D., Gilbert, D. J., Jenkins, N. A., Copeland, N. G., and Parada, L. F. (1993). trkC, a receptor for neurotrophin-3, is widely expressed in the developing nervous system and in non-neuronal tissues. *Development (Cambridge, UK)* **118**, 463–475.

Tsai, L. H., Takahashi, T., Caviness, V. S., Jr., and Harlow, E. (1993). Activity and expression pattern of cyclin-dependent kinase 5 in the embryonic mouse nervous system. *Development (Cambridge, UK)* **119**, 1029–1040.

Uemura, T., Shepherd, S., Ackerman, L., Jan, L. Y., and Jan, Y. N. (1989). *numb*, a gene required in determination of cell fate during sensory organ formation in *Drosophila* embryos. *Cell (Cambridge, Mass.)* **58**, 349–360.

Uyttendaele, H., Marazzi, G., Wu, G., Yan, Q., Sassoon, D., and Kitajewski, J. (1996). Notch4/int-3, a mammary proto-oncogene, is an endothelial cell-specific mammalian Notch gene. *Development (Cambridge, UK)* **122**, 2251–2259.

Van Doren, M., Ellis, H. M., and Posakony, J. W. (1991). The *Drosophila* extramacrochaetae protein antagonizes sequence-specific DNA binding by daughterless/achaete-scute protein complexes. *Development (Cambridge, UK)* **113**, 245–255.

Walsh, C. A. (1996). Neural development: Identical twins separated at birth? *Curr. Biol.* **6**, 26–28.

Walsh, C., and Cepko, C. L. (1992). Widespread dispersion of neuronal clones across functional regions of the cerebral cortex. *Science* **255**, 434–440.

Walsh, C., and Cepko, C. L. (1993). Clonal dispersion in proliferative layers of developing cerebral cortex. *Nature (London)* **362**, 632–635.

Wanaka, A., Milbrandt, J., and Johnson, E. M., Jr. (1991). Expression of FGF receptor gene in rat development. *Development (Cambridge, UK)* **111**, 455–468.

Ware, M. L., Fox, J. W., Gonzalez, J. L., Davis, N. M., Lambert de Rouvroit, C., Russo, C. J., Chua, S. C., Jr., Goffinet, A. M., and Walsh, C. A. (1997). Aberrant splicing of a mouse disabled homolog, mdab1, in the scrambler mouse. *Neuron* **19**, 239–249.

Ware, M. L., Reid, C. B., Tavazoie, S. F., and Walsh, C. A. (1997). Lineage analysis of early ferret cerebral cortex studied with an alkaline phosphatase retroviral library. *Soc. Neurosci. Abstr.* **23**, 580.

Williams, B. P., and Price, J. (1995). Evidence for multiple precursor cell types in the embryonic rat cerebral cortex. *Neuron* **14**, 1181–1188.

Xiang, X., Osmani, A. H., Osmani, S. A., Xin, M., and Morris, N. R. (1995). NudF, a nuclear migration gene in *Aspergillus nidulans*, is similar to the human LIS-1 gene required for neuronal migration. *Mol. Biol. Cell* **6**, 297–310.

Xu, Y., Baldassare, M., Fisher, P., Rathbun, G., Oltz, E. M., Yancopoulos, G. D., Jessell, T. M., and Alt, F. W. (1993). LH-2: A LIM/homeodomain gene expressed in developing lymphocytes and neural cells. *Proc. Natl. Acad. Sci. U.S.A.* **90**, 227–231.

Xuan, S., Baptista, C. A., Balas, G., Tao, W., Soares, V. C., and Lai, E. (1995). Winged helix transcription factor BF-1 is essential for the development of the cerebral hemispheres. *Neuron* **14**, 1141–1152.

Yue, T. L., Stadel, J. M., Sarau, H. M., Friedman, E., Gu, J. L., Powers, D. A., Gleason, M. M., Feuerstein, G., and

Wang, H. Y. (1992). Platelet-activating factor stimulates phosphoinositide turnover in neurohybrid NCB-20 cells: Involvement of pertussis toxin-sensitive guanine nucleotide-binding proteins and inhibition by protein kinase C. *Mol. Pharmacol.* **41,** 281–289.

Zhong, W., Feder, J. N., Jiang, M. M., Jan, L. Y., and Jan, Y. N. (1996). Asymmetric localization of a mammalian numb homolog during mouse cortical neurogenesis. *Neuron* **17,** 43–53.

Zhong, W., Jiang, M. M., Weinmaster, G., Jan, L. Y., and Jan, Y. N. (1997). Differential expression of mammalian Numb, Numblike and Notch1 suggests distinct roles during mouse cortical neurogenesis. *Development (Cambridge, UK)* **124,** 1887–1897.

Vertebrate Tissue Specification

37

Tissue Determination

An Introduction

Sally A. Moody
Department of Anatomy and Cell Biology
Institute for Biomedical Sciences
The George Washington University
Washington, D.C. 20037

I. Introduction
II. Tissue Specification

III. References

I. Introduction

The previous sections presented exciting discoveries in our understanding of the processes that determine cell fate. They show that several events, including cell ancestry, gene regulation, and cell-cell signaling, coordinate to determine embryonic patterns and specific cell types. These fate-determining influences have been presented in an animal-by-animal approach to emphasize the benefits of the different experimental models for answering questions about embryonic cell fate and cell lineage determination. In this section we focus on the determination of specific tissues. Nearly all multicellular animals contain similar tissues: epidermis, nervous system, digestive organs, circulatory system and blood, and contractile muscle. The following chapters present the cellular and molecular mechanisms that establish some of these tissues, combining information gathered from a number of animal models. This allows us to appreciate those mechanisms that are conserved and those that are species-specific.

II. Tissue Specification

It is an opportune time to consider tissue determination from a cross-species perspective. There have been considerable advances in this research over the past decade, in part due to homology cloning techniques, the Human Genome project, animal genome projects, transgenic mouse technology, and improved technologies for manipulating small or inaccessible embryos and their cells. Information from several animal models is converging to identify common themes and species-specific themes in the cellular and molecular mechanisms of tissue differentiation. In addition, the knowledge base for understanding cell type determination can now be applied to the next level of organization, that of the tissues and organs. Organs are derived from cells of different tissue origins that must coordinate in function, and often errors in these interactions are responsible for congenital malformations. The following chapters illustrate that our understanding of cell-cell interactions and gene regulation is now sufficient to enable us to begin to elucidate the complex interactions that occur between germ layers, cells, and tissue types to construct an organ.

In addition, these studies will lead to a better understanding of the molecular, cellular, and morphogenetic basis of birth defects. Hundreds of human congenital disorders have been described at the clinical level over the past 100 years, but methodologies for prediction, intervention, or prevention await a complete understanding of the underlying developmental causes. Currently, human clinical genetic investigation relies on two approaches. Genetic mapping of affected families compares the inheritance pattern of the disease phenotype with the inheritance pattern of identified genetic markers. Alternatively, a candidate mutant gene is identified in a similarly affected animal model, and then sought in human conditions. In both approaches a basic understanding of cell fate determination and tissue formation can help to elucidate underlying causes. For example, families with Hirschprung's disease (congenital aganglionic megacolon) were found to have mutations in different elements in a signal transduction pathway necessary for neural crest invasion of the hindgut and differentiation into enteric neurons (e.g., Angrist *et al.*, 1996; Chakravarti, 1996; Robertson *et al.*, 1997; Pingault *et al.*, 1998). Previous knowledge of the role of Hoxa13 in the formation of animal limbs led to the discovery that Hoxa13 is mutated in people with the hand-foot-genital syndrome (Mortlock and Innes, 1997). Similarly, study of the structure, function, and developmental role of the family of fibroblast growth factor (FGF) receptors contributed to the discovery that mutations resulting in their constitutive activation cause several human skeletal malformations (e.g., Horton, 1997; Tartaglia *et al.*, 1997; Wilkie, 1997).

Of course, the clinical goal is to correct life-threatening or debilitating congenital diseases in children and to promote tissue repair after traumatic or degenerative processes in adults. However, these attempts are best aided by a full understanding of the developmental processes described in this book. This section will consider the factors and interactions that regulate tissue fate determination from a multianimal viewpoint. It reviews our current understanding of the normative cell fate determinative processes in a few select tissues, and illustrates the advantages of utilizing many different animal models for addressing these problems. For example, frog is best suited for studies of the initiation of vertebrate epidermal fate, whereas mouse is best suited for studies of later differentiation events (Chapter 38 by Sargent and Morasso). In vertebrate hematopoiesis, frog is used for defining early steps in mesoderm specification, mouse for understanding fetal and adult hematopoiesis, cytokine influences, and effects of transcription factor gene deletions, and zebrafish for identifying novel blood mutants (Chapter 39 by Liao and Zon). Comparing results from *Drosophila*, frog, avians, zebrafish, and mouse reveals the relative roles of myogenic regulatory factors and signaling molecules in muscle specification (Chapter 41 by Buckingham and Tajbakhsh). The complex molecular and morphogenetic events that organize the endoderm has relied on all the vertebrate systems to reveal underlying patterning, signaling, and gene expression events (Chapter 40 by Gannon and Wright). The approaches outlined in the following chapters will not only elucidate the complexities of vertebrate tissue differentiation and organogenesis, but also lay the foundations for understanding and intervening in congenital and degenerative disorders.

References

Angrist, M., Bolk, S., Halushka, M., Lapchak, P. A., and Chakravarti, A. (1996). Germline mutations in glial cell line-derived neurotrophic factor (GDNF) and RET in a Hirschsprung disease patient. *Nat. Genet.* **14,** 341–344.

Chakravarti, A. (1996). Endothelin receptor-mediated signaling in Hirschsprung disease. *Hum. Mol. Genet.* **5,** 303–307.

Horton, W. A. (1997). Fibroblast growth factor receptor 3 and the human chondrodysplasias. *Curr. Opin. Pediatr.* **9,** 437–442.

Mortlock, D. P., and Innes, J. W. (1997). Mutation of HOXA13 in hand-foot-genital syndrome. *Nat. Genet.* **15,** 179–180.

Pingault, V., Bondurand, N., Kuhlbrodt, K., Goerich, D. E., Prehu, M. O., Puliti, A., Herbarth, B., and Hermans-Borgmeyer, I. (1998). SOX10 mutations in patients with Waardenburg-Hirschsprung disease. *Nat. Genet.* **18:** 171–173.

Robertson, K., Mason, I., and Hall, S. (1997). Hirschsprung's disease: Genetic mutations in mice and men. *Gut* **41,** 436–441.

Tartaglia, M., DiRocco, C., Lajeunio, E., Valeri, S., Velardi, F., and Battaglia, P. A. (1997). Jackson-Weiss syndrome: Identification of two novel FGFR2 missense mutations shared with Crouzon and Pfeiffer craniosynostic disorders. *Hum. Genet.* **101,** 47–50.

Wilkie, A. O. (1997). Craniosynostosis: Genes and mechanisms. *Hum. Mol. Genet.* **6,** 1647–1656.

38

Differentiation of Vertebrate Epidermis

Thomas D. Sargent
Maria I. Morasso
Section on Vertebrate Development
Laboratory of Molecular Genetics
National Institute of Child Health and Human Development
Bethesda, Maryland 20892

I. Introduction

A central problem in developmental biology is understanding how a terminally differentiated cell can be derived from the products of a cleaving fertilized egg. Epidermis is an excellent tissue in which to study this problem at the cellular and molecular levels: Epidermal specific gene expression initiates very early in the development of *Xenopus* embryos, using a tissue-autonomous mechanism that is closely tied to the specification of the central nervous system and other ectodermal derivatives. With the establishment of the stratified morphology in later embryonic stages, the epidermis is a tissue which is self-renewing throughout the life of the organism. In all terrestrial vertebrates, mature epidermis comprises mitotically active basal cells adjacent to the dermis, which undergo sequential differentiation through the spinous, granular, and finally to the cornified layers on the surface of the skin. Each of these compartments expresses a unique array of specific differentiation markers. The keratinocyte, which is the major cell type of the epidermis, is derived from the ectoderm. Cells from other embryonic origins are also found in the epidermis, for example, melanocytes of neural crest origin and Langerhans cells derived from the bone marrow.

The keratinocytes of the epidermal basal layer and hair follicle, which include the epidermal stem cell compartment (Cotsarelis *et al.*, 1990; Jones and Watt, 1993), can be isolated from neonatal mouse epidermis, cultured, and induced to differentiate *in vitro* by specific signals, such as elevated extracellular calcium concentrations. Each stage of differentiation is characterized by the appearance of specific structural proteins. Regulatory molecules have been identified that may participate in the control of the keratinocyte phenotype in each of the epidermal compartments. This chapter reviews the earliest steps in epidermal specification, which can be readily approached in the *Xenopus* system, and the regulation of epidermal cell differentiation in stratified skin and its appendages, utilizing the elegant molecular genetics of the mouse.

II. Early Development: Initiation of Epidermis in *Xenopus*

As reviewed in Chapter 19 by Dawid, the "midblastula transition" (MBT) is the point at which the zygotic genome becomes transcriptionally active in *Xenopus*. This is the first opportunity for tissue-specific gene expression in the embryo. By early gastrula stage, or about 2 hours after the MBT, a number of zygote-specific (i.e., nonmaternal) transcripts have appeared, many of which are present in localized patterns in the embryo. Among the most abundant of these RNAs are transcripts of a gene family encoding type I (XK70 and XK81) and type II (XK76) keratins, which are expressed exclusively in the ectoderm (Jamrich *et al.*, 1987). These keratin genes continue to be expressed until metamorphosis, and as early as midgastrula, their protein products are assembled into the intermediate filaments that are the major protein product of epidermal cells (Fig. 1). While several regulatory genes have been identified that activate in specific tissues at or near the MBT, the epidermal keratins are probably the first genes expressed in development that encode tissue-specific structural proteins. Therefore, epidermis can be considered to be the first biochemically differentiated tissue to appear in *Xenopus* development.

In principle, a tissue can be specified by one of two general modes: (1) inductive, in which a signaling molecule is secreted by another tissue, usually adjacent, binds to receptors on the target tissue, and elicits the appropriate developmental response, or (2) autonomous, in which no external signal is necessary. When considered as an intact tissue, *Xenopus* epidermis falls into the

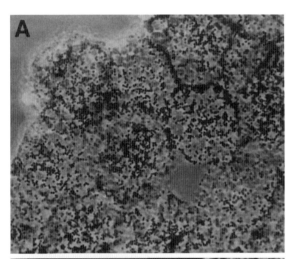
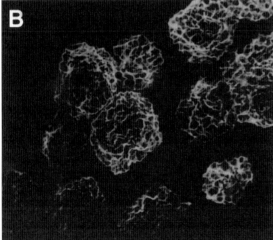

Figure 1 Midgastrula stage *Xenopus* ectoderm stained with rabbit antiserum recognizing type I keratin proteins that appear during embryogenesis. The intermediate filaments of the cytoskeleton are clearly visible, indicating that by this stage keratins have been synthesized and assembled into functional structures. Modified from Jamrich *et al.* (1987). (A) Bright field; (B) fluorescence.

second category. This has been demonstrated by explant experiments in which cells fated to become epidermis, such as those from the animal hemisphere, are dissected away from the remainder of the embryo and cultured in saline (Sudarwati and Nieuwkoop, 1971). Epidermal marker genes, such as keratins, are invariably activated in the explant at the appropriate time and levels, even when the explants are removed as early as the 8-cell stage (Jones and Woodland, 1986). In the absence of external inducers or attached mesodermal tissue, no other cell types arise. Thus, *Xenopus* ectoderm is programmed as epidermis from very early in cleavage, perhaps by maternally deposited factors present in the egg.

During gastrulation, the dorsal half of the ectoderm is diverted from epidermis and respecified as the central nervous system and associated anterior structures, such as the cement gland (Sive *et al.*, 1989; Drysdale and Elinson, 1993; Mathers *et al.*, 1995; Blitz and Cho, 1995). This process is one of the classically defined cases of embryonic induction, and depends on inducers secreted by the Spemann organizer region of the dorsal marginal zone. For many years it was assumed that, since ectoderm in isolation differentiates as epidermis, neural induction took place by means of an "instructive" mechanism. In other words, epidermis was considered the "ground state" of ectoderm, which could be overridden by a specific inductive signal, converting part of the ectoderm to nervous system. This interpretation is basically correct, but the mechanism underlying the neural–epidermal switch has turned out to be fundamentally different from what had been expected. This was first suggested by experiments in which whole blastula-stage embryos or isolated ectodermal explants were dissociated into single cells by treatment with medium lacking divalent ions, followed by reaggregation at the equivalent of the gastrula stage. Neural-specific gene expression was observed under these conditions (Sato and Sargent, 1989; Grunz and Tacke, 1989; Godsave and Slack, 1989). Since signaling from the dorsal mesoderm was prevented by the explant removal and dissociation procedures, this result was unexpected, and suggested that ectoderm might in fact be predetermined as neural, rather than epidermal. A further indication that this might be the case, and a possible molecular mechanism for regulating the epidermal–neural switch, was the observation that interference with the activin signaling pathway, using a dominant negative receptor for this transforming growth factor β (TGFβ) family member, resulted in neuralization of the ectoderm (Hemmati-Brivanlou and Melton, 1994). It has become subsequently clear that it is not activin, but members of the bone morphogenic protein (BMPs) family that are responsible for maintaining, and possibly initiating, the epidermal identity of ectoderm (Chang *et al.*, 1997).

BMP2, BMP4, and *BMP7* are all expressed in the early *Xenopus* embryo. By the midgastrula stage,

BMP4 RNA is predominantly present in ventrolateral ectoderm and mesoderm, while *BMP2* and *BMP7* are more broadly expressed (Fainsod *et al.*, 1994; Hemmati-Brivanlou and Thomsen, 1995; Thomsen, 1997). Furthermore, addition of soluble BMP4 to dissociated ectodermal cells prevents neuralization (Wilson and Hemmati-Brivanlou, 1995). Thus, the autonomous epidermal specification of ectoderm depends on a feedback loop in which the ectodermal cells produce a ligand that binds to receptors on the same cells, eliciting, or at least maintaining, epidermal gene expression. This model provides a mechanism for neural induction, in which inducers secreted by the Spemann organizer could antagonize this feedback loop, allowing the neural phenotype to emerge by default. In fact, this interpretation appears to be correct: Four proteins have been identified so far that can act to convert ectoderm into neural tissue. They are follistatin (Hemmati-Brivanlou *et al.*, 1994), Xnr3 (a TGFβ family member; Hansen *et al.*, 1997), noggin (Smith and Harland, 1992), and chordin (Sasai *et al.*, 1994). All of these are expressed in or around the Spemann organizer region at the beginning of gastrulation and potentially could induce neural tissue *in vivo*. Follistatin binds to activin, and can act as a neural inducer under artificial conditions. It has also been shown to bind BMP4, and could play a similar role in the embryo. Similarly, noggin and chordin bind BMP4 with high affinity *in vitro,* and are probably embryologically functional neural inducers, although they act via an inhibitory rather than a direct, instructive route. The mechanism of Xnr3 action is still unknown, but this molecule also appears to function by antagonizing the BMP–epidermal feedback loop.

At late blastula to early gastrula stages, at least some of the future neural ectoderm is initially epidermal, prior to coming under the influence of the organizer and the inhibitors of BMP signaling secreted by this tissue. This is manifested by the presence of epidermal keratin mRNA, visualized by *in situ* hybridization of midgastrula stage *Xenopus* embryos. RNA signal is clearly visible in ectoderm in direct contact with involuted dorsal mesoderm and hence induced to neural plate, indicating that keratin expression was under way prior to the diversion from the epidermal to a neural program (Fig. 2). This early, transient epidermal behavior of dorsal ectoderm could indicate that the BMP4 signal is not necessary for the initial specification of epidermis, since *BMP4* expression is predominantly confined to ventral ectoderm, at least by midgastrula stages (Schmidt *et al.*, 1995; Hemmati-Brivanlou and Thomsen, 1995). It is possible that BMP2 or BMP7, which are more broadly expressed in the ectoderm and mesoderm, or the low levels of maternal BMP proteins might partly account for the keratin expression in dorsal ectoderm (Clement *et al.*, 1995; Thomsen, 1997). However, it cannot be ruled out that BMP signaling is required for maintenance but not for initiation of epider-

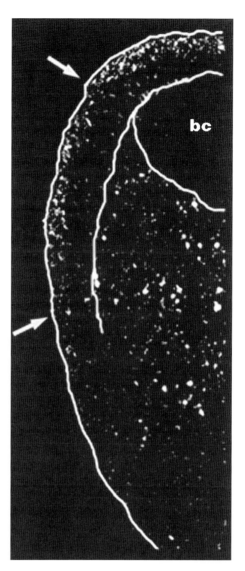

Figure 2 *In situ* hybridization of embryonic keratin probe to a thin section of a midgastrula stage *Xenopus* embryo. In this dark-field view, silver grains indicating expression of keratin mRNA can be seen in the outer layer of the ectoderm, including the area adjacent to the involuting dorsal mesoderm (bounded by the arrows). This shows that keratin gene expression has taken place, transiently, in the future neural plate region prior to inactivation of epidermal specification by neural induction. Silver grains in the endodermal region, beneath the blastocoel (bc), represent nonspecific background. The outline of the embryo is indicated in white. Modified from Jamrich *et al.* (1987).

mis, at least on the basis of existing data. It is also important to draw a distinction between the function of BMPs in epidermal specification, which is essentially cell-autonomous, and the classic concept of induction, which refers to a developmental signal transmitted from one cell to a neighboring cell in the embryo. In other words, it is not really appropriate to describe BMPs as "epidermal inducers," since the source and recipient of the signal are the same.

Nevertheless, it is clear that BMP signaling plays a central role in epidermal fate, so the signal transduction pathway associated with these factors is of considerable interest from the standpoint of skin biology. The BMP receptors are transmembrane serine/threonine kinases that function as heteromeric complexes of type I and type II molecules, which interact to bind ligand at high affinity and phosphorylate target proteins (Letsou *et al.*, 1995; Rosenzweig *et al.*, 1995). Genetic analysis has resulted in the identification of major downstream components of BMP signaling, the Smad proteins (for the *Sma* genes of *Caenorhabditis elegans* and the *Mothers Against Decapentaplegic* gene in *Drosophila*; reviewed by Massague, 1996). *Smad* genes are conserved in many animal species, and comprise a gene family with at least five related members in vertebrates. In *Xenopus*, the Smad1 protein can transduce the BMP4 signal; overexpression of this protein has the same effect on ectodermal development as activation of BMP (Wilson *et al.*, 1997). Smad proteins are broadly expressed, phosphorylated by a BMP-dependent mechanism, and translocate to the nucleus, where they presumably activate transcription of BMP-responsive genes (Graff *et al.*, 1996; Hoodless *et al.*, 1996). The mechanism by which Smads bind DNA and affect transcription is not well understood. Nor is much known about the range of genes that respond to this regulatory signal. However, one particularly interesting target gene that has been identified is the homeodomain protein Msx1.

Msx1, *Msx2*, and *Msx3* comprise a family of genes related to the *Drosophila Msh1* (muscle segment homeobox) gene, and have been implicated in the establishment and morphogenesis of structures derived from epithelial–mesenchymal interactions in vertebrate embryos, such as hair and teeth (Hill *et al.*, 1989; MacKenzie *et al.*, 1991, 1992; Jowett *et al.*, 1993; Shimeld *et al.*, 1996; see below). In *Xenopus*, *Msx1* is expressed as early as the MBT (with a minor maternal component), and by midgastrula *Msx1* RNA is localized predominantly to ventral ectoderm and mesoderm (Suzuki *et al.*, 1997). Like other ventral and epidermal genes, *Msx1* expression is blocked by disruption of the BMP signaling pathway. Furthermore, this gene is an immediate early target of BMP4; addition of soluble BMP4 protein to dissociated *Xenopus* ectoderm activates *Msx1* transcription even in the presence of protein synthesis inhibitors, indicating that the BMP signal is transduced to the level of *Msx1* gene expression by maternally inherited protein factors (Suzuki *et al.*, 1997). Such factors would presumably include, and could even be limited to, the receptors and Smads intermediates, although additional effectors could be involved. In addition, Msx1 is itself an important mediator of the epidermal fate. Injection of RNA encoding this protein into fertilized *Xenopus* eggs results in ventralization, effectively overriding the diversion from epidermis that results from disruption of BMP signaling (Suzuki *et al.*, 1997). Msx1 acts as a transcriptional repressor, both as a conventional DNA-

binding protein (Woloshin *et al.*, 1995) and as a DNA-independent transcriptional cofactor (Catron *et al.*, 1995), so it may function in *Xenopus* ectodermal development by repressing the activation of genes involved in the specification of neural tissue. No specific target genes for Msx1 have been identified. However, Msx1 has been shown to interact directly with another class of homeodomain proteins, the distal-less-related or *Dlx* factors (Zhang *et al.*, 1997).

Distal-less (*Dll*) is another gene identified by *Drosophila* molecular genetic analysis, and has important functions in patterning the embryos of this species (Cohen *et al.*, 1989). In mouse and human there are six homologs of *Dll*, which are organized as three pairs, each closely linked to the 3′ end of a *HOX* cluster (Simeone *et al.*, 1994; Nakamura *et al.*, 1996; Morasso *et al.*, 1997; Quinn *et al.*, 1997). *Dlx* homeoproteins are DNA-dependent transcriptional activators (Zhang *et al.*, 1997; J. Feledy, unpublished). In the case of *Dlx1*, *Dlx2*, and *Dlx5*, this activation is inhibited by *Msx* proteins, through specific interactions between the homeodomains of the two proteins. This prevents association of Dlx protein with target gene sequences (Zhang *et al.*, 1997). Thus *Msx* and *Dlx* genes have antagonistic functions. It seems unlikely, however, that inhibition of Dlx3 is the basis for epidermal specification. The *Dlx* gene that is active during gastrulation, *Dlx3* (originally referred to as *Xdll2* in *Xenopus*) is expressed predominantly in ventral ectoderm (Papalopulu and Kintner, 1993; Dirksen *et al.*, 1994), in a pattern similar to that of *Msx1* (Fig. 3), whereas the neural-suppression model for *Msx1* function would suggest the existence of one or more panectodermal neurogenic factors which would be inhibited by Msx1 in ventral ectoderm. Direct tests for interaction between Msx1 and Dlx3 have not been reported. However, in view of the conservation of Dlx homeodomains, it is likely that Msx1 will antagonize Dlx3 in a manner similar to its effects on Dlx1, Dlx2, and Dlx5.

III. Other Ectodermal Fates

A remarkable property of the early ectoderm in *Xenopus* is the extent of its plasticity: ectoderm can be respecified as virtually any embryonic tissue, depending on the inducers used. As outlined above, interruption of the autonomous BMP signaling pathway results in conversion from epidermal to neural tissue. The response of ectoderm to this ligand is probably more complex than a simple binary switch, however. By restoring BMP4 to dissociated ectodermal cells at variable levels, Hemmati-Brivanlou and colleagues demonstrated that the presumptively neuralized cells could be converted not just back to epidermis, but to at least one other ectodermal fate, that of cement gland (Wilson *et al.*,

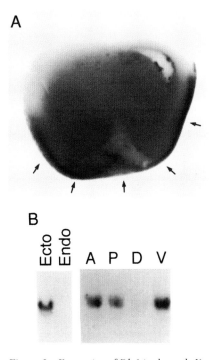

Figure 3 Expression of *Dlx3* in the early *Xenopus* embryo. (A) Whole-mount *in situ* hybridization to a midgastrula stage embryo revealing localized *Dlx3* expression (arrows) in the ventral ectoderm. Ventral is oriented toward the bottom of the figure. (B) Northern blot analysis of RNA from dissected gastrula-stage embryos. *Dlx3* mRNA is detected in ectoderm (Ecto) but not in endoderm (Endo), and in ventral (V) but not dorsal (D) halves of embryos. There is no obvious anterior (A)–posterior (P) difference in expression.

1997). In addition, Morgan and Sargent (1997) showed that intermediate levels of BMP signaling could respecify *Xenopus* ectoderm as neural crest. These findings raise the possibility that BMP4 functions as a type of morphogen in the *Xenopus* embryo, inducing different tissues depending on its effective local concentration. Such a function for BMP4 would differ somewhat from the classic definition of morphogen in that the BMP levels sensed by a given cell would depend not on a diffusion gradient of the factor emanating from a localized source in the embryo, but rather on the local concentration of the antagonist, such as chordin or noggin, secreted by the Spemann organizer.

The graded response to BMP4 is reminiscent of the effect on ectoderm of varying concentrations of the mesoderm inducer activin: high concentrations induce a more dorsal type of mesoderm (e.g., the organizer-specific homeodomain gene *Goosecoid*), while lower activin levels induce a more ventral/lateral set of mesodermal markers, such as muscle-specific actin or brachyury. Interestingly, all concentrations of activin result in the complete suppression of epidermal gene expression (keratin XK81A), but at the lowest dosage, no mesoderm is induced, as judged from the molecular markers tested (Green *et al.*, 1992). In view of the effect of dissocia-

tion alone, it would be interesting to know if these cells expressed markers of neural specification. In the absence of epidermal or mesodermal identity, such activin-treated cells could be respecified as neural, or neural crest (Kengaku and Okamoto, 1993; see below), or placed into a developmentally intermediate state which does not correspond to any of the known primordial cell types. It is important to note that in these graded-response experiments it is essential to dissociate the *Xenopus* ectoderm into single cells, or small clusters, to equalize the exposure to the soluble factor. If carried out long enough, this dissociation alone results in neuralization. However, short periods of dispersion do not interfere with epidermis, presumably because the BMP signal is restored before the transition to neural specification can take place.

In the embryo, neural crest arises at the boundary between the neural plate and surrounding epidermis, as the result of an inductive interaction between the two tissues (Mancilla and Mayor, 1996). Neural crest cells depart from the neural tube and migrate throughout the body to specific destinations where they differentiate into a complex array of tissues, including bone, neurons and glia, muscle, endocrine cells, and others (Bronner-Fraser, 1993). *Xenopus* ectoderm can be experimentally induced to form neural crest by treatment of explants with a mixture of noggin and low concentrations of fibroblast growth factor (Kengaku and Okamoto, 1993; Mayor *et al.*, 1995), or by increasing the level in ectoderm of the translation initiation factor elF4AII by RNA injection (Morgan and Sargent, 1997). This latter result implies the existence of translationally regulated RNAs encoding factors involved in neural or neural crest specification. Possible candidates for such factors are the calcium-sensitive protein kinase C isoforms (PKCα, -β, and -γ), which are up-regulated as part of neural crest induction (Morgan and Sargent, 1997). PKCα and -β have been reported to play a role in *Xenopus* neural induction (Otte and Moon, 1992), and as discussed below are also important in the regulation of epidermal cell differentiation, raising the interesting possibility of a conserved mechanism for these two different processes in ectodermal development.

IV. Formation and Differentiation of Stratified Epidermis

Early mammalian embryos are more difficult to manipulate than those of *Xenopus* at the equivalent stage. Consequently the initial phase of ectodermal differentiation in mammals is not so well understood as it is in amphibians. However, the molecular genetics tools that can be brought to bear using the murine system have been invaluable in studying the control of later aspects of epidermal development. In the mammalian embryo, the presumptive surface ectoderm envelopes the embryo during gastrulation and neurulation, forming a simple epithelium comprising a single cell layer. In the rabbit, this epithelium contains abundant keratins, indicating that in the mammal, as in the amphibian, these cytoskeletal proteins are part of the earliest manifestation of the epidermal cell phenotype (Banks-Schlegel, 1982).

Around the time at which the dermal mesenchymal cells arrive from the paraxial (dorsolateral skin) and somatopleural (ventral skin) mesoderm (Sengel, 1990), the mammalian epidermis undergoes a transition from a single-layered to a double-layered tissue. The outer epithelium, or periderm, first appears on the upper limb bud at day 9 or 10 of mouse development, then on the rest of the body, appearing last on the ventral abdominal surface, around day 12 to 14 (M'Boneko and Merker, 1988). The periderm probably arises from surface ectoderm, by a process in which cells detach from the basement membrane and proliferate, eventually covering the surface of the fetus. Peridermal cells are distinct from the epidermis they overlay, and express a partially overlapping set of keratins (Fisher, 1994; Byrne *et al.,* 1994). Around day 15.5 in mouse gestation, stratification begins, resulting from the detachment of cells from the basal lamina and their migration to a more superficial location in the epidermis. This ultimately results in a multilayered structure, bounded by the basal keratinocytes on the inside, adjacent to the dermis, and the cornified layer on the outside. The initial differentiation process is similar to the ongoing program of keratinocyte development in adult epidermis (see below), with one major exception: in the fetus the suprabasal cells do not cease dividing, but simultaneously stratify and proliferate laterally, probably to accommodate the need to rapidly expand the embryonic surface during growth (Holbrook, 1991).

From the late fetal stages onward, stratified epidermis comprises several layers of cell types situated above the dermis (Fig. 4). The bottom or basal layer is attached to the dermal basal lamina via hemidesmo-

somes, and includes a portion of the epidermal stem cell population (additional epidermal stem cells reside within the hair follicle; Cotsarelis *et al.,* 1990), which give rise to transit amplifying cells, and daughter stem cells. The transit amplifying cells divide a few times, then develop into "committed" keratinocytes which detach from the dermis and enter the suprabasal compartment. All basal keratinocytes express a specific pair of keratins, K5 and K14, while the three basal cell phenotypes can be distinguished by integrin expression patterns (Jones and Watt, 1993; Jones *et al.,* 1995). Stratification results from the detachment of basal keratinocytes from the basement membrane and migration into the spinous cell compartment. In adult epidermis this is accompanied by a cessation of cell division. Another major change is the down-regulation of K5/K14 expression and the activation of two other genes encoding keratins K1 and K10, which are specific markers for spinous cells. The next phenotypic transformation occurs when spinous cells give rise to granular cells. This differentiation step includes the down-regulation of keratins K1 and K10 and the activation of genes encoding another set of epidermal markers, some of which are constituents of the cornified cell envelope such as loricrin and profilaggrin. The final phase of keratinocyte differentiation is the formation of cornified envelopes. This process includes granular cell death, destruction of all organelles, proteolytic processing of profilaggrin into filaggrin, covalent cross-linking of cornified envelope precursors through the action of one or more transglutaminases, and attachment of lipid molecules to the cross-linked envelope. The end product of this complex differentiation program, the squame, resides on the surface of the epidermis and provides the main barrier function of the skin. Squames are continually shed and replaced by keratinocyte differentiation throughout the life of the organism. The entire differentiation process, from basal cell to squame takes about 30 days in normal human skin (Watt, 1989; Fuchs, 1990; Fuchs and Byrne, 1994).

Thus the epidermis is formed and maintained by a differentiation program based on a localized compart-

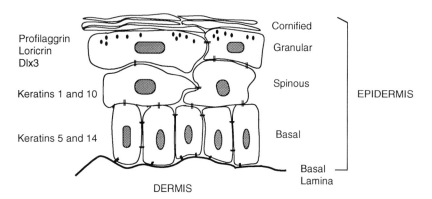

Figure 4   Schematic diagram of stratified epidermis. The layers are labeled to the right, and some key layer-specific markers to the left. Desmosomes and hemidesmosomes are indicated by double lines.

ment of stem cells and leading to a single fate. The accessibility of this tissue and the availability of numerous molecular markers, detectable by cDNA probes and antibodies, make epidermis an ideal experimental system in which to study differentiation. Considerable progress has been made in understanding how this pathway is regulated. The keratinocyte maturation process is tightly associated with a stepwise program of transcriptional regulation, as judged by the sequential induction and repression of structural and enzymatic differentiation-specific markers. This process can be achieved in mouse keratinocytes cultivated *in vitro* by increasing the calcium ion concentration in the culture medium from 0.05 to 0.12 mM, which mimics the endogenous calcium gradient present in the skin (Menon *et al.*, 1985; Yuspa *et al.*, 1989; Menon and Elias, 1991). Calcium-mediated keratinocyte differentiation is associated with increased phospholipase C activity, elevated diacylglycerol levels, and activation of protein kinase C (PKC). PKC signaling specifically in the late stages of epidermal differentiation has been demonstrated, showing that activation of PKC induces expression of late differentiation markers such as loricrin, profilaggrin, and transglutaminase, and concomitantly suppresses the spinous-specific markers K1 and K10 in cultured keratinocytes (Dlugosz and Yuspa, 1993). This reprogramming of gene expression mimics changes taking place *in vivo* as spin-

ous cells make the transition to granular cells. PKC activators such as the phorbol ester TPA also induce the expression of late differentiation markers in the absence of changes in the calcium levels. In addition, this effect is blocked by the PKC inhibitor bryostatin, so the negative effect of TPA is probably transduced by activation of PKC. Furthermore, both protein and RNA synthesis appear to be required for this effect to be achieved, suggesting that PKC activation results in the up-regulation of one or more protein factors which are then responsible for the down-regulation of K1/10 expression. Subsequent steps in the keratinocyte pathway, such as up-regulation of cornified envelope proteins, are also sensitive to bryostatin treatment, and hence are probably also regulated by PKC. Thus it is likely that the general control of the transition from spinous to granular cell phenotype is accomplished via this kinase (Dlugosz and Yuspa, 1993).

V. Epidermal Appendages

In addition to its protective function, the epidermis, in conjunction with the dermis, also gives rise to several types of accessory structures or appendages. These include, most conspicuously, the hair in mammals, and feathers and scales in other vertebrates. Eccrine

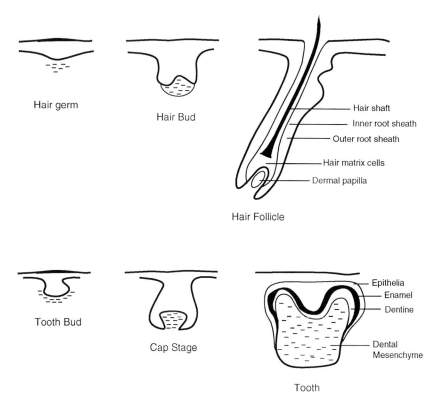

Figure 5 Schematic diagram of hair (top) and tooth (bottom) development. In hair, the underlying dermis induces the formation of the hair germ which is first visualized as the epidermal thickening, or placode. The hair germ proliferates and invaginates into the dermis, and the follicle encloses the mesenchymal cells (dashed) that will form the dermal papilla. In tooth, development also commences with thickening of surface ectoderm and condensation of mesenchymal cells (dashed). This stage is followed by formation of the dental papilla. In the final stage, the mesenchymal cells differentiate into the odontoblast pulp layer, and the adjacent epithelial cells differentiate into ameloblasts that deposit the enamel.

sweat glands, nails, teeth, and mammary glands are also epidermally derived structures. All of these appendages arise as the result of reciprocal interactions between developing epidermis and underlying mesenchyme, typically the dermis. These interactions have many similar aspects, including not only the morphological changes but also the signaling and regulatory molecules involved. In the following section we discuss two of the best understood cases of epidermal appendage development, the hair follicle and the tooth.

The first morphologically distinguishable event in hair follicle formation is the thickening of the surface ectodermal cells, to form an epidermal placode. This results from a region-specific inductive signal from the dermis (Fig. 5; Hardy, 1992). The induced epidermis invaginates into the dermal layer, eventually enclosing a small cluster of mesenchymal cells at the bottom of the developing follicle. This cluster becomes the dermal papilla and remains closely associated with the fully formed hair follicle that consists of the outer root sheath, inner root sheath, hair matrix cells, and hair shaft with its characteristic keratin-based structure. Once completed, the hair follicle undergoes cyclic growth and regeneration (Hardy, 1992).

The positional information required for the spatial patterning of body hair is presumably conveyed by the action of signaling molecules secreted by one or more of the cell types involved. One example of such a molecule is sonic hedgehog (Shh), which is expressed in the epithelial component of hair and tooth and has been proposed to act upstream of homeobox genes in a regulatory cascade that affects patterning (Bitgood and McMahon, 1995; Iseki et al., 1996). In addition, polypeptide growth factors and their receptors also are expressed during hair follicle development (reviewed by Holbrook et al., 1993). Studies using transgenic mice have provided considerable insight into the potential roles of these molecules in the regulation of the hair follicle. Keratinocyte growth factor, a member of the fibroblast growth factor (FGF) family, is expressed in the dermal papilla and affects keratinocyte proliferation and differentiation. Ectopic expression of KGF under the basal K14 promoter caused basal cells to enter the epidermal instead of the hair follicle lineage, resulting in an absence of hairs in areas of affected skin (Guo et al., 1993). Members of the TGFβ family, such as BMP2 and BMP4, are also known to be expressed in the hair follicle (Jahoda and Reynolds, 1993) in a pattern similar to that of Shh. In the mature follicle, BMP2 is restricted to the precortex cells. To test the importance of this localized BMP expression, Blessing et al., (1993) directed ectopic expression of the highly similar BMP4 to the outer root sheath by means of a tissue-specific keratin gene promoter. This resulted in a substantial reduction in the proliferation of both outer root sheath

and hair matrix cells and triggered a transition in keratin gene expression. These results are consistent with a function for BMP2 as a differentiation-promoting signal affecting the progression from hair matrix to the hair shaft.

Homeobox containing transcription factors have also been detected in localized compartments of the hair follicle: Dlx3 is expressed in the germinative epidermal cells of the hair matrix (Robinson and Mahon, 1994) and the POU-class protein skn-1 is expressed in the developing hair (Andersen et al., 1993). Another interesting transcription factor, lymphoid enhancer factor-1 (LEF-1) is also expressed in the hair follicle (Zhou et al., 1995). It is first detected in the epidermis, prior to the appearance of any morphological changes, in a temporal and spatial fashion that resembles Shh (Bitgood and McMahon, 1995). Ectopic expression of LEF-1 with a keratin K14 basal cell-specific promoter resulted in animals that had an aberrant patterning of the hair follicles (Zhou et al., 1995). Targeted inactivation of the LEF-1 gene resulted in complete absence of hair and vibrissae, demonstrating the requirement of LEF-1 for normal hair morphogenesis (van Gendersen et al., 1995).

Another structure that has been extensively studied as a developmental system is the tooth. Tooth morphogenesis also starts as a mesenchymally induced thickening of the surface ectoderm that subsequently undergoes invagination and surrounds the condensed bud of underlying mesenchymal cells. This associated mesenchyme is derived from the branchial arch neural crest, and forms the odontoblasts of the dental papilla, analogous to the dermal papilla of the hair follicle. This tooth germ continues to grow, and the odontoblast cells eventually differentiate into the dental pulp and dentine layers, while the epithelial cells become the ameloblasts that produce the dental enamel.

As with the hair follicle, a constellation of growth factors and their receptors are expressed during tooth development, many of which are utilized in both structures. BMP4 is expressed at the onset of tooth development, in the thickened presumptive dental epithelium, later shifting to the condensing mesenchyme. BMP2 is expressed in the epithelium at the bud stage and later in tooth development shifts to the mesenchyme (Vainio et al., 1993). In vitro studies have shown that BMPs are important epithelial signals that can induce the expression of the Msx1 and Msx2 genes in the mesenchyme (Thesleff et al., 1995; see below).

A series of transcription factors have been shown to be expressed at different stages of tooth morphogenesis, for example, N-myc, c-fos, LEF-1 (reviewed by Thelseff et al., 1995), as well as the homeobox-containing transcription factors Msx1, Msx2, Dlx2, and Dlx3 (Hill et al., 1989; MacKenzie et al., 1991, 1992; Robinson and Mahon, 1994). The expression of Msx1 is restricted to the mesenchyme while LEF-1 expression shifts between the epithelial and mesenchymal tissues (Kratochwil *et al.*, 1996). Targeted inactivation of *Msx1* and *LEF-1* have shown that these genes are necessary for tooth formation (Satokata and Maas, 1994; van Gendersen *et al.*, 1995). Interestingly, besides being defective in hair development (see above), LEF-1 null mice are also deficient in tooth and mammary gland development (van Gendersen *et al.*, 1995). All of these structures are derived from epithelial–mesenchymal interactions initiated from surface ectoderm, suggesting a similar regulatory role for LEF-1 in their formation. The role of Shh has not been confirmed in tooth development, but as with the hair follicle, its expression also commences and coincides with the epithelial thickening that demarcates the site of formation of the structure.

LEF-1 expression is activated by BMP4 (Kratochwil *et al.*, 1996), suggesting the identification of common molecular mechanisms that govern the cascade of signaling that may be applicable to all these epithelial–mesenchymal derived structures involving the interacting factors BMP, LEF-1, and the homeobox containing genes *Msx1* and *Msx2*.

VI. Transcriptional Control of Keratinocyte Differentiation

While RNA stability or translation are important factors in the maturation of the suprabasal keratinocyte, it is reasonably clear that the major mode of epidermal gene regulation is transcriptional (Fuchs, 1990). Considerable effort has gone into the characterization of transcription factors in epidermal cells and the analysis of their functions in development (reviewed in Fuchs and Byrne, 1994; Eckert and Welter, 1996). Evidence supporting roles in the regulation of keratinocyte gene expression has been reported for several transcription factors, including AP1, AP2, the retinoid receptor family, and the homeodomain protein Dlx3.

AP1, which comprises heterodimers of members of the c-fos, and c-jun families, is an important target for the PKC signal transduction pathway discussed earlier (Eckert and Welter, 1996). Consistent with this function, DNA binding sites for AP1 have been identified in the regulatory DNA of genes expressed in the later stages of keratinocyte differentiation. For example, the loricrin and involucrin genes, which are active in granular layer cells, depend on interactions with AP1 for normal expression (Eckert and Welter, 1996). Furthermore, AP1 DNA-binding activity rises from very low to high levels when basal keratinocytes are stimulated to differentiate by exposure to elevated extracellular calcium. The increase in AP1 is attributable to upregulation of the *Fra-1*, *Fra-2*, *c-Jun*, and *JunD* genes,

and does not involve c-Fos. This increase in AP1 activity depends on signaling by PKCα, as determined by studies with isoform-specific inhibitors (Rutberg *et al.*, 1996).

One or more of the alternatively spliced AP2 isoforms are expressed in gastrula and neurula-stage epidermis and neural crest in *Xenopus*, and at early stages of epidermal development in the mouse (Winning *et al.*, 1991; Byrne *et al.*, 1994). This factor has been implicated in the response to retinoids, phorbol esters, and cAMP (Luscher *et al.*, 1989), all of which have major effects on epidermal cells. AP2 also was identified as an important component of keratin gene expression in the *Xenopus* embryo (Snape *et al.*, 1991; Winning *et al.*, 1991) and in mouse skin (Leask *et al.*, 1991). However, the AP2 gene has been inactivated by homologous recombination in the mouse, and while this results in a host of defects, including failure to close the skin on the ventral side, formation and stratification of the epidermis is not prevented (Schorle *et al.*, 1996; Zhang *et al.*, 1996). It is conceivable that AP2-related genes might partly mask the effects on epidermis of this gene deletion, but it is more likely that AP2 plays a positive but nondeterministic role in epidermal cell differentiation (Magnaldo *et al.*, 1993).

It has been known for some time that vitamin A and retinoids can influence epidermal morphology. However, the molecular basis for this is still somewhat unclear. Increasing the level of retinoic acid *in vivo* or *in vitro* tends to inhibit the terminal differentiation of keratinocytes. For example, treatment of cultured keratinocytes with retinoic acid (RA) inhibits the expression of differentiation markers such as K1 and K10 (Kopan *et al.*, 1987). This effect is not limited to suprabasal keratins, however, as keratins K5 and K14 can also be suppressed by RA (Tomic-Canic *et al.*, 1996). Reduction or removal of RA has the opposite effect, that is, the rate of differentiation of basal keratinocytes into suprabasal cells is accelerated. However, targeted inactivation of the two principal RA receptors (RARβ and RARγ) expressed in epidermis does not appear to interfere with skin morphology (reviewed by Anderson and Rosenfeld, 1995). Very different results have been obtained with transgenic mice expressing dominant negative RA receptors. Saitou *et al.* (1995) reported a drastic inhibition of suprabasal cell formation in mice expressing an inhibitory mutant RA receptor in basal cells, driven by the keratin K14 promoter. This is surprising, since as noted above, reducing the level of RA itself has the opposite effect. When a similar dominant negative RA receptor was targeted to the suprabasal cells (using a keratin K1 promoter), the effects were more subtle, albeit lethal, primarily confined to the barrier function of the cornified layer (Imakado *et al.*, 1995). One problem with these studies is that the dominant negative molecule potentially can interfere with a spectrum of retinoid receptors, making it difficult to

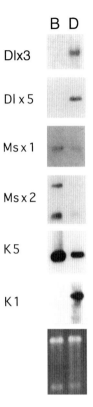

Figure 6 Expression of *Dlx* and *Msx* genes in epidermis. RNA isolated from purified preparations of either basal (B) or differentiated suprabasal (D) mouse epidermal keratinocytes was electrophoresed on a denaturing agarose gel, blotted to a nylon membrane, and probed with radiolabeled cDNA corresponding to *Dlx3*, *Dlx5*, *Msx1*, and *Msx2*. Duplicate filters were also probed for keratin K5 and K1 mRNA to monitor the fractionation (K5 is basal-specific and K1 is suprabasal-specific). Dlx3 and Dlx5 are both completely confined to the suprabasal compartment, while Msx1 and Msx2 are predominantly basal, with some expression of Msx1 in suprabasal keratinocytes. The low level of Msx2 in the suprabasal RNA is probably due to contamination with basal cells, as indicated by the K5 panel.

assign functions to particular species. In general, the functions of RA can be thought of as modulating the levels of keratinocyte-specific gene expression, as opposed to playing a determinative role in epidermal differentiation.

Like the early epidermis of the *Xenopus* embryo, stratifying epidermis of adult vertebrates express members of the Dlx and Msx homeodomain proteins. Of the six *Dlx* genes, *Dlx1*, *Dlx2*, and *Dlx6* are not detectably transcribed in epidermis (M. I. Morasso unpublished), while *Dlx3* and *Dlx5* are both expressed in the suprabasal compartment (Fig. 6). *Dlx8*, which is paired with *Dlx3* (Morasso *et al.*, 1997), has not been tested for expression in epidermis. Of the three *Msx* genes described to date, *Msx1* is apparently expressed in both basal and suprabasal epidermis, at a lower level in suprabasal, while *Msx2* is clearly basal-specific, within the limits of detection. *Msx3* has not yet been analyzed for epidermal expression.

In the interfollicular epidermis, there is reasonably convincing evidence for a role for Dlx3 in the regulation of gene expression during late keratinocyte differentiation. This homeoprotein is most abundant in the granular layer, along with cornified envelope precursor proteins such as profilaggrin and loricrin. Targeted misexpression of Dlx3 in basal cells of transgenic mice using a keratin K5 promoter fragment results in extensive alteration of epidermal morphology (Fig. 7; Morasso *et al.*, 1996). There is little if any cornified layer present, and the overall suprabasal cell population is greatly reduced. In the basal layer, the keratinocytes are spatially disorganized. Cell proliferation was also inhibited in basal cells ectopically expressing Dlx3. At the level of gene expression the situation is complex and interesting. Keratin genes are not strongly affected; K5 and K14 proteins accumulate to roughly normal levels in the disturbed basal layer, while K1 and K10 are confined to the disrupted suprabasal strata. On the other hand, the granular stage markers filaggrin and loricrin exhibit ectopic expression in the transgenic basal cells. Accumulation of these proteins in the suprabasal layer is drastically reduced, probably reflecting the premature termination of the differentiation program (Fig. 8). Therefore, expression of the granular-layer homeoprotein Dlx3 in basal cells has resulted in the partial conversion of these cells to a more differentiated, namely

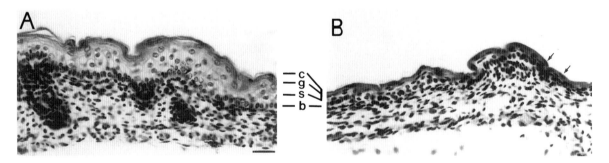

Figure 7 Effect on epidermis of ectopic expression of *Dlx3* in the basal layer of embryonic day 18.5 mice. A *Xenopus Dlx3* open reading frame cassette was expressed in transgenic mice using a keratin K5 promoter. This resulted in a marked thinning of the epidermis (B) compared to nontransgenic siblings (A). All suprabasal layers are greatly reduced. c, Cornified; g, granular; s, spinous; b, basal layers. Reproduced from Morasso *et al.* (1996).

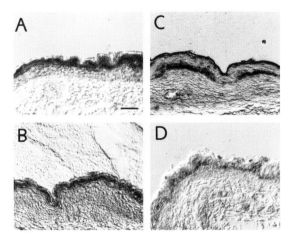

Figure 8 Ectopic expression of profilaggrin and loricrin in basal cells of transgenic mice misexpressing *Dlx3*. *In situ* hybridization of sections of ventral epidermis from normal (A, B) or transgenic mice misexpressing *Xenopus Dlx3* in basal cells (C, D). Both filaggrin (A, C) and loricrin (B, D) are reduced in level, and ectopically expressed in basal cells in affected epidermis. Reproduced from Morasso *et al.* (1996).

granular, phenotype. This evidently blocks the normal keratinocyte development sequence, but does not radically interfere with the keratin gene expression program which thus may be controlled by a Dlx3-independent mechanism.

Like Dlx2 and Dlx5, Dlx3 is a transcriptional activator (J. Feledy, unpublished). Interestingly, the profilaggrin gene, which is apparently up-regulated by Dlx3, has a moderately high affinity binding site for this regulatory protein, just downstream from its TATA element (Morasso *et al.*, 1996). It is possible, then, that profilaggrin is a direct and positively modulated target of Dlx3 homeoprotein. Targeted inactivation of the *Dlx3* gene results in early lethality, prior to epidermal stratification (M. I. Morasso and T. D. Sargent unpublished) so this approach does not shed any further light on the function of this factor in skin development.

The role, if any, for Dlx3 in the initial phase of mammalian epidermis, prior to stratification, has not been substantiated, since this gene does not activate until later in development. The first detectable expression in epidermis is at E10.5 in the apical ectodermal ridge (AER) of the mouse forelimb. The AER is not stratified, although periderm formation initiates in the vicinity of the AER at approximately the stage when Dlx3 expression commences. By E14.5, *Dlx3* is transcribed in prestratification limb epidermis (Morasso *et al.*, 1995). This is interesting, as *Dlx3* is exclusively expressed in the suprabasal cells in stratified skin. The AER epithelium interacts with underlying mesenchyme to promote limb outgrowth and differentiation (Zeller and Duboule, 1997), and other sites of *Dlx3* expression, such as hair follicles and mammary epithelium, are also associated

with interaction with mesenchyme. Thus it is possible that this homeoprotein has a function at the earliest stages of expression that is distinct from its role in later epidermis.

VII. Unanswered Questions and Future Directions

While there have been major strides in the past decade, there still remains much to be learned about the specification of ectoderm and the control of epidermal cell differentiation. Given the tendency for regulatory mechanisms to be conserved in evolution, and to be utilized repeatedly in development, it seems reasonable to postulate that the earliest regulatory events identified in the *Xenopus* embryo might be relevant to later processes such as stratification and epidermal appendage formation. For example, is the BMP–Smads–Msx1 pathway used to maintain epidermis in the *Xenopus* gastrula also functioning in these terminal differentiations? The epidermis of *Msx1* knock-out mice is not noticeably defective (Satokata and Maas, 1994), but functional redundancies often mask the role of regulatory genes in such experiments. As discussed above, BMPs are probably important in appendage development. BMP6 is induced when basal cells differentiate, and retrovirally expressed BMP6 blocks basal cell division, suggesting an effect on the earliest steps in interfollicular keratinocyte differentiation (Drozdoff *et al.*, 1994).

The coexpression of Msx1 and Dlx3 in the ventral ectoderm of *Xenopus,* and the antagonistic relationship between these two homeoprotein families suggest a control mechanism in these cells that depends on both factors. In mouse epidermis, Dlx3 is most likely a positive, differentiation-promoting regulator, which is consistent with its presence in the presumptive epidermis in the frog embryo. However, if Msx1 functions to inhibit Dlx3, then why does ectopic overexpression of Msx1 expand the epidermal field? A related question concerns the molecular basis of the "default" neuralization of *Xenopus* ectoderm; are there maternally inherited neurogenic factors in the *Xenopus* egg? If not, then what programs the expression of neural-specific genes when the epidermal identity fails? Also, as noted earlier, it is possible that the BMP signal is not needed for the initiation of epidermis, but only for its maintenance past the early gastrula stage. This could be settled by a suitable experiment with BMP signaling inhibitors, using sufficiently sensitive methods to detect abortive expression of keratins or other epidermal genes.

The role of Dlx3 in the differentiation of suprabasal keratinocytes, suggested by the ectopic expression data, raises the interesting question of target genes in epidermis for this activator protein. *Loricrin* and *profilaggrin* are candidates, based on the up-regulation of these genes in basal cells misexpressing Dlx3, but of course a less direct relationship is also possible. The role of Dlx3 in the early *Xenopus* embryo is also very much an open question, which is a problem that should be amenable to the powerful methods available to analyze gene function at the early stages of *Xenopus* development, such as RNA injection into fertilized eggs and dominant-negative approaches. To date, Dlx3 is the only transcription factor identified that seems capable of changing the identity of epidermal cells when ectopically expressed in the skin. However, since ectopic expression of Dlx3 in basal keratinocytes does not result in the activation of suprabasal keratins or the inactivation of basal keratin expression, the transition is presumably not complete. This implies the existence of other regulatory genes acting upstream of or parallel to Dlx3. The regulatory region of the *Xenopus Dlx3* gene comprises only a few hundred 5′-flanking base pairs, which should facilitate the identification of candidates for factors that control expression of Dlx3 in epidermis.

The initial specification and terminal differentiation of epidermis and its appendage structures, such as hair and tooth, utilize signal transduction pathways and transcriptional regulators that are also important in many other aspects of embryogenesis. The use of epidermis as a model system should continue to provide valuable insights into the mechanisms that control the development of this and other tissues in the vertebrate embryo.

Acknowledgments

We thank Drs. Jean-Pierre Saint-Jeannet and Andrzej Dlugosz for many helpful comments on the manuscript, and Dr. Jules Feledy for unpublished data.

References

Andersen, B., Schonemann, M. D., Flynn, S. E., Pearse, R. V., Singh, H., and Rosenfeld, M. G. (1993) Skn-1a and skn-1i: Two functionally distinct Oct-2-related factors expressed in epidermis. *Science* **260**, 78–81.

Anderson, B., and Rosenfeld, M. G. (1995). New wrinkles in retinoids. *Nature (London)* **374**, 118–119.

Banks-Schlegel, S. P. (1982). Keratin alterations during epidermal differentiation: A presage of adult epidermal maturation. *J. Cell Biol.* **93**, 551–559.

Bitgood, M. J., and McMahon, A. P. (1995). Hedgehog and Bmp are co-expressed at many sites of cell–cell interaction in the mouse embryo. *Dev. Biol.* **172**, 126–138.

Blessing, M., Nanney, L. B., King, L. E., Jones, C. M., and Hogan, B. L. (1993). Transgenic mice as a model to study the role of TGF-beta-related molecules in hair follicles. *Genes Dev.* **7**, 204–215.

Blitz, I. L., and Cho, K. W. (1995). Anterior neurectoderm is progressively induced during gastrulation: The role of the *Xenopus* homeobox gene orthodenticle. *Development (Cambridge, UK)* **121**, 993–1004.

Bronner-Fraser, M. (1993). Mechanisms of neural crest cell migration. *BioEssays* **15**, 221–230.

Byrne, C., Tainsky, M., and Fuchs, E. (1994). Programming gene expression in developing epidermis. *Development (Cambridge, UK)* **112**, 2369–2383.

Catron, K. M., Zhang, H., Marshall, S. C., Inostroza, J. A., Wilson, J. M., and Abate, C. (1995). Transcriptional repression by Msx1 does not require homeodomain DNA-binding sites. *Mol. Cell. Biol.* **15**, 861–871.

Chang, C., Wilson, P., Mathews, L., and Hemmati-Brivanlou, A. (1997). A *Xenopus* type I activin receptor mediates mesodermal but not neural specification during embryogenesis. *Development (Cambridge, UK)* **124**, 827–837.

Clement, J. H., Fettes, P., Knochel, S., Lef, J., and Knochel, W. (1995). Bone morphogenetic protein 2 in the early development of *Xenopus laevis*. *Mech. Dev.* **52**, 357–370.

Cohen, S. M., Bronner, G., Kutter, F., Jurgens, G., and Jackle, H. (1989). Distal-less encodes a homeodomain protein required for limb development in *Drosophila*. *Nature (London)* **338**, 432–434.

Cotsarelis, G., Sun, T. T., and Lavker, R. M. (1990). Label-retaining cells reside in the bulge area of pilosebaceous unit: Implications for follicular stem cells, hair cycle, and skin carcinogenesis. *Cell* **61**, 1329–1337.

Dirksen, M. L., Morasso, M. I., Sargent, T. D., and Jamrich, M. (1994). Differential expression of a Distal-less homeobox gene Xdll-2 in ectodermal cell lineages. *Mech. Dev.* **46**, 63–70.

Dlugosz, A., and Yuspa, S. (1993). Coordinate changes in gene expression which mark the spinous to granular cell transition in epidermis are regulated by protein kinase C. *J. Cell Biol.* **120**, 217–225.

Drozdoff, V., Wall, N. A., and Pledger, W. J. (1994). Expression and growth inhibitory effect of decapentaplegic Vg-related protein 6: Evidence for a regulatory role in keratinocyte differentiation. *Proc. Natl. Acad. Sci. U.S.A.* **91**, 5528–5532.

Drysdale, T. A., and Elinson, R. P. (1993). Inductive events in the patterning of the *Xenopus laevis* hatching and cement glands, two cell types which delimit head boundaries. *Dev. Biol.* **158**, 245–253.

Eckert, R. L., and Welter, J. F. (1996). Transcription factor regulation of epidermal keratinocyte gene expression. *Mol. Biol. Rep.* **23**, 59–70.

Fainsod, A., Deissler, K., Yelin, R., Marom, K., Epstein, M., Pillemer, G., Steinbeisser, H., and Blum, M. (1997). The dorsalizing and neural inducing gene follistatin is an antagonist of BMP-4. *Mech. Dev.* **63**, 39–50.

Fainsod, A., Steinbeisser, H., and De Robertis, E. M. (1994). On the function of BMP-4 in patterning the marginal zone of the *Xenopus* embryo. *EMBO J.* **13**, 5015–5025.

Fisher, C. (1994). The cellular basis of development and differentiation in mammalian keratinizing epithelia. *In* "The Keratinocyte Handbook" (I. Leigh, B. Lane, and F. Watt, eds.), pp. 131–150. Cambridge Univ. Press, London.

Fuchs, E. (1990). Epidermal differentiation: The bare essentials. *J. Cell Biol.* **111**, 2807–2814.

Fuchs, E., and Byrne, C. (1994). The epidermis: Rising to the surface. *Curr. Opin. Genet. Dev.* **4**, 725–736.

Godsave, S. F., and Slack, J. M. W. (1989). Clonal analysis of mesoderm induction in *Xenopus laevis. Dev. Biol.* **134**, 486–480.

Graff, J. M., Bansal, A., and Melton, D. A. (1996). *Xenopus* Mad proteins transduce distinct subsets of signals for the TGF superfamily. *Cell (Cambridge, Mass.)* **85**, 479–487.

Green, J. B. A., New, H. V., and Smith, J. C. (1992). Responses of embryonic *Xenopus* cells to activin and FGF are separated by multiple dose thresholds and correspond to distinct axes of the mesoderm. *Cell* **71**, 731–739.

Grunz, H., and Tacke, L. (1989). Neural differentiation of *Xenopus laevis* ectoderm takes place after dis-aggregation and delayed reaggregation without inducer. *Cell Differ. Dev.* **28**, 211–218.

Guo, L., Yu, Q. C., and Fuchs, E. (1993). Targeting expression of keratinocyte growth factor to keratinocytes elicits striking changes in epithelial differentiation in transgenic mice. *EMBO J.* **12**, 973–986.

Hansen, C. S., Marion, C. D., Steele, K., George, S., and Smith, W. C. (1997). Direct neural induction and selective inhibition of mesoderm and epidermis inducers by Xnr3. *Development (Cambridge, UK)* **124**, 483–492.

Hardy, M. H. (1992). The secret life of the hair follicle. *Trends Genet.* **8**, 55–60.

Hemmati-Brivanlou, A., and Melton, D. A. (1994). Inhibition of activin receptor signaling promotes neuralization in *Xenopus. Cell (Cambridge, Mass.)* **77**, 273–281.

Hemmati-Brivanlou, A., and Thomsen, G. A. (1995). Ventral mesodermal patterning in *Xenopus* embryos: Expression patterns and activities of BMP-2 and BMP-4. *Dev. Genet.* **17**, 78–89.

Hemmati-Brivanlou, A., Kelly, O. G., and Melton, D. A. (1994). Follistatin, an antagonist of activin is expressed in the Spemann organizer and displays direct neuralizing activity. *Cell (Cambridge, Mass.)* **77**, 283–295.

Hill, R. E., Jones, P. F., Rees, A. R., Sime, C. M., Copeland, N. G., Jenkins, N. A., Graham, E., and Davidson, D. R. (1989). A new family of mouse homeobox containing genes: Molecular structure, chromosomal location and developmental expression of Hox-7.1. *Genes Dev.* **3**, 26–37.

Holbrook, K. A. (1991). Structure and function of the developing human skin. *In* "Physiology, Biochemistry and Molecular Biology of the Skin" (L.A. Goldsmith, ed.), 2nd ed., pp. 63–110. Oxford Univ. Press, London.

Holbrook, K. A., Smith, L. T., Kaplan, E. D., Minami, S. A., Hebert, G. P., and Underwood, R. A. (1993). Expression of morphogens during human follicle development *in vivo* and a model for studying follicle morphogenesis *in vitro. J. Invest. Dermatol.* **101**(Suppl.), 39S–49S.

Hoodless, P. A., Haerry, T., Abdollah, S., Stapleton, J., O'Connor, M. B., Attisano, L., and Wrana, J. L. (1996). MADR1, a MAD-related protein that functions in BMP2 signaling pathways. *Cell (Cambridge, Mass.)* **85**, 489–500.

Imakado, S., Bickenbach, J. R., Bundman, D. S., Rothnagel, J. A., Attar, P. S., Wang, X.-J., Walczak, V. R., Wisniewski, S., Pote, J., Gordon, J. S., Heyman, R. A., Evans, R. M., and Roop, D. R. (1995). Targeting expression of a dominant-negative retinoic acid receptor mutant in the epidermis of transgenic mice results in loss of barrier function. *Genes Dev.* **9**, 317–329.

Iseki, S., Araga, A., Ohuchi, H., Nohno, T., Yoshioka, H., Hayashi, F., and Noji, S. (1996). Sonic hedgehog is expressed in epithelial cells during development of whisker, hair, and tooth. *Biochem. Biophys. Res. Commun.* **218**, 688–693.

Jahoda, C. A. B., and Reynolds, A. J. (1993). Dermal–epidermal interactions: Follicle-derived cell populations in the study of hair-growth mechanisms. *J. Invest. Dermatol.* **101**(Suppl.), 33S–38S.

Jamrich, M., Sargent, T. D., and Dawid, I. B. (1987). Cell-type-specific expression of epidermal cytokeratin genes during gastrulation of *Xenopus laevis. Genes Dev.* **1**, 124–132.

Jones, E. A., and Woodland, H. R. (1986). Development of the ectoderm in *Xenopus*: Tissue specification and the role of cell association and division. *Cell (Cambridge, Mass.)* **44**, 345–355.

Jones, P. H., and Watt, F. M. (1993). Separation of human epidermal stem cells from transit amplifying cells on the basis of differences in integrin function and expression. *Cell (Cambridge, Mass.)* **73**, 713–724.

Jones, P. H., Harper, S., and Watt, F. M. (1995). Stem cell patterning and fate in human epidermis. *Cell (Cambridge, Mass.)* **80**, 83–93.

Jowett, A. K., Vainio, S., Ferguson, M. W. J., Sharpe, P. T., and Thesleff, I. (1993). Epithelial–mesenchymal interactions are required for Msx1 and Msx2 gene expression in the developing muring molar tooth. *Development (Cambridge, UK)* **117**, 461–470.

Kengaku, M., and Okamoto, H. (1993). Basic fibroblast growth factor induces differentiation of neural tube and neural crest lineages of cultured ectoderm cells from *Xenopus* gastrula. *Development (Cambridge, UK)* **119**, 1067–1078.

Kopan, R., Traska, G., and Fuchs, E. (1987). Retinoids as important regulators of terminal differentiation: Examining keratin expression in individual epidermal cells at various stages of keratinization. *J. Cell Biol.* **105**, 427–440.

Kratochwil, K., Dull, M., Farinas, I., Galceran, J., and Grosschedl, R. (1996). LEF-1 expression is activated by BMP-4 and regulates inductive tissue interactions in tooth and hair development. *Genes Dev.* **10**, 1382–1394.

Leask, A., Byrne, C., and Fuch, E. (1991). Transcription factor AP2 and its role in epidermal-specific gene expression. *Proc. Natl. Acad. Sci. U.S.A.* **88**, 7948–7952.

Letsou, A., Arora, K., Wrana, J. L., Simin, K., Twombly, V., Jamal, J., Staehling-Hampton, K., Hoffmann, F. M., Gelbart, W. M., Massague, J., and O'Connor, M. B. (1995). *Drosophila* Dpp signaling is mediated by the punt gene product: A dual ligand-binding type II receptor of the TGF beta receptor family. *Cell (Cambridge, Mass.)* **80**, 899–908.

Luscher, B., Mitchell, P. J., Williams, T., and Tjian, R. (1989). Regulation of transcription factor AP-2 by the morphogen retinoic acid and by second messengers. *Genes Dev.* **3**, 1507–1517.

M'Boneko, V., and Merker, H. J. (1988). Development and morphology of the periderm of mouse embryos (days 9–12 of gestation). *Acta Anat. (Basel)* **133**, 325–336.

MacKenzie, A., Leeming, G. L., Jowett, A. K., Ferguson, M. W. J., and Sharpe, P. T. (1991). The homeobox gene Hox7.1 has specific regional and temporal expression patterns during early murine craniofacial embryogenesis, especially tooth development *in vivo* and *in vitro*. *Development (Cambridge, UK)* **111**, 269–285.

MacKenzie, A. K., Ferguson, M. W. J., and Sharpe, P. T. (1992). Expression patterns of the homeobox gene Hox-8 in the mouse embryo suggest a role in specifying tooth initiation and shape. *Development (Cambridge, UK)* **115**, 403–420.

Magnaldo, T., Vidal, R. G., Ohtsuki, M., Freedberg, I. M., and Blumenberg, M. (1993). On the role of AP2 in epithelial-specific gene expression. *Gene Express.* **3**, 307–315.

Mancilla, A., and Mayor, R. (1996). Neural crest formation in *Xenopus laevis:* Mechanisms of Xslug induction. *Dev. Biol.* **177**, 580–589.

Massague, J. (1996). TGF signaling: Receptors, transducers and Mad Proteins. *Cell (Cambridge, Mass.)* **85**, 947–950.

Mathers, P. H., Miller, A., Doniach, T., Dirksen, M. L., and Jamrich, M. (1995). Initiation of anterior head-specific gene expression in uncommitted ectoderm of *Xenopus laevis* by ammonium chloride. *Dev. Biol.* **171**, 641–654.

Mayor, R. M., Morgan, R. M., and Sargent, M. G. (1995). Induction of the prospective neural crest of *Xenopus. Development (Cambridge, UK)* **121**, 767–777.

Menon, G. K., Grayson, S., and Elias, P. M. (1985). Ionic calcium reservoirs in mammalian epidermis: Ultrastructural localization by ion-capture cytochemistry. *J. Invest. Dermatol.* **84**, 508–512.

Menon, G. K., and Elias, P. M. (1991). Ultrastructural localization of calcium in psoriatic and normal human epidermis. *Arch. Dermatol.* **127**, 57–63.

Morasso, M. I., Mahon, K. A., and Sargent, T. D. (1995). A *Xenopus* Distal-less gene in transgenic mice: Conserved regulation in distal limb epidermis and other sites of epithelial–mesenchymal interaction. *Proc. Natl. Acad. Sci. U.S.A.* **92**, 3968–3972.

Morasso, M. I., Markova, N. G., and Sargent, T. D. (1996). Regulation of epidermal differentiation by a Distal-less homeodomain gene. *J. Cell Biol.* **135**, 1879–1887.

Morasso, M. I., Yonescu, R., Griffin, C. A., and Sargent, T. D. (1997). Localization of human DLX8 to chromosome 17q21.3-q22 by fluorescence *in situ* hybridization. *Mamm. Genome* **8**, 302–303.

Morgan, R., and Sargent, M. G. (1997). The role in neural patterning of translation initiation factor eIF4AII; induction of neural fold genes. *Development (Cambridge, UK)* **124**, 2751–2760.

Nakamura, S., Stock, D. W., Wynder, K. L., Bollekens, J. A., Takeshita, K., Nagai, B. M., Chiba, S., Kitamura, T., Freeland, T. M., Zhao, Z., Minowada, J., Lawrence, J. B.,

Weiss, K. M., and Ruddle, F. R. (1996). Genomic analysis of a new mammalian Distal-less gene: Dlx7. *Genomics* **38**, 314–324.

Otte, A. P., and Moon, R. T. (1992). Protein kinase C isozymes have distinct roles in neural induction and competence in *Xenopus. Cell (Cambridge, Mass.)* **68**, 1021–1029.

Papalopulu, N., and Kintner, C. (1993). *Xenopus* Distal-less related homeobox genes are expressed in the developing forebrain and are induced by planar signals. *Development (Cambridge, UK)* **117**, 961–975.

Quinn, L. M., Johnson, B. V., Nicholl, J., Sutherland, G. R., and Kalionis, B. (1997). Isolation and identification of homeobox genes from the human placenta including a novel member of the Distal-less family, DLX4. *Gene* **187**, 55–61.

Robinson, G. W., and Mahon, K. A. (1994). Differential and overlapping expression domains of Dlx2 and Dlx3 suggest distinct roles for Distal-less homeobox genes in craniofacial development. *Mech. Dev.* **48**, 199–215.

Rosenzweig, B. L., Imamura, T., Okadome, T., Cox, G. N., Yamashita, H., ten Kijke, P., Heldin, C. H., and Miyazono, K. (1995). Cloning and characterization of a human type II receptor for bone morphogenetic proteins. *Proc. Natl. Acad. Sci. U.S.A.* **92**, 7632–7636.

Rutberg, S. E., Saez, E., Glick, A., Dlugosz, A. A., Spiegelman, B. M., and Yuspa, S. H. (1996). Differentiation of mouse keratinocytes is accompanied by PKC-dependent changes in AP1 proteins. *Oncogene* **13**, 167–176.

Saitou, M., Sugai, S., Tanaka, T., Shimouchi, K., Fuchs, E., Narumiya, S., and Kakizuka, A. (1995). Inhibition of skin development by targeted expression of a dominant-negative retinoic acid receptor. *Nature (London)* **374**, 159–162.

Sasai, Y., Lu, B., Steinbeissser, H., Geissert, D., Gont, L. K., and De Robertis, E. M. (1994). *Xenopus* Chordin: A novel dorsalizing factor activated by organizer-specific homeobox genes. *Cell (Cambridge, Mass.)* **79**, 779–790.

Sato, S. M., and Sargent, T. D. (1989). Development of neural inducing capacity in dissociated *Xenopus* embryos. *Dev. Biol.* **134**, 263–266.

Satokata, I., and Maas, R. (1994). Msx1 deficient mice exhibit cleft palate and abnormalities of craniofacial and tooth development. *Nat. Genet.* **6**, 348–356.

Schmidt, J. E., Suzuki, A., Ueno, N., and Kimelman, D. (1995). Localized BMP-4 mediates dorsal/ventral patterning in the early *Xenopus* embryo. *Dev. Biol.* **169**, 37–50.

Schorle, H., Meier, P., Buchert, M., Jaenisch, R., and Mitchell, P. J. (1996). Transcription factor AP2 essential for cranial closure and craniofacial development. *Nature (London)* **381**, 235–238.

Sengel, P. (1990). Pattern formation in skin development. *Int. J. Dev. Biol.* **34**, 33–50.

Shimeld, S. M., McKay, I. J., and Sharpe, P. T. (1996). The murine homeobox gene Msx3 shows highly restricted expression in the developing neural tube. *Mech. Dev.* **55**, 201–210.

Simeone, S., Acampora, D., Pannese, M., D'Esposito, M., Stornaiuolo, A., Gulisano, M., Mallamaci, A., Kastury,

K., Druck, T., Huebner, K., and Boncinelli, E. (1994). Cloning and characterization of two members of the vertebrate Dlx gene family. *Proc. Natl. Acad. Sci. U.S.A.* **91**, 2250–2254.

Sive, H. L., Hattori, K., and Weintraub, H. (1989). Progressive determination during formation of the anteroposterior axis in *Xenopus laevis. Cell (Cambridge, Mass.)* **58**, 171–180.

Smith, W. C., and Harland, R. M. (1992). Expression cloning of noggin, a new dorsalizing factor localized to the Spemann organizer in *Xenopus* embryos. *Cell (Cambridge, Mass.)* **70**, 829–840.

Snape, A. M., Winning, R. S., and Sargent, T. D. (1991). Transcription factor AP2 is tissue-specific and is closely related or identical to keratin transcription factor 1 (KTF-1). *Development (Cambridge, UK)* **113**, 283–293.

Sudarwati, S., and Nieuwkoop, P. D. (1971). Mesoderm formation in the anuran *Xenopus laevis* (Daudin). *Wilhelm Roux's Arch.* **166**, 189–204.

Suzuki, A., Ueno, N., and Hemmati-Brivanlou, A. (1997). *Xenopus* Msx1 mediates epidermal induction and neural inhibition by BMP4. *Development* **124**, 3037–3044.

Thesleff, I., Vaahtokari, A., and Partanen, A.-M. (1995). Regulation of organogenesis: Common molecular mechanisms regulating the development of the teeth and other organs. *Int. J. Dev. Biol.* **39**, 35–50.

Thomsen, G. H. (1997). Antagonism within and around the organizer: BMP inhibitors in vertebrate body patterning. *Trends Genet.* **13**, 209–211.

Tomic-Canic, M., Day, D., Samuels, H. H., Freedberg, I. M., and Blumenberg, M. (1996). Novel regulation of keratin gene expression by thyroid hormone and retinoid receptors. *J. Biol. Chem.* **271**, 1416–1423.

Vainio, S., Karavanova, I., Jowett, A., and Thesleff, I. (1993). Identification of BMP-4 as a signal mediating secondary induction between epithelial and mesenchymal tissues during early tooth development. *Cell (Cambridge, Mass.)* **75**, 45–58.

van Gerdensen, C., Okamura, R. M., Farinas, I., Quo, R.-G., Parslow, T. G., Bruhn, L., and Grosschedl, R. (1995). Development of several organs that require inductive epithelial–mesenchymal interactions is impaired in LEF-1 deficient mice. *Genes Dev.* **8**, 2691–2703.

Watt, F. M. (1989). Terminal differentiation of epidermal keratinocytes. *Curr. Opin. Cell Biol.* **1**, 1107–1115.

Wilson, P. A., and Hemmati-Brivanlou, A. (1995). Induction of epidermis and inhibition of neural fate by BMP-4. *Nature (London)* **376**, 331–336.

Wilson, P. A., Lagna, G., Suzuki, A., and Hemmati-Brivanlou, A. (1997). Concentration-dependent patterning of the *Xenopus* ectoderm by BMP4 and its signal transducer Smad1. *Development (Cambridge, UK)* **124**, 3177–3184.

Winning, R. S., Shea, L. J., Marcus, S. J., and Sargent, T. D. (1991). Developmental regulation of transcription factor AP-2 during *Xenopus laevis* embryogenesis. *Nucleic Acids Res.* **19**, 3709–3714.

Woloshin, P., Song, K., Degnin, C., Killary, A. M., Goldhamer, D. J., Sassoon, D., and Thayer, M. J. (1995). Msx1 inhibits MyoD expression in fibroblast X 10T1/2 cell hybrids. *Cell (Cambridge, Mass.)* **82**, 611–620.

Yuspa, S. H., Kilkenny, A. E., Steinert, P. M., and Roop, D. R. (1989). Expression of murine epidermal differentiation markers is tightly regulated by restricted extracellular calcium concentrations *in vitro. J. Cell Biol.* **109**, 1207–1217.

Zeller, R., and DuBoule, D. (1997). Dorso-ventral limb polarity and origin of the ridge: On the fringe of independence? *BioEssays* **19**, 541–546.

Zhang, J., Hagopian-Donaldson, S., Serbedzija, G., Elsemore, J., Plehn-Dujowich, D., McMahon, A. P., Flavell, R. A., and Williams, T. (1996). Neural tube, skeletal and body wall defects in mice lacking transcription factor AP2. *Nature (London)* **381**, 238–241.

Zhang, H., Hu, G., Wang, H., Sciavolino, P., Iler, N., Shen, M. M., and Abate-Shen, C. (1997). Heterodimerization of Msx and Dlx homeoproteins results in functional antagonism. *Mol. Cell. Biol.* **17**, 2920–2932.

Zhou, P., Byrne, C., Jacobs, J., and Fuchs, E. (1995). Lymphoid enhancer factor 1 directs hair follicle patterning and epithelial cell fate. *Genes Dev.* **9**, 700–713.

39

Conservation of Themes in Vertebrate Blood Development

Eric C. Liao[1,2]
Leonard I. Zon[1]
[1]Division of Hematology/Oncology
Children's Hospital
Department of Pediatrics and Howard Hughes Medical Institute
Harvard Medical School
Boston, Massachusetts 02215
[2]Division of Health Sciences and Technology
Massachusetts Institute of Technology
Cambridge, Massachusetts 02139

I. Introduction

Hematopoiesis provides an excellent model system for the study of cellular differentiation events. Morphological, molecular, and anatomic changes during blood development reveal key concepts of induction, patterning, specification, and differentiation that occur throughout embryogenesis. The identification and characterization of a pluripotent hematopoietic stem cell population enables one to study progressive restriction of lineage potential during hematopoiesis with *in vitro* culture assays. Significant advances in the isolation of lineage-specific cytokines and surface markers provide molecular control in studying specific cellular proliferation and differentiation (Metcalf, 1993). Furthermore, targeted gene disruptions of an increasing number of lineage-specific transcription factors in mice have led to functional assignments of these molecules along a developmental pathway (Orkin, 1995; Shivdasani and Orkin, 1996). Expression cloning strategies in *Xenopus* also have identified early mesoderm induction events that lead to blood formation (Mead *et al.,* 1996). More recently, investigation of hematopoiesis from a genetic approach is facilitated with the identification of a large number of zebrafish mutants (Ransom *et al.*, 1996; Weinstein *et al.*, 1996).

This chapter describes our current understanding of the developmental biology of hematopoiesis, and showcases several experimental approaches taken to dissect the process. We examine common themes of blood development in five vertebrate species: human, mouse, chick, *Xenopus*, and zebrafish, and highlight the unique experimental advantages of each model organism. Progress in our understanding of hematopoiesis not only reveals molecular mechanisms underlying fate determination and lineage restriction, but also has direct relevance for therapy of a number of blood disorders and cancers.

II. The Hematopoietic System

A. Mesoderm Induction

Blood arises from ventral mesoderm during gastrulation (Fig. 1). In *Xenopus*, experiments on embryonic patterning have demonstrated that factors regulating axis formation significantly affect the initiation of the blood program. For instance, many early hematopoietic-specific transcription factors are expressed in the ventral region of the embryo in late gastrula and early neurula stages. Perturbing the axis with UV irradiation leads to a radially symmetric embryo with uniform expression of these transcription factors. Dorsalization by lithium

chloride abrogates the expression of these transcription factors (Kelley *et al.*, 1994). Hence, the factors that regulate embryonic patterning are likely to have a profound effect on hematopoiesis. Because zygotic transcription in a *Xenopus* embryo occurs at stage 8 (5.5 hours postfertilization), and the hematopoietic transcription factors are expressed at stage 11 (8 hours), the induction of the hematopoietic program is likely to occur between stages 8 and 11 during gastrulation.

Mesoderm is patterned to a ventral fate by a class of factors known as bone morphogenetic proteins (BMP) (Graff, 1997). BMP-2, -4, and -7 have been shown to induce ectopic globin expression in *Xenopus* animal pole explants, where blood is not normally formed (Hemmati-Brivanlou and Thomsen, 1995; Maeno *et al.*, 1996; Zhang and Evant, 1996). Recent experiments have demonstrated that induction of blood by BMP-4 occurs after mesoderm fate has been determined (T. Huber and L. I. Zon, 1998, in press). These data support a model where mesoderm induction occurs and is subsequently patterned to a ventral fate by BMP-4. When dominant negative BMP receptors are targeted to the ventral region of the embryo, blood formation fails to occur (Maeno *et al.*, 1994). In addition, a zebrafish mutant called *swirl* has a defect in the BMP-2 gene and has a dorsalized body plan (Kishimoto *et al.*, 1998). This dorsalized mutant also fails to express hematopoietic transcription factors such as GATA-1 and GATA-2 in the ventral mesoderm. Collectively, these observations support that BMPs pattern the mesoderm toward a ventral and hematopoietic fate.

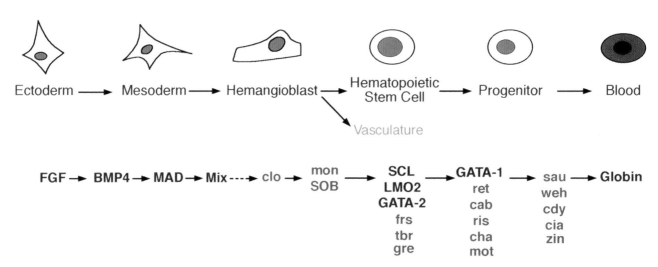

Figure 1 Developmental hematopoiesis. Early during development, an ectodermal cell is induced to become mesoderm by factors that emanate from the endodermal pole. This mesodermal cell is patterned to ventral mesoderm and ultimately forms hemangioblasts which have the potential to become either vascular or hematopoietic progenitors. The hematopoietic stem cells are able to proliferate and terminally differentiate to red blood cells. We and others have defined a cascade of events that are important in this process. This includes early mesodermal inducing factors, such as fibroblast growth factor (FGF) and other transforming growth factor β (TGFβ) related family members that induce mesoderm. Subsequently, BMP-4 signals through SMADs and Mix to lead to ventralization of the embryo. Genes such as *cloche* (clo) regulate hemangioblast formation. The transcription factors SCL, LMO2, and the GATA-binding proteins regulate the proliferation and differentiation of hematopoietic cells (known genes are shown in red). Several zebrafish mutations (in blue) are also included, which may act at the noted stages of development (recessive mutations are in lowercase and the dominant mutation *SOB* is capitalized).

In recent years, there have been several exciting developments in our understanding of BMP signaling. BMPs can homo- or heterodimerize and bind to type I and type II receptors. There is cross-regulation at a receptor level with particular ligands. In addition, there can be heterodimerization of those ligands. On cross-linking with ligands, a family of cytoplasmic signal transduction molecules called Smads become phosphorylated, homo- and heterodimerize, and then enter into the nucleus where they activate transcription (Massague, 1992; Massague, 1996). Recent evidence has suggested that Smad-1 is downstream of BMP-4 and Smad-2 is downstream of activin signaling (Graff *et al.,* 1996). Smad-1 and Smad-2 can each heterodimerize with Smad-4, which is constitutively expressed. Thus, the Smad molecules are signal transducers downstream of receptor activation.

Another class of proteins, the Mix homeodomain factors, are involved in specifying the hematopoietic mesoderm (Mead *et al.,* 1996). Four distinct Mix factors have been isolated in *Xenopus*, each expressed in distinct regions of the embryo. The Mix family of homeoproteins are responsive to activin and BMP, and are induced by the Smad proteins. The Mix.1 gene product is a potent inducer of hematopoiesis. Overexpression of Mix.1 leads to ventralization similar to that caused by BMP-4, and a dominant negative Mix.1 mutant blocks BMP-4 signaling in the early embryo. Each Mix gene is expressed in distinct but overlapping patterns. Mix.1 is expressed equally in both the mesoderm and endodermal compartments. Mix.3 and Mix.4 are expressed mostly in the endoderm, however, Mix.4 also is expressed in the mesoderm. Mix.4 expression in the mesoderm is excluded from the dorsal marginal zone. The Mix molecules can heterodimerize with each other without any distinct specificity, and each can ventralize the embryo. Recently, Whitman and colleagues defined an activin response element in the Mix.2 promoter (Chen *et al.,* 1997). The activin response element binds a novel fork head protein called FAST-1 that is in a complex with a phosphorylated form of Smad. This complex is able to activate the Mix.2 promoter. Collectively, these studies establish a signal cascade for the ventral pattern formation, involving BMP signaling to Smad, and then to Mix.1 (Fig. 1). The targets of Mix. 1, Mix.3, and Mix.4 during development remain to be defined.

The future of these studies is to define specifiers of hematopoietic tissues. Recent evidence in our laboratories demonstrated that GATA-1, SCL, and LMO2 can directly induce blood in *Xenopus* animal cap cells cultured with fibroblast growth factor (FGF). However, c-Myb does not have this activity. The induction of blood by the aforementioned factors does not require exogenous BMP-4. These studies open a new avenue for understanding organogenesis at a molecular level.

B. The Hemangioblast, a Common Precursor for Blood and Vasculature

First coined by His in 1900, the term hemangioblast has been used to describe the intimate relationship between blood and endothelial cells during early vertebrate development (Sabin, 1920). Multiple studies have suggested that this bipotential precursor cell contributes to both hematopoietic and endothelial progenitors (Fig. 1). These studies have examined expression patterns of a number of early blood and endothelial molecular markers (SCL, GATA-2, CBF, CD34, SCF/c-kit, fli-1, flk-1, Quek-1), and have found them to co-localize in distinct anatomical sites (Zon, 1995). These genes have been implicated by mouse targeted gene disruptions to function in either hematopoiesis (SCL, GATA-2, CD34, SCF/c-kit), vasculogenesis (fli-1), or both (CBF, flk-1) (reviewed in Shivdasani and Orkin, 1996). For instance, the flk-1 knock-out mouse has both hematopoietic and vascular defects (Shalaby *et al.,* 1995). This mouse mutant is reminiscent of the *cloche* mutation in zebrafish, which also affects both hematopoiesis and vasculogenesis (Stainer *et al.,* 1995). The *cloche* gene is not flk-1, based on linkage analysis (Liao *et al.,* 1997). The *cloche* mutation should define an interesting molecule which leads to the formation of both tissues.

It is not certain whether the other genes demonstrated to be necessary for hematopoiesis also act at the hemangioblast level, and therefore could play a role in vasculogenesis. It is possible that the lethal anemia in a homozygous null background of these genes masks any possible vascular defect. For example, when the SCL is replaced in the SCL−/− knockout under the action of the GATA-1 promotor, an angiogenesis defect of yolk sac vitelline vessel is observed (Visvader *et al.,* 1998). Our recent data corroborate a vascular function for SCL, where forced expression of SCL can rescue both blood and vascular differentiation defects in the cloche zebrafish mutant (Liao *et al.,* 1998). Whether the hemangioblast can be isolated as a single cell remained speculative. However, recently Choi and colleagues demonstrate that a blast colony derived from embryonic stem cell embryoid bodies (blast colony-forming cell, BL-CFC) possesses both vascular and hematopoetic characteristics (Choi *et al.,* 1998). Taken together, these results suggest that bipotential (vascular and hematopoietic) genes such as SCL, flk-1, and CBF exist, and may act in a cell similar to BL-CFC. What the BL-CFC may correspond to in an *in vivo* context remains undefined.

C. Primitive Hematopoiesis

Vertebrate blood development occurs in two distinct phases: primitive (embryonic), and definitive (adult). The two phases of hematopoiesis can be distinguished on the basis of cell morphology, time of initiation, site of development, and globin chain subtypes. For example, primitive hematopoiesis in mouse consists of nucleated erythrocytes that arise in yolk sac mesoderm around embryonic day 7.5 (E7.5), and expresses (ζ,ϵ/βH1) globin chains. Murine definitive hematopoiesis is characterized by non-nucleated erythrocytes produced from the fetal liver around E11 of gestation, which express (α,βmajor) globin chains. A distinct population of hematopoietic stem cells in the dorsal aorta, gonads, and mesonephros (AGM) also contributes to definitive hematopoiesis (Section II, D). Finally, primitive hematopoiesis is restricted to the erythroid lineage and definitive hematopoiesis produces blood cells of all lineages.

Distinct ventral sites are used by each species for the first phase of hematopoiesis. In mice, primitive hematopoiesis takes place in the yolk sac blood islands, which are derived from extraembryonic hematopoieitic mesoderm (Fig. 2C). In chicks and most birds, primitive hematopoiesis begins in the yolk sac, as blood islands arise in the posterior area opaca at the early somite stage. In amphibians such as *Xenopus*, blood is first formed in the ventral mesoderm and migrates to form a "V-shaped" hematopoietic blood island, the VBI (Fig. 2B) (Kelley *et al.*, 1994; Rollins-Smith and Blair, 1990). Hemoglobination occurs sequentially from the anterior to posterior, beginning at 36 hours postfertilization.

In teleosts (bony fish) such as zebrafish, embryonic hematopoiesis occurs in the intermediate cell mass (ICM) of the embryo (Fig. 2A). Unlike all other vertebrates that initiate primitive hematopoiesis on the yolk, teleost primitive hematopoiesis begins in the ICM, within the embryo proper. The ICM is derived from the paraxial mesoderm of the embryo beginning around the five somite stage. The ICM is analogous to the yolk sac hematopoietic tissue of higher vertebrates, as many

hematopoietic transcription factors expressed in the yolk sac of higher vertebrates such as GATA-1 and GATA-2 also are found to be expressed in the fish ICM (Detrich *et al.*, 1995). As in *Xenopus*, the hematopoietic cells of the zebrafish ICM differentiate in an anterior to posterior pattern.

We have defined subpopulations of blood progenitors within the ICM of fish. Anterior ICM (A-ICM) contains smaller cells with more differentiated morphology, exhibiting oval condensed nuclei (Detrich *et al.*, 1995), whereas the posterior ICM (P-ICM) cells have a less differentiated morphology of larger cell size and round prominent nuclei. The difference between A-ICM and P-ICM cells also is defined molecularly, where GATA-2, SCL, and LMO-2 are expressed in both A-ICM and P-ICM, but GATA-1 expression is restricted to A-ICM (Liao *et al.*, 1998; Thompson *et al.*, 1998). The P-ICM cells are therefore likely to be hematopoietic stem cells, whereas the A-ICM consists of the differentiated blood cells derived from primitive hematopoiesis. As circulation begins around 26 hours postfertilization, cells in the A-ICM travel throughout the embryo, whereas the P-ICM cells appear to remain localized to the posterior tail. We believe that the A-ICM progenitor cells home to hematopoietic niches that later will give rise to definitive blood. Furthermore, the P-ICM cells appear to reinitiate the hematopoietic program at 4 days postfertilization (dpf), where SCL and GATA-1 expression can be detected ventral to the axial vein of the posterior tail, which we term the ventral vein region (VVR) (Liao *et al.*, 1998).

D. Definitive Hematopoiesis in the Aorta–Gonads–Mesonephros Region

The main source of definitive hematopoietic stem cells long has been believed to be intraembryonic in origin

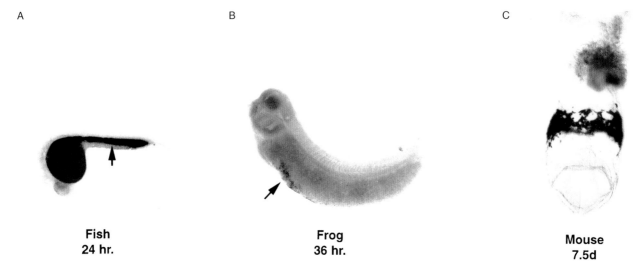

A   B   C

Fish
24 hr.

Frog
36 hr.

Mouse
7.5d

Figure 2 Primitive sites of hematopoiesis. (A) Zebrafish intermediate cell mass in a 24-hour embryo, shown by GATA-1 RNA *in situ* hybridization (arrow). (B) *Xenopus* blood islands in a 36-hour embryo, delineated with GATA-1 RNA *in situ* (arrow). (C) Murine yolk sac hematopoietic mesoderm in a 7.5-day embryo, stained with GATA-1 RNA *in situ*. (Courtesy of Stuart Orkin.)

(Cormier *et al.*, 1986; Dieterlen-Lievre, 1984; Tavian *et al.*, 1996). Definitive hematopoiesis in the chicken is exclusively intra-embryonic (Fig. 4), where the stem cells reside in the ventral wall of the dorsal aorta (Dieterlen-Lievre, 1975; Dieterlen-Lievre and Martin, 1981; Pardanaud *et al.*, 1996). The first definitive hematopoietic cells in vertebrates arise from the aorta-gonads-mesonephros (AGM) region (Figs. 3 and 4AB) (Medvinsky *et al.*, 1993; Muller *et al.*, 1994). However, a recent *in vivo* study provides evidence that definitive hematopoiesis also receives significant contribution from extraembryonic yolk sac-derived hematopoietic stem cells (Turpen *et al.*, 1997). Multipotential hematopoietic progenitors also have been detected in murine paraaortic splanchnopleura (Fig. 4B) (Dieterlen-Lievre, 1993; Godin *et al.*, 1993, 1995). However, these multipotential cells are incapable of restoring the hematopoietic system in lethally irradiated hosts.

With reciprocal transplantation experiments in *Xenopus*, Turpen *et al.* (1997) demonstrate *in vivo* that a common progenitor gives rise to primitive and definitive hematopoiesis. As mentioned earlier, the site of primitive hematopoiesis in *Xenopus* is the ventral blood island (VBI), which is equivalent to the mammalian yolk sac with respect to blood development. A second site of early hematopoiesis, the dorsal lateral plate (DLP), is analogous to the mammalian AGM and paraaortic region (Maeno, 1985; Turpen *et al.*, 1981; Weber *et al.*, 1991). It has been shown that the VBI can contribute to primitive hematopoiesis, and to a lesser extent, definitive hematopoiesis. On the other hand, DLP contributes to only definitive hematopoiesis. Turpen *et al.* demonstrate that either the ventral marginal zone or the lateral

marginal zone can each give rise to both the VBI and the DLP, whereas the dorsal marginal zone never contributes to blood. Interestingly, when the DLP is transplanted to the VBI region at stage 12.5 of development, DLP can now contribute to primitive hematopoiesis. However, if the transplantation of DLP to the VBI environment is done later at stages 15-18, the ability of DLP to contribute to primitive hematopoiesis is decreased. Collectively, these results provide evidence that the ventral marginal zone gives rise to both the primitive (VBI) and definitive (DLP) hematopoietic progenitors, and that both VBI and DLP are bipotential with respect to primitive and definitive hematopoiesis. Moreover, it appears that a ventral signal is responsible for specifying the primitive hematopoietic fate.

The definitive hematopoietic stem cells populating AGM have been characterized. Müller *et al.* (1994) demonstrated that AGM derived cells possess colony forming activity (CFU-S) and can long-term repopulate lethally irradiated recipients. Sanchez *et al.* (1996) demonstrated that the first definitive stem cells are generated in the AGM at day 10 of mouse gestation, and that these cells express the c-kit receptor. Later at day 11, hematopoietic stem cells (HSC) also are detected in the fetal liver, expressing c-kit as well. HSC from the AGM and fetal liver are also CD34+ and contain a heterogeneous population of Mac+ and Mac− cells. Mac-1 is a surface adhesion molecule previously demonstrated to mark long-term repopulating cells of the 13-day postcoitus liver (Morrison *et al.*, 1995). The dif0erential expression of Mac-1 adhesion molecule in the early hematopoietic stem cells may reflect a maturation process for these cells.

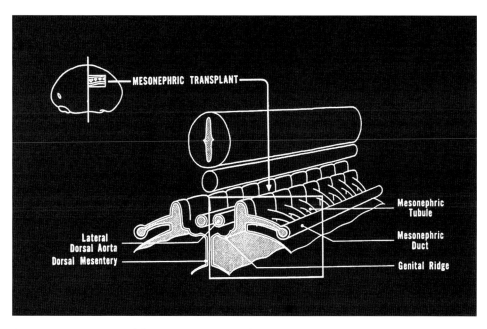

Figure 3 Cross section of the *Xenopus* aorta-gonad-mesonephros (AGM) region. Shown are the position of the definitive progenitors in most vertebrates. (Courtesy of Jim Turpen.)

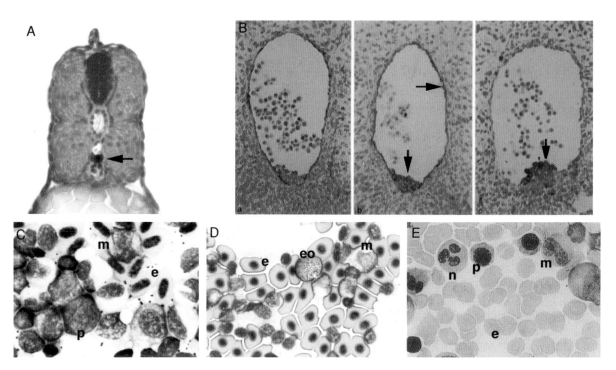

Figure 4 Adult sites of hematopoiesis. (A) Paraaortic hematopoiesis in zebrafish can be detected by c-myb RNA *in situ* in the ventral floor of the dorsal aorta (arrow). This position of definitive hematopoiesis is conserved throughout vertebrate life. (B) AGM hematopoiesis in human. In the left panel binding of the endothelial-specific lectin, eulex europeous, is to aortic endothelial cells but not to CD34+ cells clumped in the vessel wall. In the middle panel, CD31 staining is expressed in both endothelial cells (horizontal arrow) and intraaortic blood progenitors (vertical arrow). In the right panel a panleukocyte marker CD45 only stains hematopoietic cells (arrow). (Courtesy of Bruno Péault.) (C) Definitive hematopoiesis is shown by tissue smear of a zebrafish adult kidney, demonstrating proerythroblasts (p), myeloid precursors (m), and terminally differentiated erythrocytes (e). Note the similarity to the spleen tissue smear of an adult *Xenopus laevis* (D), which demonstrates myeloid precursors (m), eosinophils (eo), and terminally differentiated erythrocytes (e). (E) Definitive hematopoiesis in humans. The bone marrow sample contains terminally differentiated erythrocytes (e), proerythroblasts (p), as well as neutrophils (n) and myeloid (m) precursors. (Courtesy of Sam Lux.) Note that the mammalian erythrocytes are enucleated.

The origin of primitive and definitive erythrocytes has been investigated in cell culture studies (Kennedy *et al.*, 1997; Nakano, 1996). When D3 embryonic stem (ES) cells are cultured with OP9 stromal cells, two distinct waves of erythropoiesis result, corresponding to EryP (primitive erythropoiesis) and EryD (definitive erythropoiesis). The EryP and EryD populations differ in morphology and globin chains (EryP expressed mainly ζ and ε-globin, whereas EryD expressed α and β-globin). Also, c-kit signaling is necessary for the development and/or the maintenance of multipotential precursors in EryD, and erythropoietin is able to cause differentiation of EryP to EryD. Kennedy *et al.* (1997) working with embryoid bodies showed that EryP and EryD and other hematopoietic lineages arise from a common progenitor precursor.

In short, vertebrate definitive hematopoiesis arises from two sources: first, the yolk sac consisting non-self-renewing progenitors that contribute definitive sites of hematopoiesis; and second, the self-renewing definitive population that arises in the AGM. Recent evidence further suggest that yolk-derived progenitors contribute to definitive hematopoiesis. The *in vitro* studies may reca-pitulate different aspects of hematopoietic ontogeny during the transition from primitive to definitive hematopoiesis.

E. Fetal and Adult Hematopoiesis

After AGM hematopoiesis in mammals, definitive hematopoietic stem cells localize to the fetal liver (Capel *et al.*, 1989; Fleishman *et al.*, 1982). During fetal liver hematopoiesis, all lineages of blood arise. After birth, the site of definitive hematopoiesis switches from the fetal liver to bone marrow (Fig. 4E) (Russell, 1979; 1984). In mouse, the spleen is utilized as a major erythropoietic site, but this is not true of humans. In chickens, the fetal liver is not really utilized as a site of hematopoiesis; rather, the site of hematopoiesis shifts directly from the AGM to the bone marrow in adults. In *Xenopus*, the liver and spleen (Fig. 4D) are the major sites of adult hematopoiesis. In zebrafish, the kidney maintains adult hematopoiesis (Fig. 4C), but the larval site remains unclear. Thus, the location of adult hematopoiesis during vertebrate development is loosely conserved.

F. Pluripotent Stem Cell

Blood development is characterized by progressive lineage-restricted fate decisions. As a pluripotent hematopoietic stem cell (PHSC) gives rise to cells of a particular blood lineage, it is regulated by a host of cytokines, growth factors, signal transduction molecules, and transcription factors. The identification of the PHSC was largely based on correlation of surface marker profiles to a functional assay, such as repopulating ability in a lethally irradiated host. Purification of such cells involves fluorescence activated cell sorting (FACS) of hematopoietic populations that retained the abilities of self-renewal, maintenance, and differentiation into multiple lineages (Li and Johnson, 1995; Spangrude *et al.*, 1988; 1991). For example, primitive stem cells with long-term bone marrow repopulating ability are associated with the surface marker profile of Rhodamine-123^{lo} lineage$^-$Ly6A/E$^+$ c-kit$^+$, whereas mature stem cells (progenitors) enriched for short-term bone marrow repopulating ability has the Rhodamine-123$^{med/hi}$ lineage$^-$ Ly6A/E$^+$ c-kit$^+$ marker profile (Li and Johnson, 1992).

Stem cells and progenitors also have been characterized in terms of the cytokine receptors that they express (McKinstry *et al.*, 1997). It was found that pools of cells with stem cell activity (marker profile Rhodamine-123^{lo} lineage$^-$ Ly6A/E$^+$ c-kit$^+$) are enriched for cells that express receptors for IL-3, G-CSF, IL-6, and IL-1. In this stem cell population, no receptors were detected for M-CSF, GM-CSF, or LIF. In contrast, the progenitor cell population (marker profile Rhodamine-123$^{med/hi}$ lineage$^-$ Ly6A/E$^+$ c-kit$^+$) is enriched for cells bearing receptors for IL-3, G-CSF, IL-6, IL-1, and GM-CSF. No receptors for M-CSF were detected for the progenitor cell population. Stem cell populations defined functionally by their expressed cytokine receptors can be useful for clinical applications, in treating diseases such as thalassemia (hemoglobin defects) and severe combined immunodeficiency syndromes.

Another important marker associated with hematopoietic stem cells and early progenitors is the surface glycoprotein CD34 (Andrews *et al.*, 1992; Berenson *et al.*, 1988; 1991). In addition to hematopoietic tissues, CD34 is expressed in vascular endothelium and some fibroblasts. Clinically, the anti-CD34 antibody has been exploited widely to purify stem cells and progenitors for purposes of allogeneic and autologous bone marrow transplants. Recently, the biological function of CD34 was characterized. Targeted disruption of the CD34 locus in mice yielded only mild hematopoietic abnormality, suggesting it is not essential for hematopoietic nor vascular development (Cheng *et al.*, 1996; Suzuki *et al.*, 1996). However, the study of CD34 expression in early blood and vessel development provides new insight. A detailed analysis of CD34 expression during early mouse embryonic development confirms that CD34 is expressed on endothelial precursors (angioblasts) and hematopoietic progenitors (Wood *et al.*, 1997). Further, CD34 is an excellent marker for vasculogenesis, the process by which angioblasts differentiate and organize to form vascular channels in the embryo. CD34 is also strongly correlated with angiogenesis, the process by which new capillaries sprout from existing vessels. As expected, the study of CD34 expression pattern confirmed the existence of previously identified sites of hematopoiesis in yolk blood islands, ventral wall of the aorta, paraaortic region, the AGM region, and embryonic liver. The expression of CD34 on both hematopoietic and endothelial precursors also lends support to the existence of the hypothesized hemangioblast, a bipotential precursor that gives rise to blood and vascular progenitors.

G. Cytokines and Receptors

Hematopoiesis is governed by a number of cytokines that act on stem cells and progenitors to mediate differentiation and proliferation. Cytokines have been identified that could stimulate specific lineages of colony forming units (CFU) to develop lineage-restricted pathways: SCF (kit ligand, mast cell factor, or steel factor), IL-3, IL-6, G-CSF, GM-CSF, erythropoietin, and thrombopoietin. Further, clinical applications of many of these cytokines (erythropoietin, G-CSF, GM-CSF, thrombopoietin) have revolutionized treatment for anemias, bone marrow transplants, leukemias, and other cancers. As the field of cytokines and cytokine-receptor literature is extensive and beyond the scope of this review, we will limit our discussion to one cytokine-receptor pair as an example: the SCF ligand and c-kit receptor.

Stem cell factor (SCF), also referred to as the murine steel factor, kit ligand, or mast cell growth factor, is the gene product of the Steel (Sl) locus (Russell, 1979). SCF binds to c-kit, a receptor tyrosine kinase coded by the W locus (Besmer *et al.*, 1993; Geissler *et al.*, 1981). Null mutations in either the Sl locus or W locus of mice cause death *in utero* or in the perinatal period, due to severe macrocytic anemia. Missense mutation in the c-kit receptor leading to loss of kinase activity, thereby abolishing signal transduction, also leads to severe macrocytic anemia (Tan *et al.*, 1990). A gradient of mutant phenotypes occur within the Sl and W loci, as hypomorphs of either locus with varying strengths have been identified (Blouin and Bernstein, 1993). Since the genetic identification of SCF and c-kit receptor, much work has elucidated their function and biology. Through *in vivo* studies, biochemistry and *in vitro* culture assays, the work on SCF has shed light on early hematopoiesis and cytokine function. SCF and c-kit are expressed along migration pathways for germ cells, melanocytes, central nervous system, and hematopoietic cells. By studying SCF/c-kit expression, it was confirmed that hematopoietic stem cells can be found in the AGM and ventral wall of the aorta (Bernex *et al.*, 1996). SCF thus plays a key role in regulating hematopoiesis.

III. Transcription Factors and Knockout Mice

A. Targeted Gene Disruption in Embryonic Stem Cells, Knockout Mice, and Chimeras

The ability to disrupt specific genes of interest in murine ES cells has allowed investigators to address the loss-of-function question in a mammalian system. Embryonic stem cells are derived from the inner cell mass of the 32-cell stage blastocyst, and can be manipulated in a cell culture system. Genes to be disrupted are homologously recombined and selected in ES cells. When reimplanted, the genetically engineered ES cell can contribute to all tissues of a chimeric mouse, where germ line contribution would allow the recovery of a mouse heterozygous for the gene disruption. Mice homozygous for the gene disruption can be raised from a heterozygous mating.

Embryonic stem cells also can be studied in culture to assay any functional changes that are brought about by specific gene disruptions. In the case of hematopoiesis, disruption of certain genes (SCL, GATA-1, GATA-2) has led to altered ability of the ES cell to differentiate into certain lineages of blood cells (Orkin, 1995; Shivdasani and Orkin, 1996). This provides a powerful *in vitro* assay to look for very specific effects of gene disruptions on lineage specificity, proliferation, and differentiation.

Fate determination decisions are mediated by the regulation of transcription factors to express genes leading to differentiation along lineage restricted paths. One approach to understanding fate determination is to identify the transcription factors that act as molecular switches during blood differentiation. The study of hematopoietic transcription factors in knockout mice has led to functional assignments for a number of such factors, as we will review briefly in this section.

B. Transcription Factors Acting at the Stem Cell Level: SCL, LMO-2, and GATA-2

The basic helix-loop-helix transcription factor SCL was isolated in association with childhood T-cell acute lym-

phocytic leukemia (Begley *et al.*, 1989a; Chen *et al.*, 1990; Finger *et al.*, 1989). Similarly, the LIM domain nuclear protein LMO-2 (Rbtn-2), when dysregulated, can lead to T-cell leukemia (Begley *et al.*, 1989b; Boehm *et al.*, 1991; Rabbits, 1994). Both SCL and LMO-2 are expressed in a number of tissues, with coexpression in hematopoietic cells. Through gene targeting, it was determined that SCL and LMO-2 are required for the differentiation of the hematopoietic stem cell to all lineages of blood cells. In the knock-out mice lacking either SCL or LMO-2, embryonic red cell production does not occur, leading to embryonic lethality *in utero* (Shivdasani *et al.*, 1995; Warren *et al.*, 1994). Subsequent ES cell culture study establishes that definitive hematopoiesis also requires SCL (Porcher *et al.*, 1996). In fact, it has been shown that SCL and LMO-2 physically interact in erythroid cells (Larson *et al.*, 1996; Valge-Archer *et al.*, 1994). More recently, it was found that a novel DNA-binding complex consisting of SCL, LMO-2, GATA-1, E47, and Lbd1/NL1 is formed in erythroid cells (Wadman *et al.*, 1997). This multiprotein complex may mediate transcription activation of a number of target genes that are expressed during erythroid differentiation.

GATA-2 is a zinc-finger transcription factor that is expressed in hematopoietic progenitors, early erythroid cells, mast cells, and megakaryocytes. Targeted disruption of GATA-2 leads to embryonic anemia, causing lethality by E11 (Tsai *et al.*, 1994). *In vitro* culture of ES cells lacking GATA-2 demonstrate a proliferative function for GATA-2, as primitive and definitive erythrocyte colonies can form, but proliferation of cells in these colonies appears to be impaired.

C. Transcription Factors Essential for Erythroid and Megakaryocytic Differentiation: GATA-1 and FOG

GATA-1 is the founding member of the GATA family of zinc finger transcription factors that bind to the GATA cis-element of many erythroid-expressed genes (Evans and Felsenfeld, 1989; Tsai *et al.*, 1989). Disruption of GATA-1 leads to maturation arrest of erythroid precursors at the proerythroblast stage, followed by apoptosis of arrested cells (Pevny *et al.*, 1991, 1995; Simon *et al.*, 1992; Weiss *et al.*, 1994; Weiss and Orkin, 1995). Recently, FOG, a partner of GATA-1 has been identified by a yeast two-hybrid strategy (Tsang *et al.*, 1997). FOG is a multitype zinc finger molecule that is coexpressed with GATA-1 during hematopoietic development, and cooperates in mediating erythroid and megakaryocytic differentiation. Gene disruption of FOG leads to a proerythroblast arrest phenotype similar to the GATA-1 knockout, further supporting that GATA-1 and FOG cooperate in mediating erythroid and megakaryocytic differentiation (Tsang *et al.*, 1998).

IV. Genetic Advantages of Zebrafish Blood Mutants

A. Zebrafish as a Genetic Model System

The classic genetic approach whereby a mutant hunt is initiated in a biological process of interest has proven to be a method for the investigation of many areas: from *Saccharomyces cerevisiae* DNA polymerases to *Drosophila* eye formation. Until now, a genetic analysis of this sort for blood formation has not been possible. With the discovery of a large number of zebrafish mutants affecting hematopoiesis, genetic analysis of blood development in a vertebrate organism is now in progress (Ransom *et al.*, 1996; Weinstein *et al.*, 1996).

Zebrafish is the vertebrate model system of choice for genetic investigation because of its many unique biological attributes (see Chapter 24 by Driever). These developmental advantages of zebrafish include: external fertilization, large brood size, large optically transparent embryos, and short generation time. Embryos also can be raised in the haploid state, to screen for recessive mutations. Furthermore, gynogenetic diploids can be generated to facilitate mapping and recombination frequency analysis. Two large-scale mutagenesis screens have generated a large number of mutants with defects in early hematopoiesis (Driever *et al.*, 1996; Haffter *et al.*, 1996). In our laboratory, we are characterizing 26 complementation groups of blood mutants.

B. Zebrafish Blood Mutants

The zebrafish hematopoietic mutants can be mainly categorized into four classes: bloodless, hypochromic, decreasing blood, and photosensitive mutants. The bloodless mutants we are studying include *moonshine (mon)*, *sort-of-anemic (SOA)*, and *sort-of-bloodless (SOB)* (Fig. 1). These three mutants are bloodless at the onset of circulation, but do have some blood cells at early time points. When examined at the molecular level with GATA-1 and SCL RNA *in situ* hybridization, it appears that these three mutants do have blood progenitors, as the P-ICM cells are positive for SCL (see Section II, C). However, the A-ICM is negative for SCL and GATA-1, suggesting that differentiation of blood progenitors to contribute to primitive hematopoiesis does not occur in these mutants. Interestingly, homozygotes of most *mon* alleles are embryonic lethal, whereas *SOA* and *SOB* homozygotes can be raised to adulthood. It appears that *mon* has a blood progenitor differentiation defect in both primitive and definitive hematopoiesis, whereas *SOA* and *SOB* have a similar defect affecting only primitive hematopoiesis. Recent data also suggest that the defect in *SOA* and *SOB* affects erythropoietic, myeloid,

and lymphoid lineages. Homozygotes of *mon* also have an increase in the number of iridophores (specialized neural crest derived cells), but such a defect is not found in *SOA* nor *SOB*. Given both the neural crest and blood defects of *mon*, a possible candidate of the mutation is the SCF/c-kit signaling pathway. SCF/c-kit ligand receptor pair is involved in both blood cell differentiation/proliferation, as well as melanocyte and neural crest cell development. The zebrafish homologue of the c-kit receptor has been cloned, and is not linked to the *mon* locus. We are currently fine mapping these bloodless mutations and attempting to clone the zebrafish SCF by homology to test it as a candidate for *mon*.

There are five hypochromic mutants in our collection: *chardonnay (cdy)*, *chianti (cia)*, *sauternes (sau)*, *weißherst (weh)*, and *zinfandel (zin)* (Fig. 1). All five mutants have been fine mapped, and positional cloning efforts are in progress. These mutants exhibit hypochromic blood around 2 days postfertilization (dpf), and the number of circulating cells also are significantly reduced. The globin chains of these mutants have been analyzed, and it appears that many of the hypochromic mutants exhibit globin chain imbalance (where the level of α globin is not matched by that of β globin). The circulating red cells also appear to be smaller than wild-type cells. The hypochromic microcytic red cell phenotype is suggestive of a thalassemialike defect. In fact, the globin locus has been cloned in zebrafish, and the *zin* mutant maps to that region. (A. Brownlie, manuscript in preparation). The *sau* locus has also been positionally cloned, and represents the aminolevulinic acid synthase (ALA-S) gene, a critical enzyme in the earliest steps of globin synthesis (A. Brownlie, manuscript in preparation). The *sau* locus is one of the first zebrafish genes identified by the positional cloning approach (Section IV.C). The other hypochromic mutants may be genes involved in hemoglobination or regulation of globin chain switching. All of the hypochromic mutants are homozygous embryonic viable, except for *weh* which is embryonic lethal.

The decreasing blood mutants can be further subdivided into two classes: those with proerythroblast arrest (PEA) phenotype, and those without. The mutants with the PEA phenotype include *cabernet (cab)*, *chablis (cha)*, *merlot (mot)*, *retsina (ret)*, and *reisling (ris)* (Fig. 1). Of the PEA mutants, the lethality of the mutation is in the order of *cab > ret > cha > mot > ris*, where *cab* homozygotes always die at around 6 dpf but *ris* adults are viable and reproduce normally. The PEA mutants have normal numbers of red cells until 3-4 dpf, when the number of red cells drops off precipitously. The morphology of the embryonic red cells appears normal, but the adult red cells have a proerythroblast phenotype, similar to that seen in the GATA-1 mouse knockout. To date, all the PEA mutants have been excluded for GATA-1 by linkage analysis. All PEA mutants are similar to wild type in terms of GATA-1 and GATA-2 expression during early development. At 4 dpf, a second hematopoietic wave is initiated in the ventral-posterior tail region, as delineated by SCL, GATA-1, and c-myb RNA *in situ* analysis. In the mutant embryos, this second wave of hematopoiesis appears to occur earlier, at 3 dpf, and the number of SCL positive cells appears to be increased. Hence, in the PEA mutants there may be a compensatory homeostatic mechanism whereby the animal attempts to increase the number of blood progenitors to overcome the differentiation defect. Consistent with this is the observation that when homozygotes are raised to adulthood, they exhibit hyperplastic kidney. Kidney is the adult site of hematopoiesis in zebrafish, hence its hypercellularity is likely a consequence of increased blood production. The homozygote adults also exhibit cardiomegaly and pallor, both symptoms of severe anemia.

The decreasing blood mutants without proerythroblast arrest phenotype include *frascati (frs)*, *grenache (gre)*, and *thunderbird (tbr)* (Fig. 1). These mutants are similar to the PEA mutants in that the homeostatic compensation is also present at 3 dpf, as evident by SCL and GATA-1 RNA expression. These mutants can be raised to adulthood, with strength of lethality in the order of *frs* > *tbr* = *gre*. The embryonic and adult peripheral blood smears of these mutants exhibit wild-type morphology, but the number of cells in circulation is dramatically decreased. Like the PEA mutants, these decreasing blood mutants have hyperplastic kidney, cardiomegaly, and pallor in the adult state.

C. Genetic Analysis of Blood Mutants

To uncover the molecular defects of each of the blood mutants under investigation, we are primarily exploiting two approaches: candidate gene and positional cloning. In the case of the candidate gene approach, we have cloned many transcription factors that function in blood differentiation (GATA-1, GATA-2, SCL, LMO-2, etc.). To date, none of the blood mutants match these candidates. The process of cloning other blood genes are ongoing, and such genes are readily positioned on the zebrafish genetic map. Since the majority of the blood mutants have been mapped, a colocalization of mapped candidate gene with any mutant locus would lead us to investigate the correlation further. This candidate gene approach has been successful in zebrafish, demonstrating that BMP-2 was the molecular defect in the *swirl* mutant.

The second approach is positional cloning of the mutant locus. In brief, positional cloning involves identification of genetic markers closely linked to the mutation of interest. The closely linked markers then serve as the address by which clones of genomic sequence are identified from a number of different large insert libraries. We have constructed genomic libraries where large fragments of DNA are inserted into either yeast artificial chromosome (500 kb insert size) or P1 clones (20 kb insert size). These clones are oriented with the linked genetic markers and a chromosome walk can be initiated to identify genomic clones containing the mutant locus.

V. Conclusion

Distinct components of the hematopoietic program are conserved throughout vertebrate ontogeny. First, the hematopoietic tissue initiates from mesoderm early during development, in a ventral location. This ventral mesoderm is patterned by a BMP cascade, involving MAD and Mix molecules. The resulting cells are likely to be bipotential for their ability to form hematopoietic or vascular tissue. Subsequent events which are environmentally regulated lead to the fate decision between vasculogenesis and hematopoietic (primitive or definitive) lineages. With respect to hematopoiesis, the primitive lineage is highly favored in the yolk sac environment and only embryonic red cells are formed. Some cells in the ventral yolk sac contribute to definitive hematopoiesis, but the majority of definitive cells arise from a second site within the embryo called the aorta-gonad-mesonephros (AGM). AGM initially expresses early markers of vasculogenic and hematopoietic potential, but does not terminally differentiate *in situ*. The initial wave in the AGM likely enters the circulation and travels to the fetal liver. A second wave of AGM hematopoiesis occurs in the loose mesenchyme surrounding the dorsal aorta and ducts of Cuvier. A new site of hematopoiesis is established (such as the fetal liver), and contributes to the production of all myeloid lineages. The lymphoid lineages are derived from the AGM cells and/or fetal liver cells that migrate to the thymic epithelium, where they mature.

As fetal life continues, there is a shift to adult sites of hematopoiesis which typically include the bone marrow and occasionally the spleen. Based on the different sites of hematopoiesis seen in distinct adult vertebrates, it is likely that the homing of these hematopoietic stem cells (HSC) is regulated by a species-specific molecular address resulting from the interaction of HSC surface receptors and the cellular environment. Once hematopoietic stem cells arise in a particular environment, cytokines are available to drive proliferation and terminal differentiation of blood cells. Finally, the different lineages of blood cells are released into circulation where they execute their respective specialized functions.

Convergence of molecular and genetic methods in the different vertebrate systems will further define the hematopoietic program. We are encouraged by the zebrafish system as a powerful genetic tool toward identifying molecules involved in blood differentiation. Novel factors involved in mesoderm patterning and early hematopoiesis will also be isolated by expression cloning strategies in *Xenopus*. Targeted disruption of these genes in mice will place the function of such factors in a mammalian context. We expect that these studies will further our understanding of the mechanisms underlying cell fate determination, and further reveal conservation of key molecules and principles in vertebrate blood development.

Acknowledgments

Our thanks go to Sam Lux, Stuart Orkin, Bruno Péault, and Jim Turpen for contribution of figures. We also thank Barry Paw and Ramesh Shivdasani for review of this manuscript. E. C. L. is supported by the Howard Hughes Predoctoral Fellowship in Biological Sciences. L. I. Z. is an Associate Investigator of the Howard Hughes Medical Institute.

References

Andrews, R. G., Bryant, E. M., Bartelmez, S. H., Muirhead, D. Y., Knitter, G. H., Bensinger, B. W., Strong, D. M., and Berntein, I. D. (1992). CD34+ marrow cells devoid of T and B lymphocytes, reconstitute stable lymphopoiesis and myelopoiesis in lethally irradiated allogeneic baboons. *Blood* **80**, 1693–1701.

Begley, C. G., Aplan, P. D., Denning, S. M., Haynes, B. F., Waldman, T. A., and Kirsch, I. R. (1989a). The gene SCL is expressed during early hematopoiesis and encodes a differentition-related DNA-binding motif. *Proc. Natl. Acad. Sci. U.S.A.* **86**, 10128–10132.

Begley, C. G., Aplan, P. D., Davey, M. P., Nakahara, K., Tchorz, K., Kurtzberg, J., Hershfield, M. S., Haynes, B. F., Cohen, D. I., Waldman, T. A., and Kirsch, I. R. (1989b). Chromosomal translocation in human leukemic stem-cell line disrupts the T-cell receptor delta-chain diversity region and results in a previously unreported fusion transcript. *Proc. Natl. Acad. Sci. U.S.A.* **86**, 2031–2035.

Berenson, R. J., Andrews, R. G., Bensinger, W. I., Kalamasz, D., Knitter, G., Buckner, C. D., and Berntein, I. D. (1988). Antigen CD34$^+$ marrow cells engraft lethally irradiated baboons. *J. Clin. Invest.* **81**, 951–962.

Berenson, R. J., Bensinger, W. I., Hill, R. S., Andrews, R. G., Garcia Lopez, J., Kalamasz, D., Still, B. J., Spitzer, G., Buckner, C. D., Berntein, I. D., and Thomas, E. D. (1991). Engraftment after infusion of CD34$^+$ marrow cells in patients with breast cancer or neuroblastoma. *Blood* **77**, 1717–1722.

Bernex, F., De Sepulveda, P., Kress, C., Elbaz, C., Delouis, C., and Panthier, J. J. (1996). Spatial and temporal patterns of c-kit-expressing cells in the WlacZ/+ and Wlacz/

Wlacz mouse embryos. *Development (Cambridge, UK)* **122**, 3023–3033.

Besmer, P., Manova, K., Duttlinger, R., Huang, E. J., Packer, A., Gyssler, C., and Bachvarova, R. F. (1993). The kit-ligand (steel factor) and its receptor c-kit/W: Pleotropic roles in gametogenesis and melanogenesis. *Development* **125** (Suppl.), 125–137 Review.

Blouin, R., and Bernstein, A. (1993). "The White Spotting and Steel Hereditary Anemias of the Mouse" (S. A. Feig and M. H. Freedman, eds.), p. 157. CRC Press, Boca Raton, Florida.

Boehm, T., Foroni, L., Kaneko, Y., Perutz, M. P., and Rabbits, T. H. (1991). The rhombotin family of cycteine-rich LIM domain oncogenes: Distinct members are involved in T-cell translocations to human chromosome 11p15 and 11p13. *Proc. Natl. Acad. Sci. U.S.A.* **88**, 4367–4371.

Capel, R., Hawley, R., Covarrubias, L., Hawley, T., and Mintz, B. (1989). Clonal contributions of small numbers of retrovirally marked hematopoietic stem cells engrafted in irradiated neonatal W/Wv mice. *Proc. Natl. Acad. Sci. U.S.A.* **86**, 4564–4568.

Chen, Q., Cheng, J. T., Tasi, L. H., Schneider, N., Buchanan, G., Carroll, A., Crist, W., Ozanne, B., Siciliano, M. J., and Baer, R. (1990). The tal gene undergoes chromosome translocation in T cell leukemia and potentially encodes a helix-loop-helix protein. *EMBO J.* **9**, 415–424.

Chen, X., Weisberg, E., Fridmacher, V., Watanabe, M., Naco, G., and Whitman, M. (1997). Smad4 and FAST-1 in the assembly of activin-responsive factor. *Nature (London)* **389**, 85–89.

Cheng, J., Baumhuetter, S., Cacalano, G., Carver-Moore, K., Thibodeaux, H., Thomas, R., Broxmeyer, H. E., Cooper, S., Hague, N., Moore, M., and Lasky, L. A. (1996). Hematopoietic defects in mice lacking the sialomucin CD34. *Blood* **87**, 479–490.

Choi, H., Kennedy M., Kazarov, A., Papadimitriou, J. C., and Keller, G. (1998). A common precursor for hematopoietic and endothelal cells. *Development (Cambridge, UK)* **125**, 725–732.

Cormier, F., de Paz, P., and Dieterlen-Lievre, F. (1986). *In vitro* detection of cells with monocytic potential in the wall of the chick embryo aorta. *Dev. Biol.* **118**, 167–175.

Detrich III, H. W., Kieran, M. W., Chan, F. Y., Barone, L. M., Yee, K., Rundstadler, J. A., Pratt, S., Ransom, D., and Zon, L. I. (1995). Intraembryonic hematopoietic cell migration during vertebrate development. *Proc. Natl. Acad. Sci. U.S.A.* **92**, 10713–10717.

Dieterlen-Lievre, F. (1975). On the origin of haematopoietic stem cells in the avian embryo: An experimental approach. *J. Embryol. Exp. Morphol.* **33**, 607–619.

Dieterlen-Lievre, F. (1984). Emergence of intraembryonic blood stem cells in avian chimeras by means of monoclonal antibodies. *Comp. Immunol.* **3**, 75–81.

Dieterlen-Lievre, F. (1993). Hemopoiesis during avian ontogeny. *Poult. Sci. Rev.* **5**, 273–305.

Dieterlen-Lievre, F., and Martin, C. (1981). Diffuse intraembryonic hemopoiesis in normal and chimeric avian development. *Dev. Biol.* **88**, 180–191.

Driever, W., Solnica-Krezel, L., Schier, A., Neuhauss, S., Malicki, J., Stemple, D., Stainier, D., Zwartkruis, F., Abdelilah, S., Rangini, Z., Belak, J., and Boggs, C. (1996). A genetic screen for mutations affecting embryogenesis in zebrafish. *Development (Cambridge, UK)* **123**, 37–46.

Evans, T., and Felsenfeld, G. (1989). The erythroid-specific transcription factor Eryf1: A new finger protein. *Cell (Cambridge, Mass.)* **58**, 877–885.

Finger, L. R., Kagan, J., Christopher, G., Kurtzberg, J., Nowell, P. C., and Croce, C. M. (1989). Involvement of the TCL5 gene on human chromosome 1 in T-cell leukemia and melanoma. *Proc. Natl. Acad. Sci. U.S.A.* **86**, 5039–5043.

Fleishman, R., Custer, R., and Mintz, B. (1982). Totipotent hematopoietic stem cells: Normal self-renewal and differentiation. *Cell (Cambridge, Mass.)* **30**, 352–360.

Geissler, E. N., MacFarland, E. C., and Russel, E. S. (1981). Analysis of pleiotropism at the dominant white-spotting (W) locus of the house mouse: A description of ten new W alleles. *Genetics* **97**, 337–361.

Godin, I. E., Garcia-Porrere, J. A., Coutinho, A., Dieterlen-Lievre, F., and Marcos, M. A. R. (1993). Para-aortic splanchnopleura from early mouse embryos contains B1a cell progenitors. *Nature (London)* **364**, 67–70.

Godin, I., Dieterlen-Lievre, F., and Cumano, A. (1995). Emergence of multipotential hemopoietic cells in the yolk sac and para-aortic splanchnopleura in mouse embryos, beginning at 8.5 days post coitus. *Proc. Natl. Acad. Sci. U.S.A.* **92**, 773-777.

Graff, J. M. (1997). Embryonic patterning: To BMP or not to BMP, that is the question. *Cell (Cambridge, Mass.)* **89**, 171–174.

Graff, J. M., Bansal, A., and Melton, D. A. (1996). *Xenopus* mad proteins tranduce distinct subsets of signals for the TGFβ superfamily. *Cell (Cambridge, Mass.)* **85**, 479–487.

Haffter, P., Granato, M., Brand, M., Mullins, M. C., Hammerschmidt, M., Kane, D. A., Odenthal, J., van Eeden, F. J., Jiang, Y. J., Heisenberg, C. P., Kelsh, R. N., Furutani-Seiki, M., Vogelsang, E., Beuchle, D., Schach, U., Fabian, C., and Nusslein-Volhard, C. (1996). The identification of genes with unique and essential functions in the development of the zebrafish, *Danio rerio*. *Development (Cambridge, UK)* **123**, 1–36.

Hemmati-Brivanlou, A., and Thomsen, G. H. (1995). Ventral mesodermal patterning in *Xenopus* embryos: Expression patterns and activities of BMP-2 and BMP-4. *Dev. Genet.* **17**, 78–89.

Kelley, C., Yee, K., Harland, R., and Zon, L. I. (1994). Ventral expression of GATA-1 and GATA-2 in the *Xenopus* embryo defines induction of hematopoietic mesoderm. *Dev. Biol.* **165**, 193–205.

Kennedy, M., Firpo, M., Choi, K., Wall, C., Robertson, S., Kabrun, N., and Keller, G. (1997). A common precursor for primitive erythropoiesis and definitive haematopoiesis. *Nature (London)* **386**, 488–493.

Kishimoto, Y., Lee, K., Zon, L., Hammerschmidt, M., and Schulte-Merker, S. (1998). The molecular nature of zebrafish swirl: BMP2 function is essential during dorsal-ventral patterning. *Development (Cambridge, UK)* **124**, 4457–4466.

Larson, R. C., Lavenir, I., Larson, T. A., Baer, R., Warren, A. J., Wadman, I., Nottage, K., and Rabbitts, T. H.

(1996). Protein dimerization between LMO2 (Rbtn2) and Tal1 alters thymocyte development and potentiates T cell tumorigenesis in transgenic mice. *EMBO J.* **15**, 1021–1027.

Li, C. L., and Johnson, G. R. (1992). Rhodamine 123 reveals heterogeneity within murine Lin-, Sca-1+ hemopoietic stem cells. *J. Exp. Med.* **175**, 1443–1447.

Li, C. L., and Johnson, G. R. (1995). Murine hematopoietic stem and progenitor cells. Enrichment and biological characterization. *Blood* **85**, 1472–1479.

Liao, W., Bisgrove, B. W., Sawyer, H., Hug, B., Bell, B., Peters, K., Grunwald, D. J., and Stainier, D. Y. R. (1997). The zebrafish gene cloche acts upstream of a flk-1 homologue to regulate endothelial differentiation. *Development (Cambridge, UK)* **124**, 381–389.

Liao, E. C., Paw, B. H., Oates, A. C., Pratt, S. J., Postlethwait, J. H., and Zon, L. I. (1998). SCL/Tal-1 transcription factor acts downstream of cloche to specify hematopoietic and vascular progenitors in zebrafish. *Genes & Dev.* **12**, 621–626.

McKinstry, W. J., C. L., Rasko, J. E. C., Nicola, N. A., Johnson, G. R. and Metcalf, D. (1997). Cytokine receptor expression on hematopoietic stem and progenitor cells. *Blood* **89**, 65–71.

Maeno, M., Todate, A. and Katagiri, C. (1985). The localization of precursor cells for larval and adult hemopoietic cells of *Xenopus laevis* in two regions of embryos. *Dev. Growth Differ.* **27**, 137–148.

Maeno, M., Ong, R. C., Suzuki, A., Ueno, N., and Kung, H. F. (1994). A truncated bone morphogenetic protein 4 receptor alters the fate of ventral mesoderm to dorsal mesoderm: Roles of animal pole tissue in the development of ventral mesoderm. *Proc. Natl. Acad. Sci. U.S.A.* **91**, 10260–10264.

Maeno, M., Mead, P. M., Kelley, C., Xu, R.-H., Kung, H.-F., Suzuki, A., Ueno, N., and Zon, L. I. (1996). The role of BMP-4 and GATA-2 in the induction and differentiation of ventral mesoderm. *Blood* **88**(6), 1965–1972.

Massague, J. (1992). Receptors for the TGF-beta family. *Cell (Cambridge, Mass.)* **69**, 1067–1070.

Massague, J. (1996). TGFbeta signaling: Receptors, transducers, and Mad proteins. *Cell (Cambridge, Mass.)* **85**, 947–950.

Mead, P. E., Brevanlou, I. H., Kelley, C. M., and Zon, L. I. (1996). BMP-4-responsive regulation of dorsal-ventral patterning by the homeobox protein Mix.1. *Nature (London)* **382**, 357–360.

Medvinsky, A. L., Samoylina, N. L., Muller, A. M., and Dzierzak, E. A. (1993). An early pre-liver intraembryonic source of CFU-S in the developing mouse. *Nature (London)* **364**, 64–67.

Metcalf, D. (1993). Hematopoietic regulators: Redundancy or subtlety? *Blood* **82**, 3515–3523.

Morrison, S. J., Hoummati, H. D., Wandycz, A. M., and Weissman, I. L. (1995). The purification and characterization of fetal liver hematopoietic stem cells. *Proc. Natl. Acad. Sci. U.S.A.* **92**, 10302–10306.

Muller, A. M., Medvinsky, A., Strouboulis, J., Grasveld, F., and Dzierzak, E. (1994). Development of hematopoietic stem cell activity in the mouse embryo. *Immunity* **1**, 291–301.

Nakano, T., Kodama, H. and Honjo, T. (1996). *In vitro* development of primitive and definitive erythrocytes from different precursors. *Science* **272**, 722–724.

Orkin, S. H. (1995). Transcription factors and hematopoietic development. *J. Biol. Chem.* **270**, 4955–4958.

Pardanaud, L., Luton, D., Prigent, M., Bourcheix, L. M., Catala, M., and Dierterlen-Lievre, F. (1996). Two distinct endothelial lineages in ontogeny, one of them related to hemopoiesis. *Development (Cambridge, UK)* **122**, 1363–1371.

Pevny, L., Simon, M. C., Robertson, E., Klein, W. H., Tsai, S. F., DAgati, V., Orkin, S. H., and Costantini, F. (1991). Erythroid differentiation in chimaeric mice blocked by a targeted mutation in the gene for transcription factor GATA-1. *Nature (London)* **349**, 257–260.

Pevny, L., Lin, C. S., D'Agati, V., Simon, M. C., Orkin, S. H., and Costantini, F. (1995). Development of hematopoietic cells lacking transcription factor GATA-1. *Development (Cambridge, UK)* **121**, 163–172.

Porcher, C., Wojciech, S., Rockwell, K., Fujiwara, Y., Alt, F. W., and Orkin, S. H. (1996). The T cell leukemia oncoprotein SCL/tal-1 is essential for development of all hematopoietic lineages. *Cell (Cambridge, Mass.)* **86**, 1–20.

Rabbits, T. H. (1994). Chromosomal translocations in human cancer. *Nature (London)* **372**, 143–149.

Ransom, D. G., Haffter, P., Odenthal, J., Brownlie, A., Vogelsang, E., Kelsh, R. N., Brand, M., van Eeden, F. J. M., Furutani-Seiki, M., Granato, M., Hammerschmidt, M., Heisenberg, C. P., Jiang, Y. J., Kane, D. A., Mullins, M. C., and Nusslein-Volhard, C. (1996). Characterization of zebrafish mutants with defects in embryonic hematopoiesis. *Development (Cambridge, UK)* **123**, 311–319.

Rollins-Smith, L. A., and Blair, P. (1990). Contribution of ventral blood island mesoderm to hematopoiesis in postmetamorphic and metamorphosis-inhibited *Xenopus laevis*. *Dev. Biol.* **142**, 178–183.

Russell, E. N. (1979). Hereditary anemias of the mouse: A review for geneticists. *Adv. Genet.* **20**, 357–459.

Russell, E. S. (1984). Developmental studies of mouse hereditary anemias. *Am. J. Med. Genet.* **18**, 621–641.

Sabin, F. R. (1920). Studies on the origin of blood vessels and of red blood corpuscles as seen in the living blastoderm of chicks during the second day of incubation. *Contrib. Embryol.* **9**, 213–262.

Sanchez, M. J., Holmes, A., Miles, C., and Dzierzak, E. (1996). Characterization of the first definitive hematopoietic stem cells in the AGM and liver of mouse embryo. *Immunity* **5**, 513–525.

Shalaby, F., Rossant, J., Yamaguchi, T. P., Gertsenstein, M., Wu, X. F., Breitman, M. L., and Schuh, A. C. (1995). Failure of blood-island formation and vasculogenesis in Flk-1 deficient mice. *Nature (London)* **376**, 62–66.

Shivdasani, R. A., and Orkin, S. H. (1996). The transcriptional control of hematopoiesis. *Blood* **87**, 4025–5039.

Shivdasani, R. A., Mayer, E. L., and Orkin, S. H. (1995). Absence of blood formation in mice lacking the T-cell leukaemia oncoprotein tal-1/SCL. *Nature (London)* **373**, 432–434.

Simon, M. C., Pevny, L., Wiles, M. V., Keller, G., Costantini, F., and Orkin, S. H. (1992). Rescue of erythroid development in gene targeted GATA-1 mouse embryonic stem cells. *Nat. Genet.* **1**, 92–98.

Spangrude, G. J., Heimfeld, S., and Weissman, I. L. (1988). Purification and characterization of mouse hematopoietic stem cells. *Science* **241**, 58–62.

Spangrude, G. J., Smith, L., Uchida, N., Ikuta, K., Heimfeld, S., Friedman, J., and Weissman, I. L. (1991). Mouse hematopoietic stem cells. *Blood* **78**, 1395–1402.

Stainer, D. Y. R., Weinstein, B. M., Detrich III, W. D., Zon, L. I., and Fishman, M. C. (1995). *Cloche*, an early acting zebrafish gene, is required by both the endothelial and hematopoietic lineages. *Development (Cambridge, UK)* **121**, 3141–3150.

Suzuki, A., Andrew, D. P., Gonzalo, J. A., Fukumoto, M., Spellberg, J., Hashiyama, M., Takimoto, H., Gerwin, N., Webb, I., Molieux, G., Amakawa, R., Tada, Y., Wakeham, A., Brown, J., McNiece, I., Ley, K., Buthcer, E. C., Suda, T., Gutierrez-Ramos, J.-C., and Mak, T. W. (1996). CD34-deficient mice have reduced eosinophil accumulation after allergen exposure and show a novel crossreactive 90-kD protein. *Blood* **87**, 3550–3562.

Tan, J. C., Nocka, K., Ray, P., Traktman, P., and Besmer, P. (1990). The dominant W42 spotting phenotype results from a missense mutation in the c-kit receptor kinase. *Science* **247**, 209–212.

Tavian, M., Coulombei, L., Luton, D., San Clemente, H., Dieterlen-Lievre, F., and Peault, B. (1996). Aorta-associated CD34+ hematopoietic cells in the early human embryo. *Blood* **87**, 67–72.

Thompson, M. A., Ransom, D. G., Pratt, S. J., MacLennan, H., Kieran, M. W., Detrich III, H. W., Vail, B., Huber, T. L., Paw, B., Brownlie, A. J., Oates, A. C., Fritz, A., Gates, M. A., Amores, A., Bahary, N., Talbot, W. S., Her, H., Beier, D. R., Postlethwait, J. H., Zon, L. I. (1998). The *cloche* and *spadetail* genes differentially affect hematopoiesis and vasculogenesis. *Dev. Biol.* **197**, 248–269.

Tsai, F.-Y., Keller, G., Kuo, F. C., Weiss, M., Chen, J., Rosenblatt, M., Alt, F. W., and Orkin, S. H. (1994). An early hematopoietic defect in mice lacking the transcription factor GATA-2. *Nature (London)* **371**, 221–226.

Tsai, S. F., Martin, D. I., Zon, L. I., DAndrea, A. D., Wong, G. G., and Orkin, S. H. (1989). Cloning of cDNA for the major DNA-binding protein of the erythroid lineage through expression in mammalian cells. *Nature (London)* **339**, 446–451.

Tsang, A. P., Visvader, J. E., Turner, C. A., Fujiwara, Y., Yu, C., Weiss, M., Crossley, M., and Orkin, S. H. (1997). FOG, a multitype zinc finger protein, acts as a cofactor for transcription factor GATA-1 in erythroid and megakaryocytic differentiation. *Cell (Cambridge, Mass.)* **90**, 109–119.

Tsang, A. P., Fujiwara, Y., Hom, D. B., and Orkin, S. H. (1998). Failure of megakaryopoiesis and arrested erythropoiesis in mice lacking the GATA-1 transcription cofactor FOG. *Genes & Dev.* **12**, 1176–1188.

Turpen, J. B., Knudson, C. M., and Hoefen, P. S. (1981). The early ontogeny of hematopoietic cells studied by grafting cytogenetically labeled tissue anlagen: Localization of a prospective stem cell compartment. *Dev. Biol.* **85**, 99–112.

Turpen, J. B., Kelley, C. M., Mead, P. E., and Zon, L. I. (1997). Bipotential primitive-definitive progenitors in the vertebrate embryo. *Immunity* **7**, 325-334.

Valge-Archer, V. E., Osada, H., Warren, A. J., Foster, A., Li, J., Baer, R., and Rabbitts, T. H. (1994). The LIM protein Rbtn2 and the basic helix-loop-helix protein TAL1 are present in a complex in erythroid cells. *Proc. Natl. Acad. Sci. U.S.A.* **91**, 8617–8621.

Visvader, J. E., Fujiwara, Y., and Orkin, S. H. (1998). Unsuspected role for the T-cell leukemia protein SCL/tla-1 in vascular development. *Genes & Dev.* **12**, 473–479.

Wadman, I. A., Osada, H., Grutz, G. G., Agulnick, A. D., Westphal, H., Forster, A., and Rabbitts, T. H. (1997). The LIM-only protein Lmo2 is a bridging molecule assembling an erythroid, DNA-binding complex which includes the TAL1, E47, GATA-1, and Ldb1/NL1 proteins. *EMBO J.* **16**, 3145–3157.

Warren, A. J., Colledge, W. H., Carlton, M. B., Evans, M. J., Smith, A. J., and Rabbitts, T. H. (1994). The oncogenic cysteine-rich LIM domain protein rbtn2 is essential for erythroid development. *Cell (Cambridge, Mass.)* **78**, 45–57.

Weber, R., Blum, B., and Muller, P. (1991). The switch from larval to adult globin gene expression in *Xenopus laevis* is mediated by erythroid cells from distinct compartments. *Development (Cambridge, UK)* **112**, 1021–1029.

Weinstein, B., Schier, A., Abdelilah, S., Malicki, J., Solnica-Krezel, L., Stemple, D. L., Stainier, D. Y. R., Zwartkruis, F., Driever, W., and Fishman, M. (1996). Hematopoietic mutations in zebrafish. *Development (Cambridge, UK)* **123**, 303–309.

Weiss, M. J., and Orkin, S. H. (1995). Transcription factor GATA-1 permits survival and maturation of erythroid precursors by preventing apoptosis. *Proc. Natl. Acad. Sci. U.S.A.* **92**, 9623–9627.

Weiss, M. J., Keller, G., and Orkin, S. H. (1994). Novel insight into erythroid development revealed through *in vitro* differentiation of GATA-1 embryonic stem cells. *Genes Dev.* **8**, 1184–1197.

Wood, H. B., May, G., Healy, L., Enver, T., and Morriss-Kay, G. M. (1997). CD34 expression patterns during early mouse development are related to modes of blood vessel formation and reveal additional sites of hematopoiesis. *Blood* **90**, 2300–2311.

Zhang, C., and Evant, T. (1996). BMP-like signals are required after the midblastula transition for blood cell development. *Dev. Genet.* **18**, 267–278.

Zon, L. I. (1995). Developmental biology of hematopoiesis. *Blood* **86**, 2876–2891.

40

Endodermal Patterning and Organogenesis

Maureen Gannon
Christopher V. E. Wright
Department of Cell Biology
Vanderbilt University Medical Center
Nashville, Tennessee 37232

I. Introduction
   A. Origin of the Digestive Tube and Derivatives
   B. Pancreas Development
   C. *Pdx-1* Expression during Pancreas Development
II. Classic Studies on Epithelial–Mesenchymal Interactions
III. Relative Merits of Studying Endoderm Development in Different Systems
IV. Molecular Analysis of Endodermal Patterning
   A. Posterior Foregut Specification in Amphibian Embryos
   B. Posterior Foregut Patterning
   C. Epithelial–Mesenchymal Interactions Revisited

V. Genetic Analysis of Endoderm Development
   A. Posterior Foregut Development as a Model for Endodermal Regionalization and Organogenesis
   B. *Isl1* in Pancreas Development
   C. Additional Mouse Mutations with Relevance to Endodermal Patterning
   D. Limitations of Traditional Knockouts and Alternative Approaches
VI. Transgenic Analysis of Gene Regulation and Function
VII. Growth Factors in Lung and Pancreas Development
VIII. Major Questions and Challenges Remaining References

I. Introduction

The basic cellular interactions and molecular signaling events leading to anterior–posterior (A/P) and dorsal–ventral (D/V) patterning of the mesoderm and neuroectoderm of the vertebrate embryo have received intense scrutiny from many outstanding laboratories, as detailed in other sections of this book. However, although the subject of several elegant recombination studies performed in the 1960s and 1970s, the level of understanding regarding endodermal patterning has lagged significantly behind that of the other germ layers. With the recent identification and genetic analysis of several gene products that affect endodermal development, however, this situation is rapidly changing. In this chapter we discuss some of the advances in the molecular analysis of vertebrate endodermal patterning and organogenesis in the light of classic studies of endodermal–mesenchymal interactions. We will present the insight into endoderm development offered by the various model organisms, and list their current limitations. We

will survey some informative developmental studies performed in other endodermal organs, although we will highlight the work done in our laboratory and others analyzing posterior foregut specification and differentiation in *Xenopus* and mouse embryos, to emphasize the power of using complementary model systems to address a particular developmental problem. Finally, we will summarize some of what we believe to be the major challenges to be tackled in the near future in the field of endoderm development.

A substantial problem posed for the developmental biologist studying vertebrate endodermal differentiation and organogenesis is that all "endodermally derived" organs are actually an interdependent association of epithelial (endodermal) and mesenchymal (mesodermal) components. The study of this germ layer is also complicated by the fact that the early endoderm is very difficult to work with, consisting of a fragile unicellular layer in chick and mouse. Furthermore, studies of the earliest steps of specification in this germ layer are compromised because of the difficulties in identifying and isolating endodermal precursors prior to gastrulation, with the possible exception of the *Xenopus* embryo.

A central question for those studying endodermal development is to determine how the continuous digestive–respiratory tube becomes subdivided along the A/P and D/V axes to produce distinct regions of specialized epithelium. As we are now becoming increasingly aware, the basic mechanisms controlling initial A/P patterning display a surprisingly high level of conservation throughout animal evolution. Thus, since the majority of phyla in the animal kingdom possess some type of alimentary canal, it is reasonable to presume that the analysis of various organisms, including nonvertebrates, will generate information relevant to vertebrate endodermal differentiation. Other major areas of investigation for the endodermal biologist focus on the molecular genetic dissection of morphogenetic and differentiation processes in the formation of organs like the lung, liver and pancreas, including the role of epithelial–mesenchymal interactions and the pathways by which the coordinated differentiation of multiple cell types is achieved within the organ primordia.

A. Origin of the Digestive Tube and Derivatives

In the frog *Xenopus*, the endodermal lineage is spatially restricted to the large, yolky, vegetally located blastomeres in the early embryo (Fig. 1B). The most dorsal of these cells, which are fated to become pharyngeal endoderm, comprise part of the Nieuwkoop center, which serves the critical function of inducing the Spemann organizer in overlying blastomeres (reviewed in Kimelman *et al.*, 1992). Creation of the Nieuwkoop center results

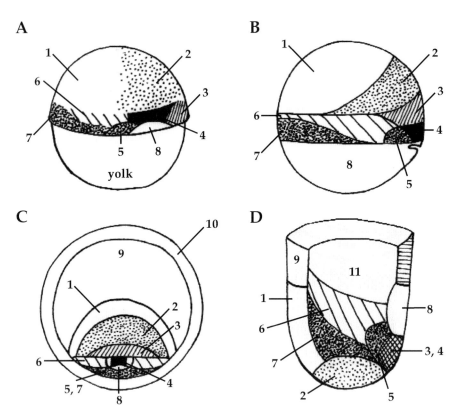

Figure 1 Simplified fate maps of early vertebrate gastrulae. At this stage, there is still considerable overlap of prospective fates, therefore differently shaded regions represent general areas of organization and not specific boundaries. (A) Zebrafish embryo at 50% epiboly. Areas of the epiblast producing different tissues are labeled according to Kimmel *et al.* (1990) and Tam and Quinlan (1996). Definitive endoderm is derived from cells at the dorsolateral margin of the epiblast. The animal pole is to the top and dorsal is to the right. (B) *Xenopus* embryo at stage 10. Cells specified to give rise to different cell types are labeled according to Keller (1976) and Kimelman *et al.* (1992). Endoderm arises from vegetally located cells. The animal pole is toward the top and dorsal (indicated by the dorsal lip) is to the right. (C) Dorsal view of the chick embryo epiblast at stage XIV. The flat epiblast is divided into the area pellucida (inner circle), which generates the embryonic tissues and extraembryonic ectoderm, and the area opaca (outer circle), which produces solely extraembryonic tissues (Torrey and Feduccia, 1979; Hatada and Stern, 1994). Definitive endoderm is derived from cells adjacent to prospective heart and notochord. Posterior is to the bottom. (D) Right half of the egg cylinder of a mouse embryo at 6.5 dpc viewed internally showing fate of different regions of the epiblast (Tam and Quinlan, 1996). The epiblast gives rise to both embryonic and extraembryonic tissues. The region of future embryonic endoderm is adjacent to the forming primitive streak (horizontal shading) at its posterior end, and to the prospective heart and notochord at its anterior end. Posterior is toward the upper right. 1, Epidermis; 2, neural ectoderm; 3, notochord; 4, prechordal plate; 5, cardiac lateral plate mesoderm; 6, paraxial mesoderm; 7, lateral plate mesoderm and blood; 8, endoderm; 9, extraembryonic ectoderm; 10, area opaca; 11, extraembryonic mesoderm.

from activation of a Wnt-like signaling pathway in an area of high concentrations of activin/Vg1-like signaling (e.g., Cui *et al.*, 1996). The situation in other well-studied vertebrates (e.g., zebrafish, chick and mouse) is quite different in that definitive embryonic mesoderm and endoderm are derived from delamination of cells within the epiblast (Fig. 1). In chick embryos, endoderm precursors are derived from the posterior third of the Eyal-Giladi and Kochav (1976) stage XIV epiblast (Fig. 1C), where they are located adjacent to prospective cardiac mesoderm (Hatada and Stern, 1994). The endodermal sheet forms from a combination of gastrulation through the primitive streak and ingression from the epiblast (Stern, 1992; Hatada and Stern, 1994). While passing through the anterior primitive streak at Hamburger and Hamilton (1951) stage 3–3+, endoderm precursors are closely associated with cells that will give rise to the notochord (Selleck and Stern, 1991), a tissue implicated in endodermal patterning, as we describe later. The mechanisms (stochastic or otherwise) involved in allocating cells within the epiblast to either the endodermal or mesodermal lineages, and the inductive effects imposed on the endoderm by their very first interactions with mesodermal precursors, are currently unknown.

In zebrafish, chicks, and mammals, the movements of gastrulation, combined with the growth and folding of the lateral and ventral body wall, result in the formation of the endodermally lined anterior and posterior intestinal portals (AIP and PIP), which produce the foregut and hindgut, respectively (Fig. 2). Further ventral growth closes the embryonic gut toward the middle of the embryo; eventually only a portion of the midgut is

continuous with the yolk in zebrafish and chick, or with the visceral yolk sac in mammals. Pharyngeal endoderm gives rise to the pharynx, and outpocketings from this region in part produce precursors for the thyroid and thymus. In this review we will concentrate mainly on those endodermal derivatives posterior to the pharynx. The foregut gives rise to the esophagus, stomach, and duodenum with evaginations from it producing the lungs, liver, gall bladder, and pancreas. The midgut develops into the jejunum and ileum of the small intestine, as well as the rostral colon, while the hindgut becomes the distal large intestine and rectum.

Endodermally derived cells generate only the epithelial lining and glands of the digestive tract and accessory organs; the splanchnic lateral plate mesoderm provides the underlying connective tissue of the organs and the smooth muscle layer responsible for peristalsis. Vertebrate gut derivatives all begin as evaginations of the endodermal epithelium that invade the surrounding mesenchyme. In the lungs, which begin as a ventral outpocketing from the esophageal region at 9.5 days post coitum (dpc) in the mouse embryo, endodermal epithelium completely lines the trachea, bronchi, and alveoli. The liver, gall bladder, and pancreas arise from the rostral duodenum, just caudal to the stomach. Within this region of the posterior foregut, the liver, gall bladder, and ventral pancreatic bud emerge ventrally, while the dorsal pancreatic bud emerges dorsally (Fig. 3). The endodermal component proliferates and branches to produce the hepatic component of the liver, and the pancreatic exocrine and endocrine cells and ductal network.

Even a fairly superficial examination reveals that endodermal and mesenchymal tissues are in intimate

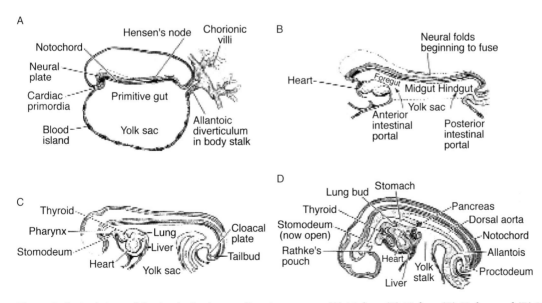

Figure 2 Sagittal views of the developing human digestive system at (A) 16 days, (B) 18 days, (C) 22 days, and (D) 28 days. (C and D) are analogous to mouse embryos at 8.5 and 10.0 days post coitum (dpc). By 10.0 dpc, the outpocketings which will form the major endodermal derivatives are apparent, including thyroid, lung, liver, and pancreas. Anterior is to the left and dorsal is to the top. (Adapted, with permission, from Gilbert, 1994.)

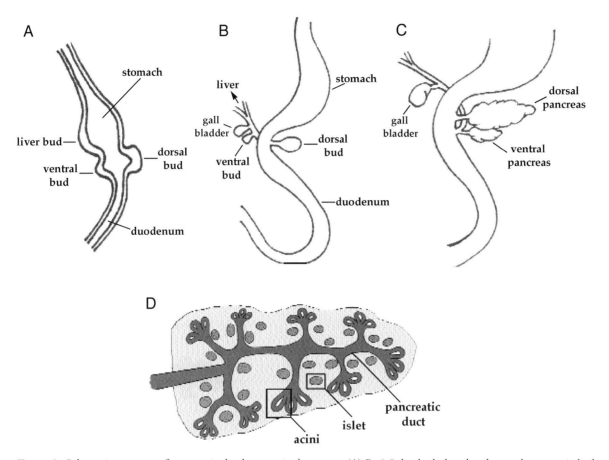

Figure 3 Schematic summary of pancreatic development in the mouse. (A) By 9.5 dpc both dorsal and ventral pancreatic buds are apparent. (B) At 10.0. dpc the ventral pancreatic bud is still close to the liver primordium and gall bladder, but by (C) 10.5 dpc, the ventral bud has contacted the dorsal bud; the two buds subsequently merge. (D) In the mature pancreas, branches from the main pancreatic duct terminate in acinar cell clusters that secrete digestive enzymes. The islets of Langerhans develop from the distal tips of growing ducts, becoming isolated within the more abundant acinar tissue as the pancreas matures. Internal organs at different stages are not drawn to scale. (Based on Gittes and Rutter, 1992.)

contact throughout embryonic development. It is therefore likely that multiple reciprocal interactions occur between these germ layers, controlling the processes of proliferation, differentiation, branching, and morphogenesis of the endodermal organs (reviewed in Haffen *et al.*, 1987). While the molecules initiating outgrowth of the endoderm into surrounding mesenchyme have not yet been identified, it is clear that dorsal–ventral differences exist in the nature of the signal(s), the source of the signal(s), and/or in the response of the epithelium to these signals. In this chapter we will present some of the experimental evidence for these differences and the source(s) of potential inductive signals.

B. Pancreas Development

The focus of our laboratory is the specification and differentiation of the posterior foregut, with particular emphasis on the molecular and cellular mechanisms leading to the formation of the pancreas. In this review, we will use pancreatic differentiation and organogenesis as

a paradigm for studies on endodermal patterning, epithelial–mesenchymal interactions, and cell fate decisions, although specific examples of key studies in other endoderm derivatives will also be discussed. Recent studies on pancreatic development have led to a better understanding of the factors regulating pancreatic growth and differentiation in general, and islet cell-specific gene expression in particular. Knowledge of the factors regulating β-cell function and insulin production will have enormous relevance to understanding the development of diabetes, and may lead to potential avenues of therapy. A brief summary of the key features of pancreatic development will now be given to facilitate the interpretation of the studies outlined later in the chapter (see Fig. 3).

In mouse embryos at 8.5 dpc, the foregut endoderm, except for the most anterior region, is still an open tube that is laterally contiguous with the extraembryonic membranes. At this stage, foregut endoderm contacts cardiac mesoderm at its anterior–ventral surface while its dorsal surface contacts the notochord. Endodermal outpockets arise from the dorsal and ventral posterior foregut. The ventral outgrowth gives rise to (in anterior–posterior order) the liver, gall bladder, and ventral pancreas. The dorsal pancreatic evagination, which produces the bulk of the mature pancreas, becomes evident at 9.0 dpc, with the definitive ventral pancreatic bud forming at 9.5 dpc. The buds continue to grow, with the dorsal bud proliferating more rapidly, until they merge into a single organ by 11.5 dpc.

There is general agreement in the field that precursor cells located within the ductal epithelium of the pancreas produce both the endocrine or exocrine cell types (see Fig. 4), despite some earlier controversy regarding an alternate origin of pancreatic endocrine cells from the neural crest, as discussed below. The majority of pancreatic cells differentiate as exocrine acinar cells that synthesize, store, and secrete digestive enzymes into the rostral duodenum via the branching pancreatic ductal network. Other cells bud off from the distal end of the growing ducts and become organized into endocrine clusters within the exocrine tissue, forming the islets of Langerhans. Newly formed or primitive islets are found close to the ducts. The four major cell types within the islets and their hormone products are α

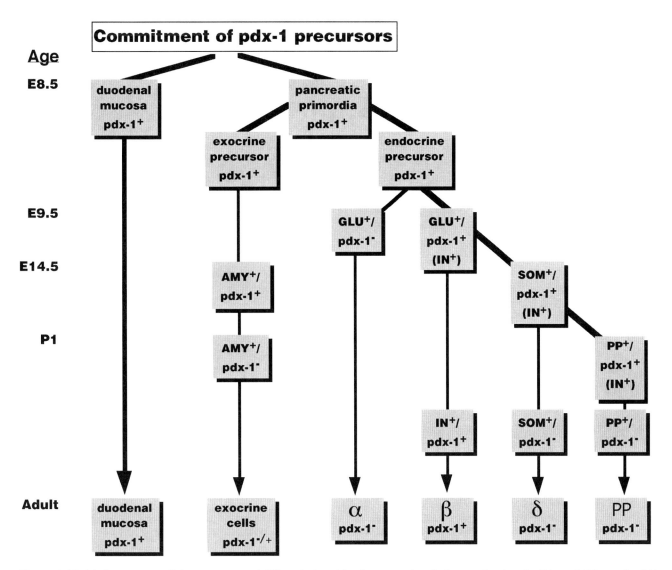

Figure 4 Model of pancreatic cell determination and differentiation with reference to the *pdx-1* expression domain. Prior to 8.5 dpc, cells within the posterior foregut become committed to express the homeobox gene *pdx-1*. *pdx-1* expressing cells give rise to the rostral duodenum and pancreas. Duodenal expression of *pdx-1* is maintained throughout life. Initially, all cells of the pancreatic buds are Pdx-1-positive. The first glucagon- (GLU) and insulin- (IN) positive cells are observed at 9.5 dpc some of which are *pdx-1*-negative. By 14.5 dpc, the majority of β cells, but only a subset of α cells, are *pdx-1* (+). Glucagon (+)/Pdx-1 (+) cells may also be insulin (+) and may represent the progenitors of mature β, δ, and PP cells. At 14.5 dpc the first amylase (AMY)-positive cells arise and many of these also contain Pdx-1. By 16.5 dpc Pdx-1 expression is restricted mainly to islets and a few ductal cells. In the adult, Pdx-1 is found in 90% of β, 15% of δ, 9% of PP, and 3% of α cells. Much lower expression is found in subsets of acinar cells (Wu *et al.*, 1997). (Adapted from Guz *et al.*, 1995.) The reader is reminded that direct lineage tracing experiments for this ontogeny are not available and that, for simplicity, not every potential connection between cell types is shown.

(glucagon), β (insulin), δ (somatostatin), and PP (pancreatic polypeptide). Transcription of endocrine pancreas-specific genes, such as glucagon and insulin, is first detected at 9.5 dpc in the pancreatic buds prior to islet morphogenesis and formation of the pancreatic epithelium (Gittes and Rutter, 1992; Teitelman *et al.*, 1993).

Phylogenetically, the presence of morphologically distinct islets occurs with the appearance of the vertebrates (reviewed in Rosenberg, 1996). In protochordates, insulin, somatostatin, glucagon, and PP cells are distributed throughout the mucosa of the digestive tract or bile duct. Tunicates and *Amphioxus* have neuroendocrine cells located in the brain and digestive tract. Among the vertebrates a distinct endocrine organ containing islets appears for the first time in agnathans (jawless fish such as the lamprey), in which insulin and somatostatin cells are clustered adjacent to the bile duct epithelium. During the course of evolution, therefore, it seems that a distinct pancreatic organ gradually arose by aggregation and migration of islet cells from the gut mucosa into localized clusters. The exocrine pancreas first makes its appearance in the chondricthyes (cartilagenous fish such as sharks and rays). In these fish, islets are closely associated with the ductal epithelium of the pancreas, with cells present for each of the four islet hormones. Teleosts (modern bony fish) show a diversity in the relationship between exocrine and endocrine pancreatic cell types. Some species contain both cell types in a discrete pancreatic organ as in higher vertebrates. In others (e.g., the rainbow trout), a remarkable situation has arisen in which endocrine cells are completely separated from exocrine tissue, lying in a "Brockmann body" that is located close to the bile duct. Within the vertebrate gut, somatostatin, glucagon, PP, and other enteroendocrine cells are dispersed throughout the epithelium, while insulin-producing cells are restricted to islets. Surprisingly, in several strains of rat, extrapancreatic islets containing insulin, glucagon, and somatostatin have been observed within the duodenum near the bile and pancreatic ducts (Bendayan and Park, 1997). These islets are detected as early as 12.0 dpc and, although their physiological role is unclear, insulin-producing cells in these islets respond in the same way as pancreatic islets under diabetic conditions, undergoing degeneration.

The origin of exocrine and endocrine cell types from a common endodermal precursor has been vigorously debated over the years; an early hypothesis being that islet cells came from neural crest cells that migrated into the developing pancreas (Pearse, 1977; Alpert *et al.*, 1988). Primarily, this idea arose because all four islet cell types arise from common multipotent precursors that coexpress multiple islet hormones and a battery of markers shared with neurons (De Krieger *et al.*, 1992; Lukinius *et al.*, 1992). Indeed, pancreatic

islet cells, and similar hormone-producing cells in the gut epithelium, are often referred to as neuroendocrine cells. During development, islet endocrine cells become restricted in their potential and eventually express a single hormone (Fig. 4) (Teitelman *et al.*, 1993), although mature islet cells continue to synthesize several neural markers such as tyrosine hydroxylase and enolase, and express transcription factors such as *Isl1*, *Pax6* and *Beta2/NeuroD* that are also found in specific neural tissues (Alpert *et al.*, 1988; Thor *et al.*, 1991; Naya *et al.*, 1995; Lee *et al.*, 1995). Despite the correlative evidence for a neural origin, the balance of the experimental evidence gathered more recently is completely consistent with a purely endodermal origin for islet cells. The work in our laboratory and others has shown that multipotent islet cells express the homeodomain-containing transcription factor, *pdx-1*, as do the duodenal epithelial cells from which the pancreatic buds arise (Fig. 4) (Ohlsson *et al.*, 1993; Guz *et al.*, 1995; Offield *et al.*, 1996). On the other hand, neurons do not express *pdx-1*. While this argument is subject to the criticism that neural crest-derived cells may activate *pdx-1* transcription on arrival at the pancreas, explant studies also support the endodermal origin. Cultured explants of mouse pancreatic endoderm plus mesenchyme (Ahlgren *et al.*, 1996, 1997) or isolated endoderm cultured under the renal capsule of adult mice (Gittes *et al.*, 1996), can produce islet cells when removed at stages prior to neural crest cell arrival at this A/P region of the embryo. In addition, preliminary studies using prepancreatic endoderm from β-galactosidase-expressing transgenic Rosa26 mice recombined with nontransgenic mesenchyme suggest that insulin-producing cells are derived from the lineagelabeled endoderm (Slack *et al.*, 1997).

C. *Pdx-1* Expression during Pancreas Development

In all vertebrates examined to date, cellular progenitors of the rostral duodenum and pancreas can be identified by their expression of *pdx-1* and its homologs (Fig. 5) (Wright *et al.*, 1989; Ohlsson *et al.*, 1993; Guz *et al.*, 1995; Gamer and Wright, 1995; Offield *et al.*, 1996; Kim *et al.*, 1997). *pdx-1* was first isolated from *Xenopus* as *XlHbox8* (Wright *et al.*, 1989); homologs were subsequently identified independently in several laboratories based on their expression in islets and ability to bind to specific DNA sequences and thereby activate insulin and somatostatin gene expression (Ohlsson *et al.*, 1993; Leonard *et al.*, 1993; Miller *et al.*, 1994). In mouse, *pdx-1* expression is first detected at 8.0 dpc, preceding insulin and glucagon expression (Fig. 6). Early in development, Pdx-1 expression marks all pancreatic cells and the intervening duodenal epithelium (Guz *et al.*, 1995). In adults, Pdx-1 is maintained in all rostral duodenum

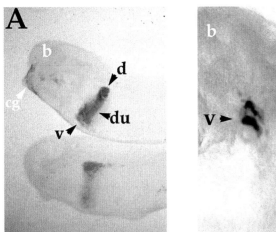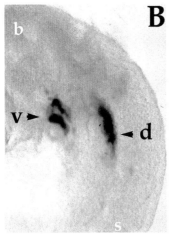

Figure 5 Pdx-1 expression in the posterior foregut is conserved in vertebrates. Lateral views of (A) frog tadpoles at stage 32 (lower) and stage 34 (upper), showing Pdx-1 (XlHbox8) expression visualized immunohistochemically at stage 32 in posterior foregut endoderm corresponding to the future pancreas and duodenum, and at stage 34 in the duodenum and the now apparent dorsal and ventral pancreatic buds. Anterior is to the left. (B) Partial view of a 9.5 dpc mouse embryo showing *pdx-1*-driven β-galactosidase expression in the dorsal and ventral pancreatic buds. The head is to the upper left; the embryo is curled with the tail at the lower left. b, Brain; cg, cement gland; d, dorsal pancreatic bud; du, duodenum; v, ventral pancreatic bud; s, somites.

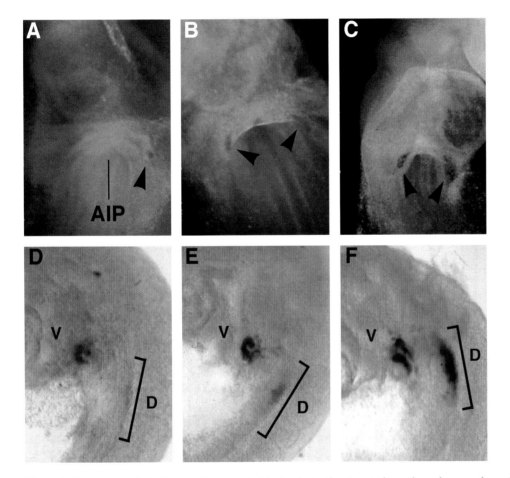

Figure 6 Time course of *pdx-1* expression as marked by β-galactosidase inserted into the endogenous locus. Ventral (A–C) or lateral (D–F) views of mouse embryos from 8.0 dpc to 9.5 dpc. (A) The earliest expression of *pdx-1* is at ~8.0 dpc (7 somites) on the left ventral side of the AIP (arrow). (B) By the 8- to 9-somite stage, expression is detected on both left and right sides ventrolateral to the AIP (arrows). (C–F) Progressing toward 9.0–9.5 dpc, the ventral expression domains fuse as the ventral foregut closes anteriorly, and then spreads. (D) At 9.0 dpc, dorsal bud expression is first detected at the bracketed region and (E, F) expands and intensifies until it becomes the predominant bud. AIP, Anterior intestinal portal; D, dorsal; V, ventral. Anterior is toward the top in A–C, and the upper left in D–F.

cells and islet β-cells, but only subsets of acinar and somatostatin-producing cells (Guz *et al.*, 1995; Gamer and Wright, 1995; Wu *et al.*, 1997). The expression pattern alone suggests that *pdx-1* plays multiple, distinct roles during development in the adult, including specification of the posterior foregut, proliferation, and differentiation of different pancreatic lineages, activation of insulin gene expression, and maintenance of β-cell function, including islet and glucose homeostasis.

II. Classic Studies on Epithelial– Mesenchymal Interactions

Early tissue recombination studies addressed the potential for reciprocal inductive interactions between mesenchyme and epithelium. A general conclusion from the resulting data is that the mesenchyme adjacent to the endoderm epithelium is involved in inducing and supporting its proliferation and morphogenesis (Goldin and Wessels, 1979).

There is evidence that the subjacent mesenchyme can to some extent control the differentiated character of the endoderm that it contacts in the lung (Alescio and Cassini, 1962; Wessels, 1970) and gastrointestinal tract (Yasugi and Mizuno, 1978; Lacroix *et al.*, 1984). In the latter case, mesenchyme transplanted from the developing intestine, to either the presumptive esophageal or gastric endoderm, induced intestine-specific epithelial morphogenesis. The reciprocal nature of the endoderm–mesenchyme interaction is revealed by the finding that either intestinal or extraembryonic endoderm can induce gastrointestinal smooth muscle differentiation in various gut mesenchymes (Kedinger *et al.*, 1990; Apelqvist *et al.*, 1997). The molecules involved in these interactions remain largely unknown but, as we discuss in sections IV,C and VII, some recent candidates have begun to emerge.

In these tissue recombination studies, an important distinction has been made between so-called instructive and permissive inductive interactions between the two germ layers. An instructive interaction involves a signal(s) passing from the inducing tissue to provoke the "naive" responding tissue to activate only one of several potential genetic programs, leading to a specific differentiation pathway being adopted by that tissue. Without the inducing signal, the responding tissue fails to differentiate. In contrast, in a permissive interaction, the responding tissue is already restricted in its developmental potential, and carries a particular cell/tissue type identity, although this remains latent until provided with the appropriate environmental stimuli (e.g., growth factors, extracellular matrix) that support growth, morphogenesis, and differentiation (discussed in Slack, 1983, and Gilbert, 1994).

The developing lung provides a particularly good example of an instructive interaction, because mesenchyme from different regions of the developing respiratory system can elicit quite different responses from the adjacent endoderm (Wessels, 1970; Goldin and Wessels, 1979). Grafting tracheal mesenchyme to more distal lung epithelium suppresses the branching that normally occurs at the end buds, while transplanting mesenchyme from the distal branching regions to the normally linear tracheal region induces its outgrowth and branching (Alescio and Cassini, 1962; Wessels, 1970; Goldin and Wessels, 1979). Recent results indicate that such induced tracheal branches are not only morphologically similar to terminal lung buds, but express markers of differentiated alveolar epithelium (Shannon, 1994). In this case, then, the instructive capacity of different regions of lung mesenchyme apparently overrides any inherent predisposition of lung endoderm along the proximodistal axis.

This kind of result from lung endoderm–mesenchyme recombinants suggested that the endoderm of the digestive tube and its evaginations could be considered naive, and that specific prepatterns within the mesenchyme controlled the regionalized differentiation of the endoderm. It was, however, unclear if this could be generalized; that is, do specific mesenchymes provide instructive signals to endoderm to elicit differentiation into lungs, stomach, small intestine, etc? An early study on chick embryos seemed to suggest that intestinal or hepatic mesenchyme could instructively signal to lung epithelium, dominantly inducing intestine or liver epithelial differentiation (reviewed in Wolff, 1968). These conclusions were based on a morphological analysis of epithelial differentiation in tissue recombinants. This approach is fairly limiting, however, and reexamination of the original data leads to the conclusion that there is little, if any, support for the hypothesis that mesenchyme from different regions of the digestive tract can instruct lung endoderm to "transdifferentiate" along pathways associated with other organs. Thus, with the exception of transdominant mesenchymal influences within the proximal–distal axis of the lung itself, the degree of plasticity associated with lung epithelium (specifically its capacity to be completely reprogrammed by different mesenchymes) remains unknown pending more precise molecular analyses.

An example of a permissive inductive interaction occurs in pancreas development (reviewed in Wessels, 1977). When presumptive pancreatic epithelium from 15-somite stage (9.0 dpc) mouse embryos was recombined with a variety of different mesenchymes, including either pancreatic or salivary gland mesenchyme, the endoderm differentiated into morphologically recognizable acini containing amylase in zymogen granules, although these studies did not report islet differentiation and/or morphogenesis. Therefore, insofar as exocrine development is concerned, it can be concluded that "prepancreatic" endoderm has already acquired a pancreatic identity by 9.0 dpc and that the surrounding mesenchyme serves merely a permissive role, secreting diffusible signaling molecules to promote outgrowth (Golosow and Grobstein, 1962; Wessels and Cohen, 1967). In addition to its role in inducing proliferation and outgrowth, the mesenchyme may also provide a substrate through which branching morphogenesis can occur (Fell and Grobstein, 1968). Overall, these studies suggest that the permissive signals are similar in different sources of mesenchyme, and that differentiation of particular endodermal cell types results from a combination of these signals with predispositions carried within the responding tissue.

Tissue recombination strategies have also been used to analyze the sequence of inductive events leading to specification and/or outgrowth of liver and pancreatic epithelium. Because chick and quail nuclei have different histological properties, LeDouarin (reviewed in LeDouarin, 1975) could show that liver development requires successive interactions between ventrally located cardiac mesoderm, which is an apparent source of instructive signal(s) for liver specification, and liver mesenchyme, which permissively promotes outgrowth and morphogenesis. However, it was still unclear from her studies whether cardiac mesoderm could provide a signal capable of inducing naive endoderm toward the liver fate, as hepatocyte differentiation was not morphologically identified in endoderm from regions anterior or posterior to the liver primordium placed into coculture with heart mesoderm.

While cardiac mesoderm seems to induce the liver fate from the posterior foregut, the tissue sources and signals that specify the fate of other regions of endoderm are generally unknown. Most likely based on spatial proximity arguments, Wessels and Cohen (1967) proposed that dorsal axial tissues (notochord) were the source of signals for the specification and outgrowth of the dorsal pancreatic bud. However, the ability of axial tissue to induce pancreatic development from naive endoderm was not tested in these studies. The source of the signal that induces the ventral budding still remains unknown (see below for more discussion on this point). However, more precise experimentation to determine if axial tissues do in fact carry "instructive" signals for pancreatic specification is now being initiated (Kim *et al.*, 1997; Hebrok *et al.*, 1998).

Nevertheless, the early explant studies were important in providing the groundwork for current studies, through the identification of potentially critical tissue interactions and specific windows of developmental time during which instructive and permissive signals might operate. However, the lack of early markers for individual cell types meant that only relatively late aspects of terminal differentiation could be used as analytical criteria. Functionally dissecting the early steps in the cell fate decision-making process had to await the identification of transcription factors and intercellular signaling molecules that mediate these tissue interactions, and that display regionally restricted expression patterns along the A/P axis of the endoderm during development. Furthermore, the earlier studies were generally limited to the "cut-and-paste" techniques of experimental embryology, analyzing the developmental effects of adding or removing relatively large groups of cells to or from the endoderm. The advent of techniques for molecular overexpression in *Xenopus* and chick, transgenic and "knockout" technology in the mouse, and saturation mutagenesis in zebrafish, will yield a much more incisive and coherent understanding of the role that each of these factor(s) plays in endodermal patterning and differentiation.

III. Relative Merits of Studying Endoderm Development in Different Systems

It has become increasingly clear in recent years that the gene networks controlling patterning and specification within the individual germ layers (ectoderm, mesoderm, and endoderm) are surprisingly conserved throughout evolution, from nematodes to flies to humans. Thus, it is a fairly safe assumption that the genes important for early endodermal patterning and regionalization are similar in most metazoans, with additional functions or family members being invented/recruited for more elaborate endodermal derivatives, such as the lungs or liver, in more advanced vertebrates. For example, the GATA family of transcription factors is expressed in the intestine of nematodes, flies, and vertebrates, and has been shown genetically to be required for the formation and differentiation of the gut in all three phyla (Arceci *et al.*, 1993; Reuter, 1994; Rehorn *et al.*, 1996; Kuo *et al.*, 1997; Molkentin *et al.*, 1997; Zhu *et al.*, 1997). Binding sites for these transcription factors have also been identified within the regulatory regions of endoderm-specific genes in species from *Caenorhabditis elegans* (MacMorris *et al.*, 1992; Fukushige *et al.*, 1996; Zhu *et al.*, 1997) to mammals (K. Zaret, 1997, personal communication). A complete understanding of endodermal patterning, differentiation, and organogenesis will therefore come from adding findings from studies of invertebrates to those obtained in vertebrates. Given that the focus of this chapter is vertebrate endoderm development, we now present a nonexhaustive listing of the advantages, available techniques, and current shortcomings of different vertebrate systems, as applied to the analysis of cell lineage–tissue relationships and genetic interactions–hierarchies in organogenesis. It should be emphasized that the high degree of conservation of molecular mechanisms between species strongly suggests that complementary studies in these model systems will provide the most rapid and comprehensive understanding of endodermal development.

1. **Mouse**. Advantages: Preexisting mutants, recent establishment of mutant screens, capacity for reverse genetics, transgene-based misexpression or overexpression, explant studies becoming more feasible, ROSA-26 mice useful for recombination/chimera studies. Drawbacks: High cost and space for housing multiple lines, long gestation period, few embryos per litter, relative inaccessibility of the embryo, generally no fast method of gene isolation once mutant is known.

2. **Avian**. Advantages: Easily accessible embryo, relatively large and flat gastrulation stage embryos facilitate explant studies, quail/chick chimeras allow lineage analysis, limited retrovirus-based molecular overexpression possible. Drawbacks: Limited genetics, current inability to make transgenics, difficult to culture embryos long-term through organogenesis.

3. **Xenopus**. Advantages: Huge numbers of easily produced embryos, relatively well-defined fate map, large embryos facilitate explant studies, transplantation, recombination and injection of RNA/expression plasmids. Drawbacks: Current inability to do genetics, transgenics being developed but not yet readily available (Slack, 1996).

4. **Zebrafish**. Advantages: Genetics/large scale mutagenesis screens possible, large numbers of transparent embryos, rapid development, injection of candidate genes to rescue mutant phenotypes, laser ablation to analyze effect of specific cells on development, cell lineage tracing and cell transplantation, recent reports of novel proviral insertion method for generating mutants and rapid cloning of genes (Gaiano *et al.*, 1996), relatively "simple" organ structure provide an easier model for analyzing the pathways patterning the vertebrate digestive system. Drawbacks: Current paucity of early markers, lack of rapid gene cloning methods once mutants are identified, difficult to get large quantities of embryos for biochemistry.

IV. Molecular Analysis of Endodermal Patterning

While the basic questions about the mechanisms underlying gut development remain the same, the methods

used to address them have become more sophisticated, pushing back the time in development when alterations in gene expression can be measured and meaningfully interpreted. Thus, we are now gaining large insights into the genetic mechanisms controlling the first steps of endodermal development. As we progress, however, our evaluation of differentiation may require the redefinition of concepts like specification, determination and differentiation to recognize that each of these broad divisions actually contains multiple, isolatable steps distributed along an apparent developmental continuum.

The number of transcriptional regulators and growth factors showing restricted expression in the digestive tube and its derivatives is ever increasing. While some are expressed in particular organs or cell types, others show germ layer restriction. Table I lists some of the prominent factors involved in endoderm specification, patterning, and differentiation. While this is necessarily only a partial list, it provides a sense of the complexity and combinatorial nature of the factors controlling endodermal development. The cataloging of endodermally expressed factors has been enhanced by techniques such as *in situ* hybridization and reverse transcription–polymerase chain reaction (RT–PCR) that allow the detection of even low levels of gene expression in sometimes vanishingly small pieces of tissue or cultured explants from very early embryos. For example, RT–PCR analysis has led to the surprising discovery that expression of islet hormones in the pancreas, and albumin and α-fetoprotein in the liver, previously thought of as late stage differentiation markers, can in fact be detected prior to overt formation of the individual organ buds (Gittes and Rutter, 1992; Gualdi *et al.*, 1996). As described below, such findings are causing a substantial reconsideration of the mechanisms of endoderm specification and differentiation.

A. Posterior Foregut Specification in Amphibian Embryos

The large blastomeres of the *Xenopus* embryo and its well-characterized fate map facilitate analyses of cell fate decisions and cell interactions (see Chapter 19 by David; Chapter 20 by Sullivan *et al.*). Early experiments (Okada, 1954a,b) showed that explanted vegetal cells, which are fated to become the embryonic gut, undergo morphogenesis toward intestinal structures only when cocultured with mesodermal cells, a result that can be considered to be consistent with the requirement for endoderm–mesenchyme interactions in other vertebrates as described in Sections II, IV,C, and VII. This connection is also indicated by the observation that mesoderm inducers like activin or fibroblast growth factor 2 (FGF2; bFGF) induce morphologically identifiable endodermal tissues from animal cap ectoderm, most likely by a secondary induction from the mesoderm in-

TABLE I

Transcriptional Regulators and Intercellular Signaling Molecules Expressed in the Developing Digestive Tract and Derivatives

Name	Type	Expression[a]	Ref.[b]
Endodermal transcription factors			
pdx-1/XlHbox8/IPF-1/STF-1/IDX-1	Divergent *Antp* homeodomain	Posterior foregut, antral stomach, rostral duodenum, whole pancreas in embryos, mainly islet β cells in adults	1
T(ebp)/Nkx2.1/TTF-1	Nkx homeodomain	Thyroid, lung (ventral forebrain)	2
Nkx 2.2	Nkx homeodomain	Islet β cells (ventral CNS)	2
Nkx2.5 and Nkx2.8	Nkx homeodomain	Foregut (cardiac mesoderm)	2
Nkx6.1	Nkx homeodomain	Early whole pancreas, becomes restricted to islet β cells	2
Isl1	LIM homeodomain	Islet cells (motor neurons)	3
HoxA and HoxD clusters	*AbdB*-related homeodomain	Nested expression of cluster members with progressively more 5' genes having more caudal anterior limits	4
cdx-1	*caudal*-like homeodomain	Small intestine, colon	5
cdx-2/3	*caudal*-like homeodomain	Duodenum, small intestine, colon, islets	5
cdx-4	*caudal*-like homeodomain	Embryonic intestine, adult islets	5
HNF1	POU-homeodomain	Lung, stomach, liver, intestine, pancreas (renal tubules)	6
HNF6	Cut-homeodomain	Liver, intestine, pancreas (dorsal root ganglia, midbrain)	7
HNF3β	Winged helix	Lung, liver, pancreas, small intestine, (node, notochord, floorplate)	8
HNF4	Steroid hormone receptor	Stomach, liver, islets, intestine, extraembryonic endoderm, (kidney)	9
Beta2/NeuroD	bHLH	Intestinal enteroendocrine cells, pancreatic endocrine cells, (differentiating neurons)	10
p48	bHLH	Exocrine pancreas	11
Pax4	*paired*-homeodomain	Islet β cells, (ventral spinal cord)	12
Pax6	*paired*-homeodomain	Islets (CNS, eye, nose)	12
Pax8	*paired*-homeodomain	Thyroid, (CNS, kidney)	12
GATA 4/5/6	GATA factor	Gut epithelium (broadly in splanchnic mesoderm)	13
Endodermal intercellular signaling molecules			
Sonic hedgehog (Shh)	—	Distal lung buds, entire gut except pancreatic buds (limb bud, notochord, floorplate)	14
Indian hedgehog (Ihh)	—	Similar pattern to Shh	14
Bone morphogenetic protein (BMP)-4	—	Distal lung buds	15
Transcription factors expressed in gut-associated mesoderm			
Isl1	LIM homeodomain	Mesenchyme of dorsal but not ventral pancreatic bud	3
Hox A and B clusters	*AbdB*-related homeodomain	Nested pattern in lung mesenchyme: 5'-most genes expressed more distally	4
Hoxd-12, Hoxd-13	*AbdB*-related homeodomain	Mesenchyme of terminal hindgut	4
Fkh-6	Winged helix	Gut mesenchyme	16
GATA 4/6	GATA factor	Intestinal smooth muscle	13
Gli, Gli2, Gli3	Zinc finger	Lung mesenchyme	17
Signaling molecules expressed in gut-associated mesoderm			
Fibroblast growth factor (FGF)-7 and FGF-10	—	Lung, duodenum, (heart, spinal cord, posterior pituitary, cervical vertebrae)	18
BMP-4	—	Early ventral mesoderm, distal lung, gut mesenchyme	15

[a]Parentheses indicate regions of expression outside the gut proper.

[b]Key to references:

1, Wright *et al.*, 1989; Ohlsson *et al.*, 1993, Leonard *et al.*, 1993; Miller *et al.*, 1994; Guz *et al.*, 1995; Offield *et al.*, 1996. 2, Guazzi *et al.*, 1990; Mizuno *et al.*, 1991; Lazzaro *et al.*, 1991; Lints *et al.*, 1993; Sussel *et al.*, 1998; Brand *et al.*, 1997. 3, Thor *et al.*, 1991; Pfaff *et al.*, 1996; Ahlgren *et al.*, 1997. 4, Dollé *et al.*, 1991; Yokouchi *et al.*, 1995; Roberts *et al.*, 1995; Mollard and Dziadek, 1997. 5, Duprey *et al.*, 1988; Gamer and Wright, 1993; Suh *et al.*, 1994; Laser *et al.*, 1996; Marom *et al.*, 1997; Walters *et al.*, 1997. 6, Tronche and Yaniv, 1992; Strandmann *et al.*, 1997. 7, Rausa *et al.*, 1997. 8, Ang *et al.*, 1993; Monaghan *et al.*, 1993; Sasaki and Hogan, 1993; Wu *et al.*, 1997; Rausa *et al.*, 1997. 9, Duncan *et al.*, 1994; Miquerol *et al.*, 1994. 10, Naya *et al.*, 1995; Lee *et al.*, 1995; Mutoh *et al.*, 1997. 11, Hagenbuchle *et al.*, 1997. 12, Plachov *et al.*, 1990; Walther and Gruss, 1991; Zannini *et al.*, 1992; Turque *et al.*, 1994; Sosa-Pineda *et al.*, 1997. 13, Arceci *et al.*, 1993; Kelley *et al.*, 1993; Laverriere *et al.*, 1994; Morrisey *et al.*, 1996, 1997. 14, Echelard *et al.*, 1993; Bitgood and McMahon, 1995; Bellusci *et al.*, 1996; Urase *et al.*, 1996; Apelqvist *et al.*, 1997; Hebrok *et al.*, 1998. 15, Bitgood and McMahon, 1995; Roberts *et al.*, 1995; Bellusci *et al.*, 1996. 16, Kaestner *et al.*, 1996. 17, Grindley *et al.*, 1997. 18, Yamasaki *et al.*, 1996; Bellusci *et al.*, 1997b.

duced first in the explants by the factors tested (Asashima *et al.*, 1991; Jones *et al.*, 1993). Thus, it is an important finding that endoderm can be specified independently of other tissues. The recently identified activin-responsive HMG domain-containing genes *Xsox17α* and *Xsox17β* (Hudson *et al.*, 1997), and the Vg1-responsive homeobox gene *Mixer* (Henry and Melton, 1998), can both independently convert ectoderm directly to endoderm without concurrent induction of mesoderm.

While it is currently difficult to undertake genetic studies in *Xenopus*, the cloning of the *pdx-1* homolog in the frog, called *XlHbox8* (Wright *et al.*, 1989), allowed standard molecular embryological techniques to be used to study the signaling pathway(s) involved in endoderm specification and regionalization. *XlHbox8* begins to be expressed very shortly after gastrulation in endodermal precursors occupying the dorsovegetal region of the embryo. The use of this very early marker to analyze regional specification in the endoderm obviated relying on more complex and much later events like epithelial ductule formation. The main findings of Gamer and Wright (1995) and Henry *et al.* (1996) were that a program for spatially restricting *XlHbox8* expression to the dorsal vegetal side was initiated and maintained in vegetal explants that were isolated prior to the onset of zygotic transcription and independent from the presence of mesoderm. This therefore provides strong evidence that endodermal regionalization is capable of occurring through molecular processes activated autonomously within the endodermal precursors.

Other experiments have identified a surprising amount of plasticity of developmental fate of amphibian endodermal precursors in response to their extracellular environment. Wylie *et al.* (1987) showed that endodermally fated vegetal blastomeres could change their fate, to mesodermal or neuroectodermal tissue, if transplanted to appropriate ectopic locations in host gastrulation stage embryos where they become exposed to different combinations of local intercellular signaling molecules and/or extracellular matrix (ECM). Adding to these concepts, some of our studies with *XlHbox8* suggested that its expression is activated in the dorsovegetally located endodermal precursors in an area where high local concentrations of activin/Vg1 signals overlap with relatively low concentrations of FGF (Gamer and Wright, 1995). These results are remarkably consistent with the effects of inhibiting transforming growth factor β (TGFβ)/activin-like signal transduction within endodermal cells, whereby they are converted first to mesodermal and then neural tissue, with increasing blockage of the signaling pathway (Hemmati-Brivanlou and Melton, 1994). Thus, the ability to manipulate preendodermal cells early in development, along with the ability to measure expression of a posterior foregut-

specific transcription factor rather than detecting much later events (tubule formation, presence of pancreatic zymogen granules), has focused attention on the cell-autonomous specification of anterior endoderm and on potential signaling molecules that may be involved. It remains to be seen to what degree autonomous endoderm specification occurs in other vertebrates, in which the endoderm progenitors are intermingled with cells destined for other fates, instead of beginning from a topologically separated location away from other germ layer precursors as in *Xenopus*.

B. Posterior Foregut Patterning

The posterior foregut produces distinct dorsal (pancreas) and ventral (pancreas, liver, and gall bladder) derivatives (Fig. 3). Recently, the specification of these fates in mammals, and the potential inductive/inhibitory influences of adjacent mesoderm, have been reanalyzed using a combination of powerful molecular biological techniques with the finesse of experimental embryological manipulations. These studies focus on many of the issues raised by earlier experimentation, but provide a substantial leap forward in defining the instructive capacity of mesodermal signal(s), the innate capacity of endoderm to respond to these signals, and the possible molecular nature of the signals.

1. Analysis of Hepatic Specification in Vitro

As described above, there is some evidence that embryonic cardiac tissue produces signals that induce the hepatic fate. This interaction was reassessed more recently using RT–PCR to detect the expression of the liver markers albumin and α-fetoprotein at very early embryonic stages without having to wait for complex morphogenetic processes (Gualdi *et al.*, 1996). In agreement with the observations of LeDouarin (1975) in the chick embryo, contact with cardiac mesoderm was shown to be important for hepatic specification in the mouse, since liver-specific gene expression could be activated only in endodermal explants that actually contacted mesodermal cells with the capacity to become beating heart tissue. These results also suggest that the signal from cardiac mesoderm is either membrane-bound, or not readily diffusible, and may be associated with the extracellular matrix.

However, a substantial advance over the previous chick studies was the finding that cardiac mesoderm could act instructively to induce liver-type gene expression patterns within more posterior endodermal epithelium (Gualdi *et al.*, 1996). "Ectopic induction" of hepatic gene expression by cardiac mesoderm was only observed when dorsal axial tissues, such as the notochord, were removed from the endodermal explant, suggesting that these tissues express dominant inhibitors of the induction of a liver fate. Extrapolating to the situation *in vivo*, one might infer that the dorsal axial mesoderm normally inhibits the formation of a liver diverticulum from the dorsal foregut. Indeed, when placed in contact with the ventral foregut endoderm, the source of endogenous hepatic precursors, the notochord substantially inhibited expression of liver markers (Fig. 7; Gualdi *et al.*, 1996). This contrasts the positive influence of the notochord in pancreas induction, as described below. The study of Gualdi *et al.*, is also noteworthy for the extraordinary technical achievement of *in vivo* footprinting of the albumin enhancer in explanted 9.5 dpc embryonic tissues, revealing DNA sequences bound by transacting factors, among which were identified an HNF3 protein (Gualdi *et al.*, 1996) and GATA4 (K. Zaret, 1997, personal communication).

How such inductive and inhibitory signals result in the fine details of posterior foregut patterning remains to be elucidated. It is possible that the domain of *pdx-1* expression (which includes the antral stomach, rostral duodenum, bile duct, and pancreas; detailed further below) endows a particular endodermal "compartment" with the potential to become pancreas, but that the interplay between opposing hepatic (cardiac) and pancreatic (notochord) influences results in the final pattern of posterior foregut derivatives. There is fascinating anecdotal evidence for a close lineage relationship between hepatic and pancreatic cell types in certain rodent models of pancreatic regeneration. Rats fed either a copper-deficient diet (Dabeva *et al.*, 1995) or a methionine-deficient diet with simultaneous daily injec-

tions of ethionine (Scarpelli and Rao, 1981) undergo pancreatic degeneration with substantial acinar atrophy. Pancreatic regeneration (in the presence of a carcinogen in Scarpelli and Rao, 1981) was associated with the appearance of cells morphologically resembling hepatocytes and expressing markers consistent with a liverlike character, including liver-enriched transcription factors such as HNF-1a, HNF-4, and C/EBP, as well as glycogen, albumin, and liver enzymes. Thus, under the right stimuli, pancreatic epithelial cells may be able to "transdifferentiate" toward a liver fate. These models may provide additional ways of estimating the degree of plasticity present in apparently terminally differentiated endodermal lineages. Moreover, an understanding of the molecular nature of these processes may prove useful in designing reagents to counteract degenerative pathophysiologies of endodermal organs.

2. Axial Signals in Pancreas Development

Wessels and Cohen (1967) suggested that dorsal axial signals that are restricted along the A/P axis might locally specify the location of the pancreatic primordia within the posterior foregut. The dorsally located notochord directly contacts the endodermal epithelium that is fated to produce the dorsal pancreatic bud throughout the time when it can be inferred that the pancreatic fate is regionally specified: at around the 13-somite stage (stage 11) in chick, or 10–12 somites (~9 dpc) in mice (Wessels and Cohen, 1967; Pictet *et al.*, 1972). If this is true, then the notochord may play two functions in patterning the endoderm: promotion of dorsal (pan-

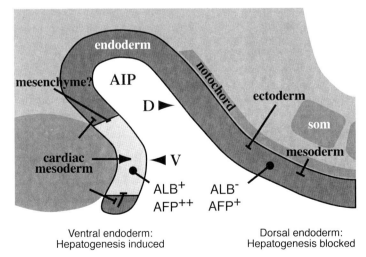

Figure 7 Positive and negative signals in hepatogenesis. Schematized 8.5 dpc mouse embryo showing tissue interactions inferred in the specification of hepatic primordia (lightly shaded region of endoderm) within the ventral endoderm. Cardiac mesoderm may induce localized hepatogenesis by positive signals (arrow) and/or by blocking (barred lines) inhibitory signals from other ventral tissues. In the dorsal–posterior region, inhibitory signals from axial mesoderm block the intrinsic capacity of the dorsal endoderm to express the liver fate. See text for further explanation. Anterior is to the left. AIP, Anterior intestinal portal; D, dorsal endoderm; V, ventral endoderm; som, somites; ALB, albumin; AFP, α-fetoprotein. (Adapted, with permission, from Gualdi *et al.*, 1996, © Cold Spring Harbor Laboratory Press.)

creatic) fates and inhibition of ventral (hepatic) fates, as already alluded to above.

Recently, the effect of extirpating the notochord on pancreas-specific gene expression and development was directly analyzed in chick embryos (Kim *et al.*, 1997). Removing it toward the end of the period of specification (stage 11) caused a complete down-regulation of *Isl1* and *Pax6* (both of which are already expressed at this stage in the chick) in the dorsal bud during the culture period and failure to activate *pdx-1* expression (which, in contrast to the mouse, is apparently activated in chick at stage 16, after *Isl1* and *Pax6*). Glucagon, insulin, and carboxypeptidase A (an exocrine marker) also were not expressed, while the general endodermal marker, HNF3β, was unaffected. Pancreatic gene expression in the ventral pancreatic bud, on the other hand, remained quite normal. Even in the absence of the notochord, the dorsal endoderm went through the first stages of its morphogenetic program, evaginating and invading the dorsal mesenchyme, and undergoing limited branching. Thus, some important aspects of endodermal development are intrinsically programmed prior to the stage at which the notochord was removed, but continued presence of the notochord (for how long is still unclear) is needed for full development to proceed, and to initiate or maintain tissue-specific gene expression. Preliminary experiments suggest that activin βB and FGF2 may participate in the endogenous notochord to endoderm signal (Hebrok *et al.*, 1998). Whatever the molecular nature of these notochord signals, they may, at least at this stage, serve only a permissive role since the notochord could not induce the expression of pancreas-specific genes within nonpancreatic endoderm (Kim *et al.*, 1997).

These results suggest that the gut endoderm of 10- to 12-somite embryos has already received some positional cues from other tissues, perhaps even as early as gastrulation, and that regionally restricted fates lie dormant within the endodermal layer. In agreement with this, we have carried out preliminary studies with chick embryos, in which newly gastrulated stage 4 (primitive streak stage) endoderm was explanted and cultured in the absence of mesoderm, and found that a region of specific Pdx-1 immunoreactivity develops within anterior but not posterior ventral endoderm (M. A. Gannon, D. Bader, and C. V. E. Wright, 1994, unpublished observations). This also suggests that regional patterns of predispositions toward particular developmental fates may be laid down surprisingly early within the endoderm in the chick, as well as in frogs, as described earlier. Moreover, our results indicate that the mesoderm may not always be required for *pdx-1* induction, in contrast to the chick dorsal pancreas, which needs axial mesoderm to induce *pdx-1* expression. An evolutionary angle on this question is provided by the observation that an endoderm–mesoderm interaction is required for

the expression of a *pdx-1* homolog, *Lox3*, in the leech (Weeden and Shankland, 1997). Future studies will undoubtedly reveal the degree to which core features of the mesoderm to endoderm signaling pathways leading to *pdx-1* induction are similar in different organisms.

C. Epithelial–Mesenchymal Interactions Revisited

As mentioned earlier, the pancreatic organ has evolved from distinct intestinal endocrine and exocrine cells into a complicated but structurally well-ordered association of cell types within a single organ in the vertebrates. In the mammalian pancreas, immature islets are associated with growing ductal tips, and mature as they bud off to form distinct bodies within the acini tissue. Islet maturation is concomitant with the distal outgrowth of the epithelium and its associated mesenchyme away from newly formed islets, lending support to the idea that islet maturation may be inhibited by close association with mesenchyme. Among the vertebrates, however, epithelial outgrowth and production of pancreatic ducts and acini is not always connected in the same way to the formation of islets, and in some cases there is good evidence that the pancreatic mesenchyme is not always required for islet formation. For example, in species like hagfish, the endocrine cells are localized to the proximal region of the pancreatic duct, rather than originating from distal ductal epithelium, indicating that the primary role of the mesenchyme is to induce exocrine outgrowth. The question of the requirement for mesenchyme in islet differentiation is similarly raised by the observation that islets can differentiate without pancreatic outgrowth in some strains of rats (Bendayan and Park, 1997).

Indeed, recent studies provide direct evidence that mesenchymal factors inhibit formation of mature islets in mammals, while promoting acinar differentiation (Gittes *et al.*, 1996). Although the endogenous factor responsible is not yet identified, it may be in the FGF family, since ectopic FGF4 expression under control of a *pdx-1* promoter/enhancer in transgenic mice results in an immature pancreatic epithelium containing very few differentiated endocrine cells (Miller *et al.*, 1997). Other descriptive studies in the mouse show that islet maturation is associated with down-regulation of *HNF-6*, which encodes a cut-homeodomain transcription factor (Rausa *et al.*, 1997). *HNF-6* and one of its targets, *HNF3β*, are expressed throughout the pancreas early in development, but their expression patterns diverge at 18 dpc, such that *HNF-6* is maintained in ductal epithelium, and *HNF3β* in mature islets and subsets of acinar cells (Wu *et al.*, 1997, Rausa *et al.*, 1997). This switch in transcription factor complement may be critical for the differentiation of mature islets from ductal epithelium. Whether *HNF-6* is truly incompatible with islet maturation might be determined by misexpression of this factor in islet cells.

V. Genetic Analysis of Endoderm Development

As indicated in Table I, the specific expression patterns of many transcriptional regulators and growth factors in the developing gut make them good candidates for playing a regulatory role in the growth and differentiation of the endoderm. The great advances made in targeted mutagenesis in the mouse have allowed genetic investigation of the role of several of these genes. For example, targeted mutagenesis of *pdx-1* shows that it is essential for pancreatic outgrowth and differentiation, but not for initial aspects of pancreatic specification (Jonsson *et al.*, 1994; Offield *et al.*, 1996; Ahlgren *et al.*, 1996). In addition, the analysis of mice homozygous for a null mutation in the LIM-homeodomain protein-encoding gene *Isl1* has revealed separate requirements for this factor in dorsal but not ventral pancreatic bud formation, and its common requirement for endocrine cell specification in both pancreatic buds (Ahlgren *et al.*, 1997). These mutations are discussed in detail below, and a model for their function in pancreatic development is presented in Fig. 11.

A. Posterior Foregut Development as a Model for Endodermal Regionalization and Organogenesis

The expression pattern of Pdx-1 in the posterior foregut and throughout pancreatic organogenesis (Figs. 6 and 9), as well as *in vitro* evidence implicating it as a potent transactivator of insulin gene transcription, identifies this factor as one of the key regulators of differentiation of this region of the endoderm. Targeted mutagenesis of the murine *pdx-1* gene was carried out in the Edlund

laboratory (Jonsson *et al.*, 1994) and our own (Offield *et al.*, 1996). Newborn *pdx-1* homozygous null mice specifically lacked a pancreas (Fig. 8) and pancreatic development was arrested at an early stage, with limited ductal branching (Jonsson *et al.*, 1994; Offield *et al.*, 1996; Ahlgren *et al.*, 1996). Mature acinar or islet cells were not formed, although transient populations of insulin-positive cells and longer-lived glucagon-positive cells were found in the partially outgrown ductular epithelium (Fig. 11). The relevance of these findings is underscored by the correlation of a null mutation in human *pdx-1* with pancreatic agenesis in a newborn (Stoffers *et al.*, 1997a). Paradoxically, despite its isolation as a putative insulin gene transactivator, the presence of insulin-positive cells in the homozygous *pdx-1* null mice clearly shows that insulin expression does not absolutely require Pdx-1. It is currently unclear whether these early *pdx-1*-independent endocrine cells represent genuine precursors of future islet cells, or if mature pancreatic islets are normally derived only from a subpopulation of *pdx-1*-expressing cells. A definitive answer to these questions awaits the development of precise, cell fate-independent lineage-tracing techniques that can be applied to the developing endoderm. Additionally, Pdx-1 expression in mature islets (particularly in β cells), may be important for coupling the level of insulin expression to stimuli such as an increased blood glucose level, or for maintaining other aspects of a β-cell fate. For example, while there is an apparently small decrease in islet size in *pdx-1* (+/−) adult mice, these mice are severely impaired in their capacity to clear a glucose load from the bloodstream (Dutta *et al.*, 1997).

In addition to the missing pancreas, other cell types within the posterior foregut that normally express *pdx-1* were also affected in *pdx-1* (−/−) mice. The pylorus and rostral duodenum were malformed, and the normal columnar epithelium of this part of the *pdx-1* expression domain was replaced by a cuboidal, bile duct-like epithelium (Offield *et al.*, 1996). The Brunner's glands that are located at the rostral end of the duode-

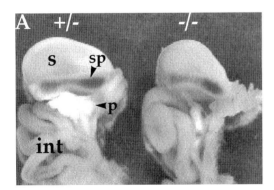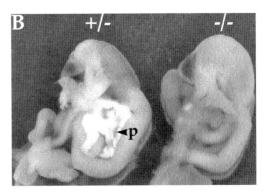

Figure 8 Pancreatic agenesis in *pdx-1* (−/−) animals. (A, B) Two different views of digestive organs from 18.5 dpc wild-type (left) and *pdx-1* (−/−) (right) embryos. The (−/−) embryos show absence of pancreatic tissue (white tissue in wild-type) and malformations at the stomach–duodenal junction not shown here (see Offield *et al.*, 1996). S, Stomach; sp, spleen; p, pancreas; int, intestines. (From Offield *et al.*, 1996, © Company of Biologists, Ltd.)

num, and which also form by outpocketing from the duodenal epithelium, were absent. In addition, the number of enteroendocrine cells expressing serotonin, secretin, and CCK in the rostral duodenal mucosa were greatly reduced. Among the antral stomach endocrine cells, subpopulations of gastrin, PYY, somatostatin, and serotonin-positive cells normally express Pdx-1. Gastrin cells are almost completely absent from the antral stomach in *pdx-1* (−/−) neonates and the number of PYY cells is greatly reduced. The corresponding large increase in serotonin cells suggests that *pdx-1* influences key points in the formation, proliferation, or maintenance of these cell types as they differentiate from a Pdx-1-positive precursor (Larsson *et al.*, 1996).

Further analysis of the *pdx-1* (−/−) mutant phenotype was facilitated by generating another mutant *pdx-1* allele in which a bacterial *lacZ* reporter cassette inserted into the locus is under the control of endogenous *pdx-1* regulatory sequences. This allowed the tracing of *pdx-1* expressing cells in heterozygous and ho-

mozygous null animals, and provided further critical insight into the role of *pdx-1* in pancreatic specification and differentiation. At 9.5 dpc, dorsal and ventral pancreatic bud formation occur on schedule in *pdx-1* null embryos (Fig. 9) (Offield *et al.*, 1996). By 11.5 dpc, however, defects in pancreatic development were conspicuous in the homozygous mutants. The ventral bud was no longer detectable, while the dorsal bud was severely arrested in its outgrowth, forming only a minimally branched ductule that remained in this state throughout embryogenesis and in newborns (Fig. 9). This limited outgrowth in the absence of terminal differentiation is similar in some respects to the degree of morphogenesis observed in the chick following notochord removal (Kim *et al.*, 1997), and provides evidence that the initial stages of pancreatic differentiation can occur without *pdx-1*.

Since the initial stages of dorsal and ventral bud formation are apparently normal in *pdx-1* (−/−) mutants, and the dorsal epithelium can initiate branching

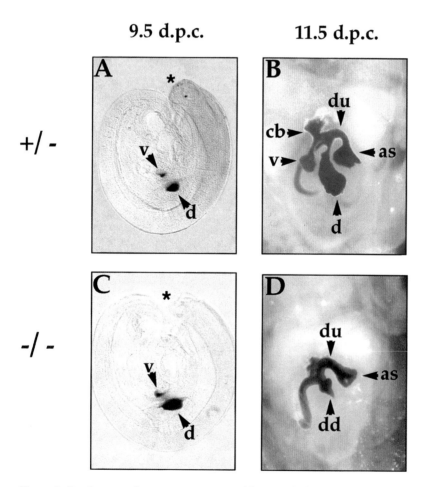

Figure 9 Development of pancreatic tissues in wild-type and *pdx-1* (−/−) mice visualized via a *pdx-1/lacZ* fusion null allele. (A, C) At 9.5 dpc, dorsal and ventral pancreatic buds, and the intervening duodenum at lower levels, express β-galactosidase in (+/−) and (−/−) embryos. Anterior is upper right (asterisk); heads were removed for genotyping. (B, D) By 11.5 dpc, the dorsal and ventral buds are larger in (+/−) embryos and *pdx-1* is expressed throughout the pancreatic buds, rostral duodenum, antral stomach, and common bile duct. In contrast, pancreatic buds are absent in (−/−) embryos and a unique ductal structure (dd) replaces the dorsal bud. The pattern of *pdx-1* expression is otherwise identical to wild type. d, Dorsal pancreatic bud; v, ventral pancreatic bud; as, antral stomach; du, duodenum; dd, dorsal ductule; cb, common bile duct.

morphogenesis and differentiation, *pdx-1* seems to be required for events subsequent to regional specification. Using epithelial–mesenchymal recombinants, Ahlgren *et al.* (1996) carried out important ancillary experiments to prove that the defect in *pdx-1* mutant embryos is restricted to the endoderm. *Pdx-1* (−/−) mesenchyme could provide normal levels of a permissive induction signal to promote morphogenesis and differentiation of wild-type endoderm, while mutant endoderm could not respond appropriately to wild-type mesenchyme. Thus, these results show clearly that the absence of *pdx-1* prevents the endoderm from responding even in the presence of the appropriate mesenchymal growth/proliferation signals, and shows that normal pancreatic differentiation is tightly linked to proliferation and morphogenesis.

B. *Isl1* in Pancreas Development

Several lines of evidence suggest that the signals inducing the formation of the dorsal and ventral pancreatic bud precursors are realized through different mechanisms. First, *pdx-1* expression is initiated first in the ventral endoderm of the still open gut tube prior to its expression in the dorsal endoderm (Fig. 6A–D). Second, the notochord is important for pancreatic morphogenesis and expression of *pdx-1* and exocrine and endocrine markers in dorsal, but not ventral, pancreas (Kim *et al.*, 1997). Third, the *Isl1* transcription factor affects differentiation of the dorsal and ventral pancreatic buds differently (Ahlgren *et al.*, 1997), as we will now describe.

    Isl1 was first identified as an islet transcription factor, although its expression pattern during embryogenesis is broader, being found also in motor neurons of the ventral neural tube, differentiating pancreatic endocrine cells (beginning at 9 dpc in the dorsal bud and 11 dpc in the ventral bud), and within the splanchnic mesenchyme adjacent to the duodenum and dorsal bud, but not in the mesenchyme contacting the ventral pancreatic bud (Fig. 10) (Thor *et al.*, 1991; Ahlgren *et al.*, 1997). *Isl1* expression coincides and colocalizes with islet hormone gene expression at the time that differentiation of each endocrine cell type can be detected, whereas expression of *Isl1* in the splanchnic mesoderm seems to be differentially regulated along its dorsal–ventral axis (Fig. 10). *Isl1* (−/−) embryos die around 9.5 dpc, most likely due to defects in vascular integrity and a disrupted dorsal aorta (Pfaff *et al.*, 1996). The effects of *Isl1* deficiency on pancreatic differentiation were analyzed in whole embryos and in cultured epithelial–mesenchymal recombinants. In *Isl1* (−/−) embryos, the dorsal pancreatic bud failed to form and this region showed much reduced levels of *pdx-1*, while ventral bud formation was unaffected. In addition, there was a depletion of mesenchyme surrounding the dorsal bud. Analysis of sectioned embryos and cultured explants

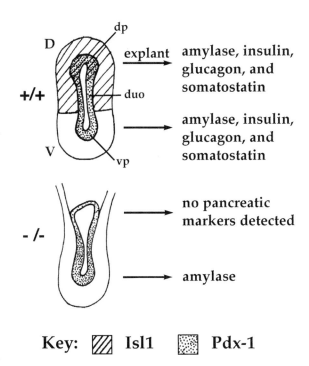

Key: ▨ Isl1   ▨ Pdx-1

Figure 10 *Isl1* (−/−) embryos lack dorsal pancreatic mesenchyme and differentiated islet cells. (*Top*) At 9.5 dpc, *Isl1* is expressed in splanchnic mesenchyme surrounding the duodenum and dorsal pancreatic bud, and within the dorsal pancreatic epithelium (hatched area). Expression of *Isl1* within the ventral pancreatic bud begins at 11.0 dpc (not shown) but is never detected in the mesenchyme surrounding the ventral bud. Pdx-1 (stippled area) is expressed throughout the endoderm of the posterior foregut. Cultured explants of either dorsal or ventral pancreatic endoderm plus mesenchyme differentiate into exocrine and endocrine cell types after 7 days.(*Bottom*) *Isl1* (−/−) embryos show an absence of mesenchyme surrounding the dorsal pancreatic bud, which has reduced levels of Pdx-1. Cultured explants of dorsal pancreatic endoderm fail to express any pancreatic markers, while explants of ventral endoderm plus mesenchyme differentiate into exocrine cells; islet endocrine cells were not detected. Recombinants of (−/−) dorsal endoderm with (+/+) mesenchyme result in exocrine cell differentiation from dorsal endoderm (not shown), but islets were never formed. These results suggest an independent requirement for Isl1 in dorsal mesenchyme and all pancreatic islets. (Based on Ahlgren *et al.*, 1997.) D, dorsal; V, ventral; dp, dorsal pancreatic bud; vp, ventral pancreatic bud; duo, duodenum.

failed to detect differentiated islets in either pancreatic bud, although the ventral outgrowth generated amylase-positive acinar cells. In cultured explants, wild-type mesenchyme could rescue the differentiation of exocrine tissue within the dorsal bud, but differentiated endocrine cells were still not formed from either bud. These studies suggest that it has two separate functions in pancreatic development (Fig. 10). First, it is required within the dorsal mesenchyme for the proper maintenance of this tissue, which signals to induce the subsequent outgrowth and differentiation of exocrine tissue from the associated dorsal endoderm. Second, *Isl1* expression within all islet endocrine cells is necessary for the differentiation of these cells in both pancreatic buds (Fig. 11).

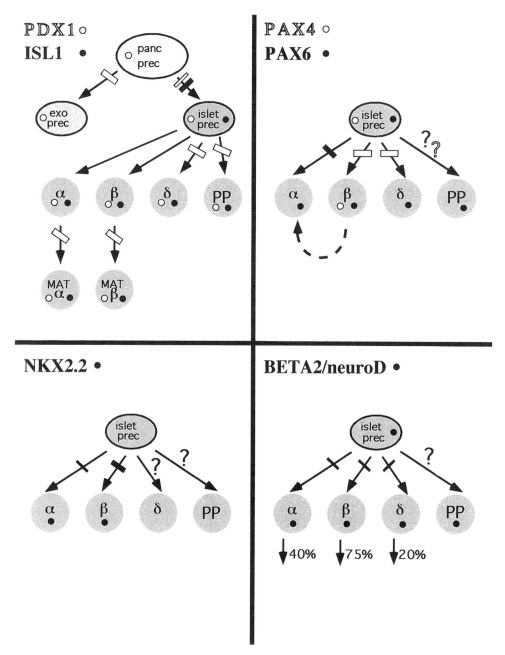

Figure 11 Role of transcription factors in islet endocrine cell differentiation. Nuclear expression of factors is marked by appropriately shaded circles in the diagrammatically represented cells. (*Top left*) *pdx-1* and *Isl1* act early within the islet cell lineage. Pdx-1 (white circles) is expressed very early within all pancreatic precursor cells. *pdx-1* (−/−) mutants have an early block in pancreatic development; exocrine cells are never detected (thick white bar) and there is a partial block in islet development (thin white bar). Although early, transient glucagon- or insulin-positive cells are seen, mature islets never develop (thick white bars). *Isl1* (black circles) is expressed in islet precursors and differentiated islets fail to form in its absence (black bar). (*Top right*) *Pax4* (white circles) and *Pax6* (black circles) are both expressed in subsets of islet precursors. *Pax4* expression is maintained in β-cells, while *Pax6* is expressed in all islet cells. Loss of Pax4 causes loss of both β- and δ-cells (white bars) and an increased number of α-cells, which may happen by conversion of β-cell precursors (dashed arrow). Loss of Pax6 specifically affects the formation of α-cells (black bar) although islet organogenesis is grossly disrupted. (*Bottom left*): Nkx2.2 is expressed in α- and β-cells (black dots). Loss of function results in an absence of β-cells (thick bar) and a reduced number of α-cells (thin bar). (*Bottom right*) *Beta2/neuroD* is expressed in islet precursors and all four islet cell types (black circles). Loss of function causes decreased numbers of differentiated islet cells (bars); the most dramatic decrease is a 75% reduction in β-cells. In each panel the thickness of the bar indicates the degree to which differentiation is blocked: wide bar: total, thin bar: partial. Question marks indicate markers that were not analyzed in these studies.

C. Additional Mouse Mutations with Relevance to Endodermal Patterning

For those genes that might have an early broad role in processes required for embryonic viability, and a later role in more specific tissues, the early lethality resulting from standard gene inactivation may preclude a genetic analysis of their later functions. On the other hand, although a gene product may normally play a role in many cells, a loss-of-function phenotype may be manifested within only a subset of these, due to the oft-cited phenomenon of functional compensation by related gene family members that are coexpressed in nonaffected tissues. Nevertheless, with this caveat in mind, several null mutations in genes encoding transcription factors show specific effects on different cell lineages in the pancreas, suggesting that the identity of each islet endocrine cell type is regulated through the combination of transcription factors expressed (summarized in Fig. 11). As indicated by the aforementioned studies on *pdx-1* (−/−) and *Isl1* (−/−) embryos, homozygous null embryos also provide a source of tissue with which to determine if and how the deleted gene is involved in the normal interactions between mesenchyme and endoderm. Next, we will briefly describe some of the recently generated mutations with endodermal phenotypes, beginning with those that affect the pancreas, although we refer the reader to the original papers and recent reviews (Sander and German, 1997; Madsen *et al.*, 1997) for deeper discussion of the implications that these studies hold for islet cell differentiation and function, possible links to diabetes, and potential therapies thereof.

1. bHLH Transcription Factors

The bHLH factor *Beta2/NeuroD* is expressed in pancreatic/intestinal endocrine cells and the brain, and can activate insulin gene transcription in cell lines (Naya *et al.*, 1995; Lee *et al.*, 1995). Beta2/NeuroD acts as a determination factor similar to the muscle-specific bHLH, MyoD, in that it can redirect early *Xenopus* ectoderm to a neural fate (Lee *et al.*, 1995). It is therefore intriguing that mice homozygous for *Beta2* deletions showed no defect in neural development (presumably due to genetic compensatory mechanisms), but instead developed severe diabetes and died a few days after birth (Naya *et al.*, 1997). The early lethality precluded an analysis of potential late-manifesting neuronal/behavioral defects. While pancreatic Pdx-1 seemed unaffected by the loss of *Beta2*, there was a 75% reduction in the number of insulin-positive cells (see Fig. 11) and islet morphogenesis was arrested, suggesting that Pdx-1 expression is not in itself enough for wild-type levels of insulin production or proper islet maturation. Intestinal secretin and CCK cells failed to develop and pancreatic acinar cells, although present and differentiated, had a disrupted apical-basal polarity.

Another bHLH gene, *p48*, is expressed specifically in the exocrine pancreas. Null mutant mice die shortly after birth, apparently with no pancreas (Hagenbuchle *et al.*, 1997). Amazingly, this deficiency was specifically confined to the exocrine pancreas, and functional endocrine cells were in fact present but mislocalized in the nearby spleen. This is the first identification of a factor specifically affecting the differentiation of acinar cells from pancreatic endoderm, and provides strong genetic evidence that mammalian endocrine cells can develop in the absence of exocrine tissue.

2. Pax Genes

Pax4 and *Pax6* encode related paired box-containing transcription factors (Pax6 also has a homeodomain) that are both expressed in the central nervous system (CNS) and developing pancreas (Walther and Gruss, 1991; Turque *et al.*, 1994; Sosa-Pineda *et al.*, 1997). Within the pancreatic epithelium, *Pax4* is probably expressed in subpopulations of islet precursor cells, subsequently becoming restricted to mature insulin-producing cells. *Pax6* is also expressed in subsets of prepancreatic epithelial cells, but its expression is maintained in all four (α, β, δ, and PP) mature islet endocrine subtypes (Sosa-Pineda *et al.*, 1997; St.-Onge *et al.*, 1997). Inactivation of *Pax4* results in increased numbers of glucagon-producing α-cells at the expense of β- and δ-cells, suggesting that *Pax4* is important for the generation of mature β- and δ-cells and regulates the α- versus β-/δ-cell lineage decision (Sosa-Pineda *et al.*, 1997). In embryos homozygous for the dominant *Pax6* mutation *Small eye* (*Sey^{Neu}*), which encodes a truncated Pax6 protein lacking the carboxy terminal transactivation domain, the numbers of all four types of endocrine cells are decreased, islet morphology is grossly disrupted, and insulin and glucagon transcription is greatly reduced (Sander *et al.*, 1997). In contrast, a conventional *Pax6* gene inactivation specifically affects the differentiation of only α-cells and the morphogenesis of mature islets (St.-Onge *et al.*, 1997). Not unexpectedly, mice lacking *Pax4* and *Pax6* fail to develop any mature endocrine cells (St.-Onge *et al.*, 1997). These results are summarized in Fig. 11.

3. Homeodomain Proteins

Several *Nkx*-like transcription factors are expressed in restricted patterns within the endoderm. *Nkx-2.2* and *Nkx-6.1* are both expressed in the ventral CNS and, within the pancreas, predominantly in insulin-producing β cells and subsets of glucagon-producing α cells (Rudnick *et al.*, 1994; Jensen *et al.*, 1996; Oster *et al.*, 1996; Sussel *et al.*, 1997). Homozygous *Nkx-2.2* mutant animals die within a few days after birth, lacking detectable insulin-producing cells and having reduced glucagon cell numbers (Fig. 11) (Sussel *et al.*, 1998). *Nkx-2.1* (also known as TTF-1 and T/ebp) was initially identified as a regulator of thyroid and lung gene expression. It is expressed at 8.5–9.5 dpc in the pharyngeal endoderm and later in the thyroid diverticulum derived from it, as well as in the bronchial epithelium and a restricted region of the forebrain, including the precursors of the posterior pituitary (Lazzaro *et al.*, 1991; Lints *et al.*, 1993). Homozygous null mutants die very shortly after birth, most likely because of the severely compromised lung development (Kimura *et al.*, 1996); mutant lungs have a very rudimentary bronchial tree and a lack of lung parenchyma. In addition, these animals completely lack both the thyroid gland and pituitary. Thus, *Nkx-2.1* most likely plays a role in specification of the thyroid and posterior pituitary primordia, transcriptional regulation in these tissues, and differentiation and organogenesis of the lung. Defects in thyroid development are also observed in mice homozygous for a null mutation in *Pax8* (Mansouri *et al.*, 1996). The relationship between Nkx2.1 and Pax8 in thyroid development is currently unclear.

Similar to the well-known nested expression patterns of HOX cluster genes in the paraxial mesoderm and neuroectoderm, the anterior expression boundaries of HOX cluster members within the lung and hindgut become progressively more posteriorly located in moving from 3′to 5′ within the cluster (Roberts *et al.*, 1995). In the lung, for example, *Hoxa-1* is expressed throughout the mesenchyme, while *Hoxa-4* is concentrated in distal mesenchyme at the tips of the buds (Mollard and Dziadek, 1997). *Hoxd-12* and *Hoxd-13*, the two most posteriorly expressed HoxD members, are both expressed in the mesenchyme of the terminal region of the hindgut, while *Hoxd-13* is also expressed in the endoderm of the rectum (Dollé *et al.*, 1991). Homozygous mutations in either gene lead to loss and/or dysmorphologies of certain rectal sphincter muscles such that this area structurally resembled more anterior regions of the large intestine (Kondo *et al.*, 1996).

The *caudal*-related homeobox genes are also expressed in the posterior of the vertebrate embryo; mutations in *Drosophila caudal* result in loss of posterior structures (Macdonald and Struhl, 1986). The three murine *caudal* homologs, *Cdx1*, *Cdx2/3*, and *Cdx4*, exhibit overlapping expression patterns in the posterior primitive streak, and also in posterior mesodermal and endodermal tissues of slightly later chick and mouse embryos (Duprey *et al.*, 1988; Gamer and Wright, 1993; Suh *et al.*, 1994; Laser *et al.*, 1996; Marom *et al.*, 1997). Although Cdx2 has been shown to interact with the promoters of intestine-specific genes (Suh *et al.*, 1994), preliminary indications are that its inactivation does not affect intestinal differentiation (perhaps due to functional compensation by other Cdx members), but causes homeotic transformations in the ribs and axial skeleton (Chawengsaksophak *et al.*, 1997). However, analysis of adult heterozygotes revealed a high incidence of intestinal polyps which occasionally became metaplastic, indicating that *Cdx2* may regulate epithelial proliferation and/or maintain the differentiated state. Homeotic alterations in axial skeletal identities were also observed in *Cdx1* homozygous null mutant mice, possibly related to its potential earlier role as a global regulator of Hox gene expression patterns (Subramanian *et al.*, 1995). There were no obvious defects in intestinal development in *Cdx1* (−/−) mutants and, so far, proliferation/differentiation defects are not reported for heterozygous adults.

4. Retinoic Acid Receptors

Compound null mutants for retinoic acid (RA) receptors α and β2 show defects in branching and severe lung hypoplasia (Mendelsohn *et al.*, 1994). In other systems, effects of RA on development have been correlated with alterations in expression of *Hox* genes (Conlon and Rossant, 1992; Kessel, 1992), and/or cell adhesion and extracellular matrix molecules (Zerlauth and Wolf, 1984; Vasios *et al.*, 1991; Jones *et al.*, 1993). Thus, the lung defects observed in RAR mutants could result from changes in *Hox* gene expression in the lung with subsequent changes in the ECM and cell adhesion.

5. GATA Factors

GATA factors, named for their preferred DNA consensus binding site, have been implicated in endodermal differentiation in *C. elegans*, *Drosophila*, and mammals. Mutations in the GATA family members *end-1* in *C. elegans*, and *serpent* in *Drosophila*, both result in the complete elimination of endodermally derived intestinal structures (Reuter, 1994; Rehorn *et al.*, 1996; Zhu *et al.*, 1997). In mammals, GATA4 is expressed in the ventral foregut as well as the developing myocardium (Arceci *et al.*, 1993). GATA4 function was recently analyzed by targeted mutagenesis (Kuo *et al.*, 1997; Molkentin *et al.*, 1997). Although cardiac myocyte differentiation occurred in mutant embryos, the heart primordia did not migrate to the ventral midline and the primitive heart tube failed to form. The mutant embryos died between 8.5 and 10.5 dpc. However, these early defects in cardiac development are thought to be secondary to the lack of growth and ventral folding of the foregut endoderm and anterior intestinal portal—a hypothesis supported by experiments using chimeric embryos composed of GATA4 (−/−) and ROSA26/GATA4 (+/+) embryonic stem (ES) cells (Narita *et al.*, 1997). Embryos in which most of the visceral and embryonic endoderm was composed of (+/+) cells, while the rest of the embryo was GATA4 (−/−), showed a rescue of heart and foregut defects, showing that endodermal expression of GATA4, rather than in the cardiac mesoderm itself, is required for ventral morphogenesis. These genetic analyses therefore support previous experimental evidence that the foregut endoderm specifically affects later aspects of heart differentiation and morphogenesis (Linask and Lash, 1988; Antin *et al.*, 1994; Sugi and Lough, 1994; Gannon and Bader, 1995), and furthermore suggest that reciprocal signaling events occur throughout development between differentiating cardiac mesoderm and prospective liver endoderm.

6. Winged Helix Transcription Factors

Fkh6 is expressed in the mesenchyme throughout the gastrointestinal tract (Kaestner *et al.*, 1996). Targeted disruption of *Fkh6* indicates that it acts to negatively regulate gastric and duodenal epithelial cell proliferation since homozygous mutants displayed a 4-fold increase in the number of dividing epithelial cells, distorting the gastric and intestinal tissue architecture (Kaestner *et al.*, 1997). In addition, *Fkh6* (−/−) mice showed slightly reduced expression of the bone morphogenic protein (BMP) genes *Bmp2* and *Bmp4* in the foregut endoderm. These results add to the *Isl1* knockout in providing clear genetic evidence that the mesenchyme plays an essential supportive role in endoderm development.

In contrast to *Fkh6*, the well-studied liver transcription factor *HNF3β* is broadly expressed in the embryo in notochord, floor plate, and endoderm (Ang *et al.*, 1993; Monaghan *et al.*, 1993; Sasaki and Hogan, 1993; Gualdi *et al.*, 1996). The *fork head* (*fkh*) gene of *Drosophila*, to which *HNF3β* is related, is essential for gut development in fly embryos (Weigel *et al.*, 1989; Hoch and Pankratz, 1996). *Fkh*-mutant embryos exhibit homeotic transformations of the ectodermally derived foregut to head ectoderm and degeneration of the endodermally derived midgut. The level of similarity between *fkh* and *HNF3β* suggested a conserved function in endoderm development, but the early embryonic lethality of mouse embryos lacking HNF3β (Ang and Rossant, 1994; Weinstein *et al.*, 1994) has so far precluded a specific study of the function of this gene in the endoderm alone. This situation will be remedied by the development of conditional gene inactivation as described below.

7. HNF4

An orphan member of the steroid hormone receptor family, HNF4, like HNF3, is enriched in adult liver and activates liver-specific genes (Duncan *et al.*, 1994), although it is expressed at lower levels in other tissues (Table I). The *Drosophila* homolog, encoded by the *HNF-4(D)* gene, is expressed in the midgut and fat bodies (the fly "liver"), suggesting a conserved function for this gene product in endoderm and liver differentiation (Zhong *et al.*, 1993). In mice, homozygous null embryos

die prior to hepatic differentiation because of loss of HNF4 function in the extraembryonic endoderm, where it is first strongly expressed (Chen *et al.*, 1994). (Interestingly, many "liver-specific" genes, including α-fetoprotein and apolipoproteins, are also first expressed in extraembryonic endoderm.) Thus, like HNF3β, the role of HNF4 in liver development remains unresolved. A role in the pancreas is suggested by the finding that mutations in this gene in humans are associated with maturity-onset diabetes of the young (MODY1; Yamagata *et al.*, 1996). In another exciting example of evolutionary conservation in endodermal development, two new members of the nuclear hormone receptor family have recently been identified in *C. elegans*. Interestingly one of these, *end-2*, also promotes endoderm development and lies within the same genomic region as the GATA factor *end-1*. The other gene is similar to HNF4 and is strongly expressed in the endoderm (J. Rothman, 1997, personal communication).

D. Limitations of Traditional Knockouts and Alternative Approaches

The data pointing to *pdx-1* as a critical regulator of pancreatic development and an insulin gene transactivator, thereby implicating it in the pathogenesis of diabetes, have stimulated enormous interest in precisely defining its function in later aspects of pancreatic development. The failure of pancreatic development in *pdx-1* (−/−) mutants prohibited the analysis of later functions for *pdx-1* in acinar and islet differentiation and function. For example, since transient insulin-positive cells were found in *pdx-1* (−/−) animals, a key issue that remains is whether and/or how *pdx-1* expression is linked to insulin gene regulation in mature bone fide β-cells.

To carry out such as analysis, one would ideally like to control the time and cell type of inactivation of *pdx-1*, or any other gene of interest. Such a fine level of control will facilitate the functional analysis of genes involved in several developmental stages or tissues, including those for which early inactivation causes early lethality or a severe block in organogenesis.

Technologies for conditional gene inactivation in the mouse have not only become a reality but are improving at a rapid pace (e.g., see Barinaga, 1994; Rajewsky *et al.*, 1996). One method for rendering a gene susceptible to future inactivation is summarized in Fig. 12. An artificial allele is generated by inserting 30-bp sequences called *lox*P sites, which are specifically recognized by bacteriophage P1 Cre recombinase (Sauer and Henderson, 1989; Sauer, 1993), into sites that do not disrupt gene function (e.g., in an intron) but flank functionally critical parts of the gene. This still physiologically normal "floxed" allele is then electroporated into ES cells and clones selected for the replacement of the endogenous locus with the floxed version. These cells

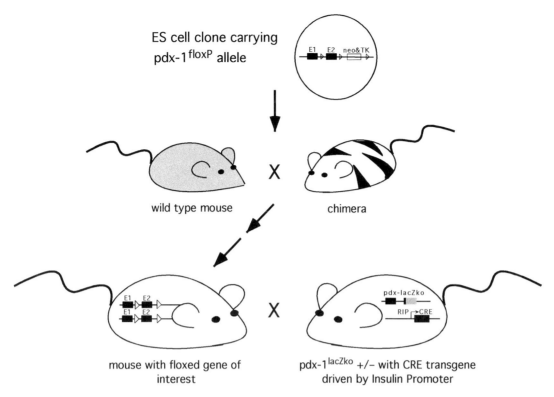

Figure 12 Strategy for conditional inactivation of the *pdx-1* gene. The gene is replaced with a "floxed" allele by homologous recombination in ES cells. Open triangles represent *lox*P sites. ES cells carrying correctly recombined alleles are then selected for removal of the neo/TK cassettes, and used to generate chimeras, which are mated to an appropriate strain allowing detection of germ-line transmission (see Hogan *et al.*, 1994). Offspring carrying the floxed allele are then crossed with specific Cre recombinase-producer transgenic lines. In this example, mice homozygous for the floxed *pdx-1* allele are crossed to a line expressing Cre under control of the rat insulin promoter (RIP). E1 and E2, exons 1 and 2; neo&TK, neomycin resistance and thymidine kinase genes for selection; pdx-lacZKO, null allele of *pdx-1* containing β-galactosidase gene.

are then used to generate mice carrying the floxed allele in all cells. The floxed mice can then be crossed with Cre-producer transgenic lines in which Cre expression is driven by different tissue-specific promoters/enhancers. Cre recombinase-dependent gene inactivation has been used successfully in the brain (Tsien *et al.*, 1996) and colon (Shibata *et al.*, 1997). Many laboratories are now generating conditional inactivation of their particular gene of interest, and as more tissue/cell type-specific promoters become available the number of such studies will undoubtedly increase.

Our laboratory, and others, have begun such studies on the *pdx-1* locus, replacing the endogenous locus by a floxed allele in which the homeodomain exon is flanked by *lox*P sites. Preliminary results using this strategy indicate that *pdx-1* functions in adult β cells to regulate insulin expression and maintain β cell identity (Ahlgren *et al.*, 1998). The ultimate goal is to assess the effect of inactivating *pdx-1* at different times during development, in each of the different pancreatic cell lineages (see Fig. 4), as well as in gastric and duodenal enteroendocrine and mucosal cells. These studies will directly address the requirement for *pdx-1* in the maintenance of islet β-cell function and insulin gene regulation, roles assigned on the basis of results in β-cell lines

(Ohlsson *et al.*, 1993; Peers *et al.*, 1994; Peshavaria *et al.*, 1994; Stoffers *et al.*, 1997b).

VI. Transgenic Analysis of Gene Regulation and Function

The identification of *pdx-1* as a very early competence factor for pancreatic differentiation, coupled with its evolutionarily conserved expression within the posterior foregut of frog, chick, and mouse, make it an ideal entry point for dissecting the mechanisms of endoderm regionalization. For example, how is *pdx-1* expression restricted to the posterior foregut, and what regulates its dynamic pattern of expression throughout development? It will be important to determine if there are separate cis-acting regulatory modules directing duodenal and pancreatic expression of *pdx-1*. However, a critical goal given its putative role as an insulin gene regulator is to identify the specific nucleotide sequences required for β-cell expression of *pdx-1*. Determining the factors regulating β-cell specification, differentiation, and insulin gene expression will have enormous implications regarding the ontogeny of diabetes and the development of therapeutic strategies.

Our analysis has concentrated on the characterization of *pdx-1* regulatory regions using *lacZ* reporter genes in transgenic mice. Beginning with large fragments of genomic DNA encompassing the *pdx-1* locus, sequences sufficient to recapitulate the endogenous *pdx-1* expression pattern were localized to a 4.5-kb segment of DNA lying in the 5′ upstream region of the gene, with no apparent contribution from the intron or 3′ sequences (L. Gamer, M. Offield, and C. V. E. Wright, 1996, unpublished observations). Working within the context of this 5′ region, it is now possible to define smaller, independent modules that direct *pdx-1* expression to different subsets of cells. Our collaborators in the laboratory of Roland Stein (Vanderbilt University), comparing β- and non-β-cell lines, have identified a 1-kb subfragment of this 5′ region that directs β-cell-specific expression of reporter genes (Wu *et al.*, 1997). The *in vivo* relevance of this fragment was demonstrated by finding that it can direct islet-specific expression in embryos and adults (see Fig. 13, and Wu *et al.*, 1997; M. Gannon and C. V. E. Wright, 1997, unpublished observations). In our experience, combining tissue culture cell studies with transient transgenic analysis is an efficient and thorough method of detecting vital regulatory elements. Further study of this islet-specific fragment will facilitate the identification and isolation of up-

stream regulatory factors. For example, we have identified HNF3β, which is expressed in islet cells *in vivo*, as a candidate factor for regulating expression by binding sequences within this region of *pdx-1* (Wu *et al.*, 1997).

Similar transgenic dissection of the intestinal fatty acid binding protein (*Fabpi*) gene, has identified sequences regulating positional information along the duodenum-to-colon and crypt-to-villus axes of the gut epithelium. *Fabpi* is expressed in an anterior–posterior gradient, being maximal in the rostral duodenum and decreasing toward the ileum and colon (Cohn *et al.*, 1992). The dissection has so far revealed the presence of region-specific activator and repressor elements, as well as cell type-specific regulatory sequences. In addition to the definition of tissue-specific regulatory elements leading to information on the trans-acting factors operating through these sequences (e.g., tissue-specific, ubiquitous, growth factor responsive), these analyses yield new promoter/enhancer cassettes to drive ectopic expression of regulatory molecules and growth factors to different subsets of cells in transgenic mice. For example, transgenes driven by cell- or germ layer-specific promoters/enhancers might test whether a regionally expressed transcription factor can alter or override the phenotype of cells in which that factor is normally not expressed. Similarly, growth factor misexpression, or

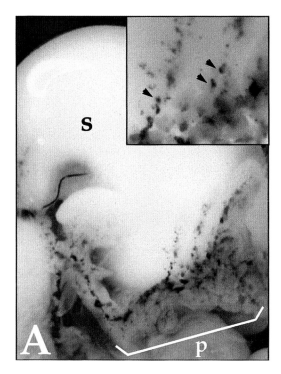
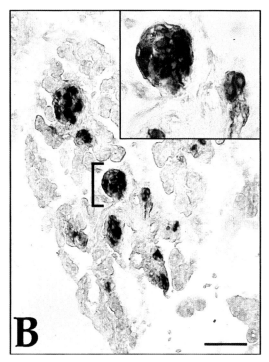

Figure 13 Transgenic analysis of *pdx-1* β-cell-specific cis-acting regulatory regions. (A) Digestive organs of 2-day-old neonatal pups carrying a β-galactosidase reporter transgene under control of 1 kb of *pdx-1* upstream sequence show reporter expression (blue staining) in islets within the pancreas (p) (bracket), but not within the exocrine cells, stomach (s), or duodenum. The inset shows islets at higher magnification (arrowheads). Expression is detected in both islets that are close to the ductal epithelium and those that are more separated from it. (B) Transgene expression colocalizes with insulin (dark brown) within the islets. The bracketed islet is shown at higher magnification in the inset. (Reprinted, with permission, from Wu *et al.*, 1997.) Bar = 45 μm.

tissue-specific inhibition of their endogenous signaling pathway using dominant negative interference, can address the involvement of these pathways in cell–cell or epithelial–mesenchymal interactions. For example, a metallothionein promoter-driven dominant-negative TGFβ type II receptor was used to block TGFβ signaling in selected epithelial cells including the liver, pancreas, intestine, and kidney. The transgenic pancreata showed increased proliferation, apoptosis, and fibrosis of acinar cells while islets were unaffected, suggesting that TGFβ signaling normally represses acinar cell proliferation and promotes their differentiation (Bottinger *et al.*, 1997). A selection of further studies in which transgenic mice were used to determine how certain factors act in the endoderm will be discussed below.

VII. Growth Factors in Lung and Pancreas Development

Several different families of intercellular signaling molecules have recently been implicated in specific aspects of lung morphogenesis, including FGFs, TGFβs, BMPs, and Sonic hedgehog (Shh) (see references in Bellusci *et al.*, 1997a,b). In mice carrying a transgene encoding a dominant-negative FGF receptor (DNFGF2R), in which expression was directed to the distal epithelium by the surfactant protein-C (SP-C) promoter/enhancer, lung morphogenesis was arrested at the stage of initial bifurcation of the tracheal tube (Peters *et al.*, 1994). A connection between FGF signaling and lung development in mammals is apparently conserved because mutations in the *breathless* or *branchless* genes, which encode an FGF receptor and ligand, respectively, result in defects in migration and branching of cells in the tracheal system of *Drosophila* (Lee *et al.*, 1996; Sutherland *et al.*, 1996). FGF10 is expressed in the distal mesenchyme of the developing mouse lung buds (Fig. 14)

(Bellusci *et al.*, 1997b) and, in agreement with the DNFGF2R experiment, elicits branching of lung epithelium *in vitro* (Bellusci *et al.*, 1997b). The potent embryonic patterning molecule, Shh, is expressed at low levels throughout the endodermal epithelium, with higher expression at the distal bud tips (Bitgood and McMahon, 1995; Bellusci *et al.*, 1996, 1997a; Urase *et al.*, 1996), while its receptor, Patched (*ptc*), and a downstream target, the zinc-finger transcription factor, *Gli*, are expressed mainly in mesenchyme adjacent to the distal buds (Bellusci *et al.*, 1997a; Grindley *et al.*, 1997). *Shh* overexpression in the distal epithelium, under control of the SP-C promoter, caused upregulated *Ptc* and *Gli* expression, and increased mesenchymal cell proliferation that prevented alveolar epithelium dilation. Thus, *Shh* may normally regulate some aspects of epithelial–mesenchymal interactions in the developing lung.

Although *Shh* is broadly expressed throughout the embryonic gut endoderm (Echelard *et al.*, 1993; Bitgood and McMahon, 1995), *Shh* null mutants, which have drastic developmental defects in several *Shh*-expressing tissues (Chiang *et al.*, 1996), lack some foregut derivatives while the midgut and hindgut appear normal (Chin Chiang, Vanderbilt University, 1997, personal communication).

In both chick and mouse, *Shh* expression is conspicuously absent from the dorsal and ventral pancreatic bud endoderm, where *pdx-1* is expressed at high levels (Apelqvist *et al.*, 1997; Hebrok *et al.*, 1998; see Fig. 15). The exclusion of Shh from both buds, but not from the intervening duodenal epithelium, led to the hypothesis that the initiation of pancreatic outgrowth requires *Shh* down-regulation. This was recently tested by using a *pdx-1* cassette to ectopically express Shh in the early pancreatic epithelium (Apelqvist *et al.*, 1997). The dramatic results of this manipulation are summarized in Fig. 15. Similar to its effect on lung endoderm, the ectopic Shh induced adjacent mesenchymal ex-

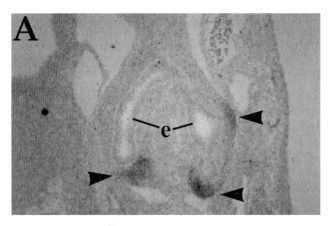
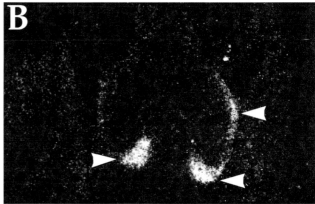

Figure 14 *FGF10* is expressed in lung mesenchyme adjacent to the distal tips of growing endodermal buds. (A, B) are bright- and dark-field views, respectively, of an 11.5 dpc embryo section in which *FGF10* expression (arrows) was detected by *in situ* hybridization. E, Endodermal epithelium.

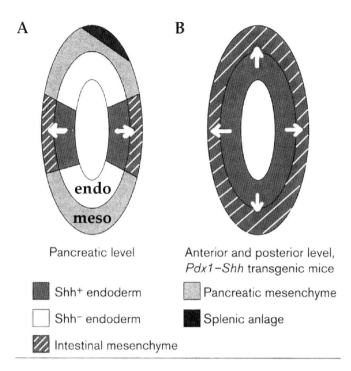

Figure 15 Model of the role of Shh in posterior foregut differentiation. (A) At the level of the pancreas, Shh expression is in the duodenal endoderm (solid dark gray) and is excluded from dorsal and ventral pancreatic buds (white). Shh in the endoderm induces (indicated by arrows) the adjacent mesenchyme to adopt an intestinal smooth muscle fate (hatched mesenchyme). Mesenchyme adjacent to Shh (−) endoderm differentiates into pancreatic (light gray) and splenic (black) mesenchymes. (B) In regions anterior and posterior to the pancreatic endoderm in wild-type mice, and now ectopically in the region of the pancreas in *pdx-1*-Shh mice, Shh is expressed throughout the endoderm, resulting in induction of intestinal smooth muscle-like differentiation (hatched mesenchyme) over a much wider area of the surrounding mesoderm (arrows). The spleen is absent in transgenic animals. (Adapted, with permission, from Apelqvist *et al.*, 1997.)

pression of *ptc*, which is normally restricted to the duodenal mesenchyme. Spleen development was blocked and pancreatic mesoderm seemed to be replaced by a smooth muscle layer resembling that around the intestinal epithelium. Of relevance to this finding is the previous observation that Shh-producing intestinal epithelium can induce the expression of smooth muscle markers in fibroblasts *in vitro* (Kedinger *et al.*, 1990). The extended expression of Shh in the early pancreatic endoderm may therefore impart at least some aspects of duodenal character to the endoderm, at least as judged by its ability to affect the differentiation of the subjacent mesenchyme. The pancreatic epithelium of *pdx*-Shh transgenics lacked properly organized islets or acinar structures, although glucagon-, insulin-, and amylase-positive cells were all present. In addition, mucin-positive cells that are normally restricted to the duodenum were found in the pancreas. It will be important to determine whether the effect of the *pdx-1*-Shh transgene on the pancreatic epithelium results directly from Shh effects within the pancreatic primordia, or is caused indirectly by alterations in the character of the associated mesenchyme.

These studies provide molecular evidence that signals from the endoderm affect the development of the mesenchyme, and again emphasize the reciprocal na-

ture of endoderm–mesoderm interactions during development of the respiratory and digestive tubes and associated organs. In addition, the up-regulation by Shh of *Ptc* in lung and pancreas, and *Gli* in the lung, demonstrates that the Shh signaling pathway in endodermal organogenesis shares features associated with other model systems of inductive interactions such as the limb (Marigo *et al.*, 1996; Vortkamp *et al.*, 1996).

VIII. Major Questions and Challenges Remaining

Our knowledge of the regulatory factors and intercellular signaling molecules expressed in the posterior foregut endoderm and/or mesenchyme has increased greatly over the last few years. However, a much larger collection of precisely characterized early markers distributed over the entire endoderm will be necessary to allow a more complete analysis of this germ layer. Some of the markers to be cloned in the future will undoubtedly add to the current list of transcription factors and signaling molecules that regulate endodermal development. But, the generation of a catalog of cell surface markers, combinations of which might form a unique identification code for particular cells, will be very useful as we strive to develop a rigorous understanding of cell lineage during normal embryogenesis, under conditions of regeneration, or after interference with developmental processes by gene inactivation. Furthermore, such markers might eventually allow the development of characterization and purification strategies leading to the isolation of, for instance, pancreatic islet precursor or stem cells (as was done for the hematopoietic stem cell). In this light, the classic (often called "risky") fishing expedition, and information from genome sequencing projects and functional screens, are likely to continue to be highly fruitful avenues in attacking endodermal specification, differentiation, morphogenesis, and organogenesis.

A major challenge ahead, with relevance not only to endodermal patterning and differentiation but also to developmental processes in general, is to determine how a multitude of dynamically changing signals received by a cell (including secreted molecules, cell surface contacts, extracellular matrix and even, in some cases, nutritional cues) are interpreted and integrated intracellularly to result in specific alterations in behavior, physiology, or architecture. This will also entail substantial progress in the understanding of the biochemistry of signal transduction, especially as it relates to a three-dimensional environment rather than under *in vitro* conditions, and how this transcriptionally or functionally activates different batteries of transcriptional regulators.

Moreover, there is currently very little understanding of the mechanisms by which cohorts of cells are programmed en masse to embark on complex morphogenetic processes (such as rolling up, budding, branching). In a complementary fashion, there is a huge amount to be learned about how specific progenitor/pioneer cells recruit surrounding cells to initiate specialized morphogenesis and differentiation programs. In this regard, it may be possible to establish cell culture models that mimic certain aspects of endodermal or pancreatic development in a reductionist manner, and to use these to study lineage choices in response to extracellular and environmental signals. Our capacity to compare rapidly the changes in gene expression profiles between different developmental stages or in response to different external stimuli will probably be revolutionized by "DNA chip technology," in which vast numbers of specific sequences arrayed in a known pattern on a small detector can be used to measure very accurately and selectively the relative abundances of mRNAs expressed in tissues or cells of interest.

The paradigm in which individual "master regulators" act to control cell or tissue identity, which was popularized with the identification of molecules like the dominant muscle determinant MyoD, although an attractive hypothesis, will probably be found to be too simplistic, at least for complex tissues like the pancreas.

It is more likely that we will have to understand how titrating the levels of various transcription factors, during prescribed windows of developmental time, acts to influence cell fate choice, and to what extent stochastic principles govern these events. Key insights into the combinatorial nature of transcriptional control and tissue identity may come from a cross-species comparison of the regulation of critical early-acting genes such as *pdx-1*, in which conserved sequences/elements might be found to bind conserved upstream trans-acting factors.

A phylogenetic analysis of the regulation of *pdx-1* and other conserved factors will also give clues to the ancestral role for these factors in endoderm patterning. Major questions here include the following. What is the positional information that is used to initiate the regionalization process, and will information relevant to vertebrates be gained from analyzing more primitive organisms? For example, while the leech lacks a pancreatic organ, it may be possible to identify the mesodermal signals that induce regionalized expression of its *pdx-1* homologs, generating information regarding the evolution of signaling mechanisms in endodermal segmentation. From this same standpoint, it could be valuable to search for *pdx-1* homologs in other forerunners of the vertebrates such as the cephalochordate, *Amphioxus*, and to consider information arising from extensive genetic analyses of lower, unsegmented oragnisms such as *C. elegans*. What sorts of information might one gain from such studies? An entirely speculative view might be that factors specifying endoderm in general or broad domains would be conserved, but that more cell type-specific factors, such as *Pax 4* and *Pax6*, may be used differently in different animals. Whatever the case, a powerful argument can be made that a broad-based approach to endoderm development will shed light on the regulatory complexities and hierarchies that were established during the evolution of the metazoan respiratory and digestive tract.

The explosion of interest in the zebrafish model for vertebrate embryogenesis may soon provide an abundance of mutational data relevant to endodermal patterning and organogenesis. For example, one recent screen resulted in the identification of nine mutations affecting the development of endodermal organs (Pack et al., 1996). Although the underlying genetic lesions are still unknown, they affect distinct endodermal tissues. Several mutations perturb foregut differentiation: one results in lack of an esophagus, two cause degeneration of the liver, and four affect pancreatic exocrine cells together with anterior intestine development, although differentiated islet endocrine cells occur at their normal location adjacent to the proximal pancreatic duct. As part of this large-scale screen, mutants with heart defects were also isolated. Two of these, *miles apart* and *bonnie and clyde* result in cardia bifida and a failure of ventral body morphogenesis (Chen et al.,

1996; Stainier *et al.*, 1996), similar to the phenotype of GATA4 (−/−) mice described earlier. It will be interesting to see if these zebrafish phenotypes also result from mutations in the GATA4 pathway. Future screens that focus on the development of the pancreas, or other endodermal organs, may uncover additional regulatory genes in an unbiased manner.

The genetic analysis possible in fish, together with techniques of laser ablation and accurate fate mapping, may also help to clarify the lineage relationships between different pancreatic cell types. The phenotype of *Pax4* homozygous null mice (missing β cells, increased α cells) suggests the possibility of a conversion between the fates of islet cells following drastic alteration of their transcription factor complements. But, it is also possible that a specific precursor cell type overproliferated in the absence of inhibitory signals from the deleted β cells. To distinguish clearly between these possibilities, strict lineage-tracing methods must be developed, preferably that will allow cells to be unambiguously identified independently of their transient or ultimate fate. This would also impact the identification of the putative pancreatic stem cell, and address the question of whether single committed cells act to initiate islet formation, or if a community effect is involved. Conditional gene inactivation promises to produce much incisive information on the role of individual factors in single cell types, as well as their effects on morphogenesis and differentiation as a whole. Nevertheless, more efficient methods of expressing Cre and achieving recombination in precisely characterized groups of cells, together with rigorous methods of tracing those cells that undergo the recombination event, will allow future experimentation to arrive at more definitive conclusions.

Finally, the ultimate challenge is to use information gathered from all the various avenues of study to design therapeutic reagents for degenerative diseases in which endoderm differentiation and/or morphogenesis is destabilized and misregulated such as diabetes, cancer, Crohn's disease, and liver fibrosis.

Acknowledgments

Due to space restrictions, we regret that we were unable to comprehensively represent the enormous number of contributions made by all researchers in the field of endoderm development. We thank Brigid Hogan, Roland Stein, Saverio Bellusci, and members of the Wright laboratory for critical reading of the manuscript. We also thank Chin Chiang, Rob Costa, Doug Melton, Joel Rothman, and Jeffrey Whitsett for communicating results prior to publication. This work was supported by NIH Grant DK42502 to C.V.E.W., and postdoctoral fellowships from the VUMC Molecular Endocrinology Training Grant (NIHT32 DK07563) and the Juvenile Diabetes Foundation International (397019) to M.G.

References

Ahlgren, U., Jonsson, J., and Edlund, H. (1996). The morphogenesis of the pancreatic mesenchyme is uncoupled from that of the pancreatic epithelium in IPF1/PDX1-deficient mice. *Development* (*Cambridge, UK*) **122**, 1409–1416.

Ahlgren, U., Pfaff, S. L., Jessell, T. M., Edlund, T., and Edlund, H. (1997). Independent requirement for Isl1 in formation of pancreatic mesenchyme and islet cells. *Nature* (*London*) **385**, 257–260.

Ahlgren, U., Jonsson, J., Jonsson, L., Simu, K., and Edlund, H. (1998). β-cell-specific inactivation of the mouse *Ipf1/Pdx1* gene results in loss of the β-cell phenotype and maturity onset diabetes. *Genes and Dev.* **12**, 1763–1768.

Alescio, T., and Cassini, A. (1962). Induction *in vitro* of tracheal buds by pulmonary mesenchyme grafted on tracheal epithelium. *J. Exp. Zool.* **150**, 83–94.

Alpert, S., Hanahan, D., and Teitelman, G. (1988). Hybrid insulin genes reveal a developmental lineage for pancreatic endocrine cells and imply a relationship with neurons. *Cell* (*Cambridge, Mass.*) **53**, 295–308.

Ang, S.-L., and Rossant, J. (1994). *HNF3β* is essential for node and notochord formation in mouse development. *Cell* (*Cambridge, Mass.*) **78**, 561–574.

Ang, S.-L., Wierda, A., Wong, D., Stevens, K. A., Cascio, S., Rossant, J., and Zaret, K. S. (1993). The formation and maintenance of the definitive endoderm lineage in the mouse: Involvement of HNF3/*forkhead* proteins. *Development* (*Cambridge, UK*) **119**, 1301–1315.

Antin, P. B., Taylor, R. G., and Yatskievych, T. (1994). Precardiac mesoderm is specified during gastrulation in quail. *Dev. Dyn.* **200**, 144–154.

Apelqvist, A., Ahlgren, U., and Edlund, H. (1997). Sonic hedgehog directs specialised mesoderm differentiation in the intestine and pancreas. *Curr. Biol.* **7**, 801–804.

Arceci, R. J., King, A. A. J., Simon, M. C., Orkin, S. H., and Wilson, D. B. (1993). Mouse GATA-4: A retinoic acid-inducible GATA-binding transcription factor expressed in endodermally derived tissues and heart. *Mol. Cell. Biol.* **13**, 2235–2246.

Asashima, M., Uchiyama, H., Nakano, H., Eto, Y., Ejima, D., Sugino, H., Davids, M., Plessow, S., Born, J., Hoppe, P., Tiedemann, H., and Tiedemann, N. (1991). The vegetalizing factor from chicken embryos: Its EDF (activin A)-like activity. *Mech. Dev.* **34**, 135–141.

Barinaga, M. (1994). Knockout mice: Round two. *Science* (*Res. News*) **265**, 26–28.

Bellusci, S., Henderson, R., Winnier, G., Oikawa, T., and Hogan, B. L. M. (1996). Evidence from normal expression and targeted misexpression that Bone Morphogenetic Protein-4 (Bmp-4) plays a role in mouse embryonic lung morphogenesis. *Development* (*Cambridge, UK*) **122**, 1693–1702.

Bellusci, S., Furuta, Y., Rush, M. G., Henderson, R., Winnier, G., and Hogan, B. L. M. (1997a). Involvement of Sonic hedgehog (*Shh*) in mouse embryonic lung growth and morphogenesis. *Development* (*Cambridge, UK*) **124**, 53–63.

Bellusci, S., Grindley, J., Emoto, H., Itoh, N., and Hogan, B. L. M. (1997b). Fibroblast growth factor 10 (FGF10) and branching morphogenesis in the embryonic mouse lung. *Development* **124**, 4867–4878.

Bendayan, M., and Park, I.-S. (1997). Extrapancreatic islets of Langerhans: Ontogenesis and alterations in diabetic condition. *J. Endocrinol.* **153**, 73–80.

Bitgood, M. J., and McMahon, A. P. (1995). *Hedgehog* and *bmp* genes are coexpressed at many diverse sites of cell–cell interaction in the mouse embryo. *Dev. Biol.* **172**, 126–138.

Bottinger, E. P., Jakubczak, J. L., Roberts, I. S. D., Mumy, M., Hemmati, P., Bagnall, K., Merlino, G., and Wakefield, L.M. (1997). Expression of a dominant-negative mutant TGF-β type II receptor in transgenic mice reveals essential roles for TGF-β in regulation of growth and differentiation in the exocrine pancreas. *EMBO J.* **16**, 2621–2633.

Brand, T., Andree, B., Schneider, A., Buchberger, A., and Arnold, H.-H. (1997). Chicken *Nkx2-8*, a novel homeobox gene expressed during early heart and foregut development. *Mech. Dev.* **64**, 53–59.

Chawengsaksophak, K., James, R., Hammond, V. E., Kontgen, F., and Beck, F. (1997). Homeosis and intestinal tumors in Cdx2 mutant mice. *Nature (London)* **386**, 84–87.

Chen, J.-N., Haffter, P., Odenthal, J., Vogelsang, E., Brand, M., Van Eden, F. J. M., Furutani-Seiki, M., Granato, M., Hammerschmidt, M., Heisenberg, C.-P., Jiang, Y.-J., Kane, D. A., Kelsh, R. N., Mullins, M. C., and Nusslein-Volhard, C. (1996). Mutations affecting the cardiovascular system and other internal organs in zebrafish. *Development (Cambridge, UK)* **123**, 293–302.

Chen, W. S., Manova, K., Weinstein, D. C., Duncan, S. A., Plump, A. S., Prezioso, V. R., Bachvarova, R. F., and Darnell, J. E., Jr. (1994). Disruption of the HNF-4 gene, expressed in visceral endoderm, leads to cell death in embryonic ectoderm and impaired gastrulation of mouse embryos. *Genes Dev.* **8**, 2466–2477.

Chiang, C., Litingtung, Y., Lee, E., Young, K. E., Corden, J. L., Westphal, H., and Beachy, P. A. (1996). Cyclopia and defective axial patterning in mice lacking *Sonic hedgehog* gene function. *Nature (London)* **383**, 407–413.

Cohn, S. M., Simon, T. C., Roth, K. A., Birkenmeier, E. H., and Gordon, J. I. (1992). Use of transgenic mice to map *cis*-acting elements in the intestinal fatty acid binding protein gene (Fabpi) that control its cell lineage-specific and regional patterns of expression along the duodenal–colonic and crypt-villus axes of the gut epithelium. *J. Cell. Biol.* **119**, 27–44.

Conlon, R.A., and Rossant, J. (1992). Exogenous retinoic acid rapidly induces anterior ectopic expression of murine *Hox-2* genes *in vivo*. *Development (Cambridge, UK)* **116**, 357–368.

Cui, Y., Tian, Q., and Christian, J. L. (1996). Synergistic effects of Vg1 and Wnt signals in the specification of dorsal mesoderm and endoderm. *Dev. Biol.* **180**, 22–34.

Dabeva, M. D., Hurston, E., and Shafritz, D. A. (1995). Transcription factor and liver-specific mRNA expression in facultative epithelial progenitor cells of liver and pancreas. *Am. J. Pathol.* **147**, 1633–1648.

De Krieger, R. R., Aanstoot, H. J., Kranenburg, G., Reinhard, M., Visser, W. J., and Bruining, G. J. (1992). The midgestational human fetal pancreas contains cells co-expressing islet hormones. *Dev. Biol.* **153**, 368–375.

Dolle, P., Izpisua-Belmonte, J.-C., Boncinelli, E., and Duboule, D. (1991). The *Hox-4.8* gene is located at the 5′ extremity of the *Hox-4* complex and is expressed in the most posterior parts of the body during development. *Mech. Dev.* **36**, 3–13.

Duncan, S. A., Manova, K., Chen, W. S., Hoodless, P., Weinstein, D. C., Bachvarova, R. F., and Darnell, J. E., Jr. (1994). Expression of transcription factor HNF-4 in the extraembryonic endoderm, gut, and nephrogenic tissue of the developing mouse embryo: HNF-4 is a marker for primary endoderm in the implanting blastocyst. *Proc. Natl. Acad. Sci. U.S.A.* **91**, 7598–7602.

Duprey, P., Chowdhury, K., Dressler, G. R., Balling, R. Simon, D., Guenet, J. L., and Gruss, P. (1988). A mouse gene homologous to the *Drosophila* gene *caudal* is expressed in the epithelial cells from the embryonic intestine. *Genes Dev.* **2**, 1647–1654.

Dutta, S., Bonner-Weir, S., Wright, C. V. E., and Montminy, M. (1997). Altered glucose tolerance in *pdx-1* deficient mice. *Nature (London)* **392**, 560.

Echelard, Y., Epstein, D. J., St.-Jaques, Y., Shen, L., Mohler, J. McMahon, J., and McMahon, A. P. (1993). Sonic hedgehog, a member of a family of putative signalling molecules, is implicated in the regulation of CNS polarity. *Cell (Cambridge, Mass.)* **75**, 1417–1430.

Eyal-Giladi, H., and Kochav, S. (1976). From cleavage to primitive streak formation: A complementary normal table and a new look at the first stages of the development of the chick. *Dev. Biol.* **49**, 321–337.

Fell, P.E., and Grobstein, C. (1968). The influence of extra-epithelial factors on the growth of embryonic mouse pancreatic epithelium. *Exp Cell Res.* **53**, 301–304.

Fukushige, T., Schroeder, D. F., Allen, F. L., Goszcynski, B., and McGhee, J. D. (1996). Modulation of gene expression in the embryonic digestive tract of *C. elegans*. *Dev. Biol.* **178**, 276–288.

Gaiano, N., Amsterdam, A., Kawakami, K., Allende, M., Becker, T., and Hopkins, N. (1996). Insertional mutagenesis and rapid cloning of essential genes in zebrafish. *Nature (London)* **383**, 829–832.

Gamer, L.W., and Wright, C.V.E. (1993). Murine *Cdx-4* bears striking similarities to the *Drosophila caudal* gene in its homeodomain sequence and early expression pattern. *Mech. Dev.* **43**, 71–81.

Gamer, L., and Wright, C. V. E. (1995). Autonomous endodermal determination in *Xenopus*: Regulation of expression of the pancreatic gene *XlHbox8*. *Dev. Biol.* **171**, 240–251.

Gannon, M., and Bader, D. (1995). Initiation of cardiac gene expression occurs in the absence of anterior endoderm. *Development (Cambridge, UK)* **121**, 2439–2450.

Gilbert, S. F. (1994). "Developmental Biology," 4th Ed. Sinauer, Sunderland, Massachusetts.

Gittes, G. K., and Rutter, W. J. (1992). Onset of cell-specific gene expression in the developing mouse pancreas. *Proc. Natl. Acad. Sci. U.S.A.* **89**, 1128–1132.

Gittes, G. K., Galante, P. E., Hanahan, D., Rutter, W. J., and Debas, H. T. (1996). Lineage-specific morphogenesis in the developing pancreas: Role of mesenchymal factors. *Development (Cambridge, UK)* **122**, 439–447.

Goldin, G. V., and Wessels, N. K. (1979). Mammalian lung development: The possible role of cell proliferation in the formation of supernumerary tracheal buds and branching morphogenesis. *J. Exp. Zool.* **208**, 337–346.

Golosow, N., and Grobstein, C. (1962). Epitheliomesenchymal interaction in pancreatic morphogenesis. *Dev. Biol.* **4**, 242–255.

Grindley, J. C., Bellusci, S., Perkins, D., and Hogan, B. L. M. (1997). Evidence for the involvement of the *Gli* gene family in embryonic mouse lung development. *Dev. Biol.* **188**, 337–348.

Gualdi, R., Bossard, P., Zheng, M., Hamada, Y., Coleman, J. R., and Zaret, K. S. (1996). Hepatic specification of the gut endoderm *in vitro*: Cell signaling and transcriptional control. *Genes Dev.* **10**, 1670–1682.

Guazzi, S., Price, M., DeFelice, M., Damante, G., Mattei, M.-G., and DiLauro, R. (1990). Thyroid nuclear factor 1 (TTF-1) contains a homeodomain and displays a novel DNA binding specificity. *EMBO J.* **9**, 3631–3639.

Guz, Y., Montminy, M. R., Stein, R., Leonard, J., Gamer, L. W., Wright, C. V. E., and Teitelman, G. (1995). Expression of murine STF-1, a putative insulin gene transcription factor, in β cells of pancreas, duodenal epithelium and pancreatic exocrine and endocrine progenitors during ontogeny. *Development (Cambridge, UK)* **121**, 11–18.

Haffen, K., Kedinger, M., and Simon-Assmann, P. (1987). Mesenchyme-dependent differentiation of epithelial progenitor cells in the gut. *J. Pediatr. Gastroenterol. Nutr.* **6**, 14–23.

Hagenbuchle, O., Krapp, A., Knofler, M., and Wellauer, P. K. (1997). The bHLH protein p48, a DNA-binding subunit of PTF1, acts as a morphogen during pancreas development. *Exp. Clin. Endocrinol. Diabetes* **105**, A9.

Hamburger, V., and Hamilton, H.L. (1951). A series of normal stages in the development of the chick embryo. *J. Morphol.* **88**, 49–67.

Hatada, Y., and Stern, C. D. (1994). A fate map of the epiblast of the early chick embryo. *Development (Cambridge, UK)* **120**, 2879–2889.

Hebrok, M., Kim, S. K., and Melton, D. A. (1998). Notochord repression of endodermal Sonic hedgehog permits pancreas development. *Genes and Dev.* **12**, 1705–1713.

Hemmati-Brivanlou, A., and Melton, D.A. (1994). Inhibition of activin receptor signalling promotes neuralization in *Xenopus*. *Cell (Cambridge, Mass.)* **77**, 273–281.

Henry, G. L., and Melton, D. A. (1998). *Mixer*, a homeobox gene required for endoderm development. *Science* **281**, 91–96.

Henry, G. L., Brivanlou, I. H., Kessler, D. S., Hemmati-Brivanlou, A., and Melton, D. A. (1996). TGF-β signals and a prepattern in *Xenopus laevis* endodermal development. *Development (Cambridge, UK)* **122**, 1007–1015.

Hoch, M., and Pankratz, M.J. (1996). Control of gut development by *fork head* and cell signaling molecules in *Drosophila*. *Mech. Dev.* **58**, 3–14.

Hogan, B., Beddington, R., Costantini, F., and Lacy, E. (1994). "Manipulating the Mouse Embryo: A Laboratory Manual," 2nd Ed. Cold Spring Harbor Laboratory, Cold Spring Harbor, New York.

Hudson, C., Clements, D., Friday, R. V., Stott, D., and Woodland, H. R. (1997). Xsox17α and β mediate endoderm formation in Xenopus. *Cell (Cambridge, Mass.)* **91**, 397–405.

Jensen, J., Serup, P., Karlsen, C., Funder, T. F., and Madsen, O. D. (1996). mRNA profiling of rat islet tumors reveals Nkx 6.1 as a β-cell specific homeodomain transcription factor. *J. Biol. Chem.* **271**, 18749–18758.

Jones, F. S., Holst, B. D., Minowa, O., DeRobertis, E. M., and Edelman, G. M. (1993). Binding and transactivation of the promoter for the neural cell adhesion molecule by HoxC6 (Hox-3.3). *Proc. Natl. Acad. Sci.* **90**, 6557–6561.

Jones, E. A., Abel, M. H., and Woodland, H. R. (1993). The possible role of mesodermal growth factors in the formation of endoderm in *Xenopus laevis*. *Roux's Arch. Dev. Biol.* **202**, 233–239.

Jonsson, J., Carlsson L., Edlund, T., and Edlund, H. (1994). Insulin-promoter-factor 1 is required for pancreas development in mice. *Nature (London)* **371**, 606–609.

Kaestner, K. H., Bleckmann, S. C., Monaghan, A. P., Sclondorff, J., Mincheva, A., Lichter, P., and Schutz, G. (1996). Clustered arrangement of winged helix genes *fkh-6* and *MFH-1*: Possible implications for mesoderm development. *Development (Cambridge, UK)* **122**, 1751–1758.

Kaestner, K. H., Silberg, D. G., Traber, P. G., and Schutz, G. (1997). The mesenchymal winged helix transcription factor Fkh6 is required for the control of gastrointestinal proliferation and differentiation. *Genes Dev.* **11**, 1583–1595.

Kedinger, M., Simon-Assmann, P., Bouziges, F., Arnold, C., Alexandre, E., and Haffen, K. (1990). Smooth muscle actin expression during rat gut development and induction in fetal skin fibroblastic cells associated with intestinal embryonic epithelium. *Differentiation* **43**, 87–97.

Keller, R. E. (1976). Vital dye mapping of the gastrula and neurula of *Xenopus laevis*. II. Prospective areas and morphogenetic movements of the deep layer. *Dev. Biol.* **51**, 118–137.

Kelley, C., Blumberg, H., Zon, L. I., and Evans, T. (1993). GATA-4 is a novel transcription factor expressed in endocardium of the developing heart. *Development (Cambridge, UK)* **118**, 817–827.

Kessel, M. (1992). Respecification of vertebral identities by retinoic acid. *Development (Cambridge, UK)* **115**, 487–501.

Kim, S. K., Hebrok, M., and Melton, D. A. (1997). Notochord to endoderm signaling is required for pancreas development. *Development (Cambridge, UK)* **124**, 4243–4252.

Kimelman, D., Christian, J. L., and Moon, R. T. (1992). Synergistic principles of development: Overlapping patterning systems in *Xenopus* mesoderm induction. *Development (Cambridge, UK)* **116**, 1–9.

Kimmel, C. B., Warga, R. M., and Schilling, T. F. (1990). Origin and organization of the zebrafish fate map. *Development (Cambridge, UK)* **108**, 581–594.

Kimura, S., Hara, Y., Pineau, T., Fernandez-Salguero, P., Fox, C. H., Ward, J. M., and Gonzalez, F. J. (1996). The T/ebp null mouse: Thyroid-specific enhancer-binding protein is essential for the organogenesis of the thyroid, lung, ventral forebrain, and pituitary. *Genes Dev.* **10**, 60–69.

Kondo, T., Dolle, P., Zakany, J., and Duboule, D. (1996).

Function of posterior *HoxD* genes in the morphogenesis of the anal sphincter. *Development* **122**, 2651–2659.

Kuo, C. T., Morrisey, E. E., Anandappa, R., Sigrist, K., Lu, M. M., Parmacek, M. S., Soudais, C., and Leiden, J. M. (1997). GATA4 transcription factor is required for ventral morphogenesis and heart tube formation. *Genes Dev.* **11**, 1048–1060.

Lacroix, B., Kedinger, M., Simon-Assman, P. M., Haffen, K. (1984). Effects of human fetal gastroenteric mesenchymal cells on some developmental aspects of animal gut endoderm. *Differentiation* **28**, 129–135.

Larsson, L.-I., Madsen, O. D., Serup, P., Jonsson, J., and Edlund, H. (1996). Pancreatic-duodenal homeobox 1-role in gastric endocrine patterning. *Mech. Dev.* **60**, 175–184.

Laser, B., Meda, P., Constant, I., and Philippe, J. (1996). The *caudal*-related homeodomain protein Cdx-2/3 regulates glucagon gene expression in islet cells. *J. Biol. Chem.* **271**, 28984–28994.

Laverriere, A. C., MacNeill, C., Mueller, C., Poelmann, R. E., Burch, J. B. E., and Evans, T. (1994). GATA 4/5/6, a subfamily of three transcription factors transcribed in developing heart and gut. *J. Biol. Chem.* **269**, 23177–23184.

Lazzaro, D., Price, M., DeFelice, M., and DiLauro, R. (1991). The transcription factor TTF-1 is expressed at the onset of thyroid and lung morphogenesis and in restricted regions of the foetal brain. *Development (Cambridge, UK)* **113**, 1093–1104.

LeDouarin, N. (1975). An experimental analysis of liver development. *Med. Biol.* **53**, 427–455.

Lee, J. L., Hollenberg, S. M., Snider, L., Turner, D. L., Lipnick, N., and Weintraub, H. (1995). Conversion of *Xenopus* ectoderm into neurons by neuroD, a basic helix loop helix transcription factor. *Science* **268**, 836–844.

Lee, T., Hacohen, N., Krasnow, M., and Montell, D. J. (1996). Regulated Breathless receptor tyrosine kinase activity required to pattern cell migration and branching in the *Drosophila* tracheal system. *Genes Dev.* **10**, 2912–2921.

Leonard, J., Peers, B., Jonson, T., Ferreri, K., Lee, S., and Montminy, M. R. (1993). Characterization of somatostatin transactivating factor-1, a novel homeobox factor that stimulates somatostatin expression in pancreatic islet cells. *Mol. Endocrinol.* **7**, 1275–1283.

Linask, K. K., and Lash, J. W. (1988). A role for fibronectin in the migration of avian precardiac cells: I. Dose-dependent effects of fibronectin antibody. *Dev. Biol.* **129**, 315–324.

Lints, T. J., Parsons, L. M., Hartley, L., Lyons, I., and Harvey, R. P. (1993). *Nkx-2.5*: A novel murine homeobox gene expressed in early heart progenitor cells and their myogenic descendents. *Development (Cambridge, UK)* **119**, 419–431.

Lukinius, A., Ericsson, J. L. E., Grimelius, L., and Korsgren, O. (1992). Electron microscopic immunocytochemical study of the ontogeny of fetal human and porcine endocrine pancreas, with special reference to colocalization of the four major islet hormones. *Dev. Biol.* **153**, 376–390.

Macdonald, P. M., and Struhl, G. (1986). A molecular gradient in early *Drosophila* embryos and its role in specifying the body plan. *Nature (London)* **324**, 537–545.

MacMorris, M., Broverman, S., Greenspoon, S., Lea, K., Madej, C., Blumenthal, T., and Spieth, J. (1992). Regulation of vitellogenin gene expression in transgenic *Caenorhabditis elegans*: Short sequences required for activation of the *vit-1* promoter. *Mol. Cell. Biol.* **12**, 1652–1662.

Madsen, O. D., Jensen, J., Petersen, H. V., Pedersen, E. E., Oster, A., Andersen, F. G., Jorgensen, M. C., Jensen, P. B., Larsson, L.-I., and Serup, P. (1997). Transcription factors contributing to the pancreatic β-cell phenotype. *Horm. Metab. Res.* **29**, 265–270.

Mansouri, A., Hallonet, M., and Gruss, P. (1996). Pax genes and their roles in cell differentiation and development. *Curr. Opin. Cell Biol.* **8**, 851–857.

Marigo, V., Johnson, R. L., Vortkamp, A., and Tabin, C. J. (1996). Sonic hedgehog differentially regulates expression of *GLI* and *GLI3* during limb development. *Dev. Biol.* **180**, 273–283.

Marom, K., Shapira, E., and Fainsod, A. (1997). The chicken *caudal* genes establish an anterior–posterior gradient by partially overlapping temporal and spatial patterns of expression. *Mech. Dev.* **64**, 41–52.

Mendelsohn, C., Lohnes, D., Decimo, D., Lufkin, T., LeMeur, M., Chambon, P., and Mark, M. (1994). Function of the retinoic acid receptors (RARs) during development (II): Multiple abnormalities at various stages of organogenesis in RAR double mutants. *Development (Cambridge, UK)* **120**, 2749–2771.

Miller, C. P., McGehee, R. E., Jr., and Habener, J. F. (1994). IDX-1: A new homeodomain transcription factor expressed in rat pancreatic islets and duodenum that transactivates the somatostatin gene. *EMBO J.* **13**, 1145–1156.

Miller, C., Williamson, M., Li, X., Arnold, G., Gimlich, R., and Wong, G. (1997). Alterations of pancreatic development by misexpression of growth and differentiation factors in transgenic mice. *Exp. Clin. Endocrinol. Diabetes* **105**, A10.

Miquerol, L., Lopez, S., Cartier, N., Tulliez, M., Raymondjean, M., and Kahn, A. (1994). Expression of the L-type pyruvate kinase gene and the hepatocyte nuclear factor 4 transcription factor in exocrine and endocrine pancreas. *J. Biol. Chem.* **269**, 8944–8951.

Mizuno, K., Gonzalez, F. J., and Kimura, S. (1991). Thyroid-specific enhancer binding protein (T/EBP): cDNA cloning, functional characterization, and structural identity with thyroid transcription factor TTF-1. *Mol. Cell. Biol.* **11**, 4927–4933.

Molkentin, J. D., Lin, Q., Duncan, S. A., and Olson, E. N. (1997). Requirement of the transcription factor GATA4 for heart tube formation and ventral morphogenesis. *Genes Dev.* **11**, 1061–1072.

Mollard, R., and Dziadek, M. (1997). Homeobox genes from clusters A and B demonstrate characteristics of temporal colinearity and differential restrictions in spatial expression domains in the branching mouse lung. *Int. J. Dev. Biol.* **41**, 655–666.

Monaghan, A. P., Kaestner, K. H., Grau, E., and Schutz, G. (1993). Postimplantation expression patterns indicate a role for the mouse *forkhead*/HNF-3 α, β, and γ genes in determination of the definitive endoderm, chordamesoderm and neuroectoderm. *Development (Cambridge, UK)* **119**, 567–578.

Morrisey, E. E., Ip, H. S., Lu, M. M., and Parmacek, M. S. (1996). GATA-6: A zinc finger transcription factor that is expressed in multiple cell lineages derived from lateral mesoderm. *Dev. Biol.* **177**, 309–322.

Morrisey, E. E, Ip, H. S., Tang, Z., Lu, M. M., and Parmacek, M. S. (1997). GATA-5: A transcriptional activator expressed in a novel temporally and spatially-restricted pattern during embryonic development. *Dev. Biol.* **183**, 21–36.

Mutoh, H., Fung, B. P., Naya, F. J., Tsai, M.-J., Nishitani, J., and Leiter, A. B. (1997). The basic helix-loop-helix transcription factor Beta2/NeuroD is expressed in mammalian enteroendocrine cells and activates secretin gene expression. *Proc. Natl. Acad. Sci. U.S.A.* **94**, 3560–3564.

Narita, N., Bielinska, M., and Wilson, D. B. (1997). Wild-type endoderm abrogates the ventral developmental defects associated with GATA-4 deficiency in the mouse. *Dev. Biol.* **189**, 270–274.

Naya, F. J., Stellrecht, C. M. M., and Tsai, M.-J. (1995). Tissue specific regulation of the insulin gene by a novel basic helix loop helix transcription factor. *Genes Dev.* **9**, 1009–1019.

Naya, F. J., Huang, H.-P., Qui, Y., Mutoh, H., DeMayo, F. J., Leiter, A. B., and Tsai, M.-J. (1997). Diabetes, defective pancreatic morphogenesis, and abnormal enteroendocrine differentiation in Beta2/NeuroD-deficient mice. *Genes Dev.* **11**, 2323–2334.

Offield M. F., Jetton, T. L., Labosky, P. A., Ray, M., Stein, R., Magnuson, M. A., Hogan, B. L. M., and Wright, C. V. E. (1996). PDX-1 is required for pancreatic outgrowth and differentiation of the rostral duodenum. *Development (Cambridge, UK)* **122**, 983–995.

Ohlsson, H., Karlsson, K., and Edlund, T. (1993). IPF-1, a homeodomain-containing transactivator of the insulin gene. *EMBO J.* **12**, 4251–4259.

Okada, T. S. (1954a). Experimental studies on the differentiation of the endodermal organs in Amphibia. I. Significance of the mesenchymatous tissue to the differentiation of the presumptive endoderm. *Mem. Coll. Sci. Univ. Kyoto Ser. B.* **21**, 1–6.

Okada, T. S. (1954b). Experimental studies on the differentiation of the endodermal organs in Amphibia. II. Differentiating potencies of the presumptive endoderm in the presence of the mesodermal tissues. *Mem. Coll. Sci. Univ. Kyoto Ser. B.* **21**, 7–14.

Oster, A., Jensen, J., Serup, P., Madsen, O. D., and Larsson, L. I. (1996). Development of the endocrine pancreas of the rat related to the homeobox gene products of PDX-1 and Nkx 6.1. "Program of the 10th International Congress of Endocrinology," International Society of Endocrinology, San Francisco.

Pack, M., Solnica-Krezel, L., Malicki, J., Neuhauss, S. C. F., Schier, A. F., Stemple, D. L., Driever, W., and Fishman, M. C. (1996). Mutations affecting development of zebrafish digestive organs. *Development (Cambridge, UK)* **123**, 321–328.

Pearse, A. E. G. (1977). The APUD concept and its implication: Related endocrine peptides in brain, intestine, pi-

tuitary, placents and anuran cutaneous glands. *Med. Biol.* **55**, 115–125.

Peers, B., Leonard, J., Sharma, S., Teitelman, G., and Montminy, M. R. (1994). Insulin expression in pancreatic islet cells relies on cooperative interactions between the helix loop helix factor E47 and the homeobox factor STF-1. *Mol. Endocrinol.* **8**, 1798–1806.

Peshavaria, M., Gamer, L., Henderson, E., Teitelman, G., Wright, C. V. E., and Stein, R. (1994). XlHbox8, an endoderm-specific *Xenopus* homeodomain protein, is closely related to a mammalian insulin gene transcription factor. *Mol. Endocrinol.* **8**, 806–816.

Peters, K., Werner, S., Liao, X., Wert, S., Whitsett, J., and Williams, L. (1994). Targeted expression of a dominant negative FGF receptor blocks branching morphogenesis and epithelial differentiation of the mouse lung. *EMBO J.* **13**, 3296–3301.

Pfaff, S. L., Mendelson, M., Stewart, C. L., Edlund, T., and Jessel, T. M. (1996). Requirement for LIM homeobox gene Isl1 in motor neuron generation reveals a motor neuron-dependent step in interneuron differentiation. *Cell (Cambridge, Mass.)* **84**, 309–320.

Pictet, R., Clark, W. R., Williams, R. H., and Rutter, W. J. (1972). An ultrastructural analysis of the developing embryonic pancreas. *Dev. Biol.* **29**, 436–467.

Plachov, D. K., Chowdhury, K., Walther, C., Simon, D., Guenet, J. L., and Gruss, P. (1990). Pax8, a murine paired box gene expressed in the developing excretory system and thyroid gland. *Development (Cambridge, UK)* **110**, 643–651.

Rajewsky, K., Gu, H., Kuhn, R., Betz, U. A. K., Muller, W., Roes, J., and Schwenk, F. (1996). Perspective series: Molecular medicine in genetically engineered animals. Conditional gene targeting. *J. Clin. Invest.* **98**, 600–603.

Rausa, F., Samadani, U., Ye, H., Lim, L., Fletcher, C. F., Jenkins, N. A., Copeland, N. G., and Costa, R. H. (1997). The cut-homeodomain transcriptional activator HNF-6 is coexpressed with its target gene HNF-3β in the developing murine liver and pancreas. *Dev. Biol.* **192**, 228–246.

Rehorn, K.-P., Thelen, A. M., Michelson, A. M., and Reuter, R. (1996). A molecular aspect of hematopoiesis and endoderm development common to vertebrates and *Drosophila*. *Development (Cambridge, UK)* **122**, 4045–4056.

Reuter, R. (1994). The gene *serpent* has homeotic properties and specifies endoderm versus ectoderm within the *Drosophila* gut. *Development (Cambridge, UK)* **120**, 1123–1135.

Roberts, D. J., Johnson, R. L., Burke, A. C., Nelson, C. E., Morgan, B. A., and Tabin, C. (1995). Sonic hedgehog is an endodermal signal inducing *Bmp-4* and *Hox* genes during induction and regionalization of the chick hindgut. *Development (Cambridge, UK)* **121**, 3163–3174.

Rosenberg, L. (1996). Pancreatic embryology, phylogeny and complexity. *In* "Cellular Inter-relationships in the Pancreas—Implications for Islet Transplantation." (L. Rosenberg and W. P. Duguid, eds.), pp. 29–53. R. G. Landes.

Rudnick, A., Ling, T. Y., Odagiri, H., Rutter, W. J., and German, M. S. (1994). Pancreatic beta cells express a diverse set of homeobox genes. *Proc. Natl. Acad. Sci. U.S.A.* **91**, 12203–12207.

Sander, M., and German, M. S. (1997). The β cell transcription factors and development of the pancreas. *J. Mol. Med.* **75**, 327–340.

Sander, M., Neubuser, A., Kalamaras, J., Ee, H. C., Martin, G. R., and German, M. S. (1997). Genetic analysis reveals that Pax6 is required for normal transcription of pancreatic hormone genes and islet development. *Genes Dev.* **11**, 1662–1673.

Sasaki, H., and Hogan, B. L. M. (1993). Differential expression of multiple fork head related genes during gastrulation and pattern formation in the mouse embryo. *Development (Cambridge, UK)* **118**, 47–59.

Sauer, B. (1993). Manipulation of tansgenes by site-specific recombination: Use of Cre recombinase. *In* "Methods in Enzymology" (P. M. Wassarman and M. L. DePamphilis, eds.), Vol. 225, pp. 890–900. Academic Press, San Diego.

Sauer, B., and Henderson, N. (1989). Cre-stimulated recombination at *loxP*-containing DNA sequences placed into the mammalian genome. *Nucleic Acids Res.* **17**, 147–161.

Scarpelli, D. G., and Rao, M. S. (1981). Differentiation of regenerating pancreatic cells into hepatocyte-like cells. *Proc. Natl. Acad. Sci. U.S.A.* **78**, 2577–2581.

Selleck, M. A. J., and Stern, C. D. (1991). Fate mapping and cell lineage analysis of Henson's node in the chick embryo. *Development (Cambridge, UK)* **112**, 615–626.

Shannon, J. M. (1994). Induction of alveolar type II cell differentiation in fetal tracheal epithelium by grafted distal lung mesenchyme. *Dev. Biol.* **166**, 600–614.

Shibata, H., Toyama, K., Shioya, H., Ito, M., Hirota, M., Hasegawa, S., Matsumoto, H., Takano, H., Akiyama, T., Toyoshima, K., Kanamaru, R., Kanegae, Y., Saito, I., Nakamura, Y., Shiba, K., and Noda, T. (1997). Rapid colorectal adenoma formation initiated by conditional targeting of the *Apc* gene. *Science* **278**, 120–123.

Slack, J. M. W. (1983). "From Egg to Embryo. Regional Specification in Early Development" (P. W. Barlow, D. Bray, P. B. Green, and J. M. W. Slack, eds.) Cambridge Univ. Press.

Slack, J. M. W. (1996). High hops of transgenic frogs. *Nature (London)* **383**, 765–766.

Slack, J. M. W., Percival, A. C., Neumann, I., and Lear, P. V. (1997). Embryological origin of beta cells. *Exp. Clin. Endocrinol. Diabetes* **105**, A18.

Sosa-Pineda, B., Chowdhury, K., Torres, M., Oliver, G., and Gruss, P. (1997). The *Pax4* gene is essential for differentiation of insulin-producing β-cells in the mammalian pancreas. *Nature (London)* **386**, 399–402.

Stainier, D. Y. R., Fouquet, B., Chen, J.-N., Warren, K. S., Weinstein, B. M., Meiler, S. E., Mohideen, M.-A. P. K., Neuhauss, S. C. F., Solnica-Krezel, L., Scheir, A. F., Zwartkruis, F., Stemple, D. L., Malicki, J., Driever, W., and Fishman, M. C. (1996). Mutations affecting the formation and function of the cardiovascular system in the zebrafish embryo. *Development (Cambridge, UK)* **123**, 285–292.

Stern, C. D. (1992). Mesoderm induction and development of the embryonic axis in amniotes. *Trends Genet.* **8**, 158–163.

Stoffers, D. A., Zinkin, N. T., Stanojevic, V. Clarke, W. L., and Habener, J. F. (1997a). Pancreatic agenesis attributable to a single nucleotide deletion in the human *IPF1* gene coding sequence. *Nature Genetics* **15**, 106–110.

Stoffers, D. A., Thomas, M. K., and Habener, J. F. (1997b). Homeodomain protein IDX-1. A Master regulator of pancreas development and insulin gene expression. *Trends Exp. Med.* **8**, 145–151.

St.-Onge, L., Sosa-Pineda, B. Chowdhury, K., Mansouri, A., and Gruss, P. (1997). *Pax6* is required for differentiation of glucagon-producing α-cells in mouse pancreas. *Nature (London)* **387**, 406–409.

Strandmann, E. P., Nastos, A., Holewa, B., Senkel, S., Weber, H., and Ryffel, G. U. (1997). Patterning the expression of a tissue-specific transcription factor in embryogenesis: HNF1α gene activation during *Xenopus* development. *Mech. Dev.* **64**, 7–17.

Subramanian, V., Meyer, B. I., and Gruss, P. (1995). Disruption of the murine homeobox gene *Cdx1* affects axial skeletal identities by altering the mesodermal expression domains of *Hox* genes. *Cell (Cambridge, Mass.)* **83**, 641–653.

Sugi, Y., and Lough, J. (1994). Anterior endoderm is a specific effector of terminal cardiac myocyte differentiation of cells from the embryonic heart forming region. *Dev. Dyn.* **200**, 155–162.

Suh, E., Chen, L., Taylor, J., and Traber, P. G. (1994). A homeodomain protein related to caudal regulates intestine-specific gene transcription. *Mol. Cell. Biol.* **14**, 7340–7351.

Sussel, L., Hartigan, D. J., Kalamaras, J., Meneses, J., Pederson, R., German, M. S., and Rubenstein, J. L. R. (1997). The Nkx-2.2 homeobox gene plays a critical role in islet cell development. *Exp. Clin. Endocrinol. Diabetes* **105**, A20.

Sutherland, D., Samakovlis, C., and Krasnow, M. A. (1996). *branchless* encodes a *Drosophila* FGF homolog that controls tracheal cell migration and pattern of branching. *Cell (Cambridge, Mass.)* **87**, 1091–1101.

Tam, P. P. L., and Quinlan, G. A. (1996). Mapping vertebrate embryos. *Curr. Biol.* **6**, 104–106.

Teitelman, G., Alpert, S., Polak, J. M., Martinez, A., and Hanahan, D. (1993). Precursor cells of mouse endocrine pancreas coexpress insulin, glucagon and the neuronal proteins tyrosine hydroxylase and neuropeptide Y, but not pancreatic peptide. *Development (Cambridge, UK)* **118**, 1031–1039.

Thor, S., Ericsson, J., Brannstrom, T., and Edlund, T. (1991). The homeodomain LIM protein Isl-1 is expressed in subsets of neurons and endocrine cells. *Neuron* **7**, 881–889.

Torrey, T. W., and Feduccia, A. (1979). Avian cleavage and germ layer formation. *In* "Morphogenesis of the Vertebrates," 4th Ed. Pp. 152–153. Wiley, New York.

Tronche, F., and Yaniv, M. (1992). HNF1, a homeoprotein member of the hepatic transcription regulatory network. *BioEssays* **14**, 579–587.

Tsien, J. Z., Chen, D. F., Gerber, D., Tom, C., Mercer, E. H., Anderson, D. J., Mayford, M., Kandel, E. R., and Tonegawa, S. (1996). Subregion- and cell type-restricted gene knockout in mouse brain. *Cell (Cambridge, Mass.)* **87**, 1317–1326.

Turque, N., Plaza, S., Radvanyi, F., Carriere, C., and Saule, S. (1994). Pax-qnr.pax-6, a paired-box-containing and homeobox-containing gene expressed in neurons, is also expressed in pancreatic endocrine cells. *Mol. Endo.* **8,** 929–938.

Urase, K., Mukasa, T., Igarashi, H., Ishii, Y., Yasagi, S., Momoi, M. Y., and Momoi, T. (1996). Spatial expression of Sonic hedgehog in the lung, epithelium during branching morphogenesis. *Biochem. Biophys. Res. Commun.* **225,** 161–166.

Vasios, G., Mader, S., Gold, J. D., Leid, M., Lutz, Y., Gaub, M.-P., Chambon, P., and Gudas, L. (1991). The late retinoic acid induction of laminin B1 gene transcription involves RAR binding to the responsive element. *EMBO J.* **10,** 1149–1158.

Vortkamp, A., Lee, K., Lanske, B., Segre, G. V., Kronenberg, H. M., and Tabin, C. J. (1996). Regulation of rate of cartilage differentiation by Indian hedgehog and PTH-related protein. *Science* **273,** 613–622.

Walters, J. R. F., Howard, A., Rumble, H. E. E., Prathalingam, S. R., Shaw-Smith, C. J., and Legon, S. (1997). Differences in expression of homeobox transcription factors in proximal and distal human small intestine. *Gastroenterology* **113,** 472–477.

Walther, C., and Gruss, P. (1991). *Pax-6,* a murine paired box gene, is expressed in the developing CNS. *Development (Cambridge, UK)* **113,** 1435–1449.

Weeden, C. J., and Shankland, M. (1997). Mesoderm is required for the formation of a segmented endodermal cell layer in the leech *Helobdella. Dev. Biol.* **191,** 202–214.

Weigel, D., Jurgens, G., Kuttner, F., Seifert, E., and Jackle, H. (1989). The homeotic gene *fork head* encodes a nuclear protein and is expressed in the terminal regions of the *Drosophila* embryo. *Cell (Cambridge, Mass.)* **57,** 645–658.

Weinstein, D. C., Ruiz I Altaba, A., Chen, W. S., Hoodless, P., Prezioso, V. R., Jessell, T. M., and Darnell, J. E., Jr. (1994). The winged-helix transcription factor *HNF-3β* is required for notochord development in the mouse embryo. *Cell (Cambridge, Mass.)* **78,** 575–588.

Wessels, N. K. (1970). Mammalian lung development: interactions in formation and morphogenesis of tracheal buds. *J. Exp. Zool.* **175,** 455–466.

Wessels, N. K. (1977). Tissue interactions and cell differentiation. *In* "Tissue Interactions in Development," pp. 105–121. Benjamin, Menlo Park, California.

Wessels, N. K., and Cohen, J. H. (1967). Early pancreas organogenesis: Morphogenesis, tissue interactions, and mass effects. *Dev. Biol.* **15,** 237–270.

Wolff, E. (1968). Specific interactions between tissues during organogenesis. *Curr. Top. Dev. Biol.* **3,** 65–94.

Wright, C. V. E., Schnegelsberg, P., and DeRobertis, E. M. (1989). XlHbox8: A novel *Xenopus* homeoprotein restricted to a narrow band of endoderm. *Development (Cambridge, UK)* **105,** 787–794.

Wu, K.-L., Gannon, M., Peshavaria, M., Offield, M. O., Henderson, E., Ray, M., Marks, A., Gamer, L. W., Wright, C. V. E., and Stein, R. (1997). Hepatocyte nuclear factor β is involved in pancreatic β-cell-specific transcription of the *pdx-1* gene. *Mol. Cell. Biol.* **17,** 6002–6013.

Wylie, C. C., Snape, A., Heasman, J., and Smith, J. C. (1987). Vegetal pole cells and commitment to form endoderm in *Xenopus laevis. Dev. Biol.* **119,** 496–502.

Yamagata, K., Furuta, H., Oda, N., Kaisaki, P. J., Menzel, S., Cox, N. J., Fajans, S. S., Signorini, S., Stoffel, M., and Bell, G. I. (1996). Mutations in the hepotocyte nuclear factor-4α gene in maturity-onset diabetes of the young (MODY1). *Nature (London)* **384,** 458–460.

Yamasaki, M. Miyake, A., Tagashira, S., and Itoh, N. (1996). Structure and expression of the rat mRNA encoding a novel member of the fibroblast growth factor family. *J. Biol. Chem.* **271,** 15918–15921.

Yasugi, S., and Mizuno, T. (1978). Differentiation of the digestive tract epithelium under the influence of the heterologous mesenchyme of the digestive tract in bird embryos. *Dev. Growth Differ.* **20,** 261–267.

Yokouchi, Y., Sakiyama, J., and Kuroiwa, A. (1995). Coordinate expression of *Abd-B* subfamily genes of the *HoxA* cluster in the developing digestive tract of chick embryo. *Dev. Biol.* **169,** 76–89.

Zannini, M., Francis-Lang, H., Plachov, D., and DiLauro, R. (1992). Pax-8, a paired domain-containing protein, binds to a sequence overlapping the recognition site of a homeodomain and activates transcription from two thyroid-specific promoters. *Mol. Cell. Biol.* **12,** 4230–4241.

Zerlauth, G., and Wolf, G. (1984). Kinetics of fibronectin release from fibroblasts in response to 12-0-tetradecanoylphorbol-13-acetate and retinoic acid. *Carcinogenesis (London)* **5,** 863–868.

Zhong, W., Sladek, F. M., and Darnell, J. E., Jr. (1993). The expression pattern of a *Drosophila* homolog to the mouse transcription factor HNF-4 suggests a determinative role in gut formation. *EMBO J.* **12,** 537–544.

Zhu, J., Hill, R. J., Heid, P. J., Fukuyama, M., Sugimoto, A., Priess, J. R., and Rothman, J. H. (1997). *end-1* encodes an apparent GATA factor that specifies the endoderm precursor in *Caenorhabditis elegans* embryos. *Genes Dev.* **11,** 2883–2896.

41

Myogenic Cell Specification during Somitogenesis

Margaret Buckingham
Shahragim Tajbakhsh
CNRS URA 1947
Department of Molecular Biology
Pasteur Institute
75724 Paris, France

I. Introduction

In many species skeletal muscle represents a major component of the body mass and its formation takes place progressively (see Hauschka, 1994). In the mouse, for example, signals from surrounding tissues acting on the mesodermal cells of the somites lead to the specification of muscle progenitor cells and formation of the myotome. Other muscle progenitor cells and migrate out of the somites to found skeletal muscle masses elsewhere in the body. An initial wave of primary myogenesis leads to the formation of primary fibers which will mainly contribute to the slow fibers of adult skeletal muscle. Subsequently a second wave of myogenesis occurs, which is characterized by the proliferation of previously quiescent myoblasts, and the formation of secondary fibers, which mature into the different fast fibers of the adult. This process is associated with the onset of innervation although not entirely dependent on it, and will continue throughout the fetal period. Muscle fibers continue to grow after birth, when the formation of mature motor end plates with slow or fast nerves leads to the adult fiber phenotype. Even in the adult, skeletal muscle remains remarkably plastic. The fiber type and hence the transcriptional status of different muscle genes is affected by physiological parameters, such as exercise, and adult muscles retain a precursor cell population, the satellite cells, which permits regeneration to take place. In this review we concentrate on the early

stages of skeletal myogenesis in the embryo, taking the mouse as a model, with other species cited when appropriate, in the context of the genes and the signaling molecules which lead to the acquisition of myogenic identity.

II. The MyoD Family of Myogenic Regulatory Factors

The formation of skeletal muscle has provided a paradigm for the study of tissue differentiation since the 1970s. The availability of mammalian muscle cell lines as well as primary cultures, in which dividing myoblasts spontaneously form muscle fibers, facilitated investigation of the battery of muscle specific genes which are activated during this process (see Buckingham, 1985). It also led to the isolation of the MyoD family of myogenic regulatory factors (MRFs) in the late 1980s. MyoD was identified by subtractive hybridization as a sequence present in dividing myoblasts but not in a related fibroblast-type cell, whereas myogenin was isolated on the basis of its accumulation in muscle cells at the onset of differentiation. Subsequently two other members of this family of basic helix–loop–helix transcription factors were isolated, Myf5 also present in dividing myoblasts and MRF4 only expressed in mature muscle fibers in culture. All four factors will activate transcription via certain E-box motifs (CANNTG) present in the promoters and enhancers of many genes expressed in differentiated skeletal muscle. These factors also have the surprising property of effecting myogenic conversion when overexpressed in different nonmuscle cell types, and it was on this basis that MyoD was described as a myogenic determination factor (see Weintraub *et al.*, 1991). A second family of transcription factors, the myocyte enhancer-binding factor 2 (MEF-2) family of RSRF/MADS box proteins, (see Olson *et al.*, 1995) clearly also play a major role in activating muscle gene transcription, acting directly through AT rich sequence motifs and also in conjunction with the MyoD family. However, it is not evident that they play a role at an earlier stage of skeletal myogenesis.

All four *Mrf* genes have been mutated in the mouse, and the results are summarized in Table I. As expected from cell culture experiments, *Myogenin* −/− embryos lack most differentiated skeletal muscle (Hasty *et al.*, 1993, Nabeshima *et al.*, 1993). Strikingly, embryos with mutations in both *Myf5* and *MyoD* genes do not form skeletal muscle and furthermore lack the precursor myoblast population (Rudnicki *et al.*, 1993). Null mutations in either gene alone do not prevent skeletal muscle formation, although in the case of *Myf5* this is delayed and the early myotome does not form (Braun *et al.*, 1992; Tajbakhsh *et al.*, 1996b), in keeping with the fact that *Myf5* only is expressed at this time. Later activation of *MyoD* takes place in the absence of *Myf5* and leads to skeletal muscle formation. In the absence of MyoD (Rudnicki *et al.*, 1992), *Myf5* expression which is usually down-regulated in older embryos, remains high. As is often the case, more detailed examination of the mutants has revealed additional facets. In *Myf5* −/− embryos, *MyoD* activation is delayed, indicating that *Myf5* acts genetically upstream of *MyoD* (Tajbakhsh *et*

al., 1997). Early muscle differentiation is indeed observed in *Myogenin* −/− embryos (Nabeshima *et al.*, 1993, Venuti *et al.*, 1995) corresponding to the time when *Mrf4* is expressed in the myotome (see Buckingham, 1994). *Myf5* null mice do not survive after birth due to respiratory problems, related to rib defects which probably reflect repercussions from the early delay in muscle formation. However *MyoD* −/− mice are viable and these animals exhibit an impairment of the satellite cell population, which affects muscle growth, and strikingly, muscle regeneration (Megeney *et al.*, 1996). The results of gene knock-out experiments on this family of MRFs has led to the conclusion that Myf5 and MyoD are important for the specification and/or maintenance of the muscle progenitor cell population acting upstream of Myogenin and MRF4, which are implicated in muscle cell differentiation.

III. The Embryology of Skeletal Muscle Formation

In order to understand how Myf5 and MyoD may act on the muscle progenitor cell population, it is necessary to review the embryology of muscle formation. Much of our current knowledge of this in higher vertebrates is based on experimental manipulations in avian embryos, particularly those using chick/quail chimeras (see Christ and Ordahl, 1995). Skeletal muscle in the body is derived from paraxial mesoderm which segments into somites following a rostral/caudal developmental gradient. The initially epithelial somite matures into a ventral sclerotomal compartment of mesenchymal cells from which the cartilage and bone of the vertebral column and ribs are derived, and a dorsal epithelium, the dermomyotome, which gives rise to the progenitor cells of skeletal muscle as well as cells which form the dermis of the back. In addition, the epithelial somite has angiogenic potential, contributing endothelial cells to the vasculature (Eichmann *et al.*, 1993, Wilting *et al.*, 1995). If the epithelial somite is rotated in the embryo to reverse its dorsoventral (Aoyama and Asamoto, 1988) or mediolateral orientation (Ordahl and Le Douarin, 1992), it will mature and give rise to correctly positioned somite derivatives, hence the important concept that cells in the somite are initially pluripotent. In support of these findings, experiments in which surrounding tissues have been ablated show that they play a critical role in specifying somite cell fate (see Christ and Ordahl, 1995, Dietrich *et al.*, 1997). This also is demonstrated by explant experiments (see Cossu *et al.*, 1996a).

The tissues involved in skeletal muscle formation are shown in Fig. 1. The axial structures (neural tube and notochord) and the dorsal ectoderm are sources of molecules which lead to the activation of myogenesis, whereas lateral mesoderm has been shown to retard or

TABLE I

Summary of Results Obtained from Gene Targeting Studies of the Myogenic Regulatory Factors

Gene(s) mutated	Myoblasts	Muscle fibers	Description
Myf5	+	+	Lethal at birth due to rib truncations; ~2.5 day delay in myotome formation; early muscle perturbations in trunk and tail
MyoD	+	+	Viable and fertile; ~30% reduction in size; defects in muscle fiber regeneration
Myogenin	+	−	Lethal at birth; almost all skeletal muscle fibers absent
Mrf4	+	+	Viable and fertile; some mild rib defects
Myf5/MyoD	−	−	Lethal at birth; total absence of muscle fibers and precursor myoblasts

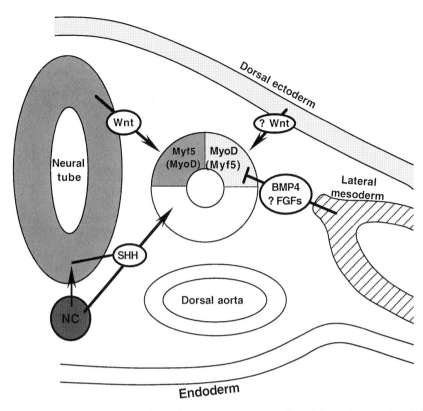

Figure 1 Local environmental signals act in concert to specify and determine muscle cell fate from the dorsal-half of the immature somite. Positive signals emanating from the dorsal neural tube (e.g., Wnts) and notochord/floor plate (via SHH) are required to specify and determine muscle progenitor cells located in the medial portion of the dorsal somite compartment, whereas dorsal ectoderm signals are required to promote myogenesis in the lateral portion of the dorsal somite. Negative signals (e.g., BMP4) produced by the lateral mesoderm negatively regulate myogenesis, as assayed by maintained *Pax3* expression and absence of *MyoD* expression in the lateral dermomyotome. Notochord/floor plate signals (via SHH) activate Pax1 expression and promote sclerotome formation in the ventral somite compartment.

inhibit its onset (see Cossu *et al.*, 1996a). The classic view has been that muscle progenitor cells are specified in response to signals, in the epaxial (medial) and hypaxial (lateral) domains of the dermomyotome from which cells will delaminate and migrate to sites of skeletal muscle formation. Progenitor cells from the epaxial domain adjacent to the axial structures will form the deep back muscles of the body, initially derived from the dorsal domain of the myotome, whereas cells from the hypaxial domain will contribute to the more laterally positioned muscles such as those of the body wall, and muscles formed as a result of longer range migration such as those of the limbs or diaphragm. The relative proportion of epaxial and hypaxial cells produced by a somite will depend on its position along the body axis, reflecting the nature of the surrounding tissues and also their developmental maturity together with that of the paraxial mesoderm when the somite is formed (see Dietrich *et al.*, 1997, and below). Head muscles, such as the extraocular muscles are derived from nonsomitic paraxial head (anterior to the first somite) and prechordal (anterior-most) mesoderm. However, the anterior-most somites also contribute to some head muscles. It was previously thought that some head muscles had a neural crest origin. This is not the case (see Christ and

Ordahl, 1995), although the neural crest probably plays an important role in their patterning (Noden, 1983). Formation of these muscles depends on the myogenic regulatory factors and their differentiation ressembles that of muscles in the body, although the contractile protein isoforms expressed and hence fiber types are often distinct. It is probable that such anterior muscle cells also are specified in response to similar types of signaling from surrounding tissues, as has been observed for the somites of the trunk, but this has not been studied.

IV. Signaling Molecules That Specify Myogenic Progenitor Cells in the Somite

Three major classes of signaling molecules have been implicated in the patterning of the somite, namely, Sonic Hedgehog, Wnts, and bone morphogenetic proteins (BMPs) (see Cossu *et al.*, 1996a).

Sonic hedgehog (SHH) is produced by the notochord and later by the floor plate of the neural tube. The notochord has been shown to have a ventralizing effect (i.e., to be necessary for sclerotome formation, marked by expression of *Pax1*), and indeed the introduction of supernumerary notochords dorsally will inhibit the onset of myogenesis and enhance the expression domain of *Pax1* and the formation of axial cartilage (Brand-Sabéri *et al.*, 1993; Pourquié *et al.*, 1993). In the absence of the neural tube the notochord alone, when positioned normally, will result in weak activation of *MyoD* in avian embryos and it has been suggested that the neural tube is necessary for the maintenance of the myogenic program (Pownall *et al.*, 1996, see Borycki *et al.*, 1997). The presence of the notochord leads to patterning of the neural tube (see Tanabe and Jessell, 1996) and thus also plays a role indirectly via an effect on the maturation of this neuroectoderm. In zebrafish mutants where the notochord is absent, the specification of the so-called adaxial muscle cells (Devoto *et al.*, 1996) which initially abut the notochord, is eliminated although later muscle progenitor cell populations are present (see Blagden *et al.*, 1997). This early muscle cell population, marked by the expression of a slow myosin heavy chain isoform, can be rescued by injection of Sonic hedgehog, normally produced by the notochord (Blagden *et al.*, 1997). In zebrafish, other members of the hedgehog family also are expressed in axial structures, Tiggy winkle hedgehog (TwHH) is in the neural tube and Echidna hedgehog (EHH) in the notochord. In higher vertebrates additional *Hedgehog* genes are also present, but only SHH has been reported in mid-line structures (see Hammerschmidt *et al.*, 1997). SHH is capable of inducing expression of *Pax1* and of suppressing the expression of *Pax3*, a marker of the dermomyotome, either in explant experiments or when expressed ectopically in the chick embryo, in keeping with observations on the ventralizing role of the notochord (Fan and Tessier-Lavigne, 1994, Johnson *et al.*, 1994). *Sonic Hedgehog* null mice lack most sclerotome derivatives, but also show perturbations in myogenesis, particularly in the extent of *Myf5* activation which may reflect a primary effect of the mutation on the patterning of the neural tube (Chiang *et al.*, 1996). However, explant experiments with unsegmented paraxial mesoderm from chick (Münsterberg *et al.*, 1995) or mouse (Tajbakhsh *et al.*, 1998) show that SHH appears to cooperate with the Wnts (see below) for initial activation of myogenesis. Interestingly, whereas the Wnts alone do not induce much myogenesis with explants of unsegmented paraxial mesoderm from young embryos [mouse, embryonic day (E8.5): 6–10 somites], this is not the case for unsegmented paraxial mesoderm from older embryos (mouse E9.5), where addition of SHH has little effect. The nature of this "maturation" phenomenon is not clear. It is interesting in this context that *Pax1* activation in the sclerotome follows somite formation and maturation throughout development, whereas the developmental kinetics of myogenesis as measured by *MyoD*

expression in quail embryos, is more complex perhaps reflecting developmental regulation of the production of axial signals, as well as the competence of somites to respond to them (Borycki *et al.*, 1997).

Involvement of the Wnts in myogenic induction was suggested by work in *Drosophila* where mutations in *wingless* affect skeletal muscle formation (Baylies *et al.*, 1995). A number of Wnts are expressed in the dorsal neural tube, and the *open brain* (*opd*) mouse mutation that affects the dorsal neural tube results in abnormal epaxial myogenesis (Spörle *et al.*, 1996). Explant experiments also point to the importance of the dorsal neural tube for myogenesis (see Cossu *et al.*, 1996a). In such experiments with unsegmented paraxial mesoderm and cells expressing different Wnts it has been shown that Wnts 1, 3a, and 4 will induce the expression of skeletal muscle markers in chick (Münsterberg *et al.*, 1995; Stern *et al.*, 1995). In the mouse, Wnts 1, 4, and 7a, and to a lesser extent Wnts 5a and 6, promote myogenesis (Tajbakhsh *et al.*, 1998). Wnts 1, 3a, 4, and 6 also have been shown to promote expression of Pax3, a marker of the dermomyotome (Fan *et al.*, 1997). Wnts 1, 4, and 3a are present in the dorsal neural tube whereas Wnts 7a, 4, and 6 have been reported to be expressed in dorsal ectoderm (Parr *et al.*, 1993; see also Tajbakhsh *et al.*, 1998). These Wnts may therefore be candidate signaling molecules for the specification of muscle progenitors in the dermomyotome.

An important question concerns potential target genes, particularly *Myf5* and *MyoD*. In *Drosophila* where there is a single *MyoD*-type gene *nautilus*, manipulation of *wingless* shows that the Wingless signal is required for the formation of *nautilus*-expressing medial muscle precursor cell clusters (Ranganayakulu *et al.*, 1996). In *Xenopus* embryos expression of a dominant negative Wnt blocks induction of *MyoD* (Hoppler *et al.*, 1996). In mouse embryo explants, signaling from axial structures preferentially activates *Myf5* (Cossu *et al.*, 1996b) and this is seen preferentially with Wnt 1 producing cells (Tajbakhsh *et al.*, 1998). The older neural tube, on the other hand, will activate MyoD, at the stage when this gene is normally activated in epaxial muscle progenitor cells *in vivo* (Tajbakhsh *et al.*, 1998). Furthermore, explants of medial- or lateral-half somite with associated structures from *Myf5* null mice revealed that *MyoD* can be activated independently in each of these domains (Tajbakhsh *et al.*, 1998). In the absence of axial structures, myogenesis via the hypaxial route still takes place (Rong *et al.*, 1992, see Christ and Ordahl, 1995). Moreover, *MyoD* is preferentially activated in explants of unsegmented paraxial mesoderm in the presence of dorsal ectoderm (Cossu *et al.*, 1996b). Dorsal ectoderm also has been shown, in chick, to exert a dorsalizing influence on the somite, with activation of *Pax3* (Fan *et al.*, 1994) and to be required for maintenance of *Pax3* (Maroto *et al.*, 1997) and *Myf5/MyoD* expression (see also Dietrich *et al.*, 1997). When grown with mouse unsegmented paraxial mesoderm, Wnt 7A producing cells mimic this preferential activation of *MyoD*, rather than *Myf5* (Tajbakhsh *et al.*, 1998). In such explant cultures, activation of one myogenic factor gene is followed by coexpression of both *MyoD* and *Myf5* in the same cells (Cossu *et al.*, 1996b). In birds, it is mainly MyoD that has been used as a myogenic marker, although *Myf5* also is up-regulated in response to signaling molecules. The temporal difference in their activation during embryogenesis is not so evident (see Pownall *et al.*, 1996), nor is it at present clear whether there is an epaxial–hypaxial bias in their expression pattern. In the mouse embryo, however, *Myf5* is expressed first in the epaxial lip of the dermomyotome, as well as the newly forming myotome. In contrast, MyoD expression is strongest in the hypaxial myotome once the cells have left the dermomyotome (Cossu *et al.*, 1996b). However in *Myf5* mutant embryos there is a delay in *MyoD* activation here, too, suggesting a requirement for the prior activation of *Myf5 in vivo* (Tajbakhsh *et al.*, 1997), which is not apparent in explant cultures (Cossu *et al.*, 1996b).

In the embryo other tissues probably influence the activation of myogenesis by dorsal ectoderm. This is the case for lateral mesoderm (Fig. 1), which in mouse explant cultures retards the onset of myogenesis (Cossu *et al.*, 1996b). In avian embryos where this effect was first observed, BMP4 expressing cells have been shown to exert a similar inhibition (Pourquié *et al.*, 1996). BMP4 is indeed present in lateral mesoderm during this critical period. Other factors such as the fibroblast growth factors (FGFs), present in this tissue, may also exert a negative effect on myogenic differentiation. It should be noted that BMPs also are present in other adjacent tissues such as the dorsal region of the neural tube. The presence of noggin in the epaxial lip of the dermomyotome may antagonize the inhibitory action of these molecules (Hirsinger *et al.*, 1997; Marcelle *et al.*, 1997). Other molecules which may influence signaling also are present here. These include other Wnts, such as Wnt 11 (Marcelle *et al.*, 1997) which may act as a relay and also the Frzb class of dominant negative Wnt receptors which potentially bind and sequester Wnts (Leyns *et al.*, 1997). In *Xenopus* embryos, Frzb, secreted by cells of the Spemann organizer, has been shown to bind and sequester Wnt 8 and block the induction of *MyoD* (Wang *et al.*, 1997) in a similar way to a dominant negative Wnt construct.

The localization of the receptors themselves and the intracellular signal transduction components, for the SHH, Wnt, and BMP families awaits detailed analysis in the somite and adjacent structures. A further level of complexity also is introduced by quantitative considerations. Indeed some of the apparently contradictory results reported with SHH or BMPs, on the activation

of myogenesis may be a reflection of this. Signal response elements in potential target genes such as *Myf5* or *MyoD* have not yet been identified precisely. However, regulatory sequences which drive transgene expression in the embryo have been described for *MyoD*. In addition to a more proximal regulatory region which will partially reproduce the pattern of *MyoD* expression (Tapscott *et al.*, 1992), the gene has an enhancer element situated at about 20 kb upstream of the Cap site which, alone with a neutral promoter, will direct appropriate transgene expression in the embryo (Goldhamer *et al.*, 1995). Regulation of the *Myf5* gene, which is organized in tandem with, and about 6 kbp downstream of, *Mrf4* has been only partially described to date. The intragenic region upstream of *Myf5* contains elements which will direct transgene expression to some of the sites in the embryo where the endogenous gene is expressed (Patapoutian *et al.*, 1993). A 5.5-kb fragment from this region driving an *nlacZ* reporter, however, is clearly not capable of reproducing qualitatively or quantitatively the expression pattern seen with an *nlacZ* targeted *Myf5* allele (Tajbakhsh *et al.*, 1996a). Mutations in the *Mrf4* gene which have repercussions on *Myf5* expression suggest that *Myf5* regulatory elements also are present in this part of the locus (see Olson *et al.*, 1996).

The role of surrounding tissues, and signaling molecules in activating the myogenic program has been discussed, however there is some evidence to suggest that derepression may be occuring in the somite, and the Notch signaling pathway has been proposed as a potential candidate (see Cossu *et al.*, 1996a). In cultured muscle cells, Notch signaling has been shown to repress myogenesis (Kopan *et al.*, 1994; Lindsell *et al.*, 1995). Mouse mutants in Notch (Conlon *et al.*, 1995) or the ligand Delta (Hrabe de Angelis *et al.*, 1997) show somite disorganization but no major effect on early muscle formation; although redundancy within this receptor family in the mouse makes it difficult to draw conclusions. In *Xenopus*, expression of a truncated form of Notch causes an increase in muscle tissue (Coffman *et al.*, 1993), and in *Drosophila* mutations in *Notch* and other neurogenic genes on this pathway lead to overproduction of *nautilus* (D-*MyoD*) expressing cells and expansion of muscle tissue (Corbin *et al.*, 1991). It is intriguing that the Notch and Wingless signaling pathways interact at the level of the signal transduction intermediate, Dishevelled (Axelrod *et al.*, 1996). Recently it has been shown that mutation of *Drosophila numb*, which acts to block Notch-mediated repression of genes in muscle progenitor cells, results in a defective muscle pattern caused by the duplication of some fates in the myogenic lineage accompanied by the corresponding loss of others. Numb is asymmetrically distributed in muscle progenitors and segregates asymmetrically as these cells divide. The implication therefore is that this determines the allocation of cell fates in muscle forming mesoderm (Gomez and Bate, 1997). In vertebrates there is some suggestive evidence for the repression of an earlier onset of myogenesis. In the mouse embryo, transitory *Myf5* expression has been detected in unsegmented paraxial mesoderm prior to somite formation (Cossu *et al.*, 1996a; Kopan *et al.*, 1994). In chick even prior to mesoderm formation a striking myogenic potential of clonally cultured epiblast cells has been demonstrated (George-Weinstein *et al.*, 1996). This phenomenon has not been reproduced with cells from early mouse embryos, but these are difficult to culture and furthermore, it has been shown that mammalian cells depend on a community effect to undergo myogenic commitment (Cossu *et al.*, 1995). The suggestion based on embryological observations that myogenesis may be a "default pathway" is not recent and this remains an intriguing if somewhat obscure consideration.

More immediately obvious in the context of signals which activate myogenesis, is the caveat about selective effects on cell proliferation or cell survival which may bias the interpretation of experiments. The neural tube has been shown to produce somite "survival" factors (see Teillet and Le Douarin, 1983) and the Wnts affect cell growth, indeed Wnt1 (formerly Int1) was first isolated as a transforming growth factor (see Nusse and Varmus, 1992). The amplification of one somitic cell type at the expense of another may well be an important contributory factor to the relative importance which different progenitor cell populations assume (see later).

V. Migration of Muscle Progenitor Cells

Intimately linked with myogenic cell specification is the question of their migration from the dermomyotome. Muscle progenitor cells which already have activated *Myf5* in the epaxial lip of the dermomyotome delaminate from this epithelium and migrate into the central compartment of the somite where the first differentiated muscle, the myotome, will form. In the absence of the myogenic factor Myf5, cells which are marked by expression of *nlacZ* targeted into the *Myf5* gene (Tajbakhsh *et al.*, 1996b) leave the dermomyotome, indicating that Myf5 is not required for this

epithelial–mesenchymal transition. However, the cells migrate aberrantly: some become localized dorsally under the surface ectoderm (Fig. 2), and others can be detected ventrally in the sclerotome compartment. Myf5 is therefore required for the correct positioning of muscle progenitor cells in the central compartment of the somite.

Normally, once the myotome begins to form, a basal lamina is laid down, apparently by the myotomal cells themselves, and this structure separates the myotome from the underlying sclerotome. In *Myf5* null mutants where muscle progenitor cells remain developmentally arrested, the basal lamina does not form. The failure of muscle progenitor cells in *Myf5* null mice to respond correctly to positional cues is most probably due to the lack of a cell surface component that is not synthesized in the absence of this transcription factor. Cells which have activated the *Myf5* gene but do not have the Myf5 protein remain multipotent (Fig. 3). According to their local environment they will become incorporated into other differentiation pathways, expressing an early dermal marker, such as the bHLH factor, Dermo1, if they have migrated under the surface ectoderm, or another bHLH factor, Scleraxis, which marks early cartilage cell differentiation if they are positioned ventrally in the sclerotome. Later, as the ribs begin to form, β-galactosidase from the targeted *Myf5-nlacZ* alleles of muscle progenitor cells is still present in the cartilage cells of the ribs, providing a striking demonstration of their change in cell fate (Tajbakhsh *et al.*, 1996b) (Fig. 4).

In the limb, *Myf5* and *MyoD* are activated at about the same time and *Myf5* null mice have no apparent phenotype in these muscles. However in double mutant *Myf5* −/− *MyoD* −/− mice which carry a *MyoD-nlacZ* transgene marking muscle precursor cells as β-galactosidase positive, some of these cells are found in the condensing bone masses of the limb (M. A. Rudnicki, 1997, personal communication). In this case, bone is derived from the mesenchymal cells of the limb (lateral mesoderm derivative); the change in cell fate adopted by muscle progenitor cells is therefore not restricted to somite derivatives (paraxial mesoderm). It is difficult in the mouse embryo at these later stages to carry out lineage tracing experiments to prove formally that β-galactosidase positive cells present in bone, for example, are

---

Figure 2 Muscle progenitor cells exit the somite and express a dermal marker in *Myf5* null embryos. (A) Muscle cells in the myotome show proper patterning in *Myf5* +/− embryos at E10.75 (42 somites). (B) Aberrant patterning of muscle progenitor cells in *Myf5* null embryos is observed in the occipital (o), cervical (C1, bar), and thoracic (T1, arrow) somites. (C, D) Phase-contrast micrographs of paraffin sections of X-gal stained heterozygous (C) and homozygous (D, E) embryos. In the homozygous embyro β-gal+ cells accumulate medially and appear under the ectoderm. *In situ* hybridization reveals that in heterozygotes (C) *Dermo1* transcripts (dark grains) normally appear under the ectoderm and remain distinct from myotomal cells (m). In the homozygous mutant, however, some cells under the ectoderm (arrowheads in E) are β-gal+ and express *Dermo1* (region indicated by bracket in D). The limit of *Dermo1* expression at this stage does not extend fully dorsally toward the neural tube to mark more dorsally located β-gal+ cells. Stars in C and D indicate the third branchial arch. nt, Neural tube; FL, forelimb. Bar: 150 μm for C, D and 25 μm for E.

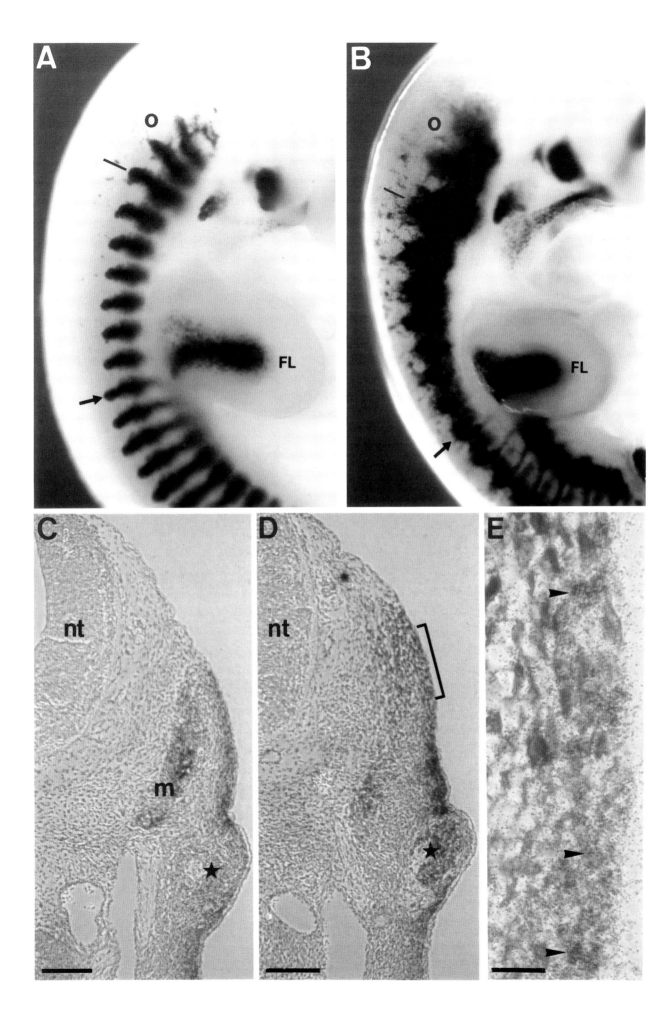

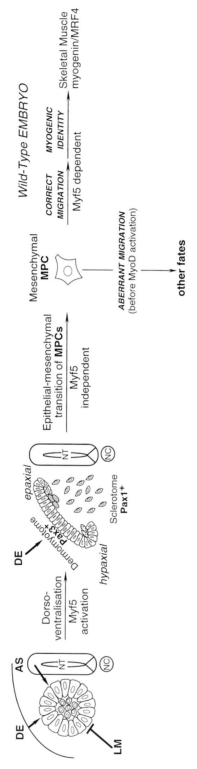

SPECIFICATION   **DETERMINATION**   **DIFFERENTIATION**

Wild-Type EMBRYO

Myf5-nlacZ NULL EMBRYO

Figure 3   Schema summarizing known key events involved in lineage restriction of muscle progenitor cells in the somite. Here, we refer to muscle specification as intra- or extracellular events which progressively restrict mesodermal cell fate to establish a muscle progenitor cell. Determination (or commitment) is also viewed as a multistep process, where the final determined state is the acquisition of myogenic identity via Myf5 (or MyoD) function. In molecular terms Myf5 (and MyoD) act in the final stages to establish myogenic identity, but how this is done remains largely unknown. In the absence of Myf5, muscle precursor cells (MPCs) remain multipotential in *Myf5-nlacZ* null embryos, in spite of the fact that instructive (determination) events activate the *Myf5* promoter, and some of these cells leave the myogenic lineage (see text). In the *Myf5-nlacZ* null, the determined state is later imposed by MyoD on β-gal+ cells which have not left the muscle lineage. Determined cells then differentiate as mononucleated myocytes in the myotome and express *Myogenin, Mrf4*, and other muscle markers. In the case of the limb, specified muscle progenitors migrate into the limb field in the absence of *Mrf* gene expression. AS, axial structures; DE, dorsal ectoderm; LM, lateral mesoderm; NC, notochord; NT, neural tube.

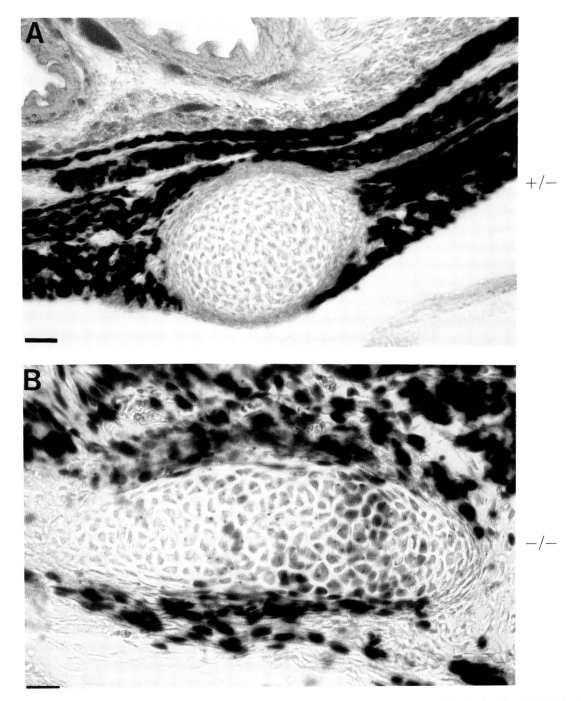

Figure 4 Muscle progenitor cells in *Myf5* null embryos are found in ribs. Cryostat sections of *Myf5* +/− (A) and *Myf5* −/− (B) E16.5 fetuses reveal β-gal⁺ cells in the rib of the *Myf5* −/− fetus. These β-gal⁺ cells are morphologically distinct from muscle cells, appearing cuboidal in shape like the adjacent cartilage cells. Note also that the muscles in this region are perturbed in *Myf5* −/− fetuses. Bar: 50 μm.

not cartilage precursors that have activated *Myf5* aberrantly. However, the fact that this is never seen in normal (heterozygous) embryos and that it occurs at different sites (trunk and limb) makes it very unlikely.

That a mutation in a gene expressed in one tissue may have secondary effects on other tissues is of course a recognized phenomenon. *Myf5* null mice have a rib phenotype, and in fact die of respiratory failure at birth

(Braun *et al.*, 1992; Tajbakhsh *et al.*, 1996b). The distal parts of the rib are missing or malformed. The incorporation of supernumerary muscle progenitor cells into the rib may be a contributory factor, as may the continuing disorganization of the surrounding intercostal muscles which is seen in the fetus, well after the activation of *MyoD* (Tajbakhsh *et al.*, 1996b). It also has been proposed that the absence of the early myotome, a potential source of growth factors, may have repercussions on the adjacent sclerotome cells, at a critical stage for rib development (Grass *et al.*, 1996).

A proportion of muscle progenitor cells derived from the lateral dermomyotome will delaminate from what has been described as the hypaxial somitic bud and form the hypaxial myotome. Other muscle progenitor cells from this part of the somite will migrate further, for example, into the limb field, under the influence of the presumptive limb somatopleure (Hayashi and Ozawa, 1995), or to sites where the muscle masses of the tongue or diaphragm will form. A gene critical for this migration, *Pax3*, has already been mentioned as a marker of the dermomyotome. *Pax3* encodes a paired/homeodomain type transcription factor belonging to the *Pax* gene family, homologues of the *Drosophila paired* gene (Tremblay and Gruss, 1994). *splotch* (*Sp*) mice, which have two mutated alleles of *Pax3*, lack skeletal muscles of the limbs (Bober *et al.*, 1994, Goulding *et al.*, 1994), diaphragm, and tongue (see Tajbakhsh *et al.*, 1997). Skeletal muscles which are derived from the hypaxial somitic bud, such as those of the body wall, are also absent (see Tajbakhsh *et al.*, 1997). Furthermore some perturbation in epaxial muscles such as those in the deep back also has been noted (Tajbakhsh *et al.*, 1997).

Pax3 is expressed in a number of mesodermal and neurectodermal derived cells including skeletal muscle precursors and neural crest. Indeed, effects in the latter which lead to aberrant pigmentation account for the name "*splotch*" (see Tremblay and Gruss, 1994). As far as skeletal muscle is concerned, *Pax3* already is expressed in unsegmented paraxial mesoderm, becoming confined to the dorsal aspect of the somite as it differentiates. Subsequently, *Pax3* transcripts become concentrated in the hypaxial somitic bud, and it has been demonstrated by using chick/quail chimeras that muscle progenitor cells which migrate from this domain are *Pax3* positive (Williams and Ordahl, 1994). In the absence of Pax3, the hypaxial somitic bud appears foreshortened but still contains some Pax3 positive cells which appear not to delaminate in the absence of Pax3 (Bober *et al.*, 1994, Daston *et al.*, 1996).

Overexpression of *Pax* genes in cultured cells leads to a transformed phenotype, favoring proliferation over differentiation (Epstein *et al.*, 1995). This suggests that Pax3 in the hypaxial somitic bud plays a role in the proliferation of muscle progenitor cells, and this may also be the case when these cells have migrated to more distal locations such as the limb. Initially, myogenic factors are not detectable in these cells and are only expressed later, immediately prior to muscle cell differentiation (Sassoon *et al.*, 1989; Tajbakhsh and Buckingham, 1994). In the limb bud, signaling molecules similar to those which play a role along the axis are present (Tickle, 1996); Sonic hedgehog in the zone of polarizing activity (ZPA) and Wnt 7a in the dorsal ectoderm may play a similar role in myogenic cell specification.

Other transcription factors such as the homeobox proteins Msx1 or Lbx also are potentially involved in retaining muscle progenitor cells in a proliferative phase. These are also expressed in the hypaxial somitic bud and in the muscle progenitor cells which migrate to the limb (Dietrich *et al.*, 1998; Houzelstein *et al.*, 1998; Jagla *et al.*, 1995). Interestingly, not all somites along the anterior–posterior axis of the embryo show expression of these genes; for example, Lbx1 expression is detectable only at limb levels. Msx1 has been shown, in *in vitro* experiments, to antagonize the expression of *MyoD* (Song *et al.*, 1992) and this observation has been extended to the homologous gene *Msx2* (Woloshin *et al.*, 1995). *Msx1* null mice do not have a muscle phenotype, possibly because of functional compensation by Msx2 (see Houzelstein *et al.*, 1997) or perhaps the Lbx proteins.

Pax3 may play a role in the proliferation of the muscle progenitor cell population. However, it also is implicated in the migration of these cells, probably acting through the tyrosine kinase receptor c-met (Epstein *et al.*, 1996; Yang *et al.*, 1996). Together with its ligand hepatocyte growth factor (HGF) or scatter factor, c-met is essential for the formation of muscle masses (which require longer range cell migration) such as those in the limb (Bladt *et al.*, 1995; Maina *et al.*, 1996). c-met is present at a number of sites in the embryo, including the dermomyotome in both the epaxial lip and hypaxial somitic bud as well as in cells which migrate from the latter (Yang *et al.*, 1996). Its expression in the hypaxial dermomyotome may be necessary for cells to undergo an epithelial–mesenchymal transition and indeed ectopic expression of scatter factor induces delamination (Brand-Saberi *et al.*, 1996a; Heymann *et al.*, 1996). It is also probably involved in the longer range migration of muscle progenitor cells, since it is expressed in these migratory mesenchymal cells as well as in the limb where the dorsal and ventral muscle masses will form. Furthermore ectopic expression of scatter factor can result in ectopic migration and muscle formation (Brand-Saberi *et al.*, 1996a, Takayama *et al.*, 1996). Other components have also been implicated in this migration, including N-cadherin (Brand-Saberi *et al.*, 1996b), and molecules associated with the connective tissue cell matrix, such as hyaluronic acid (Krenn *et al.*, 1988) and fibronectin (see Brand-Saberi *et al.*, 1996b). The notion that the muscle progenitor cells are "naive," and that their migration and final localization are determined by external factors, comes from classic embryological experiments (see Christ and Ordahl, 1995).

Unexpected regulatory relationships between Pax3, Myf5, and MyoD were uncovered when *splotch* mice were crossed with *Myf5-nlacZ* heterozygotes to obtain double mutant embryos (Tajbakhsh *et al.*, 1997).

Given the phenotypes of the individual mutants it might have been predicted that initially, as in the case of *Myf5* −/− embryos, skeletal muscles would be absent, but that once *MyoD* was activated these mice would have a splotch type phenotype, that is, most epaxial muscles would be present. In fact, these double mutant mice lack skeletal muscles in the body (Tajbakhsh et al., 1997) (Fig. 5). Essentially, muscle differentiation does not take place—*Myogenin* is not expressed—and MyoD is not activated. This finding places *Myf5* and *Pax3* upstream of *MyoD* in the genetic hierarchy that regulates myogenesis. Interestingly, anterior muscles, both those thought to be derived from the more cranial somites and those from paraxial head and prechordal mesoderm, form normally. Myf5 and MyoD are necessary for the formation of these muscles, but Pax3 does not appear to play a role. Indeed, neither Pax3 nor its orthologue Pax7 is present in paraxial head or prechordal mesoderm (see Tajbakhsh *et al.*, 1997). Presumably the processes governing muscle cell migration and possibly *MyoD* activation are different in this domain of the embryo. Indeed, in the head, the movement of sheets of mesoderm rather than individual cells, together with a potentially instructive role for neural crest cells which permeate the mesoderm, are distinguishing features (Noden, 1983).

The phenotype seen in the body of the double mutant raises a number of questions. First, it is possible that Pax3 directly regulates *MyoD*. Experiments in chick embryos, where ectopic expression of Pax3 activates *MyoD*, suggest that this is the case (Maroto *et al.*, 1997). Furthermore, in explant experiments with unsegmented paraxial mesoderm, dorsal ectoderm activates *Pax3* (Fan and Tessier-Lavigne, 1994, Maroto *et al.*, 1997), consistent with subsequent Pax3 activation of *MyoD*. An alternative view is that both Myf5 and Pax3, as discussed here, are critical for the correct migration of muscle progenitor cells, a pre-requisite for subsequent myogenesis. *Pax3*, and indeed *c-met*, are expressed in the epaxial dermomyotome lip, and may be involved in positioning these cells, prior to *MyoD* activation (Tajbakhsh *et al.*, 1997; Yang *et al.*, 1996). At present there is no evidence at the molecular level for transactivation of *MyoD* regulatory sequences by Pax 3, nor is there any direct evidence that Pax3, perhaps via c-met, plays a complementary role to Myf5 in epaxial precursor cell migration. There are preliminary indications that the *c-met* −/− muscle phenotype is not identical to that of *splotch* mice (S. Tajbakhsh, M. Buckingham, and C. Ponzetto, 1998, unpublished observations), suggesting that the role of Pax3 in myogenesis is more complex, and that it has other targets.

VI. Different Muscle Progenitor Cell Populations

Cells in the epithelial somite are multipotent. However, once signals from surrounding tissues have been interpreted, muscle progenitors become distinguishable by virtue of their position in the embryo and hence the environmental influences to which they are subject. Classically distinct skeletal muscle masses are characterized by contractile protein isoforms and metabolic enzymes that mark slow and fast fiber types, for example. However, well before this relatively late stage when innerva-

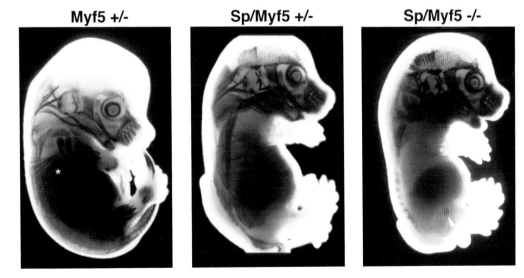

Figure 5 Skeletal muscles are missing in the body of *Sp/Myf5* −/− double homozygous embryos. E14.5 *Myf5* +/− (left), *Sp/Myf5* +/− (middle), and *Sp/Myf5* −/− (right) embryos were analyzed by X-gal staining in toto. The *Myf5* +/− embryo was stained more briefly with X-gal to distinguish muscle groups more clearly. The prominent M. latissimus dorsi (left, asterisk) is missing in Sp embyros (middle, right). Some β-gal[+] cells are reproducibly detected in the hindlimb of *Sp* mutants (right, arrowhead).

tion plays a major role in determining muscle phenotype, there are indications of muscle cell heterogeneity.

An example of early heterogeneity is provided by *Drosophila* (see Bate, 1992), where the unique MyoD-like gene *nautilus* is expressed in only a subset of muscle progenitor cells. Expression of other genes, S59 and *apterous*, encoding homeodomain proteins, and vestigial, with a product potentially involved in protein–protein interactions, marks other subsets of these cells (Abmayr *et al.*, 1995). The gene *twist* acts upstream and is critical for the formation of all somatic muscles. However, it is not specific to this cell type and is important for the subsequent specification of other mesodermal derivatives. *twist* expression is down-regulated as muscle cells begin to differentiate. At the cellular level, a founder cell hypothesis has been proposed to explain how particular myoblasts are selected to contribute to a particular muscle. The grasshopper provides a striking example of this, where neighboring cells fuse with a so-called pioneer cell and are thus recruited to myogenesis (Ho *et al.*, 1983); ablation of the muscle pioneer cell prevents formation of a muscle. There is no direct evidence for this mechanism in *Drosophila*, but expression of a gene such as S59 may mark such founder cells (Dohrmann *et al.*, 1990). The identification of the gene *myoblast city* which, when mutated, abolishes the formation of multinucleate muscles and leaves only a subset of differentiated muscle cells, provides further support for the existence of founder cells (Rushton *et al.*, 1995). The neurogenic genes affect muscle cell specification; in neurogenic mutants the domain of expression of *nautilus* or S59 is increased, suggesting either increased recruitment of surrounding cells by a founder cell or premature expression of these genes in cells which would normally express them later, and which are already "myogenic" (see Bate, 1992).

Evidence for distinct early muscle cell populations also comes from a vertebrate, the zebrafish. Adaxial muscle cells express MyoD (Weinberg *et al.*, 1996) and are marked by slow myosin expression in the paraxial mesoderm even before somitogenesis (Devoto *et al.*, 1996) and as discussed previously their specification depends on SHH (Blagden *et al.*, 1997; Currie and Ingham, 1996; Du *et al.*, 1997). These large cuboidal-shaped cells subsequently migrate from a medial position adjacent to the notochord, laterally. A subset of the adaxial cells, misleadingly named muscle pioneers (Felsenfeld *et al.*, 1991), are marked by the expression of engrailed (Hatta *et al.*, 1991) and remain more medially at the myoseptum. A later population of myogenic cells, derived from the somites form the bulk of the muscle, and can be distinguished by antibodies to fast myosin isoforms (Devoto *et al.*, 1996).

In amniotes, epaxial muscle progenitor cells are the first to be distinguished as MyoD positive (Myf5 in mouse), and these will delaminate from the epaxial der-

momyotome lip, derived from the dorsomedial quadrant of the somite, to form the early epaxial (dorsal) myotome (Denetclaw *et al.*, 1997). Later, hypaxial progenitor cells from the hypaxial somitic bud also contribute to the hypaxial (ventral) myotome. In explant experiments, as discussed, *MyoD* is first activated by the signals which specify this cell population (Cossu *et al.*, 1996b), although *in vivo* an early dependence on Myf5 is observed (Tajbakhsh *et al.*, 1997). Muscle cells in the myotome express different combinations of myogenic factors (Smith *et al.*, 1994), although it is not very clear to what extent this reflects differences in maturation, rather than differences in their origin. However, the epaxially derived deep back muscles are present in *Myogenin* null mice (Nabeshima *et al.*, 1993, Venuti *et al.*, 1995), suggesting a difference between these and other skeletal muscles, with respect to MRFs that drive differentiation initially (see Buckingham, 1994). It has been suggested that another muscle progenitor cell population may be present in the central region of the dermomyotome, marked by continuing expression of *Pax3* (Tajbakhsh *et al.*, 1997), *Sim1*, and, interestingly in the zebrafish context, *Engrailed1* (R. Spörle, 1997, personal communication). It remains to be shown that the cells expressing these genes contribute to the myotome; however, the suggestion is that they may do so when the central dermomyotome disintegrates in older somites (see Tajbakhsh and Spörle, 1998).

Once muscle progenitor cells have been specified, their developmental history in terms of temporal as well as spatial parameters will determine their characteristics. In the myotome, for example, a later population of myoblasts, derived from the lateral, or possibly more central dermomyotome, express the avian FGF receptor, FREK. This also marks replicating skeletal muscle myoblasts in the limb at later stages of development. Expression of an FGF receptor may initially prevent the cells from differentiating by the action of FGF and thus in the somite as well as the limb ensure the maintenance of a precursor population for the subsequent growth of the myotome, and of muscle masses generally (Marcelle *et al.*, 1995). In chick the timing of migration of muscle progenitor cells from the somite to the limb influences their subsequent differentiation as slow versus fast-type muscle (Van Swearingen and Lance-Jones, 1995). The intrinsic differences in the embryonic myoblast population, clearly documented for birds in terms of myosin isoforms expressed as they differentiate may also be a reflection of this phenomenon (see Stockdale, 1992). It appears clear that the myoblasts that remain latent in the mammalian and avian limb do not participate in the first wave of primary myogenesis. When they are activated later, they proliferate and then participate in the wave of secondary myogenesis, having acquired distinct properties as fetal myoblasts (see Hauschka, 1994; Stockdale, 1992). It has been suggested that expression of transforming growth factor β (TGFβ) receptors may be critical in distinguishing fetal myoblast precursors from the embryonic myoblasts (Cusella-De Angelis *et al.*, 1994). Satellite cells that remain quiescent in adult muscle and are activated on injury, permitting regeneration, also have distinct characteristics (see Hauschka, 1994; Stockdale, 1992). Interestingly in *Drosophila*, continued expression of *twist* is a feature of embryonic myoblasts that do not differentiate immediately and will contribute to the adult musculature of the fly (Bate *et al.*, 1991). Although innervation in the adult amniote is critical in determining muscle fiber type, it has been shown that unexpected heterogeneity exists in the satellite cell population. The phenotype of adult myoblasts when they differentiate *in vitro* reflects their fiber type of origin (Rosenblatt *et al.*, 1996), again demonstrating the effect of the environment, even at a late stage, on the myogenic potential of a cell.

The extent to which the location of different skeletal muscle masses in the body results in intrinsic differences in cell populations is beginning to become more apparent. The *Myf*5/*Pax*3 double mutant mouse is an example (Tajbakhsh *et al.*, 1997). However, there have been other indications. For example, in an elegant transplantation experiment, Hoh showed that myoblast cells from the masseter muscle of the cat, when transplanted into the muscle mass of the limb, will still differentiate into muscle fibers expressing characteristic superfast myosin isoforms seen only in the masseter muscle (Hoh and Hughes, 1988). Indications of differences in relation to the anterior–posterior axis were provided by a myosin transgene (Donoghue *et al.*, 1992; Grieshammer *et al.*, 1992) that is expressed at a higher level in more caudal muscles. This property is cell autonomous and points to intrinsic positional information acquired by the muscle precursor cell. Indeed, the range of genetically determined human muscle diseases, which affect some muscles and not others, also points to the existence of positional identity (see Engel and Franzini-Armstrong, 1994).

This phenomenon has begun to be dissected genetically in *Drosophila*. The homeotic genes of the *Bithorax* complex have been shown to control the identities of abdominal somatic muscles and their precursors by functioning directly in cells of the mesoderm (Michelson, 1994). It seems probable that the expression of Hox genes in different anterior–posterior domains of mesoderm in the trunk and limb may also underlie the patterning of skeletal muscle in vertebrates. Until relatively recently there was circumstantial evidence in *Drosophila* that the expression of genes such as S59 in subsets of developing muscles and their precursors may regulate their identity (see Abmayr *et al.*, 1995). However, manipulation of *Krüppel*, originally identified as a member of the gap class of segmentation genes, and encoding a zinc finger-type transcription factor, now shows that gain or loss of *Krüppel* expression in sibling founder cells is sufficient to change the identity of the muscles to which they normally give rise (Ruiz-Gomez *et al.*, 1997). Krüppel is required for the maintenance of expression of S59. It also maintains expression of a novel gene *knockout*, identified as a direct target of Krüppel, in a subset of muscles, and apparently required for their proper innervation (Hartmann *et al.*, 1997).

Dissecting genetic hierarchies during the different stages of tissue formation gives insight into how muscle identity may be set up. Many of the genes discussed in this chapter were first identified by genetic screens in *Drosophila*. It is not evident, of course, that the same proteins are players in vertebrates. S59 homologues, for example, have been identified in vertebrates, but none have yet been shown to be expressed in skeletal muscle. In *Drosophila* and in *nematode* (Krause and Weingtraub, 1992) where there is also only one *MyoD*-type gene, myogenesis will proceed in the absence of this myogenic regulatory factor, whereas the MyoD dependent pathway appears to have been selected and embroidered on as vertebrates evolved. However, even if the dominant strategy for muscle cell specification varies between species, the differences are intriguing. If one looks in greater detail, there may be more unexpected types of muscle progenitor in amniotes, which represent either vestiges of a specification strategy that is more prominent in another species or a subpopulation that contributes significantly to the myogenic strategy of the organism. These differences and similarities will provide important clues to understanding several congenital and degenerative muscle diseases.

References

Abmayr, S. M., Erickson, M. S., and Bour, B. A. (1995). Embryonic development of the larval body wall musculature of *Drosophila melanogaster*. *Trends Genet* **11**, 153–159.

Aoyama, H., and Asamoto, K. (1988). Determination of somite cells: Independence of cell differentiation and morphogenesis. *Development (Cambridge, UK)* **104**, 15–28.

Axelrod, J. D., Matsuno, K., Artavanis-Tsakonas, S., and Perrimon, N. (1996). Interaction between wingless and notch signaling pathways mediated by dishevelled. *Science* **271**, 1826–1832.

Bate, M. (1992). Mechanisms of muscle patterning in *Drosophila*. *Semin. Dev. Biol.* **3**, 267–275.

Bate, M., Rushton, E., and Currie, D. A. (1991). Cells with persistent *twist* expression are the embryonic precursors of adult muscles in *Drosophila*. *Development (Cambridge, UK)* **113**, 79–89.

Baylies, M. K., Martinez-Arias, A., and Bate, M. (1995). Wingless is required for the formation of muscle founder cells in *Drosophila*. *Development (Cambridge, UK)* **121**, 3829–3837.

Bladt, F., Riethmacher, D., Isenmann, S., Aguzzi, A., and Birchmeier, C. (1995). Essential role for the c-met receptor in the migration of myogenic precursor cells into the limb bud. *Nature (London)* **376**, 768–771.

Blagden, C. S., Currie, P. D., Ingham, P. W., and Hughes, S. M. (1997). Notochord induction of zebrafish slow muscle mediated by sonic hedgehog. *Genes Dev.* **11**, 2163–2175.

Bober, E., Franz, T., Arnold, H. H., Gruss, P., and Tremblay, P. (1994). Pax-3 is required for the development of limb muscles: A possible role for the migration of dermomyotomal muscle progenitor cells. *Development (Cambridge, UK)* **120**, 603–612.

Borycki, A.-G., Strunk, K. E., Savary, R., and Emerson, C. P., Jr., (1997). Distinct signal/response mechanisms regulate *pax1* and *QmyoD* activation in sclerotomal and myotomal lineages of quail somites. *Dev. Biol.* **185**, 185–200.

Brand-Sabéri, B., Ebensperger, C., Wilting, J., Balling, R., and Christ, B. (1993). The ventralizing effect of the notochord on somite differentiation in chick embryos. *Anat. Embryol.* **188**, 239–245.

Brand-Saberi, B., Müller, T. S., Wilting, J., Christ, B., and Birchmeier, C. (1996a). Scatter factor/hepatocyte growth factor (SF/HGF) induces emigration of myogenic cells at interlimb level *in vivo*. *Dev. Biol.* **179**, 303–308.

Brand-Saberi, B., Gamel, A. J., Krenn, V., Müller, T. S., Wilting, J., and Christ, B. (1996b). N-cadherin is involved in myoblast migration and muscle differentiation in the avian limb bud. *Dev. Biol.* **178**, 160–173.

Braun, T., Rudnicki, M. A., Arnold, H. H., and Jaenisch, R. (1992). Targeted inactivation of the muscle regulatory gene Myf5 results in abnormal rib development and perinatal death. *Cell (Cambridge, Mass.)* **71**, 369–82.

Buckingham, M. E. (1985). Actin and myosin multigene families: Their expression during the formation of skeletal muscle. *Essays Biochem.* **20**, 77–109.

Buckingham, M. (1994). Which myogenic factors make muscle? *Curr. Biol.* **4**, 61–63.

Chiang, C., Litingtung, Y., Lee, E., Young, K. E., Corden, J. L., Westphal, H., and Beachy, P. A. (1996). Cyclopia and defective axial patterning in mice lacking *Sonic hedgehog* gene function. *Nature (London)* **383**, 407–413.

Christ, B., and Ordahl, C. P. (1995). Early stages of chick somite development. *Anat. Embryol.* **191**, 381–396.

Coffman, C. R., Skoglund, P., Harris, W. A., and Kintner, C. R. (1993). Expression of an extracellular deletion of *Xotch* diverts cell fate in *Xenopus* embryos. *Cell (Cambridge, Mass.)* **73**, 659–671.

Conlon, R. A., Reaume, A. G., and Rossant, J. (1995). *Notch1* is required for the coordinate segmentation of somites. *Development (Cambridge, UK)* **121**, 1533–1545.

Corbin, V., Michelson, A. M., Abmayr, S. M., Neel, V., Alcamo, E., Maniatis, T., and Young, M. W. (1991). A role for the Drosophila neurogenic genes in mesoderm differentiation. *Cell (Cambridge, Mass.)* **67**, 311–323.

Cossu, G., Kelly, R., Donna, S. D., Vivarelli, E., and Buckingham, M. (1995). Myoblast differentiation during mammalian somitogenesis is dependent upon a community effect. *Proc. Natl. Acad. Sci. U.S.A.* **92**, 2254–2258.

Cossu, G., Tajbakhsh, S., and Buckingham, M. (1996a). How is myogenesis initiated in the embryo? *Trends Genet.* **12**, 218–223.

Cossu, G., Kelly, R., Tajbakhsh, S., Donna, S. D., Vivarelli, E., and Buckingham, M. (1996b). Activation of different myogenic pathways: myf-5 is induced by the neural tube and MyoD by the dorsal ectoderm in mouse paraxial mesoderm. *Development (Cambridge, UK)* **122**, 429–437.

Currie, P. D., and Ingham, P. W. (1996). Induction of a specific muscle cell type by a hedgehog-like protein in zebrafish. *Nature (London)* **382**, 452–455.

Cusella-De Angelis, M. G., Molinari, S., Donne, A. D., Coletta, M., Vivarelli, E., Bouche, M., Molinaro, M., Ferrari, S., and Cossu, G. (1994). Differential response of embryonic and fetal myoblasts to TGFβ: A possible regulatory mechanism of skeletal muscle histogenesis. *Development (Cambridge, UK)* **121**, 637–649.

Daston, G., Lamar, E., Olivier, M., and Goulding, M. (1996). Pax-3 is necessary for migration but not differentiation of limb muscle precursors in the mouse. *Development (Cambridge, UK)* **122**, 1017–1027.

Denetclaw, W. F., Christ, B., and Ordahl, C. P. (1997). Location and growth of epaxial myotome precursor cells. *Development (Cambridge, UK)* **124**(8); 1601–1610.

Devoto, S. H., Melançon, E., Eisen, J. S., and Westerfield, M. (1996). Identification of separate slow and fast muscle precursor cells *in vivo*, prior to somite formation. *Development (Cambridge, UK)* **122**, 3371–3380.

Dietrich, S., Schubert, F. R., and Lumsden, A. (1997). Control of dorsoventral pattern in the chick paraxial mesoderm. *Development (Cambridge, UK)* **124**, 3895–3908.

Dietrich S., Schubert F. R., Healy, C., Sharpe, P. T., and Lumsden, A. (1998). Specification of the hypaxial musculature. *Development* **125**, 2235–2249.

Dohrmann, C., Azpiazu, N., and Frasch, M. (1990). A new *Drosophila* homeo box gene is expressed in mesodermal precursor cells of distinct muscles during embryogenesis. *Genes Develop.* **4**, 2098–2111.

Donoghue, M. J., Morris-Valero, R., Johnson, Y. R., Merlie, J. P., and Sanes, J. R. (1992). Mammalian muscle cells bear a cell-autonomous, heritable memory of their rostrocaudal position. *Cell (Cambridge, Mass.)* **69**, 67–77.

Du, J. S., Devoto, S. H., Westerfield, M., and Moon, R. T. (1997). Positive and negative regulation of muscle cell identity by members of the hedgehog and TGFβ gene families, *J. Cell Biol.* **139**; 145–156.

Eichmann, A., Marcelle, C., Bréant, C., and Le Douarin, N. M. (1993). Two molecules related to the VEGF receptor are expressed in early endothelial cells during avian embryonic development. *Mech. Dev.* **42**, 33–48.

Engel, A. G., and Franzini-Armstrong, C. (1994). "Myology," 2nd Ed., Vol. 2. McGraw, New York.

Epstein, J. A., Lam, P., Jepeal, L., Maas, R. L., and Shapiro, D. N. (1995). Pax3 inhibits myogenic differentiation of cultured myoblast cells. *J. Biol. Chem.* **270**, 11719–11722.

Epstein, J. A., Shapiro, D. N., Cheng, J., Lam, P. Y., and Maas, R. L. (1996). Pax3 modulates expression of the c-Met receptor during limb muscle development. *Proc. Natl. Acad. Sci. U.S.A.* **93**, 4213–4218.

Fan, C. M., and Tessier-Lavigne, M. (1994). Patterning of mammalian somites by surface ectoderm and notochord: Evidence for sclerotome induction by a *hedgehog* homolog. *Cell* (*Cambridge, Mass.*) **79**, 1175–1186.

Fan, C.-M., Lee, C. S., and Tessier-Lavigne, M. (1997). A role for Wnt proteins in induction of dermomyotome. *Dev. Biol.* **191**, 160–165.

Felsenfeld, A. L., Curry, M., and Kimmel, C. B. (1991). The fub-1 mutation blocks initial myobrifil formation in zebrafish muscle pioneer cells. *Dev. Biol.* **148**, 23–30.

George-Weinstein, M., Gerhart, J., Reed, R., Flynn, J., Callihan, B., Mattiacci, M., Miehle, C., Foti, G., Lash, J. W., and Weintraub, H. (1996). Skeletal myogenesis: The preferred pathway of chick embryo epiblast cells *in vitro. Dev. Biol.* **173**, 279–291.

Goldhamer, D. J., Brunk, B. P., Faerman, A., King, A., Shani, M., and Emerson, C. P., Jr., (1995). Embryonic activation of the myoD gene is regulated by a highly conserved distal control element. *Development* (*Cambridge, UK*) **121**, 637–649.

Gomez, M. R., and Bate, M. (1997). Segregation of myogenic lineages in *Drosophila* requires Numb. *Development* (*Cambridge, UK*) **124**, 4857–4866.

Goulding, M., Lumsden, A., and Paquette, A. J. (1994). Regulation of Pax-3 expression in the dermomyotome and its role in muscle development. *Development* (*Cambridge, UK*) **120**, 957–971.

Grass, S., Arnold, H. H., and Braun, T. (1996). Alterations in somite patterning of Myf5 deficient mice: A possible role for FGF-4 and FGF-6. *Development* (*Cambridge, UK*) **122**, 141–150.

Grieshammer, U., Sassoon, D., and Rosenthal, N. (1992). A transgene target for positional regulators marks early rostrocaudal specification of myogenic lineages. *Cell* (*Cambridge, Mass.*) **69**, 79–93.

Hammerschmidt, M., Brook, A., and McMahon, A. P. (1997). The world according to *hedgehog. Trends Genet.* **13**, 14–21.

Hartmann, C., Landgraf, M., Bate, M., and Jäckle, H. (1997). *Krüppel* target gene *knockout* participates in the proper innervation of a specific set of *Drosophila* larval muscles. *EMBO J.* **16**, 5299–5309.

Hasty, P., Bradley, A., Morris, J. H., Edmondson, D. G., Venuti, J. M., Olson, E. N., and Klein, W. H. (1993). Muscle deficiency and neonatal death in mice with a targeted mutation in the myogenin gene. *Nature* (*London*) **364**, 501–506.

Hatta, K., BreMiller, R. A., Westerfield, M., and Kimmel, C. B. (1991). Diversity of expression of *engrailed* homeoproteins in zebrafish. *Development* (*Cambridge, UK*) **112**, 821–832.

Hauschka, S. D. (1994). The embryonic origin of muscle. *In* "Myology" (A. G. Engel and C. Franzini-Armstrong, eds.), Vol. 1, pp. 3–74. McGraw-Hill, New York.

Hayashi, K., and Ozawa, E. (1995). Myogenic cell migration from somites is induced by tissue contact with medial region of the presumptive limb mesoderm in chick embryos. *Development* (*Cambridge, UK*) **121**, 661–669.

Heymann, S., Koudrova, M., Arnold, H.-H., Köster, M., and Braun, T. (1996). Regulation and function of SF/HGF during migration of limb muscle precursor cells in chicken. *Dev. Biol.* **180**,, 566–578.

Hirsinger, E., Duprez, D., Jouve, C., Malapert, P., Cooke, J., and Pourquié, O. (1997). Noggin acts downstream of Wnt and Sonic Hedgehog to antagonize BMP4 in avian somite patterning. *Development* (*Cambridge, UK*) **124**, 4605–4614.

Ho, R. K., Ball, E. E., and Goodman, C. S. (1983). Muscle pioneers: Large mesodermal cells that erect a scaffold for developing muscles and motoneurones in grasshopper embryos. *Nature* (*London*) **301**, 66–69.

Hoh, J. F. Y., and Hughes, S. (1988). Myogenic and neurogenic regulation of myosin gene expression in cat jaw-closing muscles. *J. Muscle Res. Cell Motil.* **9**, 59–72.

Hoppler, S., Brown, J. D., and Moon, R. T. (1996). Expression of a dominant-negative Wnt blocks induction of MyoD in *Xenopus* embryos. *Genes Dev.* **10**, 2805–2817.

Houzelstein, D., Cohen, A., Buckingham, M. E., and Robert, B. (1997). Insertional mutation of the mouse *Msx1* homeobox gene by an nlacZ reporter gene. *Mech. Dev.* **65**, 123–133.

Houzelstein, D., Auda-Boucher, G., Tajbakhsh, S., Buckingham, M. E., Fontaine-Perus, J., and Robert, B. (1998). The homeobox gene Msx1 is expressed in the lateral dermomyotome of a subset of somites, and in muscle progenitor cells migrating into the forelimb. Submitted for publication.

Hrabe de Angelis, M., McIntyre II, J., and Gossler, A. (1997). Maintenance of somite borders in mice requires the *Delta* homologue Dll1. *Nature* (*London*) **386**, 717–721.

Jagla, K., Dollé, P., Mattei, M.-G., Jagla, T., Schuhbaur, B., Dretzen, G., Bellard, F., and Bellard, M. (1995). Mouse *Lbx1* and human *LBX1* define a novel mammalian homeobox gene family related to the *Drosophila lady bird* genes. *Mech. Dev.* **53**, 345–356.

Johnson, R. L., Laufer, E., Riddle, R. D., and Tabin, C. (1994). Ectopic expression of sonic hedgehog alters dorsal–ventral patterning of somites. *Cell* (*Cambridge, Mass.*) **79**, 1165–1173.

Kopan, R., Nye, J. S., and Weintraub, H. (1994). The intracellular domain of mouse Notch: A constitutively activated repressor of myogenesis directed at the basic helix-loop-helix region of MyoD. *Development* (*Cambridge, UK*) **120**, 2385–2396.

Krause, M., and Weintraub, H. (1992). CeMyoD expression and myogenesis in C. *elegans. Semin. Dev. Biol.* **3**, 277–285.

Krenn, V., Gorka, P., Wachtler, F., Christ, B., and Jacob, H. J. (1988). On the origin of cells determined to form skeletal muscle in avian embryos. *Anat. Embryol.* **179**, 49–54.

Leyns, L., Bouwmeester, T., Kim, S.-H., Piccolo, S., and De Robertis, E. M. (1997). Frzb-1 is a secreted antagonist of Wnt signaling expressed in the Spemann organizer. *Cell (Cambridge, Mass.)* **88**, 747–756.

Lindsell, C. E., Shawber, C. J., Boulter, J., and Weinmaster, G. (1995). Jagged: A mammalian ligand that activates Notch 1. *Cell (Cambridge, Mass.)* **80**, 909–917.

Maina, F., Casagranda, F., Audero, E., Simeone, A., Comoglio, P. M., Klein, R., and Ponzetto, C. (1996). Uncoupling of Grb2 from the Met Receptor *in vivo* reveals complex roles in muscle development. *Cell (Cambridge, Mass.)* **87**; 531–542.

Marcelle, C., Wolf, J., and Bronner-Fraser, M. (1995). The *in vivo* expression of the FGF receptor FREK mRNA in avian myoblasts suggests a role in muscle growth and differentiation. *Dev. Biol.* **171**, 100–114.

Marcelle, C., Stark, M. R., and Bronner-Fraser, M. (1997). Coordinate actions of BMPs, Wnts, Shh and Noggin mediate patterning of the dorsal somite. *Development (Cambridge, UK)* **124**, 3955–3963.

Maroto, M., Reshef, R., Münsterberg, A., Koester, A. E., Goulding, M., and Lassar, A. B. (1997). Ectopic Pax-3 activates MyoD and myf-5 expression in embryonic mesoderm and neural tissue. *Cell (Cambridge, Mass.)* **89**, 139–148.

Megeney, L. A., Kablar, B., Garrett, K., Anderson, J. E., and Rudnicki, M. A. (1996). MyoD is required for myogenic stem cell function in adult skeletal muscle. *Genes Dev.* **10**, 1173–1183.

Michelson, A. M. (1994). Muscle pattern diversification in *Drosophila* is determined by the autonomous function of homeotic genes in the embryonic mesoderm. *Development (Cambridge, UK)* **120**, 755–768.

Münsterberg, A. E., Kitajewski, J., Bumcrot, D. A., McMahon, A. P., and Lassar, A. B. (1995). Combinatorial signaling by Sonic hedgehog and Wnt family members induces myogenic bHLH gene expression in the somite. *Genes Dev.* **9**, 2911–2922.

Nabeshima, Y., Hanaoka, K., Hayasaka, M., Esumi, E., Li, S., Nonaka, I., and Nabeshima, Y. (1993). Myogenin gene disruption results in perinatal lethality because of severe muscle defect. *Nature (London)* **364**, 532–535.

Noden, D. M. (1983). The role of the neural crest in patterning of avian cranial skeletal, connective and muscle tissues. *Dev. Biol.* **96**, 144–165.

Nusse, R., and Varmus, H. E. (1992). *Wnt* genes. *Cell (Cambridge, Mass.)* **69**, 1073–1087.

Olson, E. N., Arnold, H.-H., Rigby, P. W. J., and Wold, B. J. (1996). Know your neighbors: Three phenotypes in null mutants of the myogenic bHLH gene *MRF4*. *Cell (Cambridge, Mass.)* **85**, 1–4.

Olson, E. N., Perry, M., and Schulz, R. A. (1995). Regulation of muscle differentiation by the MEF2 family of MADS box transcription factors. *Dev. Biol.* **172**, 2–14.

Ordahl, C. P., and Le Douarin, N. M. (1992). Two myogenic lineages within the developing somite. *Development (Cambridge, UK)* **114**, 339–353.

Parr, B. A., Shea, M. J., Vassileva, G., and McMahon, A. P. (1993). Mouse Wnt genes exhibit discrete domains of expression in the early embryonic CNS and limb buds. *Development (Cambridge, UK)* **119**, 247–261.

Patapoutian, A., Miner, J. H., Lyons, G. E., and Wold, B.

(1993). Isolated sequences from the linked *myf*-5 and *MRF4* genes drive distinct patterns of muscle-specific expression in transgenic mice. *Development (Cambridge, UK)* **118**, 61–69.

Pourquié, O., Coltey, M., Teillet, M.-A., Ordahl, C., and Le Douarin, N. M. (1993). Control of dorsoventral patterning of somitic derivatives by notocord and floor plate. *Proc. Natl. Acad. Sci. U.S.A.* **90**, 5242–5246.

Pourquié, O., Fan, C.-M., Coltey, M., Hirsinger, E., Wanatabe, Y., Bréant, C., Francis-West, P., Brickell, P., Tessier-Lavigne, M., and Le Douarin, N. (1996). Lateral and axial signals involved in avian somite patterning: A role for BMP4. *Cell (Cambridge, Mass.)* **84**, 461–471.

Pownall, M. E., Strunk, K. E., and Emerson, C. P., Jr. (1996). Notochord signals control the transcriptional cascade of myogenic bHLH genes in somites of quail embryos. *Development (Cambridge, UK)* **122**, 1475–1488.

Ranganayakulu, G., Schulz, R. A., and Olson, E. N. (1996). Wingless signaling induces *nautilus* expression in the ventral mesoderm of the *Drosophila* embryo. *Dev. Biol.* **176**, 143–148.

Rong, P. M., Teillet, M.-A., Ziller, C., and Le Douarin, N. M. (1992). The neural tube/nothocord complex is necessary for vertebral but not limb and body wall striated muscle differentiation. *Development (Cambridge, UK)* **115**, 657–672.

Rosenblatt, J. D., Parry, D. J., and Partridge, T. A. (1996). Phenotype of adult mouse muscle myoblasts reflects their fiber type of origin. *Differentiation* **60**, 39–45.

Rudnicki, M. A., Braun, T., Hinuma, S., and Jaenisch, R. (1992). Inactivation of *MyoD* in mice leads to up-regulation of the myogenic HLH gene *Myf5* and results in apparently normal muscle development. *Cell (Cambridge, Mass.)* **71**, 383–390.

Rudnicki, M. A., Schneglesberg, P. N. J., Stead, R. H., Braun, T., Arnold, H.-H., and Jaenisch, R. (1993). MyoD or myf-5 is required for the formation of skeletal muscle. *Cell (Cambridge, Mass.)* **75**, 1351–1359.

Ruiz-Gomez, M., Romani, S., Hartmann, C., Jäckle, H., and Bate, M. (1997). Specific muscle identities are regulated by *Krüppel* during *Drosophila* embryogenesis. *Development (Cambridge, UK)* **124**, 3407–3414.

Rushton, E., Drysdale, R., Abmayr, S. M., Michelson, A. M., and Bate, M. (1995). Mutations in a novel gene, *myoblast city*, provide evidence in support of the founder cell hypothesis for *Drosophila* muscle development. *Development (Cambridge, UK)* **121**, 1979–1988.

Sassoon, D., Lyons, G., Wright, W. E., Lin, V., Lassar, A., Weintraub, H., and Buckingham, M. (1989). Expression of two myogenic regulatory factors myogenin and MyoD1 during mouse embryogenesis. *Nature (London)* **341**, 303–307.

Smith, T. H., Kachinsky, A. M., and Miller, J. B. (1994). Somite subdomains, muscle cell origins, and the four muscle regulatory factor proteins. *J. Cell Biol.* **127**, 95–105.

Song, K., Wang, Y., and Sassoon, D. (1992). Expression of Hox-7.1 in myoblasts inhibits terminal differentiation and induces cell transformation. *Nature (London)* **360**, 477–481.

Spörle, R., Günther, T., Struwe, M., and Schughart, K. (1996). Severe defects in the formation of epaxial musculature in *open brain* (*opb*) mutant mouse embryos. *Development* (*Cambridge, UK*) **122**, 79–86.

Stern, H. M., Brown, A. M., and Hauschka, S. D. (1995). Myogenesis in paraxial mesoderm: Preferential induction by dorsal neural tube and by cells expressing Wnt-1. *Development* (*Cambridge, UK*) **121**, 3675–3686.

Stockdale, F. E. (1992). Myogenic cell lineages. *Dev. Biol.* **154**, 284–298.

Tajbakhsh, S., and Buckingham, M. (1994). Mouse limb muscle is determined in the absence of the earliest myogenic factor myf-5. *Proc. Natl. Acad. Sci. U.S.A.* **91**, 747–751.

Tajbakhsh, S., and Spörle, R. (1998). Somite development: Constructing the vertebrate body. *Cell* (*Cambridge, Mass.*) **92**, 9–16.

Tajbakhsh, S., Bober, E., Babinet, C., Pournin, S., Arnold, H., and Buckingham, M. (1996a). Gene targeting the myf-5 locus with LacZ reveals expression of this myogenic factor in mature skeletal muscle fibers as well as early embryonic muscle. *Dev. Dyn.* **206**, 291–300.

Tajbakhsh, S., Rocancourt, D., and Buckingham, M. (1996b). Muscle progenitor cells failing to respond to positional cues adopt non-myogenic fates in *myf-5* null mice. *Nature* (*London*) **384**, 266–270.

Tajbakhsh, S., Rocancourt, D., Cossu, G., and Buckingham, M. (1997). Redefining the genetic hierarchies controlling skeletal myogenesis: Pax-3 and myf-5 act upstream of MyoD. *Cell* (*Cambridge, Mass.*) **89**, 127–138.

Tajbakhsh, S., Vivarelli, E., Kelly, R., Papkoff, J., Duprez, D., Buckingham, M., and Cossu, G. (1998). Differential activation of Myf5 and MyoD by different Wnts in explants of mouse paraxial mesoderm and the later activation of myogenesis in the absence of Myf5. *Development* In press.

Takayama, H., LaRochelle, W. J., Anver, M., Bockman, D. E., and Merlino, G. (1996). Scatter factor/hepatocyte growth factor as a regulator of skeletal muscle and neural crest development. *Proc. Natl. Acad. Sci. U.S.A.* **93**, 5866–5871.

Tanabe, Y., and Jessell, T. M. (1996). Diversity and pattern in the developing spinal cord. *Science* **274**, 1115–1123.

Tapscott, S. J., Lassar, A. B., and Weintraub, B. (1992). A novel myoblast enhancer element mediates MyoD transcription. *Mol. Cell. Biol.* **12**, 4994–5003.

Teillet, M. A., and Le Douarin, N. M. (1983). Consequences of neural tube and notochord excision of the peripheral nervous system in the chick embryo. *Dev. Biol.* **98**, 192–211.

Tickle, C. (1996). Vertebrate limb development. *Semin. Cell Dev. Biol.* **7**, 137–143.

Tremblay, P., and Gruss, P. (1994). Pax: Genes for mice and men. *Pharmacol. Ther.* **61**, 205–226.

Van Swearingen, J., and Lance-Jones, C. (1995). Slow and fast muscle fibers are preferentially derived from myoblasts migrating into the chick limb bud at different developmental times. *Dev. Biol.* **170**, 321–337.

Venuti, J. M., Morris, J. H., Vivian, J. L., Olson, E. N., and Klein, W. H. (1995). Myogenin is required for late but not early aspects of myogenesis during mouse development. *J. Cell Biol.* **128**, 563–576.

Wang, S., Krinks, M., and M. Moos, J. (1997). Frzb, a secreted protein expressed in the Spemann organizer, binds and inhibits Wnt-8. *Cell* (*Cambridge, Mass.*) **88**, 757–766.

Weinberg, E. S., Allende, M. L., Kelly, C. S., Abdelhamid, A., Murakami, T., Andermann, P., Doerre, O. G., Grunwald, D. J., and Riggleman, B. (1996). Developmental regulation of zebrafish *MyoD* in wild-type, no *tail* and *spadetail* embryos. *Development* (*Cambridge, UK*) **122**, 271–280.

Weintraub, H., Davis, R., Tapscott, S., Thayer, M., Krause, M., Benezra, R., Blackwell, K., Turner, D., Rupp, R., Hollenberg, S., Zhuang, Y., and Lassar, A. (1991). The MyoD gene family: Nodal point during specification of the muscle cell lineage. *Science* **251**, 761–766.

Williams, B. A., and Ordahl, C. P. (1994). Pax-3 expression in segmental mesoderm marks early stages in myogenic cell specification. *Development* (*Cambridge, UK*) **120**, 785–793.

Williams, B. A., and Ordahl, C. P. (1997). Emergence of determined myotome precursor cells in the somite. *Development* (*Cambridge, UK*) **124**, 4983–4997.

Wilting, J., Brand-Saberi, B., Huang, R., Zhi, Q., Köntges, G., Ordahl, C. P., and Christ, B. (1995). Angiogenic potential of the avian somite. *Dev. Dyn.* **202**, 165–171.

Woloshin, P., Song, K., Degnin, C., McNeille Killary, A., Goldhamer, D. J., Sassoon, D., and Thayer, M. J. (1995). MSX1 inhibits MyoD expression in fibroblasts × 10T1/2 cell hybrids. *Cell* (*Cambridge, Mass.*) **82**, 611–620.

Yang, X. M., Vogan, K., Gros, P., and Park, M. (1996). Expression of the met receptor tyrosine kinase in muscle progenitor cells in somites and limbs is absent in Splotch mice. *Development* (*Cambridge, UK*) **122**, 2163–2171.

Zebrafish (*continued*)
    gastrula, 384–385
      neural, 391–394
      tail bud, 407–410
    gastrulation, 389–391
    genetic model system, 577
    head region, 402–403
    Hox cluster genes, 400–412
    mesoderm, paraxial, 404–407
    neural crest, 415–416
      cell fate restriction, 417

      cell potential restriction, 420–421
      development, 416–417
      genetic analysis of development, 418–420
      regulative interactions among cells, 421–423
    nonaxial signals, 394–395
    rostrocaudal patterning, 394-395
    tail bud, 407–412
Zeste-white genes, *see Zw2; Zw3*
Zf-T, 391
Zw2, 110
Zw3, 258-259, 261, 262, 263–264